OCEAN CIRCULATION

The circulation of the oceans is a fundamental process in the movement of energy and materials around the planet. In recent years, the interaction between ocean circulation and climate change has become one of the most active research frontiers in the Earth sciences. Ocean circulation, and the geophysical fluid dynamical principles that underpin it, are subjects taught at graduate level in many Earth science, oceanography, and atmospheric sciences departments. Ocean circulation is driven and regulated by interaction with the atmosphere (wind), by tidal dissipation, and by regional differences in the temperature and salinity, and subsequently, density, of the oceans. There are several books that deal with wind-driven ocean circulation, but few, if any, cover thermohaline-driven circulation and its energetics. This is the first advanced textbook to cover both these important aspects of large-scale ocean circulation. It is based on Rui Xin Huang's many years of teaching an advanced course at Woods Hole Oceanographic Institution and Massachusetts Institute of Technology.

This book provides a concise introduction to the dynamics and thermodynamics of the oceanic general circulation, including the thermodynamics of seawater and the energetics of the ocean circulation; an exhaustive theory of wind-driven circulation; thermohaline circulation, including water mass formation/erosion, deep circulation, and the hydrological cycle; and the interaction between the wind-driven and thermohaline circulation. Highly illustrated to help the reader establish a clear mental picture of the physical principles involved, it will be invaluable on advanced courses in ocean circulation and as a reference text for oceanographers and other Earth scientists.

RUI XIN HUANG is a Scientist Emeritus at the Department of Physical Oceanography, Woods Hole Oceanographic Institution. He has been awarded the Von Alan Clarke Jr. Chair of Excellence in Oceanography at the same institution and has served as Chair Professor, Green Card Project, Ocean University of China. He has also worked at the Institute of Mechanics, Academy of Science, China; Massachusetts Institute of Technology; Geophysical Fluid Dynamical Laboratory, Princeton; and the University of Hawaii. His research interests include physical oceanography and climate dynamics, and he has authored/co-authored over 90 scientific publications in these areas.

OCEAN CIRCULATION

Wind-Driven and Thermohaline Processes

RUI XIN HUANG
Woods Hole Oceanographic Institution

CAMBRIDGE
UNIVERSITY PRESS

CAMBRIDGE UNIVERSITY PRESS
Cambridge, New York, Melbourne, Madrid, Cape Town, Singapore,
São Paulo, Delhi, Dubai, Tokyo

Cambridge University Press
The Edinburgh Building, Cambridge CB2 8RU, UK

Published in the United Kingdom by Cambridge University Press, UK

www.cambridge.org
Information on this title: www.cambridge.org/9780521852289

First published 2010

Printed in the United Kingdom at the University Press, Cambridge

A catalog record for this publication is available from the British Library

ISBN 978-0-521-85228-9 Hardback

Contents

Preface

With great progress being made in science and technology, we are becoming more interested in finding out how the climate system, including the oceanic general circulation, works on our planet. This book is written for the general reader who is searching for knowledge about oceanic circulation and its relevance to climate and the global environment on Earth.

During the process of collecting the materials for this book, I have tried to achieve a sensible balance between the physical concepts fundamental to the oceanic circulation, well-established theories, and recent developments associated with the frontiers in our field. As its title suggests, the book is about the wind-driven and thermohaline processes in the oceans. Although many theories about the oceanic general circulation have developed over recent decades, it is clear that our understanding of the circulation remains rudimentary at best. Since this book is intended as a textbook for graduate students, I have made a major effort to describe and explain the physical aspects of the circulation without relying on the sometimes complicated mathematics. To aid the reader, I have included many diagrams illustrating the physics.

In terms of the theoretical part of the book, I have made every effort to present new theories and thoughts about the energetic theory of the oceanic general circulation. Although energetics is one of the fundamental aspects of any dynamical system, the importance of examining the energetics of the oceanic general circulation has so far not been widely appreciated. In fact, there is no reliable estimate of the fundamental terms of energy balance, in particular the balance of mechanical energy, which is now believed to play a critically important role in regulating the oceanic general circulation. Clearly, much work still needs to be done, most probably by young students who may be inspired by the fact that so many aspects of the energetics of oceanic circulation remain uncertain or barely known. Since new theories are created almost every day, the situation is rather similar to that in the computer industry, where any product you buy may already be obsolete. Thus, in publishing a book about thermohaline circulation and its energetic theory, I could find myself in a situation where the theories collected in this book may soon be out of date and will have to be replaced by new theories which will be created in the near future. Nevertheless, I will be happy if this book can serve as a learning base for young students on their journey to uncover the mysteries of oceanic circulation.

This book includes many of my own personal views. Although I have made an effort to be broad-minded, the book – like most of the books published previously – necessarily reflects a personal view of the subjects discussed. Because it is a textbook, I have also included many results, some of them very elementary and well known, and others which may be somewhat new to the community at large. I do not claim any personal credit for all the material presented in the book.

To illustrate the historical development of oceanic circulation, I have cited some of the ideas tested by pioneers in the field. As our understanding has progressed, some of these ideas have proved to be unworthy or even incorrect. I believe it is important for the young reader to learn from some of these mistakes made by our predecessors, so that they will not fall into similar traps.

I have followed a long and winding road in science. When I was young, I enjoyed a simple and happy student life until I finished my undergraduate education. During my years in school, I benefited greatly from many excellent teachers, who taught me how to think, and how to work as an honest student and a future scientist. During the so-called Cultural Revolution I lost 10 years of the most precious time in my career. Along with many other young people, I forgot what I had learned in school and did virtually no science during that period.

Life then changed to a completely different path, and a goal I had never dared to dream of came true when I entered graduate school in China in 1978. Owing to the selfless and persistent encouragement of my English teacher, Ms. Mary Van de Water, I came to the United States as a graduate student in 1980. My career in oceanography started 28 years ago when I took part in and eventually graduated from the MIT/WHOI Joint Program in Oceanography.

I had the good fortune to meet and get to know the late Hank Stommel when I came back to Woods Hole as an entrance-level scientist. Despite the great differences in our experiences, we became close friends. Over a period of more than five years, I talked to Hank every day, and his personal approach to science and to life had a profound impact on me. Most of all, I started to think about the physics of oceanic circulation, rather than the mathematical and technical details. This book is dedicated to my lifelong memory of his impact on oceanography and to his personal charm.

I also received a great deal of help from my teachers at MIT and Woods Hole during and after my student years, including Glen Flierl, Mark Cane, and Carl Wunsch. In particular, my former teacher and now close friend, Joseph Pedlosky, has given me much help and personal advice over the past two decades.

During the period of writing this book, I have received considerable help from many good friends and colleagues, including Terry Joyce, Ray Schmidt, Xiangze Jin, Wei Wang, Qinyu Liu, Ted Durland, Zijun Gan, Yuping Guan, Hua Jiang and others. In particular, Joe Pedlosky and Fu Jia read part of the draft and offered many constructive comments; Bruce Warren helped me to update the figures associated with deepwater formation and deep circulation. Many of my former students read the lecture notes I used for the graduate course "Theory of the oceanic general circulation" offered to the MIT/WHOI Joint Program students. In addition, parts of my lecture notes have been used for seminars that I presented at the Ocean University of China, South China Sea Institute of Oceanology, and other oceanographic research institutes in China. In particular, Ms. Ru Chen and Ping Zhai of the Ocean University of China read through the early draft of the book and pinpointed numerous mistakes. Yuebing Zou provided great help in drawing some figures.

Finally, I am very grateful for the mentoring help from my first graduate advisor, Ji Ping Chao, who taught me how to work as a scientist. During the initial period of my study in the USA as a graduate student, I received tremendous spiritual support from Howard and Vivian Raskin. My wife, Luping Zou, continually reminds me of my goal in writing the book; without her encouragement and support, this book would never have been finished.

My scientific research has been supported by the National Science Foundation over the past two decades. The writing of the book was made possible through generous support from the Van Alan Clark Chair of Excellence in Oceanography. Ms. Barbara Gaffron read the manuscript with great attention to detail, and made the text flow more smoothly.

Part I

Introduction

1
Description of the world's oceans

The main focus of this book is the study of large-scale circulation in the world's oceans. As a dynamical system, the circulation in the world's oceans is controlled by the combined effects of external forcing, including wind stress, heat flux through the sea surface and seafloor, surface freshwater flux, tidal force, and gravitational force. In addition, the Coriolis force should be included, because all our theories and models are formulated in a rotating framework. In this chapter, I first describe surface forcing and the distribution of physical properties. I then discuss the classification of different kinds of motion in the world's oceans, and briefly review the historical development of theories of oceanic general circulation.

1.1 Surface forcing for the world's oceans

The ocean is forced from the upper surface, including wind stress, and heat and freshwater fluxes. In addition, tidal forces affect the whole depth of the water column, and geothermal heat flux and bottom friction also contribute to the establishment and regulation of the motions in the ocean. However, the surface forces are the primary forces for the oceanic circulation, and these are the focus of this section.

1.1.1 Surface wind forcing

Wind stress is probably the most crucial force acting on the upper surface of the world's oceans. The common practice in physical oceanography is to treat the effect of wind as a surface stress imposed on the upper surface of the ocean. The sea surface wind stress is usually calculated from the geostrophic wind 10 m above the sea surface, using bulk formulae. However, the air–sea interface is actually a transition zone between the atmospheric boundary layer and the oceanic boundary layer. Most importantly, the oceanic boundary layer is a wave boundary layer in which surface waves and turbulence play vitally important roles in regulating the vertical fluxes of momentum, heat, freshwater, and gases. Strictly speaking, therefore, the so-called wind stress on the sea surface should be replaced by the radiation stress between the wave boundary layer in the upper surface and the water below. Wind stress acting on the water below should depend on many dynamical aspects of these two boundary layers, such as the stability of the atmospheric boundary layer and the age of surface waves in the upper ocean. However, the discussion in this book follows the traditional approach, and the term "wind stress" is used for simplicity.

Furthermore, the distribution of wind stress on the upper ocean should be a final product of the atmosphere–ocean coupled system, and such interaction involves very complicated dynamical processes that are the subject of air–sea interactions and are beyond the scope of this book. Thus, in this book we will treat the wind stress as an external forcing for the oceanic general circulation.

Wind stress at sea level is the surface expression of the turbulent motions in the atmosphere, which occupy rather broad spectra in both space and time. It is common knowledge that wind stress changes over different time scales, from seconds to interannual and centennial time scales. The most important cycles in wind stress include the diurnal cycle and the seasonal cycle, in addition to changes on longer time scales, from interannual to decadal. Similarly, wind stress varies on spatial scales over a very broad spectrum. However, for the theory of oceanic general circulation, wind stress is normally referred to the smoothed wind stress data for large spatial scales and long time scales.

The dominant player in setting up the global wind stress pattern is the equator–pole temperature difference. Owing to this surface differential heating, atmospheric circulation is organized in the form of "Hadley cells." The prime feature of the surface wind stress is the strong westerlies associated with the Jet Stream at mid latitudes of both hemispheres. The existence of a quasi-steady circulation requires a near balance of the surface frictional torque; therefore, easterlies should exist at low latitudes. In fact, both the equatorial Pacific and equatorial Atlantic Oceans are dominated by easterlies (also known as trades) (see Fig. 1.1).

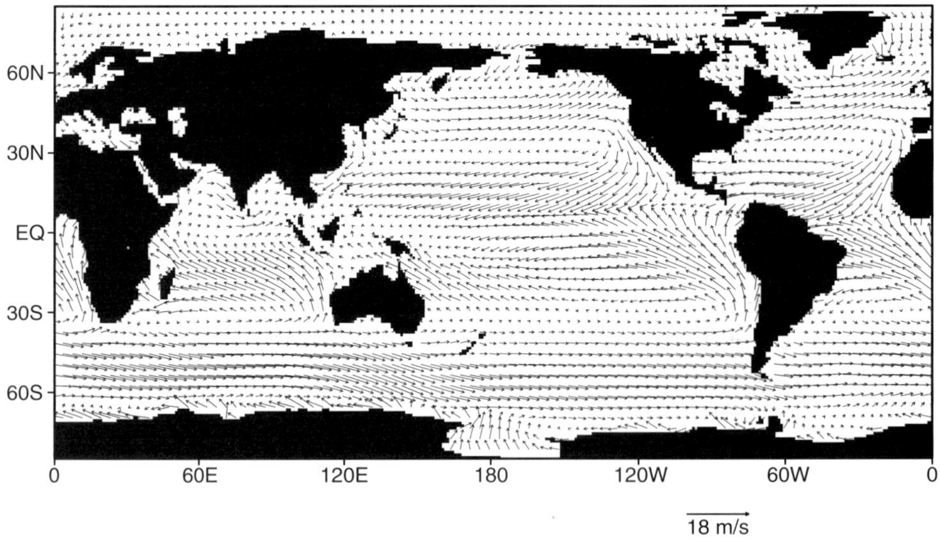

Fig. 1.1 Annual mean wind vector (in m/s) on the world's oceans, based on the European Centre for Medium-Range Weather Forecasts (ECMWF) dataset (Uppala *et al.*, 2005).

Owing to the large-scale distribution of land and ocean, wind stress on the surface is far from being zonally symmetric. Another major player in setting up the global wind stress pattern is the Earth's rotation. For example, at low latitudes the near-surface branch of the equatorward return flow of the Hadley cell is turned westward and appears as the northeast trade wind in the Northern Hemisphere and the southeast trade wind in the Southern Hemisphere. Under many such dynamical constraints, the sea surface wind stress pattern takes complicated forms. In fact, the most outstanding feature in the North Pacific Basin is the huge cyclonic wind stress pattern in the subpolar basin and the anticyclonic wind stress pattern in the subtropical basin. Similar features also exist in the Atlantic Basin and in the Southern Hemisphere, including the South Pacific, South Atlantic and South Indian Oceans.

Strong circulation is induced by wind stress in the upper kilometer of the world's oceans. The most outstanding features of wind-driven circulation include gigantic anticyclonic gyres in subtropical basins, and cyclonic gyres in subpolar basins, as shown in Figure 1.2. In addition, there is a strong circumpolar current system in the Southern Ocean, which is one of the most crucial branches of circulation in the world's oceans.

Wind stress is one of the most crucial driving forces of the oceanic circulation. As explained in Chapter 4, the westerlies at mid latitudes are responsible for the equatorward surface drift, the so-called "Ekman drift," and the easterlies at low latitudes are responsible for the poleward surface drift. The anticyclonic wind stress in the subtropical basin drives the anticyclonic circulation in the subtropical basin, and the cyclonic wind stress in the subpolar basin drives the cyclonic circulation there. In the Southern Ocean the westerly wind appears as a continuously strong belt around the whole latitudinal band; this wind

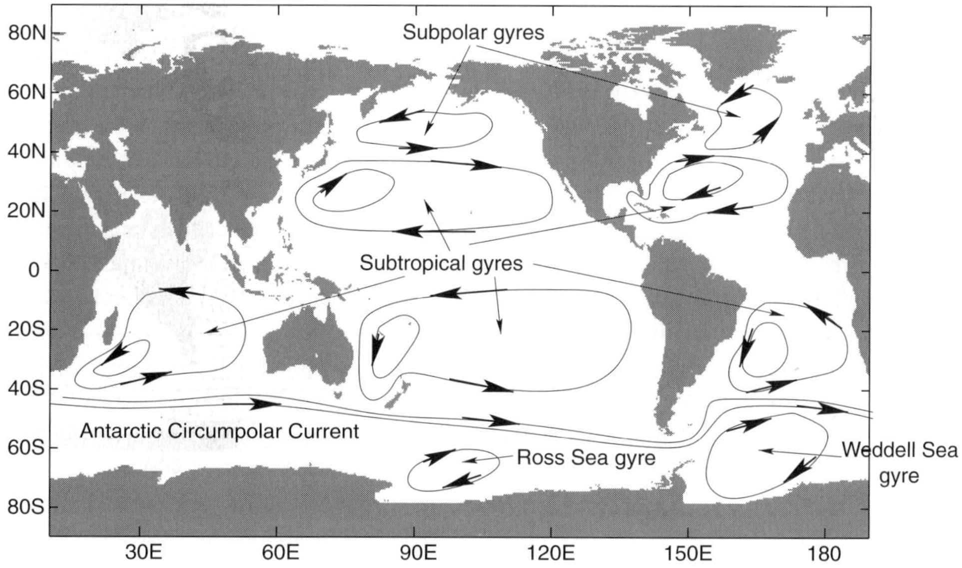

Fig. 1.2 Sketch of the major wind-driven gyres and currents in the world's oceans.

stress gives rise to the strong northward Ekman transport and upwelling, and is a direct driving force of the Antarctic Circumpolar Current (ACC).

Although a layman can observe the surface waves created by wind blowing on the sea surface, and a yachtsman can discern the surface wind drift from observation, the dynamical effects of wind stress include phenomena of huge spatial scales on the order of hundreds or thousands of kilometers – phenomena undetectable to the layman's eyes. A comprehensive understanding of the wind-driven motions in the oceans can only be achieved by systematic scientific research. In fact, the theory of the wind-driven circulation has to be developed side by side with the progress of *in situ* observations through the development of modern scientific instrumentation.

1.1.2 Surface thermohaline forcing

Heat and freshwater fluxes through the air–sea interface are the most critical forcing boundary conditions for the temperature and salinity distribution in the oceans. In addition, the oceans also receive geothermal heat from the seafloor; however, under present-day geological conditions, the amount of heat received from the seafloor is relatively small, approximately a thousand times smaller than that through the air–sea interface, so it is a rather minor contributor to the oceanic general circulation, except near the seafloor.

Surface heat flux

I first discuss the heat fluxes through the air–sea interface. The heat flux maps presented in this section are based on the NCEP-NCAR Reanalysis Project (Kistler *et al.*, 2001). In the following figures, downward heat flux into the ocean is defined as positive, and upward heat flux, leaving the oceans, is defined as negative.

The most essential forcing for the climate system on Earth is solar radiation, and this energy is in the form of short waves. Most of the energy required for maintaining the climate system can ultimately be traced back to solar radiation. Since the atmosphere is nearly transparent for solar radiation, most of it can penetrate the atmosphere and reach the lower boundary of the atmosphere, over both the land and the oceans. The amount of short-wave radiation reaching the sea surface depends primarily on the latitudinal location. Furthermore, cloudiness may be another major player in regulating the amount of solar radiation which can reach the sea surface. On the sea surface, part of the incoming short-wave radiation is reflected; thus, what the ocean receives is the net short-wave radiation, as shown in Figure 1.3. The global maxima of net short-wave radiation are closely related to the cold tongues in the eastern part of the equatorial Pacific and Atlantic Oceans.

The net short-wave radiation at each station on the sea surface is balanced by the heat transport within the ocean through advection and diffusion, plus upward heat flux through the air–sea interface. The major term in the heat flux from the ocean to the atmosphere is the latent heat flux associated with evaporation. The latent heat content for water vapor is approximately 2,500 kJ/kg; thus a relatively small amount of evaporation can transfer a large amount of heat from the ocean to the atmosphere.

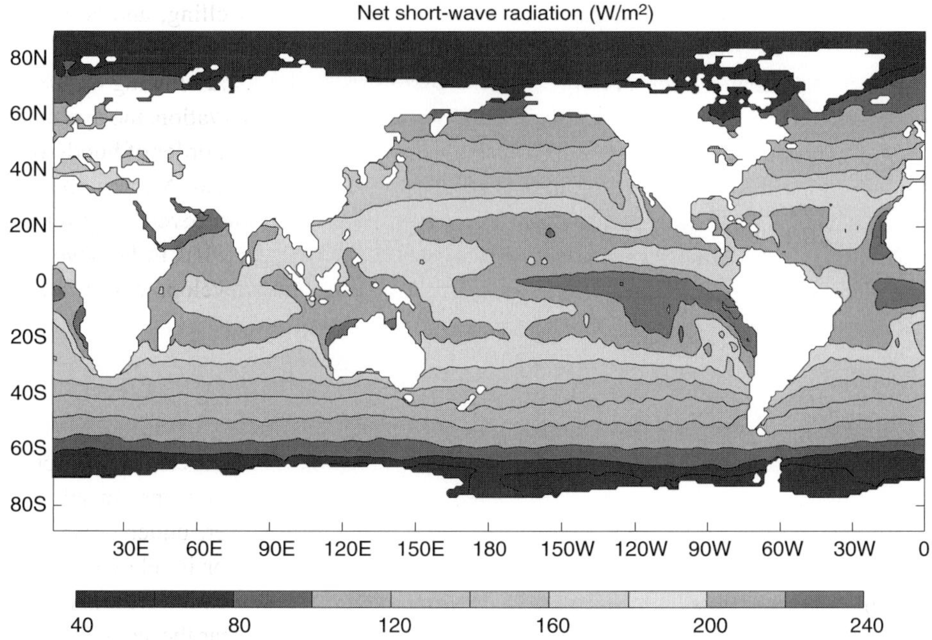

Fig. 1.3 Annual mean (NCEP-NCAR) net short-wave radiation (W/m^2). See color plate section.

There are several places where the latent heat flux is maximal (Fig. 1.4). First, the global maxima of latent heat loss exist in western boundary outflow regimes, such as the Kuroshio and the Gulf Stream, where the warm water brought by these strong western boundary currents meets cold and dry air from the continents, and strong latent heat loss is induced. These places are closely linked to a high rate of evaporation, as will be discussed shortly. Second, centers of strong latent heat loss exist at extratropics/subtropics in both hemispheres. There are areas of very low rate of latent heat flux associated with the cold tongues of surface waters in the eastern equatorial Pacific and Atlantic Oceans. Latent heat loss is generally small at high latitudes, where low sea surface temperature cannot sustain much evaporation.

The backward radiation from the ocean to the atmosphere and outer space is made up of two components: short-wave radiation reflected from the sea surface and long-wave radiation. The long-wave radiation is due to the fact that the equivalent radiation temperature of the Earth is rather low; this outgoing radiation is directly controlled by sea surface temperature and the local atmospheric conditions. The bulk formula for long-wave radiation is

$$IR \uparrow\downarrow = IR \uparrow - IR \downarrow$$

i.e., the heat flux associated with long-wave radiation is the outgoing long-wave radiation from the ocean to the atmosphere minus the backward long-wave radiation from the atmosphere to the ocean. Both these terms are proportional to the fourth power of temperature at

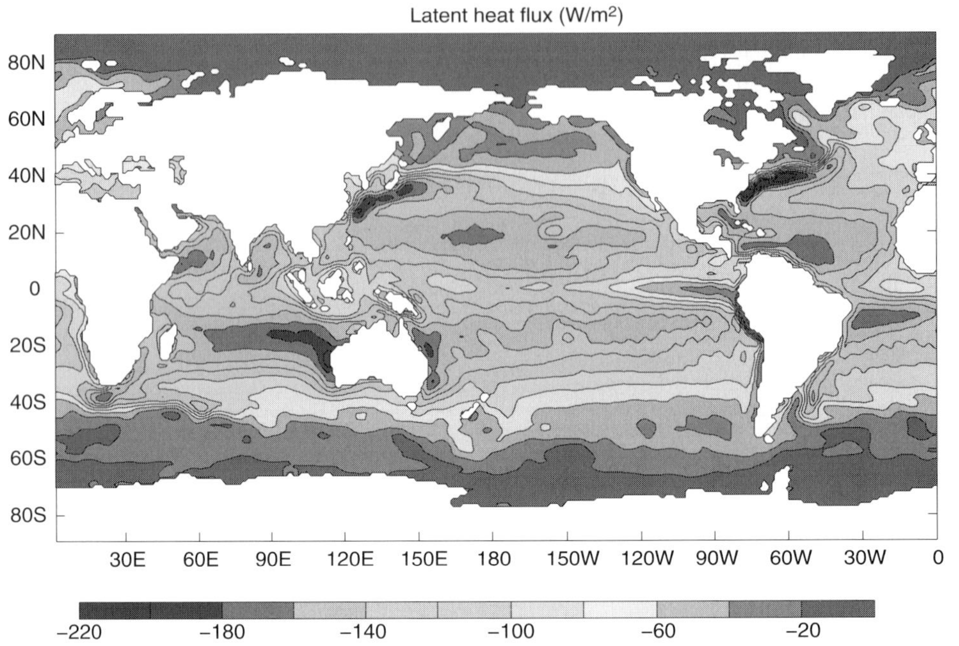

Fig. 1.4 Annual mean (NCEP-NCAR) latent heat flux due to evaporation (W/m^2). See color plate section.

the sea surface and of the atmosphere. The annual mean net heat flux of long-wave radiation is shown in Figure 1.5.

Owing to the competition of these two processes, the pattern of outgoing long-wave radiation is more complicated than other fluxes. In general, it is high near the western boundary outflow regimes in the subtropical basins of both hemispheres, especially in the Pacific Ocean. In comparison, it is much lower in the equatorial band and at high latitudes.

Sensible heat loss to the atmosphere is intimately related to the difference between the sea surface temperature and the atmospheric temperature. The most important sites of large sensible heat flux from the ocean to the atmosphere are over the Gulf Stream in the North Atlantic Ocean and the Kuroshio in the North Pacific Ocean. These high sensible heat flux regimes are clearly related to the warm water flowing in both the Gulf Stream and the Kuroshio. Note that, at high latitudes, the annual mean flux of sensible heat is actually from the atmosphere to the ocean. In particular, sensible heat flux in the Indian and Atlantic sectors of the Southern Ocean is from the atmosphere to the ocean, indicating that sea surface temperature is lower than the atmospheric temperature (Fig. 1.6). Such a low sea surface temperature is closely related to the cold deep water brought up by the strong Ekman upwelling driven by the Southern Westerlies under the current land–sea distribution.

The net air–sea heat flux, which is the sum of the four previous terms in the heat balance, is shown in Figure 1.7. As expected, there is a strong heat gain along the equatorial band, in

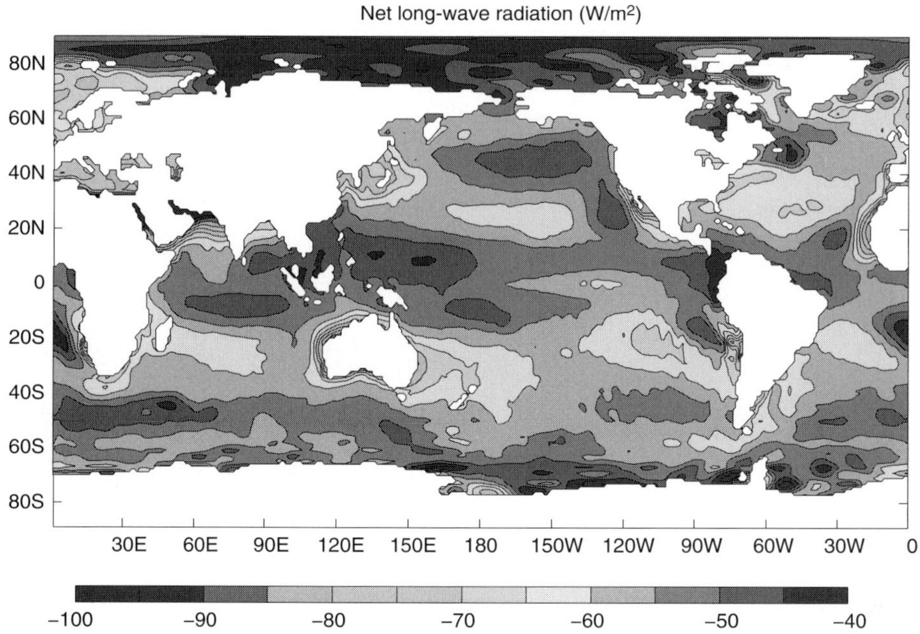

Fig. 1.5 Annual mean (NCEP-NCAR) net long-wave radiation (W/m^2). See color plate section.

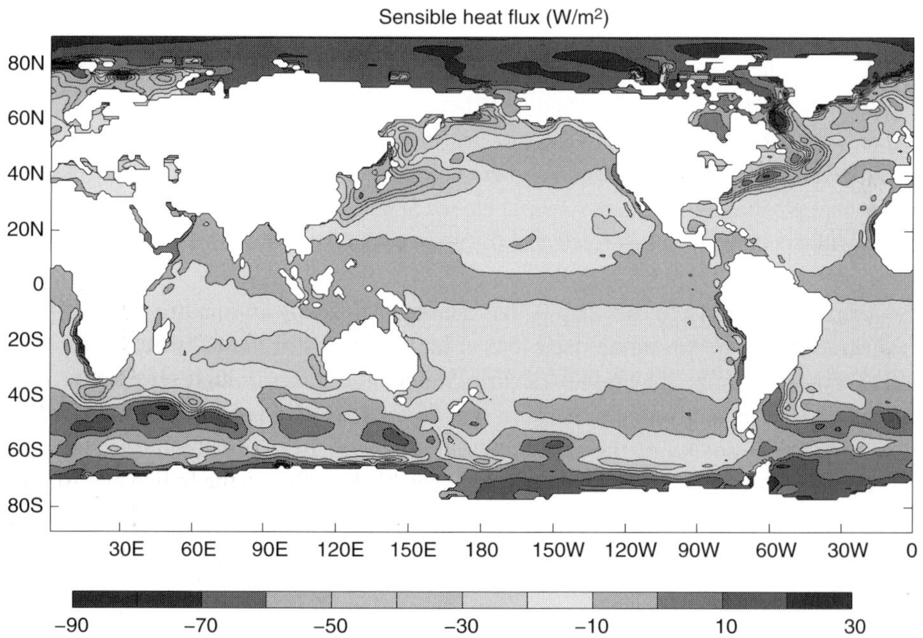

Fig. 1.6 Annual mean (NCEP-NCAR) sensible heat flux in the world's oceans (W/m^2). See color plate section.

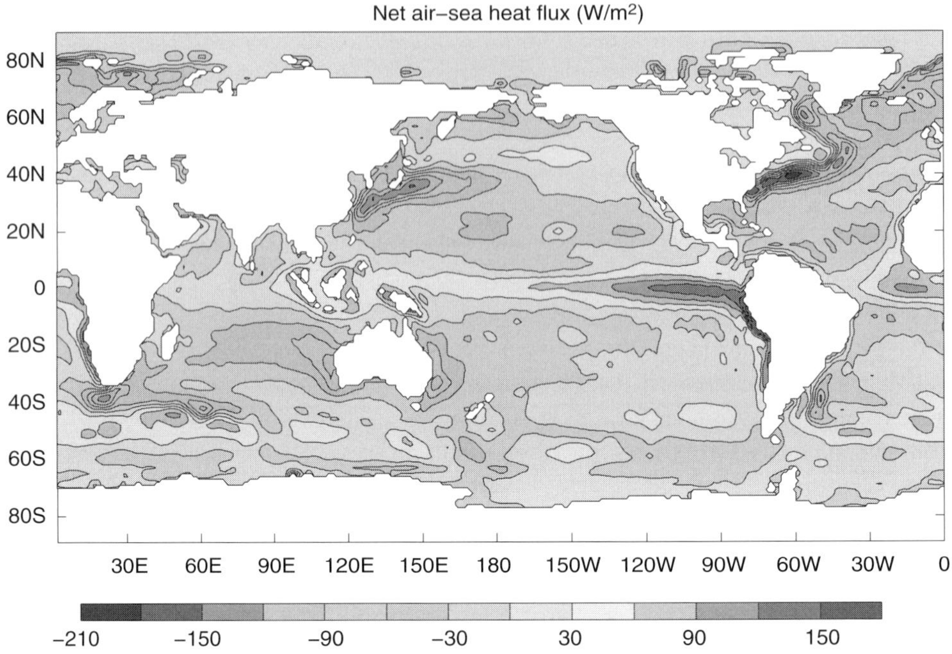

Fig. 1.7 Annual mean (NCEP-NCAR) net air–sea heat flux in the world's oceans (W/m^2). See color plate section.

particular the cold tongues in both the Pacific and Atlantic Oceans. In addition, the western coasts of South America and Africa appear as heat absorption bands, linked to the downward sensible heat flux, which is due to the low sea surface temperature associated with strong coastal upwelling. Both the Kuroshio and the Gulf Stream are major sites of heat loss in the world's oceans. The high-latitude Atlantic Ocean appears as another major site of heat loss in the world's oceans, which is related to the present-day strong meridional overturning in this basin.

Another major feature of this map is that the net heat flux is asymmetric with respect to the equator. Given the strong net heat loss at high latitudes in the Northern Hemisphere, one may expect a similar situation to occur in the Southern Hemisphere. However, a close examination reveals a different pattern. In fact, in the Indian sector and the South Atlantic sector of the Southern Ocean, the net heat flux is downward, i.e., the ocean there gains heat from the atmosphere, instead of losing heat. Comparing Figures 1.6 and 1.7, it is readily seen that these areas of net heat gain in the Southern Ocean are closely related to the downward sensible heat flux associated with the cold water upwelling driven by the strong westerlies in this latitudinal band.

The net air–sea heat flux distribution shown in Figure 1.7 implies that there is a meridional heat transport in the ocean, otherwise the underlying ocean would continuously cool or warm depending on the sign of the heat flux. In order to demonstrate the meridional heat flux, we first calculate the zonally integrated net air–sea heat flux, then integrate the zonal heat flux meridionally, starting from the South Pole, $H_f = \int_{\theta_S}^{\theta} \dot{q} a d\theta$, where a is the radius of the Earth, θ is the latitude, θ_S is the latitude of the South Pole, and $\dot{q} = \dot{q}(\theta)$ is the meridional distribution of net air–sea heat flux obtained by zonally integrating the flux shown in Figure 1.7. Accordingly, a positive slope of the curve shown in this figure indicates a downward heat flux into the ocean at the latitude of concern, and a negative slope indicates an upward heat flux at this latitude. For example, the strong positive slope over the equatorial band and the latitudinal band of 58° S–42° S indicates strong heat absorption by the ocean (Figure 1.8a).

On the other hand, a positive value of H_f indicates the northward heat flux in the oceans; thus, over the entire Northern Hemisphere, there is a poleward heat flux. As a matter of fact, in the Northern Hemisphere, the poleward heat flux reaches a maximum of nearly 2 PW (1 PW = 10^{15} W) around 15° N. The corresponding poleward heat flux in the Southern Hemisphere is much smaller and changes its sign several times. In fact, the result obtained from this approach shows a northward heat flux in the latitudinal band of 58° S–20° S; however, values of poleward heat flux obtained from other more comprehensive methods indicate that meridional heat flux in the ocean is mostly southward in the Southern Hemisphere, as discussed in Section 5.3.1. Such a large discrepancy in poleward heat flux is due to the fact that the air–sea heat flux data obtained from observations is not very accurate, especially in the Southern Ocean, where reliable *in situ* observations are sparse.

Similarly, there is a strong zonal transport of heat in the ocean. In order to demonstrate the zonal heat transport, we integrate the net air–sea heat flux, starting from the longitude of the southern tip of South America. As shown in Figure 1.8b, there is a westward heat flux in the Pacific Basin, with a peak value of 1.4 PW. The zonal heat flux is a manifestation of the zonal asymmetric nature of the thermal forcing of the oceans. This zonal heat flux is intimately linked to the oceanic currents, which are discussed in later chapters.

Surface freshwater flux

The oceans exchange freshwater with the atmosphere through evaporation and precipitation, plus river run-off. The river run-off is the result of water vapor from sea surface evaporation and precipitation on land. Freshwater exchange with the atmosphere is one of the most important forcing conditions for both the oceanic general circulation and the climate system. Evaporation is the most crucial vehicle for bringing heat from low-latitude ocean to the atmosphere, where water vapor is carried poleward. Water vapor carries a large amount of latent heat, and this is one of the vital mechanisms of poleward heat transport in the climate system. Water vapor in the atmosphere eventually condenses and releases the latent heat content, returning to the oceans or land as precipitation.

Freshwater flux through the air–sea interface plays a vital role in regulating the hydrological cycle in the ocean. In particular, freshwater flux is the key ingredient in controlling the salinity distribution in the oceans. Water density is primarily controlled by temperature and salinity, and thus freshwater flux is one of the key players in regulating the thermohaline circulation through its direct connection with the salinity distribution, which is one of the most important topics related to thermohaline circulation and climate. Meridional transport

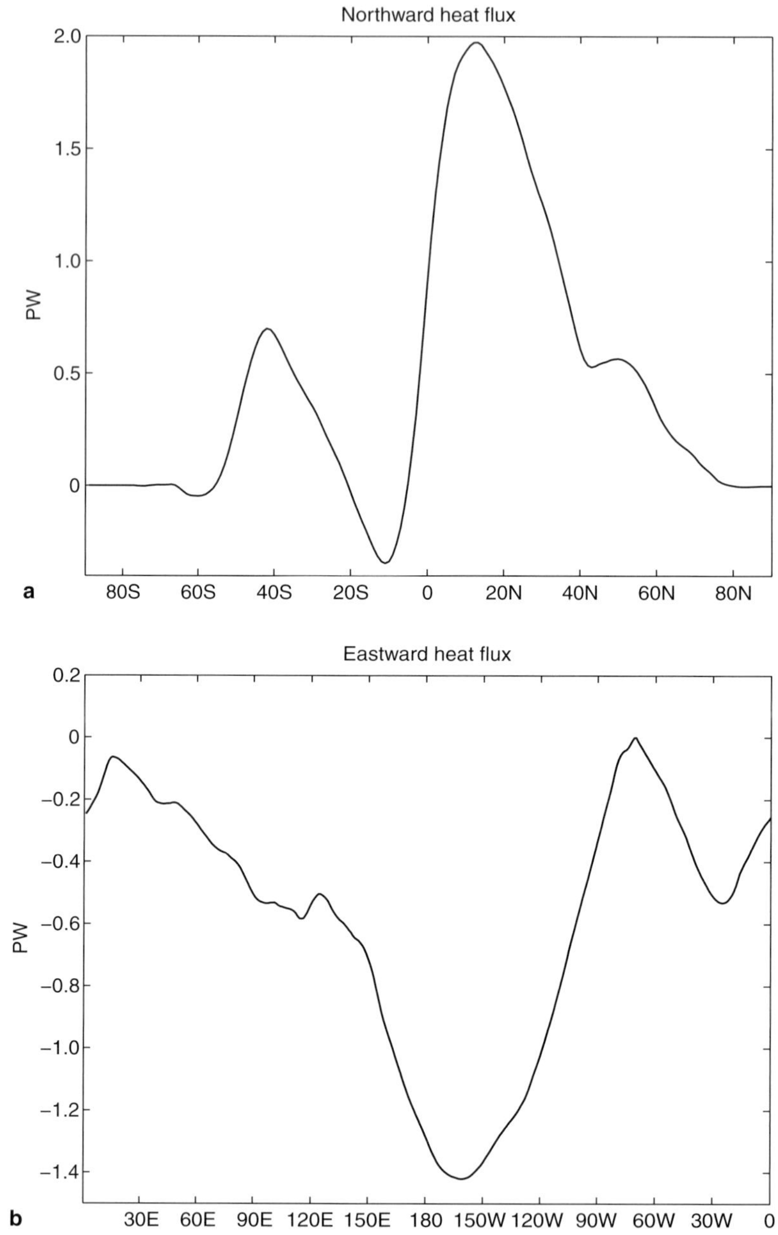

Fig. 1.8 **a** Northward and **b** eastward heat transport H_f in the world's oceans, in PW (1 PW = 10^{15} W).

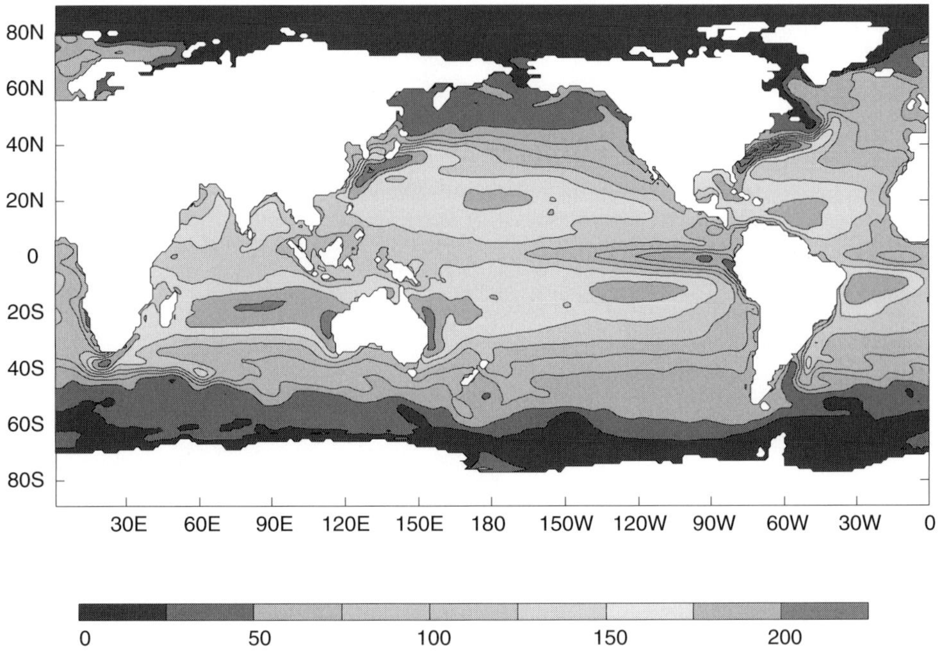

Fig. 1.9 Annual mean (NCEP-NCAR) evaporation rate in the world's oceans (cm/yr). See color plate section.

of water vapor in the atmosphere, and the associated return flow of water in the ocean, is one of the most important mechanisms responsible for the meridional heat transport in the climate system.

Evaporation is strongly related to the latent heat loss from the oceans; thus the global pattern of evaporation (Fig. 1.9) has basically the same pattern as the latent heat loss to the atmosphere (see Fig. 1.4). Here again, strong evaporation appears in the subtropics and the western boundary current system in both hemispheres. In particular, both the Gulf Stream and the Kuroshio system are the regimes with the maximal evaporation rate in the global oceans. Another strong evaporation regime is in the South Indian Ocean and both east and west of Australia.

On the other hand, the evaporation rate is very low over the cold tongues in both the eastern equatorial Pacific and Atlantic Oceans. In the subpolar basin of the North Pacific Ocean and the Labrador Sea of the North Atlantic Ocean, evaporation rate is quite low, due to the low sea surface temperature. In addition, over the vast regime of the Southern Ocean, evaporation is quite low, and this is due to the cold sea surface temperature associated with the strong local Ekman upwelling induced by the Southern Westerlies.

The oceans receive the returning water as precipitation from the atmosphere, plus river run-off. The major regimes of strong precipitation include the equatorial Pacific Ocean and the South Pacific Convergence Zone, which extends southeastward in the South Pacific

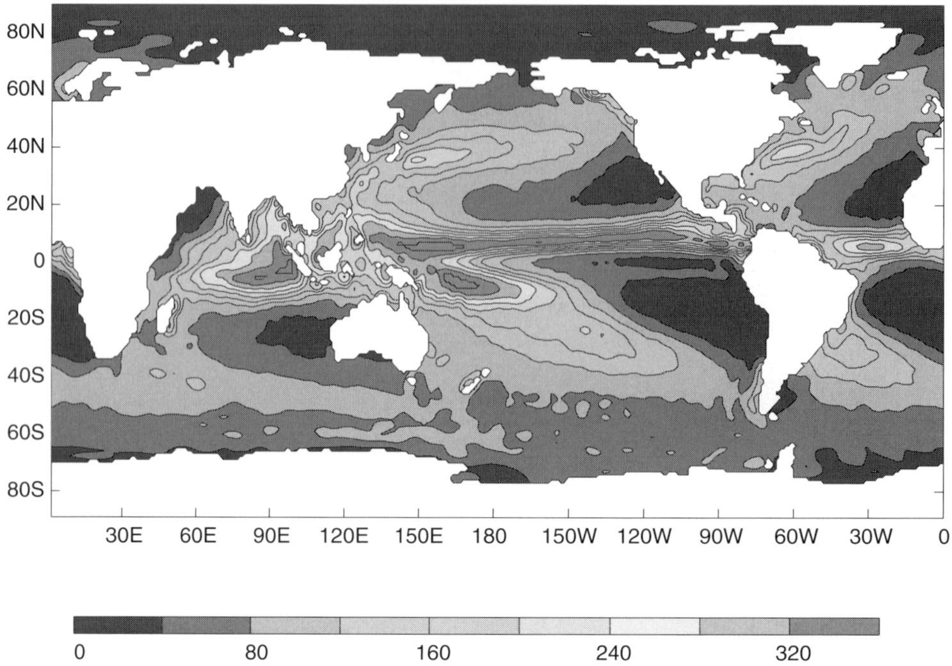

Fig. 1.10 Annual mean (NCEP-NCAR) precipitation rate (cm/yr). See color plate section.

Ocean, with an annual mean precipitation rate of more than 3.5 m/yr (Fig. 1.10). In addition, there are large regimes of strong precipitation in the equatorial Indian Ocean and equatorial Atlantic Ocean. In contrast, the eastern parts of the subtropical basins in both hemispheres appear as major sites of minimum precipitation, where the annual mean precipitation rate is on the order of only 0.5 m/yr.

The difference in evaporation and precipitation is what really affects the haline circulation in the oceans (Fig. 1.11). The pattern of the net freshwater flux across the air–sea interface is dominated by two features: the strong precipitation bands shown in Figure 1.10 and the strong and relatively narrow regimes of strong evaporation shown in Figure 1.8. Overall, the equatorial band is dominated by a net gain of freshwater, which is primarily due to the very strong precipitation shown in Figure 1.10. In particular, there is a strong net freshwater flux, on the order of 3 m/yr, into the western equatorial Pacific Ocean and the eastern equatorial Indian Ocean. This location is coincident with the Pacific–Indian Warm Pool, where the global maximal sea surface temperature is located. This is also the major ascendant branch of the Walker circulations in the atmosphere.

On the other hand, the eastern parts of the five subtropical basins in both hemispheres appear to be the "deserts" of the oceans, where the annual mean net freshwater loss to the atmosphere is on the order of 1–1.5 m/yr. As discussed in the next section, the strong freshwater loss in these vast areas is closely related to the high surface salinity in the oceans.

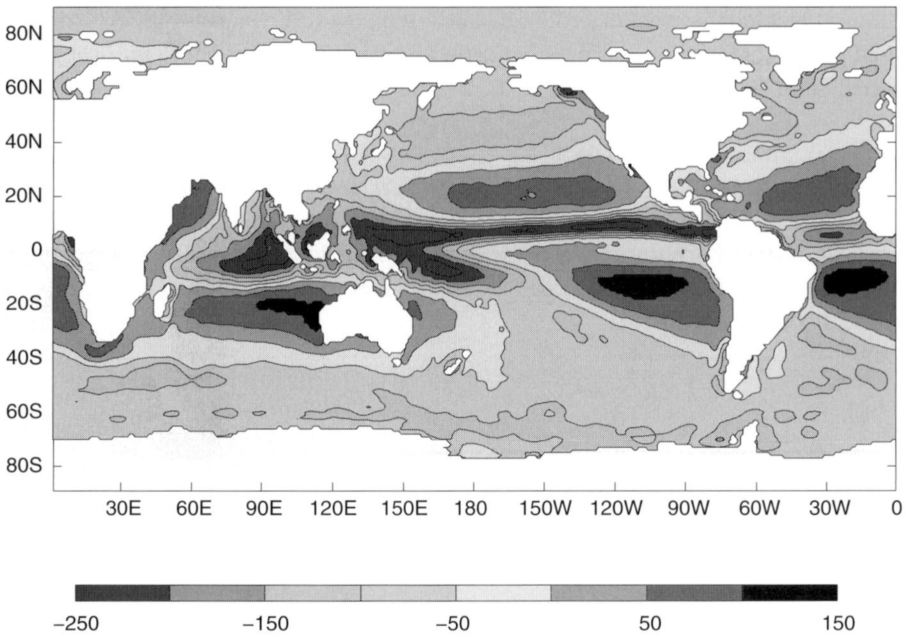

Fig. 1.11 Annual mean (NCEP-NCAR) rate of evaporation minus precipitation (cm/yr). See color plate section.

1.1.3 Other external forcing

Tidal flows are one of the major components of the large-scale motions in the oceans. According to the traditional view, tides are not considered as part of oceanic general circulation; and in the classical theoretical framework, the oceanic general circulation is mostly driven by surface forcing. It is, however, now commonly accepted that tidal dissipation is also one of the key factors in regulating the thermohaline circulation in the world's oceans. In fact, tidal dissipation is a major contributor to diapycnal mixing in the deep ocean; therefore energy sources due to tidal dissipation will be discussed in connection with the mechanical energy balance in the world's oceans.

In addition, geothermal heat flux also contributes to the thermal circulation, although it is believed that its influence is mostly limited to the deep ocean and a few sites associated with the hot plumes above the "hot spots" on the seafloor of the world's oceans. In general, new seafloor is continuously formed at the top of the mid-ocean ridge, so it is relatively shallow and is associated with very active geothermal heat flux release. As the newly created seafloor spreads outward from the mid-ocean ridge, it moves away from the active geological processes in the Earth's mantle. As a result, the geothermal heat released from the seafloor also gradually diminishes.

Owing to the technical difficulty of making heat flux measurements over the seafloor of the world's oceans, there is no reliable global dataset for the geothermal heat flux. However, to estimate the global impact of geothermal heat flux, the following empirical formula can be used (DeLaughter *et al.*, 2005; this formula is a modification from earlier studies by Stein and Stein (1992, 1994):

$$\dot{q} = \begin{cases} 481t^{-1/2} & if\, t < 44 \\ 48.8 + 97.5e^{-0.0323t} & if\, t \geq 44 \end{cases} \tag{1.1}$$

where heat flux is in mW/m^2, and t is the age of the seafloor, in Myr (million years). The age of seafloor can be estimated from the following empirical formulae, which relate the age of seafloor to its depth (m):

$$t = \begin{cases} \sqrt{\dfrac{d - 2600}{349}} & if\, 2600 \leq d < 3996 \\ -\dfrac{\ln[(5302 - d)/2190]}{0.0323} & if\, d \geq 3996 \end{cases} \tag{1.2}$$

The relations between seafloor depth, age, and geothermal heat flux, as defined in Eqns. (1.1) and (1.2), are shown in Figure 1.12.

The horizontal distribution of geothermal heat flux in the world's oceans calculated from Eqns. (1.1) and (1.2) is shown in Figure 1.13, and the total amount of geothermal heat flux is estimated as 32 TW. As expected from these formulae, geothermal heat flux is large in the vicinity of the global mid-ocean ridge system. Away from the mid-ocean ridge, the geothermal heat flux gradually declines, and the mean geothermal heat flux for the old-age seafloor is on the order of 50 mW/m^2. Thus, geothermal heat flux, in general, is a rather weak source of thermal energy. Although the strong geothermal heat flux released near the mid-ocean ridges may play an important role in driving thermal circulation near the

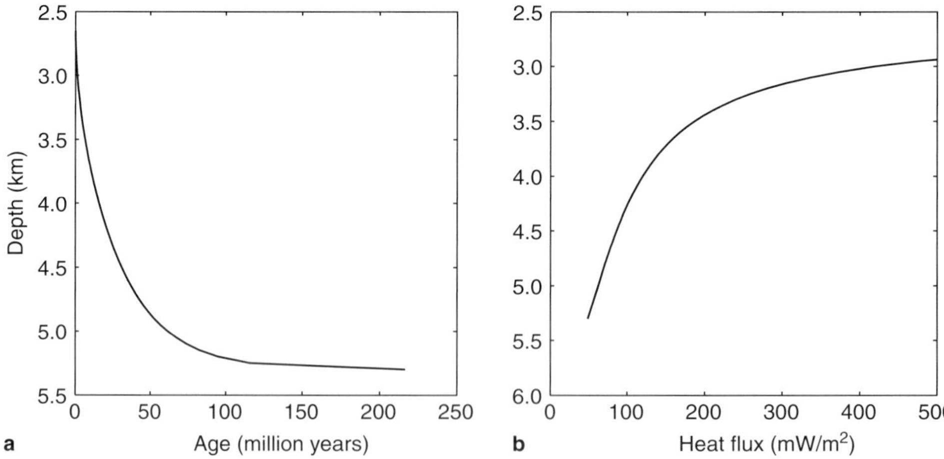

Fig. 1.12 Empirical relations between seafloor depth and **a** seafloor age, **b** geothermal heat flux.

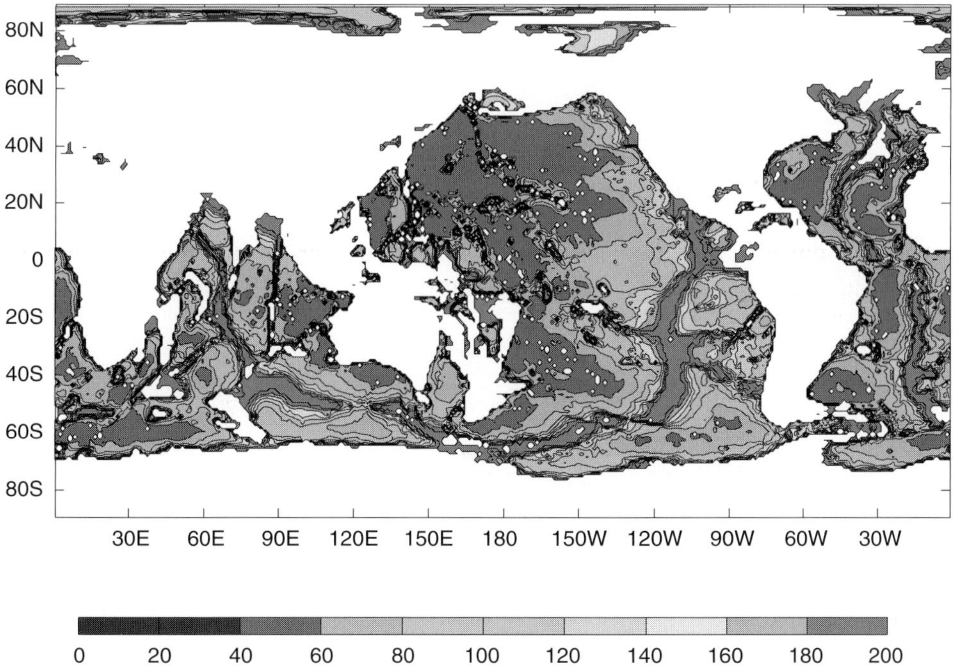

Fig. 1.13 Geothermal heat flux based on a semi-empirical formula; seafloor shallower than 2.6 km is excluded (mW/m^2). See color plate section.

ridges, away from the ridges the contribution of geothermal heat flux is mostly limited to the abyssal ocean.

1.2 Temperature, salinity, and density distribution in the world's oceans

The study of the oceanic circulation must start by building a clear mental picture of the physical conditions in the ocean, because water property distribution in the world's oceans is a result of the circulation system in response to the surface forcing described in the previous section. This section serves as a brief introduction to descriptive large-scale oceanography.

1.2.1 Surface distribution of temperature, salinity, and density

Sea surface temperature is closely related to the air–sea heat fluxes discussed in the previous section. In fact, there is a strong negative feedback between sea surface temperature and air–sea heat fluxes, i.e., a positive sea surface temperature anomaly induces more evaporation, sensible heat loss, and long-wave radiation, which tend to reduce the positive temperature anomaly. On the other hand, a negative sea surface temperature anomaly reduces evaporation, sensible heat loss, and outgoing long-wave radiation, and thus pushes sea

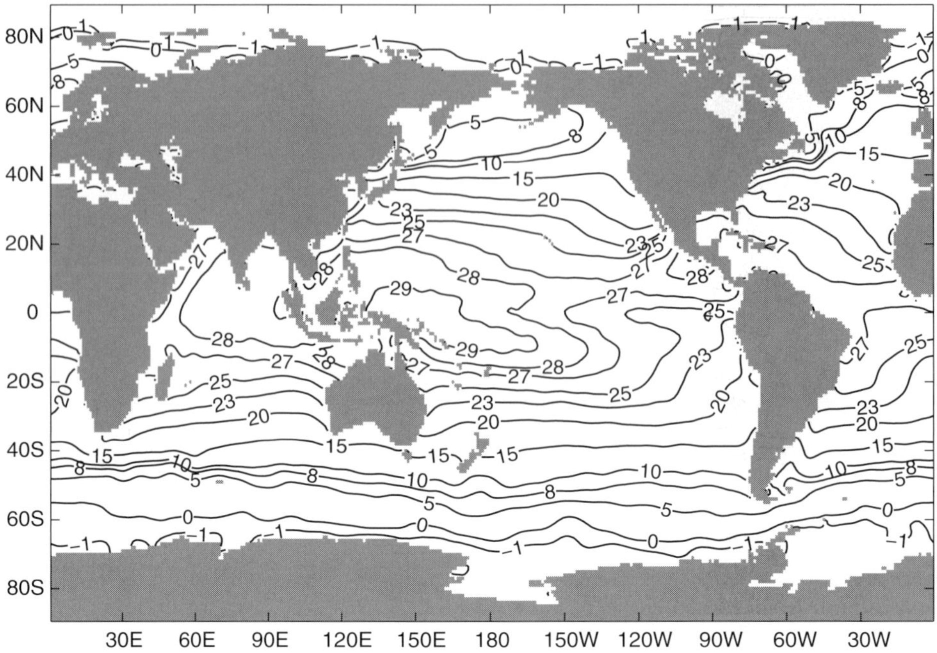

Fig. 1.14 Annual mean sea surface temperature (°C), based on WOA-01.

surface temperature back toward the normal range. In addition to the sea surface heat fluxes, advection and diffusion in the oceanic interior join together in setting up the temperature distribution. The connection between the surface thermal forcing, sea surface temperature, and the current is the major subject of the thermohaline circulation. The maps presented in this section are based on the *World Ocean Atlas 2001* (WOA-01; Conkright *et al.*, 2002).

In general, sea surface temperature is reduced in the poleward direction, with the highest temperature at the equatorial band (Fig. 1.14). In particular, sea surface temperature is maximal in the Warm Pool in the western equatorial Pacific and eastern equatorial Indian Oceans. There is a strong zonal temperature gradient along the equatorial band, with a regime of rather low temperature associated with the so-called "cold tongue" in the eastern equatorial Pacific Ocean. The existence of the cold tongue in the eastern equatorial Pacific Ocean is due to the upwelling of cold water driven by the easterlies in the equatorial band. In addition, the strong upwelling near the coast of Peru driven by the equatorward trade wind also contributes.

In subtropical basins, temperature is generally high near the western boundary and low in the eastern basin; this pattern is closely associated with the wind-driven anticyclonic gyre. In subpolar basins, the zonal temperature gradient flips sign, with the high temperature appearing in the eastern basin and the low temperature appearing in the western basin. This feature is closely related to the wind-driven cyclonic circulation there.

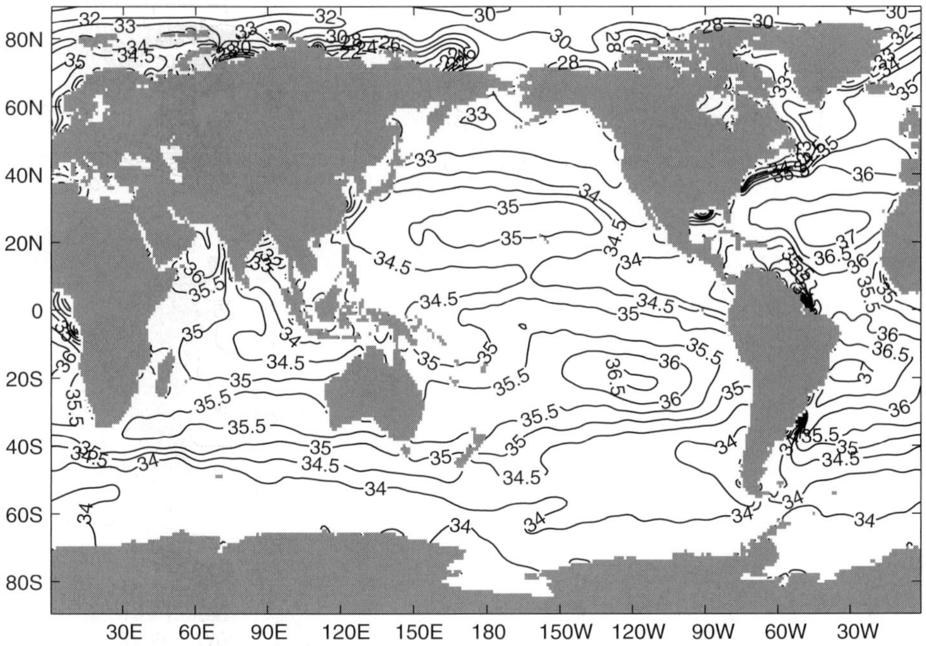

Fig. 1.15 Annual mean sea surface salinity in the world's oceans, based on WOA-01.

In the Southern Ocean, there is a very strong thermal front, especially in the Indian and Atlantic sectors around the latitudinal band of 40–50° S. This cold front is due to the strong upwelling driven by the Southern Westerlies. As discussed above, this cold front is the key factor in connection with the downward sensible heat flux through the air–sea interface. This cold front is also in close connection with the strong eastward current system and the strong velocity front. This temperature front will be discussed in detail later.

Sea surface salinity is high in the centers of the subtropical basins of both the Atlantic and Pacific Oceans (Fig. 1.15). These high-salinity areas are closely related to the high evaporation minus precipitation (E − P) rate over the five oceanic deserts, as shown in Figure 1.9.

There is a very close link between surface salinity and the local rate of evaporation minus precipitation. Assuming that the surface salinity is regulated by a one-dimensional balance of vertical advection and diffusion, strong evaporation minus precipitation should correspond to high salinity on the sea surface. This correlation between the sea surface salinity and net freshwater flux across the air–sea interface can be seen clearly from the scatter diagram in Figure 1.16. The low-salinity branch, with salinity lower than 32, primarily represents the surface water in the Arctic Ocean, where the low temperature and remote distance from the low latitudinal band limits both evaporation and precipitation. On the upper left, the branch of strong precipitation (negative E − P) corresponds primarily to the Warm Pool in the western equatorial Pacific and eastern equatorial Indian Oceans.

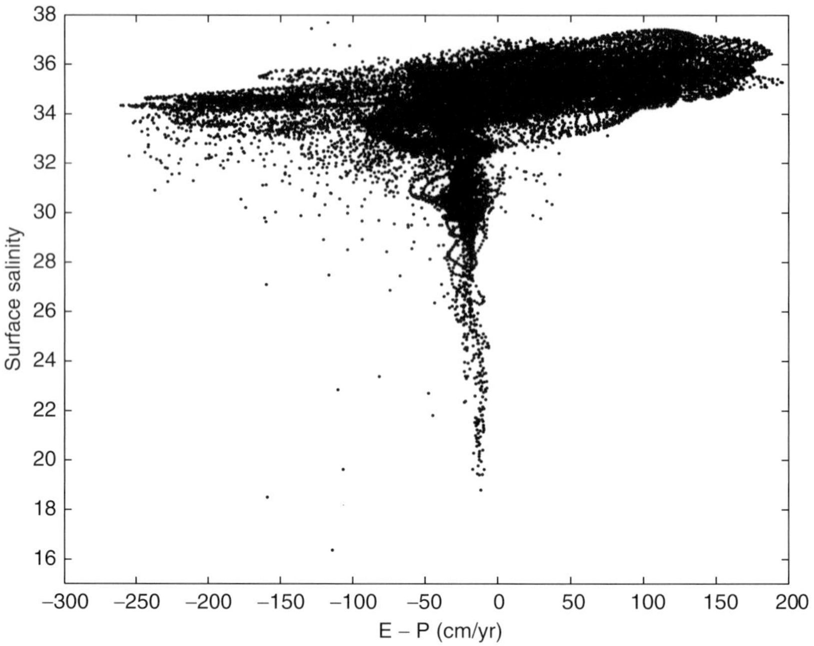

Fig. 1.16 Scatter diagram for the relation between the rate of evaporation minus precipitation and sea surface salinity in the world's oceans.

The upper branch suggests a nearly linear relationship between the local evaporation minus precipitation rate, i.e., a strong local excess of evaporation corresponds to high salinity, and a strong local excess of precipitation corresponds to low salinity. In comparison with the connection between sea surface temperature and air–sea heat fluxes, the connection between the sea surface salinity and air–sea freshwater flux is quite different because it is unidirectional, i.e., air–sea freshwater flux can directly affect the local sea surface salinity; however, sea surface salinity cannot directly affect the local air–sea freshwater flux. The lack of feedback from the local sea surface salinity to evaporation and precipitation is a unique element of the thermohaline circulation. For example, this disconnection between the surface freshwater flux and the sea surface salinity leads to the persistence of a surface freshwater anomaly at high latitudes and the so-called "halocline catastrophe," which is discussed in Chapter 5.

Of course, sea surface salinity is also controlled by other dynamical processes, including advection and diffusion in the subsurface layers; nevertheless, the local air–sea freshwater flux is one of the most important factors regulating the salinity distribution in the upper ocean.

Although both temperature and salinity are crucial factors in controlling the circulation, water density is dynamically linked to the current. Sea surface density distribution has a pattern rather similar to that of sea surface temperature because density in the upper ocean

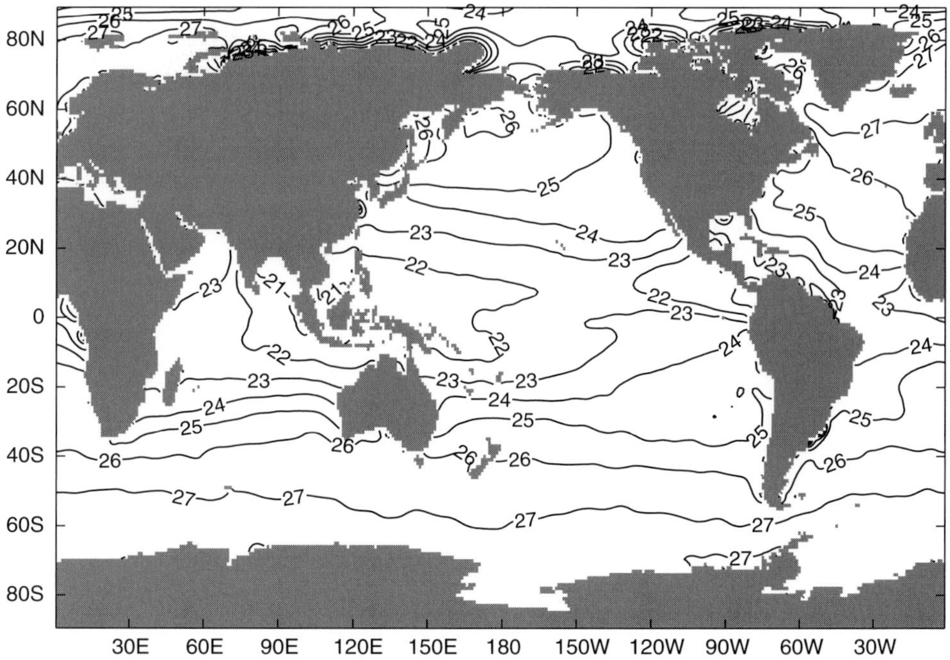

Fig. 1.17 Sea surface density (kg/m^3), based on WOA-01.

is primarily controlled by temperature, except for the high latitudes, where surface density is primarily controlled by sea surface salinity (Fig. 1.17). The high sea surface temperature, in combination with low salinity, gives rise to the lowest sea surface density in the Bay of Bengal and the Warm Pool. There is a zone of low-density water extending from the low-salinity tongues in the eastern equatorial boundary of the Pacific Ocean (Fig. 1.15). In the subtropical basins, density is relatively low in the western basins, as compared with the eastern basins, and this feature is apparently related to the surface temperature pattern shown in Figure 1.14.

One of the salient features in the surface density, temperature and salinity distribution in the world's oceans is the asymmetry in the zonally averaged meridional distribution. In fact, surface water in the Southern Hemisphere is colder and saltier than that in the Northern Hemisphere. As a result, surface water in the Southern Hemisphere is heavier than that in the Northern Hemisphere (Fig. 1.18). The asymmetric nature of meridional distribution of water properties at the sea surface reflects many physical aspects of the climate system. First, the Southern Hemisphere is mostly a water hemisphere, while the Northern Hemisphere is mostly a land hemisphere. More important, the combination of the Antarctic continent and the adjacent circumpolar water channel creates the coldest and densest surface water in the world's oceans, which sinks to the bottom of the world's oceans and dominates the abyssal circulation.

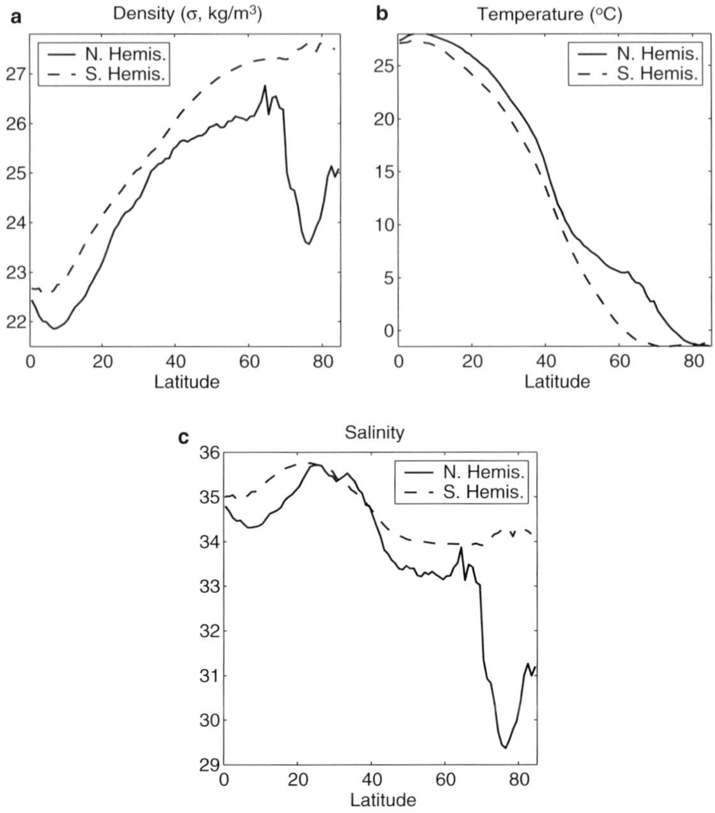

Fig. 1.18 The contrast of meridional distribution of **a** surface density (σ, kg/m^3), **b** temperature (°C), and **c** salinity in the Northern (*solid line*) and Southern (*dashed line*) Hemispheres.

To examine the distribution of surface temperature and salinity more closely, we subtract the zonal mean value and plot the deviation from the zonal mean for both the surface temperature and salinity. The most interesting features that can be identified from the deviation from the zonal mean surface temperature and salinity distribution are as follows.

First, there is a strong east–west asymmetry of temperature distribution in both the Pacific and Atlantic Basins (Fig. 1.19). This feature is closely related to the circulation driven by wind stress. The most direct connection is the coastal upwelling driven by the along-shore wind. For example, the equatorward wind at mid latitudes in both hemispheres drives coastal upwelling along the eastern boundaries, and coastal downwelling near the western boundaries. The strong upwelling driven by trade winds near the coast of Peru brings cold water rich in nutrients from deeper in the ocean to the upper surface layer and thus creates one of the largest fishing grounds in the world's oceans. There are other similar cold-water upwelling zones near the eastern coasts of North and South Atlantic Ocean. As a result, there is an outstanding east–west temperature contrast at mid latitudes in the Pacific and

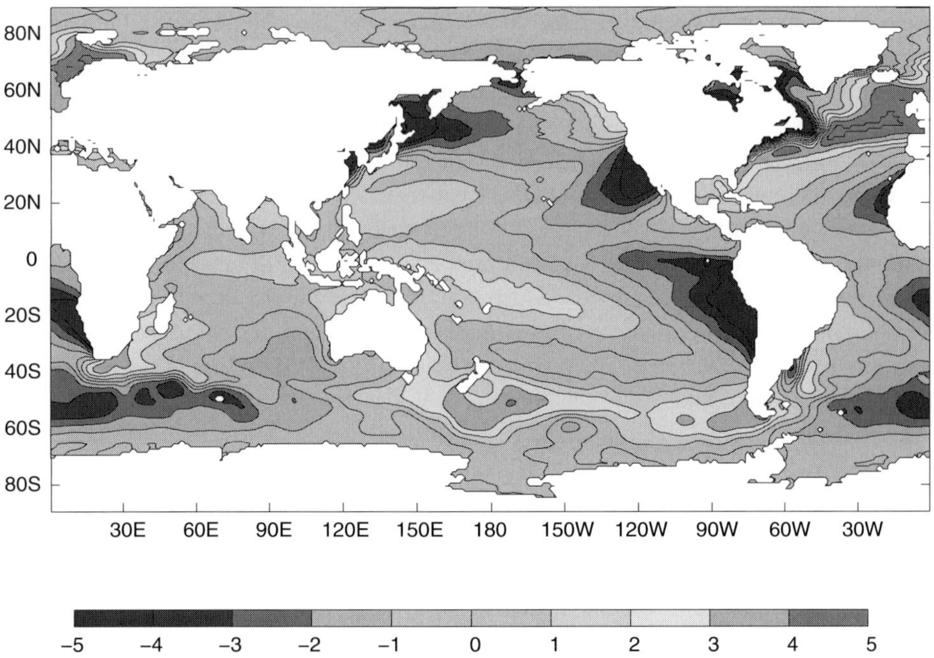

Fig. 1.19 Annual mean sea surface temperature anomaly (°C), deviation from the zonal mean. See color plate section.

Atlantic Basins. Another major factor is the wind-driven gyres. The anticyclonic gyre in the subtropical basin brings relatively cold water from the mid latitudes and thus contributes to the low temperature in the eastern basins.

Cyclonic and anticyclonic gyres, shown in Figure 1.2, are the major players in setting water properties in the upper ocean. Cyclonic gyres in the subpolar basins bring relatively warm water to the eastern part of the subpolar basins, thus maintaining a relatively warm surface temperature. On the other hand, cold water brought by the cyclonic gyres from the north creates the regime of cold surface water in the western part of the subpolar basins. In addition, the strong cold and dry air from the Eurasian continent must contribute to the low surface temperature north of Japan. Similar, there is a narrow band of low temperature around the eastern coast of Canada and northeastern coast of the USA. As a result, over the high-latitude band in both the North Pacific and North Atlantic Oceans, the sea surface temperature near the eastern boundaries is substantially warmer than that near the western boundaries (Fig. 1.19).

Second, there is a major contrast between the Atlantic and Pacific Oceans. Sea surface temperature in the eastern and middle parts of the northern North Atlantic Ocean is about 5°C warmer than the zonal mean temperature. In comparison, sea surface temperature along the eastern boundary of the northern North Pacific Ocean is only slightly warmer than the zonal mean temperature. The dramatic difference in the zonal sea surface temperature distribution is one of the major indications of the vitally important role played by the thermohaline circulation in the modern climate system. In fact, there is a strong meridional overturning circulation (MOC) in the Atlantic Ocean, with deepwater formation and sinking at high northern latitudes, while there is no such circulation in the Pacific Ocean.

The sea surface salinity deviation from the zonal mean also shows a remarkable contrast between the Atlantic Ocean and the Pacific Ocean (Fig. 1.20a). In fact, sea surface salinity in the Atlantic Ocean, especially in the North Atlantic, is much higher than in the Pacific Ocean. In addition to the difference in sea surface temperature, this big difference in salinity between the Atlantic and Pacific Oceans is one of the most critical factors in regulating the global thermohaline circulation, and thus the climate system. There are complicated dynamics related to this salinity difference, and we will discuss the relevant issues in connection with thermohaline circulation.

In addition, sea surface salinity in the Bay of Bengal is the lowest, which is related to the strong precipitation and huge amount of river run-off. Salinity is very high in the Arabian Sea, due to the strong evaporation there; however, salinity in the South Indian Ocean is the lowest of all three main basins in the Southern Hemisphere, which may be due to the horizontal advection of low-salinity water from the Bay of Bengal and the Warm Pool.

At 600 m depth, the salinity contrast between the North Atlantic and North Pacific Oceans becomes even larger (Fig. 1.20b). However, in the Southern Hemisphere, salinity in the South Indian Ocean is the highest. There is an eastward extension of this high salinity, which is due to the strong advection of the ACC.

The difference between temperature and salinity gives rise to the difference in surface density, which is a vitally important factor regulating the wind-driven circulation and thermohaline circulation in the world's oceans. Although the effect of high salinity in the Atlantic Ocean is partially compensated by the high surface temperature, surface density in the Atlantic Ocean is much higher than in the Pacific Ocean. In particular, surface density in the latitudinal band of 60–70° N in the northern North Atlantic Ocean is more than 2 kg/m^3 higher than the northern North Pacific Ocean (Fig. 1.21). This is the most crucial dynamical factor regulating the deepwater formation in the Northern Hemisphere.

The mechanisms responsible for making these two oceans so different are the subject of the theory of thermohaline circulation. What is the primary reason for these differences? One of the major forces inducing such dramatic differences may be the hydrological cycle. The differences induced by the hydrological cycle are complicated and far-reaching, and involve complicated air–sea interactions. In addition, it is also closely related to the geometry of the land–sea distribution and the topography on land.

Examining the water vapor flux in the atmosphere reveals that the easterlies at low latitudes transport about 0.36 Sv of freshwater from the Atlantic Basin to the Pacific Basin across the narrow land bridge of Central America. Although this low-latitude water vapor transport may be partially compensated by the corresponding transport at high latitudes, there is still a substantial surplus of freshwater flux received by the Pacific Ocean. As shown in Figure 1.22, the Pacific Ocean receives a net freshwater flux, but the Atlantic Ocean loses freshwater through excessive evaporation; thus, water in the North Atlantic Ocean must be

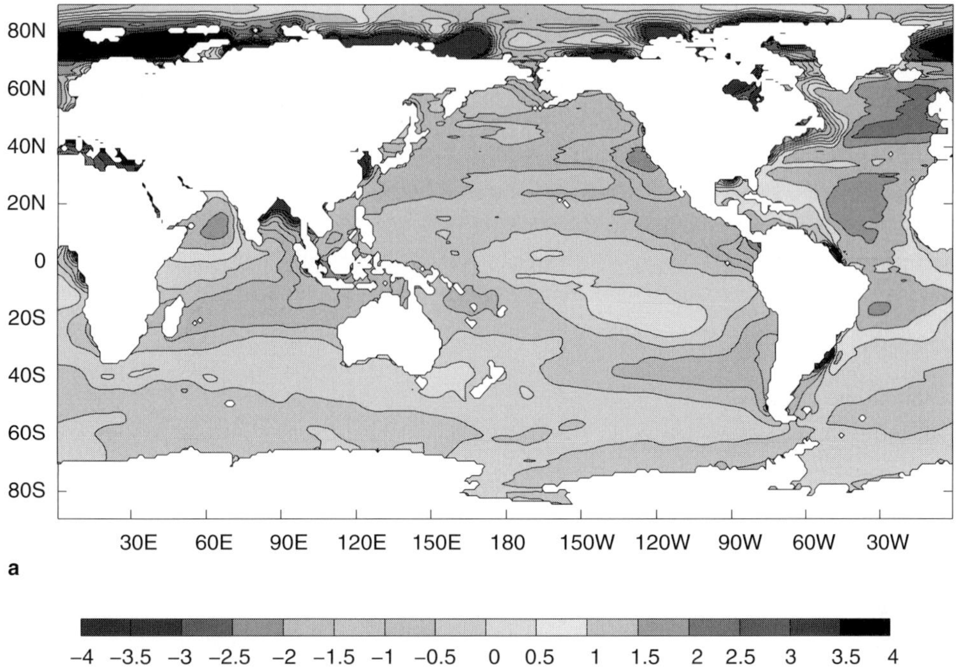

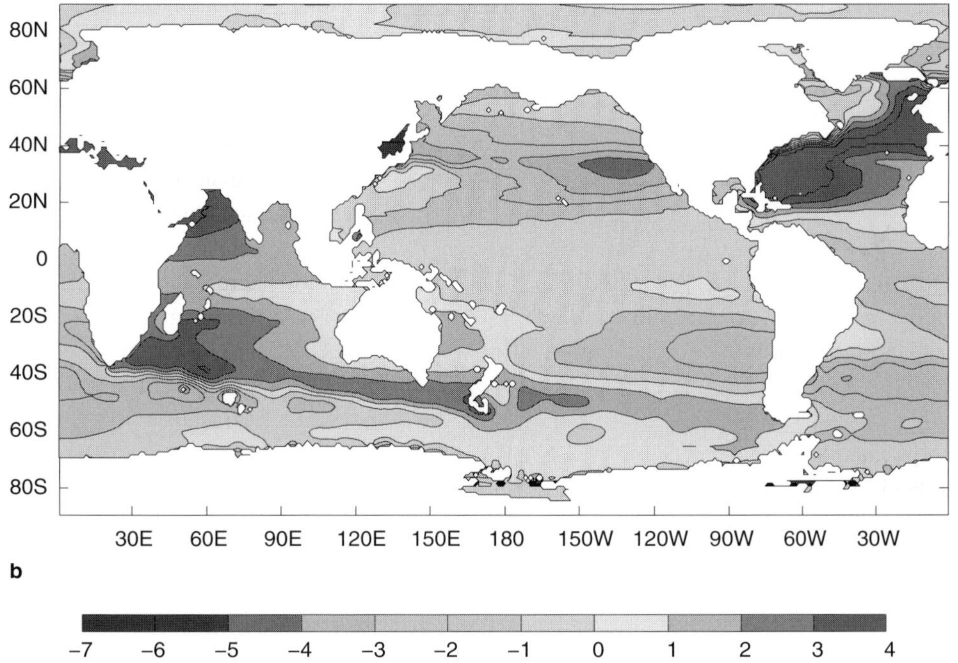

Fig. 1.20 Annual mean salinity anomaly; **a** at sea surface, **b** at depth of 600 m, deviation from the zonal mean. See color plate section.

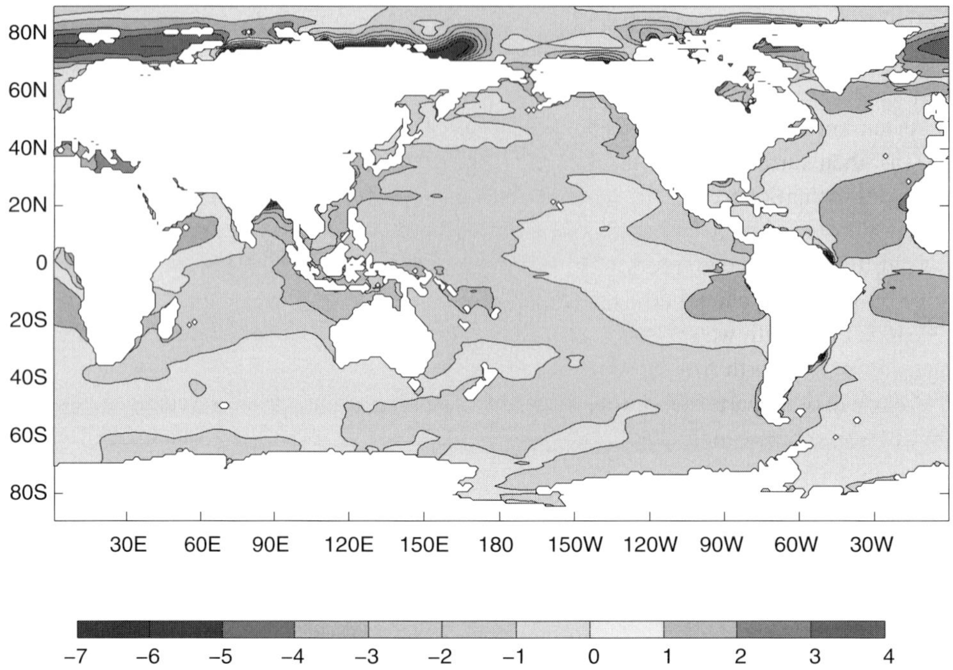

Fig. 1.21 Annual mean sea surface density anomaly, deviation from the zonal mean (kg/m^3). See color plate section.

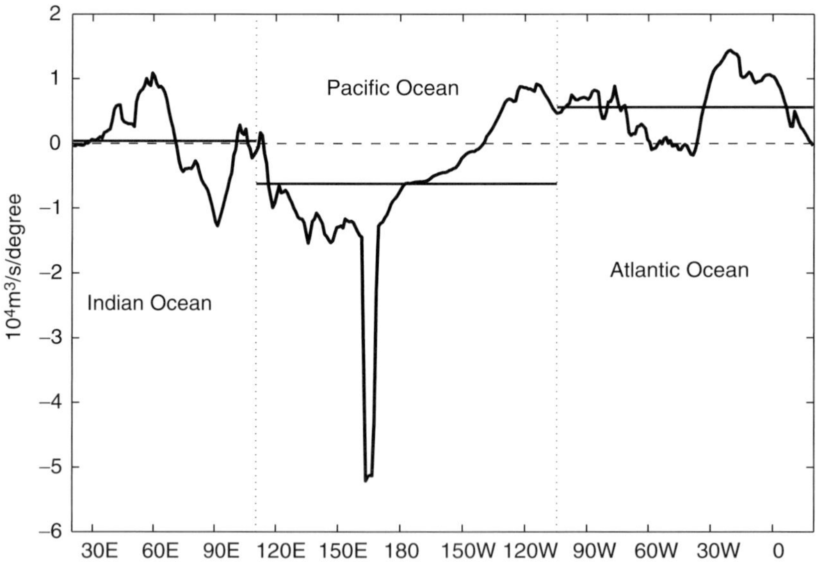

Fig. 1.22 Meridional integrated annual-mean net evaporation minus precipitation through the air–sea interface for each one-degree longitudinal band.

saltier than in the North Pacific Ocean. This difference is the most essential ingredient that makes the deepwater formation possible in the North Atlantic Ocean but not in the North Pacific Ocean.

Assuming that the sea surface temperature at high latitudes is the same in these two oceans, then surface water density is different in the two oceans. Since water is saltier in the North Atlantic Ocean, surface water becomes very dense after cooling, and it can sink to the bottom in the northern North Atlantic Ocean. On the other hand, surface water in the North Pacific Ocean is too fresh, so that when it is cooled, the corresponding density is still so light that it cannot sink to the bottom. Instead, the bottom part of the North Pacific Ocean is filled with water which is a mixture of bottom water from the Antarctic and deep water from the North Atlantic Ocean.

Although other parts of the atmosphere–ocean coupling system can be changed, the salinity difference between the Pacific and Atlantic Oceans may still exist, due to the easterlies at low latitudes under the modern land–sea distribution; therefore this effect will be a major factor in setting up any possible global circulation pattern.

1.2.2 Meridional distribution of temperature, salinity, and density

The focus of this section is the distribution of water mass properties through examination of the meridional distribution of temperature, salinity, and density in the world's oceans. Since seawater is slightly compressible, for a water parcel adiabatically moving down in the ocean its *in situ* temperature will gradually increase. Therefore, to subtract this warming effect due to compression, the so-called potential temperature Θ is widely used in the study of large-scale oceanic circulation and water properties. Potential temperature is the temperature that a water parcel would have if it were moved to the sea surface adiabatically, without exchanging heat and salt with the environment. The exact definition is discussed in Chapter 2.

In some sense the goal of physical oceanography is to describe the dynamics regulating the three-dimensional distribution of (Θ, S) (where S is salinity) in the world's oceans, and to predict the change in (Θ, S) properties associated with changes in external forcing conditions. Despite the efforts made over the last century, we are still a long way from reaching such a goal.

(Θ, S) distribution is closely related to oceanic current, climate, and global environment. As the world's oceans play a vital role in global climate changes, predicting changes in (Θ, S) distribution becomes one of the most exciting research frontiers. Scientifically, temperature and salinity distribution in the world's oceans is regulated by the wind-driven and thermohaline circulation. Water masses are primarily formed at the sea surface through air–sea interactions; subsequently, they are transported by currents in the oceans, and transformed through mixing in the oceanic interior. As the first step, we discuss the meridional distribution of (Θ, S) under the current climatic conditions.

Before we start to examine temperature and salinity distributions along meridional and zonal sections, it is important for the reader to remember that the property distributions in these vertical maps do not necessarily indicate the primary flow direction. It is necessary to emphasize that flows in the oceans are primarily in the quasi-horizontal planes in the form of nearly geostrophic currents and eddies, and that the vertical velocity with which water parcels glide up and down following the isopycnal surfaces is several orders of magnitude smaller than the horizontal geostrophic velocity. Therefore, when studying such vertical sections, the reader should be careful not to form the impression that water properties are transported along the same vertical planes.

Meridional distribution of potential temperature

The most outstanding features in the upper ocean are the bowl-shaped isothermals in the subtropical oceans around 30° off the equator and a few hundred meters below the sea surface, as can be seen clearly in the Atlantic section (Fig. 1.23), the Pacific section (Fig. 1.24), and the Indian section (Fig. 1.25). A close examination reveals the subsurface maximum of the vertical potential temperature gradient in this vicinity, which is called the main thermocline, or the permanent thermocline. The formation and maintenance of the main thermocline is the subject of the thermocline theory, which is directly related to the wind-driven circulation in the upper ocean.

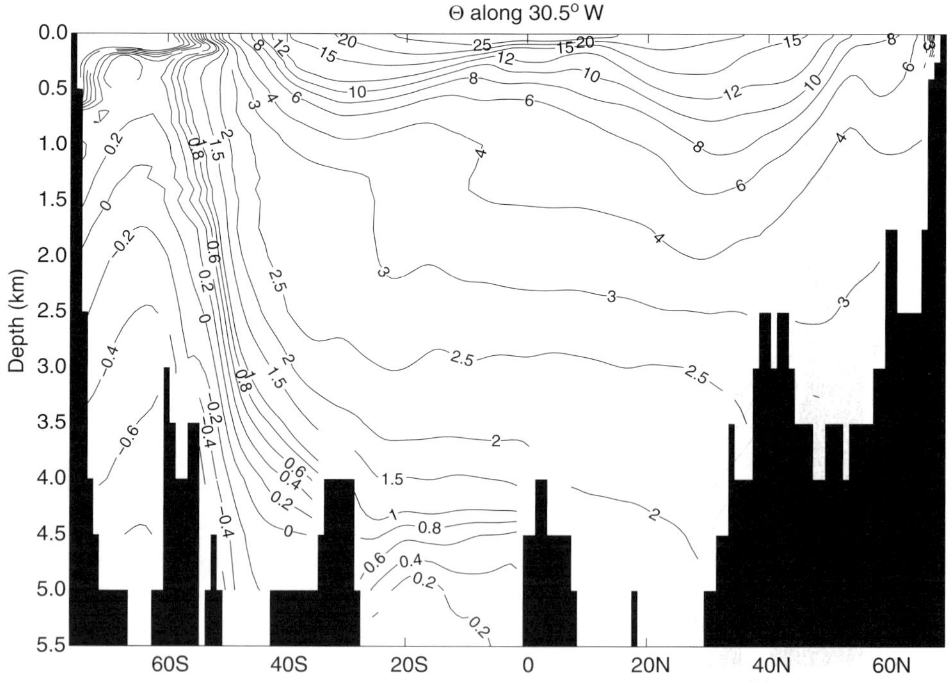

Fig. 1.23 Meridional section of potential temperature (°C) along 30.5° W (Atlantic Ocean).

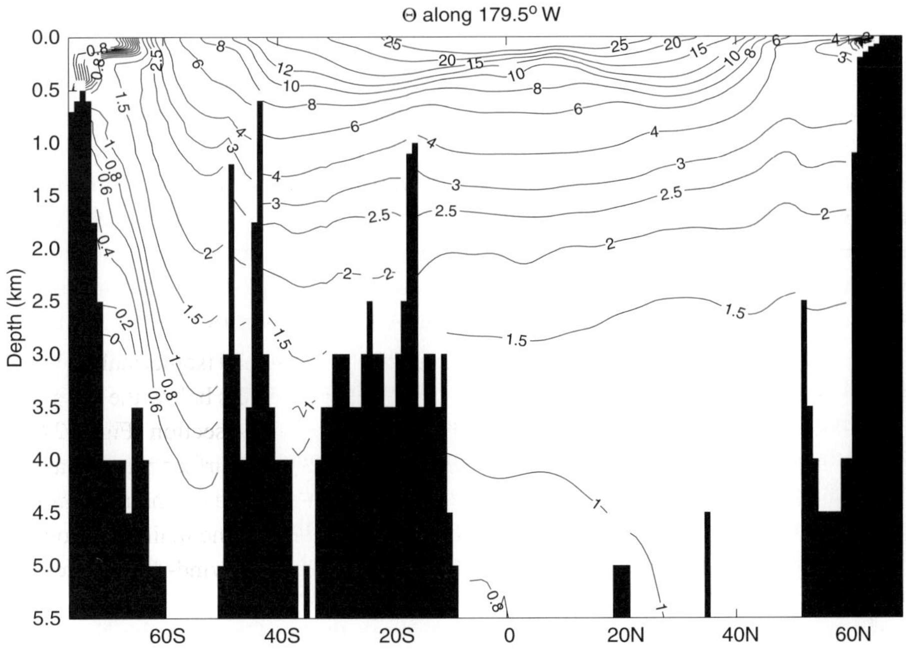

Fig. 1.24 Meridional section of potential temperature (°C) along 179.5° W (Pacific Ocean).

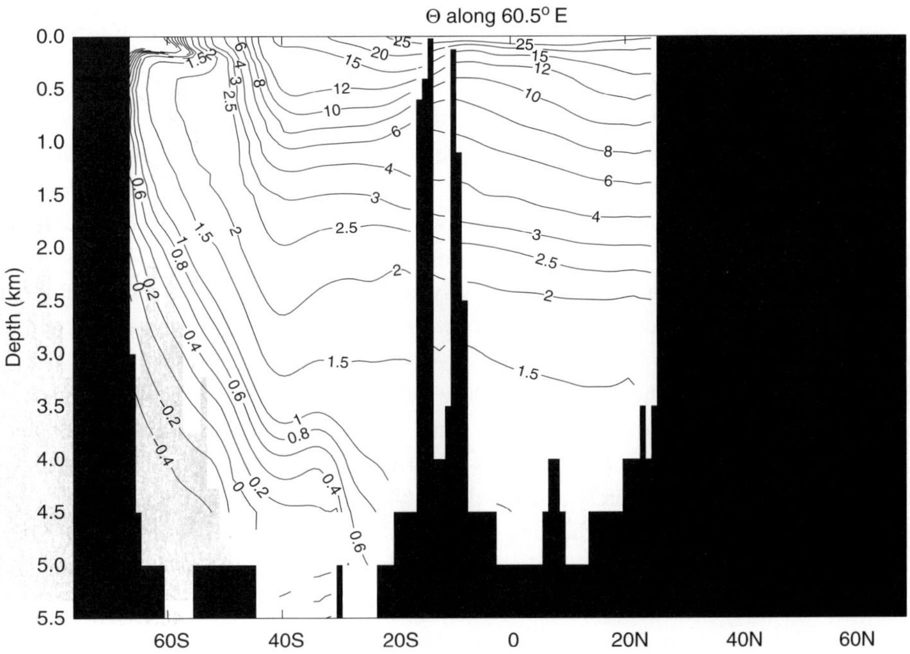

Fig. 1.25 Meridional section of potential temperature (°C) along 60.5° E (Indian Ocean).

Another major feature is the strong front from the sea surface to the deep ocean in the Southern Ocean. In the Atlantic section, this front is extremely strong and appears in the latitude band of 50–60° S (Fig. 1.23). In the Pacific section, this front appears in the latitude band of 60–70° S (Fig. 1.24). In the Indian section, this front is slightly more diffuse (Fig. 1.25). This temperature front is the most unique feature in the world's oceans under the modern land–sea distribution. There is no meridional land barrier at this latitudinal band in the Southern Ocean. As a result, a strong circumpolar current exists which acts as a barrier between the cold water from the Antarctic continent and the warm water from the north.

Cold water originating from Antarctica can be seen in the deep ocean in all three sections, with potential temperature as cold as −0.8°C in the Atlantic section and −0.4°C in the Indian section; however, deep water in the Pacific section is slightly warmer, with the coldest potential temperature being about 0°C. Cold water formed near the edge of Antarctica spreads northward as bottom water in the world's oceans; its presence can clearly be seen in the Atlantic section all the way to the equator. Of course, during this long-distance transportation deep water is gradually warmed up, and the bottom potential temperature near the equator in this Atlantic section is slightly warmer than 0°C.

In the northern North Atlantic Ocean, the upward tilting of the 3°C isothermal in latitudes higher than 50° N indicates a northern source of cold deepwater formation. In comparison, there is no sign of deepwater formation in the North Pacific Ocean or the North Indian Ocean. The formation of cold deep water at high latitudes in both hemispheres is directly linked to the thermohaline circulation in the world's oceans, and thus it is very closely linked to climate and its variability on the Earth.

Meridional distribution of salinity

Salinity is another major water property directly linked to the oceanic general circulation. In the North Atlantic Ocean, there is a tongue-like feature of high salinity, starting from the upper ocean and penetrating vertically to the depth of 2 km. At the 2 km level, the core of this high-salinity tongue extends southward and across the equator (Fig. 1.26). This salinity tongue is mostly the signature of the high-salinity Mediterranean Water. Since it is a subsurface salinity maximum, it is not directly connected to the surface salinity along this section. In fact, it is produced by a westward lateral transport of high salinity at a depth of roughly 1 km.

In the Southern Hemisphere, an outstanding tongue of low salinity, originating from the sea surface at 50° S, extends to a depth of 1 km. This low salinity is a major indicator of Antarctic Intermediate Water (AAIW). This feature extends all the way to the equator, with the core remaining approximately at 1 km depth.

In the depth range of 2–4 km, the high salinity originates from the North Atlantic Ocean, and the relatively low salinity below the level of 4 km is associated with Antarctic Bottom Water (AABW) formed near the edge of the Antarctic continent.

In some of the classical papers published early in the development of physical oceanography, these tongue-like features had been used as indicators for water movement; however,

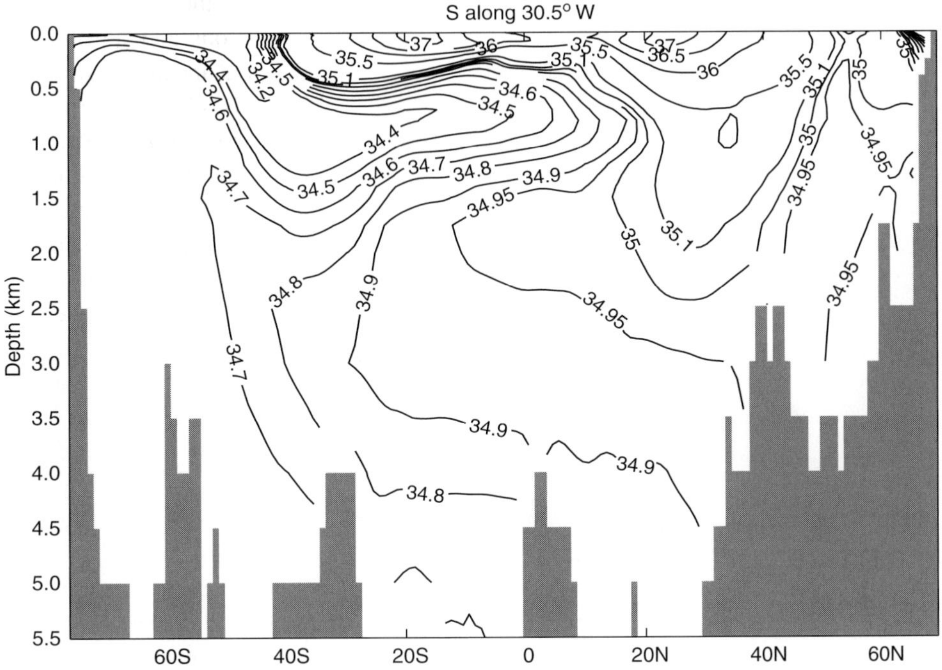

Fig. 1.26 Meridional section of salinity along 30.5° W (Atlantic Ocean).

the distribution of salinity and other tracers is the result of many complex dynamical processes, including advection/diffusion due to the mean current and meso-scale eddies. Thus, these tongue-like features cannot be interpreted as the direction of the mean flow.

In the 179.5° W section in the Pacific Ocean (Fig. 1.27), the salinity distribution in the upper 400 m in the subtropics is dominated by relatively high salinity; however, between 500 m and 1.5 km the salinity distribution is dominated by tongues of relatively low salinity from high latitudes. In comparison with the Atlantic section, the low-salinity tongue from the south is less prominent. In contrast to the Atlantic section, here the AABW appears as slightly saltier water spreading northward in the deep ocean below 4 km.

In the Indian section (Fig. 1.28), the low-salinity tongue associated with AAIW is the dominant feature in the Southern Hemisphere, similar to the Atlantic section. The relatively high-salinity intrusion from the north dominates the depth between 2 and 4 km in the Southern Hemisphere. The AABW appears as the relatively low-salinity layer extending from the Antarctic northward on the seafloor. In the North Indian Ocean, the influence of high-salinity water from the Red Sea can be seen clearly at the depth of 1 km.

Meridional distribution of density

Although temperature and salinity are most important in oceanic circulation, the density of seawater is directly related to the pressure gradient force and thus the velocity field. Since

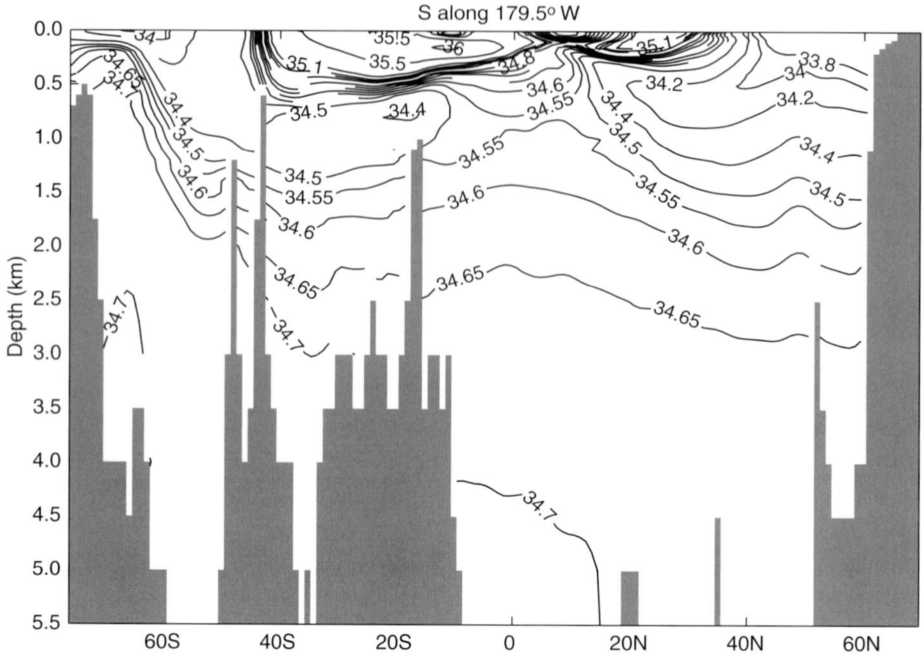

Fig. 1.27 Meridional section of salinity along 179.5° W (Pacific Ocean).

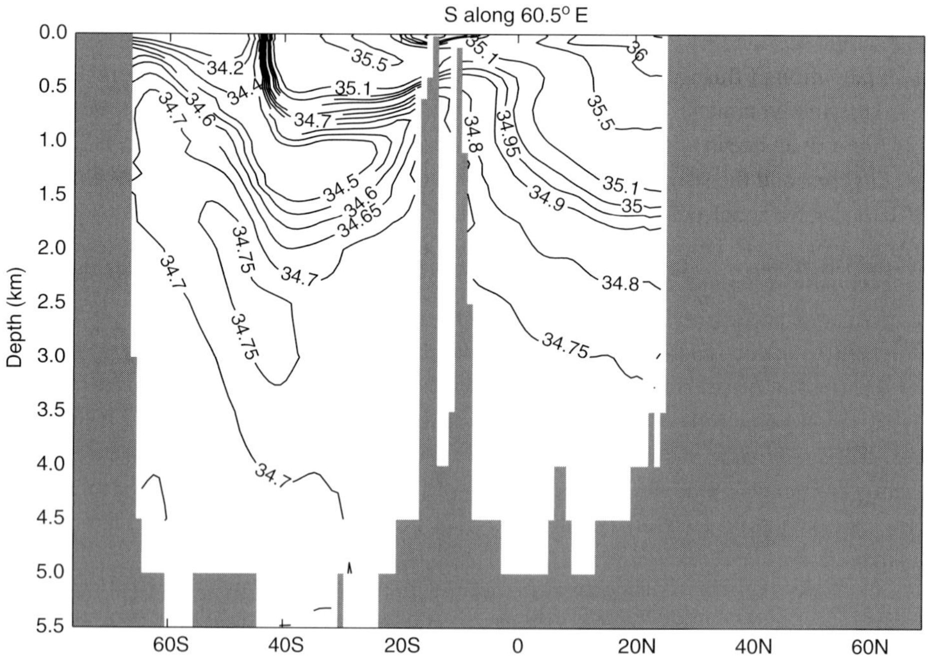

Fig. 1.28 Meridional section of salinity along 60.5° E (Indian Ocean).

seawater is slightly compressible, density increases with depth. However, a large part of the density increase in the vertical direction is dynamically inert, so that another quantity, potential density, which is quite similar to potential temperature, has been widely used in dynamical oceanography. Potential density is the density that a water parcel would have if it were moved to the sea surface adiabatically, without exchange of heat and salt with the environment; the exact definition of potential density is discussed in Chapter 2.

Before we start to examine density distribution along meridional and zonal sections, it is of utmost importance again to remind the reader that isopycnal surfaces in these maps do not necessarily mean the primary flow direction. In fact the primary form of property transportation is taking place in the nearly horizontal trajectories defined by a strong horizontal geostrophic current, with a relatively small component of vertical velocity along isopycnal surfaces, and an even smaller component of the so-called diapycnal velocity, which is the velocity going through the isopycnal surface in the perpendicular direction.

In oceanography, density or potential density is often given as an anomaly from 1,000 kg/m^3. In general, potential density is within the range of 20–28 σ units (kg/m^3). At temperatures near freezing point, the thermal expansion coefficient is nearly zero. Since the saline contraction coefficient is almost constant, at low temperature, density is primarily controlled by salinity. However, for the high temperature range, the density of seawater is primarily controlled by temperature.

The increase of potential density with depth indicates that water columns are generally stable. There is, however, a seemingly unstable feature at 3–4 km depth near the equator in the Atlantic section (Fig. 1.29). In fact, the water column is quite stable in this vicinity, and the explanation of this apparently unstable stratification will be presented in the discussion of the thermodynamics of seawater in Chapter 2.

In the upper ocean of the subtropics, the bowl-shaped isopycnals indicate the wind-driven gyres, and the strong density fronts on the northern and southern edges of the bowls indicate strong zonal currents associated with these wind-driven gyres. In the equatorial regime, isopycnals below the top 500 m and above 2.5 km are mostly flat, suggesting relatively slow zonal currents in this depth range.

There are strong density fronts in the Southern Ocean, especially in the upper 1.5 km in the Atlantic section: the left margin in Figure 1.29. However, the density front below this depth is more diffuse than the temperature front shown in Figure 1.23. The weakening of the density front at mid depth is due to partial compensation of density by the salinity front (Fig. 1.26) here. The small area at the northern end of this section with a high potential density of $\sigma_0 = 27.88$ kg/m^3 indicates the overflow of deep water formed in the Nordic seas, while the high potential density on the seafloor in the Southern Hemisphere indicates dense bottom water originating from Antarctica.

In the Pacific section, there are also clear signs of wind-driven gyres in the subtropics of both hemispheres, with the sloping isopycnals on the edges of the gyres indicating strong zonal currents on both sides of the bowl-shaped gyres (Fig. 1.30).

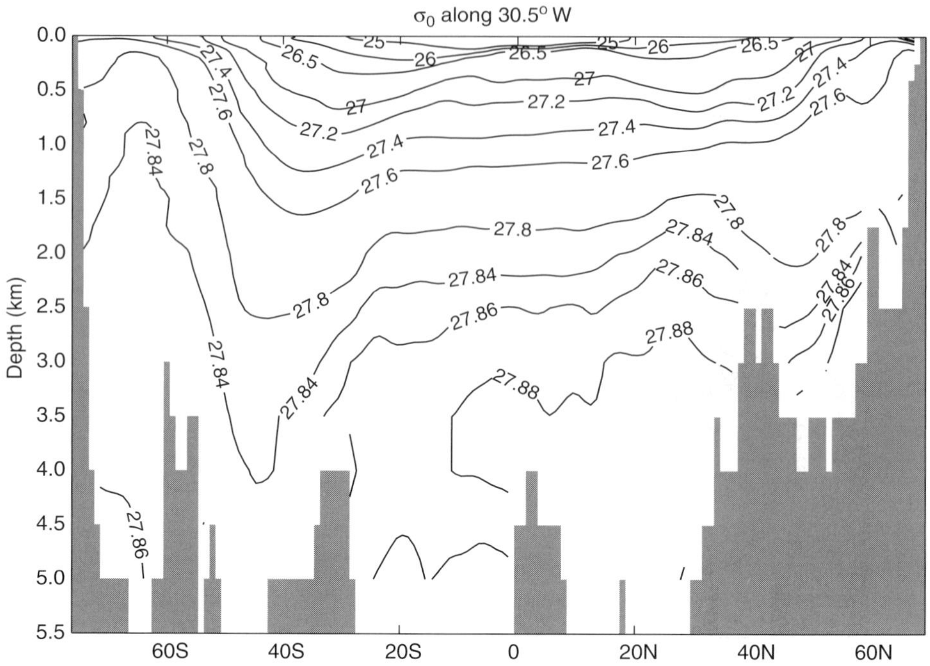

Fig. 1.29 Meridional section of potential density (kg/m^3) along 30.5° W (Atlantic Ocean).

The isopycnal front in the Southern Ocean in the Pacific section seems much stronger than that in the Atlantic section shown in Fig. 1.29. The reason for this difference is due to different salinity patterns in these two sections. In contrast to the Atlantic section, here the salinity gradient gives rise to the density gradient, which is of the same sign as that due to temperature; thus, the density front here is quite strong down to 3 km depth. From geostrophy, this strong density front should correspond to a strong zonal velocity shear at this depth.

The high density in the deep ocean in the Southern Hemisphere is directly related to the AABW. In addition, the high-density water near the equator is a clear sign of the influence of AABW from the south. The lack of dense water in the high-latitude Northern Hemisphere indicates the lack of deepwater formation in the North Pacific Ocean.

In the Indian section, there is also a clear sign of a wind-driven gyre in the subtropics of the Southern Hemisphere, with the sloping isopycnals on the edges of the gyre indicating strong zonal currents on both sides of the bowl-shaped gyres (Fig. 1.31).

The isopycnal front in the Southern Ocean in this section seems more diffuse than that in the Pacific section shown in Figure 1.30. Similar to the Atlantic section, but in contrast to the Pacific section, here the salinity gradient gives rise to a density gradient of sign opposite to that due to temperature; thus, the density front here is quite weak below 1.5 km. From geostrophy, this weak density front should correspond to a relatively weak zonal velocity shear at this depth range.

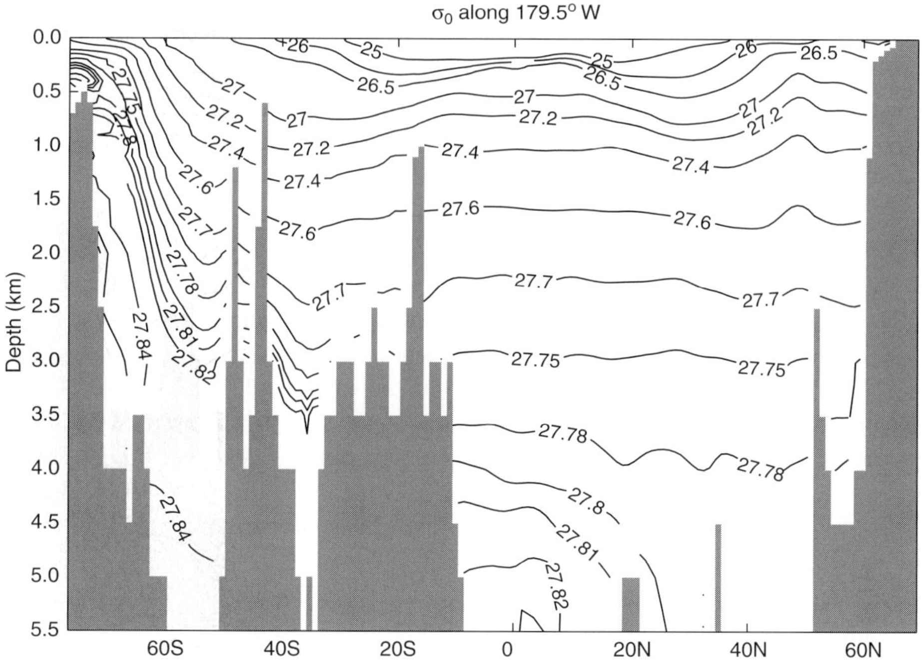

Fig. 1.30 Meridional section of potential density (kg/m^3) along 179.5° W (Pacific Ocean).

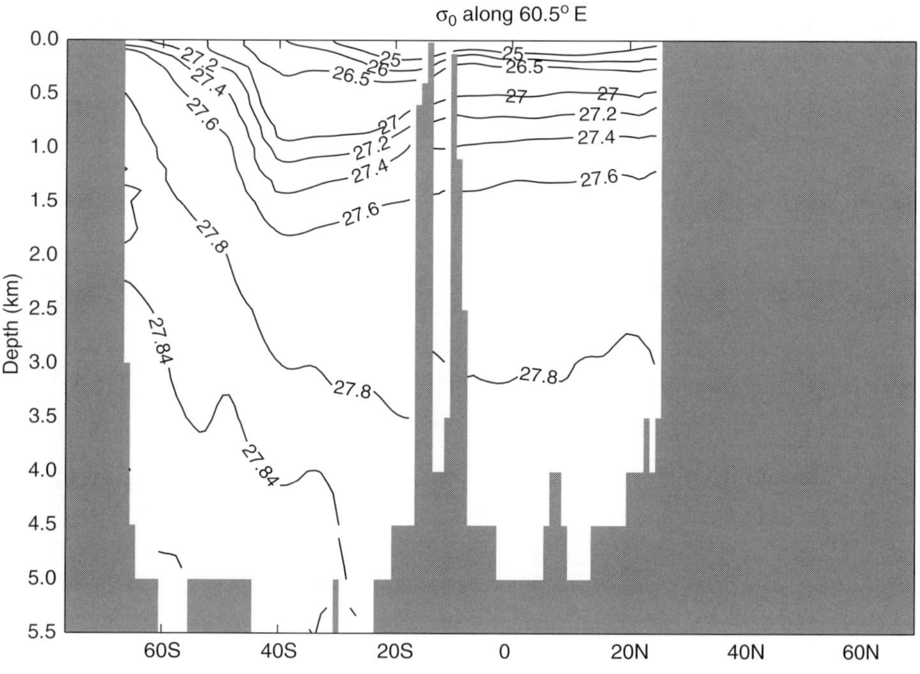

Fig. 1.31 Meridional section of potential density (kg/m^3) along 60.5° W (Indian Ocean).

1.2.3 Distribution of potential temperature, salinity, and density in the Southern Ocean

The Southern Ocean is characterized by the existence of a circumpolar current system, which separates the cold water from the south and the warm water from the north. As an example, water property distributions for a zonal section along 58.5° S are shown in the following figures. The low potential temperature and salinity in the Atlantic sector between 30° W and 30° E indicate the influence of AABW, with its major source in the Weddell Sea (located at 30–60° W) (Figs. 1.32 and 1.33).

The relatively warm water in the upper ocean near the longitudinal band of 60° E, 180° E, and 90° W implies the relatively small influence of cold water upwelling originating from North Atlantic Deep Water (NADW) in both the Pacific and Indian Oceans.

Salinity distribution along this section is dominated by freshwater from the sea surface and the relatively fresh AABW originating from the Antarctic continent (Fig. 1.33). The relatively warm and fresh water features near 60° E, 140° E, 180° W, and 90° W shown in Figures 1.2.32 and 1.33 are related to the meridional meanderings of the ACC associated with the major topographic features in the Southern Ocean.

The high-density water found in the Atlantic sector is again a clear sign of the cold dense bottom water originating from the Weddell Sea (Fig. 1.34). The most interesting features in

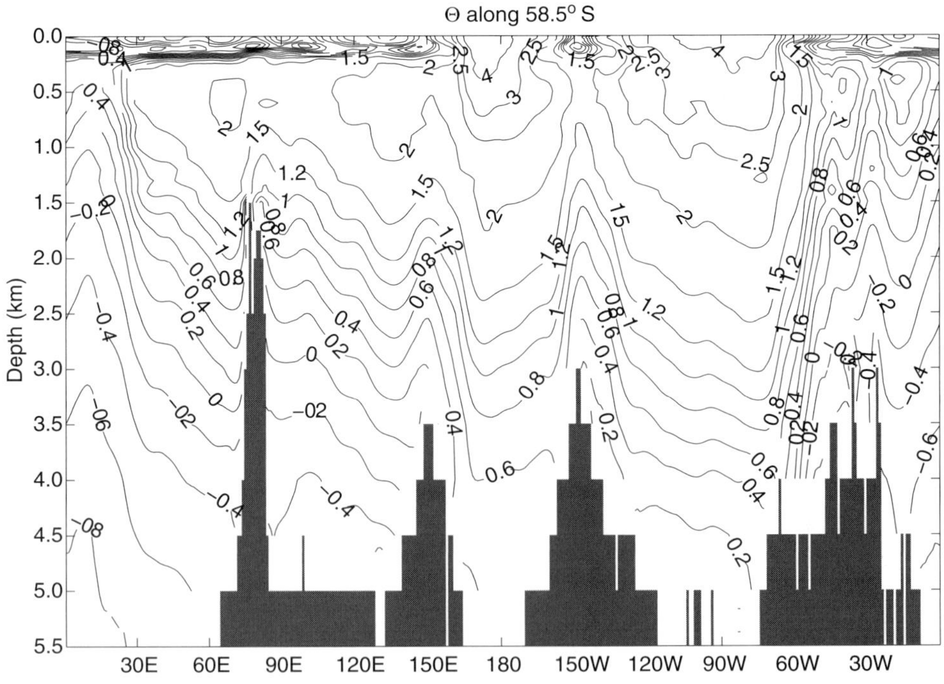

Fig. 1.32 Zonal section of potential temperature (°C) along 58.5° S (Southern Ocean).

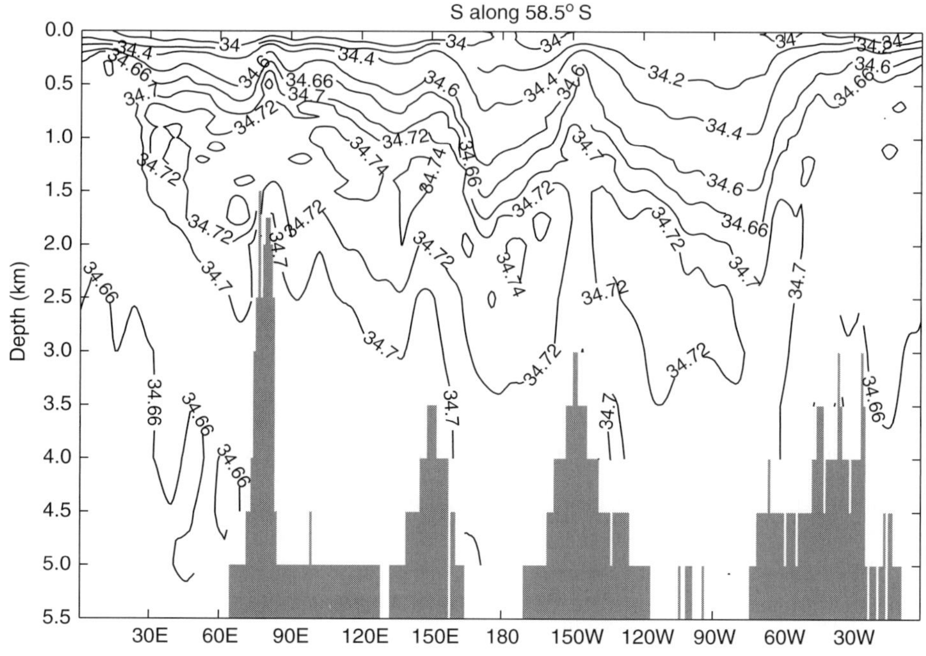

Fig. 1.33 Zonal section of salinity along 58.5° S (Southern Ocean).

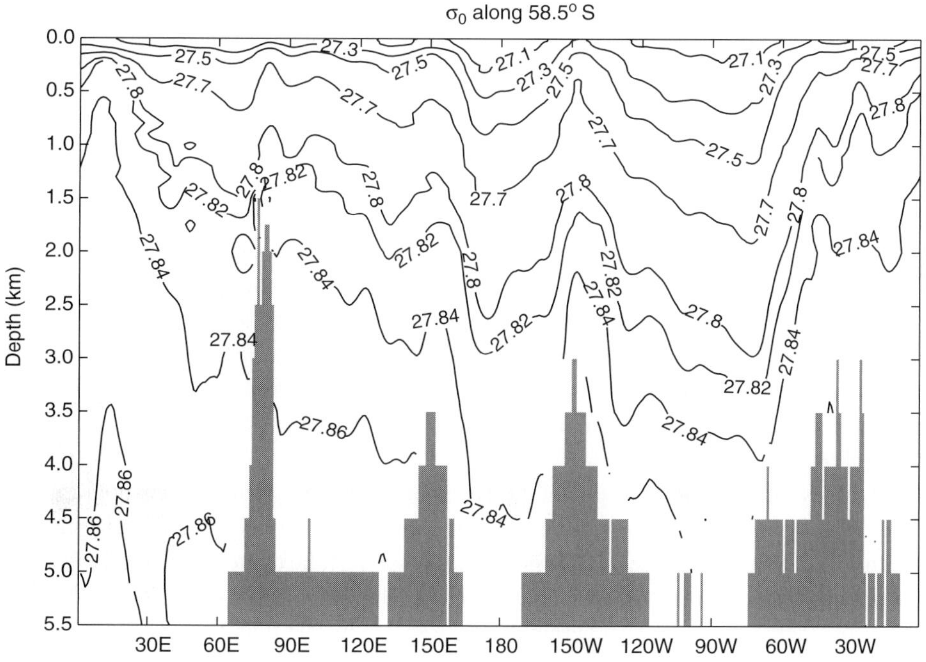

Fig. 1.34 Zonal section of potential density (kg/m^3) along 58.5° S (Southern Ocean).

the zonal density map are the strong density fronts associated with the major topographic features in the Southern Ocean. Since large-scale oceanic circulation should satisfy the geostrophy constraint, these meridional density fronts imply strong meridional velocity fronts in the Southern Ocean associated with these major topographic features. The lack of any meridional land barrier makes the establishment of a zonal pressure difference in this latitudinal band impossible. As a result, there would be no geostrophic current in the meridional direction if there were no bottom topography. However, the existence of major topographic features in the Southern Ocean induces some major, permanent, eddy-like features, which play an important role in the circulation at this latitudinal band.

1.3 Various types of motion in the oceans

1.3.1 Introduction

Motions in the world's oceans cover a tremendous range of spatial and temporal scales. These motions can be classified as turbulence, waves, and currents.

At the small-scale (down to the order of millimeters) and high-frequency end of the spectrum, motions in the oceans can be classified as turbulence, surface and internal gravity waves. Although these motions also interact with the large-scale motions, many aspects of their study differ from that of large-scale dynamics; and therefore they are not the focus of this book.

At the other end of the spectrum, there are large-scale (on the order of 1 km to 10,000 km) waves and currents, as shown in Figure 1.35. (Note that although internal waves can have a horizontal scale of more than 1 km, we do not include them among the large-scale motions discussed in this book.) These motions are further classified into four main categories.

Tides

The study of tides is probably one of the most familiar branches of physical science. Mankind started to observe tidal motions at the beginning of civilization. In general, tidal motions involve spatial scales of the whole basin. Owing to the complicated shapes of coastline and bottom topography, tidal motions can also involve spatial scales much smaller than basin-scale; however, such relatively small-scale motions are related to the basin-scale tidal motions in the open oceans. Although most components of the tides have a frequency on the order of a day, some components can have much longer periods, such as the annual and semi-annual solar tides.

The theory of tidal motion is one of the oldest branches of dynamical oceanography. The modern theory of tidal motions started from the classical work of Newton's (1687) theory of equilibrium tides and Laplace's (1775) formulation of the tidal equation. Since tidal motions are primarily controlled by astronomical forces and the shape of coastline and seafloor, they can be considered nearly time-invariant for time scales shorter than centennial. In fact, tidal tables for the world's oceans have been printed for many years into the future.

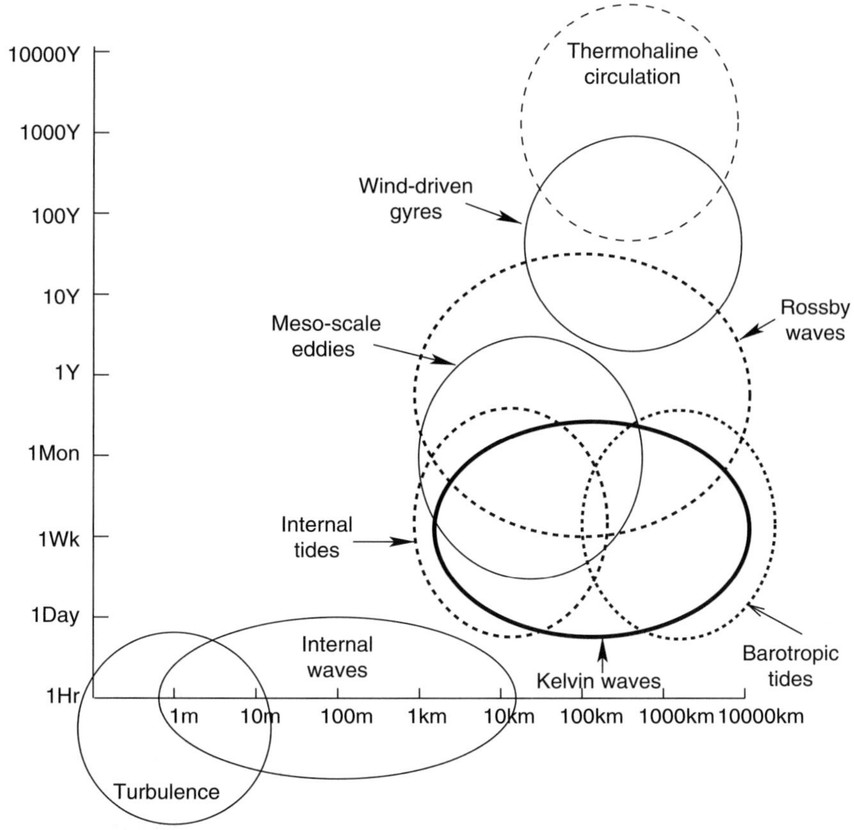

Fig. 1.35 Various types of motion in the oceans.

Large-scale and low-frequency waves

These include Rossby waves and Kelvin waves, and they play a crucial role in the time evolution of the oceanic circulation. The major difference between these waves and the commonly encountered gravity waves is that these waves are characterized by the fact that the Earth's rotation is their restoring force. The existence of these waves can be detected either through satellite measurements or by *in situ* observations.

Meso-scale eddies

These are the most energetic component of the oceanic circulation; 99 percent of the total kinetic energy in the oceans belongs to meso-scale (or synoptical scale) eddies, which can also be called geostrophic turbulence. They play an essential role in the energy cascade in the oceans. However, our knowledge about the meso-scale eddies in the oceans remains very incomplete because there are not enough data. Although satellite altimetry has provided a wealth of data for the surface expression of meso-scale eddies, observing the meso-scale eddies in the vast oceans remains a great challenge for technology. With the deployment of the Argo floats in the world's oceans, we may see much better data coverage in the near future.

Quasi-steady current systems

This category includes wind-driven circulation and thermohaline circulation. Large-scale motions in the oceans are organized into circulation systems, which are regulated by various forces. These circulation systems play an essential role in setting up the global environments and climate, and are the main focus of this book.

It is very important to recognize that the oceanic circulation is a complicated dynamical system; thus, the separation into the above four categories of motion should be treated with caution because the differences between these different types of motion can be fuzzy. In fact, all these components interact in very complicated and nonlinear ways. It is a common practice to study the theories of individual components of the system; however, such theories should be combined with study about the interactions between the different components of the circulation system.

1.3.2 Two types of circulation

By excluding motions associated with waves and eddies, large-scale flows in the oceans can be separated into three components: tidal flows, wind-driven circulation, and thermohaline circulation. Clearly, such a classification is somewhat subjective because there must be non-linear interactions between these components. Since the theory of tidal motions, especially the barotropic tides, has been well established, we will confine our discussion here to the other two components.

For example, circulation in the Atlantic Basin can be separated into three major parts: the wind-driven gyres in the upper ocean, the meridional overturning cells associated with deep/bottom water formation and spreading, and the Antarctic Circumpolar Current (ACC) system. North of 30° S, motions in the Atlantic Basin can be further separated into the following components. At the surface, there is the wind-stress-dominated Ekman layer, with wind-driven gyres below this layer. Motions at depths below the wind-driven gyres are typically classified as part of the thermohaline circulation (Fig. 1.36). The separation of wind-driven gyre and the thermohaline circulation is, of course, artificial; but such a separation can help us to simplify the problems and thus obtain a clear picture of the physical processes involved in each of these components.

The Antarctic Circumpolar Current system is a complicated combination of both the wind-driven and thermohaline circulation. However, it does not appear as a horizontal gyre; instead, it is the only circum-Earth current in the world's oceans under the current geometric and climatic conditions. It plays a unique role as the global artery, feeding bottom water to the world's oceans and receiving water at intermediate levels from other oceans.

This book focuses on the theories of wind-driven circulation and thermohaline circulation in the world's oceans. These two types of motion share some similarities, but they also differ

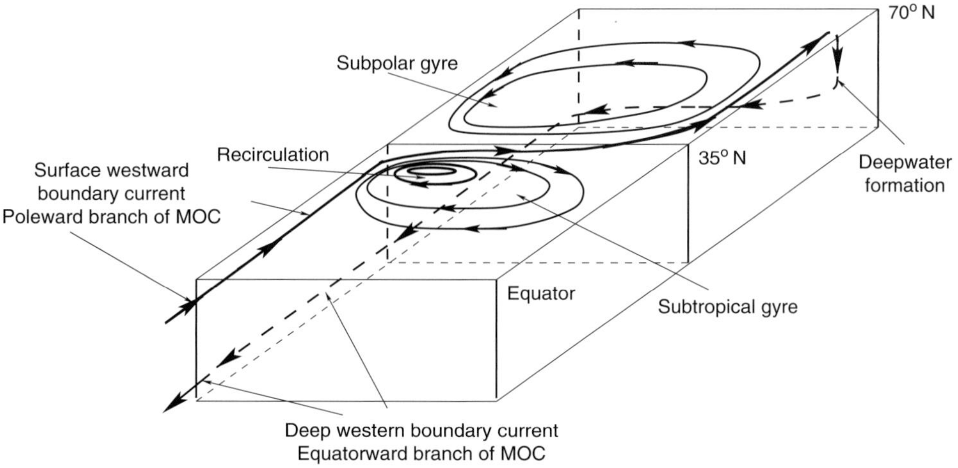

Fig. 1.36 Schematic view of the circulation in the North Atlantic Ocean, including the subtropical wind-driven gyre, the subpolar wind-driven gyre, the recirculation, and meridional overturning cell (MOC) associated with the thermohaline circulation.

in many fundamental ways. Table 1.1 shows some of the most essential characteristics of these two types of circulation. This simple table is not intended as an accurate description of the characteristics of the circulations; instead, it offers the reader concise, conceptual descriptions of these motions.

The wind-driven circulation

Wind-driven circulation generally refers to the circulation in the upper ocean ($\sim$1 km) which is primarily driven by wind stress.

Wind-driven circulation in a homogeneous ocean

Wind stress can drive horizontal circulation in a homogeneous ocean; thus, the existence of wind-driven circulation is independent of the surface thermohaline forcing. Assuming that the circulation is steady and transport in the bottom Ekman layer is negligible, the vertically integrated flow in the ocean interior should satisfy the Sverdrup constraint, which is defined more accurately in Chapters 2 and 4.

Wind-driven circulation in a stratified ocean

If diapycnal mixing across the main thermocline is negligible and the abyss is almost motionless, the vertically integrated steady horizontal circulation in the interior upper ocean should satisfy the Sverdrup constraint. There is vertical shear of the horizontal velocity, which is intimately related to the horizontal density gradient in the upper ocean. The theory relevant to the density structure and the horizontal velocity field is the so-called ventilated

Table 1.1. *Comparison between the wind-driven circulation and thermohaline circulation; volume fluxes are in Sv ($1\ Sv = 10^6 m^3/s$)*

Type of circulation		Wind-driven	Thermohaline
Primary driving forces		Wind stress on sea surface	Wind stress and tidal dissipation
Secondary driving forces (Preconditions for the circulation)		Thermohaline circulation to set up the background stratification	Surface heat and freshwater fluxes
Primary motions	Plane	Horizontal	Vertical
	Form	Horizontal gyration	Meridional overturning cell
	Volume flux	50–100 Sv	~10 Sv
Secondary motions	Plane	Vertical	Horizontal
	Driving force	Ekman pumping	Source/sink due to water mass formation
	Volume flux	~10 Sv	~10 Sv
Vertical scale		The upper 1,000 m	Over the whole depth of oceans
Lowest-order dynamics		Ideal fluid	Mixing driving
Water mass formation and erosion	Formation	Subduction in the subtropical gyre	Deepwater formation at high latitudes
	Erosion/transformation	Obduction in the subpolar gyre	Water mass erosion/transformation at low latitudes and abyss
Dynamics	Theory	Sverdrupian	To be developed
	Formulas	$\beta hv = fw_e$	Scaling laws
	Uncertainty of the theory	Small	The direction of meridional overturning can be flipped

thermocline theory. However, the ventilated thermocline theory assumes some kind of background stratification which is presumably set up by thermohaline circulation external to the wind-driven circulation. In this sense, the so-called wind-driven circulation in a stratified ocean represents the combined results of both wind stress and thermohaline circulation.

Variability

The wind-driven gyre is established through the westward propagation of Rossby waves. Since at mid latitude the first baroclinic mode takes about 10–20 years to cross the subtropical basins, wind-driven circulation can vary on decadal time scales.

How well do we understand the wind-driven circulation?

The circulation in the upper ocean has been observed over the past half-century. Since the wind-driven circulation varies on decadal time scales, there is now a major effort to assemble a large set of data which can provide useful insights into the variability of the wind-driven circulation.

The wind-driven circulation theories discussed in this book form a collection of the developments over the past decades. These theories are mostly based on ideal-fluid models, and the roles of meso-scale eddies are omitted. Therefore, these theories can only provide some conceptual framework which can help us to understand the complicated circulation systems in the oceans. We will emphasize the development of the ventilated thermocline in the 1980s, which has provided the theoretical cornerstone for the theoretical framework of the modern theory of wind-driven circulation in the oceans. In particular, the models with multiple layers or continuous stratification provide a clear dynamical picture of the wind-driven gyres. Furthermore, the ideal-fluid thermocline theory can also be used as a convenient tool for understanding the variability of the wind-driven circulation in the upper ocean.

Thermohaline circulation

The term thermohaline circulation has been widely used in oceanography and climate study. This term generally refers to the circulation associated with differences in temperature and salinity in the ocean, although the exact definition remains debatable, as will be explained shortly.

The classical definitions

The classical view of thermohaline circulation has been discussed in many books and papers, but the exact definition remains unclear. Wunsch (2002) wrote:

A reading of the literature on climate and the ocean suggests at least seven different, and inconsistent, definitions of term "thermohaline circulation":

a) the circulation of mass, heat, and salt;
b) the abyssal circulation;
c) the meridional overturning circulation of mass;
d) the global conveyor, that is, the diffusively defined gross property movements in the ocean that together carry heat and moisture from low to high latitudes;

e) the circulation driven by surface buoyancy forcing;
f) the circulation driven by density and/or pressure differences in the deep oceans;
g) the net export, by the North Atlantic Ocean, of a chemical substance such as the element protactinium.

As it turns out, none of those definitions is suitable for describing the thermohaline circulation in the oceans.

The new definition

Thermohaline circulation is a circulation driven by mechanical stirring, which transports mass, heat, freshwater and other properties in the meridional/zonal direction. Mechanical stirring is supported by external sources of mechanical energy from wind stress and tidal dissipation. In addition, surface heat and freshwater fluxes are necessary for setting up the circulation. A more refined discussion about the thermohaline circulation will be presented in subsequent sections.

- *Is there a pure thermohaline circulation?*
  Although surface thermohaline forcing alone may be able to drive the so-called pure thermohaline circulation, such a circulation is so weak that it is irrelevant to the circulation observed in the world's oceans. The meaning of this statement is implicitly included in the new definition discussed above, and this will be discussed in detail in connection with the energetics of the world's oceans and the thermohaline processes in the ocean.
- *Variability of the thermohaline circulation*
  Common wisdom may suggest that oceanic circulation does not change with time; however, oceanic circulation does vary greatly over long time scales. There is evidence from paleoproxies that thermohaline circulation and the associated water mass formation/erosion has been through great changes on decadal to centennial and millennial time scales.
- *How well do we understand thermohaline circulation?*
  The major problem associated with the thermohaline circulation is that it varies on time scales of centennial to millennial. Theories of oceanic circulation and numerical models need to be verified by observations. Thus far our understanding of thermohaline circulation under climate conditions different from the present comes mostly from incomplete sources, including paleoproxies, theories, and numerical simulations. It is clear that theories based on such an incomplete foundation are rudimentary. Despite the long-term efforts in collecting data for the circulation during the past, we do not have enough data to accurately describe thermohaline circulation under climatic conditions different from the present. This very unsatisfactory situation is due to the tremendously broad range in both temporal and spatial scales involved in the thermohaline circulation.

The unified picture

The two types of circulation discussed above are combined together in the oceans. For example, the circulation system in the North Atlantic Ocean is shown in Figure 1.36. There are the wind-driven subtropical gyre and the subpolar gyre, and the strong surface western boundary currents associated with them. In addition, in the northwest corner of the subtropical gyre there is a regime of recirculation.

Superimposed on these two gyre systems, there is the meridional overturning circulation (MOC), with a warm surface current moving across the equator and joining the surface western boundary current. At the boundary between the subtropical and subpolar gyres, they move eastward and turn northward in the eastern basin. This corresponds to the North Atlantic Current in the Atlantic Ocean. At the northeast corner, this current is cooled down and sinks, feeding the deep water. The deep water moves toward the western boundary, and from there it flows equatorward as the deep western boundary current.

The Gulf Stream is a complicated current system that combines contributions from three major components in the upper ocean. The first contribution is due to the return flow of the linear Sverdrup dynamics, about 25 Sv. The second contribution is associated with the so-called recirculation. Due to the combination of eddies and bottom pressure torque, the volumetric flux of the recirculation is on the order of 100 Sv. The third contribution is due to the northward surface branch of the meridional overturning cell, which may contribute as much as 15 Sv.

In the North Pacific Ocean, the meridional overturning cell is very weak or non-existent; thus the transport of the Kuroshio system does not include a significant contribution from the meridional thermohaline circulation.

1.4 A survey of oceanic circulation theory

1.4.1 Introduction

It is well known that oceanic circulation is driven by wind stress, heat flux, and freshwater flux associated with evaporation, precipitation, and river run-off. In addition, tidal dissipation also contributes. Although mankind started to observe the ocean circulation many centuries ago, the development of the modern dynamical theory of the oceanic circulation is relatively new. Thus far, our understanding of the oceanic general circulation remains rudimentary in comparison with the theory of atmospheric circulation, which has been developed to a much more complete stage during the last century.

In his famous monograph about atmospheric general circulation, Lorenz (1967) reviewed the historical development of the relevant theory. The Hadley cell and the Jet Stream in both the Northern and Southern hemispheres characterize circulation in the atmosphere. In a crude way, the circulation in the atmosphere can be described in terms of an axisymmetric circulation.

On the other hand, circulation in the oceans is much more complicated. Due to the existence of meridional boundaries associated with the major continents, individual basins are bounded by the coastlines. As a result, oceanic general circulation appears as separate gyres and meridional overturning cells in individual basins, with the only exception being that, under the modern geometry of land–sea distribution, there is a circum-Earth current system, the ACC, which resembles the circulation in the atmosphere.

As we will see, the existence of the eastern/western boundaries is one of the major differences between atmospheric and oceanic circulation. These meridional boundaries in

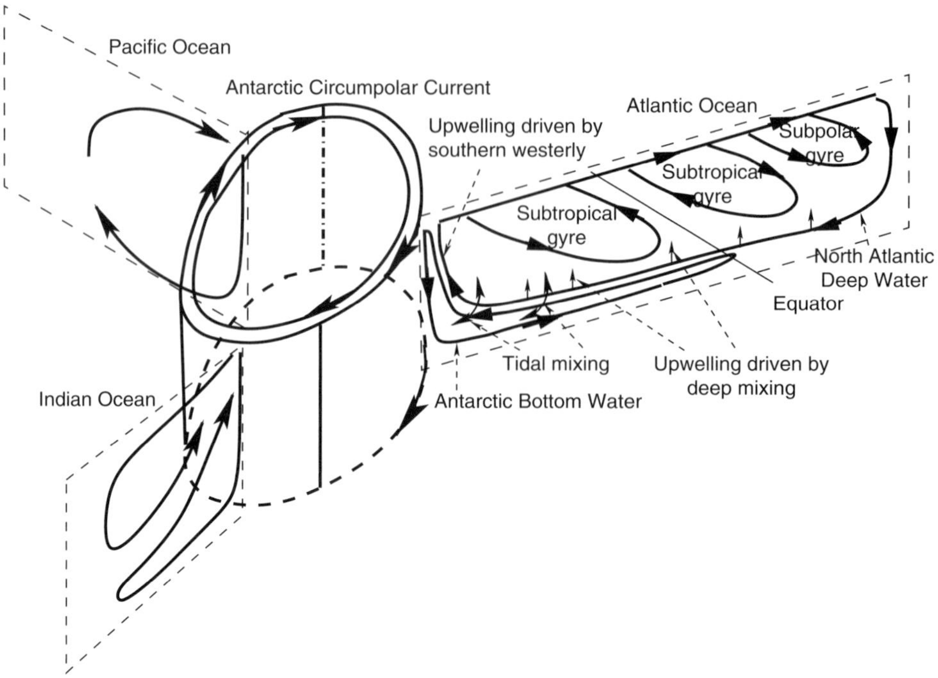

Fig. 1.37 Sketch of the circulation system in the Atlantic Basin, including three horizontal wind-driven gyres in the upper ocean, two meridional overturning cells, the Antarctic Circumpolar Current; the corresponding circulation details in the Pacific and Indian Oceans are omitted.

each basin create the east–west pressure gradient, and thus make the existence of meridional geostrophic currents in each basin possible. As an example, a sketch of the circulation in the Atlantic Basin, including the ACC, is depicted in Figure 1.37. A more comprehensive review of the general circulation in the world's oceans can be found in the excellent monographs by Schmitz (1996a, b).

The circulation system in the Atlantic Basin is a combination of meridional cells and horizontal gyres. There are three major gyres in the upper ocean: a subtropical gyre and a subpolar gyre in the North Atlantic Ocean, plus a subtropical gyre in the South Atlantic Ocean. The existence of these gyres is primarily due to wind stress on the ocean, and the structure of these gyres is the main focus of the wind-driven circulation theory. There are also small tropical gyres near the equator, but they are less well defined, so we will not discuss them here. There are two major meridional overturning cells in the Atlantic Ocean – the main cell associated with the NADW and a deep cell associated with the AABW. In the traditional theory of thermohaline circulation, these two cells are considered to be driven by NADW and AABW respectively. As will be explained later, however, deepwater formation cannot provide the mechanical energy required for sustaining the circulation against friction and dissipation. In fact, strong upwelling driven by the Southern Westerlies and deep mixing driven by tidal dissipation, as depicted in Figure 1.37, are the most important mechanisms sustaining the thermohaline circulation.

Finally, the ACC is a combination of the wind-driven circulation and the thermohaline circulation; it is the global artery and plays the most critical role in the global oceanic circulation and climate. The dynamical structure of the ACC and its maintenance are rather complicated. In simple words, this current system is a result of the interaction between the wind-driven circulation and the thermohaline circulation.

Our knowledge about the oceanic circulation has evolved at a rather slow pace, due to the difficulties in observing the oceanic general circulation, a turbulent system with extremely broad spatial and temporal scales. Although we have seen great progress in understanding the structure of the global oceans, many essential aspects of this global picture remain unclear, and it is one of the most crucial and exciting research frontiers.

In the early stages, our knowledge of oceanic general circulation was mostly observational. *The Oceans*, written by Harald Sverdrup, Martin Johnson, and Richard Fleming (Sverdrup *et al.*, 1942) is one of the classic books on oceanography. An historical article on oceanic circulation was written by Henry Stommel (1957), in which he summarized the theories of oceanic circulation at that time. During the past 50 years there have been many major breakthroughs in the theory of oceanic general circulation, and the following account is a concise history of modern theories of large-scale dynamical oceanography.

1.4.2 Thermal structure and circulation in the upper ocean

The main thermocline is one of the most outstanding features in the world's oceans, and it can be easily identified from hydrographic sections. In this section, the climatology of Levitus *et al.* (1998) is used to diagnose the thermal structure in the world's oceans. As an example, the structure of both temperature and density for the upper ocean along 158.5° E is shown in Figure 1.38. By definition, the thermocline is a thin layer where the vertical gradient of temperature is a local maximum. There are many types of thermocline, including the diurnal thermocline, the seasonal thermocline, the main thermocline, and the abyssal thermocline. The diurnal thermocline exists in the top layer of the upper ocean, and it is closely related to the diurnal cycle. The seasonal thermocline exists in the upper hundred meters of the ocean, and is closely related to the seasonal cycle in the upper ocean. The main thermocline exists within the depth range of 100–800 m. Because it is far away from the sea surface and shielded from the direct forcing in the seasonal cycle, it is also called the permanent thermocline. The abyssal thermocline exists in the deep ocean, which will be discussed in Section 5.2, which is concerned with deep circulation.

Since motions in the ocean are intimately related to density, the pycnocline, defined as a subsurface layer of local maximum of a vertical density gradient, may be more important dynamically. However, in most cases, the contribution of salinity to the density is much smaller than that due to temperature, and therefore the thermocline and the pycnocline are closely linked to each other. In many studies people have used the term thermocline when it would have been more accurate to use the term pycnocline. Since thermal (or density)

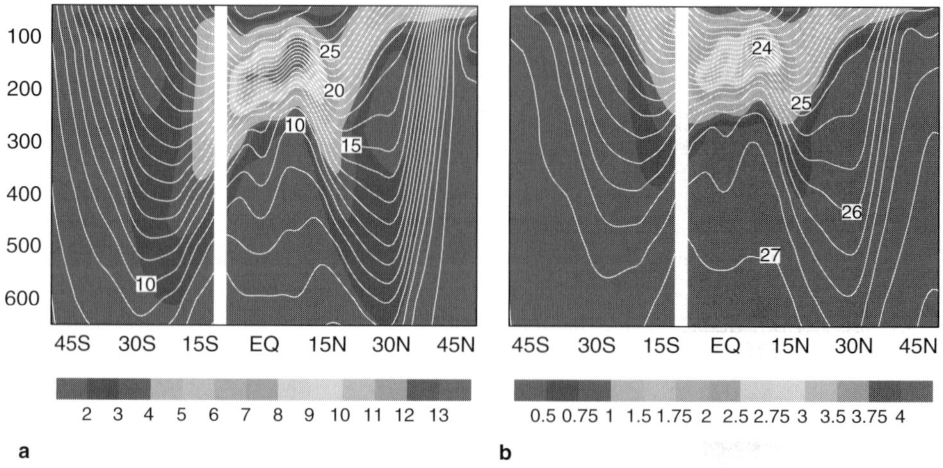

Fig. 1.38 Meridional section (along 158.5° E; vertical axis depicts depth (m)) **a**, thermal structure (contours) (°C); and **b**, stratification (contours) (kg/m^3), overlaid with the vertical gradient in the colored map. See color plate section.

structure is intimately associated with the circulation, the theory of the thermocline is also a theory for the circulation in the upper ocean.

Note that both the main thermocline and the main pycnocline are conceptual features defined for the basin scale; thus, for a given station or section, the depth of these layers may not be well defined. At this section (158.5° E), the main thermocline can be readily identified from the left panel of Figure 1.38. It is close to the 20°C isothermal surface in the equatorial ocean, and it gradually shifts to a lower temperature of 11–12°C at mid latitudes. The main thermocline is clearly asymmetric with respect to the equator, which indicates that forcing and boundary conditions for the wind-driven circulation are asymmetric. On the other hand, the main pycnocline is not clearly defined for this section; see the right panel of Figure 1.38.

The depth of the main thermocline varies greatly with geographic location, Figure 1.39. It is rather shallow near the eastern boundaries of the equatorial regimes due to equatorial upwelling driven by the easterlies at low latitudes. It is also shallow along the eastern boundaries of the Southern Hemisphere, owing to the strong coastal upwelling driven by the along-shore component of the trade wind. Thus, the shallowness of the main thermocline in these parts of the ocean is associated with the relatively cold surface temperature induced by local wind-driven upwelling, and these areas are called the "cold tongues" in the oceans.

The main thermocline in the western part of the equatorial oceans is deeper than in the eastern part because warm water is piled up under the equatorial easterly. In the Pacific Ocean this body of warm water is called the Warm Pool. Both the Warm Pool and the Cold Tongue play vitally important roles in the global climate system, especially in the dynamics of the El Niño–Southern Oscillation (ENSO).

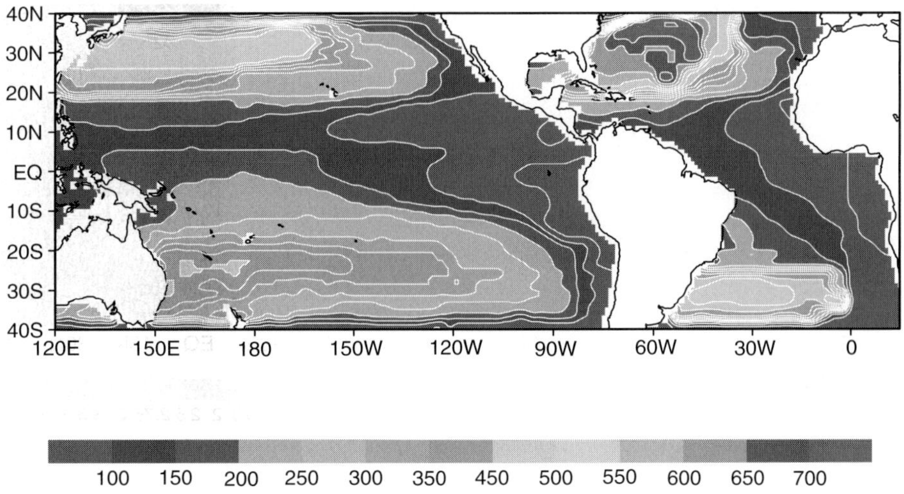

Fig. 1.39 Depth of the main thermocline (m) for the Pacific and Atlantic Oceans. See color plate section.

The thermocline is much deeper at mid latitudes, primarily due to the downward pushing associated with the negative wind stress curl in the subtropics. In the western part of the subtropical gyre of the North Atlantic Ocean (North Pacific Ocean), it can reach 800 m (550 m) depth (Fig. 1.39). The thermocline in the Southern Hemisphere is relatively shallower, and is about 500 m (450 m) for the South Atlantic Ocean (South Pacific Ocean). The difference in thermocline depth reflects the difference in wind stress forcing and the stratification in different oceans. As will be explained in the study of wind-driven circulation, the square of the thermocline depth is inversely proportional to the stratification. In the North Atlantic Ocean, stratification is relatively weak, due to high salinity induced by strong evaporation. In comparison, low surface salinity associated with excessive precipitation in the North Pacific Ocean gives rise to strong stratification. As a result, the thermocline in the North Atlantic Ocean is much deeper than in other oceans.

The isopycnal surfaces representing the main thermocline outcrop along the boundary between the subtropical and subpolar gyres in the Northern Hemisphere; thus, there is no main thermocline in subpolar basins. In the Southern Hemisphere, the main thermocline outcrops along the northern edge of the ACC. In both cases, the main thermocline (main pycnocline) appears as a strong front associated with strong current systems that are mostly zonally oriented.

The temperature of the main thermocline varies greatly, depending on location. For example, the temperature of the main thermocline is about 21°C at the western part of the equatorial Pacific Ocean, and it gradually shifts toward 18°C near the eastern equatorial Pacific Ocean where the cold tongue exists (Fig. 1.40). At mid latitudes, the main thermocline corresponds to a much lower temperature, about 8–10°C for the subtropical North Pacific Ocean and about 10–12°C for the subtropical North Atlantic Ocean.

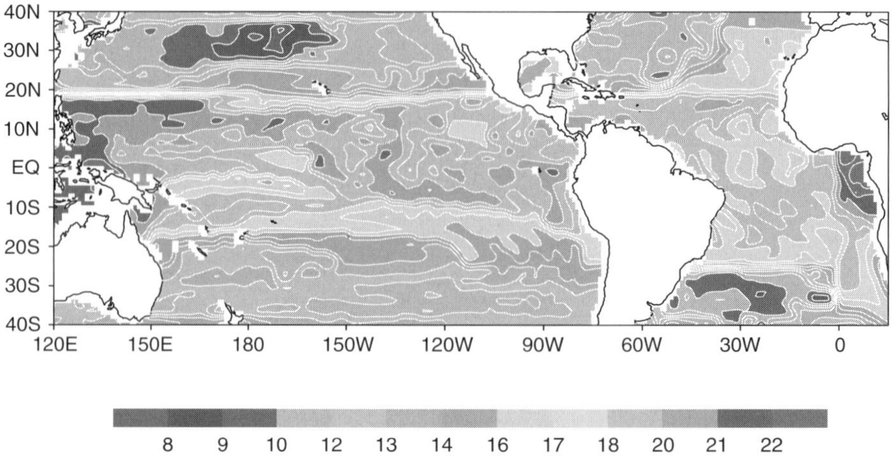

Fig. 1.40 The temperature (°C) of the main thermocline of the Pacific and Atlantic Oceans. See color plate section.

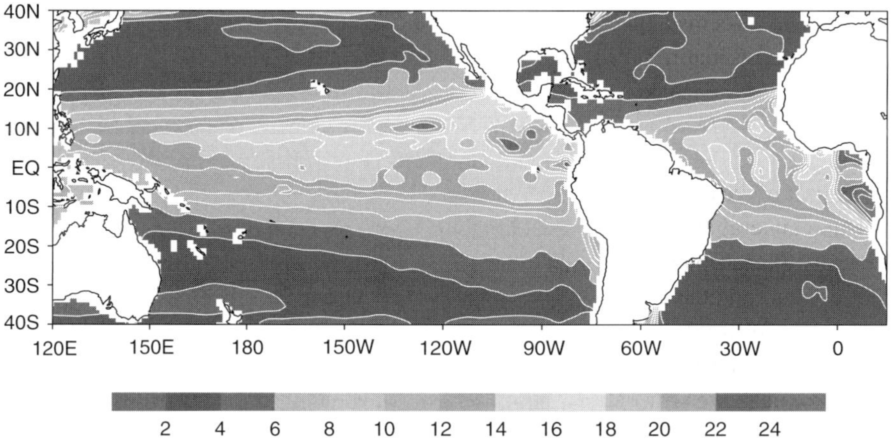

Fig. 1.41 The temperature gradient (°C/100m) of the main thermocline in the Pacific and Atlantic Oceans. See color plate section.

The vertical temperature gradient of the main thermocline also varies greatly over the world's oceans. Within the subtropical gyres, the gradient is on the order of 2–4°C/100 m; however, it is much larger in the equatorial ocean, varying from 10°C/100 m in the western part to 20°C/100m in the eastern part of the equatorial Pacific Ocean (Fig. 1.41).

1.4.3 Early theories for the wind-driven circulation

Ekman theory

In the early stages, our knowledge of oceanic general circulation was mostly observational. The first milestone was a paper by Ekman (1905), in which he discussed the structure of the wind-driven circulation in the surface boundary layer. According to his theory, the velocity in the boundary layer should have a spiral structure, the vertically integrated volume transport is $\tau/f\rho_0$, and it points 90° to the right of the wind stress (in the Northern Hemisphere). This layer is now called the Ekman layer; the flux in this layer is called the Ekman flux, forming the theoretical foundation of modern wind-driven circulation theories.

The Ekman layer and its associated spiral velocity profile in the atmospheric boundary layer can be readily observed. In fact, the beautiful Ekman spiral can be demonstrated by releasing a line with many balloons attached. It took a long time before the Ekman theory could be verified in the ocean. The major difficulty in the ocean is the presence of strong surface waves in the upper ocean. However, after a long time delay, the Ekman spiral in the upper ocean was finally confirmed through *in situ* measurements (Price *et al.*, 1987). The structure of the stratified Ekman layer is much more complicated. For the most up-to-date information, the reader is referred to Price and Sundermeyer (1999).

Before the 1940s our knowledge of oceanic circulation theory was confined to simple dynamical calculations of currents based on a level of no motion, the Ekman layer, waves, and tides. For example, *The Oceans*, written by Sverdrup *et al.* (1942), is an amazing summary of the state of the art of oceanography in the early 1940s. At the time *The Oceans* was published, it seems that a lot was known about the oceanic circulation. Thus, the book was a rather intimidating collection of knowledge, as Stommel (1984b) recalled in his autobiography.

Sverdrup's theory

Over the past 60 years there have been many major breakthroughs in our understanding of the oceanic general circulation. The second milestone in wind-driven circulation theory is the work by Sverdrup (1947), in which he established the simple relationship between the wind stress curl and the circulation in the basin interior.

In the atmosphere, both westerlies and easterlies are the necessary components of the atmospheric circulation. The westerlies at mid latitudes and the easterlies at low latitudes give rise to a negative wind stress curl in the subtropics, which drives an equatorward flow in the ocean interior. In order to find the circulation in the basin, Sverdrup integrated the wind stress curl westward, starting with a no-zonal flux condition at the eastern boundary. His solution did not include the flow near the western boundary. Apparently, at that time it was unclear how to deal with the western boundary and why the integration should be started from the eastern boundary.

Theories of western intensification

The existence of swift currents, such as the Gulf Stream, near the western boundary of the basin was recognized through frequent trade activity between Europe and the American continent several hundred years ago. However, a dynamical explanation for such swift currents was first established only in the late 1940s, when Stommel (1948) studied an idealized model for the North Atlantic Ocean, including bottom friction and the latitudinal change of the Coriolis parameter, which is now called the β effect. In fact, Stommel was not aware of Sverdrup's theory because it had been published in a journal (*Proceedings of the National Academy of Sciences of the USA*) rarely used by oceanographers.

Other dynamical explanations of the western boundary current appeared shortly afterward, including a theory of the western boundary layer with lateral friction by Munk (1950), and the inertial western boundary layer theory by Charney (1955) and Morgan (1956).

Homogeneous model or reduced-gravity model

The common feature of these models is that they all treat the wind-driven oceanic circulation as a single moving layer. Since seawater is nearly constant, the simplest way to simulate the ocean circulation is to assume that the ocean is homogeneous in density. Such an ocean would have no vertical structure. The model used by Stommel (1948) belongs to this category.

Another possible approach is to assume that the wind-driven circulation is confined to the upper layer of the ocean; thus, the circulation can be treated in terms of a single moving layer model, such as those of Sverdrup (1947) and Munk (1950). This is called the reduced-gravity model. The essence of a reduced-gravity model is to treat the main thermocline (or the pycnocline) in the oceans as a step function in density, so that density in the upper layer equals a constant ρ and density in the lower layer is $\rho + \Delta\rho$. Furthermore, the lower layer is assumed to be infinitely deep, so the pressure gradient in the lower layer is infinitely small and the corresponding volume transport negligible. As a good approximation, we can assume that the lower layer is motionless; thus we actually deal with a single moving layer. The pressure gradient in the upper layer is described by $\nabla p/\rho = g'\nabla h$, where h is the upper layer depth and $g' = g\Delta\rho/\rho$ is called a reduced gravity, which is on the order of 0.01–0.02 m/s^2.

The basic idea is demonstrated in Figure 1.42, where the structure of the water column is depicted. The seasonal thermocline (pycnocline) near the upper surface can clearly be seen. The main thermocline and the main pycnocline are located at about the same depth (800 m); this is due to the fact that the contribution of salinity to the density stratification is relatively small at this location (and many other locations). The density structure is now represented in terms of two layers of constant density, as shown by the heavy lines in Figure 1.42b. The lower layer extends all the way to the seafloor. Since this layer is very thick, the horizontal pressure gradient and velocity in this layer are very small and negligible. Thus, the reduced-gravity model can be used to capture the first baroclinic mode of the circulation and the depth of the main thermocline.

1.4.4 Theoretical framework for the barotropic circulation

The backbone of theories of the barotropic circulation is the barotropic potential vorticity constraint. Hough (1897) devoted a minor portion of his tidal study to the currents produced by a zonally distributed evaporation and precipitation, ignoring friction. He found that a uniformly accelerated system of purely east–west geostrophic currents would exist. His model had several limitations. First, there was no friction in the model, so that he was unable to obtain a steady solution, nor was he able to discern the long-term effect of evaporation and precipitation. Second, his model had no meridional boundary, which is a crucial constraint

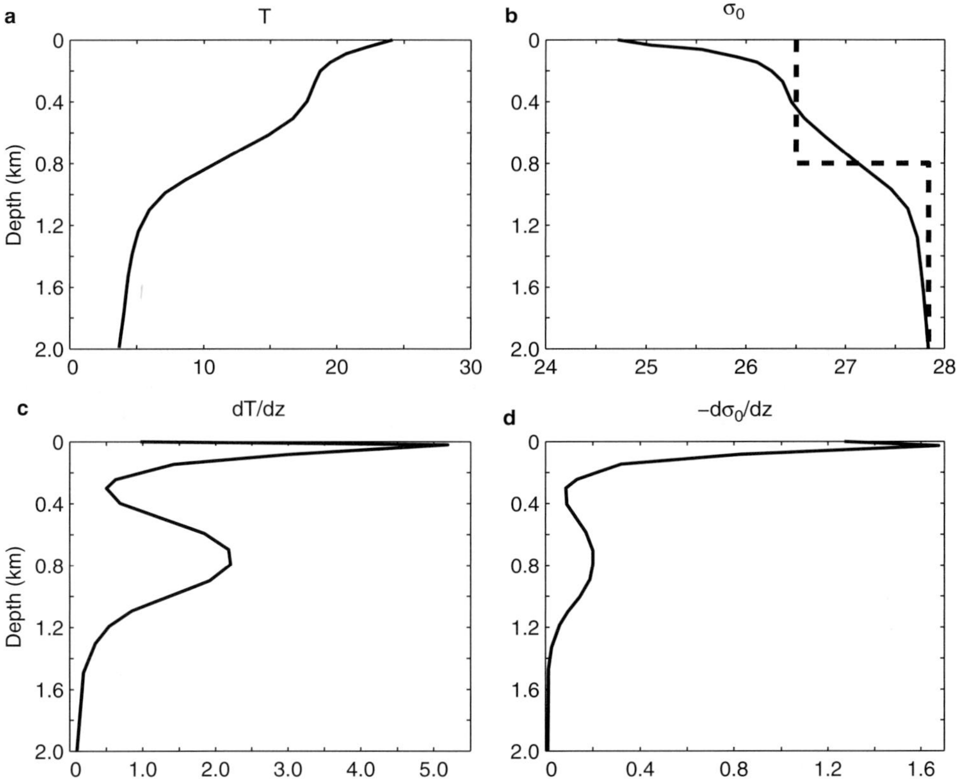

Fig. 1.42 Temperature and stratification at a station (70.5° W, 30.5° N) from annual mean climatology: **a** Temperature (T, °C); **b** Stratification (σ_0, kg/m^3), the heavy lines indicate the equivalent stratification used in a reduced-gravity model; **c** dT/dz (0.01°C/m); **d** $-d\sigma_0/dz$ (0.01 kg/m^4).

on the oceanic circulation. Hough published his results without even a figure to show the structure of the solution; thus, his solution remained unnoticed until Stommel (1957) publicized this solution with a beautiful illustration.

Goldsbrough (1933) discussed an ocean model forced by evaporation and precipitation. By choosing a rather special form of precipitation and evaporation pattern (where the zonally integrated evaporation and precipitation along each latitude vanishes), he was able to obtain a steady circulation, even though his model also had no friction. Because Goldsbrough's solution required a special form of precipitation and evaporation, his model was quite unrealistic. The major reason why Goldsbrough's theory has been largely ignored is due to the small size of the barotropic current predicted by his theory. As we discuss later, another major shortcoming in his theory was the lack of salinity in the model. If salt and mixing were added into his model, the baroclinic velocity induced by the freshwater flux could be a hundred times stronger than the barotropic velocity predicted by his original theory.

Nevertheless, the theories of Hough and Goldsbrough are more general than they first appear. The general nature of their theories can be explained more clearly with the aid of two basic mechanisms.

First, Ekman (1905) showed that the frictional stress of the wind is confined to a thin layer on the upper surface, so the motions below this thin surface layer can be treated as frictionless. Thus, for the ocean below this thin layer, the only role that wind stress plays is to create the horizontal Ekman transport in the upper ocean. The horizontal convergence of the Ekman transport gives rise to the Ekman pumping, which drives the interior flow equatorward in the subtropical basin.

In terms of potential vorticity dynamics, the lowest-order balance for the oceanic interior is as follows. In the oceanic interior, the relative vorticity is negligible, so potential vorticity is reduced to f/h, where f is the Coriolis parameter and h is the layer thickness. In order to balance the compression due to the large-scale Ekman pumping associated with the negative wind stress curl, water columns move toward lower latitudes, where f is smaller. Similarly, precipitation plays a role similar to Ekman pumping because it compresses water columns in the ocean, and thus reduces h. As a result, precipitation drives an equatorward flow in the ocean interior, very much like the effect of Ekman pumping.

Second, mass conservation requires a return flow, which is accomplished by western boundary currents. The frictional or inertial western boundary layers provide a vital dynamical component that helps to close the circulation in terms of the conservation of mass, energy, and potential vorticity. The interior flow field can therefore be combined with some kind of western boundary layer, such as the bottom friction model of Stommel (1948), the lateral friction model of Munk (1950), or the inertial model of Charney (1955) and Morgan (1956). Thus, the boundary layer theory that was developed in traditional fluid dynamics found its use in dynamical oceanography. In mathematical terms, the problem is treated by the perturbation method. In the interior ocean the flow is described by the low-order dynamics, i.e., essentially inviscid and linear. This lower-order dynamics cannot satisfy the boundary conditions at the western wall, so that within the western boundary layer, high-order terms, such as frictional or inertial terms, must be included to provide some corrections in order to match the interior solution and the western boundary conditions.

In this spirit, Stommel proposed to use western boundary currents to close the circulation driven by evaporation at low latitudes and precipitation at high latitudes. The solution he proposed overcomes the strong limitations implied in Goldsbrough's original model. Accordingly, circulation in a closed basin driven by an arbitrary pattern of wind stress, or freshwater flux across the air–sea interface due to evaporation minus precipitation, can be very well described by the theory.

Therefore, these seemingly unrelated theories for the oceanic circulation are essentially linked to each other in terms of potential vorticity balance, as shown in Figure 1.43. The dynamics of the reduced-gravity model will be discussed in detail in connection with the theory of the wind-driven circulation. In addition, the further development of three-dimensional structure of the wind-driven circulation also follows similar approaches based on potential vorticity dynamics.

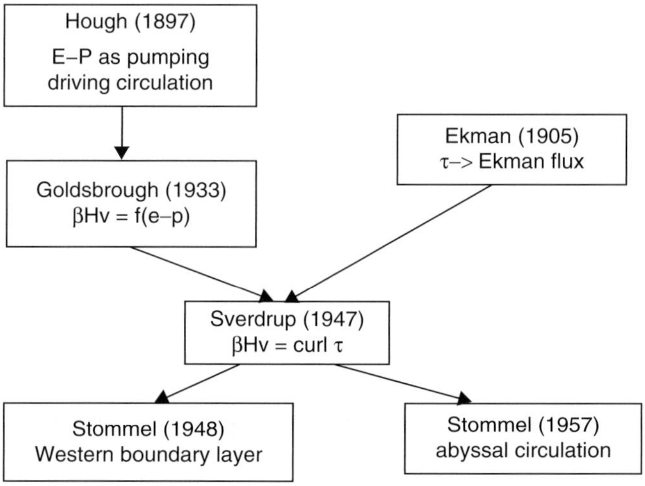

Fig. 1.43 Connections between different paths during the early development of the oceanic general circulation.

1.4.5 Theories of the baroclinic wind-driven circulation

Although single-layer models have been useful tools in describing wind-driven circulation, the theories of wind-driven circulation based on such models remained primarily two-dimensional and they could not provide the vertical structure of the gyre. In the 1950s and 1960s, people tried very hard to work out theories about the vertical structure of the oceanic circulation.

As discussed above, one of the most outstanding features in the oceans is the existence of the main thermocline, which is closely related to the velocity structure in the upper ocean. Thus, the baroclinic theory of the wind-driven circulation is also called the thermocline theory. The theory of the thermocline was first proposed in two papers published side by side in *Tellus* by Welander (1959) and Robinson and Stommel (1959). There have been many attempts to find solutions to the thermocline equations; however, most of these solutions are similarity solutions that cannot satisfy some of the essential boundary conditions. The most serious deficit of these solutions is their inability to satisfy the Sverdrup constraint. Without satisfying this constraint, these solutions are incapable of describing the basin-wide structure of the wind-driven gyre.

During the 1960s and 1970s, the development of oceanic circulation theory was relatively slow, owing to a lack of understanding of the physical processes involved in the circulation. Numerical models had been developed; however, without the physical insights obtained from observations or theoretical studies, results from numerical experiments proved as hard to understand as data from the oceans. The second phase of development of the theory of oceanic circulation began in the 1980s. This new phase is characterized by combining observations, theory, and numerical models.

Three-dimensional structure of the wind-driven circulation

As our understanding of the oceanic circulation deepened, it was realized that the ocean can be described to a very good approximation as an ideal fluid system. Recent field observations have indicated that diapycnal diffusion in the subsurface ocean interior is on the order of 10^{-5} m^2/s. A major theoretical difficulty in the 1970s was the puzzle of how the subsurface layers in an ideal-fluid model ocean could be in motion. In a rotating fluid, large-scale motions should follow the geostrophic contours, $f/\Delta h$ (where Δh is the layer thickness). Since interfacial friction is assumed to be infinitesimal, simple intuition suggests that geostrophic contours in subsurface layers should be parallel to the latitudinal circles. (However, such a simple intuition turns out to be incorrect, as discussed below.) Because of the no-flux condition at the eastern wall, all geostrophic contours in subsurface layers are blocked, and the subsurface layers should be stagnant.

This puzzle was solved by Rhines and Young (1982). Using a quasi-geostrophic model, they were able to show that closed geostrophic contours can be formed due to large interfacial deformation induced by strong forcing applied to the surface layer. As a result, there could be infinitely many possible non-stagnant solutions to the problem instead of the solution of no motion, as was thought previously. Furthermore, they showed, subject to assumptions about the effects of eddies on the mean flow, that potential vorticity should be homogenized within these closed geostrophic contours; thus the system should possess a unique solution that is stable to small perturbations. Their theory provided a theoretical background for motions in the subsurface layers.

A second way of getting the subsurface water in motion was proposed by Luyten *et al.* (1983): in their model, the isopycnal outcropping effectively bypasses the blocking due to the eastern boundary. In some sense, their model is a very nice extension of the classical conceptual model of ventilation through outcropping proposed by Iselin (1939) much earlier. Of course, such an extension included many conceptual breakthroughs, such as formulating the model in a solid dynamical framework and introducing the concepts of the ventilated zone, the pool regime, and the shadow zone.

Although Iselin proposed his conceptual model of late winter ventilation, it was not clear why the ocean should pick up only the late winter properties for ventilation. To explain this phenomenon, Stommel (1979) analyzed the physical processes involved and showed that there are indeed some processes in the oceans that select only the late winter properties for ventilation. This mechanism is now called the Stommel demon. Accordingly, in order to study the climatological mean circulation, it is possible to avoid the complexity of the seasonal cycle by simply choosing the late winter properties, such as the mixed layer depth and density. To date, the Stommel demon has remained a main theoretical backbone of the modern theory of wind-driven ocean circulation.

Another classical approach to the thermocline theory is the ideal-fluid thermocline theory proposed by Welander (1959, 1971a). His theory is basically to treat the wind-driven circulation as a perturbation to the background stratification set up by an external thermohaline circulation (which is not explicitly included in the model). Welander (1971a) showed that the ideal-fluid thermocline problem can be reduced to solving a second-order ordinary differential equation. However, his solution can satisfy only two boundary conditions in the vertical direction, and for a long time it was not clear how to improve his theory to accommodate more boundary conditions, as required by the physics.

The connections between these seemingly different approaches were unified into a theory of the three-dimensional structure of the wind-driven circulation in the continuously stratified oceans by Huang (1988a, b). It was demonstrated that the problem can be reduced to solving a free-boundary problem of a second-order ordinary differential equation in density coordinates. This theory was further extended to a model incorporating a mixed layer on top (Huang, 1990a; Williams, 1991). The model with a mixed layer can provide a more realistic description of the three-dimensional structure of the wind-driven circulation in the oceans.

The baroclinic structure of the inertial western boundary currents

Although theories of western boundary currents based on a single-moving layer have been simple, elegant, and successful, the corresponding part associated with the multiple moving layers has not. The difficulty associated with multi-layer inertial western boundary currents was first discussed by Blandford (1965). Basically, he searched for solutions with two moving layers, but failed to find any continuous solutions. Instead, he found that the solutions break down before they reach the latitude corresponding to the Gulf Stream's separation from the coast. The difficulties associated with the discontinuity of the inertial western boundary current have been discussed by Luyten and Stommel (1985) in terms of virtual control.

Using a streamfunction coordinate transformation, Huang (1990b) showed that continuous solutions for the two-moving-layer inertial western boundary currents do exist, and that these solutions can be matched to the multi-layer ventilated thermocline solution in the ocean interior (Huang, 1990c). However, the continuity of the inertial western boundary currents does impose certain dynamical constraints on the thermocline structure of the interior oceans.

1.4.6 Theory of thermohaline circulation

The Stommel–Arons theory for the deep circulation

In the early theory of the thermohaline circulation, the deep part of the circulation was idealized as a source–sink-driven circulation. In a series of seminal papers, Stommel and his colleagues developed the framework of the deep circulation (e.g., Stommel, 1957; Stommel and Arons, 1960a, b).

The basic assumptions of their models are as follows. Deep circulation in the world's oceans is assumed to be steady and it is driven by idealized point sources prescribed in an ocean with no bottom topography. The source of deep water is balanced by a uniform upwelling prescribed over the whole basin.

The most important dynamical consequence of these assumptions is as follows. First, uniform upwelling specified in the ocean drives poleward flow in the abyssal oceanic interior, as dictated by the linear potential vorticity constraint, similar to the case of wind-driven circulation discussed above. Second, in order to balance mass and potential vorticity in the abyssal ocean, deep western boundary currents are required.

In fact, the most crucial conclusion from such a simple theory is the prediction of the existence of deep western boundary currents. Almost immediately after the theory was postulated by Stommel (1957), a deep western boundary current was found off the eastern coast of the USA by Swallow and Worthington (1957). This combination of theory and *in situ* observations was hailed as one of the most important discoveries in physical oceanography of the last century.

Abyssal circulation beyond the Stommel–Arons theory

The Stommel–Arons theory of the abyssal circulation was very simple and successful, so it dominated the theoretical field of deep circulation for more than 20 years. However, their theory is essentially limited by several of the assumptions it makes. Although their theory has been very successful in predicting the deep western boundary currents, the uniform poleward flow in the abyssal ocean could not be verified. It was realized in the 1980s that in order to describe the deep circulation accurately, many of the simplifications made in their theory must be replaced with more realistic assumptions.

Departing from the steady-state assumption, Kawase (1987) studied the spin-up process of an inverse reduced-gravity model (the meaning of this term will be explained in later chapters), in which he assumed that the interfacial upwelling is linearly proportional to the interfacial displacement from the mean. His solution clearly demonstrated the critical role of the coastal Kelvin waves in setting up the deep circulation, especially the deep western boundary currents. Rhines and MacCready (1989) noted that the bottom of the oceans is far from being flat. In fact, the oceans' bottom has the shape of a bowl, like a Chinese wok. Since the horizontal area of the deep oceans increases upward, the deep circulation in the interior oceans may be clockwise, instead of counterclockwise as suggested by the classical Stommel–Arons theory.

Stommel and Arons (1960a) assumed that the upwelling velocity was uniform basin-wide; such an assumption was a way to simplify the model, but it is not necessarily true. There is much evidence to suggest that upwelling is not uniform. Using an idealized two-level model, Huang (1993a) was able to show that upwelling is very strong along the equator and the eastern boundary.

The baroclinic structure of the abyssal circulation with continuous stratification has been discussed by Pedlosky and his co-workers in a series of papers (Pedlosky, 1992; Christopher and Pedlosky, 1995). For a model with a flat bottom and given stratification, it was shown that the vertical and meridional velocity can have alternate signs because the basic equations have eigenfunctions that oscillate in both vertical and horizontal directions.

These theories are based on simple assumptions about the bottom topography and mixing. However, the situation in the oceans is much more complicated. As recent field experiments have indicated, mixing is highly inhomogeneous in both space and time. As a result of the nonlinear interaction between stratification, flow over topography, and mixing, the abyssal circulation is very complicated; it is therefore one of the most exciting research frontiers.

Conceptual models for the thermohaline circulation

The seemingly similar situation that both the ocean and the atmosphere are under strong thermal forcing has convinced many people to draw parallel comparisons between the thermal circulation in the atmosphere and in the oceans. Since the discovery of cool and dense water overlying the bottom of the world's oceans, the concept of thermally driven circulation in the ocean has been gradually formed. Because salinity should also contribute to the density distribution, and thus the oceanic general circulation, the circulation related to density differences in the ocean is called the thermohaline circulation. The advances in atmospheric science in understanding the thermally driven circulation, plus our experience from thermal engines in daily life, would have been the major impetus for the formation of early theories of thermohaline circulation.

The essential element of the early theories is that cold and dense deep water formed at high latitudes plays the role of the driving force for the thermohaline circulation in the world's oceans. A typical example is the two-dimensional conceptual model proposed by Wyrtki (1961), in which high-latitude cooling produces dense water that sinks to the seafloor. The equatorward spreading of deep water drives the meridional circulation, including the poleward branch of the return flow. The same idea was also used in the classical box model of Stommel (1961), in which he assumed that the overturning rate is linearly proportional to the meridional difference of density in the upper ocean.

However, there is an essential difference in terms of how heating/cooling applies to atmosphere and ocean. In fact, at the beginning of the twentieth century Sandstrom (1908, 1916) had postulated that thermal forcing in the ocean is incapable of driving any strong circulation. Although his postulation was also cited in the classic textbook *Physical Oceanography* by Defant (1961), it was largely ignored. Beginning in the late 1990s, the energetics of thermohaline circulation became a research frontier. According to the new paradigm, the thermohaline circulation is not driven by surface heating/cooling; instead, it is driven by external sources of mechanical energy from wind stress and tidal dissipation, as briefly discussed below.

The hydrological cycle and the haline circulation

It is common knowledge that salinity is a key factor in controlling the thermohaline circulation. Thus, freshwater flux across the air–sea interface should be one of the primary driving forces of the oceanic general circulation. Although salinity distribution in the ocean has been simulated in numerical models based on the surface salinity relaxation condition, the dynamical role of freshwater flux was not explored thoroughly in most early studies.

Before the 1990s, the role of freshwater-driven circulation was discussed in very few papers, such as those by Hough (1897), Goldsbrough (1933), and Stommel (1957, 1984a). These studies focused on the barotropic circulation driven by freshwater flux.

In a departure from the traditional salinity boundary conditions, Huang (1993b) postulated that the suitable condition for salinity balance should be the net freshwater flux through the air–sea interface. Assuming strong diapycnal mixing, numerical simulations showed that freshwater flux can drive a haline circulation. Although the barotropic component of such a haline circulation is relatively weak, the baroclinic component has a strength that is comparable to the circulation driven by heat flux or wind stress. The unique role of surface freshwater flux in maintaining and regulating the circulation is also a vital research frontier related to the thermohaline circulation, as discussed below.

The multiple equilibria and variability of the thermohaline circulation

The multiple equilibria for the thermohaline circulation were first discussed in a seminal paper by Stommel (1961). Based on a two-box model, he predicted that there should be two steady states (one stable and one unstable) in the so-called thermal mode, and a stable steady state in the haline mode. In common with much of his work, his model was thought to be too simple, and people did not appreciate its physical meaning for two decades. However, this changed rapidly in the 1980s. Due to a very strong need to understand the climate system, people started to consider possible multiple equilibrium states of the climate, including multiple solutions to the thermohaline circulation.

Major contributions include F. Bryan's (1986) work on the multiple states of the model Atlantic Ocean and the associated change in the poleward heat flux. He also introduced the use of the so-called mixed boundary conditions, which have since gained wide acceptance by numerical modelers. Manabe and Stouffer (1988) found multiple states in an air–sea coupled general circulation model, which shares some similarities to the solutions predicted by the box model of Stommel. Marotzke (1990) found many more multiple states of the thermohaline circulation, including the so-called flushing phenomenon associated with the halocline catastrophe.

The critical role of virtual salt flux (or freshwater flux) in controlling the thermohaline catastrophe and its variability on the decadal time scale was explored in numerical models. One of the most interesting topics is the thermohaline variability on decadal or longer time scales (e.g., Weaver and Sarachik, 1991).

The flux condition for the salinity balance seems to be the essential ingredient for the thermohaline variability, as shown by many studies (e.g., Weaver *et al.*, 1991). In fact, freshwater flux alone can give rise to haline oscillation on a decadal time scale (e.g., Huang and Chou, 1994). Thus, freshwater flux due to evaporation and precipitation may be the essential ingredient for climate variability.

1.4.7 Mixing and energetics of the oceanic circulation

Since the early days of oceanic general circulation modeling it was realized that the choice of sub-grid parameterizations was rather subjective and may involve great uncertainty. In order to assess the vertical diffusivity in the oceans, Munk analyzed the balance of tracers in the world's oceans; his paper "Abyssal recipes" (Munk, 1966) remains a classic. Using a one-dimensional model, he concluded that the global mean vertical diffusivity is approximately 10^{-4} m^2/s.

Mixing in the oceans is non-uniform

Many *in situ* observations carried out afterward, however, indicated that diffusivity in the oceans is highly inhomogeneous. In particular, diffusion is very weak (on the order of 2×10^{-5} m^2/s; Ledwell *et al.*, 1993) in the subsurface ocean, which is much smaller than the global mean value of 10^{-4} m^2/s. To narrow the gap between Munk's global mean diffusivity and the low diffusivity observed in the subsurface ocean, it was postulated that strong diffusion near the bottom or side boundaries of the ocean may contribute to the seemingly high global mean value of Munk. In fact, recent *in situ* observations indicate that diapycnal diffusivity can be very strong near the bottom and close to the mid-ocean ridge (on the order of 10^{-3} m^2/s; Ledwell *et al.*, 2000).

Highly non-uniform mixing in the deep ocean posed serious questions about the validity of the classical Stommel–Arons theory. It is obvious that abyssal circulation in the presence of complicated bottom topography and driven by such non-uniform mixing can be drastically different from the classical theory of Stommel and Arons (1960a). Flow induced by bottom-intensified mixing over sloping boundaries has been studied extensively. Phillips (1970) and Wunsch (1970) pointed out that a thermal insulation boundary condition applied to a sloping bottom requires that isotherms must be perpendicular to the local slope. In a rotating fluid such a density gradient near the bottom boundary thus induces an along-slope flow (the primary circulation) and an uphill flow (the secondary circulation) in the bottom boundary layer. Phillips *et al.* (1986) also showed that a tertiary flow perpendicular to the slope might exist, due to the convergence and divergence of the secondary circulation. Mixing over a sloping bottom boundary has been studied extensively (Garrett *et al.*, 1993). Recent numerical studies of abyssal circulation induced by bottom-intensified mixing demonstrate rather complicated flow patterns (e.g., Cummins and Foreman, 1998).

Energetics of oceanic circulation

Oceanic circulation requires a source of mechanical energy to overcome friction and dissipation. For a long time, the common wisdom has been that thermal forcing can provide the mechanical energy sustaining the thermohaline circulation. However, a close examination reveals the following fundamental difference between the atmospheric circulation and the oceanic circulation: the atmosphere is heated from below and cooled from the middle and upper levels, so the atmosphere can work as a thermal engine. However, the ocean is heated and cooled from the upper surface, at roughly the same level; such a forcing is now called horizontal differential heating. About 100 years ago, Sandstrom (1908, 1916) had postulated that thermal forcing in the ocean is incapable of driving a strong circulation. Thus, the oceanic circulation is not a heat engine; instead, it is a conveyor belt for heat and freshwater driven by external sources of mechanical energy.

In order to maintain the stratification observed in the ocean, the cold and dense deep water must be warmed up and returned to the surface. This process implies mass flux across isopycnal surfaces, and is called diapycnal mixing. Since isopycnals in the ocean are nearly horizontal, diapycnal mixing is often referred to as vertical mixing. Vertical mixing in a stratified fluid pushes light water downward and heavy water upward. As a result, the center of mass is moved upward owing to vertical mixing, and the total amount of gravitational potential energy is increased. Therefore, mixing requires external sources of mechanical energy. The sources of external mechanical energy and their distribution within the ocean are critically important for oceanic circulation and climate.

The mechanical energy required to sustain the oceanic circulation was first explored by Faller (1966); however, the connection between sources of external mechanical energy and oceanic general circulation remained unexplored. Although many scientists working on small-scale mixing problems knew that mixing in a stratified fluid requires mechanical energy, most numerical modelers and theoreticians for large-scale oceanic circulation completely ignored the possible link between diapycnal mixing and the external mechanical energy required to sustain the mixing. Vertical diffusivity in numerical models remains as an adjustable parameter in the models, which were tuned to fit the meridional overturning circulation to observations. For example, in the book *Physics of Climate* (Peixoto and Oort, 1992), the energetics for the oceanic circulation was mostly confined to balance of thermal energy.

The thermohaline circulation in the oceans is strongly influenced by diffusion; thus, the traditional tools based on ideal-fluid assumptions, such as conservation of tracers or potential vorticity, cannot provide complete dynamical pictures. Studying the thermohaline circulation from other viewpoints, such as the balance of mechanical energy, is necessary; the major breakthrough in the energetics of oceanic circulation came only in the late 1990s.

The dominating role of wind stress over the Southern Ocean in providing mechanical energy and sustaining the thermohaline circulation was demonstrated through numerical experiments by Toggweiler and Samuels (1993, 1998). Wunsch (1998) made the first reliable estimate of wind energy input to geostrophic currents in the world's oceans.

A somewhat surprising conceptual breakthrough was postulated by Munk. After studying tidal problems for many decades, he finally came up with the crucial idea that tidal dissipation can provide the mechanical energy sustaining diapycnal mixing in the ocean. As a result, tidal dissipation can play a key role in regulating oceanic circulation, and thus the climate, on a planet. Munk and Wunsch (1998) postulated that tidal dissipation and energy due to wind work on the surface geostrophic current are the vital sources of mechanical energy sustaining diapycnal mixing in the world's oceans. An attempt was made to establish the balance of different kinds of energy for the oceanic general circulation by Huang (1998a). In particular, the importance of external sources of mechanical energy required for sustaining diapycnal mixing and the balance of available potential energy were examined (e.g., Huang, 1998b, 1999). Many studies regarding the energetics of the oceanic circulation have appeared since then. Progress along these lines has been summarized in review papers by Wunsch and Ferrari (2004) and Ferrari and Wunsch (2009). I discuss some fundamental issues related to the energetics of the oceanic general circulation in Chapters 3, 4 and 5.

2

Dynamical foundations

2.1 Dynamical and thermodynamic laws

Since the atmospheric and oceanic circulations take place within a rather thin layer of fluid on the surface of the Earth, large-scale motions in the atmosphere and oceans can most conveniently be described in spherical coordinates. The basic equations have been discussed in many standard textbooks; we will assume that the reader is familiar with them and therefore these equations will be introduced here in a concise way. One of the best sources is the book by Holton (2004), *An Introduction to Dynamic Meteorology*.

2.1.1 Basic equations

The momentum equations

In a rotating frame of reference (Fig. 2.1), Newton's second law can be written in vector notation as

$$\frac{D\vec{u}}{Dt} = -2\vec{\Omega} \times \vec{u} - \frac{\nabla p}{\rho} + \vec{g} + \vec{F} \tag{2.1}$$

where $\vec{u} = (u\vec{i}, v\vec{j}, w\vec{k})$ is the three-dimensional velocity vector; $D\vec{u}/Dt$ is the total derivative or the "material derivative" of velocity, i.e., the rate of time change for an observer who rides with the fluid particles; $\vec{\Omega} = \omega\vec{z}$ ($\vec{z}$ is a unit vector) is the vector representing the rotation of the Earth; p and ρ are pressure and density; $\vec{g}$ is the gravity vector; and $\vec{F}$ is the friction force, such as the surface wind stress, bottom drag and internal friction.

The convention of spherical coordinates used in physical oceanography is (λ, θ, r), where $0 \leq \lambda \leq 2\pi$ is longitude, $-\pi/2 \leq \theta \leq \pi/2$ is latitude, and r is radius. In spherical coordinates, the total derivative of velocity is

$$\frac{D\vec{u}}{Dt} = \left(\frac{du}{dt} - \frac{uv\tan\theta}{r} + \frac{wu}{r}\right)\vec{i} + \left(\frac{dv}{dt} + \frac{u^2\tan\theta}{r} + \frac{wv}{r}\right)\vec{j} + \left(\frac{dw}{dt} - \frac{u^2+v^2}{r}\right)\vec{k} \tag{2.2}$$

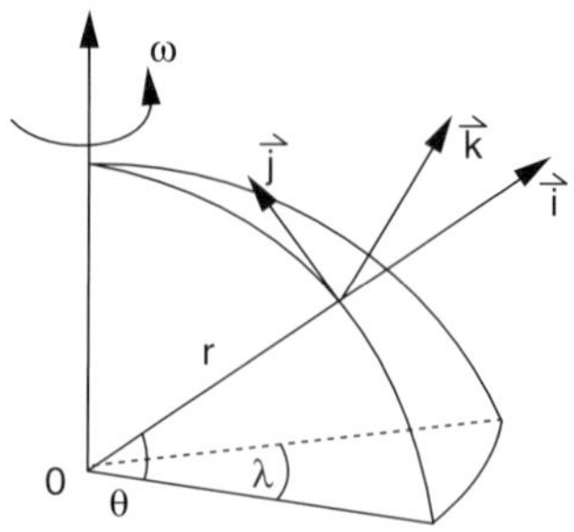

Fig. 2.1 Sketch of spherical coordinates in a rotating frame of reference.

where we have introduced a new operator

$$\frac{d}{dt} = \frac{\partial}{\partial t} + \frac{u}{\rho r \cos\theta}\frac{\partial}{\partial\lambda} + \frac{v}{\rho r}\frac{\partial}{\partial\theta} + w\frac{\partial}{\partial r}$$

In spherical coordinates, the first term on the right-hand side of Eqn. (2.1) is

$$2\vec{\Omega} \times \vec{u} = (2\omega w \cos\theta - 2\omega v \sin\theta)\vec{i} + 2\omega u \sin\theta \vec{j} - 2\omega u \cos\theta \vec{k}$$

Therefore, the momentum equations can be written in three components

$$\frac{du}{dt} - \frac{uv\tan\theta}{r} + \frac{uw}{r} - 2\omega v \sin\theta + 2\omega w \cos\theta = -\frac{1}{\rho r\cos\theta}\frac{\partial p}{\partial\lambda} + F_\lambda \tag{2.3}$$

$$\frac{dv}{dt} + \frac{u^2\tan\theta}{r} + \frac{vw}{r} + 2\omega u \sin\theta = -\frac{1}{\rho r}\frac{\partial p}{\partial\theta} + F_\theta \tag{2.4}$$

$$\frac{dw}{dt} - \frac{u^2+v^2}{r} - 2\omega u\cos\theta = -\frac{1}{\rho}\frac{\partial p}{\partial r} - g + F_r \tag{2.5}$$

The continuity equation

The continuity equation or the law of mass conservation in vector notation is

$$\frac{\partial\rho}{\partial t} + \nabla\cdot(\rho\vec{u}) = 0 \tag{2.6}$$

In spherical coordinates it becomes

$$\frac{\partial\rho}{\partial t} + \frac{1}{r\cos\theta}\left[\frac{\partial}{\partial\lambda}(\rho u\cos\theta) + \frac{\partial}{\partial\theta}(\rho v\cos\theta)\right] + \frac{\partial\rho w}{\partial z} = 0 \tag{2.6$'$}$$

where we have used the approximation $\delta r \simeq \delta z$ because of the thinness of oceanic circulation compared with the radius of the Earth. Similar approximations will apply to the momentum equations, as discussed in Sections 2.6 and 2.7.

Thermodynamic energy equation

For the basic formulation discussed in this section, only the internal energy associated with temperature is considered:

$$c_v \frac{dT}{dt} + p \frac{d\alpha}{dt} = \dot{q} \tag{2.7}$$

A more complete treatment of energy is presented in Chapter 3.

Tracer equations

Similar to the continuity equation, many tracers in the ocean obey conservation equations in the following form

$$\frac{\partial \rho X}{\partial t} + \nabla \cdot (\rho \vec{u} X) = \nabla \cdot (\kappa \nabla X) + Q(X) \tag{2.8}$$

where X is the tracer, the first term on the right-hand side is the contribution due to diffusion, and the last term $Q(X)$ is the distributed source/sink of the tracer. For example, salinity in the ocean obeys Eqn. (2.8), with no source/sink in the ocean interior; temperature distribution also obeys a similar equation, where $Q(X)$ is interpreted as being contributions due to different sources.

2.1.2 Integral properties

Equations (2.1), (2.6), (2.7), and (2.8) constitute the basic equations describing the oceanic general circulation. These differential equations are the most complete dynamical laws for the large-scale motions in the oceans, so that their solutions are rather complicated. For many applications, it is also desirable to explore the circulation in terms of the integral forms derived from these equations. These integral relations include many conservation laws, such as conservation of mass, angular momentum, potential vorticity, and different forms of energy.

Conservation of angular momentum

Since linear momentum is not conserved in rotating coordinates, we consider the conservation of angular momentum. Multiplying Eqn. (2.3) by a factor of $r\cos\theta$, and using definitions $dr/dt = w$ and $rd\theta/dt = v$, the following conservation relation can be obtained

$$\frac{d}{dt}\left[r\cos\theta\,(u + \omega r\cos\theta)\right] = r\cos\theta\left(-\frac{1}{\rho r\cos\theta}\frac{\partial p}{\partial\lambda} + F_\lambda\right) \tag{2.9}$$

In the study of the world ocean general circulation, a more convenient form is the balance of the zonally average angular momentum, which can be obtained by taking the zonal average of Eqn. (2.9)

$$\frac{d\overline{M}^\lambda}{dt} = \overline{r\cos\theta F_\lambda}^\lambda \tag{2.10}$$

where

$$\overline{M}^{\lambda} = \overline{r\cos\theta\,(u + \omega r\cos\theta)}^{\lambda} \tag{2.11}$$

is the zonally averaged angular momentum. The physical meaning of Eqns. (2.9) and (2.10) is that the total angular momentum of the atmospheric/oceanic general circulation must be conserved.

During its long existence, the atmospheric general circulation shows neither substantial slowing down nor speeding up; thus the total frictional torques received from the atmospheric planetary boundary layer must be approximately zero. A straightforward application of the angular momentum conservation is as follows. There are strong westerlies at mid latitudes in both hemispheres. Since the total frictional torques over the surface must be zero for a steady state of atmospheric circulation, there must be easterlies on the Earth to offset the friction torques due to westerlies. In fact, easterlies exist near the equator and at high latitudes. The coexistence of westerlies at mid latitudes and easterlies at low and high latitudes poses essential constraints over the wind-driven circulation in the world's oceans, as discussed in Chapter 4. The most important application of this conservation principle in dynamical oceanography is the study of the basic structure of the Antarctic Circumpolar Current (ACC).

Mechanical energy conservation

Note that the main focus in this book is the study of the large-scale motions in the oceans, including the wind-driven circulation and the thermohaline circulation. The study of wind-driven circulation is based on the theoretical framework in which motions in the upper ocean are treated in terms of adiabatic motions, so that the dynamical equations are separated from the thermodynamic equations. For the study of wind-driven circulation, therefore, mechanical energy conservation is one of the most crucial dynamical constraints.

Multiplying Eqns. (2.3), (2.4), and (2.5) by $(u,\ v,\ w)$ and summing up lead to the conservation of total mechanical energy

$$\frac{dE}{dt} = uF_{\lambda} + vF_{\theta} + wF_{r} \tag{2.12}$$

where

$$E = \frac{1}{2}\left(u^2 + v^2 + w^2\right) + \frac{p}{\rho} + g(r - r_0) \tag{2.13}$$

is the total mechanical energy. In this derivation, we ignored the $\partial p/\partial t$ term and assumed that the density is approximately constant. Equation (2.12) states that change in the total mechanical energy is balanced by the loss due to friction.

On the other hand, the study of thermohaline circulation is closely related to thermodynamics, and thus we will use the thermodynamics in combination with dynamics equations. However, mechanical energy balance in the ocean also plays a vitally important role in controlling of the oceanic general circulation, as discussed in connection with the basic theory of the thermohaline circulation in Chapter 3, where a more comprehensive treatment of energy conservation is presented.

2.2 Dimensional analysis and nondimensional numbers

Dimensional analysis is one of the basic tools in mechanics. It has been extensively used in a variety of mechanical problems, especially in complex systems where multiple factors exist and analytical relations are difficult to find.

The basic idea of dimensional analysis is the principle that the mathematical form of the functional relation between different physical quantities should be independent of the units used for these quantities.

There exist four fundamental physical quantities: length, time, mass, and temperature. We call them fundamental owing to the fact that each of them cannot be expressed as functions of others without introducing other quantities, while all other physical quantities have dimensions that can be defined or expressed in terms of these four fundamental quantities.

Temperature can be defined in two ways. When we treat a system as a collection of its microscopic particles, the system needs to be described in terms of statistical physics. Accordingly, the thermal state of a material can be treated using the statistics of the molecules' movements. Temperature can be defined in terms of the root-mean-square of the kinetic energy of the molecules; thus, temperature should have a dimension as energy in statistical physics. On the other hand, if the system is treated as a macroscopic object, temperature is defined as a primary dimension. Since our focus is on the macroscopic phenomena in the oceans, we treat temperature as a primary dimension.

Any measurable or observable physical variables can be classified into two categories: dimensional or nondimensional. Quantities whose values depend on the system of measurement units are called dimensional; quantities whose values are independent of the system of measurement units are called nondimensional. Typical dimensional quantities are *mass, length, time, energy, velocity*, etc. Typical nondimensional quantities are *angle, the ratio of the inertial force to the viscosity force,* etc. Although, traditionally, an angle can be expressed in terms of different units, such as degree or radian, it is more convenient to treat the angle as a nondimensional quantity.

Any dimensional physical quantity is defined in terms of dimension: a unit expresses magnitude of the dimension. For example, the depth of the basin has a dimension of length, and can be expressed in meters, centimeters, or kilometers.

Various unit systems have been used in oceanography in the past; we will primarily use the International System of Units adopted since 1985, which is also called the MKS system because it is based on meter–kilogram–second–degree. Another popular system commonly used in scientific research is the CGS unit, which is based on centimeter–gram–second–degree.

2.2.1 Dimensions of the commonly used variables in physical oceanography

We will use notation [x] to denote the dimensions of a given physical quantity x. For example, velocity c can be defined as length divided by time, and has a dimension $[c] = L/T$.

According to Newton's second law, $Force = mass \times acceleration$; thus, force has a dimension $[ML/T^2]$. Pressure is one of the most frequently used variables in oceanography and has a dimension $[p] = M/LT^2$.

Table 2.1. *Primary dimensions*

Quantity	Dimension	Unit
Mass	[M]	kg (kilogram)
Length	[L]	m (meter)
Time	[T]	s (second)
Temperature	[Θ]	K

Table 2.2. *Secondary dimensions*

Variable	Dimension	Unit
Velocity	[L/T]	m/s
Acceleration	[L/T^2]	m/s^2
Force	[ML/T^2]	N (newton)
Energy	[ML^2/T^2]	J (joule)

Table 2.3. *Dimensions of the most frequently used quantities in physical oceanography*

Quantity	Dimension	Unit
Temperature	[Θ]	K
Salinity	None	‰[a]
Pressure and wind stress	[M/LT^2]	Pa or bar
Velocity	[L/T]	m/s
Mixing coefficient	[L^2/T]	m^2/s

[a] Salinity is defined in fraction of mass, or in parts per thousand, so it is a nondimensional unit. This unit is sometimes called the practical salinity unit, or psu; however, it is more appropriate not to include psu after the value for salinity.

Tables 2.1 and 2.2 list the dimensions of the most commonly used physical variables, including four primary dimensions (Table 2.1) and four secondary dimensions (Table 2.2).

The most frequently used dimensional variables in physical oceanography are listed in Table 2.3.

The commonly used dimensional units in physical oceanography are included in Table 2.4. Note that the commonly used pressure unit is db (decibar), and for each 1 m of depth the *in situ* pressure increases by approximately 1 db.

Table 2.4. *Commonly used dimensional units in physical oceanography*

Physical quantity	Unit	Equivalent	
Time	d (day)	86,400 s	
	yr (year)	31,558,000 s	
Force	Dyn	1g cm/s^2	10^{-5}N
	N	1 kg m/s^2	
Pressure and Stress	dyn/cm^2	1 g/cm/s^2	0.1 N/m^2 (Pa)
	Pa	1 kg/m/s^2	
	Bar	10^5 N/m^2	10^5 Pa
	db (decibar)	10^4 N/m^2	10^4 Pa
Energy	Erg	10^{-7} J	
	Cal	4.184 J	
Volume flux	Sv[a] (sverdrup)	10^6 m^3/s	
Latent heat	cal/g	4,184 J/kg	

[a] Although Sv has been commonly used as a volumetric flux, it can also be used as a mass flux (see the discussion in Section 5.3.1). Because water density is in the range of 1,020–1,050 kg/m^3, when it is used as a unit of mass transport, 1 Sv $\approx 10^9$/kg/s.

2.2.2 Dimensional homogeneity

All terms appearing in any equation describing a physical relation should have the same dimension, that is, dimensional homogeneity. The homogeneity of dimension can be used as a powerful tool in dimensional analysis. For example, the depth of the thermocline D obeys a cubic equation

$$D^3 - aD = b \tag{2.14}$$

Since D has a dimension of L, the dimensional homogeneity of this equation requires that a should have a dimension of L^2, and b should have a dimension of L^3. Accordingly, if we introduce the following depth scales $d_w = \sqrt{a},\ d_\kappa = b^{1/3}$, this equation can be rewritten as

$$D^3 - d_w^2 D = d_\kappa^3 \tag{2.15}$$

As we show shortly, by comparing the magnitude of d_w^2 and d_k^3, this equation can be further simplified by introducing a nondimensional depth.

It is an important practice to check the dimensions of each term in an equation to see whether they are dimensionally homogeneous. ***Although an equation in which all terms are of the same dimension is not necessarily a correct equation, an equation with different dimensions for some of its terms is definitely in error.***

2.2.3 The nondimensional parameters

Since an equation should be dimensionally homogeneous, it can always be reduced to a nondimensional form by dividing by its dimensions. Therefore, any equation describing the physical processes can be reduced to a general form

$$f(\Pi_1, \Pi_2, \Pi_3, \ldots) = 0 \tag{2.16}$$

where Π_i are nondimensional parameters of the dynamical system.

2.2.4 A few simple applications of dimensional analysis

Thermocline depth

As the first example, we discuss the depth of the thermocline. A simple mass balance leads to an equation determining the depth of the thermocline D, as shown in Eqn. (2.15), where the two depth scales are defined as follows

$$d_w = \left(-\frac{w_e f \rho_0 A}{c g \Delta\rho}\right)^{1/2}, \quad d_\kappa = \left(\frac{\kappa f \rho_0 A}{c g \Delta\rho}\right)^{1/3} \tag{2.17}$$

and these are the scaling depth of the thermocline due to wind stress and mixing; w_e is the Ekman pumping rate, f the Coriolis parameter, ρ_0 the mean reference density, A the horizontal area of the basin, c a constant coefficient, g the gravity, $\Delta\rho$ the density difference across the main thermocline, and κ the diapycnal mixing coefficient.

In most frequently met cases, the diapycnal mixing coefficient is rather small, and the thermocline is primarily controlled by wind-stress-induced Ekman pumping; thus, $d_\kappa \ll d_w$, so the equation discussed above can be reduced to the following nondimensional form

$$d^3 - d = \varepsilon^3 \tag{2.18}$$

where $d = D/d_w$ is the nondimensional thickness, and $\varepsilon = d_\kappa/d_w \ll 1$ is a small nondimensional parameter, which can be used to expand the solution of Eqn. (2.18) in a power series. Thus, the sensitivity of the thermocline depth can be explored. Therefore, nondimensionalization of dynamical equations can often help us to identify the most crucial dynamical balances in complicated equations. Such insight can lead to a better understanding of complicated physical processes in a dynamical system.

A practical example

From a numerical model describing the meridional overturning cell in the Atlantic basin, the depth of the main thermocline D was described in terms of the following equation

$$\frac{cg'}{L_y^n} D^3 + \frac{A_I L_x}{L_y^s} D^2 - \frac{\tau L_x}{f \rho_0} D = \kappa A \tag{2.19}$$

where $g\prime$ is the reduced gravity, L_y^n and L_y^s are meridional width of the layer depth fronts in the Northern and Southern hemispheres, L_x is the zonal width of the basin, $A$is the horizontal area of the basin, A_I and κ are eddy and diapycnal mixing coefficients respectively, and τ is the wind stress. A simple check of the consistency of this equation reveals: the dimension of the first term on the left-hand side of Eqn. (2.19) is $\left[cg'D^3/L_y^n\right] = L^3/T^2$, the dimension of the term on the right-hand side of Eqn. (2.19) is $[\kappa A] = L^4/T$, the dimension of the second and third term on the left-hand side is also L^4/T. Therefore, it is clear that the dimension of the first term on the left-hand side is wrong. In order to have a dimensionally correct equation, there should be an additional factor in this term, which has a dimension of *TL*. A careful re-examination of the derivation of this equation reveals that the first term in Eqn. (2.19) should have an additional factor of β in the denominator:

$$\frac{cg'}{\beta L_y^n}D^3 + \frac{A_I L_x}{L_y^s}D^2 - \frac{\tau L_x}{f\rho_0}D = \kappa A \tag{2.20}$$

A more interesting example of dimensional analysis is to introduce the nondimensional variable as the following. Equation (2.20) can be rewritten as

$$D^3 + d_I D^2 - d_w^2 D - d_\kappa^3 = 0 \tag{2.21}$$

where d_I, d_w, and d_k all have a dimension of length. This equation can be solved by using one of these length scales and introducing a nondimensional depth. The details of such an application are discussed in Section 5.5.1.

The atomic bomb explosion

Dimensional analysis can provide very useful information, especially when the system has a simple geometry. One of the classic examples used to demonstrate the power of dimensional analysis is the case of the strong shock waves associated with a point source explosion.

There are three key physical quantities; the amount of energy released, $[E] = ML^2/T^2$, the density of air, $[\rho] = M/L^3$, and the time $[T] = T$. This system has a very simple geometry. Since there is no intrinsic length scale imposed, the distance of the shock wave propagation must depend on a power combination of the three basic quantities, i.e., energy, density, and time $d = E^\alpha \rho^\beta T^\gamma$. Since d has a dimension of L, the only possible combination is $\alpha = 1/5$, $\beta = -1/5$, and $\gamma = 2/5$; thus we have the equation for the propagation of the shock waves

$$d = (E/\rho)^{1/5} t^{2/5} \tag{2.22}$$

The most beautiful example of the power of dimensional analysis was demonstrated by G.I. Taylor. After the first test of the atomic bomb, he was invited to watch a movie recorded during the experiment. Although all the information was classified and thus not available, using the scaling analysis outlined above, G.I. Taylor was able to infer the amount of energy released from the atomic explosion by substituting the time evolution of the shock front he saw in the movie. The reader who is interested in the details of dimensional analysis and the exciting story of G.I. Taylor is referred to the classic book on dimensional analysis by Sedov (1959).

2.2.5 Important nondimensional numbers in dynamical oceanography

Ekman number: $E = \dfrac{\nu}{2\Omega L^2}$

where υ is the viscosity, Ω is the angular velocity of the Earth, and L is the horizontal length scale. The Ekman number is the ratio of frictional force and the Coriolis force. This nondimensional number is introduced in the study of the frictional boundary layer in a rotating frame. Within this boundary layer the frictional force is balanced by the Coriolis force, but outside this boundary layer the frictional force is negligible in comparison with the Coriolis force.

Froude number: $Fr = \dfrac{U}{\sqrt{gh}}$

where U is the velocity scale, g is the gravity acceleration, and h is the depth of the water. The Froude number is the ratio of the advection velocity and the speed of the surface gravity waves. When the Froude number is equal to one, the advection velocity is equal to the phase velocity of the gravity waves, and the flow is called critical. When the Froude number is larger than one, the perturbation signals cannot propagate upstream, so the flow is called supercritical.

Grasshof number: $G_T = \dfrac{g\alpha\Delta Th^3}{\nu^2}$, $G_S = \dfrac{g\beta\Delta Sh^3}{\nu^2}$

where $\alpha\Delta T$ and $\beta\Delta S$ are the density perturbations associated with the temperature and salinity difference, h is the height of the container, and υ is the viscosity. The Grasshof number is the ratio of the buoyancy force due to temperature or salinity difference and the frictional force.

Peclet number: $Pe = \dfrac{UL}{\kappa}$

where U and L are the velocity and length scales, and κ is the temperature diffusivity. The Peclet number is defined as the ratio of advection to diffusion; therefore a large Peclet number indicates that advection is more important than diffusion.

Prandtl number: $Pr = \dfrac{\nu}{\kappa}$

where υ is the viscosity and κ is the temperature diffusivity. The Prandtl number is used to describe the ratio of momentum dissipation and tracer mixing. At the level of molecular dissipation and mixing, the Prandtl number for seawater is approximately equal to 8; however, owing to strong eddy and turbulent activity, the equivalent Prandtl number for the large-scale geostrophic turbulence may be quite different from this value.

Rayleigh number: $Ra = \dfrac{g\alpha \Delta T h^3}{\kappa \nu} = G_T \cdot Pr$

where ΔT is the scale of temperature difference in connection with the convection, and h is the height scale. The Rayleigh number is one of the most crucial nondimensional numbers used in the study of thermal circulation. When the Rayleigh number is higher than certain critical values, the character of the convection switches from one regime to another regime.

Reynolds number: $Re = \dfrac{UL}{\nu}$

where U and L are the velocity and length scale of the motion, and v is the viscosity. The Reynolds number is the ratio of the inertial force to the viscosity force. When *Re* is small, the friction force is important, i.e., it is comparable with the inertial force, and the flow is laminar. When *Re* is large, the frictional force is less important, i.e., it is much smaller than the inertial force, so that the frictional effect is confined within very thin boundary layers near the solid boundary of the flow field. The flow under a large Reynolds number is turbulent.

Richardson number: $Ri = -\dfrac{g d\rho/dz}{\rho\,(du/dz)^2}$

The Richardson number is the ratio of the stratification and the square of the vertical velocity shear. A strong stratification can inhibit the instability, but a strong velocity shear is in favor of the instability; thus, the Richardson number is widely used as an index for instability. A necessary condition for instability is $Ri < 1/4$. However, this does not necessarily mean that instability will actually occur.

Rossby number: $Ro = \dfrac{U}{2\Omega L}$

where U is the velocity, Ω is the angular velocity of the Earth's rotation, and L is the horizontal scale of the flow. The Rossby number is the ratio of the inertial force and the Coriolis force. When *Ro* is much smaller than one, the inertial terms in the momentum equation can be ignored. The most commonly used geostrophic approximation is obtained under several assumptions, including the assumption that *Ro* is much smaller than one. On the other hand, when *Ro* is close to one or larger than one, the inertial terms in the momentum equations must be retained.

Schmidt number: $Sc = \dfrac{\nu}{\kappa_S}$

where v is the viscosity and κ_S is the salinity diffusivity. The Schmidt number is the ratio of the relative importance of momentum dissipation and tracer diffusion.

2.3 Basic concepts in thermodynamics

In the study of oceanic circulation, both dynamics and thermodynamics are very useful tools. The following sections contain a rather concise summary of thermodynamics, which is essential for the description and understanding of large-scale oceanic circulation and its energetics.

In thermodynamics, variables are classified into the so-called state variable and non-state variables. State variables are variables describing the state of a thermodynamic system, including temperature, pressure, volume, density, different forms of energy, entropy, etc. Non-state variables include heat and mechanical energy flux exchanged with the environment.

2.3.1 Temperature

Temperature is a very old concept in science. Observations indicate that two bodies placed together for a long time eventually reach a thermal equilibrium. When such a final state is reached, they will have the same temperature. Thus, we have:

The zeroth law of thermodynamics: *If two systems are in thermal equilibrium with a third system, these two systems are in thermal equilibrium with each other.*

In order to establish a temperature scale, fixed points of reference must be defined. Different choices of the reference point and the corresponding values assigned to these points give rise to different temperature scales. The most frequently used temperature scales are:

- ***The Fahrenheit scale (°F)*:** The freezing temperature of pure water at one atmospheric pressure is assigned the number 32, and the corresponding boiling temperature is assigned number 212. Although this scale is still used in countries like the USA, it is no longer used in scientific studies.
- ***The Celsius scale (°C)*:** The freezing temperature of pure water at one atmospheric pressure is assigned the number 0, and the corresponding boiling temperature is assigned the number 100.
- ***The Kelvin scale (K)*:** In 1854, Kelvin suggested using the triple point of pure water (solid, liquid, and vapor coexisting in equilibrium) as a more accurate reference point. In addition, the zero point of the scale is set at the so-called absolute zero, i.e., the lowest possible temperature at which there is virtually no thermal movement of molecules. In the Kelvin scale, the freezing point of pure water under one atmospheric pressure is approximately 273.16 K.
- For oceanographic applications an ***International Temperature Scale*** (ITS-90) has been adopted, which is very close to the Celsius scale, with minor differences.

2.3.2 Energy

The concept of energy was first introduced by Newton, who postulated two types of energy: kinetic and potential energy. Energy is one of the most important scalar quantities for the description of natural phenomena. Energy is linked with mass in the relativistic theory, according to Einstein's famous equation $E = mc^2$. However, when the velocity is much smaller than the speed of light, the Newtonian system is accurate enough for almost every practical purpose. In a Newtonian system that has no net mass exchange with its environment, the total mass of the system is conserved, i.e., there is no transformation between mass and energy within the system.

Energy can exist in different forms, such as mechanical energy, thermal energy, chemical energy, electrical energy, and atomic energy, etc. Most energy sources in the climate system, including the atmosphere, the oceans, and other components, can be traced back to the radiative heat flux from the Sun, except for tides and geothermal heat flux. Energy can be transformed from one form to another. However, the extent of energy conversion may be complete or partial, depending on the forms of energy involved. For example, mechanical energy and chemical energy can be converted 100 percent into thermal energy. On the other hand, thermal energy can be only partially converted to mechanical energy. Therefore, caution must be exercised when we discuss the energy and its conversion of a particular system. Energy relevant to large-scale oceanic circulation can be transferred in various forms of energy flux, the most common forms of which include mechanical work, heat flux, kinetic energy, and gravitational potential energy.

Work

Work (or mechanical work) is a form of energy exchange between two systems. Since the concept of work was first introduced in mechanical systems, its exact definition is also based on simple mechanical devices. Work is defined as the interaction between two systems, if the sole effect external to each system can be reduced to the change in the vertical position of a weight. Although a weight is referred to in this definition, it is not necessary to have a weight involved in the actual process, and the equivalence of the weight lift is what is conceptually required by the definition. Unless stated otherwise, the convention is to define the work as positive when it is done on the system by its environment, and negative if the work is done by the system on its environment. Energy has a dimension of force times length.

When an external force $\vec{F}$ is applied to a mass moving through a distance $\vec{dx}$, the mechanical work done by the force is $dW = \vec{F} \cdot \vec{dx}$. Assuming the specific volume of a parcel of air or water obtains an increment $dv > 0$, the work done by the external pressure force P is negative, $dW = -Pdv < 0$. For the case shown in Figure 2.2a, $dv > 0$ during the process from state A to state B, so the total work done by the external force is negative

$$\Delta W = -\int_{v_A}^{v_B} P dv < 0$$

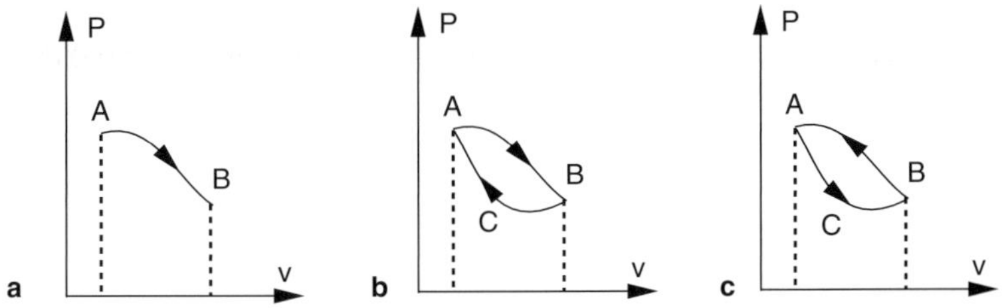

Fig. 2.2 Work done due to volumetric change.

Note that work done during a process depends on how pressure varies from state A to state B, so work is not a state variable. For example, in the case shown in Figure 2.2b, where the system returns from state B back to state A via path BCA, the net work done within this cycle is negative, i.e., there is a net work done to the environment because $\Delta W = -\oint_{ABCA} P dv < 0$. On the other hand, if the direction of the process is reversed, as in the case shown in Figure 2.2c, the net work is positive, $\Delta W = -\oint_{ACBA} P dv > 0$.

Heat

Heat is the energy transferred between two parcels of different temperature and the heat received by the system is

$$dQ = mcdT$$

where m is the mass of a parcel, c is the specific heat, and dT is the temperature change.

First law of thermodynamics

The total amount of energy of a system is conserved.

This law can be written in a differential form

$$dE = dQ + dW \tag{2.23}$$

i.e., the change in internal energy, dE, of the system is equal to the heat received by the system plus the work done to the system.

2.3.3 Entropy

Reversible processes and irreversible processes

A reversible process is one in which the system is very close to its equilibrium at all times, so that the process can be reversed with no changes in the system and its environment. In an irreversible process, the system moves away from the equilibrium state, so that *the system and its environment* cannot come back to their initial states. The adjective "irreversible" does not mean that the system cannot come back to its initial state; all it means is that some irreversible changes must occur in the environment if the system is to be brought back to its initial state.

Change of entropy

Entropy is an essential thermodynamic variable of a system, which was introduced in the study of thermodynamics of irreversible processes. Change of entropy of a system is defined as

$$d\eta \geq dq/T \tag{2.24}$$

where the equal sign is valid for reversible processes only, and dq is the heat received by the system. For irreversible processes there is an additional increase of entropy, as indicated by the "greater than" sign. For a reversible process, the system and its environment remain unchanged after a complete cycle, i.e.,

$$\oint d\eta = \oint dq/T|_{reversible} = 0 \tag{2.25}$$

Although calculating entropy for an ideal gas is quite easy, the calculation of seawater entropy was not easy because there was no simple and reliable formula. However, standard formulae are now available. Accordingly, entropy is defined as a thermodynamic state variable of seawater, and for given temperature, salinity and pressure, entropy can be calculated from standard formulae, as discussed later in this section.

2.3.4 The second law of thermodynamics

Entropy change of a system due to heat exchange with the environment must obey $d\eta \geq dq/T$.

The equals sign is valid only for reversible processes. In general cases, the inequality applies, i.e., the total entropy of the system and its environment increases due to irreversible processes.

Clausius inequality: The total entropy of an isolated system cannot decrease, i.e., $\Delta\eta_{total} \geq 0$. By definition, an isolated system includes the system and its environment.

The Carnot cycle

In order to extract the maximal work from a heat source, an idealized cycle, called the Carnot cycle, based on one kilogram of ideal gas, is designed as follows (Fig. 2.3). An imaginary perfect engine works between a heat source at temperature T_1 and a cooling source at temperature $T_2 < T_1$. The equation of state for one kilogram of ideal gas is

$$Pv = RT \quad \text{and} \quad R = c_v(\gamma - 1) \tag{2.26}$$

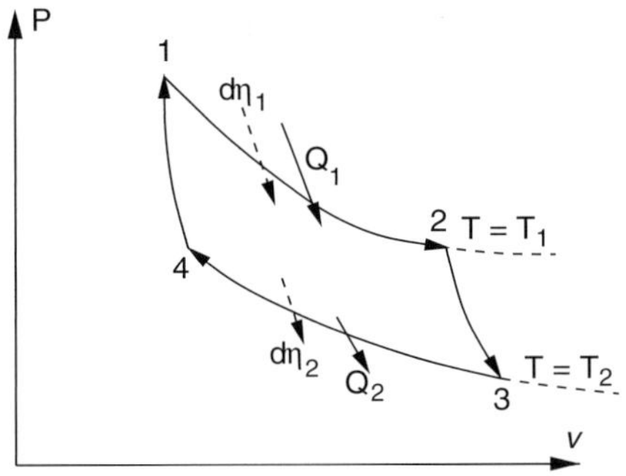

Fig. 2.3 A cycle in the idealized Carnot engine.

where c_v is the specific heat under constant volume, and γ is a power index which is defined for the reversible adiabatic process. During a reversible adiabatic process temperature and specific volume obey

$$Tv^{\gamma-1} = const. \tag{2.27}$$

The idealized Carnot cycle consists of the following four stages, as shown in Figure 2.3:

1. **From state** 1 $\rightarrow$ 2. This is an isothermal expansion process, during which the internal energy remains constant and the work exported equals the heat Q_1 absorbed from the heat source T_1;

$$Q_1 = W_{1\rightarrow 2} = \int_{v_1}^{v_2} P dv = RT_1 \int_{v_1}^{v_2} dv/v = RT_1 \ln(v_2/v_1) > 0 \tag{2.28}$$

   Here the exported work by the system is defined as positive, because our focus is on the work exported to the environment.

2. **From state** 2 $\rightarrow$ 3. This is an adiabatic expansion process, during which the work exported equals the decrease of internal energy of the system:

$$W_{2\rightarrow 3} = c_v (T_1 - T_2) > 0 \tag{2.29}$$

3. **From state** 3 $\rightarrow$ 4. This is an isothermal compression process, during which the work imported equals the heat flux from the system to the cooling source T_2:

$$-Q_2 = W_{3\rightarrow 4} = RT_2 \ln(v_4/v_3) < 0 \tag{2.30}$$

   where the negative sign in the front of heat flux Q_2 indicates that the system loses heat to the cooling source T_2.

4. **From state** 4 $\rightarrow$ 1. This is an adiabatic compression process, during which the work imported equals the increase of internal energy of the system:

$$W_{4\rightarrow 1} = -c_v (T_1 - T_2) < 0 \tag{2.31}$$

Therefore, the total work exported from the system during the complete cycle is the sum of work from each stage

$$W = W_{1\rightarrow 2} + W_{2\rightarrow 3} + W_{3\rightarrow 4} + W_{4\rightarrow 1} = RT_1 \ln(v_2/v_1) + RT_2 \ln(v_4/v_3) \tag{2.32}$$

Using Eqn. (2.27), the volume at these four states satisfies

$$v_4/v_3 = v_1/v_2 \tag{2.33}$$

Thus

$$W = R \ln(v_2/v_1)(T_1 - T_2) = Q_1 (T_1 - T_2)/T_1 \tag{2.34}$$

If we define the efficiency of a heat engine as $\varepsilon = W/Q_1$, the efficiency of the Carnot engine is

$$\varepsilon_{efficiency} = 1 - T_2/T_1 \tag{2.35}$$

Therefore, theoretically, the crucial way to increase the efficiency of a thermal engine is to raise the temperature of the heat source and to reduce the temperature of the cooling source.

Implications of the second law of thermodynamics

While the first law of thermodynamics is a statement of energy conservation that all forms of energy are equivalent, the second law of thermodynamics distinguishes energy in two categories: thermal energy, and mechanical energy (including kinetic energy and gravitational potential energy) and its equivalent. While mechanical energy can be converted into thermal energy 100 percent, thermal energy cannot be converted into mechanical energy 100 percent. The theoretical limit of this conversion rate is set by the Carnot efficiency $\varepsilon = 1 - T_2/T_1$, where T_1 and T_2 are the temperature of the hot and cold reservoirs of the ideal Carnot engine. This upper limit of thermal efficiency is not attainable in practice, and in order to raise efficiency various engineering designs or arrangement are required. Such designs and special arrangements do not exist in nature, so the efficiency of many thermal processes in both the atmosphere and oceans is very low; sometimes it is almost zero.

One of the most vital applications of the second law of thermodynamics in dynamical oceanography is the study of the thermohaline circulation in the ocean. Although the ocean is forced by heating and cooling at the sea surface, there is no efficient way to convert such thermal energy into mechanical energy. Thus, without additional mechanical energy to overcome mechanical energy loss due to friction and dissipation, there would be no strong meridional overturning cells in the world's oceans. This issue will be elucidated in connection with the Sandstrom theorem associated with the thermohaline circulation.

The concept of a heat engine has been widely adopted in studies about climate, atmospheric and oceanic general circulation. However, the mechanisms involved in these systems can be very complicated and differ in fundamental ways from the everyday thermal engines in the automobile and airplane, which we are mostly familiar with.

The atmosphere is a heat engine. If we apply the Carnot formula to the atmosphere and choose the mean temperature at sea level, the heat source is at the equator, so $T_1 = 300$ K, and the high latitude is the cooling source, $T_2 = 200$ K; thus we have $\varepsilon = 1 - T_2/T_1 \approx 33\%$. However, the real efficiency of the atmospheric heat engine is much lower than this theoretical upper limit. The total heat flux through radiation is about $Q_1 = 238$ W/m^2, and the rate of work produced by the atmospheric heat engine is about 2 W/m^2. Thus, the efficiency of the atmospheric heat engine is only about 0.8%. Although the ocean has been called a heat engine in many papers and books, as will be discussed shortly, the ocean is actually not a heat engine; instead, it is a heat-transporting machine, driven by external mechanical energy due to wind stress and tidal dissipation.

Entropy balance in the Carnot engine

Since entropy balance of a system is crucial for understanding how the system works, let us now examine the entropy fluxes associated with a system. The entropy fluxes associated with the reversible heat exchanges are defined in Eqn. (2.24) as $d\eta = dq/T$. The question

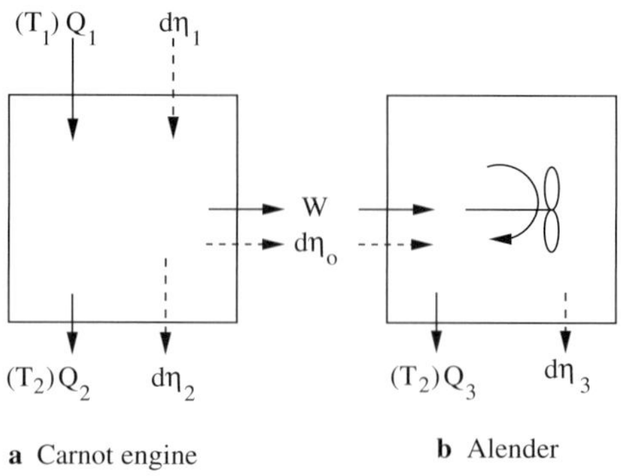

Fig. 2.4 Balance of energy and entropy in **a** an ideal Carnot engine and **b** a blender.

is how to define the entropy flux of an object associated with the exchange of mechanical energy with the environment.

We start with the balance of heat and entropy for an ideal Carnot engine, as shown in Figs. 2.3 and 2.4a. From state 1 to state 2, the incoming entropy flux is

$$d\eta_1 = Q_1/T_1 \tag{2.36}$$

and the outgoing entropy flux during transition from state 3 to state 4 is

$$d\eta_2 = Q_2/T_2 \tag{2.37}$$

However, for this ideal heat engine, we have

$$Q_2 = Q_1 - W = Q_1 T_2/T_1 \tag{2.38}$$

and this leads to

$$d\eta_2 = d\eta_1 \tag{2.39}$$

The processes in an ideal Carnot engine are completely reversible; thus, there should be no entropy production. As a result, the mechanical energy exchange with the environment is entropy-free, i.e.

$$d\eta_o = 0$$

The balance of entropy for a blender

To illustrate the idea that mechanical energy exchange is entropy-free, we examine the entropy balance for a blender which receives mechanical energy output from a Carnot engine. We denote the mechanical energy input as W (Fig. 2.4b). Assuming that no incoming heat flux exits to the blender, that the internal blades convert all the mechanical energy into thermal energy, and that the blender has a net heat flux output which is associated with an entropy flux output, then

$$Q_3 = W, \quad d\eta_3 = W/T_2 \tag{2.40}$$

Inside the blender, the blades act as an internal source of entropy, which is equal to W/T_2. Here again, the entropy balance for the blender requires that the external source of mechanical energy should be an entity without entropy attached to it.

Entropy balance in Carnot and non-Carnot engines

For these cases, whatever sign and magnitude we assign to the entropy output from a Carnot engine, it should be consistent with the incoming source of entropy for the blender. Therefore, the mechanical energy exchanges between a system and its environment should contribute no entropy flux to the system. This argument can be understood as follows: if we regard mechanical as equivalent thermal energy, its equivalent temperature would be infinite; then we have

$$d\eta_o = W/T_{equiv} = W/\infty = 0 \tag{2.41}$$

All non-thermal forms of energy exchange with the environment, including electromagnetic energy or chemical energy, can be equivalent to mechanical energy and treated in the same way as described above.

The balance of entropy for an ideal Carnot engine is a perfect cycle, as illustrated by the solid arrows in Figure 2.5. During the isothermal transition from state 1 to state 2, the engine is at the high temperature and heat flows into the system with the associated entropy flux, as shown by the solid arrow in the upper part of Figure 2.5. During the adiabatic transition from state 2 to state 3, the process is reversible and without heat flux, so that entropy of the system remains constant. During the isothermal transition from state 3 to state 4, heat and entropy flow from the engine to the cooling source. Under the assumptions of an ideal cycle, the entropy flux out of the engine at this stage is exactly the same as that of the incoming entropy flux during the transition from state 1 to state 2. During the adiabatic

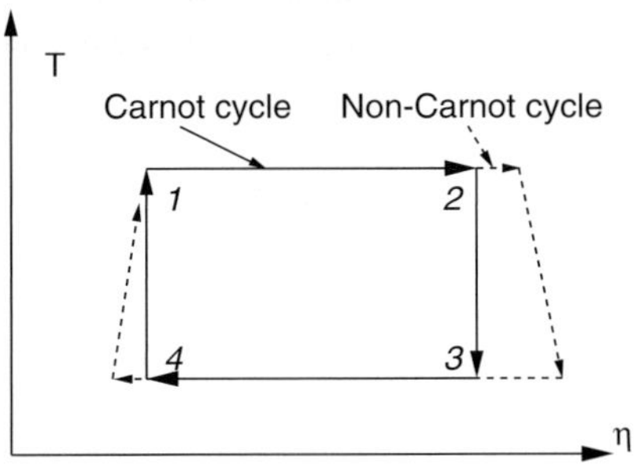

Fig. 2.5 Entropy cycles for an idealized Carnot engine and a non-Carnot engine.

transition from state 4 to state 1, entropy of the system remains constant. Therefore, there is no net change of entropy during the complete cycle. An important point to note is that the amount of entropy increase due to the transition from state 1 to 2 is exactly balanced by the transition from state 3 to 4; thus there is no net entropy production in the system and the environment.

On the other hand, if the engine is not a perfect engine, there will be an extra amount of entropy production associated with each step in the cycle. As indicated by the dashed arrows in Figure 2.5, irreversible processes in the engine make the entropy cycle different from the simple one denoted by the solid arrows. In order to make the system return to the same initial state 1 as that of the perfect Carnot cycle, the entropy produced must be removed and sent to the environment, indicated by the slightly longer arrow denoting the entropy flux for the transition from state 3 to 4, as compared with that during the transition from state 1 to 2. As a result, the total amount of entropy for the isolated system, i.e., the engine and its environment, increases due to these irreversible processes.

2.3.5 Energy versus entropy

Although energy and entropy are two very common quantities in physics, there is something worth examining more closely. These two thermodynamic quantities are closely linked to each other. According to the first law of thermodynamics, the same amount of heat transformation at different temperatures is considered to be equal. However, according to the second thermodynamic law, thermal energy exchange between the system and its environment may have different qualities, so that the same amount of heat flux at different temperatures is considered to have different effects on the system. In fact, heat transport at high temperature is considered to be energy flux of high quality, whereas heat transport at low temperature is considered to be of low quality.

The quality of energy can be illustrated in the following imaginary experiment. Assume that two perfect Carnot engines are driven by the same amount of heat flux from heat reservoirs with different temperatures $T_2 > T_1$, and they are cooled by the same cold reservoir with temperature T_0. According to the discussion above, the engine associated with heat flux with higher temperature should have a higher efficiency than the engine working with heat flux with lower temperature. The difference between these two engines is due to the difference in the quality of heat fluxes they received.

In addition, a simple example related to our daily life is as follows. Two pots each containing a sunflower are exposed to the same conditions, such as the pot soil, air temperature, the amount of water and fertilizer, except for the light. One pot is in the sunlight, but the other one is under a fluorescent light. The sunlight is radiation with high equivalent temperature on the order of several thousands of degrees; on the other hand, the fluorescent light has a much lower equivalent temperature. Although the amount of light energy for these two pots is maintained at the same level, the sunflower in the sunlight should grow much better than that under the fluorescent light because of the difference in the quality of energy in the light.

In summary, entropy can be used as a quality index for the energy transported between systems. As a thermodynamic system, the oceanic circulation should be subject to both the first and second laws of thermodynamics, i.e., balance of both energy and entropy should be essential for the study of the oceanic general circulation.

2.4 Thermodynamics of seawater

Seawater is a mixture of pure water and many different kinds of chemical component, and the mass fraction of each component is denoted as m_i. Other essential thermodynamic variables include temperature, salinity, and pressure; these state variables are denoted as (T, S, P) in this book; the lower-case letter p will be reserved for pressure in the dynamic analysis.

2.4.1 Basic differential relations of thermodynamics

Thermodynamic relations of seawater can be established in different ways. For example, they can be defined in terms of entropy or the Gibbs function. We first introduce thermodynamics from entropy. We will then use the Gibbs function to establish a unified system of defining the thermodynamics of seawater.

Basic differential relations for a multiple-component system

Thermodynamics of a multiple-component system can be established from the definition of specific entropy η. Two other crucial variables of a system, including temperature and specific chemical potential, can be defined as follows

$$\frac{1}{T} = \left(\frac{\partial \eta}{\partial e}\right)_{v,m_i}, \quad \mu_i = -T\left(\frac{\partial \eta}{\partial m_i}\right)_{e,v}, \quad i = 1, 2, \ldots, n \tag{2.42}$$

where e is the specific internal energy, v is the specific volume, m_i and μ_i are the mass fraction and chemical potential for the i-th component of the seawater.

During a reversible adiabatic process (which means that neither heat exchange with the environment nor change in entropy exist there), change in the specific internal energy is balanced by the pressure work, i.e.

$$de = -P dv$$

where P is pressure; this relation can be rewritten as

$$\left(\frac{\partial e}{\partial v}\right)_{\eta,m_i} = -P \tag{2.43}$$

A convenient way to derive some of these thermodynamic relations is to use the Jacobian expression

$$\frac{\partial(f,g)}{\partial(x,y)} = \frac{\partial f}{\partial x}\frac{\partial g}{\partial y} - \frac{\partial g}{\partial x}\frac{\partial f}{\partial y}$$

For example, combining Eqn. (2.43) with Eqn. (2.42) leads to

$$\left(\frac{\partial \eta}{\partial v}\right)_{e,m_i} = \frac{\partial(\eta,e)}{\partial(v,e)} = \frac{\partial(\eta,e)}{\partial(\eta,v)}\frac{\partial(\eta,v)}{\partial(v,e)} = -\left(\frac{\partial e}{\partial v}\right)_{\eta,m_i}\left(\frac{\partial \eta}{\partial e}\right)_{v,m_i} = \frac{P}{T} \tag{2.44}$$

Using Eqns. (2.42), (2.43), and (2.44), the specific entropy satisfies the following ***Gibbs relation***

$$d\eta = \frac{1}{T}de + \frac{P}{T}dv - \sum_{i=1}^{n}\frac{\mu_i}{T}dm_i \tag{2.45}$$

Another very useful thermodynamic function is the ***Gibbs function***

$$\mathsf{g} = e + Pv - T\eta \tag{2.46}$$

Differentiating Eqn. (2.46) leads to

$$d\mathsf{g} = -\eta dT + v dP + \sum_{i=1}^{n}\mu_i dm_i \tag{2.47}$$

Since T and P are intensive variables, they are homogeneous everywhere in an equilibrium system. g is an extensive variable of state of the system; thus it should be a linear function in the mass fractions

$$\mathsf{g}(T,P,m_j) = \sum_{i=1}^{n}\frac{\partial \mathsf{g}}{\partial m_i}dm_i$$

From Eqn. (2.47), we have the ***Euler relation***

$$e + Pv - T\eta = \sum_{i=1}^{n}\mu_i m_i \tag{2.48}$$

Taking the derivative of Eqn. (2.48) and comparing it with Eqn. (2.47), we obtain the ***Gibbs–Duhem equation***

$$\eta dT - v dP + \sum_{i=1}^{n} m_i d\mu_i = 0 \tag{2.49}$$

Seawater as a two-component system

In general, the thermodynamics of a multi-component system is quite complicated. However, many aspects of these complications are not essential for the study of dynamical oceanography; thus it is desirable to simplify the thermodynamics of seawater. One of the common practices in dynamical oceanography is to assume that the ratios of different chemical components remain constant in the world's oceans. And seawater can be treated as an equivalent two-component system, comprising salt and water.

For such a two-component system, we will use the following notations for the mass fractions of salt and water in seawater

$$m_s = s, \ m_w = 1 - s \quad \text{and} \quad dm_w = -ds \tag{2.50}$$

The common expression of salinity is the practical salinity unit (psu), which is defined in parts per thousand (by weight), i.e., $S = 1,000 * s$; however, in this section we will use s for consistency of notation. Thus, Eqns. (2.48) and (2.45) are reduced to

$$T\eta = e + Pv - (1 - s)\,\mu_w - s\mu_s \tag{2.51}$$

$$d\eta = \frac{1}{T}de + \frac{P}{T}dv - \frac{\mu_s - \mu_w}{T}ds = \frac{1}{T}de + \frac{P}{T}dv - \frac{\mu}{T}ds \tag{2.52}$$

where $\mu = \mu_s - \mu_w$ is the specific chemical potential of seawater, and μ_s and μ_w are the partial chemical potential for the salt and water in seawater.

Three types of energy

In addition to the internal energy, there are two commonly used thermodynamic functions: specific enthalpy is defined as

$$h = e + Pv \tag{2.53}$$

and specific free enthalpy (Gibbs function), defined in Eqn. (2.46) as

$$\mathsf{g} = e + Pv - T\eta$$

Using Eqn. (2.50), this relation is reduced to

$$\mathsf{g} = s\mu_s + (1 - s)\,\mu_w \tag{2.46$'$}$$

Thus, there are three types of energy frequently used, including the specific internal energy e, specific enthalpy h, and specific free enthalpy g (Gibbs function). The meaning and connections between these forms of energy are discussed at the end of this section.

In the above approach we build the seawater thermodynamics on the basis of specific entropy. Another way to set up the seawater thermodynamics is to start from the Gibbs function. For example, both specific entropy and specific chemical potential can be defined in terms of the Gibbs function

$$\eta = -\left(\frac{\partial \mathrm{g}}{\partial T}\right)_{s,P}, \quad \mu = \left(\frac{\partial \mathrm{g}}{\partial s}\right)_{T,P} \tag{2.54}$$

The specific chemical potential of pure water is negative infinite by definition, so it is cumbersome to deal with. However, as will be shown later, the term really relevant to problems in dynamical oceanography is the partial chemical potential for the water in seawater, μ_w.

Assuming that the second derivatives are continuous, the order of partial differentiation can be exchanged; cross-differentiation of Eqn. (2.54) leads to the following relation

$$\left(\frac{\partial \mu}{\partial T}\right)_{s,P} = -\left(\frac{\partial \eta}{\partial s}\right)_{T,P}$$

Using specific entropy, specific volume, and mass fraction of salt as independent variables, Eqns. (2.51), (2.52), and (2.53) lead to the following differential relations

$$de = Td\eta - Pdv + \mu ds \tag{2.55}$$

$$dh = Td\eta + vdP + \mu ds \tag{2.56}$$

$$d\mathrm{g} = -\eta dT + vdP + \mu ds \tag{2.57}$$

From the above discussion, we have the following **Maxwell relations**:

$$\eta = -\left(\frac{\partial \mathrm{g}}{\partial T}\right)_{P,s}, v = \left(\frac{\partial \mathrm{g}}{\partial P}\right)_{T,s}, \quad \mu = \left(\frac{\partial \mathrm{g}}{\partial s}\right)_{T,P}, \quad c_p = \left(\frac{\partial h}{\partial T}\right)_{P,s} = T\left(\frac{\partial \eta}{\partial T}\right)_{P,s} \tag{2.58}$$

where c_p is specific heat capacity. Assuming that the second derivatives are continuous, the order of partial differentiation can be exchanged; and we have the following relations

$$\frac{\partial^2 \mathrm{g}}{\partial T \partial T} = -\frac{\partial \eta}{\partial T} = -\frac{c_p}{T} \tag{2.59a}$$

$$\frac{\partial^2 \mathrm{g}}{\partial T \partial P} = \frac{\partial v}{\partial T} = -\frac{\partial \eta}{\partial P} \tag{2.59b}$$

$$\frac{\partial^2 \mathrm{g}}{\partial T \partial s} = -\frac{\partial \eta}{\partial s} = \frac{\partial \mu}{\partial T} \tag{2.59c}$$

$$\frac{\partial^2 \mathrm{g}}{\partial P \partial s} = \frac{\partial \mu}{\partial P} = \frac{\partial v}{\partial s} \tag{2.59d}$$

2.4.2 Basic relations for seawater thermodynamic functions

Specific internal energy

For a two-component system, the specific energy consists of two parts, and each of them is proportional to its mass fractions

$$e = se_s + (1 - s)\, e_w = T\eta - Pv + s\mu_s + (1 - s)\, \mu_w \tag{2.60}$$

The partial internal energy for salt and pure water in seawater is defined as

$$e_s = e + (1 - s)\frac{\partial e}{\partial s}, \quad e_w = e - s\frac{\partial e}{\partial s} \tag{2.61}$$

Using T, P, s as independent variables, we have the differential forms

$$d\eta = \frac{\partial \eta}{\partial T}dT + \frac{\partial \eta}{\partial P}dP + \frac{\partial \eta}{\partial s}ds \tag{2.62}$$

$$dv = \frac{\partial v}{\partial T}dT + \frac{\partial v}{\partial P}dP + \frac{\partial v}{\partial s}ds \tag{2.63}$$

Substituting these relations into Eqns. (2.55) and using Eqns. (2.59a) and (2.59c) leads to

$$de = \left(c_p - P\frac{\partial v}{\partial T}\right)dT + \left(T\frac{\partial \eta}{\partial P} - P\frac{\partial v}{\partial P}\right)dP + \left(\mu - T\frac{\partial \mu}{\partial T} - P\frac{\partial v}{\partial s}\right)ds \tag{2.64}$$

Specific enthalpy

For a two-component system, the specific enthalpy also contains two parts proportional to the corresponding mass fractions. Using Eqn. (2.60), we have the following relation

$$h = sh_s + (1 - s)\, h_w = T\eta + s\mu_s + (1 - s)\, \mu_w \tag{2.65}$$

The partial enthalpy for salt and pure water in seawater is defined as

$$h_s = h + (1 - s)\frac{\partial h}{\partial s}, \quad h_w = h - s\frac{\partial h}{\partial s}, \quad h_s - h_w = \frac{\partial h}{\partial s} \tag{2.66}$$

The differential relation Eqn. (2.56) can be rewritten as

$$\begin{aligned} dh &= T\frac{\partial \eta}{\partial T}dT + \left(v + T\frac{\partial \eta}{\partial P}\right)dP + \left(\mu + T\frac{\partial \eta}{\partial s}\right)ds \\ &= c_p dT + \left(v - T\frac{\partial v}{\partial T}\right)dP + \left(\mu - T\frac{\partial \mu}{\partial T}\right)ds. \end{aligned} \tag{2.67}$$

Specific free enthalpy (Gibbs function)

For a two-component system, the specific free enthalpy function similarly contains two parts

$$\mathsf{g} = s\mathsf{g}_s + (1 - s)\, \mathsf{g}_w = s\mu_s + (1 - s)\, \mu_w \tag{2.68}$$

The partial free enthalpy for salt and pure water in seawater is defined as

$$\mathsf{g}_s = \mathsf{g} + (1 - s)\frac{\partial \mathsf{g}}{\partial s} = \mu_s \tag{2.69}$$

$$\mathsf{g}_w = \mathsf{g} - s\frac{\partial \mathsf{g}}{\partial s} = \mu_w \tag{2.70}$$

where the partial chemical potentials satisfy the following relation

$$\mu_s - \mu_w = \mu = \frac{\partial \mathsf{g}}{\partial s} \tag{2.71}$$

2.4.3 Density, thermal expansion coefficient, and saline contraction coefficient

Seawater density is a nonlinear function of state variables

$$\rho = \rho\,(S, T, P) \tag{2.72}$$

which can be calculated from the commonly used standard subroutine. In dynamical oceanography, density is often given as an anomaly from 1,000 kg/m^3, and this is sometimes called the σ unit (kg/m^3).

Density is low for warm and fresh water, but it is high for cold and salty water. For the normal range of temperature (-2 to 30°C) and salinity (30–37), surface density is within the range of 20–30 kg/m^3 (Fig. 2.6a). Since cold water appears in high latitudes only, where salinity is normally low (in most cases, it is lower than 35), surface density is no larger than 28 kg/m^3.

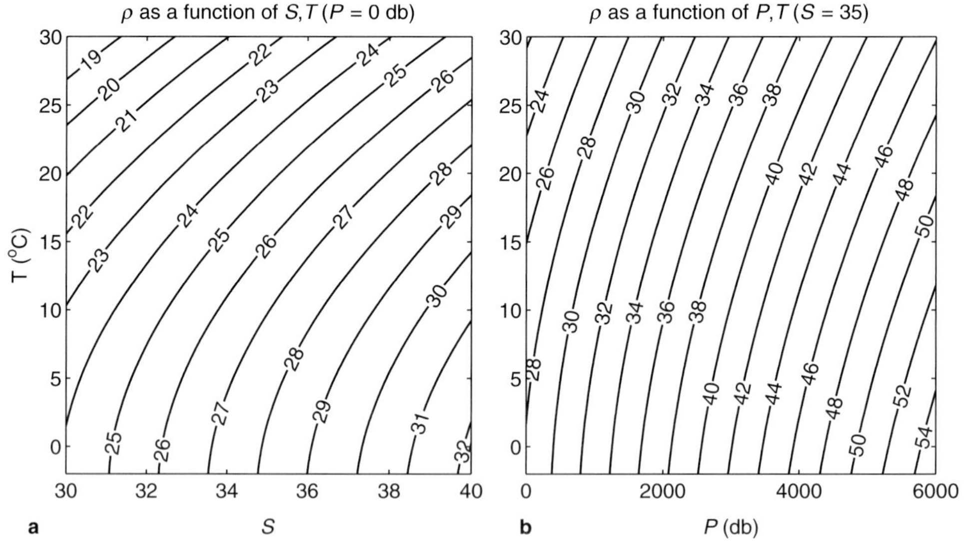

Fig. 2.6 Seawater density (kg/m^3): **a** as a function of (S,T); **b** as a function of (P,T).

Density increases with depth, owing to the compression of water. At great depth, density can reach high values. For example, at 6,000 db level, and 0°C, density is about 54 kg/m^3. However, a large part of the density increase with depth in the world's oceans is dynamically inert; thus the so-called potential density is introduced in oceanography, and its definition is discussed shortly.

Changes in density can be described in terms of the thermal expansion coefficient and the saline contraction coefficient

$$\alpha = -\frac{1}{\rho}\left(\frac{\partial \rho}{\partial T}\right)_{P,s} \tag{2.73}$$

$$\beta = \frac{1}{\rho}\left(\frac{\partial \rho}{\partial S}\right)_{T,P} \tag{2.74}$$

The thermal expansion coefficient increases with temperature and salinity almost linearly (Fig. 2.7a). Since salinity in the ocean varies within a rather narrow range of 30–37, changes in thermal expansion coefficient induced by changes in salinity are small. It is also well known that the thermal expansion coefficient for freshwater at near freezing temperature is very small and can be negative. The thermal expansion coefficient also increases with the *in situ* pressure (Fig. 2.7b). This is linked to the thermobaric effect, which is discussed shortly.

The saline contraction coefficient declines very slightly with increases in temperature, salinity, and pressure (Fig. 2.8). Thus, in theoretical studies, the saline contraction coefficient can be treated as approximately constant.

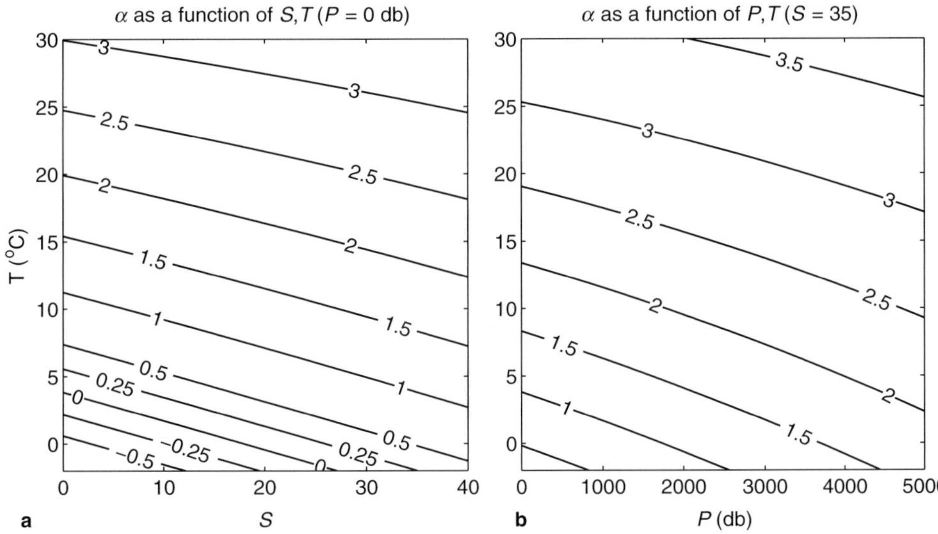

Fig. 2.7 Thermal expansion coefficient (10^{-4}/°C): **a** as a function of (S,T); **b** as a function of (P,T).

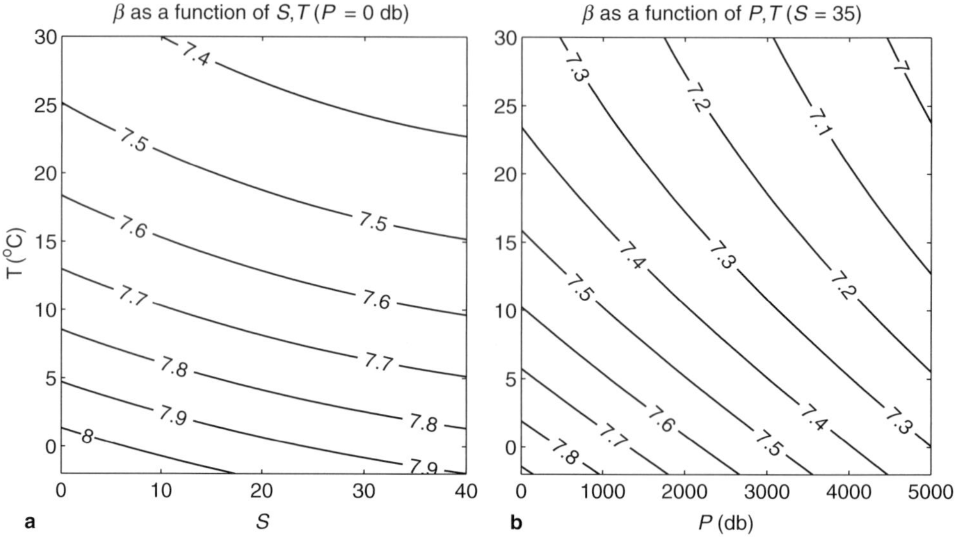

Fig. 2.8 Haline contraction coefficient (10^{-4}/psu): **a** as a function of (S,T); **b** as a function of (P,T).

2.4.4 Specific heat capacity

Specific heat capacity differentiates in two different cases. First, if heating/cooling takes place under a constant pressure, we have

$$c_p = \left(\frac{\partial h}{\partial T}\right)_{P,s} = T\left(\frac{\partial \eta}{\partial T}\right)_{P,s} \tag{2.75}$$

which is the specific heat capacity under constant pressure.

Second, if heating/cooling takes place with a constant volume, we have

$$c_v = \left(\frac{\partial e}{\partial T}\right)_{v,s} = T\left(\frac{\partial \eta}{\partial T}\right)_{v,s} \tag{2.76}$$

which is the specific heat capacity with constant volume.

These two types of specific heat are related through the following relation

$$c_p = c_v + \frac{T\alpha^2}{\rho K_T} \tag{2.77}$$

where

$$K_T = \frac{1}{\rho}\left(\frac{\partial \rho}{\partial P}\right)_{T,s} \tag{2.78}$$

is the isothermal compressibility of the fluid. Although seawater is almost incompressible, nevertheless its density can change, and therefore maintaining a constant specific volume is not a good assumption. As a result, the specific heat capacity under constant pressure is most commonly used.

2.4.5 Compressibility and adiabatic temperature gradient

The compressibility of a fluid can be defined in two ways. First, assuming that temperature remains constant during the compression, we obtain the compressibility under constant temperature, K_T. Second, assuming that the entropy of the fluid remains constant, i.e., the process is adiabatic and reversible, we obtain the compressibility under constant entropy, K_η. Thus

$$K_T = \frac{1}{\rho}\left(\frac{\partial \rho}{\partial T}\right)_{T,s}, \quad K_\eta = \frac{1}{\rho}\left(\frac{\partial \rho}{\partial P}\right)_{\eta,s} \tag{2.79}$$

The compressibility is closely related to the speed of sound waves $c = \sqrt{(\partial P/\partial \rho)_\eta}$.

$$\left(\frac{\partial v}{\partial P}\right)_\eta = \frac{\partial(v,\eta)}{\partial(P,\eta)} = \frac{\partial(v,\eta)}{\partial(T,P)}\frac{1}{\partial(P,\eta)/\partial(T,P)} = -\left(\frac{\partial v}{\partial T}\frac{\partial \eta}{\partial P} - \frac{\partial \eta}{\partial T}\frac{\partial v}{\partial P}\right)\frac{1}{\partial\eta/\partial T}$$

or

$$\left(\frac{\partial v}{\partial P}\right)_\eta = \left[\left(\frac{\partial v}{\partial T}\right)^2 + \frac{c_p}{T}\left(\frac{\partial v}{\partial P}\right)_T\right]\frac{T}{c_p} \tag{2.80}$$

Using Eqn. (2.77), this leads to a simple relation between these two compressibility coefficients

$$K_\eta = \frac{c_v}{c_p}K_T \tag{2.81}$$

Owing to the nonlinearity of the equation of state, compressibility changes with state variables. In particular, compressibility is larger for colder water (Fig. 2.9). Thus, cold water formed at high latitudes can sink to the bottom of the ocean because it is more compressible than the warm and salty water formed at low latitudes, such as the Mediterranean overflow. This nonlinearity of seawater has very important implications for the thermohaline circulation, as discussed in Section 5.1.3.

2.4.6 Adiabatic lapse rate

The adiabatic vertical temperature gradient Γ is defined as the temperature change per unit pressure change in a reversible adiabatic process. From thermodynamic relations, Γ can be calculated from the following equation

$$\Gamma = \left(\frac{\partial T}{\partial P}\right)_\eta = \frac{\partial(T,\eta)}{\partial(P,\eta)} = \frac{\partial(T,\eta)}{\partial(T,P)}\frac{1}{\partial(P,\eta)/\partial(T,P)} = \frac{T}{c_p}\frac{\partial v}{\partial T} = \frac{\alpha T}{\rho c_p} \tag{2.82}$$

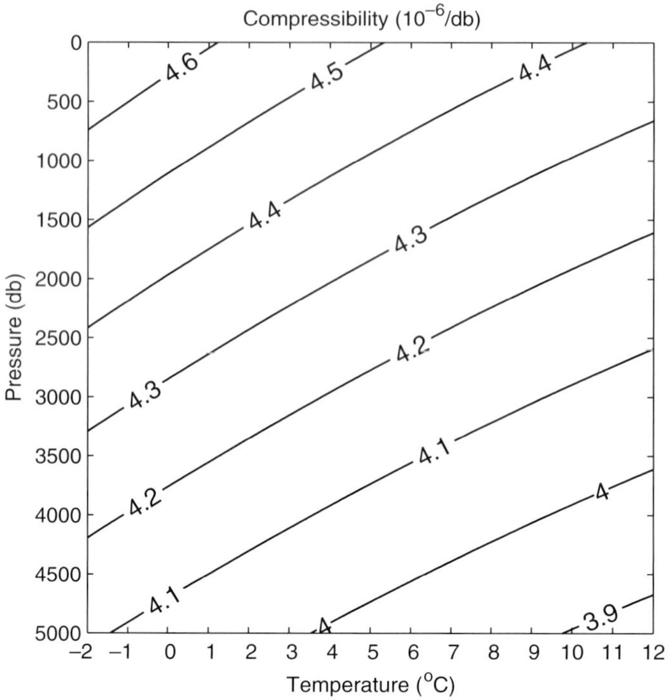

Fig. 2.9 Compressibility under constant temperature (K_T) of seawater as a function of temperature and pressure, with $S = 35$.

This quantity will be used to introduce potential temperature and potential density. Since all three variables (T, ρ, c_p) in this expression vary less than 10 percent for seawater, while the thermal expansion coefficient α can vary over a much wider range, Γ is mostly controlled by α. There are two interesting situations. First, as shown in Figure 2.7a, α increases with temperature almost linearly, and Γ also increases with temperature (Fig. 2.10a). As a result, if two water parcels with the same salinity but different temperatures are brought downward adiabatically and without exchanging salt with the environment, the warm-water parcel is warmed up more than the cold-water parcel; this will be further discussed shortly.

Second, α is negative for freshwater at temperatures near freezing. Therefore, the corresponding adiabatic lapse rate is negative (Fig. 2.10a); this means that if a parcel of freshwater of near-freezing temperature is brought downward without exchanging heat or salt with the environment, its temperature could drop, and this is the opposite of the warming up of salty water discussed above.

Adiabatic lapse rate in the deep ocean is larger than at the sea surface. Because α increases with pressure, as shown in Figure 2.7b, Γ also increases slightly with pressure (Fig. 2.10b).

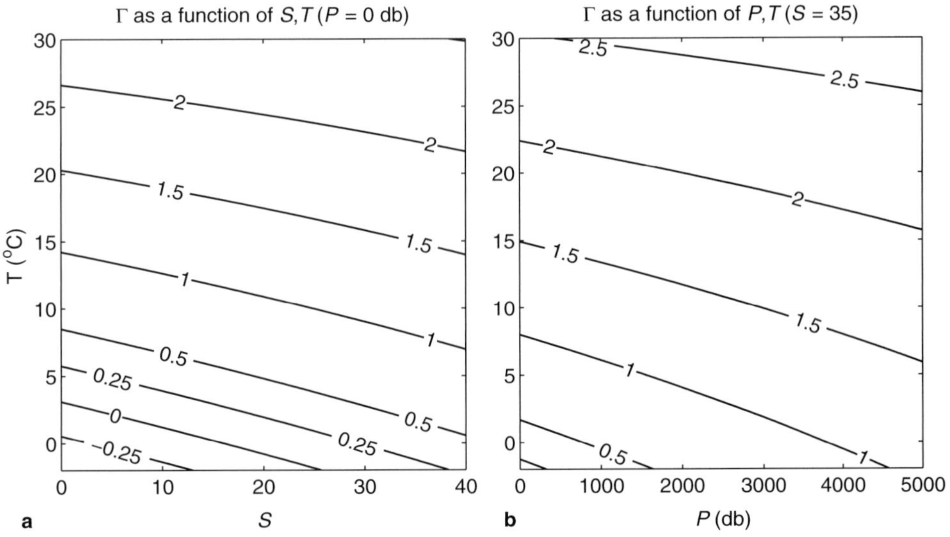

Fig. 2.10 Adiabatic lapse rate (10^{-4}°C/db): **a** as a function of (S,T), under atmospheric pressure (0 db); **b** as a function of (P,T), with $S = 35$.

2.4.7 Potential temperature

Now let us examine the following case. When a water parcel is adiabatically moved downward in a water column without changing its salinity, its temperature is increased owing to compression. From Eqn. (2.62)

$$\delta T = -\frac{\partial \eta / \partial P}{\partial \eta / \partial T} \delta p = \Gamma \delta P \tag{2.83}$$

For example, if $T = 283$ K, $\alpha = 1.77 \times 10^{-4}$/°C, $\delta p = 5,500$ db, $c_p \simeq 4200$ J/kg/K, then $\delta T \simeq 0.64$°C. Therefore, if a water parcel is moved from the sea surface to a depth of 5.5 km (the equivalent pressure change is very slightly larger than 5,500 db), its temperature increases about 0.64°C. Such an increase in temperature is due to compression and receipt of mechanical energy from the environment; therefore, the *in situ* temperature of a water parcel adiabatically moving in the ocean is not a conserved quantity. For the study of oceanic circulation, however, it is highly desirable to label the water parcel with some conserved quantity, i.e., a property without such an energy exchange. The concept of potential temperature was thus introduced for such a purpose.

Potential temperature has been widely used in physical oceanography; it means the temperature of a seawater parcel during vertical movement, without including the effects of compression or salinity exchange. Equation (2.83) can be used to make a rough estimate of the change of temperature due to compression; for the exact change, the exact equation of state should be used. For example, when a water parcel with salinity 35 and temperature 0°C (10°C) at the sea surface sinks to the 5,500 db level, slightly less than 5 km below sea level, its temperature becomes 0.461°C (10.821°C). Thus, temperature increase during the downward movement depends on the temperature of the water parcel. In fact, cold water warms up less than warm water, as shown in this example, and this is due to the relation between the adiabatic lapse rate and the thermal expansion coefficient, as discussed in the previous subsection.

Potential temperature was first introduced and defined by Helland-Hansen (1912). The current definition is slightly modified from the original definition by explicitly specifying the reference pressure. Thus, potential temperature $\Theta(S, T, P, P_r)$ is defined as the temperature of a seawater parcel if it is moved adiabatically from an initial pressure P to a reference pressure P_r with no change in its salinity

$$\Theta\,(S_0, T_0, P_0, P_r) = T_0 + \int_{P_0}^{P_r} \Gamma(S_0, \Theta(S_0, T_0, P_0, P), P)dP \tag{2.84}$$

For a water parcel with salinity $S = 34.85$ at the surface and $T = 0$°C, we have $\Gamma =$ 0.0355°C/1,000 db; if $T = 20$°C, $\Gamma = 0.1841$°C/1,000 db; at the depth of 5,000 db and $T = 0$°C, $\Gamma = 0.118$°C/1,000 db. As discussed in the previous section, Γ increases with pressure, and this is primarily due to the increase of the thermal expansion coefficient with pressure. From the sea surface to 4 km depth, the typical adiabatic temperature range is approximately 0.6°C; however, the *in situ* temperature range is about 20°C. At depths below 4 km, the amplitude of the adiabatic temperature lapse rate and the vertical gradient of *in situ* temperature are comparable.

Why is Γ so much different from the vertical gradient of the real temperature? This difference is primarily due to the fact that large-scale vertical motion is dynamically forbidden; thus, water masses in the deep ocean do not come through the local vertical movement. Instead, water masses are formed at high latitude and move down along isopycnal surfaces. Thus, the large vertical temperature gradient observed at low or mid latitudes is the result of lateral motions on a global scale.

2.4.8 Potential density

Seawater is almost incompressible; nevertheless, its density can change slightly due to pressure change. In fact, density in the ocean generally increases with depth; however, a major part of the vertical density gradient is due to the increase of pressure alone. In the study of dynamical oceanography, it is often desirable to identify density changes due to other physical factors, such as temperature and salinity. Thus, potential density was introduced and has been widely used.

Potential density is defined as the density of a seawater parcel if it is moved adiabatically and without changing its salinity from an initial pressure P to a reference pressure P_r. Its

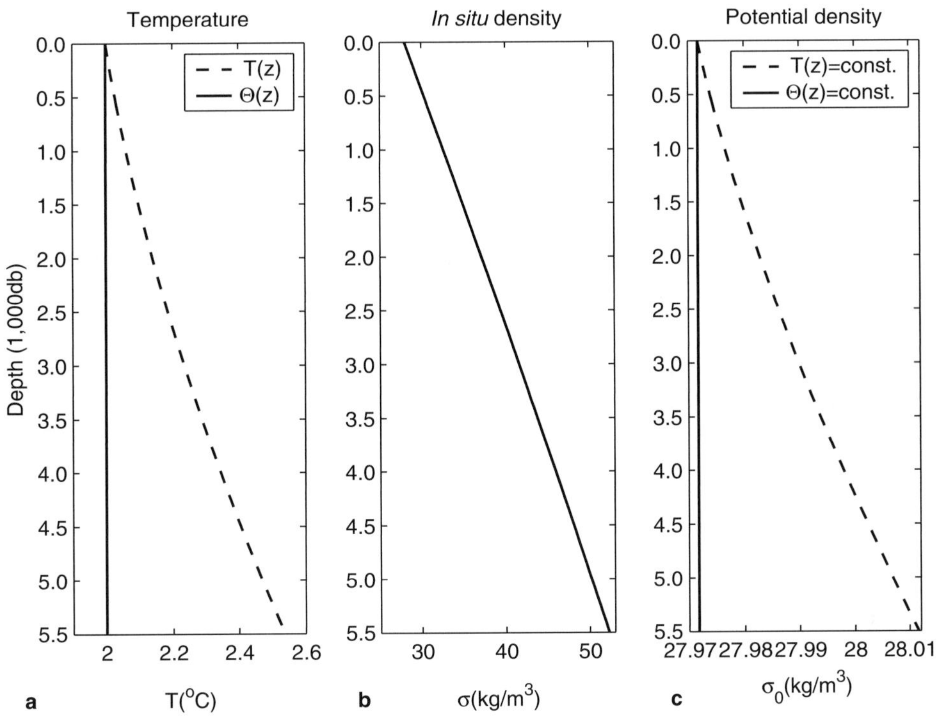

Fig. 2.11 Idealized vertical profiles of **a** *in situ* temperature and potential temperature, **b** *in situ* density and **c** potential density.

definition can be written as

$$\sigma(S, T, P, P_r) = \rho(S, \Theta(S, T, P, P_r), P) = \rho(S, \Theta, P) \tag{2.85}$$

There are commonly accepted subroutines to calculate both potential temperature and potential density in both FORTRAN and Matlab, so such a calculation can easily be carried out for practical purposes. A simple example is shown in Figure 2.11c, where the heavy solid line indicates potential density for the case of a constant potential temperature profile; on the other hand, the dashed line indicates the potential density profile for the case when the *in situ* temperature profile is constant.

Potential density has been widely used in the study of water column stability and relevant issues, such as in the definition of the Rossby deformation radius. The most frequently used potential density is based on using the sea surface as the reference level. Owing to the nonlinearity of the equation of state of seawater, however, stratification inferred from potential density defined in this way, denoted as σ_0 here, can be unstable in the Atlantic Ocean. For example, the potential density, σ_0, distribution along the 30.5° W section for the Atlantic shows that it is not monotonic in the vertical section (Fig. 2.12). Such a seemingly

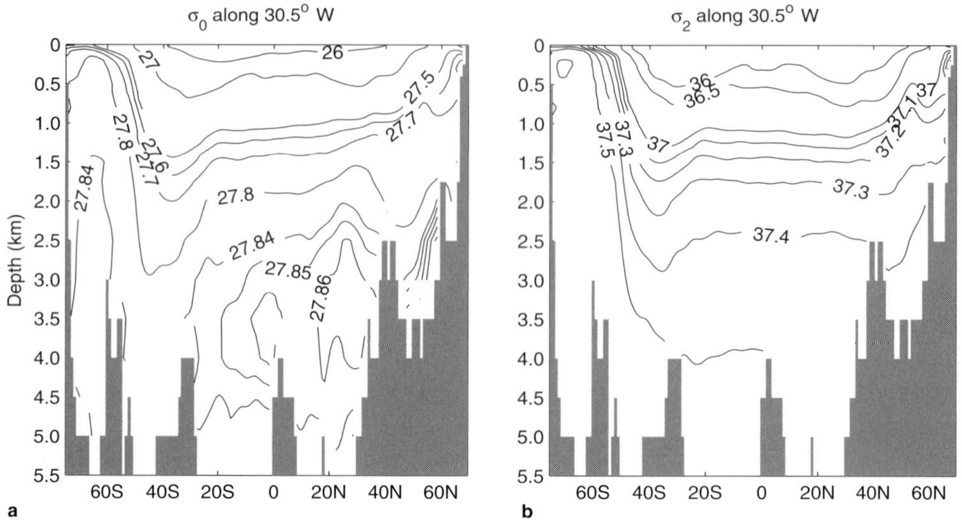

Fig. 2.12 Meridional distribution of potential density along 30.5° W (kg/m^3), using two different reference levels.

unstable stratification is artificial. Were the equation of state linear, there would be no such artificial problems.

Apparently, the vertical gradient of σ_0 changes signs at about 4 km depth. Thus, the water column at this depth range seems gravitationally unstable. A column of water taken from depth range 4–4.5 km would be unstable if it were adiabatically brought to the sea surface. Under the adiabatic assumption, at the sea surface, water from 4 km depth would be heavier than water from 4.5 km depth. However, the water column at 4–4.5 km is actually quite stable to small perturbations. The only problem for such a seemingly unstable case is that such extremely large perturbations do not happen in the ocean. As a matter of fact, water below 4.5 km belongs to the Atlantic Bottom Water formed near Antarctica, while water at depth 4 km belongs to the Atlantic Deep Water formed near the Greenland/Norwegian Sea.

In fact, the water column in the depth range of 4–4.5 km is stable for small perturbations in the vertical direction, and the seemingly unstable situation inferred from the vertical gradient of σ_0 is an artifact due to the nonlinearity of the equation of state. To overcome this difficulty, we often use potential density defined at different levels. For example, σ_0 (or σ_Θ) is used for the upper ocean; σ_2 (using 2,000 db as the reference pressure, which corresponds to the *in situ* pressure at the depth of slightly less than 2 km) is used for the mid-depth circulation; and σ_4 (using 4,000 db as the reference pressure) is used for analyzing the deep circulation. If we want to study a process that takes place over a large depth range, the neutral surface can be used, as discussed in the next subsection.

Note that using σ_2 improves the situation, but there may arise some minor problems near the sea floor; thus, for abyssal circulation, it is better to use σ_4. A simple approach relying on choosing a simple reference pressure does not seem to work for the whole depth range of the ocean, and this is a technical difficulty which oceanic general circulation models based on density coordinates must overcome. To avoid the potential problems associated with such artificial instabilities, the commonly used MICOM (Miami Isopycnal Coordinate Ocean Model) model is based on σ_2 as a compromise.

Using potential temperature we can compare the density variation due to contributions from both the pressure and potential temperature changes. As we have seen above, the range of adiabatic temperature is about 0.6°C within the top 4 km; the corresponding density change due to this effect is about 0.10 kg/m^3. For the same range of depth change, the density increases from 27 kg/m^3 to 45 kg/m^3, i.e., an increase of 18 kg/m^3. A large part of the vertical density variance in the oceans is due to change in pressure. This part of density variance is dynamically inactive, except for keeping the water column stable. This is an important reason why potential density is so widely used in the study of oceanic circulation.

As an example, we examine the case of a water parcel at the sea surface with a temperature of 2°C. If this water parcel is moved downward, without exchanging heat and salt with the environment, its potential temperature is kept constant; however, its *in situ* temperature increases with pressure, indicated by the dashed line in Figure 2.11a. The *in situ* density also increases (Fig. 2.11b); such an increase in density is due to the compressibility of seawater, as explained above. The corresponding potential density does not change, as indicated by the solid line in Figure 2.11c; however, if the *in situ* temperature were vertically constant, $T \equiv 2°\text{C}$, the corresponding potential density would increase with depth, as indicated by the dashed line in Figure 2.11c.

2.4.9 Thermobaric effect

In order to separate the effects of density change due to temperature and pressure, in this subsection water density is defined as a function of potential temperature, salinity, and pressure. Since density can change solely due to pressure change, using potential temperature, instead of temperature, is more convenient in the discussion of thermobaric effect. Accordingly, the commonly used thermal expansion coefficient is defined as $\alpha = (\partial\rho/\partial\Theta)_{P,S}$. To the first-order approximation, seawater density is linearly proportional to changes in Θ, S, P; in a Taylor expansion, we have

$$\frac{\rho}{\rho_0} - 1 = -\alpha\delta\Theta + \beta\delta S + \gamma\delta P + \frac{\partial^2\rho}{\partial\Theta\partial P}\delta\Theta\delta P + \cdots \qquad (2.86)$$

where $\gamma = K_\eta$. The second-order term on the right-hand side is the so-called ***thermobaric effect***, i.e., density changes due to the combination of temperature and pressure. Owing to the contribution from the second-order term in Eqn. (2.86), α increases when pressure is enhanced. This will be shown in the following discussion.

As an example, changes in specific volume and specific internal energy of two water parcels under adiabatic compression are shown in Figure 2.13. It is readily seen that cold

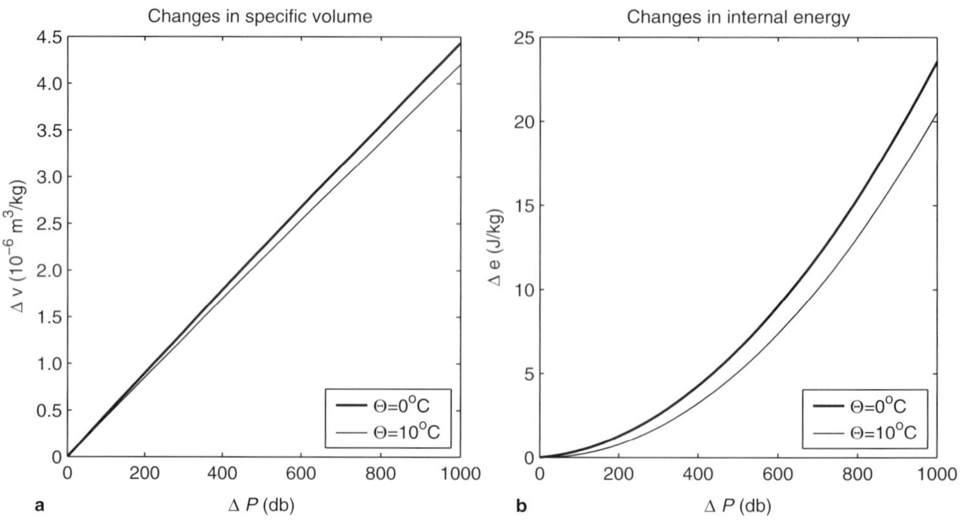

Fig. 2.13 Changes in **a** specific volume and **b** internal energy as a function of potential temperature and pressure.

water is more compressible, and thus the internal energy of a cold-water parcel also increases more quickly with the increase of pressure than a warm-water parcel.

The different behaviors of compressibility γ and thermal expansion coefficient α can best be illustrated by following the two examples shown in Figure 2.14. By definition

$$\gamma = \frac{\partial\,(\ln \rho)}{\partial P}, \qquad \alpha = -\frac{\partial\,(\ln \rho)}{\partial \Theta}$$

thus

$$\frac{\partial \gamma}{\partial \Theta} = -\frac{\partial \alpha}{\partial P} \tag{2.87}$$

It is well known that α is very small at low temperature. For example, $\alpha \simeq 0.5 \times 10^{-4}/°\text{C}$ at $\Theta = 0°\text{C}$ and $S = 35$. As T increases, α increases almost linearly (Fig. 2.14). The fact that the thermal expansion coefficient is very small at nearly freezing temperature has crucial dynamical implications for high-latitude oceanography. Since the haline contraction coefficient β is nearly constant, a very small α means that salinity is the dominating factor in regulating the stratification and flow there.

Another critical fact is that α increases with pressure, i.e., $\partial\alpha/\partial P > 0$, as shown in Figures 2.14a and 2.15a. According to Eqn. (2.87), this is consistent with the fact that $\partial\gamma/\partial\Theta < 0$, as shown in Figure 2.14b. Furthermore, the derivative $\partial\alpha/\partial\Theta$ declines with pressure and temperature (Fig. 2.15b). Thus, cold water is more compressible than warm water: the so-called thermobaric effect. Although sea surface temperature in most parts of the world's oceans is relatively warm, water in the subsurface ocean and deep ocean is

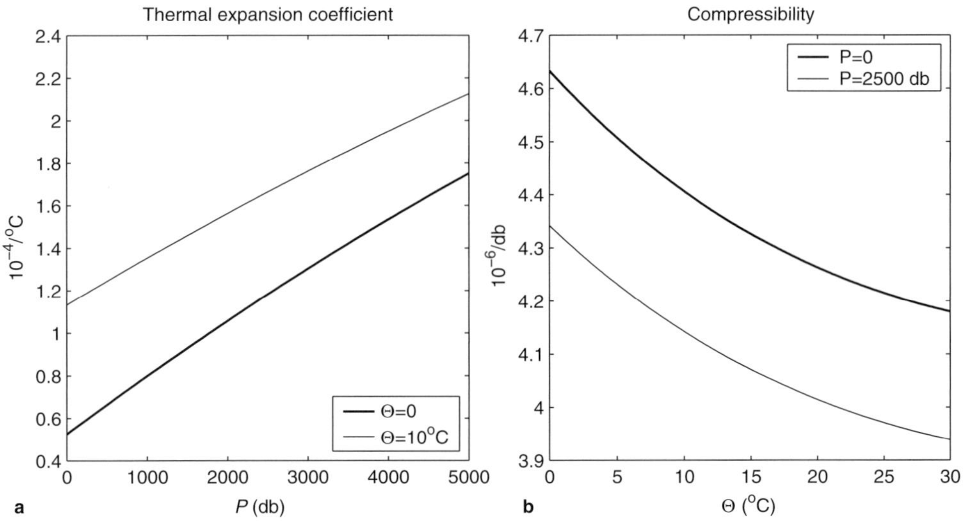

Fig. 2.14 **a** Pressure dependence of the thermal expansion coefficient α (10^{-4}/°C); **b** temperature dependence of the compressibility γ (10^{-6}/db), for two water parcels with $S = 35$.

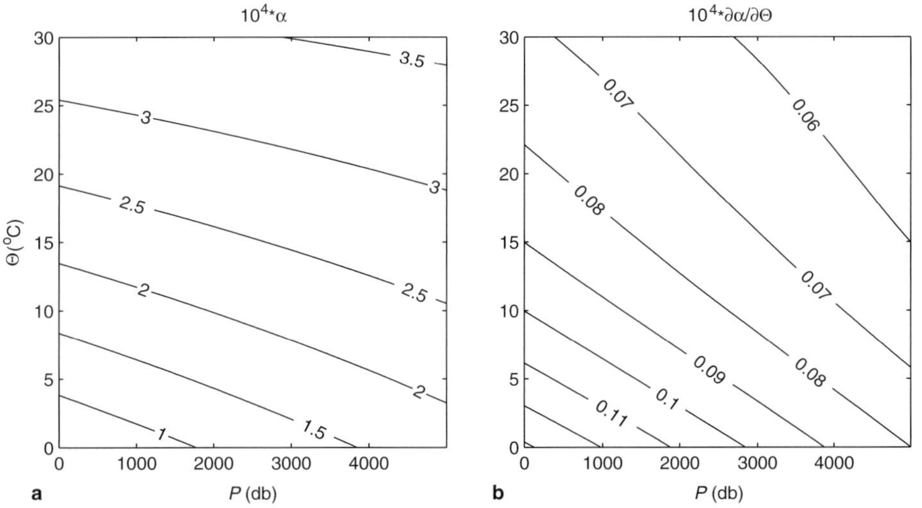

Fig. 2.15 **a** Thermal expansion coefficient and **b** its derivative as a function of potential temperature and pressure, with $S = 35$.

quite cold. In fact, water temperature in the world's oceans is mostly below 5°C; thus, only this part of the curves in Figure 2.15 is most relevant to the circulation at middle and deep levels. This is a very important physical property of seawater; it will be discussed in detail later in connection with deepwater formation and circulation.

2.4.10 Cabbeling

When two water parcels with the same mass but different temperature and salinity mix together, the newly generated water parcel may have a density greater than the mean density of the original water parcels. In particular, if the original water parcels have the same density, then the newly formed water parcel may have a density greater than the mean density of the original parcels. As a result, the newly formed water parcel will sink. This process, called cabbeling in oceanography, is due to the nonlinearity of the equation of state of seawater, especially the increase of the thermal expansion coefficient with temperature.

As an illustration, if two water parcels, A and B, with the same volume and same density ($\sigma_0 = 26\ \mathrm{kg/m^3}$) mix, then the newly formed water parcel should have the mean temperature and salinity; thus, it is represented by point C in the $T - S$ diagram (Fig. 2.16). It is clear that water parcel C is heavier than both A and B, so it will sink to the density layer below in the ocean. Cabbeling plays an essential role in forming the dense water sinking into the deep ocean, and is intimately connected with the formation of North Atlantic Deep Water (NADW) and Antarctic Bottom Water (AABW).

2.4.11 Neutral surface and neutral density

To overcome the problem associated with potential density discussed above, we can use the local pressure as the reference pressure to define the potential density, σ_r, where the subscript r indicates the local reference pressure r. Therefore, the so-called neutral surface is defined as a surface whose normal is in the direction of $-\alpha\nabla\Theta + \beta\nabla S$, where α and β are the coefficients of thermal expansion and saline contraction. If we can construct a surface

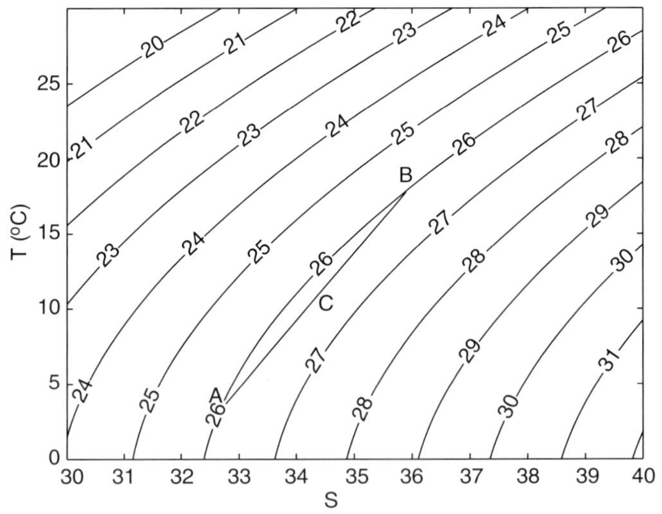

Fig. 2.16 Density at the sea surface as a function of temperature (T) and salinity (S).

whose tangent is everywhere perpendicular to $-\alpha\nabla\Theta + \beta\nabla S$ in the world's oceans, such a surface would be the neutral surface we need. Unfortunately, such a surface turns out to be a helical surface; and this means that a trajectory beginning from one station (x, y) at depth z and going around the basin cannot come back to the same depth, even if it reaches the same location (x, y). In general, the discrepancy between the starting and ending depths is small, on the order of a few meters. Nevertheless, this helicity renders the exact definition of neutral surfaces in the ocean impossible.

To overcome this problem, an algorithm has been developed by Jackett and McDougall (1997). For any given location in the world's oceans, from the *in situ* observed temperature, salinity, and pressure, this algorithm provides the corresponding neutral density variable γ^n. The surfaces defined in this way are approximately neutral and they stay within a few tens of meters of an ideal surface anywhere in the world.

2.4.12 Spiciness

Since a thermodynamic state of seawater can be uniquely determined by three variables: temperature (or potential temperature), salinity, and pressure, plotting potential temperature and salinity distribution on an isopycnal surface is redundant. In fact, potential temperature and salinity on an isopycnal surface are not independent. Thus, to represent the thermodynamic properties on an isopycnal surface more concisely, another thermodynamic function which is "perpendicular" or "orthogonal" to density is desirable. This thermodynamic variable is called "spiciness" (Munk, 1981). Since its introduction in oceanography, spiciness has been used as a tool in the study of double diffusion and other processes in the oceans.

There are many different ways to define such a thermodynamic function. A scale-invariant constraint for defining spiciness is to require that, at any given point in the (Θ, S) space, the slopes of the isopycnals and spiciness isopleths are equal and of opposite sign (Fig. 2.17). The spiciness can be defined in terms of a polynomial (Flament, 2002):

$$\pi\,(\Theta, S) = \sum_{i=0}^{5}\sum_{j=0}^{4} b_{ij}\Theta^i\,(S - 35)^j \tag{2.88}$$

where b_{ij} are the coefficients determined through a least-square fitting to the constraints.

2.4.13 Stability and Brunt–Väisälä frequency

Assume that a water parcel at level z is displaced to a slightly higher level $z + \zeta$. Its density will be different from that of the new environment. Whether such an initial perturbation is stable depends on the vertical distribution of water properties. We will assume that the processes are adiabatic and reversible, so the new density of the parcel at the new location will be

$$\rho' = \rho + (\partial\rho/\partial z)_{\eta,S}\,\zeta \tag{2.89}$$

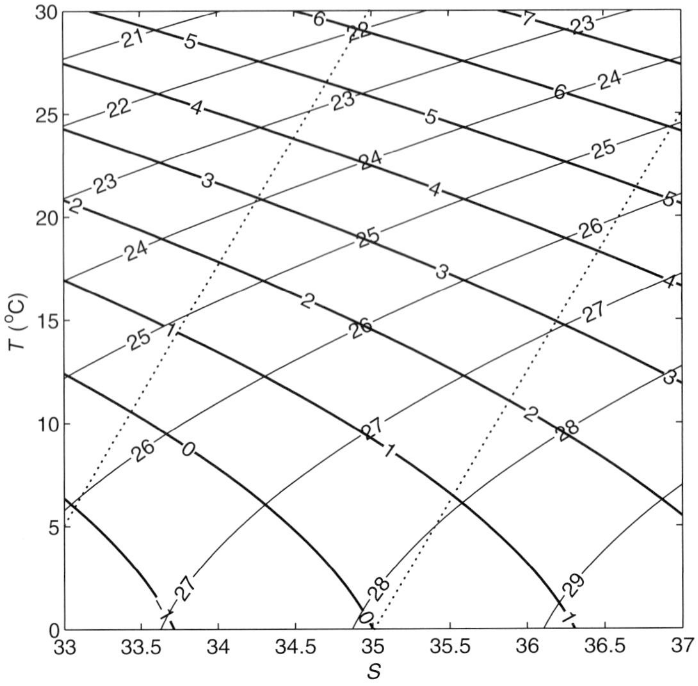

Fig. 2.17 Isopycnals (*thin lines*) and spiciness isopleths (*heavy lines*) at sea surface; the *dotted lines* indicate the domain of minimizing error in least-square fitting (kg/m^3) (redrawn from Flament, 2002).

where $(\partial\rho/\partial z)_{\eta,S}$ denotes the vertical gradient of density under the assumptions of constant entropy and salinity.

The density of the new environment is $\rho_{envir} = \rho + d\rho/dz \cdot \zeta$. Thus, the buoyancy force acting on the parcel will be

$$(g/\rho)\left(\rho_{envir} - \rho'\right) = (g/\rho)\left[d\rho/dz - (\partial\rho/\partial z)_{\eta,S}\right] \cdot \zeta \tag{2.90}$$

where g is gravity.

If the water parcel is heavier than its environment, the buoyancy force is negative, so the water parcel is pushed back toward its original position by buoyancy force. When the water parcel is pushed toward its original position, it will overshoot because of inertia. Thus, small perturbations can exist as stable oscillations. In most cases, water columns observed in the oceans are stably stratified. The corresponding frequency of oscillations is called the buoyancy frequency or the Brunt–Vaisala frequency

$$N^2 = -(g/\rho)\left[d\rho/dz - (\partial\rho/\partial z)_{\eta,S}\right] \tag{2.91}$$

Since

$$(\partial \rho / \partial z)_{\eta,S} = (\partial \rho / \partial P)_{\eta,S} \cdot \partial P / \partial z \tag{2.92}$$

where $(\partial \rho / \partial P)_{\eta,S} = 1/c^2$ (c is the speed of sound); the hydrostatic equilibrium gives $\partial P / \partial z = -\rho g$. Thereby, the buoyancy frequency is reduced to

$$N^2 = -\frac{g}{\rho}\frac{d\rho}{dz} - \frac{g^2}{c^2} \tag{2.93}$$

2.4.14 Thermodynamics of seawater based on the Gibbs function

Due to the nonlinear property of the seawater, many thermodynamic quantities of seawater had to be determined through experimental measurements, and some other quantities had to be introduced through different thermodynamic relations. The use of different means of measurement and formulae can give rise to some inconsistency within the thermodynamic quantities calculated from different means. Fofonoff (1962) proposed to overcome such problems by defining one set of unified formulae in terms of the Gibbs function; thereby all other thermodynamic variables can be defined from the combination of the Gibbs function and its derivatives. About 30 years later, his idea was carried out by Feistel (1993, 2003) and Feistel and Hagen (1995). By a least-square fitting to all available seawater thermodynamic properties, the Gibbs function has been expressed in terms of a power series of (T, S, P) with 100 double-precision coefficients. Since all thermodynamic functions are derived from the same set of polynomials, the calculated thermodynamic variables are self-consistent within the error bounds. This formulation is now available in both FORTRAN and Matlab (Feistel, 2005); thus, by using the standard functions, one can calculate all the thermodynamic properties, such as density, specific heat, specific enthalpy, and specific entropy. Table 2.5 lists some of the most commonly used thermodynamic functions defined through the Gibbs function. Here s denotes the mass fraction of salt, as it is the notation used for introducing the thermodynamics functions.

2.4.15 Entropy of seawater

Although entropy is one of the fundamental thermodynamic variables of seawater, it has not been widely used in descriptive or theoretical studies of physical oceanography. Since no reliable formulae for calculating the entropy of seawater were available, there were very few studies related to entropy.

This situation changed considerably thanks to the recent work of Feistel and his colleagues, e.g., Feistel (1993, 2003). Using standard subroutines based on the Gibbs function, all thermodynamic functions, in particular entropy, can now be calculated.

According to the formula, entropy depends on the temperature almost linearly, and it increases slightly with salinity (Fig. 2.18a). In fact, a linear function, $\eta = c_1 T + c_2$, fits the more accurate calculation based on the standard subroutine within 1.5%, depicted as the thin

Table 2.5. *Thermodynamics functions/variables in terms of Gibbs function*

Thermodynamic functions	Formula based on Gibbs function
Specific enthalpy (heat content)	$h = \mathsf{g} - T\dfrac{\partial \mathsf{g}}{\partial T}$
Specific free energy (Helmholtz free energy, available work)	$f = \mathsf{g} - P\dfrac{\partial \mathsf{g}}{\partial P}$
Specific internal energy	$e = \mathsf{g} - T\dfrac{\partial \mathsf{g}}{\partial T} - P\dfrac{\partial \mathsf{g}}{\partial P}$
Specific chemical potential	$\mu = \dfrac{\partial \mathsf{g}}{\partial s}$
Specific chemical potential of water in seawater	$\mu_w = \mathsf{g} - \mu \cdot s$
Specific chemical potential of salt in seawater	$\mu_s = \mu - \mu_w$
Specific entropy	$\eta = -\dfrac{\partial \mathsf{g}}{\partial T}$
Specific volume	$v = \dfrac{\partial \mathsf{g}}{\partial P}$
Specific heat capacity (at constant pressure)	$c_p = \left(\dfrac{\partial h}{\partial T}\right)_{p,s} = -T\dfrac{\partial^2 \mathsf{g}}{\partial T^2}$
Haline contraction coefficient	$\beta = \dfrac{1}{\rho}\left(\dfrac{\partial \rho}{\partial s}\right)_{P,T} = -\dfrac{1}{v}\dfrac{\partial^2 \mathsf{g}}{\partial s \partial P}$
Thermal expansion coefficient	$\alpha = -\dfrac{1}{\rho}\left(\dfrac{\partial \rho}{\partial T}\right)_{P,s} = \dfrac{1}{v}\dfrac{\partial^2 \mathsf{g}}{\partial T \partial P}$
Isothermal compressibility	$K_T = \dfrac{1}{\rho}\left(\dfrac{\partial \rho}{\partial P}\right)_{T,s} = -\dfrac{1}{v}\dfrac{\partial^2 \mathsf{g}}{\partial P^2}$
Speed of sound	$c^2 = v^2\dfrac{\partial^2 \mathsf{g}}{\partial T^2}\left(\dfrac{\partial^2 \mathsf{g}}{\partial T \partial P} - \dfrac{\partial^2 \mathsf{g}}{\partial T^2}\dfrac{\partial^2 \mathsf{g}}{\partial P^2}\right)$
Adiabatic lapse rate	$\Gamma = -\dfrac{\partial^2 \mathsf{g}}{\partial T \partial P}\dfrac{\partial^2 \mathsf{g}}{\partial T^2}$
Vertical stability (Brunt–Vaisala frequency)	$N^2 = -\dfrac{g^2}{v^2}\left(\dfrac{\partial^2 \mathsf{g}}{\partial T \partial P}\right)^2 \dfrac{\partial^2 \mathsf{g}}{\partial T^2}$

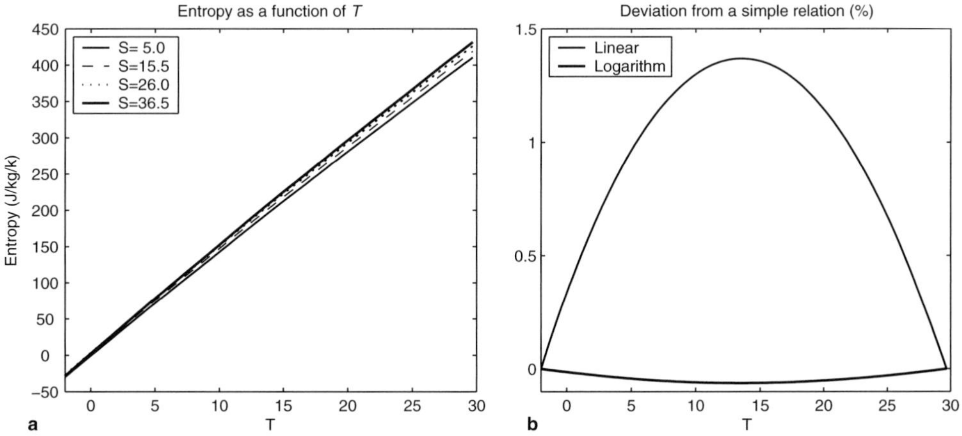

Fig. 2.18 **a** Entropy as a function of temperature (T) and salinity (S) at a standard atmospheric pressure; **b** deviation from a simple linear function, and a logarithmic function, defined as percentage of the maximal value of entropy in this range (425 J/kg/K).

line in Figure 2.18b. Furthermore, a function, $\eta = d_1 \ln T + d_2$, fits the accurate definition with errors less than 0.1%, depicted as the heavy line in Figure 2.18b. Thus, isentropic analysis is very close to analysis on a potential temperature surface, and using isentropic coordinates in oceanography is very close to using potential temperature as the vertical coordinate. Some preliminary studies have been carried out with isentropic analysis, e.g., Gan *et al.* (2007).

2.4.16 Relation between internal energy, enthalpy, and free enthalpy

For the study of physical oceanography, three forms of energy exist, and they can be used to explore the physics of circulation under different conditions:

Specific internal energy *e*

This form of energy can be used for a system with a fixed volume, and thus no exchange of pressure work with its environments.

Specific enthalpy *h*

This form of energy plays a vital role in the study of fluid motion, where the pressure work through volume change is included as part of the energy transfer between the system and the environments.

Specific free enthalpy (Gibbs function) g

As the name suggests, this function indicates that part of enthalpy of the system which is free or available. There is another function, the specific free energy, defined as $f = \mathrm{g} - Pv$;

however, for problems involving the motion of water, specific free enthalpy is the most suitable function which can be used to describe the thermodynamics of the oceanic circulation.

Broadly speaking, free enthalpy is directly linked to the amount of energy which is convertible to mechanical energy. For example, if the enthalpy of a system is increased, it does not necessarily mean that the system is capable of producing more mechanical work. In fact, if both the enthalpy and entropy of a system increase, the free enthalpy of the system may decline due to a large increase in entropy. As a result, the capability of the system to produce mechanical energy is reduced.

The difference between enthalpy and internal energy is the term Pv, which is relatively small in general. At sea level, this term is very small because sea-level pressure is equal to one standard atmospheric pressure, giving $Pv \simeq 100$ J/kg. As a result, internal energy and enthalpy are virtually equal. As we already know, both internal energy and enthalpy increase almost linearly with temperature (Fig. 2.19). As discussed above, entropy also increases with temperature almost linearly. On the other hand, the free enthalpy declines with temperature. This fact thus shows that the increase in internal energy or enthalpy does not necessarily mean that the system has more energy that is available or "free." The decline of free enthalpy with the increase in temperature is due to the rapid increase of entropy. In the abyss, where the pressure term makes a sizable contribution to enthalpy, enthalpy is much larger than internal energy (Fig. 2.19b).

The value of different forms of energy in the ocean also sensitively depends on the salinity (Fig. 2.20). It is readily seen (Fig. 2.20a) that internal energy, enthalpy, and entropy are convex functions of salinity, i.e., $\partial^2 e/\partial S^2 < 0$, $\partial^2 h/\partial S^2 < 0$, and $\partial^2 \eta/\partial S^2 < 0$. On the other hand, free enthalpy is a concave function of salinity, i.e., $\partial^2 \mathsf{g}/\partial S^2 > 0$. These

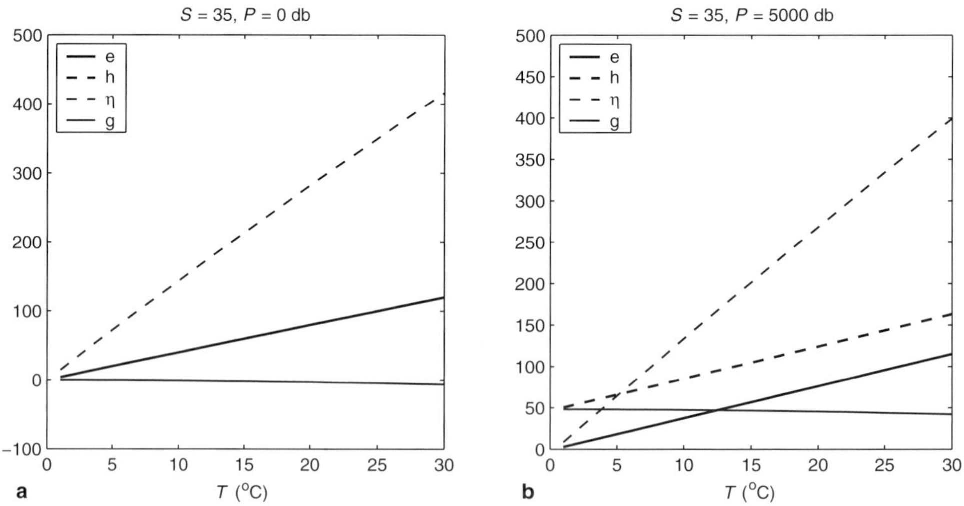

Fig. 2.19 Temperature-dependence of internal energy (e), enthalpy (h), free enthalpy (g) (all in kJ/kg), and entropy (η) (in J/kg/K).

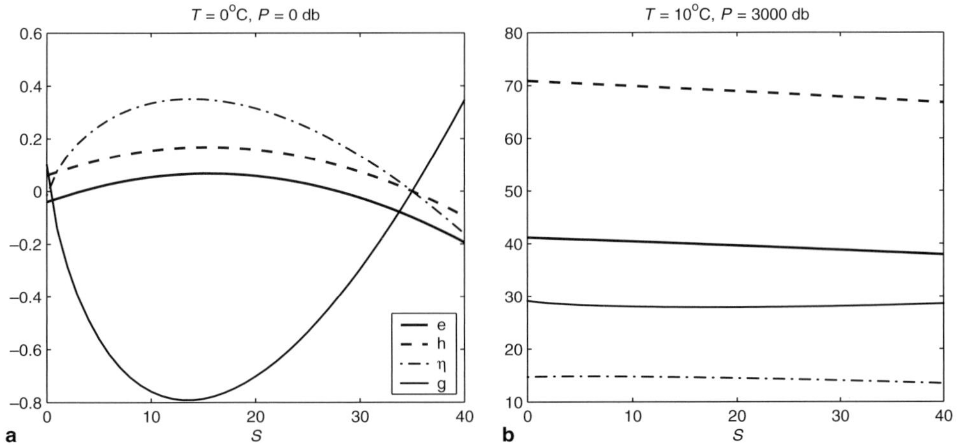

Fig. 2.20 Salinity-dependence of internal energy (e), enthalpy (h), free enthalpy (g) (all in kJ/kg), and entropy (η) (in J/kg/K).

thermodynamic features remain true in general; however, at higher temperature it is less obvious (Fig. 2.20b). The characteristics of these thermodynamic functions have important implications for mixing, as will be seen later.

Although the difference in internal energy and enthalpy is negligible at the sea surface, their difference increases with depth. In fact, at a constant temperature, the Pv term increases from a value of approximately 100 J/kg at the sea surface to a value of approximately 30,000 J/kg at the 3,000 db level (Fig. 2.20b). Since the contribution from the entropy term at low temperature is small, enthalpy and free enthalpy increase with pressure. On the other hand, at higher temperature, free enthalpy is much smaller than enthalpy due to the negative contribution associated with the large value of entropy (Fig. 2.21).

As a further example, we examine the changes in different forms of energy associated with the mixing of two seawater parcels. Assume that we have two parcels with the same temperature (10°C) and pressure (sea-level pressure), and each of them has one kilogram of seawater, with salinity of 25 and 35. Now let these two water parcels mix under constant temperature (10°C) and pressure (sea-level pressure). The final product is a parcel of two kilograms and salinity 30. Seawater properties for salinity at 25, 30, and 35 are shown in Table 2.6.

As discussed above, internal energy, enthalpy, and entropy are convex functions of salinity. Therefore, the final values of internal energy, enthalpy, and entropy after mixing are larger than the mean of the initial values. On the other hand, free enthalpy is a concave function of salinity. As a result, the total amount of free enthalpy after mixing is lower than the sum of the initial values. This characteristic of free enthalpy indicates that after mixing the system has less energy available, although both the internal energy and enthalpy of the system are increased. The increase of entropy after mixing overshadows the increase of enthalpy, leading to the reduction of free enthalpy.

Table 2.6. *Changes in thermodynamic properties of seawater after mixing. The difference between the final state and the initial states is defined as* $\Delta = 2f_2 - f_1 - f_3$, *where* f *is a seawater thermodynamic property, based on the formulae of Feistel (2003)*

	Parcel property			
Parcel	1	2	3	Δ
S	25	30	35	
e (J/kg)	40,487	40,141	39,781	15.0442
h (J/kg)	40,586	40,240	39,879	15.0419
η (J/kg/K)	147.8896	145.7726	143.3879	0.2678
g (J/kg)	$-1,288.8$	$-1,035.4$	$-7,211$	-60.7814
v (10^{-3} m^3/kg)	0.9812	0.9775	0.9738	-0.00002287

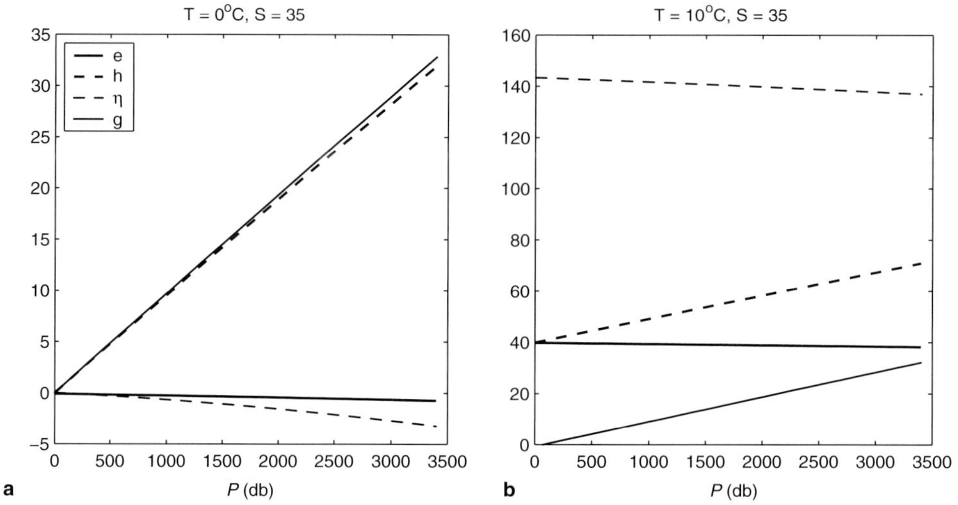

Fig. 2.21 Pressure-dependence of internal energy (e), enthalpy (h), free enthalpy (g) (all in kJ/kg), and entropy (η) (in J/kg/K).

In this case, due to the reduction of volume, the system has actually received mechanical work from the environment. The fact that both the internal energy and enthalpy increase after mixing implies that the system must receive thermal energy from the environment as well. However, the amount of free enthalpy is actually reduced, which is a strong indication that the amount of energy available from the system is reduced. A more rigorous definition of energy available for the oceanic general circulation is presented at the end of Section 3.7.

2.5 A hierarchy of equations of state for seawater

2.5.1 Introduction

As is well known, the ideal gas law plays a vitally important role in the theoretical study of atmospheric circulation. Accordingly, it would be highly desirable if we could find a similar equation of state for seawater, thereby promoting the theoretical study of oceanic general circulation. Many simple versions of the equation of state have been proposed and used in the past. Unfortunately, however, in most cases these equations of state have not been linked to the other thermodynamic properties of seawater, such as enthalpy, entropy, and chemical potential. Therefore, using such simplified equations of state may introduce some inconsistency between the dynamics and thermodynamics of seawater. It is therefore desirable to have a simplified equation of state which comes with a set of thermodynamic functions that are internally consistent with the equation of state. However, the practical approach seems unclear.

One way to overcome such problems is to use the Gibbs function. Since the Gibbs function has the merit that once it is defined (in terms of T, S, P), all thermodynamic properties are derived from this function accordingly. Such a unification of seawater thermodynamics was first suggested by Fofonoff (1962). As an example, we proceed to discuss a simple choice in the following section, and the point is to examine carefully whether the thermodynamic properties derived from such a Gibbs function can be used to approximate the equation of state of seawater. This means that all the important thermodynamic properties derived from this Gibbs function should be compared with those from the more accurate expressions accepted by the research community, with small and tolerable errors.

2.5.2 Simple equations of state

The simplest possible choice is to have an equation of state that has a thermal expansion coefficient linearly dependent on both temperature and pressure. We begin with the following relations

$$\alpha = -\frac{1}{\rho}\frac{\partial \rho}{\partial T} = \alpha_0 + \alpha_1 T + \gamma_1 P \tag{2.94}$$

$$\beta = \frac{1}{\rho}\frac{\partial \rho}{\partial S} \tag{2.95}$$

$$\gamma = \frac{1}{\rho}\frac{\partial \rho}{\partial P} = \gamma_0 - \gamma_1 T \tag{2.96}$$

The equation of state which satisfies these constraints is

$$\rho = \rho_0 e^{-(\alpha_0 + 0.5\alpha_1 T + \gamma_1 P)T + \beta S + \gamma_0 P} \tag{2.97}$$

The most crucial step is to introduce the Gibbs function, so that all thermodynamic variables can be derived from the same Gibbs function consistently. As discussed in Section 2.4, the specific volume is related to the Gibbs function in the following way:

$$v = \frac{\partial \mathrm{g}}{\partial P}$$

Therefore, the desirable Gibbs function can be determined by solving the following differential equation

$$\frac{\partial \mathrm{g}}{\partial P} = \frac{1}{\rho_0} e^{(\alpha_0 + 0.5\alpha_1 T + \gamma_1 p)T - \beta S - \gamma_0 P} \tag{2.98}$$

The solution is

$$\mathrm{g} = \frac{e^{(\alpha_0 + 0.5\alpha_1 T + \gamma_1 P)T - \beta S - \gamma_0 P}}{\rho_0 (\gamma_1 T - \gamma_0)} + C(T, S) \tag{2.99}$$

where the constants α_0, α_1, γ_0, γ_1 and β can be chosen by best-fitting the UNESCO equation of state, $C(T, S)$ as an arbitrary function. This equation of state represents a thermal expansion coefficient that increases with temperature and pressure. The same coefficient γ_1 also reflects the special thermodynamic property of seawater, the so-called thermobaric effect: cold seawater is more compressible than warm seawater. On the other hand, the saline contraction coefficient is constant, which is a good approximation for seawater.

Basic relations from this Gibbs function are

$$\frac{\partial \mathrm{g}}{\partial T} = \frac{(\gamma_1 T - \gamma_0)(\alpha_0 + \alpha_1 T + \gamma_1 p) - \gamma_1}{\rho_0 (\gamma_1 T - \gamma_0)^2} e^{(\alpha_0 + 0.5\alpha_1 T + \gamma_1 P)T - \beta S - \gamma_0 P} + \frac{\partial C(T, S)}{\partial T} \tag{2.100}$$

$$\frac{\partial \mathrm{g}}{\partial s} = \frac{-\beta}{\rho_0 (\gamma_1 T - \gamma_0)} e^{(\alpha_0 + 0.5\alpha_1 T + \gamma_1 P)T - \beta S - \gamma_0 P} + \frac{\partial C(T, S)}{\partial s} \tag{2.101}$$

From this equation of state, one can derive all the thermodynamic properties of seawater consistently (Feistel and Hagen, 1995; Feistel, 2003); thus, these thermodynamic functions can be used to study the thermodynamics of seawater under a consistent framework of thermodynamics:

Enthalpy (heat content)

$$h = \mathrm{g} - T\frac{\partial \mathrm{g}}{\partial T} = \frac{2\gamma_1 T - \gamma_0 - (\gamma_1 T - \gamma_0)(\alpha_0 + \alpha_1 T + \gamma_1 P) T}{\rho_0 (\gamma_1 T - \gamma_0)^2} e^{(\alpha_0 + 0.5\alpha_1 T + \gamma_1 P)T - \beta S - \gamma_0 P} + C(T, S) - \frac{\partial C(T, S)}{\partial T} \tag{2.102}$$

Free energy (Helmholtz free energy, available work)

$$f = \mathrm{g} - P\frac{\partial \mathrm{g}}{\partial P} = \frac{1 - P(\gamma_1 T - \gamma_0)}{\rho_0 (\gamma_1 T - \gamma_0)} e^{(\alpha_0 + 0.5\alpha_1 T + \gamma_1 P)T - \beta S - \gamma_0 P} + C(T, S) \tag{2.103}$$

Internal energy

$$
\begin{aligned}
e &= \mathrm{g} - T\frac{\partial \mathrm{g}}{\partial T} - P\frac{\partial \mathrm{g}}{\partial P} \\
&= \frac{2\gamma_1 T - \gamma_0 - (\gamma_1 T - \gamma_0)(\alpha_0 + \alpha_1 T + \gamma_1 P)\,T - P(\gamma_1 T - \gamma_0)^2}{\rho_0 (\gamma_1 T - \gamma_0)^2} e^{(\alpha_0 + 0.5\alpha_1 T + \gamma_1 P)T - \beta S - \gamma_0 P} \\
&\quad + C(T, S) - \frac{\partial C(T, S)}{\partial T}
\end{aligned} \tag{2.104}
$$

Chemical potential

$$
\mu = \frac{\partial \mathrm{g}}{\partial s} = \frac{-\beta}{\rho_0 (\gamma_1 T - \gamma_0)} e^{(\alpha_0 + 0.5\alpha_1 T + \gamma_1 P)T - \beta S - \gamma_0 P} + \frac{\partial C(T, S)}{\partial S} \tag{2.105}
$$

Entropy

$$
\begin{aligned}
\eta = -\frac{\partial \mathrm{g}}{\partial T} &= -\frac{(\gamma_1 T - \gamma_0)(\alpha_0 + \alpha_1 T + \gamma_1 P) - \gamma_1}{\rho_0 (\gamma_1 T - \gamma_0)^2} e^{(\alpha_0 + 0.5\alpha_1 T + \gamma_1 P)T - \beta S - \gamma_0 P} \\
&\quad - \frac{\partial C(T, S)}{\partial T}
\end{aligned} \tag{2.106}
$$

Other thermodynamic functions can be obtained similarly.

For simplicity, we list the thermodynamic functions for the case when the equation of state is a simple linear function of temperature, salinity, and pressure. Omitting the arbitrary function $C(T, S)$ introduced in Eqn. (2.99), the Gibbs function takes the form

$$
\mathrm{g} = -\frac{e^{\alpha T - \beta S - \gamma P}}{\rho_0 \gamma} \tag{2.107}
$$

where α, β, γ and ρ_0 are all consistent. From the Gibbs function, we can derive the following list of thermodynamic variables which are consistent with each other.

$$
\frac{\partial \mathrm{g}}{\partial T} = -\frac{\alpha}{\rho_0 \gamma} e^{\alpha T - \beta S - \gamma P}, \quad \frac{\partial \mathrm{g}}{\partial s} = \frac{\beta}{\rho_0 \gamma} e^{\alpha T - \beta S - \gamma P} \tag{2.108}
$$

$$
h = \mathrm{g} - T\frac{\partial \mathrm{g}}{\partial T} = \frac{-1 + \alpha T}{\rho_0 \gamma} e^{\alpha T - \beta S - \gamma P} \tag{2.109}
$$

$$
f = \mathrm{g} - P\frac{\partial \mathrm{g}}{\partial P} = -\frac{1 + \gamma P}{\rho_0 \gamma} e^{\alpha T - \beta S - \gamma P} \tag{2.110}
$$

$$
e = \mathrm{g} - T\frac{\partial \mathrm{g}}{\partial T} - P\frac{\partial \mathrm{g}}{\partial P} = \frac{-1 + \alpha T - \gamma P}{\rho_0 \gamma} e^{\alpha T - \beta S - \gamma P} \tag{2.111}
$$

$$
\mu = \frac{\partial \mathrm{g}}{\partial s} = \frac{\beta}{\rho_0 \gamma} e^{\alpha T - \beta S - \gamma P} \tag{2.112}
$$

$$
\eta = -\frac{\partial \mathrm{g}}{\partial T} = \frac{\alpha}{\rho_0 \gamma} e^{\alpha T - \beta S - \gamma P} \tag{2.113}
$$

Note that under the assumption of a linear equation of state, all basic thermodynamic functions have a factor which is an exponential function of temperature, salinity, and pressure. In particular, entropy increases with temperature exponentially. As a result, the Gibbs function (the free enthalpy) declines with the increase in temperature, although both the internal energy and enthalpy increase with temperature.

2.6 Scaling and different approximations

The partial differential equations describing oceanic motions constitute a very complicated equation system. These equations can describe thousands of kinds of phenomena having time scales from seconds to thousands of years, and length scales from millimeters to thousands of kilometers. Thus, trying to find the "complete solution" or "general solution" of such a system is unrealistic. In order to understand the real-world oceanic motions by means of such an equation system, we have to limit ourselves to certain scales, thereby simplifying the general equation system.

The basic principle of scaling is that we want to focus our study on some specific phenomenon which has certain time and space scales. Therefore, we will use these scales to estimate and compare the magnitude of all terms in the equations. We keep only the "important terms" and discard the smaller terms, assuming that they are not as important for studying the phenomenon in which we are really interested. It is crucial to emphasize that scale analysis simplifies equations and helps to highlight the physics important for our study; however, the result of scaling is implicitly defined in the basic assumptions of the scaling. Different scales will certainly lead to quite different equations, which describe different dynamical processes.

Note that scaling is an art for dealing with complicated systems, so it is not a foolproof method. There are indeed many cases in which some terms that are much smaller than other terms may play crucial roles in the overall balance of vorticity and energy, so they cannot be discarded. For example, terms associated with small-scale turbulence dissipation/mixing often have magnitudes much smaller than other terms; however, the interaction between large and small scales is one of the major players that ultimately regulates the global-scale wind-driven circulation and thermohaline circulation. Since we will derive equations in the local Cartesian coordinates, the following notations are used in this section: $dx = r\cos\theta d\lambda$, $dy = rd\theta$, and $dz = dr$.

2.6.1 Hydrostatic approximation

Scaling of the vertical momentum equation

For basin-scale motions in the oceans the magnitude of terms in the vertical momentum equation can be estimated as follows

$$
\begin{array}{llllll}
 & \dfrac{dw}{dt} & -2\omega u\cos\theta & -\dfrac{u^2+v^2}{a} & = -\dfrac{1}{\rho}\dfrac{\partial p}{\partial z} & -g \\
\textit{Scale} & \dfrac{UW}{L} & f_0 U & U^2/a & P_0/\rho_0 H & g \\
\textit{Magnitude} & 10^{-13} & 10^{-5} & 10^{-8} & 10 & 10
\end{array}
\qquad (2.114)
$$

We assume that the horizontal and vertical scales of the motions are of the same order of magnitude as the width and depth of the basin, $L \simeq 10^6 m$, $H \simeq 5 \times 10^3 m$; the horizontal and vertical velocity scales are $U \simeq 0.1\, m/s$, $W \simeq 10^{-6} m/s$; the time scale of horizontal advection, U/L, is used as the time scale; $\rho_0 = 10^3$ kg/m^3 is the mean reference density; and $a = 6,370$ km is the radius of the Earth; $P_0 \simeq 5 \times 10^7 N/m^2$, $f_0 \simeq 10^{-4}/s$. Thus, to a very high accuracy, the vertical momentum equation can be replaced with the so-called ***hydrostatic approximation***

$$\frac{\partial p}{\partial z} = -\rho g \tag{2.115}$$

Further simplification

Equation (2.115) can be simplified further because the horizontally varying pressure component is also in hydrostatic equilibrium with the horizontally varying density field. Define standard mean density and pressure profiles $\bar{\rho}(z)$ and $\bar{p}(z)$ that satisfy

$$\frac{1}{\bar{\rho}}\frac{d\bar{p}}{dz} = -g \tag{2.116}$$

The pressure and density fields can be separated into two parts, i.e., the standard profiles and perturbations

$$p(x, y, z, t) = \bar{p}(z) + p'(x, y, z, t)$$
$$\rho(x, y, z, t) = \bar{\rho}(z) + \rho'(x, y, z, t)$$

where we have assumed that

$$p' \ll \bar{p}, \quad \rho' \ll \bar{\rho}$$

Substituting these into Eqn. (2.115) and expanding the denominator in Taylor series leads to

$$\begin{aligned}-\frac{1}{\rho}\frac{\partial p}{\partial z} - g &= -\frac{1}{\bar{\rho}(1 + \rho'/\bar{\rho})}\frac{\partial(\bar{p} + p')}{\partial z} - g \\ &= -\frac{1}{\bar{\rho}}\left[\frac{d\bar{p}}{dz} + \frac{\partial p'}{\partial z} - \frac{\rho'}{\bar{\rho}}\frac{d\bar{p}}{dz} - \frac{\rho'}{\bar{\rho}}\frac{\partial p'}{\partial z}\right] - g.\end{aligned}$$

The first term on the right-hand side is canceled by the gravity term; the fourth term is a product of two small terms, so it is small and can be ignored. Thus, the hydrostatic approximation relation (Eqn. (2.115)) is reduced to the following simplified equation

$$\frac{\partial p'}{\partial z} + \rho' g = 0 \tag{2.117}$$

Note that in Eqn. (2.117) the perturbations of pressure and density are used, instead of the total variables. In the ocean, we have the estimate: $|\rho'/\bar{\rho}| \leq 0.01$, so that the magnitude of these two terms is

$$\frac{\Delta p}{\rho_0 H} \simeq \frac{\rho' g}{\rho_0} \simeq 10^{-1} m/s^2$$

Thus, using Eqn. (2.117), we have increased the accuracy 100 times with the same simple hydrostatic relation.

Introduction of the hydrostatic relation substantially simplifies the vertical momentum equation. Now the pressure can easily be calculated by a simple integration. It is important to remember that the hydrostatic relation is derived under the assumption of large-scale motions; thus, if the scale of the motions is very small, the vertical acceleration term cannot be ignored.

To discover the conditions under which the hydrostatic approximation breaks down, we examine the vertical momentum equation, including the time-dependent term and using the simplified terms in Eqn. (2.117) derived above

$$\frac{dw}{dt} = -\frac{1}{\rho_0}\frac{\partial p'}{\partial z} - g\frac{\rho'}{\rho_0} \tag{2.118}$$

Assuming that the time scale is set by horizontal advection, $T \simeq U/L$, this leads to the following condition for the hydrostatic approximation:

$$w \ll \frac{gL\rho'}{U\rho_0} \tag{2.119}$$

The continuity equation can be reduced to the following form

$$\frac{d\rho'}{dt} + \frac{\rho_0}{g}N^2 w = 0 \tag{2.120}$$

where $N^2 = -\frac{g}{\rho_0}\frac{\partial \overline{\rho}}{\partial z}$ is the buoyancy frequency. Equation (2.120) leads to an estimate of w:

$$w \simeq \frac{gU}{L\rho_0 N^2}\rho' \tag{2.121}$$

Thus, combining Eqns. (2.119) and (2.121) leads to a criterion for the validity of the hydrostatic approximation:

$$\gamma^2/R_i \ll 1 \tag{2.122}$$

where $\gamma = H/L$ is the aspect ratio, $R_i = N^2h^2/U^2$ is the Richardson number, and H is the vertical scale.

It is evident that, if the stratification is strong and the flow is weak (large R_i), then the hydrostatic approximation is valid, even if γ is not small; whereas, for weakly stratified conditions on a small horizontal scale (~1 km), the hydrostatic approximation may break down. For example, the hydrostatic approximation is not valid for motions on the oceanic convective scales, and the vertical momentum equation (2.114) should be used to predict the time evolution of the vertical velocity.

2.6.2 The traditional approximation

First let us take note of the fact that when the hydrostatic approximation is used, the $\omega w \cos\theta$ term in the zonal momentum equation (2.3) has to be dropped so that the physical constraint that the Coriolis force does no work will not be violated. (The Coriolis force is a virtual force introduced in rotating coordinates; thus, with non-rotating coordinates, it does not exist. A virtual force does not do work.) In fact, at mid latitudes we have the following estimate

$$\frac{v \sin\theta}{w \cos\theta} = \delta^{-1} \tan\theta \gg 1 \tag{2.123}$$

so that the $\omega w \cos\theta$ term can be neglected.

Without making the assumption of hydrostatic equilibrium, we can also achieve the same approximation (e.g., Phillips, 1966). Using $1/a$ to simplify the metric terms in the zonal momentum equation, we have

$$\frac{du}{dt} = -\frac{1}{\rho} p_x + F_x + \left(2\omega + \frac{u}{a\cos\theta}\right)(v \sin\theta - w\cos\theta) \tag{2.124}$$

Then, to conserve the angular momentum, we must drop the $w \cos\theta$ term; thus the angular momentum conservation is

$$\frac{d}{dt}\left[a\cos\theta\,(u + \omega a \cos\theta)\right] = a\cos\theta\left(F_x - \frac{1}{\rho} p_x\right) \tag{2.125}$$

Now, to conserve energy the Coriolis force term in the z-equation has to be dropped. The simplified set of equations is

$$\frac{du}{dt} - \frac{uv\tan\theta}{a} + \frac{uw}{a} = -\frac{1}{\rho_0}\frac{\partial p'}{\partial x} + 2\omega v \sin\theta + F_x \tag{2.126}$$

$$\frac{dv}{dt} + \frac{u^2\tan\theta}{a} + \frac{vw}{a} = -\frac{1}{\rho_0}\frac{\partial p'}{\partial y} - 2\omega u \sin\theta + F_y \tag{2.127}$$

$$\frac{dw}{dt} - \frac{u^2 + v^2}{a} = -\frac{1}{\rho_0}\frac{\partial p'}{\partial z} - g\frac{\rho'}{\rho_0} + F_z \tag{2.128}$$

2.6.3 Scaling of the horizontal momentum equations

For large-scale motions in the oceans we have the scales shown in Table 2.7. In this table, the scale of horizontal pressure difference is based on a sea surface elevation difference of

Table 2.7. *Fundamental scales for oceanic circulation*

	Oceanic gyres	Gulf Stream
U (m/s)	0.1	1
W (m/s)	10^{-6}	10^{-6}
L (m)	10^{6}	10^{5}
D (m)	10^{4}	10^{3}
L/U (s)	10^{7}	10^{5}
$\Delta p/\rho$ (m^2/s^2)	10	10

1 m. Using these basic scales, we can scale the horizontal momentum equations as

	$\frac{du}{dt}$	$-2\omega v\cos\theta$	$+2\omega w\cos\theta$	$+\frac{uw}{a}$	$-\frac{uv\tan\theta}{a}$	$=-\frac{1}{\rho_0}\frac{\partial p}{\partial x}$	$+F_x$
Scale	U^2/L	f_0U	f_0W	UW/a	U^2/a	$\Delta P/\rho_0 L$	
Gyres	10^{-8}	10^{-5}	10^{-10}	10^{-14}	10^{-9}	10^{-5}	
Gulf Stream	10^{-5}	10^{-4}	10^{-10}	10^{-13}	10^{-7}	10^{-4}	

(2.129)

For large-scale motions in the atmosphere and oceans, the spherical curvature terms associated with factor $1/a$ are negligible, and the Coriolis force term associated with the vertical velocity is negligible. Therefore, the horizontal momentum equations are reduced to the following forms

$$\frac{du}{dt} - 2\omega v\sin\theta = -\frac{1}{\rho_0}\frac{\partial p}{\partial x} + F_x \tag{2.130}$$

$$\frac{dv}{dt} + 2\omega u\sin\theta = -\frac{1}{\rho_0}\frac{\partial p}{\partial y} + F_y \tag{2.131}$$

β-plane approximation

Many practical problems studied in dynamical oceanography are associated with meridional scales much smaller than the radius of the Earth. For such problems, the spherical geometry can be approximately represented by a set of local Cartesian coordinates whose horizontal surface is tangential to the local spherical surface. In this new coordinate system, the Coriolis parameter is approximated by its Taylor expansion around the origin of the local Cartesian coordinates

$$f = 2\omega\sin\theta \simeq 2\omega\sin\theta_0 + 2\omega\cos\theta_0\,(y - y_0)/a$$

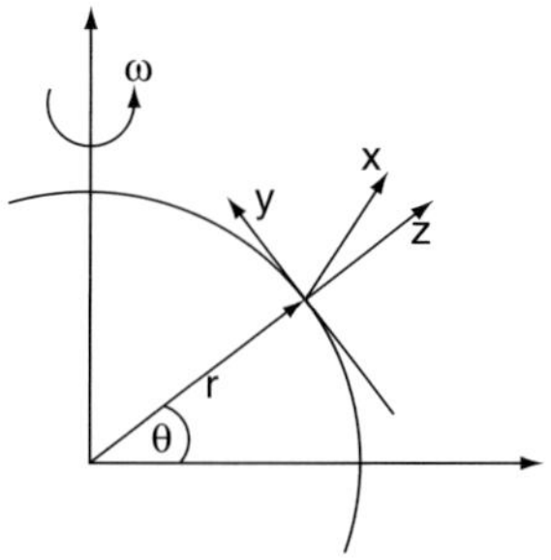

Fig. 2.22 Sketch of the spherical coordinates and a locally defined β-plane.

or

$$f = f_0 + \beta (y - y_0), \quad \textit{for } |y - y_0| \ll a \tag{2.132}$$

where $\beta = 2\omega \cos \theta_0 / a$ is an approximation of the meridional gradient of the planetary vorticity. Using local Cartesian coordinates (Fig. 2.22), we have

$$f = f_0 + \beta y \tag{2.132$'$}$$

Thus, under the β-plane approximation, the corresponding momentum equations are reduced to

$$\frac{du}{dt} - f v = -\frac{1}{\rho_0} \frac{\partial p}{\partial x} + F_x \tag{2.133}$$

$$\frac{dv}{dt} + f u = -\frac{1}{\rho_0} \frac{\partial p}{\partial y} + F_y \tag{2.134}$$

Note that the Coriolis parameter is a linear function of y; thus, the meridional gradient of f is equal to β. The introduction of the β-plane greatly reduces the complexity of spherical geometry and the associated dynamics, making many simple analytical studies feasible. On the other hand, the β-plane is an approximation, which is valid under the constraint of Eqn. (2.132); thus, for problems with horizontal scales not satisfying Eqn. (2.132), the original equations in spherical coordinates may have to be used.

f-plane approximation

For motions of even smaller scale, the change in the Coriolis parameter can be totally neglected; thus, the basic equations (2.133) and (2.134) are further simplified to

$$\frac{du}{dt} - f_0 v = -\frac{1}{\rho_0} \frac{\partial p}{\partial x} + F_x \tag{2.135}$$

$$\frac{dv}{dt} + f_0 u = -\frac{1}{\rho_0} \frac{\partial p}{\partial y} + F_y \tag{2.136}$$

Table 2.8. *Comparison of three sets of coordinates*

	Spherical coordinates	β-plane	f-plane
f	$2\omega \sin\theta$	$f = f_0 + \beta y$	f_0
$\beta = df/dy$	$2\omega \cos\theta / a$	$\beta = \beta_0$ (constant)	0
dx	$a \cos\theta d\lambda$	dx	dx

Introducing the β-plane and f-plane provides very simple and useful tools for the study of the atmosphere and oceans without getting into the algebraic complexity of spherical geometry. However, it is important to remember that some essential elements of the problems appear in slightly different forms in these coordinates (see Table. 2.8). Using the local Cartesian coordinates simplifies the equation, but it may also introduce small distortions of the dynamical picture.

2.6.4 Geostrophy and the thermal wind relation

For basin-scale circulations, the time-dependent terms are often much smaller than the Coriolis force terms and this can be argued as follows. Let us introduce the ***Rossby number***, which is defined as the ratio of the time-dependent term and the Coriolis force term

$$Ro = \frac{U^2/L}{fU} = \frac{U}{fL} \tag{2.137}$$

As shown in Eqn. (2.129), $Ro \simeq 0.001$ for the gyre scale circulation, and $Ro \simeq 0.1$ for the Gulf Stream. Therefore, for basin-scale circulation, the Rossby number is much smaller than one, so that the time-dependent term, or the inertial term, can safely be ignored in the horizontal momentum equations. Furthermore, for basin-scale motions in the subsurface layers, horizontal friction is also negligible. As a result of these approximations, for large-scale motions in the oceans, geostrophy applies:

$$-fv = -\frac{1}{\rho_0}\frac{\partial p}{\partial x} \tag{2.138}$$

$$fu = -\frac{1}{\rho_0}\frac{\partial p}{\partial y} \tag{2.139}$$

For gyre-scale motions, geostrophy is a very good approximation. Therefore, large-scale motions in the ocean follow the pressure contours, in dramatic contrast to the down-pressure-gradient motions which are more familiar to our daily experience. Taking the partial derivative with respect to z and using the hydrostatic approximation gives rise to the

thermal wind relation:

$$f\frac{\partial v}{\partial z} = -\frac{g}{\rho_0}\frac{\partial \rho}{\partial x} \tag{2.140}$$

$$f\frac{\partial u}{\partial z} = \frac{g}{\rho_0}\frac{\partial \rho}{\partial y} \tag{2.141}$$

The thermal wind relation links the vertical shear of horizontal velocity with the horizontal density gradient. This relation suggests that wherever there is a horizontal density gradient, there is vertical shear of the horizontal velocity.

2.7 Boussinesq approximations and buoyancy fluxes

In classical fluid dynamics, comprehensive theories have been developed for the homogeneous fluid. Because of the constant density of the fluid, the momentum equations are entirely separated from the thermodynamic equations. In the atmosphere and the oceans, density is not entirely uniform. In fact, the buoyancy force due to small density deviations is one of the main forces driving the circulation. Because density appears as the denominator of the pressure gradient force term, it is readily seen that density field is very closely tied to the velocity field.

Water density varies from 1,020 kg/m^3 at the surface in the tropics to 1,070 kg/m^3 in the deepest ocean trenches. Hence, if density appears as just a multiplying coefficient, we can use an average density with an error of no more than 4%. However, if the density appears in the buoyancy force term, such a replacement will throw away the forcing term and change the dynamics entirely. The Boussinesq approximations have been developed in order to simplify the dynamical equations in a consistent way. In fact, under the Boussinesq approximations both the atmosphere and the oceans can be treated as if they were incompressible fluid, while the buoyancy force due to small density deviations is still retained.

2.7.1 Boussinesq approximations

Approximation in the continuity equation

For large-scale motions, the density field can be separated into a mean reference density and the perturbation

$$\rho = \rho_0\left(1 + \frac{fUL}{gH}\rho'\right) = \rho_0\left(1 + RoF\rho'\right) \tag{2.142}$$

where ρ_0 is the fixed-value mean reference density, f the Coriolis parameter, U the horizontal velocity, L and H the horizontal and vertical length scales, ρ' is the nondimensional density, $Ro = U/fL$ is the Rossby number, and $F = f^2L^2/gH$. The continuity equation (2.6) can be reduced to the following nondimensional form

$$RoF\frac{D\rho'}{Dt'} + \left(1 + RoF\rho'\right)\nabla'\cdot\vec{u}' = 0 \tag{2.143}$$

where $t' = tU/L$, $\nabla' = L\nabla$, and $\vec{u}' = \vec{u}/U$ is the nondimensional velocity vector. In the case where RoF is extremely small or equal to zero, the assumption of volume conservation is a good approximation; thus, the approximate form of continuity in dimensional form is

$$\nabla\cdot\vec{u} = 0 \tag{2.144}$$

In spherical coordinates, the corresponding equation is

$$\frac{1}{r\cos\theta}\left[\frac{\partial}{\partial\lambda}(u\cos\theta) + \frac{\partial}{\partial\theta}(v\cos\theta)\right] + \frac{\partial w}{\partial z} = 0 \tag{2.145}$$

In local Cartesian coordinates, it is reduced to

$$\frac{\partial u}{\partial x} + \frac{\partial v}{\partial y} + \frac{\partial w}{\partial z} = 0 \tag{2.146}$$

For global-scale motions, $L = a = 6,400$ km is the radius of the Earth, $f = 2\omega$, the horizontal velocity scale is $U = 0.1$ m/s, and the depth scale of the motion is $H = 800$ m. The corresponding nondimensional numbers are $Ro \simeq 10^{-4}$ and $F \simeq 100$. Thus, errors in replacing the mass conservation with the volume conservation for global-scale motions are approximately 1%.

On the other hand, if RoF is not very small, the errors introduced by the volume conservation approximation may not be entirely negligible. Since $RoF = fUL/gH$, we expect that the errors introduced by neglecting the density change terms can be relatively large for large-scale heating/cooling of the thin layer in the upper ocean.

It is very important to emphasize that the volume conservation is only an approximation to the continuity equation; it does not mean that the density of the fluid is constant. In fact, both temperature and salinity of water parcels can change with time due to surface forcing and diffusion. The common practice in dynamical oceanography is that the density of seawater is calculated from the equation of state, after both temperature and salinity of a given water parcel are determined. Therefore, the following set of equations is used:

$$\frac{dT}{dt} = \frac{\partial T}{\partial t} + \nabla\cdot(\vec{u}T) = Q_T \tag{2.147a}$$

$$\frac{dS}{dt} = \frac{\partial S}{\partial t} + \nabla\cdot(\vec{u}S) = Q_S \tag{2.147b}$$

$$\rho = \rho(T, S, p) \tag{2.147c}$$

where Q_T and Q_S are the sources of heat and salt due to diffusion and surface fluxes.

Approximations in the momentum equations

Since density is approximately constant, we will neglect density variation in the horizontal momentum equations, keeping the buoyancy contribution due to density variation only in the vertical momentum equation, giving

$$\frac{d\vec{u}}{dt} + 2\vec{\Omega} \times \vec{u} = -\frac{1}{\rho_0}\nabla p' - \vec{g}\frac{\rho'}{\rho_0} \tag{2.148}$$

where $p' = p - \bar{p}(z)$, $\rho' = \rho - \bar{\rho}(z)$, $\bar{\rho}(z)$ and $\bar{p}(z)$ are the mean density and pressure profiles, and ρ_0 is the mean reference density ($\rho_0 \simeq 1,035\ kg/m^3$ for the world's oceans).

Using Taylor series to expand the right-hand-side of Eqn. (2.148):

$$R.H.S. \simeq -\frac{1}{\rho}\nabla p' - \vec{k}g\left[\frac{1}{\rho}\left(\frac{\partial\rho}{\partial\Theta}\right)_{p,S}\theta' + \frac{1}{\rho}\left(\frac{\partial\rho}{\partial S}\right)_{p,\Theta}S' + \frac{1}{\rho}\left(\frac{\partial\rho}{\partial p}\right)_{\Theta,S}p'\right] \tag{2.149}$$

where Θ is potential temperature. Note that $\left(\frac{\partial\rho}{\partial p}\right)_{\Theta,S} = \frac{1}{c^2} \simeq \left(2\times10^3\right)^{-2}$; hence in the vertical momentum equation we have the following estimation for the ratio between the vertical pressure gradient and the buoyancy contribution from the pressure perturbation (after subtracting the standard pressure!):

$$\frac{\frac{1}{\rho_0}\frac{\partial p'}{\partial z}}{\frac{g}{\rho_0}\left(\frac{\partial\rho}{\partial p}\right)_{\Theta,S}p'} = \frac{c^2}{Hg} \simeq \frac{4\times10^6}{5\times10^3\times10} = 80 \gg 1 \tag{2.150}$$

where we assume that $H = 5,000$ m. Thus the last term on the right-hand side of Eqn. (2.149) can be neglected. Physically, when the Boussinesq approximations apply, the fluid is regarded as incompressible with infinite sound speed; hence the above treatment is self-consistent.

The Boussinesq approximations

Combining the analysis above, the oceanic circulation problem can be solved in terms of a basic state consisting of an adiabatic state of no motion and a dynamical perturbation part to the basic state. Thus, the dynamical structure of the oceanic circulation can be solved as follows.

1. *The adiabatic basic state*

   In the basic state, density and pressure are in hydrostatic equilibrium:

$$\nabla\bar{p} = -\vec{g}\bar{\rho}, \quad \bar{\rho} = \bar{\rho}\left(T, \bar{p}, S_0\right), \quad S_0 = const. \tag{2.151}$$

2. *The perturbation state*

$$\frac{d\vec{u}}{dt} + 2\vec{\Omega} \times \vec{u} = -\frac{1}{\rho_0}\nabla p' - \vec{g}\frac{\rho'_\Theta}{\rho_0} \tag{2.152}$$

where

$$\frac{\rho'_\Theta}{\rho_0} = -\alpha\Theta' + \beta S' \tag{2.153}$$

Both $\alpha = -\frac{1}{\rho}\left(\frac{\partial\rho}{\partial\Theta}\right)_{p,S}$ and $\beta = \frac{1}{\rho}\left(\frac{\partial\rho}{\partial S}\right)_{p,\Theta}$ are assumed to be constant.

Equations (2.152) and (2.153) combined with Eqns. (2.144) and (2.147) constitute a complete set of basic equations of a dynamical system under the so-called Boussinesq approximations, i.e., the following simplifications are assumed:

- Using the volume conservation to replace the mass conservation
- Retaining the buoyancy forcing associated with small density deviation
- Using volume conservation in the tracer (temperature and salinity) prognostic equations.

2.7.2 Potential problems associated with Boussinesq approximations

Boussinesq approximations have been widely used in the study of atmospheric and oceanic circulations. In particular, using the volume conservation to replace the continuity equation filters out the sound waves and greatly simplifies the dynamics. The momentum equation is linearized and is therefore much easier to handle.

However, these approximations can also cause some problems, which require caution:

- In reality neither α nor β is constant, depending nonlinearly upon temperature and pressure.
- Mass conservation is violated; thus, gravitational potential energy is not conserved, owing to the existence of an artificial source/sink of gravitational potential energy in the model.
- Sea surface elevation and bottom pressure due to surface heating/cooling and freshwater fluxes are incorrectly simulated in such models.
- The accurate calculation for both temperature and salinity should be based on mass conservation, as stated in Eqn. (2.8); thus, using the volume conservation approximation in the advection term, such as Eqns. (2.147a) and (2.147b), can introduce errors on the order of a few percent.

2.7.3 Buoyancy fluxes

Since Boussinesq approximations have been widely used in oceanography, the equivalent dynamical effects of surface heat and freshwater fluxes are simulated in terms of two related fluxes: density and buoyancy fluxes. Density flux is defined as

$$F_\rho = -\rho_0\left(\alpha F_T - \beta F_S\right), \quad \alpha = -\frac{1}{\rho_0}\frac{\partial\rho}{\partial T}\Big|_{p,S}, \quad \beta = \frac{1}{\rho_0}\frac{\partial\rho}{\partial S}\Big|_{p,T} \tag{2.154}$$

where $F_T = Q/\rho_0 c_p$, Q is the net heat flux into the ocean, c_p is the heat capacity of water; $F_S = (E-P)S/(1-S)$, where $E-P$ is evaporation minus precipitation.

Buoyancy is defined as $b = -g\Delta\rho/\rho_0$; thus, light water is more buoyant. In terms of the air–sea fluxes, therefore, buoyancy flux is defined as

$$F_b = g\left(\alpha F_T - \beta F_S\right) \tag{2.155}$$

and its units are m^2/s^3.

Although the concepts of both density flux and buoyancy flux are very useful in describing dynamical effects due to surface thermohaline forcing, these fluxes are artificial or conceptual and are applicable to Boussinesq models only. We need to be clear-headed about the possible problems.

First, density flux associated with surface thermal forcing is an artifact associated with the Boussinesq approximations. When water is cooled down, its density increases, so the water parcel should shrink, but the total mass should be the same as before. In a Boussinesq model, density is increased; however, owing to the volume conservation approximation, the increase in density gives rise to more mass in the ocean, and thus an artificial density flux from the atmosphere to the ocean.

Second, density flux associated with evaporation and precipitation through the sea surface is also incorrect. In the ocean, precipitation and evaporation add a mass flux of

$$F_{\rho,real} = \rho_{fresh}(P - E) \tag{2.156}$$

Owing to precipitation, surface salinity tends to decline and the corresponding density decreases. This decline of density due to precipitation in a Boussinesq model gives rise to a loss of mass because the total volume is assumed to be unchanged; thus precipitation is interpreted in terms of a negative density flux through the sea surface. According to Eqn. (2.154), the dynamical effect of precipitation is interpreted as the equivalent density flux

$$F_{\rho,Bouss} = \rho_0\beta\frac{E - P}{1 - S}S \simeq \rho_0\left(E - P\right)\beta S \tag{2.157}$$

Since $\beta S \simeq 0.02$, the equivalent density flux in a Boussinesq model is 50 times smaller than the real density flux, and the sign is opposite to the real density flux.

2.7.4 Pitfalls of using the buoyancy flux to diagnose energetics of the oceanic circulation

If the equation of state is linear in temperature, salinity, and pressure, then buoyancy is linearly proportional to density, so the balance (or the transport) of buoyancy, the temperature, the density, or the gravitational potential energy are equivalent to each other. However, if the equation of state is nonlinear, such a simple relation is no longer valid.

Since the lateral and bottom surfaces are assumed to be insulated, the total heat flux through the sea surface in a steady state must be in balance; however, the buoyancy flux through the upper surface is generally not balanced. To illustrate this problem, we discuss the buoyancy flux balance in a meridional strip of ocean in a steady state (Fig. 2.23).

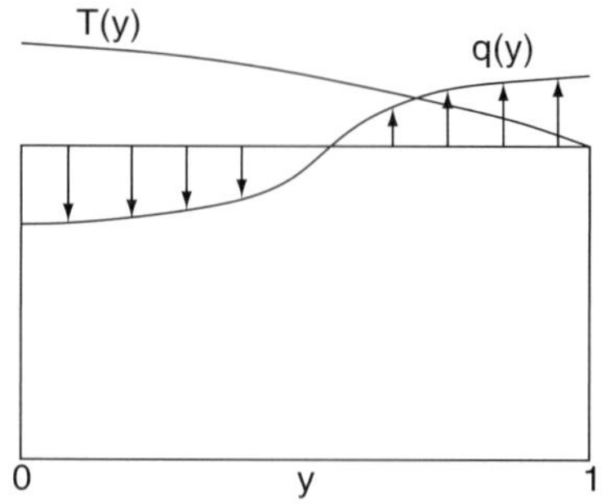

Fig. 2.23 Sketch of a model basin (in nondimensional length), subjected to the air–sea heat flux $q(y)$, and with the sea surface temperature profile $T(y)$.

In the following analysis, we define heat flux into the ocean as positive. Assuming that the heat flux into the ocean is q (in W/m^2), the corresponding buoyancy flux is $F_b = \alpha g q / \rho_0 c_p$, where α is the thermal expansion coefficient, and c_p is the specific heat under constant pressure. Although $\overline{\iint q dx dy} = 0$ for a steady state, $\overline{\iint F_b dx dy} \neq 0$, because $\alpha = \alpha(T)$ is a function of temperature. For simplicity, we assume that sea surface temperature is a linear function of y, $T = T_0(1 - y)$, and the thermal expansion coefficient is a linear function of temperature, $\alpha = \alpha_0 T = \alpha_0 T_0 \,(1 - y)$. Then we have the total buoyancy gain and loss:

$$B^+ = \frac{g}{\rho_0 c_p} \int_0^{1/2} \alpha q dy, \quad B^- = \frac{g}{\rho_0 c_p} \int_{1/2}^{1} \alpha q dy \qquad (2.158)$$

It is clear that buoyancy flux is not balanced, even if the heat flux is balanced for a steady state. Similarly, if we treat the air–sea flux at a station in terms of a one-dimensional model, then the annual mean buoyancy flux is non-zero, even though the annual mean heat flux is zero for a steady state. In particular, it is important to note that although buoyancy has been widely used as a tool in diagnosing the structure of the thermal circulation and its energetics, its success is limited to the case when the equation of state is linear. For the case with a nonlinear equation of state, ***buoyancy is not a conserved quantity***, and using buoyancy transport as a tool in diagnosing the circulation in the ocean may introduce some artifacts; therefore, such a method should be used with caution.

2.7.5 Balance of buoyancy in a model with a nonlinear equation of state

To show the potential problems of using buoyancy as a diagnostic tool, we examine the imbalance of buoyancy from the models. For the simple model in Figure 2.23 and the world's oceans, the buoyancy imbalance is listed in Table 2.9, in which the imbalance of the model is defined as

$$Imbalance = (1 - |Loss/Gain|) \times 100\% \qquad (2.159)$$

Table 2.9. *Buoyancy flux balance for three idealized model oceans and the world's oceans*

Buoyancy Unit	One-dimensional model (Fig. 2.22) Nondimensional unit			World oceans $10^8 m^4/s^3$	
Buoyancy profile	$q = \begin{Bmatrix} 1, \ y < 0.5 \\ -1, \ y > 0.5 \end{Bmatrix}$	$q = 0.5 - y$	$q = \cos(\pi y)$	Annual mean	Monthly mean
Gain	5/48	3/8	$\frac{1}{2\pi} + \frac{1}{\pi^2}$	2.16	6.15
Loss	−1/48	−1/8	$-\frac{1}{2\pi}$	−2.37	−5.82
Unbalance	80%	66%	$\frac{2}{\pi + 2} = 39\%$	−10%	5%

For the simple one-dimensional model, three different meridional profiles of buoyancy flux in nondimensional units are included. For the first case, the heat flux profile is a Heaviside function, and the imbalance in buoyancy flux is 80%. For the second case, a linear profile is assumed, and the imbalance is 66%. For the third case, the sinusoidal profile is assumed, but the imbalance is also quite large.

For the world's oceans, we discuss the surface buoyancy flux associated with heat flux alone, which is defined as $F_b = -g\delta\dot{\rho}/\rho_0 = g\alpha q/\rho_0 c_p$, where q is the surface heat flux, in W/m^2. The calculation is based on the climatology of temperature and salinity, and the heat flux is based on the NCEP data averaged from 1948 to 2001. The heat flux field has been modified in order to make sure that the net heat flux through the air–sea interface is exactly zero, i.e., the heat flux is balanced to meet the assumption of a steady state. Based on NCEP data, the total downward buoyancy flux in the world's oceans is approximately 6.15×10^8 m^4/s^3, and the outward buoyancy flux is about -5.82×10^8 m^4/s^3; the imbalance is about 5%. Thus, the net downward buoyancy flux at the surface is approximately 3.3×10^7 m^4/s^3, which must be gradually reduced as we move downward. However, if the buoyancy flux calculation is based on the annual mean heat flux, the downward buoyancy flux is 10% smaller than the upward buoyancy flux.

The major difference between these two calculations is in the mean thermal expansion coefficient. Because the thermal expansion coefficient $\alpha = -\partial\rho/\rho\partial T$ is sensitively dependent on the mean temperature, the annual mean buoyancy flux can be quite different from the monthly mean buoyancy flux. Thus, we can see that although the mean air–sea heat flux can be well defined from observations, the mean air–sea buoyancy flux cannot be uniquely defined, owing to the nonlinear nature of the equation of state.

2.8 Various vertical coordinates

In meteorology and oceanography, various coordinate systems have been used in theoretical studies and numerical models. A classic reference is Kasahara (1974). Although the physical problems are the same in different coordinates, the mathematical formulation can be quite different. Using special coordinates can help us to simplify the formulation and thus make the presentation of the relevant physics clear and easy.

2.8.1 Vertical coordinate transformation

Assume that $\zeta = \zeta(x, y, z, t)$ is a single-valued monotonic function of z. For any dependent variable we have the following relation: $A(x, y, z, t) = A(x, y, z(x, y, \zeta, t), t)$. Note that the requirement of a single-value monotonic function may be violated. For example, if we choose the potential density σ_0 as a vertical coordinate, the transformation is not monotonic in the Atlantic Ocean, as discussed in Section 2.4.

Relation for partial derivatives

For partial derivatives other than the vertical derivative we have

$$\left(\frac{\partial A}{\partial s}\right)_\zeta = \left(\frac{\partial A}{\partial s}\right)_z + \frac{\partial A}{\partial z}\left(\frac{\partial z}{\partial s}\right)_\zeta, \quad \text{where } s = x, y, t \tag{2.160}$$

The relation for the vertical derivative

We have the following relation

$$\frac{\partial A}{\partial \zeta} = \frac{\partial A}{\partial z}\frac{\partial z}{\partial \zeta}, \quad \text{or} \quad \frac{\partial A}{\partial z} = \frac{\partial A}{\partial \zeta}\frac{\partial \zeta}{\partial z} \tag{2.161}$$

Substituting Eqn. (2.161) into Eqn. (2.160) gives

$$\left(\frac{\partial A}{\partial s}\right)_\zeta = \left(\frac{\partial A}{\partial s}\right)_z + \frac{\partial A}{\partial \zeta}\frac{\partial \zeta}{\partial z}\left(\frac{\partial z}{\partial s}\right)_\zeta \tag{2.162}$$

The two-dimensional operators

For convenience we introduce the following horizontal gradient operators

$$\nabla_z = \left(\vec{i}\frac{\partial}{\partial x} + \vec{j}\frac{\partial}{\partial y}\right)_z, \quad \nabla_\zeta = \left(\vec{i}\frac{\partial}{\partial x} + \vec{j}\frac{\partial}{\partial y}\right)_\zeta$$

so that

$$\nabla_\zeta A = \nabla_z A + \frac{\partial A}{\partial \zeta}\frac{\partial \zeta}{\partial z}\nabla_\zeta z \tag{2.163}$$

$$\nabla_\zeta \cdot \vec{B} = \nabla_z \cdot \vec{B} + \frac{\partial \vec{B}}{\partial \zeta}\frac{\partial \zeta}{\partial z} \cdot \nabla_\zeta z \tag{2.164}$$

The total derivative

In the new coordinates, the total derivative, or the so-called material derivative, is defined as

$$\frac{dA}{dt} = \left(\frac{\partial A}{\partial t}\right)_\zeta + \vec{v} \cdot \nabla_\zeta A + \dot{\zeta}\frac{\partial A}{\partial \zeta} \tag{2.165}$$

where $\dot{\zeta} = \frac{d\zeta}{dt}$ is the "vertical velocity" in ζ-coordinates. Note that this quantity may have a dimension other than LT^{-1}.

Horizontal pressure force

Using Eqn. (2.163), we obtain

$$-\frac{1}{\rho}\nabla_z p = -v\nabla_\zeta p + v\frac{\partial p}{\partial z}\nabla_\zeta z = -v\nabla_\zeta p - \nabla_\zeta \varphi \tag{2.166}$$

where $v = 1/\rho$ is the specific volume, and $\varphi = gz$ is the geopotential.

2.8.2 Commonly used vertical coordinates in oceanography

Many different vertical coordinates have been used in oceanography. The horizontal pressure gradient term has different forms of expression in these coordinates. For numerical modeling or data analysis, it is desirable that pressure terms can be reduced to a gradient of a streamfunction defined on the coordinate surfaces. On the other hand, if the pressure term is expressed in terms of the small difference between two large terms, large numerical errors may be introduced into the numerical analysis.

z-coordinate, $\zeta = z$

This is the commonly used vertical coordinate, and the convention is $z = 0$ at the mean sea surface and $z > 0$ upward. In this coordinate, the last term in Eqn. (2.166) vanishes; thus, p is the streamfunction for the horizontal mass transport $\rho f \vec{u}$.

Pressure coordinate: $\zeta = p$

In this coordinate, the first term on the right-hand side of Eqn. (2.166) vanishes. Thus, the geopotential $\varphi = gz$ is a streamfunction. However, the absolute geopotential for a pressure surface is unknown because the free elevation is uncertain. This can be overcome by using the geopotential relative to a reference level as follows. If p_0 is a deep reference pressure level, i.e., geostrophic motions are negligible at pressure surface p_0, then the hydrostatic approximation leads to

$$\phi - \phi_0 = -\int_{p_0}^{p} dp/\rho \tag{2.167}$$

Introduce the specific volume anomaly

$$\delta = v\,(S, T, p) - v\,(35, 0, p) = v - \tilde{v} \qquad (2.168)$$

Since the integral of $\tilde{v}$ between $p = [p_0, p]$ is a function of p and p_0 only, it is independent of x,y, and thus makes no contribution to the geostrophic velocity. As a result, the dynamic height $-\int_{p_0}^{p} \delta dp$ is the streamfunction of the horizontal volume transport $f\vec{u}$.

The major advantages of using pressure coordinates are the following. First, the equation of state of seawater is defined in terms of pressure, so that density calculation based on a pressure coordinate is straightforward. Second, the mass-conserving model can be quite conveniently formulated in the pressure coordinate.

The common practice of using the z-coordinate in oceanic research comes from historical tradition. However, most data collected from the open ocean are based on the *in situ* pressure, not the geopotential height. On the other hand, many datasets and most numerical models are based on the z-coordinate. Since equation of state is defined in terms of pressure, not geopotential height, the application of these data and models involves conversion between these two vertical coordinates. If the model is based on the pressure coordinate, there is no need for the coordinate conversion; however, this has not been common practice in oceanography.

Steric anomaly coordinate

The pressure effect on the seawater density can be subtracted as, in a family of pressure-corrected specific volume anomaly,

$$\delta_p = v - f\,(p) \qquad (2.169)$$

where $f\,(p)$ is a function of pressure only. For example, the traditional definition of steric anomaly is

$$\delta = v\,(S, T, p) - v\,(35, 0, p) \qquad (2.170)$$

Other choices for $f\,(p)$ can be used as well. Using Eqn. (2.167), the right-hand side of Eqn. (2.166) is reduced to

$$-v\nabla_{\delta_p} p + \nabla_{\delta_p} \int_{p_0}^{p} v dp = p\nabla_{\delta_p} v - \nabla_{\delta_p}\,(v_0 p_0) - \nabla_{\delta_p} \int_{v_0}^{v} p dv \qquad (2.171)$$

After manipulation, this is reduced to

$$-\nabla_{\delta_p}\pi = -\nabla_{\delta_p}\left(\delta_0 p_0 + \int_{\delta_0}^{\delta} p d\delta_p\right) \qquad (2.172)$$

where $\delta_0 = \delta_0\,(S, T, p_0)$ varies on pressure surface p_0. Therefore, the “acceleration potential,” π, which was first introduced by Montgomery (1937), is a streamfunction for the volume transport $f\vec{\mathbf{u}}$.

Density coordinate

Using the *in situ* density as a vertical coordinate, the corresponding terms in Eqn. (2.166) are reduced to

$$-\nabla_\rho (p/\rho) + \nabla_\rho \int_{p_0}^{p} dp/\rho \tag{2.173}$$

Therefore, the corresponding streamfunction is

$$p/\rho - \int_{p_0}^{p} dp/\rho \tag{2.174}$$

Similarly, a whole family of density coordinates, called orthobaric density, can be defined as

$$\rho_p (S,T,p) = \rho (S,T,p) - f (\rho,p) \tag{2.175}$$

where $f (\rho,p)$ is a function of density and pressure only. It is readily shown that an exact streamfunction exists for such a density coordinate. For example, if $f (\rho,p) = f (p)$, after some manipulations Eqn. (2.166) is reduced to

$$\begin{aligned} -v\nabla_z p &= -v\nabla_{\rho_p} p + \nabla_{\rho_p} \int_{p_0}^{p} v dp \\ &= -\nabla_{\rho_p} (p_0 v_0) + p\nabla_{\rho_p} v - \nabla_{\rho_p} \int_{\rho_0}^{\rho} p dv \\ &= -\nabla_{\rho_p} \left\{ \frac{p_0}{\rho_{p,0} + f (p_0)} - \int_{\rho_{p,0}}^{\rho_p} \frac{p d\sigma}{[\sigma + f (p)]^2} \right\} \end{aligned} \tag{2.176}$$

where $\rho_{p,0} = \rho_0 - f (p_0)$ is the reference value.

Global pressure-corrected density coordinate

A special choice of the orthobaric density can be defined by choosing the correction obtained by the global-averaged adiabatic compressibility

$$\sigma_g (S, \Theta, p) = \rho (S, \Theta, p) - f (p) \tag{2.177}$$

$$f (p) = \int_{p_0}^{p} \overline{\rho K_\eta}^{\lambda,\Theta} dp \tag{2.178}$$

where K_η is the adiabatic and isohaline compressibility of seawater. Note that the potential temperature is used in the definition because it is a much better conservable quantity. σ_g can be used as a vertical coordinate and provides a relatively easy tool for diagnosing the water properties associated with the global-scale circulation. In particular, as demonstrated above, an exact streamfunction exists for flow on the σ_g surfaces; thus, it can be used as a vertical coordinate in numerical models for the oceanic general circulation.

Potential density coordinate

If we choose $\zeta = \sigma$, i.e., the potential density is used as the vertical coordinate, Eqn. (2.166) is reduced to

$$-v\nabla_z p = -v\nabla_\sigma p + \nabla_\sigma \int_{p_0}^{p} v dp \tag{2.179}$$

Since $v = \delta + v_0(p)$ and

$$\nabla_\sigma [pv_0(p)] - \nabla_\sigma \int_{p_0}^{p} v_0(p)\, dp - p\nabla_\sigma [v_0(p)] = 0 \tag{2.180}$$

Eqn. (2.166) is reduced to

$$-v\nabla_z p = -\left[\nabla_\sigma \left(\delta p - \int_{p_0}^{p} \delta dp\right) - p\nabla_\sigma \delta\right] \tag{2.181}$$

It turns out that the last term on the right-hand side of Eqn. (2.179) is of the same order as the first two terms, so this term cannot be omitted; thus, there is no exact streamfunction for the potential density coordinate.

Other vertical coordinates

There are many other thermodynamic variables which can be used as vertical coordinates, such as potential temperature, neutral density, entropy, and others. However, in most of such vertical coordinates. the horizontal pressure gradient terms cannot be expressed in terms of a streamfunction.

2.9 Ekman layer

In the upper ocean there exists a thin boundary layer below the sea surface where the wind stress is balanced by frictional force due to vertical-shear-induced turbulence and pressure force. The earliest study of this layer can be traced back to the classical work by Ekman (1905); however, it turns out that the structure of the Ekman layer in the ocean is much more complicated, due to the existence of surface waves, wave breaking, turbulence, and other dynamical processes, such as the Stokes drift and the Langmuir cell.

Since Ekman postulated the theory of this frictional layer for the ocean more than 100 years ago, the Ekman layer has become one of the fundamental elements of the oceanic general circulation machinery. Despite the pivotal position of the Ekman layer in oceanic dynamics and much effort devoted to observing the Ekman layer in the ocean, it was not until the mid-1980s that a clear picture of the Ekman layer was observed in the upper ocean through instrumentation (Price *et al.*, 1987). Even today, many questions remain unclear regarding the structure of the Ekman layer in the ocean. In this section, we limit discussion to the most elementary theory of the Ekman layer only, and leave many of the details for the reader to discover by reading the recently published literature.

2.9.1 Classical theory of Ekman layer below a free surface

For the study of oceanic general circulation, the most important application of the Ekman theory is the case where wind stress is imposed on the free surface of the ocean. In this section, we start with a model with a simple assumption about the vertical diffusivity, and a more complete analysis will be discussed afterward. For the following analysis, we assume that seawater density is homogeneous and the ocean is in a quasi-steady state.

Model formulation

The model is formulated for a relatively thin surface layer, on the order of a few tens of meters, where the vertical shear of turbulence force is a dominating factor in the dynamic balance. For large-scale motions in the oceanic interior, away from the coast and equator, the horizontal momentum equations are

$$-fv = -\frac{1}{\rho_0}\frac{\partial p}{\partial x} + \frac{\partial}{\partial z}\left(A\frac{\partial u}{\partial z}\right) \quad (2.182a)$$

$$fu = -\frac{1}{\rho_0}\frac{\partial p}{\partial y} + \frac{\partial}{\partial z}\left(A\frac{\partial v}{\partial z}\right) \quad (2.182b)$$

where ρ_0 is a constant reference density. The solution is subject to a boundary condition that velocity vanishes at large depths below the sea surface and the vertical stress on the sea surface should match the wind stress imposed on the ocean, i.e.

$$(u, v) \to 0, \quad as\ z \to -\infty \quad (2.183)$$

$$A\frac{\partial u}{\partial z} = \tau^x/\rho_0, \quad A\frac{\partial v}{\partial z} = \tau^y/\rho_0 \quad at\ z = 0 \quad (2.184)$$

where A is an eddy viscosity.

Equations (2.182a) and (2.182b) can be simplified by decomposing the velocity into two parts

$$u = U_g + u_e = -\frac{1}{\rho_0}\frac{\partial p}{\partial y} + u_e \quad (2.185a)$$

$$v = V_g + v_e = -\frac{1}{\rho_0}\frac{\partial p}{\partial x} + v_e \quad (2.185b)$$

where U_g and V_g are the components of the geostrophic velocity, which is in balance with the pressure gradient. We will assume that the vertical scale of the Ekman layer is very small (on the order of 30 m), so that the vertical shear of the geostrophic velocity can be neglected. This velocity decomposition leads to a simplified set of equations for the ageostrophic velocity:

$$-fv_e = \frac{\partial}{\partial z}\left(A\frac{\partial u_e}{\partial z}\right) \tag{2.186a}$$

$$fu_e = \frac{\partial}{\partial z}\left(A\frac{\partial v_e}{\partial z}\right) \tag{2.186b}$$

This system can easily be solved; however, there are some simple properties which can be derived without even solving the system. In fact, integrating Eqn. (2.186a) and using boundary conditions of Eqns. (2.183, 2.184), we obtain

$$-f\int_{-\infty}^{0} v_e dz = A\frac{\partial u_e}{\partial z}|_{z=0} = \tau^x/\rho_0 \tag{2.187}$$

Similarly, we can obtain the second relation for u_e; thus, for the volume flux integrated over the Ekman layer, we have

$$-fV_e = \tau^x/\rho_0, \quad fU_e = \tau^y/\rho_0 \tag{2.188}$$

These two relations can be combined and written in the following vector form

$$\vec{U}_e = -\vec{z} \times \vec{\tau}/f\rho_0 \tag{2.189}$$

where $\vec{U}_e$ is the total volume flux integrated over the whole depth of this surface boundary layer. Therefore, wind stress in the upper ocean drives a volume flux, the so-called Ekman transport, which is 90° to the right of wind-stress direction (in the Northern Hemisphere). The volume flux is linearly proportional to the wind stress and inversely proportional to the Coriolis parameter. Most interestingly, the Ekman transport $\vec{U}_e$ is independent of A, even if A may not be a constant.

Assuming $\tau = 0.1$ N/m^2, $f = 10^{-4}$/s, so that $V_e = \tau/f\rho \simeq 1$ m^2/s. If the frictional layer is 20 m thick, then the horizontal velocity is $v_e \simeq 0.05$ m/ s. Since the North Atlantic basin is approximately 6,000 km wide, the total equatorward volume flux driven by the mid-latitude westerlies is $V_eL = 6 \times 10^6$m^3/s $= 6$ Sv.

The Ekman spiral

To pursue the vertical structure of velocity within this boundary layer, we need to know more about the eddy diffusivity A. This turns out to be a very complicated issue. Because of waves and turbulent motions in the upper ocean, the equivalent eddy diffusivity is difficult to measure and is unlikely to be constant. However, for simplicity of the analysis, we will assume that the eddy viscosity is constant.

One way to solve the equation system is to introduce a complex velocity, in much the same way as in the theory of two-dimensional incompressible fluid mechanics:

$$M = u_e + iv_e \tag{2.190}$$

The corresponding equations for two unknown variables (u_e, v_e) are reduced to an equation of the single unknown variable M:

$$\frac{d^2M}{dz^2} - i\frac{f}{A}M = 0 \tag{2.191}$$

The general solution of Eqn. (2.191) is

$$M = c_1 e^{\lambda z} + c_2 e^{-\lambda z}, \quad \lambda = \frac{\sqrt{2}}{2}(1+i)\sqrt{\frac{f}{A}} \tag{2.192}$$

Applying the lower boundary condition, $M \to 0$ $as\, z \to -\infty$, leads to $c_2 = 0$.

At the sea surface, the corresponding boundary condition is

$$\left.\frac{dM}{dz}\right|_{z=0} = \lambda c_1 = \frac{1}{\rho_0 A}\left(\tau^x + i\tau^y\right) \tag{2.193}$$

Thus, $c_1 = \frac{1-i}{\sqrt{2fA\rho_0}}(\tau^x + i\tau^y)$, and the final solution is

$$u_e + iv_e = \frac{1-i}{\sqrt{2fA\rho_0}}(\tau^x + i\tau^y)Exp\left[\frac{\sqrt{2}}{2}(1+i)\sqrt{\frac{f}{A}}z\right] \tag{2.194}$$

This solution is in the form of a spiral called the Ekman spiral. In the Northern Hemisphere, the surface velocity vector is rotated 45° to the right of the wind-stress direction, and the velocity vector rotates clockwise with depth, as shown in Figure 2.24.

As an example, the horizontal velocity of the classical Ekman spiral calculated from Eqn. (2.194) is shown as the solid line in Figure 2.25. The other two curves in this figure are for the case of anisotropic vertical eddy diffusivity, which is discussed shortly.

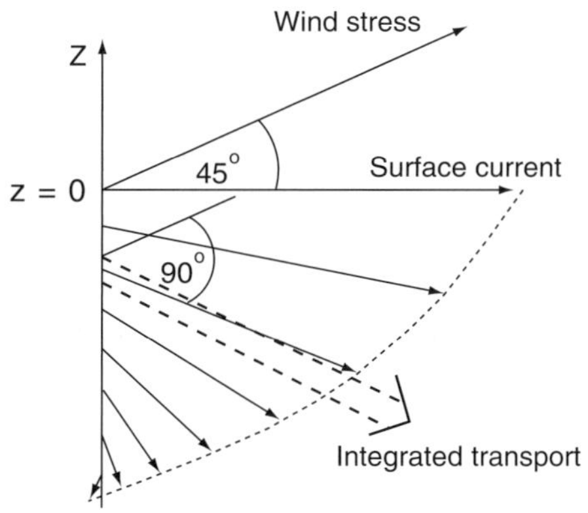

Fig. 2.24 Sketch of the Ekman spiral in the Northern Hemisphere.

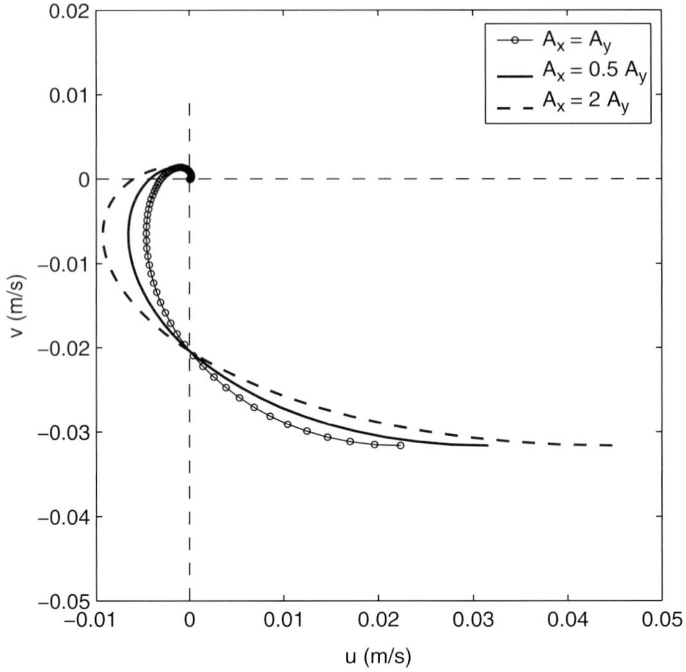

Fig. 2.25 Typical profiles for the Ekman spiral: the *solid line* for the case with isotropic vertical diffusivity; the *circled line* for the case with the diffusivity in the downwind direction double that in the crosswind direction; and the *dashed line* for the case with diffusivity in the crosswind direction double that in the downwind direction.

Ekman pumping

Owing to the spatial variation of both τ and f, there is a non-zero horizontal convergence (divergence) of the Ekman transport, the so-called Ekman pumping (upwelling). The rate of Ekman pumping can be calculated from the continuity equation

$$u_x + v_y + w_z = 0 \tag{2.195}$$

Using the upper boundary condition of $w = 0$ at $z = 0$, the vertical velocity at the base of the Ekman layer is

$$w_e = \int_{-H}^{0} (u_x + v_y)\, dz = \frac{\partial}{\partial x} \int_{-H}^{0} u_e dz + \frac{\partial}{\partial y} \int_{-H}^{0} v_e dz \tag{2.196}$$

where we have assumed that the background geostrophic velocity is uniform in space and H is constant. The pumping velocity is related to the wind stress as follows:

$$w_e = \frac{\partial}{\partial x}\left(\frac{\tau^y}{f\rho_0}\right) - \frac{\partial}{\partial y}\left(\frac{\tau^x}{f\rho_0}\right) = \frac{1}{f\rho_0}\left[\frac{\partial \tau^y}{\partial x} - \frac{\partial \tau^x}{\partial y}\right] + \frac{\beta \tau^x}{f^2 \rho_0} \tag{2.197}$$

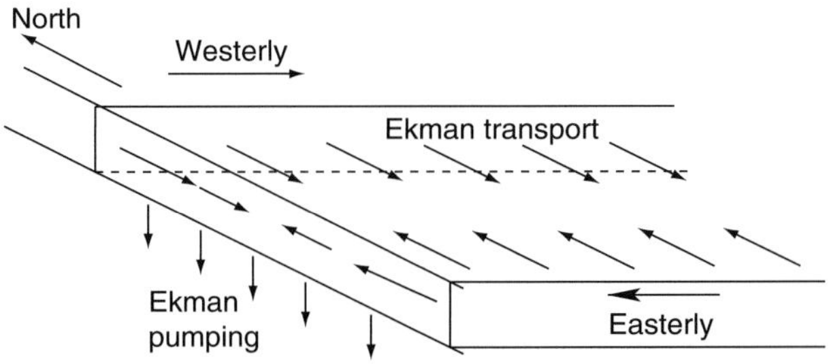

Fig. 2.26 Sketch of the Ekman pumping in a subtropical basin.

Therefore, the Ekman pumping rate consists of two parts: the first part is due to the wind-stress curl, and the second part is due to the beta effect. Assuming that the scale of wind stress is $\tau \simeq 0.1\,\mathrm{N/m^2}$, and the length scale is 1,000 km, the pumping velocity is estimated as $w_e \simeq -10^{-6}\,\mathrm{m/s}$. Over the world's oceans, westerly winds prevail at mid latitudes and easterlies at low latitudes. The equatorward Ekman transport driven by westerlies at mid latitudes and the poleward Ekman transport driven by easterlies at low latitudes converge in the subtropical basin. As a result, Ekman pumping prevails over the subtropical basin (Fig. 2.26). Similarly, strong westerlies at mid latitudes and weak westerlies or easterlies at high latitudes give rise to Ekman upwelling for the subpolar basin. Thus, wind-stress curl in the world's oceans gives rise to basin-scale Ekman pumping/upwelling. As will be discussed in relation to the wind-driven circulation theory, Ekman pumping is the primary driving force for the upper ocean circulation.

There are also other mechanisms which can induce upwelling/downwelling. As shown in the last term of Eqn. (2.197), the change in the Coriolis parameter, the so-called "beta effect," can give rise to Ekman upwelling/downwelling. In addition, strong along-shore winds can drive strong upwelling/downwelling in coastal areas. For example, near the coast of California, strong equatorward trade winds drive strong offshore Ekman transport in the upper ocean. Since there is no mass flux across the coastline, water must be drawn from depth near the coast in order to sustain the strong offshore Ekman flux. The upwelling/downwelling near coastal areas is a critically important component of the water mass cycle in the world's oceans.

2.9.2 Ekman spiral with inhomogeneous diffusivity

Although the Ekman theory has been widely regarded as the backbone of modern dynamical oceanography, the Ekman spiral predicted by classical theory was not verified through oceanic observations for a long time. According to the theory, for a steady wind stress, the velocity vector on the sea surface from classical theory is 45° to the right of wind stress (in the Northern Hemisphere), and the velocity vector rotates in the form of a spiral in a vertical direction. Observations, however, indicate that the angle between surface wind stress and surface drift velocity vector is in the range between 5° and 20° (Cushman-Roisin, 1994).

Recent observations also indicate that the surface velocity lies at more than the predicted 45° to the right of the wind. More importantly, the observed current amplitude decreases at a faster rate than it turns to the right, i.e., the observed velocity profiles in the Ekman layer seem "flat" (Chereskin and Price, 2001).

Despite great efforts to find solutions that fit the observations, most models within the framework of the classical laminar theory (with a diffusivity which is spatially isotropic and independent of time) fail to produce the observed "flat" spiral. It seems that other essential dynamical processes, such as buoyancy flux through the air–sea interface, stratification, the diurnal cycle, or even the Stokes drift, may have to be included in order to explain the observed structure of the Ekman layer (Price and Sundermeyer, 1999).

Owing to complicated dynamical processes in the upper ocean, parameterization of turbulent dissipation in the upper ocean remains a great challenge. Observations indicate that in a thin surface layer immediately below the sea surface, waves and turbulent activities are rather strong, so dissipation is nearly constant or increases slightly with depth; however, the dissipation rate declines with depth below this shallow layer. Direct observations in the California Current indicate that turbulent diffusivity declines exponentially below 20 m depth (Chereskin, 1995). Terray *et al.* (1996) carried out field observations and found that the dissipation rate is higher and roughly constant in a near-surface layer, but below this layer the dissipation rate decays as $|z|^{-2}$. This is further refined as a -2.3 power law, i.e., dissipation rate decays as $|z|^{-2.3}$, by Terray *et al.* (1999).

A crude model that can incorporate this complexity is a two-layer model with a power law of vertical diffusivity in each layer. In addition, the same approach can be used to find solutions for the case where the vertical diffusivity changes with depth exponentially. In such cases the solutions are in form of Bessel functions. A simple linear profile $A = \alpha|z|$ was used in previous studies, e.g., Madsen (1977). Such a profile is questionable because it is inconceivable that turbulent diffusivity is zero at the sea surface (N.E. Huang, 1979). Therefore, it seems more reasonable to choose a linear profile in the surface layer, $n_1 = 1$, starting with a finite diffusivity on the sea surface. For the second layer, it is found that an inverse power profile with $n_2 = -0.7$ has a best fit for the diffusivity diagnosed by Chereskin (1995). The application of this two-layer model is referred to in the study by Wang and Huang (2004a).

Another way to narrow the gap between observations and the classical Ekman spiral is to relax the isotropic assumption of vertical diffusivity. Although vertical diffusivity in the previous example is assumed to be isotropic, it is possible that, due to surface waves and other processes, the turbulence diffusivity may be non-isotropic. Assuming that the turbulent motions in the Ekman layer are non-isotropic, the structure of the Ekman spiral can easily be derived as follows. For simplicity, we will assume that vertical diffusivity is non-isotropic and remains constant in the vertical direction, i.e., $A_x = const.$, $A_y = const.$, but $A_x \neq A_y$. The corresponding momentum equations for the ageostrophic velocity components are

$$-f v_e = \frac{\partial}{\partial z}\left(A_x \frac{\partial u_e}{\partial z}\right) \tag{2.198a}$$

$$fu_e = \frac{\partial}{\partial z}\left(A_y \frac{\partial v_e}{\partial z}\right) \tag{2.198b}$$

Integrating Eqn. (2.198a) and using boundary conditions, we obtain

$$-f\int_{-\infty}^{0} v_e dz = A_x \frac{\partial u_e}{\partial z}|_{z=0} = \tau^x/\rho_0 \tag{2.199}$$

Introducing a coordinate transform:

$$z = z'\sqrt{A/f} \tag{2.200}$$

where $A = \sqrt{A_x A_y}$ is the geometric mean of the vertical diffusivity. Parallel to the previous approach, we can introduce the complex velocity

$$M = Ru_e + iv_e, \quad R = \sqrt{A_x/A_y} \tag{2.201}$$

Eqations (2.198a, 2.198b) are reduced to

$$\frac{d^2M}{dz'^2} - M = 0 \tag{2.202}$$

The solution which satisfies the condition of no flow as $z \to -\infty$ is

$$M = Ru_e + iv_e = c \cdot Exp\left[(1+i)\sqrt{\frac{f}{2A}}z\right] \tag{2.203}$$

Applying the stress boundary condition at the sea surface leads to

$$c = \frac{1-i}{\rho_0}\sqrt{\frac{1}{2fA}}\left(\tau^x + iR\tau^y\right) \tag{2.204}$$

The structure of the Ekman spiral under the assumption of non-isotropic vertical diffusivity is shown in Figure 2.24. It is clear that if the diffusivity is spatially non-isotropic, the Ekman spiral should have a structure quite different from the classical one.

2.10 Sverdrup relation, island rule, and the β-spiral

Many aspects of the fundamental structure of the wind-driven circulation in the ocean have been successively described by some classical theories, as discussed in many introductory textbooks about dynamical oceanography. In this section, we give a brief presentation of the dynamical laws. As discussed shortly, these laws are essentially various forms of the potential vorticity balance in the upper ocean.

2.10.1 Sverdrup relation

For large-scale steady wind-driven circulation in the ocean interior, the nonlinear advection terms are negligible, so the time-dependent term in Eqns. (2.133) and (2.134) can be omitted. Therefore, in a β-plane it can be described in terms of depth-integrated equations

$$-fV = -P_x/\rho_0 + \tau^x/\rho_0 + F_x \tag{2.205}$$

$$fU = -P_y/\rho_0 + \tau^y/\rho_0 + F_y \tag{2.206}$$

$$U_x + V_y = 0 \tag{2.207}$$

where (U, V) are the depth-integrated volume flux, (τ^x, τ^y) are the wind-stress components, (F_x, F_y) are the forces due to bottom or lateral friction, which are assumed to be negligible except within the narrow western boundary, and (P_x, P_y) are the pressure gradient terms, which can be calculated from the density structure. As discussed in Section 2.6, the hydrostatic relation can be reduced to a relation between the perturbations of pressure and the density.

Assuming that both the bottom and lateral friction are negligible, cross-differentiating and subtraction of Eqns. (2.205) and (2.206) leads to

$$\beta V = -\left(\partial\tau^x/\partial y - \partial\tau^y/\partial x\right)/\rho_0 \tag{2.208}$$

where $\beta = df/dy$ is the gradient of planetary vorticity; this is called the ***Sverdrup relation***. This relation represents the basic vorticity balance in the wind-driven circulation, i.e., in the ocean interior the planetary vorticity gradient term associated with meridional advection is balanced by the wind-stress curl. This is valid for the ocean interior only where other terms, such as the nonlinear advection terms, the bottom or lateral friction terms, are much smaller than the planetary vorticity and wind-stress torque terms; thus, they can be omitted for the lowest-order dynamic balance in the oceanic interior. For example, in the subtropical basin, the westerlies at mid latitudes and easterlies at low latitudes give rise to negative wind-stress torque. Under this negative vorticity drive, water parcels move equatorward and thus create the anticyclonic gyres in the world's oceans. The application of the Sverdrup relation is discussed in Chapter 4.

2.10.2 The island rule

The Sverdrup relation discussed above applies to circulation in the open ocean. The same set of equations can apply to the circulation around a big island in the middle of the ocean, such as Australia, or any other large island (Godfrey, 1989).

Integrating the momentum equations along line CBA (Fig. 2.27), we obtain

$$P_C - P_A = \int_{ABC} \tau^l dl + \rho_0 f_B T_0 \tag{2.209}$$

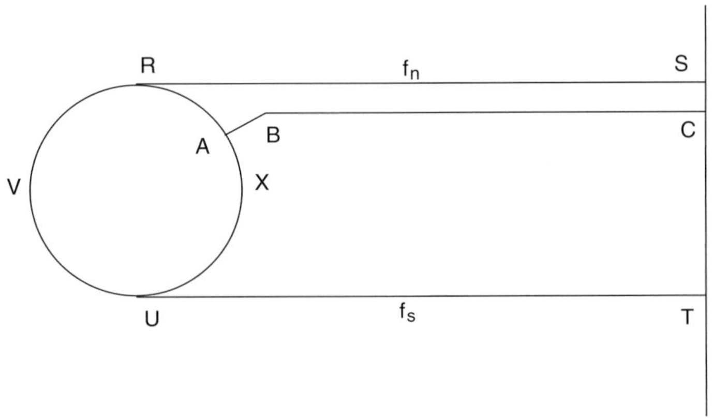

Fig. 2.27 Sketch of the island rule.

where the integral on the right-hand side is an along-path integral of the wind-stress component τ^l and T_0 is the total volume flux across this section, including both the interior flow and the western boundary current. In deriving this relation, we have used the assumptions that friction is negligible for the ocean interior, and that semi-geostrophy applies to the western boundary current. Since the width of the western boundary is much smaller than the size of the basin, the above formulae can be applied to any cross-section, without specifically mentioning the inclusion of the western boundary segment.

Since no mass flux can cross the eastern boundary, a similar integral along the eastern segment *ST* gives

$$P_S - P_T = \int_{TS} \tau^l dl \tag{2.210}$$

Applying the above argument to line *RSTU*, we have

$$P_R - P_U = \int_{UTSR} \tau^l dl + \rho_0 (f_s - f_n) T_0 \tag{2.211}$$

where f_n and f_s are the Coriolis parameters at the northern and southern segments (Fig. 2.27). Similar to Eqn. (2.209), we can obtain an equation for the western coast of the island; thus, the final expression for the volume flux circulating the island is

$$T_0 = \frac{1}{\rho_0 (f_n - f_s)} \oint_{RSTUV} \tau^l dl \tag{2.212}$$

where the integral is calculated along a complete loop along the path *RSTUVR*.

Note that $f_R - f_U = \beta \Delta y$ and $\oint \tau^l dl = curl\tau \cdot \Delta x \Delta y$, so this relation can be reduced to the classical Sverdrup relation as follows:

$$T_0 = \frac{curl\tau}{\rho_0 \beta} \Delta x \tag{2.213}$$

One of the most successful applications of the island rule is to predict the volume flux of the Indonesian Throughflow. From simple intuition, one may think that the volume flux of the throughflow is mostly controlled by local forces, such as wind stress and tides. However, the island rule makes a solid link between the large-scale wind stresses and the volume transport through the relatively narrow channel.

It is also amazing to see that the island rule compares favorably with observations, even for the case of time-dependent flow and with friction. There are many big islands in the world's oceans, and the island rule can be a very powerful tool in understanding the circulation there.

2.10.3 Vertical structure of the horizontal velocity field

Large-scale motions are predominantly horizontal, and the vertical structure of the horizontal velocity field depends on the horizontal density structure through the thermal wind relation.

Taylor–Proudman theorem

For large-scale steady circulations in the ocean interior, the time-dependent term, the frictional terms, and the inertial terms are negligible; thus, the dynamics is regulated by geostrophy and the hydrostatic relations:

$$fv = p_x/\rho_0 \tag{2.214}$$

$$fu = -p_y/\rho_0 \tag{2.215}$$

$$0 = p_z + \rho g \tag{2.216}$$

From these relations we can derive the thermal wind relation:

$$fu_z = g\rho_y/\rho_0 \tag{2.217}$$

$$fv_z = -g\rho_x/\rho_0 \tag{2.218}$$

The Taylor–Proudman theorem can be obtained under different assumptions.

1. For homogeneous fluid, $\rho = const.$
   In this case, the horizontal density gradient vanishes, so that

$$u_z = v_z = 0 \tag{2.219}$$

Thus, fluid columns move as a rigid body, a phenomenon which can be demonstrated through laboratory experiments.

2. For stratified fluid under the f-plane approximation, i.e., $f = const.$
   For this case we add on two more equations of continuity and density conservation:

$$u_x + v_y + w_z = 0 \tag{2.220}$$

$$u\rho_x + v\rho_y + w\rho_z = 0 \tag{2.221}$$

Cross-differentiating the momentum equations and using the continuity equation lead to

$$w_z = 0 \tag{2.222}$$

Since vertical velocity is zero at the sea surface,

$$w = 0, \; at \; z = 0 \tag{2.223}$$

Eqn. (2.222) leads to $w \equiv 0$ for the whole depth of a water column. At the sea floor, the no-penetration condition is

$$w = -\vec{v} \cdot \nabla h, \quad at \; z = -h \tag{2.224}$$

Therefore, zero vertical velocity is equivalent to

$$\vec{u} \cdot \nabla h = 0, \; at \; z = -h \tag{2.225}$$

i.e., flow must follow the depth contours. Since $w = 0$, using density conservation and thermal wind relation equations (2.217) and (2.218), we finally obtain

$$-uv_z + vu_z = 0, \quad i.e., \; \vec{k} \cdot \vec{u} \times \vec{u}_z = 0 \tag{2.226}$$

This relation means that flow stays in the same direction over the whole depth of the water column.

This is essentially the conclusion of the classical "law of parallel solenoids" (Neumann and Pierson, 1966), which has ruled the theoretical framework of physical oceanography for a long period of time, between 1920 and 1970. As this book says:

For non-accelerated, frictionless ocean currents the flow must always be parallel, not only to the isobars but also to the isopycnals. If this condition were not fulfilled, the distribution of mass would be altered by the fluid motion. It also follows that the isobars and isopycnals at one level must be parallel to those at other levels if the density stratification is continuous and if the above-mentioned assumptions hold (p. 90).

Although the conclusion of this "law of parallel solenoids" was obviously in contradiction to observations, for a long time it was not easy to see how to move away from such a persuasive theory. A real breakthrough from this seemingly solid theoretical argument of the parallel solenoid flow was due to the pioneering work of Stommel and Schott (1977), which is discussed in the next section.

Before the discussion of the β-spiral, it is worthwhile to explore what went wrong with this seemingly solid argument. To begin with, we start from the following momentum equations based on the *in situ* density:

$$-fv = -p_x/\rho \tag{2.227}$$

$$fu = -p_y/\rho \tag{2.228}$$

Cross-differentiating Eqns. (2.227) and (2.228), then using the continuity equation Eqn. (2.220), we obtain

$$\beta v = f w_z + \frac{p_y \rho_x - p_x \rho_y}{\rho^2} \tag{2.229}$$

Using Eqns. (2.227), (2.228), and (2.221), this is reduced to

$$\beta v = f w_z + f w \rho_z / \rho \tag{2.230}$$

The magnitude of each term in this equation can be estimated as follows. Assume the following scales: $\beta \simeq 10^{-11}/\text{s/m}$, $v \simeq 10^{-2}\text{m/s}$, $f \simeq 10^{-4}/\text{s}$, $\text{w} \simeq \Delta w \simeq 10^{-6}\text{m/s}$, $\Delta z \simeq 10^3\text{m}$, $\Delta\rho/\rho \simeq 10^{-3}$; thus, the scale of each term is: 10^{-13}, 10^{-13}, and 10^{-16} (s^{-2}). This scaling analysis indicates that the convergence of the vertical velocity is primarily due to the planetary β effect. On the other hand, the contribution due to non-constant density in the momentum equation is really negligible.

By going through the above analysis carefully, Eqn. (2.222) is the most questionable step and, for the case of stratified flow, the assumption of an f-plane approximation seems to be at the core of the problem of the "law of parallel solenoids."

β-spiral

Before the wide application of modern instrumentation, shipboard measurements were mostly limited to hydrographic data, such as temperature, salinity, and pressure. Although the thermal wind relation can be used to calculate the vertical shear of horizontal geostrophic velocity, it cannot provide the total velocity. In order to calculate the absolute velocity, the baroclinic velocity field has to be combined with knowledge of the absolute velocity at a certain depth. Before the development of current measurement based on the Doppler effect, it was very difficult to determine such a reference velocity. The common practice was to select a so-called reference level which was relatively deep. Assuming the flow at such a reference level is very slow and negligible, the absolute velocity can be obtained through vertically integrating the baroclinic velocity inferred from the thermal wind relation, starting from such a reference level. The reference level is also called the level of no motion. This approach has been widely used; however, the choice of the reference level remains somewhat arbitrary, and the search for such a level of no motion was a crucial issue in large-scale oceanography. Through this search, an important discovery was made: that of the β-spiral.

In this section we will show that if ρ is not constant, the horizontal velocity vector rotates with depth. We begin with the general form of density balance in a steady state:

$$u\rho_x + v\rho_y + w\rho_z = A_H \nabla^2 \rho + A_v \rho_{zz} - \alpha \dot{q}/c_p \tag{2.231}$$

where the right-hand-side terms are contributions from horizontal mixing with a diffusivity of A_H, vertical mixing with a diffusivity of A_v, and local heating/cooling $\dot{q}$. Defining the angle of horizontal velocity vector $\theta = \tan^{-1}(v/u)$, Eqns. (2.217), (2.218), and (2.220) lead to an expression for the rate of rotation with depth:

$$\frac{d\theta}{dz} = \frac{\gamma}{u^2+v^2}\left(w\rho_z + \alpha\dot{q}/c_p + A_H\zeta_z/\gamma - A_v\rho_{zz}\right), \quad \gamma = \frac{g}{f\rho_0} \tag{2.232}$$

Assuming that mixing and thermal forcing are negligible in the ocean interior, this equation is reduced to

$$\frac{d\theta}{dz} = \frac{\gamma}{u^2+v^2}w\rho_z \tag{2.233}$$

Using the density coordinate, this expression is reduced to the following relation:

$$\frac{d\theta}{d\rho} = \frac{\gamma w}{u^2+v^2} \tag{2.234}$$

This means that if the vertical velocity is zero, the horizontal velocity should point in the same direction, which is vertically constant. Accordingly, density surfaces at different depths for a water column should also slope in the same direction: essentially a conclusion from the "law of parallel solenoids."

However, in a subtropical gyre due to wind-driven forcing, the Ekman flux has a non-zero divergence that pushes a downward motion in the subtropical basin, so that$w$ is non-zero and negative. Since the background stratification satisfies $d\rho/dz < 0$, $d\theta/dz$ is positive, meaning a right-hand (clockwise) spiral in the Northern Hemisphere; while in a subpolar basin, wind-stress curl is positive, so w is positive; thus, $d\theta/dz$ is negative, implying a left-hand (anticlockwise) spiral. As a result, density surfaces at different depths of a water column should slope in directions that are rotated vertically.

The β-spiral implies that both u and v must cross zero at a certain depth; thus, a "level of no motion" for each component of the horizontal velocity would exist. Accordingly, the absolute velocity can be calculated. The β-spiral was first used by Stommel and Schott (1977) to calculate the absolute velocity from the density field observations.

Assuming that water is an ideal fluid and the motions obey geostrophy, the thermal wind relation equations (2.217, 2.218) can be rewritten as

$$u_z = -\gamma h_y\rho_z; \quad v_z = \gamma h_x\rho_z; \quad \gamma = g/f\rho_0 \tag{2.235}$$

where h is the height of a given density surface, and h_x, h_y are the slopes of the density surface in the x and y directions. In addition, we have the vorticity equation:

$$\beta v = f w_z \tag{2.236}$$

For an ocean with a flat bottom, vertical velocity at the bottom is zero. If there was no β effect, the vertical velocity in the whole water column would be zero, and according to Eqn. (2.222) there would be no spiral in the horizontal velocity. Thus, this type of velocity spiral in the ocean is called the β-spiral.

In isopycnal coordinates, the vertical velocity satisfies

$$w = uh_x + vh_y \tag{2.237}$$

Differentiating Eqn. (2.237) with respect to z and using Eqns. (2.235, 2.236) leads to

$$uh_{xz} + v\left(h_y - \beta z/f\right)_z = 0 \tag{2.238}$$

From the climatology of density distribution, h_{xz} and h_{yz} can be calculated. At certain depths these coefficients, h_{xz} or $\left(h_y - \beta z/f\right)_z$, can vanish, so either the u or v component vanishes at such depths. In combination with the thermal wind relation, these zero-velocity levels can be used as the reference level to calculate the absolute velocity in the oceans. Examples of the β-spiral are shown in Figs. 2.28 and 2.29.

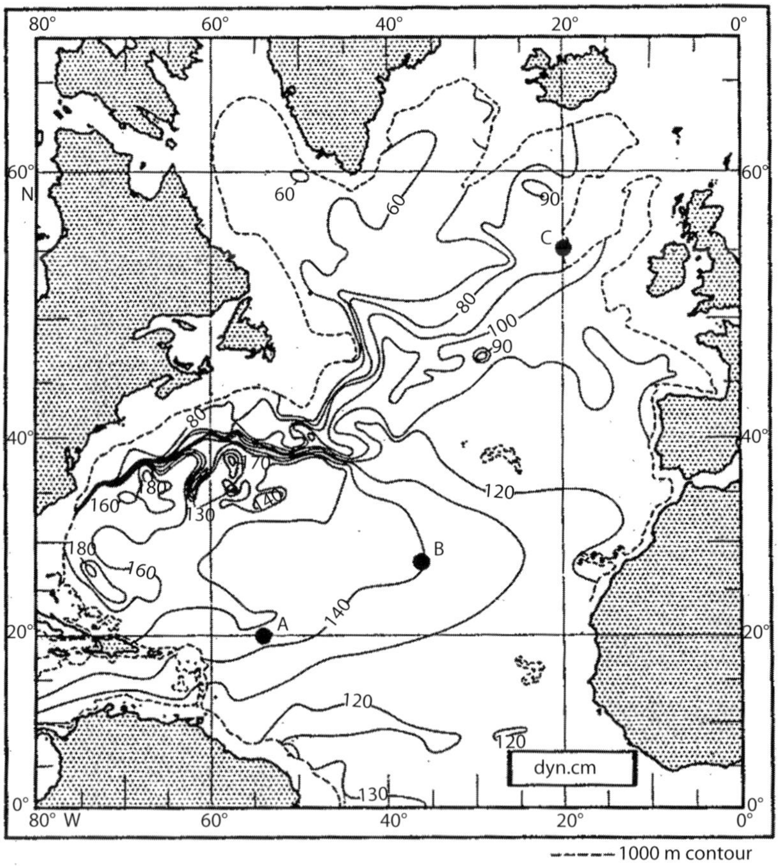

Fig. 2.28 Locations of the stations (A, B, and C) where the β-spiral was calculated on a map of dynamical topography of the 100-dbar level relative to 1,500 dbar for the North Atlantic Ocean (Schott and Stommel, 1978).

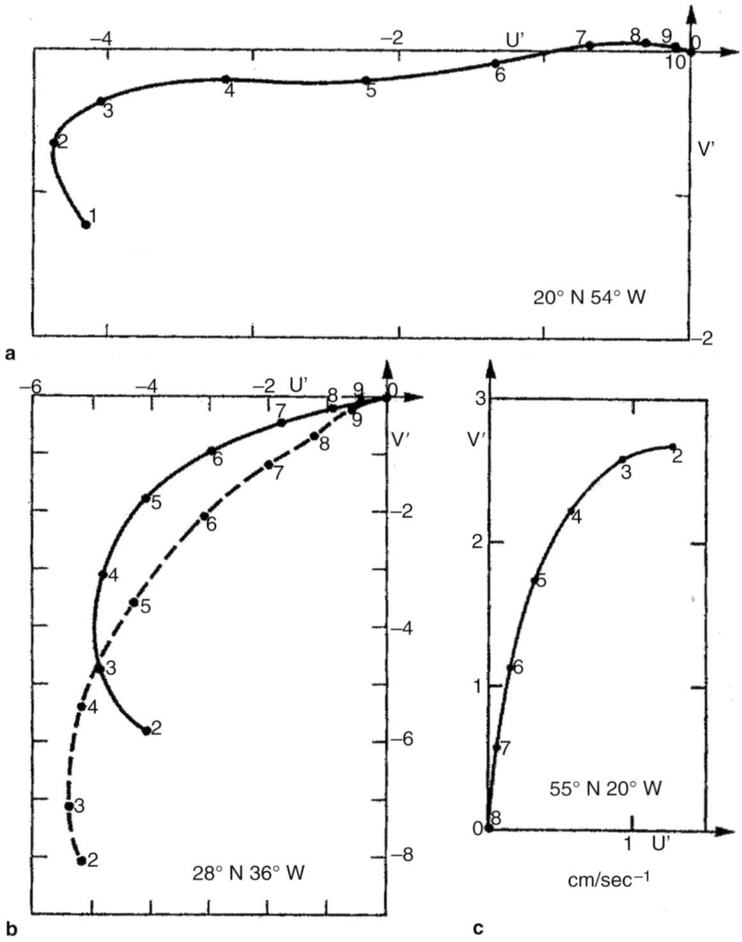

Fig. 2.29 β-spiral diagnosed from observations at the three stations located in Fig. 2.28, depth in 100 m (Schott and Stommel, 1978).

The β-spiral can be calculated from a simple analytical model of the ideal-fluid thermocline (Fig. 2.30). Velocity profiles were calculated from the analytical model (Huang, 2001), which will be discussed in Section 4.2, for the subtropical gyre from 15° N to 45° N and 0° to 60° N; four stations labeled with nondimensional coordinates for a rectangular basin $[x, y] = [0{:}1, 0{:}1]$. It is readily seen that horizontal velocity rotates in the form of a right-hand spiral.

Cooling spiral

Near the western boundary or in a subpolar basin, cooling may become a dominating factor. Accordingly, cooling ($\dot{q} < 0$) induces a spiral which rotates anticlockwise in the

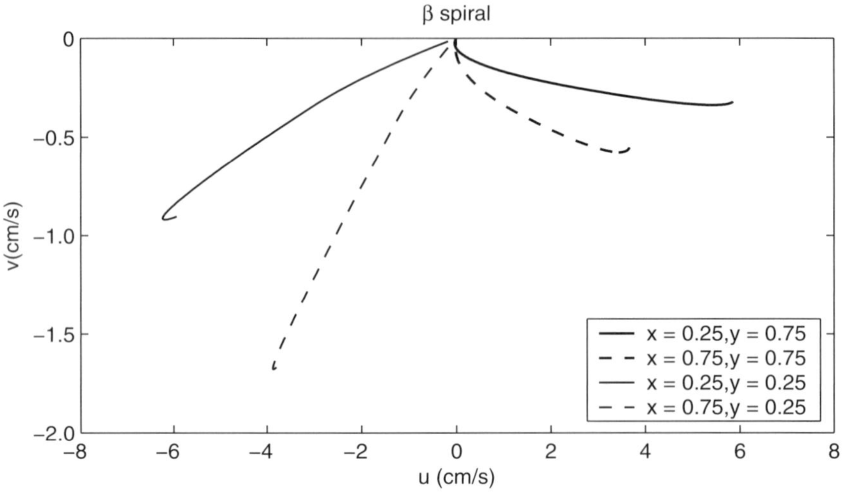

Fig. 2.30 β-spirals diagnosed for four stations in a subtropical basin interior from an ideal-fluid thermocline model; *heavy* (*thin*) lines for the northern part of the subtropical gyre where the zonal velocity is positive (negative).

downward direction. The ratio of the vertical advection term $w\rho_2$ and the cooling term $\alpha\dot{q}/c_p$ is $R = w\rho_z/\left(\alpha\dot{q}/c_p\right)$. Within the Gulf Stream recirculation, we have the following scales: $w \sim 3\times10^{-7}$ m/s, $\rho_2 \sim 3\times10^{-3}$ kg/m^4, $\alpha \sim 2\times10^{-3}$/K, $c_p = 4,200$ J/kg/K. Assuming the cooling rate is 100 W/m^2 for the 500-m-thick layer, we have the following estimate: $\dot{q} \sim 0.2\text{W/m}^3$; the corresponding ratio is $R \approx 0.009 \ll 1$. Therefore, within the recirculation regime, heat loss to the atmosphere overpowers the contribution due to stratification, leading to a cooling spiral rotating in the anticlockwise sense, opposite the clockwise β-spiral in the subtropical gyre interior (Spall, 1992).

Inferring velocity from hydrographic data

The large-scale motions in the ocean interior are described by the geostrophic, hydrostatic equations of motion, continuity, and conservation of density:

$$\vec{f}\times\rho_0\vec{u} = -\nabla p - g\rho\vec{k} \tag{2.239}$$

$$\nabla\cdot(\rho\vec{u}) = 0 \tag{2.240}$$

$$\vec{u}\cdot\nabla\rho = 0 \tag{2.241}$$

From these equations we can derive the conservation of the Bernoulli function, $B = p + \rho g z$, and the potential vorticity $Q = f\rho_2$ along the streamlines:

$$\vec{u}\cdot\nabla B = \vec{u}\cdot\nabla Q = 0 \tag{2.242}$$

Therefore, the streamline must be along the intersection of these two surfaces of B and Q, i.e., the velocity should be in the following form:

$$\vec{u} = A(x, y, z)\vec{P},\ \vec{P} = \frac{\nabla\rho \times \nabla Q}{|\nabla\rho \times \nabla Q|} \tag{2.243}$$

From hydrographic data, $\vec{P}$ can be calculated. The constant $A(x, y, z)$ can be determined by applying the relation (Eqn. (2.243)) to two adjacent levels and by using the thermal wind relation. Taking two different depths z_k and z_m, we can write down two sets of equations:

$$A^{(k)}P_x^{(k)} - A^{(m)}P_x^{(m)} = \Delta u_{km} \tag{2.244}$$

$$A^{(k)}P_y^{(k)} - A^{(m)}P_y^{(m)} = \Delta v_{km} \tag{2.245}$$

where

$$\Delta u_{km} = \frac{g}{f\rho_0}\int_{z_m}^{z_k} \frac{\partial\rho}{\partial y} dz' \tag{2.246}$$

$$\Delta v_{km} = -\frac{g}{f\rho_0}\int_{z_m}^{z_k} \frac{\partial\rho}{\partial x} dz' \tag{2.247}$$

are the velocity shear from the thermal wind relation. If the determinant of Eqns. (2.244, 2.245) is non-zero, the coefficient at these two levels, $A^{(k)}$ and $A^{(m)}$, can be determined,

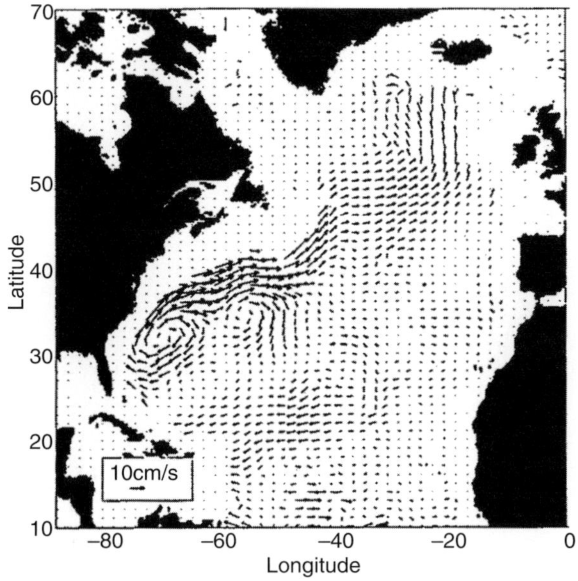

Fig. 2.31 A map of horizontal velocity diagnosed from the P-vector method for the North Atlantic (Chu, 1995).

so the horizontal velocity can be determined at these two levels. This method is called the P-vector method (Chu, 1995), and the application of this method to the North Atlantic produced maps of the horizontal velocity at different levels (Fig. 2.31).

The P-vector method is quite similar to the β-spiral method in that they are both based on many important assumptions, including geostrophy and conservation of potential vorticity. To determine the coefficient A, we have to calculate the determinant of Eqns. (2.232, 2.233); thus, the second-order derivatives of the density field are included in the calculation.

A major substantial difficulty in applying the β-spiral method or the P-vector method is that such calculations involve the second derivatives of the isopycnal slope. Owing to internal waves and meso-scale eddies in the ocean, hydrographic datasets contain a rather high level of noise. Therefore, using the β-spiral method or other similar extensions to infer the absolute velocity from hydrographic data has not been very successful. With the rapid development in modern instrumentation, there are more accurate ways of determining the absolute velocity directly from observations.

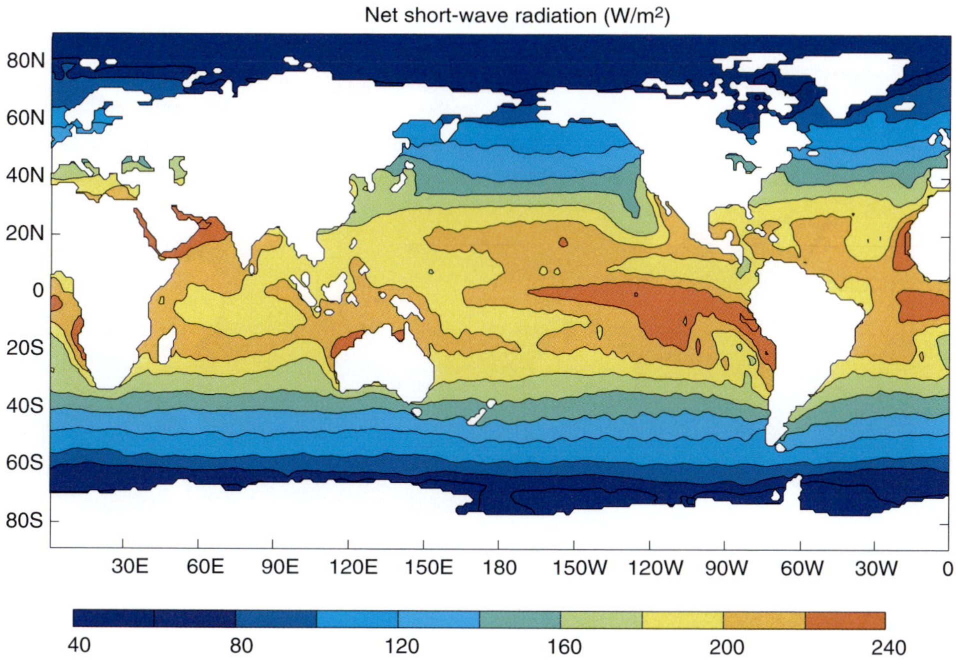

Fig. 1.3 Annual mean (NCEP-NCAR) net short-wave radiation (W/m^2).

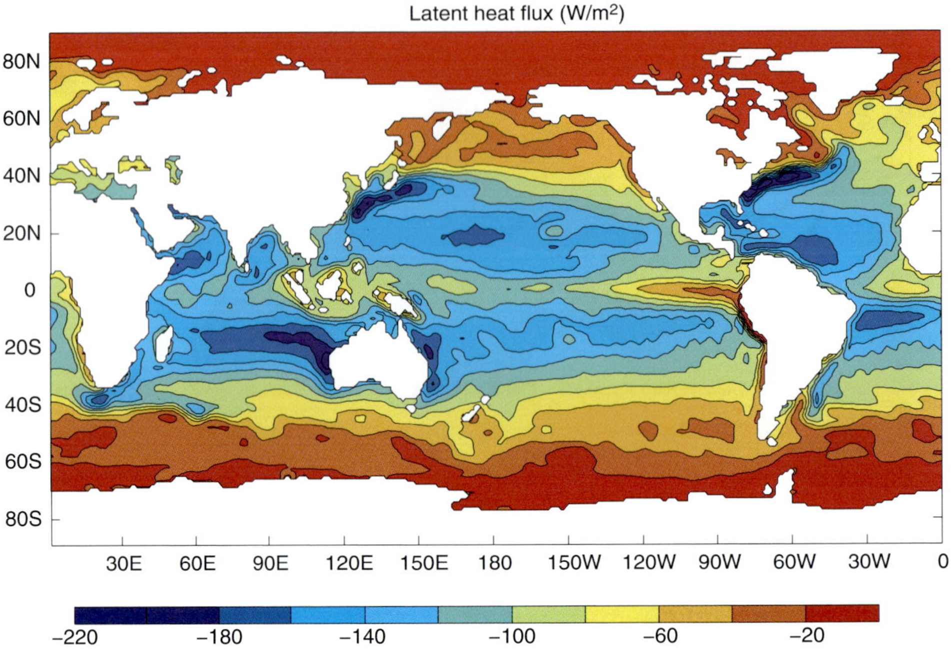

Fig. 1.4 Annual mean (NCEP-NCAR) latent heat flux due to evaporation (W/m^2).

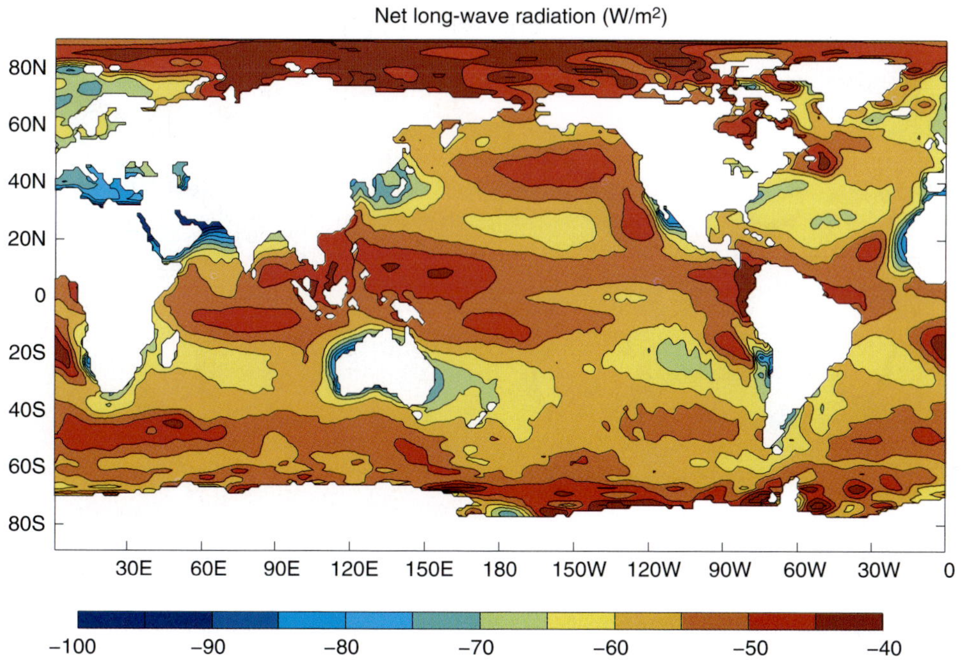

Fig. 1.5 Annual mean (NCEP-NCAR) net long-wave radiation (W/m^2).

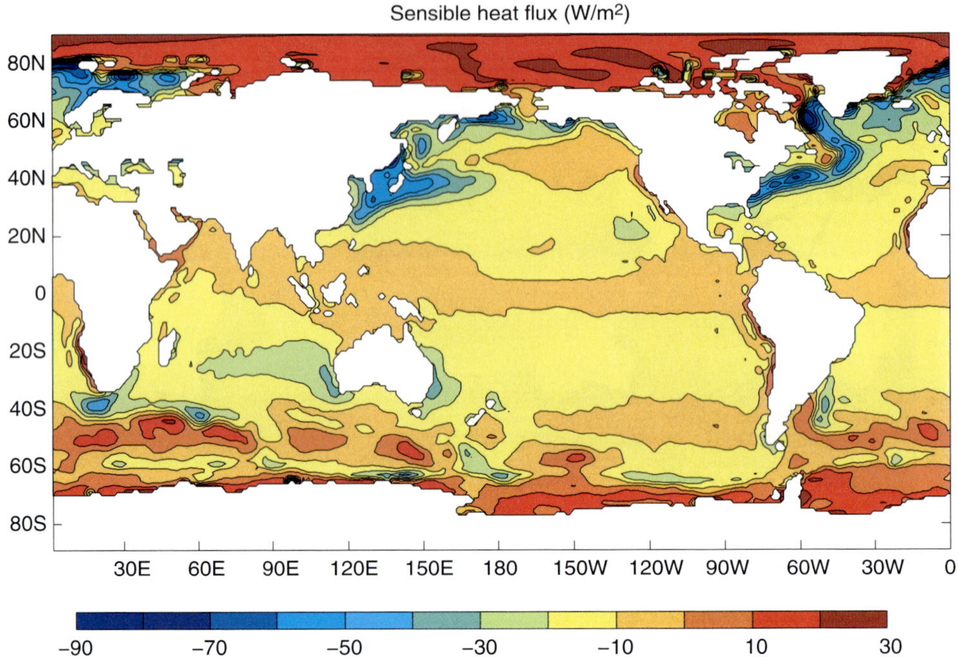

Fig. 1.6 Annual mean (NCEP-NCAR) sensible heat flux in the world's oceans (W/m^2).

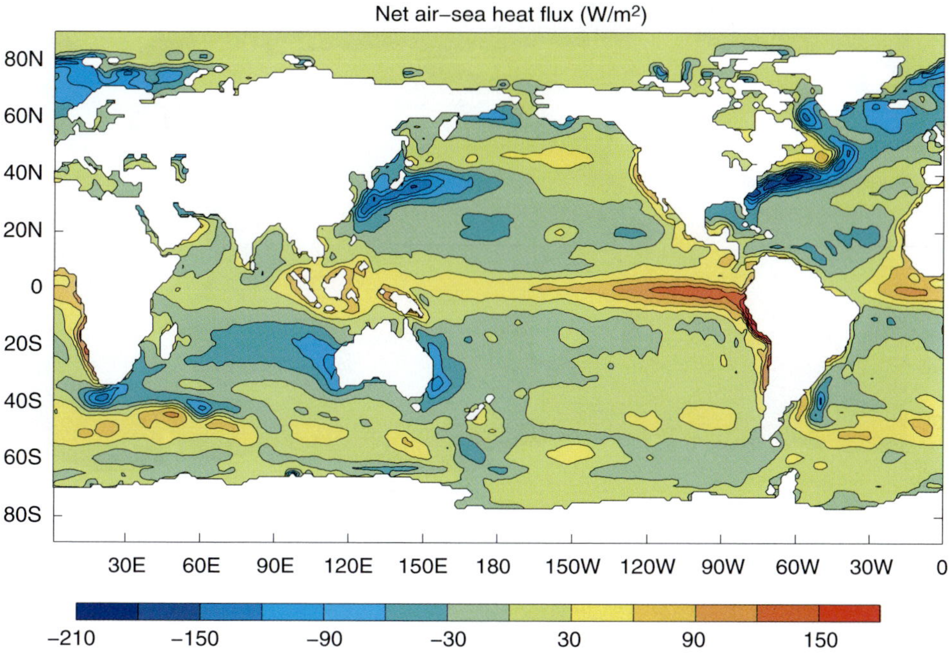

Fig. 1.7 Annual mean (NCEP-NCAR) net air–sea heat flux in the world's oceans (W/m^2).

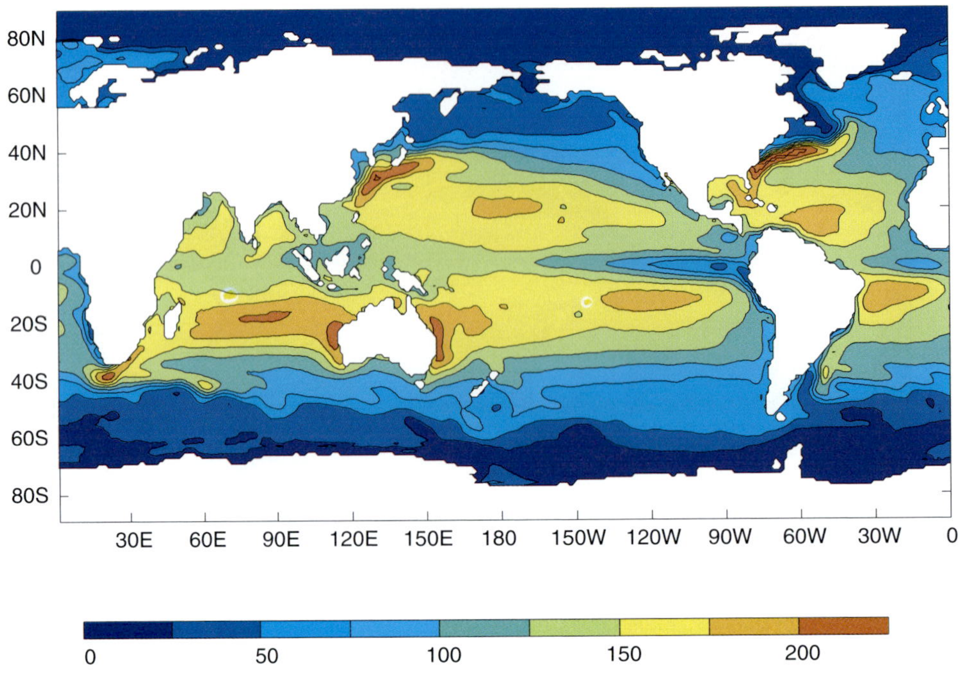

Fig. 1.9 Annual mean (NCEP-NCAR) evaporation rate in the world's oceans (cm/yr).

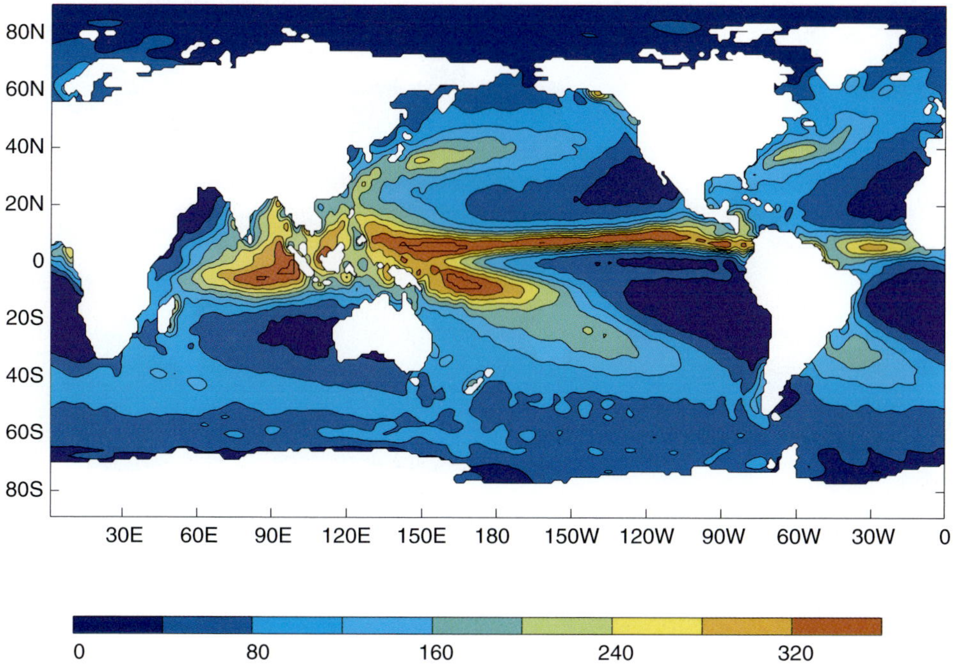

Fig. 1.10 Annual mean (NCEP-NCAR) precipitation rate (cm/yr).

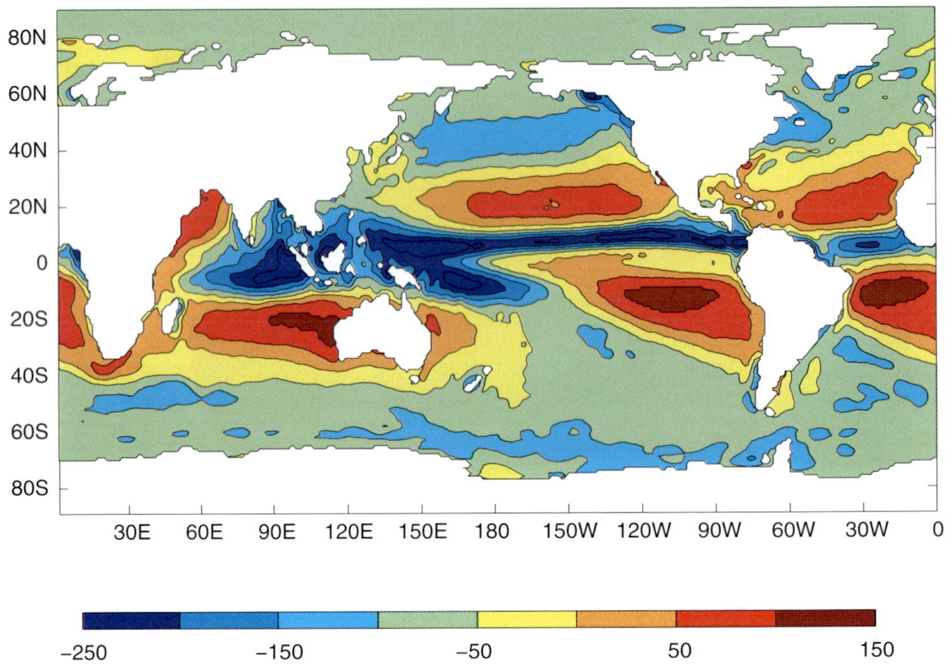

Fig. 1.11 Annual mean (NCEP-NCAR) rate of evaporation minus precipitation (cm/yr).

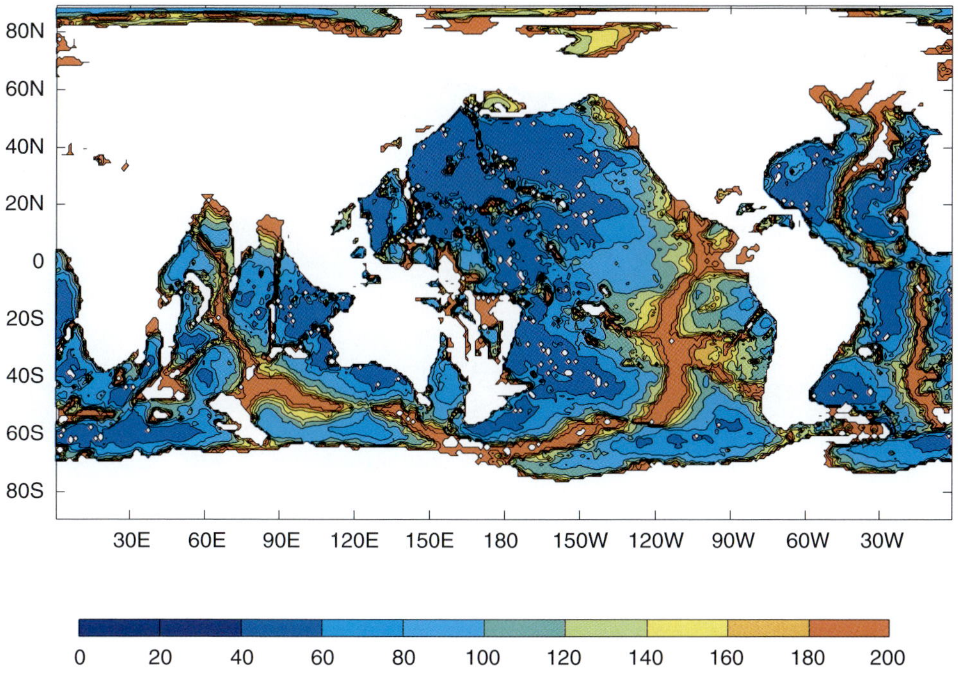

Fig. 1.13 Geothermal heat flux based on a semi-empirical formula, seafloor shallower than 2.6 km is excluded (mW/m^2).

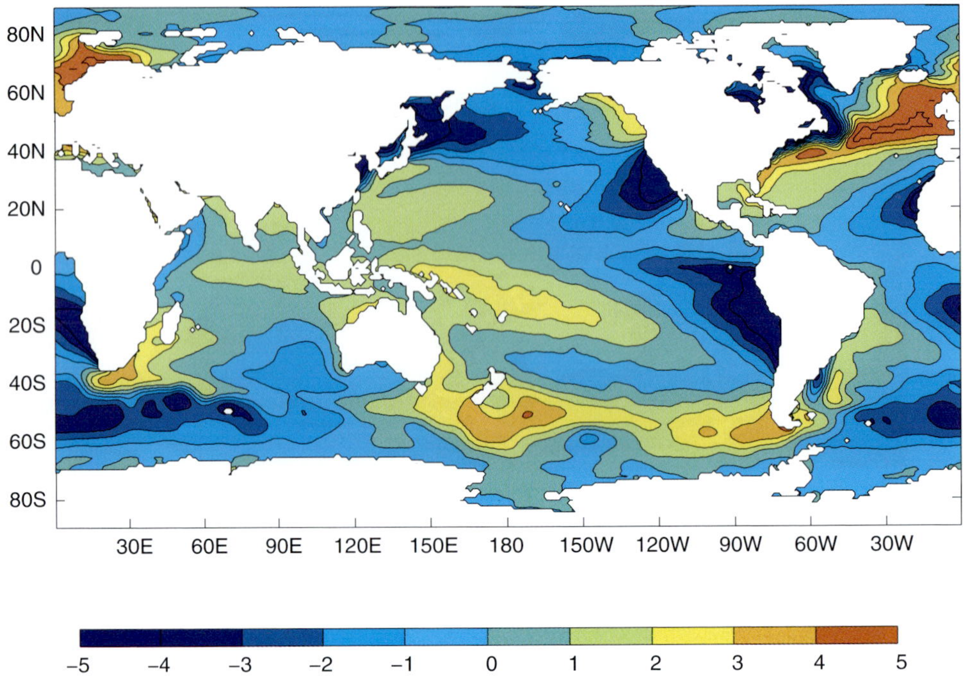

Fig. 1.19 Annual mean sea surface temperature anomaly (°C), deviation from the zonal mean.

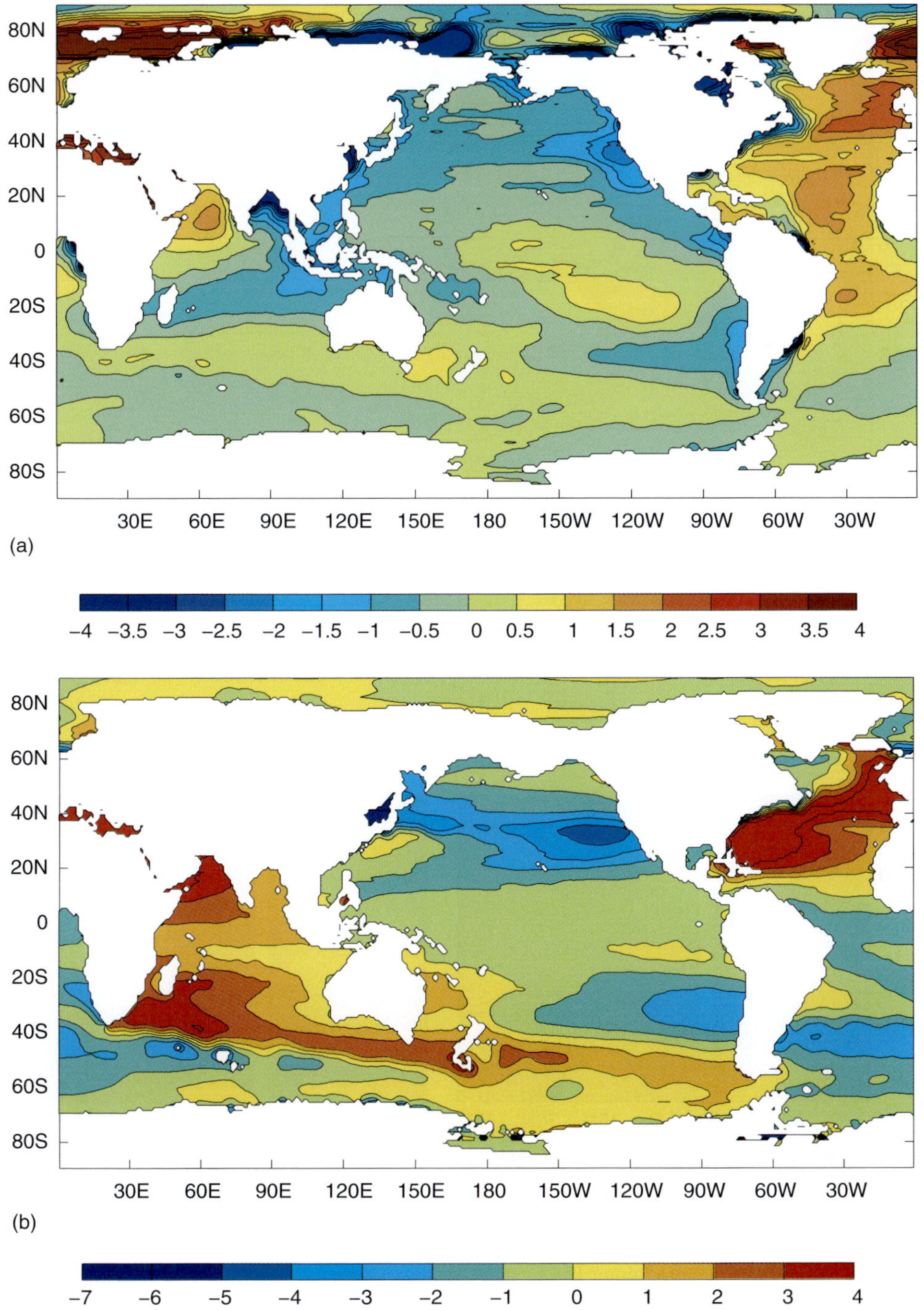

Fig. 1.20 Annual mean salinity anomaly; **a** at sea surface, **b** at depth of 600 m, deviation from the zonal mean.

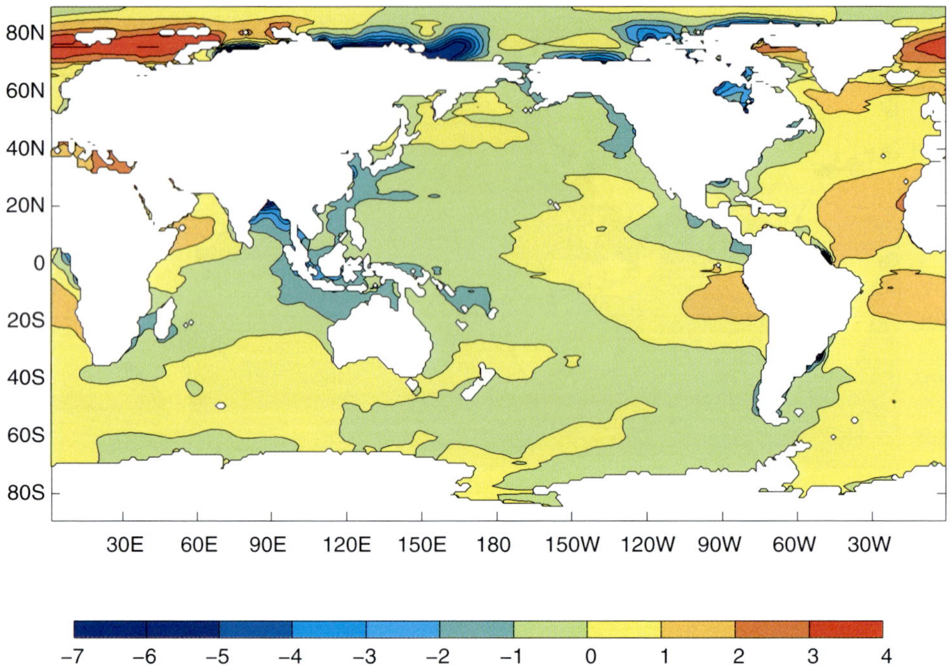

Fig. 1.21 Annual mean sea surface density anomaly, deviation from the zonal mean (kg/m^3).

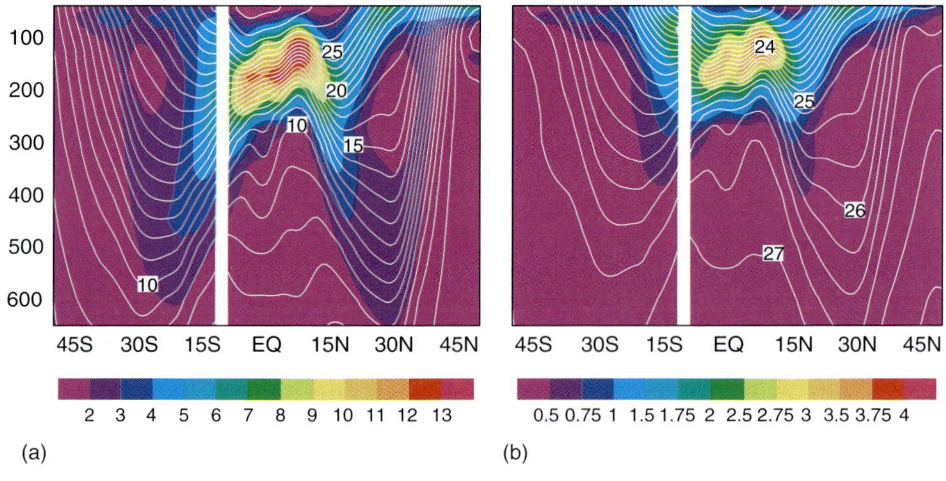

Fig. 1.38 Meridional section (along 158.5° E; vertical axis depicts **a** depth (m) of thermal structure (contours) (°C), and **b** stratification (contours) (kg/m^3), overlaid with the vertical gradient in the colored map.

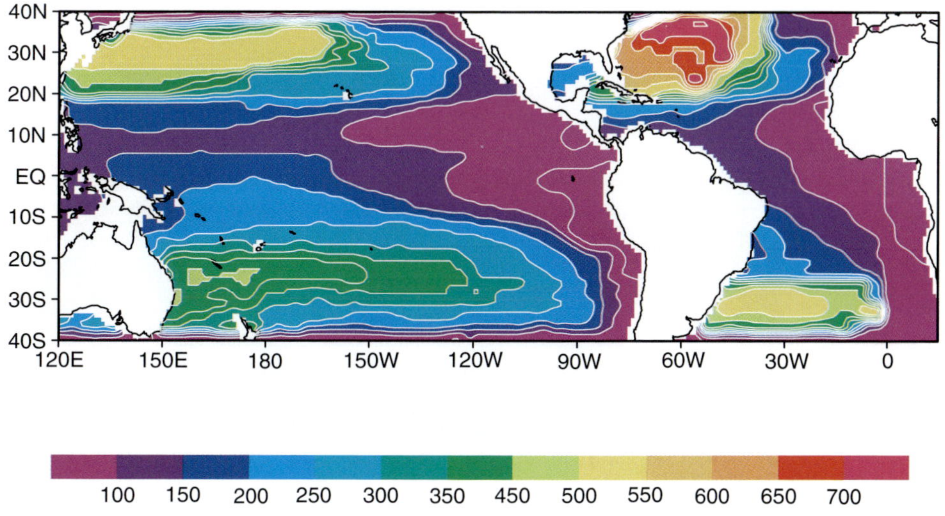

Fig. 1.39 Depth of the main thermocline (m) for the Pacific and Atlantic Oceans.

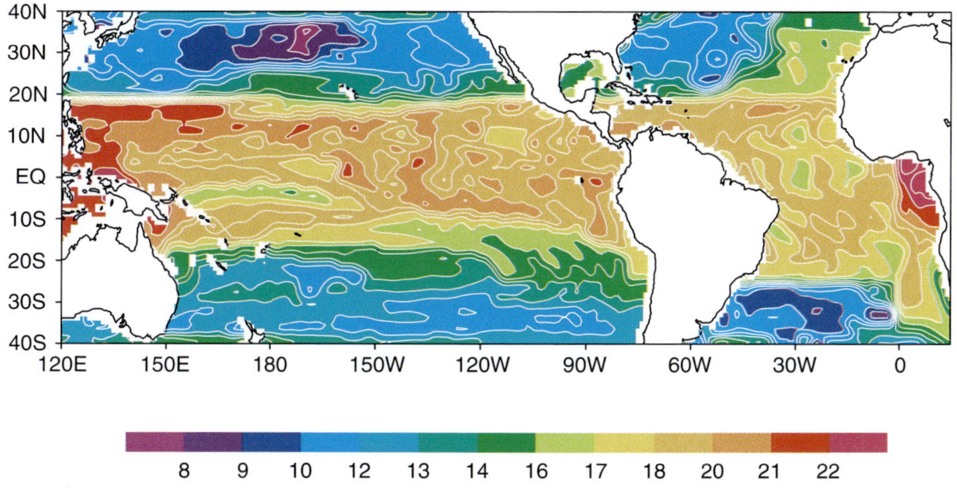

Fig. 1.40 The temperature (°C) of the main thermocline of the Pacific and Atlantic Oceans.

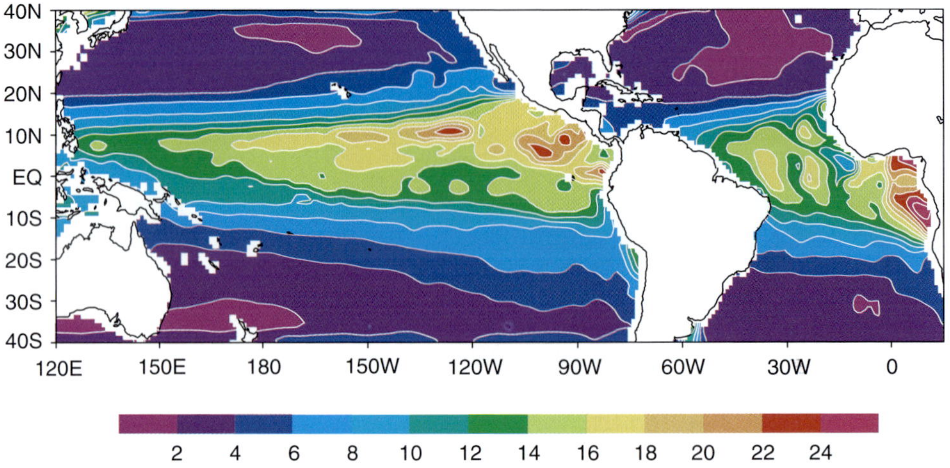

Fig. 1.41 The temperature gradient (°C/100m) of the main thermocline in the Pacific and Atlantic Oceans.

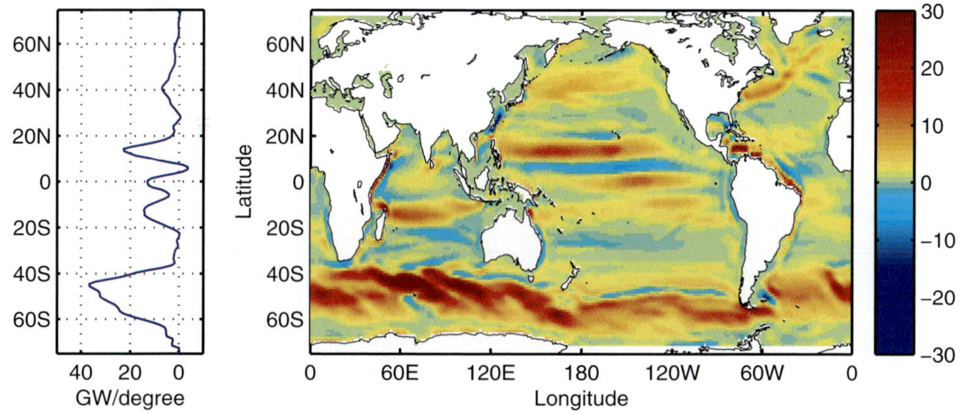

Fig. 3.10 Distribution of wind energy input through surface currents (*right panel*), in mW/m^2, and its latitudinal distribution (*left panel*) (Huang *et al.*, 2006).

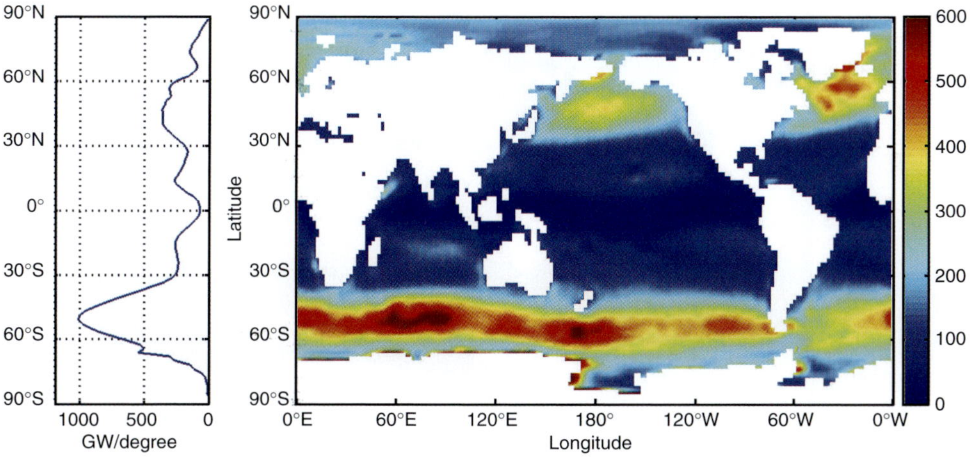

Fig. 3.11 Distribution of wind stress work on surface waves (*right panel*) (mW/m^2), and its latitudinal distribution (*left panel*) (Wang and Huang, 2004b).

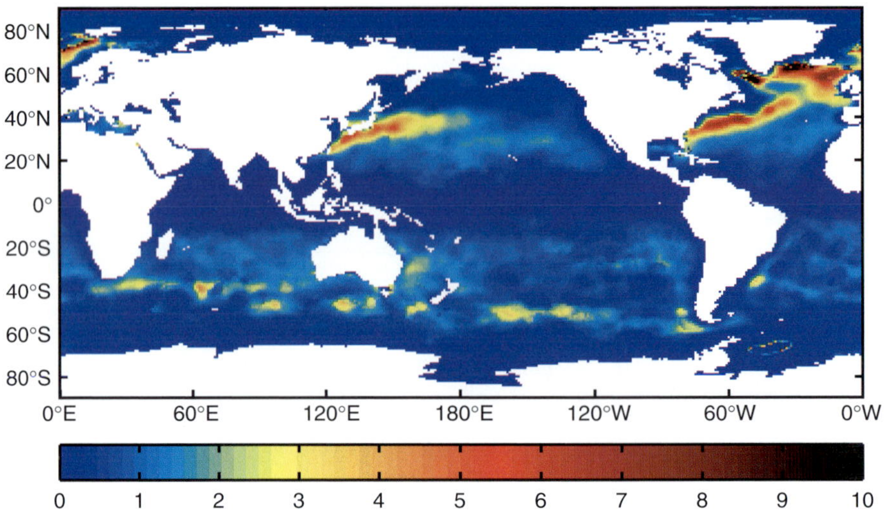

Fig. 3.14 Annual mean loss of GPE to convective adjustment (mW/m^2) (Huang and Wang, 2003).

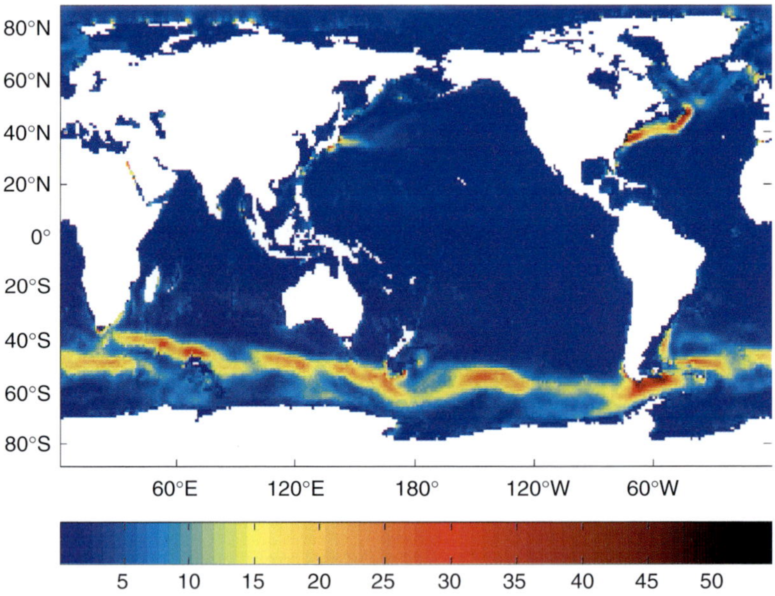

Fig. 3.16 Conversion rate of mean GPE to eddy GPE through baroclinic instability based on the empirical eddy parameterization of Gent and McWilliams (1990) (mW/m^2) (Huang and Wang, 2003).

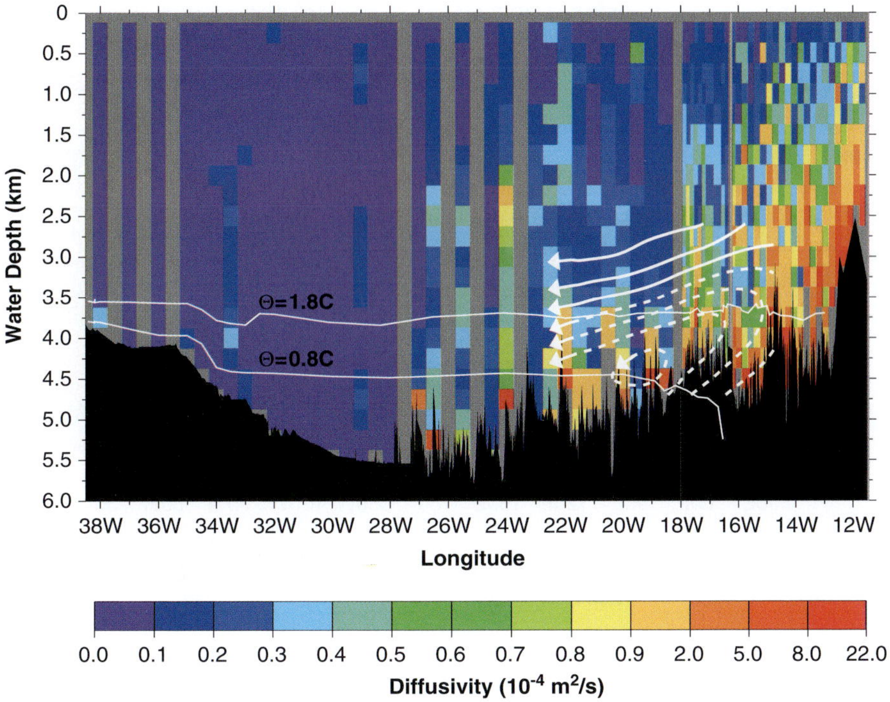

Fig. 3.17 Depth–longitude section of cross-isopycnal diffusivity, K_v, in the Brazil Basin inferred from velocity microstructure observations (Polzin *et al.*, 1997); with additional data from a later cruise (Ledwell *et al.*, 2000). The *thin white lines* mark the observed depths of the 0.8°C and 1.8°C isotherms; the *thicker white lines with arrows* are a schematic representation of the stream function estimated from an inverse calculation (St Laurent *et al.*, 2001) (adapted from Mauritzen *et al.*, 2002).

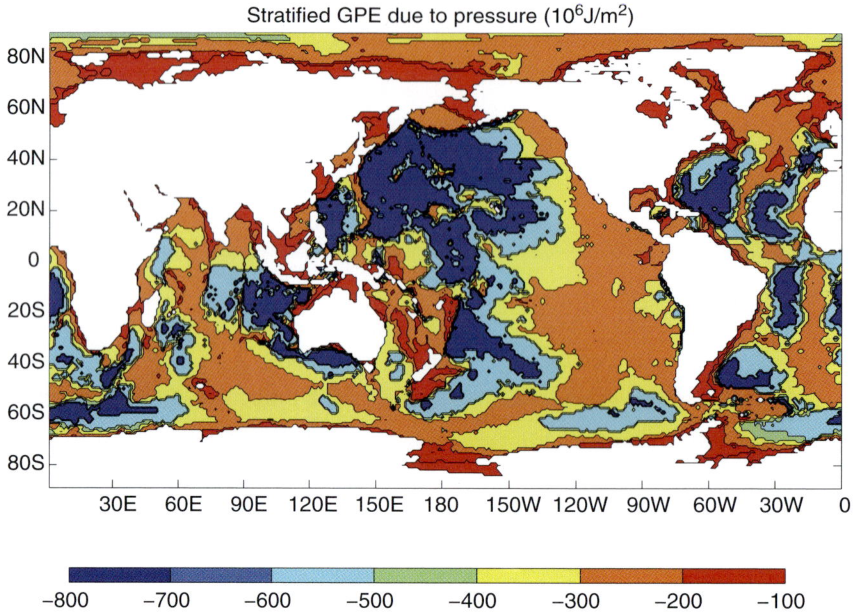

Fig. 3.23 Contribution of pressure to stratified gravitational potential energy (SGPE) (in 10^6 J/m^2).

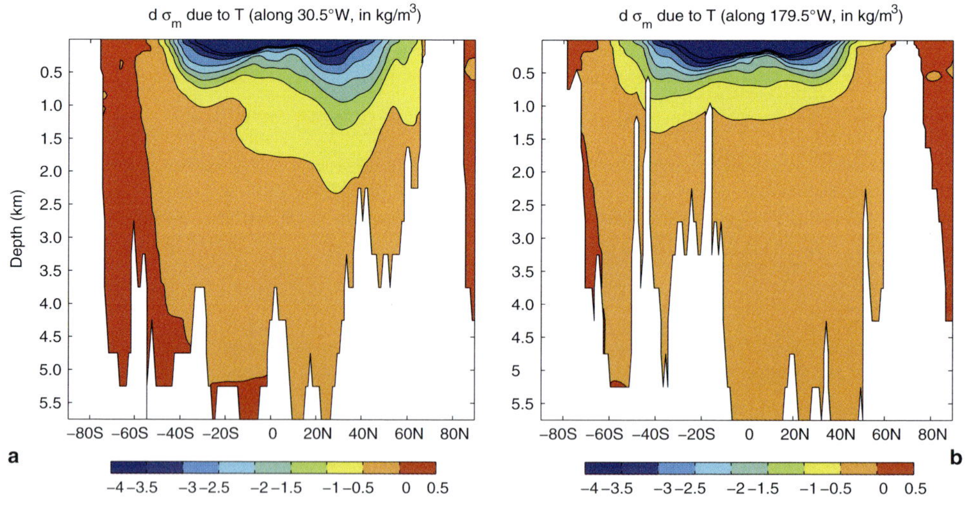

Fig. 3.24 **a**, **b** Density anomaly due to temperature contribution for two meridional sections.

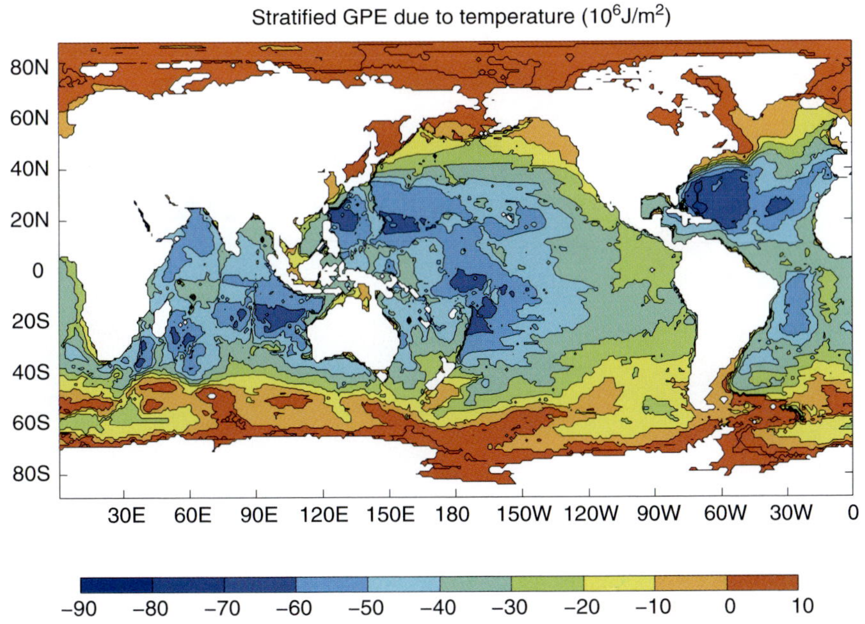

Fig. 3.25 Contribution of temperature to SGPE (in 10^6 J/m^2).

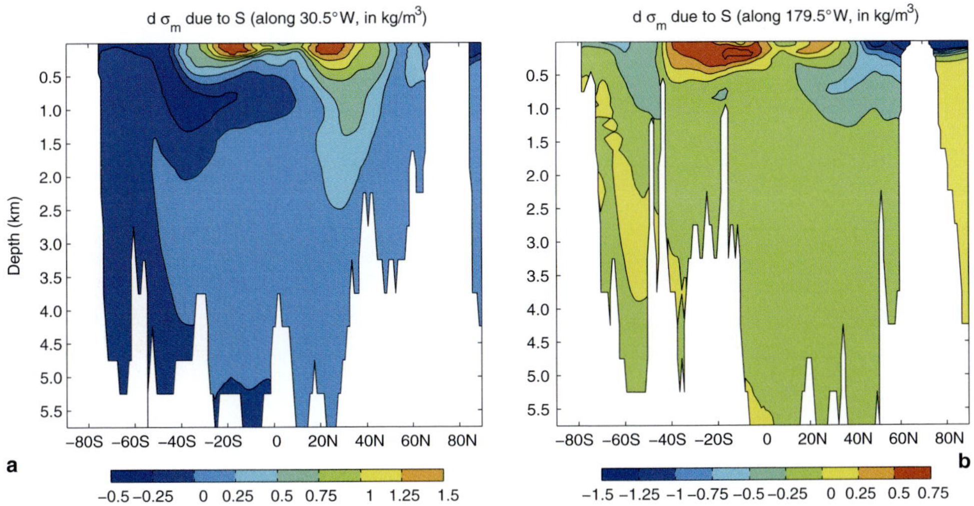

Fig. 3.26 Density difference due to temperature effect for two sections: **a** through the Atlantic Ocean, and **b** the Pacific Ocean (kg/m^3).

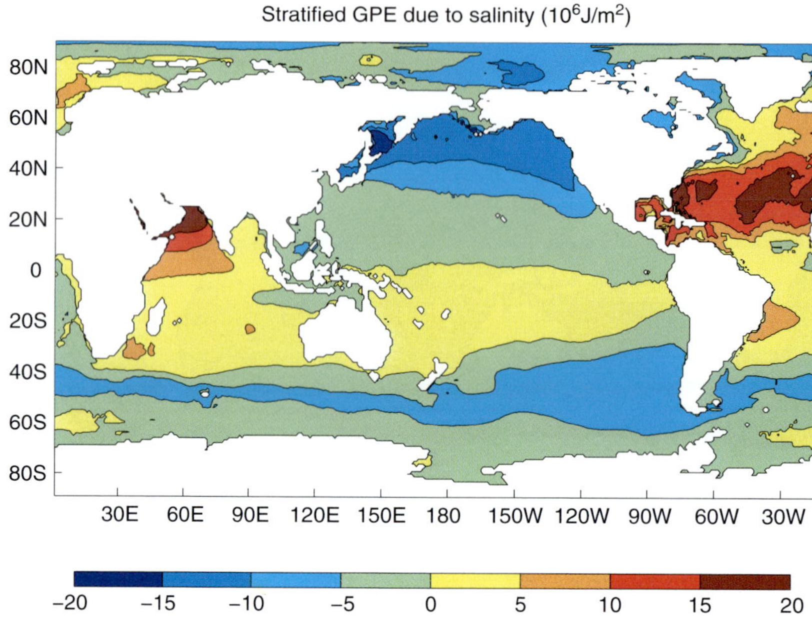

Fig. 3.27 Contribution of salinity to SGPE (in 10^6 J/m^2).

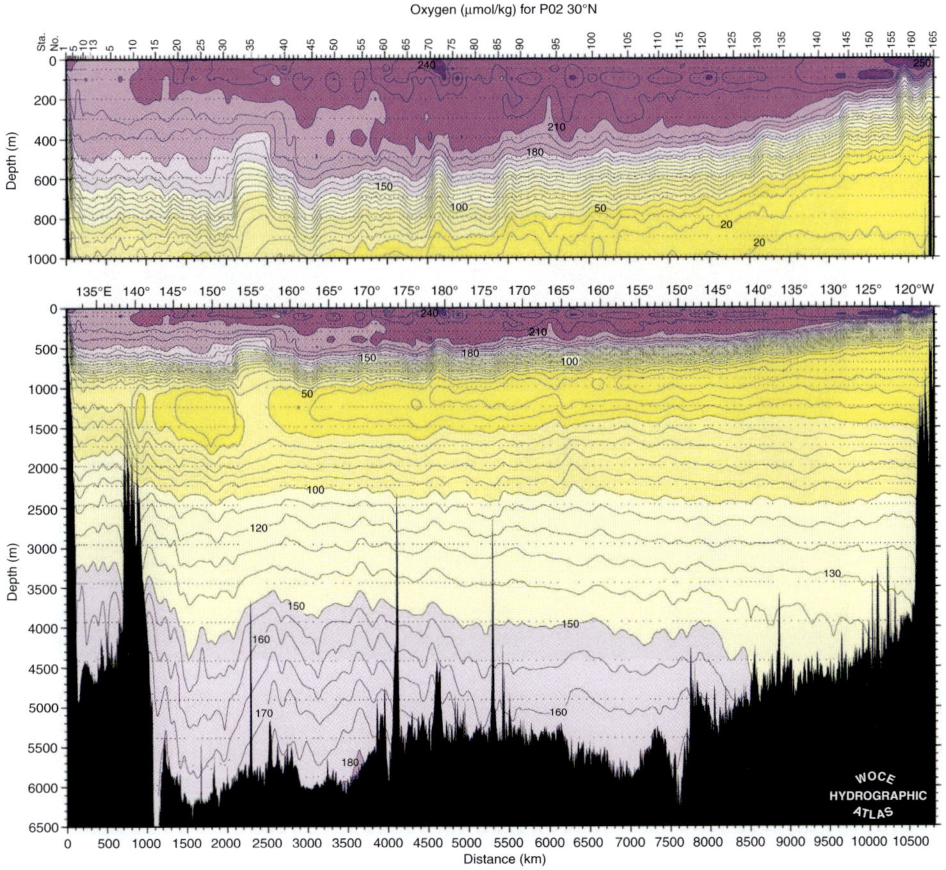

Fig. 4.29 Oxygen content for a 30° N zonal section through the center of the subtropical basin in the North Pacific (Talley, 2007).

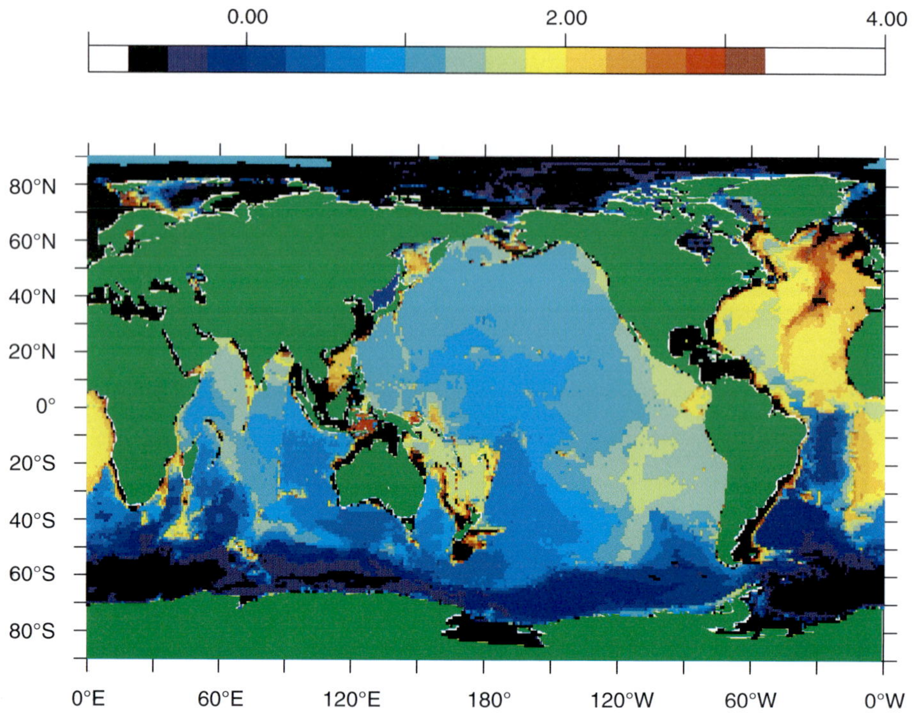

Fig. 5.1 Potential temperature at the bottom of the world's oceans based on Levitus *et al.*'s (1998) Climatology. Note that the sea floor is rather shallow along the mid-ocean ridge in the Atlantic Basin, so bottom water over the ridge is relatively warm.

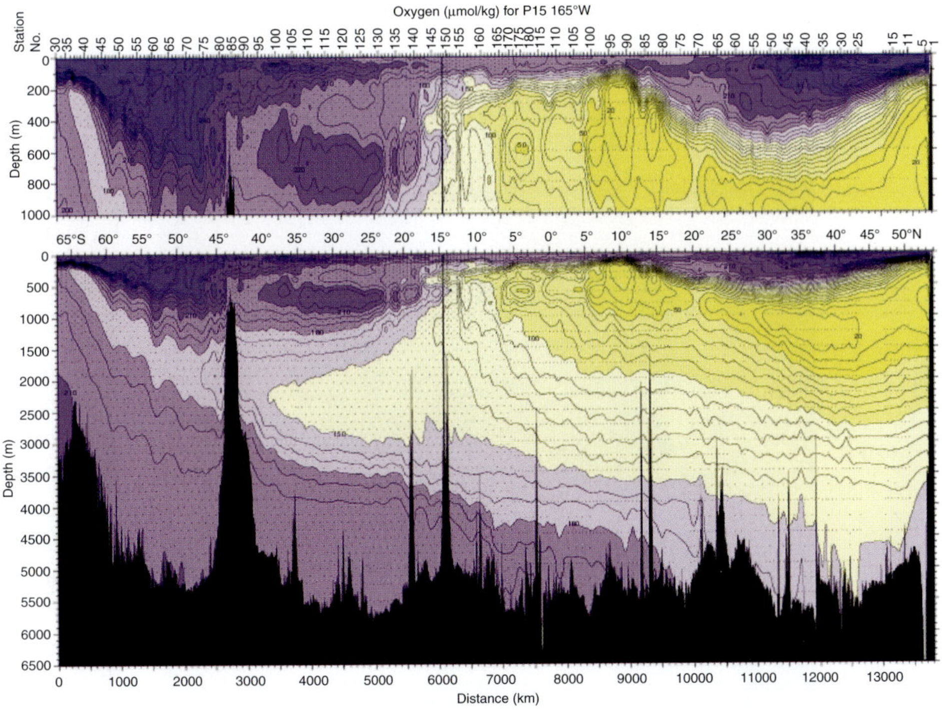

Fig. 5.2 Oxygen concentration P15 section (approximately along 165° W) from WOCE. The contour increment is 10 μmol/kg, and the concentration level between the yellow and light purple color is 150 μmol/kg. The concentration in the bottom layer in the Southern Hemisphere is above 190 – 200 μmol/kg (Talley, 2007).

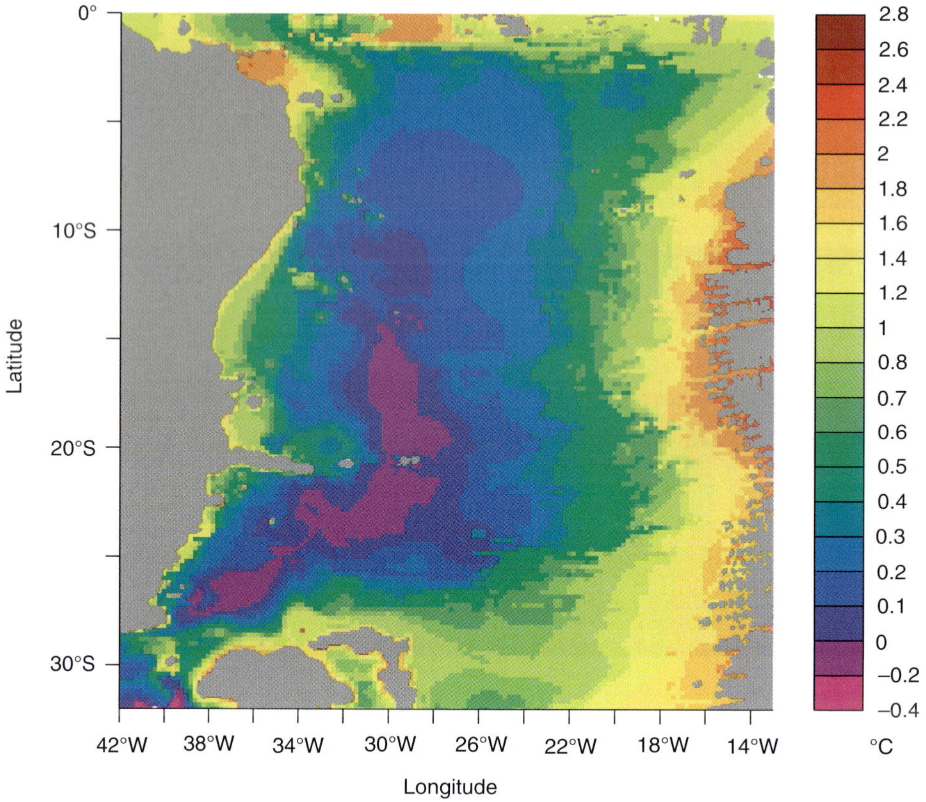

Fig. 5.11 Potential temperature on the sea floor in the Brazil Basin (Morris *et al.*, 2001).

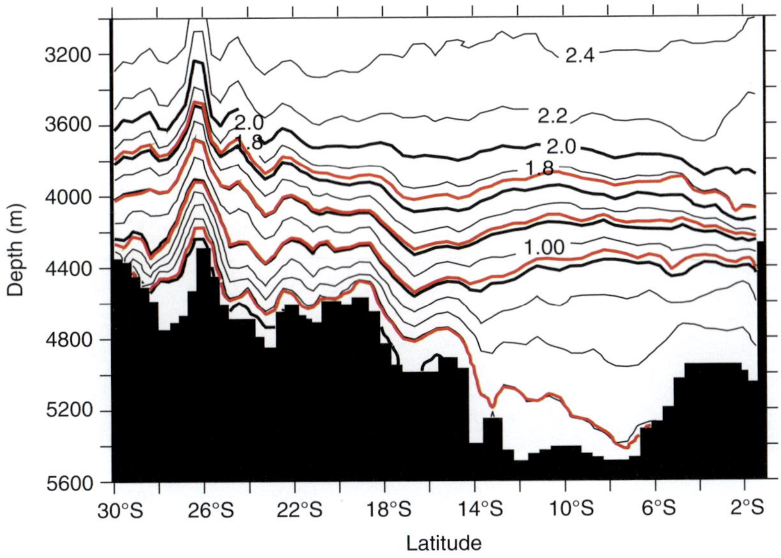

Fig. 5.12 A vertical section of potential temperature (*black*) and neutral density (*color*) from the western Brazil Basin, indicating the downward flow of water (over the seafloor from left to right) through the channel (Morris *et al.*, 2001).

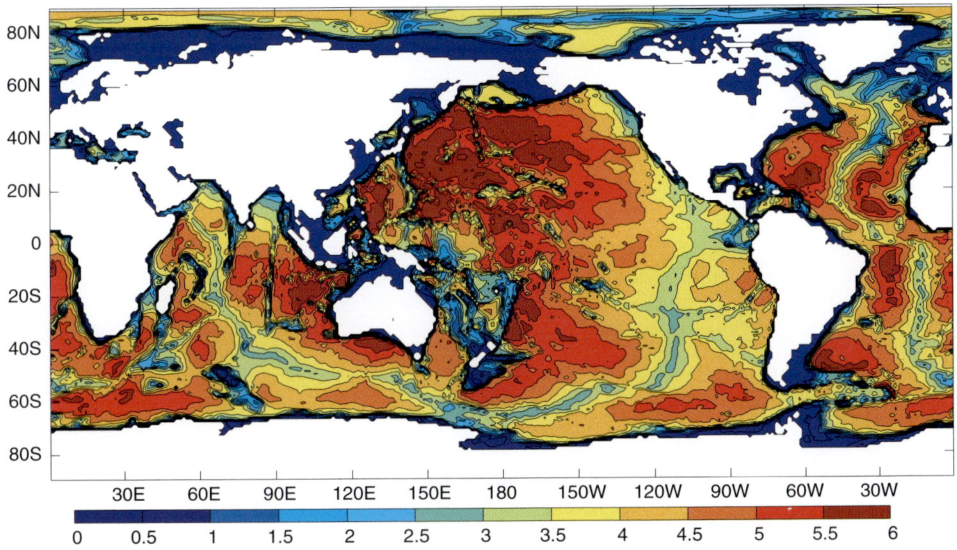

Fig. 5.60 Sea floor topography (in km) based on NOAA topographic dataset.

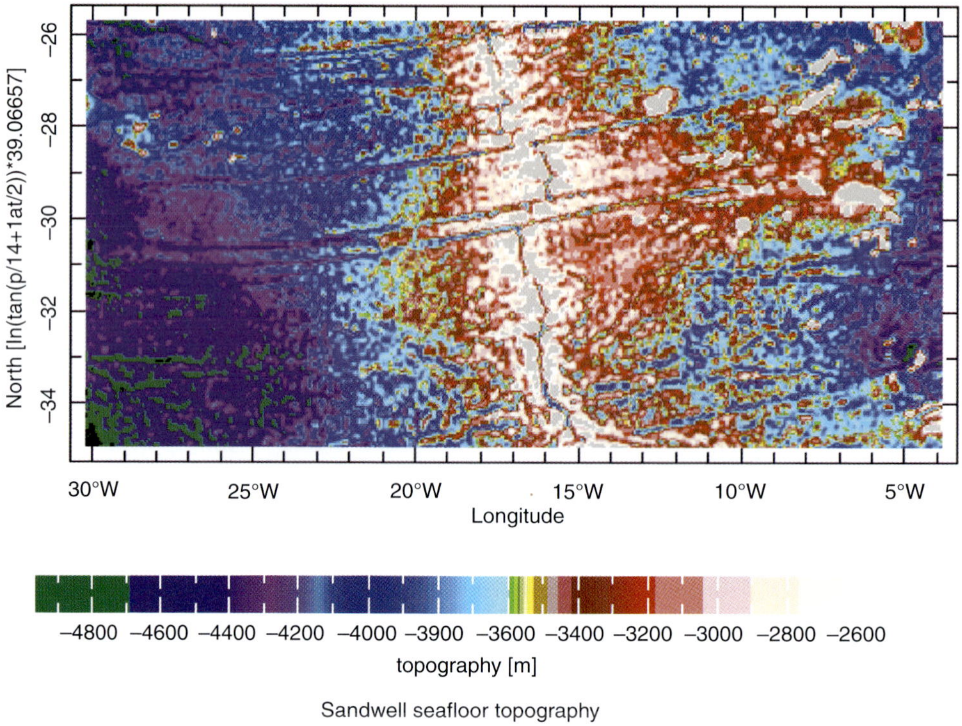

Fig. 5.76 Fine structure of sea floor topography (based on Smith and Sandwell, 1997).

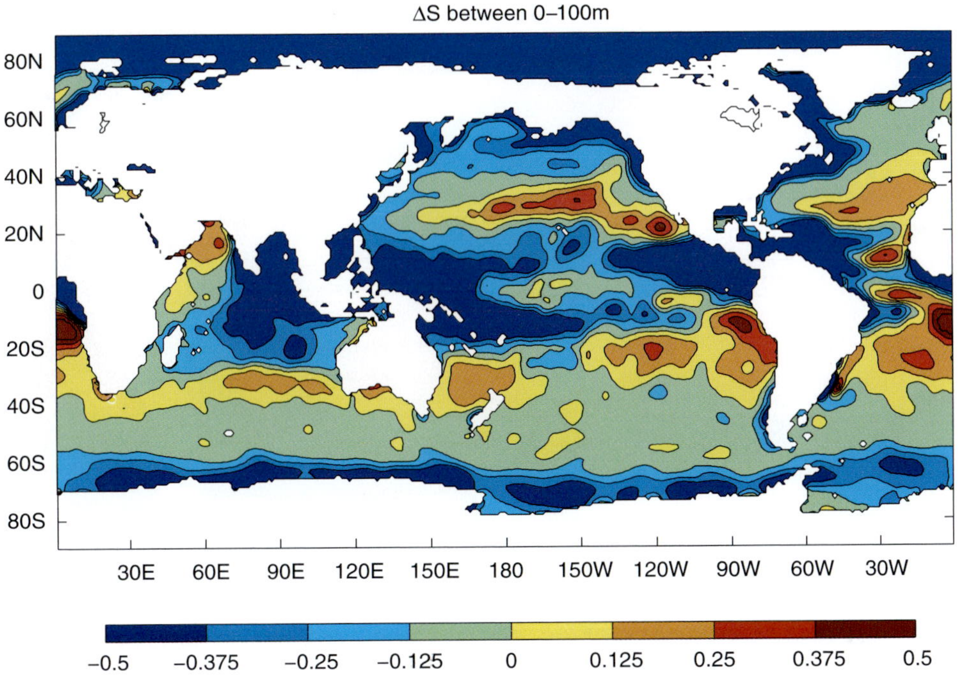

Fig. 5.88 Annual mean salinity difference between sea surface and 100 m depth.

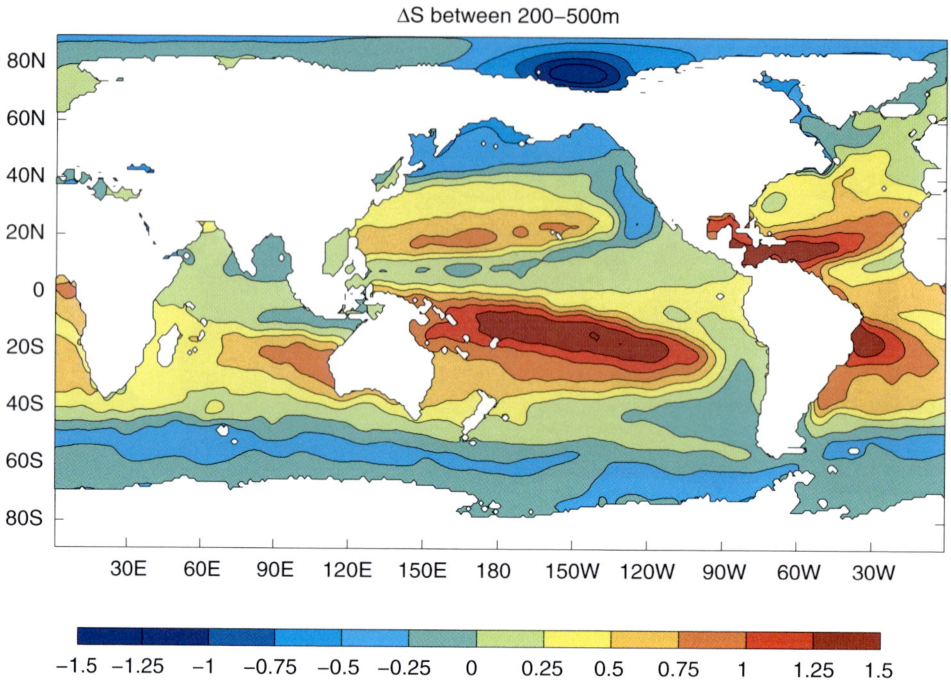

Fig. 5.89 Annual mean salinity difference between 200 and 500 m depth.

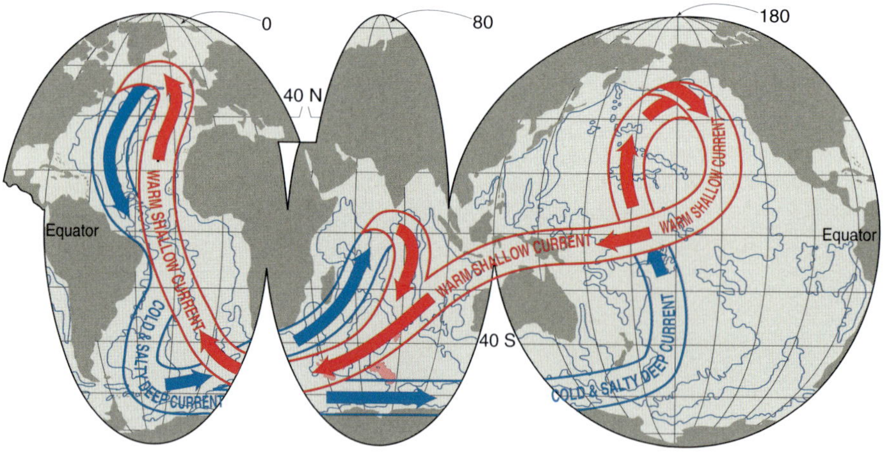

Fig. 5.152 A two-layer thermohaline conveyor belt originally proposed by Broecker (1991), modified and redrawn by Schmitz (1995) (adapted from Schmitz, 1996a).

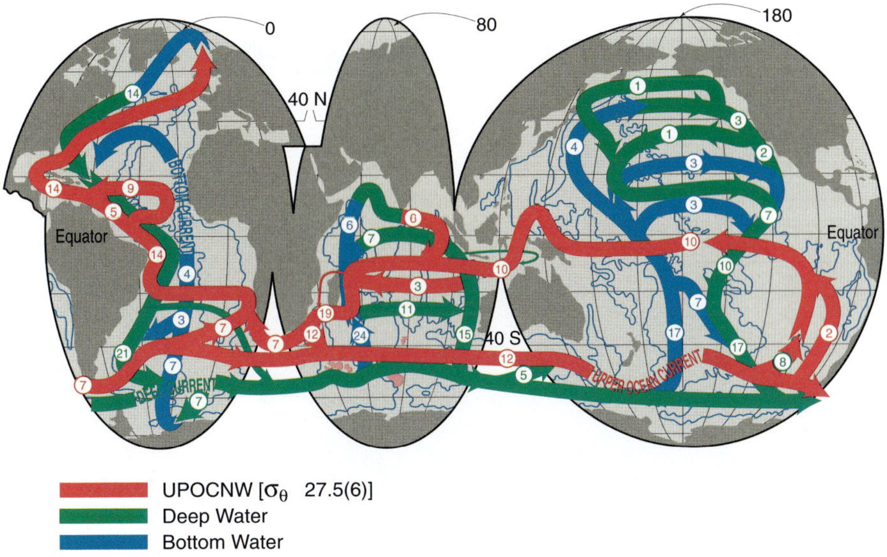

Fig. 5.153 Water mass transports (in Sv) in the world's oceans; the *blue lines* for bottom water, *green lines* for deep water, and *red* for the upper ocean (Schmitz, 1996b).

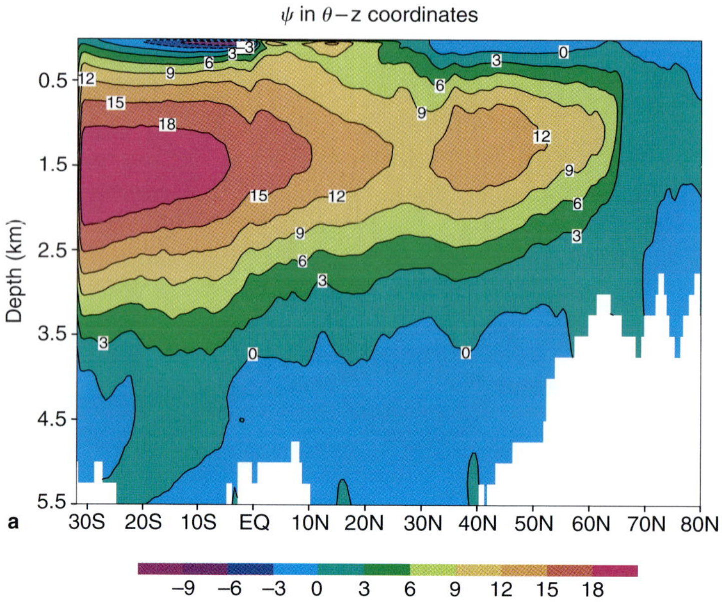

a

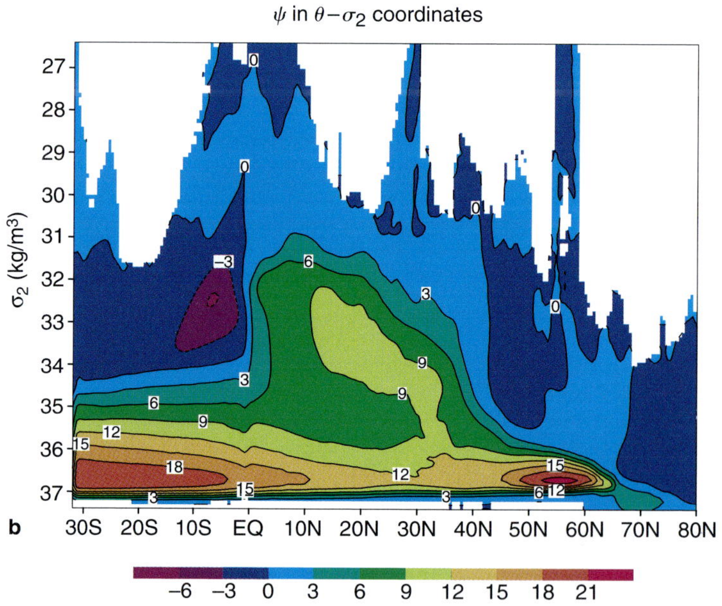

b

Fig. 5.178 **a** Annual-mean meridional overturning circulation of the Atlantic Ocean, θ–z coordinates, in Sv; **b** annual-mean meridional overturning circulation of the Atlantic Ocean, θ–σ coordinates, in Sv; **c** annual-mean meridional overturning circulation of the Atlantic Ocean, θ–Θ coordinates, in Sv; **d** annual-mean meridional overturning circulation of the Atlantic Ocean, θ–S coordinates, in Sv.

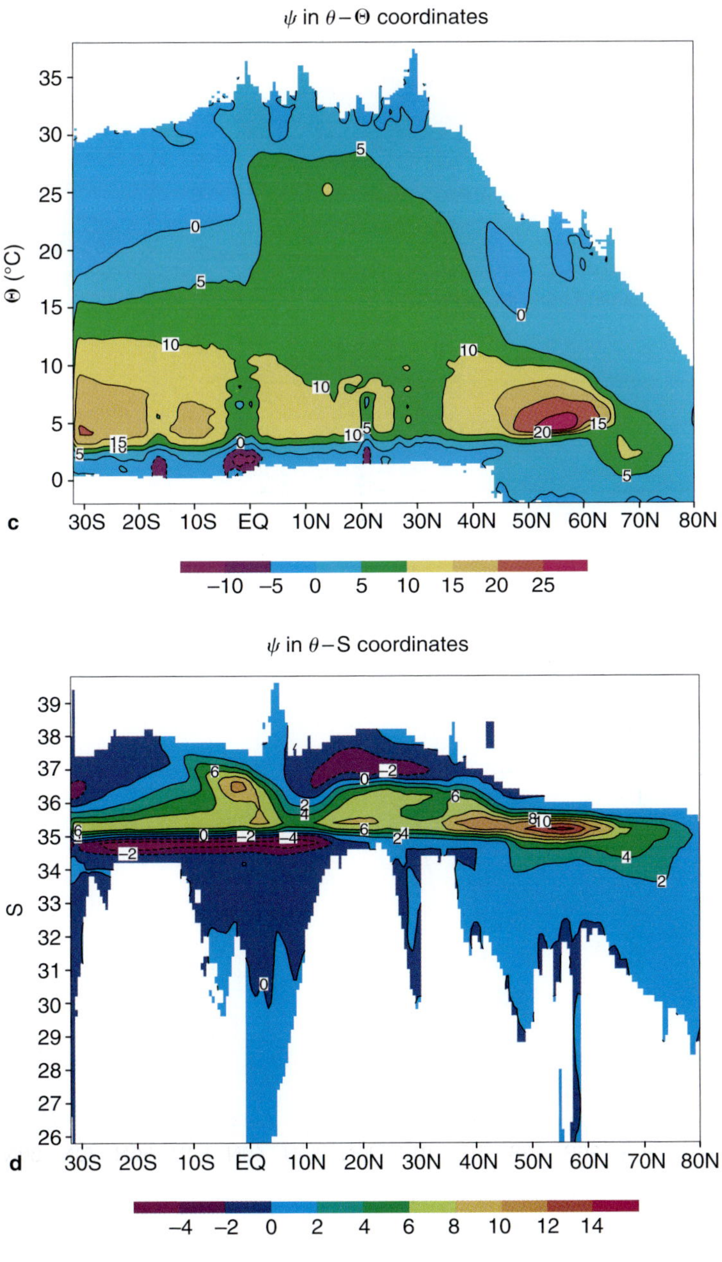

Fig. 5.178 Continued

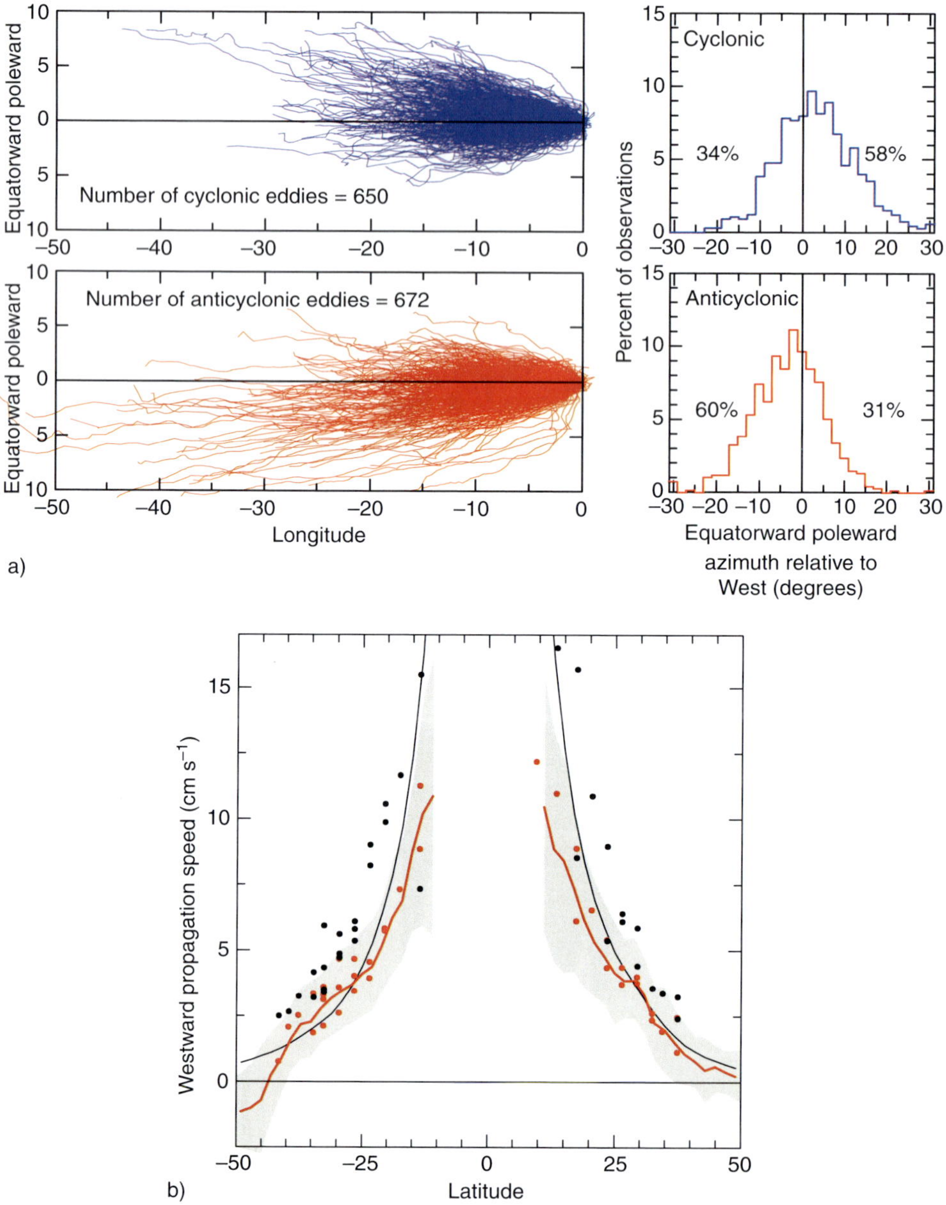

Fig. 5.191 **a** The global propagation of eddies with lifetimes ≥12 weeks; *left panels* show changes in relative position, *right panels* are histograms of the mean propagation angle relative to due west; **b** the latitudinal variation of the westward zonal propagation speeds of large-scale SSH (*black dots*) and small-scale eddies (*red dots*); the *red line* indicates the zonal average of the propagation speeds of all eddies with lifetimes ≥12 months, the *gray shading* indicates the central 68% of the distribution in each latitudinal band, and the *black line* indicates the propagation speed of the non-dispersive baroclinic Rossby waves (Chelton *et al.*, 2007).

3

Energetics of the oceanic circulation

3.1 Introduction

Energetics is one of the fundamental aspects of the climate system. Over the past decades many studies have been devoted to the energetics of the oceanic circulation. Although a more general definition of the oceanic general circulation may include wind-driven circulation, thermohaline circulation, and tides, the commonly used definition of oceanic general circulation is confined to the wind-driven circulation and thermohaline circulation.

Wind-driven circulation is a direct consequence of wind stress applied to the sea surface; therefore the energetics of wind-driven circulation must be closely linked to wind stress energy input. However, the cause of the thermohaline circulation seems complicated and remains the subject of hot debate. As explained shortly, the nature of the thermohaline circulation may depend on the viewpoint of the person who studies the problem. Therefore, research into the energetics of the oceanic circulation is often focused on the causes of the thermohaline circulation.

3.1.1 Energetic view of the ocean

Most previous studies on the energetics of the thermohaline circulation have been focused on the balance of thermal energy, in particular the air–sea heat fluxes and the meridional transport of thermal energy. A typical example is the book, *Physics of Climate* (Peixoto and Oort, 1992), in which the balance of thermal energy and its transformation have been discussed in great detail.

For a long time the importance of the mechanical energy balance, in particular energy sources from wind stress and tidal forces, failed to be recognized in the study of the thermohaline circulation. However, there were some discussions on the sources of mechanical energy and their potential role in maintaining the oceanic general circulation. As a matter of fact, a preliminary report, "Sources of energy for the ocean circulation and a theory of the mixed layer," by Faller (1966) was published in the proceedings of the Fifth US National Congress of Applied Mechanics. Unfortunately, little attention was paid to this important issue till the late 1990s; but since then the balance of mechanical energy has become a hot topic among scientists. Departing from the traditional view of energetic analysis of the oceanic general circulation, we primarily focus on the balance of mechanical energy and its role in maintaining the oceanic general circulation.

3.1.2 Different views of the oceanic circulation

People study the thermohaline circulation from different points of view; therefore, there are quite a few theories regarding the fundamental mechanism controlling the thermohaline circulation. However, our understanding on this issue remains incomplete and any specific law or theory for the thermohaline circulation can explain only some aspects of the problems, leaving others unresolved.

Surface buoyancy theory

The thermohaline circulation is driven by pressure gradient force due to the meridional difference in density set up by the meridional differential surface buoyancy forcing.

Although the rotation of the Earth makes the movement of water bodies much more complicated, the thermohaline circulation is essentially linked to the meridional pressure difference. Accordingly, variations in surface thermohaline forcing should lead to changes in the thermohaline circulation. In particular, strong meridional differential buoyancy forcing should lead to strong meridional overturning and poleward heat flux.

Mechanical energy theory

Thermohaline circulation is driven by external sources of mechanical energy provided by wind stress and tidal dissipation.

As a mechanical system, the circulation requires sources of mechanical energy to overcome friction and dissipation. However, surface thermohaline forcing cannot provide such mechanical energy efficiently; thus, external sources of mechanical energy are needed to maintain the circulation. Accordingly, changes in strength and distribution of the external sources of mechanical energy should affect the thermohaline circulation. In this theory, the surface thermohaline forcing is treated as the precondition which is necessary for setting up the thermohaline circulation.

Entropy theory

Thermohaline circulation in the oceans is an orderly dissipation system which is not in a thermodynamic equilibrium state. As a dissipation system, the oceans continuously generate entropy due to internal dissipation, including momentum dissipation, heat transfer, and freshwater mixing. In order to maintain the circulation in a quasi-steady state, the entropy produced in the system must be removed as the so-called negative entropy flux.

In fact, solar insolation brings in a large amount of thermal energy with very low entropy into the oceans. This low-entropy energy flux is used to drive many critically important processes in the ocean, in particular in the upper ocean near the air–sea interface. For example, low-entropy energy flux associated with the light penetrates into the upper ocean and becomes the key source of energy sustaining photosynthetic processes in the ocean, which represent one of the major components of the ecological system on the Earth. Overall, heat flux from the ocean to the atmosphere is associated with a much lower temperature and thus a high entropy flux. The difference between low incoming and high outgoing entropy fluxes through the air–sea interface leads to a huge negative entropy flux, which is responsible for maintaining the orderly dissipation system in the ocean, including the circulation of water, and the chemical, biological, and ecological cycles. Thus, it is clear that the study of entropy balance in the ocean is a crucial way of revealing the structure of the oceanic circulation.

The commonly accepted theories of the thermohaline circulation discussed in most published papers and books belong to the surface buoyancy type, i.e., it is believed that thermohaline circulation is driven by surface thermohaline forcing. Regarding energetics, the total energy flux is nearly balanced for a quasi-steady state, i.e., the first law of thermodynamics is satisfied.

However, different forms of energy have different qualities: according to the second law of thermodynamics, thermal energy is not the same as mechanical energy. Thus, a new paradigm of thermohaline circulation is emerging. The new theory claims that the thermohaline circulation is driven by external sources of mechanical energy, not by the thermal energy associated with surface thermal forcing. Although the total energy flux is balanced, the input mechanical energy is of high quality and this is turned into thermal energy of low quality through internal dissipation in the oceans.

The entropy theory is based on the exact balance of entropy, including its sources from input, internal generation, and sinks in the form of output. The entropy theory has not received enough attention at present; however, this may change in the near future with a better understanding of the complicated machinery of the oceanic general circulation. This chapter is dedicated to the energetics of the oceanic general circulation; in particular the balance of mechanical energy in the world's oceans.

3.2 Sandstrom's theorem

3.2.1 The oceanic circulation as a thermodynamic cycle

Is the ocean a heat engine?

The atmosphere and ocean work together as a heat engine, if we neglect the small contribution of tidal energy to the oceanic circulation. The atmosphere itself can also be considered as a heat engine, which is driven by heating from below and cooling from the middle and top levels, with an efficiency of 0.8% (the corresponding Carnot efficiency is about 33%). Although the ocean is also subject to thermal forcing, both heating and cooling are applied at the sea surface, which is nearly on the same geopotential level. It turns out that the ocean is not a heat engine at all. In fact, differential heating is only a precondition for the thermohaline circulation, and not the driving force of the circulation. The driving force for the oceanic circulation is the wind stress and tides which contribute the mechanical energy required for maintaining the quasi-steady circulation in the ocean against friction and dissipation. Thus, the ocean is a mechanical conveyor driven by external mechanical energy that transports thermal energy, freshwater, carbon dioxide, and other tracers.

The inability of surface thermal forcing to drive the oceanic circulation was recognized a long time ago. Sandstrom discussed this fundamental issue 100 years ago; his postulation is known as "Sandstrom's theorem" in the literature.

Sandstrom's theorem

Sandstrom (1908, 1916) considered the mechanical energy balance of a steady circulation in the ocean. His original papers are in German and not easily accessible, but a concise description of his idea can be found in the book by Defant (1961).

Assuming the circulation is steady, the circulation along a closed streamline should not change with time. This is valid for circulation in moving coordinates or in absolute coordinates, i.e., the rotation of the Earth does not contribute. Therefore, omitting the rotation term in Eqn. (2.1) and integrating it along a closed streamline, we have

$$0 = \frac{dC}{dt} = \frac{d}{dt}\oint_s \vec{u} \cdot d\vec{x} = -\oint_s v dp + \oint_s \vec{F} \cdot d\vec{x} + \oint_s \vec{g} \cdot d\vec{x} \tag{3.1}$$

Since gravity is the gradient of the geopotential, the last term vanishes. As friction always leads to the loss of mechanical energy, this equation is reduced to

$$-\oint_s v dp = \oint_s p dv = -\oint_s \vec{F} \cdot d\vec{x} > 0 \tag{3.2}$$

where v and p are the specific volume and pressure, respectively, and the integration is taken along a closed streamline s. Thus, Sandstrom postulated that, in order to overcome friction, the system should be able to generate a net production of mechanical energy over each cycle defined by the closed streamline.

Following the approach to the idealized Carnot cycle discussed in Section 2.3, Sandstrom simplified the oceanic circulation by adopting the following assumptions regarding each of the four idealized stages within a cycle of the engine (Fig. 3.1).

1. Heating-induced expansion is under a constant pressure ($1 \to 2$).
2. Transition from the heating source to the cooling source is adiabatic ($2 \to 3$).
3. Cooling-induced contraction is under a constant pressure ($3 \to 4$).
4. Transition from the cooling source to the heating source is adiabatic ($4 \to 1$).

We define the downward direction as positive for the pressure coordinate, which is an upside-down flip of Figure 2.2. In addition, in the following discussion, work is defined as positive if the system does work to its environment. When the system goes through a clockwise cycle, as shown in Figure 3.1a (corresponding to Fig. 2.2c), the work done by the system is

$$-\oint_s v dp = -\int_{p_2}^{p_3} (v_{23} - v_{41})\, dp < 0 \tag{3.3}$$

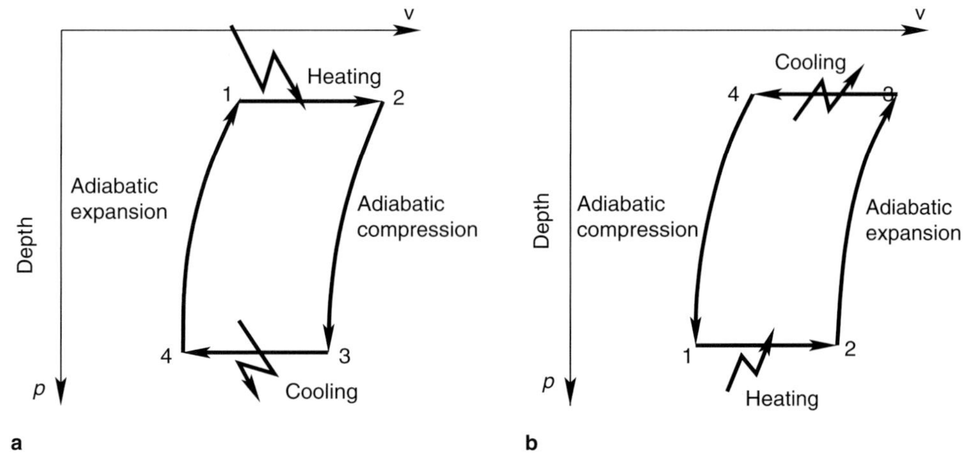

Fig. 3.1 **a, b** Two types of idealized Carnot cycle for the oceanic thermohaline circulation, postulated by Sandstrom (1916).

thus, the system cannot generate mechanical energy to sustain the circulation.

If the system goes through an anticlockwise cycle, i.e., the heating source is located at the high-pressure side and the cooling source is located at the low-pressure side, as shown in Figure 3.1b (corresponding to Figure 2.2b), the work done by the system is

$$-\oint_s v dp = -\int_{p_2}^{p_3} (v_{41} - v_{23})\, dp > 0 \tag{3.4}$$

i.e., the system can generate mechanical energy to sustain the circulation by itself. Thus, Sandstrom came to the conclusion that a closed steady circulation can be maintained in the ocean only if the heating source is situated at a level lower than the cooling source.

Sandstrom also carried out laboratory experiments to demonstrate his postulation. In the first experiment, the heating source was put at a level lower than the cooling source. Strong circulation was observed between the levels of heating and cooling. In the second experiment, the heating source was put at a level higher than the cooling source. He reported that no circulation occurred, and a stable stratification was observed between the heating and cooling levels.

Sandstrom's theorem can be concisely stated as follows:

A closed steady circulation can be maintained in the ocean only if the heat sources are situated at a lower level than the cold sources (Defant, 1961, p. 491).

Sandstrom's theorem was questioned by Jeffreys (1925), who pointed out that any horizontal density (temperature) gradients must induce a circulation. By including diffusion terms in the density balance equation, Jeffreys concluded that a circulation should be induced even if the heating source were put at a level higher than the cooling source.

The application of Sandstrom's theorem to the oceanic circulation poses a serious puzzle. The ocean is mostly heated and cooled from the upper surface. Due to thermal expansion, sea level at low latitudes, where heating takes place, is about 1 m higher than the sea level at high latitudes, where cooling takes place. According to Sandstrom's theorem, there should be no convectively driven circulation. The existence of a strong overturning circulation in the oceans is seemingly in conflict with Sandstrom's theorem.

Since heating and cooling apply to the sea surface, actually to the same level, circulation will be extremely weak if the system has no additional mechanical energy for supporting the circulation. However, the circulation observed in the world's oceans is very strong due to the fact that the effective depth of heating is moved downward owing to tidal and wind mixing.

It can be shown that heating/cooling an infinitely thin layer on the upper surface does not generate much gravitational potential energy (GPE). On the other hand, cooling induces an unstable stratification, and convective overturning ensues, which leads to a loss of GPE.

Although the discrepancy between Sandstrom's theorem and the situation in the oceans had been known for many years, the problem remained unsolved for a long time. Some researchers tried to find a mechanism which makes heating penetrate much deeper than cooling, without yielding any satisfactory results because cooling-induced convection at high latitudes can easily penetrate to 1 km (or deeper) below the surface, while no one has yet been able to identify any mechanism that would make heating penetrate even deeper.

The difficulty associated with Sandstrom's theorem may stem from the fact that ***his argument was entirely based on thermodynamics, without a more rigorous fluid dynamic analysis***. His model involves highly idealized circulation physics, especially as it completely excludes diffusion, internal friction, and wind stress. Diapycnal diffusivity in the ocean is about 1,000 times stronger than the molecular diffusivity. Including diffusion can substantially change the model's behavior, in particular the energetics of the circulation.

However, vertical mixing in a stratified fluid requires mechanical energy to push light (heavy) fluid downward (upward) during mixing; thus, mechanical energy is required for the maintenance of the stratification. Consequently, one of the key issues for the thermohaline circulation study is the energy source for mixing, and its spatial distribution and temporal evolution.

Mixing-induced circulation in a tube model

For extreme simplicity, thermal circulation in the ocean can be studied using a tube model. The model consists of a closed rectangular loop in the vertical plane (Fig. 3.2). Assume that the tube has a uniform cross-section of unit area, and that temperature and velocity are uniform across each section. The cooling source locates at a distance D from the origin (the central point of the low arm), while the heating source locates at the directly opposite position of the upper arm. When $D > B$, the cooling source locates at a vertical level of $D - B$ in the right arm. When $D > B + H$, the cooling source locates in the upper arm. Temperature at the cooling and the heating sources is maintained constant. The fluid is Boussinesq; in other words, the heating/cooling does not change the volume of the fluid.

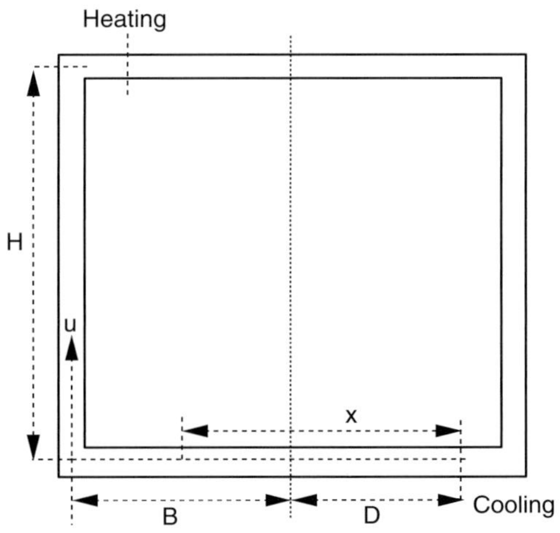

Fig. 3.2 An idealized tube model for the ocean (Huang, 1999).

Density distribution in a steady state is governed by the following one-dimensional balance between advection and diffusion:

$$u\rho_x = \kappa\rho_{xx} \tag{3.5}$$

where x is the along-tube coordinate, and κ is the mixing coefficient. We introduce the following nondimensional numbers: $s = \frac{x}{L}$, $b = \frac{B}{L}$, $d = \frac{D}{L}$, $h = \frac{H}{L}$, where $L = 2B + H$. The solution to the left (right) of the cooling source is denoted with a superscript *left* (*right*):

$$\rho^{left} = \rho_2 - \Delta\rho R_1 \left(e^{\alpha s} - 1\right), \quad \rho^{right} = \rho_2 - \Delta\rho R_2 \left(e^{-\alpha s} - 1\right) \tag{3.6}$$

where ρ_2 is the density at the cooling source, ρ_1 is the density at the heating source, and $\Delta\rho = \rho_2 - \rho_1$ is the density difference:

$$R_1 = \left(e^{\alpha} - 1\right)^{-1}, \quad R_2 = \left(e^{-\alpha} - 1\right), \quad \alpha = \frac{uL}{\kappa} \tag{3.7}$$

For an application to the ocean, model parameters are chosen as follows: $L = 1$ km, $b = 0.25$, $h = 0.5$. In the steady state, pressure torque is balanced by friction

$$\oint_s \rho\vec{g} \cdot d\vec{s} = r\rho_0 u \tag{3.8}$$

where $\vec{g}$ is gravitational vector, and r is the friction parameter.

The solution shows that the strength of the circulation strongly depends on the location of the cooling source (Fig. 3.3). The solutions for $d < 0.5$ (where the heating source is higher than the cooling source) are shown in the left-hand panel. Since diffusivity in the ocean is normally lower than 10^{-4} m^2/s, only the left part of this panel is relevant. For any given d,

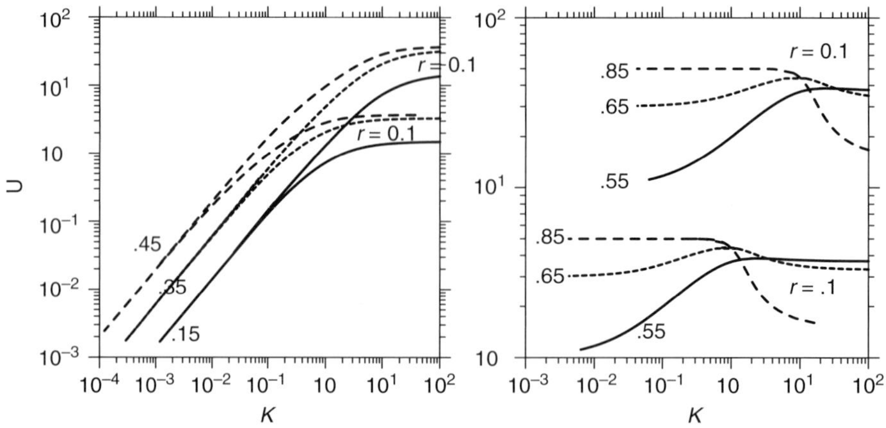

Fig. 3.3 Efficiency of the tube model: dependency of the circulation rate (U, in 10^{-7} m/s) on mixing coefficient κ (in 10^{-4} m^2/s). Numbers on the *left-hand side* of each panel indicate the nondimensional relative position d of heating/cooling source; numbers in the *right-hand side* of each panel indicate the non-dimensional frictional parameter r (Huang, 1999).

the circulation is almost linearly proportional to the mixing coefficient κ but is insensitive to the friction parameter r. Thus, the circulation is mixing-controlled, and resembles in some respects the meridional circulation in the ocean.

To obtain a circulation on the order of 10^{-7} m/s, we need a mixing coefficient on the order of 10^{-4} m^2/s. When mixing is very weak, on the order of 10^{-7} m^2/s (molecular diffusion), the nondimensional velocity will be on the order of 10^{-3} (10^{-10} m/s or 3 mm/yr, dimensionally), which is very difficult to observe. ***Thus, we can say that although both Sandstrom's theorem and Jeffreys' argument are correct on their own terms, both of their statements are incomplete and inaccurate in some way***.

For the cases of $d > 0.5$ (where the cooling source is higher than the heating source) shown in the right-hand panel of Figure 3.3, the circulation is insensitive to the diffusivity. Here again, our discussion is focused on the realistic range of diffusivity in the oceans, $\kappa \leq 10^{-4}$ m^2/s. When the value of κ varies 1,000 times, the corresponding circulation rate changes only slightly. However, the circulation rate is very sensitive to the friction parameter r. For example, an increase of one order of magnitude of r will give rise to the same order of change in the circulation. Thus, we can say that when the cooling source is at a level higher than the heating source, the circulation is frictionally controlled.

3.2.2 Where does Sandstrom's theorem stand?

Pure thermally driven circulation

A closer examination shows that circulation driven solely by thermal forcing can be classified into the following three types.

- **Type 1:** The heating source is located at a pressure level higher than the cooling source; it is well known that in this case there is a strong circulation. The Rayleigh–Benard thermal convection discussed in many textbooks belongs to this category.
- **Type 2:** The heating source is located at a pressure level lower than the cooling source. Jeffreys (1925) argued that wherever there is a horizontal density difference, there should be circulation; but he did not state how fast the circulation is. Although Type 2 thermal circulation can be weak, it is detectable, as will be discussed shortly.
- **Type 3:** Heating and cooling sources located at the same pressure level. This is also referred to as horizontal differential heating or horizontal convection. This type of heating/cooling resembles the situation in the ocean, where heating/cooling takes place primarily at the upper surface, neglecting the penetration of solar radiation and geothermal heating. This type of thermal circulation is very weak, as will be seen shortly.

The Paparella–Young theorem

Paparella and Young (2002) discussed horizontal convection in a rectangular model ocean based on the Boussinesq approximations. Their approach can be extended to the case of flow governed by two-dimensional, non-Boussinesq equations, including conservation of mass, momentum, and thermal energy, plus the equation of state (Wang and Huang, 2005). The basic equations (the continuity equation, the momentum equations, the thermal energy equation, and the equation of state) are:

$$\frac{\partial \rho}{\partial t} + \nabla \cdot (\rho \vec{u}) = 0 \tag{3.9}$$

$$\rho \frac{D\vec{u}}{Dt} = -\nabla p + \rho \vec{g} + \mu \nabla^2 \vec{u} \tag{3.10}$$

$$\rho c_p \frac{DT}{Dt} = \kappa \rho c_p \nabla^2 T - p \nabla \cdot \vec{u} + \Phi \tag{3.11}$$

$$\rho = \rho_0 \left[1 - \alpha \left(T - T_0\right)\right] \tag{3.12}$$

where ρ (ρ_0) is density (mean density) of the fluid, g is gravitational acceleration, T is temperature, T_0 is a constant reference temperature, p is pressure, c_p is the specific heat, α is the thermal expansion coefficient, and Φ is the dissipation function

$$\Phi = \mu \left[2\left(\frac{\partial u}{\partial x}\right)^2 + 2\left(\frac{\partial w}{\partial z}\right)^2 + \left(\frac{\partial w}{\partial x} + \frac{\partial u}{\partial z}\right)^2 \right] + \lambda \left(\frac{\partial u}{\partial x} + \frac{\partial w}{\partial z}\right)^2 \tag{3.13}$$

where $\lambda = -2\mu/3$ is commonly accepted as a parameter for most fluids.

Taking the dot product of Eqn. (3.10) with the velocity vector leads to the conservation equation of mechanical energy

$$\frac{\partial}{\partial t} \left[\rho E_k + \rho g z\right] + \nabla \cdot \left[\left(\rho E_k + \rho g z + p\right) \vec{u}\right] = -p \nabla \cdot \vec{u} + \mu \nabla^2 E_k + \mu \left\|\nabla \vec{u}\right\|^2 \tag{3.14}$$

where $E_k = \frac{1}{2}\vec{u} \cdot \vec{u}$ is the kinetic energy per unit mass and $\|\nabla \vec{u}\|^2 \equiv \nabla u \cdot \nabla u + \nabla w \cdot \nabla w$ is the deformation of velocity.

In a steady state, no mechanical energy enters/leaves the system through the boundaries; thus, averaging Eqn. (3.14) over the volume leads to a simple balance between the pressure work and dissipation

$$\langle p\nabla \cdot \vec{u}\rangle - \mu \left\langle \|\nabla \vec{u}\|^2 \right\rangle = 0 \tag{3.15}$$

where $\langle\ \rangle$ denotes the ensemble average.

Assume $\alpha = const, p = p_0 + p', \ p_0 = -\rho_0 g z$. To a very good approximation, Eqn. (3.9) can be reduced to the following relation:

$$\nabla \cdot \vec{u} = -\frac{1}{\rho_0}\frac{D\rho}{Dt} = \alpha \frac{DT}{Dt} = \alpha\kappa\nabla^2 T \tag{3.16}$$

thus

$$\langle p\nabla \cdot \vec{u}\rangle = -\rho_0 g \alpha\kappa \left\langle z\nabla^2 T\right\rangle \tag{3.17}$$

Using the Green formulae, $\left\langle z\nabla^2 T\right\rangle$ is reduced to

$$\left\langle z\nabla^2 T\right\rangle = \frac{1}{HL}\iint z\left(\frac{\partial^2 T}{\partial x^2} + \frac{\partial^2 T}{\partial z^2}\right) dxdz = -\frac{\bar{T}_{top} - \bar{T}_{bottom}}{H} \tag{3.18}$$

where H and L are the depth and width of the model ocean, and $\bar{T}_{top}$ and $\bar{T}_{bottom}$ are the mean temperatures at the top and bottom boundaries, respectively. For the case of heating/cooling from the upper surface with a linear reference temperature profile, the mean temperature at the upper boundary is $\bar{T}_{top} = (T_h + T_c)/2$, where T_h and T_c are the temperatures at the hot and cold ends; the bottom temperature cannot be lower than T_c, i.e., $\bar{T}_{bottom} \geq T_c$; thus, Eqn. (3.18) is reduced to

$$\left\langle z\nabla^2 T\right\rangle \geq -\frac{\Delta T}{2H} \tag{3.19}$$

where $\Delta T = T_h - T_c$, and the mean mechanical energy generation rate is

$$\frac{1}{\rho_0}\langle p\nabla \cdot \vec{u}\rangle \leq \frac{\kappa\alpha g\Delta T}{2H} = \frac{\kappa g'}{2H} \tag{3.20}$$

This relation provides an upper limit for the mechanical energy convertible from thermal energy: $\langle p\nabla \cdot \vec{u}\rangle HL/\rho_0 \leq \kappa g' L/2$, which is independent of the water depth. From Eqn. (3.15), the mean dissipation rate within the model ocean should be

$$\nu \left\langle \|\nabla \vec{u}\|^2 \right\rangle \leq \frac{\kappa g'}{2H} \tag{3.21}$$

where $\nu = \mu/\rho_0$.

A direct application of the Paparella–Young theorem provides an upper limit for the mechanical energy convertible from surface thermal forcing in the world's oceans. Assume that the equator–pole temperature difference is $\Delta T \simeq 30°C$, $\overline{H} \simeq 3.75$ km is the mean depth of the world's oceans, $\alpha \simeq 2 \times 10^{-4}/°C$, $\kappa \simeq 1.5 \times 10^{-7} m^2/s$, and the energy conversion rate is $\langle p\nabla \cdot \vec{u} \rangle \sim \rho_0 g \alpha \kappa \Delta T / 2\overline{H} \simeq 1.2 \times 10^{-9}$ W/m^3. The total volume of the ocean is 1.3×10^{18} m^3, so the total conversion rate is estimated at less than 1.5×10^9 W. This is 1,000 times smaller than the rate of tidal dissipation. Since the total amount of heat flux going through the oceans is approximately 2×10^{15} W, the efficiency of the ocean as a heat engine is estimated at less than 7×10^{-7}, which is an extremely low value! If we use the total heat flux associated with the incoming short-wave radiation from the Sun to the upper ocean, the equivalent efficiency of the oceanic heat engine is twenty times smaller.

The extension of the Paparella–Young theorem

Some of the key assumptions made in the Paparella–Young theorem are worth further exploration.

- *Molecular mixing value near the upper/lower boundary.* This may not be true in general. For example, if the ocean is heated from below and cooled from above, it is well known that the thermal boundary layer on the seafloor may be turbulent when the Rayleigh number is larger than the critical value. On the other hand, the surface mixed layer is actually a boundary layer dominated by strong surface waves and turbulence. Even if the ocean is heated and cooled from the upper surface, due to the huge horizontal dimensions of the ocean and the extreme low molecular viscosity, the Reynolds number corresponding to the real ocean is on the order of 10^{10} or more; thus, the circulation forced by such thermal boundary conditions may be in the regime of turbulence. The circulation for such a conceptual ocean, without forcing of wind or tides, remains a theoretical challenge.
- *Constant thermal expansion coefficient.* In reality, this is not constant. Modification due to the nonlinearity of the equation of state remains unknown.
- *Hydrostatic approximation* within the ascending (descending) plume above (below) the heating (cooling) source. This assumption may not be valid and the correction due to this factor remains unclear.

3.2.3 Laboratory experiments testing Sandstrom's theorem

Results from previous laboratory and numerical experiments

Most previous results from laboratory and numerical experiments suggested that circulation induced by horizontal differential heating occupies the whole depth of the tank; such a pattern of circulation is called the fully penetrating flow (Rossby, 1965, 1998).

New experimental results

The basic ideas of Sandstrom's theorem have recently been tested by means of laboratory experiments, and circulation driven by horizontal differential heating was very carefully examined, using a double-walled Plexiglas tank ($20 \times 15 \times 2.5$ cm^3) filled with salt water (Wang and Huang, 2005). The vacuum inside the double walls provides the best heat insolation. In contrast with most previous results of experimental and/or numerical modeling concerning the flow induced by heating/cooling from above and below, the new findings show that circulation appears as a shallow cell adjacent to the boundary of thermal forcing, instead of a cell penetrating to the full depth of the tank.

The typical quasi-steady mean velocity field for the following four cases was measured and is shown in Figure 3.4:

- *Case 1:* Heating/cooling from below, with heating on the left-hand side
- *Case 2:* Heating/cooling from above, with heating on the right-hand side
- *Case 3:* Heating/cooling from a sloping bottom, with heating on the left-hand side and at a level lower than the cooling source
- *Case 4:* Heating/cooling from a sloping bottom, with heating on the right-hand side and at a level higher than the cooling source.

The temperature difference of the thermal sources in these cases was kept the same, $\Delta T = 18.5°C$. It can be seen that the flow pattern produced by heating/cooling from the top is a mirror image of heating/cooling from the bottom. The pattern in Figure 3.4c, d is quite different from that in Figure 3.4a, b. In general, circulation for Case 3 is much stronger than for Case 1; the circulation cell is rather tall near the heating end, but is shallow and intense near the cooling end. Case 4 is quite different from the other cases (Fig. 3.4d). The circulation cell for this case is confined to the right half of the tank and the rest of the tank is left undisturbed. In addition, the circulation is separated from the bottom boundary, except near the heating end. Most noticeably, the strength of the circulation is greatly reduced. Nevertheless, circulation in all cases is undoubtedly visible to the naked eye.

The results indicate that there always exists a relatively stable steady circulation which occurs within a shallow depth of the fluid, or a so-called partial cell circulation. We then see that although no external mechanical energy input exists in the system, a circulation driven by horizontal convection does exist in the experimental situations. Thus, Sandstrom's theorem is inaccurate, although the circulation driven by horizontal convection is quite weak. If the results from the laboratory experiments can be extended and applied to the oceans, then we may come to the conclusion that surface thermal forcing alone is capable of driving a circulation that is so weak that it may not be able to penetrate to the deep ocean. Of course, caution needs to be exercised: there are major differences between the laboratory and the ocean, such as the huge difference in the Rayleigh number, the Reynolds number, and the effect of rotation.

The experimental results for the cases with horizontal differential thermal forcing (Cases 1 and 2) fit the classical scaling 1/5-power law of Rossby (1965), i.e., the nondimensional streamfunction maximum obeys $\Psi \sim Ra_L^{1/5}$, where the horizontal Rayleigh number is defined as $Ra_L = g'L^3/\nu\kappa$, where $g' = g\alpha\Delta T$, and κ and ν are molecular diffusivity and viscosity of the fluid, respectively. The nondimensional streamfunction maximum is defined as $\Psi = \tilde{\Psi}/\kappa$, where $\tilde{\Psi}$ is the maximum of the dimensional streamfunction defined as $\tilde{\Psi} = max|\int_0^\delta \bar{u}(z)dz|$. Alternatively, one can also use the vertical Rayleigh number $Ra_\delta = g'\delta^3/\nu\kappa$, where δ is the thickness of the velocity boundary layer adjacent to the heating/cooling surface. However, for the cases with a slanted bottom, the circulation seems to obey different power laws (Fig. 3.5). All these results suggest that circulation driven by

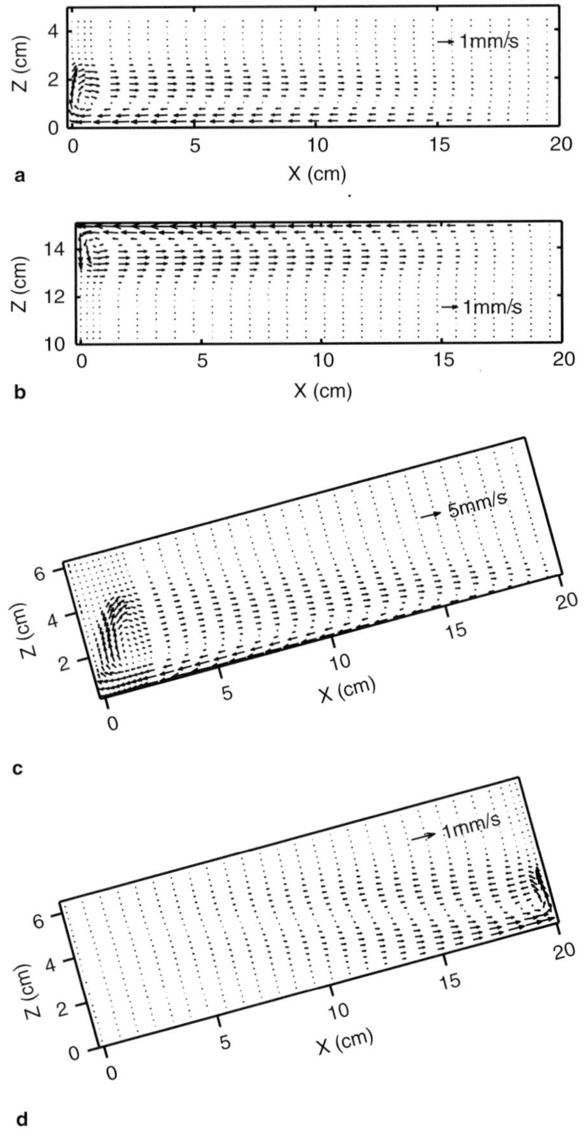

Fig. 3.4 **a–d** Time-mean circulation driven by purely thermal forcing obtained from laboratory experiments (Wang and Huang, 2005).

differential heating can be very complicated, and is sensitive to the slope of the boundary where heating/cooling applies.

Challenges posed by laboratory experiments

Laboratory experiments often pose exciting questions. In the experiments described above, results turn out quite differently from previous results and raise many questions:

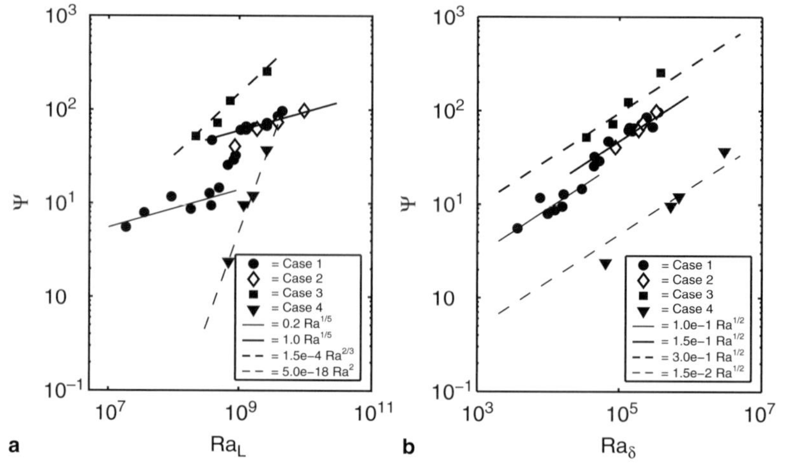

Fig. 3.5 The relationship between nondimensional streamfunction maximum, Ψ, and Rayleigh number: **a** horizontal Rayleigh number (Ra_L); **b** vertical Rayleigh number (Ra_δ) (Wang and Huang, 2005).

- Why a partial-penetration circulation, which differs from previous results? Under what conditions would the circulations penetrate fully?
- What is the scaling law for the case with a sloping bottom?
- What is the effect of the size of the laboratory experiments; namely, if the size of the tank is enlarged (i.e., the Rayleigh number is much larger) will the results be qualitatively changed?

The essential limitation of Sandstrom's theorem

A severe limitation of Sandstrom's theorem stems from its omission of wind stress. Wind energy is the most important contributor to the mechanical energy balance in the world's oceans. Wind stress drives strong circulation in the world's oceans, including Ekman transport in the upper ocean and the ACC (Antarctic Circumpolar Current). Any model or theory of the oceanic general circulation that does not include the contribution from wind is incomplete, and its implications should be interpreted with caution.

3.3 Seawater as a two-component mixture

Seawater contains many different chemical components. In the study of dynamical oceanography, however, it is highly desirable to simplify the system. The common practice is to treat the thermodynamics of seawater in terms of a two-component system. Such an assumption implies that the ratios of different chemical components in seawater are fixed, and the dynamical effects of separate components are combined and treated in terms of a single component, namely the salt concentration in seawater. When seawater is treated as a two-component mixture, each component – the salt and pure water – obeys the corresponding continuity equation

$$\frac{\partial \rho_s}{\partial t} + \nabla \cdot (\rho_s \vec{u}_s) = 0 \tag{3.22}$$

$$\frac{\partial \rho_w}{\partial t} + \nabla \cdot (\rho_w \vec{u}_w) = 0 \tag{3.23}$$

where $\rho_s = \rho s$, $\rho_w = \rho(1-s)$, and s is the mass fraction of salt in seawater, which is related to the commonly used salinity S through the relation $s = S/1,000$; $\vec{u}_s$ and $\vec{u}_w$ are the densities and velocities of salt and pure water components in seawater. Note that velocities in the expressions are regarded as the velocity of the macroscopic parcels of salt and pure water, respectively. Velocity defined in this way includes the contribution from small-scale turbulent motions at a given time and location, so it is different from the ensemble mean velocity used in many studies.

3.3.1 *Description in coordinates moving with the center of mass*

To specify the system more clearly, we introduce a coordinate system moving with its center of mass. The velocity of the center of mass is

$$\vec{u} = \frac{\rho_w \vec{u}_w + \rho_s \vec{u}_s}{\rho_w + \rho_s} \tag{3.24}$$

The corresponding continuity equation in terms of $\vec{u}$ and $\rho = \rho_w + \rho_s$ is obtained by adding Eqns. (3.22) and (3.23):

$$\frac{\partial \rho}{\partial t} + \nabla \cdot (\rho \vec{u}) = 0 \tag{3.25}$$

The continuity of salt and pure water can be rewritten as follows

$$\frac{\partial \rho_s}{\partial t} + \nabla \cdot \left(\rho_s \vec{u} + \vec{J}_s\right) = 0 \tag{3.26}$$

$$\frac{\partial \rho_w}{\partial t} + \nabla \cdot \left(\rho_w \vec{u} + \vec{J}_w\right) = 0 \tag{3.27}$$

where $\rho_s \vec{u}$ and $\rho_w \vec{u}$ are the advective transports of salt and pure water associated with the movement of the center of mass, and

$$\vec{J}_s = \rho_s (\vec{u}_s - \vec{u}), \quad \vec{J}_w = \rho_w (\vec{u}_w - \vec{u}) \tag{3.28}$$

are diffusive fluxes relative to the center of mass. Note that diffusive fluxes of salt and pure water are of the same magnitude and opposite signs, i.e.

$$\vec{J}_s + \vec{J}_w = 0 \tag{3.29}$$

Using the definition $\rho_s = \rho s$ and the continuity equation (3.25), one can rewrite Eqn. (3.26) in the following form

$$\rho \frac{ds}{dt} = -\nabla \cdot \vec{J}_s \tag{3.30}$$

3.3.2 Natural boundary condition for salinity balance

In many previous studies, the relaxation condition or the virtual salt flux condition has been used for salinity balance in the ocean. Although such boundary conditions are convenient to use, and models under such boundary conditions can provide nice-looking solutions, these boundary conditions are unphysical. In this section, we present the genuine suitable boundary conditions for salinity at the sea surface. The details of different types of surface boundary condition for salinity balance are discussed in Section 5.3.2.

Assume that the sea surface is specified by the following equation

$$F\left(\vec{x}_s, t\right) = 0 \tag{3.31}$$

Taking the time derivative of this relation leads to

$$\frac{\partial F}{\partial t} + \vec{u}_F \cdot \nabla F = 0, \quad \vec{u}_F = \frac{d\vec{x}_s}{dt} \tag{3.32}$$

The physical boundary condition at the sea surface is the following: the rate of freshwater leaving the ocean is equal to the rate of evaporation (minus precipitation) ω, and there is no salt leaving the ocean. Although there can be a very small amount of salt carried out by surface winds, this is so small that it can be ignored for the study of dynamical oceanography. Therefore, the upper boundary conditions are

$$\rho_w \left(\vec{u}_w - \vec{u}_F\right) \cdot \vec{n} = \rho_w \omega \tag{3.33a}$$

$$\rho_s \left(\vec{u}_s - \vec{u}_F\right) \cdot \vec{n} = 0 \tag{3.33b}$$

where $\vec{n} = \nabla F / |\nabla F|$ is the outward normal vector of the sea surface.

When the sea surface does not move, $\vec{u}_F \cdot \vec{n} = 0$, and Eqn. (3.33b) implies $\vec{u}_s = 0$ on the sea surface. This condition should be interpreted as follows: the vertical component of turbulent velocity of salt must be zero at the surface. Near the sea surface we can treat the process in terms of a one-dimensional model in the vertical direction. Since $\vec{u}_s = 0$ on the upper surface, $\vec{u}_s \equiv 0$ holds even below the surface. Conceptually, one can treat the salt as stagnant in space, i.e., salt is like a fixed grid in space. On the other hand, freshwater moves through the salt grid. One can imagine the movement of freshwater associated with precipitation and evaporation as water percolating through this grid of salt, very much like hot water percolating through the coffee grounds in a coffee-making machine.

Adding the two equations in Eqn. (3.33) leads to

$$\left(\vec{u} - \vec{u}_F\right) \cdot \vec{n} = \omega \rho_w / \rho \tag{3.34}$$

Using Eqn. (3.32), this is reduced to

$$\frac{\partial F}{\partial t} + \vec{u} \cdot \nabla F = \frac{\rho_w}{\rho} \omega |\nabla F| \tag{3.35}$$

Equation (3.33b) can be rewritten as

$$(\vec{u}_s - \vec{u}) \cdot \vec{n} = -(\vec{u} - \vec{u}_F) \cdot \vec{n} \tag{3.36}$$

Using Eqns. (3.34), (3.36), and the definition of $\vec{J}_s$, this gives rise to a formula for the diffusive salt flux at the sea surface

$$\vec{J}_s \cdot \nabla F = -\omega \rho_w s |\nabla F| \tag{3.37}$$

However, it is important to notice that this expression holds only below the sea surface. Above the sea surface, there is no salt and salt flux is identically zero.

Our discussion here excludes the water–sea ice interface. Since sea ice formation and melting involves the movement of both pure water and salt through the water–ice interface, the corresponding boundary conditions are more complicated. The detailed discussion of the boundary condition applied to the water–ice interface can be found in an article by Huang and Jin (2007).

3.3.3 A one-dimensional model with evaporation

To gain a better understanding of the boundary condition for salinity, we examine a one-dimensional model (Fig. 3.6). The basic idea can best be illustrated by adding on an imaginary, infinitely thin, layer of freshwater with $S = 0$ on the top.

Suppose the ocean is subject to evaporation of ω. In a one-dimensional model, mass conservation requires the balance of pure water, i.e., the vertical flux of water should be continuous. Thus, over the whole depth of the water column, freshwater, with density

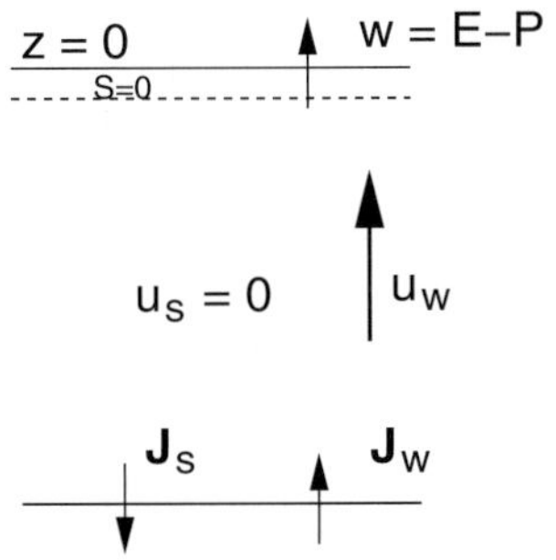

Fig. 3.6 A one-dimensional balance of transport in the upper ocean, including an imaginary infinitely thin layer of pure water on the top. All fluxes are defined in the coordinates moving with the center of mass.

Table 3.1. *Balance of water and salt flux in a steady one-dimensional model*

	Diffusive flux relative to center of mass	Advective flux	Sum
Salt	$-\rho s\,(1-s)\,\omega$	$\rho s\,(1-s)\,\omega$	0
Water	$\rho s\,(1-s)\,\omega$	$\rho\,(1-s)^2\,\omega$	$\rho\,(1-s)\,\omega$
Sum	0	$\rho\,(1-s)\,\omega$	$\rho\,(1-s)\,\omega$

$\rho_w = \rho(1-s)$, must move upward with the same vertical velocity $u_w = \omega$; however, the other component, salt (density $\rho_s = \rho s$), is stagnant, i.e., $u_s = 0$ (Fig. 3.6). The velocity of the center of mass is

$$u = \frac{\rho_w u_w + \rho_s u_s}{\rho_w + \rho_s} = \frac{\rho_w}{\rho} u_w = (1-s)\,\omega \tag{3.38}$$

The corresponding diffusive mass flux, relative to the center of mass, is

$$J_w = \rho_w\,(u_w - u) = \rho_w s\omega = \rho s\,(1-s)\,\omega > 0 \tag{3.39}$$

$$J_s = -J_w = -\rho s\,(1-s)\,\omega < 0 \tag{3.40}$$

Note that the center of mass moves upward with a velocity of $u = (1-s)\omega$, slower than the rate of evaporation because the salt component is stagnant; the diffusive salt flux defined here is the diffusive flux relative to the center of mass. The advective salt flux associated with the center is $u\rho s = J_w$, which is equal to the rate of diffusive salt flux, but with an opposite sign, i.e., $J_w = J_s$. ***Thus, the advective salt flux is exactly cancelled by the diffusive salt flux.*** The balance of water and salt fluxes for this model is shown in Table 3.1.

The salt balance in fixed standard Eulerian coordinates is

$$I_{s,Adv} = \omega\rho_s \tag{3.41}$$

Since the salt is in an exact balance, there is a salt flux which is opposite to the advective salt flux,

$$I_{s,Dif} = -\omega\rho_s \tag{3.42}$$

Therefore, the diffusive salt flux defined in the coordinates moving with the center of mass equation (3.40) is slightly (about 3.5%) smaller than the diffusive salt flux defined in the fixed Eulerian coordinates (see Eqn. (3.42)). This difference is due to the fact that the velocity of the center of mass is (1 - s)ω, which is slower than the evaporation rate ω.

The discussion above applies to regime below a thin layer of pure water. Within a thin layer of pure water, the velocity of center of mass is exactly the same as the velocity of pure water because there is no salt within this layer. As a result, both the salt flux and water flux relative to the center of mass are identically zero. The introduction of an imaginary thin layer of pure water on the top of the ocean has therefore helped us to clarify the dynamical processes associated with evaporation and precipitation, and such a procedure will be detailed in the analysis of entropy balance in the world's oceans at the end of this chapter.

3.4 Balance of mass, energy, and entropy

As a dynamical system, oceanic circulation must satisfy several balance laws, which also act as the most important theoretical tools in the study of oceanic dynamics. In most cases these laws are also called the conservation laws; however, there is an exception for entropy, because it is a peculiar thermodynamic variable which is not conserved and never decreases for any macroscopic thermodynamic systems. Thus, we will adopt the terminology of "balance laws."

The fundamental conservation/balance laws of dynamics and thermodynamics are extensively discussed in various standard textbooks. We here present a concise derivation of these laws in the form most suitable for studies of wind-driven circulation and thermohaline circulation. In this section the standard notation of tensor will be utilized. In particular, the summation convention for tensors will be used. For example, $\partial u_j/\partial x_j$ means $\sum_{i=1}^{3} \partial u_i/\partial x_i = \partial u_1/\partial x_1 + \partial u_2/\partial x_2 + \partial u_3/\partial x_3$.

3.4.1 Mass conservation

$$\frac{\partial \rho}{\partial t} + \frac{\partial \rho u_j}{\partial x_j} = 0 \tag{3.43}$$

where $\{x_j\} = (x_1, x_2, x_3)$ are the coordinates, ρ is the density, and $\{u_j\} = (u_1, u_2, u_3)$ is the velocity.

3.4.2 Momentum conservation

$$\frac{\partial \rho u_i}{\partial t} + \frac{\partial \rho u_i u_j}{\partial x_j} = -2\varepsilon_{i,j,k}\Omega_j u_k - \frac{\partial p}{\partial x_i} - \rho\frac{\partial \phi_0}{\partial x_i} + \frac{\partial \tau_{ij}}{\partial x_j} - \rho\frac{\partial \phi_T}{\partial x_i}, \quad i = 1, 2, 3. \tag{3.44}$$

where ε_{ijk} is the three-dimensional permutation symbol; the geopotential is separated into two parts:

$$\phi = \phi_0 + \phi_T = gz + \phi_T \tag{3.45}$$

where ϕ_T is the tidal potential and $\partial\phi_T/\partial x_i = [0,\ 0,\ g_T(x, y, t)\vec{k}]$ is the tidal force, which is treated as a body force specified *a priori* and remains the same after the ensemble averaging. All forces act as sources of momentum changes. In fluid dynamics, the stress tensor is defined as

$$\sigma_{ij} = \tau_{ij} + \delta_{ij} p \tag{3.46a}$$

where the first term is called the viscosity stress tensor, and for isotropic viscosity is in the form derived by Landau and Lifshitz (1959, p. 48):

$$\tau_{ij} = \mu \left[\left(\frac{\partial u_i}{\partial x_j} + \frac{\partial u_j}{\partial x_i} \right) - \frac{2}{3}\delta_{ij}\frac{\partial u_k}{\partial x_k} \right] + \mu' \delta_{ij} \frac{\partial u_k}{\partial x_k} \tag{3.46b}$$

In many oceanographic studies, the seawater is treated in terms of a Boussinesq fluid; thus, the divergence terms vanish, and the corresponding viscosity stress is reduced to the following form

$$\tau_{ij} = \mu \left(\frac{\partial u_i}{\partial x_j} + \frac{\partial u_j}{\partial x_i} \right) \tag{3.46b$'$}$$

3.4.3 Gravitational potential energy conservation

Multiplying the continuity equation (3.4.1) by $\phi = gz + \phi_T$, we obtain the GPE conservation

$$\frac{\partial \rho \phi}{\partial t} + \frac{\partial \rho u_j \phi}{\partial x_j} = \rho u_j \frac{\partial \phi}{\partial x_j} + \rho \frac{\partial \phi_T}{\partial t} \tag{3.47}$$

3.4.4 Kinetic energy conservation

From the momentum equations, we can derive the kinetic energy (KE) balance equation by the inner product of the velocity and the momentum equation. Since individual terms of the momentum equation can be interpreted as forces, the rate of doing work is the scalar product of forces times the velocity:

$$\frac{\partial \rho K}{\partial t} + \frac{\partial \rho u_j K}{\partial x_j} = -u_j \frac{\partial p}{\partial x_j} - \rho u_j \frac{\partial \phi_0}{\partial x_j} + u_i \frac{\partial \tau_{ij}}{\partial x_j} - \rho u_j \frac{\partial \phi_T}{\partial x_j} \tag{3.48}$$

where $K = u_i u_i / 2$ is the KE per unit mass. Note that the Coriolis force term makes no contribution to the energetics.

3.4.5 Internal energy conservation

A second energy conservation equation can be obtained from the thermodynamic conservation of energy. For a water parcel with volume V and lateral boundary S, the sum of the internal and kinetic energy must change at a rate equal to the sum of the work done by the individual forces plus internal sources, i.e.,

$$\begin{aligned} \frac{\partial}{\partial t} \int_V \rho(e + K) dV = & - \int_S \rho(e + K) u_j n_j dA - \int_S p u_j n_j dA + \int_S u_i \tau_{ij} n_j dA - \int F_j n_j dA \\ & - \int_V \rho u_j \frac{\partial \phi_0}{\partial x_j} dV - \int_V \rho u_j \frac{\partial \phi_T}{\partial x_j} dV, \end{aligned} \tag{3.49}$$

where

$\int_S \rho(e+K)u_j n_j dA$ is the energy flux due to mass exchange through boundaries,

$\int_S p u_j n_j dA$ is due to pressure work on boundaries,

$\int_S u_i \tau_{ij} n_j dA$ is due to work by frictional stress on boundaries,

$\int_V \rho u_j \frac{\partial \phi_0}{\partial x_j} dV$ is the work by gravity on the whole water parcel,

$\int_V \rho u_j \frac{\partial \phi_T}{\partial x_j} dV$ is due to energy input from tides on the whole water parcel,

$\int_S F_j n_j dA$ is due to diffusion of heat and mass through the boundaries, and flux F_i includes heat flux, enthalpy flux due to salt water and freshwater mixing across the boundaries.

Using Eqns. (3.29) and (2.66), this flux can be rewritten as

$$F_i = q_i + h_s J_{s,i} + h_w J_{w,i} = q_i + (h_s - h_w) J_{s,i} = q_i + \frac{\partial h}{\partial s} J_{s,i} \tag{3.50}$$

where h_s and h_w are the specific partial enthalpy of salt and pure water, and q_i and $J_{s,i}$ are the i-th component of the interfacial heat and salt fluxes.

Since Eqn. (3.49) is valid for any infinitesimal volume, we obtain the differential equation for conservation of kinetic and internal energy:

$$\frac{\partial \rho\,(e+K)}{\partial t} + \frac{\partial \rho u_j\,(e+K)}{\partial x_j} = -\frac{\partial p u_j}{\partial x_j} - \rho u_j \frac{\partial \phi_0}{\partial x_j} - \rho u_j \frac{\partial \phi_T}{\partial x_j} + \frac{\partial u_i \tau_{ij}}{\partial x_j} - \frac{\partial F_j}{\partial x_j} \tag{3.51}$$

Subtracting Eqn. (3.48) from Eqn. (3.51) leads to the conservation of internal energy:

$$\frac{\partial \rho e}{\partial t} + \frac{\partial \rho u_j e}{\partial x_j} = -p \frac{\partial u_j}{\partial x_j} + \rho \varepsilon - \frac{\partial F_j}{\partial x_j}, \tag{3.52}$$

where

$$\rho \varepsilon = \tau_{ij} \frac{\partial u_j}{\partial x_i} = \frac{1}{2} \tau_{ij} \left(\frac{\partial u_i}{\partial x_j} + \frac{\partial u_j}{\partial x_i} \right) \tag{3.53}$$

is the dissipation rate per unit volume. The kinetic conservation law can be rewritten as

$$\frac{\partial \rho K}{\partial t} + \frac{\partial \rho u_j K}{\partial x_j} = \frac{\partial \tau_{ij} u_i}{\partial x_j} - u_j \frac{\partial p}{\partial x_j} - \rho u_j \frac{\partial\,(\phi_0 + \phi_T)}{\partial x_j} - \rho \varepsilon \tag{3.54}$$

Using the continuity equation, the corresponding equation for internal energy (Eqn. (3.52)) is reduced to

$$\rho \frac{de}{dt} = -p \frac{\partial u_j}{\partial x_j} + \rho \varepsilon - \frac{\partial F_j}{\partial x_j} \tag{3.55}$$

In summary, now we have the following differential forms of conservation of kinetic energy, internal energy, and GPE

$$\frac{\partial \rho K}{\partial t} + \frac{\partial \rho u_j K}{\partial x_j} - \frac{\partial \tau_{ij} u_i}{\partial x_j} + \frac{\partial p u_j}{\partial x_j} = p\frac{\partial u_j}{\partial x_j} - \rho u_j \frac{\partial\left(\phi_0 + \phi_T\right)}{\partial x_j} - \rho\varepsilon \tag{3.56}$$

$$\frac{\partial \rho e}{\partial t} + \frac{\partial \rho u_j e}{\partial x_j} + \frac{\partial}{\partial x_j}\left(q_j + \frac{\partial h}{\partial s} J_{s,j}\right) = -p\frac{\partial u_j}{\partial x_j} + \rho\varepsilon \tag{3.57}$$

$$\frac{\partial \rho \phi}{\partial t} + \frac{\partial \rho u_j \phi}{\partial x_j} = \rho u_j \frac{\partial \phi}{\partial x_j} + \rho\frac{\partial \phi_T}{\partial t} \tag{3.58}$$

3.4.6 *Entropy balance*

From thermodynamics, the differential relation for entropy is Eqn. (2.52):

$$d\eta = \frac{1}{T} de + \frac{p}{T} dv - \frac{\mu}{T} ds$$

From the continuity equation (3.43):

$$\rho\frac{dv}{dt} = \frac{\partial u_j}{\partial x_j}$$

Using Eqn. (3.55) and the relation between diffusive salt flux and rate of salt change, Eqn. (3.30), we obtain the balance equation of entropy:

$$\rho\frac{d\eta}{dt} = \frac{1}{T}\left(\rho\varepsilon - \frac{\partial F_j}{\partial x_j}\right) + \frac{\mu}{T}\frac{\partial J_{s,j}}{\partial x_j} \tag{3.59}$$

Using the continuity equation, this can be rewritten as

$$\frac{\partial \rho\eta}{\partial t} + \frac{\partial \rho u_j \eta}{\partial x_j} + \frac{\partial}{\partial x_j}\frac{q_j}{T} + \frac{\partial}{\partial x_j}\frac{\partial h/\partial s - \mu}{T} J_{s,j} = \frac{\rho\varepsilon}{T} + q_j\frac{\partial}{\partial x_j}\frac{1}{T} + J_{s,j}\left(\frac{\partial h}{\partial s}\frac{\partial}{\partial x_j}\frac{1}{T} - \frac{\partial}{\partial x_j}\frac{\mu}{T}\right) \tag{3.60}$$

An important point to note is that mechanical energy exchange crossing the interfaces is entropy-free, as discussed in Section 2.3.3. From the discussion in Section 3.3, the salt flux is exactly zero at the air–sea interface, so the contribution for the global oceans due to the last term on the left-hand side of Eqn. (3.60) vanishes. Thus, entropy changes are due to three sources: exchange of heat fluxes with the environment, the production of entropy due to internal dissipation, and heat and freshwater mixing. A detailed analysis of each kind of entropy production is discussed in Section 3.8.

3.5 Energy equations for the world's oceans

3.5.1 Three types of time derivative for the property integral in the ocean

For a given function $F(\vec{x}, t)$ and a control volume V, there exist three different types of time derivative for the volume integral of $F(\vec{x}, t)$.

Partial derivative

$$\partial_t I_{V_0} = \frac{d}{dt} \iiint_{V=V_0} F(\vec{x}, t)\, dv = \lim_{\delta t \to 0} \frac{1}{\delta t} \left(\iiint_{V_0} F(\vec{x}, t + \delta t)\, dv - \iiint_{V_0} F(\vec{x}, t)\, dv \right) \tag{3.61}$$

This is the time differentiation with a fixed control volume $V = V_0$ in Eulerian coordinates. Since space and time are separate variables in Eulerian coordinates, the differentiation and integration are interchangeable, i.e.,

$$\partial_t I_{V_0} = \frac{d}{dt} \iiint_{V=V_0} F(\vec{x}, t)\, dv = \iiint_{V=V_0} \frac{\partial}{\partial t} F(\vec{x}, t)\, dv \tag{3.62}$$

Material derivative

$$\begin{aligned} \partial_t I_{V(m_0)} &= \frac{D}{Dt} \iiint_{V(m_0)} F(\vec{x}, t)\, dv \\ &= \lim_{\delta t \to 0} \frac{1}{\delta t} \left(\iiint_{V(m_0)} F(\vec{x}, t + \delta t)\, dv - \iiint_{V(m_0)} F(\vec{x}, t)\, dv \right) \end{aligned} \tag{3.63}$$

where $V(m_0)$ indicates that the integral is taken over the same material volume $V(m) = V(m_0)$. This is the time differentiation with the control volume which moves with the fluid parcel. Note that:

- Although the control volume may change with time in Eulerian coordinates, it always consists of the same mass elements; thus, the control volume is fixed in Lagrangian coordinate space.
- The differentiation and integration are ***not interchangeable***. In fact, there is the following relation, discussed in many classical textbooks (e.g., Kundu (1990, p. 77)):

$$\frac{D}{Dt} \iiint_V F dv = \iiint_V \left(\frac{dF}{dt} + F \nabla \cdot \vec{u} \right) dv = \iiint_V \frac{\partial F}{\partial t} dv + \oint\!\!\oint_S F \vec{u} \cdot \vec{n} ds \tag{3.64}$$

- The material derivative is meaningless for seawater owing to salt diffusion. With salt diffusion, the material surface used in the discussion above is no longer definable. Such problems associated with the inaccuracy of the material derivative and material surface may exist for all multi-component fluids, owing to the included effect of diffusion present.

Control derivative

$$\partial_t I_c = \frac{\delta}{\delta t} \iiint_{V(t)} F(\vec{x}, t)\, dv$$
$$= \lim_{\Delta t \to 0} \frac{1}{\Delta t} \left(\iiint_{V(t)} F(\vec{x}, t + \Delta t)\, dv - \iiint_{V(t)} F(\vec{x}, t)\, dv \right) \quad (3.65)$$

This is the time differentiation for the property integrated over a control volume $V(t)$, which can change with time and exchange mass with the environment. It is readily seen that the control derivative depends on the specification of $V(t)$. The most relevant example for oceanography is the property integrated over the world's oceans, defined as the water mass below the free surface. Note that:

- Owing to evaporation, precipitation, river run-off, and ice formation/melting, this control volume has mass exchange with the atmosphere/land/ice, including water properties, such as pure water and salt associated with evaporation/precipitation, river run-off and ice melting/formation. Thus, the free surface in the ocean is a moving surface in both the Eulerian and Lagrangian coordinates, and the corresponding control volume changes in space.
- The differentiation and integration are ***not interchangeable***.
- The time rate defined in this way is the rate most relevant to the balance of the oceanic properties in the world's oceans, including mass, KE, GPE, internal energy, and entropy.

Examples

As an illustration, let us examine the mass balance of a bucket of water with a unit of horizontal area and an initial volume V_0 and density ρ_0. The control volume $V(t)$ will be defined as the total volume under the free surface. Two types of forcing will be discussed: heating and precipitation.

Heating

At time $t = t_0$, the total mass is $m_0 = V_0 \rho_0$. With heating alone, water temperature increases, but the total mass remains the same: $m_1 = m_0$. At time $t_0 = \delta t$, temperature increases by δT, the total volume expands, and the density is $\rho = \rho_0(1 - \alpha \delta T)$. Set $F(\vec{x}, t) = \rho$, and for these three time derivatives we have

$$\partial_t \iiint_{V_0} \rho dv < 0, \quad \partial_t \iiint_{V(m_0)} \rho dv = 0, \quad \partial_t \iiint_{V(t)} \rho dv = 0 \quad (3.66)$$

The first time derivative is negative, due to the fact that heat expansion gives rise to smaller density and thus less mass in the original volume. Therefore, for this special case, the control derivative and the material derivative are equal because there is no mass exchange with the environment, so the material surface is well-defined in principle.

Precipitation

First, assume there is no salt: $\rho = \rho_0$. At time $t = t_0$, the total mass is $m_0 = V\rho_0$. At time $t_0 + \delta t$, $m_1 = (V_0 + \omega \delta t)\rho_0$, where $\omega = p$ is the (constant) rate of precipitation. Choose

$F = \rho_0$, and for the three time derivatives we have

$$\partial_t \iiint_{V_0} \rho_0 dv = 0, \quad \partial_t \iiint_{V(m_0)} \rho_0 dv = 0, \quad \partial_t \iiint_{V(t)} \rho_0 dv = \omega \rho_0 > 0 \tag{3.67}$$

Second, assume there is salt in the water, and the initial salinity equals S_0. Density is a linear function of salinity S, $\rho = \rho_0(1+\beta S)$. Thus, the density associated with precipitated water is $\rho = \rho_0$. The total mass is

$$m_0 = V_0 \rho_0 (1 + \beta S_0) \text{ at } t = t_0 \tag{3.68}$$
$$m_1 = m_0 + \delta m = \rho_0 [V_0 (1 + \beta S_0) + \omega \delta t] \text{ at } t_0 = \delta t$$

Thus, $\dfrac{\delta m}{m_0} = \dfrac{\omega \delta t}{V_0 (1 + \beta S_0)}$, and the new salinity is $S = \dfrac{m_0 S_0}{m_0 + \delta m} \approx \left(1 - \dfrac{\delta m}{m_0}\right) S_0$; the density change is $\delta \rho = -\rho_0 \beta \dfrac{\delta m}{m_0} S_0 < 0$.

By choosing $F(\vec{x}, t) = \rho$, we have

$$\partial_t \iiint_{V_0} \rho dv = -\rho_0 \frac{\beta S_0}{1 + \beta S_0} \omega \delta t < 0 \tag{3.69}$$

The negative value of this derivative is due to the dilution of water. For this case the material derivative is

$$\partial_t \iiint_{V(m_0)} \rho dv = ?? \tag{3.70}$$

i.e., this term is meaningless, because $V(m_0)$ has lost its meaning due to the mass exchange (mixing of salt in the water) between the control volume and the precipitated freshwater. This simple case clearly demonstrates the potential problems related to the material derivative for multi-component fluids with diffusion. The corresponding control derivative is positive because precipitation increases the total mass under the free surface:

$$\partial_t \iiint_{V(t)} \rho dv = \omega \rho_0 > 0 \tag{3.71}$$

3.5.2 *The generalized Leibnitz theorem and generalized Reynolds transport theorem*

Now we proceed to establish a general transport theorem in order to grasp the balance relationship of various oceanic properties. Let us begin with recalling the very fundamental theorem in calculus – Leibnitz's theorem: in the one-dimensional case we have

$$\frac{d}{dt} \int_{a(t)}^{b(t)} \phi(x, t)\, dx = \int_{a(t)}^{b(t)} \frac{\partial}{\partial t} \phi(x, t) dx + \phi[b(t), t] \frac{db}{dt} - \phi[a(t), t] \frac{da}{dt} \tag{3.72}$$

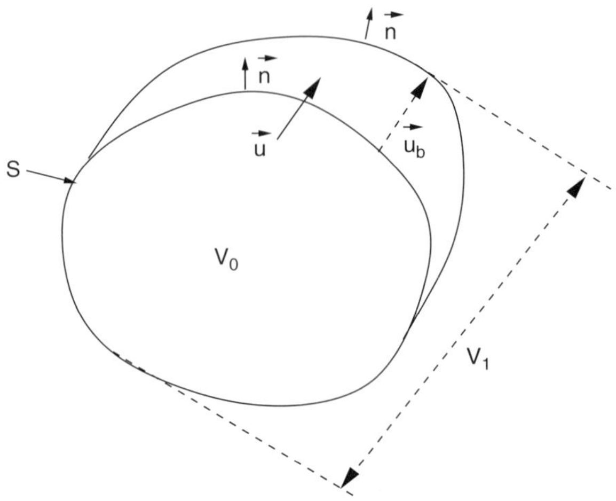

Fig. 3.7 Sketch of the time evolution of the control volume $V(t)$ and the surface S: at time t_0 the control volume is V_0, and its outer surface expands with a velocity of $\vec{u}_s$, so that at time $t_0 + \delta t$, the new volume is V_1.

In the three-dimensional case, the corresponding generalized Leibnitz theorem (Kundu, 1990, p. 75) is

$$\frac{\delta}{\delta t} \iiint_{V(t)} \varphi(\vec{x}, t)\, dv = \iiint_{V(t)} \frac{\partial}{\partial t} \varphi(\vec{x}, t)\, dv + \oiint_S \varphi(\vec{x}, t)\, \vec{u}_s \cdot \vec{n} dS \tag{3.73}$$

where the time derivative is defined as in Eqn. (3.65), $\vec{u}_s$ is the velocity of the moving boundary of the control volume $V(t)$, which changes in time; and $\vec{n}$ is the outward normal vector of the boundary S (Fig. 3.7).

Assume that a property q satisfies the following equation

$$\frac{\partial \varphi}{\partial t} + \nabla \cdot (\vec{u} F) = q \tag{3.74}$$

where F (F can be different from ϕ) is the property transported by the velocity and q is the spatial distribution of the source. The volumetric integration of Eqn. (3.74) leads to the following relation

$$\iiint_{V(t)} \frac{\partial \varphi}{\partial t} dv + \oiint_S F \vec{u} \cdot \vec{n} ds = \iiint_V q dv \tag{3.75}$$

Using Eqn. (3.73), Eqn. (3.75) is reduced to the generalized Reynolds transport theorem:

$$\frac{\delta}{\delta t} \iiint_{V(t)} \varphi dv = \oiint_S (\varphi \vec{u}_s - F \vec{u}) \cdot \vec{n} ds + \iiint_V q dv \tag{3.76}$$

Therefore, the total time rate of the property integral within a control volume changing with time consists of two terms: the first term on the right-hand side is the net transport through the moving surface, and the second term is the contribution due to the distributed source in the interior; the net transport through the moving surface is the transport induced by the moving boundary after subtracting the transport associated with the fluid velocity. The generalized Reynolds transport theorem can be used for the case of a moving control volume, which includes the possible mass exchange with the environment.

3.5.3 Energetics of the barotropic tides

Tidal dissipation is a source of the mechanical energy sustaining the oceanic general circulation, and its contribution will be discussed in detail shortly. Before discussing dissipation of the three-dimensional tidal motion, it is helpful to begin with the barotropic tide. We start from Laplace's tidal equations:

$$\partial_t u - f v = -g \partial_x (\zeta - \Gamma_T / g) + F_r^x / \rho_0 H \tag{3.77}$$

$$\partial_t v + f u = -g \partial_y (\zeta - \Gamma_T / g) + F_r^y / \rho_0 H \tag{3.78}$$

$$\partial_t \zeta + \partial_x (uH) + \partial_y (vH) = 0 \tag{3.79}$$

where ζ is the free surface height, $\Gamma_T = \Gamma_T(x, y, t)$ is the barotropic tide-generating potential (the corresponding tidal force is $\vec{F}_T = -\nabla \cdot \Gamma_T$), $\left(F_r^x, F_r^y\right) = \vec{F}_r$ is tidal friction, ρ_0 is the density of seawater (assumed constant for the barotropic tide problem), $H = h + \zeta$ is the total thickness of the water, and h is the depth of the ocean (assumed to be constant for simplicity).

Multiplying Eqns. (3.77) and (3.78) by $\rho_0 H u$ and $\rho_0 H v$, and adding the two equations, leads to

$$\frac{1}{2} \rho_0 D \partial_t \left(u^2 + v^2\right) = -\rho_0 g H \left(u \partial_x \zeta + v \partial_y \zeta\right) + \rho_0 H \left(u \partial_x \Gamma_T + v \partial_y \Gamma_T\right) + u F_r^x + v F_r^y \tag{3.80}$$

Multiplying Eqn. (3.79) by $\rho_0 g \zeta$, we obtain

$$\rho_0 g \partial_t \left(\zeta^2 / 2\right) = -\rho_0 g \zeta \left[\partial_x (uH) + \partial_y (vH)\right] \tag{3.81}$$

Adding Eqns. (3.80) and (3.81), we have

$$\frac{\rho_0}{2} \partial_t \left[H \left(u^2 + v^2\right) + g \zeta^2\right] = \rho_0 \left(\frac{u^2 + v^2}{2}\right) \partial_t \zeta - \rho_0 g \nabla_h \cdot (\vec{u} D \zeta) + \rho_0 H \vec{u} \cdot \nabla_h \Gamma_T + \vec{u} \cdot \vec{F}_r \tag{3.82}$$

where $\nabla_h = \left(\vec{i} \frac{\partial}{\partial x} + \vec{j} \frac{\partial}{\partial y}\right)$ is the two-dimensional gradient operator. Note that when integration of Eqn. (3.82) is over the world's oceans, the divergence term involving the horizontal velocity vanishes due to the zero-normal velocity condition at the lateral boundaries or the periodic boundary condition. Averaging Eqn. (3.82) over one tidal period leads to

$$0 = \rho_0 \overline{\left(\frac{u^2 + v^2}{2}\right) \partial_t \zeta} + \rho_0 \overline{H \vec{u} \cdot \nabla_h \Gamma_T} + \overline{\vec{u} \cdot \vec{F}_r} \tag{3.83}$$

Note that $\overline{\vec{u} \cdot \vec{F}_r} < 0$ because friction is in the direction opposite to velocity; thus, this is a sink of mechanical energy. The first and second terms on the right-hand side represent energy generated by tidal forcing. Using the continuity equation (3.79), the energy source terms in Eqn. (3.83) can be rewritten as

$$\rho_0 \overline{\left(\frac{u^2 + v^2}{2}\right) \partial_t \zeta} + \rho_0 \overline{H \vec{u} \cdot \nabla_h \Gamma_T} = \rho_0 \overline{(K + \Gamma_T)\, \partial_t \zeta}$$

where $K = (u^2 + v^2)/2$ is the kinetic energy. Therefore, Eqn. (3.83) is reduced to

$$\rho_0 \overline{(K + \Gamma_T)\, \partial_t \zeta} = -\overline{\vec{u} \cdot \vec{F}_r} > 0 \tag{3.84}$$

Since tidal dissipation is a sink of mechanical energy, the source on the left-hand side of Eqn. (3.84) should be positive, i.e., the vertical velocity should be positively correlated with the total mechanical energy of tides (the tidal potential and tidal kinetic energy).

The phase relation between the total mechanical energy (tidal potential plus kinetic energy) and vertical velocity is similar to the case of a swing. When children want to swing higher and higher, they sit up when the swing moves upward, and lean down low when the swing moves downward.

Note that in the discussion above, tidal potential and tidal motions are assumed to be a perfect periodic function of time, and so are the tidal motions. In reality, tidal potential, tidal velocity, and tidal amplitude decline slowly over the long geological time of the Earth. Although such slow declines are omitted in the discussion above, they may not be negligible if we examine the energetics of the oceanic general circulation from different angles. For example, if we treat the gravitational potential of both the tidal motion and the non-tidal motions (traditionally called the oceanic general circulation) as a single quantity, we can no longer ignore the decline of tidal potential, even if we are talking about motion on a time scale of days to months. This is explained in the next section.

3.5.4 Energy equations for the oceans

The basic balance equations for a fixed fluid parcel, Eqns. (3.56), (3.57), and (3.58), can be rewritten in the following forms:

$$\frac{\partial}{\partial t}(\rho K) + \nabla \cdot [(p + \rho K)\vec{u} - \mu \vec{u} \Pi] = -\rho \vec{u} \cdot \nabla \phi + p \nabla \cdot \vec{u} - \rho \varepsilon \tag{3.85}$$

$$\frac{\partial(\rho e)}{\partial t} + \nabla \cdot \left[\rho e\vec{u} + \vec{q} + \frac{\partial h}{\partial s}\vec{J}_s\right] = -p\nabla \cdot \vec{u} + \rho\varepsilon \quad (3.86)$$

$$\frac{\partial(\rho\phi)}{\partial t} + \nabla \cdot (\rho\phi\vec{u}) = \rho\vec{u} \cdot \nabla\phi + \rho\frac{\partial\phi_T}{\partial t} \quad (3.87)$$

where $\vec{u}\Pi$ is a product of the velocity vector and the stress tensor $\Pi = \pi_{ij}$, so that $(\vec{u}\Pi)_i = u_j\pi_{ij}$; $\vec{q}$ is the heat flux across the boundary of the parcel, including both the radiation heat flux from the incoming solar radiation in the upper ocean and the heat flux due to eddies and turbulence; $\vec{J}_s$ is the diffusive salt flux relative to the center of mass, as discussed in the previous section. (This diffusive salt flux may be slightly different from the traditional diffusive salt flux defined in numerical models based on fixed Eulerian coordinates.) We separate the gravitational potential into two parts:

$$\phi = \phi_0(z) + \phi_T(x, y, z, t) \quad (3.88)$$

where the first part is the gravitational potential independent of time; the second part is associated with tidal forcing which is defined as a function of time and three-dimensional coordinates, so it is slightly different from the barotropic tidal potential discussed in the previous section.

Using the generalized Reynolds transport theorem (Eqn. (3.76)), these equations can be rewritten as follows:

$$\frac{\delta}{\delta t}\iiint_V \rho K dv = \oiint_S \rho K(\vec{u}_s - \vec{u}) \cdot \vec{n} ds + \oiint_S (-p\vec{u} + \mu\vec{u}\Pi) \cdot \vec{n} ds - P + C - D \quad (3.89)$$

$$\frac{\delta}{\delta t}\iiint_V \rho e dv = \oiint_S \rho e(\vec{u}_s - \vec{u}) \cdot \vec{n} ds - \oiint_S \left(\vec{q} + \frac{\partial h}{\partial S}\vec{J}_s\right) \cdot \vec{n} ds - C + D \quad (3.90)$$

$$\frac{\delta}{\delta t}\iiint_V \rho\phi dv = \oiint_S \rho\phi(\vec{u}_s - \vec{u}) \cdot \vec{n} ds + \iiint_V \rho\frac{\partial\phi_T}{\partial t} dv + P \quad (3.91)$$

where

$$P = \iiint_V \rho\vec{u} \cdot \nabla\phi dv \quad (3.92)$$

is the interchange rate between GPE and KE,

$$C = \iiint_V p\nabla \cdot \vec{u} dv \quad (3.93)$$

is the interchange rate between internal energy and KE, and

$$D = \iiint_V \rho\varepsilon dv \quad (3.94)$$

is the dissipation rate, i.e., the internal energy converted from KE. Owing to evaporation and precipitation, fluid velocity at the boundary and the boundary velocity can be slightly different:

$$(\vec{u}_s - \vec{u}) \cdot \vec{n} = precipitation - evaporation \tag{3.95}$$

Thus, the hydrological cycle can contribute to the energetics of the world's oceans, through the first terms on the right-hand side of Eqns. (3.89), (3.90), and (3.91).

Adding up these equations, we find that all the conversion terms, P, C, and D, cancel out exactly. This leads to the conclusion that the energy contribution of tides comes from the $\iiint_V \rho \frac{\partial \phi_T}{\partial t} dv$ term in the GPE balance equation (3.91). Neglecting density variations of seawater, one might argue that the volume integration of the time derivative of the tidal potential is zero, and thus come to the conclusion that the energy contribution of tides is very small and negligible.

However, a closer examination suggests that there may be a misconception in this argument. Let us separate the density into two components, $\rho = \rho_0 + \rho'$, where ρ_0 is the mean reference density. The tidal contribution term can be rewritten as

$$\iiint_{V(t)} \rho \partial_t \phi_T dv = \rho_0 \iiint_{V(t)} \partial_t \phi_T dv + \iiint_{V(t)} \rho' \partial_t \phi_T dv \tag{3.96}$$

Note the fact that the tidal potential is a function primarily of the horizontal position, with a small percentage change in the vertical direction, so the volume integration of the time derivative of the tidal potential is equal to

$$\rho_0 \iiint_{V(t)} \partial_t \phi_T dv \simeq \rho_0 \iint_A h \partial_t \phi_T dxdy + \rho_0 \iint_A \zeta \partial_t \phi_T dxdy \tag{3.97}$$

Integrating Eqn. (3.97) over a tidal cycle, the first term associated with the time-invariant depth of the ocean (h) vanishes, so that only the free surface elevation term remains. After integration by parts, the corresponding tidal contribution averaged over a tidal cycle is $-\rho_0 \overline{\partial_t \zeta \phi_T}$. As discussed above, to maintain the tidal circulation, the tidal potential and the vertical velocity should be positively correlated, so this term is negative. As a result, the tidal contribution in this formulation turns out to be negative.

This formulation is thus incapable of providing a clear explanation for the role of tidal energy in the global circulation. This failure is not a surprise: and a closer examination reveals the problem associated with this formulation. In Eqn. (3.91), tidal potential is included as a part of the total energy of the oceanic circulation system. It is well known that the total tidal energy declines due to tidal dissipation; thus the negative sink term on the right-hand side of Eqn. (3.91) reflects the decline of tidal potential over longer time scales. The exact formulation of tidal potential energy involves complicated dynamics of the gravitational field and orbital motions of the solar system: this is beyond the scope of our discussion.

It is interesting to note that the first term on the right-hand side of Eqn. (3.83) does not explicitly appear on the right-hand side of the energy balance equations in this formulation. To dig for this term, we reformulate the problem as follows. First, we notice that, in the previous formulation, the GPE includes contributions associated with the tidal potential and non-tidal motions. It is well known that tidal dissipation is associated with loss of GPE in the Earth–Moon–Sun system. In the following analysis we will divide both GPE and KE into two parts.

The time-invariant part of GPE ϕ_0 and the tidal part of GPE ϕ_T satisfy the following equation

$$\frac{\partial(\rho\phi_0)}{\partial t} + \nabla\cdot(\rho\phi_0\vec{u}) = \rho\vec{u}\cdot\nabla\phi_0 \quad (3.98)$$

$$\frac{\partial(\rho\phi_T)}{\partial t} + \nabla\cdot(\rho\phi_T\vec{u}) = \rho\vec{u}\cdot\nabla\phi_T + \rho\phi_{T,t} \quad (3.98)$$

Changes in the total GPE can be rewritten as

$$\frac{\delta}{\delta t}\iiint_V \rho\phi dv = \frac{\delta}{\delta t}\iiint_V \rho\phi_0 dv + \frac{\delta}{\delta t}\iiint_V \rho\phi_T(t)\,dv = \frac{\delta}{\delta t}\Phi_0 + \frac{\delta}{\delta t}\Phi_T \quad (3.99)$$

Note that the second term is associated with tidal potential Φ_T, which should decline with time. If the tidal potential term is moved from the right-hand side of Eqn. (3.99) to its left-hand side, we can see that for a balance of GPE of the non-tidal motions, tidal dissipation should appear as a source of energy for the non-tidal motions.

Similarly, if we assume that the tidal velocity and non-tidal velocity are uncorrelated, then we have

$$\frac{\delta}{\delta t}\iiint_V \rho K dv = \frac{\delta}{\delta t}\iiint_V \rho K_0 dv + \frac{\delta}{\delta t}\iiint_V \rho K_{tides} dv \quad (3.100)$$

It is worth noting that the postulation of a complete uncorrelation between the tidal and non-tidal motions is only conceptual, but not strictly rigorous. In fact, one of the most important issues in energetic theory of the oceanic general circulation is that tidal dissipation can affect the non-tidal motions through mixing sustained by tidal dissipation.

Nevertheless, the separation of tidal and non-tidal motions can help us to understand the contribution of tidal dissipation to the non-tidal motions; thus, we adopt this postulation in the following discussion.

The corresponding energy equations for the non-tidal motions are

$$\frac{\delta}{\delta t}\iiint_V \rho K_0 dv = \oiint_S \rho K(\mathbf{u}_s - \vec{u})\cdot\vec{n}ds + \oiint_s(-p\mathbf{u} + \mu\vec{u}\Pi)\cdot\vec{n}ds - P_0 + C - D$$
$$- \frac{\delta}{\delta t}\iiint_V \rho K_{tides} dv - \iiint_V \rho\vec{u}\cdot\nabla\phi_T dv \quad (3.101)$$

$$\frac{\delta}{\delta t}\iiint_V \rho e dv = \oiint_S \rho e(\vec{u}_s - \vec{u})\cdot\vec{n}ds - \oiint_S\left(\vec{q} + \frac{\partial h}{\partial S}\vec{J}_s\right)\cdot\vec{n}ds - C + D \quad (3.102)$$

$$\frac{\delta}{\delta t}\iiint_V \rho\phi_0 dv = \oiint_S \rho\varphi_0(\vec{u}_s - \vec{u})\cdot\vec{n}ds + P_0 \quad (3.103)$$

where

$$P_0 = \iiint_V \rho\vec{u} \cdot \nabla\phi_0 dv = g \iiint_V \rho w dv \tag{3.104}$$

If Eqns. (3.101), (3.102), and (3.103) are added together, all exchange terms, such as P_0, C, and D, cancel exactly. Therefore, the tidal forcing terms for the non-tidal motions are the last two terms in Eqn. (3.101), and the last term can be rewritten as

$$-\iiint_V \rho\vec{u} \cdot \nabla\phi_T dv = -\iiint_V \rho\vec{u}_{non-tidal} \cdot \nabla\phi_T dv - \iiint_V \rho\vec{u}_{tidal} \cdot \nabla\phi_T dv \tag{3.105}$$

The sign of the first term on the right-hand side of Eqn. (3.105) is unclear; however, tidal flows are driven by tidal force, so that the last term in Eqn. (3.105) is positive. Thus, the last two terms in Eqn. (3.101) represent a positive contribution to the non-tidal motions, and they correspond to similar terms associated with the barotropic tides discussed in Section 3.5.3.

3.5.5 Interpretation of energy integral equations

Wind energy sources for the kinetic energy

In addition to the contribution from tides as discussed above, the KE of oceanic motions comes primarily from surface forcing. Averaged over a long period, the surface integral in Eqn. (3.89) is reduced to

$$\overline{\iint (-p\vec{u} + \mu\vec{u}\Pi) \cdot \vec{n}ds} \approx \overline{\iint [-pw + (u\tau^x + v\tau^y)]_{z=-H}^{z=\zeta} dxdy} \tag{3.106}$$

Thus, the surface integral consists of a kinetic energy source from the upper surface and a kinetic energy sink from the seafloor.

The kinetic energy source on the upper surface is further separated into contributions from two categories: large spatial scale and low frequency versus small spatial scale and high frequency.

The energy contribution due to large-scale and low-frequency atmospheric pressure perturbations is the work done associated with the so-called atmospheric loading. The global sum of this term is small, estimated as 0.04 TW by Wang *et al.* (2006). Wind energy input to the surface currents can be classified as: (1) wind stress work on the surface geostrophic currents, estimated as 0.88 TW (Wunsch, 1998); and (2) wind stress work on the surface ageostrophic currents (the Ekman layer), estimated as 3 TW (Wang and Huang, 2004a).

In the meantime the contribution due to small-scale and high-frequency atmospheric pressure perturbations and wind stress is the energy input into the surface waves, estimated as 60 TW (Wang and Huang, 2004b).

The kinetic energy sink from the seafloor boundary represents energy dissipation due to bottom friction and form drag, whose value remains uncertain.

Sources of internal energy

In the internal energy balance equation, the first term on the right-hand side of Eqns. (3.90) or (3.102) constitutes only a minor source of internal energy due to evaporation and precipitation. The major source/sink of internal energy comes from the exchange with the atmosphere above and geothermal heat from the seafloor below, as expressed in the surface integral $-\oint\!\!\oint_S \left(\vec{q} + \frac{\partial h}{\partial S}\vec{J}_s\right) \cdot \vec{n} ds$, where the fluxes can be parameterized as

$$\vec{q} + \frac{\partial h}{\partial S}\vec{J}_s = \vec{F}_{rad} - \rho c_p \kappa_T \nabla T - \frac{\partial h}{\partial S}\rho \kappa_S \nabla S \tag{3.107}$$

where the first term on the right-hand side is the solar radiation; the second term is the turbulent flux, which is normally parameterized as a downward temperature gradient flux. Similarly, one can parameterize the diffusive salt flux as a downward salinity gradient. Strictly speaking, the diffusive salt flux used in the formulae should be defined in the coordinates moving with the center of mass, and this can be in a form slightly different from the traditional definition of salt flux used in numerical models. In addition, a salinity (temperature) gradient can induce temperature (salinity) diffusion. However, the diffusion parameterization in currently used numerical models is too crude to take into account such physics.

Although salt diffusion may contribute to internal energy for each individual grid cell in the ocean interior, there is no salt flux across the air–sea interface. Thus, the corresponding flux term is zero at the sea surface, assuming that freshwater flux across the air–sea interface is balanced and has the same temperature. In this discussion, the salt flux across the water–sea ice interface is excluded.

Of note is the fact that the sum of the internal energy fluxes through the system is not self-balanced, even though the system reaches a quasi-steady state. If all these equations are added up, the exchange terms in Eqns. (3.101, 3.102, 3.103) cancel out exactly. Thus, for a quasi-steady state, the energy balance is to be set up among the inputs, including internal energy from the sea surface and seafloor, plus wind and tidal energy inputs. As required by the first law of thermodynamics, energy should be conserved, i.e., the total amount of energy input should be equal to the amount of energy output. The only way for the ocean to lose energy is in the form of heat loss to the atmosphere; therefore the total heat loss to the atmosphere is the sum of the following terms: the heat absorbed by the ocean, the heat flux from geothermal heating (32 TW), and the net mechanical energy received by the oceans, including the mechanical energy source due to wind (64 TW) and tidal dissipation (3.5 TW). The net heat loss to the atmosphere, i.e., the difference between the outgoing heat flux and the incoming heat flux, is estimated as 99.5 TW.

The balance of GPE

The GPE plays an important role in the balance of mechanical energy in the ocean. From the balance of GPE (see Eqn. (3.103)), the only direct source term is that due to mass exchange through the air–sea interface, namely that from precipitation and evaporation;

however, this term turns out to be extremely small and negligible, as will be shown later. The contribution from wind stress manifests itself in the conversion term P_0, which represents the transfer between the potential and kinetic energy. There are two pathways for the ocean to receive GPE.

1. *Surface wind stress.* As explained in detail in the first subsection of Section 3.6.1, wind stress energy input to the surface geostrophic current is converted into GPE in the world's oceans through Ekman pumping. This is a major source of GPE in the ocean. However, strong baroclinic instability associated with strong zonally oriented currents, such as the ACC, is a major sink of GPE.
2. *Surface thermohaline forcing* has to undertake a zigzag route to be converted into GPE. First it is converted into internal energy of water parcels, then it is converted into kinetic energy through expansion/contraction, and finally it can be converted to the gravitational potential energy through the P_0 term.

For a steady state, the total vertical mass transport averaged over a period of tides should be zero at any fixed level, so that $\bar{P}_0$ vanishes, i.e., there is no net exchange between KE and GPE. However, for an ocean with a seasonal cycle, the GPE balance also goes through a similar cycle, in which the late winter cooling-induced convection appears as a major sink of GPE and the diapycnal mixing and other dynamical processes constitute the sources of GPE. As an example, a detailed analysis of GPE balance for the case with a regular seasonal cycle is discussed in Section 3.7.3.

If we neglect any contribution due to evaporation and precipitation, integration of Eqn. (3.103) leads to a statement of $0 = 0$. Thus, one seems to be led to the conclusion that the balance of GPE is a trivial problem, and the study of GPE for the large-scale circulation is neither interesting nor useful.

However, such a seemingly trivial relation may be one of the most important equations for the study of oceanic energetics. The term $\bar{P}_0$ reflects synthetically quite a few physical processes. Averaging over a period of tides, this term can be rewritten as

$$g\overline{\rho w} = g\bar{\rho}\bar{w} + g\overline{\rho' w'} \tag{3.108}$$

This equation can be interpreted in at least two possible ways, as follows.

First, the overbars are interpreted as the ensemble mean of the product of density and vertical velocity at each point, and the primes indicate the deviation from the ensemble mean due to turbulence and internal waves. Then the first term on the right-hand side of Eqn. (3.108) is the energy transfer between the GPE and KE of the mean state, while the second term is due to the turbulence and internal waves. When the second term is examined in further detail, we find two cases to be differentiated. Within the deep ocean interior with stable stratification, vertical mixing pushes heavy water upward (downward), so that ρ' and w' are both positive (negative); therefore this term is positive. This means that mixing due to turbulence and internal waves in the stably stratified ocean increases the GPE of the mean state. On the other hand, within the convective overturning zone, instantaneous stratification is unstable before the convective overturning takes place. During convective adjustment, dense and heavy water parcels ($\rho' > 0$) move downward ($w' < 0$); thus, ρ' and w' are of opposite signs, and convective mixing reduces the GPE of the mean state.

Second, the overbars are interpreted as the time and horizontal mean, and the primes stand for the deviation from such a mean. As shown in Section 3.7.3, the perturbation term can be interpreted as the meridional transport of GPE associated with advection.

In the above argument, we exclude the time-varying component of gravity. In fact, in a slightly different formulation, we can include the time-dependent gravity, and the relation corresponding to Eqn. (3.108) is

$$\overline{g\rho w} = g_0\overline{\rho w} + g_0\overline{\rho' w'} + \overline{g_T \rho w} \tag{3.109}$$

The last term is an additional term due to tidal force, and here it appears as a net source of energy conversion.

3.5.6 An energy diagram for the world's oceans

World ocean circulation encompasses an immense range of spatial and time scales. Many of the basic elements of the energetics of the circulation remain unclear owing to the severe technical challenges of collecting data over such a huge dimension and under the extreme conditions in the open ocean.

Energy in the ocean can be roughly classified into four categories: GPE, KE, internal energy, and chemical potential (Fig. 3.8). In the study of the oceanic general circulation we further separate the GPE and KE into two parts: the energy contained in the mean state and that in meso-scale eddies, turbulence, and internal waves.

The oceans receive a huge amount of thermal energy through solar insolation, and exchange heat with the atmosphere in the forms of short-wave radiation, latent heat flux, sensible heat flux, and long-wave radiation. The total amount of internal energy in the ocean is immense, estimated as 20 YJ (2×10^{25} J). Internal energy can be converted into KE through expansion/compression, $p\nabla \cdot \vec{v}$. However, the ocean's capability of converting internal energy into KE is very limited, as discussed in previous sections; thus, we will not discuss the balance of internal energy in this book. Readers who are interested in this subject can find much useful information in the book *Physics of Climate* by Peixoto and Oort (1992).

We include chemical potential as a form of energy, in addition to mechanical energy and internal energy. However, chemical potential in the ocean is directly linked to evaporation induced by the solar radiation. The ocean has chemical potential because seawater is a multi-component mixture. It is the existing concentration difference of salt in seawater that gives rise to the chemical potential. The total amount of chemical potential in the oceans is estimated as 3.6×10^{24} J. However, we should be aware of the fact that only the difference in chemical potential between different water parcels is dynamically active. The hydrological cycle in the form of evaporation and precipitation gives rise to salinity difference in the oceans. In fact, evaporation (precipitation) extracts (inputs) pure water from

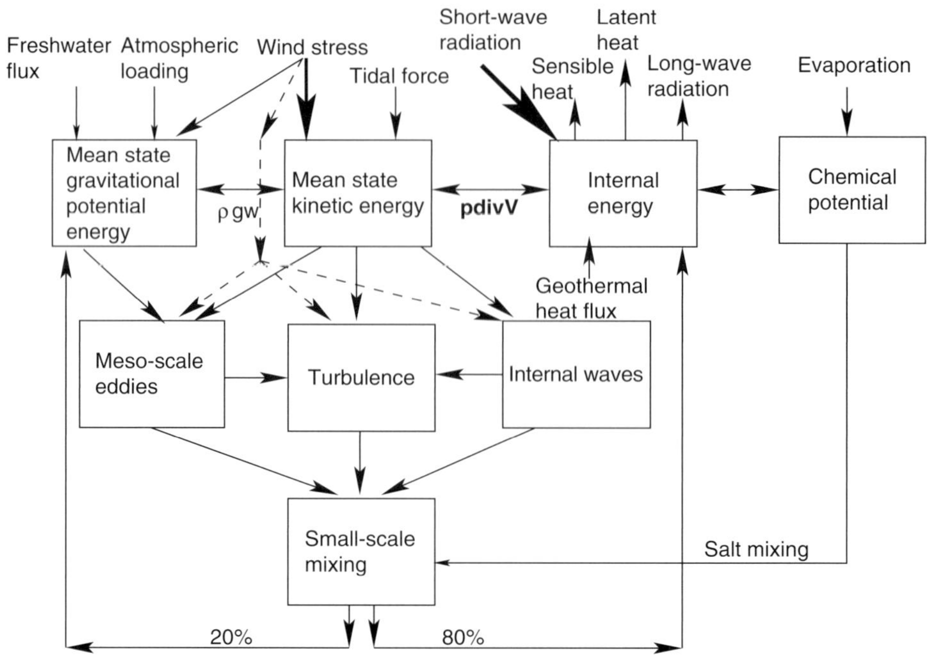

Fig. 3.8 Energy diagram for the world's oceans.

(to) the seawater, acting as the source of chemical potential energy. A difference in salinity induces a difference in chemical potential, which is the driving force of salt mixing at the molecular level. As a matter of fact, however, salt mixing in the oceans is primarily regulated by turbulence and internal waves, which are much stronger than the molecular mixing.

The most important forms of energy of the oceanic circulation are kinetic and potential energy. The sources of KE of the mean state include tidal force and wind stress. However, wind stress also makes direct contributions to meso-scale eddies, turbulence, and internal waves, mostly through surface waves and other dynamical processes in the upper ocean.

We note that oceanic circulation takes place in the environment of a gravity field, and gravitational force is one of the major forces regulating the stratification and circulation. The total amount of GPE is immense; however, only a very small proportion of this energy is dynamically active. We will delay this discussion until the end of this chapter. GPE and KE can be converted into each other through vertical motions, which can be expressed as $\rho g w$.

A major part of the mechanical energy in the oceans is contained in meso-scale eddies, turbulence, and internal waves. In fact, it is estimated that the total energy associated with meso-scale eddies is two orders of magnitude larger than that associated with the mean currents. Unfortunately, there is no reliable estimate of the energy stored in oceanic motions in these forms.

At the smallest spatial and temporal scale, mixing plays a vital role in maintaining the stratification due to temperature and salinity. One of the major characteristics of stratified fluid is the following: vertical mixing in a stratified ocean pushes light water downward and heavy water upward, resulting in an increase in GPE of the mean state. Therefore, turbulent mixing in the oceans (due to internal wave breaking and small-scale turbulence) does not turn all mechanical energy into the so-called dissipation heat; instead, a small fraction of the turbulent KE, estimated as approximately 20%, is fed back to the mean state on a global scale. It is imaginable that the efficiency of mixing varies greatly, depending upon the environmental conditions; however, its dynamical specification remains at the frontiers of research. This energy transform is indicated by the arrows at the bottom of Figure 3.8.

3.6 Mechanical energy balance in the ocean

3.6.1 Mechanical energy sources/sinks in the world's oceans

It has been shown in the previous section that mechanical energy sources and sinks play vital roles in maintaining/regulating the oceanic circulation; in this section we focus on the mechanical energy balance in the world's oceans. A sketch of mechanical energy distribution in the oceans (Fig. 3.9) shows that wind stress and tidal dissipation are the most important sources of mechanical energy driving the oceanic general circulation.

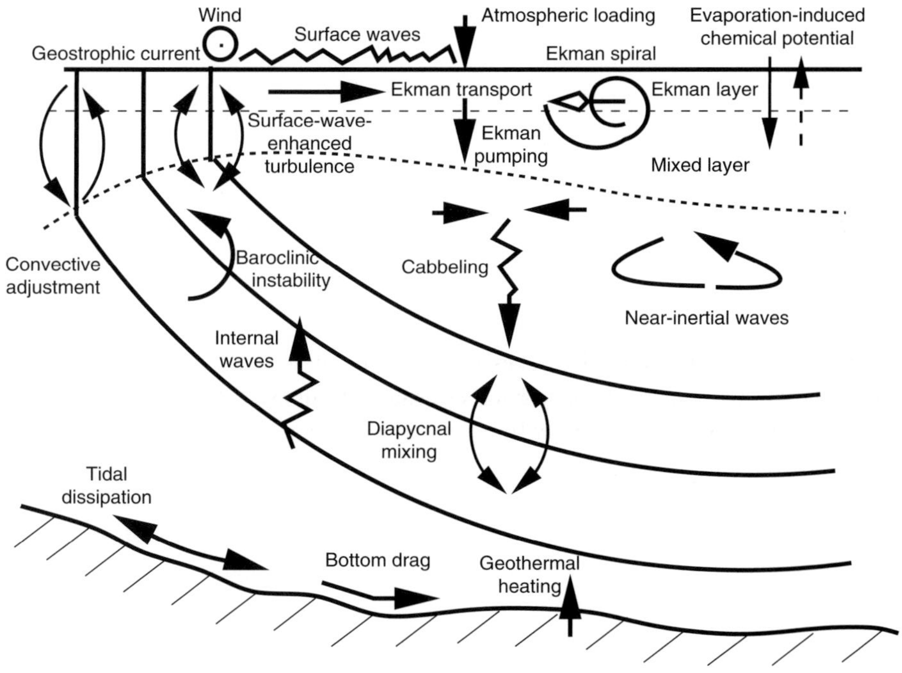

Fig. 3.9 Mechanical energy diagram for the ocean circulation.

Wind energy input

Wind stress on the surface of the ocean drives both surface currents and waves. The mechanical energy transferred from wind into the ocean, W_{wind}, is defined as $W_{wind} = \overline{(\sigma_{ij} \cdot \vec{u}) \cdot \vec{n}}$, where the stress tensor includes both the viscosity stress tensor and the pressure: $\sigma_{ij} = \pi_{ij} + \delta_{ij}p$. Thus, energy input from wind includes contributions from wind stress and the pressure. Although wind stress is more frequently discussed in relation to large-scale circulation, sea-level atmospheric pressure and its perturbations also affect the large-scale circulation.

It is difficult to separate wind stress from sea-level atmospheric pressure because they are closely related to each other; however, wind stress energy input can be roughly divided into the following components

$$W_{wind} = \overline{(\sigma_{ij} \cdot \vec{u}) \cdot \vec{n}} = \vec{\tau} \cdot (\vec{u}_{0,g} + \vec{u}_{0,ag}) + \overline{\vec{\tau}' \cdot \vec{u}_0'} + \overline{p' w_0'} + \overline{p w_0} \tag{3.110}$$

where the overbar indicates average in space and time; the perturbations are defined in terms of the spatial and temporal scales of the surface waves; $\vec{\tau}$, $\vec{u}_{0,g}$, $\vec{u}_{0,ag}$ are the spatially and temporally averaged tangential stress, surface geostrophic and ageostrophic velocity, respectively; $\vec{\tau}'$ and $\vec{u}_0'$ are the perturbations; p' and w_0' are perturbations of the surface pressure and velocity component normal to the surface; and $\bar{p}$ and $\bar{w}_0$ are the sea surface pressure and vertical velocity averaged over time scales much longer than the typical wave periods.

The first and second terms on the right-hand side of Eqn. (3.110) are the wind stress work on the quasi-steady currents on the surface, and quasi-steadiness is defined in terms of the time scale of the typical surface waves. The third term is work done by wind stress on surface waves, and the last two terms are the work done by atmospheric pressure. Energy contributions from these terms are discussed separately as follows.

Work input to surface geostrophic currents

Wind stress energy input through the surface geostrophic current is

$$W_{wind,geo\,curr} = \vec{\tau} \cdot \vec{u}_{0,g} \tag{3.111}$$

where surface geostrophic currents can be calculated from $\vec{u}_{0,g} = g\vec{k} \times \nabla\eta/f$ (where η is the sea surface height inferred from either satellite altimeter data or numerical models), except near the equator. The total amount of energy input is estimated as 0.88 TW (Wunsch, 1998). Although the wind energy input is positive around 40° N, the major energy input is through the Southern Ocean and the equatorial band (Fig. 3.10). In addition, the North Equatorial Counter Current (NECC) is a place of energy sink because the eastward current is against the easterlies there.

By means of numerical modeling, energy input to surface currents can be calculated, and this energy input for the period 1993–2003 is estimated as 1.16 TW (Huang *et al.*, 2006). Away from a narrow band (within ±3° of the equator), the surface currents can be separated

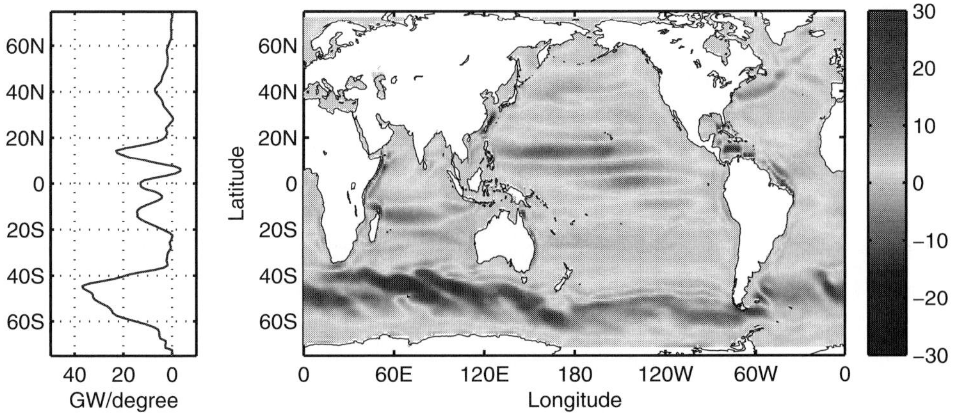

Fig. 3.10 Distribution of wind energy input through surface currents (*right panel*), in mW/m^2, and its latitudinal distribution (*left panel*) (Huang *et al.*, 2006). See color plate section.

fairly well into the geostrophic and ageostrophic components. The energy input into the geostrophic current calculated from a numerical model for this period is 0.87 TW; almost the same as that deduced from satellite data.

Energy input through the surface geostrophic current can be fed directly into the large-scale current, and can thereby be efficiently turned into GPE through the vertical velocity conversion term.

Work input to surface ageostrophic currents

Surface ageostrophic currents can be described in terms of the Ekman theory. The horizontal momentum equations, including the time-dependent term and the geostrophic flow, are

$$u_t - fv = -p_{s,x}/\rho_0 + (Au_z)_z, \quad v_t + fu = -p_{s,y}/\rho_0 + (Av_z)_z \tag{3.112}$$

where $u = u_g + u_e, \quad v = v_g + v_e$ are the sum of geostrophic velocity and ageostrophic velocity in the Ekman layer, ρ_0 is the reference density, and $\rho_s = \rho_s(x, y)$ is the surface pressure associated with large-scale circulation. The geostrophic velocity satisfies $\vec{u}_g = \vec{k} \times \nabla p_s / f\rho_0$. The corresponding boundary conditions are

$$Au_{e,z}|_{z=0} = \tau^x/\rho_0, \quad Av_{e,z}|_{z=0} = \tau^y/\rho_0, \quad (u_e, v_e) \to 0, \text{ at } z \to -\infty \tag{3.113}$$

where the lower limit is to be understood as the base of the Ekman layer, and within the Ekman layer the vertical shear of the geostrophic velocity is negligible.

Multiplying these two equations by u and v, respectively, and integrating the result over the depth of the Ekman layer, we obtain

$$E_t = S - P - D \tag{3.114}$$

where

$$E = \rho_0 \int_{-\infty}^{0} 0.5(u^2 + v^2)dz \tag{3.115}$$

$$S = S_g + S_e, \quad S_g = \tau^x u_g + \tau^y v_g, \quad S_e = \tau^x u_e(0) + \tau^y v_e(0) \tag{3.116}$$

$$P = \vec{U}_e \cdot \nabla p_s/\rho_0, \; D = \rho_0 \int_{-\infty}^{0} A(u_{e,z}^2 + v_{e,z}^2)dz \tag{3.117}$$

where E is the total KE of the Ekman layer, S the rate of wind energy input, P the rate of pressure work by the current integrated over the Ekman layer, and D the rate of dissipation integrated over the Ekman layer. Note that $\vec{U}_e = \rho_0 \int_{-\infty}^{0} \overrightarrow{u_e} dz = -\vec{k} \times \vec{\tau}/f$, so

$$P = \vec{U}_e \cdot \nabla p_s/\rho_0 = \vec{\tau} \cdot \vec{u}_g = S_g \tag{3.118}$$

i.e., pressure work done by the Ekman transport is exactly the same as the wind stress work on the surface geostrophic currents. The pressure work done by the Ekman transport is also related to the GPE generated by Ekman pumping:

$$\iint_A PdA = \iint_A \vec{U}_e \cdot \nabla p_s/\rho_0 dA = -\iint_A w_e p_s/\rho_0 dA + \oint \vec{U}_e \cdot \vec{n}dl \tag{3.119}$$

where A is the area of the ocean. Thus, Eqn. (3.114) can be further reduced to

$$E_t = S_e - D \tag{3.120}$$

Therefore, we conclude that wind stress energy input to the surface geostrophic currents is equal to the increase in GPE in the world's oceans via Ekman pumping, including coastal upwelling/downwelling. On the other hand, wind stress energy input to the surface ageostrophic current is used to maintain the Ekman spiral via the vertical turbulent dissipation in the Ekman layer.

The exact amount of wind stress energy input through the Ekman spiral is estimated as 2.4 TW (Wang and Huang, 2004a) for frequencies greater than 1/(2 days). The spatial pattern of this energy source is quite similar to that of the energy input to surface waves, detailed in the next subsection. In addition, there exists a large amount of energy input through the near-inertial waves, which is due to the resonance at a frequency of $\omega = -f$. Alford (2003) used a slab model and obtained an estimate of 0.47 TW. However, Plueddemann and Farrar (2006) argued that a slab model may give rise to a near-inertial energy input that is twice as much as that obtained from a more accurate mixed-layer model. Thus, a more realistic value might be 0.23 TW.

The above estimate is based on the smoothed wind stress data, excluding the contribution from strong nonlinear events such as hurricanes and typhoons. Using tropical cyclone data from 1984 to 2003, Liu *et al.* (2008) estimated that there is an additional contribution due to hurricanes/typhoons of 0.1 TW to the surface currents, including 0.03 TW to the near-inertial waves.

The convergence of the Ekman flux gives rise to a pumping velocity at the base of the Ekman layer w_e, which is responsible for pushing the warm water into the subsurface ocean and thus forming the main thermocline in subtropical oceans. Ekman pumping sets up wind-driven circulations and an associated bow-shaped main thermocline in the world's oceans. During this process GPE is increased, which is a very efficient way indeed of converting KE into GPE.

Work input to surface waves

Another major channel of energy input from wind and sea-level atmospheric pressure into the ocean is through surface waves. Wind stress drives surface waves in the oceans, and this energy input can be treated by means of the form drag of the atmospheric boundary layer. Wind energy input to surface waves can be estimated as

$$W_{waves} = \tau\bar{c} = \rho_a u_{*a}^2 \bar{c} \approx \rho_w u_{*w}^2 \bar{c} \tag{3.121}$$

where $\bar{c}$ is the effective phase speed, $u_{*a} = \sqrt{\tau/\rho_a}$ and $u_{*w} = \sqrt{\tau/\rho_w}$, ρ_a and ρ_w are the density of air and water, and τ is the stress at the air–sea interface. The corresponding value of $\bar{c}$ can be determined from field experiments. Observation data fit the formula $W_{waves} = 3.5\rho_a u_{*a}^3$. Since u_{*a} depends on both wind and sea state, the effects of waves on W_{waves} is implied here. To have the best fitting for the observational data, an empirical formula is proposed:

$$W_{waves} = A\rho_a u_{*a}^3 \tag{3.122}$$

where A is the empirical coefficient representing the energy flux factor, $\bar{c}/u_{*a}$:

$$A = \begin{cases} 0.5c_{p*}, & \text{as } c_{p*} \le 11 \\ 12c_{p*}^{-1/3}, & \text{as } c_{p*} > 11 \end{cases} \tag{3.123}$$

where $c_{p*} = c_p/u_{*a}$ is the wave age, and c_p is the phase velocity at the wind sea peak frequency.

Using the smoothed NCEP (National Centers for Environmental Prediction) wind stress data, the total contribution to the global oceans is estimated as 60 TW (Wang and Huang, 2004b) (Fig. 3.11). How this huge amount of energy is distributed and dissipated in the oceans remains unclear at this time. It is roughly estimated that 36 TW of this energy is dissipated locally through wave breaking, turbulence, and internal wave generation in the upper ocean; about 20 TW is propagated to remote areas in the form of long waves, which gradually dissipate their energy. Eventually, these long waves break and dissipate the rest of their energy in the shallow/marginal seas and along the beaches of the world's oceans.

The strong nonlinear events in the atmosphere are smoothed out in the low-resolution wind stress dataset. The total contribution of subtropical cyclones, hurricanes, and typhoons to surface waves is estimated as 1.6 TW (Liu *et al.*, 2008). Since the total amount of 60 TW mentioned above is the result of rounding up from a number slightly lower than 60 TW, we

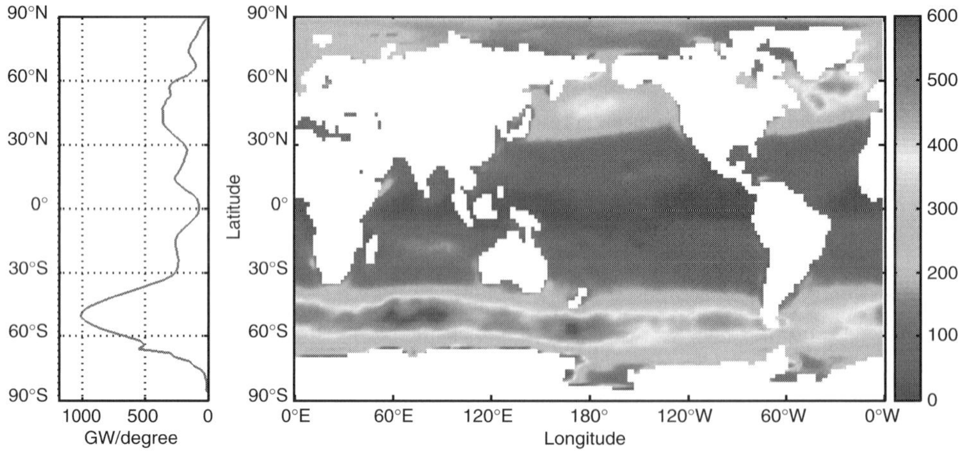

Fig. 3.11 Distribution of wind stress work on surface waves (*right panel*) (mW/m^2), and its latitudinal distribution (*left panel*) (Wang and Huang, 2004b). See color plate section.

tentatively use 60 TW for the total amount of energy in the surface waves, including the contribution due to hurricanes/typhoons.

Work by atmospheric loading

When sea-level atmospheric pressure varies with time, the sea surface moves vertically in response. Therefore, changes in sea-level atmospheric pressure can transport mechanical energy into the oceans. The relevant process is called atmospheric loading. A common practice in satellite altimetry data processing is to assume that the ocean responds to changes in sea level instantaneously, and the corresponding changes in sea surface height are removed from the original satellite data through the so-called inverse barometer effect (Wunsch and Stammer, 1997).

In reality, the response of sea level is not instantaneous, and changes in sea-level atmospheric pressure induce vertical motions of sea surface; thus, there is mechanical energy transported from the atmosphere to the oceans. The total amount of energy due to this source for the world's ocean circulation remains unclear. Based on data concerning sea surface height from satellites and data regarding atmospheric pressure, a preliminary estimate for the work done by atmospheric loading for the world's oceans is about 0.04 TW (Wang *et al.*, 2006).

The implications for numerical models

The above discussion on the wind energy input based on the currently available wind stress datasets is of rather low spatial resolution. The effects of eddies and turbulence, with scales on the order of 1° or less are smoothed out in such datasets. In particular, contributions from strong nonlinear events, such as tropical cyclones, hurricanes, and typhoons, are excluded.

It is reasonable to expect that, when using wind stress data with better spatial and temporal resolutions, the estimation of the wind stress energy input may vary substantially.

It is more important to bear in mind that sea surface wind stress is the manifestation of the atmospheric motions, which have broad spatial and temporal scales, on the sea surface. It is obvious that wind energy input to the ocean, as expressed in forms of Eqn. (3.110), can hardly be described simply by specifying a wind stress on the upper surface of the ocean. In particular, the large amount of mechanical energy input to the turbulence motions in the upper ocean needs to be parameterized in forms different from the low-resolution wind stress data used in current numerical models.

Tidal dissipation

The primary source of mechanical energy supporting deep mixing very likely comes from tidal dissipation in the deep ocean. Owing to tidal dissipation, the parameters related to the orbits of the Earth and Moon change continuously. In fact, the distance between the Earth and the Moon has gradually increased over the long geological past. Tidal dissipation can be inferred from changes in the Moon's orbit; such changes can be measured very accurately through laser tracking. The total amount of tidal dissipation in the world's oceans is 3.5 TW (Fig. 3.12). However, the spatial distribution of tidal energy dissipation is still not accurately known.

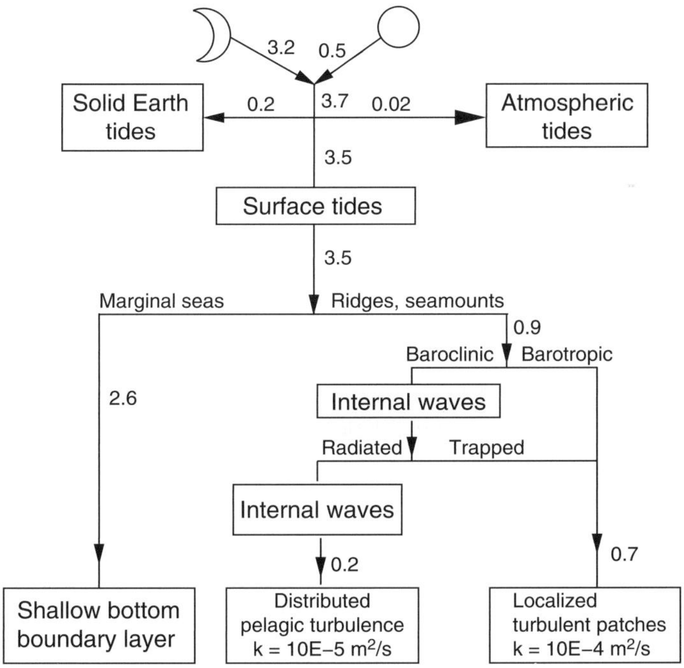

Fig. 3.12 Tidal dissipation diagram (TW) (modified from Munk and Wunsch, 1998).

Based on barotropic tidal modeling with satellite altimeter data assimilation, the energy dissipation estimation is as follows: 2.6 TW is dissipated within the shallow seas of the world's oceans, with the remaining 0.9 TW dissipation being distributed in the deep ocean (Figs. 3.9 and 3.12) (Munk and Wunsch, 1998). Tidal dissipation in the deep ocean most probably takes place over rough topography, where barotropic tidal energy is converted into energy for internal tides and internal waves that sustain the strong bottom-intensified diapycnal mixing on the order of 10^{-3} m^2/s. The conversion of KE to GPE through internal waves and turbulence is of rather low efficiency, typically around 20% (Osborn, 1980), which is sometimes referred to as the mixing coefficient; thus, the amount of GPE generated by tidal dissipation in the open ocean is around 0.18 TW.

In the above discussion we mainly concerned ourselves with mixing in the deep ocean, so that the potential role of tidal dissipation in the shallow seas or marginal seas, which accounts for about 2.6 TW, was excluded. Actually, tidal dissipation in marginal seas, such as the South China Sea or the Sea of Japan, can certainly play a very important role in regulating the local circulation and water mass modification. Furthermore, these marginal seas may exchange water masses with the open ocean and thus contribute to the water mass formation and balance of the Pacific Ocean. However, the connection between marginal seas and open oceans remains unclear for the time being.

Source/sink of GPE due to thermohaline forcing at the air–sea interface

Thermohaline forcing at the air–sea interface is usually treated in terms of the buoyancy flux. Awareness is needed, however, of the fact that adopting the concept of buoyancy flux may lead us to overlook the important physical fact that, along with the freshwater flux going through the air–sea interface, fluxes of other physical quantities, such as mass, enthalpy, and entropy, also exist.

In the following we separate the GPE source/sink due to surface heat flux and freshwater flux into two parts: (1) fluxes that change the surface density and thus induce a change in the total GPE of the water column, so that the GPE calculation is based on a mass conserving model; (2) the mass flux associated with surface freshwater flux, which is counted as a different item in the total budget.

Source/sink of GPE associated with surface density change due to surface thermohaline fluxes

The total amount of GPE source/sink due to surface heating and cooling is

$$\frac{d\chi}{dt} = \frac{g}{2c_p} \iint\limits_S \dot{q}\left(\alpha_{heat} h_{heat} - \alpha_{cool} h_{cool}\right) dxdy \tag{3.124}$$

where $\dot{q}$ is the local heat flux rate, h_{heat} (h_{cool}) is the depth of heating (cooling) penetration and α_{heat} (α_{cool}) is the thermal expansion coefficient during heating (cooling). The derivation of Eqn. (3.124) is dealt with in the Appendix. Assuming that heating/cooling penetration depth is very thin, Eqn. (3.124) suggests that heating/cooling on the surface does not contribute much to GPE at all. The same argument applies to the surface buoyancy flux as well.

Physical processes in this regard need to be considered in detail. First, solar radiation can penetrate into the first 10–20 m of water, and this penetration depth depends on the radiation frequency and the sea state, including the abundance of phytoplankton (Morel and Antoine, 1994). In any case, the effective penetration depth h_{heat} is finite. The GPE source due to the penetration of solar radiation is affected by many physical processes, including atmospheric conditions, sea state, and biological activity in the upper ocean. It is estimated that the global contribution is on the order of 0.01 TW (Huang and Wang, 2003), so it is negligible for the theory of oceanic circulation. However, the penetration of solar radiation may be one of the key controlling factors in the marine ecological system.

As a result of actions of surface waves and turbulence, the effective depth of heating penetration is the depth of the mixed layer. We emphasize that GPE generated in this way is not a direct contribution from heat flux; instead, the GPE source here is due to the KE conversion from surface waves and turbulence.

Another vital process taking place in the mixed layer is the convective adjustment due to cooling and salinification. During the cooling process, the density of seawater increases. Sea level declines and the center of mass moves down (Fig. 3.13a). In addition, density at the sea surface becomes heavier than the water below; thus, a gravitationally unstable stratification appears. Due to this unstable stratification, a rapid convective adjustment process takes place and density becomes nearly homogenized in the upper part of the water column (Fig. 3.13b). During this process, GPE is converted into KE for turbulence and internal waves. The GPE of the mean state declines in both stages.

Therefore, the effective center of cooling is not at the sea surface; instead it locates halfway down the well-mixed layer's depth. Because of this asymmetry associated with

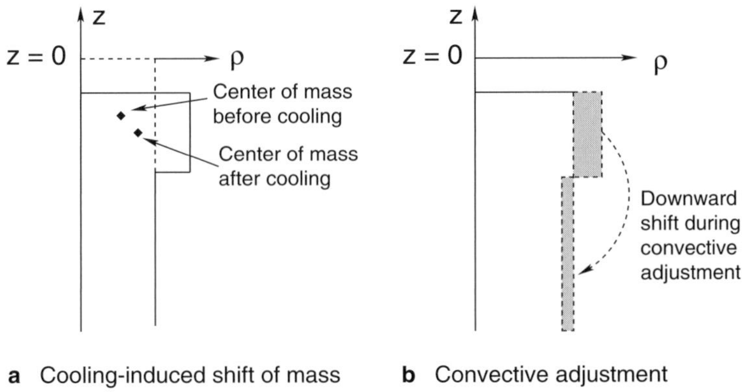

Fig. 3.13 Gravitational potential energy (GPE) loss due to **a** cooling-induced shift of mass, **b** convective adjustment.

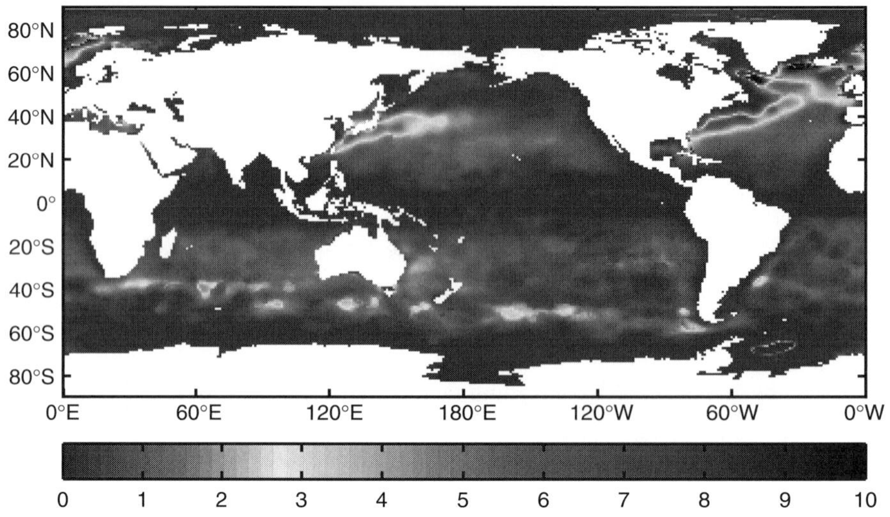

Fig. 3.14 Annual mean loss of GPE to convective adjustment (mW/m^2) (Huang and Wang, 2003). See color plate section.

cooling/salinification, surface buoyancy forcing gives rise to a sink of GPE. The quantitative amount of energy loss through this process remains unclear. A preliminary estimate based on the monthly mean climatology for this sink term is about 0.24 TW (Fig. 3.14). Because such an estimate is based on monthly mean climatology, it may be exceeded by estimations based on a calculation in which the diurnal cycle is resolved.

The corresponding gain of GPE associated with buoyancy gain is estimated as 0.13 TW (Huang and Wang, 2003). Note that a gain of GPE is converted from KE associated with surface waves and turbulence, so this should not be counted again as a separate source of mechanical energy for the world's oceans.

Source/sink of GPE due to surface mass flux

Freshwater flux through the air–sea interface is actually a mass flux. As discussed in Section 3.5, the GPE balance of the world's oceans is

$$\frac{\delta}{\delta t}\iiint_V \rho\phi dv = \oiint_S \rho\phi\left(\vec{\mathbf{u}}_s - \vec{\mathbf{u}}\right)\cdot\vec{\mathbf{n}}ds + \iiint_V \rho\frac{\partial\phi_T}{\partial t}dv + P \tag{3.125}$$

The first term on the right-hand side is the GPE source/sink due to surface mass exchange associated with the hydrological cycle. The global sum of this term is small, estimated as −0.007 TW (Huang, 1998a). The negative sign is due to the fact that freshwater is transported to the atmosphere from low latitudes, where sea level is high, and returns to the oceans at high latitudes, where sea level is low.

In addition, precipitation also carries kinetic energy associated with the water droplets. The total contribution is estimated as 0.4 TW (Faller, 1966); however, this energy is mostly dissipated through the turbulence motions in the skin layer on the sea surface, and is negligible for the mechanical energy balance of the large-scale circulation in the world's oceans.

Loss of GPE through baroclinic instability

Meso-scale eddies in the oceans

Meso-scale eddies are among the most salient features of the ocean. It is estimated that the amount of eddy KE is about 100 times larger than the KE of the time-mean flow. Despite great efforts expended on carrying out field observations and numerical simulations, no reliable estimate for the total amount of eddy KE in the world's oceans is available. However, satellite observational data have provided a global distribution of both the mean and eddy KE on the sea surface. The estimated ratio from such a data analysis of these two forms of KE is on the order of 100, which is consistent with the theoretical prediction based on scaling, as discussed in Section 4.1.3.

Conversion of GPE from the mean state to eddies

The steep isopycnal surface in the oceans, such as that along the meridional edges of the wind-driven gyre, is unstable owing to baroclinic instability. Baroclinic instability is one of the most critical mechanisms transforming GPE from the mean state to eddies, and the parameterization of eddies has constituted a major focus since the 1990s. Owing to baroclinic instability, the isopycnal slope is reduced. As shown in Figure 3.15, the flattening of the isopycnal surface is equivalent to moving the wedge of dense water from left to right; thus moving the center of mass downward and letting GPE in the mean state be released. The released GPE in the mean state is transformed into the kinetic and potential energy of meso-scale eddies. Owing to baroclinic instability in the world's oceans, a large amount of GPE of the mean state can be converted into eddy kinetic/potential energy.

The parameterization proposed by Gent and McWilliams (1990) is widely accepted and has been applied in non-eddy-resolving numerical simulations with comparative success. The basic idea is to parameterize the baroclinic instability in terms of isopycnal layer thickness diffusion, and its formulation can be better explained in terms of the energy

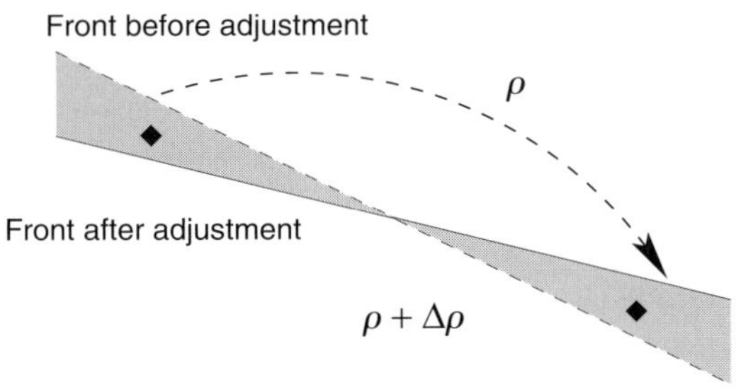

Fig. 3.15 Density front adjustment due to baroclinic instability.

conversion. Accordingly, the conversion rate of mean GPE to eddy GPE is governed by the following equation

$$\frac{D^*}{Dt}(g\rho z) = \nabla \cdot (g\rho\kappa_{th}L) + g\rho w + g\kappa_{th}\frac{\nabla\rho \cdot \nabla\rho}{\rho_z} \tag{3.126}$$

where D^*/Dt is the substantial derivative that advects with the effective transport velocity, including the Eulerian mean velocity and the eddy transport velocity; κ_{th} is the thickness diffusivity; $L = -\nabla\rho/\rho_z$, and ρ is the potential density. The first term on the right-hand side integrates to zero, so it does not contribute. The second term is the conversion from KE to GPE. The third term is a sink, ρ_z being negative, associated with the eddies. Accordingly, the total conversion rate in the world's oceans can be calculated by integrating this term over the total volume

$$\dot{P}_{e,bi} = g\iiint \kappa_{th}\frac{\nabla\rho \cdot \nabla\rho}{\rho_z}dxdydz \tag{3.127}$$

The potential density gradient is calculated using the center of each layer as the reference level for the potential density, and a central difference scheme in space.

As suggested by Gent *et al.* (1995), $\kappa_{th} = 1,000$ m^2/s, and the density field is calculated from temperature and salinity climatology. This scheme is a good approximation that applies for the case of relatively small isopycnal slopes. The common practice is to set a limit for the isopycnal slope. For cases with isopycnal slope limit $|\nabla\rho/\rho_z| \leq 0.01$, the total rate of GPE transformation is estimated as 0.9 TW. The horizontal distribution of this GPE conversion rate is shown in Figure 3.16. It is clear that most energy transformation takes place in the ACC, the Gulf Stream and, to a much smaller extent, in the Kuroshio.

However, numerical simulations with eddy-permitting resolution revealed that such eddy parameterization schemes are not very accurate. Thus, any result based on such parameterization should be treated as an approximate estimate, and a more accurate value may need to be calculated from the next generation of numerical models with much finer resolution. The most up-to-date estimate currently available is the in the range of 0.3–0.8 TW (Ferrari and Wunsch, 2009).

Diapycnal and along-isopycnal mixing

Tracers, including temperature and salinity, mix in the oceans as results of internal wave breaking and turbulence. Mixing can be classified into diapycnal mixing and along-isopycnal mixing. Since along-isopycnal mixing involves the least amount of GPE, it is the dominant form of mixing at large scales.

In a stably stratified fluid, any vertical mixing increases GPE because, in such mixing, light fluid is pushed downward while heavy fluid is pushed upward. Molecular diffusion can also cause mixing, but at a negligible rate compared with other mixing mechanisms in the ocean; thus it is not the dominant form of mixing for large-scale motions in the oceans.

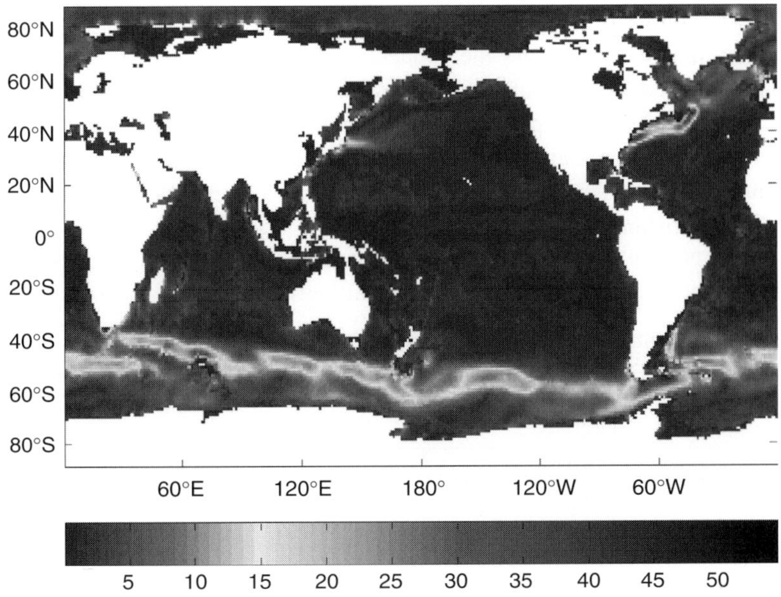

Fig. 3.16 Conversion rate of mean GPE to eddy GPE through baroclinic instability based on the empirical eddy parameterization of Gent and McWilliams (1990) (mW/m^2) (Huang and Wang, 2003). See color plate section.

The energy sources supporting diapycnal mixing in the ocean include:

- *Wind stress* input through the sea surface and tidal dissipation, which was discussed in the previous section.
- *Turbulent mixing* associated with fast currents, especially that associated with flow over a sill and/or the down-slope flow afterward, can provide a strong source of energy that supports mixing. There is plenty of observational evidence indicating that the coefficient of such mixing can be on the order of 10^{-3}–10^{-1} m^2/s. For example, the diffusivity can be as large as 0.1 m^2/s within the Romanche Fracture Zone, where water drops more than 500 m within 100 km of a downward flow (Polzin *et al.*, 1996).
- *Internal lee waves* generated by flows over topography, such as the ACC, are probably one of the most important contributors.

The total amount of energy supporting diapycnal mixing in the oceans remains unclear, because mixing is highly non-uniform in space and may also vary over time. According to the energetic theory of thermohaline circulation, the meridional overturning rate is directly controlled by the strength and distribution of the external mechanical energy available for supporting mixing in the ocean, including mixing in the mixed layer and mixing in the subsurface layers. Therefore, understanding the physics related to the spatial and temporal distribution of mixing is one of the most important research frontiers in physical oceanography.

In the ocean interior, diapycnal mixing is closely related to the dissipation of turbulent kinetic energy. In order to discuss the connection between the energetics of mixing and dissipation associated with turbulence and internal waves, we separate the flow into two components, the mean and the perturbations. Of course, we need to bear in mind that such a separation is only conceptual for our understanding the dynamics associated with mixing, and a distinct separation is impossible. Accordingly, the velocity, density, and pressure are separated into the mean and perturbations:

$$u_i = \bar{u}_i + u_i', \quad \rho = \bar{\rho} + \rho', \quad p = \bar{p} + p' \tag{3.128}$$

In addition, we assume that the Boussinesq approximations hold and the tidal force is omitted. Multiplying Eqn. (3.44) by u_i' and averaging leads to the equation for the turbulent kinetic energy:

$$\frac{\partial}{\partial t}K' + \bar{u}_j \frac{\partial}{\partial x_j}K' = \overline{u_j'\alpha'}\frac{\partial \bar{p}}{\partial x_j} - \overline{(\bar{\alpha} + \alpha')\, u_j' \frac{\partial p'}{\partial x_j}} + \nu \frac{\partial}{\partial x_j}\left(u_i'\sigma_{ij}'\right) - \varepsilon - \overline{\left(u_i'u_j'\right)}\frac{\partial \bar{u}_i}{\partial x_j} \tag{3.129}$$

where

$$\sigma_{ij}' = \frac{\partial u_i'}{\partial x_j} + \frac{\partial u_j'}{\partial x_i}, \quad \varepsilon = \nu \sigma_{ij}' \sigma_{ji}' \tag{3.130}$$

Scaling analysis indicates that for steady-state flows the basic balance is between the turbulence production, the dissipation, and the buoyancy work (Turner, 1973; Osborn, 1980):

$$\overline{u_i'u_j'}\frac{\partial \bar{u}_i}{\partial x_j} = -\varepsilon - \frac{\overline{u_3'\rho' g}}{\bar{\rho}} \tag{3.131}$$

Defining the flux Richardson number as the ratio of the buoyancy flux to the turbulent production:

$$R_f = \frac{\overline{u_3'\rho' g}}{\bar{\rho}} \bigg/ \left[-\overline{u_i'u_j'}\frac{\partial \bar{u}_i}{\partial x_j}\right] \tag{3.132}$$

the eddy density diffusivity is defined as

$$\kappa_\rho = g\overline{u_3'\rho'} \big/ \bar{\rho} N^2 \tag{3.133}$$

Then, the balance of turbulent dissipation in Eqn. (3.131) is reduced to

$$\kappa_\rho = \frac{R_f \varepsilon}{\left(1 - R_f\right) N^2} \tag{3.134}$$

If $R_f \leq R_{f,critical} = 0.15$, then we have

$$\kappa_\rho \leq 0.2 \frac{\varepsilon}{N^2} \tag{3.135}$$

Thus, among the turbulent kinetic energy dissipated in the ocean, a small proportion of it can be converted back to the GPE of the large-scale mean flow. This is a very peculiar and important upscale energy cascade in the stratified turbulence. This means that not all the kinetic energy of turbulence is dissipated into heat, and a small percentage of such dissipation is actually fed back to the large-scale flow.

However, this feedback involves many complicated processes, and the mixing efficiency should not simply be taken as a constant. In fact, the efficiency of mixing varies significantly. Peltier and Caulfield (2003) presented the most up-to-date review of the mixing efficiency in stratified shear flows.

In situ observations indicate that in the upper ocean, below the mixed layer, diapycnal diffusivity is on the order of 10^{-5} m^2/s, which is much larger than the molecular diffusivity. In other places in the oceans, diffusivity can be much higher than this background rate (Fig. 3.17). The strong diapycnal (or vertical) mixing in the oceans is driven by strong internal wave breaking and turbulence.

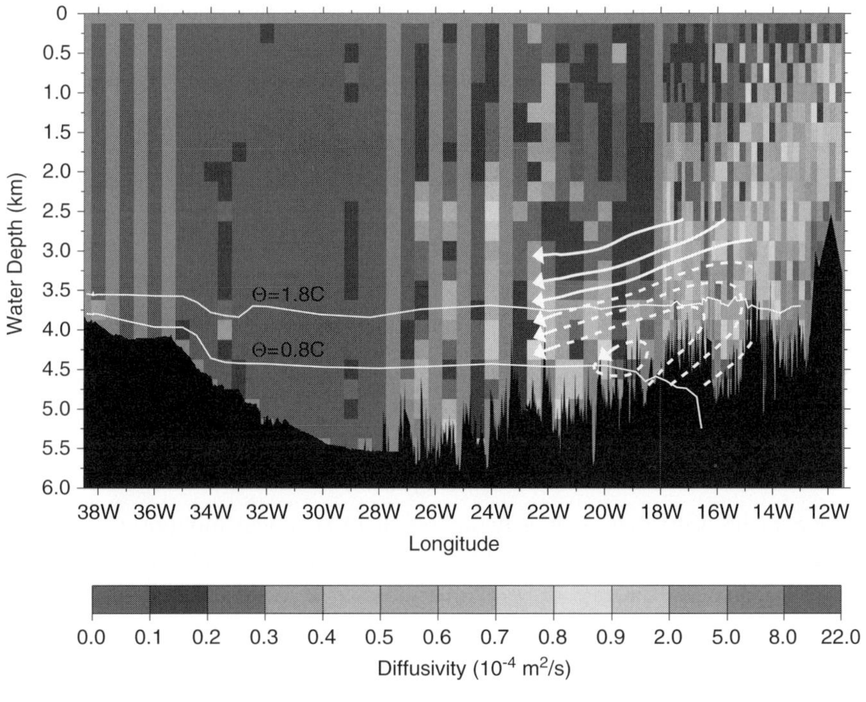

Fig. 3.17 Depth–longitude section of cross-isopycnal diffusivity, K_v, in the Brazil Basin inferred from velocity microstructure observations (Polzin *et al.*, 1997); with additional data from a later cruise (Ledwell *et al.*, 2000). The *thin white lines* mark the observed depths of the 0.8°C and 1.8°C isotherms; the *thicker white lines with arrows* are a schematic representation of the stream function estimated from an inverse calculation (St Laurent *et al.*, 2001) (adapted from Mauritzen *et al.*, 2002). See color plate section.

Note that although there is a large amount of external mechanical energy, in particular that from the wind stress input to surface waves and surface currents, the pathways from these sources to internal waves and turbulence supporting diapycnal mixing remain unclear at present. Despite many decades of research effort, our understanding of the role of mixing in the thermohaline circulation remains rudimentary, and it is fair to say that there is no simple solution.The study related to these pathways is beyond the scope of this book, and interested readers can find the most up-to-date information in the book *The Turbulent Ocean* by Thorpe (2005) and a review article by Ferrari and Wunsch (2009).

Geothermal heating

In addition to energy sources described above, geothermal heat and hot plumes provide a total heat flux of 32 TW to the oceans. Though being three orders of magnitude smaller than the heat flux across the air–sea interface, this may be a significant component of the driving force for the abyssal circulation. Since geothermal heating occurs at great depth and the corresponding cooling takes place at the sea surface, geothermal heat can be more efficiently converted into GPE. The source of GPE due to geothermal heat can be calculated using Eqn. ((3.A8)) in the Appendix. The global rate of this conversion is about 0.05 TW (Huang, 1999), which is a minor term in comparison with other major terms; but nevertheless is not negligible, especially for the abyssal circulation and temperature distribution in the abyss.

Bottom drag

Oceanic currents moving over rough bottom topography must overcome bottom drag or form drag, as already shown in Figure 3.9. The total amount of mechanical energy dissipation through bottom drag for the world's ocean circulation remains unclear. The preliminary estimate for the open oceans was about 0.4 TW (Wunsch and Ferrari, 2004). More recently, this has been updated by Sen *et al.* (2008) as 0.2–0.8 TW. It is obvious that bottom drag remains one of the major unknowns in oceanic circulation theory. There is no reliable observational data about bottom drag, and most numerical models cannot resolve the deep circulation accurately.

Double diffusion

Seawater contains salt; thus, it is a two-component chemical mixture. On the level of molecular mixing, heat diffusion is 100 times faster than salt diffusion; however, in a turbulent environment, the rate of salt diffusion can be comparable to the rate of heat diffusion. The difference in heat and salt diffusion for a laminar fluid versus turbulent fluid environment constitutes one of the most important aspects of the thermohaline circulation in the oceans.

Double diffusion in the oceans primarily manifests itself in two forms: salt fingers and diffusive convection. Salt fingers appear when warm and salty water lies over cold and fresh water. The mechanism sustaining salt fingers is the release of GPE due to the salt fingers' downward motion driving an instability. Let us imagine a vertical pipe which connects the cold and fresh water in the lower part of the water column with the warm and salty water in the upper part of the water column. The wall of the pipe is very thin, so it allows a rather efficient heat flux into the pipe, thus warming up the cold water. At the same time, the low salt diffusivity preserves the freshness of the ascending water parcel. Therefore, the buoyancy difference drives the upward motion of water inside the pipe, and a self-propelled fountain can be built in the ocean.

The subtropical gyre interior is a salt finger's favorite place, where strong solar insolation and excessive evaporation lead to warm and salty water above the main thermocline. In this regime, warm and salty fingers move downward while cold and fresh plumes move upward. Since heat is 100 times more easily diffused between the salt fingers and the environment, salt fingers lose their buoyancy and continue their movement downward. In this way the GPE released is used to drive the mixing. According to new *in situ* observations in the upper thermocline of the subtropical North Atlantic, the diffusivity for the temperature (salinity) is on the order of 4×10^{-5} m^2/s (8×10^{-5} m^2/s) and the equivalent density diffusivity is negative (Schmitt *et al.*, 2005).

The other possible double diffusive process is the diffusive convection that takes place if cold and fresh water overlies warm and salty water. The system is relatively stable and allows an oscillatory instability.

Based on climatological data of temperature and salinity in the world's oceans, the release of GPE due to salt fingers can be calculated using the double diffusive flux associated with salt finger parameterization (Zhang *et al.*, 1998) (Fig. 3.18). The global sum of GPE release is small, in the order of 26.8 GW. Since GPE is increased due to the thermal diffusion associated with salt fingers, with a global total of 18.8 GW, the net value is about 8 GW. Details of this estimate are discussed in Section 5.3.4.

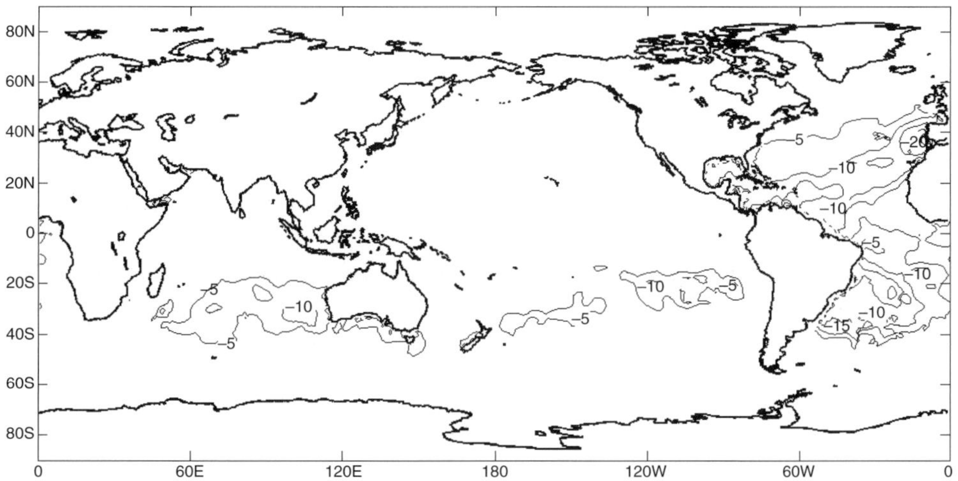

Fig. 3.18 Global distribution of GPE release due to salt diffusion associated with salt fingering (mW/m^2) (the background diffusivity is set to $K^{\infty} = 0$).

Note that: (1) this energy should not be considered as an external mechanical energy input because salt fingers can only release GPE from the stratification which is set up by the large-scale circulation, and the circulation itself is driven by external mechanical energy; (2) although GPE release from salt fingers seems small compared with other sources of energy, it may play a vital role in some parts of the world's oceans, such as the subtropical North Atlantic.

Mechanical energy input due to biomixing

Although biological activity in the ocean has been treated as a passive response to the oceanic environment, there is a possibility that biological activity in the ocean may contribute to the large-scale circulation. The total amount of chemical power produced through biological activity in the world's oceans, commonly called the primary production rate, is estimated as 62.7 TW. Based on several approaches, about 1% of this huge amount of power may contribute to the oceanic circulation; thus, the global sum of mechanical energy input to large-scale circulation from biological sources is estimated as 0.6 TW (Dewar *et al.*, 2006). It is clear that more accurate estimations should be available in the future.

Cabbeling

Owing to the nonlinearity of the equation of state, water density increases during both diapycnal and isopycnal mixing and the newly formed water with higher density sinks; this is the process called cabbeling, as already shown in Figure 3.9. As a result, GPE is converted into energy for internal waves and turbulence. The total amount of GPE loss associated with cabbeling in the oceans remains unknown.

3.6.2 Source of chemical potential energy

Seawater contains salt, and therefore it possesses a certain amount of energy in the form of chemical potential. Chemical potential differs from mechanical energy in the ordinary sense; however, it is also different from internal energy, or the so-called thermal energy. In fact, chemical potential energy in the ocean can be converted into equivalent mechanical energy, either through the osmotic pump or the generation of electricity through the utilization of the salt concentration difference in the ocean.

As detailed in Section 3.8, the maintenance of the oceanic hydrological cycle and the salinity distribution requires that a large amount of entropy produced through salt and pure water mixing must be extracted through the air–sea interface. From the point of view of mechanical energy balance, the separation of pure water from seawater requires a huge amount of equivalent work. For a unit mass of evaporation, the amount of equivalent mechanical energy is equal to $\mu_w(0) - \mu_w(S)$, and for the global oceans, the total amount is estimated as 32 TW. This means that some source of energy is required for the balance of chemical potential in the world's oceans. Note that this is a huge amount of energy, much greater than that due to tidal dissipation. Thus, the maintenance of the hydrological cycle is a very important part of the global thermohaline circulation; however, this subject remains largely unexplored.

It is reasonable to propose that a major portion of this energy is dissipated within the surface layer as soon as water from precipitation is mixed with the seawater, and the remaining portion may be used to drive mixing of salt and pure water in the oceanic interior; however, a precise picture of the balance of the energy associated with chemical potential remains elusive.

3.6.3 A tentative scheme for balancing the mechanical energy in the ocean

The tentative balance of mechanical energy, including both the KE and GPE, is shown in Figure 3.19. As discussed in the previous section, we have also included the contribution to the chemical potential, because this is the energy maintaining the hydrological cycle and the salinity stratification in the oceans.

It seems clear that at this time we do not yet know the balance of mechanical energy, even to the lowest order of precision. We know that a great deal of KE is constantly input into the ocean, but do not know how such energy is distributed and eventually dissipated in the oceans. Similarly, although we know that there are two major sinks of GPE, it is not clear how energy is transported and supplied to such energy sinks. It is fair to say that most quantities for the energy fluxes listed in Figure 3.19 are accurate to a factor of two only. In some cases, the uncertainty may be even greater. The more accurate description of energy pathways and the production of better estimates require further study.

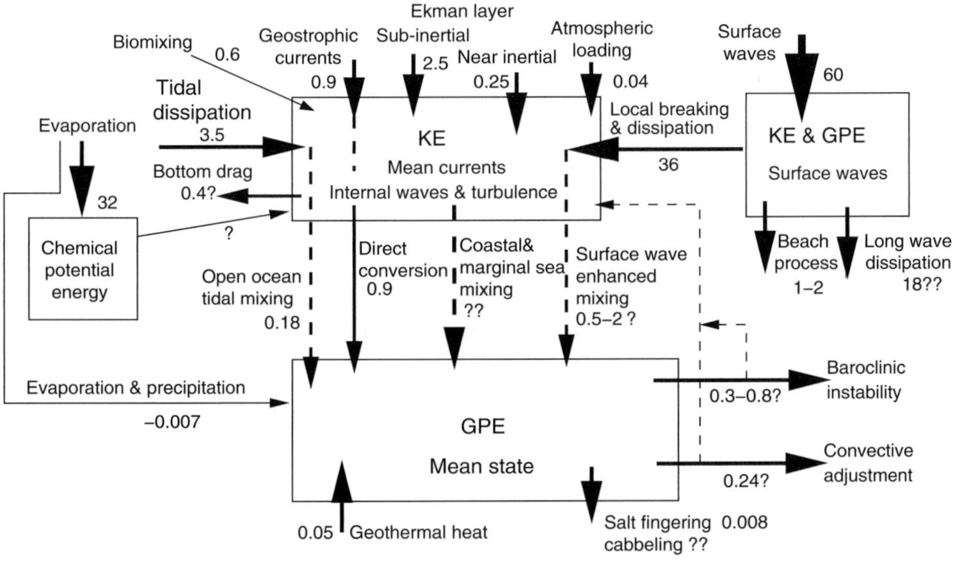

Fig. 3.19 An attempt to balance mechanical energy for the world's oceans (TW).

3.6.4 Remaining challenges in the energetics of the world's oceans

Wind is the main driving force of oceanic circulation and climate changes

Over the past decade, the role of tidal dissipation in driving mixing in the open ocean has received a lot of attention. In contrast, the role of wind stress in sustaining mixing and the thermohaline circulation has not received the attention due to it. Enthusiasm for the study of tidal mixing may partly be due to the fact that the tidal problem is a very well-formulated one and *in situ* measurements are relatively easy to carry out. In contrast, surface waves and turbulence are much tougher to deal with, both in terms of theoretical/numerical studies and field observations. However, the reality is that wind stress is actually the most important driving force for the thermohaline circulation and climate change, for the following reasons:

- Wind stress energy input is the dominant factor in controlling the meridional overturning circulation in the Atlantic Ocean under current climatic conditions. This has been demonstrated through numerical simulations based on an oceanic general circulation model, e.g., Toggweiler and Samuels (1998).
- Wind stress energy input changes greatly over a broad range of time scales, from interannual, decadal, centennial, to other longer time scales. In fact, data analysis has shown that wind energy input to geostrophic currents (Fig. 3.20) and to surface waves and the Ekman layer (Fig. 3.21) has increased about 15–20% over the past two decades. In the meantime, tidal dissipation can be regarded as nearly constant for time scales shorter than centennial.
- Wind stress energy input affects the upper ocean directly, making it the most important dynamic zone in various application areas, including weather, fisheries, transportation, and the environment.

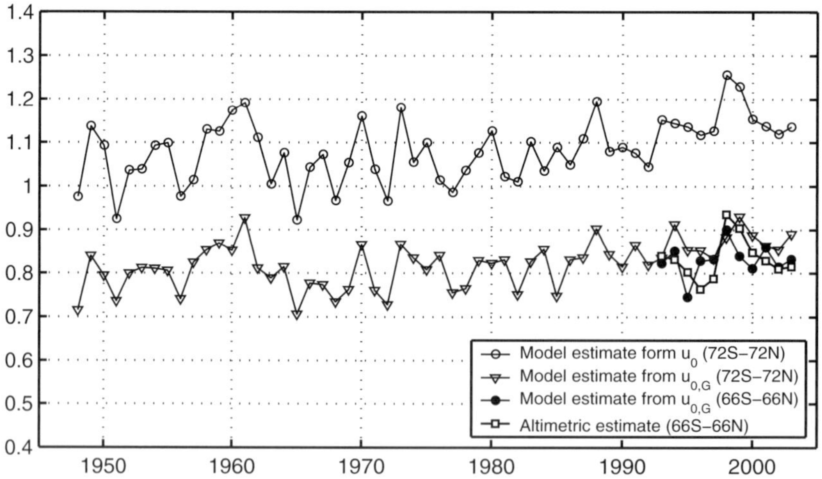

Fig. 3.20 Annual mean wind energy input to the surface geostrophic current diagnosed from the numerical model, and the results from altimetry data (TW) (Huang *et al.*, 2006).

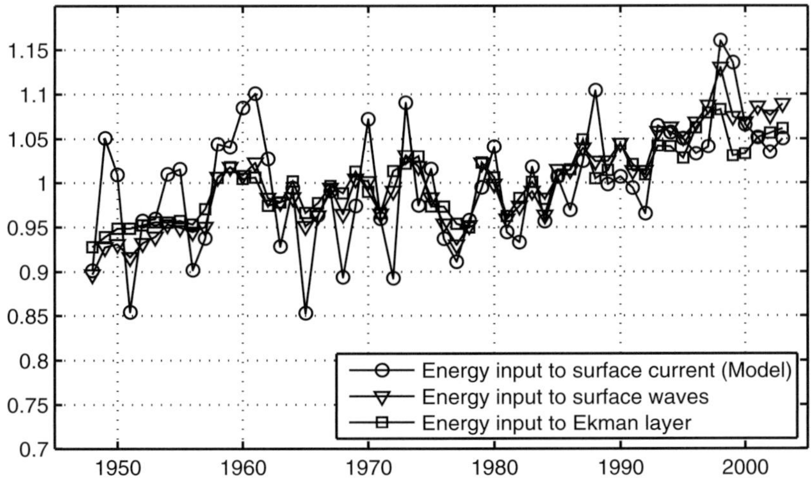

Fig. 3.21 The time evolution of the normalized wind energy input through the geostrophic current (*circles*), surface waves (*inverted triangles*), and Ekman layer (*squares*) (Huang *et al.*, 2006).

Changes in wind stress energy input

Perhaps the most important point deserving serious attention is that wind stress energy input to surface geostrophic current varies greatly from year to year, and has been increasing over the past 50 years, as shown in Figure 3.20.

Similarly, wind energy input to surface waves and the Ekman layer has also increased noticeably over the past decades (Fig. 3.21). Variations in wind energy input to the oceans have originated from different sources. First, due to global warming, wind and other nonlinear dynamical processes may be more energetic. Second, the ozone hole which first appeared near the South Pole is capable of intensifying the southern polar vortex. Through wave–wave interactions in the atmosphere, the intenstification of the polar vortex propagates downward and manifests as the intensification of the south-westerly and enhances the surface west wind over the Southern Ocean (see, e.g., Yang *et al.*, 2007).

As shown in these two figures, we can see, thanks to the availability of reliable measurements, including satellite data, that energy input to the oceans, through geostrophic currents, ageostrophic currents, and surface waves, has kept increasing steadily since the 1980s. Thus, it is natural that the oceanic general circulation should adjust in response to such changes in energy input.

Changes in tidal dissipation

Many factors regulate tidal dissipation, including the shape of the basin, bathymetry, and sea level. These aspects of boundary conditions change over a wide range of time scales.

Global-scale changes in land–sea distribution due to continental drift

These changes have a millennial or longer time scale. Due to changes in the size and shape of the basins, tidal dissipation has varied greatly over the geological past; thus, the energy supporting diapycnal mixing may vary as well (Fig. 3.22). Note that this result was based on a numerical model with a flat bottom. It is well known that bottom topography plays a vitally important role in controlling the barotropic tidal flow and dissipation; however, reconstruction of the paleotopography may be very challenging. Thus, the results from a simple tidal model are quoted here to demonstrate the basic idea that tidal dissipation may have changed tremendously over the geological past.

If we go further back into the geological past, the change in the gravitational field of the Earth–Moon system and the rotation rate of the Earth may also have to be taken into consideration. Paleo records indicate that the Earth's rotation has slowed down over the past 900 million years; this should have affected the tidal flow and dissipation rate.

Owing to the movement of the mantle, continents have drifted greatly over the geological past. Accordingly, the shape of the world's oceans has also changed dramatically. It is a grand challenge to map out the position and bathymetry of the oceans and the associated tidal flows.

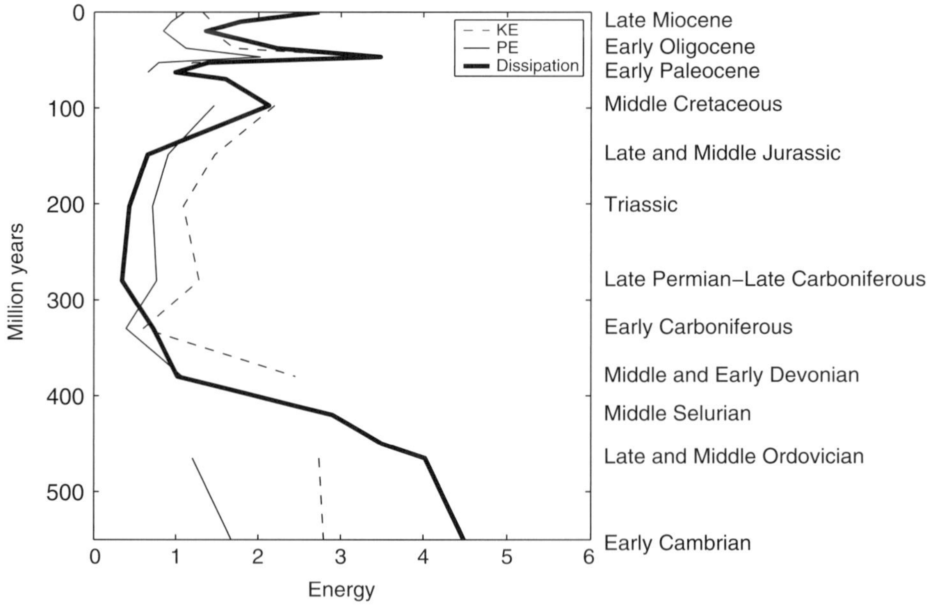

Fig. 3.22 Changes in the M_2 tidal dissipation (*thick line*, in 10^{12} W), KE (*dashed line*, in 10^{17} J), and potential energy (*thin line*, in 10^{17} J), over the past 550 million years (based on data from Kagan and Sundermann, 1996).

Global sea-level change

On time scales shorter than millennial, the mean sea level of the world's oceans can change due to the glacial–interglacial cycles. For example, during the Last Glacial Maximum (LGM), great changes took place, resulting in a tidal dissipation and thermohaline circulation that were substantially different, including the following:

- Sea level was more than 100 m lower than at present. Tidal dissipation in shallow seas was much reduced. As a result, tidal flow in the deep oceans was much faster. Hence, global tidal dissipation during the LGM was 50% higher than at present (Egbert *et al.*, 2003).
- The meridional temperature difference in the atmosphere was larger, so the wind was much stronger; thus wind energy input into the ocean was much stronger.

Thus, it would be an excellent project to simulate the oceanic circulation and climate during the LGM with a new parameterization of diapycnal mixing which changes with climate.

3.7 Gravitational potential energy and available potential energy

3.7.1 Gravitational potential energy

Gravitational potential energy is one of the most important components in the energy balance of the oceanic general circulation. In order to make full use of the mass conservation rule, the definition of GPE can be conveniently rewritten in the mass coordinate, using $\rho dv = dm$, so that

$$\Phi = g \iiint \rho z dv = g \iiint z dm \tag{3.136}$$

The density of GPE is defined as

$$\chi = g \iiint \rho z dv \Big/ \iiint dv \tag{3.137}$$

Note that the value of GPE depends on the choice of reference level. For example, using the sea surface as a reference level will give rise to a negative value for GPE. Using the mean seafloor depth of the ocean, $\bar{D} = -3750$ m, as the reference level, the total amount of GPE in the world's oceans is estimated as 2.1×10^{25} J ($\chi \simeq 1.4 \times 10^7$J/m^3).

The density of seawater varies only slightly. Therefore, a major part of GPE in the ocean is dynamically inert, so that only a small part of this energy associated with the density deviation from the mean value is dynamically active. Different ways of differentiating between the dynamically active and non-active components of GPE exist, such as the conceptions of stratified GPE (SGPE) and available potential energy (APE), which are discussed in the following sections.

Because seawater is almost constant, mass conservation is replaced by volume conservation in the commonly used Boussinesq approximations. However, replacing mass conservation by volume conservation can induce an artificial source/sink of mass and GPE in the models. In order to avoid such problems, one should use a model based on mass conservation for studying the GPE balance.

Definition of stratified gravitational potential energy

The basic idea of SGPE is to separate GPE into two parts: the part due to the basin mean density and the part due to the density anomaly. We define the global mean depth, density, and reference pressure as

$$\bar{z} = \iiint z dv/V, \quad \bar{\rho} = \iiint \rho dv/V, \quad \bar{p}_r = \iiint p dv/V, \quad V = \iiint dv \quad (3.138)$$

For the world's oceans $\bar{z} = -2365$ m, $\bar{\rho} = 1038.43$ kg/m^3, $\overline{p_r} = 2455$ db. Accordingly, the integrand of GPE in Eqn. (3.136) can be separated into two parts $\overline{\rho z} = \overline{\bar{\rho} \bar{z} + \rho' z'}$, because the other terms, $\bar{\rho} z'$ and $\rho' \bar{z}$, make no contribution to the global integral. Therefore, the total GPE can be separated into two parts

$$\Phi = \Phi^0 + \Phi' \quad (3.139)$$

where

$$\Phi^0 = g \bar{\rho} \bar{z} V, \quad \Phi' = g \iiint \rho' z' dv \quad (3.140)$$

are the GPE associated with the mean density and the SGPE associated with the stratification. While Φ^0 is reference-level dependent, Φ' is not and its density is estimated as $\chi' = \Phi'/V \simeq -1.0 \times 10^5 J/m^3$ for the world's oceans. χ' is negative because ρ' and z' are negatively correlated. Apparently, SGPE reflects the energy associated with the vertical stratification; however, most of such energy cannot be converted into KE, and thus it is not an effectively usable energy source for driving the oceanic circulation. By definition, if water density is homogeneous, SGPE should be zero.

To examine the contributions to SGPE from pressure, temperature, and salinity, we introduce the following decomposition

$$\rho' z' = [(\rho - \sigma_m) + (\sigma_{m0} - \overline{\sigma_m}) + (\overline{\sigma_m} - \bar{\rho}) + (\sigma_m - \sigma_{m0})] (z - \bar{z}) \quad (3.141)$$

where ρ is the *in situ* density; $\sigma_m = \sigma_m (T, S, p, \overline{p_r})$ is the potential density, using $\overline{p_r} =$ 2455 db as the reference pressure; and $\sigma_{m0} = \sigma_m \left(T, \bar{S}, p, \overline{p_r}\right)$; $\bar{S} = 34.718$ is the mean salinity; $\overline{\sigma_m} = \overline{\iiint_V \sigma_m dv/V} = 1038.70$kg/m^3, so the difference between $\bar{\rho}$ and $\overline{\sigma_m}$ is small.

From the definition of $\bar{z}$, the global integration of the third term in Eqn. (3.141) vanishes, i.e., $\iiint_v (\overline{\sigma_m} - \bar{\rho})(z - \bar{z})dv = 0$. Thus, SGPE can be separated into three components

$$\Phi' = \iiint_v \rho' z' dv = \Phi_p + \Phi_T + \Phi_S \tag{3.142}$$

Now let us look at these terms more closely. The first term, $\Phi_p = \iiint_v (\rho - \sigma_m)(z - \bar{z})dv$, is due to the density anomaly associated with pressure difference. Since the compressibility of seawater is roughly constant, $\rho - \sigma_m$ is a nearly linear function of z, with a zero value at the reference level, $(\rho - \sigma_m)(z - \bar{z}) \leq 0$ for the whole water column. Because the compressibility of seawater varies with change in salinity and temperature only very slightly, this term is roughly proportional to the square of the ocean depth at each location (Fig. 3.23).

The second term in Eqn. (3.142), $\Phi_T = \iiint_v (\sigma_{m0} - \overline{\sigma_m})(z - \bar{z})dv$, reflects the density difference due to temperature in the ocean. In the subtropics, warm water in the upper ocean gives rise to a large difference of $\sigma_{m0} - \overline{\sigma_m} \leq 0$. In the Atlantic sector, the density anomaly is positive only for the small part where the Antarctic Bottom Water (AABW) dominates (see Fig. 3.24a). Beginning with the large negative value in the warm water pool, the contribution due to temperature remains negative everywhere in the Pacific sector, except for small areas near the Antarctic and in the Arctic Ocean (see Fig. 3.24b).

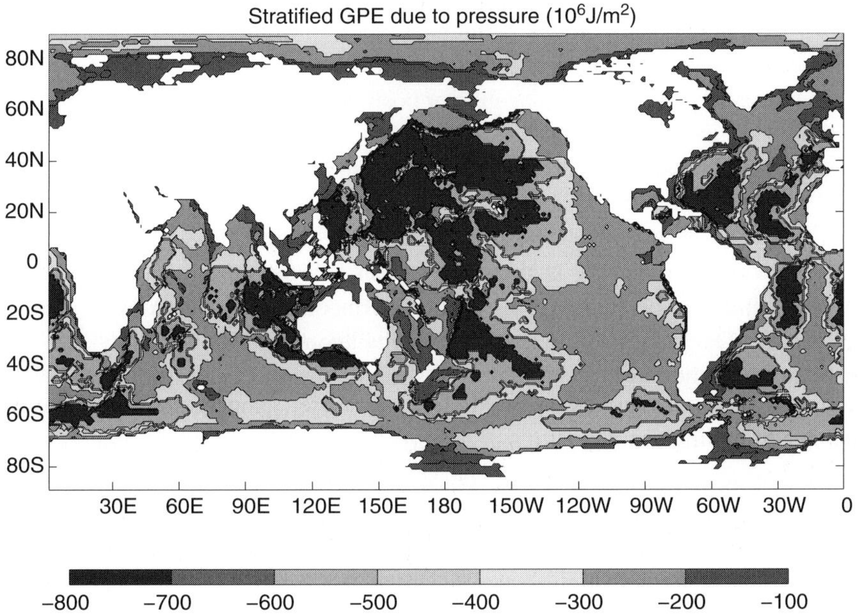

Fig. 3.23 Contribution of pressure to stratified gravitational potential energy (SGPE) (in 10^6 J/m^2). See color plate section.

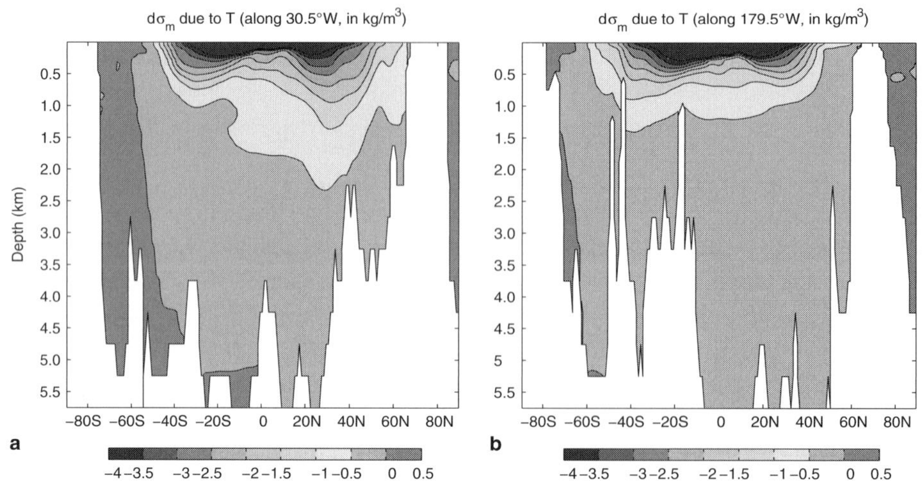

Fig. 3.24 **a, b** Density anomaly due to temperature contribution for two meridional sections. See color plate section.

The contribution of temperature can be seen clearly through the meridional section of density (Fig. 3.25). In the subtropics, warm water in the upper ocean gives rise to a large difference in $\sigma_{m0} - \overline{\sigma_m} \leq 0$, while in the abyssal ocean the cold water temperature leads to a large positive $\sigma_{m0} - \overline{\sigma_m} \geq 0$ term. Therefore the contribution of this term is negative everywhere except in polar latitudes. At high latitudes, low temperature in the upper ocean gives the term $\sigma_{m0} - \overline{\sigma_m}$ a much smaller value, which is sometimes positive in the upper ocean. Thus, this term has a small negative or positive value at high latitudes (Fig. 3.25).

The third term in Eqn. (3.142), $\Phi_S = \iiint_v (\sigma_m - \sigma_{m0})(z - \bar{z})dv$, represents an integrated measure of the salinity contribution to SGPE. In the North Atlantic Ocean, especially in the subtropical basin, salinity is much higher than the mean value in the world's oceans, which is thus $\sigma_m - \sigma_{m0} \geq 0$ for most of the Atlantic section. This leads to a domain of relatively high positive density in the upper ocean. On the other hand, in the Southern Ocean, low salinity associated with Antarctic Intermediate Water (AAIW) and Antarctic Deep Water (AADW) gives rise to negative density anomaly (Fig. 3.26a).

The situation in the North Pacific Ocean is the opposite. Owing to the relatively low salinity in this basin, the density anomaly is positive for two small areas: a shallow layer in the South Pacific subtropical gyre and a relatively salty band in the core of the Antarctic Circumpolar Water; everywhere else the density anomaly due to salinity is negative.

In the North Atlantic Ocean, especially in the subtropical basin, salinity is much higher than the mean salinity value in the world's oceans, which is thus $\sigma_m - \sigma_{m0} \geq 0$ in the upper ocean, leading to a relatively high energy density in this basin. Another area of high value, in the Arabian Sea, is also due to the high salinity there. In comparison, salinity is low in the North Pacific Ocean, the Southern Ocean, and the Arctic Ocean; thus, the contribution to SGPE is negative in these regions (Fig. 3.27).

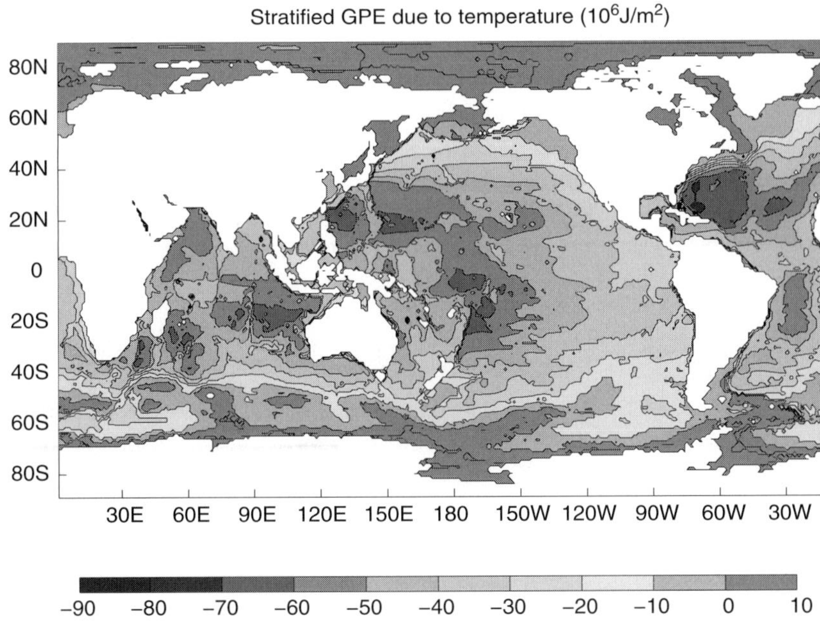

Fig. 3.25 Contribution of temperature to SGPE (in 10^6 J/m^2). See color plate section.

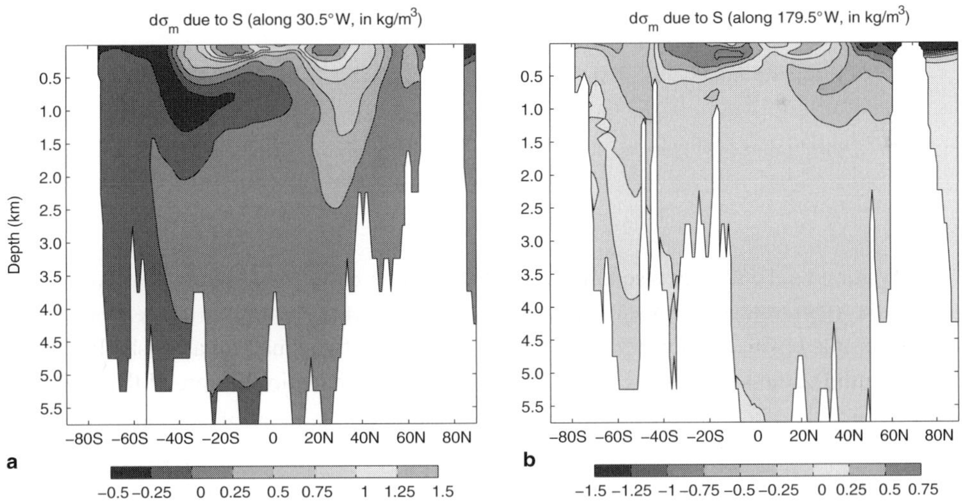

Fig. 3.26 Density difference due to temperature effect for two sections: **a** through the Atlantic Ocean, and **b** the Pacific Ocean (kg/m^3). See color plate section.

Table 3.2. *Contributions to stratified gravitational potential energy (SGPE), global energy density in 10^3 J/m^3*

Term	Pressure (ϕ_p)	Temperature (ϕ_T)	Salinity (ϕ_S)	Sum
Mean energy density	−93.17	−7.27	−0.086	−100.5

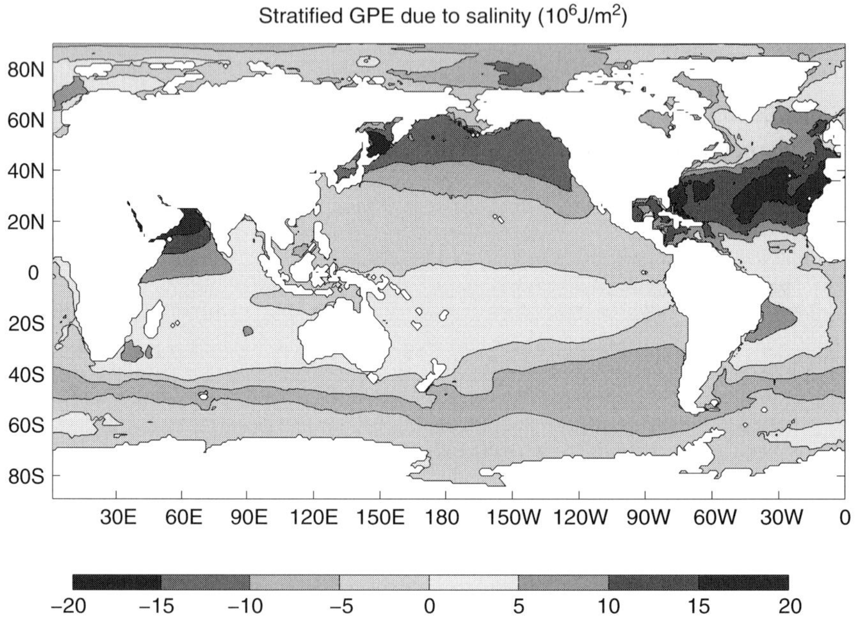

Fig. 3.27 Contribution of salinity to SGPE (in 10^6 J/m^2). See color plate section.

The global sum of these terms is listed in Table 3.2. It is clear that the pressure term dominates, the temperature term is ten times smaller, and the salinity term is 1,000 times smaller.

By definition, the magnitude of SGPE associated with temperature and salinity is an integral measure of the strength of the thermal and haline components of the circulation. With changes in these components, the corresponding SGPE components change accordingly. For example, if the hydrological cycle diminishes greatly, the SGPE due to salt should also decline. In the classical theory of oceanic general circulation, the dynamical role of compressibility of seawater is ignored in most discussions. As shown above, on the other hand, compressibility of seawater is the dominant player in creating SGPE. The meaning of these terms remains unclear and should be explored in the future.

3.7.2 Available potential energy

Another way of separating the GPE into active and non-active parts is based on identifying GPE associated with horizontal density differences, which are directly linked to the overturning circulation. The basic idea, first postulated by Margules (1905), is to find a reference state which has the minimal GPE, and the difference in GPE between the physical state and the reference state is defined as available potential energy (APE) for the physical state. However, the application of APE to atmospheric dynamics appeared primarily after Lorenz (1955) introduced an approximate definition of APE.

A simple example

The general definition of APE in oceanography has two parts, including contributions from GPE and internal energy. First, available GPE (AGPE) is defined as

$$\chi = g \int (z - Z)\, dm \tag{3.143}$$

where g is gravity, z (Z) are the geopotential height of a mass element in the physical (reference) state. The reference state is defined as a state with minimal GPE, a state attainable through adiabatic adjustment and redistribution of mass.

As an example, we examine the AGPE in a two-layer model in a two-dimensional space, assuming it is of unit length in the third dimension. The thermocline, represented by a solid curve in Figure 3.28a, separates water of density $\rho + \Delta\rho$ and ρ. The reference state is shown in Figure 3.28b, with $\overline{h} = \int_0^L h dx / L$.

Using $z = -H$ as the reference level for GPE, the total GPE of a water column in the physical and reference states is

$$\chi^0 = \frac{g(H-h)^2}{2}\Delta\rho + \frac{gH^2}{2}\rho, \quad \chi^r = \frac{g(H-\overline{h})^2}{2}\Delta\rho + \frac{gH^2}{2}\rho$$

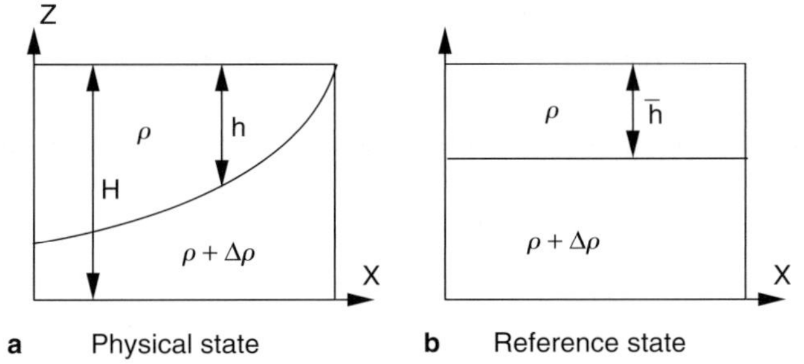

Fig. 3.28 Definition of the available gravitational potential energy (AGPE) for a two-layer model: **a** physical state, **b** reference state.

The AGPE of the system is

$$\Pi^a = \int_0^L \chi^0 dx - \int_0^L \chi^r dx = \frac{g\Delta\rho}{2}\int_0^L \left[(H-h)^2 - \left(H-\bar{h}\right)^2\right] dx$$

For the case of a linear profile $h = H(1 - x/L)$, the AGPE is $\Pi^a = g\Delta\rho H^2 L/24$.

Taking the warm water pool in the Pacific Ocean as an example, where $H = 100$ m, $L = 10,000$ km, and $\Delta\rho = 30$ kg/m^3, the meridional width of the pool is approximately 200 km, so the total amount of AGPE is estimated as $\Pi^a_{wwp} \simeq 2.5 \times 10^{17}$J. The density of AGPE is 125 J/m^3; much larger than the KE associated with the currents. If all this energy is converted to KE, this can accelerate a barotropic current from a speed of zero to 0.5 m/s.

Meso-scale APE

Owing to the nonlinear equation of state of seawater and the complicated bottom topography, ascertaining the state of minimal GPE in the ocean is difficult. There are several ways to overcome such difficulties. First, seawater is nearly incompressible, hence the compressibility of seawater can be ignored. Under such an approximation, the reference state of minimal GPE can be found through a simple sorting program (Huang, 1998b). Second, the horizontal-mean density profile can be used as the reference state as a compromise. This approach overcomes the difficulty in searching for the reference state using the nonlinear equation of state. In fact, in most studies of the basin-scale oceanic circulation APE are calculated, according to Oort *et al.* (1989, 1994) and Reid *et al.* (1981), as

$$\Pi^a_{MS} = -g \iiint_V \frac{[\rho - \bar{\rho}(z)]^2}{2\bar{\rho}^h_{\Theta,z}} dv \tag{3.144}$$

where the reference state $\bar{\rho}(z)$ is defined by horizontal averaging of the *in situ* density, and $\bar{\rho}^h_{\Theta,z}$ is the horizontal-mean vertical gradient of potential density. This definition has been widely used for the diagnosis of AGPE in basin-scale (or global-scale) circulation problems. Since it is adapted from meso-scale dynamics (Pedlosky, 1987a), we will classify it as meso-scale AGPE (MS AGPE). As shown shortly, this definition can be refined further, depending on the horizontal scale used to obtain the reference state.

This definition is simple and easy to use, but it has some problems as well. First, although it is a sound approximation for the study of meso-scale eddies and baroclinic instability, it is inaccurate for the study of the thermohaline circulation. It has been found that the application of this traditional definition of AGPE and its sources can lead to substantial errors, as shown in the case of an incompressible ocean (Huang, 1998b). Second, the contribution due to internal energy is not clearly defined in this formulation. Third, horizontal mixing of the density field is inevitably required during adjustment from the physical state to the reference state; thus the definition is not truly based on adiabatic processes.

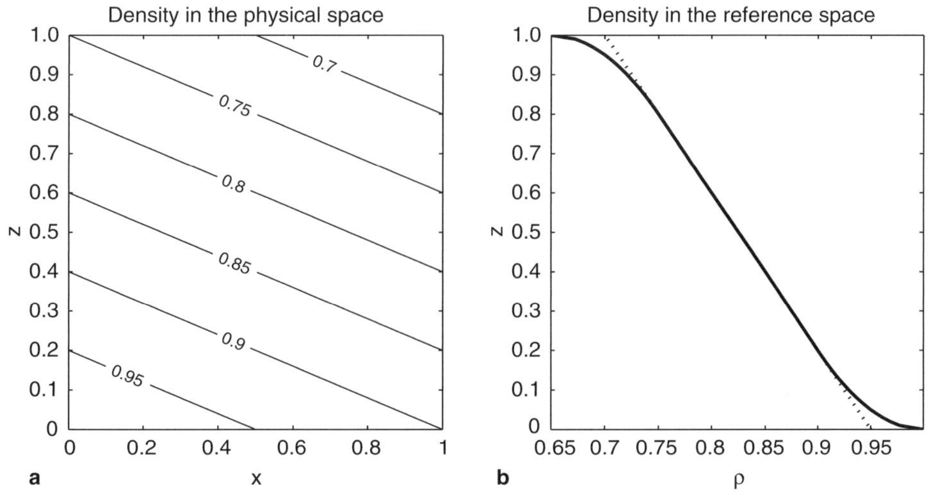

Fig. 3.29 Stratification in the **a** physical and **b** reference state, for the case $a = 0.1$ and $b = 0.25$. *Solid* (*dotted*) line in (**b**) indicates stratification in the reference state under the exact (meso-scale) definition (Huang, 2005a).

The pitfalls of the MS AGPE can be illustrated by the following example. Let us consider a simple stratification (Fig. 3.29a):

$$\rho = \frac{\rho^*}{\rho_0} = 1 - ax - bz, \quad 0 \leq x = \frac{x^*}{L} \leq 1, \quad 0 \leq z = \frac{z^*}{H} \leq 1 \tag{3.145}$$

where the superscript * denotes dimensional variables. By definition, stratification in the reference state used in the MS AGPE is the horizontal mean density $\overline{\rho} = \overline{\rho}^x(z)$, depicted by the dotted straight line (Fig. 3.29b), while stratification in the reference state used in the exact definition is the solid curve.

It is clear that water with a density lighter than 0.7 and heavier than 0.95 does not exist in the reference state defined in the MS AGPE. Thus, the transition from the physical state to the reference state must involve horizontal mixing, i.e., the reference state defined in the MS AGPE cannot be reached adiabatically.

Using the MS AGPE, it is readily seen that

$$\chi_0 = g \iint \rho z dx dz = g \int \overline{\rho}^x dz = \chi_{ref} \tag{3.146}$$

That is to say, based on the MS definition of APE, the total amount of GPE in the physical state and the reference state is exactly the same. In fact, the adjustment process implied in the MS definition involves only horizontal shifting and mixing of water masses, so the total amount of GPE should not change. However, using the formulae from the MS definition,

AGPE in this system is equal to

$$\Pi_{MS} = \frac{a^2}{24b} g \rho_0 H^2 L \tag{3.147}$$

On the other hand, the amount of AGPE from the exact definition can be calculated analytically. For the case with $a = 0.1$ and $b = 0.25$, $\Pi^a_{MS} = 0.00167 g \rho_0 H^2 L$ and $\Pi^a_{real} = 0.00153 g \rho_0 H^2 L$; thus, AGPE from the MS definition is about 10% larger than that from the exact definition.

Available potential energy in the compressible ocean

The exact definition of available potential energy

Using the general definition, the available potential energy Π^a can be defined as

$$\Pi^a = g \int (z - Z)\, dm + \int \left(e - e^r\right) dm + p_s \left(V - V^r\right) \tag{3.148}$$

where z (Z) is the vertical position in the physical (reference) state, with $z = 0$ defined as the deepest point on the bottom; e is the internal energy calculated in terms of the Gibbs function as discussed in Section 2.4.1, superscript r indicates the reference state, $dm = \rho dv$ is the mass element, and V (V^r) is the volume of the ocean in the physical (reference) state. Since the calculation is carried out in the mass coordinate, the choice of reference level of gravitational potential energy does not matter. The mass of each water parcel is assumed to be conserved during the adjustment from the physical to the reference state. In this calculation we omit the air–sea exchange of mass, heat, and freshwater fluxes; however, the atmospheric pressure force is retained, and this is assumed to be constant.

Thus, APE consists of three parts: the available GPE (AGPE), the available internal energy (AIE), and the pressure work (PW). Although the total amount of GPE is dependent on the choice of the reference level, the total amount of AGPE is not.

Finding the reference state

Using an iterative computer program, one can search for the reference state that is stably stratified and has a global minimal GPE, although we are unable to guarantee that such a state is also a state of global minimum of total potential energy.

The chief procedure is an iterative search program. There are two inner loops, and both of them are based on sorting according to the potential density of each grid box. The first inner loop moves upward from the deepest point. Since the maximal depth defined in the climatological data is 5,750 m, we will use 5,880 db as the highest reference pressure to begin the iteration process. Density in all grid boxes is sorted out, with the heaviest water parcel sitting on the bottom. In this process, bottom topography enters the calculation of the thickness of the individual layer $dz = dm_{ijk}/\rho_{ijk}/S(z)$, where dm_{ijk} and ρ_{ijk} are the mass and density of the grid box ijk, and $S(z)$ is the horizontal area at level z. To improve the stability of the water column at all depths, repeat the sorting process as follows. All water parcels lying above the pressure of 5,800 db will be re-sorted by using a smaller reference pressure of 5,750 db. A new array of density is formed, including the new density calculated from the new reference pressure (5,750 db) and the old density calculated from the previous reference pressure (5,880 db). This new density array is accordingly sorted again. Similarly, using pressures with intervals of 100 db, i.e., using reference pressures of 5,650 db, 5,550 db, ... to re-sort, will guarantee the stability of the whole water column at any given depth.

A second inner loop moves from the sea surface downward with similar pressure increments. This loop is used to provide correction to *in situ* pressure and density. A simple computer program can be written to sort out the stratification for any given number of reference pressure levels. Thus, the reference state, which is stably stratified at any given level, can be constructed to any reasonable desired degree of accuracy.

APE in the world's oceans

The above search program was applied to the annual-mean climatology of temperature and salinity, with "realistic" topography, from the Levitus 1998 dataset (Levitus *et al.*, 1998). For simplicity we ignore the bottom-topography-induced blocking and separation of basins at different levels. Thus, the world's oceans are treated as a single basin, whose horizontal area equals that of the world's oceans at the corresponding levels.

AGPE has been calculated based on Eqn. (3.148) for the world's oceans and three major basins separately, and the results are based on 5,800 reference levels, with a reference pressure increment of 1 db.

The calculated AGPE represents a theoretical upper limit of energy that can be converted into KE if all the forcing of the system is withdrawn. Under such an idealized situation, water parcels with large densities sink to the bottom and push the relatively lighter water parcels upward. This process is illustrated by the potential contribution of GPE through adiabatic adjustment (Figs. 3.30 and 3.31). It is readily seen that the primary contributor of AGPE is dense water in the Southern Ocean, which is associated with steep isopycnal surfaces sustained by a strong wind stress forcing.

During the adjustment, dense water from high latitudes sinks and spreads as horizontal layers to the bottom of the world's oceans; and the relatively light deep water at low latitudes is pushed upward. Thus, deep water at high latitudes contributes to AGPE positively, but deep water at low latitudes contributes to AGPE negatively.

In the calculation of APE for individual basins, the boundary between the Atlantic and Indian Oceans is set along 20° E, the boundary between the Indian and Pacific Oceans is set along 156° E, and the boundary between the Atlantic and Pacific Oceans is set along 70° W. The Arctic Ocean is separated from the North Pacific Ocean along the Arctic circle and AGPE is calculated for the Atlantic and the Arctic together. The results are summarized in Table 3.3, including the cases with/without the Mediterranean. After the adjustment, the sea level drops 4.2 cm; thus, the sea surface atmospheric pressure input is 1.43×10^{16} J of work on the oceans, amounting to an energy source of 0.01 J/m^3, which is negligible

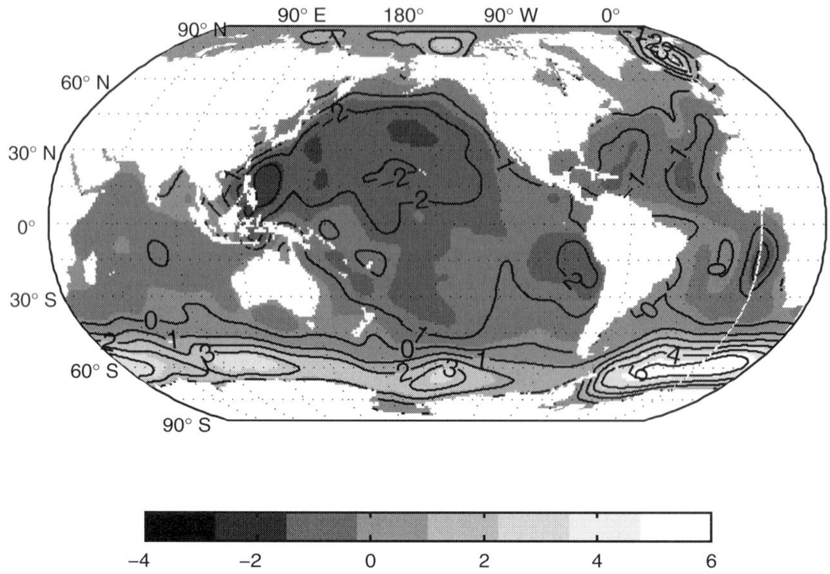

Fig. 3.30 Global contribution to AGPE through adiabatic adjustment (in 10^{11} J/m^2) (Huang, 2005a).

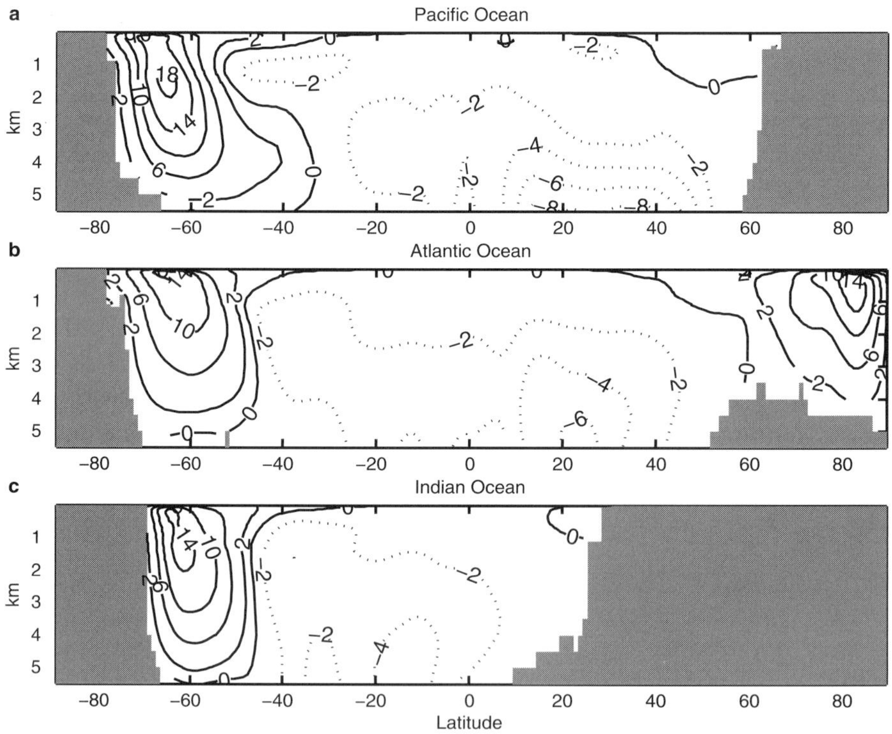

Fig. 3.31 Contribution to AGPE through adiabatic adjustment in **a** the Pacific Ocean, **b** the Atlantic Ocean, and **c** the Indian Ocean (in 10^{13} J/m^2) (Huang, 2005a).

Table 3.3. *Available potential energy (APE) for different basins in the world's oceans (including or excluding the Mediterranean Sea), in J/m*^{3}

Basin		AGPE	AIE	APE
Atlantic	With Mediterranean	2,316.4	$-1,608.4$	708.0
	Without Mediterranean	2,338.1	$-1,699.3$	638.8
Pacific		970.7	-489.0	481.7
Indian		1,235.4	-762.7	472.7
World's oceans	With Mediterranean	1,463.8	-799.4	664.4
	Without Mediterranean	1,474.5	-850.3	624.2

Table 3.4. *Available gravitational potential energy (AGPE) dependence on the equation of state, in J/m*^{3}

Equation of state	UNESCO	Linear in T, S	Linear in T, S, P	Boussinesq approximations
AGPE	1,474.5	793.4	1,316.0	904.6

compared with other terms. Comparatively speaking, AGPE density in the Atlantic Ocean is the highest, while in the Pacific Ocean it is the lowest.

The effect of nonlinearity of the equation of state of seawater is of vital importance in calculating AGPE; this can be further illustrated by processing the same climatological dataset, but with different equations of state. A linear equation of state is defined as $\rho = \rho_0 [1 - \alpha (T - T_0) + \beta (S - S_0) + \gamma P]$, where $\rho_0 = 1,036.9$ kg/m^3, $\alpha = 0.1523 \times 10^{-3}$/°C, $\beta = 0.7808 \times 10^{-3}$, and $\gamma = 4.462 \times 10^{-6}$/db. It is interesting to note that an equation of state with a linear dependence on pressure can substantially improve the calculation of the AGPE.

The calculation above is based on mass coordinates; thus the mass of each grid box is conserved during the adjustment. A traditional Boussinesq model does not conserve mass, so that the meaning of APE inferred from such a model is questionable. As an example, APE based on the Boussinesq approximations can be calculated and is noticeably smaller than that calculated from the truly compressible model (Table 3.4). (In such a calculation the volume of a water parcel remains unchanged after adjustment, even if its density is adjusted to the new pressure.) It is speculated that a smaller AGPE may affect the model's behavior during the transient state.

An interesting and potentially very important point is that AIE in the world's oceans is negative. For reversible adiabatic and isohaline processes, changes in internal energy obey $de = -pdv$. Since cold water is more compressible than warm water, during the exchange of water parcels the increase in internal energy associated with the cold water parcel is larger than the internal energy decline associated with the warm water parcel (Fig. 2.13). Thus,

AIE associated with the adjustment to the reference state is negative. In contrast, warm air is more compressible than cold air; therefore the corresponding AIE in the atmosphere is positive.

Release of AGPE

Although a special term (available GPE) has been given to the difference in GPE between the physical state and the reference state, such energy may not be completely released, owing to the existing geostrophic constraint in the ocean. Consider a two-dimensional, two-layer model ocean, with density ρ and $\rho + \Delta\rho$ (Fig. 3.32). The slope of the interface is $\mathbf{S}$, and the interface height is $b = \mathbf{S}x + H/2 - \mathbf{S}L/2$. The total amount GPE in the physical state is

$$\chi_0 = \frac{g\rho}{2}H^2L + \frac{g\Delta\rho}{24}\left(3H^2L + \mathbf{S}^2L^3\right) \tag{3.149}$$

and the AGPE of the system is

$$\Pi = \frac{g\Delta\rho}{24}L^3\mathbf{S}^2 \tag{3.150}$$

During an adjustment, the slope of the interface declines to a new value, $\mathbf{S}_n$, and AGPE is converted into KE. Under the conditions of geostrophy and stagnancy of the lower layer, velocity and total kinetic energy are:

$$u = \frac{g'\mathbf{S}_n}{f}, \quad E_k = \frac{HL}{4}\rho u^2 \tag{3.151}$$

where $g' = g\Delta\rho/\rho$ is the reduced gravity, and f is the Coriolis parameter.

The total amount of GPE of the system is reduced:

$$\Delta\chi = \frac{g\Delta\rho}{24}L^3\left(\mathbf{S}_l^2 - \mathbf{S}_n^2\right) \tag{3.152}$$

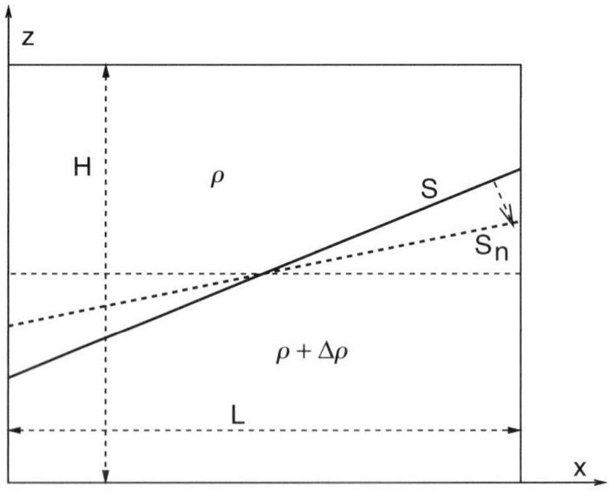

Fig. 3.32 Sketch of the adjustment of a front in a two-layer ocean.

Assuming that the process is reversible and adiabatic, the total energy is therefore conserved:

$$E_k = \Delta \chi \tag{3.153}$$

From Eqns. (3.151), (3.152), and (3.153), we obtain

$$\frac{\mathbf{S}_n^2}{\mathbf{S}_l^2 - \mathbf{S}_n^2} = \frac{1}{6}\left(\frac{L}{\lambda}\right)^2 = r^2, \quad \text{or} \quad \mathbf{S}_n^2 = \frac{r^2}{1+r^2}\mathbf{S}^2 \tag{3.154}$$

where $\lambda = \sqrt{g'H}/f$ is the deformation radius. The fraction of APE converted into KE is thus

$$\eta = \frac{\mathbf{S}^2 - \mathbf{S}_n^2}{\mathbf{S}^2} = \frac{1}{1+r^2} \tag{3.155}$$

For a circum-global current system, with $L = 1,000$ km, $H = 1,000$ m, and $g' \simeq 0.01$ m/s^2, at mid latitudes $f \simeq 10^{-4}$/s, $r^2 \simeq 167$, $\eta \simeq 1/168 \simeq 0.006$; while for an equatorial channel, $f \simeq 10^{-5}$, $r^2 \simeq 10/6$, $\eta \simeq 3/8$. Thus, the APE of such a system can be quite efficiently converted into KE. However, the conversion rate for a mid- or high-latitude channel is much smaller. This result is consistent with the scaling of the wind-driven gyre. As discussed in Section 4.1.3, the ratio between kinetic energy of the mean flow, kinetic energy of eddies, and APE of the mean state is $K_{mean} : K_{eddies} : \Pi = 1 : 100 : 1000$. Therefore, owing to the geostrophic constraint, only a small portion of the APE can be converted into kinetic energy. We should also mention that the velocity field is associated with huge amount of APE.

Redefinition of meso-scale AGPE

Although problems do occur when the MS AGPE is applied to the study of basin-scale dynamics, such a concept remains a very powerful tool in the study of meso-scale dynamics, in particular for understanding the essence of baroclinic instability. It is well known that the GPE of the mean state is the quantity that is released and converted into KE and potential energy of meso-scale eddies; therefore, it is highly desirable to be able to find a way to keep using this commonly accepted concept of AGPE. One way is to limit its usage to problems with a horizontal scale on the order of the deformation radius. In the following discussion, Eqn. (3.144) is applied to the world's oceans with a grid based on different horizontal resolutions.

First (Case A), we apply Eqn. (3.144) to each $1^\circ \times 1^\circ$ grid cell in the world's oceans, i.e., density averaged over the four corners of each $1^\circ \times 1^\circ$ grid cell is used as the reference density and stratification in Eqn. (3.144). Assuming that the density distribution within this grid cell is a bi-linear function of the local coordinates, Eqn. (3.144) is reduced to a simple finite difference form. The results of this calculation will be referenced as MS_1 AGPE.

Second (Case B), Eqn. (3.144) is applied to the world's oceans, with a $2^\circ \times 2^\circ$ resolution and similar finite difference scheme as for MS_1 AGPE. This case is termed MS_2 AGPE.

Table 3.5. *Global sum of AGPE of all cases, in EJ (10^{18} J). Values in the last column are the total amount of AGPE and the net APE, calculated from the exact definition of APE*

Case	A (MS_1)	B (MS_2)	C (MS)	Exact
AGPE	1.07	8.22	1,277	1,880(810)

Third (Case C), Eqn. (3.144) is applied to the entire world's oceans: we will see that the density profile $\overline{\rho}(z)$ is obtained by horizontal averaging over the global oceans, the corresponding global mean vertical potential density gradient is expressed as $\bar{\rho}^{h}_{\Theta,z}$. The results obtained are referred to as MS AGPE.

The total amount of AGPE and its horizontal distribution corresponding to these three cases, as listed in Table 3.5, are different. Most strikingly, we see that the total amount of AGPE increases dramatically as the horizontal scale, which is used to calculate the reference density $\overline{\rho}$, increases. The large differences in the total amount of AGPE calculated for these three cases can easily be explained by Eqn. (3.144). Assuming that density is a linear function in space, AGPE calculated from Eqn. (3.144) is linearly proportional to the square of the grid size. Therefore, as the horizontal scale of density averaging increases from 1° (Case A) to 2° (Case B), to basin scale or global scale (Case C), the total amount of AGPE increases 1,000 times (Table 3.5).

MS_1 AGPE gives a global sum of 1.07 EJ, which is ten times smaller than the total amount of meso-scale energy in the world's oceans, 13 EJ, as estimated by Zang and Wunsch (2001). Since meso-scale eddies obtain their energy primarily from AGPE, this estimated amount of AGPE (1.07 EJ) may not be large enough to support the meso-scale activity.

By doubling the grid size, the global integrated MS_2 AGPE is 8.22 EJ, which is comparable with the total KE of meso-scale eddies, 13 EJ. Major density fronts in the world's oceans mostly locate at mid latitudes, where the first baroclinic radius of deformation is about 40 km. The typical wavelength of a meso-scale eddy is estimated as 2π times the Rossby deformation radius, which is approximately 200 km at mid latitudes; thus, a $2° \times 2°$ grid is the optimal size to catch GPE available through baroclinic instability.

Note that the reference state used in this calculation is a density field which is piecewise constant within each grid cell. Such a reference state is not a state of minimal GPE for the world's oceans, because additional adjustment of the density field is possible and the GPE can be still lower. Nevertheless, the horizontal distribution of MS_1 AGPE or MS_2 AGPE provides important information regarding the potential energy source supporting baroclinic instability on the scale of the deformation radius.

The most remarkable feature of the MS_1 AGPE distribution in the world's oceans (Fig. 3.33) is the high density of MS_1 AGPE associated with strong density fronts, including the western boundary currents, such as the Kuroshio and Gulf Stream between the latitudes of 20° N and 40° N, and the ACC.

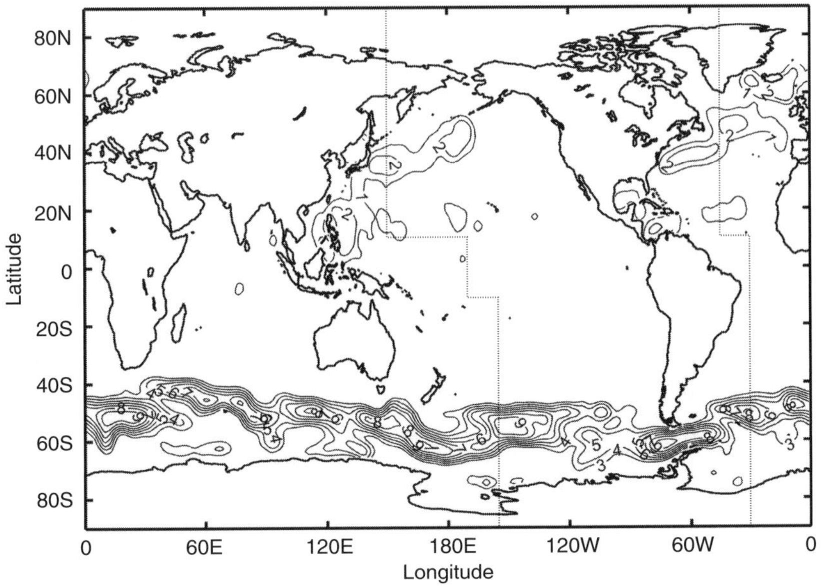

Fig. 3.33 Horizontal distribution of the vertically integrated MS_1 AGPE in the world's oceans (in $10^3 J/m^2$) (Feng *et al.*, 2006).

Meridional distribution of AGPE

The distribution of MS_1 AGPE for a composed meridional section, indicated by the faint line in Figure 3.33, is shown in Figure 3.34. All major currents and frontal structures in the upper ocean can be clearly identified, including the cores of ACC appearing in the latitude of 50° S, the Gulf Stream at latitudes of 40° N. The equatorial current system can be unmistakably linked to the frontal structure in the upper 500 m.

Therefore, two distinctly different definitions of AGPE exist and both can be used as diagnostic tools in the study of oceanic circulation. MS_1 AGPE can be used to diagnose the AGPE associated with strong density fronts and currents in the system. MS_1 AGPE is available for meso-scale activity, such as the baroclinic instability. In fact, MS_1 AGPE is directly linked to the transformation rate of GPE from the mean state to eddies, which is commonly parameterized in terms of the Gent *et al.* (1995) scheme and other similar variations.

On the other hand, the exact definition of AGPE can be used as a powerful tool in diagnosing the energetics of the basin-scale circulation. For example, the energetics of the basin-scale circulation over decadal or centennial time scales can be examined, using the exact definition of AGPE.

Relationships between different forms of GPE

Different forms of GPE have been discussed above and their relations are summarized in Table 3.6. GPE depends on the total mass of water in the oceans and the geometry of the

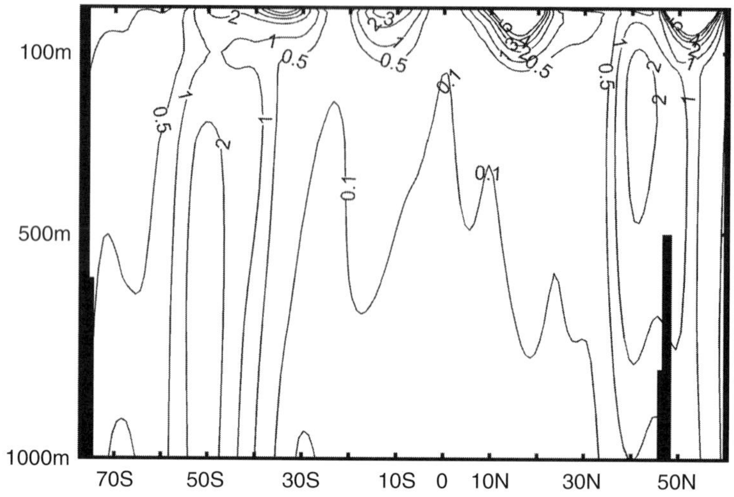

Fig. 3.34 Meridional section of MS_1 AGPE density distribution in the Atlantic Ocean (J/m^3) (Feng *et al.*, 2006).

seafloor. Over long geological time scales, the shapes of the basins in the world's oceans change and sea level can also change, owing to glacial–interglacial cycles. However, for the time scales relevant to dynamical oceanography, most of the GPE is dynamically inert.

Stratified GPE represents part of GPE which is associated with the density deviation from the global mean. Changes of SGPE are related to variations of stratification, induced by the departures of the mean temperature and salinity, on very long time scales.

The more relevant term is the APE. The definition of APE depends on the choice of reference state. The commonly used definition of APE is now redefined as the MS_1 APE, with typical length scale on the order of 100 km, which applies to meso-scale problems in the oceans. This term can be used for the study of the energetics of meso-scale motions in the oceans.

Thermohaline circulation has a length scale equal to the basin scale, to which the traditional definition of APE is not applicable. Thus, the original definition of available potential energy is used. For problems with centennial time scales, the combination of APE and MS_1 APE may be useful.

Comparison with kinetic energy

The total amount of kinetic energy in the world's oceans remains poorly known. It is argued that most of the kinetic energy is associated with meso-scale eddies; however, there is no reliable estimate, owing to the technical difficulty in measuring velocity in the ocean with the high resolution relevant to meso-scale eddies. On the other hand, numerical model simulations are far from being able to resolve all the energetic eddies in the ocean. Thus, our discussion here is based on some rough estimates. The currently available estimates

Table 3.6. *Different forms of gravitational potential energy (GPE)*

Energy form	Global sum	Length scale	Type of motion	Processes	Time scale
GPE	2.1×10^{25} J	Global scale	Thermohaline circulation	Seafloor variation	>Myr
				Sea level variation	Kyr to Myr
SGPE	-1.4×10^{23} J	Basin scale	Thermohaline circulation	Stratification changes due to changes in mean T and S	Kyr to Myr
APE	810 EJ	Basin scale	Thermohaline/wind-driven circulation	Wind stress, tides, and thermohaline forcing	Decadal to millennial
MS_1 AGPE	1–8 EJ	100 km	Wind-driven circulation, meso-scale eddies	Wind stress and thermohaline forcing	Seasonal to decadal

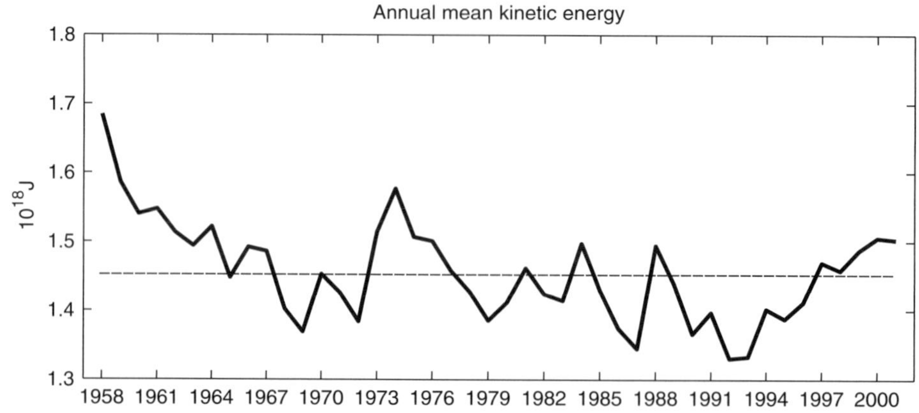

Fig. 3.35 The annual mean kinetic energy of the world's oceans diagnosed from Simple Ocean Data Assimilation (SODA) (Carton and Giese; 2008), *dashed line* indicates the mean value (1.46 EJ).

are: the total amount of energy for eddies is 13 EJ, among which the total kinetic energy is 2.6 EJ (Ferrari and Wunsch, 2009).

A diagnosis of the Simple Ocean Data Assimilation (SODA) data (Carton and Giese, 2008) returns a similar value. The SODA data have a moderate horizontal resolution of $0.5^\circ \times 0.5^\circ$ and 40 vertical levels. For example, the mean kinetic energy averaged over the period 1958–2001 is about 1.46 EJ. The total kinetic energy, as diagnosed from this dataset, went through large-amplitude, interannual oscillations (Fig. 3.35). In general, it is comparable with the value of 2.6 EJ as postulated by Ferrari and Wunsch (2009). It is speculated that the total kinetic energy of the world's oceans may be even larger than the value diagnosed from the current SODA data, if model data with much finer spatial and temporal resolution were available.

The total kinetic energy goes through a noticeable annual cycle, as shown in Figure 3.36. It is also interesting to note that two-thirds of the total kinetic energy (1.05 EJ) is due to the zonal velocity, in particular the fast zonal flow associated with the equatorial jets and the ACC, as shown in Figure 3.37. On the other hand, if the annual mean velocity is used in the calculation, the total kinetic energy is reduced to 0.91 EJ, as shown by the dot-dash line in Figure 3.36.

3.7.3 Balance of gravitational potential energy in a model ocean

Many traditional oceanic circulation models are based on Boussinesq approximations, and thereby artificial sources and sinks of energy are induced inevitably in such models. Furthermore, owing to the replacement of mass conservation with volume conservation, GPE is not conserved in such models; so diagnosing the GPE balance is difficult, if not impossible. To solve this kind of problem, a mass-conserving oceanic general circulation model, PCOM (Pressure Coordinates Ocean Model), was developed by Huang *et al.* (2001).

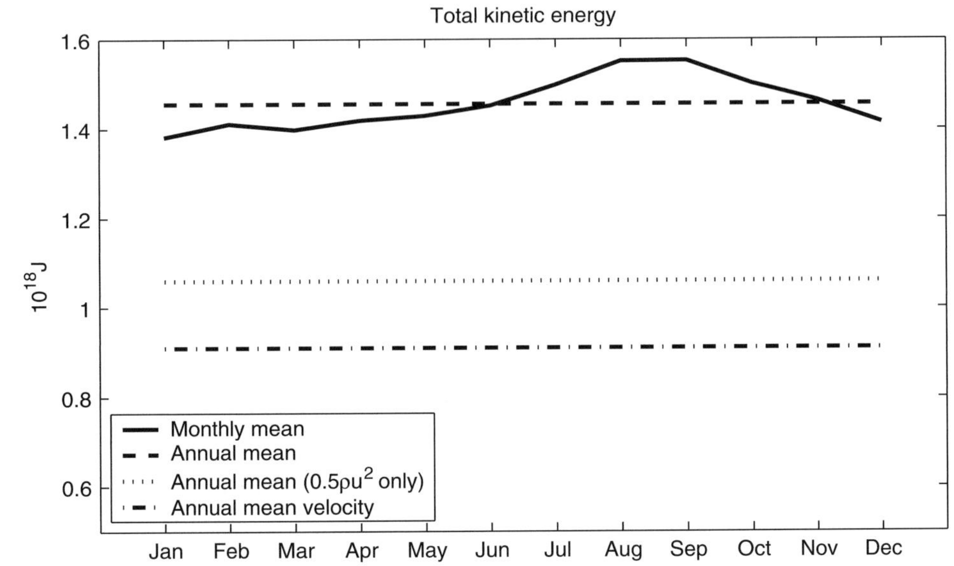

Fig. 3.36 Total kinetic energy of the world's oceans over the course of a year diagnosed from SODA (Carton and Giese, 2008).

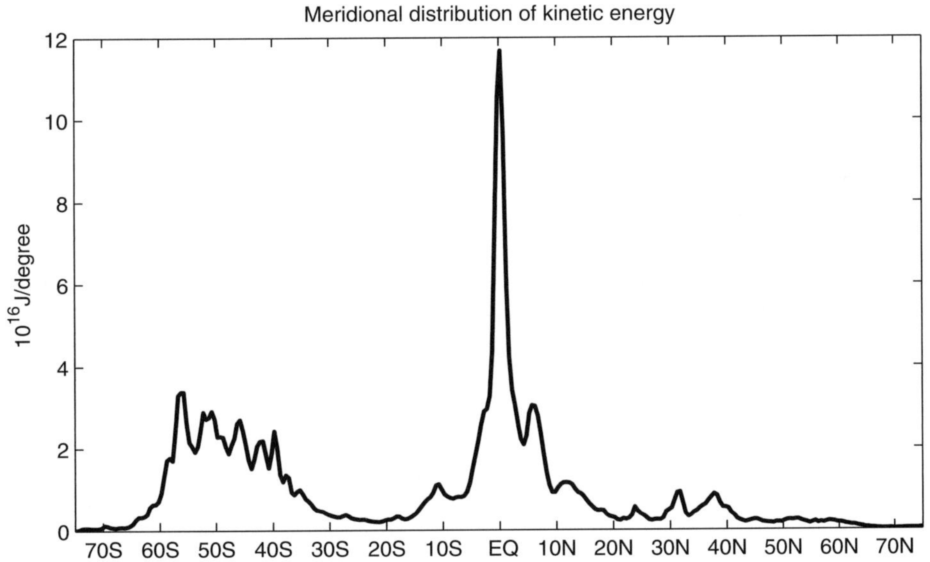

Fig. 3.37 The meridional distribution of zonal-mean kinetic energy of the world's oceans diagnosed from SODA (Carton and Giese, 2008).

Buoyancy work has been used as a tool in studying the energetics of thermohaline circulation. For the case of a linear equation of state, buoyancy work is quite close to the GPE balance diagnosed from a mass-conserving model. However, for the case with a nonlinear equation of state, buoyancy is no longer a conserved quantity, and energy analysis based on buoyancy work may not reflect the real physics.

Theoretically, we may seek a procedure of establishing a theory for the total energy conservation of the ocean, containing a small parameter whose lowest order is the Boussinesq approximations with zero divergence and the next order the total energy, including the GPE, that should be conserved. Nevertheless, using a mass-conserving model can provide a much clearer picture of GPE balance in the ocean.

A numerical model that will be discussed here is a pressure-η coordinate model, which is a slight modification of the pressure-σ coordinate model of Huang *et al.* (2001). η-coordinates are defined as $\eta = (p - p_t)/r_p$, $r_p = (p_b - p_t)/p_B$, where p_b is the bottom pressure, p_t is the sea-level atmospheric pressure, and $p_B = p_B(x, y)$ is the time-invariant reference bottom pressure. The pressure-η coordinate is fairly close to the commonly used z-coordinate model. In the following discussion, we will set $p_t = 0$ and p_B equal to the bottom pressure p_b of the initial state of rest.

The model ocean is a $60° \times 60°$ basin that is 5 km deep. The southern boundary of the model ocean is the equator, and the northern boundary is at 60° N. The horizontal resolution is $2° \times 2°$, and it has 30 layers stretching from a thin layer of 30 m on the top to about 300 m on the bottom. The actual layer thickness is in the pressure unit. The model starts from an initial state of rest with a uniform temperature of 10°C; thus a 30 m layer in the upper ocean corresponds to a pressure increment of 30.8 db. The layer thickness, in terms of pressure or equivalent thickness in geopotential height units, will slightly evolve with time in the present pressure-η coordinates. However, most results will be presented in terms of geometric height.

The equation of state adopted in the model has two versions:

1. In the case of a linear equation of state,
$$\rho = \rho_0 \left(1 - \alpha T\right) \tag{3.156}$$
where $\alpha = 0.0001523/°\text{C}$ is a constant thermal expansion coefficient.
2. For the case of a nonlinear equation of state a cubic polynomial expression for the temperature dependency is used:
$$\rho = 1028.106 + 0.7948(S - 35.0) - 0.05968T - 0.0063T^2 + 3.7315 \times 10^{-5}T^3 \tag{3.157}$$
With $S = 35$, this equation gives a temperature dependence of density quite close to that of seawater.

The model ocean is subject to a thermal relaxation boundary condition for the surface temperature: a reference temperature of 0°C at the northern boundary linearly increasing in the meridional direction to 25°C at the southern boundary. In addition, no wind stress or haline forcing exists in the model.

For all cases considered, a constant vertical diffusivity of 0.3×10^{-4} m^2/s and a horizontal diffusivity 10^3 m^2/s are used. A bottom friction parameterization is applied, and a nondimensional bottom friction parameter of $c_0 = 2.6 \times 10^{-3}$ is adopted. In order to highlight the essential physics of GPE balance in the model ocean, neither mixing tensor rotation nor eddy transport is included.

Four cases will be discussed here:

- *Case A:* surface heating at low latitudes and cooling at high latitudes, with a linear equation of state
- *Case B:* the same as Case A, except with a nonlinear equation of state
- *Case C:* bottom heating at high latitudes and cooling at low latitudes, with a linear equation of state
- *Case D:* the same as Case C, except with a nonlinear equation of state.

The reason for placing heating at high latitudes for Cases C and D is as follows. If the model ocean were heated along the equator from below, strong convection would develop there. Thereby, a circulation structure quite different from that of high latitudes would develop owing to the zero Coriolis parameter along the equator, such as in Cases A and B. By placing heating at a high-latitude seafloor, the circulation patterns are almost mirror images of the cases where there is heating from the surface along the equator. In addition, the vertical layer thickness arrangement is flipped, so that the fine vertical resolution near the bottom is suitable for resolving the strong stratification associated with the thin cold bottom boundary layer for Cases C and D.

The GPE balance

In the mass coordinate GPE is defined as $\chi = \iiint \phi r_p/g \, d\eta dx dy$, where $\phi = gz$, the geopotential, is a diagnostic variable $\phi = gz_b + r_p \int_\eta^{p_B} 1/\rho \, d\eta$. For steady circulations the time-dependent term vanishes, resulting in a five-term balance of GPE, i.e.

$$ADV + HM + VM + SF + CA = 0 \tag{3.158}$$

where *ADV* is advection, *HM* and *VM* are horizontal and vertical mixing, *SF* is surface forcing, and *CA* is convective adjustment. Note that the effect of cooling is separated into two parts: *SF* and *CA*. Thus, the loss of GPE due to convective adjustment is counted as a separate item.

GPE balance for Case A

For Case A, i.e., forcing from above and with a linear equation of state, the primary GPE balance is between the source due to vertical mixing (19.8 GW) and the loss due to convective adjustment (17.1 GW) (Fig. 3.38a). There is a small portion of GPE (2.7 GW) that is converted to KE. Since the model is subject to thermal forcing only, the conversion from GPE to KE is the only source maintaining the circulation against friction.

There is also a tiny source of GPE due to the surface thermal forcing. The smallness of this source term is due to the fact that heating and cooling are exactly balanced for a steady state, so the contribution to GPE is nearly canceled, as detailed in the Appendix. If

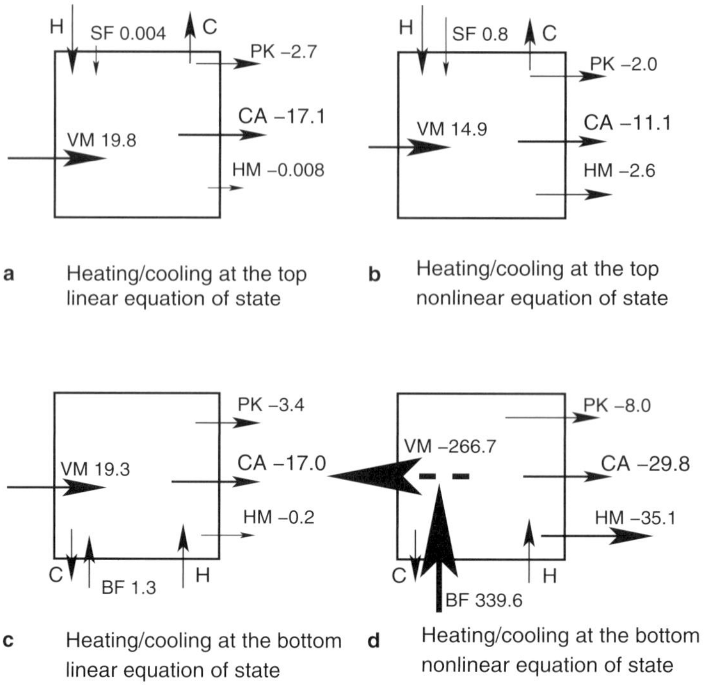

Fig. 3.38 **a–d** GPE balance (GW): *CA* is the GPE sink due to convective adjustment *VM*, *HM* is the GPE source due to vertical (horizontal) mixing, *PK* is the sink due to the conversion from GPE to KE, *SF* is the source of GPE due to surface thermal forcing, and *BF* is the source of GPE due to bottom thermal forcing (Huang and Jin, 2006).

the surface layer thickness were further reduced, this source term would diminish further. Since most solar radiation can penetrate only to a depth of less than 10 m, the source of GPE due to surface heating/cooling is negligible for the world's oceans. (As explained in Section 3.6, a rough estimate of GPE generated by solar radiation penetration in the world's oceans is on the order of 0.01 TW.) Furthermore, induced convective adjustment leads to a large loss of GPE and is the major term in the balance of GPE in the present case.

Note that horizontal mixing appears as a tiny sink of GPE. Since the equation of state is linear, there is no cabbeling. However, horizontal mixing of temperature may affect the geopotential height of water columns and thus change the total amount of GPE for the model ocean.

GPE balance for Case B

In Case B, i.e., with heating/cooling on the upper surface and a nonlinear equation of state (Fig. 3.38b), the source/sink of GPE associated with vertical mixing and convective adjustment is smaller than that with a linear equation of state. On the other hand, the most noticeable contribution of cabbeling is the loss of GPE (2.6 GW) due to horizontal mixing in the model. Because the equation of state is now nonlinear, cabbeling takes place, reducing both the amount of GPE generated by vertical mixing and the loss of GPE due to convective adjustment.

GPE balance for Case C

In Case C, i.e., heating/cooling on the seafloor and a linear equation of state, the balance of GPE is rather similar to that in Case A (heating/cooling from the upper surface). The major difference is that surface thermal forcing now appears as a source of GPE, with a value of 1.3 GW; much larger than that in Case A. This is related to the fact that now heated water must push the whole water column above it, and thus convert a much larger amount of internal energy into GPE.

GPE balance for Case D

Now we turn to Case D, i.e., heating/cooling on the seafloor and a nonlinear equation of state. In this case, heating/cooling appears as a large source of GPE. In addition, vertical mixing becomes a large sink of GPE, rather than a source. This fact is due to the effect of the strong cabbeling appearing over the bottom of the model ocean, which turns the source of GPE due to vertical mixing into a sink. It is worth noting that every term in the GPE balance in this case is greatly enhanced; therefore, the circulation seems more energetic. The implications of these modeling results remain unexplored.

Distribution of source/sink of GPE in a pure thermal circulation

The sources/sinks of GPE are non-uniformly distributed in a meridional direction (Fig. 3.39a). Convective adjustment is primarily confined to high latitudes, while vertical mixing linearly increases toward low latitudes, where stratification is stronger. Surface thermal forcing produces a small amount of GPE at low latitudes; however, this production of GPE due to heating at low latitudes is offset by the loss of GPE due to cooling at high latitudes, and the contribution from vertical mixing at low and mid latitudes is the main contributor to the generation of GPE in the model basin.

In a steady state, GPE is balanced in each local box. At lower latitudes, the primary balance of GPE is between the source due to vertical mixing and the flux divergence of GPE due to advection. At high latitudes, the balance is primarily between the divergence of GPE transport and the sink due to convective adjustment (Fig. 3.39a). There is a poleward transport of GPE, shown in Figure 3.39b, whose construction will be discussed shortly.

The distribution of the sources/sinks of GPE due to vertical mixing and to convective adjustment are quite different from each other. Vertical mixing generates GPE primarily in the upper ocean at low latitudes, where stratification is strong (Fig. 3.40a). On the other hand, convective adjustment dissipates GPE primarily at high latitudes. In the present case, convective adjustment works all the way to the bottom near the northern boundary (Fig. 3.40b).

A vertical profile of GPE sources/sinks is shown in Figure 3.41. In the upper 500 m, a strong source of GPE exists, primarily due to diapycnal mixing in the main thermocline,

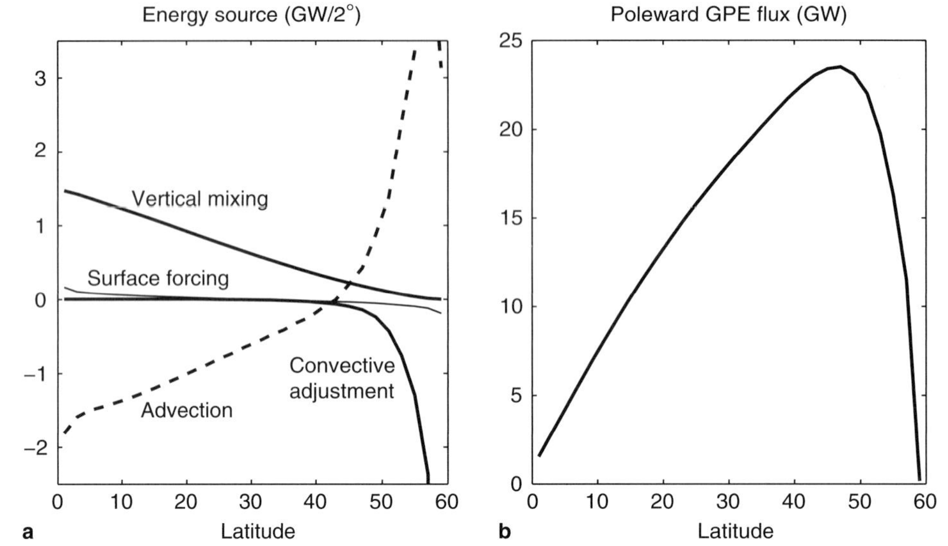

Fig. 3.39 GPE sources/sinks and balance for Case A: **a** meridional distribution of GPE sources/sinks due to vertical mixing, surface forcing, convective adjustment, and advection; **b** poleward transport of GPE (Huang and Jin, 2006).

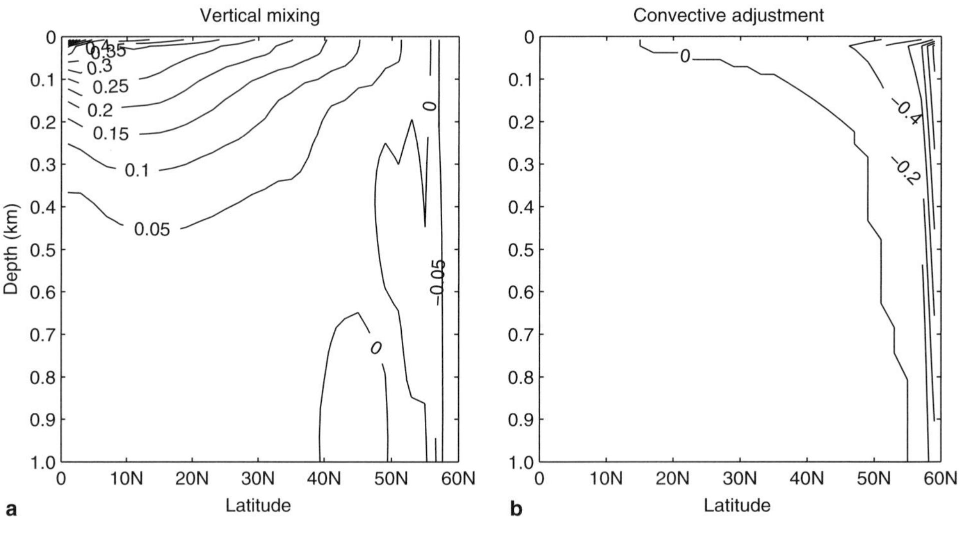

Fig. 3.40 **a** Zonally integrated source of GPE due to vertical mixing ($\kappa = 0.3 \times 10^{-4}$ m^2/s); **b** zonally integrated sink of GPE due to convective adjustment (all in GW/100 m for each 2° latitude band).

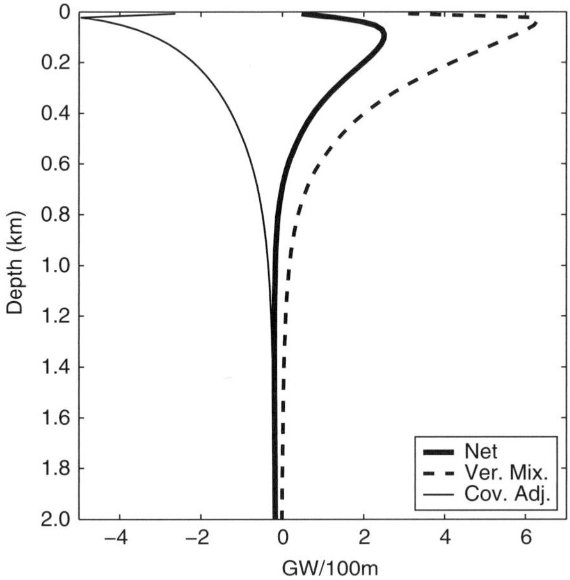

Fig. 3.41 Vertical profile of GPE source due to vertical mixing (*dashed line*) and sink due to convective adjustment (*thin line*) and net GPE source (*heavy line*) (in GW/100 m).

while convective adjustment is a major sink of GPE in the model ocean; however, convective adjustment takes place over the whole depth range, functioning as a large sink of GPE at depths below 600 m.

It is obvious that vertical mixing overpowers the loss of GPE due to convective adjustment in the upper ocean. On the other hand, the loss of GPE due to convective adjustment overpowers vertical mixing in the deep ocean (Fig. 3.41).

Since this model has no wind stress forcing, the energy required for sustaining circulation against friction must come from the conversion of GPE to KE. Recalling Eqn. (3.108), the rate of conversion from KE to GPE is

$$g\overline{\rho w} = g\overline{\rho}\overline{w} + g\overline{\rho' w'}$$

We can interpret the overbar as the basin mean and the perturbations as

$$\rho' = \bar{\rho}^{x,z} - \bar{\rho}^{x,y,z}$$
$$w' = \bar{w}^{x,z} - \bar{w}^{x,y,z}$$

Vertical velocity perturbation is positive in most places except near the northern boundary, where it is large and negative, corresponding to the sinking branch of the meridional circulation. Density perturbation is negative in the southern basin, but it is positive in the northern basin (Fig. 3.42a).

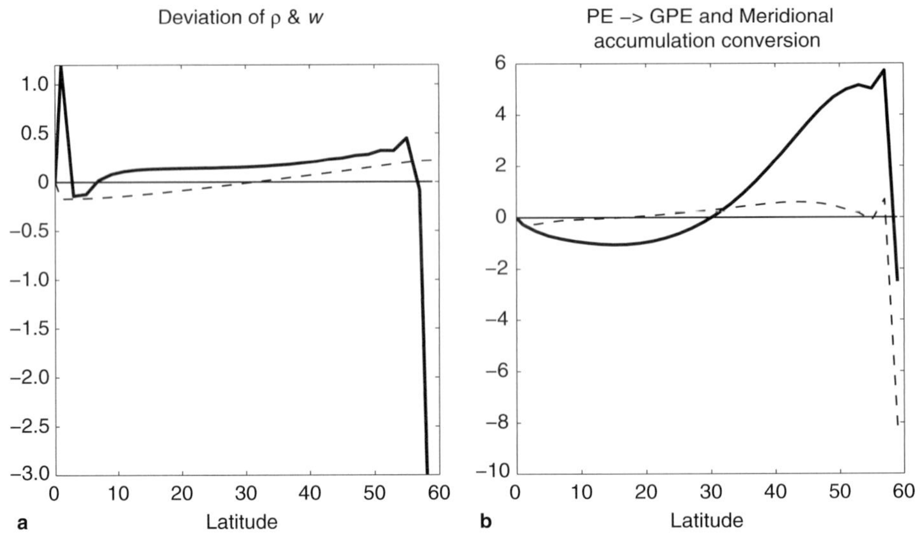

Fig. 3.42 GPE sources/sinks and balance for Case A: **a** meridional distribution of the zonal and vertical mean density (deviation from the basin mean (*dashed line*, kg/m^3) and vertical velocity (*heavy line*, 10^{-6} m/s); **b** KE→GPE conversion ($c(y)$ (*dashed line*, GW per 2° latitude band) and meridional accumulated conversion of PE ($C(y)$ (*heavy line*, GW).

Here we can also interpret the overbars as the horizontal mean at different vertical levels; thus the KE to GPE conversion term is

$$C(y) = \int_{ys}^{y} c(y)dy, \quad c(y) = g \iint \left[\rho - \overline{\rho(k)}^{x,y}\right]\left[w - \overline{w(k)}^{x,y}\right]dxdz \tag{3.159}$$

where $c(y)$ is the conversion rate of KE to GPE at each meridional band, and $C(y)$ is the corresponding meridional accumulated rate, starting from the equator. Both $c(y)$ and $C(y)$ are shown in Figure 3.42b. It can readily be seen that $c(y)$ is negative near the southern and northern boundaries, indicating a conversion from GPE to KE, while it is positive at mid latitudes, indicating that KE is converted back to GPE. The total rate of conversion from GPE to KE is 2.6 GW, as shown by the −2.6 value at the northern boundary.

Since convective adjustment is a large sink of GPE, the balance of GPE requires a large amount of poleward transport of GPE in the model ocean (Fig. 3.40b). Such meridional transport of GPE is produced by a meridional integration of the source of GPE due to vertical mixing, surface forcing, and the sink due to convective adjustment, as shown in Figure 3.39a. In addition, the KE conversion to GPE is shown as the dashed line in Figure 3.42b. The meridional accumulation of these source and sink terms gives rise to the meridional transport profile in Figure 3.39b.

In summary, the balance of GPE in this model ocean is as follows. GPE is generated at low latitudes primarily owing to vertical mixing, and carried by the meridional overturning circulation to high latitudes where it is used to support convective adjustment, and on its migration northward there is a continuous transformation between GPE and KE. It must be borne in mind that this conversion is not unidirectional, as shown in Figure 3.42b.

Seasonal cycle

In a time-dependent problem, GPE varies with time; however, changes in GPE need to be balanced by sources/sinks at each time step. (In numerical schemes, such as the leap-frog scheme used in this model, GPE is not exactly conserved for a time-dependent problem. However, the errors contained are small, and it is worth analyzing the balance of GPE in order to reveal the physical processes involved.)

As an example, we examine the annual cycle of GPE fluxes of the same model, but subject to a simple sinusoidal cycle of relaxation temperature

$$T^* = 25\left(1 - \frac{\theta}{\theta_N}\right) + 5\frac{\theta}{\theta_N}\left(1 - \sin 2\pi t\right) \tag{3.160}$$

where $\theta_N = 60°$. In this case, convective adjustment is active for the winter only (Fig. 3.43).

Similarly, the surface forcing is now a sink of GPE during the cooling season and a source of GPE during the heating season. On the other hand, sources of GPE due to vertical mixing and sinks of GPE due to conversion to KE remain nearly constant all year round. As a result, GPE has a pronounced seasonal cycle, as shown by the dashed line in Figure 3.43. During the spring, summer, and fall seasons, GPE is accumulated primarily through vertical mixing, with a minor contribution due to surface heating. This accumulated GPE

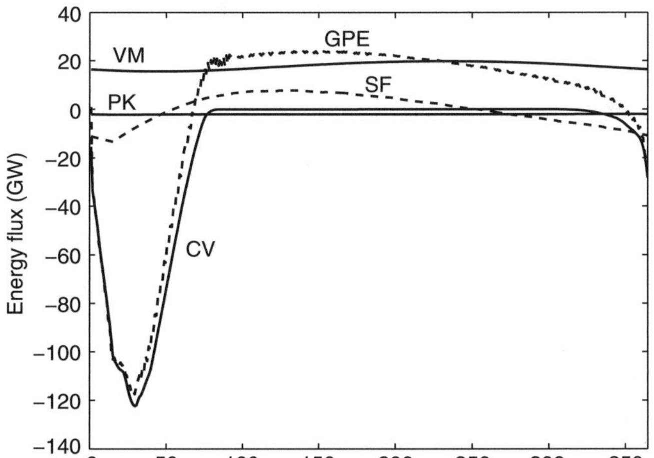

Fig. 3.43 Annual cycle of energy fluxes: *CA* is the GPE sink due to convective adjustment, *VM* is the GPE source due to vertical mixing, *PK* is the sink due to the conversion from potential energy to KE, *SF* is the source of GPE due to surface thermal forcing, and *GPE* is the total amount of GPE for the model ocean (Huang and Jin, 2006).

is quickly lost to the convective adjustment during the winter season. This seasonal cycle reinforces the basic idea that ***cooling on the surface does not create mechanical energy; it can transfer GPE to KE only***.

3.7.4 Balance of GPE/AGPE during the adjustment of circulations

GPE and its balance are one of the most important aspects of the oceanic circulation theory. Discussions in previous sections have been focused on the balance of GPE and the spatial distribution of AGPE for steady circulations. At shorter time scales the oceanic circulation, including GPE and AGPE, may undergo a transition. In particular, the total amount of mechanical energy may change with time, i.e., sources and sinks are not in exact balance all the time. This is, in some ways, much like the oxygen deficit experienced during a short-distance run. You run so fast that a large amount of oxygen is consumed, which is much larger than the amount of oxygen you can take in through breathing. As a result, your oxygen is unbalanced during the short duration of the run.

Similarly, for the oceanic circulation, a strong cooling at high latitudes on short time scales, such as interannual or decadal, can release a large amount of GPE and convert it to KE, thus giving rise to a strong circulation over a relatively short time scale. It is notable that the total amount of GPE of the mean state is not conserved for such short time scales because cooling does not create GPE or KE; instead, cooling can only release the large amount of GPE originally stored in the system and convert it to KE. Thus, strong circulation can be induced by strong cooling on decadal or shorter time scales, and such a strong circulation is not inconsistent with the theory of energetics of oceanic circulation.

In this section we discuss the balance of GPE and AGPE during a time evolution process of the circulation. This is also an area which has not received much attention so far. As an example, we examine the balance of GPE and AGPE for the case of a circulation with purely surface thermal forcing (i.e., without wind stress and freshwater flux).

The model is the same as used in Section 3.7.3 with a fixed vertical mixing coefficient of $\kappa = 0.3 \times 10^{-4}$ m^2/s. The model is subject to an upper surface relaxation condition, with a linear relaxation temperature profile; 25°C at the equator, linearly declining to 0°C at the northern boundary (60° N).

The model starts from an initial state of uniform temperature of 10°C, and runs for 5,000 years in order to reach a quasi-equilibrium. At the end of this spin-up process, the mean sea level is −1.70 m below the original level of $z = 0$. The drop of mean sea level in the model is due to the reduction of volume induced by cooling. Thereafter, the model is re-started at time $t = 0$ from the quasi-equilibrium state, and run for 2,000 years under two slightly different relaxation temperature profiles:

1. *The Cooling Case:* the surface relaxation temperature is 25°C at the equator and linearly declines to −2°C at the northern boundary (60° N).
2. *The Warming Case:* the surface relaxation temperature is 25°C at the equator and linearly declines to 2°C at the northern boundary (60° N).

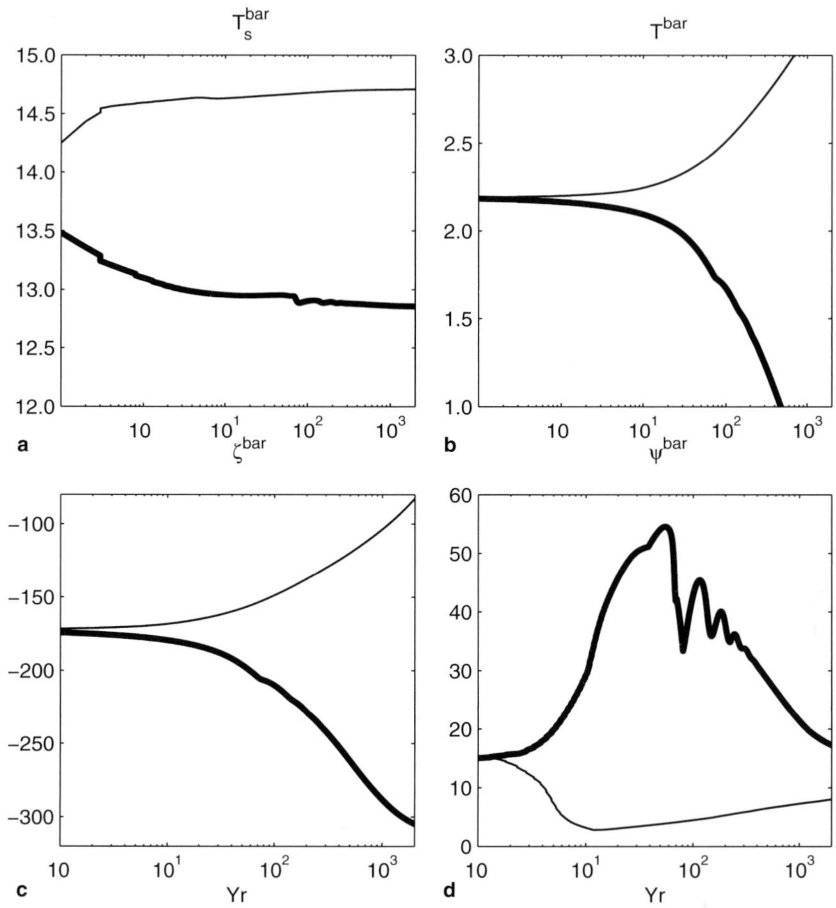

Fig. 3.44 Time evolution of the thermohaline circulation after switching on cooling (*heavy lines*) or warming (*thin lines*): **a** mean surface temperature; **b** basin mean temperature; **c** mean free surface elevation (all in cm); **d** the meridional overturning rate (Sv).

In the Cooling Case, the basin-mean temperature of the model declined quickly; the mean free surface of the model ocean went down from -1.7 m to -3.0 m. With such declines in the temperature and the free surface height, the meridional overturning rate went up greatly and reached a maximum value of more than 56 Sv; subsequently, it gradually declined (heavy lines in Fig. 3.44).

In the Warming Case, the basin-mean temperature of the model increased; the mean free surface of the model ocean went up from -1.7 m to -0.9 m. With such increases in temperature and free surface height, the meridional overturning rate went down greatly and reached a minimum value of slightly larger than 3 Sv; subsequently, it gradually recovered (thin lines in Fig. 3.44).

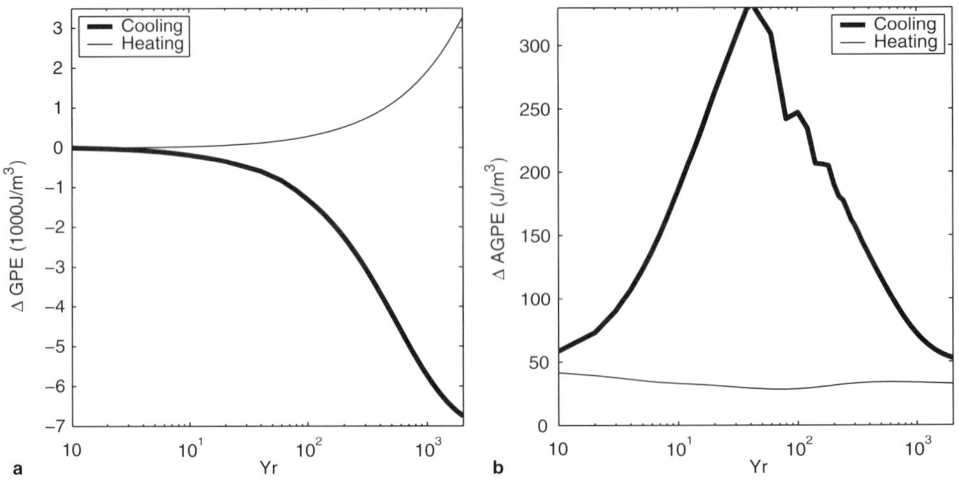

Fig. 3.45 Time evolution of **a** GPE and **b** AGPE diagnosed from the Cooling Case and the Warming Case.

Changes in circulation are closely linked to changes in GPE in the system. In the Cooling Case, GPE of the model ocean went down greatly (Fig. 3.45). On the other hand, GPE in the Warming Case increased. The trend of GPE is opposite to that of AGPE: in the Cooling Case, AGPE increased quickly, and this indicated that more GPE was now available and could be released and converted into KE, thus giving rise to a strong circulation.

We emphasize that the strong circulation appearing in the Cooling Case is due to the release of GPE originally stored in the ocean. Although sudden cooling of the ocean can give rise to strong meridional circulation on decadal–centennial time scales, such a strong circulation cannot be maintained without a continuous supply of mechanical energy from external sources.

In fact, strong cooling does not create mechanical energy; instead, cooling can only convert GPE originally stored in the ocean into KE. As shown in Figure 3.46, for the Cooling Case (heavy line) there is a strong loss of GPE during the sudden onset of cooling (Fig. 3.46a), and this loss of GPE is primarily associated with convective adjustment (Fig. 3.46b). In this case, surface cooling appears as a weak sink of GPE, not a source (Fig. 3.46c).

Note that although a large amount of GPE is lost through convective adjustment, only a small amount of GPE can be converted into KE. Comparing Figure 3.46b and d, it is readily seen that the amount of GPE converted into KE is much smaller than the amount of GPE lost through convective adjustment.

During this period of rapid change, the source of GPE due to vertical mixing is much smaller than that of the steady state, because the model has not yet reached the quasi-equilibrium state (Fig. 3.46e).

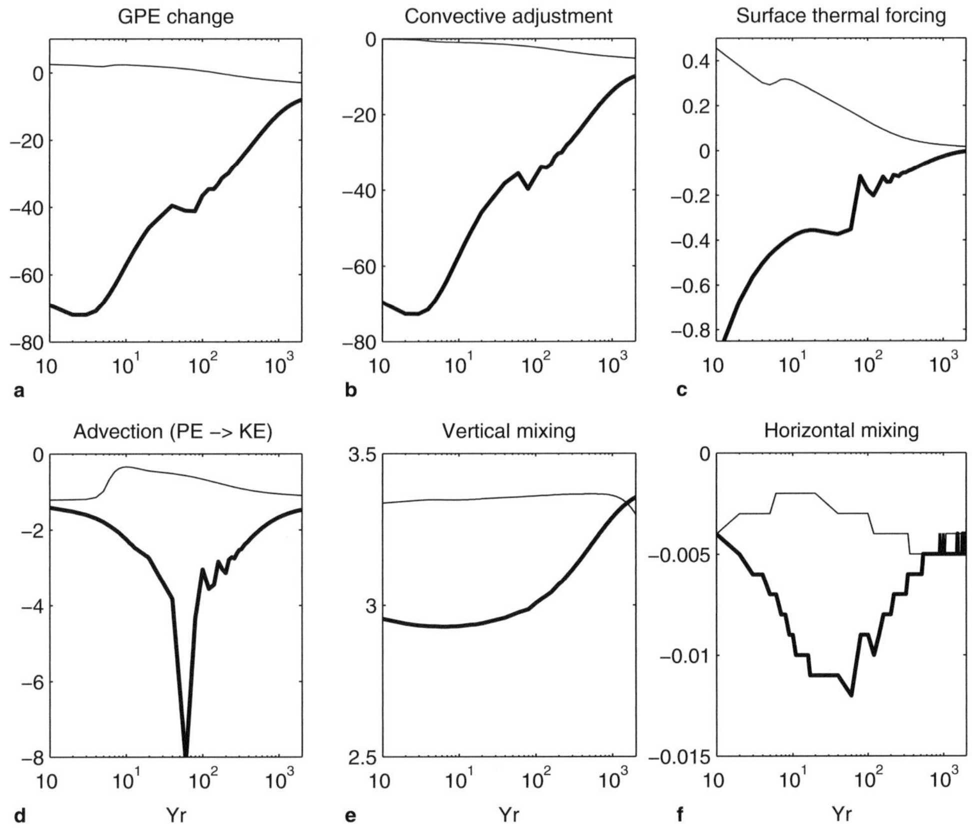

Fig. 3.46 **a** Time evolution of GPE and **b** the source/sink for GPE due to convective adjustment; **c** surface thermal forcing; **d** advection; **e** vertical mixing; **f** horizontal mixing. *Heavy lines* for the case with cooling and *thin lines* for the case with heating (GW).

Comparatively speaking, in the Warming Case (thin line) GPE increases very slowly. This slowness of increase in GPE reflects the fact that the stratification in this case is relatively stable; thus, the downward penetration of heat and the associated increase in volume and GPE of the model ocean are very slow processes. At the beginning of the experiment, the convective adjustment was almost interrupted (thin line in Fig. 3.46b). Surface heating is now working as a weak source of GPE during the spin-up phase. Overall, the conversion of potential energy to KE is relatively small because the GPE loss to convection is so much smaller than before.

Although our discussion here is limited to a simple model with only surface thermal forcing included, leaving other important forcing conditions untouched, the results obtained from this diagnosis have a much broader implication. Further study along the lines of GPE balance in the ocean should be very promising. In particular, a model including wind stress forcing could provide vitally important information about the balance of GPE in the ocean, and such knowledge will broaden our view of the physics of oceanic general circulation.

3.8 Entropy balance in the oceans

Entropy balance is one of the most fundamental thermodynamic laws governing the universe. Examining the thermohaline circulation from the viewpoint of entropy balance may bring about some new insights, though such approach has seldom been pursued, with a few exceptions. In order to simplify the problem, our discussion here excludes the potential contribution from sea ice.

3.8.1 Entropy production due to freshwater mixing

Dilution heat associated with seawater

When two water parcels of the same temperature but different salinity are adiabatically mixed under constant pressure, the final temperature of the mixture may be different from the original, and the heat loss/gain is called the dilution heat. Depending on the parameters, such as temperature, salinity, and pressure, dilution heat may be positive or negative. Assuming that each parcel has 1 kg of seawater, and using the Taylor expansion, enthalpy of the two water masses with slightly different salinity is

$$h_1 + h_2 = h\,(T, S + \Delta S, p) + h(T, S - \Delta S, P) = 2h(T, S, P) + \Delta S^2 \partial_{SS} h$$

For seawater over a rather wide range of parameter values, the second derivative is negative, $\partial_{SS} h < 0$, thus we have $h_1 + h_2 < 2h_0$. Seawater is almost incompressible, so that if temperature is maintained constant, changes in enthalpy are approximately equal to the heat received from the environment, and this is called the dilution heat. However, if mixing is adiabatic, mixing water parcels with different salinity generally leads to cooling.

Enthalpy and entropy changes associated with mixing of seawater

To explore the process of mixing and transport of freshwater in the ocean, we set up the following virtual experiment under a constant temperature of 15°C and sea-level pressure. A series of boxes numbered by N (where $N = 1, \ldots, 37$) are filled up with seawater, with the salinity (S) in the N-th box being $S = N - 1$. At the left-hand end, box $N = 1$ has no salt; it receives 1 kg of pure water and exports it to the right, and this water mixes with 1 kg of salty water from box 3 with a salinity of 2. The end product of mixing is 2 kg of water with a salinity of 1 (Fig. 3.47a). Similarly, the mass balance in each box N ($2 < N < 35$) is illustrated in Figure 3.47b. For each box the total fluxes of mass, water, and salt are balanced. Across the boundaries between two boxes, there is a net water flux moving to the right, but there is no net salt flux between boxes. This can be shown as follows. Across the boundary between boxes with salinity $S = N$ and $S = N + 1$ (Fig. 3.47b), the mass balance

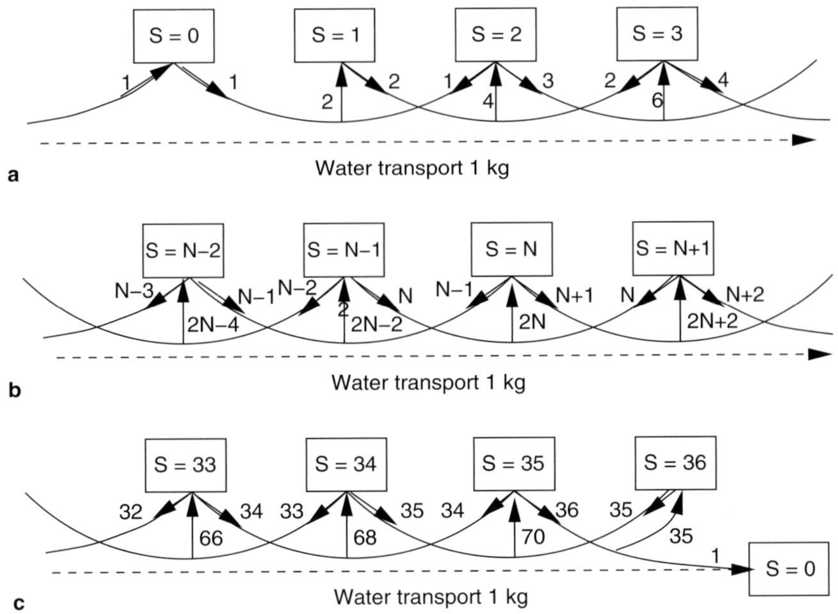

Fig. 3.47 The box train; all mass fluxes are in kg: **a** the beginning of the train, **b** the middle of the train, **c** the end of the train where the pure water is separated from the seawater with salinity 35.

of pure water is

$$(N+1)\cdot(1-0.001\cdot N)-N\cdot[1-0.001\cdot(N+1)]=1$$

i.e., there is 1 kg of water moving through the interface to the right.

In each box, the balance of enthalpy and entropy due to mixing of water from the upstream and downstream boxes under constant temperature and pressure are as follows:

$$h\,(0)+h\,(2)-1.84=2\cdot h\,(1)\quad (J) \tag{3.161a}$$

$$\eta\,(0)+\eta\,(2)+0.339-0.006=2\cdot\eta\,(1)\quad (J/K) \tag{3.161b}$$

$$\ldots$$

$$10\cdot h\,(9)+10\cdot h\,(11)+2.54=20\cdot h\,(10)\quad (J) \tag{3.161c}$$

$$10\cdot\eta\,(9)+10\cdot\eta\,(11)+0.241+0.009=20\cdot\eta\,(10)\quad (J/K) \tag{3.161d}$$

$$\ldots$$

$$35\cdot h\,(34)+35\cdot h\,(36)+18.28=70\cdot h\,(35)\;(J) \tag{3.161e}$$

$$35\cdot\eta\,(34)+35\cdot\eta\,(36)+0.263+0.063=70\cdot\eta\,(35)\quad (J/K) \tag{3.161f}$$

where $h(S)$ is the specific enthalpy for seawater with salinity S, $\eta(S)$ is the specific entropy of seawater, and the number in the front of enthalpy and entropy is the mass associated with each component involved in mixing.

The last term on the left-hand side of the enthalpy balance is the amount of heat absorption; a negative sign indicates heat released. For the entropy balance, the third number on the left-hand side indicates the irreversible entropy increase due to mixing, and the fourth number indicates (reversible) entropy increase due to heat transport from the environment; a negative value indicates entropy reduction due to loss of heat.

For example, when 1 kg of freshwater is mixed with 1 kg of seawater with a salinity of 2, 1.84 J of thermal energy is released. During this process, the irreversible mixing process induces an entropy increase of 0.339 J/K, while the heat release induces reduction of entropy, −0.006 J/K.

Near the end of this box train model, two processes take place in Box 36 with salinity 35. First, 35 kg of seawater with salinity 34 is mixed with 35 kg of seawater with salinity 36. The product is 70 kg of seawater with salinity 35, and 18.28 J of heat is absorbed (Fig. 3.48a). Similarly, entropy is produced during mixing, including an irreversible entropy increase of 0.263 J/K and 0.063 J/K of reversible increase due to heat absorption (Fig. 3.49a).

Next, of this 70 kg of seawater with salinity 35, 34 kg is used for mixing with 34 kg of seawater with salinity 33; thus, 36 kg of water with salinity 35 is left, which has to be unmixed, and the end products of this unmixing are 35 kg of seawater with salinity 36 and

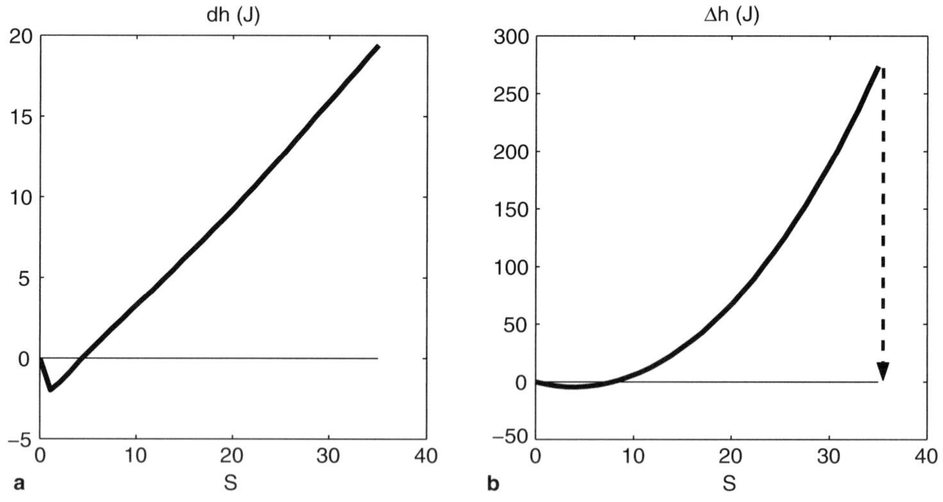

Fig. 3.48 Change of enthalpy for 1 kg of pure water going through the box train: **a** enthalpy increase during mixing under constant temperature and pressure; **b** the accumulated change in enthalpy; the *dashed arrow* indicates the reduction of enthalpy required at the end of the box train.

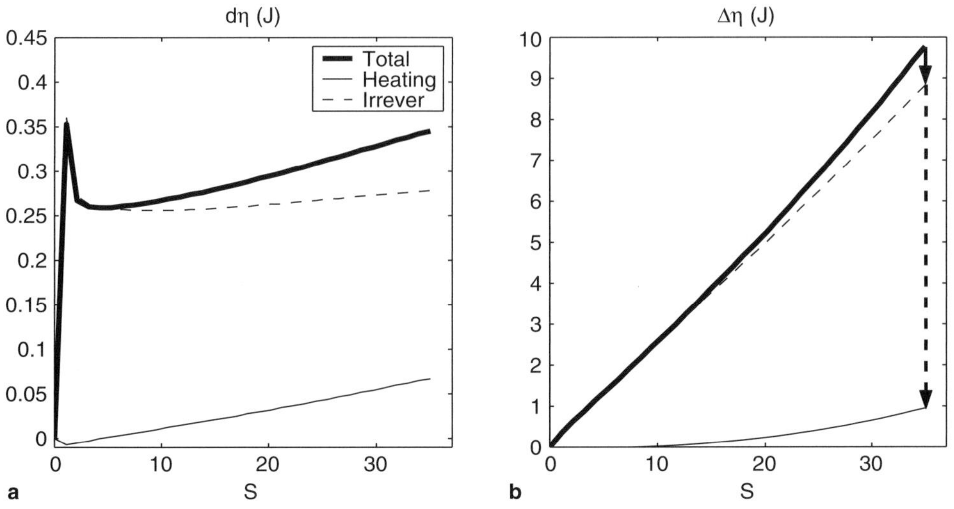

Fig. 3.49 Change of entropy for 1 kg of pure water going through the box train: **a** entropy increase during mixing under constant temperature and pressure; **b** the accumulated change in entropy; the *solid arrow* indicates the decline of entropy associated with decline in enthalpy during unmixing (which can be considered as a reversible process), and the *dashed arrow* indicates required removal of entropy at the end of the box train.

1 kg of pure water (Fig. 3.47c)

$$36 \cdot h\,(35) - 273.14 = 35 \cdot h\,(36) + h\,(0)\;(J) \tag{3.162a}$$

$$36 \cdot \eta\,(35) - (8.81 + 0.95) = 35 \cdot \eta\,(36) + \eta\,(0)\;(J/K) \tag{3.162b}$$

The total amount of enthalpy increase during mixing in this box train is 273.14 J, and at the end of the box train this amount of heat is released back to the environment (Fig. 3.48b). This process takes place within the shallow surface layer, where the solar radiation is absorbed and other outgoing heat fluxes are sent out. In addition, our discussion is based on the assumption of constant temperature and pressure. Therefore, such a heat release may not be clearly separated from the otherwise much larger fluxes through the air–sea interface.

This amount of heat release is associated with a reduction in entropy of 0.95 J/K, indicated by the solid arrow in Figure 3.49b. Assuming that the heat release is through a reversible process, this amount of entropy is considered to be reversible entropy production. However, in order to bring the system back to the original state, the net production of entropy due to mixing, 8.81 J/K, has to be removed, indicated by the dashed arrow in Figure 3.49b. This part of entropy production is through a mixing process, so it is counted as irreversible entropy production.

In general, the removal of the entropy due to mixing is required for the system to be maintained; this job may be accomplished with either an external mechanical energy input or a net negative entropy flux exchange with the atmosphere by the combination of short-wave solar radiation into the upper ocean, which brings low-income entropy, and the long-wave radiation from ocean to atmosphere, which removes high entropy produced in the ocean. This completes the whole process of transporting freshwater through the internal mixing and the unmixing near the surface associated with evaporation.

Theoretical limits of changes in enthalpy and entropy

When the number of boxes is increased further, the finite difference calculation discussed above fails to give accurate changes in total enthalpy and entropy; however, the theoretical limit of these values can be derived as follows. When the number of boxes becomes very large, the second box from the right-hand end of the box train should have salinity $S = 35$ (the mean salinity in the ocean). The corresponding salinity in the last box of the train model is $S(1 + 1/x)$, where $x = (M - 2) \to \infty$ and M is the number of boxes in the train. The mass flux required to go through the salt separation is $(1 + x)$; thus, the enthalpy balance of the salt separation is

$$(1 + x)\, h(S) = x \cdot h(S + S/x) + h(0) + \Delta h$$

or

$$\Delta h = (1 + x)\, h(S) - x \cdot h(S + S/x) - h(0) \tag{3.163}$$

As $x \to \infty$, this leads to the following relation

$$\Delta h = h_a\,(S) - h_a\,(0) \tag{3.164}$$

where

$$h_a\,(S) = h\,(S) - S\left[\frac{\partial h\,(S)}{\partial S}\right]_{T,P} \tag{3.165}$$

and this will be called the apparent specific enthalpy of seawater. Δh is the enthalpy released during this separation, or unmixing. Note that this is different from the relative apparent specific enthalpy (Lewis and Randall, 1961; Bromley, 1968; Millero and Leung, 1976; Feistel and Hagen, 1995):

$$\Phi_L\,(S) = [h(S) - h\,(0) - S\partial h\,(0)\,/\partial S]\,/S \tag{3.166}$$

Their definition of the relative apparent specific enthalpy is introduced for the case of mixing two water parcels with different salinity; however, the definition introduced above is for the case of continuous transportation of pure water through the salinity field, which increases from 0 to $S > 0$. Apart from the fact of $1/S$, their definition is related to the first derivative of $\partial h/\partial S$ at $S = 0$ (pure water); while the new definition introduced above involves $\partial h/\partial S$ at $S > 0$. The difference is readily seen because the first derivative of h varies with increase in S, as shown in Figure 2.20.

Similarly, we can introduce the apparent specific entropy, defined as

$$\eta_a = \eta(S) - S\left[\frac{\partial \eta(S)}{\partial S}\right]_{T,P} \tag{3.167}$$

thus, in this box model the theoretical limit of entropy changes due to mixing is

$$\Delta\eta = \eta_a(S) - \eta_a(0) \tag{3.168}$$

Therefore, during the freshwater transport from entering the ocean with $S = 0$ to the place with $S = 35$, and the separation associated with evaporation, the amounts of enthalpy release and entropy reduction required for each kilogram of pure water are defined by the apparent specific enthalpy and specific entropy. The parameter dependence of the entropy change is shown in Figure 3.50.

Note that a small part of the entropy increase during mixing under a constant temperature can be treated as a reversible process, and the net irreversible increase in entropy is

$$\delta\eta = \Delta\eta - \frac{\Delta h}{T} = -\frac{1}{T}\left[\mathrm{g}(S) - \mathrm{g}(0) - S\partial_S \mathrm{g}(S)\right] \tag{3.169}$$

Using Eqn. (2.70) from Chapter 2, this leads to a simple relation

$$\delta\eta = -\frac{\Delta\mu_w}{T} \tag{3.170}$$

where

$$\Delta\mu_w = \mu_w(S) - \mu_w(0) \tag{3.171}$$

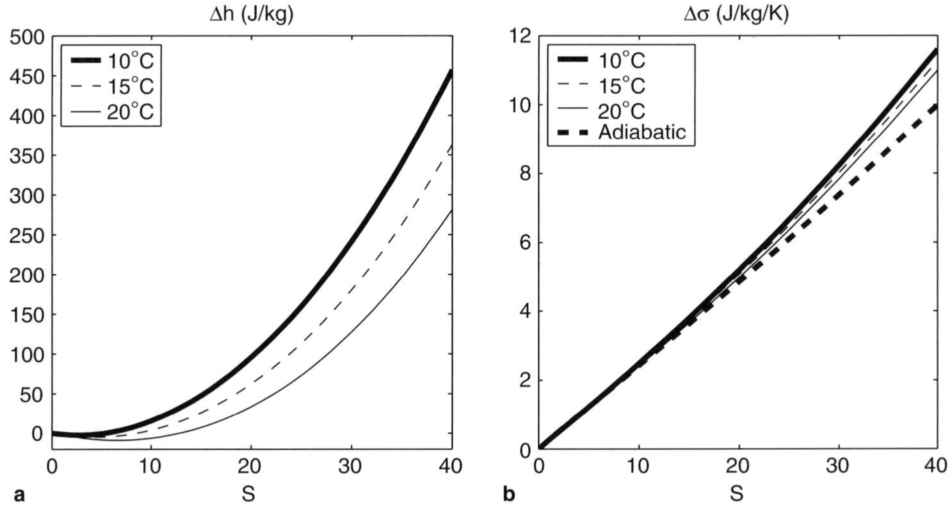

Fig. 3.50 Theoretical limit of specific mixing **a** enthalpy and **b** entropy; the *heavy dashed line* in (**b**) indicates entropy increase under the assumption of adiabatic mixing.

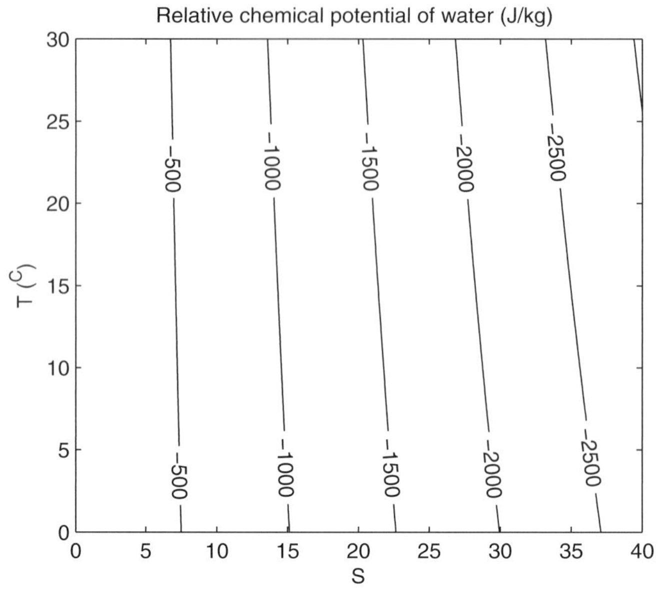

Fig. 3.51 Relative chemical potential of seawater (J/kg).

is called the relative chemical potential, and $\mu_w(S)$ is the chemical potential of water in seawater. The relative chemical potential decreases with the salinity (more negative) and slightly increases (less negative) as the temperature increases (Fig. 3.51).

During evaporation, water is separated from the water–salt mixture, i.e., the chemical potential must be increased from that corresponding to salinity S to that for the freshwater; thus the amount of equivalent external mechanical energy required is equal to

$$E_{mech} = \rho_{freshwater}\omega_{ev}\left[\mu_w(0) - \mu_w(S)\right] > 0 \tag{3.172}$$

where ω_{ev} is the evaporation rate.

Applying the same analysis for the free enthalpy, we obtain the following relation for the release of free enthalpy during the unmixing

$$\Delta \mathsf{g} = \mathsf{g}_a(S) - \mathsf{g}_a(0), \quad \mathsf{g}_a(S) = \mathsf{g}(S) - S\left(\frac{\partial \mathsf{g}}{\partial S}\right)_{T,P} \tag{3.173}$$

According to Eqn. (3.169), this leads to

$$\Delta \mathsf{g} = \Delta\mu_w < 0 \tag{3.174}$$

Therefore, owing to evaporation, change in Gibbs function (the free enthalpy) is equal to change in chemical potential, and the irreversible increase in entropy and the decrease in free enthalpy during this mixing process obey the following relation

$$\delta\eta = -\Delta \mathsf{g}/T \quad (3.175)$$

Equivalent mechanical energy required for entropy removal

The removal of entropy required for separating pure water from seawater is equivalent to applying mechanical work to push the pure water through a semi-permeable membrane against osmotic pressure. The concept of osmotic pressure is illustrated in Figure 3.52. A container is divided into two parts separated by a semi-permeable membrane.

Assuming a column of seawater of salinity 35 and 20°C, the mean density of seawater with salinity 35 and temperature of 20°C at a depth between the sea surface and a pressure level of 250 db is approximately 1,025.3 kg/m^3. The osmotic pressure under the surface pressure is approximately equal to $p_{osm} = 24.8$ atmospheric pressure. This is equivalent to the pressure at depth of $D = 247$ m.

Thus, if the surface level of pure water is exactly 247 m below the sea level, there is an equilibrium between the seawater and pure water (Fig. 3.52b). If the surface level of pure water is higher than -247 m, freshwater should move to the left part of the container through the semi-permeable membrane. Theoretically, this phenomenon could be utilized as a means of generating electricity at the mouth of a river, by building a deep freshwater reservoir near the river mouth and utilizing such a large height difference. Of course, owing to the potentially large impact on the coastal environment, this method of generating electricity is not practical.

On the other hand, if the surface level of pure water is deeper than 247 m below sea level, pure water should move to the right part of the container through the semi-permeable membrane. This phenomenon can be used to extract freshwater from seawater.

By definition, the osmotic pressure $\pi(S,T,P)$ satisfies the following equilibrium condition that the chemical potential of water on both sides of the membrane should be equal

$$\mu_w(0,T,P) = \mu_w(S,T,P+\pi) \quad (3.176)$$

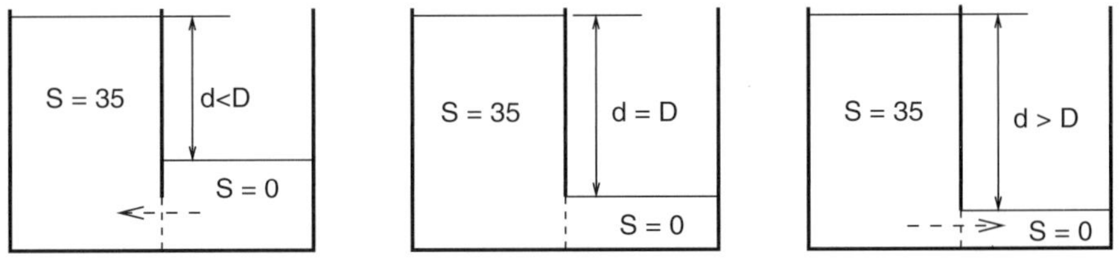

a d<D, Freshwater moves toward the seawater. **b** d=D, Equilibrium, with no motion. **c** d>D, Freshwater is separated from seawater.

Fig. 3.52 **a–c** A sketch illustrating osmotic pressure.

By definition from Eqn. (2.70), $\mu_w = \mathsf{g} - S\mathsf{g}_S$. Using Eqns. (2.58) and (2.59d),

$$\frac{\partial \mu_w}{\partial P} = \frac{\partial \mathsf{g}}{\partial P} - S\frac{\partial^2 \mathsf{g}}{\partial P \partial S} = v\,(1 + \beta S)$$

Expanding the right-hand side of Eqn. (3.176) in powers of π and retaining the linear term in π leads to the following linearization

$$\mathsf{g}\,(0) = \mathsf{g}\,(S) - S\mu\,(S) + \pi v\,(1 + \beta S) \tag{3.177}$$

or

$$\pi v = -\left[\mathsf{g}\,(S) - S\partial_S \mathsf{g}\,(S) - \mathsf{g}\,(0)\right] / \,(1 + \beta S) \simeq -\left[\mathsf{g}\,(S) - S\partial_S \mathsf{g}\,(S) - \mathsf{g}\,(0)\right] \tag{3.178}$$

Using Eqn. (3.174), this leads to

$$\pi v = -\Delta \mu_w \tag{3.179}$$

Owing to this input of mechanical energy, water is unmixed or equivalently entropy is removed from the system, and the entropy of the system is thus reduced

$$\delta \eta = -\pi v / T = \Delta \mu_w / T < 0 \tag{3.180}$$

Therefore, entropy production inferred from these two approaches is the same as in the linear approximation.

3.8.2 Balance of entropy in the world's oceans

We have discussed mechanical energy balance in the world's oceans, and now our attention turns to the entropy balance. This is a much more difficult topic, because entropy production depends on some fine details of the circulation system which are not currently available. Thus, our discussion in this section will be limited to some of the preliminary estimates of entropy production. In particular, we try to give an estimate of the entropy fluxes between the oceans and the atmosphere and outer space. In addition, we also give some lower-bound estimates of the entropy production due to freshwater and heat fluxes through the air–sea interface. Entropy production below the sea surface involves many complicated dynamical and thermodynamic processes; such a topic is beyond the scope of this book and is left for further study.

Align of balance of entropy

The balance of entropy for a water parcel, Eqn. (3.60), can be rewritten as

$$\frac{\partial \rho \eta}{\partial t} + \nabla \cdot \left(\rho \eta \vec{u} + \frac{\partial h / \partial s - \mu}{T} \vec{J}_s \right) = \frac{\rho \varepsilon}{T} - \frac{\nabla \cdot \vec{q}}{T} + \vec{J} \cdot \left(\frac{\partial h}{\partial s} \nabla \frac{1}{T} - \nabla \frac{\mu}{T} \right) \tag{3.181}$$

Following the approach adopted in Section 3.4, this equation can be transformed, using the generalized transport theorem. The corresponding rate of total entropy in the world's oceans is

$$\frac{\delta}{\delta t}\iiint_V \rho\eta dv = \oiint_S \rho\eta\left(\vec{u}_s - \vec{u}\right)\cdot\vec{n}ds + \mathrm{H}_{thermal} + \mathrm{H}_{fw,mix} + \mathrm{H}_{diss} \tag{3.182}$$

where the contribution due to air–sea mechanical energy flux is zero because it is entropy-free, and the corresponding term due to air–sea salt flux is zero because of zero salt flux. The entropy production associated with evaporation and precipitation will be discussed in terms of separating freshwater from the salty seawater. The first term on the right-hand side of Eqn. (3.182), $\oiint_S \rho\eta\left(\vec{u}_s - \vec{u}\right)\cdot\vec{n}ds$, is the flux of entropy due to mass exchange through the air–sea interface. Assuming that evaporation and precipitation are balanced for a quasi-steady state, and water leaving and entering the sea surface has the same temperature, this term vanishes.

$$\mathrm{H}_{fw,mix} = \iiint_V J_s\cdot\left(\frac{\partial h}{\partial s}\nabla\frac{1}{T} - \nabla\frac{\mu}{T}\right)dv \tag{3.183a}$$

is entropy production due to freshwater mixing,

$$\mathrm{H}_{diss} = \iiint_V \frac{\rho\varepsilon}{T}dv \tag{3.183b}$$

is entropy production due to internal dissipation,

$$\mathrm{H}_{thermal} = -\iiint_V \frac{\nabla\cdot\vec{q}}{T}dv \tag{3.183c}$$

is entropy flux and generation due to heat transport, including entropy fluxes through the air–sea interface and internal entropy generation through thermal mixing. In order to analyze this term we divide the ocean into two layers: an upper layer (about 10–15 m, denoted as V_1) and a lower layer (denoted as V_2) (Fig. 3.53).

On the sea surface, there are five heat fluxes: net short-wave radiation q_{sw}, net long-wave radiation q_{lw}, latent heat q_{lh}, sensible heat q_{sh}, and additional heat flux $q_{add} = q_{geo} + q_{dis}$ (geothermal heat q_{geo} plus dissipation heat associated with the external mechanical energy q_{dis}). Ignoring the horizontal heat flux within the upper layer, there are two heat fluxes across the interface between the upper and lower layers: $|\vec{q}_{net}| = q_{net} = q_{sw} + q_{lw} + q_{lh} + q_{sh}$ and $|\vec{q}_{add}| = q_{add}$. Thus, Eqn. (3.183c) is reduced to

$$\mathrm{H}_{thermal} = -\oiint_S \frac{\vec{q}}{T}\cdot\vec{n}ds + \oiint_{S_2}\frac{\vec{q}_{net}}{T}\cdot\vec{n}ds + \mathrm{H}_{V_1} + \mathrm{H}_{V_2} + \iiint_{V_2}\frac{\nabla\vec{q}_{net}}{T}dv \tag{3.184}$$

where $\mathrm{H}_{V_1} = \iiint_{V_1}\vec{q}\cdot\nabla\frac{1}{T}dv$ is entropy production within the upper layer, which is where the short-wave radiation is absorbed and turned into the internal energy of the water. As the equivalent temperature drops from $T_{Sun} \simeq 5.777\text{K}$ to $\bar{T}_s = 291.3\text{K}$, there is a huge

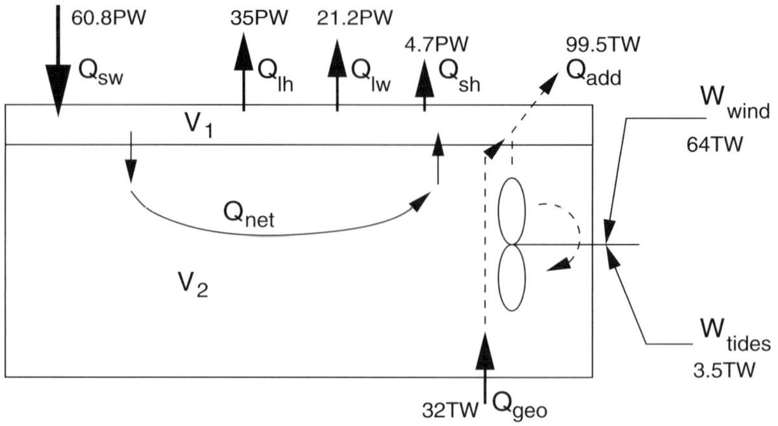

Fig. 3.53 Energy balance for the world's oceans.

production of entropy in this thin layer. Because such entropy production is not directly related to the oceanic general circulation, the corresponding negative entropy flux through the air–sea interface will be called a non-active (negative) entropy flux. The details of this layer remain unclear at this time.

$\mathrm{H}_{V_2} = \iiint_{V_2} \vec{q}_{add} \cdot \nabla \frac{1}{T} dv$ is the entropy production associated with additional heat flux in the ocean. The calculation of this term requires detailed information about the circulation, and hence is left for further study. The last term in Eqn. (3.184) represents the entropy production due to the internal heat transport and mixing, which remains unclear because its value depends on the detailed information of $\nabla \cdot \vec{q}_{net}$, which is not available from observations at this time.

We will therefore focus on the first two terms in (3.184). Since the contribution due to geothermal heat flux is unknown, this small term will be omitted. There is no lateral flux term due to periodic condition. Thus, the surface heat flux term is reduced to the flux on the sea surface (SS) $\mathrm{H}_{surf.heat} = -\iint_{SS} (\vec{q}/T) \cdot \vec{n} ds$ and entropy production through heat transport in the oceanic interior $\mathrm{H}_{heat\ mixing} = \oiint_{S_2} (\vec{q}_{net}/T) \cdot \vec{n} ds \simeq -\iint_{SS} q_{net}/T_s ds$.

Entropy fluxes through the air–sea interface

The global sum of the air–sea heat flux includes the following four terms: Q_{sw} (net short-wave radiation), Q_{lw} (net long-wave radiation), Q_{lh} (latent heat), and Q_{sh} (sensible heat). Here heat flux received by the ocean is defined as positive. Entropy flux associated with radiation is different from that of ordinary heat flux. According to the laws of Planck and Stefan–Boltzmann (Kittel and Knoemer, 1980; Yan *et al.*, 2004) energy and entropy flux associated with black body radiation obey

$$Q_r = \frac{\pi^2}{15} \frac{T^4}{(\hbar c)^3}, \quad \eta_r = \frac{4\pi^2}{45} \frac{T^3}{(\hbar c)^3}, \quad \eta_r = \frac{4}{3} \frac{Q_r}{T} \qquad (3.185)$$

Table 3.7. *Entropy balance for the world's oceans, modified from Yan et al. (2004)*

Sources and sinks	Short-wave radiation	Long-wave radiation	Latent heat	Sensible heat	Sum
E (TW)	60.8	−21.2	−35.0	−4.7	0
e (W/m^2)	163.34	−56.90	−93.96	−12.48	
$\dot{\mathrm{H}}$ (TW/K)	14.0	−97.0	−120.1	−16.0	−219.1
$\dot{\eta}$ (mW/K/m^2)	37.7	−260.4	−322.6	−42.9	−588.2

where $\hbar$ is Planck's constant, and c is the speed of light. This formula should be treated as an approximation because radiation from the Sun is not really black-body radiation.

Thus, the entropy flux associated with the air–sea heat flux is

$$\mathrm{H}_{surf.heat} = \frac{4}{3}\left(\frac{Q_{sw}}{T_{Sun}} + \frac{Q_{lw}}{\bar{T}_s}\right) + \left(\frac{Q_{lh}}{\bar{T}_s} + \frac{Q_{sh}}{\bar{T}_s}\right) \tag{3.186}$$

where $\bar{T}_s = 291.3$K is the annual mean sea surface temperature averaged over the world's oceans. The net long-wave radiation from the ocean is treated as black-body radiation, so the horizontal mean temperature is used.

The oceans receive incoming energy, including solar radiation, geothermal heat flux, and mechanical energy from wind and tidal dissipation; but the outgoing energy is in the form of heat flux. In an equilibrium state of climate, the total outgoing heat flux should include the contribution due to geothermal heat and mechanical energy input from wind and tides. For technical reasons, air–sea heat fluxes from measurements are inaccurate and not exactly balanced. Thus, we adjust the air–sea heat fluxes slightly to obtain an exactly balanced air–sea heat flux. The corresponding results are included in Table 3.7. The huge amount of negative entropy flux has been linked to the maintenance of the orderly circulation in the world's oceans. The entropy flux due to outgoing heat flux (or entropy flux), due to thermal heat, wind stress energy, and tidal dissipation as separate items, is discussed in the next section.

Entropy production due to heat mixing

Heat transported from regions of high temperature to regions of low temperature is one of the important sources of internal entropy production. The horizontal distribution of net air–sea heat flux implies the meridional and zonal transport of heat on the order of 1.5–2 PW (Fig. 1.7a, b). The entropy production due to oceanic heat transport is estimated as

$$\mathrm{H}_{heat\,mixing} = \iint_A \frac{-q_{net}}{T_s(\lambda,\theta)} r^2 \cos\theta d\lambda d\theta = 0.68 \times 10^{12} \ \ (W/K) \tag{3.187}$$

or equivalent to 1.84 mW/K/m^2. It is important to note that the entropy production calculated above may serve as the theoretical lower bound of entropy production due to heat mixing in the world's oceans only.

Entropy production due to freshwater mixing

Equation (3.183a) is rather inconvenient to apply because it involves the distribution of salt flux and other variables. It is more straightforward to estimate entropy production due to freshwater mixing, using the analysis discussed in the first subsection of Section 3.8.1. Accordingly, the rate of entropy production due to freshwater mixing in the oceans is

$$\mathrm{H}_{fw,mix} = \iint_{SS} \frac{\rho\omega_{ev}}{T} \left(\mu_w(0) - \mu_w(S)\right) dxdy = 0.11 \times 10^{12}\ (W/K) \tag{3.188}$$

or equivalent to 0.30 mW/K/m^2, where ω_{ev} is the rate of evaporation and ρ, S, and T are density of seawater, salinity, and temperature at the sea surface (Fig. 3.54). This estimate may serve as the theoretical lower bound of entropy production due to hydrological processes in the world's oceans.

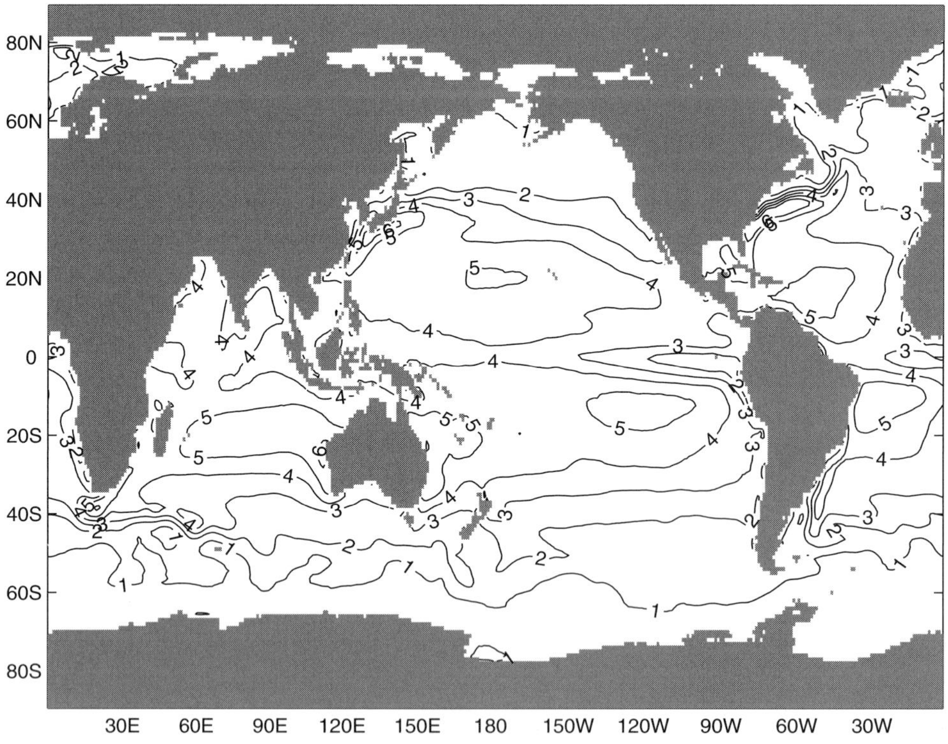

Fig. 3.54 Annual mean entropy production rate due to freshwater mixing in the world's oceans (0.1 mW/K/m^2).

The removal of entropy is implicitly part of air–sea heat exchange, including the incoming short-wave solar radiation of low entropy and the high-entropy fluxes associated with outgoing heat fluxes. Thus, this net entropy flux plays a vitally important role in maintaining the ocean circulation against the accumulation of entropy associated with salt diffusion in the ocean.

Entropy production due to mechanical energy dissipation

If the oceans were a heat engine, mechanical energy could be made from internal energy due to surface thermohaline forcing. However, the new paradigm of thermohaline circulation claims that all mechanical energy involved in thermohaline circulation comes from the external sources of wind and tides. Thus, the total mechanical energy dissipation rate is equal to the rate of input from the external sources of mechanical energy, and the conversion of energy from internal to mechanical is negligible. As a result, the total entropy production due to momentum dissipation in the world's oceans is

$$\mathrm{H}_{diss} = \iiint_V \frac{\rho\varepsilon}{T} dv \simeq \frac{W}{\bar{T}_s} \tag{3.189}$$

where $W = 67.5$ TW is the total mechanical energy from wind energy input and tidal dissipation.

Balance of entropy for the world's oceans

The most important items in the balance of entropy are shown in Figure 3.55. The incoming entropy includes two terms: that due to the incoming short-wave radiation from the Sun, and that due to the incoming geothermal heat flux. The short-wave radiation has a very high radiation temperature, so it is high-quality energy with a very low rate of entropy, $H_{sw} = 14$ TW/K. The geothermal heat flux is associated with relatively high temperature.

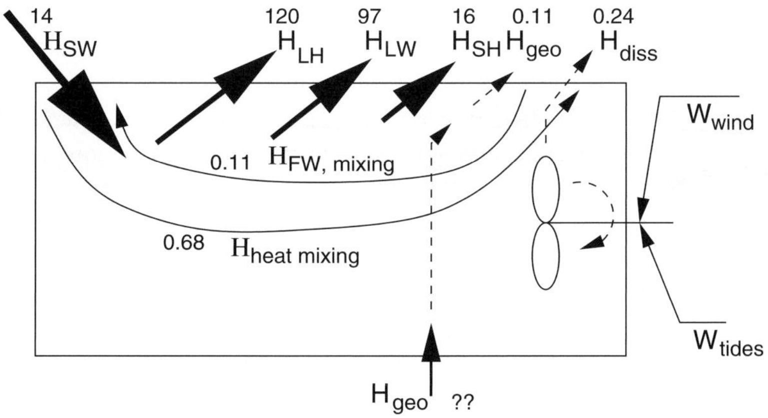

Fig. 3.55 Entropy balance for the world's oceans (in 10^{12} W/K).

Sometimes the geothermal plumes can have temperatures on the order of a few hundreds of degrees; however, the temperature may be much lower in general. The exact amount of entropy due to geothermal heat flux is much smaller than that associated with the air–sea heat flux, so it is omitted in Figure 3.55.

The outgoing entropy fluxes include the following contributions: $H_{lh} = 120$ TW/K, $H_{lw} = 97$ TW/K, and $H_{sh} = 16$ TW/K. All the mechanical energy input due to wind and tides is eventually dissipated and exported in the form of dissipated heat, $H_{diss} = 0.24$ TW/K. In addition, both the heat transport and freshwater mixing are associated with an internal source of entropy of 0.68 TW/K and 0.11 TW/K, respectively. The geothermal heat flux goes through the sea surface, $H_{geo} = 0.11$ TW/K. The entropy produced through these sources must be removed from the system in order to maintain the orderly circulation and dissipation in the world's oceans. Although these fluxes should be included as parts of the outgoing entropy fluxes related to heat flux discussed above, we list them as small items standing alone.

Active and non-active negative entropy flux

As stated above, a major portion of the negative entropy flux through the air–sea interface may not have any direct impact on the circulation. For example, if the solar radiation is received by a rock, this energy may be transferred back to the atmosphere in the form of long-wave radiation. In a steady state, there should be a big negative entropy flux; however, such entropy flux does not produce any orderly motions in the rock at all. In fact, the rock is in a thermal equilibrium, with no internal dissipation, and the only effect of solar radiation on the rock is to maintain a relatively high state of thermal equilibrium with more energetic random thermal motions of molecules in the rock.

Therefore, in order to explore the dynamic role of negative entropy flux, we need to differentiate two kinds of negative entropy flux: the ***active negative entropy flux*** and the ***non-active negative entropy flux***. The active negative entropy flux is the part of entropy flux which is directly related to the circulation, and the remaining part is called the non-active negative entropy flux.

Since the total entropy flux through the air–sea interface is approximately 588 mW/K/m^2, the ratio of active to total entropy flux is 0.36% (Table 3.8); therefore, the major part of the negative entropy flux is non-active. More importantly, the oceanic general circulation is not driven by surface thermohaline forcing from the viewpoint of energy; instead the oceanic circulation is driven by the external sources of mechanical energy, as discussed in the previous section of this chapter.

All the mechanical energy input is eventually dissipated, implying an entropy production term exported as part of the negative entropy flux. All entropy produced by the system has to be exported to the environment in the form of negative entropy flux. However, it is not a simple dissipation process, since any imposition of external mechanical energy induces the oceanic general circulation and the related mixing of freshwater and heat. The oceanic general circulation thus gives rise to additional entropy flux. According to our definition, such fluxes are active. The existence of such additional active entropy flux means

Table 3.8. *Partition of active entropy flux (mW/K/m²)*

Total entropy flux	Active entropy flux				
	Driving mechanical energy	Induced active entropy flux			
		Heat mixing	Freshwater mixing	Sum	Sum
−588.2	−0.6	−1.84	−0.30	−2.14 (0.36%)	−2.74 (0.47%)

the production of entropy associated with mechanical energy input is amplified, with an *amplification factor* of $2.14/0.6 = 3.57$.

Evaporation/precipitation idealized as a thermo-chemical engine

The salinity field in a steady state is time-invariant because salt advection has to be exactly balanced by salt diffusion. Conceptually, thus, we can treat salt as a stagnant grid fixed in space, through which pure water from precipitation moves around. As a parcel of water originating from precipitation moves through the ocean, its salinity and entropy gradually increase through mixing. This water parcel eventually comes back to the sea surface, where pure water is extracted (unmixed) from the ocean and the extra enthalpy and entropy are removed through evaporation.

To analyze the evaporation process, we add an imaginary thin layer of pure water on the top of the ocean, and separate evaporation into two steps. In the first step, pure water is extracted from seawater below this thin layer, and the entropy generated through mixing is removed. This can be achieved by the entropy removal process associated with solar insolation or, equivalently, by pushing the freshwater through the semi-permeable membrane placed at this imaginary interface. In the second step, pure water in this thin layer is transformed into water vapor, which is carried into the atmosphere by wind.

At the sea surface in subtropical oceans, solar radiation with low entropy and heat flux back to atmosphere with high entropy play the role of a conveyor that takes up the entropy produced in the ocean by freshwater mixing, and this entropy removal is equivalent to a specific amount of mechanical energy required for extracting pure water from seawater. The osmotic pressure of seawater of salinity 35 and temperature 20°C is equal to $p_{osm} = 248$ db. The global evaporation is approximately $\iint_S \omega_{ev} dS = 4.0 \times 10^{14} \mathrm{m}^3/\mathrm{y}$. The energy required for sustaining the hydrological cycle is $p_{osm} \iint_S \omega_{ev} dS = 31.6$ TW. The equivalent work is $W = \mathrm{H}_{fw,mixing} \cdot \overline{T_s} \simeq 32.3$ TW; about 2% larger than that calculated from the osmotic pressure formula. This error may be due to errors in the seawater property routine.

In the first part of this chapter, it was claimed that the ocean is not a heat engine. However, the possible contribution due to chemical potential was not included at that stage of the discussion. As discussed above, to maintain the hydrological cycle in the world's oceans, 32 TW equivalent of mechanical energy must be actually available. This huge amount of

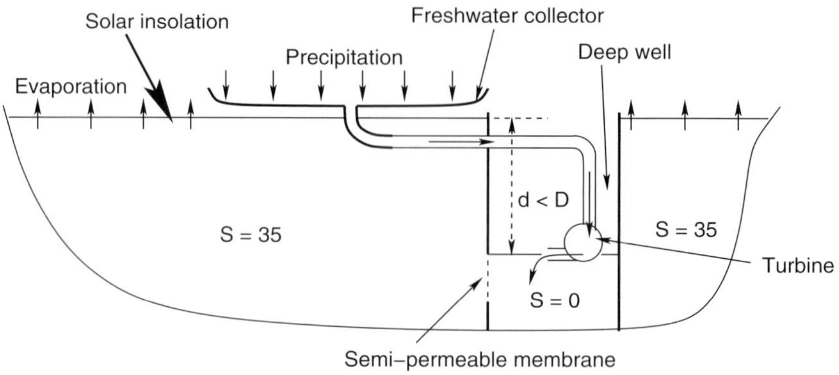

Fig. 3.56 An imaginary thermo-chemical engine driven by the hydrological cycle (evaporation and precipitation) induced by solar insolation in the world's oceans. A deep well contains the freshwater collected from precipitation, with the water level approximately $D = 246$ m below sea level.

energy has been overlooked so far. It is readily seen that such a source of energy might be utilized and converted into mechanical energy, at least in the following conceptual model.

In this conceptual model, precipitation in the ocean is collected and horizontally transported to deep wells, where the level of freshwater is maintained at approximately 246 m below sea level (Fig. 3.56). Mechanical energy in the form of electricity can be extracted from the geopotential difference between the sea surface and this deep level. Freshwater in the deep well can move through the semi-permeable membrane, which allows the throughflow of pure water only. As discussed above, the total amount of energy output is 32 TW. Assuming the total thermal energy input carried by the solar insolation is approximately 65 PW, the efficiency of this thermo-chemical energy is approximately 0.05%. This is a very low efficiency indeed. Nevertheless, in contrast with the previous claim that the ocean is not a thermal engine, the ocean is a thermo-chemical engine from the maintenance of the hydrological cycle. In principle, such a thermo-chemical engine may work in the oceans, where a strong thermohaline circulation is set up by sources of external mechanical energy. It is not clear whether or how this system will work without other sources of external mechanical energy.

Appendix: Source/sink of GPE due to heating/cooling

In the ocean, thermal forcing applies to both the sea surface and seafloor; thus, GPE change due to such forcing is a vital component of the energetics. The rate of GPE generation related to such forcing is discussed in this Appendix.

Case 1: For a water column on the upper surface

The amount of GPE for a water column with unit horizontal area from the sea surface to a depth of h is

$$\chi_0 = mgh_{cen} \tag{3.A1}$$

where g is gravity, h_{cen} is the center of mass relative to the reference level for GPE, and $m = \int_{-h}^{0} \rho_0(z)dz \simeq \bar{\rho}h$ is the total mass of the water column, with $\rho_0(z)$ as the density profile in the water column, and $\bar{\rho}$ the mean reference density. When this water column receives an amount of heat Q, its temperature increases

$$\delta T = Q/\bar{\rho}c_p h \tag{3.A2}$$

and water column height increases $\delta h \simeq \alpha Q/\bar{\rho}c_p$, where α is the thermal expansion coefficient, and c_p is the specific heat under constant pressure. Thus, after heating, the center of mass moves upward $\delta h/2$ and the total GPE of this water column is

$$\chi_1 = mg(h_{cen} + \delta h/2) \tag{3.A3}$$

The net change of GPE is

$$\Delta\chi = \frac{g\alpha Q}{2c_p}\frac{\int_{-h}^{0}\rho_0(z)dz}{\bar{\rho}} \simeq \frac{g\alpha h Q}{2c_p} \tag{3.A4}$$

Assuming that the rate of heating and cooling is balanced, the total GPE source/sink due to surface heating/cooling is

$$\frac{d\chi}{dt} = \frac{g}{2c_p}\iint_S \dot{q}\,(\alpha_{heat}h_{heat} - \alpha_{cool}h_{cool})\,dxdy \tag{3.A5}$$

where $\dot{q}$ is the local heat flux rate and the integration is taken over the global sea surface.

Equation (3.A5) can be easily extended to the general case with buoyancy flux

$$\frac{d\chi}{dt} = \frac{g}{2}\iint_S \left((\dot{b}h)_{buoy.\,gain} - (\dot{b}h)_{buoy.\,loss}\right) dxdy \tag{3.A6}$$

where $\dot{b} = \frac{\alpha\dot{q}}{c_p} + \beta(P - E)$ is the surface buoyancy flux, and P - E is the rate of precipitation minus evaporation. Since a mass coordinate is used, the results are independent of the choice of the reference level for GPE.

Case 2: For a water parcel below the sea surface

Considering a water parcel sitting at a depth h below the sea surface, the amount of GPE of a water column with unit area and initial thickness of Δh is again $\chi_0 = mgh_{cen}$, where h_{cen} is the center of mass related to the reference level for GPE, and $m = \int_{-h}^{0}\rho_0(z)dz = \bar{\rho}h$ is the total mass of the water column, with $\rho_0(z)$ as the density profile in the water column. Assume that this water column receives an amount of heat Q. After heating, the temperature increases $\delta T = Q/\bar{\rho}c_p\Delta h$, and the water column height increases $\delta h \simeq \alpha Q/\bar{\rho}c_p$. As a result, the whole water column above is pushed upward for a distance of δh, and the total GPE of this water column is

$$\chi_1 = mg(h_{cen} + \delta h) \tag{3.A7}$$

The net change in GPE for the water column after heating a water parcel is

$$\Delta\chi = \frac{g\alpha Q}{c_p}\frac{\int_{-h}^{0}\rho_0(z)dz}{\bar{\rho}} \simeq \frac{g\alpha hQ}{c_p} \tag{3.A8}$$

Note that the thickness Δh of the water parcel is assumed to be much smaller than h, so changes in GPE of this water parcel are neglected. From this formula, it is evident that, in order to become an efficient source of energy, heating needs to occur at deep levels. Geothermal heating is such an example.

Note that, for a steady state, the total amount of heat through heating and cooling should be balanced. Assuming that specific heat is constant, GPE generated from thermal forcing should be

$$\Delta\chi = \frac{gQ}{c_p}(\alpha_h h_h - \alpha_c h_c) \tag{3.A9}$$

where $\alpha_h(\alpha_c)$ and $h_h(h_c)$ are the thermal expansion coefficient and geometric height for the heating (cooling) source.

Cautions

- A mass-conserving coordinate system has been adopted in all discussions of GPE. If a Boussinesq model is used instead, in which mass conservation is replaced by volume conservation, the results are incorrect. In general, models based on Boussinesq approximations may have artificial sources/sinks of mass and GPE.
- In the above analysis, it was assumed that heating energy is entirely converted to internal energy, as $\delta T = Q/\rho_0 c_p h$ (see Eqn. (3.A2)). In reality, part of the input thermal energy would be converted into GPE; thus the corresponding temperature change should include a correction $\delta T = \varepsilon\delta T$. However, such a correction term can be shown to be much smaller than the first term, $|\varepsilon\delta T| \ll |\delta T|$, and thus negligible, even for the case of heating and cooling from the seafloor.

Part II

Wind-driven and thermohaline circulation

4
Wind-driven circulation

4.1 Simple layered models

4.1.1 Pressure gradient and continuity equations in layered models

The concept of layered models

The simplest way to simulate the ocean circulation is to assume that the ocean is homogeneous in density. Such a model has no vertical structure. As discussed in Section 1.4, there is a prominent main thermocline/pycnocline in the oceans. The subsurface maximum of the vertical density gradient can be idealized as a step function, and a natural way of simulating the ocean circulation is to treat the ocean as a two-layer fluid, using the main thermocline as the interface. The lower layer lies below the main thermocline; it is very thick and water in this layer moves much slower than that above the main thermocline. As a good approximation, one can assume that fluid in the lower layer is nearly stagnant. Such a model has one active layer only; this is called a reduced-gravity model. The advantage of a reduced-gravity model is its ability to capture the first baroclinic mode of the circulation and the depth of the main thermocline. Adding one more layer to the standard reduced-gravity model, one obtains a $2\frac{1}{2}$-layer model, which is also discussed in this chapter. The comparison of these models is outlined in Figure 4.1.

In a sense, a reduced-gravity model is equivalent to using just two grids in the density coordinate. Similarly, multi-layer models are highly truncated models in the density coordinate. Most theoretical models used in the study of large-scale oceanic circulation are layered models. The reason for favoring layered models over level models is the concept of along-isopycnal mixing.

Strictly speaking, mixing takes place predominately along the neutral surface; however, neutral surfaces cannot be defined for the global oceans. In fact, approximately neutral surfaces are defined in the least-square sense (see Jackett and McDougall, 1997; Eden and Willebrand, 1999). In comparison with a neutral density/surface, an isopycnal is conceptually simple and easy to deal with; thus, it is still widely used.

In the oceans, mixing is predominately along isopycnals because along-isopycnal mixing requires the least amount of work. In reality, mixing is not always along isopycnals. For example, baroclinic instability gives rise to cross-isopycnal mixing, which is associated with the conversion from large-scale gravitational potential energy (GPE) to synoptical-scale

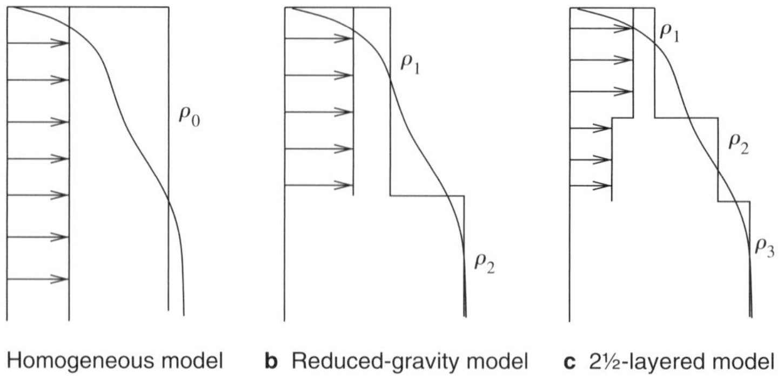

Fig. 4.1 Sketch of simple models in density coordinates, including the vertical structure of horizontal velocity: **a** homogeneous model, **b** reduced gravity model, **c** $2^1/_2$-layer model.

GPE and kinetic energy. As a result, baroclinic instability is a sink of large-scale GPE, and it requires an external source of mechanical energy for support.

The standard terminology is as follows. If the model has two layers, and both of them are in motion, it is called a two-layer model. If only the upper layer is in motion, it is called a reduced-gravity model, or a $1\frac{1}{2}$-layer model. For a multi-layer model, if both the first and the second layers are in motion, but the layers below are stagnant, it is called a $2\frac{1}{2}$-layer model, and so on. In addition, some of the common features of the most frequently used layered models include the following.

- *Most layered models assume that the water in each layer is immiscible.* Some models allow mass exchange between layers. It is readily seen that interfacial mass flux can drive the subsurface layer in motion, without interfacial friction. In fact, including the interfacial mass flux in layered models is a way of simulating the thermohaline circulation in multi-layer models.
- *Most layered models assume that layer thicknesses are always non-zero.* Some models allow zero thickness in the upper layers, i.e., outcropping. Outcropping is a strong nonlinear phenomenon, associated with surface fronts. Handling outcropping properly requires special care. In analytical models, the outcropping line is both a streamline and a zero-depth contour line. The technique of dealing with the outcropping line is discussed in Section 4.1.4. In numerical models, handling the outcropping line requires special numerical schemes that guarantee the layer thickness is always non-negative; such schemes are called positive-definite schemes (Zalesak, 1979; Smolarkiewicz, 2006).

Pressure gradients in multi-layer models

A convenient feature of layered models is that pressure gradient terms in the horizontal momentum equations are directly linked to layer thicknesses. Therefore, simple and elegant solutions in analytical form can be obtained in many cases, as discussed in this chapter.

Using the rigid-lid approximation

The essence of the rigid-lid approximation is a linearization of the upper boundary condition in which the upper boundary of the model ocean is moved from the free surface $z = \zeta$ to a flat

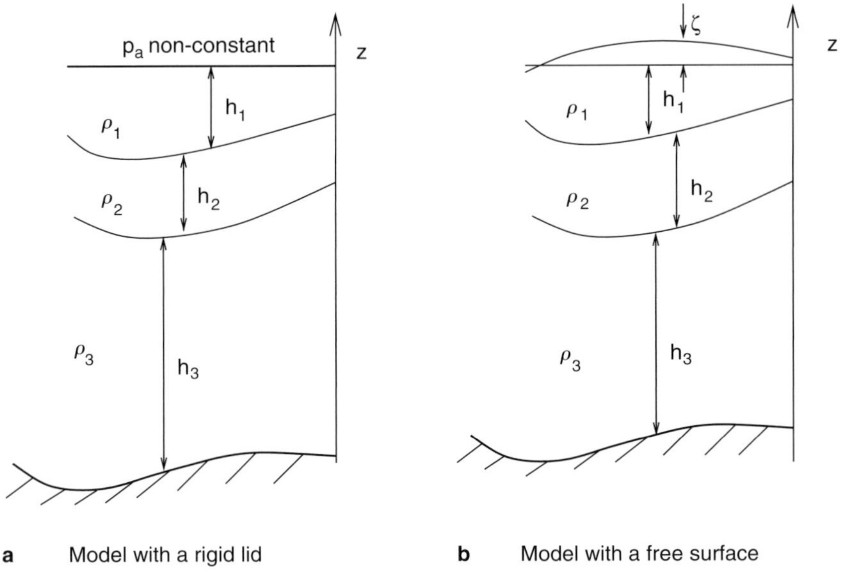

Fig. 4.2 Multi-layer models **a** with a rigid lid or **b** with a free surface.

surface $z = 0$, so the original problem of a moving boundary is reduced to one with a fixed boundary. Owing to the existence of a non-zero sea surface level $\zeta \neq 0$ at the flat surface $z = 0$, the equivalent hydrostatic pressure $p = p_a$ is not constant. Using the hydrostatic relation, one can calculate the pressure in the layers beneath. Starting from $p = p_a$ at $z = 0$ and integrating the hydrostatic relation downward, we obtain the hydrostatic pressure in the upper layer (Fig. 4.2a),

$$p_1 = p_a - \rho_1 g z \tag{4.1}$$

Note that $p_a \neq const$ is an unknown pressure at $z = 0$. In fact, p_a is equivalent to $p_{a,0} + \rho g \zeta$, where $p_{a,0}$ is the sea-level atmospheric pressure and ζ is the unknown free surface elevation. At the base of the upper layer, pressure is $p_2 = p_1 = p_a + p_1 g h_1$. Below the interfaces, pressure in the second and third layers is

$$p_2 = p_a + \rho_1 g h_1 - \rho_2 g\,(z + h_1) \tag{4.2a}$$

$$p_3 = p_a + \rho_1 g h_1 + \rho_2 g h_2 - \rho_3 g (z + h_1 + h_2) \tag{4.2b}$$

Applying the horizontal gradient operator ∇_h to Eqn. (4.2b) leads to

$$\nabla_h p_3 = \nabla_h p_a + \rho_1 g \nabla_h h_1 + \rho_2 g \nabla_h h_2 - \rho_3 \nabla_h (h_1 + h_2) \tag{4.3}$$

Assuming that the third layer is very thick and motionless; thus $\nabla_h p_3 = 0$, and we obtain

$$\nabla_h p_a = (\rho_3 - \rho_1) g \nabla_h h_1 + (\rho_3 - \rho_2) g \nabla_h h_2 \tag{4.4}$$

Taking the horizontal gradient of Eqns. (4.1) and (4.2a) and using Eqn. (4.4) leads to the pressure gradient terms in the first and second layers

$$\nabla_h p_1 = (\rho_3 - \rho_1) g \nabla_h h_1 + (\rho_3 - \rho_2) g \nabla_h h_2 \tag{4.5a}$$

$$\nabla_h p_2 = (\rho_3 - \rho_2) g \nabla_h h_1 + (\rho_3 - \rho_2)\, g \nabla_h h_2 \tag{4.5b}$$

After simple manipulations, the pressure gradients can be rewritten as

$$\frac{1}{\rho_1} \nabla_h p_1 \simeq (g_1' + g_2') \nabla_h h_1 + g_2' \nabla_h h_2 \tag{4.6a}$$

$$\frac{1}{\rho_2} \nabla_h p_2 \simeq g_2' \nabla_h h_1 + g_2' \nabla_h h_2 \tag{4.6b}$$

where we use the approximation $\rho_1 \simeq \rho_2 \simeq \rho_0$ for the denominators on the right-hand side, and $g_1' = g\,(\rho_2 - \rho_1)\,/\rho_0$, $g_2' = g\,(\rho_3 - \rho_2)\,/\rho_0$ are reduced gravity, on the order of 10^{-2} m/s^2. For a $1\frac{1}{2}$-layer model, we have

$$\nabla_h p_2 = 0, \quad \frac{1}{\rho_1} \nabla_h p_1 = g_1' \nabla_h h_1 \tag{4.7}$$

The essential step in this derivation is to assume no motion in the deep layers, so the unknown pressure p_a, which represents the contribution from the atmospheric pressure and the free surface elevation, can be eliminated. Without assuming a stagnant deep layer, the rigid-lid pressure p_a will remain as part of the pressure expression. In such cases one may have to use other approaches to eliminate p_a. For example, in numerical models based on the rigid-lid approximation, p_a can be eliminated by cross-differentiating the horizontal momentum equations, and the problem is reduced to solving an elliptic equation for the barotropic streamfunction.

Including the free surface elevation

The same expressions can be derived for a model including the free surface explicitly (Fig. 4.2b). We start from the sea surface $z = \zeta$ and integrate the hydrostatic relation downward

$$p_1 = p_{a,0} - \rho_1 g (z - \zeta)$$

$$p_2 = p_{a,0} + \rho_1 g (\zeta + h_1) - \rho_2 g (z + h_1)$$

$$p_3 = p_{a,0} + \rho_1 g (\zeta + h_1) + \rho_2 g h_2 - \rho_3 g (z + h_1 + h_2)$$

where $p_{a,0}$ is the sea-level atmospheric pressure. Assuming $p_{a,0} = const$, the pressure gradient in each layer can be written as

$$\nabla_h p_1 = \rho_1 g \nabla_h \zeta$$

$$\nabla_h p_2 = \rho_1 g \nabla_h \zeta - (\rho_2 - \rho_1) g \nabla_h h_1$$

$$\nabla_h p_3 = \rho_1 g \nabla_h \zeta - (\rho_3 - \rho_1) g \nabla_h h_1 - (\rho_3 - \rho_2) g \nabla_h h_2$$

Dividing by ρ_i and using the reduced-gravity notation, we obtain

$$\frac{1}{\rho_1}\nabla_h p_1 = g\nabla_h \zeta \tag{4.8a}$$

$$\frac{1}{\rho_2}\nabla_h p_2 \simeq g\nabla_h \zeta - g_1'\nabla_h h_1 \tag{4.8b}$$

$$\frac{1}{\rho_3}\nabla_h p_3 \simeq g\nabla_h \zeta - (g_1' + g_2')\nabla_h h_1 - g_2'\nabla_h h_2 \tag{4.8c}$$

These expressions include the gradient of the free surface, which may be part of the unknown; however, this unknown can be eliminated by assuming that the lowest layer is very thick, so that the pressure gradient in the lowest layer is negligible. If we assume $\nabla_h p_3 = 0$, $\nabla_h \zeta$ can be written in terms of $\nabla_h h_1$ and $\nabla_h h_2$; therefore the pressure gradients in the first two layers are

$$\frac{1}{\rho_1}\nabla_h p_1 \simeq g\nabla_h \zeta = (g_1' + g_2')\nabla_h h_1 + g_2'\nabla_h h_2 \tag{4.9a}$$

$$\frac{1}{\rho_2}\nabla_h p_2 \simeq g_2'\nabla_h h_1 + g_2'\nabla_h h_2 \tag{4.9b}$$

These expressions are the same as we have derived above. Thus, the "rigid-lid approximation" is a linearization of the boundary condition at the sea surface. The rigid-lid assumption replaces the free surface with a fixed boundary at $z = 0$; however, the pressure effect due to the non-constant free surface elevation is retained in terms of the equivalent pressure p_a, which is not constant.

The continuity equation

For the i-th layer, mass conservation of an incompressible fluid is

$$u_{ix} + v_{iy} + w_{iz} = 0 \tag{4.10}$$

Assuming that u_i and v_i are independent of z within each layer, integrating over each layer leads to

$$h_i(u_{ix} + v_{iy}) + w_i^{top} - w_i^{bot} = 0 \tag{4.11}$$

where w_i^{top} and w_i^{bot} are the vertical velocity at the upper and lower interfaces of the i-th layer (Fig. 4.3).

Assuming that there is no interfacial mass exchange, i.e., the diapycnal velocity is zero, $w^* = 0$, the velocity vector should be parallel to the interface, as indicated at the upper interface in Figure 4.3. Denote the interface as $H_i\ (t, x, y)$, then the vertical velocity at this interface is

$$w_i^{top} = -\left(\frac{\partial H_i}{\partial t} + u_i\frac{\partial H_i}{\partial x} + v_i\frac{\partial H_i}{\partial y}\right) \tag{4.12}$$

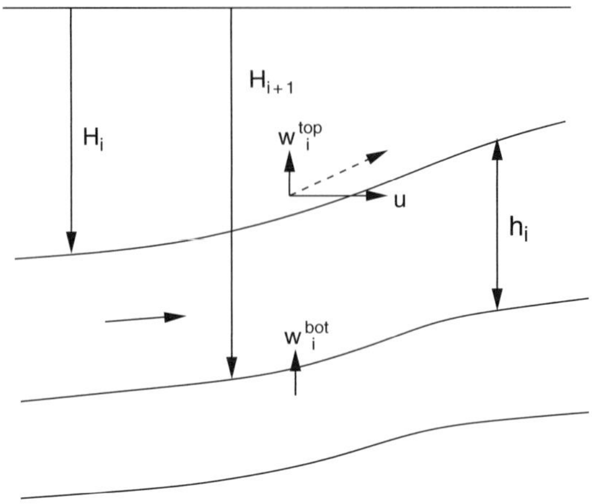

Fig. 4.3 Sketch of a layer model.

Vertical velocity at the interface H_{i+1} (t, x, y) has a similar expression, and the difference in the vertical velocity at the top and bottom of this layer is

$$w_i^{top} - w_i^{bot} = \frac{\partial}{\partial t}h_i + u_i\frac{\partial}{\partial x}h_i + v_i\frac{\partial}{\partial y}h_i, \quad h_i = H_{i+1} - H_i \tag{4.13}$$

Therefore, the final expression of the continuity equation for each layer is

$$\frac{\partial}{\partial t}h_i + \frac{\partial}{\partial x}(h_i u_i) + \frac{\partial}{\partial y}(h_i v_i) = 0 \tag{4.14}$$

Note that the thickness of the uppermost layer h_1 includes the contribution due to the free surface elevation.

4.1.2 Reduced-gravity models

The wind-driven circulation has been described in terms of the quasi-geostrophic model derived from the shallow water equation in many textbooks. In the world's oceans, wind-driven gyres in mid latitudes have horizontal scales on the order of thousands of kilometers: much larger than the synoptic scale assumed in the quasi-geostrophic approximation. For example, in the north–south direction the vertical displacement of isopycnal surfaces is of the same order of magnitude as the layer depth. As a result, one of the basic assumptions in the traditional quasi-geostrophic approximation, that deviations from the mean stratification are small, is no longer valid. Although the quasi-geostrophic theory remains a useful tool for describing the circulation, the strong nonlinearity due to the meridional change of stratification can be handled much more accurately by using simple reduced-gravity models.

The reduced-gravity model is probably one of the most useful tools in oceanic circulation simulation. For example, the reduced-gravity model can be used to simulate the free surface elevation and its time evolution under realistic wind stress forcing. In addition, the reduced-gravity model can provide the depth of the main thermocline under realistic conditions, which is impossible to obtain from the quasi-geostrophic model. This section focuses on theories for the steady wind-driven circulation based on single-moving-layer reduced-gravity models.

Formulation of a reduced-gravity model

The essence of constructing a reduced-gravity model is to treat the main thermocline (or the pycnocline) in the oceans as a step function in density, so that density in the upper layer equals a constant ρ_0 and the density in the lower layer is $\rho_0 + \Delta\rho$; furthermore, the lower layer is assumed to be infinitely deep, so the pressure gradient in the lower layer is infinitely small and the pressure gradient terms in the upper layer have simple forms, as discussed in the previous section.

The fluid is assumed to be incompressible. The dynamical effects of meso-scale eddies and turbulence are parameterized in terms of lateral friction and "bottom friction" in the horizontal momentum equations. Since the model has two layers, the so-called bottom friction is actually the interfacial friction.

In order to present analytical solutions in concise forms, the reduced-gravity model in this section is formulated in a β-plane. The corresponding model in spherical coordinates can be readily derived, and this is left as an exercise for the reader.

Under these assumptions, the momentum and continuity equations for a reduced-gravity model are

$$hu_t + h(uu_x + vu_y) - fhv = -g'hh_x + \tau^x/\rho_0 + A_h\nabla_h^2(hu) - Ru \tag{4.15a}$$

$$hv_t + h(uv_x + vv_y) + fhu = -g'hh_y + \tau^y/\rho_0 + A_h\nabla_h^2(hv) - Rv \tag{4.15b}$$

$$h_t + (hu)_x + (hv)_y = 0 \tag{4.15c}$$

where h is the layer thickness, (u, v) are the horizontal velocity, f is the Coriolis parameter, $g' = g\Delta\rho/\rho_0$ is the reduced gravity, which is on the order of 10^{-2} m/s^2, (τ^x, τ^y) are the wind stress, A_h is the coefficient of lateral friction, and R is the coefficient of interfacial friction. Note that wind stress is treated here as a body force for the whole layer, a concept that was first postulated by Charney.

In contrast to quasi-geostrophic models, where changes in layer thickness are assumed to be much smaller than the mean layer thickness, the essence of the reduced-gravity model is allowing layer thickness h to vary greatly, including the case of vanishing layer thickness. The ability to handle finite amplitude perturbations in layer thickness is one of the most important advantages of using reduced-gravity models, in comparison with models based on the quasi-geostrophic approximation.

Wind-driven circulation in the ocean interior can be described in terms of the balance between the Coriolis force, the pressure gradient, and the wind stress. However, such an interior solution cannot satisfy all boundary conditions in a closed basin. In order to describe the closed circulation in a basin, other high-order terms have to be added in, such as the nonlinear advection terms, the lateral friction terms, and the interfacial friction terms. In fact, within the western boundary regime, the potential vorticity balance is different from that in the ocean interior. For example, assuming that the friction is in the form of interfacial friction gives rise to the potential vorticity balance between the planetary vorticity advection and the interfacial friction torque. This is the case for the classical Stommel boundary layer. Similarly, assuming that the vorticity balance is between planetary vorticity advection and lateral friction or relative vorticity gives rise to the Munk boundary layer or the inertial boundary layer.

For simplicity, analysis in this section is limited to the case of a steady state, so that the time-dependent terms in the momentum and continuity equations are omitted in the following analysis.

Conservation laws

Energy balance

Multiplying Eqn. (4.15a) by u and Eqn. (4.15b) by v, and adding the results, leads to

$$h\vec{u} \cdot \nabla_h \left(\frac{u^2 + v^2}{2} + g'h \right) = W \tag{4.16}$$

where

$$W = \frac{1}{\rho_0}(u\tau^x + v\tau^y) - R(u^2 + v^2) + A_h[u\nabla_h^2(hu) + v\nabla_h^2(hv)] \tag{4.17}$$

the first term on the right-hand side is work done by the wind stress, the second term is dissipation due to interfacial friction, and the third term is dissipation due to lateral friction.

In the present case the total energy is the sum of kinetic energy and GPE. Equation (4.16) states that along streamlines changes in total energy are balanced by the source due to wind work and sinks due to interfacial and lateral friction. Since the Coriolis force is a virtual force and is always perpendicular to the velocity, it does no work on water parcels. In addition, internal energy does not appear in the balance of energy because, in such a simple reduced-gravity model, thermodynamics and dynamics are separated.

Potential vorticity balance

Dividing Eqns. (4.15a) and (4.15b) by h, then cross-differentiating and subtracting, leads to

$$\vec{u} \cdot \nabla_h q + q(u_x + v_y) = C \tag{4.18}$$

where

$$q = f + v_x - u_y \tag{4.19}$$

is the planetary vorticity plus the relative vorticity, and

$$C = \left(\frac{\tau^y}{\rho_0 h} - \frac{Rv}{h} + \frac{A_h}{h}\nabla_h^2(hv) \right)_x - \left(\frac{\tau^x}{\rho_0 h} - \frac{Ru}{h} + \frac{A_h}{h}\nabla_h^2(hu) \right)_y \tag{4.20}$$

is the source of potential vorticity due to wind-stress curl and the vorticity sinks due to interfacial and lateral friction.

Using the continuity equation (4.15c), we obtain a concise form of the potential vorticity equation

$$h\vec{u} \cdot \nabla_h \frac{f + v_x - u_y}{h} = C \tag{4.21}$$

where $Q = (f + v_x - u_y)/h$ is potential vorticity for the reduced-gravity model. A more accurate definition of potential vorticity is to include the density increment $\Delta\rho$, i.e., $Q' = (f + v_x - u_y)\Delta\rho/\rho_0 h$. When the layer thickness h is greatly reduced, the density increment is also reduced, i.e., $\Delta\rho \to 0$, as $h \to 0$. Thus, at the limit of infinitesimal layer thickness, the corresponding potential vorticity is reduced to

$$Q' = -\rho_0^{-1}\left(f + v_x - u_y\right) \partial\rho/\partial z \tag{4.22}$$

Equation (4.21) states that potential vorticity advection is balanced by potential vorticity sources and sinks, including wind stress, interfacial and lateral friction. This equation also applies to an individual layer in a multi-layer model. The relative vorticity, $v_x - u_y$, is widely used in the study of non-rotating homogeneous fluids; however, in a stratified rotating fluid, potential vorticity is used because potential vorticity combines the relative vorticity with the dynamical effects of rotation and stratification.

Solution free of forcing and dissipation

Neglecting the forcing and dissipation terms in Eqns. (4.16) and (4.21) leads to simple conservation laws for energy and potential vorticity

$$B = \frac{u^2 + v^2}{2} + g'h = F(\psi) \tag{4.23}$$

$$Q = \frac{f + v_x - u_y}{h} = G(\psi) \tag{4.24}$$

where ψ is the streamfunction introduced for a steady flow, through the following definition

$$\psi_x = hv, \quad \psi_y = -hu \tag{4.25}$$

Equations (4.23) and (4.24) state that both energy (the Bernoulli function) and potential vorticity are conserved along streamlines. As discussed shortly, in the subsurface ocean away from the western boundary regime and when the layer is shielded from direct forcing, potential vorticity is conserved along streamlines because under such circumstances mixing and dissipation along streamlines is negligible.

Functions F and G are not independent, as shown by the following: taking the gradient of Eqn. (4.23) leads to

$$\nabla B = \frac{dF}{d\psi} \nabla\psi \tag{4.26}$$

Using Eqns. (4.23) and (4.15a) without the time-dependent term, wind stress, and friction terms, we obtain the following differential relation between functions F and G:

$$\frac{dF}{d\psi} = G(\psi), \quad or \quad Q\nabla\psi = \nabla B \tag{4.27}$$

Dynamical roles of the western boundary current

From both the energy and potential vorticity equations, it is clear that a purely inertial western boundary current cannot satisfy the energy and potential vorticity balance in a closed basin. No matter how small the friction is, it plays an essential role in balancing the energy and potential vorticity in a closed basin by dissipating the potential vorticity and energy input from the wind stress. This is an extremely important point, and will be discussed in detail.

The interior solution

Formulation based on wind stress

In the ocean interior, frictional and inertial terms are negligible. For simplicity, we will assume that wind stress takes a simple form: $\tau^x = \tau^x(y)$, $\tau^y = 0$; thus, the momentum equations are reduced to

$$-fhv = -g'hh_x + \tau^x/\rho_0 \tag{4.28a}$$

$$fhu = -g'hh_y \tag{4.28b}$$

In addition, the model satisfies the continuity equation

$$(hu)_x + (hv)_y = 0 \tag{4.28c}$$

Cross-differentiating and subtracting Eqns. (4.28a) and (4.28b) leads to the vorticity equation

$$\beta hv = -\tau^x_y/\rho_0 \tag{4.29}$$

which is a simplified form of Eqn. (4.21). This equation is called the *Sverdrup relation*. Substituting Eqn. (4.29) into Eqn. (4.28a) gives rise to a first-order ordinary differential equation

$$hh_x = -\frac{f^2}{g'\rho_0\beta}\left(\frac{\tau^x}{f}\right)_y \tag{4.30}$$

Integrating this equation leads to the interior solution. However, which boundary should be used to start the integration? Although either the eastern or western boundary could be the choice, the balance of vorticity in the basin requires that the eastern boundary is the only choice for starting the integration, as will be shown in the next section. The zonal integration gives rise to the layer thickness solution

$$h^2 = h_e^2 + \frac{2f^2}{g'\rho_0\beta}\left(\frac{\tau^x}{f}\right)_y (x_e - x) \tag{4.31}$$

Using Eqn. (4.25), the Sverdrup relation (4.29) can be rewritten as $\beta\psi_x = -\tau_y^x/\rho_0$.

The eastern boundary is a streamline with $\psi = 0$; thus, the streamfunction solution is

$$\psi = \frac{1}{\rho_0\beta}\tau_y^x(x_e - x) \tag{4.32}$$

This volume transport is called the *Sverdrup transport*. The structure of wind-driven gyres described by Eqns. (4.31) and (4.32) and the associate boundary layer structure are illustrated graphically later.

Note that easterlies prevail in low latitudes and westerlies prevail in mid latitudes, i.e., τ^x is negative near the equator and reaches a positive maximum along the inter-gyre boundary; therefore, the wind-stress curl satisfies $curl\,\tau = -\tau_y^x < 0$. According to Eqn. (4.29), this negative wind-stress curl drives an equatorward flow in the interior. The depth of the main thermocline increases westward, as indicated by Eqn. (4.31).

Right at the equator, the Coriolis force vanishes, so the argument presented above does not apply. However, easterlies can also push water westward, and a balance between the wind and pressure gradient force gives rise to a higher sea level in the western part of the equatorial ocean, such as the Warm Pool in the Pacific.

In this formulation, the flow within and below the Ekman layer is considered as a single layer. At low latitudes, these two layers move in opposite directions: the Ekman flux is poleward, owing to the easterly, but the geostrophic flow driven by the Ekman pumping is equatorward, as discussed below. *Although this seems to be a disadvantage for such a formulation, this formulation also has an advantage because the Sverdrup transport ψ remains finite near the equator. On the other hand, near the equator both the Ekman flux and the geostrophic flow below the Ekman layer become unbounded and their algebraic sum is poorly defined.*

Since the Ekman volume flux is $\tau^x/f\rho_0$, the thermocline depth is controlled by the Ekman pumping, i.e., the horizontal convergence of the Ekman flux $(\tau^x/f\rho_0)_y$, while the streamfunction is controlled by the wind-stress curl equation (4.32). However, the situation near the boundary between the subtropical and subpolar gyres is slightly different from other places. Since the convergence of the Ekman flux is $\left(\frac{\tau^x}{f}\right)_y = \frac{\tau_y^x}{f} - \frac{\beta\tau^x}{f^2}$ near the inter-gyre boundary, the second term on the right-hand side may dominate and change the sign of the Ekman pumping contribution. In fact, the layer thickness may decline westward near the inter-gyre boundary. When wind stress is strong enough, the layer thickness calculated from Eqn. (4.31) may be non-positive. In such a case, the interface outcrops, and the model used above has to be modified in order to reconcile the outcropping phenomenon, which is discussed in Section 4.1.4.

Formulation based on Ekman pumping

In the discussion above, the wind stress is treated as a body force to the entire upper layer. There is another approach, in which the dynamical role of the wind stress is explored in terms of the Ekman pumping due to the Ekman flux convergence. Thus, the Ekman layer and the horizontal mass flux within it are separated from the geostrophic flow below the Ekman layer. The corresponding momentum equations are reduced to

$$-fhv = -g'hh_x \tag{4.33a}$$

$$fhu = -g'hh_y \tag{4.33b}$$

and the continuity equation is

$$(hu)_x + (hv)_y = -w_e \tag{4.33c}$$

where the Ekman pumping rate is related to the wind stress through

$$w_e = -\left(\tau^x / f\rho_0\right)_y \tag{4.34}$$

Cross-differentiating Eqns. (4.33a) and (4.33b) and using the continuity equation (4.33c) leads to the vorticity equation

$$\beta hv = f w_e = -f\left(\frac{\tau^x}{f\rho_0}\right)_y \tag{4.35}$$

Substituting Eqn. (4.35) into Eqn. (4.33a) and zonally integrating, we recover Eqn. (4.31).

Since there is a source of water coming from above, the horizontal flow field in the moving layer cannot be described in terms of a streamfunction alone. Although we can still integrate the meridional velocity by starting from the eastern boundary, what we obtain should be called the meridional volume flux rate (or the so-called *Sverdrup function*):

$$m = \frac{f}{\rho_0\beta}\left(\frac{\tau^x}{f}\right)_y (x_e - x) \tag{4.36}$$

In this formulation, the layer thickness has the same expression obtained from the previous formulation based on wind stress; but the integrated meridional volume flux is different by the amount equal to the Ekman flux, because the volume flux of the Ekman layer is not included in the geostrophic formulation. This difference is reflected in the difference in the inter-gyre boundary in these two formulations.

Any two-dimensional vector field can be separated into two parts:

$$hu = \varphi_x - \psi_y; \quad hv = \varphi_y + \psi_x \tag{4.37}$$

The decomposition is, however, not unique. For example, the solution can have an additional component ϕ', as long as $\nabla^2\varphi' = 0$. In addition, the boundary conditions for ϕ and ψ are not unique. Even if we construct such a streamfunction, it does not represent the streamlines accurately because the velocity has another component – $\nabla\varphi$.

As a compromise, we can use the pressure field or the layer thickness field to plot the streamlines. Since the pressure or the layer thickness square has a dimension different from the volume flux, we can use a modified quantity, called the virtual streamfunction, which is defined as

$$\psi^* = \frac{g'}{2f_0}(h^2 - h_e^2) = \frac{f^2}{f_0 \rho_0 \beta}\left(\frac{\tau^x}{f}\right)_y (x_e - x) \tag{4.38}$$

where f_0 is the reference latitude where the meridional volume flux is equal to ψ^*; away from the reference latitude, ψ^* has a value different from that at the reference latitude. The difference between ψ^* and m is due to the Ekman pumping in the oceanic interior. The construction of m is discussed in Section 4.7 about the communication between the subtropical and tropical oceans.

Reduced-gravity models have some essential features which are important for understanding the structure of the thermocline in multi-layer models or models with continuous stratification:

- *The stratification parameters have to be specified a priori*, such as layer thickness along the eastern boundary h_e and density jump across the interface, $\Delta\rho$. These parameters are controlled by external processes – the thermohaline circulation. Thus, the reduced-gravity model is essentially a perturbation approach, i.e., treating the wind-driven circulation as a perturbation to a given stratification profile in density coordinates.
- *The layer depth square is inversely proportional to the density jump.* As a result, strong stratification leads to a shallow thermocline and vice versa.

Common features of a western boundary layer

Within the western boundary current, scaling analysis indicates that to a very good approximation the cross-stream pressure gradient is in balance with the Coriolis force associated with the downstream velocity,

$$fhv = g'hh_x \tag{4.39}$$

However, the downstream momentum is in an ageostrophic balance, i.e., the downstream momentum balance must include downstream pressure, the Coriolis force associated with the cross-stream velocity, and other terms, such as friction or the inertial term. Thus, the boundary layer is said to be in semi-geostrophic balance. Within the western boundary regime, the zonal integration of Eqn. (4.39) leads to a simple relation between the streamfunction and the layer thickness

$$\psi = \psi_I + \frac{g'}{2f}\left(h^2 - h_I^2\right) \tag{4.40}$$

where the subscript I indicates the interior solution at the outer edge of the western boundary current:

$$h_I^2 = h_e^2 + \frac{2f^2}{g'\rho_0\beta}\left(\frac{\tau^x}{f}\right)_y (x_e - x_w) \tag{4.41}$$

$$\psi_I = \frac{1}{\rho_0\beta}\tau_y^x (x_e - x_w) \tag{4.42}$$

Like the eastern boundary, the western boundary should be a streamline, i.e., $\psi = 0$ at the western wall, and layer thickness along the western wall is

$$h_w^2 = h_I^2 - \frac{2f}{g'}\psi_I = h_e^2 - \frac{2}{g'\rho_0}\tau^x (x_e - x_w) \tag{4.43}$$

There are two important features in this solution: first, because $\psi_I > 0$ in a subtropical gyre, within the western boundary layer, layer thickness declines toward the wall. The steep layer thickness slope next to the western boundary, $h_I^2 - h_w^2 = 2f\psi_I/g'$, is caused by the strong western boundary current required for balancing the model's mass, vorticity, and energy. Second, differentiating Eqn. (4.43) with respect to y gives rise to

$$\frac{\partial h_w^2}{\partial y} = -\frac{2}{g'\rho_0}(x_e - x_w)\frac{\partial \tau^x}{\partial y} < 0 \tag{4.44}$$

Therefore, layer thickness along the western wall decreases northward.

Note that layer thickness in the reduced-gravity model can be interpreted as either pressure or free surface elevation. There are two very important dynamical consequences implied in the structure of the subtropical gyre circulation.

First, the meridional pressure at the outer edge of the western boundary is low–high–low, which is set up by the wind-driven anticyclonic gyre in the interior. This can be seen clearly from the structure of the solution at the western edge in Figure 4.4b.

Second, as described in Eqn. (4.44), the meridional pressure force along the wall is always northward along the whole western boundary. As an example, the layer thickness obtained from the Stommel boundary layer (as discussed in the next section) demonstrates this feature (Fig. 4.4a). This meridional pressure gradient is the result of the geostrophic constraint across the western boundary layer, so this feature holds for all types of western boundary layer, regardless of the specific dynamic balance within the boundary layer, i.e., it is valid for the Stommel boundary layer, the Munk boundary layer, or the inertial boundary layer.

Such a northward pressure force is an important far-field background for the coastal circulation of much smaller cross-shore scale. In a subtropical basin, this meridional pressure gradient can push a poleward coastal current along the western boundary of the subtropical basin. On the other hand, in the subpolar basin the corresponding pressure gradient is equatorward, so it can play a crucial role in setting up the equatorward current along the western boundary of the subpolar basin. In addition, due to changes in climatic conditions

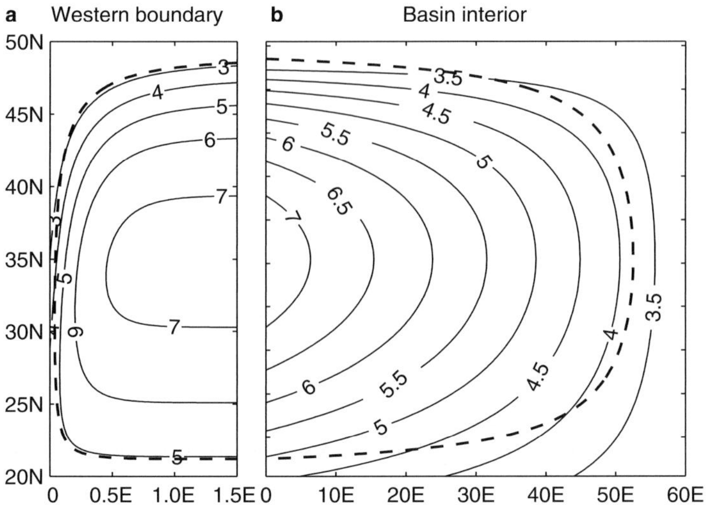

Fig. 4.4 Thermocline depth (in 100 m) and a streamline (*heavy dashed line*) in a model with a Stommel boundary layer.

over the whole basin, such a large-scale pressure field should change over decadal time scales, bringing changes to the coastal circulation.

Stommel boundary layer

Stommel (1948) made the assumption that interfacial friction is in the form of a linear drag law. Thus, within the western boundary layer the downstream momentum equation is in an ageostrophic balance

$$fhu = -g'hh_y - Rv \tag{4.45}$$

For the model with zonal wind only, the potential vorticity equation is reduced to

$$\beta hv = -\frac{\tau_y^x}{\rho_0} + R\left(u_y - v_x\right) \tag{4.46}$$

Within the western boundary layer, the contribution due to wind-stress curl is negligible compared with the contributions from the interfacial frictional torque and planetary vorticity gradient, so the potential vorticity equation is further reduced to

$$\beta hv = -Rv_x \tag{4.46'}$$

Integrating this equation across the western boundary current leads to

$$\beta\left(\psi - \psi_I\right) + \kappa g' h_x / f = 0$$

Using Eqn. (4.40), we obtain

$$h_x + \frac{\beta}{2R}\left(h^2 - h_I^2\right) = 0 \tag{4.47}$$

subject to the following boundary condition

$$h(0) = h_w \tag{4.48}$$

The corresponding solution is

$$h = h_I \frac{1 - Be^{-\eta}}{1 + Be^{-\eta}}, \quad B = \frac{h_I - h_w}{h_I + h_w} \tag{4.49}$$

where

$$\eta = \frac{h_I \beta}{R} x \tag{4.50}$$

is the stretched boundary layer coordinate. The structure of the interior solution and the western boundary layer is shown in Figure 4.4. The scale width of this boundary layer is

$$\delta_S = \frac{R}{h_I \beta} \tag{4.51}$$

From Eqns. (4.40, 4.49), the streamfunction within the boundary layer is (Fig. 4.5a):

$$\psi = \psi_I - \frac{2g' Be^{-\eta}}{f\left(1 + Be^{-\eta}\right)^2} h_I^2 \tag{4.52}$$

The global structure of the boundary layer solution depends on the choice of parameters used in the model, in particular the choice of R. From observations, the width of the Gulf Stream is about 50 km, thus we choose $\delta_s \simeq 25$ km. Assuming $\beta \simeq 2 \times 10^{-11}$/m/s, $h_I \simeq 400$ m, then the suitable choice is $R \simeq \beta h_I \delta_s = 2 \times 10^{-4}$ m/s^2.

As an example, we choose a model mimicking the North Atlantic Ocean, with parameters: $h_e = 300,\ g' = 0.015$ m/s^2, and subject to the following wind stress

$$\tau^x = -0.15 \cos\left(\frac{y - y_s}{y_n - y_s}\pi\right) \text{ (in N/m}^2\text{)} \tag{4.53}$$

Although similar maps of the Sverdrup function can be obtained from a quasi-geostrophic model, the map of thermocline depth can be obtained from a reduced-gravity model only. This is one of the most important advantages of the reduced-gravity model.

The dynamic balance of the circulation is best illustrated in terms of the potential vorticity change along closed streamlines. Since the relative vorticity is very small, the potential vorticity can be approximated by $(\Delta\rho/\rho_0)\,(f/h)$. In the basin interior the wind-stress curl is a sink of potential vorticity; thus, potential vorticity declines downstream (Fig. 4.6b).

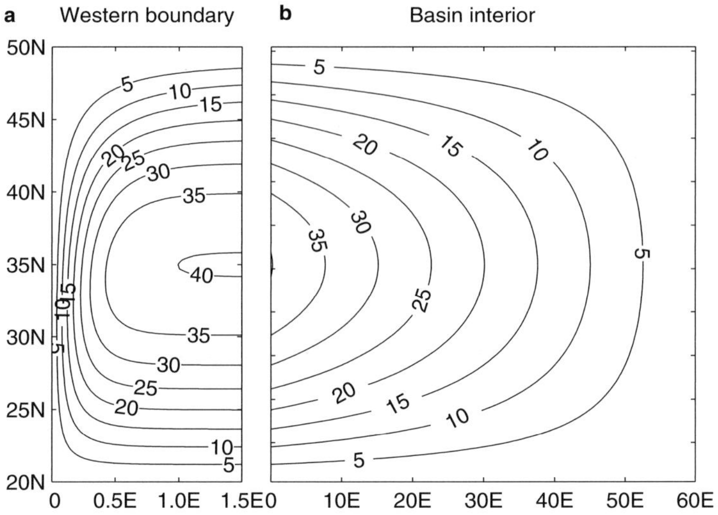

Fig. 4.5 Sverdrup transport in a model with a Stommel boundary layer (Sv).

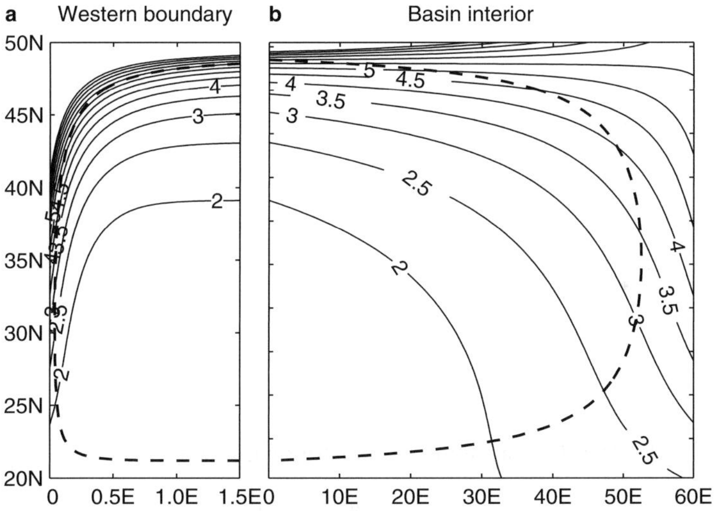

Fig. 4.6 Potential vorticity (in 10^{-10}/s/m) and a streamline (*heavy dashed line*) for a model with a Stommel boundary layer.

In the western boundary, the interfacial friction torque is a source of positive potential vorticity; thus, potential vorticity of a water parcel increases along the pathway (Fig. 4.6a).

Potential vorticity and its meridional gradient are very high within the northern part of the basin, but they are rather small elsewhere. If an active second layer is added below this top layer, the corresponding meridional gradient of potential vorticity in the second layer is

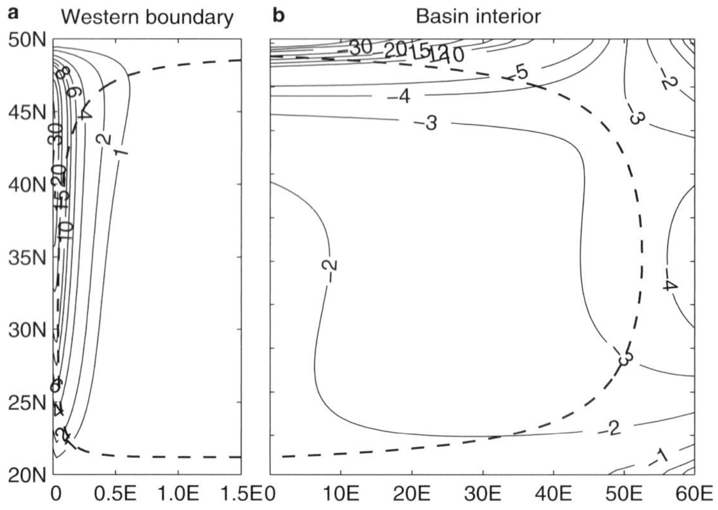

Fig. 4.7 Source and sink of potential vorticity due to wind stress and interfacial friction, including a *heavy dashed line* indicating a streamline: **a** within the western boundary layer (in $10^{-11}/s^3$); **b** in the basin interior (in $10^{-13}/s^3$).

reversed within this regime; thus, baroclinic instability is mostly active in the northern part of the basin (Pedlosky, 1987a). On the other hand, the non-zonal flow in the southeast part of the basin is associated with a rather weak potential vorticity gradient, so the baroclinic instability there should be relatively weak.

Potential vorticity balance in the basin can be diagnosed from the model, using Eqn. (4.21). In the basin interior, the interfacial friction torque is negligible, and the wind-stress curl works as a sink of potential vorticity (Figure 4.7b). Within the western boundary layer, the wind stress torque is negligible, and the frictional torque associated with the meridional velocity produces a strong positive source of potential vorticity Figure 4.7a). Note that the units used in plotting the right panel are 100 times larger than those used for the left panel, because the area of the western boundary regime is about 100 times smaller than the basin interior; the strong source of potential vorticity multiplied by a much smaller area gives rise to the exact balance of potential vorticity in the whole basin.

Similarly, the energy balance along the streamlines consists of two stages. First, in the basin interior, wind stress imposes mechanical energy into the circulation, primarily near both the northern and southern boundaries, where the zonal velocity and wind stress are large (Fig. 4.8b). This external mechanical energy input is balanced by an energy sink due to interfacial friction. Along the western boundary, the strong interfacial friction due to the strong meridional current plays the role of the mechanical energy sink in the model (Fig. 4.8a).

The boundary layer solution matches the interior solution at $\eta \to \infty$. However, at a finite distance from the wall they do not match exactly, as can be clearly seen from the mismatch

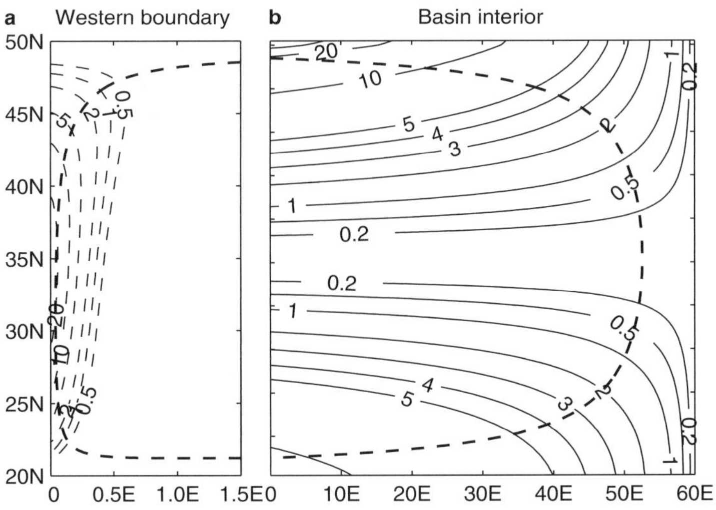

Fig. 4.8 Source and sink of mechanical work due to wind stress and interfacial friction, including a *heavy dashed line* indicating a streamline: **a** within the western boundary layer (in 10^{-4} m^3/s^4); **b** in the basin interior (in 10^{-6} m^3/s^4).

of the streamline between the interior solution and the western boundary solution. A slightly elaborate boundary solution can match with the interior solution gradually, but this is beyond the scope of discussion here.

Munk boundary layer

Munk (1950) postulated that lateral friction works as the major dissipation mechanism, and in the reduced-gravity model the lateral friction is parameterized in terms of $\nabla_h^2(h\vec{u})$, where $\nabla_h^2 = \frac{\partial^2}{\partial x^2} + \frac{\partial^2}{\partial y^2}$ is the horizontal Laplace operator. Therefore, the basic momentum equations are as follows

$$-fhv = -g'hh_x + A_h\nabla_h^2(hu) + \tau^x/\rho_0 \tag{4.54a}$$

$$fhu = -g'hh_y + A_h\nabla_h^2(hv) \tag{4.54b}$$

Based on scaling analysis, within the western boundary the momentum equations are reduced to the following form

$$-fhv = -g'hh_x \tag{4.55a}$$

$$fhu = -g'hh_y + A_h\nabla_h^2(hv) \tag{4.55b}$$

By cross-differentiation and subtraction, we obtain the vorticity equation

$$\beta\psi_x = A_h\nabla_h^2\psi_{xx} \tag{4.56}$$

where ψ is the streamfunction. The meridional derivatives are much smaller than the zonal derivatives; thus, this equation can be further reduced to

$$\beta \psi_{B,x} = A_h \psi_{B,xxxx} \tag{4.57}$$

where ψ_B is the boundary layer solution. The scale width of the western boundary layer is

$$\delta_M = (A_h/\beta)^{1/3} \tag{4.58}$$

Introducing a stretched coordinate

$$\eta = x/\delta_M \tag{4.59}$$

the vorticity equation is reduced to

$$\psi_{B,\eta\eta\eta\eta} - \psi_{B,\eta} = 0 \tag{4.60}$$

This equation is subject to the following boundary conditions. First, the boundary layer solution should be finite and it should match the interior solution at "infinity" (the outer edge of the boundary layer)

$$\psi_B \to \psi_I, \quad \eta \to \infty \tag{4.61}$$

Second, the western boundary is a streamline,

$$\psi_B = 0, \; at \; \eta = 0 \tag{4.62}$$

In addition, two types of boundary condition may apply:

a) No-slip condition: $\psi_B, \eta = 0$, at $\eta = 0$ (4.63)

b) Slippery condition: $\psi_{B,\eta\eta} = 0$, at $\eta = 0$ (4.64)

The general solution for Eqn. (4.60) is

$$\psi_B = c_1 + c_2 e^{\eta} + c_3 e^{-\frac{\eta}{2}} \cos\left(\frac{\sqrt{3}}{2}\eta\right) + c_4 e^{-\frac{\eta}{2}} \sin\left(\frac{\sqrt{3}}{2}\eta\right)$$

a) Applying the condition at infinity: $c_1 = \psi$, $c_2 = 0$.
b) Applying the boundary condition of $\psi = 0$ at the wall: $c_3 = -\psi_I$.
c) If the no-slip boundary condition applies, $\psi_{B,\eta} = 0$, $\eta = 0$; thus, $c_4 = c_3/\sqrt{3}$.

The final solution is

$$\psi_B = \psi_I \left\{ 1 - e^{-\frac{\eta}{2}} \left[\cos\left(\frac{\sqrt{3}}{2}\eta\right) + \frac{1}{\sqrt{3}} \sin\left(\frac{\sqrt{3}}{2}\eta\right) \right] \right\} \tag{4.65}$$

d) If the slippery boundary condition applies, $\psi_{B,\eta\eta} = 0$, at $\eta = 0$; thus, $c_4 = -c_3/\sqrt{3}$.

The final solution is

$$\psi_B = \psi_I \left\{ 1 - e^{-\frac{\eta}{2}} \left[\cos\left(\frac{\sqrt{3}}{2}\eta\right) - \frac{1}{\sqrt{3}} \sin\left(\frac{\sqrt{3}}{2}\eta\right) \right] \right\} \tag{4.66}$$

e) Structure of the western boundary current: the layer thickness of the boundary current can be obtained from the semi-geostrophic relation

$$h^2 = 2f\psi_I + h_w^2, \quad h_w^2 = h_I^2 - 2f\psi_I/g' \tag{4.67}$$

The corresponding velocity field can be calculated accordingly as $v = \dfrac{\psi_x}{h} = \dfrac{\psi_{B,\eta}}{\delta_M h}$.

Inertial western boundary current

The existence of an inertial western boundary current was postulated first by Stommel (1954). His basic idea is that within the western boundary the inertial term associated with horizontal advection balances the planetary vorticity term. In this way the ambiguity of the friction parameters used in models with interfacial friction or lateral friction can be avoided. The accurate formulation of this problem was first presented by Charney (1955) and Morgan (1956).

General solution

The basic equations for this case include the nonlinear advection terms, but the interfacial and lateral friction terms are omitted

$$h(uu_x + vu_y) - fhv = -g'hh_x + \tau^x/\rho_0 \tag{4.68}$$

$$h(uv_x + vv_y) + fhu = -g'hh_y \tag{4.69}$$

$$(hu)_x + (hv)_y = 0 \tag{4.70}$$

Scaling analysis leads to a simpler set of equations. In particular, the x-momentum equation is reduced to geostrophy in the cross-stream direction

$$fhv = g'hh_x \tag{4.71}$$

Combining Eqns. (4.71), (4.69), and (4.70) leads to

$$B = \frac{1}{2}v^2 + g'h = F(\psi), \quad \textit{energy conservation} \tag{4.72}$$

$$Q = \frac{f + v_x}{h} = G(\psi), \quad \textit{potential vorticity conservation} \tag{4.73}$$

where $F(\psi)$ and $G(\psi)$ are functions completely determined from the interior solution at the outer edge of the western boundary current.

$$F(\psi_I) = g'h_I(Y) \tag{4.74}$$

$$G(\psi_I) = \frac{f(Y)}{h_I(Y)} \tag{4.75}$$

where Y is the meridional coordinate at the outer edge of the western boundary layer. From Eqn. (4.42):

$$\psi_I = \frac{x_e}{\rho_0 \beta} \tau_y^x(Y) \tag{4.76}$$

Assuming this function is invertible, we can write

$$Y = Y(\psi_I) \tag{4.77}$$

Thus, both F and G are completely determined from the interior solution.

The one-to-one inversion of Eqn. (4.76) breaks down at the latitude where $\psi_I(Y)$ reaches its maximum; this is the northern limit of the purely inertial boundary layer. North of this limit, other mechanisms are needed in order to maintain a steady boundary current.

The semi-geostrophic relation leads to the relation between the streamfunction and the layer depth, the same as in the case with interfacial friction, Eqn. (4.40):

$$\psi = \psi_I + \frac{g'}{2f}\left(h^2 - h_I^2\right)$$

or

$$h^2 = h_I^2 + \frac{2f}{g'}(\psi - \psi_I) \tag{4.40$'$}$$

The meridional velocity can be calculated using the Bernoulli law

$$v = \sqrt{2\left[F(\psi) - g'h\right]} \tag{4.78}$$

In this way both the layer thickness and meridional velocity are determined in the streamfunction coordinates. Solution in the physical coordinates can be obtained through coordinate transformation

$$x = \int_0^{\psi} \frac{d\psi}{hv} \tag{4.79}$$

where hv is the function of ψ, defined in Eqns. (4.40′) and (4.78). The transformation from the physical coordinates x to the streamfunction coordinates is the well-known von Mises (1927) transformation used in fluid mechanics. This coordinate transformation was first used by Charney (1955) to solve the inertial western boundary current.

The structure of the solution obtained from a model with the same parameters (for the interior solution) as shown in the previous model with a Stommel boundary layer (Fig. 4.6) is shown in Figures 4.9 and 4.10, including the inertial western boundary current in the southern half of the western boundary.

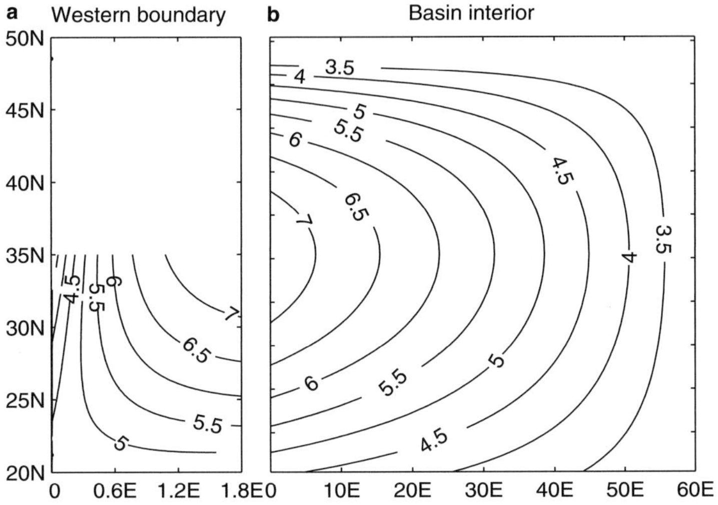

Fig. 4.9 Thermocline depth in a model with an inertial western boundary current (in 100 m) and contour interval 50 m.

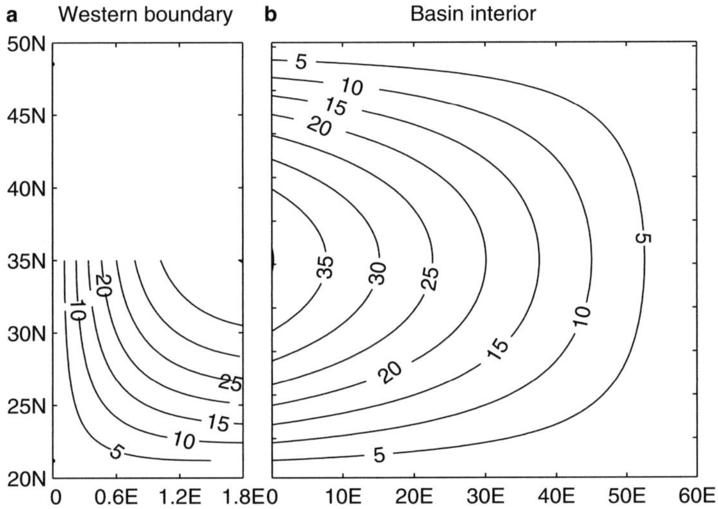

Fig. 4.10 Streamfunction in the model with an inertial western boundary current (Sv).

The inertial western boundary current has features quite different from frictional boundary layers. First, a purely inertial western boundary current is allowed for the southern half of the western boundary only, but in the northern half of the basin, a purely inertial western boundary current is not a valid solution.

Second, the width of the boundary currents may depend on parameters used in the models. In general, for the parameters suitable for simulating the ocean, the inertial western boundary current is wider than the frictional boundary layer. Simulating the western boundary current in terms of a pure inertial boundary current or a purely interfacial frictional boundary layer is idealization only. In reality, the inertial terms, the interfacial frictional terms, and the lateral frictional terms should all contribute to the dynamic balance of the western boundary current. Since the inertial boundary layer is wider than the frictional boundary layer, the western boundary layer can be separated into different regimes dominated by different dynamical processes. The outer part of the boundary layer is dominated by the inertial terms, while the frictional effect is mostly limited to the relatively narrow sub-layer near the wall (Pedlosky, 1987a).

Special case when $G(\psi)$ is constant

The inertial western boundary current has simple analytical form when potential vorticity is a constant, $G(\psi_I) = f/h_I = const$. In such a case the vorticity equation is reduced to

$$v_x + f = \frac{f_\infty h}{h_\infty} \tag{4.80}$$

Using the semi-geostrophic condition, this leads to

$$h_{xx} - \frac{ff_\infty}{g' h_\infty} h = -\frac{f^2}{g'} \tag{4.81}$$

The general solution of this equation is

$$h = a \cdot e^{-x/\lambda} + b \cdot e^{x/\lambda} + h_I$$

where

$$\lambda = \sqrt{g' h_\infty / ff_\infty} \tag{4.82}$$

is the radius of deformation. Assuming $h_\infty \simeq 400$ m and $g' \simeq 0.015$ m/s^2, we have $\delta_I \simeq \lambda \simeq 250$ km; thus, the scale width of the inertial western boundary is wider than the frictional boundary layer discussed in the previous subsection. Under the following conditions

$$h(0) = h_w, \quad h(\infty) = h_I \tag{4.83}$$

the solution is

$$h = (h_w - h_I) e^{-x/\lambda} + h_I \tag{4.84}$$

where h_w is calculated from Eqn. (4.40′).

Limitation and extension of the reduced-gravity models

Limitation of the reduced-gravity models

The major assumption used in the layered reduced-gravity models is the existence of a stagnant lower layer. The major advantages of such an assumption are as follows. First, these models filter out the external gravity modes, with phase speed $\sqrt{gH}$, where H is the depth of the ocean. Instead, these models retain only internal gravity waves, whose wave speed is $\sqrt{g'h}$, where h is the thickness of the upper layer. Since the upper layer is typically a few hundreds of meters, $h \ll H$. In addition, $g'/g \simeq 0.001$, so that large time steps can be used in reduced-gravity models. Second, the reduced-gravity models can avoid the complication due to flow over bottom topography.

The reduced-gravity models are based on the assumption that flow underneath the main thermocline is very slow and negligible. These models can be good tools for the study of circulation in the upper ocean; however, they cannot represent the complicated three-dimensional circulation associated with flows in the abyssal ocean very accurately. For example, in the study of thermohaline circulation or coastal circulation, flows in the bottom layer are an important component of the whole circulation system; thus, models ignoring bottom circulation are not suitable.

For problems associated with time scales shorter than the time scale for the first baroclinic Rossby waves to move across the basin at the latitude of interest, the errors implied by the reduced-gravity formulation may not be totally negligible. Nevertheless, reduced-gravity models have been used to simulate the seasonal cycle, such as the free surface elevation and other properties, for the subtropical basin. Furthermore, reduced-gravity models have been extensively used to study equatorial circulation on seasonal to interannual time scales. For a more accurate description of the time evolution of the sea surface, a two-layer model can be used. In such a model, the lower layer is in motion, so that the contribution due to fast-moving Rossby waves can be included in the calculation as well (Qiu, 2002b).

Layer outcropping and the Parsons model

A major difference between the reduced-gravity models discussed above and the linearized layer model used in many previous studies is that we allow the layer thickness to vary greatly. When the forcing is strong, the interface can outcrop. A layered model with outcropping requires careful treatment both analytically and numerically. Parsons (1969) discussed such a model and made the connection between the Gulf Stream and the outcropping of the main thermocline in a subtropical basin. The application of models with outcropping isopycnals to the world's oceans has been discussed in many papers, e.g., Veronis (1973), Huang and Flierl (1987). We discuss the Parsons model in Section 4.1.4.

4.1.3 The physics of wind-driven circulation

The discussion of wind-driven circulation in the previous section is based on detailed dynamical analysis which can be complicated for some cases. For a better understanding of the circulation, it is more revealing to build up a physical picture of the circulation without involving the mathematical details. Thus, in this section we will try to interpret the basic structure of the circulation in terms of the fundamental physics.

The interior solution

Meridional flow driven by Ekman pumping

The wind-driven circulation in the ocean interior can be illustrated as follows (Fig. 4.11). In order to maintain a relatively steady rotation of the Earth, the globally integrated frictional torque exerted by the atmosphere on the solid Earth and oceans should be zero; thus, both the westerlies and easterlies are necessary components of the atmospheric circulation, i.e., the prevailing westerlies at mid latitudes coexist with easterlies at low latitudes and polar regimes.

This wind stress pattern drives poleward Ekman flows at both low and high latitudes, but drives an equatorward Ekman flow at mid latitudes. As a result, the meridional convergence of Ekman flux in the upper ocean gives rise to the Ekman pumping and upwelling below the base of the Ekman layer. In the subtropical basin, a downward Ekman pumping induces a compression of the water column. In the basin interior, the relative vorticity is negligible, so potential vorticity for a water column is f/h. Ekman pumping at the base of the Ekman layer compresses the water column height h. In order to conserve potential vorticity f/h, the individual water column moves toward the equator, where the Coriolis parameter f is smaller. Thus, Ekman pumping in the subtropical basin drives an equatorward flow in the

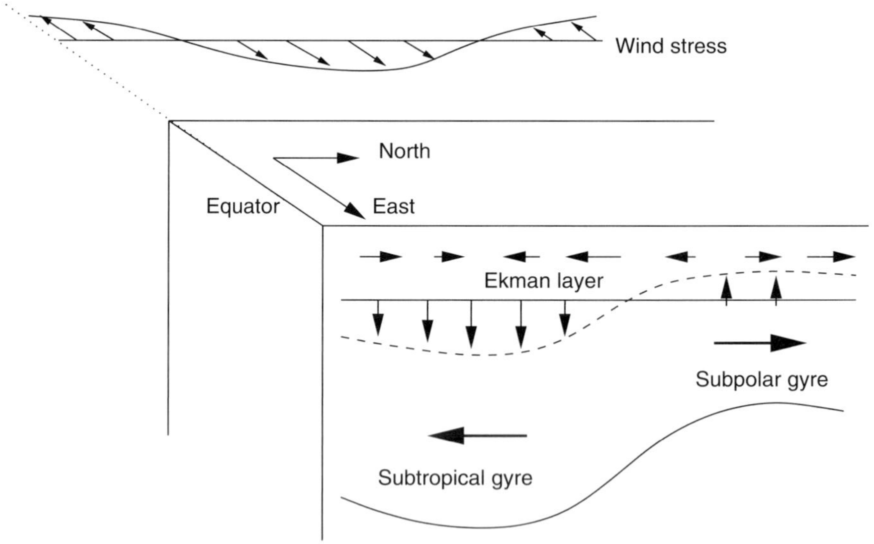

Fig. 4.11 Sketch of the wind-driven circulation. Under the Ekman pumping (upwelling), water moves equatorward (poleward) in order to conserve potential vorticity f/h.

ocean interior. Similarly, the Ekman upwelling in the subpolar basin drives a poleward flow in the ocean interior.

Closing the circulation with the western/eastern boundary layers

The above argument applies to the direction of the meridional flow in the ocean interior only, and we still do not know how to close the circulation. Obviously, in order to have a steady circulation in the subtropical basin, we must find a way to transport the water mass poleward. Since our model is two-dimensional, mass conservation requires a poleward flow in the form of either a western or an eastern boundary current. However, as will be shown, only the western boundary current can play the role of completing the vorticity balance in a closed basin.

An integral potential vorticity constraint for the steady circulation

In the discussion about the interior solution, we do not pay much attention to the possible roles played by friction because frictional force is negligible in the basin interior. However, for the circulation in a closed basin, the frictional force may play a vitally important role; this can be seen clearly through the potential vorticity balance integrated over the whole basin.

We integrate the potential vorticity equation (4.21) over area A_ψ defined by a closed streamline C_ψ (Fig. 4.12). Using the continuity equation $\nabla_h \cdot (h\vec{u}) = 0$, the integration of the left-hand side gives

$$I = \iint_{A_\psi} h\vec{u} \cdot \nabla_h \frac{f + v_x - u_y}{h} dxdy = \iint_{A_\psi} \nabla_h \cdot \left[\vec{u}\,(f + v_x - u_y)\right] dxdy \qquad (4.85)$$

Since the lateral boundary is a streamline, i.e., $\vec{u} \cdot \vec{n} = 0$, the divergence theorem leads to

$$I = \oint_{C_\psi} u_n(f + v_x - u_y)ds = 0 \qquad (4.86)$$

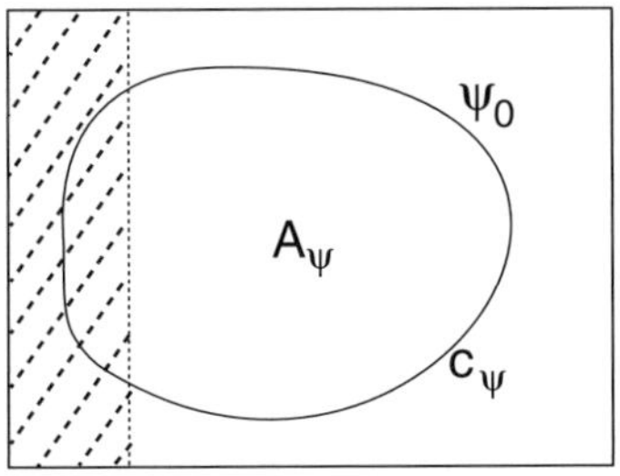

Fig. 4.12 A line integral going through the interior and the western boundary regime of the model basin.

Thus, the final balance is

$$\oint_{c\psi} \frac{\vec{\tau}}{\rho_0 h} \cdot d\vec{s} = R\oint_{c\psi} \frac{\vec{u}}{h} \cdot d\vec{s} - A_h \oint_{c\psi} \frac{\nabla^2(h\vec{u})}{h} \cdot d\vec{s} \tag{4.87}$$

The potential vorticity balance for a closed streamline (or the whole basin) is between the frictional torque (mostly generated along the edge of the basin where the friction is non-negligible) and the wind stress torque imposed over the basin. Thus, no matter how small the frictional force is, it is essential for maintaining the basin-wide vorticity balance between wind stress input and frictional output. Furthermore, a purely inertial model is physically impossible, because vorticity input from wind stress would not be balanced by friction.

Vorticity dynamics of boundary layers

In Section 4.1.2, the interior solution is obtained by starting the integration from the eastern boundary, but why not start from the western boundary? As will be shown shortly, there is no eastern boundary current, so the interior solution is valid all the way up to the eastern wall; thus, the solution in the interior can be obtained by starting the integration from the eastern boundary. On the other hand, there is a western boundary layer, so we would not be able to obtain the interior solution by integration from the western boundary.

Vorticity balance in a closed basin

Let us concentrate on a subtropical basin. Water continuously gains negative vorticity from the upper boundary. The circulation in the ocean interior is very slow, and the relative vorticity is negligible; thus, water parcels move southward to a place where the planetary vorticity is smaller. To close the circulation, water parcels with low vorticity have to move northward along the eastern/western boundary and eventually rejoin the interior flow where the water parcels should have high vorticity. That is to say, water parcels must give up their negative vorticity somewhere in order to maintain a steady circulation.

Take a control volume in the western boundary regime between two zonal sections through the western boundary, one in the south and one in the north. In the southern part of the control volume, there is an influx of low potential vorticity by the incoming flow from the ocean interior; in the northern part of the control volume, there is an outflow of high potential vorticity going to the ocean interior. In order to balance potential vorticity, water must gain positive vorticity either through the lateral boundary or the lower boundary. Thus, no matter what kind of model we use, there should always be a place where positive vorticity is generated through either interfacial or lateral friction to counterbalance the vorticity input from wind-stress curl in the interior.

Vorticity balance in the Stommel boundary layer

The frictional torque needed for balancing the negative wind stress torque imposed on the basin interior is generated along the western boundary where the current is the strongest, as will be shown shortly. The strong boundary current produces the strongest friction near the

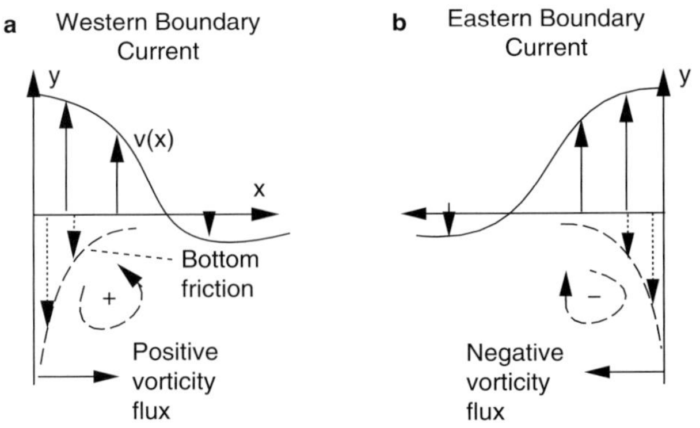

Fig. 4.13 Vorticity flux produced by a meridional boundary current with interfacial friction; **a** model with a western boundary current; **b** model with an eastern boundary current.

wall, and hence the positive source of potential vorticity (Fig. 4.13a). On the other hand, if there were a boundary current near the eastern wall, the corresponding strong current near the wall would generate a negative potential vorticity; therefore, an eastern boundary current cannot play the role of balancing potential vorticity in a closed basin.

Inertial western boundary layer as a means of a partial closing

In a single-moving-layer model, to balance the mass of the interior southward flow a northward boundary current is required. Assuming that, within the boundary current, potential vorticity contributions from wind stress and friction are negligible, potential vorticity must be conserved along streamlines, i.e., $(f + v_x)/h = G(\psi)$. Since f increases northward, the potential vorticity balance within the boundary layer requires an increasingly large negative relative vorticity. This can be achieved by horizontal convergence in the southern part of the western boundary regime.

Similar to the cases of frictional boundary currents, only an inertial western boundary current can produce the necessary negative relative vorticity. Here again, the vorticity balance rules out the possibility of closing the wind-driven gyre circulation with an inertial eastern boundary current.

In the northern part of the western boundary, flow must separate from the boundary because it is divergent. As a result, the relative vorticity declines along streamlines; thus, a purely inertial boundary current in this part of the basin is unable to compensate for the further increase in planetary vorticity f. Consequently, the purely inertial western boundary current does not work for the northern part of the western boundary. Some other mechanisms must be added into the model, such as the time-dependent terms. In fact, in this regime the meso-scale eddies, which are excluded from our simple model discussed above, are vitally important. The structure of the western boundary layer is shown schematically in Figure 4.14.

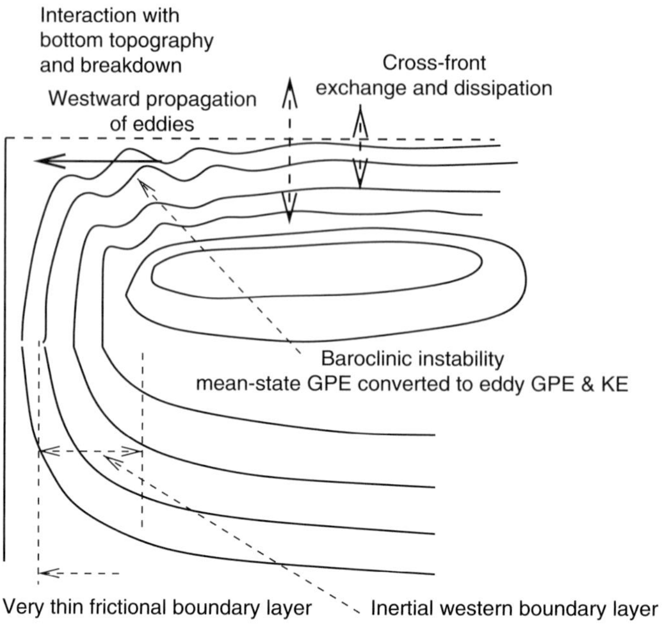

Fig. 4.14 Different dynamical zones within the western boundary regime.

In addition, a model based on a purely inertial boundary layer cannot balance the potential vorticity in the whole basin. This is because within the framework of a purely inertial boundary current there is no communication of vorticity between the current and the solid boundary, such as the bottom and the lateral walls, so the model cannot get rid of the negative vorticity imposed in the basin interior. Thus, there must be places where other mechanisms of exporting negative vorticity exist and can establish the vorticity balance in the whole basin.

Energetics of the wind-driven circulation

Gravitational potential energy in the reduced-gravity model

For a simple reduced-gravity model, the free surface elevation ζ is linked with the layer thickness $h : \zeta = h\Delta\rho/\rho$ (Fig. 4.15). Therefore, the center of mass is at a depth $(h/2 - \delta h)$ where $\delta h = \zeta/2 = h\Delta\rho/2\rho$. The value of GPE depends on the choice of a reference level. For the present case we choose $z = -h/2$ as the reference level, and the GPE for the warm water above the thermocline is

$$E_p = \delta h \cdot \rho g h = \rho g' h^2/2 \tag{4.88}$$

where we have used the relation $\rho h = (\rho - \Delta\rho)(h + \zeta)$, which is a statement of zero pressure gradient below the thermocline. In comparison, if density is equal to ρ uniformly everywhere, the corresponding free surface elevation is zero and the GPE is zero.

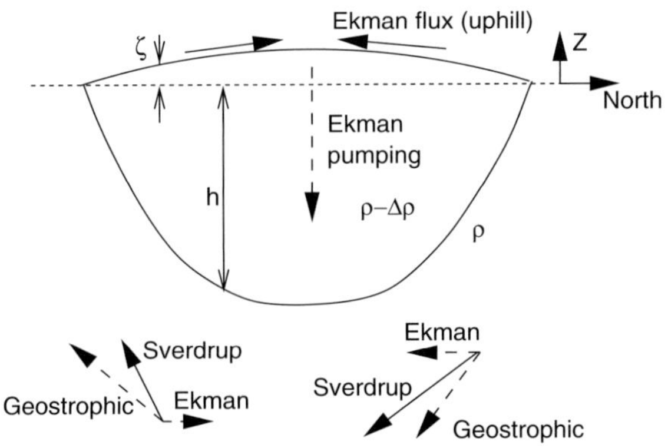

Fig. 4.15 Sketch of the velocity field and the movement of water parcels in a subtropical gyre interior for a reduced-gravity model, including (*top panel*) a meridional view of the subtropical gyre and (*lower panel*) a horizontal view of the velocity vectors. Note that the geostrophic current should follow the h-contours.

Therefore, in a reduced-gravity model, the increase of h indicates the increase of the free surface elevation ζ, and this means an increase of GPE. As discussed in Section 3.6, the amount of mechanical energy received by surface geostrophic currents from the wind stress is equal to the amount of energy required to push the Ekman flux uphill (Fig. 4.15). This uphill volume flux from both the southern and northern boundaries of the wind-driven gyre pushes water toward the gyre center with higher free surface elevation; the convergence of this volume flux in the middle of the gyre pushes warm water downward into the pool of warm thermocline water in the subtropical basin. The increase in ζ and h leads to the increase of GPE stored in the warm water pool.

Balance of GPE in a wind-driven gyre

Based on the Stommel model discussed above, the equation for the balance of mechanical energy, Eqn. (4.16), is reduced to

$$\vec{u} \cdot \nabla_h \left(g'h\right) = \vec{u} \cdot \vec{\tau}/\rho - R(u^2 + v^2) \tag{4.89}$$

Interior solution

In the basin interior, wind stress does work on the geostrophic velocity, $\vec{u} \cdot \vec{\tau} > 0$; thus, $\vec{u} \cdot \nabla_h h > 0$, h and GPE increase downstream, following the anticyclonic direction. In other words, wind stress inputs energy and pushes water toward the regime of high GPE, depicted by the arrows on the top of the surface in Figure 4.15.

This figure also can help us understand why a cyclonic gyre in the subtropical basin is impossible. For this basin-scale wind stress pattern with westerlies at mid latitudes and easterlies at lower latitudes, wind stress work in the interior of a cyclonic gyre is negative; thus, wind stress cannot drive a cyclonic circulation.

Note the minor difference in the direction of the flow for the southern half of the subtropical gyre illustrated in Figure 4.15. The streamlines shown in Figure 4.4 are the Sverdrup streamfunction, so they include both the Ekman flux and the geostrophic flow below the Ekman layer. Although these two components have similar flow direction for the northern half of the subtropical gyre, their directions are opposite in the equatorward half of the subtropical gyre. As indicated in the lower part of Figure 4.15, the Ekman flux flows toward the pole and the subsurface geostrophic current moves toward the equator. The surface velocity is the vector sum of these two components, and it receives energy from wind stress. Energy input into the Ekman layer is stored in the subtropical basin through the Ekman pumping as warm and light water is pushed down into the bow-shaped main thermocline, as indicated by the downward arrow in Figure 4.15.

In the western boundary

The wind stress work term is negligible, so the balance here is between the planetary vorticity advection and the friction term that is negative. Thus, during their northward movement, water parcels must lose their GPE, i.e., move toward a regime of lower surface elevation and shallow thermocline depth, as shown by the heavy dashed line in Figure 4.4a.

Partition of energy associated with the wind-driven gyre

The wind-driven gyres in the upper ocean are the most energetic components of the world's ocean circulation. Although the basin-scale structure of these gyres can be fairly well described using the theories discussed above, the contribution due to eddies is excluded from our discussion. In order to understand the importance of eddies in the overall dynamical framework, we need to have at least rough estimates for the partition of all the relevant forms of energy, including the potential and kinetic energy of the mean flow and the meso-scale eddies. Scaling analysis can provide us with the following rough estimate (see Gill *et al.*, 1974).

First, the density of total kinetic energy of the mean flow is

$$e_{k,mean} = \frac{1}{2}\overline{\rho h(u^2 + v^2)} = \frac{1}{2}\rho g'^2 \overline{h(h_x^2 + h_y^2)/f^2} \approx \frac{1}{2}\rho g'^2 \bar{h}\overline{h'^2}/L_y^2 f^2 \tag{4.90}$$

where the overbar indicates the horizontal average over the whole gyre, h' is the thickness perturbation, and $L_y \sim 1000\ km$ is the north–south length scale of the wind-driven gyre. In this estimate the contribution due to the east–west gradient is small, so it is ignored, i.e., the kinetic energy associated with the meridional velocity is omitted in this estimate.

The density of available gravitational potential energy of the mean flow is defined as

$$e_{p,mean} = \frac{1}{2}\rho g' \left(\overline{h^2} - \bar{h}^2\right) \approx \frac{1}{2}\rho g' \overline{h'^2} \tag{4.91}$$

Therefore, the ratio of these two types of energy is

$$\frac{E_{p,mean}}{E_{k,mean}} = \frac{e_{p,mean}}{e_{k,mean}} = \frac{L_y^2}{\lambda^2} \approx 1000 \tag{4.92}$$

where $\lambda = \sqrt{g'\bar{h}}/f \approx 30$ km is the radius of deformation. Thus, the kinetic energy of the mean flow in a wind-driven gyre is associated with a huge amount of available potential energy.

On the other hand, the potential and kinetic energy for the meso-scale eddies can be estimated as follows. If all the available potential energy can be converted into the energy of eddies, the total energy of meso-scale energy should be the same as the available potential energy, i.e.,

$$E_{eddy} \simeq E_{p,mean} \tag{4.93}$$

Assume that the typical scale of the meso-scale eddies is k^{-1}. In general, the eddy scale is greater than the deformation ratio $k^{-1} > \lambda$, and eddy energy is mostly in the form of potential energy. Similarly to the estimate of kinetic energy of the mean flow, eddy kinetic energy density can be estimated as

$$e_{k,eddy} \approx \frac{1}{2}\rho g'^2 \overline{h}\overline{h'^2} k^{-2}/f^2 \tag{4.94}$$

Thus, the ratio of eddy potential energy to eddy kinetic energy is

$$\frac{E_{p,eddy}}{E_{k,eddy}} = \frac{e_{p,eddy}}{e_{k,eddy}} = \frac{k^{-2}}{\lambda^2} \approx 10 \tag{4.95}$$

And the ratio of the eddy kinetic energy to the mean flow kinetic energy is

$$E_{k,eddy}/E_{k,mean} \simeq (k\lambda)^2 \approx 100 \tag{4.96}$$

Therefore, the eddy kinetic energy is 100 times the mean flow kinetic energy.

Physical interpretation of the interfacial friction

Although the Stommel model has sometimes been referred to as a model with "bottom friction," in reality the friction used in a reduced-gravity model is actually an interfacial friction. This interfacial friction is assumed to be linearly proportional to the velocity in the upper layer. Since the lower layer in a reduced-gravity model is assumed to be stagnant, this is in fact a crude parameterization that interfacial friction is proportional to the velocity shear across the interface. Such a parameterization in the Stommel model is therefore a crude parameterization of baroclinic instability.

Therefore, GPE loss to the so-called interfacial friction in the Stommel model can be interpreted as a loss to baroclinic instability within the western boundary. Most importantly, the mechanical energy loss through the so-called interfacial friction is not converted to internal energy; instead, this energy is converted to the GPE and kinetic energy sustaining meso-scale eddies. However, baroclinic instability in the oceans seems not to play a vital role along the western boundary of the basin. In fact, it takes place primarily within the outflow regime, such as the Gulf Stream Extension or the Recirculation; thus, the physical meaning of such a parameterization remains the subject of debate.

4.1.4 The Parsons model

Introduction

In most wind-driven circulation models, the stratification along the eastern boundary is specified *a priori*, and the solution in the ocean interior is sought under such an assumption. There is another way of looking at the same problem – instead of specifying the layer thickness along the eastern boundary, one can specify the total amount of warm water in the ocean, and find out the structure of the thermocline subject to the Sverdrup relation. In fact, that is exactly the same approach as is used in most numerical models for the wind-driven circulation – we specify the amount of water in each density category and "spin up" the model to find the solution consistent with other dynamical constraints.

Parsons model

Parsons (1969) formulated a very elegant model in an attempt to explain the Gulf Stream separation. By including friction and ageostrophic terms, his model predicts a very clear picture of the circulation in a subtropical basin. Thus, like the Stommel and Munk models, this is an analytical model that works within a closed basin.

Although the Parsons model has been extended in many studies, the same basic assumptions apply:

- Steady circulation
- Two immiscible layers, with a fixed amount of water in the upper layer
- Lower layer is infinitely deep and motionless
- Interfacial friction is proportional to the velocity in the upper layer.

The model is formulated on a β-plane, and the corresponding basic momentum and continuity equations are

$$h(\vec{u} \cdot \nabla_h)\vec{u} + h\vec{f} \times \vec{u} = -g'h\nabla h + \frac{\vec{\tau}}{\rho_0} - R\vec{u} \tag{4.97}$$

$$\nabla_h \cdot (h\vec{u}) = 0 \tag{4.98}$$

where $\vec{u} = (u, v)$ is the horizontal velocity, ∇_h is a two-dimensional operator, and

$$f = f_0 + \beta y, \quad f_0 = 2\omega \sin\theta \tag{4.99}$$

is the Coriolis parameter. Introduce the nondimensional variables

$$(x, y) = L(x', y'), \quad h = Hh', \quad \vec{\tau} = W\vec{\tau}', \quad \vec{u} = \frac{g'H}{L^2\beta}\vec{u}',$$
$$f = L\beta f', \quad f' = f_0 + y', \quad f_0 = \frac{a}{L}\tan\theta_0 - 0.5 \tag{4.100}$$

where L and H are the horizontal and vertical scales of the model ocean, W is the scale of wind stress, a is the radius of Earth, and θ_0 is the central latitude of the model ocean. After dropping off the primes, the nondimensional equations are

$$R_0 h(\vec{u} \cdot \nabla_h)\vec{u} + f\vec{k} \times h\vec{u} = -h\nabla_h h + \mu\vec{\tau} - \varepsilon\vec{u} \tag{4.101}$$

$$\nabla_h \cdot (h\vec{u}) = 0 \tag{4.102}$$

where

$$Ro = \frac{g'H}{L^4\beta^2} \ll 1, \quad \varepsilon = \frac{R}{\beta LH} \ll 1, \quad \mu = \frac{LW}{g'\rho_0 H^2} \tag{4.103}$$

are the nondimensional parameters of the model. In this model μ is a key nondimensional forcing parameter. Because there is no source or sink, a streamfunction can be introduced,

$$hu = -\psi_y, \quad hv = \psi_x \tag{4.104}$$

The boundary conditions for ψ are

$$\psi = 0, \text{ at rigid boundaries} \tag{4.105a}$$

$$\psi = 0, \text{ at } h(x, y) = 0 \text{ (the outcrop line)} \tag{4.105b}$$

The special feature of this model is the existence of an outcropping line which is both a streamline and a zero-depth line. A careful handling of these boundary conditions along the outcrop line is essential for describing the outcropping phenomenon accurately.

The continuity equation can be interpreted as an integral constraint $\frac{\partial}{\partial t}\iint hdxdy = 0$, i.e., the total amount of water in the upper layer is fixed. In the nondimensional form, this constraint is

$$\iint h(x, y)dxdy = 1 \tag{4.106}$$

For simplicity, we make use of the following additional assumptions:

- Wind stress has only the zonal component $\vec{\tau} = (\tau^x(y), 0)$.
- Inertial terms are negligible for the interior flow, i.e., $Ro \to 0$.

However, we must keep the ε term, because the circulation in a basin cannot be closed without friction.

Solutions without outcropping

Under these assumptions, the horizontal momentum equations are

$$-f\psi_x = -hh_x + \frac{\varepsilon}{h}\psi_y + \mu\tau^x \tag{4.107}$$

$$-f\psi_y = -hh_y - \frac{\varepsilon}{h}\psi_x \tag{4.108}$$

Interior solution

Neglecting the friction terms, the momentum equations are reduced to

$$-f\psi_x = -hh_x + \mu\tau^x \tag{4.109}$$

$$-f\psi_y = -hh_y \tag{4.110}$$

Cross-differentiating these two equations and subtracting leads to the vorticity equation

$$-\psi_x = \mu\tau_y^x \tag{4.111}$$

Using the vorticity equation and integrating from the eastern boundary leads to the interior solution

$$\psi_I = \mu(1-x)\tau_y^x \tag{4.112}$$

$$h_I^2 = h_e^2 + 2\lambda(1-x)f^2(\tau^x/f)_y \tag{4.113}$$

where $h_e = const$ is the nondimensional layer depth at the eastern boundary.

Western boundary layer

We introduce a stretched boundary-layer coordinate $\eta = x/\varepsilon$. The momentum equations become

$$-f\psi_\eta = -hh_\eta + \frac{\varepsilon^2}{h}\psi_y + \varepsilon\mu\tau^x \tag{4.114}$$

$$-f\psi_y = -hh_y - \frac{1}{h}\psi_\eta \tag{4.115}$$

The last two terms in Eqn. (4.114) can be neglected compared with other terms; thus, we have a geostrophic balance in the cross-stream direction; while Eqn. (4.115) indicates that momentum balance in the downstream direction must be ageostrophic. In this case the system is in a semi-geostrophic balance. Semi-geostrophy allows us to obtain useful analytical solution within strong boundary currents. The corresponding boundary conditions are

$$\psi = 0 \; at \; \eta = 0, \quad \psi \to \psi_I(0, y) \; at \; \eta \to \infty \tag{4.116}$$

where the subscript I denotes the interior solution discussed above. Dropping the small terms in Eqn. (4.114) and integrating from the western boundary leads to

$$\psi = \frac{h^2 - h_w^2}{2f} \tag{4.117}$$

where h_w is the layer thickness at the western wall. Letting $\eta \to \infty$ and using boundary conditions from Eqn. (4.116), we obtain

$$h_w^2 = h_I^2 - 2f\psi_I(0, y) \Rightarrow h_w^2 = h_e^2 - 2\mu\tau^x(0) \tag{4.118}$$

Cross-differentiating Eqns. (4.114, 4.115) leads to

$$\psi_\eta + \left(\frac{1}{h}\psi_\eta\right)_\eta = 0$$

Integrating it once gives

$$\psi + \frac{1}{h}\psi_\eta = const$$

Using boundary conditions that $\eta \to \infty$, $\psi_\eta = 0$, we obtain $h^2 + 2h_\eta = h_I^2$. Thus, the final solution is

$$h = h_I \frac{1 - Be^{-h_I\eta}}{1 + Be^{-h_I\eta}}, \quad B = \frac{h_I - h_w}{h_I + h_w} \tag{4.119}$$

Boundary layer separation

The outcropping line

When the wind forcing is strong enough, the upper layer separates from the western boundary. For example, assuming a simple sinusoidal wind stress, $\tau^x = -\tau_0 \cos(\pi y)$, the minimal layer thickness is at the northwest corner of the basin, and this minimal thickness calculated from Eqn. (4.118) may be negative

$$\min(h_w^2) = h_e^2 - 2\mu\pi\tau_0 < 0 \tag{4.120}$$

One may identify the $h = 0$ line obtained from the interior solution as the outcrop line; however, a close examination reveals that this line is not a streamline. In order to satisfy constraint (4.105), a careful handling of the outcrop line is as follows.

Since an outcropping line is not a straight line, we introduce local coordinates (r, s), where $\vec{s}$ is along the outcropping line, and $\vec{r}$ is pointed to the right-hand side of $\vec{s}$. We also assume that the radius of curvature is much larger than the boundary layer width, so that the curvature terms can be neglected. Introducing a stretched boundary layer coordinate $\eta = r\varepsilon$, the momentum equations in the new coordinates are

$$-f\psi_\eta = -hh_\eta + \frac{\varepsilon^2\psi_s}{h} + \varepsilon\mu\tau^r \tag{4.121}$$

$$-f\psi_s = -hh_s - \frac{\psi_\eta}{h} + \mu\tau^s \tag{4.122}$$

subject to the corresponding boundary conditions

$$\text{a)} \quad h = 0, \psi = 0, \ \text{at } \eta = 0 \tag{4.123}$$

$$\text{b)} \quad h \to h_I(s), \ \psi \to \psi_I(s), \ \ \text{as } \eta \to \infty \tag{4.124}$$

From Eqn. (4.121), by using boundary conditions and neglecting the small terms $O(\varepsilon)$ and higher-order $o(\varepsilon)$, we obtain

$$\psi = \frac{h^2}{2f} \tag{4.125}$$

Using the interior solution found above which satisfies

$$\psi_I = \frac{h_I^2}{2f}, \quad at\ x = X(y) \tag{4.126}$$

we have

$$\lambda(1-x)\tau_y^x = \frac{h_e^2}{2f} + \mu(1-x)f\left(\frac{\tau^x}{f}\right)_y \tag{4.127}$$

Therefore

$$X(y) = 1 - \frac{h_e^2(\mu)}{2\mu\tau^x} \tag{4.128}$$

is the equation satisfied by the outcropping line.

In order to calculate the outcropping line we make use of the integral constraint

$$\iint_\Omega h_I(x, y, \mu)dxdy = 1 \tag{4.129}$$

Substituting the solution found above,

$$1 = \iint_\Omega \sqrt{h_e^2 + 2\mu(1-x)f^2\left(\frac{\tau^x}{f}\right)_y}\, dxdy \tag{4.130}$$

where the integral domain Ω is defined by the boundaries of the basin and the outcropping line $X(y)$ defined in Eqn. (4.128) (Fig. 4.16). In this calculation, the interior solution is used up to the outcrop line; thus, the solution involves a small error on the order of $O(\varepsilon)$.

Interior boundary layer structure

Equations (4.121) and (4.122) are very similar to Eqns. (4.114 and 4.115), so the structure of the interior boundary current can be found by following a similar approach, i.e., deriving the vorticity equation, integrating the equation with respect to η once, and applying the matching boundary condition. The solution is

$$h = h_I \frac{1 - e^{-f' h_I \eta}}{1 + e^{-f' h_I \eta}}, \quad \text{where } f' = \frac{\partial f}{\partial s} \tag{4.131}$$

The structure of the solution is shown in Figure 4.17, including the interior solution, the interior boundary layer, and the western boundary layer. The interior boundary layer

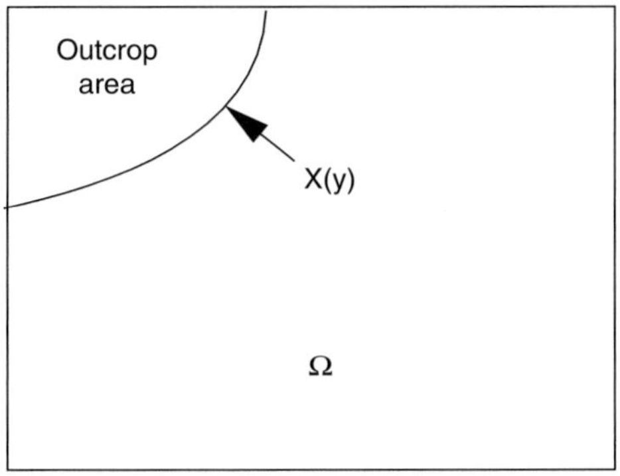

Fig. 4.16 A subtropical basin with the outcropping area.

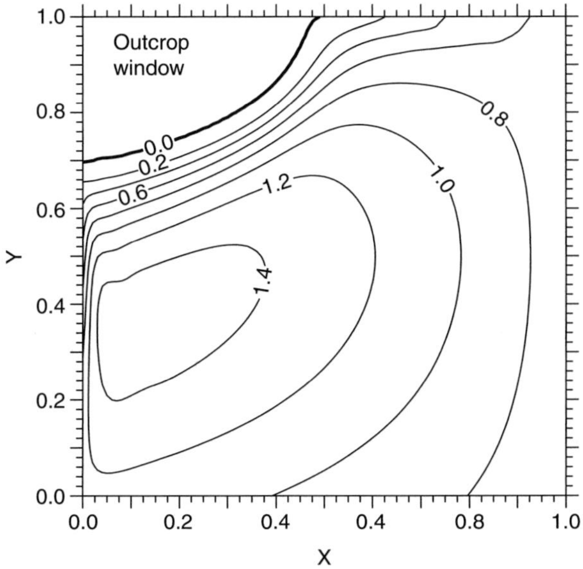

Fig. 4.17 The nondimensional upper layer thickness obtained for parameters: $\varepsilon = 0.03$, $\mu = 0.1$.

solution obtained from this model resembles the Gulf Stream after separation from the coast. In fact, the Gulf Stream is associated with a very strong density front; thus a simple reduced-gravity model, as discussed above, can provide a clear physical insight for the Gulf Stream separation and other related phenomena.

Outcropping and boundary layer separation in a two-gyre basin

The outcropping phenomenon discussed above is more pronounced for a two-gyre basin. The dynamical details have been discussed by Veronis (1973) and Huang and Flierl (1987).

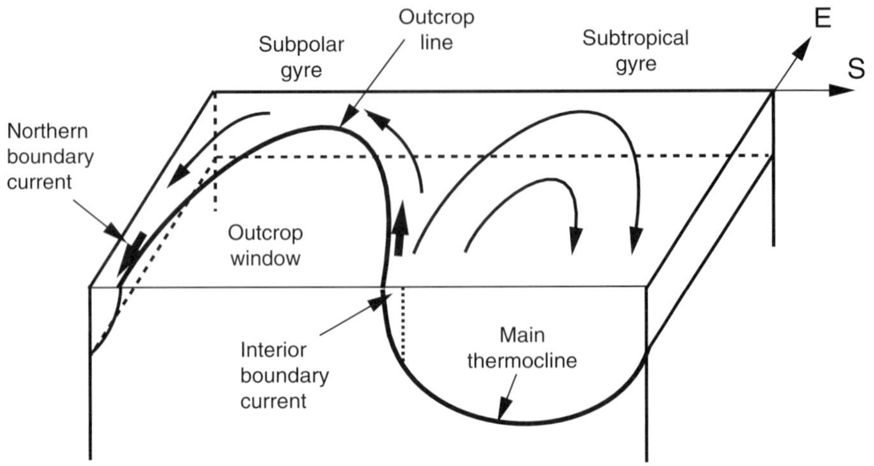

Fig. 4.18 Side view of a two-gyre basin circulation, including the strong interior boundary current and the northern boundary current.

Strong Ekman upwelling in the subpolar basin drives more prominent isopycnal outcropping. The global structure of the solution is shown in Figure 4.18, including the structure of different boundary currents for a subtropical–subpolar basin, such as the isolated northern boundary current and the isolated western boundary current. The structure of these boundary currents can be found in Huang and Flierl (1987).

4.1.5 The puzzles about motions in the subsurface layers

In the early studies, the theory of wind-driven circulation was mostly confined to modeling the homogeneous ocean or the reduced-gravity model. Although a solution from the reduced-gravity model can be interpreted as the first baroclinic mode, these models cannot provide the vertical structure of the circulation; thus, much effort has been devoted to the development of the theory of baroclinic circulation, but progress was very slow for decades. In order to really understand the new theories developed in the 1980s, we will review the knowledge of the oceanic circulation at the end of the 1970s. By learning about the history of the science, we hope to become smarter next time we face a new challenge. Sometimes it seems easy to understand a theory that has already been proved, but it is always much harder to find and prove a new theory.

Wind-driven circulation in layered models

Homogeneous model and reduced-gravity model

The interior flow of the wind-driven circulation in a basin can be described by Sverdrup (1947) dynamics. This is the famous Sverdrup relation, and it is a vorticity balance. Note that reduced gravity does not appear in the Sverdrup relation, so it does not really matter whether the model is a homogeneous model, as used by Sverdrup (1947), or a reduced-gravity model as used in Section 4.1.2. Within a subtropical gyre, wind-stress curl is negative, so the interior flow must move southward where planetary vorticity is lower.

The interior flow must be closed by some kind of boundary layer. We have the famous theories of the western boundary current, including the Stommel (1948) boundary layer and the Munk (1950) boundary layer.

Can the eastern boundary block geostrophic flow in the subsurface layers?

The no-throughflow boundary condition at the eastern boundary is a formidable obstacle to getting the subsurface layers in motion. Rooth *et al.* (1978) postulated a seemingly convincing argument based on a potential vorticity constraint applied to the subsurface layers. In fact, their argument was inspired by the intriguing results obtained from the following numerical simulation based on a three-layer model by Suginohara (1973).

The essence of the numerical model can be interpreted in terms of a $2\frac{1}{2}$-layer model, for which the basic equations are:

$$-fh_1v_1 = -g'h_1(\gamma h_{1x} + h_{2x}) + \tau^x \tag{4.132a}$$

$$fh_1u_1 = -g'h_1(\gamma h_{1y} + h_{2y}) \tag{4.132b}$$

$$(h_1u_1)_x + (h_1v_1)_y = 0 \tag{4.133}$$

$$-fh_2v_2 = -g'h_2(h_{1x} + h_{2x}) \tag{4.134a}$$

$$fh_2u_2 = -g'h_2(h_{1y} + h_{2y}) \tag{4.134b}$$

$$(h_2u_2)_x + (h_2v_2)_y = 0 \tag{4.135}$$

where h_1 and h_2 are the layer thicknesses, and $g' = g\,(\rho_3 - \rho_2)\,/\bar{\rho}$ is the reduced gravity, $\gamma = (\rho_3 - \rho_1)\,/\,(\rho_2 - \rho_1)$. As explained later, this system has a first integral, i.e., the barotropic solution, $h_1\vec{u}_1 + h_2\vec{u}_2$, can be found out from the vorticity balance. However, we need one more constraint to determine the baroclinic structure of the solution.

Suginohara made an attempt to solve this problem numerically by reducing it to a set of two ordinary differential equations, using x as the independent variable. Cross-differentiating the momentum equations (4.132a, 4.132b) and (4.134a, 4.134b), we obtain the vorticity equations, and further manipulation leads to two ordinary differential equations

$$g'\left(\gamma h_1 + \frac{f}{\beta}h_{2y}\right)h_{1x} + g'\left(h_1 - \frac{f}{\beta}h_{1y}\right)h_{2x} = \tau^x - \frac{f}{\beta}\tau^x_y \tag{4.136}$$

$$g'\left(h_2 - \frac{f}{\beta}h_{2y}\right)h_{1x} + g'\left(h_2 + \frac{f}{\beta}h_{1y}\right)h_{2x} = 0 \tag{4.137}$$

Assuming that boundary conditions are specified along the eastern boundary – say h_1 and h_2 are constant along the eastern boundary – these two equations can be rewritten in finite difference form. By marching from the eastern boundary westward, one may find solutions with the lower layer in motion and with wind-stress imposed on the upper surface only. In this way, the subsurface layers can be in motion, as suggested by some numerical experiments.

A model ocean with subsurface layers in motion was a very intriguing result, and this motivated Rooth and his colleagues to explore further the question of whether the subsurface layers can be set in motion in an ideal-fluid model. Rooth *et al.* (1978) started from a theoretical argument based on a potential vorticity equation. It is readily shown that the potential vorticity equations for these two layers are

$$\vec{u}_1 \cdot \nabla_h \left(\frac{f}{h_1} \right) = -\frac{1}{h_1} \left(\frac{\tau^x}{h_1} \right)_y \tag{4.138}$$

$$\vec{u}_2 \cdot \nabla_h \left(\frac{f}{h_2} \right) = 0 \tag{4.139}$$

Thus, potential vorticity in the upper layer changes along the streamlines owing to the direct force of Ekman pumping; however, the second layer is shielded from the direct force of Ekman pumping. As a result, potential vorticity in the second layer is conserved along geostrophic streamlines, i.e., streamlines must follow the geostrophic contours.

However, the existence of the eastern boundary blocks all geostrophic contours in the subsurface layers. Since all possible flows in the second layer must follow potential vorticity isolines, the blockage of geostrophic contours by the eastern boundary implies that all the possible flows in the subsurface layers are blocked as well.

The two approaches discussed above are completely opposite to each other, regarding the issue of whether there are motions in the subsurface layers. Now we can see some flaws in the first approach. Near the eastern boundary, the second layer is stagnant. If the second layer is in motion somewhere away from the eastern boundary, then there will be a boundary separating these two regions. Thus, simple westward marching may violate the physical law by crossing the boundary separating the geostrophic contours originating from the eastern boundary and those originating elsewhere. In mathematical terms, the problem is of the hyperbolic type, so it is better to use finite differencing in characteristic coordinates; otherwise, the numerical scheme may give rise to faulty solutions because the scheme violates the signal law governing the hyperbolic equations.

The eastern boundary blocking of the geostrophic contours seems to be a simple and very appealing argument. From such a powerful argument, one can jump to the conclusion that there would be no motion in the second layer because all these geostrophic contours are blocked by the eastern boundary, so there would be no motion in all these subsurface layers.

Inertial run-away solutions of the homogeneous ocean model

There is another puzzle related to the homogeneous ocean model. For the past several decades, people have studied numerical solutions of homogeneous ocean models, and found that when the solutions enter the highly nonlinear regime, the solutions look very unrealistic. For example, Ierley and Sheremet (1995) studied the following cases. The barotropic

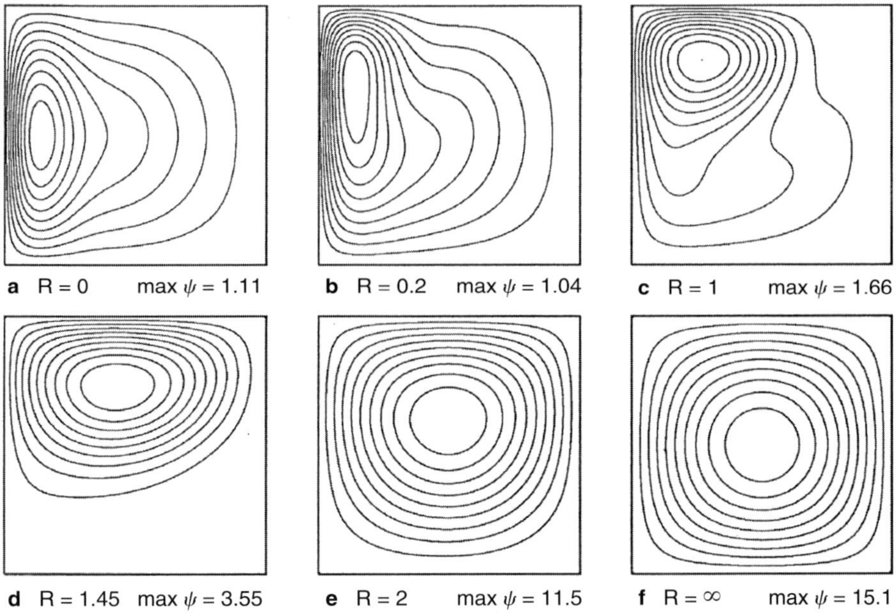

Fig. 4.19 A sequence of the streamfunction illustrating the monotonic transition from the linear Munk solution ($R = 0$) to the highly nonlinear gyre solution ($R = \infty$) for fixed viscosity, $\delta_M = 0.06$, and increasing Reynolds number R (Ierley and Sheremet, 1995).

vorticity equation for a homogeneous ocean model is

$$\nabla_h^2 \psi_t + \delta_I^2 J(\psi, \nabla_h^2 \psi) + \psi_x = \delta_M^3 \nabla_h^4 \psi + Curl\, \tau \tag{4.140}$$

where $J(\psi, \zeta) = \psi_x \zeta_y - \psi_y \zeta_x$, $\zeta = \nabla_h^2 \psi$, $u = -\psi_y$, $v = \psi_x$, $\delta_I = \sqrt{U/\beta}/L$ is a parameter based on the width of the inertial boundary layer, and $\delta_M = (A_L/\beta)^{1/3}/L$ a parameter based on the width of the viscous boundary layer. The Reynolds number is defined as $R = \delta_I^3/\delta_M^3$.

As R increases, the solution goes through a gradual transition from a solution that looks like what happens in the real oceans (for $R = 0$ to 1) to solutions that do not look like anything we observe in the oceans (for the case of $R = 3.55$ to ∞) (Fig. 4.19). This is called the run-away problem in homogeneous ocean models. It indicates that the model formulation does not represent the real oceans properly; but what is wrong with the model?

Spin-up of a linear quasi-geostrophic model ocean

Spin-up of the wind-driven circulation in a stratified ocean is closely associated with Rossby wave propagation in a basin. This problem was first discussed in terms of the quasi-geostrophic framework by Anderson and Gill (1975). Using the same formulation, Young (1981) discussed a seemingly crazy puzzle connected with the spin-up of a continuous stratified ocean without the western boundary. Within the quasi-geostrophic theory, a problem related to the general circulation involves solving a linear spin-up problem:

$$[\nabla_h^2 \psi + \beta y + (F\psi_z)_z]_t + \beta\psi_x = 0 \tag{4.141}$$

with boundary conditions

$$\psi|_{t=0} = 0; \quad \psi|_{x=a} = 0; \quad w|_{z=0} = w_e(y)\theta(t); \quad w|_{z=-H} = 0$$

where $F = F(z)$ is the nondimensional stratification, $w = -F\psi_{zt}/f_0$, $w_e\,(y)$ is the Ekman pumping imposed on the upper surface of the model ocean, $\theta(t) = 0$ for $t < 0$ and $\theta(t) = 1$ for $t \geq 0$ is a step function in time.

This problem can be solved by expanding in eigenfunctions defined by

$$(FC_z)_z = -\lambda^2 C, \quad C_z = 0, \ at\ z = 0, -H \tag{4.142}$$

There are an infinite number of eigenvalues and eigenfunctions ($\lambda_n > 0$, $C_n(z)$, $n = 0, 1, 2, \ldots$). Thus, the streamfunction ψ can be formally expanded as $\psi = \sum_0^\infty \phi_n(x, y, t)C_n(z)$, and the problem can be solved by the Galerkin method. The modal equations are

$$\nabla_h^2 \varphi_{nt} - \lambda_n^2 \varphi_{nt} + \beta\varphi_{nx} = \frac{f_0}{H} C_n(0) w_e(y)\theta(t), \quad n = 0, 1, 2, \ldots \tag{4.143}$$

Using the Laplace transform, we obtain the solution

$$\phi_n = \begin{cases} \dfrac{f_0}{H} C_n(0) w_e \theta(t) \dfrac{x-a}{\beta}, & \text{if } t > \dfrac{\lambda_n^2}{\beta}(a-x) \\ \dfrac{f_0}{H} C_n(0) w_e \theta(t) \left(-\dfrac{t}{\lambda_n^2}\right), & \text{if } t < \dfrac{\lambda_n^2}{\beta}(a-x) \end{cases} \tag{4.144}$$

The solution to Eqn. (4.144) is a Rossby wave that starts from the eastern boundary, and propagates westward with a speed of λ_n^{-2}, where the critical time $\lambda_n^2\,(a-x)\,/\beta$ is the time for the Rossby wave signals to reach a point x. Since $H\delta(z) = \sum_0^\infty C_n(0)C_n(z)$, the sum of these waves is $\lim_{t\to\infty} \psi = f_0 w_e \frac{x-a}{\beta}\delta(z)$. Therefore, the wind-driven flow is a δ function at $z = 0$, i.e., it is concentrated near the upper surface in the form of a surface-trapped jet, and there would be no flow below the surface (Fig. 4.20). Therefore, a multi-layer quasi-geostrophic model subjected to switched-on Ekman pumping does not give rise to flow in the subsurface layer.

Remarks

As we look back today, the progressive steps made in the 1980s are very closely related to attacking these seemingly difficult problems. We emphasize that, in common with other

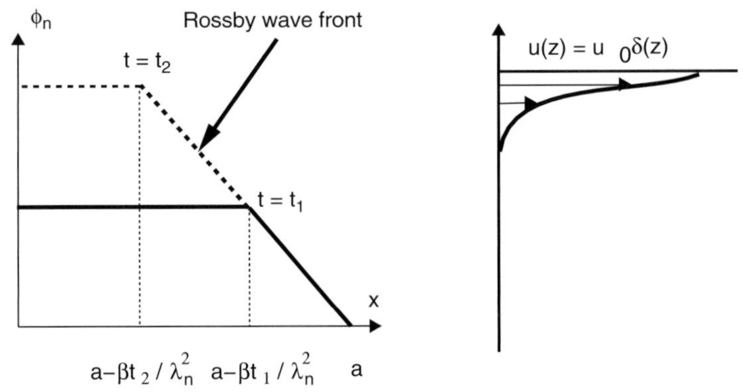

a Wave solution for a single mode

b Surface-trapped jet for a linear stratified Q–G model

Fig. 4.20 Sketch of the spin-up process in a quasi-geostrophic model with continuous stratification. The *left panel* (**a**) shows the time evolution of a single mode, and the *right panel* (**b**) shows the superposition of all vertical modes – it becomes a δ-function-like jet trapped near the sea surface (modified from Young, 1981).

scientific discoveries, each step forward in our understanding of the thermocline structure was made through very careful study of the physics. Although these new theories seem so simple, they were the result of very hard work. Now, when we look at these problems, some very simple pitfalls can be seen.

Surface-trapped jet and the run-away problem

For the quasi-geostrophic model, a surface-trapped jet is baroclinically unstable and therefore instability would develop, so the subsurface water would be driven into motion. Furthermore, in the original picture the dynamical roles of inertial terms and the western boundary were totally neglected. Adding these important components will certainly change the dynamics.

The run-away problem of the homogeneous ocean models is also related to this issue. As discussed later, for strong forcing the isopycnals move upward, and eventually outcrop. When layers become thinner and thinner, or eventually outcrop, there will be very strong velocity shear. As a result, baroclinic instability develops, transporting momentum to other layers below. Therefore, the run-away problem in homogeneous ocean models may be due to the artificial assumption of a single moving layer. The run-away problem remains a challenge for theorists of ocean circulation. This problem may also be related to the parameterization of friction in the numerical model. For example, Fox-Kemper and Pedlosky (2004) explored the connection between the run-away problem and the removal of vorticity by eddies in a barotropic model.

No solution vs. an infinite number of solutions

Close examination reveals that when the forcing is strong enough, the interface in the layer model would be deformed so much that the geostrophic contour f/h_i in the subsurface layers would have closed, typically within the northwestern corner of the basin. If there are closed geostrophic contours, flow would be free to travel along these geostrophic contours. Therefore, instead of a case of no motion, we are facing the problem of deciding which flow pattern is physically reasonable among an infinite number of solutions.

Outcropping is the most crucial nonlinearity

In the oceans, isopycnals outcrop, so numerical models should be able to incorporate the outcropping fronts. In fact, outcropping indicates a strong nonlinearity of order one in the layer thickness (continuity) equation, which is much larger than the nonlinearity associated with the inertial terms in the momentum equation. Let us begin with the continuity equation in a layered model

$$h_t + (hu)_x + (hv)_y = 0 \tag{4.145}$$

Introduce the nondimensional variables

$$h = H(1 + dh'), \quad (u, v) = U(u', v'), \quad t = t'H/U \tag{4.146}$$

where H and U are the mean layer depth and velocity, and $d = \Delta H/H$ is the nondimensional parameter indicating the nonlinearity associated with layer depth change. The nondimensional equation is

$$h'_{t'} + (u'_x + v'_y) + d(u'h'_x + v'h'_y) = 0 \tag{4.147}$$

Since layer depth can vary greatly within a basin, ΔH is of the same order as H. For the case of layer outcropping, $\Delta H = -H$, i.e., $d = -1$. In comparison, the Rossby number associated with the inertial terms in the momentum equations seldom exceeds 0.3 for large-scale circulation problems. Thus, the nonlinearity associated with change in stratification or layer thickness is the most important nonlinearity for large-scale dynamics. This is also the reason why using the quasi-geostrophic dynamics is not really suitable for studying the basin-scale general circulation; in this book we will use the primitive equations most of the time.

In reality, isopycnal outcropping is a common phenomenon in the oceans. The major problem of the seemingly unsolvable puzzles discussed above is that the nonlinearity of density advection has not been treated properly. In a sense, all these models are confined to the linear region of parameter space. Were nonlinearity included, the picture would be quite different. As we review the progress made two decades ago, one wonders why such a simple problem had not been solved much sooner. Many students think that solutions to all the easy problems have been found, and there is not much that they can do. This is not true. As you can see, there are many challenging problems waiting for us to solve. You have to look around very carefully and choose something to try for yourself.

The old way of treating the ocean in terms of models that have no outcropping does not seem appropriate; a sensible way to formulate a layered model is to take into consideration the final layer thickness change. Layer outcropping brings some new technical challenges; it also brings new and exciting physics which we will discuss later.

To illustrate the basic concept, let us examine a reduced-gravity model on a β-plane more carefully. We will assume that there are many layers vertically. At the beginning we assume that only the first layer is in motion. From Eqn. (4.31) in Section 4.1.2, we can find that the layer thickness satisfies

$$h_I^2 = H_1^2 + \frac{2f^2}{g'\rho_0\beta}\left(\frac{\tau}{f}\right)_y (x_e - x) \tag{4.148}$$

where H_1 is the upper layer thickness along the eastern boundary. The model is driven by a wind stress $\tau^x = -\cos\left(\pi y/L_y\right)$. Assume that $\beta = 2 \times 10^{-11}$ m/s, $L_x = 6{,}000$ km, $L_y = 3{,}000$ km, $g' = 0.015$ m/s^2, and $H_1 = 400$ m. The upper layer thickness map is shown in Figure 4.21a. The geostrophic contours in the second layer $q_2 = \frac{f}{h_2} = \frac{f}{H_1+H_2-h_1}$ depend on the average thickness of the second layer, H_2. If the second layer is rather thick, there would be no closed geostrophic contours, so the reduced model is self-consistent. However, if the second layer is not so thick, the strong deformation of the interface between the first and second layers would give rise to closed geostrophic contours, as shown in Figure 4.21. When closed geostrophic contours appear in the second layer, water is free to move along these contours, so the basic assumption of the reduced-gravity model is no longer valid. We discuss later how to find the solutions that are dynamically self-consistent.

4.1.6 Theory of potential vorticity homogenization

As discussed in previous sections, the most challenging question was how to bring the subsurface layers into motion in a stratified and weakly dissipated ocean. This historical puzzle was solved by Rhines and Young (1982a, b) in a series of highly original studies.

The basic ideas are as follows. Assume that the ocean can be separated into multiple layers. To begin with we assume that all subsurface layers are motionless. Strong wind forcing in the uppermost layer induces a large deformation of the interface; thus, closed geostrophic contours appear in the layer below the surface layer. Within the framework of the ideal fluid, geostrophic flow is free to move along these closed geostrophic contours. Therefore, there are infinite possible solutions and our original assumption of no motion in the subsurface layer is inconsistent.

We want to find a dynamically consistent solution, including a moving subsurface layer. However, the lowest-order dynamic balance, geostrophy, cannot provide a unique solution. For a case with two layers in motion, the Sverdrup constraint can provide only one constraint, so in order to find a solution for a two-layer model we need an additional constraint. As will be shown shortly, potential vorticity in the second layer is homogenized toward the planetary vorticity along the northern boundary of the model. Therefore, potential vorticity

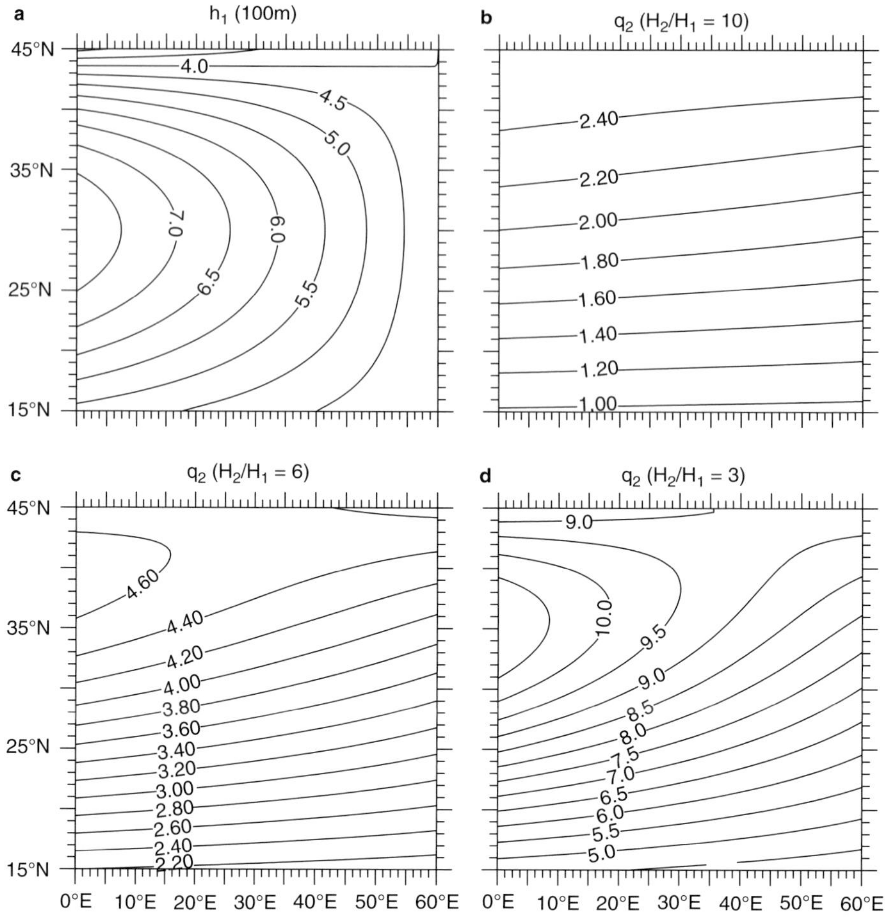

Fig. 4.21 Upper layer thickness (in 100 m) and potential vorticity in the second layer, $q_2 = f/h_2$ (in 10^{-8}/m/s) from a $1\frac{1}{2}$-layer model with $h_{1e} = 400$ m and for different layer thickness ratios (defined as the undisturbed layer thickness along the eastern boundary).

within the closed geostrophic contours is a given constant. This vorticity constraint and the Sverdrup constraint for the vertically integrated meridional volume flux combine and give rise to a solution for the two-layer model ocean, which is dynamically consistent with all dynamical constraints.

A two-layer model

On a β-plane the steady circulation of a two-layer quasi-geostrophic model can be described in terms of the potential vorticity equations (Rhines and Young, 1982a, b)

$$J(\psi_1, q_1) = w_0 - \nabla_h \cdot \Phi_1 \tag{4.149a}$$

$$J(\psi_2, q_2) = -\nabla_h \cdot \Phi_2 - D\nabla_h^2 \psi_2 \tag{4.149b}$$

where $J(g, h) = g_x h_y - g_y h_x$ is the nonlinear Jacobian term, q_1 and q_2 are potential vorticity in the upper and lower layers

$$q_1 = \beta y + F(\psi_2 - \psi_1) \tag{4.150a}$$

$$q_2 = \beta y + F(\psi_1 - \psi_2) \tag{4.150b}$$

$$w_0 = \frac{\nabla \times \vec{\tau} \cdot \hat{z}}{\rho_0 f_0} \tag{4.151}$$

is the Ekman pumping rate,

$$F = \frac{f_0^2}{g'H} = \lambda_r^{-2} \tag{4.152}$$

where g' is the reduced gravity, H is the undisturbed layer thickness, and λ_r is the Rossby radius of deformation. We have neglected relative vorticity in these equations because it is negligible for basin-scale motions. The $F(\psi_i - \psi_{i-1})$ terms are the contribution due to interface deformation, also called the stretching term, noting that the interface height is proportional to the streamfunction difference. The $D\nabla_h^2 \psi_2$ term is the bottom friction, and D is a small parameter.

The interfacial friction is parameterized in terms of Φ_i. As a crude way of mimicking baroclinic instability, the interfacial friction is assumed to be linearly proportional to the velocity shear

$$\Phi_1 = R\nabla_h(\psi_1 - \psi_2) \tag{4.153a}$$

$$\Phi_2 = R\nabla_h(\psi_2 - \psi_1) \tag{4.153b}$$

where R is a small parameter.

Steady and frictionless flow

Assuming that R and D are of the same order and very small, Eqns. (4.149a) and (149b) are reduced to

$$J(\psi_1, \beta y + F\psi_2) = w_0 + O(R) \tag{4.154a}$$

$$J(\psi_2, \beta y + F\psi_1) = O(R) \tag{4.154b}$$

where $O(R)$ denotes small terms on the order of R. From these two equations we obtain the barotropic solution

$$\psi_B = \psi_1 + \psi_2 = -\frac{1}{\beta} \int_x^{x_e} w_0 dx' \tag{4.155}$$

where $x_e = x_e(y)$ is the eastern boundary of the model. This solution is called the *barotropic solution*. For simplicity, we will assume that the Ekman pumping velocity is identically zero

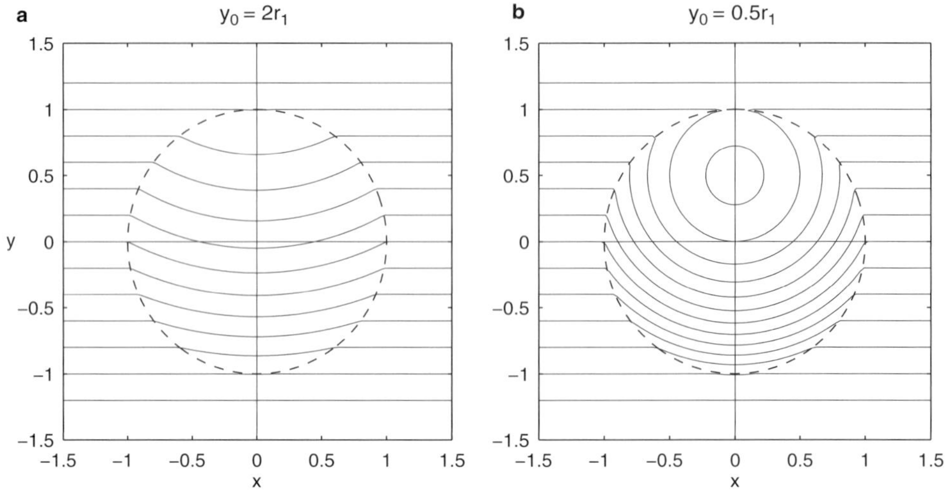

Fig. 4.22 Contours of $\hat{q}_2$. The *dashed circle* is $r = r_1$, i.e., the bounding contour of the barotropic streamfunction: **a** the forcing is weak, so $\hat{q}_2$ is dominated by the planetary vorticity term βy, and all the contours are open; **b** the forcing is stronger, so there is a region of closed contours. Flows in this region are shielded from the blocking geostrophic contours started from the "eastern" boundary (the right edge of the forcing domain where $\psi_B = 0$).

outside a circle (Fig. 4.22). We further assume the fluid to be motionless at infinity; thus, the eastern half of this circle can be chosen as the effective eastern boundary of the model. For a given Ekman pumping rate distribution, w_0, the barotropic streamfunction can be calculated accordingly. Note that the Jacobian term has a very useful property:

$$J(\psi_1, \psi_1) = J(\psi_2, \psi_2) = 0$$

Using the barotropic streamfunction ψ_B, the nonlinear equations (4.154a) and (4.154b) can be rewritten as

$$J(\psi_1, \beta y + F\psi_B) = w_0 + O(R) \qquad (4.156a)$$

$$J(\psi_2, \beta y + F\psi_B) = O(R) \qquad (4.156b)$$

These equations are linear because ψ_B is a given function from Eqn. (4.155). These equations are first-order partial differential equations in the characteristic form, with a quantity that behaves like a "barotropic potential vorticity" $\hat{q}_2 = \beta y + F\psi_B$ as the characteristics. Under the assumption of infinitesimal friction, potential vorticity in the second layer is conserved along streamlines. According to Eqn. (4.156b), potential vorticity contours in the second layer are the same as the "barotropic potential vorticity" contours.

To study the effect of strong forcing on the geostrophic contours in subsurface layers, Rhines and Young chose a very idealized Ekman pumping function

$$w_0 = -\alpha x, \ \text{for } r < r_1; \quad w_0 = 0, \ \text{for } r > r_1 \tag{4.157}$$

where $r = \sqrt{x^2 + y^2}$. Therefore, the barotropic streamfunction is

$$\psi_B = \begin{cases} \frac{\alpha}{2\beta}(r_1^2 - x^2 - y^2) & \text{for } r < r_1, \\ 0 & \text{for } r \geq r_1. \end{cases} \tag{4.158}$$

The $\hat{q}_2$ contours are circles or arcs of circles if $r \leq r_1$ (Fig. 4.22); outside the circle of non-zero Ekman pumping rate they are just straight zonal lines

$$\hat{q}_2 = \begin{cases} \frac{\alpha F}{2\beta}\left[r_1^2 + y_0^2 - x^2 - (y - y_0)^2\right], & \text{if } r < r_1, \\ \beta y, & \text{if } r \geq r_1. \end{cases} \tag{4.159}$$

where $y_0 = \frac{\beta^2}{\alpha F}$.

It can readily be seen that if forcing is weak, the barotropic potential vorticity is controlled by the planetary term βy, so vorticity contours are close to straight lines and there are no closed vorticity contours. For such cases, the eastern boundary blocks all possible flows in the second layer, as pointed out by Rooth *et al.* (1978). However, if forcing is strong enough, the second term, $F\psi_B$, dominates the value of $\hat{q}_2$, and there can be closed vorticity contours (Fig. 4.22).

For this given type of Ekman pumping rate distribution, closed contours will appear, if

$$r_1 > y_0, \ \text{or equivalently,} \ \alpha r_1 > \beta^2/F \tag{4.160}$$

In general, the closed contours will appear near the northern boundary of the forcing field, where the negative meridional gradient of the barotropic streamfunction can cancel the positive meridional gradient of the planetary vorticity.

A crucial technique used in their model is to assume a special Ekman pumping pattern which satisfies

$$\int_{-\infty}^{+\infty} w_0\left(x'\right) dx' = 0 \tag{4.161}$$

By using such a pattern, they were able to avoid the complications associated with the western boundary layer and find dynamically consistent solutions. Note that the same technique was first employed by Goldsbrough for the evaporation/precipitation-driven circulation. As we discuss later, adding the western boundary current is not an easy exercise. In fact, due to the strong friction/dissipation within the western boundary layer, potential vorticity cannot be homogenized in the subsurface layer (Ierley and Young, 1983).

When closed geostrophic contours appear in the second layer, the number of possible solutions for an ideal fluid is infinite; these solutions are in the form of

$$\psi_2 = A_2(\hat{q}_2) \tag{4.162}$$

where A_2 is an arbitrary function. In order to find solutions that are physically meaningful, one has to include the next-order terms. Typically, by working on some integrals for the next-order terms, one can find constraints on the lowest-order dynamics, and this will eventually lead to unique solutions. ***A most important attribute of this solution is that it is stable to small perturbations***, as demonstrated by Rhines and Young (1982a).

This example leads to an interesting observation. A potential vorticity field dominated by the strong planetary vorticity, combining with the eastern boundary, blocks the potential ideal-fluid flow in the subsurface interior ocean; however, strong Ekman pumping creates closed geostrophic contours in the oceanic interior, overcoming the blockage and making free solutions possible there.

Determination of the flow inside the closed geostrophic contours

For convenience, we assume that $D = R$. Integrating Eqn. (4.149b) along a closed contour, leads to

$$R\oint(2\vec{u}_2 - \vec{u}_1)\cdot d\vec{s} = 0 \tag{4.163}$$

where $\vec{u}_1 = \vec{k} \times \nabla\psi_1$ and $\vec{u}_2 = \vec{k} \times \nabla\psi_2$ are horizontal velocity in the upper and lower layers, and $\vec{k}$ is a unit vector in the vertical direction. Note that the Jacobian term vanishes identically! Equation (4.163) can be rewritten as

$$\oint\vec{u}_2 \cdot d\vec{s} = \frac{1}{3}\oint\vec{u}_B \cdot d\vec{s} \tag{4.164}$$

where $\vec{u}_B = \vec{k} \times \nabla\psi_B = \vec{u}_1 + \vec{u}_2$ is the barotropic velocity. Using Eqn. (4.162), the term on the left-hand side of Eqn. (4.164) can be rewritten as

$$\oint\vec{u}_2 \cdot d\vec{s} = \oint A_2'(\hat{q}_2)(\vec{k} \times \nabla_h\hat{v}_2)\cdot d\vec{s} = A_2'(\hat{q}_2)\oint(F\vec{u}_B - \beta\vec{x})\cdot d\vec{s} \tag{4.165}$$

where $A_2' = dA_2/d\hat{q}_2$. Thus, from Eqns. (4.164) and (4.165) we finally obtain a relation

$$A_2' = 1/3F \tag{4.166}$$

Using Eqn. (4.162), the final solution is

$$\psi_2 = \frac{1}{3F}\hat{q}_2 + const = \frac{1}{3}\psi_B + \frac{1}{3}\frac{\beta y}{F} + const \tag{4.167a}$$

$$\psi_1 = \psi_B - \psi_2 \tag{4.167b}$$

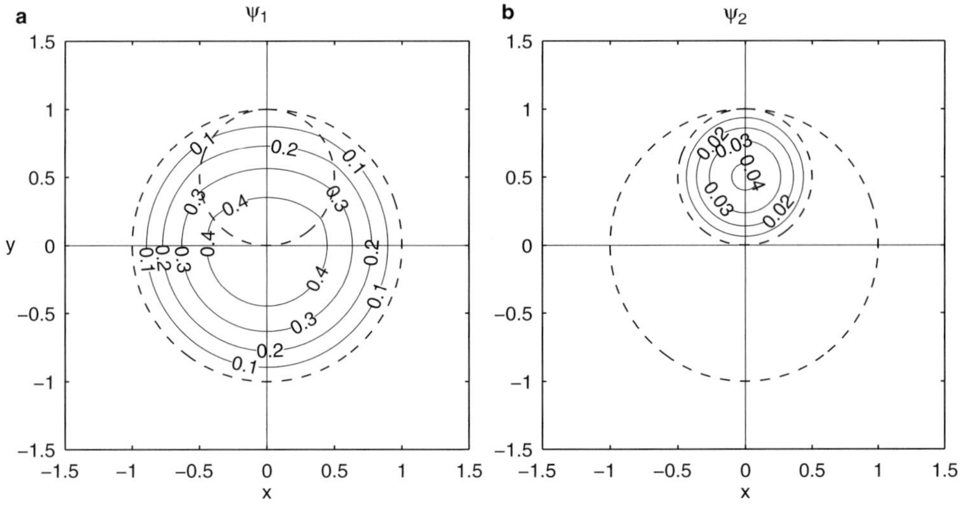

Fig. 4.23 The streamfunction maps for the **a** upper and **b** lower layer when $y_0 = 0.5r_1$ (same parameter as in Fig. 4.22b). The *small dashed circle* inside is the outermost contour within which potential vorticity is homogenized.

For the case discussed above, $w_0 = -\alpha x$, $r \le r_1$, so that flow in the lower layer is

$$\psi_2 = \begin{cases} -\dfrac{\alpha}{6\beta}(x^2 + (y - y_0)^2) + const. & \text{for closed } \hat{q}_2 \\ 0, & \text{elsewhere} \end{cases} \tag{4.168}$$

The solution is shown in Figure 4.23. In the lower layer, flow is confined to the outermost closed geostrophic contour. Outside this boundary the lower layer is stagnant, so flow in this regime is entirely confined to the upper layer. Since the lower layer is in motion within the closed geostrophic contours, flow in the upper layer is reduced as the Sverdrup constraint applies to the sum of the volumetric transport in these two moving layers.

When $D \neq R$, the corresponding solution is $q_2 = (D/2R + D)\,\hat{q}_2 + const$; thus, in the limit of $D \ll R$, potential vorticity in the second layer becomes homogenized within the closed streamlines. This is an example of the more generalized potential vorticity homogenization theory discussed by Rhines and Young (1982a, b). This example of a two-layer model, satisfying the constraint $D \ll R$, shows that potential vorticity homogenization in the subsurface layer can be realized only if that layer is shielded from strong forcing or dissipation. Therefore, a natural choice to demonstrate the idea of potential vorticity homogenization is to use a three-layer model, in which the second layer is shielded from surface forcing and bottom friction.

Three-layer model

For a model with three moving layers, an analysis similar to that presented above demonstrates that, in the second layer, potential vorticity within the closed geostrophic contours is homogenized toward its value along the northern boundary of the gyre. Since potential vorticity outside the big circle is controlled by the planetary vorticity alone, its contours are zonally oriented; thus, potential vorticity homogenization within the big circle implies an expulsion of potential vorticity contours toward the edge of the big circle. As a result, there is a sharp potential vorticity front adjacent to the outermost closed geostrophic contours, as illustrated in Figure 4.24.

Another important phenomenon related to the three-moving-layer model is that the center of the wind-driven gyre in each layer is gradually moved northward from the topmost layer to the deeper layers, as shown in Figure 4.25. This is called the northern intensification of the subtropical gyre, and this phenomenon can be identified from hydrographic data or a model with continuous stratification, as will be discussed later. Note that volume flux in the upper layer is reduced over the domain where the lower layers are in motion.

The potential vorticity homogenization theory discussed in this section represents a beautiful combination of many vital dynamic concepts related to the vertical structure of wind-driven gyres in the oceans.

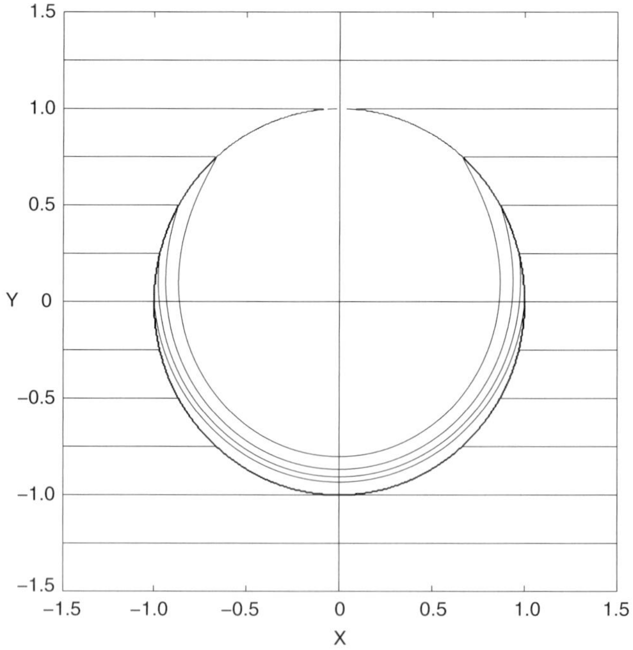

Fig. 4.24 A map of potential vorticity, illustrating the potential vorticity homogenization and expulsion of potential vorticity with non-zero Ekman pumping.

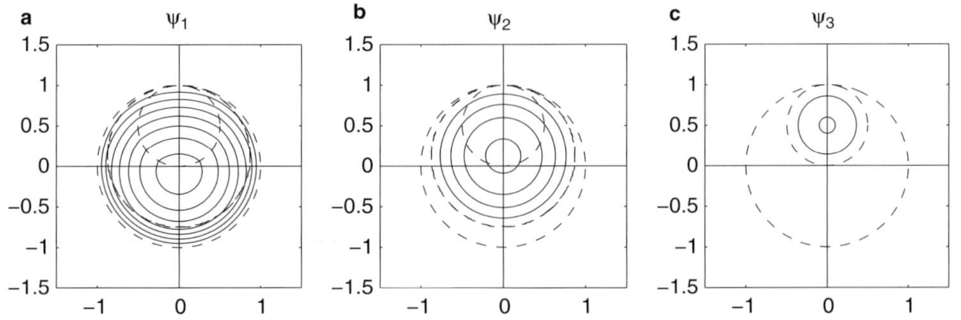

Fig. 4.25 Streamfunction maps for a three-layer model for the case with $y_0 = r_1/8$. In the *middle panel*, the *smallest dashed circle* indicates the outermost closed geostrophic contour in the lower layer, and the *middle-sized dashed circle* indicates the outermost closed geostrophic contour in the middle layer, while the *outermost dashed circle* indicates the domain with non-zero Ekman pumping.

- Strong surface forcing can induce closed geostrophic contours in the subsurface layers.
- There is an infinite number of possible solutions; however, unique solutions stable to small perturbations can be found.
- Potential vorticity is homogenized within the closed contours when the dissipation is parameterized as lateral potential vorticity diffusion.
- Potential vorticity is homogenized toward its value along the northern boundary.

These are very useful when we want to find a solution for the wind-driven circulation in the case of a multi-layer model or a model with continuous stratification.

4.1.7 The ventilated thermocline

Introduction

There is a prominent layer of steep vertical temperature gradient in the world's oceans, the so-called main thermocline, the structure of which was discussed in Section 1.4. Since density is very closely related to temperature, the structure of the thermocline is very closely related to currents in the upper ocean. The vertical shear of the horizontal velocity field is related to the horizontal density gradient through the thermal wind relation; therefore, solving the density structure is equivalent to finding the structure of the wind-driven circulation. For historical reasons the relevant theory is called thermocline theory; however, it is somewhat equivalent to a theory for the wind-driven circulation in the upper ocean.

The fundamental questions at the heart of the thermocline theory are as follows.

- First, the existence problem: Why is there the main thermocline in the ocean?
- Second, the direct problem for the thermocline: Given the surface forcing conditions, including wind stress, heat and freshwater fluxes, how is the stratification, or the potential vorticity, set up in the subsurface ocean?

From the very beginning of the development of the theory, there have been two approaches to this problem. In 1959, two papers were published side by side in *Tellus*. Robinson and Stommel (1959) proposed a theory of the thermocline in which the vertical diffusion plays a vital role; Welander (1959) proposed an ideal-fluid theory for the thermocline.

According to the theory of Robinson and Stommel (1959), the main thermocline is viewed as an internal density front or internal thermal boundary layer; thus, the vertical diffusion term should be retained as an essential component of the dynamical description. The latest advance along this line of thought is presented by Salmon (1990, 1991). The most challenging difficulty associated with this approach is the fact that nobody knows how to formulate suitable boundary value problems for the corresponding nonlinear equation system, much less how to find the solutions to this problem. In order to overcome such a difficulty, similarity solutions have been sought. The early developments along this line of research were summed up in a comprehensive review by Veronis (1969).

Similarity solutions which satisfy a given differential equation can be found through a systematic approach based on group-invariant solutions under the infinitesimal group transformation, i.e., the Lie group. Filippov (1968) applied the Lie group theory to the thermohaline equations and discussed the group-invariant solutions. More up-to-date mathematical tools can be found in books by Oliver (1986) and Rogers and Ames (1989). Similarity solutions to the oceanic thermocline have been discussed by Salmon and Hollerbach (1991) and Hood and Williams (1996); time-varying similarity solutions have been discussed by Edwards (1996).

The major disadvantage of the similarity solutions is as follows. Although they do satisfy the thermocline equations, they do not satisfy some essential boundary conditions, such as the Sverdrup constraint. A solution that does not satisfying the Sverdrup constraint cannot describe the global structure of the wind-driven circulation accurately. More importantly, the function relation for the potential vorticity in the thermocline is prescribed *a priori*; thus, such solutions cannot provide a clear answer to the direct problem of the thermocline: how potential vorticity in the thermocline is set up for given surface forcing conditions.

In the other *Tellus* paper, Welander (1959) proposed a quite different approach; he argued that the main thermocline can be treated in terms of an ideal-fluid theory. Accordingly, the structure of the main thermocline can be interpreted without the vertical and horizontal diffusion terms.

These two approaches are based on different simplifications of the same set of dynamical equations. From the beginning, the basic equations for the thermocline seemed so naively simple that many people believed they could be solved easily, so most people did not want to spend time working on the apparently incomplete ideal-fluid thermocline theory, except for Welander, who also published another very influential paper on the ideal-fluid thermocline (Welander, 1971a).

The most critical point in the ideal-fluid thermocline theory is the smallness of the vertical diffusivity in the upper ocean (below the mixed layer). Recent field observations confirm that diapycnal diffusivity above the main thermocline is indeed quite weak, on the order of 10^{-5} m^2/s (e.g., Ledwell *et al.*, 1993). Thus, flow associated with the main thermocline can be treated in terms of an ideal fluid. As explained in Section 5.4.5, the vertical diffusion term is not the most crucial ingredient in formulating the main thermocline. As a matter of fact, the main thermocline in the subtropical ocean is primarily due to the downward squeeze of the Ekman pumping induced by the negative wind-stress curl.

A convenient way to study the ideal-fluid thermocline is to formulate the problem in terms of density coordinates and the Bernoulli function $B = p + \rho g z$, where p is pressure, ρ is density, g is gravity, and z is the vertical coordinate. Assuming that the potential vorticity $Q = f\rho_z$ is a linear function of both the density and the Bernoulli function, Welander (1971a) was able to find the first analytical solution for the thermocline.

Although this is the first elegant solution for the ideal-fluid thermocline, and thus has been cited in many textbooks, it has some major defects: the solution satisfies neither the Sverdrup relation nor the eastern boundary condition, and the lower boundary is set at $z = -\infty$. Since the Sverdrup constraint is the most important constraint, thermocline solutions that do not satisfy the Sverdrup relation are less meaningful dynamically.

During the 1960s, similarity solutions remained the mainstream in thermocline theory. However, the limitations of the similarity approach became clear. Beginning in the early 1980s there were major breakthroughs in the theory of the dynamical structure of the wind-driven gyre, including the potential vorticity homogenization theory by Rhines and Young (1982a) and the ventilated thermocline by Luyten *et al.* (1983). The historic events that led to the discovery of the new thermocline theory have been vividly described by Pedlosky (2006).

According to these new theories, the wind-driven gyre includes several regions with distinctively different dynamics: the *ventilated thermocline* where potential vorticity is set at the surface; the *unventilated thermocline* where potential vorticity is homogenized; the *shadow zone* near the eastern boundary; and the *pool region* near the western boundary. These new theories were combined and extended to form a theory of the wind-driven gyre in the continuously stratified oceans by Huang (1988a, b). The dynamical role of the mixed layer, excluded from the early theories, was incorporated into the thermocline theory (e.g., Huang, 1990a; Pedlosky and Robbins, 1991; Williams, 1989, 1991). The progress made during this period was reviewed by Huang (1991a) and Pedlosky (1996).

A major defect of thermocline theories is the lack of the western boundary layer and recirculation. It is clear that in order to explain the structure of the wind-driven circulation, numerical models must be used in which the dynamical effects of vertical/horizontal mixing, the western boundary layer, and the recirculation are explicitly included. From the physical point of view, mixing must play a critical role in setting up the global structure of the thermocline; thus, the ocean can be classified into regimes of different dynamics. In particular, as Welander (1971b) pointed out, "the thermocline may not be a diffusive boundary layer, but rather an ideal-fluid regime imbedded between diffusive regimes."

More recent studies based on numerical simulation provide a more dynamically complete picture. For example, Samelson and Vallis (1997) studied the thermocline structure in a closed basin, and showed that by using a small diapycnal mixing rate in the ocean interior, the thermocline does appear in two dynamical regimes, i.e., the ventilated thermocline for the water entering from the surface layer in the subtropical basin due to Ekman pumping, and the diffusive thermocline over the density range corresponding to the subpolar basin. Vallis (2000) went through a series of carefully designed numerical experiments based on a primitive equation model and showed that stratification below the ventilated thermocline is a result of global dynamics, involving the effect of wind forcing, geometry of the world's oceans, and diffusion. In particular, his study indicated that the geometry of the ACC (Antarctic Circumpolar Current) plays a subtle but important role in setting up the stratification at the mid depth of the world's oceans.

Physical foundation of the ventilated thermocline

The modern theory of the ventilated thermocline is based on two cornerstones, i.e., the ventilation, as postulated by Iselin, and the "Stommel demon." These two concepts are discussed here before we introduce the formulation of the layered ventilated thermocline.

Iselin's conceptual model

A major conceptual difficulty in getting subsurface layers in motion is that they are not directly in contact with the atmospheric forcing. One way to get these subsurface layers in motion is the close geostrophic contours induced by strong forcing upon the surface layer. There is also another way, called ventilation, through which the subsurface layer can be put into motion. Many isopycnals outcrop in poleward parts of the oceans. When a layer outcrops, it is exposed to the atmospheric forcing directly. Thus, the outcropping layer is in motion under the wind stress forcing, and it should continue its motion even after it has been subducted under the other layers.

Iselin (1939) made a link between the T–S relation found in a vertical section and the wintertime mixed layer at higher latitudes. His schematic picture for this ventilation process is shown in Figure 4.26. The speculated motion is indicated by arrows. Iselin's model was the first prototype for water mass formation in the oceans; however, it is surprising that such a simple and important dynamical idea was not pursued further in the ensuing decades.

In modern terminology the basic idea is that, within the subtropical gyre, water is pushed downward into the thermocline by Ekman pumping and then downwells along isopycnals as it moves southward, induced by the Sverdrup dynamics. The motion of the particles after their rejection from the base of the mixed layer is confined within the corresponding isopycnal surfaces, because mixing is relatively weak below the mixed layer and above the rough bottom topography. The weakness of mixing in the upper ocean and below the mixed layer has been confirmed by observations, such as the recent tracer-release experiments. The process described by Iselin is now called "ventilation by Ekman pumping." Thus,

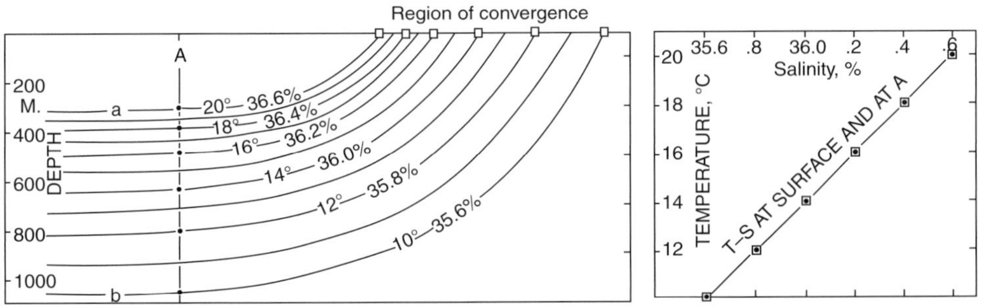

Fig. 4.26 Schematic representation of water mass formation due to water sinking along isopycnal surfaces (Iselin, 1939).

outcropping functions as windows for water in the subsurface layers to go through, escaping the blockage by the eastern boundary.

There are also other types of ventilation. For example, a water mass can be ventilated through mode water formation or deepwater formation, where convective overturning induced by cooling or salt rejection due to ice formation leads to ventilation to great depth, on the order of a few hundreds of meters or all the way to the sea floor. A water mass can also be ventilated through the western or eastern boundary currents.

The Stommel demon

One of the conceptual difficulties in building up a model for the annual mean wind-driven circulation is the strong seasonal cycle in wind stress, heat and freshwater fluxes at the sea surface. As a result, mixed layer properties vary greatly over the seasonal cycle; in particular, they change quite rapidly in late winter, as shown in Figure 4.27. It seems obvious that the seasonal variability in the surface conditions may affect the wind-driven circulation in the form of waves and time-varying currents; thus, formulating a simple analytical model to describe such complicated phenomena is a great challenge.

In a highly innovative paper, Stommel (1979) postulated that a demon is working so that only the late-winter properties are selected, i.e., water that actually enters the permanent thermocline is the water formed at the time when the mixed layer is deepest and the density is the highest. It is indeed a highly simplified idealization of the complicated processes taking place in the upper ocean. It is such a bold assumption that Henry Stommel himself was not sure, even after his paper had been published. (He told me that he destroyed all the reprints because he thought the paper was wrong.) Since then, Stommel's idea of selecting the late-winter properties as those for the water mass formed during wintertime remains the backbone of the theory of wind-driven circulation, including ventilation, subduction, and obduction.

Therefore, if we want to simulate the annual mean wind-driven circulation, we should select the late-winter thermohaline boundary conditions at the sea surface, including mixed

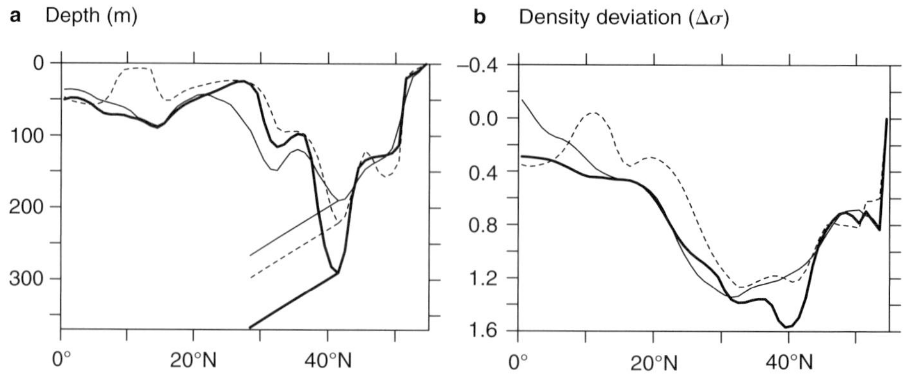

Fig. 4.27 Mixed-layer properties along 160.5° E section for the North Pacific: **a** mixed-layer depth and trajectories of water particles (*straight lines*); **b** surface density deviation from the annual mean; the *heavy line* for March, the *light line* for February, and the *dashed line* for April.

layer density and depth; however, the wind stress forcing, or the Ekman pumping rate, should be the annual mean because we are concerned about the annual mean movement of the particles in the thermocline.

For example, if we use the annual mean temperature and salinity conditions on the surface, the model will be unable to form the right type of deepwater or mode water (the meaning of mode water is explained in Chapter 5); the annual mean density on the surface is much lower than that of wintertime (such as late March), so that the deepwater formed in the model is too light.

Stommel's idea can be illustrated as follows. Assuming that at the beginning of each month a water parcel is released at the base of the mixed layer, with a constant equatorward velocity of 10^{-2} m/s and vertical velocity of $w = -0.5 \times 10^{-6}$ m/s, the trajectories of these parcels can be calculated (Fig. 4.27). As shown in this figure, water parcels released at 40° N in February will be overtaken by the mixed layer on its way down south in March. However, water parcels released in March and April can enter the permanent thermocline, and thus can be counted as effective detrainment, which contributes to the annual mean subduction and water mass formation.

Tracer-release experiments in the 1990s provided strong support for the ideal-fluid thermocline theory, because it was found that diapycnal diffusivity is very small in the main thermocline, on the order of 10^{-5} m^2/s. Thus, to a very good approximation, the structure of the thermocline and the wind-driven circulation can be treated in terms of an ideal-fluid model. It is also very important to note that wind stress is one of the crucial sources of energy supporting mixing in the ocean interior; thus, although diapycnal mixing is very small, it is not infinitesimal. Thus, running a numerical model with an extremely small and non-zero mixing coefficient may not produce results comparable with observations in the oceans.

Layer model for the ventilated thermocline

Rhines and Young (1982a, b) made a great breakthrough with their theory of potential vorticity homogenization. Inspired by their study, Luyten, Pedlosky and Stommel tried to solve the classical puzzle of motion in the subsurface layers.

In a most innovative and remarkable paper, Luyten, Pedlosky and Stommel (hereafter LPS) (Luyten *et al.*, 1983) formulated a multi-layer model for the ventilated thermocline and applied this model to the North Atlantic. The model is formulated for the ocean interior, excluding the western boundary region and the Ekman layer. The upper ocean is divided into several layers of constant density which outcrop at different latitudes (Fig. 4.28). Remember that layer models are based on density coordinates, although they are highly truncated for most cases. The uppermost layer is directly driven by Ekman pumping. After subduction, water particles retain their potential vorticity and continue their southward motion. Along the eastern boundary, the model predicts that water moving in a subsurface layer has to depart from the wall in order to maintain its potential vorticity.

Although the ventilated thermocline can be formulated in terms of the β-plane, our discussion here is based on spherical coordinates because this system provides more accurate solutions. In addition, our notations are based on dimensional variables, so the figures shown are also in dimensional units, which give the reader a clear view of the circulation calculated from the model, in comparison with the oceans.

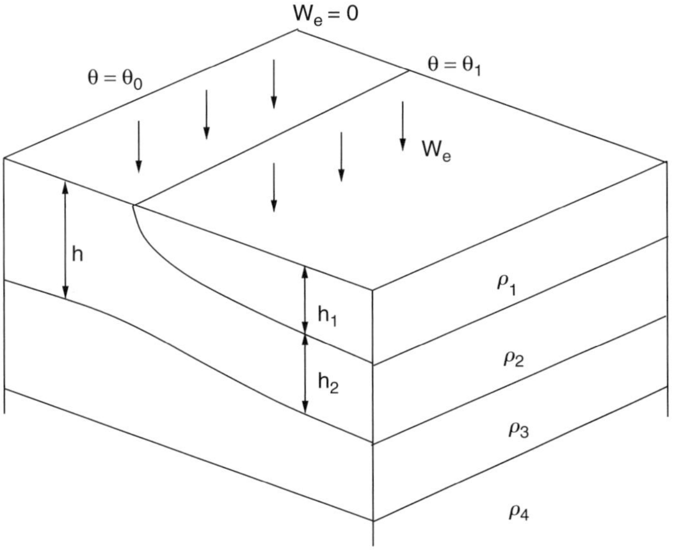

Fig. 4.28 Sketch of a multi-layer model of the thermocline in the subtropical basin. The Ekman pumping vanishes at $\theta = \theta_0$ and is downward for $\theta < \theta_0$; h is the depth of the bottom of the second layer.

The ventilated zone

In the ocean interior, the basic equations are: geostrophy for the horizontal momentum equations, hydrostatic approximation in the vertical direction, and mass conservation. In this section, we will assume that water below the second layer is stagnant. Thus, north of θ_1, only the second layer is in motion and its momentum equations are

$$2\omega \sin\theta v_2 = g' \frac{\partial h}{a\cos\theta\, \partial\lambda} \tag{4.169}$$

$$2\omega \sin\theta u_2 = -g' \frac{\partial h}{a\partial\theta} \tag{4.170}$$

where $g' = g\rho/\rho_0$, $\Delta\rho = \rho_3 - \rho_2$, and $h = h_2$ for $\theta > \theta_1$. The mass conservation equation is

$$\frac{1}{a\cos\theta}\left[\frac{\partial}{\partial\theta}(h_2 v_2 \cos\theta) + \frac{\partial}{\partial\lambda}(h_2 u_2)\right] + w_e = 0 \tag{4.171}$$

Substituting Eqns. (4.169) and (4.170) into Eqn. (4.171) gives the vorticity equation

$$\cos\theta h_2 v_2 = a \sin\theta w_e \tag{4.172}$$

This equation corresponds to Eqn. (4.35) which is formulated in a β-plane, and it can be rewritten as

$$h_2\vec{u}_2 \cdot \nabla_h (f/h_2) = f w_e/h_2 \tag{4.173}$$

Equations (4.169) and (4.170) state that h contours are streamlines, and Eqn. (4.173) states that potential vorticity changes along streamlines due to Ekman pumping.

Substituting this relation into Eqn. (4.169) and integrating in the zonal direction gives

$$h_2^2 = h_{2e}^2 + D_0^2 \tag{4.174}$$

where h_{2e} is the constant layer thickness along the eastern boundary and

$$D_0^2 = -\frac{4\omega a^2 \sin^2\theta}{g'} \int_{\lambda}^{\lambda_e} w_e\left(\lambda', \theta\right) d\lambda' \tag{4.175}$$

is the layer thickness (square) deviation from that at the eastern boundary. Note that the thickness of each layer has to be constant along the eastern boundary of the model basin in order to satisfy the boundary condition that there is no geostrophic flow across the boundary. Within the formulation of an ideal-fluid thermocline, we cannot determine h_{2e}. In fact, h_{2e} is an external parameter of the ventilated thermocline model, and different values of h_{2e} would give rise to different solutions. The basic philosophy of the ideal-fluid thermocline is to assume that h_{2e} is set up by some external processes, such as the thermohaline circulation, not directly simulated in the model. When h_2 is known, both u_2 and v_2 can be determined geostrophically.

As water moves southward, crossing latitude θ_1, the second layer continues to flow southward under the first layer. South of the outcropping line both layers are in motion; the momentum equations for the second layer have the same form as Eqns. (4.169) and (4.170), although now $h = h_1 + h_2$; thus we have

$$2\omega \sin\theta v_2 = g' \frac{\partial(h_1 + h_2)}{a\cos\theta\, \partial\lambda} \tag{4.176}$$

$$2\omega \sin\theta u_2 = -g' \frac{\partial(h_1 + h_2)}{a\partial\theta} \tag{4.177}$$

$$\frac{1}{a\cos\theta}\left[\frac{\partial}{\partial\theta}(h_2 v_2 \cos\theta) + \frac{\partial}{\partial\lambda}(h_2 u_2)\right] = 0 \tag{4.178}$$

Cross-differentiating Eqns. (4.176) and (4.177), subtracting and using the continuity equation (4.178) leads to a potential vorticity equation for layer 2 after subduction:

$$h_2\vec{u}_2 \cdot \nabla_h (f/h_2) = 0 \tag{4.179}$$

This equation applies after subduction because now the second layer is shielded from the Ekman pumping; thus, along a streamline, a water parcel maintains its potential vorticity. This is different from the case before subduction, when the second layer is directly forced by Ekman pumping, so its potential vorticity is not conserved, as stated by Eqn. (4.173). According to Eqns. (4.176) and (4.177), the flow in the second layer follows the constant h line, i.e., the h contours are streamlines. Using the law of potential vorticity conservation, Eqn. (4.179), we come to the conclusion that the h contours are also potential vorticity contours.

Note that potential vorticity in the second layer, $\sin\theta/h_2$, should be a function of h only, which we denote as $G(h)$, i.e., $\sin\theta/h_2 = G(h)$. This function $G(h)$ is set along the line $\theta = \theta_1$ (the outcrop line for layer 1 and also the subduction line for the second layer):

$$G(h) = \left.\frac{\sin\theta}{h}\right|_{\theta=\theta_1} = \frac{\sin\theta_1}{h|_{\theta=\theta_1}} = \frac{\sin\theta_1}{h} \tag{4.180}$$

The last equals sign above is due to the fact that h is constant along trajectories in the second layer after subduction. Thus, following each streamline started from the outcropping line, the potential vorticity conservation law gives us the following relation:

$$\frac{\sin\theta}{h_2} = G(h) = \frac{\sin\theta_1}{h} \tag{4.181}$$

Accordingly, the thicknesses for these two moving layers obey

$$h_2 = \frac{\sin\theta}{\sin\theta_1}h = \frac{f}{f_1}h \tag{4.182}$$

$$h_1 = \left(1 - \frac{\sin\theta}{\sin\theta_1}\right)h = \left(1 - \frac{f}{f_1}\right)h \tag{4.183}$$

Note that layer thickness ratio is entirely determined by the planetary potential vorticity. This simple relation gives rise to a simple analytical solution for the ventilated thermocline. Were the outcrop line not zonal, the solution would be in a much more complicated form, and this is the beauty of the ventilated thermocline model, as proposed by Luyten *et al.* (1983).

To determine the still unknown total layer depth h in this zone, we can use the Sverdrup relation for the barotropic mass flux. The momentum equations for the first layer are

$$2\omega\sin\theta v_1 = g'\frac{\partial}{a\cos\theta\partial\lambda}(\gamma h_1 + h_2) \tag{4.184}$$

$$2\omega\sin\theta u_1 = -g'\frac{\partial}{a\partial\theta}(\gamma h_1 + h_2) \tag{4.185}$$

where $\gamma = \frac{\rho_3-\rho_1}{\rho_3-\rho_2}$. The mass conservation equation is

$$\frac{1}{a\cos\theta}\left[\frac{\partial}{\partial\theta}(h_1 v_1\cos\theta) + \frac{\partial h_1 u_1}{\partial\lambda}\right] + w_e = 0 \tag{4.186}$$

Following the same approach in deriving the vorticity equation (4.172), substituting Eqns. (4.176, 4.177) into Eqn. (4.178), and Eqns. (4.184, 4.185) into Eqn. (4.186) leads to two relations and, adding these two relations, we obtain the Sverdrup relation for the barotropic volume flux:

$$\cos\theta(h_1 v_1 + h_2 v_2) = a\sin\theta w_e \tag{4.187}$$

Substituting Eqns. (4.176) and (4.184) into Eqn. (4.187) and integrating leads to an equation for the layer thicknesses:

$$(\gamma - 1)h_1^2 + h^2 = D_0^2 + (\gamma - 1)h_{1e}^2 + h_e^2 \tag{4.188}$$

Along the eastern boundary the no-zonal flow condition requires both h_1 and h_2 to be constant. Since $h_1 = 0$ north of θ_1, h_1 must be identically zero along the entire eastern boundary. Similarly, $h_2 = h_{2e}$ is constant along the entire eastern boundary. Note that this is one of the major difficulties in the ventilated thermocline theory that will be discussed later.

Using Eqn. (4.183), we obtain the sum of layer thickness:

$$h^2 = \frac{D_0^2 + h_{2e}^2}{1 + (\gamma - 1)(1 - f/f_1)^2}, \quad \textit{for } \theta \le \theta_1 \tag{4.189}$$

The shadow zone

The no-zonal flux condition requires the thickness of the subsurface layers to be constant along the eastern boundary. The constant layer thickness implies that the potential vorticity f/h_i is non-constant along the eastern boundary because f varies along the meridional wall. Since potential vorticity should be conserved along streamlines in subsurface layers, the eastern boundary cannot be a streamline in these layers. Thus, the region next to the eastern boundary should be a shadow zone for the subsurface layers where fluid is stagnant, and motions must be confined to the uppermost layer that is directly exposed to the Ekman pumping. Because this layer is directly forced, potential vorticity is not conserved along streamlines. For the present model, the upper layer has a zero thickness along the eastern boundary, so all motions are confined into a singular line at the junction between the upper surface and the eastern boundary. This is a weakness in the original ventilated thermocline model, and it can be improved by including a mixed layer in the upper ocean, as shown by Pedlosky and Robbins (1991). This problem can be overcome in a model of the thermocline with continuous stratification, as discussed in Section 4.2.2.

The existence of such a shadow zone near the eastern boundary is consistent with the observation that oxygen content in the depth range of 600–800 m near the eastern boundary in the subtropical basin is the lowest in the whole basin (Fig. 4.29). The source of oxygen in water parcels is attributed to recent contact with the atmosphere through the mixed layer, or photosynthesis in the upper 100 m of the ocean. Away from such sources, oxygen concentration tends to decline gradually through consumption due to biological activity. Thus, low oxygen concentration indicates water of old age, and the oxygen minimum near the eastern boundary at a depth range of 700 m indicates that water there is poorly ventilated.

To determine the boundary between the ventilated thermocline and the shadow zone, we note that the total depth h should be constant along streamlines in the second layer. Thus, Eqn. (4.189) can be used to describe this boundary in the following parametric form:

$$\frac{D_0^2 + h_{2e}^2}{1 + (\gamma - 1)(1 - f/f_1)^2} = const \tag{4.190}$$

Since the boundary between the shadow zone and the ventilated thermocline passes the boundary point (λ_e, θ_1) where $f = f_1$ and $D_0^2 = 0$, this boundary line $\lambda = \Lambda(\theta)$ is determined by

$$D_0^2(\Lambda, \theta) = h_{2e}^2(\gamma - 1)(1 - f/f_1)^2 \tag{4.191}$$

This boundary can be identified in numerical calculation as follows. Note that h is flat within the shadow zone, so $h = h_{2e}$ is valid within the shadow zone up to its western boundary. Therefore, the western boundary of the shadow zone can be identified as the place where the total depth h of this two-moving-layer solution, calculated from Eqn. (4.189), is equal to the thickness of the second layer along the eastern boundary, h_{2e}.

It can readily be seen that layer thickness is continuous across this line; however, tangential velocities are discontinuous. Within the shadow zone the second layer is stagnant

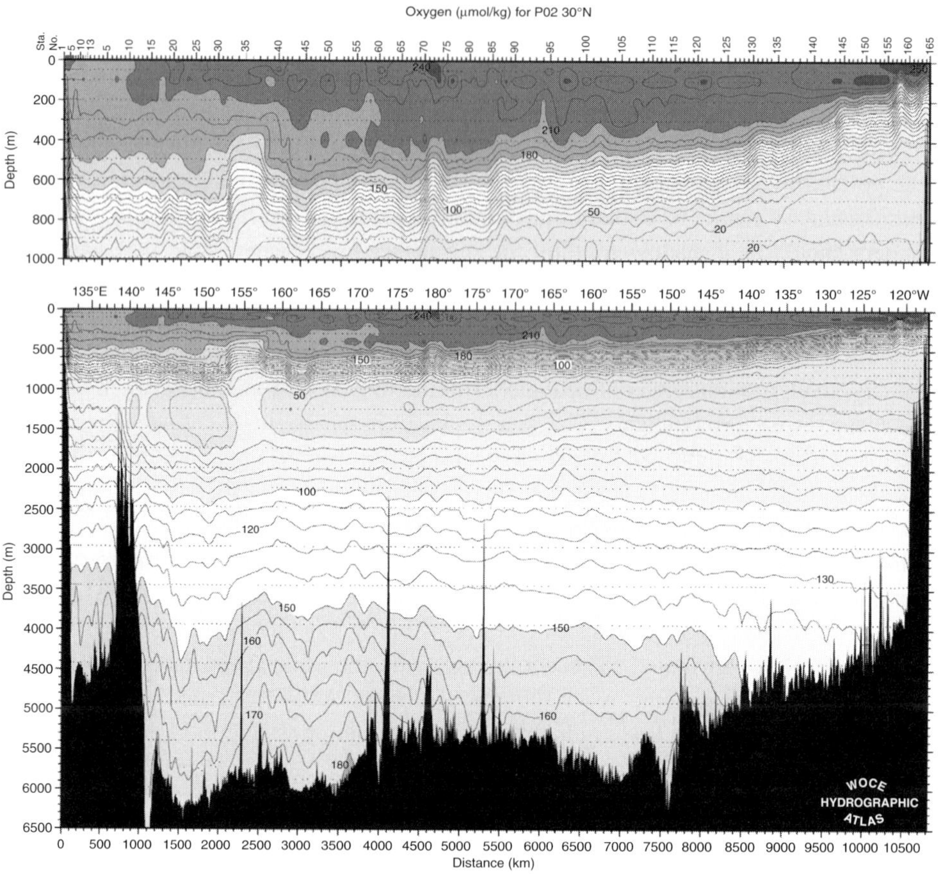

Fig. 4.29 Oxygen content for a 30° N zonal section through the center of the subtropical basin in the North Pacific (Talley, 2007). See color plate section.

and the Sverdrup transport is concentrated in the first layer. Thus, the meridional velocity can be described by

$$\cos\theta h_1 v_1 = a \sin\theta w_e \tag{4.192}$$

and the layer thickness is

$$h_1 = \left(\frac{\rho_3 - \rho_2}{\rho_2 - \rho_1}\right)^{1/2} D_0(\lambda, \theta) \tag{4.193}$$

The basic structure of the ventilated thermocline

The typical structure of the solutions of the ventilated thermocline model is shown in Figure 4.30. The model is driven by a simple sinusoidal Ekman pumping $w_e = w_0 \sin(\pi(y - y_s)/(y_n) - y_s)$, $w_0 = 10^{-6}$ m/s, $g' = 0.01$ m/s^2, and $\rho_3 - \rho_2 = \rho_2 - \rho_1$

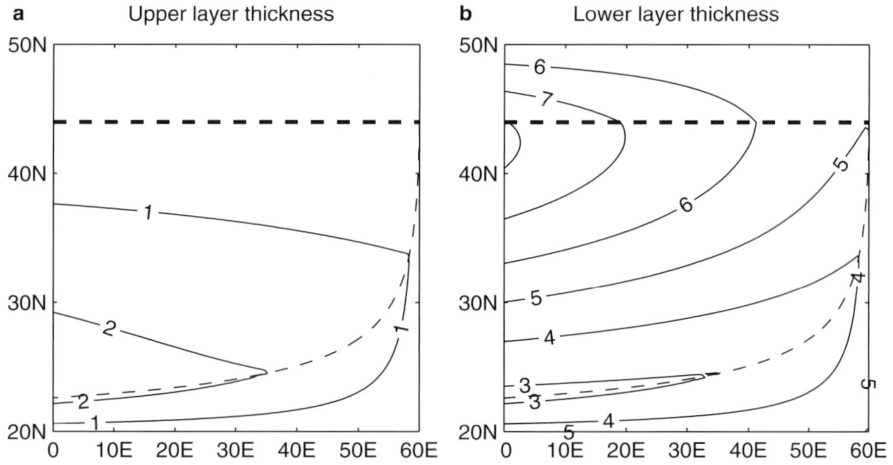

Fig. 4.30 Layer thickness (in 100 m) of a two-layer ventilated thermocline. The upper layer outcrops along 44° N, indicated by the *heavy dashed line*, and the shadow zone is depicted by the *thin dashed line* in the southeast part of the basin.

(so that $\gamma = 2$); the lower layer thickness along the eastern boundary is 500 m, and the upper layer outcrops along 44° N. Note that south of the outcrop line and near the western boundary there is water coming out from the western boundary currents whose potential vorticity cannot be determined by the interior model alone, and the region occupied by such water is called the pool zone. Strictly speaking, we do not really know the solution within the pool region. The solution drawn in Figure 4.30 is obtained under the assumption that south of the outcrop line and in the pool zone, the potential vorticity function in the second layer remains in the same form as that east of the pool zone. In fact, such a solution can be obtained if one extends the model basin beyond the current western boundary and uses part of the solution for this enlarged basin, which falls within the boundary of the current model basin.

According to the model, therefore, there are different dynamical regions in the second layer: the region north of the outcrop line, where water in the second layer is directly exposed to the Ekman pumping; south of the outcrop line there are three regions in the second layer: the pool region, where water comes from the western boundary region; the ventilated region, where water continues its southward motion, started from the open window north of the outcrop line where the second layer is directly exposed to Ekman pumping; and an eastern region, which consists of a shadow zone below the moving uppermost layer.

Note that although thickness contours in both layers are continuous, their slope is discontinuous across the boundary of the shadow zone, as shown in Figure 4.30; however, streamlines in the upper layer are continuous across this line, indicated by $2h_1 + h_2$ (because $\gamma = 2$ for the present case) contours in Figure 4.31, but the velocity is discontinuous across this line. Discontinuities in layer slope and velocity across the boundary of the shadow zone indicate changes in dynamics across this boundary.

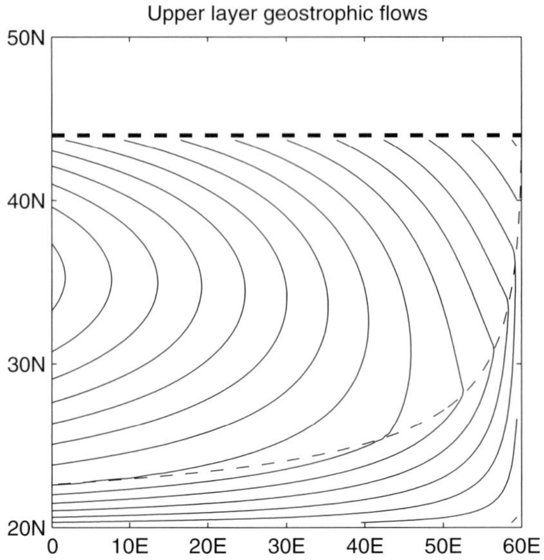

Fig. 4.31 Geostrophic flows in the upper layer, with the outcrop line and the boundary of the shadow zone depicted by *dashed lines*.

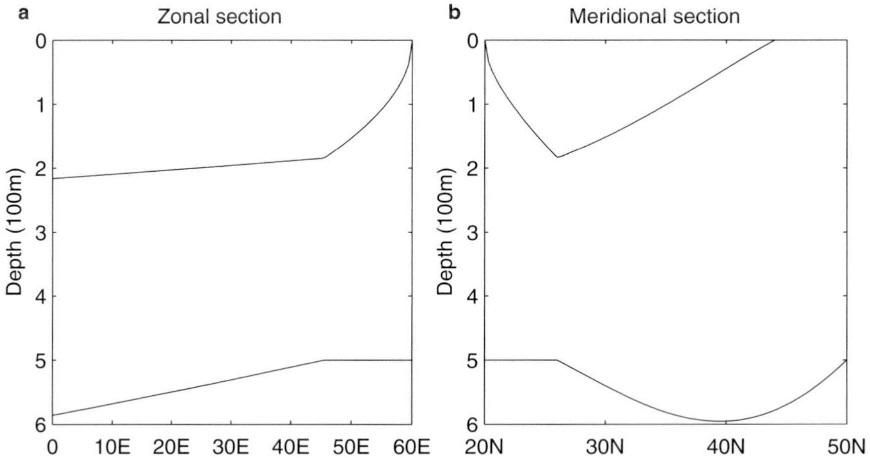

Fig. 4.32 Layer thickness along **a** a zonal section along 26° N and **b** a meridional section along 45° E of the two-ventilated-layer model. The shadow zones are indicated by a flat lower interface at the left edge of the zonal section and at the southern edge of the meridional section.

The vertical structure of solution is shown in Figure 4.32, where the shadow zone, corresponding to the places in the southeast part of the basin where the lower interface is flat, is clearly seen.

Potential vorticity maps for unventilated layers

We have assumed that the third layer is stagnant in the analysis above. Although a motionless third layer is a consistent assumption, there is a possibility that the third layer is in motion yet all the dynamical constraints are not violated. In the past, studies based on layer models have been concentrated on cases when the interfacial deformation is small and can be treated as a small perturbation. Consequently, potential vorticity contours in the subsurface layers are dominated by planetary vorticity, the β term. In the thermocline theory what we are dealing with is characterized by large layer thickness change; thus, the interfacial deformation is no longer negligible.

Let us use the same Ekman pumping field and assume that along the eastern boundary both the second and third layer thicknesses are equal to 500 m. If we assume that the third layer is still motionless and calculate the potential vorticity contours in the third layer (Fig. 4.33), then

$$\Pi_3 = \frac{\sin\theta}{h_{2e} + h_{3e} - \sqrt{D_0^2 + h_{2e}^2}} \tag{4.194}$$

Near the eastern boundary $D_0^2 \to 0$, so potential vorticity is controlled by the planetary vorticity $\sin\theta / h_{3e}$. As a result, some potential vorticity contours emerge from the eastern

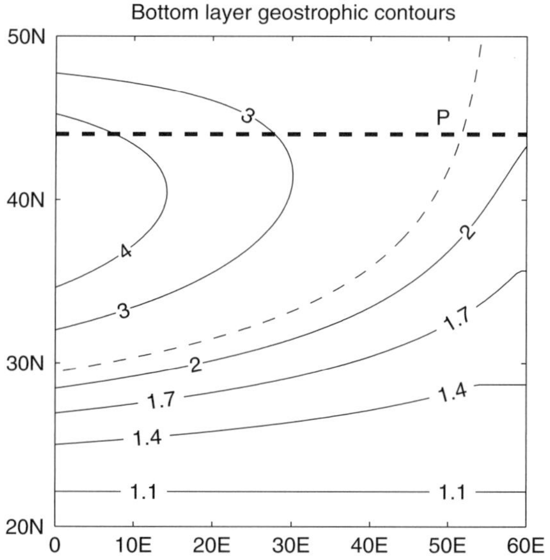

Fig. 4.33 Isolines of potential vorticity in the bottom layer, with the *heavy dashed line* indicating the outcropping line of the upper layer. Although east of the *thin dashed line* potential vorticity contours are blocked by the eastern boundary, west of the *thin dashed line* potential vorticity contours disconnected from the eastern boundary are closed through the western boundary.

boundary, along which geostrophic motion is forbidden. However, there are contours stemming from the western boundary as well. For simplicity, we assume that these contours are closed through the western boundary without being affected by friction. According to the ideal-fluid theory, water is free to travel along these contours.

The essential feature of this vorticity map is that the meridional vorticity gradient has different signs on the two sides of intersecting point P between these two dashed lines, whose location is determined by

$$\frac{d\Pi_3}{d\theta} = 0, \quad at\ \theta = \theta_0 \tag{4.195}$$

Note that near the inter-gyre boundary, $\sqrt{D_0^2 + h_{2e}^2} \approx h_{2e}$. After some algebraic manipulations, one obtains

$$\lambda_e - \lambda_p = \frac{\cos\theta_0 g' h_{2e} h_{3e}}{4\omega a^2 \sin^3\theta_0 \left(\frac{\partial w_e}{\partial \theta}\right)_{\theta=\theta_0}} \tag{4.196}$$

Point P is sometimes called the Rossby repellor, where characteristics from the eastern boundary and the western boundary meet; it is a very important singular point in the thermocline theory. The characteristic starting from P separates the basin into two regions: the eastern region where vorticity contours start from the eastern boundary and geostrophic motion is forbidden, and the western region where vorticity contours start from the western boundary and fluid is free to travel along these contours, assuming no friction in the western boundary current – an idealization widely used to construct a solution in the ocean interior in ideal-fluid thermocline theory.

Note that the existence of the western region depends on the smallness of h_{2e} and h_{3e} and strong forcing w_e. If the layers are too thick and forcing is not so strong, *P* would be located to the west of the western wall, so there would be no closed geostrophic contours. This is the case which has been explored many times before. It is also interesting to observe that for a given forcing w_e and h_{2e}, we can always choose a h_{3e} so small that the Rossby repellor is located within the basin interior. Therefore, closed potential vorticity contours always exist if the layer thicknesses in the model are chosen properly.

Potential vorticity homogenization for the pool regime and unventilated layers

The pool regime

The pool region for a ventilated layer is defined as the regime where streamlines start from the outer edge of the western boundary, instead of the outcrop line. Thus, potential vorticity in the pool regime cannot be determined by tracing backward along the streamline to the outcrop line. A more accurate approach would be to include the complicated dynamical processes in the western boundary regime. The dynamics within the pool regime was left untouched in the original ventilated thermocline model by Luyten *et al.* (1983). However,

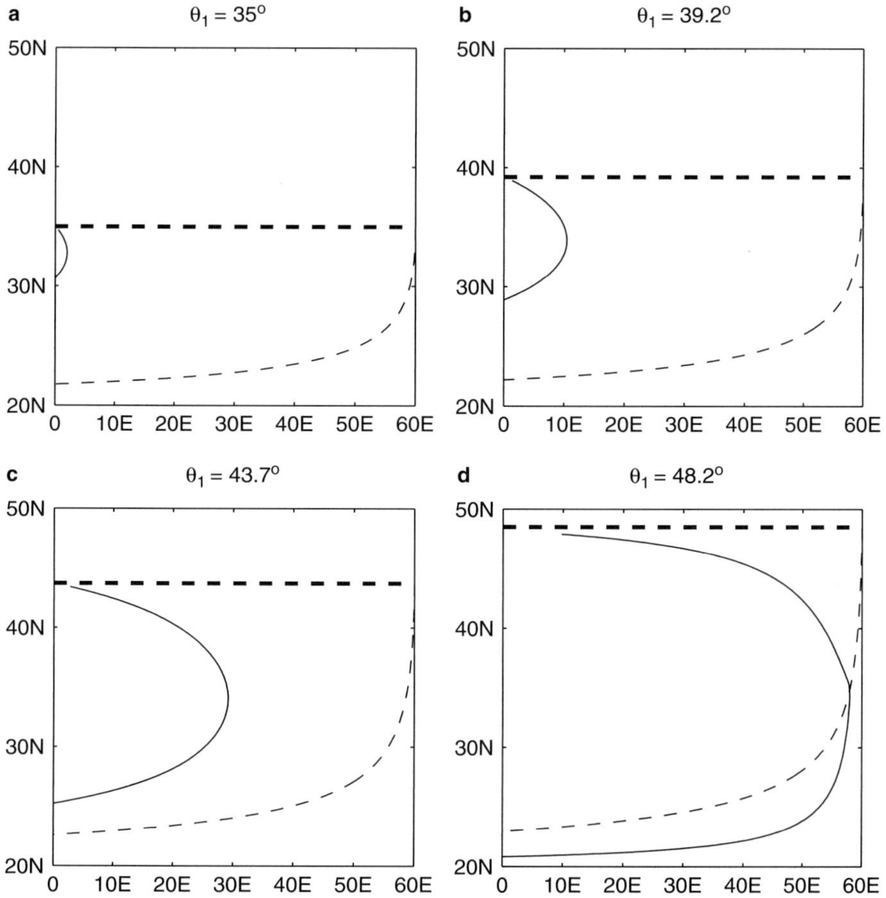

Fig. 4.34 The expansion of the shadow zone *(thin dashed line* is its boundary) and the pool region (west of the *solid line*) obtained from the ventilated thermocline model with two moving layers, as the outcrop line (*thick dashed line*) moves northward. The northern boundary of the model is set at 50° N, but the first outcrop line θ_1 is gradually moved toward the northern boundary.

it can be shown easily that as the outcrop line moves toward the inter-gyre boundary, the area occupied by the pool region also increases quickly, as shown in Figure 4.34. For the case with $\theta_1 = 48.2°$, the pool regime occupies most of the basin, and we cannot ignore this vast regime. To obtain a dynamically consistent solution for the whole basin, it is thus desirable to include the dynamical theory of the pool regime.

Potential vorticity homogenization

The common feature for the pool regime in the ventilated layer and the regime within the closed geostrophic contours in the unventilated layer is that streamlines in these regimes do not start from the outcrop lines. In the second case, streamlines are closed on their own, and we can extend the potential vorticity homogenization theory postulated by Rhines and Young (1982a, b). In the first case, streamlines are actually started from the outer edge of the western boundary. A comprehensive treatment of such a problem is difficult. As a compromise, one can assume that potential vorticity in the pool regime obeys certain laws. Two of the most frequently used assumptions are the following: first, one can leave the solution in the pool regime undetermined; second, one can assume that potential vorticity in the pool regime is also homogenized very much like it is in the case of the unventilated thermocline.

In addition, the pool region in the western part of the basin can also be treated as a ventilated pool, i.e., all the water in this pool is ventilated water from the uppermost layer which is directly in contact with surface forcing (e.g., Dewar, 1986; Dewar *et al.*, 2005). None of these approaches is perfect, but there are no simple alternatives for such a difficult problem.

The potential vorticity homogenization theory presented below follows the approach by Pedlosky (1996). For the case of a model with three moving layers, as shown in Figure 4.28, the potential vorticity balance in the third layer can be written as

$$\nabla_h \cdot (\vec{u}_3 h_3 \Pi_3) = D_3 \tag{4.197}$$

where D_3 is the vorticity dissipation term. Integrating Eqn. (4.197) over a closed area gives

$$\iint_{A_3} D_3 r \cos\theta d\lambda d\theta = 0 \tag{4.198}$$

because the advection terms integrate to zero. We also assume a special form of vorticity dissipation

$$D_3 = \nabla_h \cdot F_3 \quad and \quad F_3 = -\kappa \nabla_h \Pi_3 \tag{4.199}$$

and thus the integral (4.198) is reduced to

$$\oint_{C_3} \frac{\partial \Pi_3}{\partial n} dl = 0 \tag{4.200}$$

Since dissipation is assumed to be very weak, vorticity is nearly conserved along streamlines. Thus, $\Pi_3 = \Pi_3(p_3)$, and Eqn. (4.200) is further reduced to

$$\frac{\partial \Pi_3}{\partial p_3} \oint k \vec{u}_{3H} \cdot \vec{dl} = 0 \tag{4.201}$$

where $\vec{u}_{3H} \cdot \vec{dl}$ indicates the horizontal velocity projection onto the tangent unit vector of contour C_3. If the fluid is in motion, the integral is non-zero, so

$$\frac{\partial \Pi_3}{\partial p_3} = 0, \quad on \quad C_3 \tag{4.202}$$

By repeating the same argument for all closed contours, one comes to the conclusion that potential vorticity is uniform within the domain defined by the outermost closed potential vorticity contour. Finally, we note that if the water does not move, vorticity is not homogenized.

Note that the potential vorticity homogenization depends on the assumption of small down-gradient dissipation. The solutions with an unventilated layer where potential vorticity is homogenized can be found easily, as follows. Denote the western boundary of the shadow zone in the third layer as Λ_p. East of Λ_p the third layer is motionless, so there is only one moving layer. West of Λ_p potential vorticity is homogenized in the third layer. Since potential vorticity should be a constant along the line started from P, this constant is equal to the potential vorticity along the northern boundary. Accordingly, the third layer thickness is

$$h_3 = \frac{f}{f_0} h_{3e} \tag{4.203}$$

The zonal momentum equations for the second and third layers are

$$2\omega \sin\theta v_2 = g_3' \frac{\partial}{a\cos\theta \partial\lambda} (\gamma_3 h_2 + h_3) \tag{4.204}$$

$$2\omega \sin\theta v_3 = g_3' \frac{\partial}{a\cos\theta \partial\lambda} (h_2 + h_3) \tag{4.205}$$

where

$$g_3' = g\frac{\rho_4 - \rho_3}{\rho_0}, \quad \gamma_3 = \frac{\rho_4 - \rho_2}{\rho_4 - \rho_3} \tag{4.206}$$

The Sverdrup relation for these two layers is

$$\cos\theta (h_2 v_2 + h_3 v_3) = a \sin\theta w_e \tag{4.207}$$

Combining these three equations, one obtains an integral relation for the layer thickness

$$(\gamma_3 - 1)h_2^2 + (h_2 + h_3)^2 = D_0^2 + (\gamma_3 - 1)h_{2e}^2 + (h_{2e} + h_{3e})^2 \tag{4.208}$$

Combining with Eqn. (4.203), one can calculate the layer thickness h_2 and h_3. The velocity field can be calculated from geostrophy.

To simplify the problem, we discuss the case when layer 1 outcrops south of the southern boundary of the model basin; thus, there is actually no layer 1 within our model domain. In the following discussion, we will call layer 2 the upper layer, which now covers the whole model basin, and layer 3 will be called the lower layer. The thicknesses of both layers along the eastern boundary are set to 500 m. The density stratification and Ekman pumping field remain the same as discussed above.

The structure of the solution is shown in Figure 4.35. It can readily be seen that the solution consists of two dynamical domains separated by the eastern boundary of the potential vorticity homogenization region in the lower layer, which can be easily identified as the domain

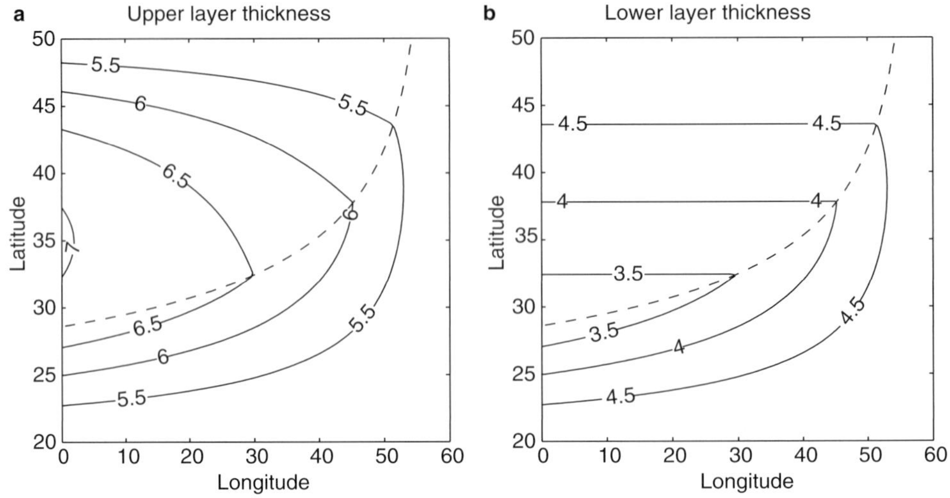

Fig. 4.35 **a** Upper and **b** lower layer thickness (in 100 m) for a $2^1/_2$-layer ventilated thermocline model.

where the lower layer thickness is independent of latitude (northwest part of Fig. 4.35b). The construction of the solution is as follows. Using Eqns. (4.203) and (4.208), we can calculate the solution within the domain of potential vorticity homogenization in the lower layer. Using this method, we can formally calculate the distribution of total layer thickness $h = h_2 + h_3$, and the eastern boundary of the domain of potential vorticity homogenization is the line where $h = h_{2e} + h_{3e}$ because east of this line the lower layer is stagnant, so the lower interface should be flat.

When the third layer is in motion, part of the Sverdrup flux is now distributed into the third layer. Consequently, over the potential vorticity homogenization zone in the third layer, the volume flux in the second layer is reduced, and so is the interfacial deformation. For example, in the present case, layer 2 has a thickness of approximately 460 m at 46° N and near the western boundary (Fig. 4.35a); however, the corresponding layer thickness for the case with no closed contour in layer 3 is slightly larger than 700 m (Fig. 4.30b). East of the closed geostrophic contours in the lower layer, layer thicknesses in these two layers are exactly compensating each other, indicating a flat interface below layer 2.

Basic features of the ventilated thermocline and beyond

The ventilated thermocline theory, combined with the potential vorticity homogenization theory, provides a rigorous framework for wind-driven circulation in the mid-latitude oceans. It has become the theoretical cornerstone for the modern theory of wind-driven circulation. The most important features of the theory and the development beyond its original form are summarized as follows.

The basic structure of the ventilated thermocline

The wind-driven circulation in the upper kilometer of the ocean consists of the following major components:

- *The Ekman layer* on the top, which is directly in touch with atmospheric forcing, including wind stress, heat flux, and freshwater flux. This layer plays the role of forming water parcels with suitable density and pumping them down into the geostrophic flow regime below.
- *The geostrophic regime* below the Ekman layer consists of several regimes which have distinctly different dynamics:
  - *The ventilated thermocline*, where water parcels are exposed to direct atmospheric forcing within the outcrop window for each isopycnal in late winter. South of the outcrop line, water parcels in the ventilated layer are subducted into the subsurface ocean where they continue their movement dictated by the Sverdrup dynamics. In particular, the potential vorticity of an individual water parcel is conserved along its trajectory.
  - *The shadow zone*, where water is stagnant within the framework of the ventilated thermocline. Of course, water there can be set in motion as part of the thermohaline circulation in the world's oceans, but this is beyond the scope of the ventilated thermocline.
  - *The pool regime* or *the potential vorticity homogenized regime*, where potential vorticity is set by processes other than ventilation through Ekman pumping.

Beyond the simple ventilated thermocline:

- **Sea surface density ρ_s is specified *a priori* in terms of the late-winter value.** ***Comment:*** There is no seasonal cycle in the model. This assumption is based on the Stommel demon, i.e., the model is forced by late-winter mixed-layer properties. An alternative is coupling a ventilated thermocline model to a mixed layer with the seasonal cycle so that ρ_s can be determined as part of the solution at the same time (e.g., Marshall and Nurser, 1991; Liu and Pedlosky, 1994). However, some complicated nonlinear processes, such as the convective adjustment in late winter, are difficult to deal with in an analytical model, and this remains one of the most challenging questions.
- **Ekman pumping is specified as the upper boundary condition**, so that the Ekman flux is counted as a separate component of the wind-driven circulation. ***Comment:*** One may choose to specify the wind stress as the upper boundary condition. In this way, the Ekman flux is counted as part of the upper layer flow, so the solution may look different from what is obtained from the previous formulation.
- **The fluid is assumed to be an ideal fluid.** Therefore density, potential vorticity, and Bernoulli function are conserved along the trajectories after subduction. ***Comment:*** Data analysis based on hydrographic observations has indicated that mixing across streamlines in the oceans renders the density, potential vorticity, and Bernoulli function non-conservative. However, the results from an ideal-fluid-based model can provide useful insights into the physical mechanisms regulating the circulation.
- **No interfacial mass flux was allowed in the original model.** ***Comment:*** Interfacial mass flux can be included, and its inclusion is a simple way to study the interaction between the wind-driven circulation and the thermohaline circulation. However, this problem is intimately related to the mechanical energy required to sustain vertical mixing in the stratified oceans, which involves complicated dynamical processes, as will be discussed in later sections.

- **No mixed layer is included in the original model.** In fact, the Ekman layer is assumed to be infinitely thin and the upper surface of the ventilated thermocline model is put at $z = 0$. ***Comment*:** This is a major deficit in the original model because all ventilated layers are forced to outcrop along the eastern boundary and have zero thickness there. As a result, the eastern boundary appears as a singular line for the model. In addition, the assumption of a zero-thickness mixed layer can substantially distort the structure of the wind-driven gyre. However, a mixed layer of finite thickness can be added and specified *a priori* as the upper boundary condition. This will be discussed for the case of a continuously stratified model.
- **Stratification along the eastern boundary must be specified. *Comment*:** This is one of the basic assumptions of the ideal-fluid thermocline theory. For the case of a reduced-gravity model with a single moving layer, one needs to specify the layer thickness as an external parameter; thus, it is conceivable that a model with n moving layers should require the specification of thickness of n layers along the eastern boundary.
- **The shadow zone near the eastern boundary is a unique feature in the ventilated thermocline model.** The boundary of the shadow zone is a characteristic of the fluid stemming from the intersection between the outcrop line and the eastern boundary. Each time this characteristic meets a new outcrop line, it splits into two characteristics, separating the flow into zones where the potential vorticity function has different structures. The number of zones with different dynamics (in terms of the potential vorticity function) doubles each time a characteristic meets a new outcrop line. Thus, as the number of ventilated layers increases, the complexity of the ventilated thermocline becomes exponential.
- **The domain where the model is applicable in the oceans is limited to the ocean interior**, away from the western boundary and the recirculation regime. The fundamental limitation of the model is that the model is based on linear Sverdrup dynamics, without including frictional effects and the nonlinear advection terms.
- **The model cannot be used for the western boundary regime** where other high-order dynamics should be included. ***Comment:*** Simple frictional boundary layers cannot be matched to the ventilated thermocline model to produce a closed circulation in a basin, because complicated dynamical and thermodynamic processes taking place in the mixed layer are omitted from the model; thus, streamlines from the outer edge of the western boundary are disconnected from the streamlines that outcrop in the mid ocean and away from the western boundary.
- **The model is not valid in the recirculation regime** where the nonlinear advection term becomes dominantly important. ***Comment*:** All models based on linear Sverdrup dynamics fail within the recirculation regime. The dynamics of the recirculation regime remains one of the major challenges for dynamical oceanography.

4.1.8 Multi-layer inertial western boundary currents

Inertial western boundary current with two moving layers

Following the success in solving the one-layer inertial western boundary current, an attempt was made to find solutions for the two-moving-layer inertial western boundary currents; however, efforts in this direction went through quite a struggle. Blandford (1965) found analytical solutions for the case when potential vorticity was constant in both moving layers; however, his solution exhibited an unexpected separation at a latitude much lower than that from a single-moving-layer model.

Basically, he found that the equation system for the two-moving-layer inertial western boundary current layers is ill-conditioned (sometimes also called a "stiff equation system"). The difficulty in dealing with the so-called ill-conditioned equation is as follows. As will be explained shortly, when potential vorticity is constant, the general solutions for inertial western boundary layers are in the form of exponential function of the x-coordinate, such as $\exp(\pm\lambda_i x)$. Since the boundary-layer solutions have to be found out by matching them to the interior solution at infinity, exponential solutions with large λ_i can grow very quickly and are therefore difficult to handle numerically. Although he used the best numerical packages available at that time, he was unable to find smooth solutions which separate from the western boundary around the latitude expected from the single-moving-layer model. This puzzle remained unsolved for two decades.

Luyten and Stommel (1985) explained the paradoxical situation in terms of the virtual control first found in the one-dimensional hydraulic problems by Woods (1968). The Luyten–Stommel model is based on an assumption of zero-potential vorticity which is not realistic. Since the relative vorticity is negligible for the interior solution, potential vorticity is reduced to the form of f/h; a zero-potential-vorticity solution can be traced back upstream to a place with either $f = 0$ (the equator) or $h = \infty$ (infinite layer depth). As discussed in this section, however, the difficulties associated with the inertial western boundary layer can be solved for the general case of non-zero potential vorticity. In fact, using the stream-function coordinate transformation reduces the ill-conditioned differential equation system into a well-behaved system, which can be solved readily.

Model formulation

Similar to the case of an inertial western boundary current with a single moving layer, the basic equations for a model with $2\frac{1}{2}$ layers include the geostrophic condition for the cross-stream momentum equations, the ageostrophic condition for the downstream momentum equations, and the continuity equations for both layers. On a β-plane these equations are

$$-fv_1 = -g'(\gamma h_{1x} + h_{2x}) \tag{4.209}$$

$$u_1 v_{1x} + v_1 v_{1y} + fu_1 = -g'\left(\gamma h_{1y} + h_{2y}\right) \tag{4.210}$$

$$(h_1 u_1)_x + (h_1 v_1)_y = 0 \tag{4.211}$$

$$-fv_2 = -g'(h_{1x} + h_{2x}) \tag{4.212}$$

$$u_2 v_{2x} + v_2 v_{2y} + fu_2 = -g'\left(h_{1y} + h_{2y}\right) \tag{4.213}$$

$$(h_2 u_2)_x + (h_2 v_2)_y = 0 \tag{4.214}$$

where $g' = g(\rho_3 - \rho_2)/\bar{\rho}$ and $\gamma = (\rho_3 - \rho_1)/(\rho_3 - \rho_2)$; and ρ_1, ρ_2, and ρ_3 are the densities of the first layer (on the top), the second layer, and the lowest layer, which is assumed to be very thick and motionless.

The system is in semi-geostrophy because the downstream velocity is geostrophically balanced by the cross-stream pressure gradient, but the cross-stream velocity obeys the ageostrophic constraint. The potential vorticity equation in both layers can be derived by cross-differentiating and subtracting the corresponding momentum equations:

$$Q_1 = (v_{1x} + f)/h_1 = F_1(\psi_1) \tag{4.215}$$

$$Q_2 = (v_{2x} + f)/h_2 = F_2(\psi_2) \tag{4.216}$$

Similarly, the Bernoulli function in each layer is also conserved:

$$B_1 = v_1^2/2 + g'(\gamma h_1 + h_2) = G_1(\psi_1) \tag{4.217}$$

$$B_2 = v_2^2/2 + g'(h_1 + h_2) = G_2(\psi_2) \tag{4.218}$$

where ψ_1 and ψ_2 are streamfunctions defined from the continuity equations

$$\psi_{ix} = h_i v_i; \quad \psi_{iy} = -h_i u_i, \quad i = 1, 2 \tag{4.219}$$

Furthermore, adding Eqns. (4.209) and (4.212) and zonally integrating leads to a simple integral constraint connecting the barotropic streamfunction in both layers with the layer thicknesses

$$(\psi_{1\infty} + \psi_{2\infty}) - (\psi_1 + \psi_2) = \frac{g'}{f}\left[\frac{\gamma\left(h_{1\infty}^2 - h_1^2\right)}{2} + h_{1\infty}h_{2\infty} - h_1 h_2 + \frac{h_{2\infty}^2 - h_2^2}{2}\right] \tag{4.220}$$

where subscript ∞ indicates the interior solution. Using the potential vorticity relations, the downstream momentum equations can be rewritten as a pair of second-order ordinary differential equations for layer thicknesses h_1 and h_2:

$$\gamma h_{1xx} + h_{2xx} - \frac{f}{g'}F_1(\psi_1)h_1 = -\frac{f^2}{g'} \tag{4.221}$$

$$h_{1xx} + h_{2xx} - \frac{f}{g'}F_2(\psi_2)h_1 = -\frac{f^2}{g'} \tag{4.222}$$

These two equations consist of a high-order system of ordinary differential equations. Unfortunately, this is a stiff equation system over the semi-infinite domain $[0, \infty]$. For example, when F_1 and F_2 are constant, Eqns. (4.221) and (4.222) possess solutions in the form of $\exp(\pm\lambda_i x)$, $i = 1, 2$, where $\{\lambda_i\}$ are the eigenvalues of the system. When $\{\lambda_i\}$ is large, the system is difficult to solve numerically, so it is called a stiff (or ill-conditioned) equation system.

Blandford used this formulation and some software packages specially designed for solving stiff equations; however, most of the solutions he found unexpectedly separated from the western boundary and he could not find solutions that were continuous. Note that numerical integration of such a stiff system is rather difficult, and this is probably the very reason why Blandford could not find smooth solutions.

Streamfunction coordinate transformation

Blandford's failure to find continuous solutions for multi-layer inertial western boundary currents may be partially related to the numerical technique he used. Mathematically, some stiff equations may become much easier to solve under certain coordinate transformations. In the present case, when dealing with an inertial western boundary current, Charney's (1955) paper has been quoted very often; however, the technique used by Charney has seldom been used in other studies. In fact, what Charney used was a streamfunction coordinate transformation which is well known in classical textbooks on fluid dynamics and which can be traced back to von Mises (1927). In the case of a one-moving-layer model, the streamfunction transformation gives the solution in closed analytical forms for a general potential vorticity profile. For the case of two moving layers, we can introduce the following transformation

$$d\psi_1 = h_1 v_1 dx \tag{4.223}$$

In the new streamfunction coordinate ψ_1, Eqns. (4.215) and (4.219) are reduced to a set of two first-order ordinary differential equations plus three algebraic relations (all in nondimensional form):

$$\frac{dv_1^2}{d\psi_1} = 2\left[F_1(\psi_1) - \frac{f}{h_1}\right] \tag{4.224}$$

$$\frac{d\psi_2}{d\psi_1} = h_r \frac{h_2 v_2}{h_1 v_1} \tag{4.225}$$

$$\frac{\gamma}{2}h_1^2 + h_r h_2 h_2 + \frac{h_r^2}{2}h_2^2 = \frac{\gamma}{2} + h_r + \frac{h_r^2}{2} - (\psi_{1\infty} + \psi_{2\infty} - \psi_1 - \psi_2) \tag{4.226}$$

$$v_1^2/2 + \gamma h_1 + h_r h_2 = G_1(\psi_1) \tag{4.227}$$

$$v_2^2/2 + h_1 + h_r h_2 = G_2(\psi_2) \tag{4.228}$$

subject to the following boundary conditions:

$$h_1(\psi_{1\infty}) = h_2(\psi_{2\infty}) = 1 \tag{4.229}$$

$$v_1(\psi_{1\infty}) = v_2(\psi_{2\infty}) = 0 \tag{4.230}$$

$$\psi_2(0) = 0, \quad \psi_2(\psi_{1\infty}) = \psi_{2\infty} \tag{4.231}$$

where $h_r = h_{2\infty}/h_{1\infty}$ is the ratio of the dimensional layer thickness of the interior solution. This system is a well-behaved system and can be solved easily, and continuous solutions can also be found. For example, two sections of a continuous solution are shown in Figure 4.36. In the physical coordinate x, the solution exponentially matches the interior solution at the outer edge of the western boundary. On the other hand, this matching condition appears in the form of a linear function at a finite distance from the western boundary in the streamfunction coordinate. Since the solution of the inertial western boundary has to be sought through a

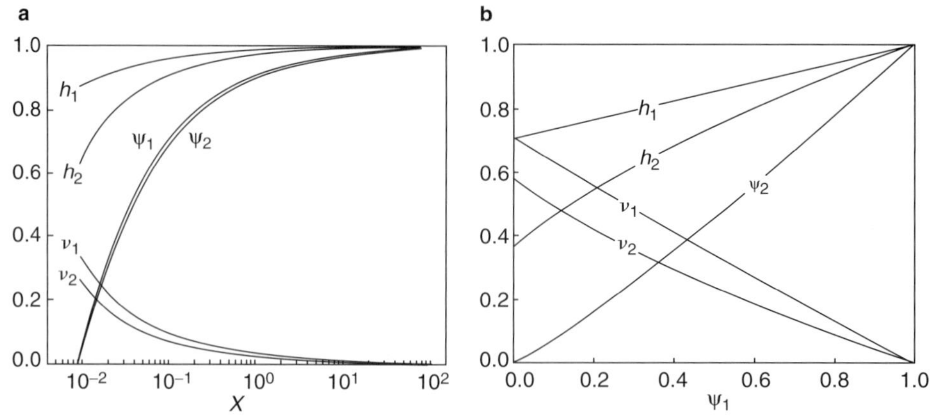

Fig. 4.36 Structure of the inertial western boundary current with two moving layers at two sections, including layer thickness, meridional velocity, and streamfunction; **a** solutions in the physical coordinate x at $y = 0.25$; **b** solutions in the streamfunction coordinate at $y = 0.55$ (Huang, 1990b).

shooting method, using the streamfunction coordinate has a great advantage in numerical calculations.

Matching the inertial western boundary current with the mid-ocean thermocline

Motivation

We have discussed thermocline structure in the ocean interior and an inertial western boundary current with two moving layers. It has been argued that mixing/dissipation within the southern part of the western boundary regime is negligible, so we would like to match the thermocline solution for the ocean interior with some kind of western boundary current. This is one more step toward constructing a unified picture for the circulation in a closed basin.

Model formulation

The model ocean (Fig. 4.37) consists of three layers of constant density, and the lowest layer is very thick and assumed motionless. The subtropical gyre interior is divided into three domains that have slightly different dynamics.

The dynamics of the model is basically the same as the classical ventilated thermocline by Luyten *et al.* (1983). However, we will assume that potential vorticity in the second layer is uniform for all the subducted water. This is not inconsistent with the observation that potential vorticity in the deep part of the Gulf Stream is practically homogenized. Potential vorticity in the uppermost layer is not uniform because it is directly exposed to surface forcing. Our assumption gives rise to a rather elegant solution for the outcropping line, the western boundary of the shadow zone, and layer thickness.

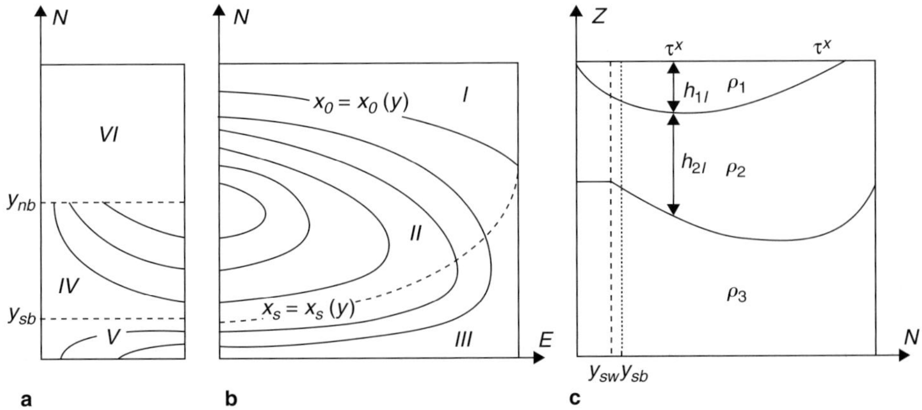

Fig. 4.37 Sketch of a model ocean for a subtropical gyre: **a** the inertial western boundary current; **b** the interior ocean with two moving layers, where $x_0(y)$ is the outcrop line for the upper layer, and $x_s(y)$ is the boundary of the shadow zone for the second layer; **c** a meridional section of the model ocean with three layers (Huang, 1990c).

Domain I

North of the outcrop line $x_0 = x_0(y)$, the first layer vanishes, so only the second layer is in motion. The wind stress is treated as a body force for the upper layer, and the basic equations are

$$-fh_2v_2 = -\gamma_2 h_2 h_{2x} + \tau^x/\rho_0 \tag{4.232}$$

$$fh_2u_2 = -\gamma_2 h_2 h_{2y} \tag{4.233}$$

$$(h_2u_2)_x + (h_2v_2)_y = 0 \tag{4.234}$$

where $\gamma_2 = g(\rho_3 - \rho_2)/\rho_0$. Cross-differentiating gives the Sverdrup relation

$$\beta h_2 v_2 = -\tau_y^x/\rho_0 \tag{4.235}$$

and the layer thickness satisfies

$$h_2^2 = h_e^2 + \frac{2f^2}{\rho_0\beta\gamma_2}\left(\frac{\tau^x}{f}\right)_y (x_e - x) \tag{4.236}$$

where h_e is the constant layer thickness along the eastern boundary of the basin.

Since potential vorticity is constant along the outcrop line,

$$f/h_2 = f_0/h_e, \quad along\ x_0\,(y) \tag{4.237}$$

Accordingly, the outcrop line satisfies

$$x_e - x_0\,(y) = \rho_0\beta\gamma_2 \frac{(f/f_0)^2 - 1}{2f^2\,(\tau^x/f)_y} h_e^2 \tag{4.238}$$

Domain II

South of $x_0(y)$ and north of $x_s(y)$ both layers are in motion, so the basic equations are

$$-fh_1v_1 = -h_1\left[(\gamma_1+\gamma_2)h_{1x} + \gamma_2 h_{2x}\right] + \tau^x/\rho_0 \tag{4.239}$$

$$fh_1u_1 = -h_1\left[(\gamma_1+\gamma_2)h_{1y} + \gamma_2 h_{2y}\right] \tag{4.240}$$

$$(h_1u_1)_x + (h_1v_1)_y = 0 \tag{4.241}$$

$$-fh_2v_2 = -h_2\left[\gamma_2 h_{1x} + \gamma_2 h_{2x}\right] \tag{4.242}$$

$$fh_2u_2 = -h_2\left[\gamma_2 h_{1y} + \gamma_2 h_{2y}\right] \tag{4.243}$$

$$(h_2u_2)_x + (h_2v_2)_y = 0 \tag{4.244}$$

where $\gamma_1 = g(\rho_2 - \rho_1)/\rho_0$. After some manipulations, the layer thicknesses satisfy

$$\gamma_2(h_1+h_2)^2 + \gamma_1 h_1^2 = \gamma_2 h_e^2 + \frac{2f^2}{\rho_0\beta}\left(\frac{\tau^x}{f}\right)_y (x_e - x) \tag{4.245}$$

We further assume that the outcrop line $x_o(y)$ has a special shape, so that potential vorticity in the second layer is constant after subduction. This assumption is supported by observations that potential vorticity in the Gulf Stream is nearly constant (Iselin, 1940; Huang and Stommel, 1990). Thus, the layer thickness is

$$h_1 = \frac{h_e}{\gamma_1+\gamma_2}\left(-\gamma_2\frac{f}{f_0} + \Delta^{1/2}\right) \tag{4.246}$$

where

$$\Delta = \left(\gamma_2\frac{f}{f_0}\right)^2 - (\gamma_1+\gamma_2)\left\{\gamma_2\left[(f/f_0)^2 - 1\right] - \frac{2f^2}{\rho_0\beta h_e^2}\left(\frac{\tau^x}{f}\right)_y (x_e - x)\right\} \tag{4.247}$$

Domain III

South of line $x_s(y)$, the second layer is stagnant, so only the first layer is in motion. The upper layer thickness satisfies

$$h_1^2 = \frac{2f^2}{\rho_0\beta\gamma_1}\left(\frac{\tau^x}{f}\right)_y (x_e - x) \tag{4.248}$$

Since the second layer is stagnant, its thickness satisfies $h_2 = h_e - h_1$ along the western edge of this domain; thus, this boundary is determined by the following equation

$$x_e - x_s(y) = \rho_0\beta\gamma_1 h_e^2\left(1 - \frac{f}{f_0}\right)^2\left[2f^2\left(\frac{\tau^x}{f}\right)_y\right]^{-1} \tag{4.249}$$

Domain IV

This is the southern part of the western boundary regime where two moving inertial boundary currents exist, which can be calculated by the streamfunction coordinate transformation discussed in the previous section.

The interaction between western boundary layer and the interior flow

Similar to the case discussed in the previous section, the western boundary layers are continuous only if the interior solution satisfies some implicit constraints. The model is forced by wind stress

$$\tau^x = -\tau_0 \frac{f}{f_0} \cos\left(\pi y / L_y\right) \tag{4.250}$$

where $\tau_0 = 0.075$ N/m^2. The horizontal dimension of the model basin is $L_x = 6,000$ km, $L_y = 3,000$ km, the stratification parameters are $\gamma_1 = \gamma_2 = 0.015$ m/s^2, $h_e = 500$ m. The solution with continuous western boundary layers is shown in Figures 4.38 and 4.39.

However, if the external parameters (such as the location where the outcrop line intercepts the eastern boundary, the lower layer thickness along the eastern boundary, the stratification parameter γ_1, or the wind stress) are changed, the western boundary layers are interrupted and system behavior can be described in terms of a saddle point in the phase space (Fig. 4.40).

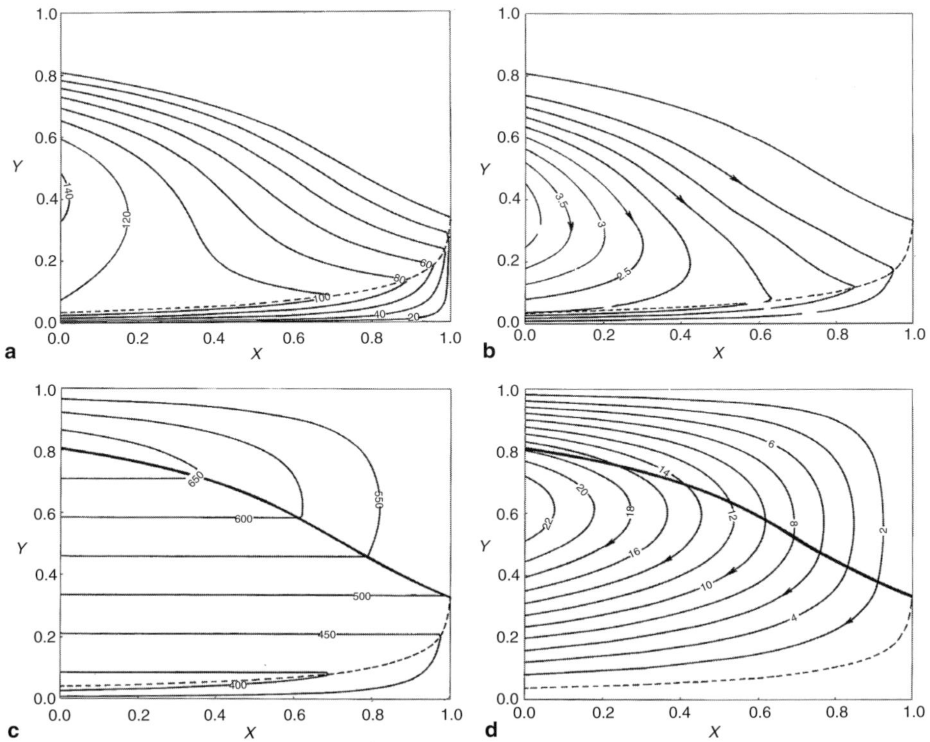

Fig. 4.38 Structure of the interior solution with external parameters $h_e = 500$ m, $y_0 = 0.334925$. The *heavy curve* indicates the outcrop line, and the *dashed curve* the boundary of the shadow zone (Huang, 1990c).

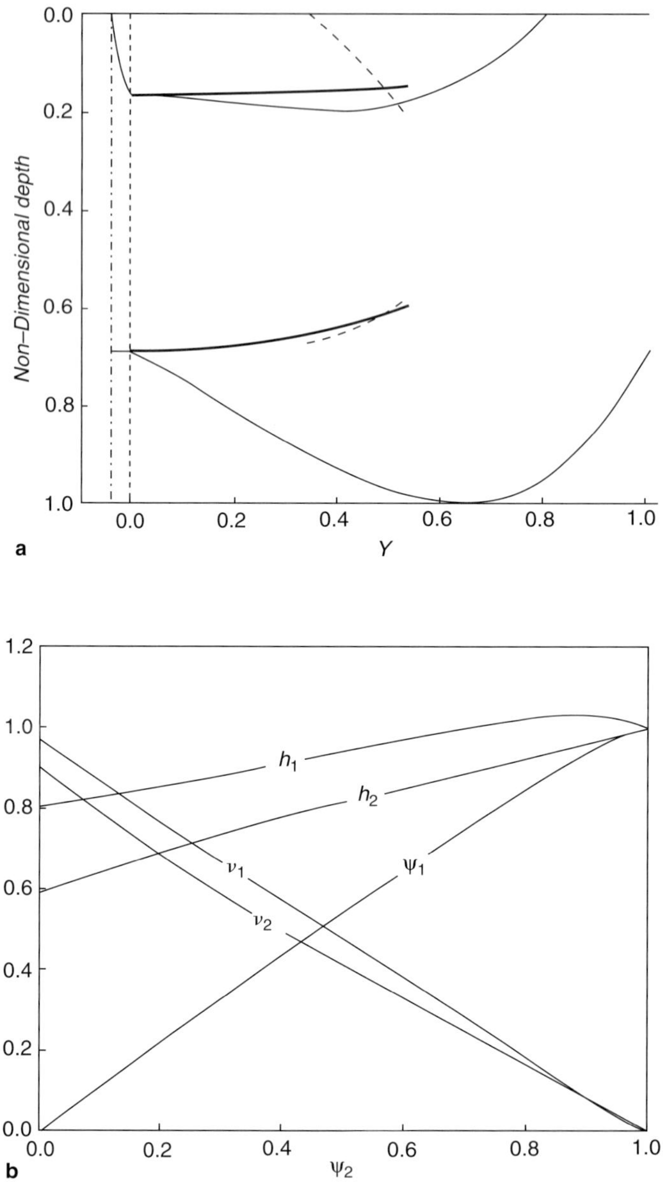

Fig. 4.39 Structure of the western boundary currents: **a** meridional section of the inertial western boundary currents. Y represents the nondimensional meridional coordinates, *thin curves* are layer interfaces at the outer edge of the western boundary, *heavy curves* are interfaces at the western wall, and the *dashed curves* are the non-physical solutions; **b** boundary layer structure in the streamfunction coordinates (Huang, 1990c).

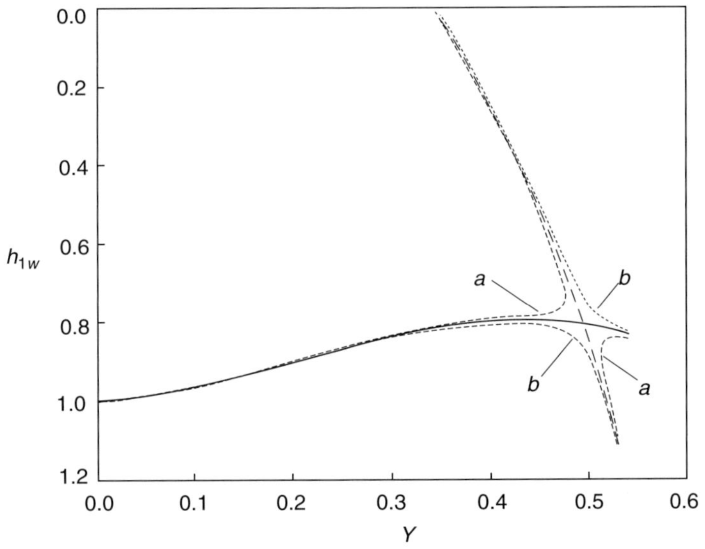

Fig. 4.40 Dependence of the boundary layer structure on the lower layer thickness at the eastern wall. The *solid line* indicates the continuous solution, the *long-dashed line* represents the spurious solution, and the *short-dashed* lines indicate branches of the solutions (Huang, 1990c).

In the classical theories of wind-driven circulation with a single moving layer, the western boundary layer has been assigned a passive role – closing the mass flux and dissipating the extra vorticity. The interior solution is uniquely determined by the wind-stress curl. For a stratified model, the solution also depends on the stratification and surface density distribution, as discussed in the previous sections.

It was surprising that Blandford (1965) failed to find continuous solutions for a two-moving-layer inertial western boundary layer model. Here we have shown that continuity of the western boundary layer imposes a constraint over the mid-ocean thermocline. Although there are many degrees of freedom for the interior thermocline, each time we add on a new piece of the circulation, the system loses some freedom; as more and more pieces are added on, there will be only few possible solutions remaining. These implicit constraints reflect the interaction between the interior flow and other parts of the circulation.

Remarks on closing the ventilated thermocline with western boundary currents

Since the ventilated thermocline theory was first proposed, an inevitable question has been whether such beautiful solutions can be closed along the western boundary. It is, however, very clear that no ideal-fluid model can accomplish such a job because, to close the circulation, we need to include friction/dissipation so that the budget of potential vorticity and energy can be closed for any closed streamlines. Matching an inertial western boundary current with multiple moving layers represents an effort in pushing this issue to the theoretical limit. The circulation obtained from an ideal-fluid model without mixing/dissipation cannot be closed. It is more convenient to use a numerical model, in which the dynamical effect of small diapycnal mixing can be incorporated, to study the circulation in a closed basin. Samelson and Vallis (1997) gave a clear picture of the linkage between the nearly ideal-fluid thermocline in the ocean interior and other parts of the basin-scale circulation through the boundary layers. However, including mixing/dissipation will render the solution analytically rather difficult to solve, and the beauty of an analytical solution hard to achieve.

4.1.9 Thermocline theory applied to the world's oceans

Potential vorticity maps in the oceans

Talley (1985) applied the LPS model to the north subtropical Pacific and successfully explained the observed shallow salinity minimum. The circulation in the Pacific can be seen clearly through maps of potential vorticity on different isopycnal surfaces (Fig. 4.41). Among many other features, potential vorticity $\rho^{-1} f \Delta\rho / \Delta z$ in the upper layers is non-homogenized, which is apparently due to the strong potential vorticity source/sink in these shallow layers; however, potential plateaus can be clearly identified for deeper isopycnal surfaces $\sigma_\theta = 26.0\ \text{kg/m}^3$ and $26.2\ \text{kg/m}^3$. These potential vorticity plateaus appear to be not inconsistent with the potential vorticity homogenization theory. However, the existence of such potential vorticity plateaus in the world's oceans may be due to quite different dynamical processes.

Potential vorticity homogenization diagnosed from numerical models

Primitive equation numerical experiments have been carried out to simulate the circulation dynamics predicted by the ventilated and unventilated thermocline theories. Potential vorticity maps generated from the Geophysical Fluid Dynamics Laboratory's primitive equation model are shown in Figure 4.42. The dashed lines are the Bernoulli contours that can be used as the streamlines for the respective isopycnal surfaces, assuming the dissipation is relatively weak. Within the major part of the subtropical gyre interior, potential vorticity is nearly homogenized.

From Figures 4.41 and 4.42, one can identify all the dynamical regions, such as the pool, the ventilated region, and the shadow zone, proposed by Rhines and Young (1982b) and Luyten *et al.* (1983). On the other hand, we will notice that there are some important discrepancies between the theories and numerical experiments or observations. A common shortcoming of these two theories is that the dynamical effects of strong eddy mixing are ignored for the sake of analytical simplicity. For example, as seen from Figure 4.42, potential vorticity is advected downstream in the form of low/high potential vorticity tongues, consistent with the potential vorticity conservation used in the LPS model. However, potential vorticity is not exactly conserved, as assumed in the ventilated thermocline theory by LPS. In fact, low/high potential vorticity tongues gradually lose their identity owing to cross-stream eddy mixing, as seen from Figures 4.41 and 4.42. There are pools of almost uniform potential vorticity in both Figures 4.41 and 4.42; however, the reason for

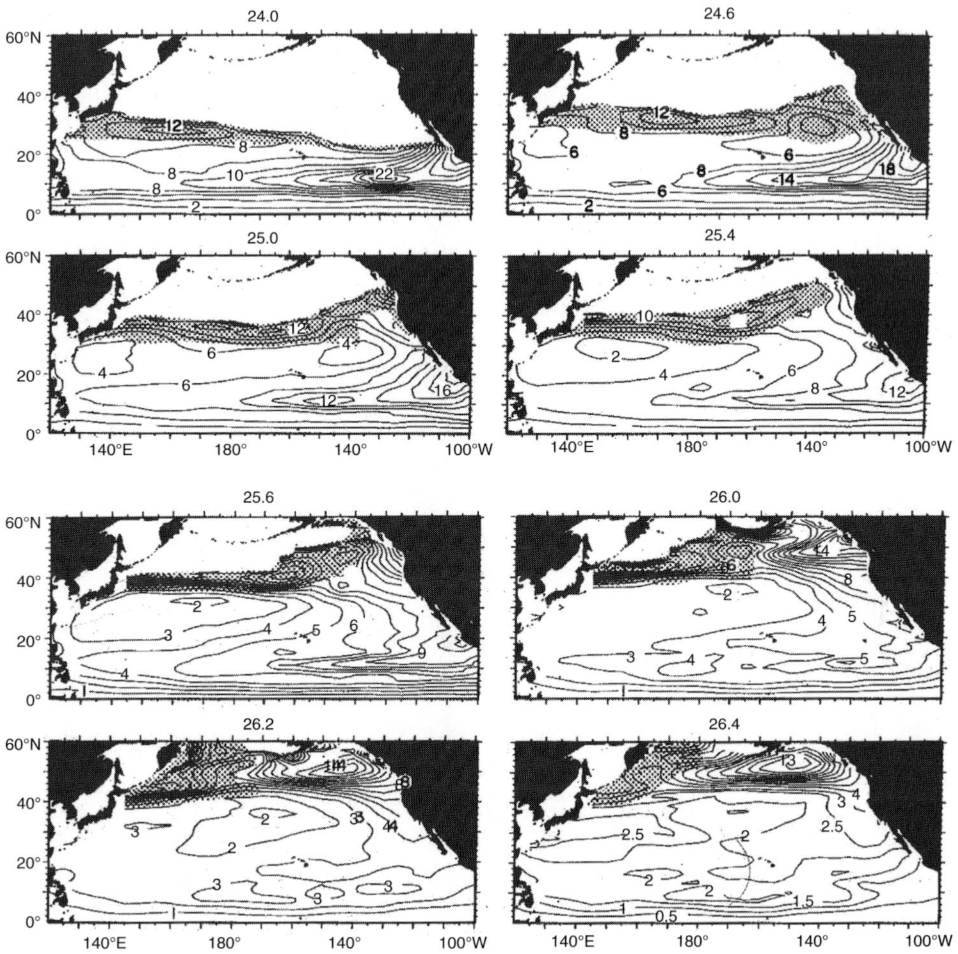

Fig. 4.41 Maps of potential vorticity on selected isopycnals for the North Pacific (in 10^{-10}/m/s). *Shaded regions* are seasonal; their southern edge is the winter sea–surface outcrop (Talley, 1988).

potential vorticity uniformity seems quite different from the original theory by Rhines and Young (1982b).

First, smearing vorticity gradients on the isopycnal surface at the base of the thermocline is relatively fast compared with the circulation time scale. In fact, potential vorticity gradients cause baroclinic instability that tends to homogenize potential vorticity on isopycnal surfaces within just part of the trajectory around a subtropical gyre.

The potential vorticity homogenization process was simulated by an eddy-resolving primitive equation model with idealized topography and forcing (Cox, 1985) (Fig. 4.43). The results from the non-eddy-resolving and eddy-resolving models gave rise to the same basic

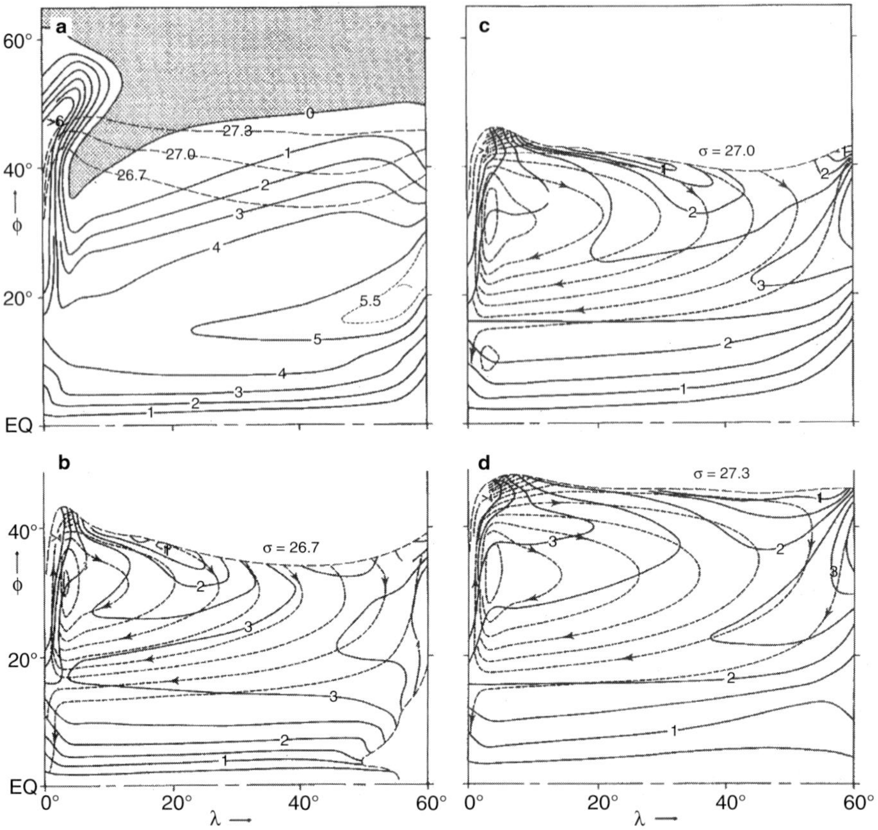

Fig. 4.42 Potential vorticity ($-f\sigma_z$) at $z = -95$ m (**a**), and on the 26.7, 27.0, and 27.3 σ surfaces (**b**, **c**, **d**) in 10^{-10}/m/s (*solid lines*). The Bernoulli function with contour interval 4 cm (equivalent vertical displacement) is shown as *short dashed lines with arrows* in **b**, **c**, **d**. The intersection of the sigma surfaces with $z = -95$ m is shown with *long dashed lines* (Cox and Bryan, 1984).

flow pattern. (It would be more accurate to call such models "non-eddy-permitted" and "eddy-permitted", because a model with $\frac{1}{3}$ degree resolution cannot really resolve eddies of higher baroclinic modes in the ocean.) However, the mixing process is quite different in these two cases. Mixing by eddies in the westward-flowing sector of the subtropical gyre is quite effective in homogenizing the potential vorticity. Whereas previous theory predicted homogenization of potential vorticity on long time scales only within a recirculating gyre, in an eddy-resolving model homogenization takes place on a much shorter time scale across recirculating/ventilating flow boundaries. In fact, anomalous potential vorticity that is advected into the thermocline from isopycnal outcrops by ventilated flow causes changes in the sign of the meridional gradient of local potential vorticity, which in turn gives rise to baroclinic instability.

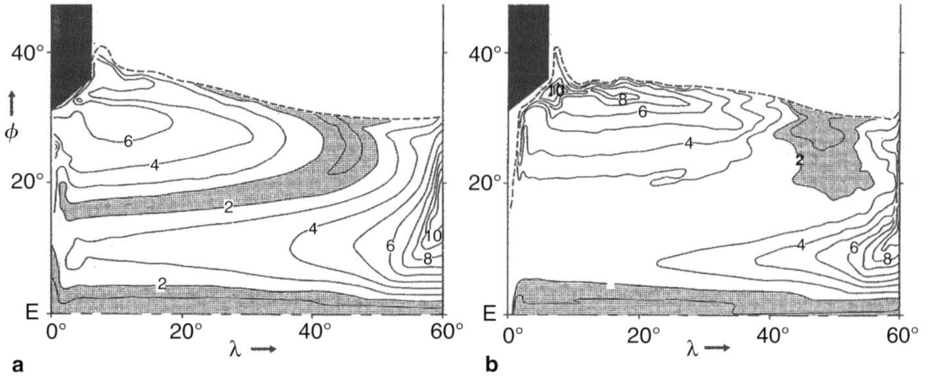

Fig. 4.43 Potential vorticity distribution on the $\sigma = 26.0$ surface obtained from a numerical model, in $10^{-10}/\mathrm{m/s}$ for **a** the case with 1° resolution, and **b** the case with 1/3° resolution (Cox, 1985).

Second, atmospheric cooling plays a crucial role in creating mode water that has nearly homogenized potential vorticity and other properties (McCartney, 1982). As shown much earlier in Figure 1.7, there is a lot of heat lost to the atmosphere near the recirculation region in both the North Atlantic and North Pacific. This strong cooling is closely associated with mode water formation, which takes place primarily south of the Gulf Stream system. A key index for mode water formation is the local maximum of mixed layer depth in late winter.

As discussed in detail in Section 5.1.5 about mode water formation, late-winter cooling leads to a deep mixed layer and a thick column of water with vertically almost homogenized properties. Springtime warming in the upper ocean induces a rapid shoaling of the mixed layer, and the newly formed mode water is sealed off. As a result, a large amount of mode water with very low potential vorticity is formed within the recirculation regime south of the Gulf Stream and the Kuroshio. As an example, mode water formed in the upper ocean can be seen clearly through a map of potential vorticity along the 65° W section in the North Atlantic Ocean (Fig. 4.44).

The same idea can be demonstrated by numerical models (Huang and Bryan, 1987). Their numerical experiments showed that at the outer edge of the western boundary outflow, cooling induces a strong mass flux from the warm and light upper layers to the cold and dense lower layers. In a four-layer model, there is a strong mass flux from the second layer to the third layer at the latitude corresponding to where the Gulf Stream leaves the western boundary (Fig. 4.45).

When water leaves the western boundary, it carries high potential vorticity; however, this high-potential vorticity tongue is rapidly transformed into quasi-uniform, low-potential vorticity water through convective overturning (Fig. 4.46). Thus, the thermodynamic processes in the mixed layer and the upper ocean could be the primary machinery generating these huge masses of quasi-homogenized potential vorticity.

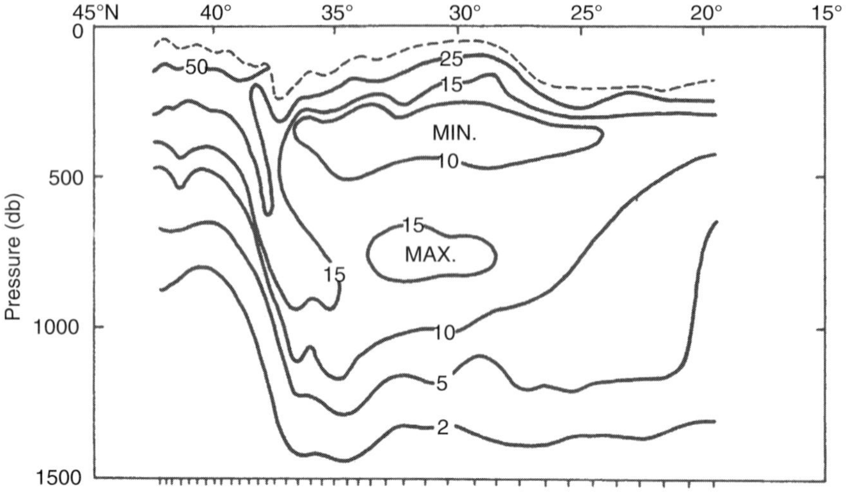

Fig. 4.44 Meridional section of potential vorticity (in 10^{-11}/m/s) along the 65° W IGY (International Geophysical Year) section. The broad shallow minimum ($q < 10$) is the 18°C water formed by convective cooling at the surface and subsequently capped off by a seasonal pycnocline. The q maximum at 800 m is the permanent pycnocline (McDowell *et al.*, 1982).

4.2 Thermocline models with continuous stratification

4.2.1 Diffusive versus ideal-fluid thermocline

Diffusive thermocline

Even though multi-layer models of the wind-driven circulation were not very successful, people have tried to attack the continuously stratified model. They simply could not resist the temptation of trying to solve such a simple-looking equation set:

$$-fv = -p_x/\rho_0 \tag{4.251}$$

$$fu = -p_y/\rho_0 \tag{4.252}$$

$$0 = -p_z - \rho g \tag{4.253}$$

$$\nabla \cdot \vec{u} = 0 \tag{4.254}$$

$$\vec{u} \cdot \nabla \rho = \kappa \rho_{zz} \tag{4.255}$$

This equation set looks really simple because all the equations are linear except the density equation (4.255). This simple-looking equation, however, turns out to be very difficult to solve. Since these equations are very complicated nonlinear partial differential equations, it was not clear how to formulate suitable boundary value problems and how to solve them. For a long time, the only way that people could solve these equations was by using similarity solutions.

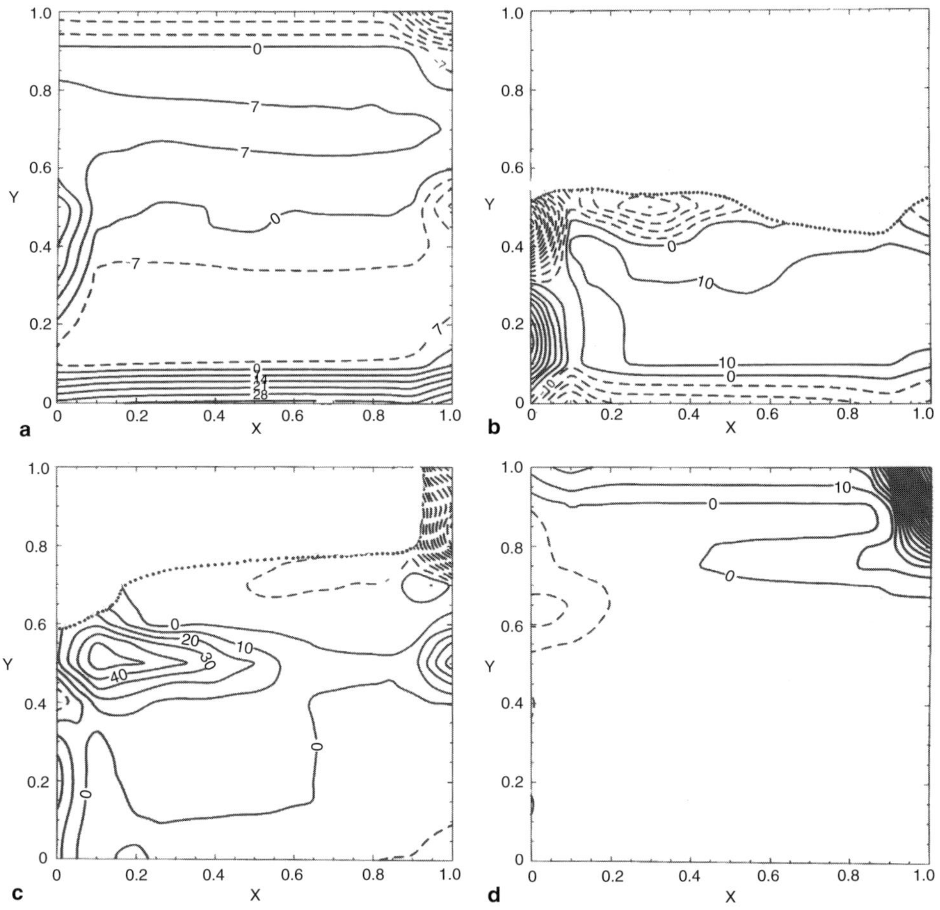

Fig. 4.45 Source of water for a four-moving-layer model (in 10^{-7}/m/s); *solid* (*dashed*) contours for source (sink): **a** the mixed layer; **b** the second layer; **c** the third layer; **d** the fourth layer. *Dotted lines* in **b** and **c** are the outcrop lines (Huang and Bryan, 1987).

For example, Needler (1967) proposed a trial solution of the form

$$p = m(x, y)e^{\kappa(x,y)z} \tag{4.256}$$

Substituting Eqn. (4.256) into the original equations leads to a solution whose pressure depends on the vertical coordinate exponentially, $p = \frac{\rho_s}{\kappa} e^{\kappa z}$; other dynamical variables have similar features.

Although similarity solutions give some nice-looking solutions, they are based on some special assumptions about balance in the original equations, and these assumptions are not always physically correct. Eventually the search for similarity solutions was largely abandoned, and people are now more interested in non-similarity solutions that can satisfy

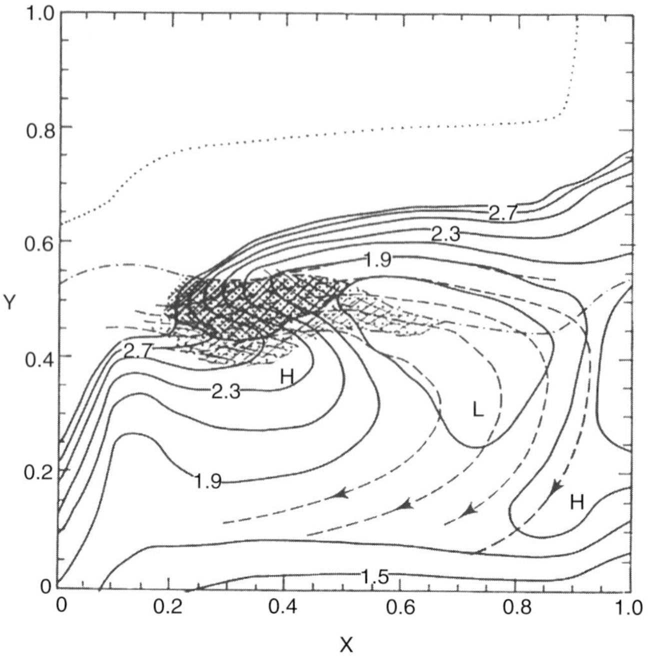

Fig. 4.46 Potential vorticity (*solid lines*) and streamfunction (*dashed lines*) of the third layer in a multi-layer model. *Light* (*heavy*) *cross-hatched* and *stippled* area for the source region with intensity weaker (stronger) than 2×10^{-6} m/s (Huang and Bryan, 1987).

the essential boundary conditions. However, there were many difficulties to overcome before we could see light in the darkness. In the following sections of this introduction we will describe some of these historical puzzles.

Ideal-fluid thermocline

Basic equations

The so-called ideal-fluid thermocline equations look very simple. They consist of geostrophy in the horizontal direction, hydrostatic approximation in the vertical direction, the incompressibility, and density conservation equations:

$$-fv = -p_x/\rho_0 \tag{4.257}$$

$$fu = -p_y/\rho_0 \tag{4.258}$$

$$0 = p_z + \rho g \tag{4.259}$$

$$\nabla \cdot \vec{u} = 0 \tag{4.260}$$

$$u\rho_x + v\rho_y + w\rho_z = 0 \tag{4.261}$$

All equations are linear except for the density conservation equation; however, the nonlinearity associated with the advection term is very strong. Since this equation system was first formulated by Welander in 1959, the challenge had been to formulate suitable boundary value problems for this equation system and to solve them.

Conservation quantities

By cross-differentiating, subtracting Eqns. (4.257) and (4.258), and applying Eqn. (4.260), one obtains the vorticity equation

$$\beta v = f w_z \tag{4.262}$$

Taking the z derivative of Eqn. (4.261) yields

$$\vec{u}_z \cdot \nabla \rho + \vec{u} \cdot \nabla \rho_z = 0 \tag{4.263}$$

Because of the thermal wind relation, the vertical shear of the horizontal velocity is perpendicular to the horizontal density gradient, so the first term on the left-hand side is reduced to

$$\vec{u}_z \cdot \nabla \rho = w_z \rho_z = \frac{\beta}{f} v \rho_z \tag{4.264}$$

Thus, Eqn. (4.263) can be rewritten as

$$\vec{u} \cdot \nabla (f \rho_z) = 0 \tag{4.265}$$

This is the potential vorticity conservation law, following a streamline.

Multiplying the momentum equations by v and u, and adding together, we obtain the mechanical energy balance, noting that the Coriolis force does no work,

$$0 = u p_x + v p_y = \vec{u} \cdot \nabla p - w p_z = \vec{u} \cdot \nabla p + w \rho g \tag{4.266}$$

Since $w = \vec{u} \cdot \nabla z$, we obtain the *Bernoulli conservation law*

$$\vec{u} \cdot \nabla (p + \rho g z) = 0 \tag{4.267}$$

Thus, density, potential vorticity, and the Bernoulli function are all conserved along streamlines. In the oceans these quantities are not exactly conserved; however, deviations from the conservation laws are small enough, so the theory of ideal-fluid thermocline has been rather successful.

Simple solutions

Reducing to a single ordinary differential equation

Welander (1971a) made another very important contribution to the ideal-fluid thermocline theory by applying the vorticity conservation and Bernoulli conservation laws to the thermocline problem. His original formulation was based on the z-coordinate. It is more convenient to formulate the problem in density coordinates and to use the *Bernoulli function*

$$B = p + \rho g z \tag{4.268}$$

as the dependent variable in the density coordinates. The first and second derivatives of the Bernoulli function with respect to the density are

$$B_\rho = gz \tag{4.269}$$

$$B_{\rho\rho} = g z_\rho = \frac{fg}{q(B,\rho)} \tag{4.270}$$

where

$$q = f\rho_z \tag{4.271}$$

is the potential vorticity. As stated above, both the Bernoulli function B and the potential vorticity q are conserved along streamlines.

Welander made a vital step in observing that if the functional form of $q(B, \rho)$ is given, the equation can be solved either analytically or numerically. In particular, he discussed several cases.

If potential vorticity is a function of the density alone

The equation can be solved easily (Welander, 1959). In fact, a double integration leads to an analytical solution

$$B = B_0 + B_{\rho,0} - \iint d\rho' \frac{fg}{q(\rho')} \tag{4.272}$$

Assume a linear function $q(B,\rho) = f\rho_z = a\rho + bB + c$

This equation can be easily integrated. First, differentiating this linear function with respect to z and applying the hydrostatic relation leads to

$$f\rho_{zz} = (a + bgz)\rho_z \tag{4.273}$$

Integrating Eqn. (4.273) twice leads to

$$\rho_z = C_1(\lambda,\theta)\, e^{\frac{az+0.5bgz^2}{f}} \tag{4.274}$$

$$\rho = \rho_0(\lambda,\theta) + C(\lambda,\theta)\int_0^z e^{-\frac{(z'+z_0)^2}{D^2 f}}\, dz' \tag{4.275}$$

where

$$C = C_1 e^{-\frac{a^2}{bg}}, \quad z_0 = \frac{a}{bg}, \quad D = \left(-\frac{2}{bg}\right)^{1/2}$$

Welander added the following boundary conditions: density at the sea surface should match the observed annual-mean surface density $\rho_s(\lambda, \theta)$, and at great depth it should approach a uniform value $\rho_{-\infty}$.

Welander also discussed the general case where q is a function that depends on a linear combination of B and ρ, say $q(B, \rho) = F(a\rho + bB + c)$. Although Welander laid the foundation for solving the thermocline equation, some very challenging difficulties remained to be overcome.

The most difficult issue had again been the conceptual difficulty associated with the boundary conditions. Welander's formulation reduces the thermocline problem to a second-order ordinary differential equation in density coordinates. A second-order ordinary differential equation can normally satisfy two boundary conditions only; however, a solution for the thermocline structure in a basin has to satisfy many boundary conditions. For example, the Ekman pumping condition requires the solution to fit a two-dimensional array, which seemed to be almost impossible within the original approach proposed by Welander. It is essential to satisfy the Ekman pumping condition, and the model can be modified by including such a constraint. In addition, specifying potential vorticity for the whole water column turns out to be unnecessary. As will be discussed in the following sections, the new theory shows that the problem is overdetermined, so it cannot satisfy other dynamically important boundary conditions.

Today, as we review the advances in the past decade, we realize that progress in solving the thermocline equation has been made step by step through the deepening of our understanding of the thermocline structure. The two contributions of Welander came primarily from some physical insights: ignoring diffusion and introducing conservation of the Bernoulli function and potential vorticity. The improvement of this solution requires additional physical insights. It was only after the innovative study of potential vorticity homogenization by Rhines and Young and the ventilated thermocline by Luyten, Pedlosky, and Stommel that we arrived at a better understanding of the physics of the thermocline. As we discuss later, similar to the situation for the multi-layered model of the thermocline, potential vorticity has to be calculated as part of the solution; one can only specify the potential vorticity for the unventilated thermocline. Therefore, Welander's suggestion of specifying the form of the potential vorticity function for all moving water requires too much information, and actually gives rise to an overdetermined system. As a result, the solution cannot satisfy more than two boundary conditions.

An analytical solution for the ideal-fluid thermocline

An analytical solution for the ideal-fluid thermocline can be found by using assumptions slightly different from Welander's. Instead of assuming a linear function of the potential vorticity, we assume that:

a) In the ventilated thermocline the potential thickness (the inverse of the potential vorticity) is a linear function of the Bernoulli function, i.e., $D = -z_\rho/f = \alpha^2 B$, where α is a given parameter;

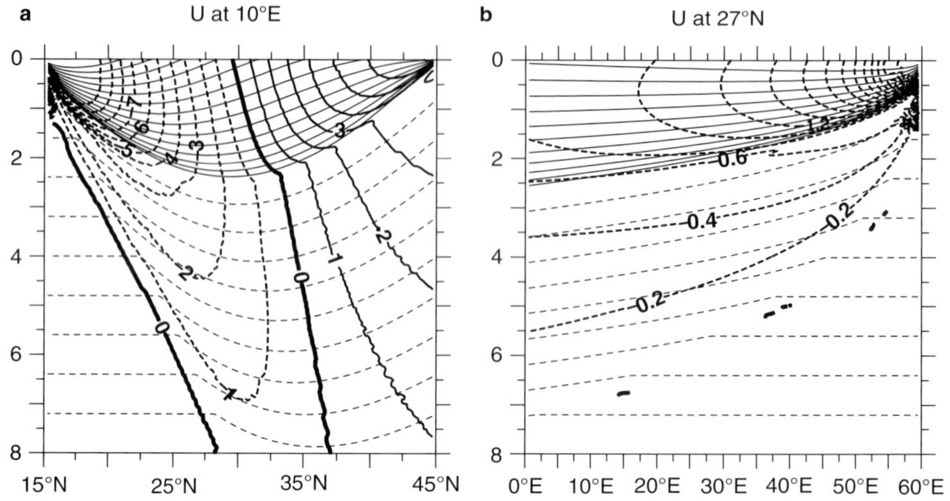

Fig. 4.47 Structure of an analytical solution for the ideal-fluid thermocline. *Solid lines* are ventilated thermocline, *dashed lines* are unventilated thermocline, and the *thick lines* are the velocity contours in 10^{-2}/m/s.

Thus, the basic equation for the ventilated thermocline is

$$B_{\rho\rho} + \left(\sqrt{fg\alpha}\right)^2 B = 0 \tag{4.276}$$

b) In the unventilated thermocline, potential thickness is constant.

The solution of Eqn. (4.276) is in the form of $B = a \cos b(\rho - \rho_s)$, where $b = \sqrt{fg\alpha}$, and $a = a(x, y)$ can be determined by the boundary conditions, including the Sverdrup constraint. After simple manipulations, the Sverdrup constraint is reduced to a single transcendental equation (Huang, 2001). An example for a model basin covering (0°–60° E, 15°–45° N), mimicking the North Atlantic, is shown in Figure 4.47. This solution includes the ventilated thermocline, the unventilated thermocline, and the shadow zone (as indicated by the flat isopycnals near the southern and eastern boundaries). Most importantly, the solution satisfies the Sverdrup constraint, so it provides a rather complete dynamical picture of the thermocline and the associated three-dimensional wind-driven gyre.

It is clear that the solution resembles the subtropical gyres observed in the oceans. In some previous studies it was speculated that the main thermocline may appear in the form of a density discontinuity in a truly continuous model. Such a density front is, however, not necessary. In fact, the simple solution shown in Figure 4.47 has a truly continuous structure in three-dimensional space, including weak discontinuities in potential vorticity. In some sense, this solution is also a similarity solution because the potential vorticity function is specified *a priori*. However, this solution is quite different from previously studied similarity solutions in two aspects. First, the solution satisfies all essential boundary conditions, and in particular the Sverdrup relation. Second, the solution includes weak discontinuity in potential vorticity, but is continuous in density structure.

This model belongs to the category of solutions in which potential vorticity in the ventilated thermocline is set as a constant *a priori*, and the surface density distribution that matches the solution is found as part of the solution. In the oceans, potential vorticity of the ventilated thermocline is set by the basin-wide circulation, including the upper boundary conditions, such as the Ekman pumping rate, the surface density, and mixed layer depth. The mixed layer depth is assumed to be zero in the case presented here; however, it is easy to modify the formulation to include a mixed layer with a non-zero depth.

4.2.2 Models with continuous stratification

How to improve the ventilated thermocline models

The ventilated thermocline theory by Luyten *et al.* (1983) laid the foundation for the ideal-fluid thermocline. In principle, their theory can be extended into a model with many more moving layers. Thus, such a model may be able to provide useful information about the thermocline structure in the ocean with continuous stratification.

However, such a multiple-layer model may encounter some difficulties when the number of layers increases. For example, from the work by Luyten *et al.* (1983), it is clear that the number of shadow zones with different dynamics increases exponentially as the number of layers is increased. In addition, when the number of layers is very large, it is a rather tedious job to derive and calculate the analytical solution. Thus, it is desirable to formulate the continuous model in the spirit of the ventilated thermocline.

The singularity in the models

One of the most critical problems in the early ventilated thermocline models is due to the lack of a mixed layer. In fact, the upper surface of the models is set to $z = 0$. This choice is apparently made to simplify the models. There are several problems associated with such an upper boundary condition.

First, the mixed layer is the major buffer between the atmosphere and the permanent thermocline. Mixed-layer depth reaches the annual maximum in late winter, to the order of 200–400 m. Volume flux in the mixed layer constitutes a substantial part of the total volume flux in the wind-driven circulation. Since density is almost vertically uniform in the mixed layer, its dynamics is quite different from that in the ocean interior. Including the mixed layer is a vital step toward a more realistic wind-driven circulation.

Second, this upper boundary forces all isopycnals to outcrop at the same depth, $z = 0$. As a result, all these models have singularity along the eastern, northern, and southern boundaries.

Third, this boundary condition excludes the contribution to the subduction rate due to the mixed-layer depth gradient. It will be shown in Section 5.1.5 that including a mixed layer with horizontally varying depth can give rise to a subduction rate that is substantially larger than the rate of Ekman pumping. Thus, including the mixed layer is an essential step in bringing the thermocline models closer to reality.

As will be shown in this section, including the mixed layer is actually not very difficult at all, and it has substantially improved the thermocline models. For example, coupling the mixed layer helps to overcome one of the major problems in thermocline models, i.e., the singularity along the eastern, northern, and southern boundaries.

The eastern boundary condition

The eastern boundary has not received enough attention. In earlier theoretical models of the wind-driven circulation with one moving layer, the eastern boundary is simply a place to start the integration. The importance of suitable eastern boundary conditions for a stratified model was first encountered in the Luyten–Pedlosky–Stommel model. A major feature of this model is that all ventilated layers have zero thickness along the eastern boundary. This is apparently inconsistent with observations that all layers have a finite thickness along the eastern boundary. Furthermore, since the calculations have to be started from the eastern boundary, it seems clear that if ventilated layers have non-zero thickness along the eastern boundary, the entire solution might change.

The special nature of the eastern boundary condition can be appreciated through an interesting argument by Killworth (1983b). Assuming that the flow can be described by the ideal-fluid thermocline equation, then the suitable kinematical condition is: at the eastern boundary

$$u = 0, \quad \text{at } x = 0 \tag{4.277}$$

so that $u_z \equiv 0$. The geostrophy equations (4.257) and (4.258) imply

$$p_y = 0, \text{ or } p = p(z), \rho = \rho(y), \quad \text{at } x = 0 \tag{4.278}$$

Along the eastern boundary, the density conservation equation (4.261) is reduced to

$$w\rho_z = 0, \quad \text{at } x = 0 \tag{4.279}$$

Now, if $\rho_z \neq 0$ at the wall, i.e., the fluid is stratified, then

$$w = 0, \quad \text{at } x = 0 \tag{4.280}$$

From the Sverdrup relation $\beta_v = f w_z$, we obtain

$$v = 0, \quad \text{at } x = 0 \tag{4.281}$$

By the thermal wind relation, this leads to

$$\rho_x =, \quad \text{at } x = 0 \tag{4.282}$$

Differentiating the density conservation equation and the Sverdrup relation repeatedly, we can show that for variables u, v, w, ρ_x, ρ_y, ρ_z, the first, second, and all higher-order x-derivatives are zero at the eastern boundary.

If the thermocline solution can be expanded in Taylor series from the eastern boundary, then the solution should be zero everywhere in the basin. Although water in the vicinity of the eastern boundary is stagnant and thus has homogeneous properties, water in the basin interior can have quite different properties. As we discussed in the previous section about the ventilated thermocline, water subducted from the basin interior can have dynamical properties completely different from the stagnant water adjacent to the eastern boundary. In fact, the system is of a hyperbolic nature; thus, along the interfaces which separate water with different dynamical properties, these quantities may be non-differentiable, and the expansion in Taylor series is not valid.

To explore suitable eastern boundary conditions for stratified models, Pedlosky (1983) studied a model of two moving layers. A new eastern boundary condition was used, which required only the vertically integrated zonal volume flux to be zero and allowed ventilated layers to have non-zero thickness along the eastern wall. Since the model has only two moving layers, the stratification along the eastern wall could be specified *ad hoc* and the solution in the interior can be calculated accordingly. This eastern boundary condition gives rise to the eastern boundary ventilated thermocline and alters the global structure of the gyre circulation.

The generalized eastern boundary conditions have been extended to the case of a continuously stratified model by Huang (1989a), where it is shown that the stratification along the eastern boundary can no longer be specified *ad hoc*; instead, it should be calculated as a part of the unified gyre-scale circulation. Thus, the eastern boundary conditions are closely tied to the gyre-scale circulation and cannot be specified arbitrarily.

There are some difficulties involved in using the generalized eastern boundary conditions. First, the model requires an unknown eastern boundary layer that can transfer water vertically. Second, the stratification along the eastern boundary implies some extra freedom of the system, so the system becomes highly underdetermined. The question is how to find a physically meaningful solution.

In searching for an answer to the general questions concerning the suitable boundary conditions at the eastern wall for continuously stratified models, Young and Ierley (1986) studied a thermohaline circulation model with vertical diffusion. They used a family of similarity solutions to explore the physical meaning of suitable eastern boundary conditions for the ideal-fluid thermocline. By examining the solutions obtained when the vertical diffusivity approached zero, they came to the conclusion that the ideal-fluid thermocline equation has weak solutions, i.e., solutions that have a density discontinuity, which they interpreted as a thermocline. However, Huang (1988a, 2001) showed that truly continuously stratified solutions of the equations do exist, although potential vorticity would be discontinuous across the base of the moving water and there would be some singularity along the boundaries of a basin. Thus, it is possible to construct a solution that has a smooth density field in the interior ocean, although the eastern boundary always involves some kind of singularity.

In a model including a mixed layer of horizontally varying density and depth, Huang (1990a) returned to the old eastern boundary condition of zero zonal velocity below the base of mixed layer. An implicit assumption of the model is that *the onshore geostrophic flow in the seasonal thermocline is exactly balanced by the offshore Ekman flux due to the southward along-shore wind stress. This new formulation successfully overcomes the artificial singularities existing in many previous theoretical models.* Due to the finite depth of the mixed layer, meridional velocity is finite everywhere. The application of this eastern boundary condition eliminates the potential vorticity singularity along the eastern, northern, and southern boundaries in previous models, and gives rise to shadow zones in the ventilated thermocline.

It is fair to say that the problem with the eastern boundary condition has not been completely solved. Since the local offshore Ekman flux may not always exactly balance the onshore flux in the seasonal thermocline, there is again some singularity involved that requires further study. In fact, there is a gap between coastal oceanography and basin-scale oceanography. In coastal oceanography, stratification in the interior ocean is assumed given, and for most cases this stratification is taken to be independent of latitude; while for basin-scale oceanography, the stratification along the eastern boundary is assumed given, presumably determined by some coastal circulation processes. These two parts should be related through the general circulation in a closed basin.

Coupling with a mixed layer of variable depth

In the early stages of development, the mixed layer was neglected in most ideal-fluid thermocline models for simplicity, and the upper boundary conditions of these models were that of specifying w_e and ρ^s at $z = 0$. Neglecting the mixed layer leads to many problems in the models. A close examination reveals that a model with a mixed layer is a fairly easy extension of the previous models whose upper surface was artificially put at the sea surface.

The mixed layer

Our concern is primarily the dynamics of the main thermocline below the mixed layer. In addition, it is very difficult to formulate a simple analytical model to incorporate the mixed-layer thermodynamics with the dynamics in the main thermocline. A major challenge involved in such a goal is the handling of the seasonal cycle in the mixed layer, with the Rossby waves interacting with each other and the mean currents. Thus, we will exclude the thermodynamics of the mixed layer; instead, we will prescribe the thermodynamic parameters of the mixed layer; the velocity in the mixed layer is, however, part of the solution. In addition, late-winter mixed-layer properties are chosen as the forcing functions, according to Stommel's (1979) suggestion.

In our idealization, the Ekman layer is treated as an infinitely thin layer on the surface of the ocean, where water is collected horizontally by Ekman drift and from which it descends owing to Ekman pumping. Since density is assumed to be vertically homogenized within the whole depth of the mixed layer, the Bernoulli function

$$B = p + \rho g z \tag{4.283}$$

is vertically constant within the mixed layer. At the sea surface, the Bernoulli function is the same as the pressure, so the horizontal pressure gradient in the mixed layer is

$$\nabla_h p = \nabla_h B^s - gz\nabla_h \rho^s \tag{4.284}$$

where the superscript s indicates the sea surface. The first term on the right-hand side is a barotropic term; the second term is a baroclinic term due to the horizontal density gradient in the mixed layer.

The horizontal pressure gradient induces a geostrophic flow in the mixed layer

$$u^g = -\frac{p_y}{f\rho_0}, \quad v^g = \frac{p_x}{f\rho_0} \tag{4.285}$$

where $f = 2\Omega \sin\theta$ is the Coriolis parameter, ρ_0 is the reference density, and subscripts x and y indicate the partial derivatives in spherical coordinates

$$\frac{\partial}{\partial x} = \frac{\partial}{a\cos\theta\partial\lambda}, \quad \frac{\partial}{\partial y} = \frac{\partial}{a\partial\theta} \tag{4.286}$$

The vertically integrated volume flux in the mixed layer is

$$\int_{-h}^{0} u dz = -\frac{1}{f\rho_0}\left(B_y^s h + \frac{1}{2} g \rho_y^s h^2\right) \tag{4.287}$$

$$\int_{-h}^{0} v dz = \frac{1}{f\rho_0}\left(B_x^s h + \frac{1}{2} g \rho_x^s h^2\right) \tag{4.288}$$

Along the eastern boundary of the model basin, the first term on the right-hand side of Eqn. (4.287) indicates an eastward flow into the eastern boundary, while the second indicates a westward flux due to the north–south density gradient. We will assume that along the eastern boundary all the thermocline layers below the base of the mixed layer are stagnant (Fig. 4.48). In reality, along the eastern boundary, water below the mixed layer is in motion, and a solution including such an eastern boundary ventilated thermocline can be obtained in a multi-layer model or a continuously stratified model as discussed by Pedlosky (1983) and Huang (1989a). As shown in these studies, although introducing the eastern boundary ventilated thermocline modified the solution, especially in the southeastern part of the basin, changes in the rest of the basin are relatively small. In addition, solutions involving the eastern boundary ventilated thermocline are not uniquely defined. Thus, to ease the analysis, we will use a simple no-flow boundary condition below the mixed layer and neglect the eastern boundary ventilated thermocline.

Accordingly, the eastern boundary condition for our model is that the zonal velocity is identically zero for water in the thermocline

$$u \equiv 0, \text{ at } x = x_e \tag{4.289}$$

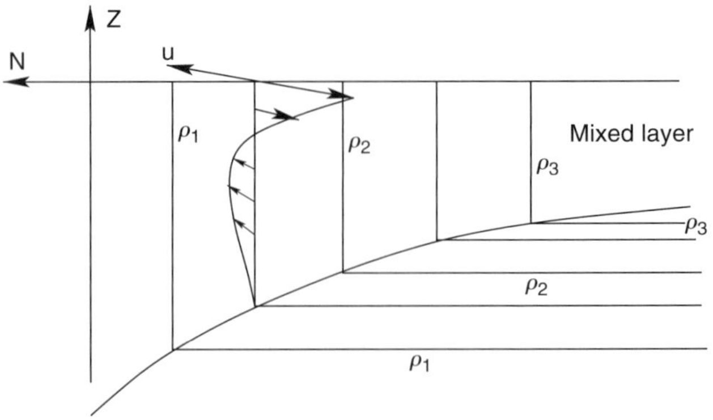

Fig. 4.48 The dynamical structure of the eastern boundary. Isopycnals are vertical within the mixed layer, and horizontal below the mixed layer, implying no motion below the mixed layer.

This implies

$$p_y = 0 \text{ for } z \leq -h \tag{4.290}$$

Therefore, the volume flux going into the eastern boundary is

$$M^g = \int \frac{gh^2}{2f\rho_0} \rho_y^s dy \tag{4.291}$$

Assuming $h = 100$ m, $f = 10^{-4}$/s, $\Delta\rho = 2$ kg/m^3, we obtain $M^g \simeq 1\,Sv$. This onshore flux of water feeds the upwelling along the eastern boundary. Upwelling along the eastern boundary is connected with the current system along the eastern boundary, details of which are excluded from our discussion here.

The actual eastern boundary of the model is slightly tilted, so the eastern boundary condition is $\vec{u} \cdot \vec{n} = 0$, *along the eastern boundary*.

The mixed layer plays a vital role in setting the stage for subduction. In a steady circulation, the annual subduction rate (the exact definition of subduction rate is discussed in Section 5.1.5) is defined as

$$S = -(w_m + \vec{u} \cdot \nabla_h h) \tag{4.292}$$

where the subscript m indicates the base of the mixed layer; the first term on the right-hand side is due to vertical pumping at the base of the mixed layer. Because of the geostrophic meridional flux in the mixed layer, the vertical pumping at the base of the mixed layer is less than the Ekman pumping

$$w_m = w_e - \frac{\beta}{f} \int_{-h}^{0} v dz = w_e - \frac{\beta}{f^2 \rho_0} \left(B_x^s + \frac{1}{2} g \rho_x^s h^2 \right) \tag{4.293}$$

The second term on the right-hand side of Eqn. (4.292) is due to lateral induction. Because the base of the mixed layer is tilted, horizontal flow gives rise to volume flux into the main thermocline. In fact, lateral induction is a major contributor to subduction into the main thermocline. This can be shown by a very simple estimation as follows.

a) Vertical pumping: $\int dx \int w_m \simeq 6 \times 10^8 \times 10^{-4} \times 3 \times 10^8 \simeq 18$ Sv.
b) Horizontal induction: $\int dx \int v dh \simeq 6 \times 10^8 \times 3 \times 10^4 \simeq 18$ Sv.

Thus, the contributions from these two terms are comparable. This estimation will be confirmed by calculation from our 3-D model.

It can readily be seen that calculating the subduction rate requires several variables, including the mixed-layer geometry (the depth), the mixed-layer density, and the Bernoulli function. The horizontal velocity in the mixed layer can be calculated from these variables. Although the thermodynamic variables of the mixed layer, such as ρ^s and h, are specified, the dynamical variables, such as B^s and $\vec{u}$, are unknown and they are part of the solution we are looking for.

Note that in this section all partial derivatives are defined in z-coordinates. Because density is vertically uniform, density coordinates are meaningless within the mixed layer. However, our analysis in the following sections is based on density coordinates, so we need to convert Eqn. (4.293) into density coordinates. When transferring between z-coordinates and density coordinates,

$$\nabla_z B = \nabla_\rho B + B_\rho \rho_x|_z$$

Thus, at the base of the mixed layer

$$B_x^s|_{\rho^s} = B_x^s|_z + gh\rho_x^s$$

Note that the terms on the right-hand side are calculated in z-coordinates because these are defined in the mixed layer, where density coordinates do not really work. Therefore, in density coordinates, the vertical velocity at the base of the mixed layer is

$$w_m = w_e - \frac{\beta h}{f^2 \rho_0} \left(B_x^s|_\rho + \frac{1}{2} B_\rho \frac{d\rho^s}{dx} \right) \tag{4.293'}$$

where we have used the relation $B_\rho = -gh$.

A free boundary value problem for the ideal-fluid thermocline

Although the ventilated thermocline model by Luyten *et al.* (1983) provided a simple and elegant solution for wind-driven gyres, it is limited to cases with a few moving layers. An extension to the case with continuously stratified ocean is much more desirable and, in order to find such solutions, we need suitably formulated models. The basic formulation of boundary value problems for the ideal-fluid thermocline was proposed by Huang (1988a, b). This formulation was modified to include the mixed layer with horizontally varying mixed-layer density and depth by Huang (1990a). The formulation is based on (potential) density coordinates ρ. It is important to note that, conceptually, the ventilated thermocline theory by Luyten *et al.* (1983) is based on a highly truncated model in density coordinates. As the number of layers is increased greatly, their model should converge to produce a model with continuous stratification.

In density coordinates, the horizontal momentum equations are reduced to

$$f\rho_0 v = B_x, \quad f\rho_0 u = -B_y \tag{4.294}$$

and the vertical momentum equation is reduced to the hydrostatic relation

$$B_\rho = gz \tag{4.295}$$

where the subscript ρ indicates a partial derivative with respect to density ρ. The linear vorticity equation, or the equivalent Sverdrup relation, is in the form

$$\beta v z_\rho = f w_\rho \tag{4.296}$$

Using Eqns. (4.294) and (4.295) to eliminate v and z, this leads to

$$B_{\rho\rho} B_x = \frac{g\rho_0 f^2}{\beta} w_\rho \tag{4.297}$$

Integrating this equation over $[\rho^s, \rho^b]$, where ρ^s is the given mixed-layer density distribution, and ρ^b is the unknown free boundary separating the moving water from the stagnant abyssal water, gives

$$\int_{\rho^s}^{\rho^b} B_{\rho\rho} B_x d\rho = -\frac{g\rho_0 f^2}{\beta} w_m \tag{4.298}$$

Using Eqn. (4.293′) and the upper boundary condition that $B_\rho = -gh$ at the base of the mixed layer, the right-hand side of Eqn. (4.298) can be rewritten as

$$-\frac{g\rho_0 f^2}{\beta} w_m = -\frac{g\rho_0 f^2}{\beta} w_e - B_\rho B_x^s - \frac{1}{2} B_\rho^2 \frac{d\rho^s}{dx} \tag{4.299}$$

Integrating by parts, the left-hand side of Eqn. (4.298) can be rewritten as

$$\int_{\rho^s}^{\rho^b} B_{\rho\rho} B_x d\rho = B_\rho B_x|_{\rho^b} - B_\rho B_x|_{\rho^s} - \int_{\rho^s}^{\rho^b} B_\rho B_{\rho x} d\rho \tag{4.300}$$

where the first term on the right-hand side is zero due to the matching boundary condition that horizontal velocity at the base of the wind-driven gyre vanishes. The last integral can be converted into another form

$$\frac{d}{dx}\int_{\rho^s}^{\rho^b} B_\rho^2 d\rho = B_\rho^2 \frac{d\rho^b}{dx}|_{\rho^b} - B_\rho^2 \frac{d\rho^s}{dx}|_{\rho^s} + 2\int_{\rho^s}^{\rho^b} B_\rho B_{x\rho} d\rho \tag{4.301}$$

Combining equations (4.298, 4.299, 4.300, and 4.301), we obtain

$$\frac{d}{dx}\int_{\rho^s}^{\rho^b} B_\rho^2 d\rho - B_\rho^2 \rho_x^b = \frac{2g\rho_0 f^2}{\beta} w_e \tag{4.302}$$

Integrating Eqn. (4.302) over domain $[x, x_e]$ leads to

$$-\int_{\rho^s}^{\rho^b} B_\rho^2 d\rho + \int_{\rho^s}^{\rho^b} B_\rho^{e^2} d\rho + \int_{\rho^s}^{\rho^b} B_\rho^{a^2} d\rho = \frac{2g\rho_0 f^2}{\beta}\int_x^{x_e} w_e dx \tag{4.303}$$

where B^e is the Bernoulli function specified along the eastern boundary, which is a given function derived from the stratification along the eastern boundary, and B^a is the specified stratification in the abyss, representing the background stratification.

This equation is a generalization of the Sverdrup relation often used for a reduced-gravity model. Since we will make use of an additional assumption that along the eastern boundary the base of moving water is the same as the base of the mixed layer, the second term on the left-hand side vanishes. As demonstrated by Huang (1990a) and Huang and Russell (1994), the calculation of the ideal-fluid thermocline is reduced to repeatedly solving the following free boundary value problem in density coordinates

$$B_{\rho\rho} = \frac{fg}{Q(B, \rho)} \tag{4.304}$$

with the following constraints

$$B_\rho = -gh(x, y) \text{ at } \rho = \rho^s \tag{4.305}$$

$$B = B^a,\ B_\rho = B_\rho^a, \text{ at } \rho = \rho^b \ (\rho^b \text{ is unknown}) \tag{4.306}$$

$$-\int_{\rho^s}^{\rho^b} B_\rho^2 d\rho + \int_{\rho^{be}}^{\rho^b} B_\rho^{a^2} d\rho = \frac{2g\rho_0 f^2}{\beta}\int_x^{x_e} w_e dx \tag{4.307}$$

Although this approach seems to be a simple extension of the early work by Welander (1971a), there are subtle differences between the new formulation and Welander's old formulation. In Welander's formulation, it was assumed that potential vorticity $Q(B, \rho)$ in Eqn. (4.304) is a given function of B and ρ, and this equation is subject to boundary conditions specified at the upper and lower boundaries. In particular, Welander did not consider the potential role of the mixed layer, i.e., he implicitly assumed $h(x, y) = 0$. In addition, he assumed that the lower boundary is fixed at $\rho = \rho_{-\infty}$. Since the ordinary differential equation (4.304) with a given function $Q(B, \rho)$ can satisfy only two boundary conditions, it was not clear how to find a solution satisfying additional boundary conditions, such as the Sverdrup constraint, Eqn. (4.307), which are essential for describing the wind-driven circulation in the ocean.

As discussed above, the major problem with Welander's original formulation is the assumption that potential vorticity is a given function for the entire thermocline. It took a long time and much effort before it was realized that the thermocline consists of many regions which are regulated by different dynamics, such as the ventilated zone, the unventilated thermocline, the shadow zone, and the pool zone. It was shown that potential vorticity in the unventilated thermocline is fairly well homogenized (Rhines and Young, 1982a; McDowell *et al.*, 1982). Owing to strong wind forcing, potential vorticity in the ventilated zone is not homogenized in general. Thus, any *a priori* assumption about the form of the potential vorticity function in the ventilated zone is artificial, and the most important progress made in the early 1980s was the realization of such an essential limit in previous approaches and the creation of a new approach which allows us to calculate potential vorticity in the ventilated zone as a part of the solution, as demonstrated by Luyten *et al.* (1983).

Another major difference in the new approaches is that the base of moving water is no longer a constant-density surface; instead, the boundary between the moving part of the wind-driven gyre and the stagnant water below is a free boundary which is calculated as part of the solution. In the model of continuous stratification, the shadow zone next to the eastern boundary in the multi-layer ventilated thermocline discussed in the original LPS model is now replaced by the regime of stagnant water with continuous stratification. Thus, the technical difficulty in dealing with the exponentially increasing number of different shadow zones can easily be avoided in the model with continuous stratification.

In light of these new discoveries, the early model of Welander may be classified as some kind of similarity solution. The continuously stratified model has incorporated these new features. As one of the major differences from Welander's model, Q in Eqn. (4.304) is a given function only for the unventilated thermocline, and is unknown for the ventilated thermocline. The special nature of the boundary value, including the facts that the lower boundary is a free boundary and that function $Q(B, \rho)$ is not completely specified, gives rise to a unique problem in which a second-order ordinary differential equation is subject to four constraints.

This free boundary value problem is solved with a shooting method by starting from a first guess of the bottom of the moving water ρ^s. Integrating upward (toward lower density) to the base of the mixed layer, we can determine the potential vorticity of the uppermost ventilated layer as $q = f\Delta\rho/\Delta h$, where $\Delta\rho$ is the density increment, and Δh is the thickness of the uppermost layer. The generalized Sverdrup relation is then checked. If it is not satisfied, the base of the moving water, ρ^b, is adjusted until the integral constraint is met.

The overall structure of a continuously stratified model is shown in Figure 4.49. For the subtropical wind-driven gyre, the model integration starts from the inter-gyre (between the subpolar and subtropical gyres) boundary in the north. In the vertical direction there are four dynamical regimes. On the top there is the mixed layer where density is vertically constant. Both the mixed-layer density and depth are specified from late-winter properties. Below the mixed layer, there is the ventilated thermocline where potential vorticity is unknown and is calculated as part of the solution. Each ventilated layer is subducted at the next outcrop line. South of this new outcrop line, the newly subducted layer continues its southward motion underneath, keeping the potential vorticity formed during the subduction process. Below

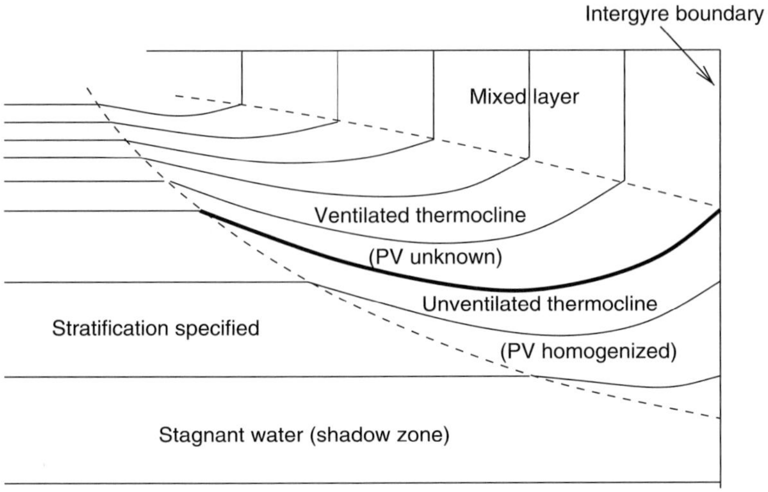

Fig. 4.49 Sketch of the ideal-fluid thermocline in a subtropical basin.

the ventilated thermocline there is the unventilated thermocline where potential vorticity is specified. Although any reasonably chosen forms of potential vorticity can be used, it is more convenient to assume that potential vorticity in the unventilated thermocline is homogenized toward the planetary vorticity along the northern boundary of the model. The lowest part of the model is the stagnant water in the abyss, where stratification is specified. Since water there does not move, a constant stratification in each layer implies that potential vorticity there is a function of latitude, although the concept of potential vorticity for stagnant water does not have much dynamical meaning.

Application to the North Pacific

This model was applied to both the North Atlantic and the North Pacific, with both h and ρ^s specified functions of geographical location, taken from the climatological mean density and depth datasets. The model ocean is divided into $m \times n$ grids, and the calculation of the three-dimensional structure of the wind-driven subtropical gyre is reduced to repeatedly solving this second-order ordinary differential equation at each station along the individual outcropping line. Typical isopycnal outcropping lines in late winter for the North Pacific are shown in Figure 4.50.

Along the northernmost outcrop line, starting from the first station next to the eastern boundary (or the northern boundary), we solve the free boundary value problem at each station. We assume that potential vorticity is a given function of density for the unventilated thermocline; thus, the solution at each station gives us the base of moving water at this station, and the Bernoulli function at the sea surface, B_s. After the completion of calculation along this outcrop line, we have a functional relation between potential vorticity and the Bernoulli function for this density ρ_1, and this function is stored in the form of a data array

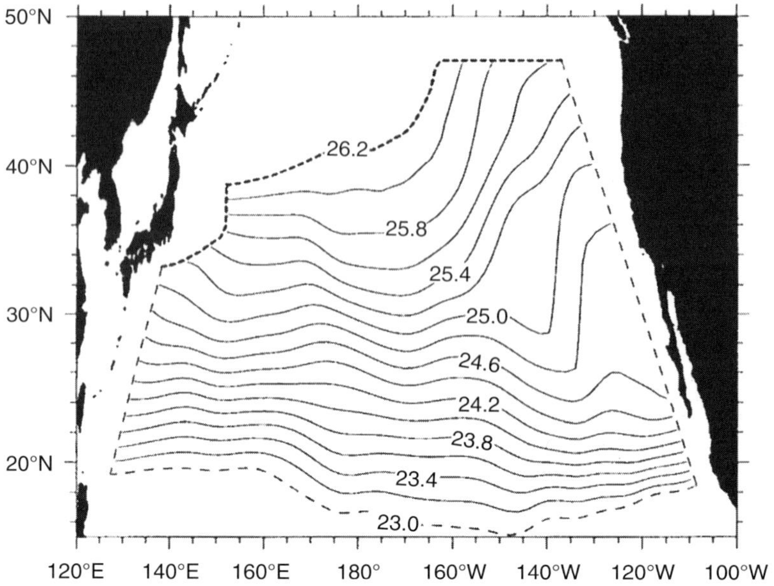

Fig. 4.50 Late-winter mixed-layer density distribution in the North Pacific, in σ (kg/m^3) (Huang and Russell, 1994).

in the computer. For outcrop lines southward, we can use this functional relation from the computer's data storage to solve the free boundary value problem along the next outcrop line with density lighter than ρ_1, and this process continues until we reach the southern boundary of the model basin.

This process produces the horizontal distribution of Bernoulli function on each isopycnal surface, and the application of the geostrophic condition gives rise to the horizontal velocity on each isopycnal surface. For example, streamlines on four isopycnal surfaces in the subtropical gyre of the North Pacific are shown in Figure 4.51. The shaded areas next to the eastern boundary depict the stagnant water on each isopycnal surface, which corresponds to the shadow zone in the multi-layered ventilated thermocline model. According to the model results, most of the wind-driven circulation in the North Pacific is ventilated.

The most prominent feature of the model is the strong ventilation due to the inclusion of a mixed layer of finite, horizontally varying depth. The southern shoaling of the late-winter mixed-layer depth gives rise to a strong lateral induction and subduction rate (Fig. 4.52c, d).

Three factors contribute to the volume flux in the ventilated thermocline: the vertical pumping from the Ekman layer convergence (Fig. 4.52a, b), the lateral induction (Fig. 4.52c), and the inter-gyre boundary outflow due to the northeast–southwest orientation of the zero-Ekman-pumping line (Fig. 4.53). In fact, these three are equal contributors. The southern shoaling of the mixed layer and the induced lateral induction from the mixed layer into the main thermocline have a very important impact on the wind-driven circulation

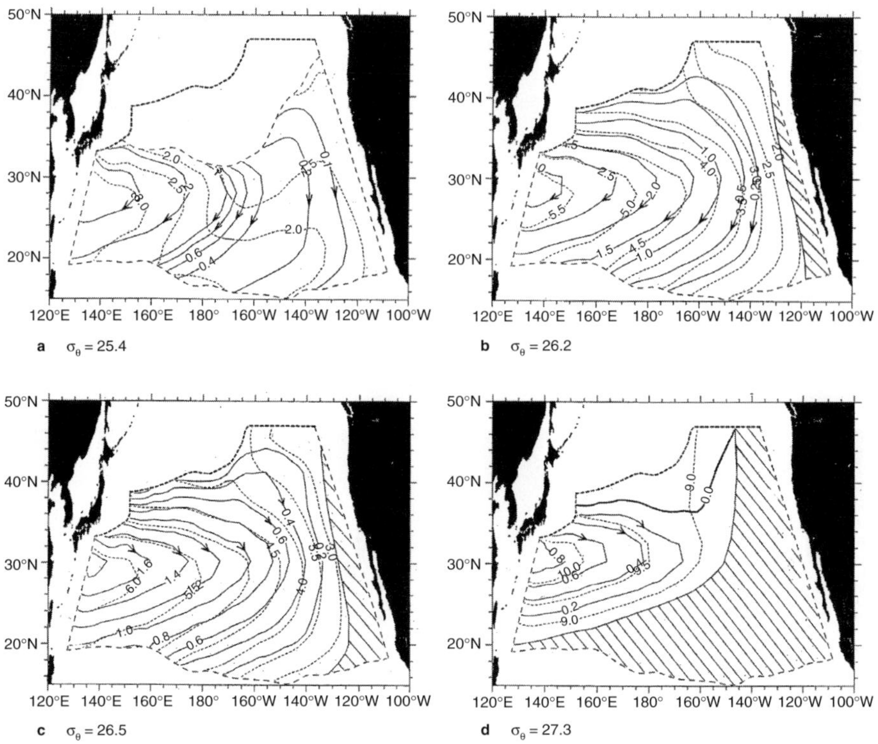

Fig. 4.51 Circulation on four different isopycnal surfaces of the wind-driven subtropical gyre in the North Pacific; *solid lines with arrows* for the layer-integrated streamlines and *numbers* for the layer-integrated volume flux (Sv), *dotted lines* for the water age in years since subduction; the shadow areas (*shaded*) for the stagnant water on each isopycnal surface (Huang and Russell, 1994).

and climate; these issues will be discussed again in Section 5.1.5 specially dedicated to water mass formation through subduction. These fluxes can be identified from either a model of continuous stratification (Fig. 4.53a) or a diagnostic calculation from historical hydrographic data (Fig. 4.53b). The numbers on the left edge of each panel indicate the mass exchange with the western boundary or the inter-gyre boundary. Since the inter-gyre boundary has a large meridional distance, most of the influx from the western boundary actually comes from the inter-gyre boundary. As shown in Figure 4.53, inter-gyre mass flux is thus a major contributor to the flux on each isopycnal layer.

4.3 Structure of circulation in a subpolar gyre

4.3.1 Introduction

Wind-driven circulation is a dominant component of the current system in the upper ocean. The simplest model is the widely used reduced-gravity model. In such a model, the specification of boundary conditions is quite similar for both the subtropical and subpolar

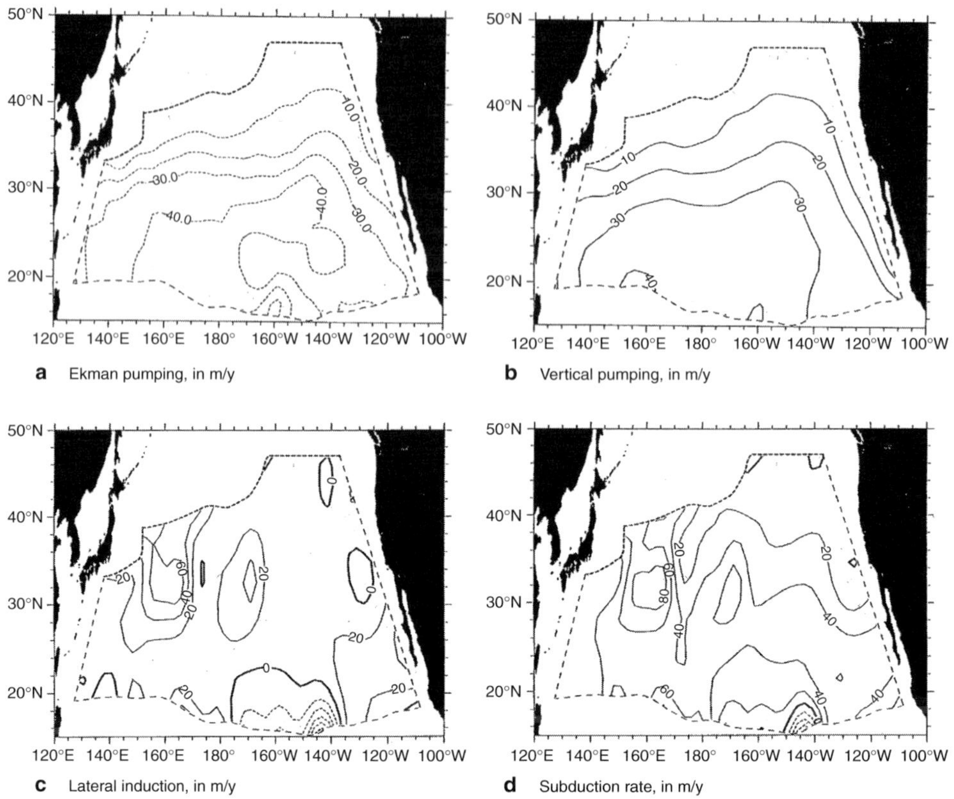

Fig. 4.52 Vertical pumping and subduction rate for the North Pacific, based on the ideal-fluid thermocline model (Huang and Russell, 1994).

gyres; therefore, the discussion in Section 4.1.2 about the reduced-gravity model with a single moving layer for the subtropical gyre also applies to the description of the subpolar gyre, with the only difference being that the Ekman pumping in the subpolar basin is positive. As a result, the geostrophic flow in the subpolar gyre interior flows poleward under the Ekman suction owing to the positive wind-stress curl, and isopycnal surfaces are dome-shaped. The corresponding western boundary current in the subpolar basin moves equatorward. The corresponding theories of the Stommel layer, the Munk layer, and the inertial western boundary layer all work in a way similar to those in the subtropical basin.

However, the wind-driven circulation in a subpolar basin is different from that in a subtropical basin in the following ways. First, Ekman pumping is upward. Although this does not seem to matter much for a model with a single moving layer, it changes the dynamics and formulation of models with multiple layers or continuous stratification. The ideal-fluid thermocline model belongs to the so-called hyperbolic system in mathematics. For a

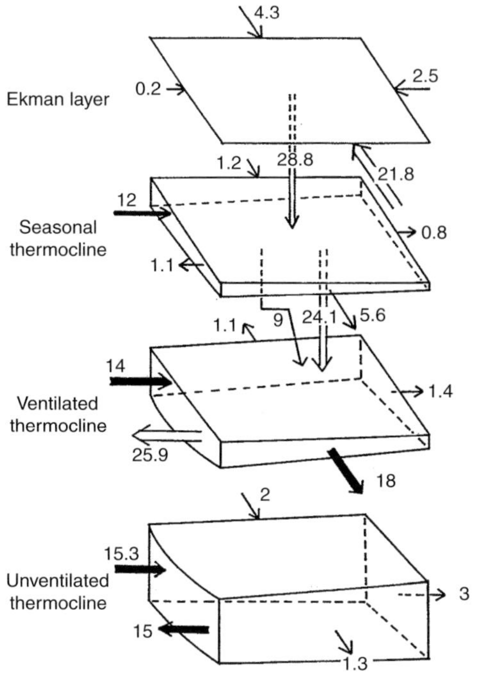

Fig. 4.53 Structure of the wind-driven circulation for the North Pacific (numbers indicate volume flux in S_V), diagnosed from **a** dynamical calculation (Huang and Qiu, 1994) and **b** the ideal-fluid thermocline model (Huang and Russell, 1994).

hyperbolic system, boundary conditions are normally specified at the upstream boundary, not the downstream boundary. In the subtropical basin, boundary conditions, such as the mixed-layer density and depth, are specified on the upper boundary of the model; however, in the subpolar basin, the sea surface is the "downstream boundary," so that mixed-layer density and depth cannot be specified as boundary conditions, as will be discussed in detail in this section.

Second, stratification in the subpolar basin is much weaker than that in the subtropical basin; thus, for the same strength of wind-stress curl, the wind-driven gyre in the subpolar basin penetrates much deeper. Thus, the possibility for the wind-driven gyre to interact with bottom topography is much greater than that in the subtropical basin. However, for simplicity, our discussion in this section will be limited to the case without interaction with bottom topography. The relevant dynamical problems were discussed by Luyten, Stommel, and Wunsch (Luyten *et al.*, 1985).

This section is devoted to the structure of the subpolar gyre, including a simple $2\frac{1}{2}$-layer model, a model with constant potential vorticity for the unventilated thermocline, and a model for the case with non-uniform potential vorticity in the unventilated thermocline. In addition, we also discuss the water mass formation and erosion in the subpolar basin.

These materials should be read after the reader has become familiar with the concepts of water mass formation in Section 5.1.5. However, these materials for the subpolar basin are included in this chapter to make the theoretical framework of wind-driven circulation complete.

4.3.2 A $2\frac{1}{2}$-layer model

The circulation in a subpolar basin is first examined using a $2\frac{1}{2}$-layer model. The major difference between circulation in a subpolar basin and in a subtropical basin is that all subsurface moving layers in a subpolar gyre are unventilated, i.e., streamlines in these layers cannot be traced back to the mixed layer. In fact, the mixed layer is at the downstream end of streamlines in the upper ocean. Thus, potential vorticity of all moving layers in the subpolar basin is assumed to be set at the outflow region of the western boundary, and in our simple model must be specified *a priori*; a simple choice is to assume that potential vorticity in these layers is homogenized. Parallel to Rhines and Young (1982a), we will assume that potential vorticity in all unventilated thermocline layer is homogenized toward their value along the inter-gyre boundary, which is assumed to be along a constant latitude.

In the pioneering work on the ventilated thermocline, Luyten, Pedlosky and Stommel (Luyten *et al.*, 1983) briefly discussed the circulation in the subpolar basin. In fact, the spirit of ventilation can easily be extended into the case of the subpolar basin, with the minor difference that now all subsurface layers are unventilated. Parallel to the discussion related to the boundary condition along the eastern wall for the subtropical basin, it is readily shown that there is a shadow zone along the eastern boundary for all the subsurface layers in a subpolar basin.

The western edge of the shadow zone is determined by the constraint that potential vorticity of the moving water in the second layer is conserved, i.e.,

$$h_2 = \frac{f}{f_0} h_{2e} \tag{4.308}$$

where f_0 is the Coriolis parameter at the southern boundary of the subpolar basin, and h_{2e} is the second-layer thickness along the eastern boundary and the inter-gyre boundary. From this relation, layer thickness increases, following the streamline because the Coriolis parameter is larger than its value at the inter-gyre boundary. Note that we here ignore the possible inter-gyre communication discussed by Pedlosky (1984). This assumption makes the formulation of the model much simpler, because stratification along the inter-gyre boundary and the eastern boundary is the same.

We first summarize the basic equations for a simple reduced-gravity model:

$$-2\omega \sin\theta h v = -\frac{g'}{a\cos\theta} h h_\lambda + \frac{\tau^\lambda}{\rho_0} \tag{4.309}$$

$$2\omega \sin \theta hu = -\frac{g'}{a} hh_\theta + \frac{\tau^\theta}{\rho_0} \tag{4.310}$$

$$(hu)_\lambda + (hv \cos \theta)_\theta = 0 \tag{4.311}$$

where g' is the reduced gravity. From these equations, the layer thickness satisfies

$$h^2 = h_e^2 + \frac{2a}{g'} \int_\lambda^{\lambda_e} \left[\tan \theta \left(\tau_\lambda^\theta - \tau_\theta^\lambda \cos \theta \right) - \tau^\lambda / \cos \theta \right] dx \tag{4.312}$$

Thus, the upper layer thickness over the whole basin can be calculated from

$$h^2 = h_e^2 + \frac{2a}{g'} \int_\lambda^{\lambda_e} P_r d\lambda \tag{4.313}$$

where

$$P_r = -2a\omega \sin^2 \theta w_e, \; w_e = \frac{1}{2\omega\rho_0 a \sin \theta} \left[\frac{1}{\cos \theta} \tau_\lambda^\theta - \tau_\theta^\lambda + \frac{\tau^\lambda}{\sin \theta \cos \theta} \right] \tag{4.314}$$

is the pumping rate.

The wind-driven circulation for a $2\frac{1}{2}$-layer model in a subpolar basin can be classified into three regions (Fig. 4.54).

Region I, where the second layer is stagnant, i.e., this is the shadow zone for the second layer. Since the upper layer is the only active layer, the solution is

$$h_1^2 = h_{1e}^2 + \frac{2a}{g'} \int_\lambda^{\lambda_e} P_r d\lambda \tag{4.315a}$$

$$h_2 = h_{1e} + h_{2e} - h_1 \tag{4.315b}$$

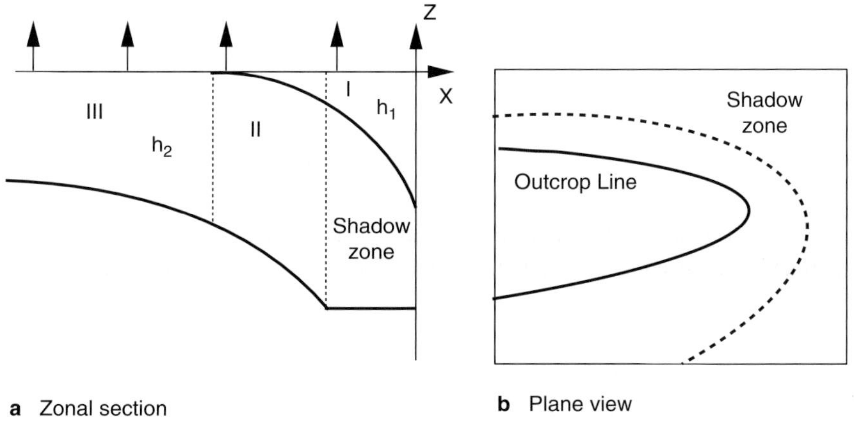

Fig. 4.54 Sketch of the wind-driven circulation in a $2^1/_2$-layer ideal-fluid model for the subpolar gyre. The *dashed line* depicts the western edge of the shadow zone in the second layer and the *solid line* for the outcrop line of the upper interface.

Using Eqn. (4.308), the western edge of this region can be defined as the place where the following constraint is satisfied:

$$h_1 = h_{1e} + \left(1 - \frac{f}{f_0}\right) h_{2e} \tag{4.316}$$

This constraint is based on the assumption that the upper-layer thickness is larger than zero, i.e., we assume that the western boundary of the shadow zone is east of the outcrop line. If the upper-layer thickness along the eastern boundary is too small, within a certain part of the basin the upper-layer outcrop line would coalesce with the western edge of the shadow zone. It is dynamically inconsistent to have the outcrop line appear east of the shadow zone boundary, because the second layer will be directly exposed to the surface forcing within the outcrop zone.

Region II, where both the first and second layers are in motion, and the solution is

$$2h_1^2 + 2h_1h_2 + h_2^2 = 2h_{1e}^2 + 2h_{1e}h_{2e} + h_{2e}^2 + \frac{2a}{g'}\int_{\lambda}^{\lambda_e} P_r d\lambda \tag{4.317a}$$

$$h_2 = \frac{f}{f_0} h_{2e} \tag{4.317b}$$

The western edge of this region is defined by the outcrop line of the upper layer

$$h_1 = 0 \tag{4.318}$$

Region III, where the first layer vanishes and the second layer is the only active layer, so the solution is

$$h_2^2 = 2h_{1e}^2 + 2h_{1e}h_{2e} + h_{2e}^2 + \frac{2a}{g'}\int_{\lambda}^{\lambda_e} P_r d\lambda \tag{4.319}$$

The shape of these dynamical regimes will be shown clearly in the following discussion about the subpolar gyre obtained from a model with continuous stratification. In particular, we will show the horseshoe shape of the shadow zone or the stagnant water below the wind-driven gyre.

4.3.3 A continuously stratified model

The three-dimensional structure of the wind-driven gyre in the subpolar basin for a model with continuous stratification can be described in terms of the Bernoulli function

$$B = p + \rho g z \tag{4.320}$$

The derivation of the basic equations discussed below parallels that for the subtropical basin. For large-scale circulation, geostrophy is a good approximation, so the horizontal momentum equations are reduced to

$$f\rho_0 v = B_x, \quad f\rho_0 u = -B_y \tag{4.321}$$

and the vertical momentum equation degenerates to the hydrostatic relation

$$B_\rho = gz \tag{4.322}$$

where the subscript ρ indicates a partial derivative. The linear vorticity equation is

$$\beta v z_\rho = f w_\rho \tag{4.323}$$

Using Eqns. (4.321, 4.322, 4.323) to eliminate v and z leads to

$$B_{\rho\rho} B_x = \frac{g\rho_0 f^2}{\beta} w_\rho \tag{4.324}$$

Integrating this equation over $[\rho^s, \rho^b]$ (where ρ^s is the mixed layer density and ρ^b is the unknown free boundary separating the moving water from the stagnant abyssal water) gives

$$\int_{\rho^s}^{\rho^b} B_\rho^2 d\rho + \int_{\rho^{se}}^{\rho^{be}} B_\rho^{e^2} d\rho + \int_{\rho^{be}}^{\rho^b} B_\rho^{a^2} d\rho = \frac{2g\rho_0 f^2}{\beta} \int_x^{x_e} w_e dx \tag{4.325}$$

From these equations, the calculation of the wind-driven gyre in the subpolar basin is reduced to solving the following free-boundary value problem in density coordinates

$$B_{\rho\rho} = \frac{fg}{Q(B, \rho)} \tag{4.326}$$

with the constraints

$$B_\rho = -gh(x, y) \text{ at } \rho = \rho^s \text{ } (\rho^s \text{ is unknown}) \tag{4.327}$$

$$B = B^a,\ B_\rho = B_\rho^a(\rho),\ \text{at } \rho = \rho^b \text{ } (\rho^b \text{ is unknown}) \tag{4.328}$$

$$-\int_{\rho_s}^{\rho_b} B_\rho^2 d\rho + \int_{\rho_s}^{\rho_b} B_\rho^{a^2} d\rho = \frac{2g\rho_0 f^2}{\beta} \int_x^{x_e} w_e dx \tag{4.329}$$

The major differences between the subtropical and subpolar gyres are: the upper boundary ρ^s is fixed in the subtropical gyre, but it is a free boundary in the subpolar basin. On the other hand, the potential vorticity function $Q(B, \rho)$ is not completely given for the subtropical gyre; however, it is completely specified for the subpolar gyre because all moving layers in the subpolar gyre are unventilated.

As a result, there is a simple relation between ρ^s and ρ^b. The depth of the isopycnal surface ρ^b along the eastern boundary (or the inter-gyre boundary) is $h_0^b = h(\rho_b)$. Note that at this station $f = f_s$, and h_0^b is also the layer thickness between isopycnals ρ^s and ρ^b. Since potential vorticity is constant within an unventilated isopycnal layer, if we trace back to the inter-gyre boundary, the water column height between ρ^s and ρ^b should be

$$\Delta h_0 = h_0^b - h_0^s = h_0^b \frac{f_0}{f_s} \tag{4.330}$$

where h_0^s is the depth of the ρ^s isopycnal surface at the inter-gyre boundary $f = f_0$. Thus,

$$h_0^s = h_0^b \frac{f_s - f_0}{f_s} \tag{4.331}$$

Using the function relation between density and depth, we have an implicit relation between ρ^b and ρ^s

$$\rho^s = F(\rho^b) \tag{4.332}$$

Therefore, this problem with two free boundaries is reduced to a problem with a single free boundary, and we can use a shooting method to solve the problem.

The case with constant potential vorticity for the unventilated thermocline

The solution in this case is in a closed analytical form. Assuming the constant stratification for the abyssal water is ρ_z^a, the depth of the wind-driven gyre is defined as

$$H = -\frac{\Delta\rho}{\rho_z^a}, \quad \Delta\rho = \rho_b - \rho_e \tag{4.333}$$

where ρ^b is the density at the base of the wind-driven gyre, and ρ^e is the density along the eastern boundary. In the gyre interior, potential vorticity is constant, equal to $-f_0\rho_z^a$; thus, the density difference between the bottom and surface of the wind-driven gyre is

$$\rho_b - \rho_s = \frac{f_0}{f}\Delta\rho \tag{4.334}$$

The depth of an isopycnal surface ρ is

$$h(\rho) = -\frac{f}{f_0\rho_z^a}(\rho - \rho_s) = \frac{f}{f_0\rho_z^a}\left(\rho - \rho_e - \frac{f - f_0}{f}\Delta\rho\right) \tag{4.335}$$

Therefore, the first integral in Eqn. (4.329) is

$$-\int_{\rho_s}^{\rho_b} B_\rho^2 d\rho = -\int_{\rho_s}^{\rho_b} (gh(\rho))^2 d\rho = -\frac{f_0}{3f}\left(\frac{g}{\rho_z^a}\right)^2 \Delta\rho^3 \tag{4.336}$$

Similarly, the second integral in Eqn. (4.329) is

$$\int_{\rho_s}^{\rho_b} B_\rho^{a^2} d\rho = \frac{1}{3}\left(\frac{g}{\rho_z^a}\right)\Delta\rho^3 \tag{4.337}$$

The Sverdrup relation is thus reduced to a simple algebraic equation in $\Delta\rho$

$$\left(\frac{g}{\rho_z^a}\right)^2 \left(\frac{f - f_0}{f}\right)\Delta\rho^3 = \frac{6g\rho_0 f^2}{\beta} w_e \Delta x \tag{4.338}$$

This equation can be used to estimate the depth of the wind-driven gyre in the subpolar basin. Using the definition $H = -\Delta\rho/\rho_z^a$, we obtain the following estimate

$$H^3 = -\frac{1}{g\rho_z^a}\left(\frac{f}{f-f_0}\right)\frac{6\rho_0 f^2}{\beta}w_e\Delta x \tag{4.339}$$

Assuming that $g = 9.8\,\text{m/s}^2$, $\rho_z^a = -0.333\times10^{-3}\,\text{kg/m}^4$, $\beta = 10^{-10}$/s/m, $f = 1.5\times10^{-4}$/s, $f_0 = 1.3\times10^{-4}$/s, $w_e = 10^{-6}$ m/s, and $\Delta x = 5,000$ km, then the depth of the wind-driven gyre is about 3 km; thus, the wind-driven gyre in the subpolar basin is quite deep. In comparison, owing to the relative strong stratification in the subtropics, most wind-driven gyres in the subtropical basin are on the order of 2 km in depth.

The case with a non-constant potential vorticity for the unventilated thermocline

As an example, we discussed the structure of the subpolar gyre circulation for a case where the potential vorticity in the unventilated thermocline is a function of density. The solution is calculated from a simple numerical model based on a shooting method for solving the nonlinear equation. The subpolar basin interior, excluding the western boundary region, is divided into $m \times n$ stations. The solution is calculated by solving the above free boundary value problem, Eqns. (4.327, 4.328, 4.329), station by station.

The model ocean is a rectangular basin of $60° \times 20°$ ($0° - 60°$ E, $45° - 65°$ N), mimicking the northern North Atlantic. Input data required for the calculation include the Ekman pumping rate and the background stratification (or the potential vorticity for all the moving water). The Ekman pumping forcing has a simple form

$$w_e = 1.0 \times 10^{-6}\sin\left(\frac{\theta-\theta_s}{\Delta\theta}\pi\right) m/s \tag{4.340}$$

where θ_s is the southern boundary of the model basin, and $\Delta\theta = 20°$ is the meridional range of the model basin (Fig. 4.55b).

As discussed above, the major difference between subtropical and subpolar gyres is that isopycnals in the subpolar gyre are all unventilated, because these isopycnals come from the deeper part of the ocean and move upward in the cyclonic gyre. As a result, the formulation of the ideal-fluid thermocline model is different for the subtropical and subpolar gyres. In the subpolar gyre, we cannot specify the density and depth of the mixed layer, but the potential vorticity for all isopycnals is specified.

Winter cooling at the sea surface is one of the most essential aspects of the circulation in a subpolar basin. In order to account for the cooling effect, Huang (1988b) proposed a model in which cooling is treated in terms of a weakly convective adjustment to the thermal structure set up by the model without considering the cooling. While the approach provides useful information about the dynamical structure of the circulation in the subpolar basin, the cooling pattern is subject to certain dynamical constraints. As a result, there may not be any solution for a given cooling pattern.

In this section, we adopt a different approach. First, we will find a steady solution with a mixed layer with zero depth for the wind-driven circulation in a subpolar basin, without

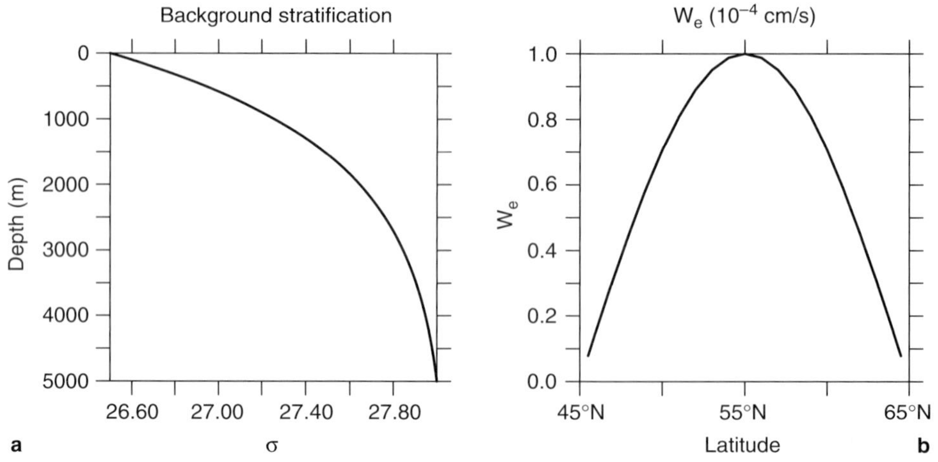

Fig. 4.55 **a** Background stratification (kg/m^3), **b** Ekman pumping velocity as a function of latitude.

imposing any density condition on the sea surface. Second, we will treat the winter cooling in terms of convection events, i.e., we will calculate the density structure at a given station as a one-dimensional process. After the convection event takes place in each station, the horizontal velocity and pressure gradient in the basin will not be in geostrophic balance. Consequently, geostrophic adjustment will take place afterward. However, the details of such geostrophic adjustment will not be discussed here. Readers who are interested in the geostrophic adjustment problem are referred to other studies, e.g., Dewar and Killworth (1990). The goal in this section will be confined to the following question: How are deep-water formation and its properties controlled by the climate conditions in the subpolar basin?

The background stratification used in this section is calculated by assuming a simple advection–diffusion balance

$$w\rho_z = \kappa\rho_{zz} \tag{4.341}$$

where $w = 1.0 \times 10^{-7}$ m/s is the upwelling velocity set to equal the Ekman suction rate specified at the sea surface, and the vertical diffusivity κ is varied from 0.01×10^{-4} to 3×10^{-4} (m^2/s). The ratio $H = \kappa/w$ is the scale height of the stratification. The example shown in this section is obtained for the case of $\kappa = 1.5 \times 10^{-4}$ m^2/s, which corresponds to a scale height of 1.5 km. This equation yields an exponential density profile. Using a density boundary condition of $\rho = 1,023$ kg/m^3 at $z = 0$ and $\rho = 1,028$ kg/m^3 at the sea floor $z = -5$ km, the density profile can be calculated accordingly (Fig. 4.55a). In addition, this profile will also be used as the resting level along the eastern boundary for all layers.

Using these forcing conditions, we obtain a cyclonic gyre in the subpolar basin, which has a dome shape. Typical density sections and velocity distribution through the middle and western parts of the gyre are shown in Figure 4.56. It is clear that wind-driven circulation

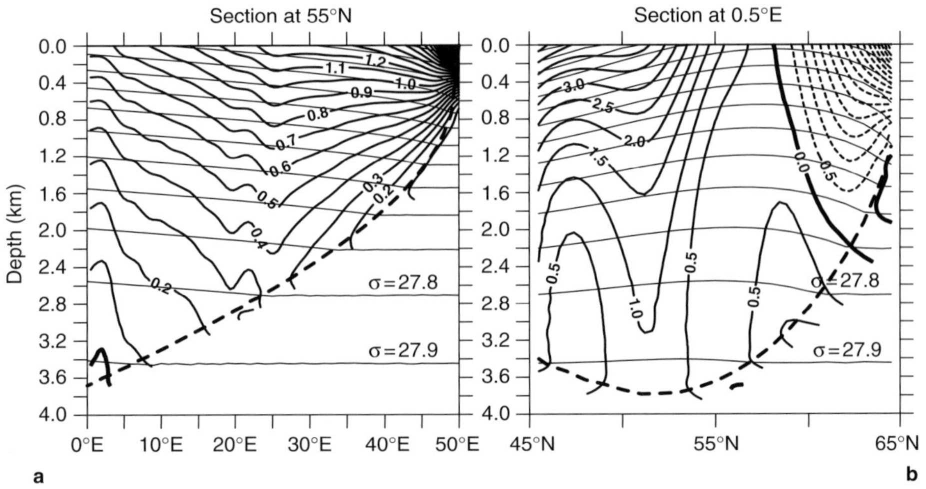

Fig. 4.56 Density stratification (*thin lines*) and horizontal velocity (*heavy lines*, in 10^{-2}/m/s) at two sections; the *dashed lines* indicate the base of the wind-driven gyre with stagnant water below, with the *heavy dashed lines* indicating the base of moving water in the wind-driven gyre: **a** zonal section taken along 55° N; **b** meridional section taken along the outer edge of the western boundary.

penetrates to a great depth in the subpolar basin. Since the wind-driven circulation is so deep, it seems rather difficult to separate the wind-driven circulation from the thermohaline circulation there; thus, the purely wind-driven circulation discussed in this section is only an idealization, and care should be taken when applying it to the oceanic situation.

Note that in this section the base of the wind-driven gyre reaches the maximal depth at latitude 46° N; thus, water belonging to this deepest part does not come from the southern boundary of the subpolar basin; instead, it comes from the western boundary at latitudes higher than 45° N. Therefore, potential vorticity for this part of the thermocline may not be homogenized toward the southern boundary of the subpolar basin. As a working assumption, we assume that potential vorticity on such an isopycnal surface is homogenized toward the planetary vorticity at the outer edge of the western boundary; thus, it is not $-f_0\rho_z^a$: instead it is $-f_b\rho_z^a$, where f_b is the Coriolis parameter for this specific water mass to enter the wind-driven gyre from the western boundary.

The sea surface elevation map clearly shows the cyclonic gyre, with a maximum sea-level depression of 45 cm in the middle of the western boundary (Fig. 4.57a). The sea surface density map clearly demonstrates the outcropping of isopycnal surfaces, with the heaviest density surface outcrops in the middle of the western boundary (Fig. 4.57b). The $\sigma = 27$ kg/m^3 surface has a dome shape, and the shadow zone along the eastern edge is too narrow to be shown in this panel (Fig. 4.57c).

The wind-driven gyre penetrates to a maximum depth of more than 3.5 km in the southern part of the western boundary region (Fig. 4.57d). Since most high-latitude oceans are not very deep, the bottom topography is likely to interact with the wind-driven gyre; thus, the

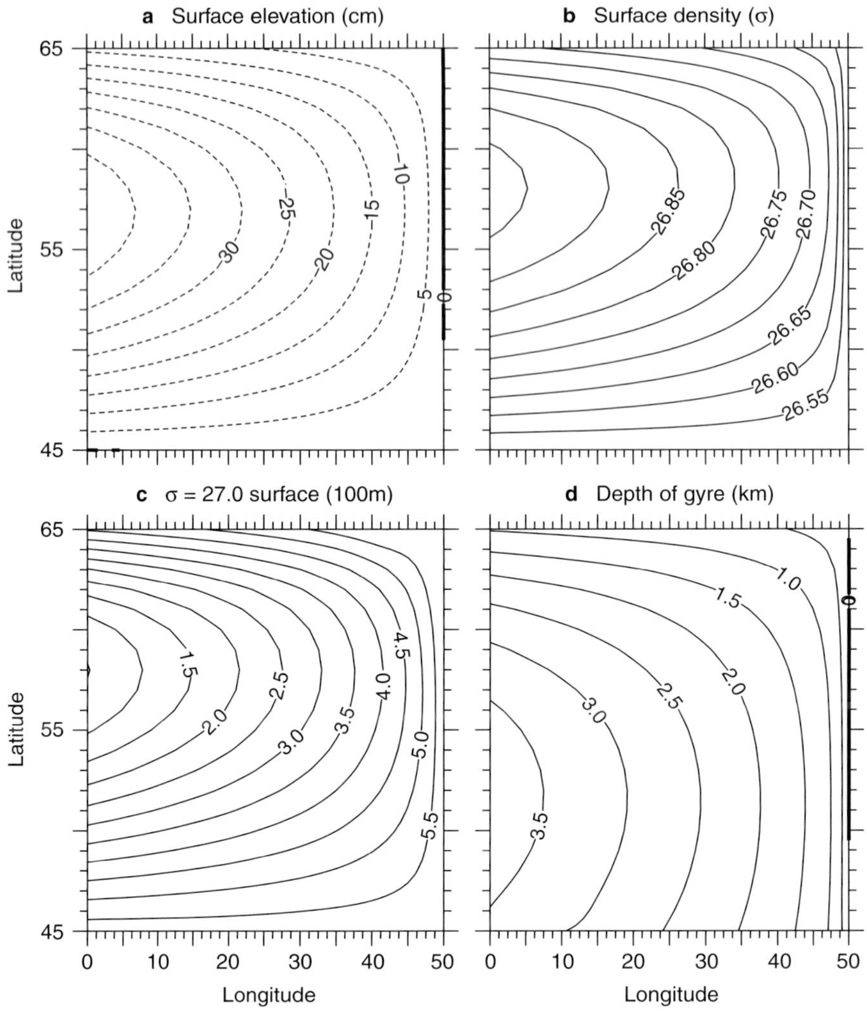

Fig. 4.57 Basic structure of the wind-driven circulation in a subpolar basin interior: **a** surface elevation (cm); **b** surface density, in σ units (kg/m^3); **c** depth of $\sigma = 27.0$ isopycnal surface (in 100 m); **d** depth of the wind-driven gyre (km).

simple Sverdrup relation for a purely wind-driven circulation may need some modification. On the other hand, any current below 2 km is rather weak, so the solution obtained from the model may still give useful information about the structure of the subpolar gyre.

Water mass formation and erosion

The subpolar gyre is dominated by Ekman upwelling and the associated water mass transformation from the thermocline to the surface mixed layer. The water mass transformation

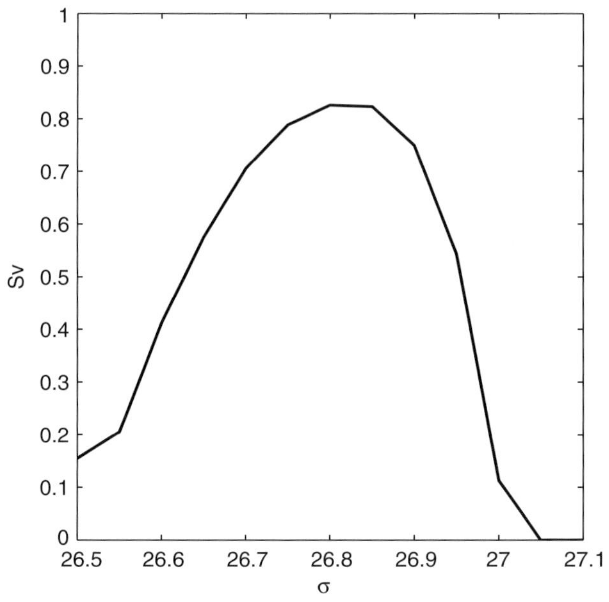

Fig. 4.58 Obduction rate distribution in density coordinates (in Sv per 0.5 kg/m^3).

rate can be defined in terms of the obduction rate (Fig. 4.58), which is discussed in detail in Section 5.1.5. Since the mixed-layer depth is set to zero in this model, the obduction rate is equal to the Ekman pumping rate. A basin-integrated budget gives rise to its distribution in density coordinates, with a peak at $\sigma = 26.8$ kg/m^3. The total amount of water flowing from the thermocline into the surface layer is equal to the total amount of the Ekman upwelling rate, 5.90 Sv.

There is another process in the subpolar basin that is the opposite to obduction, i.e., the formation of subpolar mode water, which takes place over the late-winter cooling period when strong surface cooling leads to deep convection that can reach to more than 2 km. In order to understand the dynamical factors that dominate the subpolar mode water formation, we study several experiments in which the ocean is cooled within a patch near the western boundary, in the form of a Gaussian profile:

$$Q = Q_0 \exp\left[\frac{(\theta - \theta_0)^2 + (\lambda - \lambda_0)^2}{\Delta\lambda^2}\right] \tag{4.342}$$

where $Q_0 = 5 \times 10^9$ J/m^2, which is equivalent to cooling the surface at a rate of 600 W/m^2 over 100 days. Note this is a strong heat loss to the atmosphere, while the realistic range of cooling rate observed in the Labrador Sea is about 300 W/m^2.

Owing to the cooling, a convective adjustment takes place; this is treated as a one-dimensional process. The whole water column above a certain depth will have the same temperature and density. This convection depth can be calculated by setting the cooling heat loss equal to the total amount of heat storage over the depth of the convective water column. The structure of the convective region is shown in Figure 4.59. Note that the convection can reach a depth of 2 km.

During subpolar mode water formation through such convective processes, the amount of mode water formation and its properties depend on several dynamical forcing factors. Most

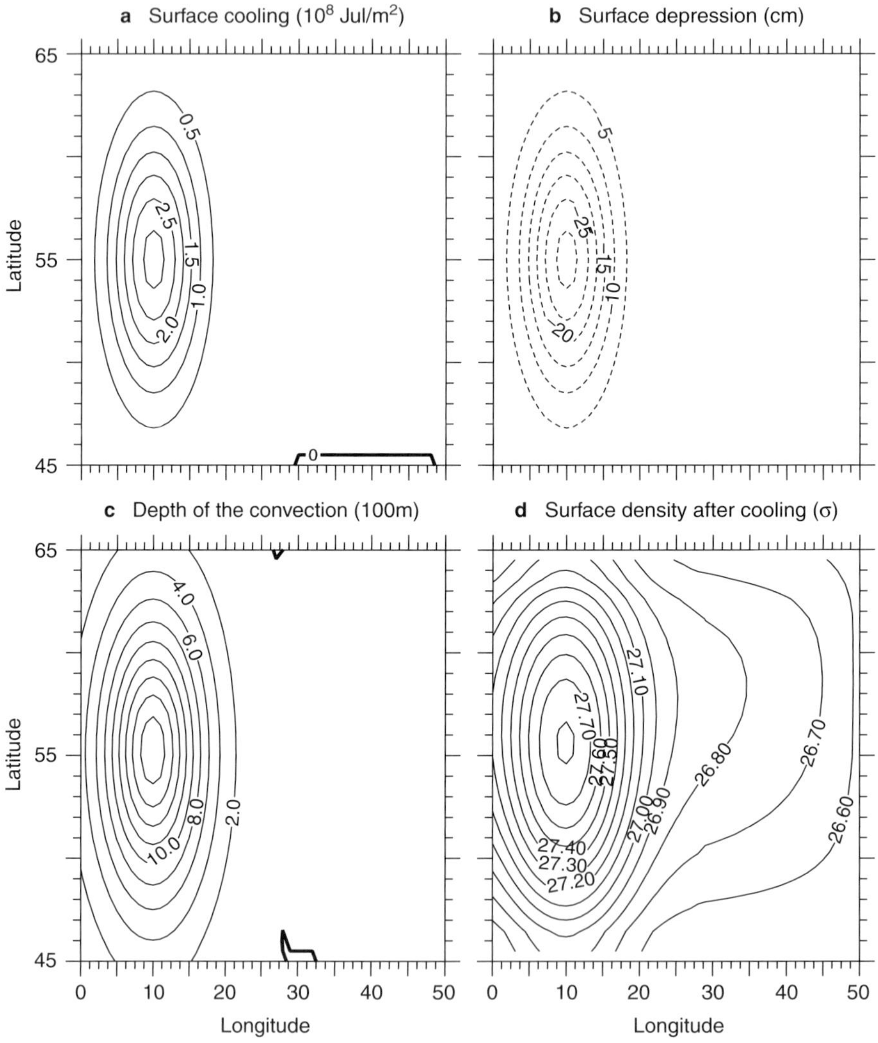

Fig. 4.59 Structure of the convective overturning zone: **a** surface cooling (in 10^8 J/m^2); **b** surface depression due to cooling (cm); **c** depth of the convection (in 100 m); **d** surface density after cooling (kg/m^3).

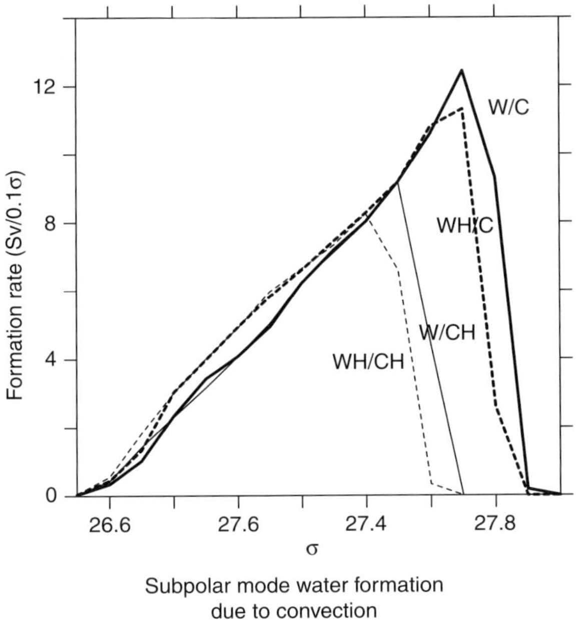

Fig. 4.60 Distribution of subpolar mode water formation rate in density coordinates, in Sv per each 0.1 kg/m^3 density interval. *Heavy solid line labeled W/C* indicates full Ekman pumping and cooling; *heavy dashed line labeled WH/C* indicates half Ekman pumping and full cooling; *thin solid line labeled W/CH* indicates full Ekman pumping and half cooling; *thin dashed line labeled WH/CH* indicates half Ekman pumping and cooling.

importantly, the amount of mode water formation is roughly proportional to the strength of cooling (Fig. 4.60). In addition, it also depends on the strength of the Ekman pumping. Ekman pumping plays the role of preconditioning the dome-shaped thermocline in the basin interior. Since stratification in the deep ocean is weaker than that on the shallow density layers, strong Ekman pumping can create a dome with weaker stratification in the middle. As a result, the same amount of cooling generates more mode water; however, the effect due to Ekman pumping is not dominant. On the contrary, the strength of cooling is the dominating factor in subpolar mode water formation.

Perturbations due to climate variability

Using this model we can also study the variability of the circulation induced by anomalous forcing. Within the framework of the ideal-fluid thermocline, a major difference between the subtropical and subpolar gyres is that potential vorticity in the ventilated thermocline changes with the alterations in the upper boundary conditions, but potential vorticity in the unventilated thermocline is specified *a priori*. As a result, the potential vorticity function in the subpolar basin does not change with the upper boundary forcing conditions.

For example, in the subtropical basin, when the outcrop lines are non-zonal, anomalous Ekman pumping can create perturbations in the form of the second baroclinic mode, which propagate along characteristics. The existence of such perturbations is due to changes in the potential vorticity function in the ventilated thermocline. However, in this ideal-fluid thermocline model, potential vorticity in the thermocline in a subpolar basin is specified *a priori*; thus, anomalous Ekman pumping can only change the flow field to the west of the perturbation source.

In order to demonstrate this point, we show the results from an experiment in which the subpolar gyre is driven by the same Ekman pumping velocity plus a small perturbation, i.e., the total Ekman pumping is in the form:

$$w_e = 10^{-6} \cos\left(\frac{\theta - \theta_0}{\Delta\theta}\pi\right) + 0.5 \times 10^{-6} \exp\left[\frac{(\theta - \theta_0)^2 + (\lambda - \lambda_0)^2}{\Delta\lambda^2}\right] \tag{4.343}$$

where $\theta_0 = 55°$, $\lambda_0 = 40°$, $\Delta\lambda = \Delta\theta = 5°$ (Fig. 4.61).

Under such anomalous Ekman pumping velocity, perturbations propagate westward from the source of Ekman pumping anomaly (Fig. 4.62). Perturbations induced by the anomalous Ekman pumping are in the form of the barotropic mode, with no baroclinic structure. This is a dramatic difference from the rich baroclinic structure induced by surface forcing conditions, as in the case of the subtropical basin discussed in Section 4.9.

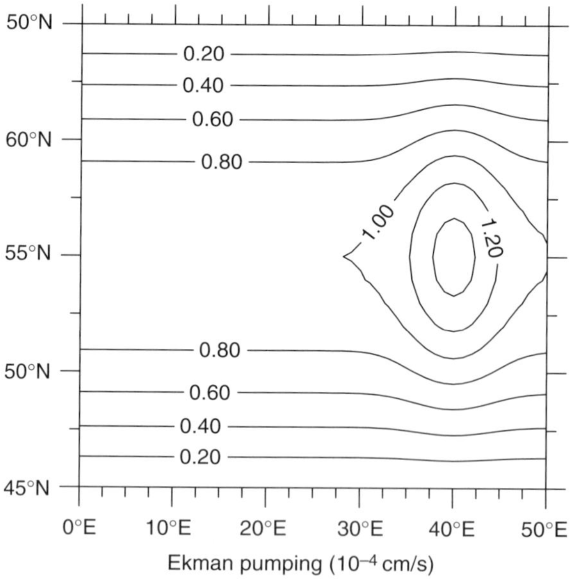

Fig. 4.61 Ekman upwelling velocity (in 10^{-6} m/s), including a patch of strong upwelling in the middle of the basin.

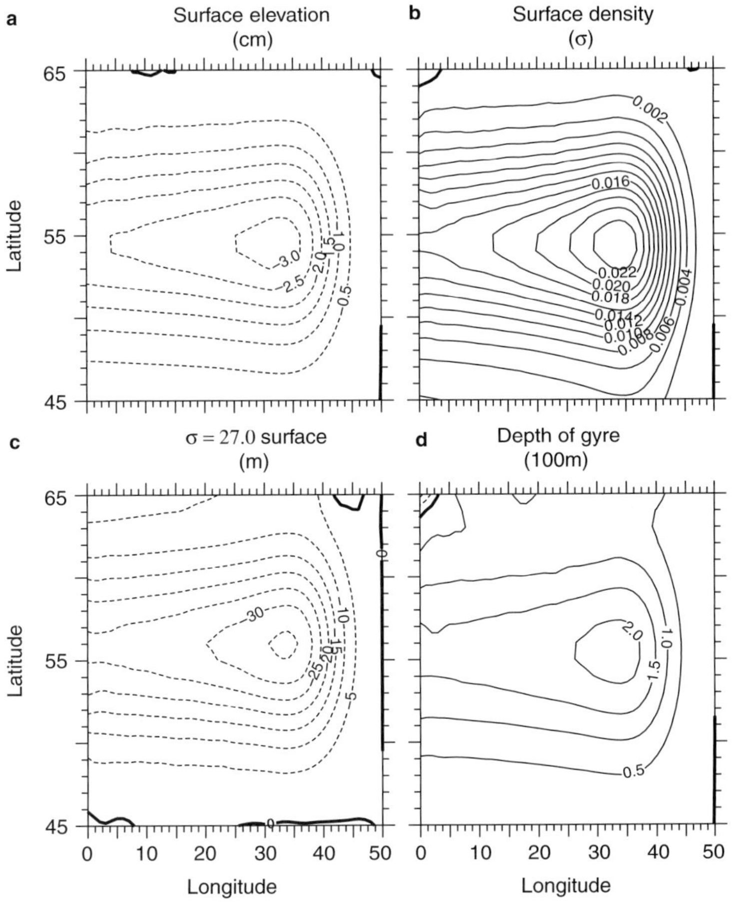

Fig. 4.62 Perturbations due to the anomalous Ekman pumping, shown in Fig. 4.61: **a** surface elevation (cm); **b** surface density (kg/m^3); **c** depth of $\sigma = 27.0$ kg/m^3 isopycnal surface (m); **d** depth of the wind-driven gyre (in 100 m).

4.4 Recirculation

4.4.1 Motivation

Our discussion has been limited to models based on the Sverdrup dynamics in the interior and western boundary currents matched to the interior solution along the western boundary, including the case with an inertial western boundary current matched to the interior solution for the southern half of the western boundary region. Within this theoretical framework, the maximal streamfunction is totally determined by a zonal integration of the Ekman pumping velocity, started from the eastern boundary of the basin. Observations, however, indicate that the maximal volume flux in the Gulf Stream is about 150 Sv, which is several times larger than the value calculated from the Sverdrup relation (Fig. 4.63).

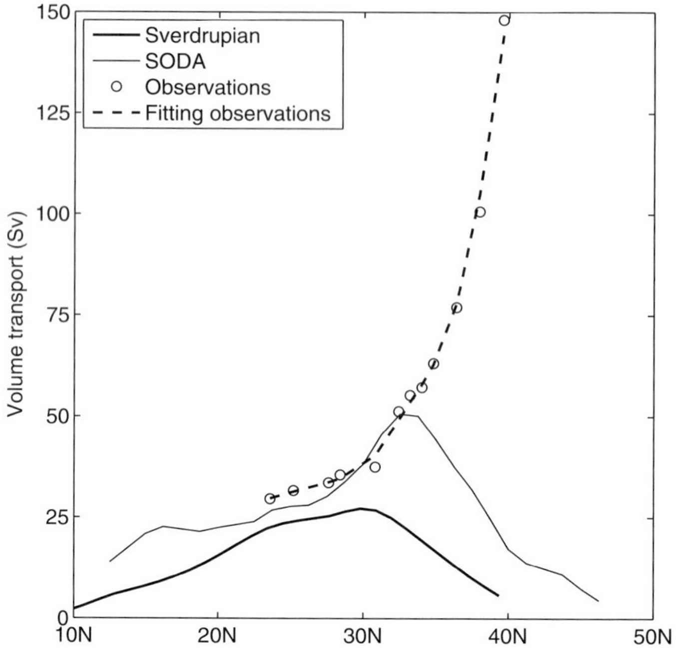

Fig. 4.63 Comparison of the Gulf Stream transport from observations (*circles*), fitted observations (*heavy dashed line*), a low-resolution model (SODA) (*thin solid line*), and the Sverdrup transport computed from wind stress (*heavy solid line*).

The linear Sverdrup dynamics suggests that the volume transport of the gyre should reach the maximal value at the latitude of the Florida Strait, and then gradually decline for sections at higher latitude. On the other hand, the observed transport is much larger than that from the Sverdrup dynamics. Even at the latitude of the Florida Strait, the observed transport is larger than that of the Sverdrup dynamics. Near 40° N, the observed transport reaches the maximal value of 150 Sv (Gill, 1971), but the transport predicted by the Sverdrup dynamics is nearly zero because this is close to the zero wind-stress curl latitude. There is a similar phenomenon in the North Pacific Ocean, where the Kuroshio Extension and its dynamics and variability have been studied extensively (for a comprehensive review, see Qiu, 2002a).

The discrepancy between the linear theory and observations is due to two factors. First, the inertial terms in the horizontal momentum equations have been neglected for analytical simplicity in the commonly used linear theories. As discussed above, inertial terms are crucial within strong boundary currents, such as the Gulf Stream, the Kuroshio, and the ACC. There are recirculation regions in the northwestern (southwestern) corners of the subtropical gyre in the Northern (Southern) Hemisphere where the circulation is strongly nonlinear. Second, there is a strong interaction with stratified flow over topography, which is called bottom pressure torque or the JEBAR (joint effect of baroclinicity and bottom relief) term, and this is discussed in detail in subsection 4.4.5, "The role of bottom pressure torque."

The contribution due to the JEBAR term can be seen clearly from the western boundary current transport as diagnosed from a $1^\circ \times 1^\circ$ low-resolution model (SODA). Such a model cannot simulate the effect of eddies. Nevertheless, the transport of the anticyclonic gyre is more than 50 Sv around 34° N.

In this section we discuss preliminary theories of the recirculation. Most of the published theoretical work relating to the role of meso-scale eddies in the recirculation is based on the quasi-geostrophic theory. Furthermore, the role of bottom pressure torque can be explained without explicitly invoking eddies; thus, the meaning of bottom pressure torque will be explained using results from a simple non-eddy-resolving model. The potential shortcomings of such models will be commented on at the end of this section.

4.4.2 Fofonoff solution

For a model with a single moving layer, the quasi-geostrophic vorticity equation is

$$J\left(\psi, \nabla_h^2 \psi + \beta y\right) = \frac{f_0 w_e}{H} - R \nabla_h^2 \psi \tag{4.344}$$

where $R \ll O(1)$ is a frictional parameter. Fofonoff (1954) suggested looking for solutions free of forcing and friction. Therefore, we set both the Ekman pumping and friction at zero ($R = 0$); since potential vorticity is conserved, such a solution should satisfy the vorticity conservation law

$$\nabla_h^2 \psi + \beta y = F(\psi) \tag{4.345}$$

where $F(\psi)$ is an arbitrary function of ψ.

There are many different solutions, as first discussed by Fofonoff (1954). However, many of these solutions are not linked to the circulation in the ocean interior, so it is desirable to find solutions which satisfy Eqn. (4.344), and are thus valid solutions for the whole basin.

Integrating Eqn. (4.344) over an area A_ψ within a closed streamline leads to an integral constraint

$$\iint_{A_\psi} \frac{f_0 w_e(x, y)}{H} = R \oint_{C_\psi} \vec{u} \cdot \vec{dl} \tag{4.346}$$

This constraint can be used to find a solution for the problem; however, it is not an easy job. One way to find a solution for this problem is to assume some simple form of this function $F(\psi)$, and thus a solution consistent with this constraint (Niiler, 1966). As an example, we assume that this function is in the form

$$\nabla_h^2 \psi + \beta y = \beta (\psi + y_0), \quad where\ \beta \gg 1 \tag{4.347}$$

Interior solution

In the ocean interior, the inertial term, i.e., the relative vorticity, can be neglected; thus, from Eqn. (4.347) the solution is

$$\psi_I = y - y_0 \tag{4.348}$$

which represents a westward flow with a uniform velocity. For convenience we choose the southern boundary as $y_0 = 0$, so there is no boundary layer along the southern boundary of the model basin.

Boundary layers along other boundaries of the model basin

It is clear that along other boundaries of the basin, the streamfunction calculated from the interior solution is not zero, $\psi_I \neq 0$. In order to satisfy the boundary condition of zero streamfunction, there must be boundary layers along all other boundaries.

At the western boundary $x = 0$. Let us separate the streamfunction into two parts:

$$\psi = \psi_I + \psi_W \tag{4.349}$$

Substituting Eqns. (4.348, 4.349) into Eqn. (4.347), we obtain

$$\nabla_h^2 \psi_W - \beta \psi_W = 0 \tag{4.350}$$

within the western boundary, the y-derivative term is much smaller than the x-derivative terms, so this equation is reduced to an ordinary differential equation in x

$$\frac{d^2}{dx^2}\psi_W - \beta \psi_W = 0 \tag{4.351}$$

subject to boundary conditions

$$\psi_W(0) = -\psi_I, \quad \text{and } \psi_W \to 0, \text{ at } x \to \infty \tag{4.352}$$

whose solution is

$$\psi_W = -ye^{-\sqrt{\beta}x} \tag{4.353}$$

Boundary layer solutions along the boundary at $x = 1$ and $y = 1$ have a similar structure; therefore, the complete solution is

$$\psi = y\left[1 - e^{-\sqrt{\beta}x} - e^{-\sqrt{\beta}(1-x)}\right] + e^{-\sqrt{\beta}(1-y)} \tag{4.354}$$

Two solutions are shown in Figure 4.64.

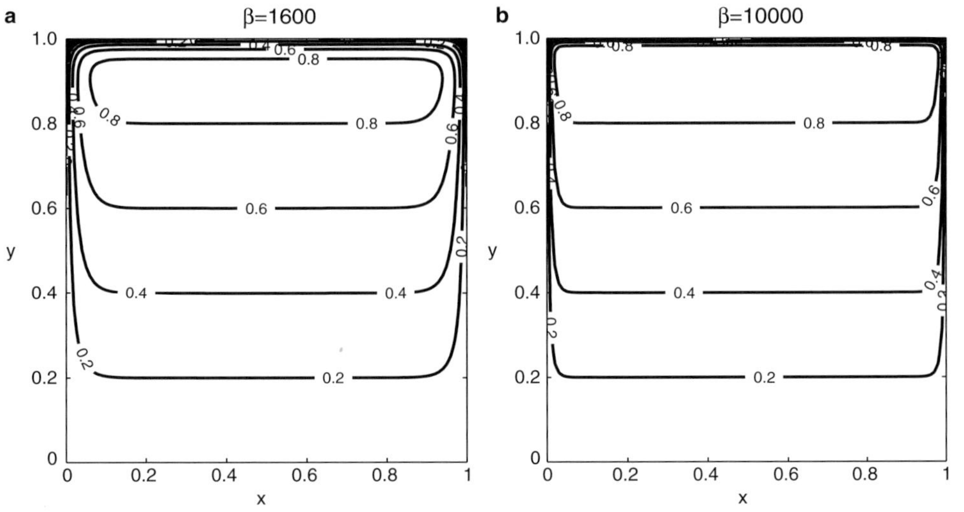

Fig. 4.64 An inertial solution in a model basin for two values of β.

4.4.3 Veronis model

The dynamical effects due to the nonlinear advection term can be explored through a model based on the shallow-water equation (Veronis, 1966). In his study, the model is formulated for a single-layer ocean under wind stress forcing and a linear bottom friction. The complete momentum equation also includes the time-dependent term and the nonlinear advection term

$$\frac{\partial}{\partial t}\vec{u} + \vec{u}\nabla_h \cdot \vec{u} + \vec{f} \times \vec{u} = -\frac{\nabla_h p}{\rho} - R\vec{u} + \frac{\vec{\tau}}{H} \tag{4.355}$$

where $H = const$ is the layer thickness. The next crucial assumption implicitly made in this model is that layer thickness perturbations remain much smaller than the undisturbed layer thickness. As a result, the continuity equation is

$$u_x + v_y = 0$$

and we can introduce a streamfunction

$$u = -\psi_y, \quad u = \psi_x \tag{4.356}$$

The corresponding vorticity equation is

$$\zeta_t + \vec{u} \cdot \nabla_h \zeta + \beta v = -R\zeta + \frac{\tau_x^y - \tau_y^x}{H} \tag{4.357}$$

where

$$\zeta = v_x - u_y \tag{4.358}$$

is the relative vorticity. The solution is subject to boundary condition $\psi = 0$ along the boundaries of the basin.

In order to find solutions in analytical form, the wind stress is assumed to have a special form

$$\tau^x = -\frac{W}{2} \sin\frac{x}{L} \cos\frac{y}{L}, \quad \tau^y = \frac{W}{2} \cos\frac{x}{L} \sin\frac{y}{L} \tag{4.359}$$

Thus, the wind-stress curl averaged over the whole layer is

$$\frac{\tau_x^y - \tau_y^x}{H} = -\frac{W}{HL} \sin\frac{x}{L} \sin\frac{y}{L} \tag{4.360}$$

Introduce the following nondimensional variables:

$$(x, y) = L\left(x', y'\right), \ t = t'/\beta L, \ \psi = \frac{W}{H\beta}\psi' \tag{4.361}$$

then the nondimensional vorticity equation, after dropping primes, is

$$\zeta_t + RoJ\,(\psi, \zeta) + \psi_x = -\varepsilon\zeta - \sin x \sin y \tag{4.362}$$

where two small parameters are introduced

$$Ro = \frac{W}{H\beta^2 L^3} \ll 1, \quad \varepsilon = \frac{R}{\beta L} \ll 1 \tag{4.363}$$

Steady, linear solution in the ocean interior

In this case, we drop the time-dependent term and the nonlinear advection terms, and the vorticity equation is reduced to

$$\varepsilon \nabla_h^2 \psi + \psi_x = -\sin x \sin y \tag{4.364}$$

The corresponding solution is

$$\psi = \frac{1}{1 + 4\varepsilon^2} \left\{ 2\varepsilon \sin x + \cos x + \frac{1}{e^{\pi D_1} - e^{\pi D_2}} \left[\left(1 + e^{\pi D_2}\right) e^{D_1 x} - \left(1 + e^{\pi D_1}\right) e^{D_2 x} \right] \right\} \sin y$$

where

$$D_1 = \frac{-1 - \sqrt{1 + 4\varepsilon^2}}{2\varepsilon}, \quad D_2 = \frac{-1 + \sqrt{1 + 4\varepsilon^2}}{2\varepsilon} \tag{4.365}$$

Numerical solutions

The nonlinear solutions can be found by a perturbation method when *Ro* is very small. By expanding in terms of a Taylor series in *Ro*, we can find the lowest-order solution with $Ro = 0$ in analytical forms. However, the general and more accurate solutions have to be calculated through numerical integration.

As the inertial term becomes more important, the circulation develops a recirculation region in the northwest corner. However, for the case of very strong nonlinearity, the circulation becomes virtually symmetric in the east–west direction, and there is no longer the western intensification, described in terms of the inertial run-away problem discussed in Section 4.1.5. As discussed in previous sections, pushing the shallow-water model to the limit of strong forcing and weak dissipation, but without considering other vital dynamical effects such as the outcrop phenomenon or the formation of closed geostrophic contours in the subsurface layers, may not be a good way to simulate the oceanic circulation.

4.4.4 Potential vorticity homogenization applied to recirculation

Numerical experiment

Solving a two-layer quasi-geostrophic model for the wind-driven circulation, we can obtain solutions that are characterized by the existence of strong recirculation in the northwestern corner of the model basin. An example is shown in Figure 4.65.

A barotropic model in a β-plane box

In order to examine the dynamics of recirculation, we will focus on a box in the northwestern corner of the model basin ($-L/\alpha < x < L/\alpha$, $-L < y < L$) where α is the aspect ratio of the model basin. The quasi-geostrophic vorticity equation is a balance between advection and dissipation

$$J(\psi, q) = R\nabla_h^2 q \tag{4.366}$$

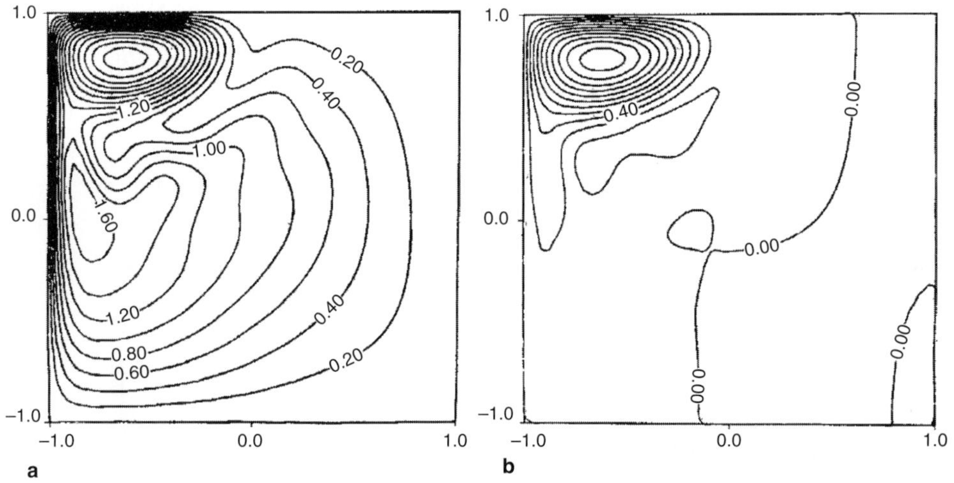

Fig. 4.65 Steady circulation (streamfunction) obtained from a two-layer quasi-geostrophic model for the wind-driven circulation; **a** upper layer and **b** lower layer (Cessi *et al.*, 1987).

The corresponding boundary conditions are $\psi = 0$ and $q = q_B(s)$, along the boundary of the basin, where s is the arc length along the boundary. The essential point is that circulation of the model ocean studied here is forced by a prescribed source of potential vorticity $q_B(s)$, which is different from the background planetary vorticity βy. The physical processes which set up such a potential vorticity forcing are not explicitly discussed here, although they may be related to advection of potential vorticity originating from lower latitudes, eddy activity, or cooling/heating. This problem can be solved by a Newton method.

For example, we can choose the following form of a potential vorticity boundary condition:

$$q_B = (Q_n - Q_s)\frac{y - L}{2L} + Q_n \tag{4.367}$$

Although q_B is required to be defined along the boundary only, here its definition is extended to the interior, i.e., $q_B = q_B(x, y)$ as defined in Eqn. (4.367). The numerical solution indicates that potential vorticity is homogenized within the interior, except for a very thick boundary layer along the edge of the box. An example is shown in Figure 4.66, with $Q_n = -\beta L/3$, $Q_s = -\beta L$, $L = 300$ km, $\beta = 2 \times 10^{-11}$/s/m, $\alpha = 0.3$.

Application of potential vorticity homogenization

Multiplying Eqn. (4.366) by ψ and integrating over the whole box gives

$$\iint q\nabla_h^2 \psi dxdy = \oint q_B \vec{u} \cdot \vec{dl} \tag{4.368}$$

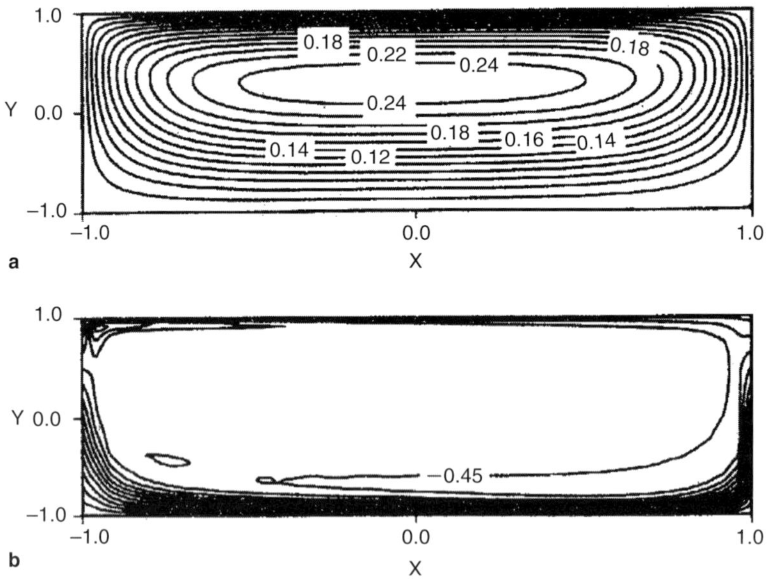

Fig. 4.66 **a** Streamfunction (in $\beta L^3 = 5.4 \times 10^5$ m^2/s) and **b** potential vorticity for the steady state solution (Cessi *et al.*, 1987).

Since potential vorticity is virtually homogenized within the box, we have

$$\iint q\nabla_h^2\psi dxdy \simeq \overline{q}\iint \nabla_h^2\psi dxdy = \overline{q}\oint \vec{u}\cdot \vec{dl} \tag{4.369}$$

Therefore, the constant value of potential vorticity within the box is

$$\overline{q} \simeq \frac{\oint q_B \vec{u}\cdot \vec{dl}}{\oint \vec{u}\cdot \vec{dl}} \tag{4.370}$$

It is interesting to note that potential vorticity is a velocity-weighted average of the vorticity source along the boundary because wherever the velocity is faster, the streamlines are closer to each other, so the cross-streamline diffusion is more effective, giving more weight to the average vorticity. This velocity-weighted average is quite counter-intuitive to the conjuncture that the places where fluid spends most time contribute least to the value of $\overline{q}$.

In order to solve Eqn. (4.370), we can introduce a new function g defined as

$$\nabla_h^2 g = 0, \text{ in the interior, and } g = q_B \text{ on the boundary} \tag{4.371}$$

Note that $g = q_B$ is a possible choice of such a function. Using the Green theorem, we have

$$\oint q_B \times \nabla_h\psi \cdot \vec{n} dl = \iint g\nabla_h^2\psi dxdy \tag{4.372}$$

Thus, Eqn. (4.370) is reduced to

$$\overline{q} = \frac{\iint q_B (\overline{q} - \beta y)\, dxdy}{\iint (\overline{q} - \beta y)\, dxdy} \tag{4.373}$$

Substituting q_B, the solution is

$$\overline{q} = \frac{Q_n + Q_s}{4} \pm \left[\frac{(Q_n + Q_s)^2}{16} - \frac{\beta L\,(Q_n - Q_s)}{6}\right]^{1/2} \tag{4.374}$$

The approach was extended to the case with a two-layer model and similar features were obtained by Cessi (1988).

A similar idea was explored by Jayne *et al.* (1996). They proposed that a strong eastward-moving jet can be unstable. If the jet is narrow and the cross-jet potential vorticity gradient changes sign, barotropic instabilities occur. Eddies generated through the instability can create zones of homogenized potential vorticity and recirculation on both the southern and northern sides of the jet.

4.4.5 The role of bottom pressure torque

The role of bottom topographic torque in enhancing the wind-driven circulation, in particular the recirculation, can be explained as follows. Our discussion here follows Greatbatch *et al.* (1991). For basin-scale motions it is more accurate to use spherical coordinates, and the basic equations include the horizontal momentum equations, the hydrostatic relation, and volume conservation

$$-fv = -\frac{1}{a\rho_0 \cos\theta}\frac{\partial p}{\partial \lambda} + \frac{\partial \tau_{\lambda z}}{\rho_0 \partial z} \tag{4.375}$$

$$fu = -\frac{1}{a\rho_0}\frac{\partial p}{\partial \theta} + \frac{\partial \tau_{\theta z}}{\rho_0 \partial z} \tag{4.376}$$

where $\tau_{\lambda z}$ and $\tau_{\theta z}$ are turbulent stress:

$$\frac{\partial p}{\partial z} = -\rho_0 b, \quad \text{where } b = g\frac{\rho - \rho_0}{\rho_0} \tag{4.377}$$

$$\frac{1}{a\cos\theta}\left[\frac{\partial u}{\partial \lambda} + \frac{\partial}{\partial \theta}(v\cos\theta)\right] + \frac{\partial w}{\partial z} = 0 \tag{4.378}$$

Vertical integration of the continuity equation leads to

$$\frac{1}{a\cos\theta}\left[\frac{\partial U}{\partial \lambda} + \frac{\partial}{\partial \theta}(V\cos\theta)\right] = 0 \tag{4.379}$$

where $(U, V) = \left(\int_{-H}^{0} u dz,\ \int_{-H}^{0} v dz\right)$ is the vertically integrated volume transport. From this equation, a layer-integrated streamfunction can be introduced:

$$aU = -\Psi_\theta, \quad aV\cos\theta = \Psi_\lambda \tag{4.380}$$

Integrating the horizontal momentum equations leads to

$$-\frac{fV}{H} = -\frac{1}{a\rho_0 H\cos\theta}\int_{-H}^{0}\frac{\partial p}{\partial \lambda}dz + \frac{\tau_\lambda}{\rho_0 H} \tag{4.381}$$

$$\frac{fU}{H} = -\frac{1}{a\rho_0 H}\int_{-H}^{0}\frac{\partial p}{\partial \theta}dz + \frac{\tau_\theta}{\rho_0 H} \tag{4.382}$$

where τ^λ and τ^θ are the surface wind stress, and bottom friction is neglected, $H(\lambda, \theta)$ is the depth of the ocean. Using the hydrostatic relation and integrating by parts, the pressure integration is reduced to

$$\int_{-H}^{0} p dz = pz|_{-H}^{0} - \int_{p_B}^{0} z dp = p_B H + \rho_0 \Phi \tag{4.383}$$

where

$$\Phi = \int_{-H}^{0} bz dz \tag{4.384}$$

is the gravitational potential energy for the water column. If there is no stratification, this term is zero. A constant density is subtracted from this energy term because what really matters is the horizontal gradient term, not the total gravitational potential energy itself. Using this equation, the pressure gradient terms in Eqns. (4.381) and (4.382) can be converted into the following forms:

$$\frac{1}{H}\int_{-H}^{0} p_\lambda dz = \frac{1}{H}\left(\frac{\partial}{\partial\lambda}\int_{-H}^{0} pdz - H_\lambda p_B\right) = p_{B,\lambda} + \frac{\rho_0}{H}\Phi_\lambda \tag{4.385}$$

$$\frac{1}{H}\int_{-H}^{0} p_\theta dz = \frac{1}{H}\left(\frac{\partial}{\partial\theta}\int_{-H}^{0} pdz - H_\theta p_B\right) = p_{B,\theta} + \frac{\rho_0}{H}\Phi_\theta \tag{4.386}$$

Cross-differentiating Eqns. (4.381) and (4.382) and subtracting leads to

$$J\left(\Psi, \frac{f}{H}\right) = J\left(\Phi, \frac{1}{H}\right) + \frac{a}{\rho_0}\left[\frac{\partial}{\partial\lambda}\left(\frac{\tau_\theta}{H}\right) - \frac{\partial}{\partial\theta}\left(\frac{\cos\theta\,\tau_\lambda}{H}\right)\right] \tag{4.387}$$

The first term on the right-hand side $J(\Phi, 1/H)$ is called the JEBAR term, and the second term on the right-hand side is the contribution due to wind-stress curl. A more comprehensive discussion of the JEBAR term is presented by Huthnance (1984). Equation (4.386) states that changes of the barotropic streamfunction along the f/H contours are due to contributions from these two terms. For given wind stress and hydrographic data, this equation can be solved using the f/H contours as the characteristic coordinates. A special numerical method can be used to solve this equation for the domain inside the closed contours of f/H (Greatbatch *et al.*, 1991).

The JEBAR term is associated with bottom pressure torque, and is the result of interaction between the stratification and bottom topography. It can readily be seen that, for the case of a flat bottom, this term vanishes.

Greatbatch *et al.* (1991) analyzed the circulation in the North Atlantic Ocean, using a model with $1° \times 1°$ resolution, plus climatological wind stress and hydrographic data. In this calculation, contributions due to wind-stress curl and the JEBAR term were calculated because, in Eqn. (4.387), their contributions are linearly separated. It can readily be seen that the contribution due to wind-stress curl reaches the maximum of 20 Sv near the Florida Strait, and gradually declines north of the strait (Fig. 4.67). South of Greenland, the contribution from wind stress is approximately 12 Sv only.

On the other hand, the contribution due to the JEBAR term reaches nearly 50 Sv in the subtropical gyre and 20 Sv in the subpolar gyre; thus the total contribution to the Gulf Stream circulation at this latitude is about 70 Sv. Remember that the total transport of the Gulf Stream is on the order of 150 Sv; it consists of three factors: the linear Sverdrup transport is approximately 10 Sv, the bottom pressure torque contributes approximately 70 Sv, and the rest is due to eddies. Therefore, the most crucial contributor to the volume transport maximum in the recirculation regime is the JEBAR term. However, the increase of the JEBAR term is closely related to the oceanic circulation. In particular, one can see this as a result of the interaction of flow, including eddies and topography. Nevertheless, the JEBAR term can be used as a tool for diagnosing the circulation.

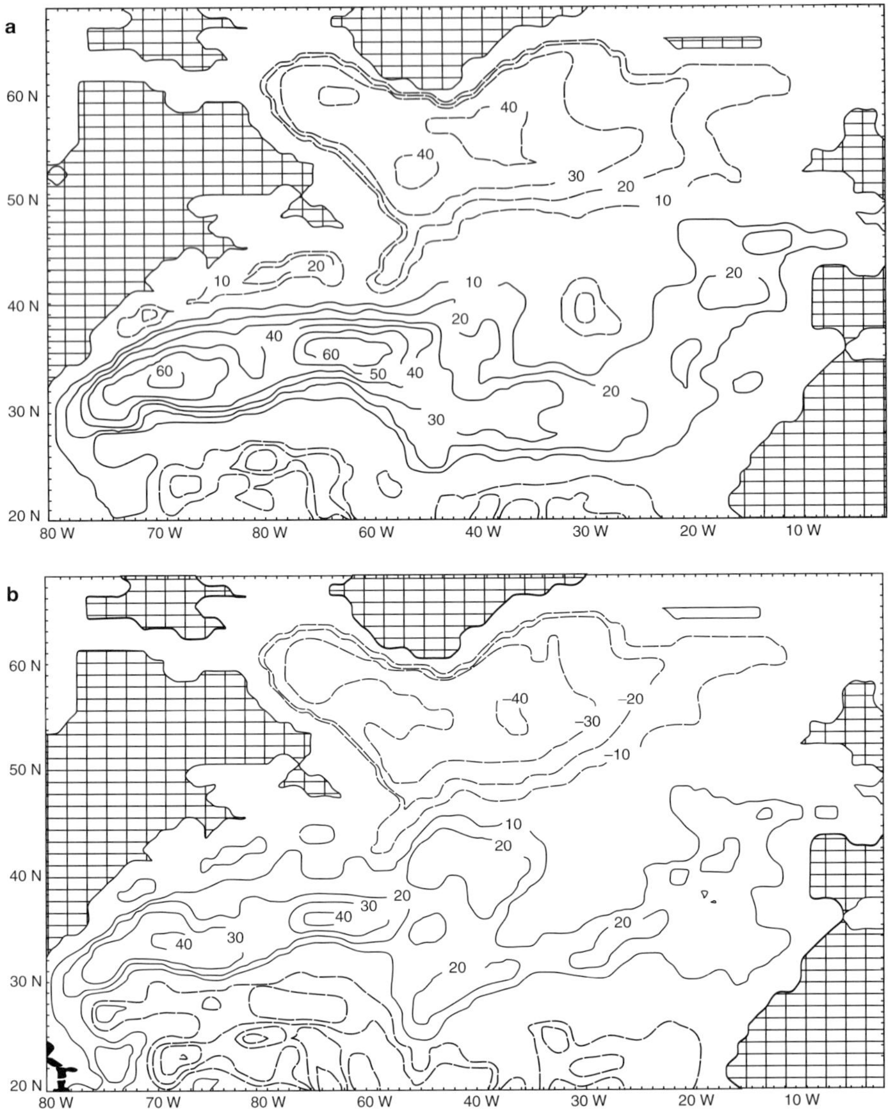

Fig. 4.67 Barotropic streamfunction maps (in Sv) obtained from numerical experiments for the North Atlantic Ocean, identifying the role of each forcing term: **a** wind stress term only; **b** JEBAR term only (Greatbatch *et al.*, 1991).

4.4.6 Final remarks

Although these theories provide some physical insight into the maintenance of recirculation, our understanding of the recirculation remains rudimentary. In fact, the recirculation in the corner of the subtropical gyres in the world's oceans shares many similar features with the strong ACC. The dynamical nature of the ACC is very complicated, involving many forcing factors and dynamical/thermodynamic processes, such as wind stress, wind-stress curl, surface heat flux and freshwater flux, interaction with bottom topography, and above all, the complex role of meso-scale eddies. In some sense, the recirculation in the subtropical gyres may be even more difficult to unravel because the existence of a meridional boundary eliminates the zonal symmetric nature of the problem, and renders it more difficult to understand.

Owing to the limitations of the theories of recirculation discussed in this section, several critical dynamical issues remain unsolved and require further exploration:

- First, deformation of isopycnal surfaces in the cross-stream direction is assumed to be small in the quasi-geostrophic models discussed above; however, observations in the oceans indicate extremely strong density fronts associated with the strong currents in the recirculation regime. The exclusion of such important physical elements may render the quasi-geostrophic theory much less useful in the recirculation regime.
- Second, the strong heat flux associated with the air–sea interaction in connection with a strong density front, such as the Gulf Stream, the Kuroshio, or the ACC, is not included in the dynamical formulation in such models. As discussed above, the potential vorticity balance within the recirculation regime is treated in terms of some adiabatic processes. On the other hand, the commonly accepted theory of subtropical mode water formation includes surface cooling as a critical factor in regulating the potential vorticity balance.
- Third, there are strong eddies observed from both *in situ* measurement and satellite altimetry, which may not be easily simulated through the simple theoretical models discussed above. Recent numerical modeling experiments indicate that results from the eddy-permitting models may sensitively depend on the resolution of the models (Hallberg and Gnanadesikan, 2001).

Therefore, despite much effort devoted to the study of the recirculation, its basic structure remains a great challenge, and a comprehensive theory of recirculation may require a new set of models based on density coordinates to handle these problems more accurately.

4.5 Layer models coupling thermocline and thermohaline circulation

4.5.1 Introduction

Most of the models discussed above are formulated for the wind-driven circulation; however, buoyancy flux is also crucial for the large-scale circulation. The difficulty of incorporating the buoyancy flux divergence into layer models lies in the nonlinearity associated with the Jacobians of interface depths (or the twisting terms, as will be discussed shortly) appearing in the vorticity equation. Luyten and Stommel (1986) went through a very careful analysis of the equations. Their contributions are the following. First, they recognized that the equations describing the problem can be put in the characteristic form. Second, they introduced the concept of the Rossby repellor and separated the basin into the western and eastern regions, where different dynamics dominate.

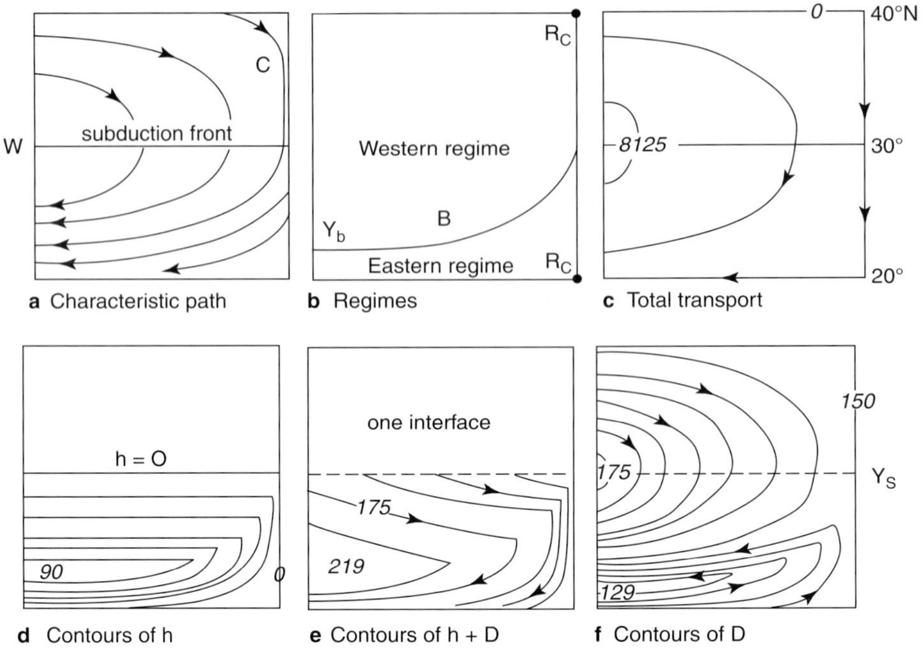

Fig. 4.68 A typical solution in a subtropical basin: **a** characteristics; **b** two regimes; **c** Sverdrup transport function $h^2 + D^2$ (m^2) with contour increments of 5 m^2; **d**, **e**, and **f** for h, $h + D$, and D, with contour increments of 15, 5, and 4 m (Luyten and Stommel, 1986).

4.5.2 A $2\frac{1}{2}$-layer model

The model includes two active layers and a stagnant layer below (Fig. 4.68) (Luyten and Stommel, 1986). The continuity equations for the active layers are

$$(hu_1)_x + (hv_1)_y = w_s - w_e \tag{4.388a}$$

$$[(D-h)\,u_2]_x + [(D-h)\,v_2]_y = -w_s \tag{4.388b}$$

where h and D are the depth of the upper and lower interfaces, w_e is the Ekman pumping velocity applied to the upper surface, and w_s is the interfacial volume flux, which is perpendicular to the interface. An important concept for a multi-layer model is the so-called diapycnal velocity. Diapycnal velocity indicates the next volume flux between two layers, so it is different from the vertical velocity at the interface. For an ideal fluid with no diffusion, diapycnal velocity should be zero. On the other hand, vertical velocity at the interface may be non-zero, even if there is no diffusion across the density interface. The non-zero vertical velocity at the interface is due to the projection of horizontal velocity on the sloping interface.

The horizontal velocity in each layer satisfies the geostrophy conditions

$$u_2 = -g' D_y / f, \quad v_2 = g' D_x / f \tag{4.389a}$$

$$u_1 = -g' \left(h_y + D_y \right) / f, \quad v_1 = g' \left(h_x + D_x \right) / f \tag{4.389b}$$

Substituting Eqn. (4.389) into the continuity equations (4.388a, 4.388b) leads to the potential vorticity equations for both layers:

$$\frac{g'}{f} \left(-h_x D_y + h_y D_x \right) - \frac{\beta g'}{f^2} h \left(D_x + h_x \right) = w_s - w_e \tag{4.390a}$$

$$-\frac{g'}{f} \left(-h_x D_y + h_y D_x \right) - \frac{\beta g'}{f^2} \left(D - h \right) D_x = -w_s \tag{4.390b}$$

The first terms on the left side are called the "twisting terms" or the Jacobi terms, the second terms are the β terms. Adding the two equations, the twisting terms cancel, and we obtain the barotropic equation

$$\frac{\beta g'}{f^2} \left(h h_x + D D_x \right) = w_e \tag{4.391}$$

Assuming that w_e is independent of x, this can be rewritten as

$$\frac{1}{2} \left(h^2 + D^2 \right) = \frac{1}{2} \left(h_e^2 + D_e^2 \right) + W_e x \tag{4.392}$$

where $W_e = \frac{f^2 w_e}{\beta g'}$, h_e, and D_e are the corresponding values at the eastern boundary $x = x_e$.

Differentiating Eqn. (4.392) with respect to x and y leads to two equations:

$$h h_x + D D_x = W_e \tag{4.393a}$$

$$h h_y + D D_y = W_{e,y} x \tag{4.393b}$$

Substituting these relations into Eqn. (4.390b), one gets

$$\left(\frac{g' W_e}{f D} \right) D_y + \left[-\frac{g' x W_{e,y}}{f D} - \frac{\beta g' h \left(D - h \right)}{f^2 D} \right] D_x = -h \frac{w_s}{D} \tag{4.394}$$

This equation is in the characteristic form, which can be solved numerically by standard methods following the characteristics. It is very important to recognize the mathematical property of this equation, because it carries a lot of crucial physical meaning. Practically, numerical schemes not based on characteristic coordinates may violate this intrinsic property and lead to unphysical solutions. As we will see shortly, Luyten and Stommel's success in solving this problem is based on recognizing this property and analyzing the special physical characters of the problem.

The two components of the characteristic velocity are:

$$\frac{dx}{ds} = u_c = -\frac{g' W_{e,y} x}{fD} - \frac{\beta g' h (D-h)}{f^2 D} \tag{4.395a}$$

$$\frac{dy}{ds} = v_c = \frac{g' W_e}{fD} \tag{4.395b}$$

In the characteristic coordinates, Eqn. (4.394) can be rewritten in terms of the total derivative

$$\frac{dD}{ds} = -\frac{h}{D} w_s \tag{4.396}$$

The characteristic velocity is not the same as the particles' velocity. In fact, these two components can be rewritten as

$$v_c = \frac{h v_1 + (D-h) v_2}{D} \tag{4.397a}$$

$$u_c = \frac{h u_1 + (D-h) u_2}{D} - \frac{\beta g' h (D-h)}{f^2 D} \tag{4.397b}$$

Luyten and Stommel applied the model to a subtropical basin and a subpolar basin; in the subtropical basin the upper layer outcrops along the mid latitude, 30° N (Fig. 4.68). The parameters used in this case are: $w_e = 0.03$ m/day, $w_s = -3$ cm/day, $x_w = -2,000$ km, midlatitude $= 30°$ N, $h_0 = 0$ m, $D_0 = 150$ m, and $g' = 0.01$ m/s^2. To make the model self-consistent, the interfacial volume flux is set to zero wherever $h = 0$.

Note that the characteristics divide the basin into the western and eastern regions separated by critical characteristic C. In the western regime all characteristics come from the western boundary. On the other hand, all characteristics in the eastern regime start from the eastern boundary. Another interesting feature is the cyclonic sub-gyre in the second layer in the southeastern corner, where a shadow zone would appear if there were no interfacial volume flux.

Note that this approach can be extended to the case of more than two moving layers; however, it is very difficult to eliminate the Jacobi terms. Pedlosky (1986) discussed a case with three moving layers; however, his solution is based on a special assumption about layer thickness (or something equivalent). Thus, the general case with multiple moving layers remains a challenge.

A key assumption made in this model is the specification of the interfacial mass flux. In reality, this should be parameterized as part of the solution, i.e., the interfacial upwelling should be determined from the internal dynamic balance. Huang (1993a) discussed such a model and demonstrated that upwelling is not uniform basin-wide; instead, it is strong along the southern and eastern boundaries.

In addition, the Rossby repellor indicates that the zonal propagation of the signal is blocked by this repellor; however, for the time-dependent problem, the first baroclinic Rossby wave can move westward, regardless of such a repellor for the steady state.

4.6 Equatorial thermocline

4.6.1 Introduction

Discussions in previous sections have focused on the circulation in the subtropical and subpolar basins, including the dynamics of western boundary layers. Now we move toward the equator, where a most unique situation is the disappearance of the Coriolis parameter. The fundamental structure of the equatorial current system and associated structures of temperature and salinity are shown in Figure 4.69.

The dynamics of the equatorial current system is much more complicated than that at mid latitudes. The most outstanding feature is the existence of the Equatorial Undercurrent (EUC), which is located within a latitudinal band of 2° around the equator. Vertically, the EUC is at a depth of 100–250 m below the surface, and there is a strong westward surface current above the core of the eastward undercurrent. In addition, there is another westward current below the eastward undercurrent. There are many other complicated currents near the equator; however, in this section our discussion will be confined to the EUC, including the associated equatorial thermocline. Another major feature is the equatorial thermocline and the thermocline ridge near 10° N, which is associated with the special feature of wind-stress curl near the Inter-Tropical Convergence Zone (ITCZ) and will be explained in the next section on the communication between the subtropics and tropics.

As discussed in previous sections on the ventilated thermocline, warm water is subducted in the subtropical gyre interior. These water masses move equatorward within the subtropical gyre interior; some of the subducted water moves back poleward via the western boundary currents, but part of it moves toward the equator in two ways, i.e., through the low-latitude western boundary or through the interior communication window. (The interior communication window is a subsurface channel which connects the subtropics with the tropics without going through either the eastern or western boundary layer, and this will be discussed in detail in Section 4.7.) These links have been identified through many observational studies. For example, Wyrtki and Kilonsky (1984) identified the source of the water mass in the equatorial current system using the Hawaii to Tahiti Shuttle Experiment; Gouriou and Toole (1993) studied the mean circulation of the upper layers of the western equatorial Pacific Ocean. These studies showed that the water mass in the equatorial current system comes from the subtropics. In particular, Gouriou and Toole (1993) pointed out that the water mass at the beginning of the undercurrent comes primarily from the Southern Hemisphere.

A zonal view of the equatorial thermocline indicates a strong tilting in both the Pacific and Atlantic Oceans, which is closely related to the prevailing easterlies in these two oceans (Fig. 4.70). As a result, there are outstanding cold tongues in the eastern parts of these oceans, where the main thermocline outcrops. The Indian Ocean is quite different because the prevailing wind is westerly. As a result, the equatorial thermocline in the Indian Ocean has a structure quite different from that in the Pacific and Atlantic Oceans.

Since the Coriolis parameter vanishes near the equator, geostrophy is invalid; thus, higher-order dynamics, such as the inertial terms and friction, must be included. In a series of

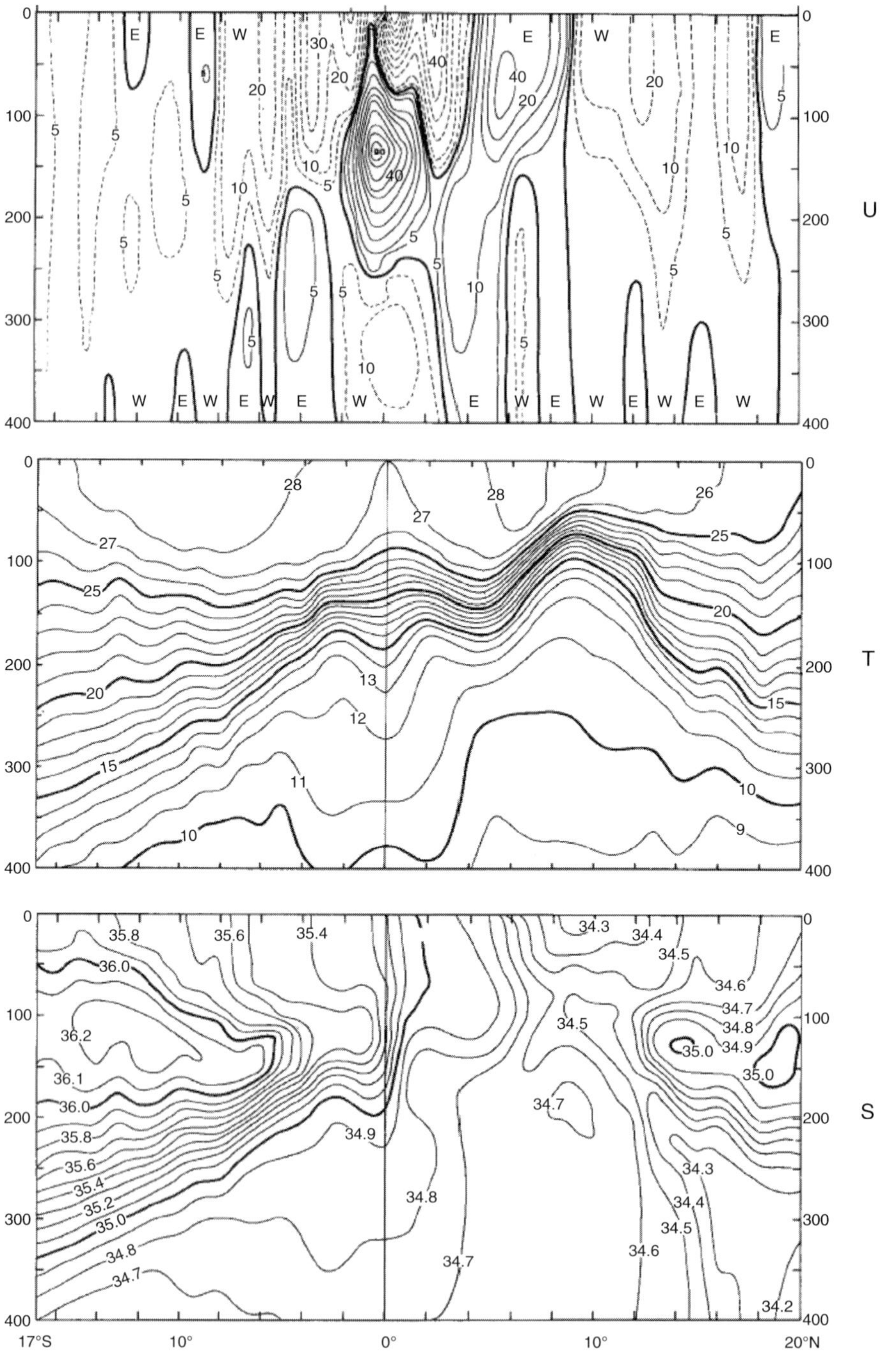

Fig. 4.69 Structure of the equatorial thermocline and circulation (Wyrtki and Kilonsky, 1984).

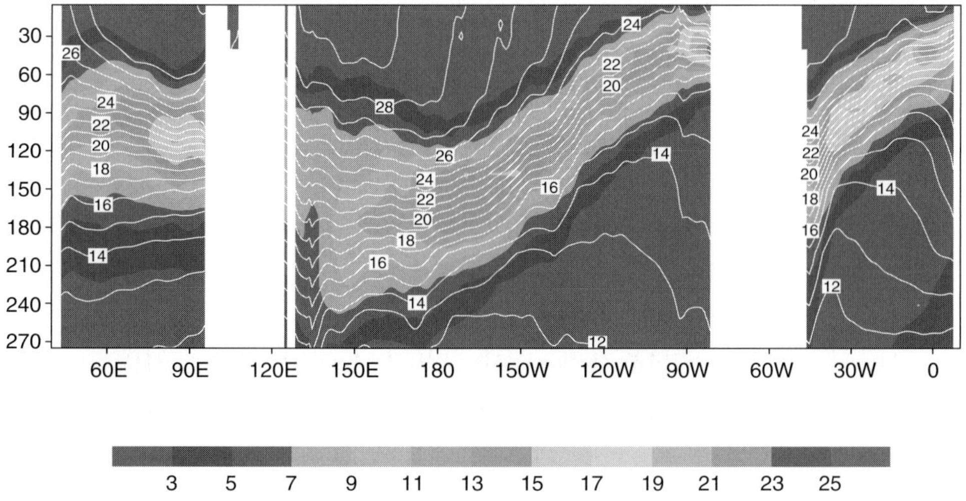

Fig. 4.70 Thermal structure of the equatorial thermocline (vertical axis depicts depth in m): contour labels are temperature (in °C), overlaid with the vertical gradient (in °C/100 m).

insightful studies, Pedlosky (1987b, 1996) developed a theory for the dynamical connection between the subtropical thermocline and the Equatorial Undercurrent, and our discussion below is based on his theory. The essence of the theory is that, to the lowest order, the Equatorial Undercurrent can be treated in terms of the ideal-fluid model. In order to conserve potential vorticity, the inertial term is retained near the equator.

First, there is a small latitudinal band near the equator where geostrophy breaks down. To replace geostrophy, we search for a new balance between the inertial terms and other terms. The outer edge of this special zone can be defined as the latitude where the Coriolis term is equal to the inertial term. The commonly used Rossby number is defined as

$$Ro = U/fL \tag{4.398}$$

Near the equator, the Coriolis parameter is approximately

$$f = \beta y \tag{4.399}$$

Approaching the equator, f declines; therefore the Rossby number $Ro = U/\beta y^2$ increases. Thus, within a distance of d_f,

$$d_f = \sqrt{U/\beta} \tag{4.400}$$

the advection term, or the relative vorticity, will be dynamically important. From observations, $U \approx 1$ m/s, $\beta = 2.28 \times 10^{-11}$/s/m; thus, $d_f \approx 208$ km. Accordingly, within 2 degrees off the equator, we expect that the inertial terms in the momentum equation will be non-negligible.

In this section we study the equatorial thermocline and current in the framework of inertial current, i.e., our study will focus on the dynamical effect of inertial terms and the linkage of the Equatorial Undercurrent, as an inertial current, with currents in the off-equatorial regimes. Study in this section will be complemented by the study of communication between the subtropics and tropics in the next section. Since the Equatorial Undercurrent is linked to the ventilated thermocline at mid latitudes, we will first describe the ventilated thermocline in the extra-equatorial regime.

4.6.2 The extra-equatorial solution

The equatorial thermocline is studied as an extension from the framework of the ventilated thermocline by Luyten *et al.* (1983). The ocean is simulated in terms of a $2\frac{1}{2}$-layer model formulated on an equatorial β-plane (Fig. 4.71). As shown in Figure 4.70, the main thermocline outcrops in the eastern part of the Pacific and Atlantic Oceans, so we set the layer thickness to zero along the eastern boundary.

The solution off equator is described by the ventilated thermocline model discussed in Section 4.1.7.

$$h^2 = \frac{D_0^2}{1 + \gamma\left(1 - f/f_1\right)^2} \tag{4.401}$$

$$h_2 = \frac{f}{f_1} h; \; h_1 = \left(1 - \frac{f}{f_1}\right) h \tag{4.402}$$

where $\gamma = g_1'/g_2', g_1' = g\left(\rho_2 - \rho_1\right)/\rho_0$, and $g_2' = g\left(\rho_3 - \rho_2\right)/\rho_0$ are the reduced gravity for layers 1 and 2, and f_1 is the Coriolis parameter at $y = y_1$, function D_0^2 is the forcing, and

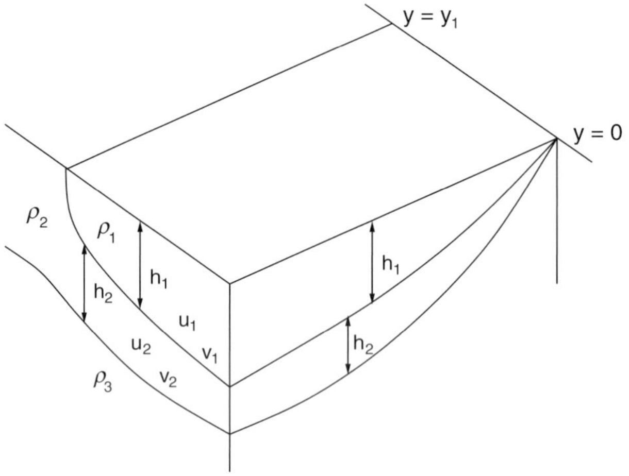

Fig. 4.71 Sketch of a $2^1/_2$-layer ventilated thermocline model for the equatorial ocean.

for the case with zonal wind it is reduced to the following form:

$$D_0^2 = -\frac{2\,(x_e - x)}{\rho_0 g_2'}\left(-\tau + \frac{f}{\beta}\frac{\partial \tau}{\partial y}\right) \tag{4.403}$$

In an equatorial β-plane we have $f = \beta y$. We will match the interior solution with an equatorial solution at $y = 1$, with the assumption that $l \ll L$. Since the wind stress varies on the planetary scale, the ratio of the last two terms in Eqn. (4.403) is $O(l/L)$; the last term can be neglected near the equator. Note that semi-geostrophy holds near the equator, so the zonal velocity is balanced by the meridional pressure gradient, or $g_2' H/l \approx \beta l U$. Thus, the characteristic velocity is linked to the layer thickness through

$$H = U\beta l^2/g_2' \tag{4.404}$$

Introducing the nondimensional variables

$$h' = h/H, \ \left(x', y'\right) = (x/L, y/L) \tag{4.405}$$

Eqn. (4.401) is reduced to

$$h^2 = \frac{g_2' \tau_0 L}{\beta^2 l^4 U^2}\frac{2\left(x_e' - x\right)(-\tau)}{1 + \gamma(1 - y/y_1)^2} \tag{4.406}$$

This equation should have an $O(1)$ balance, so we have the scales

$$l = \left(\frac{g_2' \tau_0 L}{\rho_0 \beta^4}\right)^{1/8}, \quad H = \left(\frac{\tau_0 L}{g_2' \rho_0}\right)^{1/2}, \quad U = \left(\frac{g_2' \tau_0 L}{\rho_0}\right)^{1/4} \tag{4.407}$$

The corresponding scale for the streamfunction is $\Psi = HUl$. For the Atlantic Ocean, $\tau = 0.1$ N/m^2, $g_2' = 0.01$ m/s^2, $\beta = 2 \times 10^{-11}$/m/s, and $L = 3{,}000$ km, so that $l = 280$ km, $H = 125$ m, $U = 1.57$ m/s, and $\Psi = 58$ Sv. For the Pacific Ocean, $L = 14{,}000$ km, so that $l = 339$ km, $H = 265$ m, $U = 2.3$ m/s, and $\Psi = 255$ Sv. Note that the undercurrent has fractional thickness, on the order of $0.1H$, so the volume flux contribution to the undercurrent from each hemisphere is on the order of 5 Sv for the Atlantic and 20 Sv for the Pacific. These scales are quite close to the observed Equatorial Undercurrent in the oceans.

Assuming $x_e = L$, the matching condition for the equatorial solution is that at the outer edge of the equatorial boundary layer, the solution should approach

$$h = \sqrt{\frac{-2\,(1 - x)\,\tau}{1 + \gamma\,(1 - y/y_1)^2}} \tag{4.408}$$

In order to find the solution we can use the potential vorticity and Bernoulli conservation laws. Far from the equator, conservation of potential vorticity and Bernoulli function leads to

$$\frac{y}{h_2} = Q_2\,(B_2) = Q_2\,(h) \tag{4.409}$$

because the $h = const$ or $B_2 = const$ lines are streamlines in the second layer. Tracing back to the latitude of outcropping, we have

$$\frac{y}{h_2} = \frac{y_1}{h} \tag{4.420}$$

Thus,

$$Q_2(h) = \frac{y_1}{h} \tag{4.421}$$

so

$$Q_2(B_2) = \frac{y_1}{B_2} \tag{4.422}$$

Note that h remains finite approaching the equator; however, geostrophy must break down near the equator because the Coriolis parameter vanishes at the equator. Thus the interior solution is not valid near the equator, and the equatorial region must be treated as a boundary layer.

4.6.3 The Equatorial Undercurrent as an inertial boundary current

Near the equator, the boundary current is in semi-geostrophy:

$$u_2\frac{\partial u_2}{\partial x} + v_2\frac{\partial u_2}{\partial y} - yv_2 = -\frac{\partial h}{\partial x} \tag{4.423}$$

$$yu_2 = -\frac{\partial h}{\partial y} \tag{4.424}$$

From these two equations we can derive the conservation laws, i.e., potential vorticity and Bernoulli function:

$$Q_2 = \frac{y - \partial u_2/\partial y}{h_2}, \quad B_2 = h + \frac{u_2^2}{2} \tag{4.425}$$

are conserved along the streamlines. Note that the contribution due to the zonal gradient of the meridional velocity is neglected in the relative vorticity. Thus, the solution of the equatorial boundary layer can be obtained by solving two equations:

$$\frac{\partial u_2}{\partial y} = y - y_1\frac{h - h_1}{h + u_2^2/2} \tag{4.426}$$

$$\frac{\partial h}{\partial y} = -yu_2 \tag{4.427}$$

These two equations are solved over the domain of $y = [0, y_1]$. There are three crucial points: first, the solution should match the interior solution at the outer edge of the equatorial boundary layer. Note that the layer thickness along the matching latitude is

$$H_0^2 = \frac{2\,(1-x)}{1+(1-y_n/y_1)^2} \tag{4.428}$$

The corresponding thicknesses for the upper and lower layers are

$$h_1 = H_0\,(1-y_n/y_1)\,, \quad h_2 = H_0 y_n/y_1 \tag{4.429}$$

In particular, the thickness of the upper layer must be matched with the interior solution at the matching latitude, so $h \to h_{\text{interior}}\,(y_1)\,, \, as\, y \to y_1$.

Second, there is an additional boundary condition at the equator. We assume that the equator is a streamline, so that $B_2 = B_{20}$ is a constant, which is set to be the Bernoulli function at the outer edge of the western boundary layer at some specified latitude.

Third, the upper layer thickness h_1 is unknown. For simplicity, one can choose to have $h_1 = h_1 y_n$, where y_n is the latitude where the boundary layer solution matches with the interior solution. The solution can be found by a shooting method, as described by Pedlosky (1996). This model can be improved in several aspects; however, this is beyond the focus of this section.

4.6.4 The asymmetric nature of the Equatorial Undercurrent in the Pacific

In the oceans there are western boundary currents in both hemispheres, and they all feed the Equatorial Undercurrent; thus there is competition between them. For the case of a symmetric forcing, the solution is reduced to the case discussed above. For the case of asymmetric forcing, the western boundary current from the hemisphere with stronger forcing overshoots the equator, where the two western boundary currents merge, and forms an undercurrent that is asymmetric with respect to the equator. Layer thicknesses are continuous across the matching streamline, but zonal velocity can be discontinuous.

Flow across the equator has been examined by many investigators, e.g., Anderson and Moore (1979), Killworth (1991), and Edwards and Pedlosky (1998). Since the Coriolis parameter changes sign across the equator, it works as a potential vorticity barrier for cross-equator flow. Killworth (1991) showed that cross-equator flow is confined within a few deformation radii of the equator. In order to move beyond such a range, a frictional force is required to alternate the potential vorticity of the water parcels. The problem associated with frictional flow across the equator has been studied by many investigators, such as Edwards and Pedlosky (1998).

A further complication arises with the bifurcation of the western boundary currents. At low latitudes, the westward-flowing current bifurcates when it meets the western boundary, thus forming the poleward and equatorward western boundary currents. The best example is the northward Kuroshio and the equatorward Mindanao Current in the North Pacific.

The bifurcation of the western boundary current involves complicated dynamics in three-dimensional space, including eddies and flow over complicated topography. However, a barotropic model may provide a simple solution that can be used as a crude estimate.

Model formulation

Throughout, we assume that the circulation is steady and can be treated in terms of an ideal-fluid model; thus both potential vorticity and the Bernoulli function are conserved along streamlines

$$\vec{u}_n \cdot \nabla_h q_n = 0, \quad \vec{u}_n \cdot \nabla_h B_n = 0 \tag{4.430}$$

where the subscript n labels the layer. We make such assumptions for both the thermocline and the EUC (Equatorial Undercurrent) in the nature of a null hypothesis with regard to vertical mixing. That is, we attempt to see how much of the structure of the mid-latitude and equatorial thermocline can be explained on the basis of the ideal-fluid theory. Of course this is not the only legitimate point of view. Numerical experiments (e.g., Blanke and Raynaud, 1997; Lu *et al.*, 1998) require mixing for numerical reasons and suggest the possible importance of mixing for the dynamics.

Note that the upper layer thickness is another unknown. In order to carry out the boundary layer calculation, we also need to specify an additional constraint on the upper layer thickness. Pedlosky (1987b) first assumed that h_1 is independent of latitude within the equatorial boundary layer and found some interesting solutions. Another choice is to assume that the upper layer thickness is compensated within the equatorial region (Pedlosky, 1996). This choice is based on the following idea: the geostrophic velocity in the upper layer is much smaller than that in the second layer; thus, to the lowest-order approximation, the meridional pressure gradient in the upper layer is negligible compared with that in the second layer. This approximation implies that near the equator the current in the upper layer is dominated by the local wind stress.

Of course this specification of the depth of the upper layer is entirely arbitrary. In the original theory, as described by Pedlosky (1996), the two extremes of specification of the upper layer thickness, i.e., either independent of y with no vertical shear across the interface, or completely compensated, had little effect on the undercurrent or the depth of the equatorial thermocline. In the interests of keeping our model as simple and comprehensible as possible, we retain this arbitrary, and admittedly deficient, element of the theory. Thus, we will use the following additional constraint on the upper layer thickness:

$$h_1(x, y) = h_1(x, y_m) + h(x, y_m) - h(x, y) \tag{4.431}$$

where y_m is the latitude where the equatorial thermocline solution is matched with the interior thermocline solution. This is the compensated solution. Similarly to the solution discussed in the previous section, as long as B_0 is specified, the solution can be found by a shooting method, as described by Huang and Pedlosky (2000a).

A major assumption implicitly made in the model discussed above is that the solution is symmetric with respect to the equator. This implicit assumption does not hold exactly in the world's oceans. Owing to the asymmetric nature of the atmospheric general circulation, wind stress near the equator is not symmetric. It can readily be seen that a straightforward application of this model will lead to discontinuity of the solution for the equatorial thermocline.

A natural requirement is that the solution should be continuous across the equator. Thus, the minimum requirement is that the thickness of both the upper and lower layers should be continuous across the equator. Since the Bernoulli function carried by the corresponding western boundary currents in the two hemispheres is not the same, these two boundary currents should not match exactly along the equator. In other words, the matching latitude at each longitudinal section should be a free boundary determined from the interior dynamics.

Thus, the suitable boundary value problem for the branch originating from the Northern Hemisphere branch is

$$h^n = H_0^n \text{ at } y = y_m \tag{4.432}$$

Second, the Bernoulli function should be a constant along the matching line near the equator

$$h^n + u^2/2 = B_0^n \text{ at } y = y_{sep} \tag{4.433}$$

where B_0^n is the Bernoulli function of the western boundary current at the bifurcation latitude in the Northern Hemisphere, determined by the barotropic circulation in the subtropical basin interior. In addition, we will use the following additional constraint on the upper layer thickness:

$$h_1^n(x, y) = h_1^n(x, y_m) + h^n(x, y_m) + \Delta h^{ns} \frac{y_m - y}{2y_m} - h^n(x, y) \tag{4.434}$$

where $\Delta h^{ns} = h_1^s(x, -y_m) + h^n(x, -y_m) - \left[h_1^n(x, y_m) + h^n(x, y_m)\right]$.

The suitable boundary value problem for the branch originating from the Southern Hemisphere branch is

$$h^s = H_0^s, \; at \; y = -y_m \tag{4.435}$$

In addition, the Bernoulli function should be a constant along the matching line near the equator

$$h^s + u^2/2 = B_0^s, \text{ at } y = y_{sep} \text{ and} \tag{4.436}$$

$$h_1^s(x, y) = h_1^s(x, -y_m) + h^s(x, -y_m) + \Delta h^{ns} \frac{y + y_m}{2y_m} - h^s(x, y) \tag{4.437}$$

Note that for the case with symmetric forcing, our formulation is reduced to the solution discussed by Pedlosky (1996), with the equator as the separating line of these two branches, i.e., $y_{sep} = 0$.

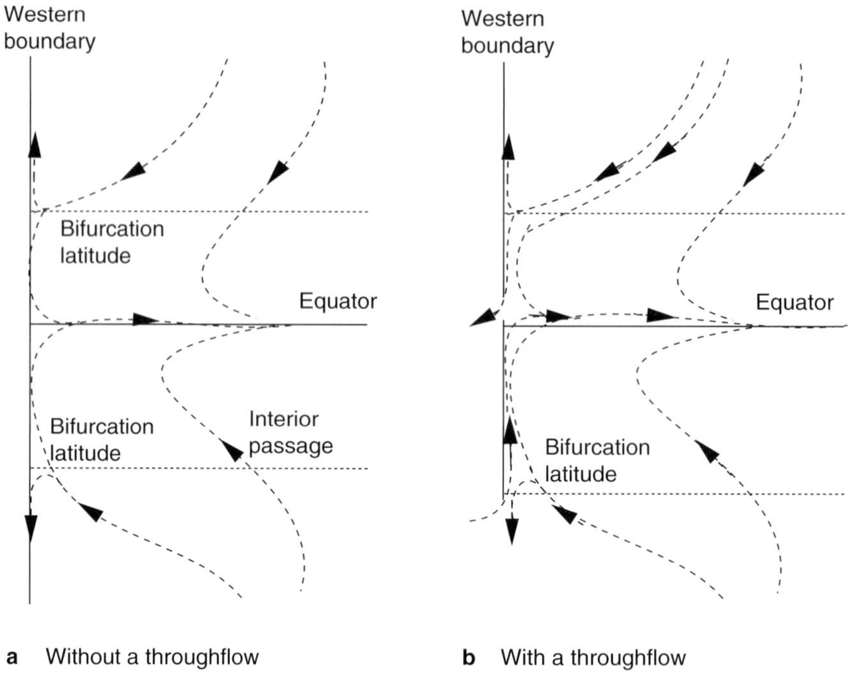

Fig. 4.72 Sketch of the Equatorial Undercurrent in a two-hemisphere basin: **a** without a throughflow, **b** with a throughflow.

Since the Bernoulli function plays a role similar to the pressure head, it is expected that the western boundary current that has the larger Bernoulli function should overshoot the equator and invade the other hemisphere, as shown in the sketch in Figure 4.72. In addition, there may be a small difference between the interior solutions at the matching latitude in the two hemispheres; thus, there is a small pressure gradient in the upper layer, resulting from $\Delta h^{ns} \neq 0$. However, this may be a small contribution to the solution, so we will assume that $\Delta h^{ns} = 0$ for the following analysis.

The boundary value problem consists of two sets of ordinary differential equation systems. Each set is subjected to the matching boundary conditions at the matching latitude y_m; these two solutions match along a free boundary y_{sep}, which is part of the solution. Across this matching boundary, the upper and lower layer thicknesses are continuous, as required. However, the zonal velocity may be discontinuous because the Bernoulli function from the two hemispheres can be different. A solution with $B_0^s = 1.25 B_0^n$ is shown in Figures 4.73 and 4.74.

The solution is very similar to the case with a symmetric forcing, and is slightly asymmetric with respect to the equator (Fig. 4.73). The thermocline depth clearly has a dumbbell shape in the equatorial regime, resembling the gross structure of the equatorial thermocline observed in the oceans. The most outstanding feature of this solution is the discontinuity of

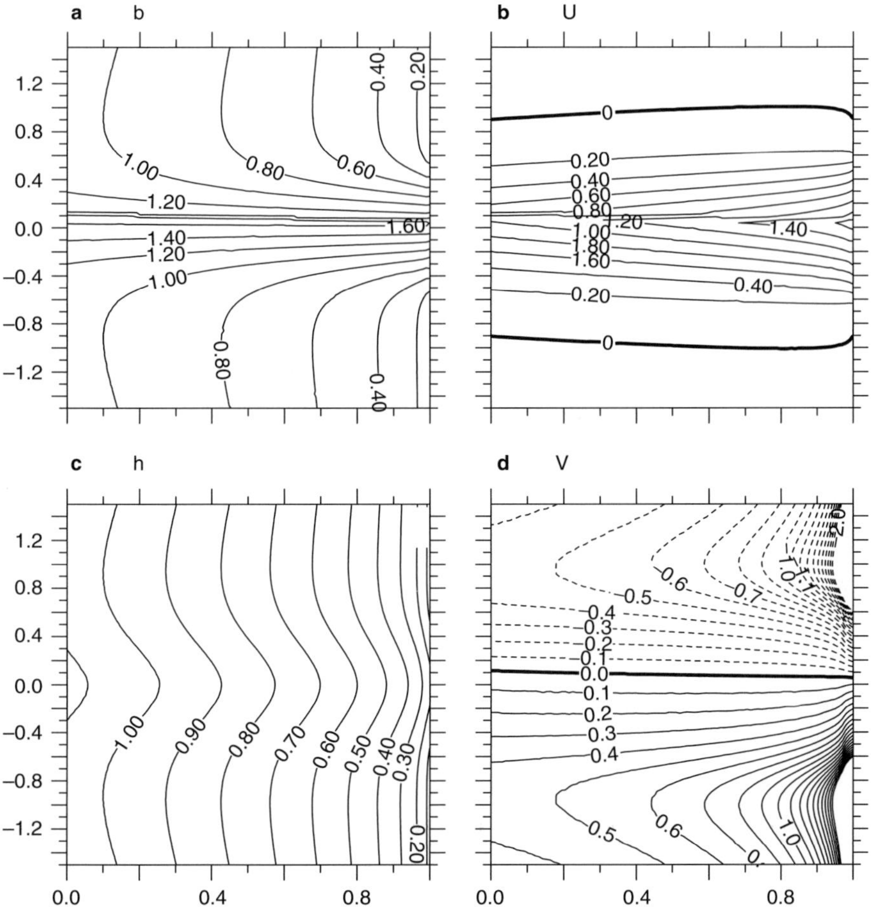

Fig. 4.73 Structure of the Equatorial Undercurrent in a two-hemisphere ocean, under the asymmetric forcing, $B_0^s = 1.25B_0^n$, in nondimensional units: **a** the Bernoulli function; **b** the zonal velocity; **c** the thermocline depth; **d** the meridional velocity (Huang and Jin, 2003).

the zonal velocity across the separating streamline; this discontinuity can be seen clearly in the meridional sections shown in Figure 4.74.

The western boundary currents from the two hemispheres meet near the equator and form the eastward-moving Equatorial Undercurrent. Owing to the differences in the Bernoulli head, the zonal velocity is discontinuous across the separating streamline, as shown in Figure 4.74a. As the undercurrent moves eastward, the zonal velocity core becomes stronger, and the separating streamline moves slightly toward the equator, as shown by the line labeled y_{sep} in the lower part of Figure 4.74a.

The meridional structure of the undercurrent can be represented by the meridional section taken along the western boundary, as shown in Figure 4.74b, c, d. As required by the

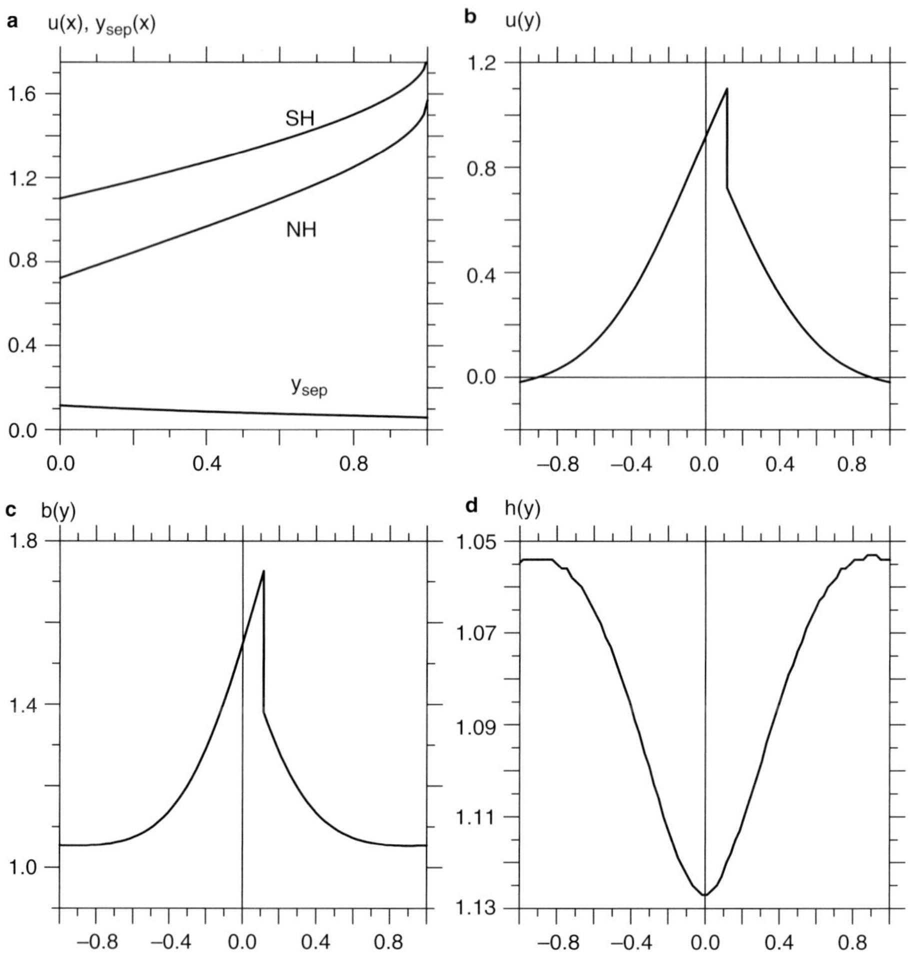

Fig. 4.74 The zonal and meridional structure of the Equatorial Undercurrent,, under the asymmetric forcing, $B_0^s = 1.25B_0^n$, in nondimensional units: **a** the zonal variation of the zonal velocity along the matching streamline: *SH* (*NH*) indicates the water particle originating from the Southern (Northern) Hemisphere, y_{sep} indicates the zonal variation of the separation point; **b** the zonal velocity at the western boundary; **c** the Bernoulli function along the western boundary; **d** the thermocline depth along the western boundary (Huang and Jin, 2003).

matching boundary condition, the thermocline depth is continuous (Fig. 4.74d). However, both the zonal velocity and the Bernoulli function are discontinuous across the separation line (Fig. 4.74b, c).

It is important to note that this ideal-fluid solution may not be very stable. In fact, the solution consists of a rather narrow band north of the equator, where a patch of negative potential vorticity originating from the Southern Hemisphere is adjacent to the positive

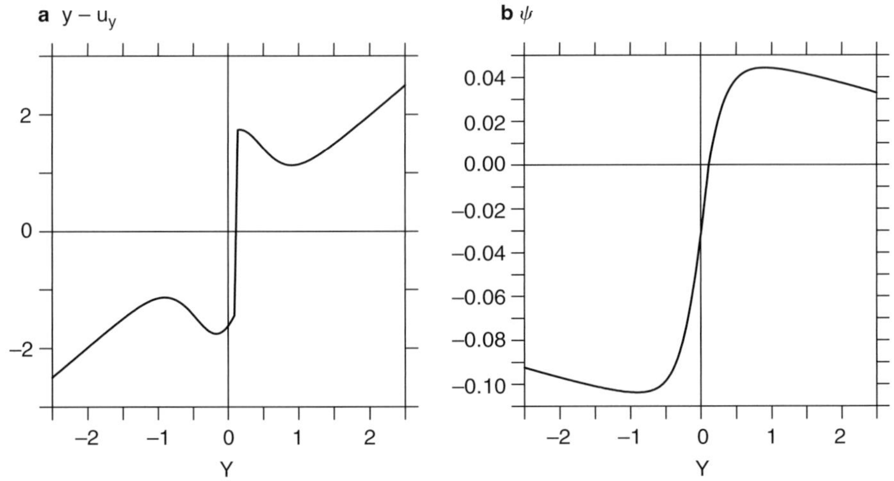

Fig. 4.75 The meridional profile of **a** the total vorticity and **b** the streamfunction at the western boundary (Huang and Jin, 2003).

potential vorticity in the Northern Hemisphere (Fig. 4.75a). The coexistence of potential vorticity with opposite signs indicates that symmetric instability may occur and thus modify the solution.

Furthermore, it is readily seen that the zonal velocity profile satisfies the necessary condition for barotropic instability. The barotropic instability is likely to smooth out the velocity cusp near the equator. Even for the hemispherical symmetric solution first obtained by Pedlosky (1987b), it can readily be seen that the zonal profile of the solution satisfies the necessary condition for barotropic instability.

The fact that symmetric instability may arise from the hemispheric asymmetric solution may also not be a significant deficiency. To some extent, it may be a realistic property. It was suggested recently by Hua *et al.* (1997) that the observed equatorial mean circulation may marginally satisfy the condition for symmetric instability. Using a numerical model, they further demonstrated that for a basic state flow with this kind of instability, the nonlinear equilibrated state of the equatorial ocean circulation exhibits some secondary flows which resemble the observed multiple equatorial jets underneath the Equatorial Undercurrent.

Application to the Pacific Ocean

Evidence of off-equator shift of the Equatorial Undercurrent core

Direct field measurements have also indicated the asymmetrical nature of the zonal velocity profile in the Equatorial Undercurrent. Hayes (1982) studied the zonal geostrophic velocity profile along two sections in the eastern equatorial Pacific. Geostrophic velocity calculated from hydrographic data is consistent with that obtained from the free-fall acoustically tracked velocimeter (TOPS). The geostrophic velocity south of the equator (at 0.5° S and

1° S) was clearly stronger than that north of the equator. Wyrtki and Kilonsky (1984) calculated the geostrophic zonal velocity from the data collected during the Hawaii to Tahiti Shuttle Experiment. Their results indicated that the center of the Equatorial Undercurrent is located at 0.5° S.

However, in the western part of the equatorial Pacific Ocean, the core of the undercurrent is located north of the equator. Tsuchiya *et al.* (1989) made a detailed analysis of the water mass properties collected during the Western Equatorial Pacific Ocean Circulation Study (WEPOCS). Their analysis clearly showed that the major portion of the water in the Equatorial Undercurrent at its beginning north of Papua New Guinea is supplied from the south by a narrow western boundary undercurrent (New Guinea Coastal Undercurrent). In fact, the water mass north of the equator can be traced back to its source in the south.

Gouriou and Toole (1993) analyzed the mean circulation in the western equatorial Pacific Ocean. As indicated in their Fig. 8B, there is a patch of negative potential vorticity north of the equator, adjacent to the positive potential vorticity in the environment. Such a patch of negative potential vorticity must come from the Southern Hemisphere.

Joyce (1988) argued that the wind stress on the equatorial ocean can force a cross-equatorial flow in the upper ocean. By applying a generalized Sverdrup relation to the equatorial oceans, he inferred that there is a southward cross-equatorial flow in the eastern equatorial Pacific Ocean, and a northward cross-equatorial flow in the western equatorial Pacific Ocean. His calculation, however, did not include the contribution due to the relative vorticity term, nor the connection with the western boundary currents.

The total volume flux going through the western boundary at the beginning of the undercurrent can be estimated from the streamfunction profile shown in Figure 4.75b. For example, the volume flux from the Southern and Northern Hemispheres is about 0.1 and 0.05 in nondimensional units. For the Pacific Ocean, the scale of the streamfunction is about 206 Sv; thus, the volume flux contribution from the western boundary currents is about 20 Sv and 10 Sv. The cross-equatorial flux, ψ_0, is about 0.03 nondimensional units, which corresponds to about 6 Sv in dimensional units. Tsuchiya *et al.* (1989) estimated that the volume flux of 5 Sv is fed to the eastward interior circulation between 3° S and the equator. If there were no Indonesian Throughflow, this would be the volume flux fed to the undercurrent. This flux rate is rather close to the 6 Sv estimated in the discussion above.

Determination of the cross-equatorial flow in the source regime of the undercurrent

The dynamical factor that controls the position of the separating streamline is the strength of the Bernoulli function at the latitude of the western boundary bifurcation. In reality, the bifurcation of the western boundary is a complicated three-dimensional phenomenon, which is determined by many dynamical factors, such as the large-scale forcing fields and the resulting pressure field, the shape of the coastline, bottom topography, and eddies. However, a barotropic bifurcation latitude may provide a simple and useful first step to this complicated problem.

The total poleward Ekman flux across a latitudinal section is

$$M_E = -\int_0^L \frac{\tau^x}{f\rho_0} dx \tag{4.438}$$

and the equatorward geostrophic flux is

$$M_G = \int_0^L \frac{f w_e}{\beta} dx \tag{4.439}$$

where w_e is the Ekman pumping velocity calculated from wind stress. The total meridional volume flux involved in the wind-driven gyre across a latitudinal section of the basin interior is the sum of these two fluxes, i.e., $M_{Sv} = M_E + M_G$. The return flow into the western boundary is M_{Sv}. The bifurcation of the western boundary current is a nonlinear phenomenon; however, here we assume that the bifurcation latitude of the western boundary current is the latitude where M_{Sv} vanishes; this can be determined from the wind stress distribution in the basin. For the climatological mean wind stress, this gives the separation latitudes at 14.5° N for the North Pacific and 15° S for the South Pacific: values which are close to the bifurcation points estimated from a numerical model (e.g., B. Huang and Liu, 1999).

For a given amount of wind stress, the depth of the main thermocline can be calculated using a simple reduced-gravity model. The corresponding Bernoulli function can be calculated according to the definition: $B = g'H$, where g' is the reduced gravity, chosen as $g' = 2$ cm/s^2 for the Pacific, and H is the depth of the main thermocline.

This simple approach is, however, unsatisfactory because wind stress is slightly stronger in the Northern Hemisphere and the corresponding Bernoulli head in the Northern Hemisphere is larger than that in the Southern Hemisphere. As a result, this simple approach would lead to a conclusion that water from the Northern Hemisphere western boundary regime should invade the Southern Hemisphere. However, observations indicate that most water forming the Equatorial Undercurrent does come from the south, as discussed above. Thus, in order to have the water from the Southern Hemisphere overshoot the equator, another mechanism is needed. A strong contender is the Indonesian Throughflow. With the throughflow, the western boundary current system changes and the source of the undercurrent also changes accordingly, as shown in Figure 4.73b.

Australia can be treated as a big island, so the simple island rule can apply and yield a circulation around it that is a simple function of wind stress, as discussed in Chapter 2. The island rule has been discussed in many papers, and the most essential physical mechanisms associated with this rule have been discussed thoroughly by Godfrey (1989, 1993).

Although volume flux associated with the throughflow predicted from the island rule can be large, on the order of 17 Sv, other physical processes reduce the flux to about 10 Sv. In our simple purely inertial model, water from the Southern Hemisphere could not penetrate into the Northern Hemisphere or feed the throughflow without the potential vorticity changing sign. The western boundary current from the Southern Hemisphere has to go through the undercurrent first. Over the eastward trajectory this volume flux upwells into the Ekman layer and then moves poleward in the form of Ekman transport. Thus, volume flux from the Southern Hemisphere becomes part of the Sverdrup volume flux in the ocean interior of the Northern Hemisphere. Within the simple ideal-fluid formulation of our model, the only option is to assume that the throughflow is fed from the western boundary current from the Northern Hemisphere; hence, we will simply assume that, indeed, about 10 Sv of water leaves the basin and forms the Indonesian Throughflow.

In the Northern Hemisphere, therefore, the separation latitude is not affected by the throughflow; volume flux in the western boundary current is zero at the separation latitude. However, the effective western boundary current that feeds the undercurrent actually comes from the ocean interior. To find the Bernoulli head of this western boundary layer, one has to search eastward, starting from the western boundary at the separation latitude, for a place where the equatorward flux, integrated from the eastern boundary, satisfies

$$\psi_{\textit{effective wbc}}^{\textit{North}} = \psi_{wbc} + \Delta\psi \tag{4.440}$$

where $\psi_{wbc} = -M_{Sv}$ is the Sverdrupian flux at the outer edge of the western boundary; and $\Delta\psi = 10$ Sv is the volumetric contribution to the throughflow from the wind-driven circulation.

In the Southern Hemisphere, the separation latitude is now determined by the constraint that the total volume flux in the western boundary layer

$$\psi_{\textit{effective wbc}}^{\textit{South}} = -M_{Sv} + \Delta\psi \tag{4.441}$$

vanishes. Without the throughflow, the separation latitude in the Southern Hemisphere is near 15° S; with the throughflow, however, the separation latitude is pushed southward to 17° S.

4.7 Communication between subtropics and tropics

4.7.1 Introduction

The subtropical and tropical cells

The linkage between the subtropics and tropics constitutes a major component of the global oceanic circulation and climate system. The circulation in the subtropical-equatorial ocean involves complicated dynamical processes and pathways. To simplify the picture, this system can be described in terms of two cells (McCreary and Lu, 1994). The division into two cells is conceptual only because the circulation is essentially three-dimensional. There is no clear boundary between these two cells, but we will use the so-called choking latitude, which will be defined shortly, as the boundary separating these two cells (Fig. 4.76a).

The tropical and subtropical cells are closely related to the following processes:

- *In the subtropical basin interior*, the horizontal convergence of Ekman flux leads to Ekman pumping that drives the anticyclonic gyre in the subtropical basin and gives rise to subduction of subtropical

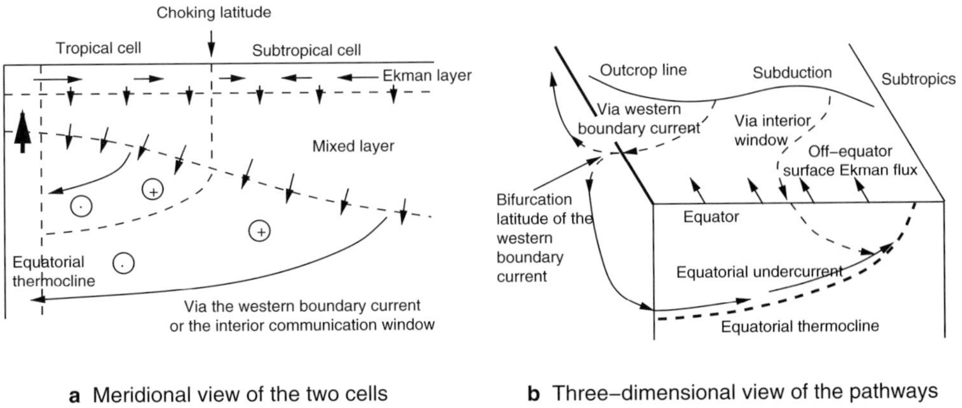

Fig. 4.76 **a** Meridional view of the tropical and subtropical cells; **b** sketch of the pathways from the subtropics to the equator.

water masses. These water masses are transported through the equatorward geostrophic flows in the subtropical thermocline, where there are three pathways for them.

- First, water masses subducted in the subtropical basin can reach the tropics through the interior communication window and the subsequent equatorward flow on isopycnal surfaces. Gu and Philander (1997) proposed that such a linkage plays a vital role in climate variability on decadal time scales (Fig. 4.76b).
- Second, water masses subducted in the subtropical basin can reach the tropics as follows. These water masses move westward and reach the western boundary where the flow bifurcates, and part of the water flows in the form of the equatorward western boundary current and eventually the eastward Equatorial Undercurrent.
- Third, at the bifurcation point of the western boundary, some of the subducted water is returned to mid latitudes via the poleward western boundary current. This part of the subducted water is not counted as the subtropical cell; on the other hand, water masses moving through the first and second pathways constitute the subtropical cell.

- *In the equatorial band*, the subsurface water mass joins the eastward Equatorial Undercurrent, and eventually it is lifted to the surface via the equatorial upwelling driven by the easterlies along the equator. It is important to emphasize that the upward motion during this phase is mostly adiabatic, without much diapycnal mixing (Bryden and Brady, 1985). As discussed in the previous section, the undercurrent can be described as an inertial boundary current in an ideal-fluid model.
- *The off-equator Ekman flux* brings water back to mid latitudes. Off equator the local convergence of Ekman flux gives rise to Ekman pumping, which pushes water down into the thermocline, and thus completes the loop.

As discussed below, not all subtropical water pumped down and going through the base of the mixed layer can reach the equator. There is a choking latitude where an interior communication window exists between the subtropics and tropics; only part of the subtropical water subducted into the subtropical thermocline can pass through this window and thus contribute to the subtropical cell. The remaining part of the subducted subtropical water turns westward and reaches the western boundary, where the current bifurcates. Only the equatorward branch of this western boundary current can contribute to the subtropical cell; on the other hand, its poleward branch is unable to reach the equatorial band and makes no contribution to the subtropical cell.

On the equatorward side of the choking latitude, the equatorward transport continues to increase owing to the equatorward geostrophic flow induced by the local Ekman pumping. This additional transport of water also goes through the two pathways similar to those of the subtropical cell, and it eventually returns poleward in the form of the off-equator Ekman transport, thus completing the cell. Accordingly, this additional transport should be classified as part of the tropical cell, as depicted in Figure 4.76a.

The mechanism described above applies to both the Pacific and Atlantic Basins, where easterlies prevail in the equatorial band; however, in the Indian Basin, westerlies prevail in the equatorial band, and the dynamics of equatorial circulation is different from that discussed in this section.

Communication window identified from tritium data in the tropical Pacific

The linkage between the subtropical and tropical oceans can be identified from tracer distribution in these regimes. In fact, such links were first discovered through tracer studies. Tracer observations indicated that in the upper layer, tritium in the North Equatorial Current has decreased since 1974, while tritium south of the North Equatorial Current increased from 1965 through 1979. Thus, the flow in the subtropical cell communicating between mid-latitude and equatorial oceans is characterized with a time scale of about 10 years. Since tritium has a half-life of 12.4 years, it can be used as a good tracer for climate change on a decadal time scale.

The interior communication window was identified through tracer studies in the 1970s and 1980s. Fine and her colleagues (Fine *et al.*, 1981, 1987; McPhaden and Fine, 1988) analyzed the tritium data and found a local tritium maximum around 140° W along the equator (Figs. 4.77 and 4.78), which they rightly attributed to the ventilation of the subtropical water via subduction. Since this local maximum of tritium is not linked to any high concentration of tritium in the western part of the equatorial band, it is readily seen that the existence of such a local maximum in tritium cannot be attributed to the Equatorial Undercurrent.

In this section, we discuss the volume flux through the interior communication window, using a simple index that is based on wind stress data only. This index can be used to illustrate the asymmetric nature of the interior communication between the tropics and the subtropics. In addition, the index can be used to infer the decadal variability of the interior communication.

The ITCZ and communication between subtropical and tropical oceans

One of the major features in wind stress near the equator is the existence of the Inter-Tropical Convergence Zone (ITCZ) in the Northern Hemisphere. The existence of a ridge-like feature in the meridional section indicates that the Ekman pumping rate is positive near the ITCZ. Such a ridge in the density section can be interpreted as a potential vorticity barrier which

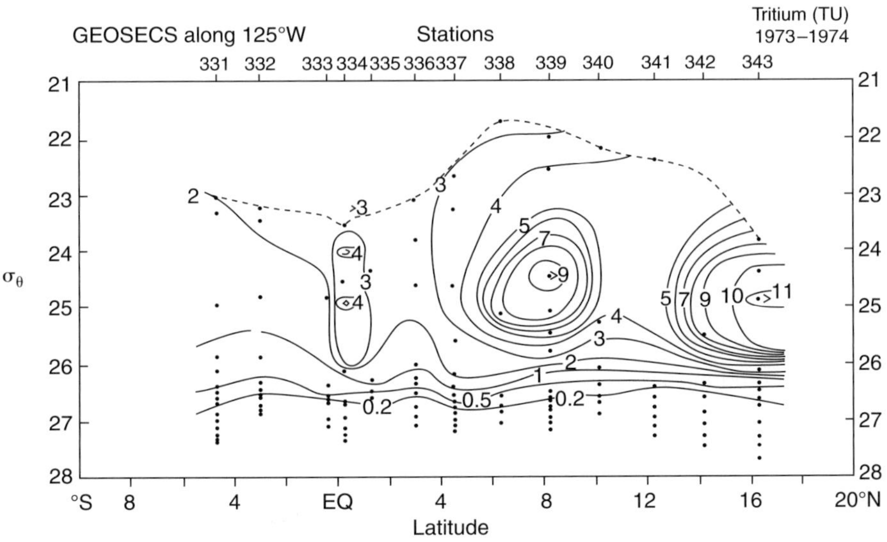

Fig. 4.77 A north–south vertical section of tritium (TU) vs. σ_θ along the eastern GEOSECS track at 125° W (Fine *et al.*, 1987).

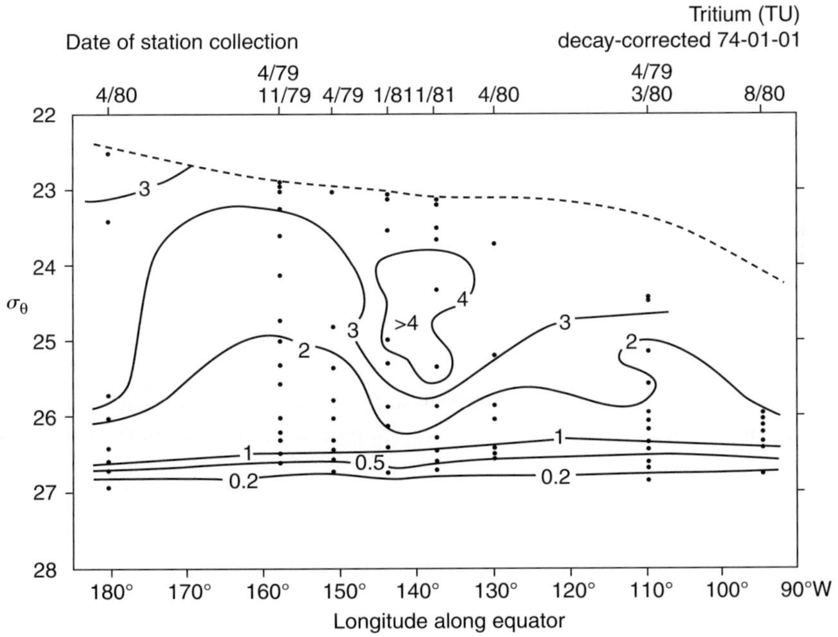

Fig. 4.78 An east–west vertical section of tritium (TU) vs. σ_θ along the equator (Fine *et al.*, 1987).

blocks the local communication between the subtropics and tropics. The position of the ITCZ can be seen clearly from the thermal structure in the ocean (Fig. 4.69). The latitude band of the ITCZ is closely linked to the choking latitude and the communication window discussed below. In fact, the pathway and the blockage of the communication have been interpreted in terms of the high potential vorticity ridge in the eastern basin (e.g., Lu and McCreary, 1995; Johnson and McPhaden, 1999).

4.7.2 Interior communication window between subtropics and tropics

Communication window inferred from data and numerical models

The interior communication associated with the circulation in the subtropical–tropical regime can be explored using analytical and numerical models (e.g., Liu, 1994; McCreary and Lu, 1994). Analysis of tracer and hydrographic data leads to a better estimate of the communication rate (e.g., Wijffels, 1993; Johnson and McPhaden, 1999). It is estimated that this communication rate is about 5 Sv for the North Pacific and 16 Sv for the South Pacific. The communication rate in the Atlantic was estimated by Fratantoni *et al.* (2000) as 1.8 Sv for the North Atlantic and 2.1 Sv for the South Atlantic. The communication window can be clearly demonstrated through analysis of a numerical model for the oceans, and pathways for the Pacific obtained from a numerical model are shown in Figure 4.79.

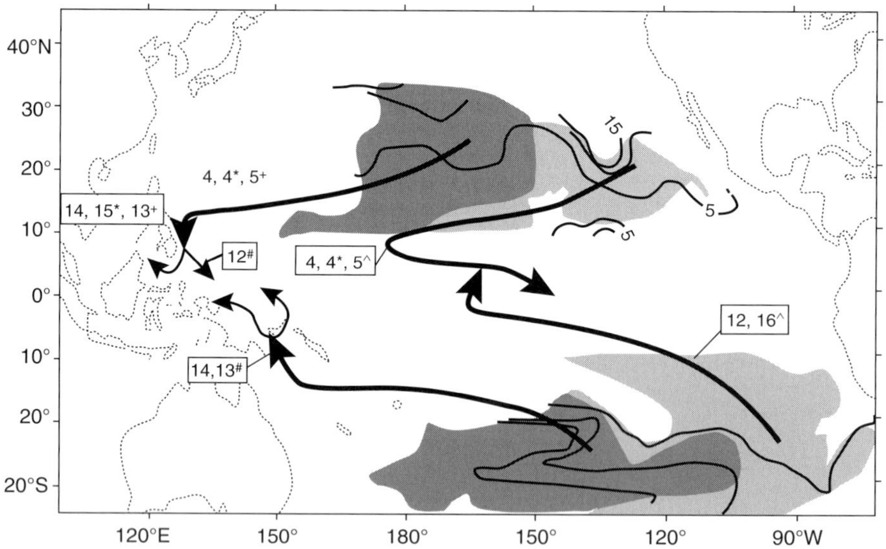

Fig. 4.79 Pathways from the subtropics to the tropics (Liu and Philander, 2001); the *thin arrows* indicate the western boundary windows and the *thick arrows* indicate the interior communication windows, identified by tracking the particles that are subducted at 50 m depth in the NCEP model data (B. Huang and Liu, 1999); *numbers in boxes* are the times in years for the particles to reach 5° latitude.

A sketch of the communication window

One outstanding feature in the subtropical-tropical ocean is the positive Ekman pumping areas within the ITCZ (or the South Pacific Convergence Zone in the South Pacific). The existence of a positive Ekman pumping rate gives rise to a small cyclonic gyre in the equatorial ocean, which is rather strong near the eastern boundary. In addition, there might be a second cyclonic gyre near the western boundary, as indicated by the heavy dashed lines in Figure 4.80.

Although the circulation involving the cyclonic gyres may seem complicated, it is, however, quite straightforward to deal with a model ocean including a patch of positive Ekman upwelling within the otherwise negative Ekman pumping. If the easterlies near the equator relax, the Ekman pumping rate will become more negative, and the western boundary of the eastern cyclonic gyre will extend far more westward; thus, the communication window will become narrower. If the western boundary of the eastern cyclonic gyre joins with the eastern boundary of the western cyclonic gyre, the interior communication window will be closed. Such a possibility can be explored, using wind stress datasets and results from an oceanic data assimilation system.

The situation can be illustrated in terms of a two-moving-layer model of the ventilated thermocline (Fig. 4.80). The lower layer outcrops in the subtropics and the upper layer covers the upper ocean in the tropics where the layer is stagnant in the shadow zone. (Within the cyclonic gyres, the lower layer is in motion because it is directly exposed to the Ekman upwelling.) However, the lower layer is in motion west of the shadow zone. In this two-layer model, the communication window is represented by potential flow in the

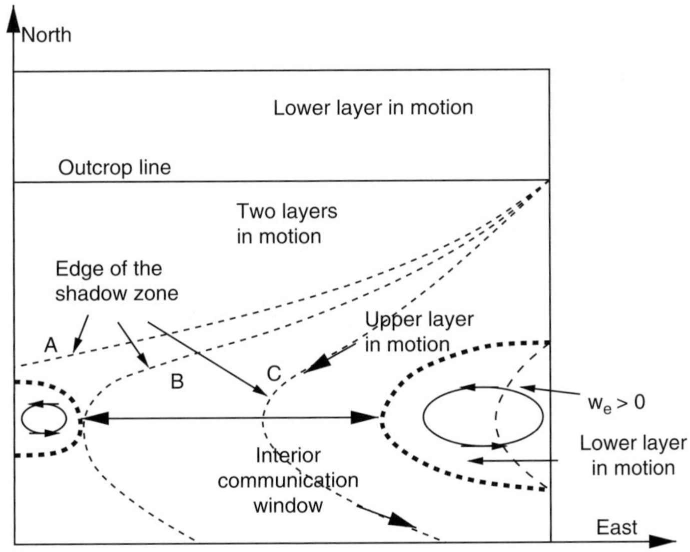

Fig. 4.80 Sketch of the communication window between the subtropics and tropics.

lower layer between the eastern cyclonic gyre and the western cyclonic gyre, depicted by the heavy dashed half ellipses in Figure 4.80.

The area covered by the shadow zone depends on the choice of layer thickness along the eastern boundary. As will be shown shortly, when the lower layer is very thick, the shadow zone is so large that its western boundary meets the western boundary north of the western cyclonic gyre, depicted by curve A. For such a case, there is no interior communication between the subtropics and tropics.

As the lower layer thickness is reduced, the boundary of the shadow zone moves toward the equator. Curve B indicates the critical case when the boundary of the shadow zone is tangential to the edge of the western cyclonic gyre. For smaller lower layer thickness, the boundary of the shadow zone is west of the western cyclonic gyre, depicted by curve C, and there is a communication window opening up between curve C and the western cyclonic gyre. The maximum width of the communication window is limited to the gap between the eastern and western cyclonic gyres.

A $2^1/_2$-layer model for an idealized ocean

The existence of the communication between the subtropics and tropics can be seen clearly using a ventilated thermocline model. The essential part of the model formulation is to simulate the effect of ITCZ by including a small positive Ekman pumping zone near the eastern boundary. We denote the bottom (and motionless) layer as layer 0, the lowest moving layer as layer 1, and the layers above have numbers increasing upward. The thickness and the depth of the i-th layer are denoted as h_i and H_i. The northernmost outcrop line is labeled as f_1.

The Sverdrup relation for an n-layer model is

$$\sum_{i=1}^{n} g_{i1} H_i^2 = D_0^2 + H_0^2 \tag{4.442}$$

where $g_{i1} = g'_i/g'_1$, and g'_i is the reduced gravity across the i-th interface. Using this relation, the Sverdrup function can be defined as

$$\psi = -\int_x^{x_e} hv dx = -\frac{f}{\beta}\int_x^{x_e} w_e dx \tag{4.443}$$

and $w_e(x, y) = -(\tau/f)_y$ is the Ekman pumping rate. North of the outcrop line f_1 there is one moving layer only, and the solution is

$$H_1 = \sqrt{D_0^2 + H_0^2} \tag{4.444}$$

South of the outcrop line f_1 there are two moving layers, and the solution is

$$H_1 = \left(\frac{D_0^2 + H_0^2}{G^2}\right)^{1/2}, \quad G^2 = 1 + g_{21}\left(1 - \frac{f}{f_1}\right)^2 \tag{4.445}$$

$$h_1 = \frac{f}{f_1}H_1, \quad h_2 = \left(1 - \frac{f}{f_1}\right)H_1 \tag{4.446}$$

However, south of f_1 there is also a shadow zone in the lower layer. Within the shadow zone, the lower layer is stagnant and the lower interface is at a constant level equal to the undisturbed depth of the lower interface set along the eastern boundary, H_0. The western boundary of the shadow zone, S_B, can be calculated by following the streamline in the lower layer starting from the eastern boundary at the outcrop latitude. However, it is more convenient to identify this boundary as the line where the lower interface depth H_1, calculated from Eqn. (4.445), is equal to H_0.

East of the shadow zone boundary S_B there can exist two dynamical regions: a shadow zone S and a cyclonic gyre CG. The boundary between the shadow zone S and the cyclonic gyre CG is determined by the zero barotropic streamline $\psi_0 = 0$. Within the shadow zone S, only the upper layer is in motion. Assuming that the reduced gravity across both the lower and upper interfaces is the same, the solution in the shadow zone is

$$h_2 = \sqrt{D_0^2 + H_0^2}, \; h_1 = H_0 - h_2 \tag{4.447}$$

Within the cyclonic gyre CG, layer 2 (the upper layer) vanishes, so layer 1 is the only moving layer, and the solution here is $h_1 = H_1 = \sqrt{D_0^2 + H_0^2}$.

We apply a $2^1/_2$-layer model to study the circulation in the subtropical-tropical ocean. The lower layer thickness is 150 m along the eastern boundary. The model is forced by a simplified wind stress profile (Fig. 4.81a), and the corresponding Ekman pumping rate is shown in Figure 4.81b.

As a departure from the practice of using a simple wind stress pattern independent of the zonal coordinate, we will also include a small positive Ekman upwelling regime near the eastern boundary of the model basin (Fig. 4.82b).

As shown in Figure 4.82, including a small regime of positive Ekman upwelling near the eastern boundary in the forcing field produces a small cyclonic gyre nearby, which is a critical component of the subtropical-tropical circulation.

A $2^1/_2$-layer model for the North Pacific

To make our discussion more practical, we apply this $2^1/_2$-layer model to the North Pacific, forced by the Ekman pumping field calculated from the Hellerman and Rosenstein (1983) wind stress data. The disadvantage of the multi-layer model is that the solution is rather sensitive to the choice of the outcrop line, the reduced gravity across the interfaces, and the

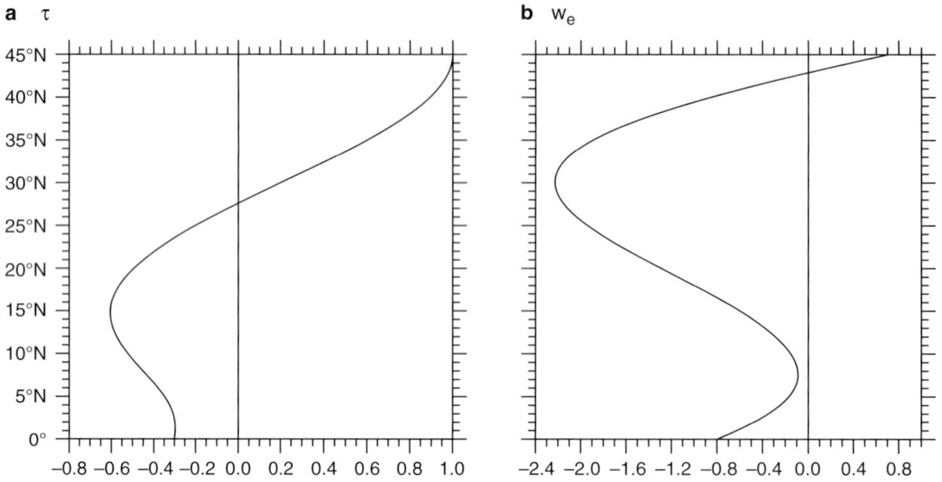

Fig. 4.81 **a** Idealized wind stress (in 0.1 N/m^2) and **b** Ekman pumping (in 10^{-6} m/s) profiles used in the analytical model.

lower layer thickness along the eastern boundary. In this section, we will simply assume that the upper layer outcrops along 18° N, and the reduced gravity for the upper and lower interfaces is 0.02 m/s^2. The only free parameter in this section is the lower layer thickness along the eastern boundary, h_{1e}.

For the case with $h_{1e} = H_0 = 150$ m, the solution is shown in Figure 4.83. The upper layer covers the area south of the outcrop line and west of the cyclonic gyre defined by the zero streamline, as shown in Figure 4.83a. Within the cyclonic gyre, there is the lower layer only, and its thickness declines toward the center of the cyclonic gyre (Fig. 4.83b). The lower interfacial depth clearly shows the bow-shaped structure of the anticyclonic gyre in the subtropics and the dome-shaped cyclonic gyre near the eastern boundary (Fig. 4.83c). The total equatorward volume flux in the upper layer going through the choking latitude, which is 9.5° N as dictated by the Sverdrup function, is more than 3 Sv (Fig. 4.83d).

The meridional volume transport in the lower layer depends on its thickness along the eastern boundary (Fig. 4.84). Note that south of the outcrop line, the lower layer is shielded from Ekman pumping, so the meridional volume transport function is also the streamfunction, except within the cyclonic gyre where the lower layer is subject to Ekman pumping directly. Three circulation regimes exist south of the outcrop line: the ventilation zone, where both the upper and lower layers are in motion; the shadow zone, where the lower layer is stagnant; and the cyclonic gyre, where the upper layer vanishes and the lower layer is in motion (indicated by dashed contours). The shadow zone is located between the two zero streamlines depicted by the heavy lines in this figure. For example, there is no shadow zone for the case with $H_0 = 100$ m, but the shadow zone is prominent in Figure 4.84c, d. As the lower layer thickness along the eastern boundary increases, the

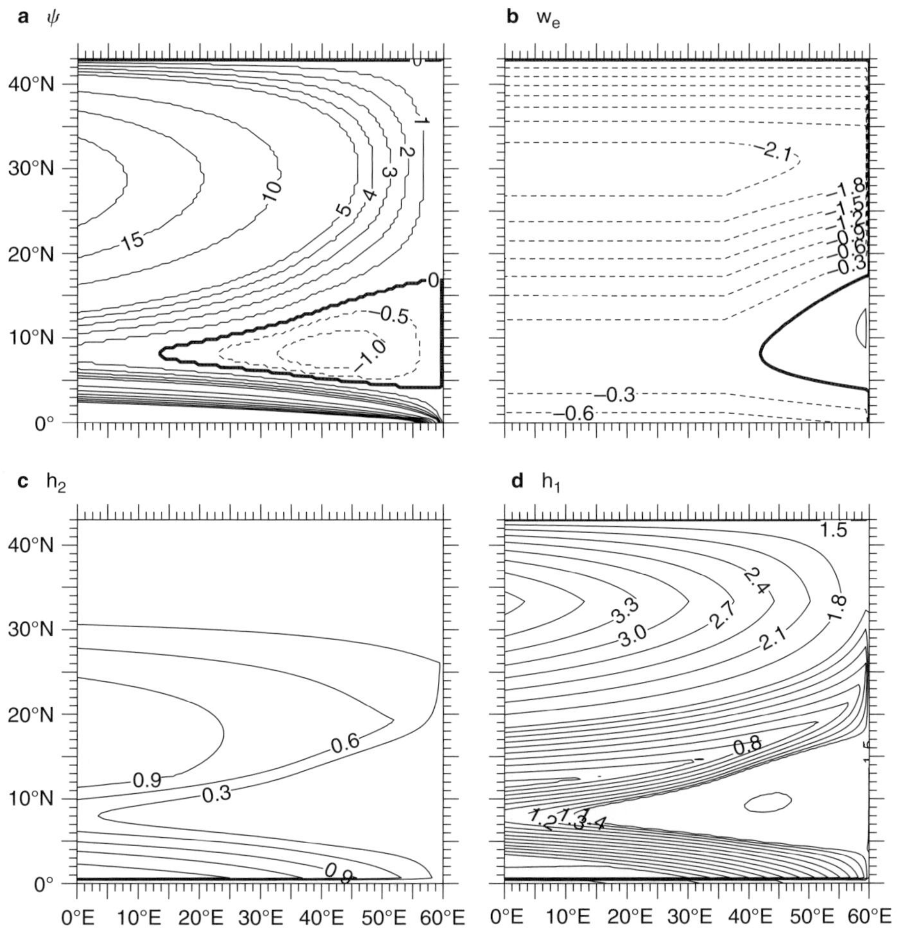

Fig. 4.82 Circulation driven by a slightly modified Ekman pumping field, with a small positive Ekman upwelling rate near the eastern boundary of the model ocean: **a** barotropic Sverdrup transport (Sv); **b** Ekman pumping rate (in 10^{-6}/m/s); **c** upper layer thickness (in 100 m); **d** lower layer thickness (in 100 m).

shadow zone expands westward, and the volume flux going through the choking latitude in the lower layer gradually declines from 2 Sv ($H_0 = 100$ m) to 0.5 Sv ($H_0 = 200$ m), and finally becomes zero for the case with $H_0 = 250$ m. Since this $2^1/_2$-layer model satisfies the Sverdrup constraint exactly, the net volume flux in the upper layer is the difference between the Sverdrup flux and the volume flux in the lower layer.

The volume flux going through the choking latitude is, thus, sensitively dependent on the choice of the lower layer thickness along the eastern boundary, as shown in Figure 4.85. When h_{1e} is larger than 240 m, the communication pathway in the lower layer is entirely blocked.

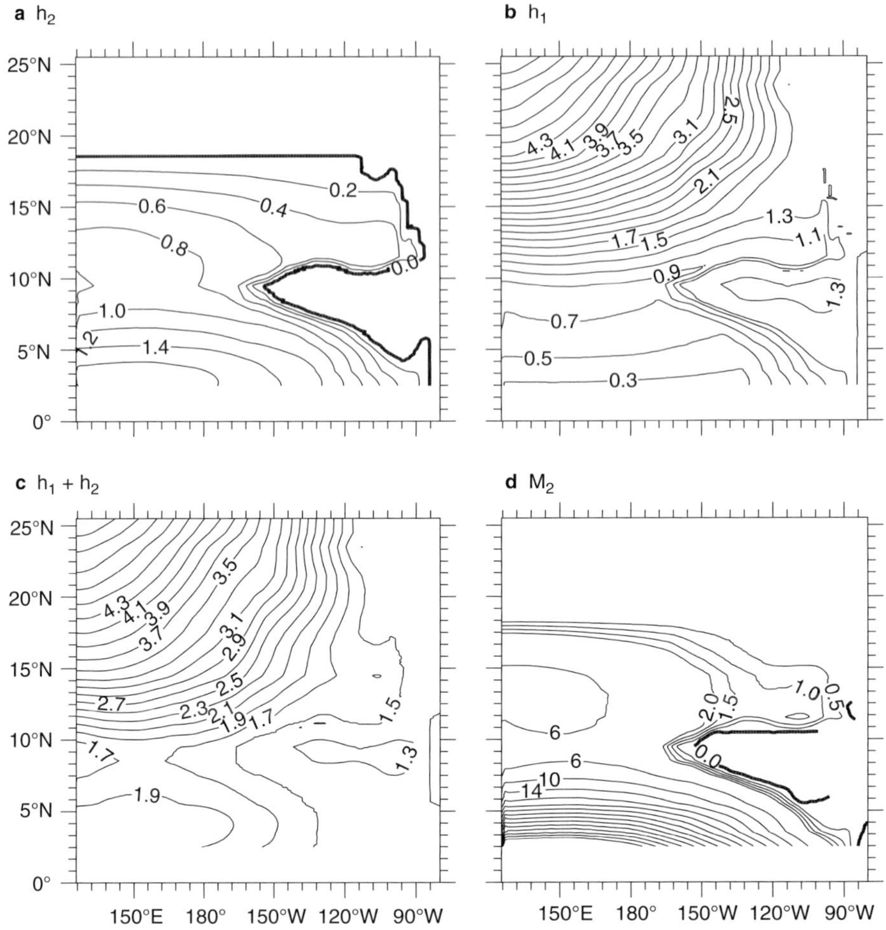

Fig. 4.83 Circulation of a $2^1/_2$-layer model ocean mimicking the North Pacific with realistic Ekman pumping; the lower layer thickness along the eastern boundary is $h_{1e} = H_0 = 150$ m, and the outcrop line is at 18° N. Panels **a**, **b**, and **c** for layer thickness (in 100 m); panel **d** is the meridional volume flux in the upper layer (Sv).

4.7.3 Communication windows in the world's oceans

It is obvious that using a simple two-layer model to decide whether there is an interior communication between the subtropics and tropics in the lower layer is rather subjective. It is desirable to calculate the interior communication rate independently of the choice of the layer thickness along the eastern boundary.

Meridional transport in the world's oceans

We begin with the barotropic flow in the world's oceans driven by the annual Ekman pumping rate obtained from the annual mean wind stress data of Hellerman and Rosenstein

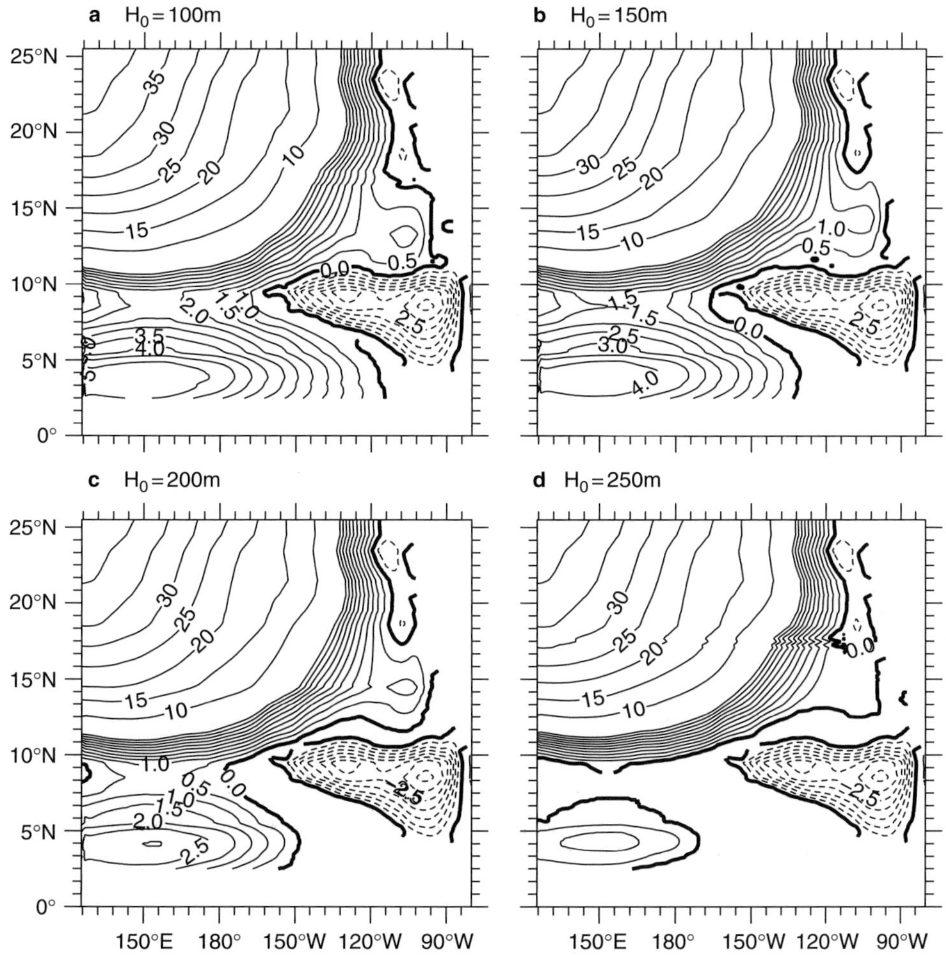

Fig. 4.84 Circulation in the lower layer for a $2^1/_2$-layer model ocean mimicking the North Pacific and driven by realistic Ekman pumping, for four different lower layer thicknesses along the eastern boundary.

(1983). The Ekman pumping rate in spherical coordinates is calculated as

$$w_e = \frac{1}{a\rho_0 2\omega \sin\theta}\left(\frac{1}{\cos\theta}\frac{\partial \tau^\theta}{\partial \lambda} - \frac{\partial \tau^\lambda}{\partial \theta} + \frac{\tau^\lambda}{\sin\theta\cos\theta}\right) \tag{4.448}$$

where a is the radius of the Earth, ω the angular velocity of the Earth's rotation, θ the latitude, λ the longitude, and τ^θ and τ^λ the meridional and longitudinal components of the wind stress. The meridional transport function (or the Sverdrup function) is defined as the zonal

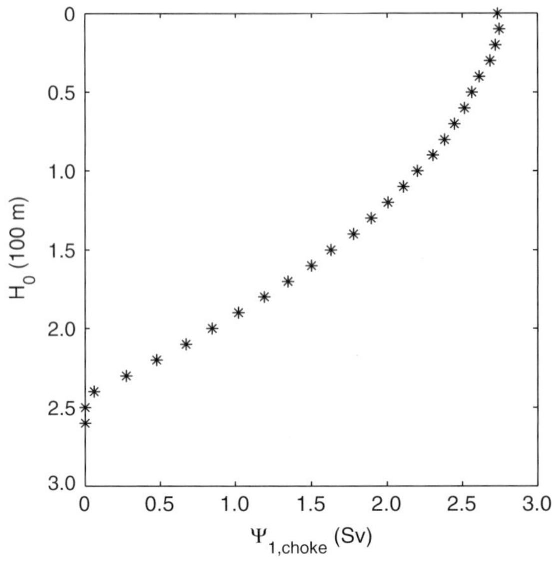

Fig. 4.85 Lower layer streamfunction at the choking latitude as a function of the layer thickness specified along the eastern boundary.

integration of the meridional volume flux within the wind-driven gyre:

$$M = -\int_{\lambda}^{\lambda_e} hva\cos\theta d\lambda = -\frac{f}{\beta}\int_{\lambda}^{\lambda_e} w_e a\cos\theta d\lambda \tag{4.449}$$

where $f = 2\omega\sin\theta$ is the Coriolis parameter and $\beta = 2\omega\cos\theta/a$. Note that the meridional volume flux in a steady state is the same for either a homogeneous ocean model with a flat bottom or a reduced-gravity model. The Sverdrup function is not a streamfunction, because the barotropic flow includes the Ekman pumping as a source in the subtropical basin. However, the barotropic streamlines can be identified from the pressure field. We introduce a "virtual streamfunction":

$$\psi^* = \frac{f}{f_0}M = -\frac{f^2}{f_0\beta}\int_{\lambda}^{\lambda_e} w_e a\cos\theta d\lambda \tag{4.450}$$

where f_0 is the reference latitude where the meridional volume flux M is equal to ψ^*. Note that ψ^* is essentially the layer-integrated pressure field, and it has the same dimension as the streamfunction. Using this virtual streamfunction, the streamlines are contours with a constant value of ψ^*. The volume flux between two streamlines satisfies $\Delta M = f_0\Delta\psi^*/f$, so away from the reference latitude ΔM is quite different from the value of $\Delta\psi^*$. At latitudes south of $f = f_0$, ΔM is larger than $\Delta\psi^*$. This difference is due to the fact that the Ekman pumping is downward, so it continuously adds on mass between two streamlines; thus the total volume flux between two streamlines, ΔM, is larger than $\Delta\psi^*$, the total volume flux between these two streamlines at the reference latitude.

The volume flux here is defined as the geostrophic flow below the Ekman layer, so we will call the gyres calculated in this way the subsurface gyres, because these gyres do not include the Ekman flux within the surface Ekman layer. In some classic papers on the wind-driven circulation, the total volume flux, including the Ekman flux, is calculated (e.g., Munk, 1950). At low latitudes the volume flux in the Ekman layer flows off the equator; however, the subsurface current may flow equatorward if the Ekman pumping rate is negative. Thus, the barotropic meridional transport function based on the total volume flux does not represent the subsurface meridional flow correctly. Even if the Ekman pumping rate is positive, the barotropic transport function based on the total volume flux tends to have a cyclonic gyre in the tropics, which has poleward volume flux stronger than that calculated from the subsurface geostrophic flow.

A simple index for the interior communication rate

As discussed above, the cyclonic gyres at the latitudes of the ITCZ are defined by the subsurface geostrophic volume flux, excluding the poleward Ekman flux in the surface layer; thus they are much weaker than the cyclonic gyres calculated from the total volume flux (including the Ekman flux). Most importantly, the cyclonic gyres from models based on total volume flux may completely block the interior passage between the subtropics and tropics, but the subsurface cyclonic gyres, as shown in Figure 4.86, do not entirely block the interior communication between the subtropical gyre and the tropics.

For the annual mean flow field, there is a window in the middle or the western basin that allows the interior communication between the subtropics and tropics. Note that there can be

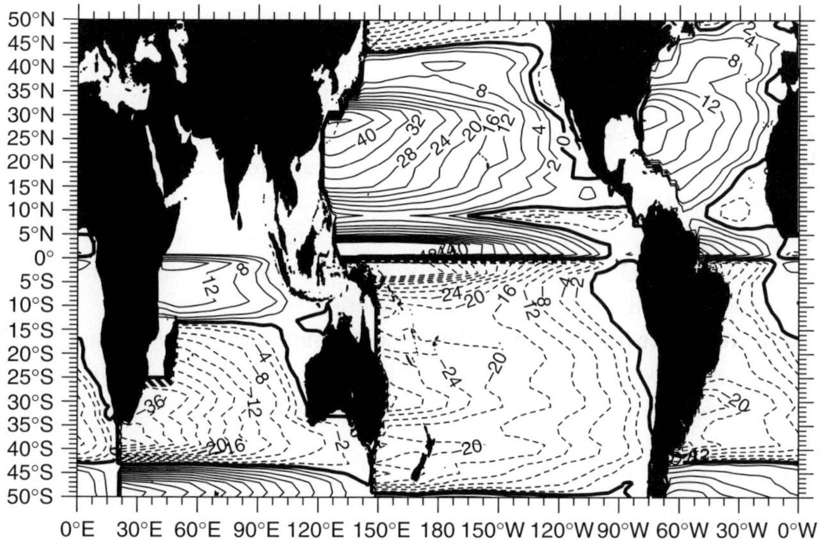

Fig. 4.86 Annual-mean barotropic meridional volume flux below the Ekman layer based on the Hellerman and Rosenstein wind stress data (Sv) (Huang and Wang, 2001).

two small cyclonic gyres at the latitude of the ITCZ; one near the eastern boundary and one near the western boundary around 9° N in the Pacific Ocean, as shown in Figure 4.86. Simply taking the meridional minimum of the Sverdrup function value at the western boundary may miss the communication window in the basin interior. Thus, for the Northern Hemisphere, the following quantity can be used as the interior communication rate (ICR):

$$I_c = \max\left[0, \min_y\left(\max_x M\right)\right] \tag{4.451a}$$

First, the maximum of M at each latitude gives the maximum equatorward volume flux at that latitude; second, taking the meridional minimum of this maximum (typically searching for the maximum between 5° and 20° off the equator) gives the choking latitude where volume flux reaches a minimum. ICR defined in this way is non-negative. The corresponding ICR for oceans in the Southern Hemisphere is defined similarly:

$$I_c = -\min\left[0, \max_y\left(\min_x M\right)\right] \tag{4.451b}$$

Since the Sverdrup function is defined in terms of the Ekman pumping rate, this index is directly related to subduction in the subtropical basin. Thus, the ICR defined above is a large-scale, wind-stress-related index for the equatorward volume flux due to subduction in the subtropical gyre interior and its subsequent equatorward flow.

The climatological mean circulation

The wind-driven circulation has quite different structure in the two hemispheres. In fact, the cyclonic tropical gyre near the western coast of South America is very weak, and its effect on the basin-scale circulation is almost negligible; thus, the interior communication in the South Pacific is totally unblocked. The cyclonic tropical gyre in the South Atlantic is also weak, and thus the interior communication is largely free.

On the other hand, the cyclonic gyre in the South Indian Ocean extends all the way to the western boundary of the basin, so there is no interior communication between the subtropics and tropics. Therefore, the equatorial currents in the South Indian Ocean have no direct connection with the subtropical gyre interior.

The ICRs obtained from the Hellerman and Rosenstein wind stress data are 4.56 Sv (19.16 Sv) for the North (South) Pacific, which are comparable with the ICR inferred from hydrographic data: 5 ± 2 Sv (16 ± 2 Sv) for the North (South) Pacific (Johnson and McPhaden, 1999).

As discussed above, the Sverdrup function is not a streamfunction for the subsurface geostrophic flow, and the exact shape of the communication window should be inferred from the streamlines of the barotropic flow. Using the virtual streamfunction introduced above, the interior communication window can be identified clearly (Fig. 4.87). For both the North Pacific and North Atlantic the communication window is a narrow passage wound

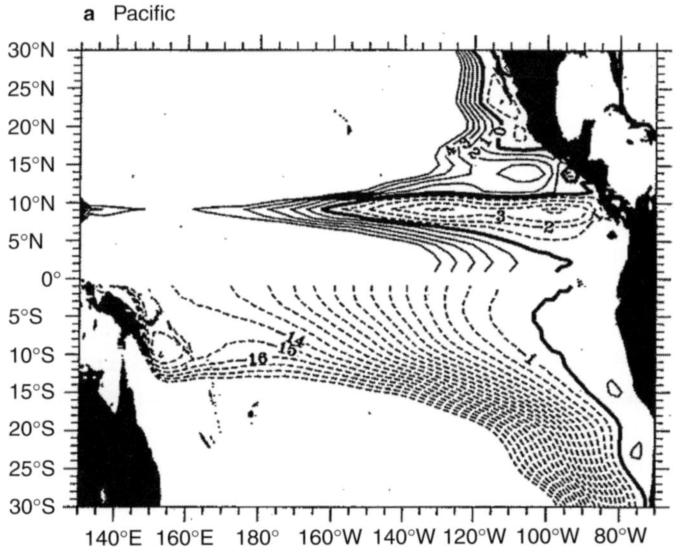

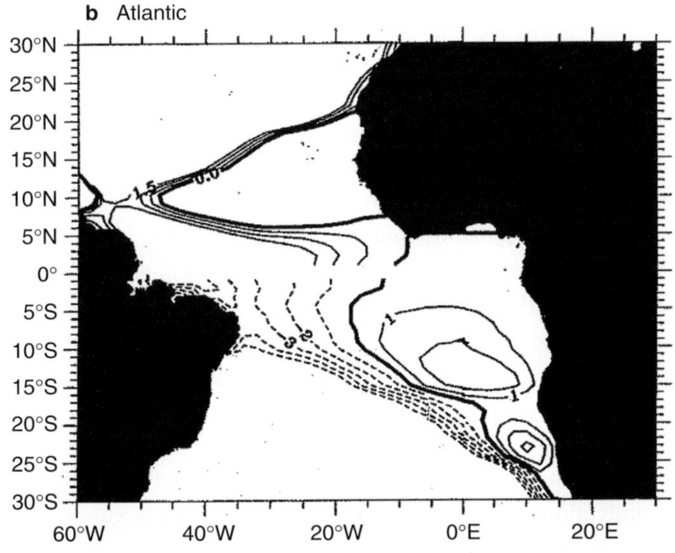

Fig. 4.87 Interior communication window identified from the potential streamfunction (Sv) in **a** the Pacific Ocean and **b** the Atlantic Ocean (Huang and Wang, 2001).

around the western edge of the cyclonic gyre and the eastern boundary. The center of the communication window in the North Pacific is near 170° E at 9° N. In particular, these pathways move eastward south of the choking latitudes. Of course, within the vicinity of the equator, other dynamical processes must be taken into consideration, such as inertia and mixing.

The communication window in the North Atlantic is rather narrow, as shown in Figure 4.87b. Thus, the interior connection between the subtropics and tropics is relatively weak, and decadal climate signals that can pass through this choking latitude and propagate to the equator may be relatively weak too.

The communication pathways in the Southern Hemisphere have a different structure; see the lower parts of Figure 4.87a, b. In both the South Pacific and South Atlantic, the communication pathway moves northwestward, without much winding around the cyclonic gyre in the eastern basin. The different features of the pathway in the two hemispheres are clearly related to the asymmetric nature of the wind stress pattern in the two hemispheres. In contrast to the ITCZ in the North Pacific and North Atlantic, the corresponding structure of the subtropical-tropical atmospheric circulation in the Southern Hemisphere is quite different. For example, the South Pacific Convergence Zone appears slanted in the South Pacific Ocean.

As discussed above, in both the North Pacific and North Atlantic, the interior communication window appears as a narrow corridor winding around the western edge of the cyclonic gyre near the eastern boundary, and it extends poleward along the eastern boundary of the basin. Thus, the interior communication in the Northern Hemisphere basins affects the eastern part of the equatorial thermocline, but with virtually no effect on the western part of the equatorial thermocline. On the other hand, the interior communication windows in the South Pacific and South Atlantic open up over the whole width of the basin, so changes in the extratropics can be transported toward the equator and affect the thermocline there. This asymmetry of the interior communication windows gives rise to vital dynamical consequence for the structure of the equatorial thermocline.

Recall that contours of virtual streamfunction indicate the streamlines only, and the volume flux within the pathways can be calculated by $\Delta M = f_0 \Delta\psi^* / f$, shown by the heavy lines in Figure 4.88 for the Pacific and Atlantic, while the maximal meridional volume flux at the corresponding latitude is indicated by thin lines. (For the Northern Hemisphere, the communication window is at the latitude where the meridional volume flux is minimal.) The total volume flux within the communication window declines poleward from the choking latitude, but it increases equatorward. This meridional change in the volume flux is due to the contribution of Ekman pumping. For example, in the North Pacific, the total volume flux within the communication pathway increases from 4.56 Sv at the choking latitude (9° N) to 13.6 Sv at 3° N; in the South Pacific, the total volume flux within the communication passage increases from 19.16 Sv at the choking latitude (13° S) to 82.37 Sv at 3° S. Such an increase in volume flux comes from the local Ekman pumping, and it constitutes the equatorward branch of the tropical cell.

4.7.4 Communication and pathways on different isopycnal surfaces

As discussed above, a simple barotropic index can be used to describe the interior communication rate, and the virtual streamfunction can be used to plot the pathways clearly.

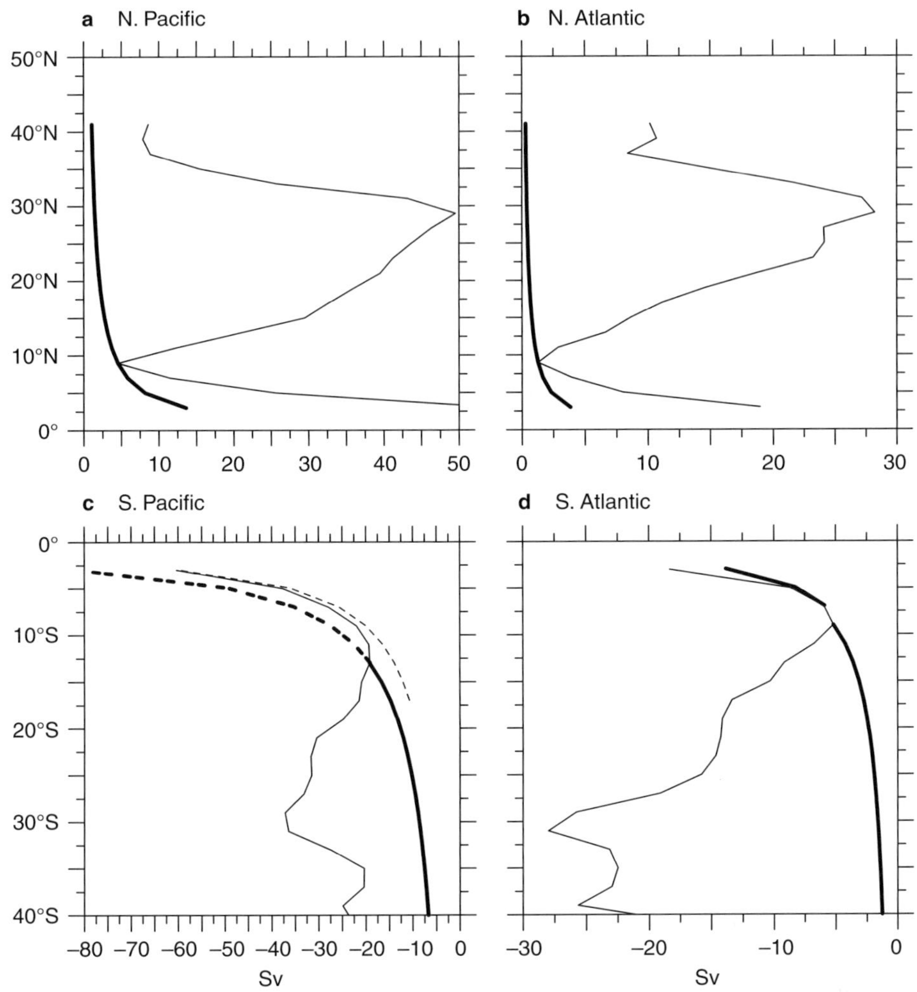

Fig. 4.88 Meridional profiles for the volume flux in the interior communication windows (*heavy lines*), the maximal meridional volume flux at the corresponding latitude (*thin lines*), and the volume flux actually reaching the equator (*thin dashed line*), in Sv (Huang and Wang, 2001).

The situation for the baroclinic circulation is more complicated, but a similar tool can be introduced.

For a steady flow with weak friction in the ocean interior, regardless of whether there are cross-isopycnal fluxes, potential vorticity conservation is reduced to

$$\nabla \cdot [\vec{u}_n h_n q_n] = 0 \quad \text{or} \quad \nabla \cdot [\vec{u}_n f] = 0 \tag{4.452}$$

where $\vec{u}_n$ is the horizontal velocity in the n-th layer (Pedlosky, 1996). Equation (4.452) can be seen as a direct result of geostrophic balance on isopycnal surfaces, and allows us to introduce a virtual streamfunction ψ_n for flow on the n-th isopycnal layer

$$u_n \sin\theta = -\frac{\partial \psi_n}{a\partial\theta} \tag{4.453a}$$

$$v_n \sin\theta = \frac{\partial \psi_n}{a\cos\theta\partial\lambda} \tag{4.453b}$$

A zonal integration of Eqn. (4.453b) leads to

$$\psi_n(\lambda,\theta) - \psi_n(\lambda_0,\theta) = \int_{\lambda_0}^{\lambda} v_n a \sin\theta\cos\theta d\lambda \tag{4.454}$$

If $\lambda(\theta)$ and $\lambda_0(\theta)$ are streamlines, the integral on the right-hand side of Eqn. (4.454) should be a constant. In particular, if we choose the western and eastern boundaries of the pathway as such streamlines: $\lambda(\theta) = \lambda_{Wn}(\theta)$ and $\lambda_0(\theta) = \lambda_{En}(\theta)$, then Eqn. (4.454) leads to

$$\psi_n(\lambda_{En},\theta) - \psi_n(\lambda_{Wn},\theta) = \int_{\lambda_{Wn}(\theta)}^{\lambda_{En}(\theta)} v_n(\lambda,\theta)\sin\theta a\cos\theta d\lambda = const \tag{4.455}$$

Thus, both interior pathway transport (IPT) per unit thickness and western boundary pathway transport (WBPT) per unit thickness satisfy

$$\int_{\lambda_{Wn}(\theta)}^{\lambda_{En}(\theta)} v_n(\lambda,\theta) a\cos\theta d\lambda = \frac{const}{\sin\theta} \tag{4.456}$$

Therefore, IPT (WBPT) per unit thickness should have a functional form $F_n(\theta) = \frac{const(n)}{\sin\theta}$. Since the velocity field obtained from models or inferred from observations does not satisfy Eqn. (4.452) exactly, IPT (WBPT) per unit thickness satisfies Eqn. (4.456) approximately.

It can readily be seen from Figure 4.89 that away from the equator and the western boundary and subduction area, IPT (WBPT) per unit thickness is approximately the functional relation $F_n(\theta)$. Thus, in mid-ocean, IPT (WBPT) per unit thickness multiplied by $\sin\theta$ is approximately constant.

Pathways from the subtropics to the tropics can be identified from data generated from numerical models, using either the virtual streamfunction or tracing the layer-integrated streamlines. For example, the interior pathway and the western boundary pathways for four isopycnal surfaces identified from the SODA data are shown in Figure 4.90. It is clear that the time taken for signals to reach the equatorial band is much longer for deep isopycnal surfaces. The interior pathways in the South Pacific are much wider than the corresponding pathways in the North Pacific, consistent with the relation for the barotropic pathways.

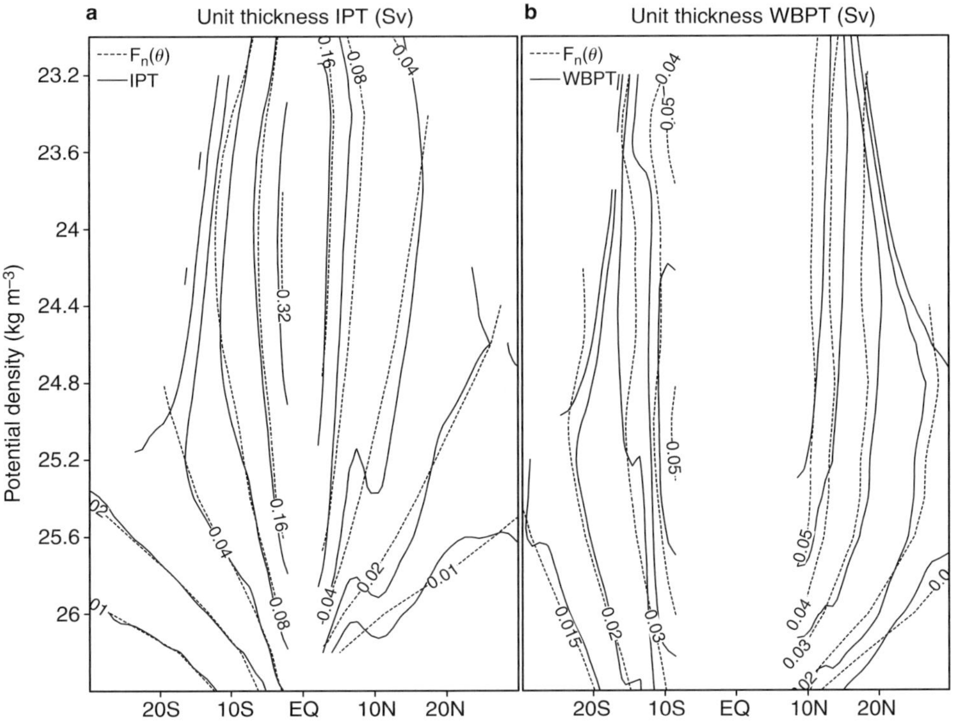

Fig. 4.89 Climatological-mean interior pathway transport (IPT) per unit thickness and western boundary pathway transports (WBPT) per unit thickness (*solid lines*); *dashed lines* indicate $F_n(\theta) = \frac{const(n)}{\sin\theta}$ (Q. Wang and Huang, 2005).

4.8 Adjustment of thermocline and basin-scale circulation

4.8.1 Geostrophic adjustment

Large-scale motions in the atmosphere and oceans are almost geostrophic, i.e., the velocity and pressure fields satisfy the geostrophic condition approximately. However, geostrophic motions are degenerated such that they cannot evolve with time. In reality, deviations from geostrophy, or the ageostrophic components of the circulation, drive the system in evolution. Geostrophic adjustment is the study of the time evolution of the system from an initial state of non-geostrophic balance to a final state in geostrophic balance. There are many beautiful examples.

Rossby (1938) first formulated a prototype model of the geostrophic adjustment. His model includes an initially stagnant and homogeneous ocean, and an initial eastward velocity jet is suddenly imposed on a strip of the ocean (Fig. 4.91).

Since there was no pressure gradient initially, the Coriolis force drove water within the jet to move southward. Thus, south of the jet, water piled up and created a high-pressure center. Similarly, the Coriolis force induced motions creating a low-pressure center to the

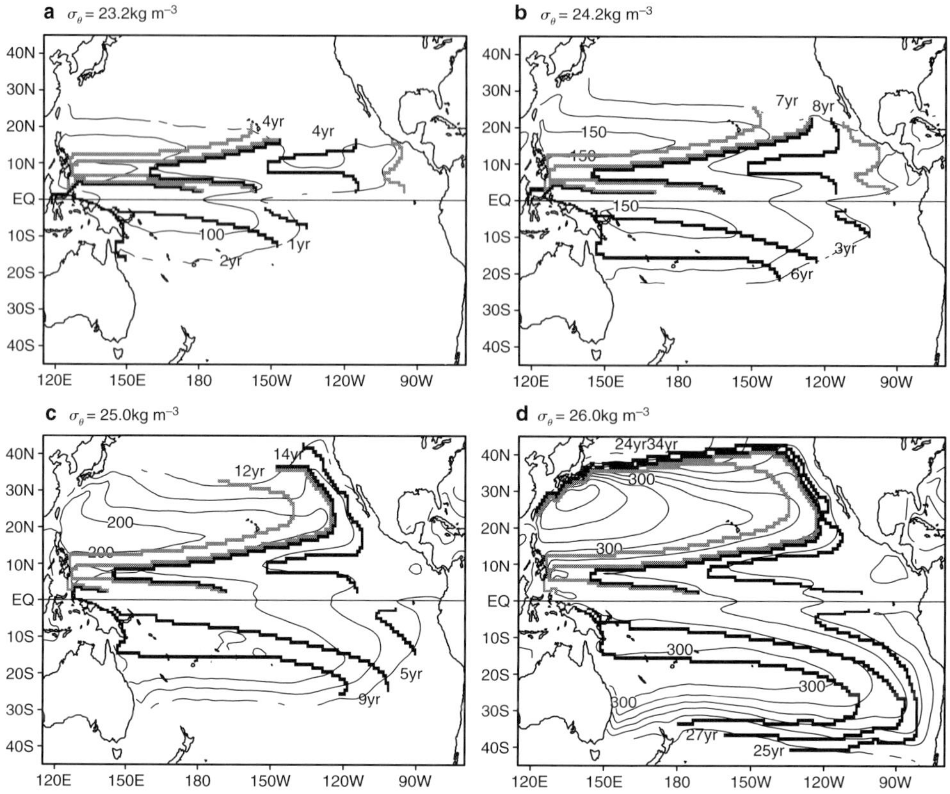

Fig. 4.90 Climatological-mean circulation on four isopycnal surfaces: **a** $\sigma_\theta = 23.2$; **b** 24.2; **c** 25.0; and **d** 26.0 kg/m^3. *Gray lines* indicate depths (m); *black and gray thick lines* indicate pathways. The ventilation time for the pathways (in years) is indicated by a *pair of numbers*: the *left* for the western boundary pathway and the *right* for the interior pathway. *Gray thick lines* near the eastern boundary in **a** and **b** indicate a narrow local pathway, a short-cut, from the subtropics to the equatorial regime (Q. Wang and Huang, 2005).

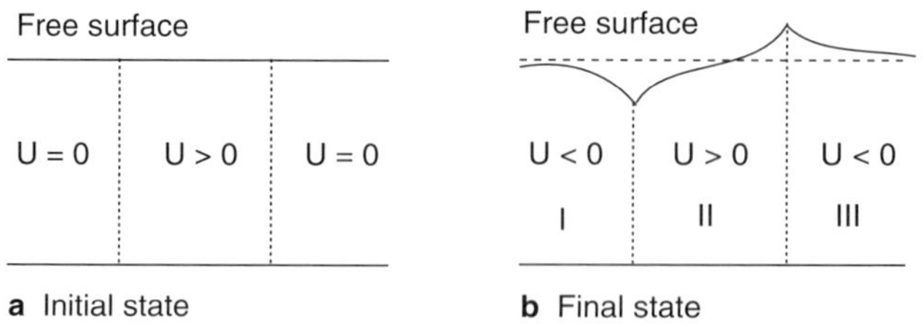

Fig. 4.91 The Rossby problem for a homogeneous ocean with an initial velocity jet (looking toward the east): **a** initial state, **b** final state.

north. For water within the jet, the southward motion further induced a westward velocity. As a result, the initially imposed eastward velocity declined and so did the total kinetic energy. Cahn (1945) analyzed the model and pointed out that the adjustment was carried out through the dispersion of inertial gravity waves.

The basic ingredients in geostrophic adjustment are the Coriolis force, which produces circulatory motions, and deviation from geostrophy, which creates waves. Rotation makes energy transfer between vortex motions and wave motions possible. As we will see later, the Rossby radius of deformation is an intrinsic length scale of geostrophic adjustment problems.

If a stone is thrown into an infinite, non-rotating, resting ocean, the gravitational oscillations will radiate all the energy to infinity and leave the ocean motionless. However, if a stone is thrown into an infinite rotating ocean, radial motions will induce circulatory motions; thus, part of the initial energy will remain in the form of vortex motions.

Rossby adjustment problem

To illustrate the basic idea of geostrophic adjustment, we discuss the classical case of Rossby (1938) adjustment. As will be demonstrated below, the final state of the adjustment can be determined from the potential vorticity conservation without solving the time evolution of the wave process. Our approach here follows Mihaljan (1963). Consider a homogeneous ocean of constant depth H. At time $t = 0$, there is a uniform current of speed $U_0 > 0$ (moving eastward). The current has a width of $2b$. We assume no variations in the x direction. The following notations will be used: the initial solution will be denoted with upper-case characters (Y, U). Since water parcels conserve some of their properties, such as potential vorticity, during the adjustment, these Lagrangian coordinates are quite useful; the solution after adjustment will be in new coordinates, denoted with lower-case characters, such as (y, u).Because there is no pressure gradient in the x direction, the x-momentum equation is

$$du/dt = fv \tag{4.457}$$

and integration leads to

$$u - U = f(y - Y) \tag{4.458}$$

The continuity condition requires that the volume of each individual water parcel should be the same in both the old and new coordinates:

$$h = H\,dY/dy \tag{4.459}$$

The final state is in geostrophy, thus

$$u = -\frac{g}{f}\frac{dh}{dy} \tag{4.460}$$

From these equations we can derive a single equation for Y:

$$\frac{d^2Y}{dy^2} - \frac{Y}{\lambda^2} = -\frac{y}{\lambda^2} - \frac{U}{f\lambda^2} \tag{4.461}$$

where $\lambda = \sqrt{gH}/f$ is the Rossby radius of deformation. This is a potential vorticity equation in terms of the Lagrangian coordinate Y. Since both the layer depth and the velocity shear are consistent, the potential vorticity is piece-wise constant, which in turn allows analytical solutions.

For the general case, the nonlinearity associated with the momentum and continuity equations is difficult to deal with in analytical study because layers can outcrop and the relative vorticity is not negligible. The advantage of this approach is the corresponding equation (4.61), which is linear. The solution is found by matching three pieces of solution over two locations, as shown in Figure 4.91. Since the solution must be finite at infinity, the suitable forms of solution are

$$Y^{II} - y = Ae^{y/\lambda} + Be^{-y/\lambda} + U_0/f \tag{4.462a}$$

$$Y^{I} - y = Ce^{y/\lambda} \tag{4.462b}$$

$$Y^{III} - y = De^{-y/\lambda} \tag{4.462c}$$

For simplicity, the center of region I is chosen as the origin of the coordinate; thus these three branches have to be matched at $y = \pm b$, with the matching conditions that both Y and h are continuous. The final solution is

$$\begin{aligned} Y^{II} &= U_0\left[1 - e^{-b/\lambda}\cosh(y/\lambda)\right]/f + y \\ Y^{I} &= U_0 e^{y/\lambda}\sinh(b/\lambda)/f + y \\ Y^{III} &= U_0 e^{-y/\lambda}\sinh(b/\lambda)/f + y \end{aligned} \tag{4.463}$$

Using Eqns. (4.459, 4.460), the corresponding velocity is

$$\begin{aligned} u^{II} &= U_0 e^{-b/\lambda}\cosh(y/\lambda) \\ u^{I} &= -U_0 e^{y/\lambda}\sinh(b/\lambda) \\ u^{III} &= -U_0 e^{-y/\lambda}\sinh(b/\lambda) \end{aligned} \tag{4.464}$$

The solution is reduced to a simple forms for two asymptotical cases:

a) If $b/\lambda \ll 1$

we have the approximate solution

$$u^{II} = U_0\left[1 - \frac{b}{\lambda} + \frac{b^2}{2\lambda^2}\left(1 + \frac{y^2}{b^2}\right)\right] + o\left(\frac{b}{\lambda}\right)^2 \approx U_0 \tag{4.465}$$

Thus, velocity is basically unchanged if the initial width of the jet is much smaller than the radius of deformation.

b) If $b/\lambda \gg 1$

the approximate solution is in the following forms

$$u(0) = U_0 e^{-b/\lambda} \approx 0 \tag{4.466a}$$

$$u\left(-b^-\right) = u\left(b^+\right) = -U_0/2 \tag{4.466b}$$

$$u\left(b^-\right) = u\left(-b^+\right) = U_0/2 \tag{4.466c}$$

Thus, if the initial velocity jet is much wider than the deformation radius, the velocity field is totally destroyed. As will be shown later, the final state after geostrophic adjustment depends on the horizontal scale of the initial perturbations.

This approach has been extended to multi-layer models, such as the geostrophic adjustment associated with density fronts and currents. In such cases, there are the barotropic mode and the baroclinic modes; each mode has its own phase speed and plays different roles in the geostrophic adjustment.

Application to a finite step in free surface

Another interesting application of this technique is the case with an initial step in the sea surface elevation and an ocean at rest (Fig. 4.92). Since the initial velocity is zero, Eqn. (4.461) is reduced to

$$\frac{d^2 Y_{II}}{dy^2} - \frac{Y_{II}}{\lambda_{II}^2} = -\frac{y}{\lambda_{II}^2} \tag{4.467}$$

where $\lambda_{II} = \sqrt{gH}/f$ is the deformation radius for domain II. The solution that is finite at $y \to \infty$ is

$$Y_{II} = y + Ae^{-y/\lambda_{II}} \tag{4.468}$$

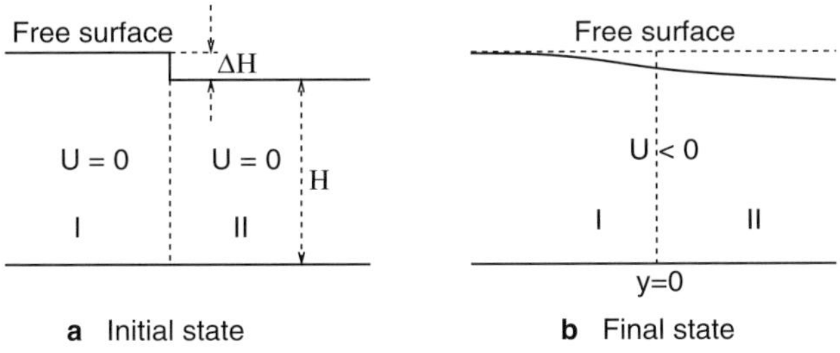

Fig. 4.92 Geostrophic adjustment induced by an initial step in sea surface elevation: **a** initial state, **b** final state.

Similarly, the solution for domain I is

$$Y_I = y + Be^{y/\lambda_I} \tag{4.469}$$

where $\lambda_I = \sqrt{g\,(H + \Delta H)}/f$ is the deformation radius for domain I.

There are two matching conditions; both Y and h are continuous at $y = 0$:

$$Y_I = Y_{II}, \quad (H + \Delta H)\, dY_I/dy = HdY_{II}/dy, \ \ at\ y = 0 \tag{4.470}$$

The first matching condition leads to $A = B$, and the second matching condition leads to

$$\frac{H + \Delta H}{H}\left(1 + \frac{B}{\lambda_I}\right) = 1 - \frac{A}{\lambda_{II}} \tag{4.471}$$

From these two relations, we obtain

$$\frac{\Delta H}{H} = -B\left(\frac{1}{\lambda_{II}} + \frac{H + \Delta H}{H}\frac{1}{\lambda_I}\right) = -B\frac{f}{\sqrt{g}}\left(\frac{1}{\sqrt{H}} + \frac{\sqrt{H + \Delta H}}{H}\right) \tag{4.472}$$

Thus, the constant is

$$B \approx -\frac{\sqrt{gH}}{f}\frac{\Delta H}{2H}\left(1 - \frac{\Delta H}{4H}\right) \tag{4.473}$$

Let $\Delta H = 2\zeta_0$, then the free surface perturbation for domain II is

$$\begin{aligned}\zeta_+ &= H\frac{dY_{II}}{dy} - \left(H + \frac{\Delta H}{2}\right) = H\left(1 - \frac{A}{\lambda_{II}}e^{-y/\lambda_{II}}\right) - \left(H + \frac{\Delta H}{2}\right)\\ &= \zeta_0\left[-1 + \zeta_0\left(1 - \frac{\Delta H}{4H}\right)e^{-y/\lambda_{II}}\right]\end{aligned} \tag{4.474}$$

For domain I , it is

$$\begin{aligned}\zeta_- &= (H + \Delta H)\frac{dY_I}{dy} - \left(H + \frac{\Delta H}{2}\right) = (H + \Delta H)\left(1 + \frac{B}{\lambda_I}e^{y/\lambda_I}\right) - \left(H + \frac{\Delta H}{2}\right)\\ &= \zeta_0 - \zeta_0\sqrt{1 + \Delta H/H}\left(1 - \frac{\Delta H}{4H}\right)e^{y/\lambda_I} \approx \zeta_0\left[1 - \left(1 + \frac{\Delta H}{4H}\right)e^{y/\lambda_I}\right]\end{aligned} \tag{4.475}$$

This example demonstrates the dynamical meaning of the Rossby radius of deformation. As shown in Eqns. (4.474) and (4.475), free surface perturbations in the final state are confined within the order of the Rossby radius of deformation.

Barotropic adjustment

The discussion in the previous section is limited to a one-dimensional model with two fronts. In this section, we discuss the temporal evolution of barotropic adjustment in a two-dimensional f-plane.

Basic equations

We use the linearized equations on an f-plane:

$$\frac{\partial u}{\partial t} - fv = -\frac{\partial \chi}{\partial x} \tag{4.476a}$$

$$\frac{\partial v}{\partial t} + fu = -\frac{\partial \chi}{\partial y} \tag{4.476b}$$

$$\frac{\partial \chi}{\partial t} + c_0^2\left(\frac{\partial u}{\partial x} + \frac{\partial v}{\partial y}\right) = 0 \tag{4.476c}$$

where $\chi = gh'$ is the geopotential perturbation, and $c_0 = \sqrt{gH}$. For the atmosphere, $H = RT_0/g = P_0/\rho g$ is the scale height. From Eqn. (4.476), we can derive the vorticity and divergence equations

$$\frac{\partial \zeta}{\partial t} + fD = 0 \tag{4.477a}$$

$$\frac{\partial D}{\partial t} - f\zeta + \nabla_h^2 \chi = 0 \tag{4.477b}$$

$$\frac{\partial \chi}{\partial t} + c_0^2 D = 0 \tag{4.477c}$$

where D is the velocity divergence, and $\nabla_h^2 = (\partial^2/\partial x^2) + (\partial^2/\partial y^2)$. By introducing velocity potential ϕ and streamfunction ψ, the horizontal velocity can be decomposed into two components:

$$u = -\psi_y + \phi_x, \quad v = \psi_x + \phi_y \tag{4.478}$$

Thus

$$\zeta = \nabla_h^2 \psi, \quad D = \nabla_h^2 \phi \tag{4.479}$$

Equation (4.477) is reduced to

$$\nabla_h^2\left(\frac{\partial \psi}{\partial t} + f\phi\right) = 0 \tag{4.480a}$$

$$\nabla_h^2\left(\frac{\partial \phi}{\partial t} - f\psi + c_0^2 \pi\right) = 0 \tag{4.480b}$$

$$\frac{\partial \pi}{\partial t} + \nabla_h^2 \phi = 0 \tag{4.480c}$$

where

$$\pi = \chi/c_0^2 = h'/H \tag{4.481}$$

is the nondimensional pressure field. Eliminating ϕ from Eqns. (4.480a) and (4.480c) gives

$$\frac{\partial}{\partial t}\left(\nabla_h^2\psi - f\pi\right) = 0 \tag{4.482}$$

That means potential vorticity is conserved, i.e.,

$$\nabla_h^2\psi - f\pi = q(x,y) = \nabla_h^2\psi_0 - f\pi_0 \tag{4.483}$$

Equations (4.480a) and (4.480b) are Laplace equations. Under the boundary conditions of finiteness at infinity, the only possible solutions are

$$\frac{\partial\psi}{\partial t} + f\phi = 0 \tag{4.484a}$$

$$\frac{\partial\phi}{\partial t} - f\psi + c_0^2 = 0 \tag{4.484b}$$

Eliminating ψ and π, we can find an equation for ϕ; thus we have a system for the linear barotropic adjustment process:

$$\frac{\partial}{\partial t}\left(\nabla_h^2\psi - f\pi\right) = 0 \tag{4.485a}$$

$$\frac{\partial^2\phi}{\partial t^2} = c_0^2\left(\frac{\partial^2\phi}{\partial x^2} + \frac{\partial^2\phi}{\partial y^2}\right) - f^2\phi \tag{4.485b}$$

$$\frac{\partial\pi}{\partial t} + \nabla_h^2\phi = 0 \tag{4.485c}$$

where the second equation is a wave equation involved with ϕ only, so that it can be solved for $\phi(x,y,t)$. It is important to remember that the velocity field has been separated into two parts, $\psi(x,y,t)$ and $\phi(x,y,t)$. When $\phi(x,y,t)$ is determined, $\psi(x,y,t)$ and $\pi(x,y,t)$ can be calculated using Eqn. (4.484).

The final state of geostrophy

From the potential vorticity equation (4.483) we can find out the final state without even solving the time evolution process. Thus

$$\nabla_h^2\psi_\infty - f\pi_\infty = q(x,y) \tag{4.486}$$

where $q(x,y)$ is the potential vorticity distribution calculated from the initial state. Since the final state is geostrophic, we also have

$$\phi_\infty = 0, \quad f\psi_\infty = c_0^2\pi_\infty \tag{4.487}$$

From these equations we find a single equation for ψ_∞:

$$\nabla_h^2\psi_\infty - \lambda^{-2}\psi_\infty = q(x,y) \tag{4.488}$$

where $\lambda = c_0/f$ is the radius of deformation. This is a Helmholtz equation; its Green function is the modified Bessel function K_0, thus

$$\psi_\infty(x, y) = -\frac{1}{2\pi}\iint_{-\infty}^{\infty} q(x, y) K_0(\rho/\lambda)\, dx d\eta \tag{4.489}$$

where $\rho = \sqrt{(\xi - x)^2 + (\eta - y)^2}$. Using polar coordinates, this can be rewritten as

$$\begin{aligned}
\psi_\infty(x, y) &= -\frac{1}{2\pi}\int_0^{2\pi}\int_0^\infty q(x\rho\cos\theta, y\rho\sin\theta) K_0\left(\frac{\rho}{\lambda}\right)\rho d\rho d\theta \\
&= -\frac{1}{2\pi}\int_0^{2\pi}\int_0^\infty \left(\nabla_h^2\psi_0 - f\pi_0\right) K_0\left(\frac{\rho}{\lambda}\right)\rho d\rho d\theta \\
&= -\frac{1}{2\pi}\int_0^{2\pi}\int_0^\infty \left(\nabla_h^2\psi_0 - \lambda^{-2}\psi_0\right) K_0\left(\frac{\rho}{\lambda}\right)\rho d\rho d\theta \\
&\quad -\frac{1}{2\pi}\int_0^{2\pi}\int_0^\infty \left(\lambda^{-2}\psi_0 - f\pi_0\right) K_0\left(\frac{\rho}{\lambda}\right)\rho d\rho d\theta.
\end{aligned} \tag{4.490}$$

Note that the first term is exactly ψ_0; thus

$$\psi_\infty(x, y) = \psi_0 - \frac{1}{2\pi}\int_0^{2\pi}\int_0^\infty \left(\lambda^{-2}\psi_0 - f\pi_0\right) K_0\left(\frac{\rho}{\lambda}\right)\rho d\rho d\theta \tag{4.491}$$

Relation between the length scales of initial perturbations.

For simplicity, let us assume that the initial perturbation is confined inside a circle of $R = L$, and both ψ_0 and π_0 are nearly constant. Thus at the origin (0, 0)

$$\psi_\infty = \psi_0 - \left(\lambda^{-2}\psi_0 - f\pi_0\right)\int_0^L K_0(\rho/\lambda)\rho d\rho \tag{4.492}$$

Using relation $d[xK_1(x)]/dx = -xK_0(x)$, we obtain

$$\begin{aligned}
\psi_\infty &= \psi_0 - (\lambda^{-2}\psi_0 - f\pi_0)\lambda^2 [-\rho/L_0 \cdot K_1(\rho/L_0)]\Big|_{\rho=0}^{\rho=L} \\
&= \psi_0 - (\psi_0 - c_0^2\pi_0/f)[-\mu \cdot K_1(\mu)],
\end{aligned} \tag{4.493}$$

where $\mu = L/\lambda$.

When $\mu = L/\lambda \ll 1$, $\mu K_1(\mu) \approx 1$, we have

$$\psi_\infty \approx \psi_0, \quad \pi_\infty \approx f\psi_0/c_0^2 \tag{4.494}$$

In this case the streamfunction, and thus the velocity field, basically remain unchanged; however, the pressure field is adjusted toward a state set up by the initial velocity field.

When $\mu = L/\lambda \gg 1$, $\mu K_1(\mu) \approx 0$, we have

$$\psi_\infty \approx c_0^2\pi_0/f, \quad \pi_\infty \approx \pi_0 \tag{4.495}$$

In this case the pressure field remains nearly unchanged; however, the streamfunction, and thus the velocity field, are adjusted toward a state set up by the initial pressure field.

In his seminal paper, Rossby (1938) concluded that the pressure field adjusted toward the velocity field. However, further studies indicated that both the velocity and pressure fields adjust, depending on the initial horizontal scale of the perturbations. For example, Yeh (1957) studied the analytical solutions of the geostrophic adjustment and pointed out that the direction of geostrophic adjustment depends on the initial horizontal scale, in comparison with the radius of deformation.

If the initial perturbation has a small horizontal scale, wave processes can disperse the energy within a time scale shorter than $1/f$. Within such a short time, the vorticity field (or the velocity field) remains basically unchanged; however, the pressure field is altered to be in geostrophic balance with the velocity.

On the other hand, if the initial horizontal scale is large, the adjustment processes take a much longer time, so a new velocity field is established that is in balance with the pressure gradient, and changes in the pressure field stop. As a result, the initial pressure perturbations can mostly remain unchanged.

An interesting application of geostrophic adjustment is the adjustment of free surface elevation and bottom pressure due to surface heating and precipitation (Huang and Jin, 2002b). Surface heating creates a surface elevation anomaly which induces a pressure perturbation in the upper ocean. This perturbation is baroclinic in nature. Since the horizontal scale of surface heating (on the order of a few hundred kilometers or larger) is much larger than the first deformation radius, signals in connection with surface elevation associated with surface heating can survive.

On the other hand, precipitation induces surface elevation and bottom pressure anomalies, which are barotropic perturbation. Precipitation has a horizontal scale on the order of a few hundred kilometers, which is much smaller than the barotropic radius of deformation (on the order of 2,000 km); thus, the initial perturbations in surface elevation and bottom pressure cannot be maintained during geostrophic adjustment. As a result, there is very little signal left behind after the geostrophic adjustment. In fact, signals associated with precipitation cannot be identified from satellite altimetry data.

An example of a vortex

Obukhov (1949) discussed a case in which there was an axi-symmetric velocity field initially, but with no pressure gradient:

$$\pi_0 = 0, \quad \phi_0 = 0 \tag{4.496}$$

$$\psi_0 = \psi_0(r) = -A\left(2 + \mu^2 - \xi^2\right)e^{-\xi^2/2} \tag{4.497}$$

where $\mu = R/L_0, \xi = r/L, r = \sqrt{x^2 + y^2}$. The solution of Eqn. (4.491) is

$$\psi_\infty(r) = -A\left(2 - \xi^2\right)e^{-\xi^2/2}, \quad \pi_\infty(r) = -\frac{A}{fL_0^2}\left(2 - \xi^2\right)e^{-\xi^2/2} \tag{4.498}$$

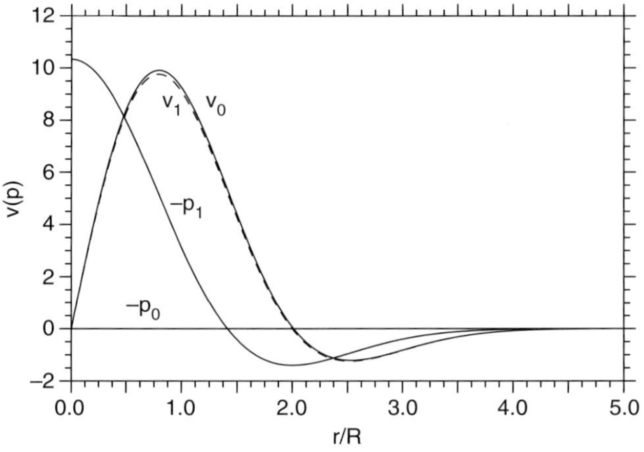

Fig. 4.93 Velocity and pressure profiles before (with subscript 0) and after geostrophic adjustment (with subscript 1).

The azimuthal velocity before and after adjustment are

$$v_0 = \frac{A}{R}\xi\left(4 + \mu^2 - \xi^2\right)e^{-\xi^2/2}, v_\infty = \frac{A}{R}\xi\left(4 - \xi^2\right)e^{-\xi^2/2} \tag{4.499}$$

Thus, the velocity ratio is

$$\Gamma = \frac{v_\infty}{v_0} = \frac{4 - \xi^2}{4 + \mu^2 - \xi^2} \tag{4.500}$$

It can readily be seen that for small-scale perturbation ($\mu = R/L_0 \ll 1$), $\Gamma \approx 1$; thus, the velocity field is almost unchanged. For large-scale perturbation ($\mu \gg 1$), $\Gamma \ll 1$; thus, the velocity field must be changed substantially.

If we choose the following values, $A = 2.5 \times 10^6$ m^2/s, $R = 5 \times 10^5$ m, and $L_0 = 2.2 \times 10^6$ m, the result is shown in Figure 4.93. For the present case, the velocity basically remains unchanged, but there is a new pressure field, which is geostrophically balanced with the velocity field, being built up during the adjustment process.

4.8.2 Basin-scale adjustment

The time evolution of the thermocline in a closed basin, either the spin-up of the wind-driven circulation or the perturbations of the existing circulation, consists of several key processes:

- *In the basin interior*, perturbations generated by anomalous forcing, such as wind-stress curl and heating/cooling, propagate westward in the form of Rossby waves.
- *At the western boundary*, Rossby waves reflect and disperse. As a result, coastal Kelvin waves are generated.

- *Coastal Kelvin waves* move equatorward and turn into equatorial Kelvin waves, which propagate eastward along the equatorial wave guide.
- *At the eastern boundary*, the equatorial Kelvin waves go through reflection and disperse, creating poleward Kelvin waves moving along the eastern boundary of the closed basin. The poleward Kelvin waves continuously send westward Rossby waves on their poleward pathways. These Rossby waves move westward in the ocean interior, completing a cycle of the adjustment, although the complete adjustment may take many cycles to complete. In addition, other processes, such as friction and dissipation, are also important in the basin-scale adjustment.

These dynamical processes are very complicated and, due to limits on space in this book, we present a concise description of some of the key processes in the following sections.

Westward propagation of Rossby waves

A good example of basin-scale adjustment is the spin-up of wind-driven circulation in a basin, and this is intimately related to the westward propagation of long Rossby waves. The classical study was published by Anderson and Gill (1975). For small perturbations of the basic state of a resting ocean, the vertical structure can be described in terms of normal modes, each of which behaves independently. For a given mode, the equations governing the perturbations are

$$u_t - fv = -\frac{1}{a\cos\theta}p_\lambda + \tau^x/H \tag{4.501}$$

$$v_t + fu = -\frac{1}{a}p_\theta + \tau^y/H \tag{4.502}$$

$$p_t + \frac{c^2}{a\cos\theta}\left[u_\lambda + (v\cos\theta)_\theta\right] = 0 \tag{4.503}$$

where a is the radius of Earth, p is the perturbation pressure divided by the density, both H and c (the speed of long internal waves) depend on the mode in question. As an example, a two-layer model ocean has two modes. The barotropic mode has

$$u = \frac{H_1 u_1 + H_2 u_2}{H_1 + H_2}; \quad p = g\zeta, \;\; H = H_1 + H_2; \quad c^2 = gH \tag{4.506}$$

where ζ is the free surface elevation. The baroclinic mode has

$$u = u_1 - u_2; \quad p = g'h; \quad H = \frac{H_1 H_2}{H_1 + H_2}; \quad c^2 = g'H \tag{4.507}$$

where $g' = (\rho_2 - \rho_1)g/\rho_2$ and h is the elevation of the interface. The typical value of the barotropic wave speed is $c = 200$ m/s, while for the baroclinic mode it is $c = 2$ m/s.

The spin-up process can be examined using the quasi-geostrophic potential vorticity equation, and the discussion below follows the notation of Hendershott (1987). Projecting the wind stress force into a sinusoidal Fourier series, then for the n-th component of the wind-stress curl the quasi-geostrophic potential vorticity equation is

$$\nabla_h^2 \psi_t - \psi_t/R^2 + \beta\psi_x = -\frac{n\pi}{\rho_0 D_0 b} \sin\left(\frac{n\pi y}{b}\right) G_n(x) \quad (4.508)$$

where $R = c/f$ is the corresponding Rossby radius of deformation. The initial condition is $\psi = 0$ everywhere, and the boundary condition is $\psi = 0$ at the boundaries $x = (-L, L)$ and $y = 0, b$.

The solution of this problem is in the form

$$\psi = \frac{n\pi}{\rho_0 D_0 b} \sin\left(\frac{n\pi y}{b}\right) \phi_n(x, t) \quad (4.509)$$

where the unknown function satisfies

$$\phi_{n,xxt} - \left(\frac{n^2\pi^2}{b^2} + \frac{1}{R^2}\right) \phi_{n,t} + \beta\phi_{n,x} = -G_n(x) \quad (4.510)$$

where the first term represents the contribution due to short waves. The exact definition of so-called short waves varies, depending on whether these are barotropic or baroclinic waves. However, let us omit the first term, and thus the equation is reduced to a first-order partial differential equation, whose solution is the sum of two parts – the *stationary part* (S) and the *transient part* (T):

$$\phi_n(x, t) = \phi_n^S(x) + \phi_n^T(x, t) \quad (4.511)$$

The initial condition of a resting ocean requires

$$\phi_n^T(x, 0) = -\phi_n^S(x) \quad (4.512)$$

The steady part of the solution is the well-known Sverdrup solution

$$\phi_n^S = -\frac{1}{\beta} \int_L^x G_n(x')\, dx' \quad (4.513)$$

The transient part obeys

$$\phi_{n,t}^T + c_n \phi_n^T = 0, \quad c_n = -\beta \left(\frac{n^2\pi^2}{b^2} + \frac{1}{R^2}\right)^{-1} < 0 \quad (4.514)$$

where c_n is the phase speed of the long waves (long waves are defined in terms of the x direction, and the short waves will be discussed below in connection with the western boundary). If we assume the wind-stress curl is independent of x, $G_n(x) = \Gamma_n$, then the transient solution, which is defined in Eqn. (4.511) and satisfies the initial condition, Eqn. (4.512), is

$$\varphi_n(x, t) = -\frac{\Gamma_n}{\beta} \left[(x - L) - H(L + c_n t - x)(x - c_n t - L)\right] \quad (4.515)$$

$$v = \psi_x = -\frac{n\pi}{\rho_0 D_0 b} \sin\left(\frac{n\pi y}{b}\right) \frac{\Gamma_n}{\beta} \left[1 - H(L + c_n t - x)\right] \quad (4.516)$$

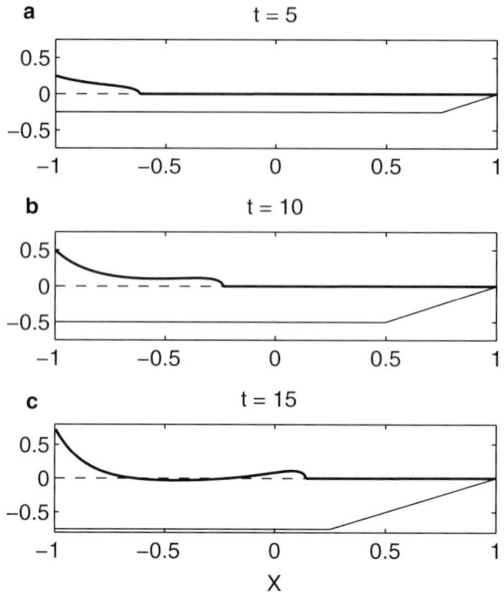

Fig. 4.94 Analytical barotropic wave solution for the first few time steps (time in nondimensional units, **a** t = 5, **b** t = 10, **c** t = 15). The *heavy solid lines* for the Rossby waves reflected from the western boundary, and the *thin solid line* for the sum of the transient part and the stationary part, as defined in Eqn. (4.511).

where $H(x)$ is the Heaviside function (or the step function): $H(x < 0) = 0, H(x > 0) = 1$. This phenomenon can be seen in Figure 4.94.

At any given station x, the Sverdrup flow is established by time $(x - L)/c_n$. Until this time, i.e., if $L + c_n t - x > 0$, the flow is still evolving with

$$\phi_n(x,t) = -\frac{\Gamma_n}{\beta} c_n t \tag{4.517a}$$

$$v = \psi_x = 0 \tag{4.517b}$$

$$u = -\psi_y = \frac{n^2\pi^2}{\rho_0 D_0 b^2} \cos\left(\frac{n\pi y}{b}\right) \frac{\Gamma_n}{\beta} c_n t \tag{4.517c}$$

i.e., before the Sverdrup flow is established, the wind stress primarily accelerates the zonal flow. Since $c_n < 0$, this solution corresponds to a westward flow at low latitude and eastward flow at mid latitudes.

In order to satisfy the no-flow condition along the western boundary, we need to use the full equation, including the third-order derivative in the potential vorticity equation. This is related to the short wave reflection from the western boundary. The approximate solution of this problem can be found from the following equation by omitting the forcing and assuming long waves in the y direction

$$\phi^R_{n,xxt} - \phi^R_{n,t}/R^2 + \beta\phi^R_{n,x} = 0 \tag{4.518}$$

subject to

$$\phi^R_n(-L,t) = -\phi^S_n(-L) - \phi^T_n(-L,t) \tag{4.519}$$

Using the transform method (Anderson and Gill, 1975; Hendershott, 1987) the solution is

$$\phi^R_n(x,t) = \frac{\Gamma_n c_n}{\beta^2}\left(\frac{\beta t}{x+L}\right)^{1/2} J_1\left[2\sqrt{\left(\beta t - \frac{x+L}{R^2}\right)(x+L)}\right] \tag{4.520}$$

The solution of the spin-up problem is the sum of these three components

$$\phi_n = \phi^S_n + \phi^T_n + \phi^R_n \tag{4.521}$$

Introducing the nondimensional length and time scales

$$x = Lx', \quad t = t'/\beta L$$

After dropping the primes and rescaling the forcing term on the right-hand side to one unit, Eqn. (4.510) is reduced to

$$\phi_{xxt} - \Lambda\phi_t + \phi_x = 1 \tag{4.522}$$

where $\Lambda = (n\pi L/b)^2 + (L/R)^2 \gg 1$ is the only parameter in the equation. We choose $L = 2{,}500$ km, $b = 3{,}000$ km, so that the model basin mimics the North Atlantic. In the ocean, the difference between the barotropic and baroclinic modes is so large that the graphic comparison between these cases is difficult. Thus, parallel to Anderson and Gill (1975), for illustrating the basic structure of these waves, we choose the following set of parameters. For the barotropic case, we choose $R = 890$ km, then $\Lambda = 20$; for the baroclinic case, we choose $R = 102$ km, then $\Lambda = 600$. Omitting the high-order dispersion term and the forcing term, the free wave solution to Eqn. (4.522) is in the form $\phi(t - \Lambda x)$. The across-basin time for the free waves to move from the eastern boundary $x = 1$ to the western boundary $x = -1$ is 2Λ; thus the corresponding time is 40 (for the barotropic waves) and 1,200 (baroclinic waves) in nondimensional time units.

As shown in Figure. 4.94, the Sverdrup solution is established after the wave front, started from the eastern boundary, passes through the station of concern. As time progresses, the zone of established Sverdrup flow expands westward. On the other hand, there are Rossby waves reflected from the western boundary, described by Eqn. (4.520), which travel eastward.

The complete physical processes involved in the adjustment can be found by solving Eqn. (4.522). Results obtained from the analytical solution and numerical solution are shown in

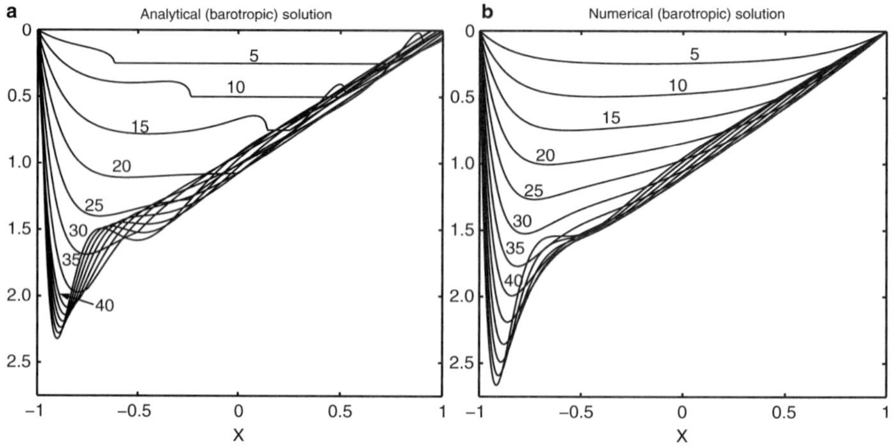

Fig. 4.95 Time evolution (with snapshots taken at equal time intervals from top to bottom of the panels) of the barotropic mode obtained from **a** analytical and **b** numerical solutions, the *numbers* indicate nondimensional time; the *eighth line* corresponds to the time of the westward wave arriving at the western boundary.

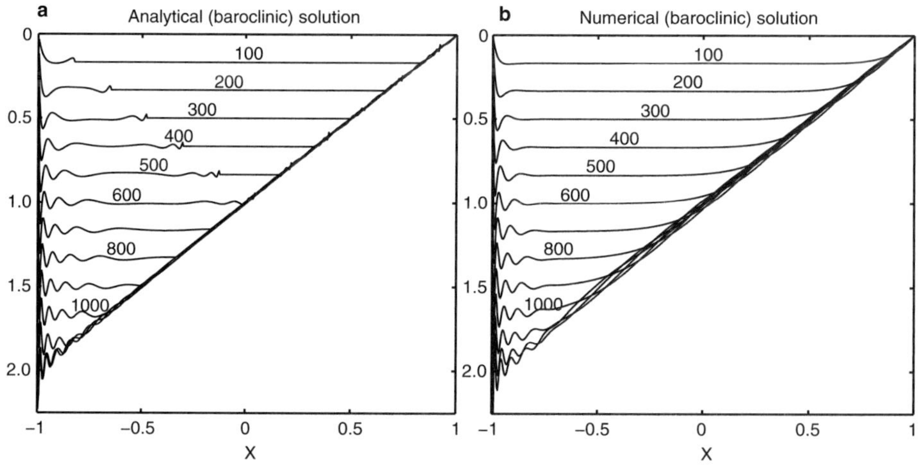

Fig. 4.96 Time evolution (with snapshots taken at equal time intervals from top to bottom of the panels) of the first baroclinic mode obtained from **a** analytical and **b** numerical solutions. The *numbers* indicate nondimensional time; the *twelfth line* corresponds to the time of the westward wave arriving at the western boundary.

Figures 4.95 and 4.96. It can easily be seen that the solutions obtained from these two approaches are quite similar, with two noticeable differences.

First, the numerical solutions are smoother. In particular, the sharper wave fronts, shown in Figure 4.94, are smoothed out owing to the dispersion term in the equation. Second, the analytical solutions do not satisfy the no-flow condition at the eastern boundary, as clearly seen in both figures. As the Rossby waves from the western boundary reach the eastern boundary, the eastern boundary condition of no-zonal flow requires new western-bound Rossby waves to be emitted from the eastern boundary. Therefore, the analytical solution discussed above, without eastern boundary correction, is no longer valid, and the non-zero value of the solution near the eastern boundary indicates the incompleteness of this analytical solution.

Note that the discussion here is focused on the westward propagation of Rossby waves. As these waves reach the western boundary, the structure of the thermocline is almost set up; however, the completion of the wind-driven circulation involves the transport of the signals back to the eastern boundary and circulating through the pathway repeatedly in the form of a combination of Rossby waves and coastal/equatorial Kelvin waves. Kelvin waves are produced when the Rossby waves reach the western boundary, and they move equatorward along the western boundary, then eastward in the form of equatorial Kelvin waves. At the eastern boundary on the equator, equatorial Kelvin waves bifurcate into two poleward-moving coastal Kelvin waves, which shed westward-moving Rossby waves on their way poleward. These processes are discussed in terms of boundary pressure perturbation in the following section.

Evolution of boundary pressure in a closed basin

When long Rossby waves impinge on the western boundary, both short Rossby waves and coastal trapped Kelvin waves are generated. A primitive shallow-water equation can be used to show the evolution of boundary pressure in a rectangular basin. The initial monopole vortex is shown upper left in Figure 4.97 (Milliff and McWilliams, 1994).

The speed of the coastal Kelvin waves is $c = \sqrt{g'h}$. For a model ocean with $h = 1{,}000$ m and $g' = 0.081\ \mathrm{m/s^2}$, $c = 9$ m/s; therefore these waves are very fast, taking 16 days to move around the model basin. The first part of the time evolution is complicated and unimportant for the understanding of the subsequent evolution, so we will focus on the time evolution after the leading edge of the monopole vortex arrives at the western boundary on day 40 (upper left in Fig. 4.97b). At this time, east of the vortex center, a Rossby wave wake has formed a negative lobe. A new positive lobe can also be identified further east.

At day 80, the monopole interacts with the western boundary, and positive signals can be identified along the entire boundary of the model basin; these signals are due to the propagation of Kelvin waves. Note that the scale width of signals along the northern boundary is slightly narrower than that along the southern boundary because of the difference in the local radii of deformation, $r(y) = \sqrt{g'h}/f(y)$. Along the eastern boundary, signals appear to spread to the western boundary, with a phase speed approximately equal to $c_l(y) = \beta r_l^2(y)$; thus, at the southern end of the eastern boundary, signals move westward much faster.

The generation of Rossby waves along the eastern boundary is due to the fact that the Kelvin waves correspond to a balance between the offshore confinement within a radius of deformation and the along-shore geostrophic velocity. As the wave signals move northward along the eastern boundary on a β-plane, the radius of deformation gradually declines, and

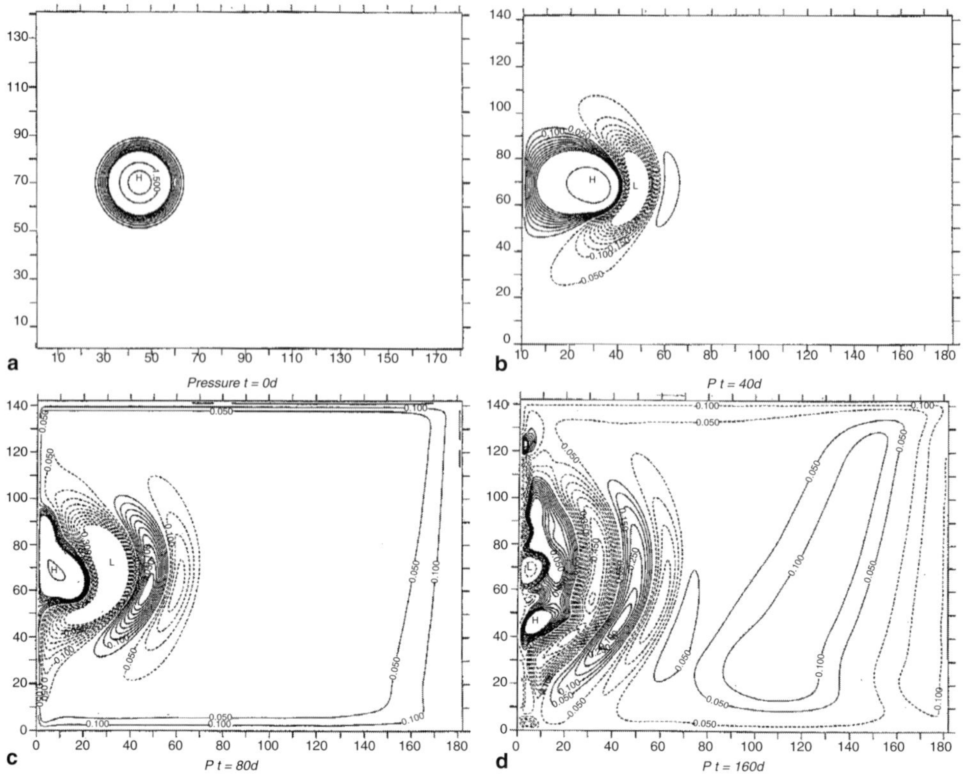

Fig. 4.97 Dynamical pressure in the solution of the shallow-water equation, at **a** day 0, **b** day 40, **c** day 80 and **d** day 160 (Milliff and McWilliams, 1994).

thus the wave amplitude along the coast must increase. The adjustment of the wave package is accomplished by continuously shedding the anomalous mass. The detailed description of Kelvin wave dispersion along the eastern boundary involves solving the wave equation subject to the boundary conditions for the whole basin; thus this is not discussed here, and the reader is referred to the work of Miles (1972) and McCalphin (1995).

At day 160, the new Rossby wave package generated from the eastern boundary moves to the basin interior, and the first cycle of the adjustment is nearly complete.

4.9 Climate variability inferred from models of the thermocline

Climate variability on decadal time scales can be examined by perturbing the steady solution of the thermocline in a multi-layer model. Perturbations induced by a surface cooling anomaly propagate downstream within the characteristic cone defined by the streamlines stemming from the western and eastern edges of the cooling source. The vertical structure of the response to the cooling anomaly depends on the structure of the thermocline circulation in which it is embedded.

In particular, we will show that anomalies produced by a localized cooling excite a fan-shaped array of characteristics which broadens out as the perturbations move southward. The disturbance produces a chain-reaction of potential vorticity anomalies that, because of the beta-spiral, constantly expand the zone of influence of a single disturbance. We now examine the climate variability for a model with multiple moving layers.

4.9.1 Multi-layer model formulation

In this section, the notation is slightly different from that used in Section 4.1.7. We denote the bottom motionless layer as layer 0, and the lowest moving layer as layer 1, and the layers above have numbers increasing upward (Fig. 4.98).

The thickness and the depth of the i-th layer are denoted as h_i and H_i, respectively; thus, for the top layer n, $H_n = h_n$. The northernmost outcrop line is labeled as f_1, and the outcrop lines southward have numerical indexes that increase accordingly. This notation is more convenient when one has to deal with an arbitrary number of layers.

Pressure gradient in a multi-layer model

First, we derive a relation between the pressure gradient in two adjacent layers. Using the hydrostatic relation, the pressure gradient in two adjacent layers satisfies

$$\frac{\nabla_h p_i}{\rho_i} = \frac{\nabla_h p_{i-1}}{\rho_i} + \gamma_i \nabla_h H_i \tag{4.523}$$

where $\gamma_i = g\left(\rho_{i-1} - \rho_i\right)/\rho_i$ is the reduced gravity across the interface. This relation applies to any pair of layers. In addition, one can assume that the lowest layer with subscript 0 is

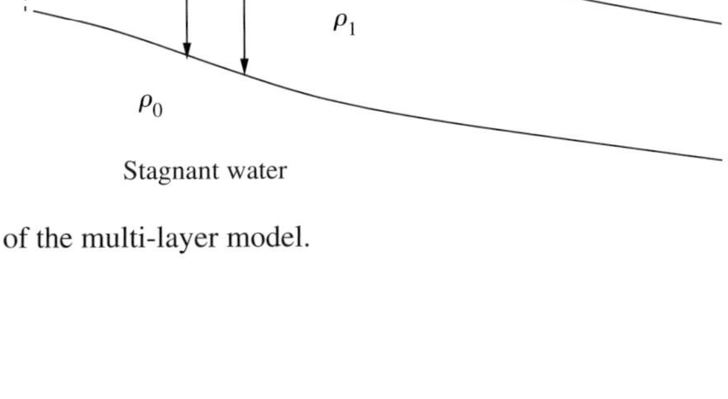

Fig. 4.98 A sketch of the multi-layer model.

very deep, so the pressure gradient there is approximately zero. Thus, the pressure gradient in the m-th layer ($m \leq n$) is

$$\frac{\nabla_h p_m}{\rho_0} = \sum_{i=1}^{m} \gamma_i H_i \tag{4.524}$$

The Sverdrup relation

By taking the curl of the momentum equations in each layer, the vorticity balance is obtained. Multiplying by the layer depth, summing up over all the moving layers, and integrating over $[x, x_e]$, one obtains the Sverdrup relation for the multi-layer reduced-gravity model:

$$\sum_{i=1}^{n} \gamma_{i1} H_i^2 = D_0^2 + H_0^2 \tag{4.525}$$

where

$$D_0^2 = -\frac{4\omega a^2 \sin^2\theta}{\gamma_1} \int_{\lambda}^{\lambda_e} w_e(\lambda', \theta) d\lambda' \tag{4.526}$$

$$\gamma_{i1} = \frac{\gamma_i}{\gamma_1}; \quad H_0^2 = \sum_{i=1}^{n} \gamma_{i1} H_{ie}^2$$

Perturbing the Sverdrup relation gives, to the lowest order,

$$\sum_{i=1}^{n} \gamma_{i1} H_i \delta h_i = \frac{1}{2} \delta D_0^2 \tag{4.527}$$

Here we assume that $\gamma_{i1} (i = 1, \ldots, n)$ remains unchanged and surface buoyancy forcing is parameterized in terms of the meridional shifting of the outcrop lines. Since anomalous buoyancy forcing makes no contribution to the right-hand side of Eqn. (4.527), climate variability due to surface temperature (or freshwater flux) perturbations must appear as internal modes, with a depth-weighted zero mean. We call such modes, which satisfy the homogeneous form of Eqn. (4.527), *dynamical thermocline modes* (DTM), since the modal structure will be a function of the background thermocline structure, which varies laterally. These modes have a three-dimensional structure; they are distinct from, although reminiscent of, the standard geostrophic one-dimensional vertical normal modes of a resting ocean, as discussed in classical quasi-geostrophic theory (Pedlosky, 1987a).

On the other hand, anomalous wind stress forcing can lead to the first baroclinic mode (or the "barotropic" mode, because in this section all the perturbations discussed are confined to the wind-driven circulation in the upper ocean).

Region II

We start our discussion by describing the anomalies produced in layers 1 and 2 in the region $f_2 < f < f_1$, shown as region II in Figure 4.99. South of the first outcrop line y_1, potential

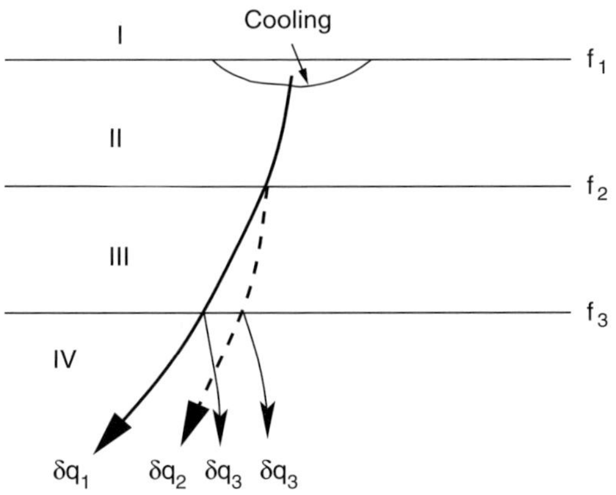

Fig. 4.99 Potential vorticity anomalies and streamlines tracing the original outcrop lines. $\delta \mathbf{q}_1$ is the anomaly generated due to cooling at outcrop line f_1; $\delta \mathbf{q}_2$ is the anomaly generated when $\delta \mathbf{q}_1$ moves crossing outcrop line f_2, and this is the bifurcation. Similarly, $\delta \mathbf{q}_3$ are the anomalies generated when $\delta \mathbf{q}_1 (\delta \mathbf{q}_2)$ moves, crossing the outcrop line f_3.

vorticity in layer 1 is conserved along the streamline, $h_1 = H_1 = const.$

$$Q_1(H_1) = f/h_1 = const. \tag{4.528}$$

At $f = f_1$, $h_1 = H$; therefore, the functional form is $Q_1(x) = f_1/x$, and the first layer thickness is

$$h_1 = fH_1/f_1 \tag{4.529}$$

Substituting Eqn. (4.529) into the Sverdrup relation, Eqn. (4.525), leads to the solution in region II:

$$H_1 = \left(\frac{D_0^2 + H_0^2}{G_2} \right)^{1/2} \tag{4.530}$$

where

$$G_2 = 1 + \gamma_{21} \left(1 - \frac{f}{f_1} \right)^2 \tag{4.531}$$

We introduce the fractional layer thickness, F, which is defined as the layer thickness divided by the total depth of the ventilated layers; thus, for region II we have

$$F_1^{II} = \frac{h_1}{H_1} = \frac{f}{f_1} \tag{4.532}$$

$$F_2^{II} = \frac{h_2}{H_1} = 1 - \frac{f}{f_1} \tag{4.533}$$

where the superscript II indicates that the definition applies to region II with two moving layers, and the subscripts indicate the individual layers.

It is important to note that f_1 in these relations is not necessarily a constant. In fact, we can write it in the form $f_1(H_1)$, which indicates that f_1 depends on the latitude of the outcrop line by tracing backward along the streamline $H_1 = const$. Therefore, in our discussion hereafter, we can write alternately $f_1 \Rightarrow f_1 + \delta f_1$, where we imagine that δf_1 represents the outcrop line perturbations induced by a heating or cooling anomaly. If $\delta f_1 < 0$, this implies that the outcrop line moves equatorward from its constant value of f_1, representing a cooling anomaly.

Since, along the first outcrop line,

$$H_1^2 = -\frac{4\omega a^2 \sin^2\theta}{\gamma_1}\int_\lambda^{\lambda_e} w_e(\lambda', \theta) d\lambda' + H_e^2 \tag{4.534}$$

it can readily be seen that $H_1^2(x) = H_{10}^2(x) + O(\delta f_1)$, where $H_{10}(x)$ is the total layer thickness along the unperturbed outcrop line. Thus, to the lowest-order approximation, the functional relation between H_1 and x remains unchanged. Using this relation we can calculate x and $\delta f_1(x)$ for a given h_1 in order to complete the necessary inversion to determine $f_1(H_1)$ to the lowest order. Accordingly, the solution in region II can be calculated either by solving the nonlinear equations (4.530) and (4.531) or by using their linearized versions.

Climate variability induced by surface cooling

Cooling can be represented by a equatorward perturbation of the first outcrop line,

$$\delta f_1 < 0 \tag{4.535}$$

$$\delta F_1^{II} = -\frac{f}{f_1^2}\delta f_1 > 0 \tag{4.536}$$

$$\delta F_2^{II} = -\delta F_1^{II} < 0 \tag{4.537}$$

$$\delta G_2 = 2\gamma_{21}(1 - F_1^{II})\frac{f}{f_1^2}\delta f_1 < 0 \tag{4.538}$$

Therefore, the cooling-induced changes in the layer depth are

$$\delta H_1 = -\frac{H_1}{2G_2}\delta G_2 > 0 \tag{4.539}$$

$$\delta h_1 = F_1^{II}\delta H_1 + H_1 \delta F_1^{II} = \frac{H_1}{G_2}\left[1 + \gamma_{21}\left(1 - F_1^{II}\right)\right]\delta F_1^{II} > 0 \tag{4.540}$$

$$\delta h_2 = \delta H_1 - \delta h_1 = -\frac{H_1}{G_2}\delta F_1^{II} < 0 \tag{4.541}$$

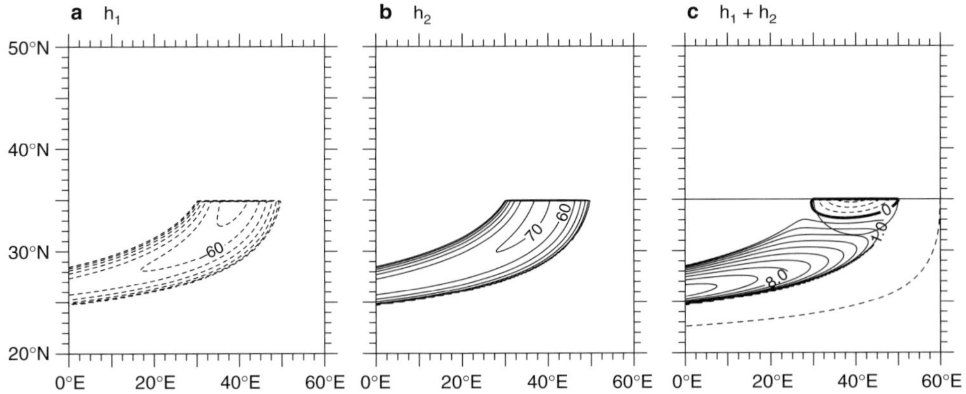

Fig. 4.100 Perturbations in layer thickness (m) in response to a regional cooling (represented by an equatorward deflection of the outcrop line, $\Delta\theta = -3.5$), represented by a southward shift of the outcrop line, indicated by the *thin half circle* in *panel* **c** (Huang and Pedlosky, 1999).

For example, assuming a localized cooling, the outcrop line moves slightly southward; as a result of the cooling, the upper layer becomes thinner, and the lower layer thicker. Furthermore, the increase in the lower-layer thickness more than compensates for the thinning of the upper layer; thus, the base of the wind-driven circulation moves downward (Fig. 4.100).

The most interesting phenomenon is that the perturbations appear in the form of the so-called second baroclinic mode, i.e., the upper interface moves upward (indicating cooling of the upper layer), but the lower interface moves downward (indicating warming of the lower layer). Surface cooling can induce warming of the lower part of the thermocline: this is a little counter-intuitive. However, a simple analysis based on the reduced-gravity model can explain such a phenomenon. As discussed in Section 4.1.2, the layer depth in a simple reduced-gravity model on a β-plane obeys

$$h^2 = h_e^2 + \frac{2f^2}{g'\rho_0\beta}\left(\frac{\tau^x}{f}\right)_y (x_e - x) \tag{4.542}$$

Assuming that the wind stress and the layer thickness along the eastern boundary do not change, taking the variations of this equation leads to

$$h\delta h = -\frac{f^2}{g'^2\rho_0\beta}\left(\frac{\tau^x}{f}\right)_y (x_e - x)\delta g' \tag{4.543}$$

Cooling the upper ocean is equivalent to reducing the reduced gravity $\delta g' < 0$; therefore, $\delta h > 0$, i.e., the depth of the wind-driven gyre, the thermocline, increases in response. This mode is called the *second dynamical thermocline mode*. Since this mode has a three-dimensional structure, it is different from the classic (one-dimensional) second baroclinic mode inferred from quasi-geostrophic theory.

Cooling creates potential thickness perturbation signals in the lowest layer that propagate along streamlines in this layer; streamlines are characteristics in an ideal-fluid thermocline. Cooling also creates a thickness anomaly in the layer above; however, this upper layer is exposed to the Ekman pumping directly, so the potential thickness of the upper layer is not conserved within this regime.

South of f_2, layer 2 is subducted, so the potential thickness anomaly of this layer is now conserved along the streamlines of this layer. Thus, south of f_2 there are two sets of characteristics, the primary and the secondary, both of which carry their own signal that propagates in different directions, as shown in Figure 4.99. More interestingly, when the third outcrop line f_3 is crossed, these two characteristics will bifurcate again, resulting in a total of four characteristics.

Climate variability due to changes in the mixed layer depth

If the mixed layer depth has a small perturbation, $\delta d < 0$, there will be perturbations of the thermocline, with the upper layer becoming thinner and the lower layer thicker, i.e., the perturbation due to the decrease of the mixed layer depth is similar to the case of cooling.

Climate variability due to changes in the Ekman pumping rate

It can easily be shown that if the outcrop line is zonal, a change in the Ekman pumping rate at a confined location can create perturbations in the form of the first baroclinic mode that propagate westward. There is no perturbation south of the latitudinal band of the Ekman pumping anomaly. On the other hand, if the outcrop lines are non-zonal, a small perturbation of the Ekman pumping rate at a given location can create perturbations in the form of the second dynamical thermocline mode that propagate southwest of the Ekman pumping anomaly (Fig. 4.101). The analysis of such a phenomenon can be found in the work of Huang and Pedlosky (1999).

Region III

South of f_2 there are two subsurface layers in which the streamlines are $H_1 = const.$ and $H_1 + \gamma_{21} H_2 = const.$ Along these streamlines, potential vorticity is conserved. Note that when the potential vorticity anomaly carried by the streamlines in layer 1 (labeled as δq_1 in Fig. 4.99) crosses the second outcrop line f_2, a new characteristic is created, and south of f_2 the potential vorticity anomaly in layer 2 propagates along streamlines in layer 2 (labeled as δq_2 in Fig. 4.99). Thus the zone influenced by the cooling, the so-called region of influence, broadens in a fan-like manner as the fluid moves southward.

Region IV

The solution in region IV is much more complicated because two new characteristics, each carrying a potential vorticity anomaly in layer 3, are created when the trajectories carrying the primary and second potential vorticity anomaly cross f_3 (Fig. 4.99). Thus, *there are four potential vorticity anomaly trajectories, including the primary, the secondary, and the tertiary*. The solution for this region was discussed by Huang and Pedlosky (2000b).

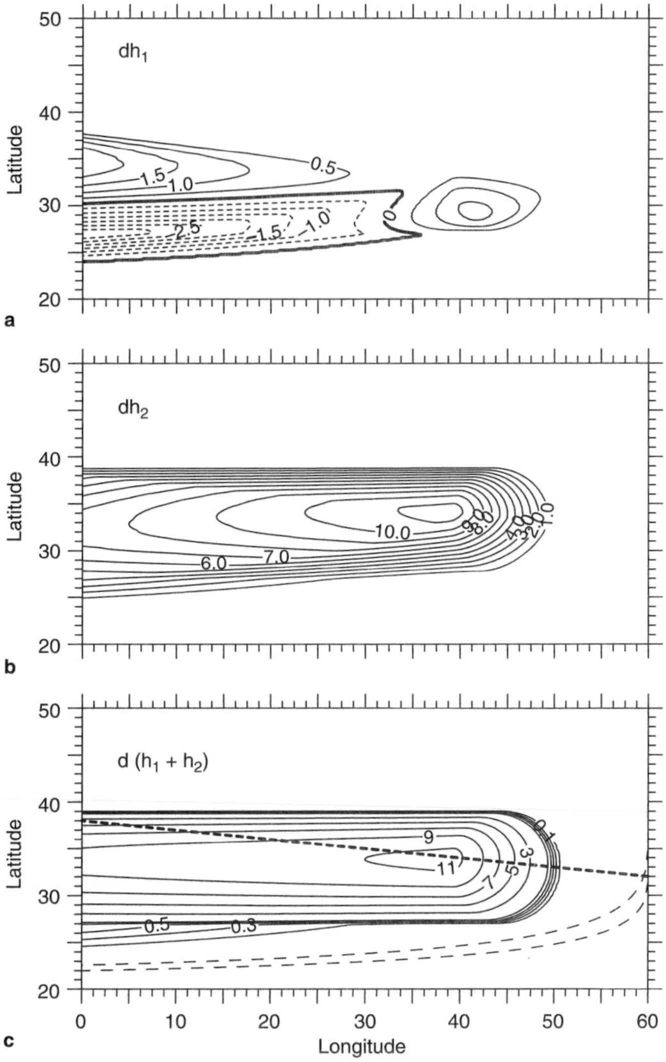

Fig. 4.101 Perturbations in layer thickness (m) in response to an increase of Ekman pumping, the outcrop line has a negative slope, as depicted by the *heavy dashed line* in *panel* **c** (Huang and Pedlosky, 1999).

In general, if there are n outcrop lines, there will be 2^{n-1} characteristics, and each of them carries different signals of potential thickness anomaly.

Climate variability

We primarily focus on the climate variability induced by anomalous buoyancy forcing imposed along the outcrop lines. The model basin is a rectangular basin 60° wide and covers the region 20° N to 15° N, with three outcrop lines at $\theta_1 = 45.5°$ N, $\theta_2 = 41°$ N, and $\theta_3 = 35°$ N. The first layer thickness along the eastern boundary is set to 300 m, and so the complete solution includes a shadow zone. We carefully choose our forcing to limit its region of influence to the ventilated zone for the sake of simplicity. The Ekman pumping rate is $w_e = 10^{-6}\sin\{[(\theta - \theta_s)/\Delta\theta]\pi\}$ (m/s), where $\theta_s = 20°$ and $\Delta\theta = 30°$.

The discussion presented here is for the case with cooling imposed on f_1; cooling/heating imposed on other outcrop lines can be treated in a similar way. A cooling anomaly is imposed in terms of a southward migration of the outcrop line within a small patch

$$\delta\theta_1 = \left[1 - \left(\frac{\lambda - \lambda_0}{\Delta\lambda}\right)^2\right]^{1/2}, \quad \text{if } \lambda_0 - \Delta\lambda \le \lambda \le \lambda_0 + \Delta\lambda; \qquad (4.544)$$
$$= 0, \textit{elsewhere}$$

where $\lambda_0 = 20°$. We choose a small perturbation in order to illustrate the fundamental structure of the variability induced by a point source of buoyancy forcing, so $d\theta_1 = -0.01°$ and $\Delta\lambda = 4°$ for the first experiment. Of course, since the perturbation solution is linear, the case of anomalous heating simply corresponds to a change of sign of all perturbation variables.

Climate variability is defined as the difference between the unperturbed solution and the perturbed solution. Surface cooling along f_1 leads to an increase of h_1 in the region of surface cooling. South of the first outcrop line, this primary potential vorticity anomaly propagates along ψ_1 and induces a decline in h_2 in region II, as shown in Figure 4.102a, b. Thus, a potential vorticity anomaly is created in layer 2.

South of f_2, the second layer is subducted, so the potential vorticity anomaly in the second layer is preserved and propagates along ψ_2. Thus, in region III we have two characteristic cones consisting of potential vorticity anomaly trajectories – the primary potential vorticity anomaly δq_1 that propagates along ψ_1, and the secondary potential vorticity anomaly δq_2 that propagates along ψ_2, as shown in Figure 4.102a, b, c.

South of f_3, the third layer is subducted, so the tertiary potential vorticity anomalies are created in the third layer, induced by the primary potential vorticity anomaly in the first layer and the secondary potential vorticity anomaly in the second layer. Thus, there are two new characteristic cones in region IV.

From the discussion above, we argue that the number of the characteristic cones carrying the perturbation information doubles each time that they cross a new outcrop line. The exponential growth of the number of characteristic cones is a major difficulty in dealing with the multi-layer model.

Within these characteristic cones, the amplitude of the layer thickness perturbations varies over two orders of magnitude. The horizontal pattern of the layer thickness perturbations is shown in Figure 4.102. Even with just three outcrop lines, the pattern of the perturbations is rather complicated, and it is clear that the sign of layer thickness perturbations alternates in the horizontal plane.

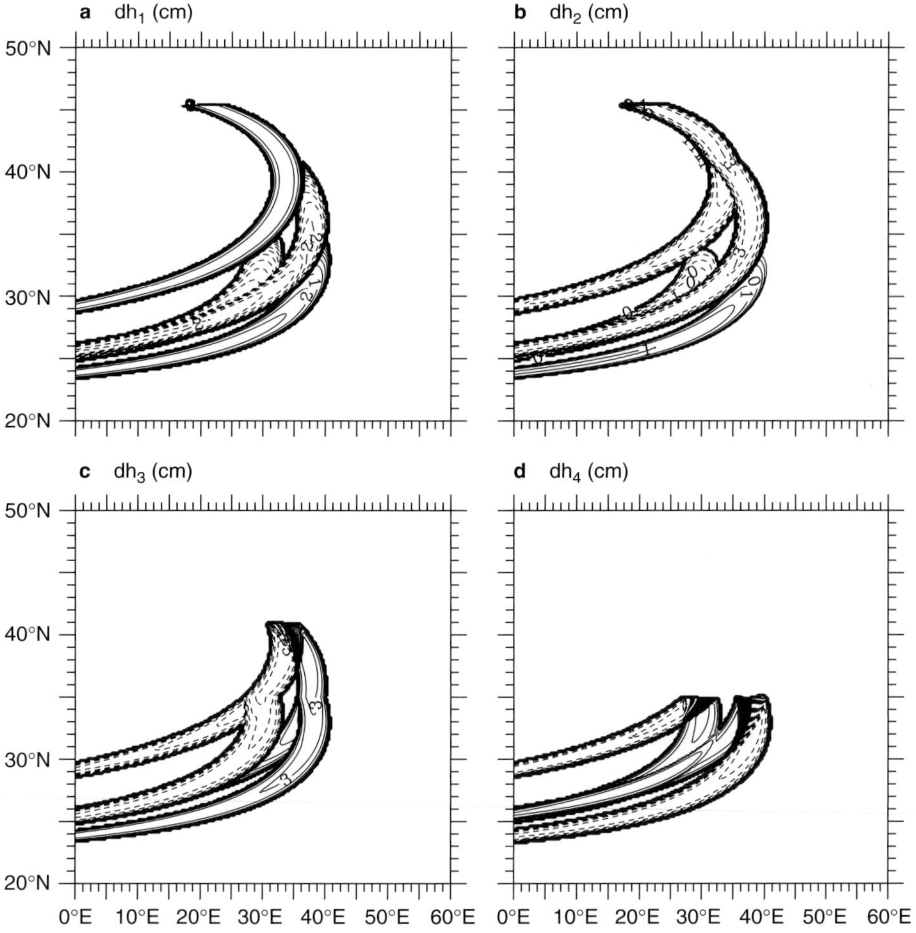

Fig. 4.102 Layer thickness perturbation maps, generated by cooling within a small area with $dy_1 = -0.01°$ and $\Delta x = 4°$. Layer thickness anomalies dh (in $10–^4$m) are presented as follows: for positive anomaly, $x = \max[\log(dh), 0]$, for negative anomaly, $x = \min[-\log(-dh), 0]$ (Huang and Pedlosky, 2000b).

As the number of moving layers increases, the vertical structure of the perturbations becomes increasingly complicated. As shown in Eqn. (4.527), variability induced by buoyancy forcing must appear in the form of internal modes, the DTM. In region II, it appears in a second baroclinic DTM, M_2^1, where the subscript 2 indicates the number of moving layers and the superscript 1 indicates the layer where the driving potential vorticity source is located.

In region III, the vertical structure of the solution in the two branches appears in a different form. Along the P-branch (the primary potential vorticity anomaly), the perturbation is in the form of M_3^1 mode. This mode is induced by a large positive dh_1. Along the S-branch

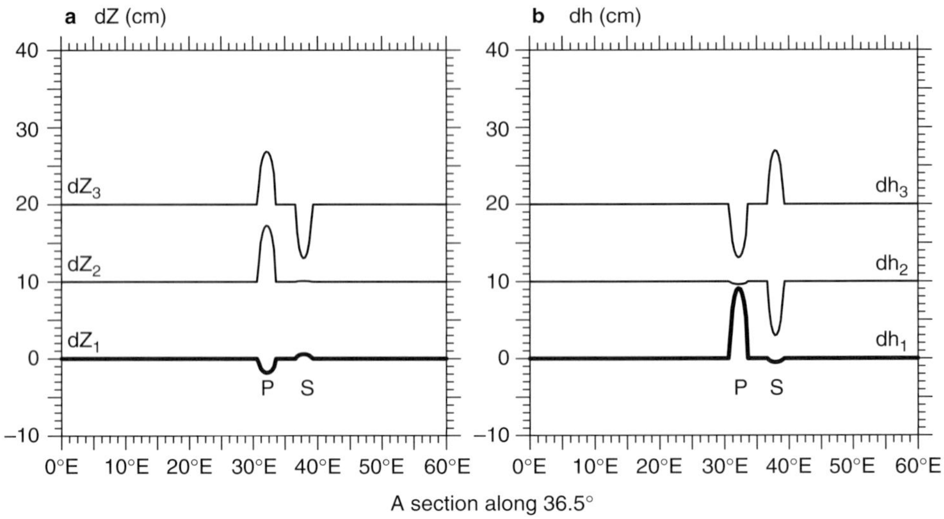

Fig. 4.103 Perturbations of interface depth (**a**) and layer thickness (**b**) along 36.5° N, generated by cooling within a small area with $dy_1 = -0.01°$ and $\Delta x = 2°$ (Huang and Pedlosky, 2000b).

(the secondary potential vorticity anomaly), the perturbation is in the form of M_3^2 mode, which is induced by a negative dh_2 (Fig. 4.103). Note that $|dh_2| < |dh_1|$ because the former is a secondary perturbation.

Note that within the S-branch there are small potential vorticity (or layer thickness) perturbations in layer 1. Such perturbations are not directly related to the surface forcing on layer 1 at the outcrop line; instead, they are induced by the secondary potential vorticity anomaly in layer 2 in the following way. A potential vorticity anomaly in layer 2 induces a slight shift of streamlines in layer 1, and thus induces a change in the potential vorticity there. Similarly, there are small potential vorticity perturbations in layer 2 within branch P.

The vertical structures of M_3^1 and M_3^2 are quite different, because they represent the thermocline's response to potential vorticity perturbations imposed on the first layer and the second layer, respectively. In addition, we note that the perturbation of the depth of the second interface has a very small negative value, due to the almost perfect compensation of the thickness perturbation of layers 2 and 3.

If there were many outcrop lines, all the perturbations would be confined within the so-called characteristic cone. The western edge of the characteristic cone is defined by the streamline on the isopycnal surface of the original cooling source. The eastern edge of the characteristic cone is more complicated. Each time that a new outcrop line is crossed, a new characteristic is created. Owing to the β-spiral, the velocity vector in the upper layer is always to the right of the velocity vector in the subsurface layer. Therefore, the eastern edge of the perturbations is represented by the streamline in the uppermost layer. In the limit, therefore, the eastern edge of the characteristic cone is defined by the envelope of the

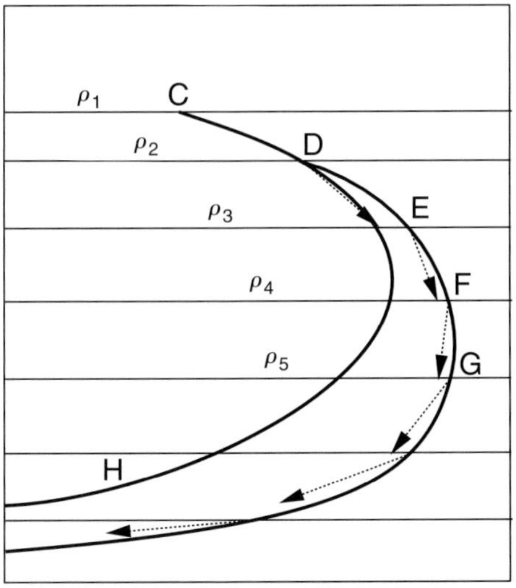

Fig. 4.104 The definition of the edge of the characteristic cone.

streamlines on the uppermost layer. Thus, in a continuous model, the eastern edge of the characteristic cone should be defined as the streamline on the sea surface, stemming from the point source of cooling (Fig. 4.104).

4.9.2 Continuously stratified model

The model is set up such that the background stratification is provided by a one-dimensional advection–diffusion balance $w\rho_z = \kappa\rho_{zz}$, where $w = 10^{-7}$ m/s is the constant upwelling velocity, and $\kappa = 10^{-6} - 60 \times 10^{-6}$ m^2/s is the vertical diffusivity. Using density boundary conditions of $\rho = 1{,}023$ kg/m^3 at $z = 0$ and $\rho = 1{,}028$ kg/m^3 at the seafloor $z = -5$ km, the density profile can be calculated. The sea surface density is a linear function of latitude $\sigma = 25 + 2(\theta - \theta_s)/(\theta_n - \theta_s)$ kg/m^3. Assuming that density is vertically homogenized within the mixed layer, this horizontal density distribution also gives the depth of the mixed layer at each location. The model is under a simple sinusoidal Ekman pumping force $w_e = -10^{-6} \sin[\pi\,(\theta - \theta_s)\,/\,(\theta_n - \theta_s)]$ m/s.

Dependence on the diffusivity κ

Examples in Figure 4.105 show the structure of the thermocline as the vertical mixing rate changes. It is clear that the ideal-fluid thermocline model with $\kappa \approx 10^{-5} - 3 \times 10^{-5}$ m^2/s produces a thermocline structure resembling the main thermocline in the ocean. Most importantly, these examples demonstrate that the ideal-fluid thermocline model can simulate the case with a step-function-like sharp density front associated with the main thermocline, although such a limit case is not realizable in the oceanic circulation under the current climate conditions.

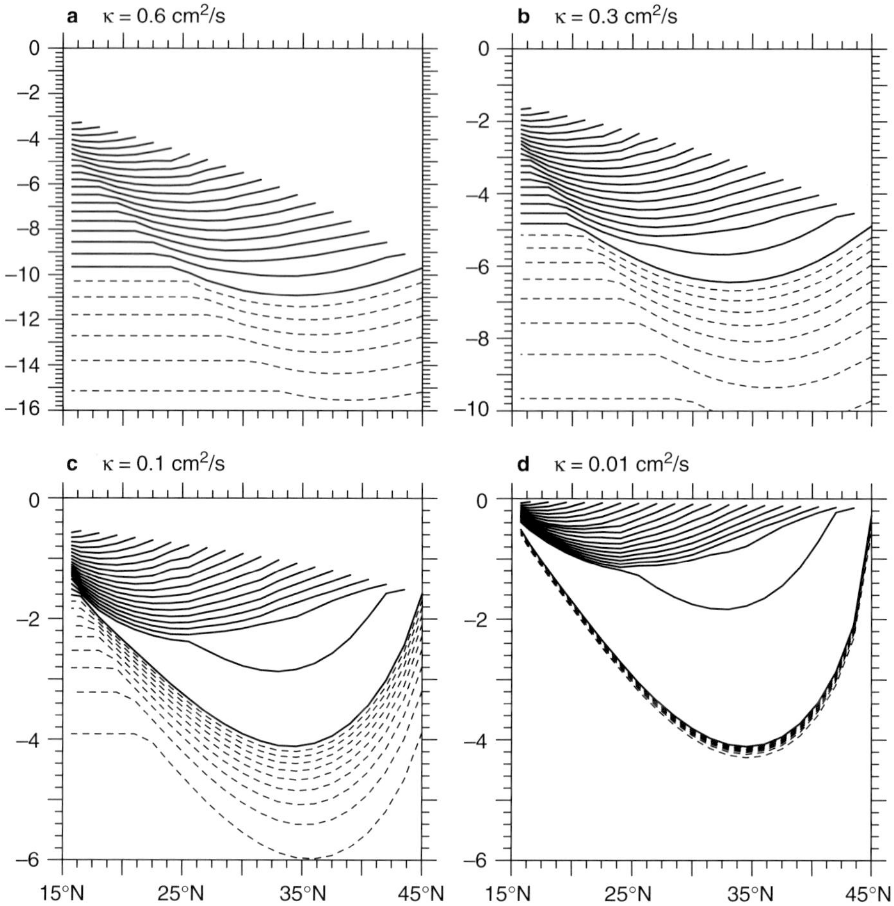

Fig. 4.105 Isopycnal sections taken along the outer edge of the western boundary (5° away from the wall), with the *solid lines* depicting ventilated thermocline and the *dashed lines* depicting the unventilated thermocline, contour interval $\Delta\sigma = 0.1$ kg/m^3. The *blank portion* in the upper ocean indicates the mixed layer where density is vertically uniform (Huang, 2000a).

One of the most outstanding features in these solutions is the low potential vorticity water formed near the northern edge of the subtropical gyre. This is a fundamental issue regarding mode water formation: Why is the potential vorticity of mode water so low? Despite much research effort, this problem remains unresolved.

Shifting of atmospheric conditions

On a decadal time scale or longer, the position of the Gulf Stream can move meridionally, and the sea surface density can increase or decline. Under such different forcing boundary conditions, the thermocline variability is shown in Figure 4.106. One of the most outstanding

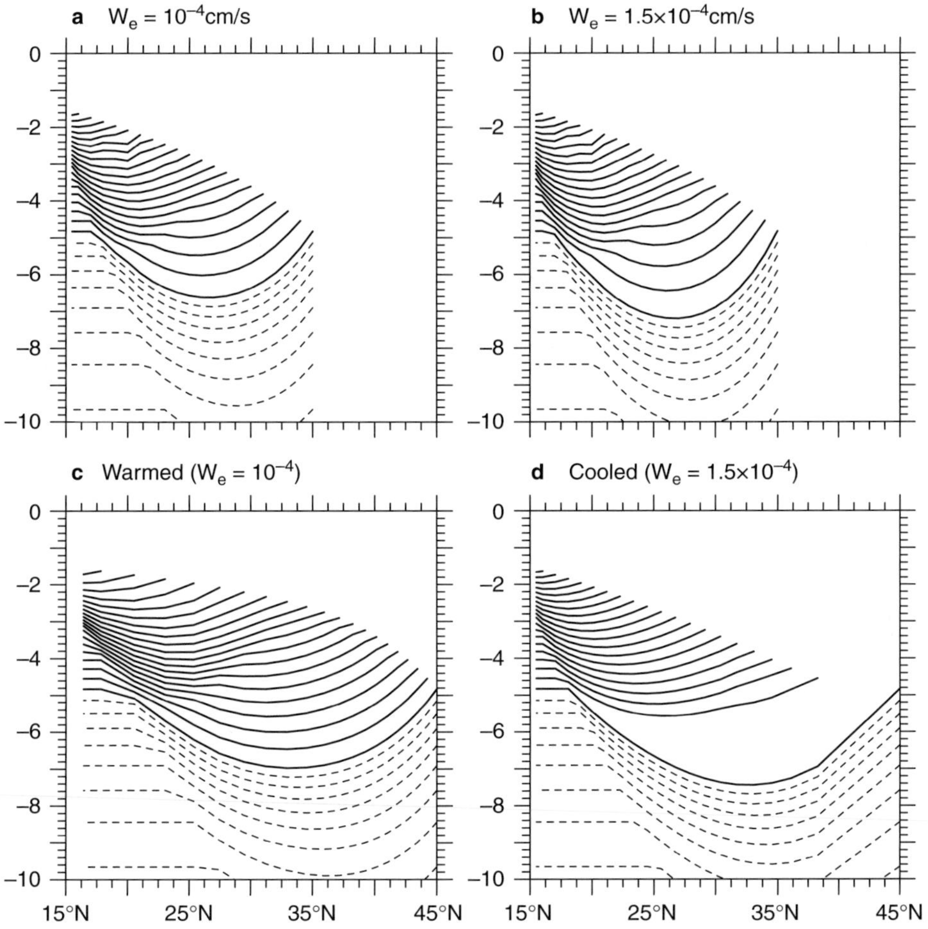

Fig. 4.106 Isopycnal sections taken along the outer edge of the western boundary (5° away from the wall): **a** the inter-gyre boundary moved to 35° N, with the maximum Ekman pumping rate unchanged; **b** the inter-gyre boundary moved to 35° N, with the maximum Ekman pumping rate increased 50% in order to have the same total amount of Ekman pumping; **c** surface warming; **d** surface cooling (Huang, 2000b).

features is the very low potential vorticity mode water created for the case of anomalous surface cooling (Fig. 4.106b, d).

The difference between these cases can be seen clearly from profiles of stratification and potential vorticity taken at a station (26° N, 5° E), which is near the western boundary and at a location corresponding to the Bermuda station in the Atlantic. Most interestingly, all perturbations appear in the form of the second dynamical thermocline mode-like features (Fig. 4.107a).

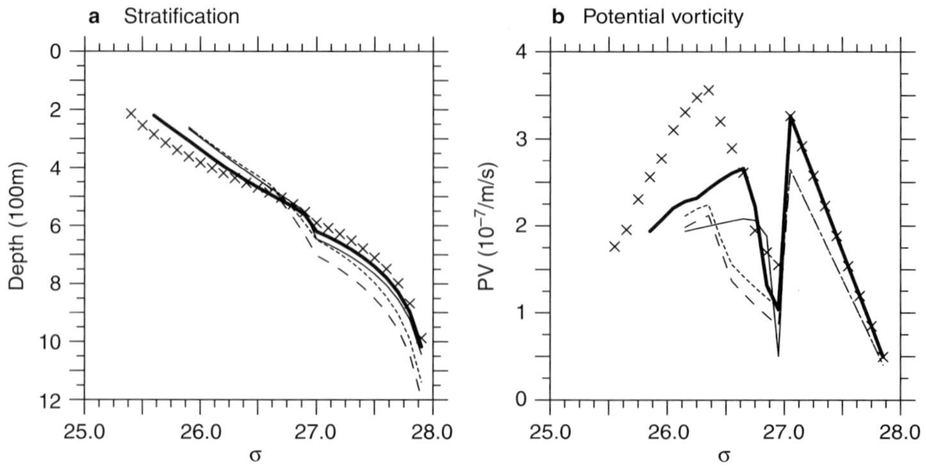

Fig. 4.107 **a** Density and **b** potential vorticity profiles at a station near the western boundary. The *solid line* is for the standard case; the *crosses* are for the case with surface warming; the *thin solid line* is for the case with surface cooling; the *thin dotted line* is for the case with inter-gyre boundary moved to 35° N, and same Ekman pumping rate as the standard case; the *long-dashed line* depicts the case with the inter-gyre boundary moved to 35° N, and the Ekman pumping rate increased 50% (Huang, 2000b).

Localized cooling

The effect of localized cooling can be found through a numerical experiment, with a cooling patch near 40° E and $y_0 = 30°$ N. This cooling induces a depression of the sea surface elevation (Fig. 4.108a) and a shoaling of the isopycnal surface $\sigma = 26.3$ kg/m^3 about 30 m; however, not only does the deep isopycnal $\sigma = 27$ kg/m^3 move down 30 m but the base of the wind-driven gyre also moves down more than 70 m. Here again, the perturbations appear in forms of the second dynamical thermocline mode. The vertical structure of the perturbation can be seen through a vertical profile taken at the center of the perturbations (Fig. 4.109).

Point source of cooling

A very interesting question is what happens when the size of the cooling anomaly gradually shrinks to a δ-function-like point source. Within the relatively crude horizontal resolution of this model, several numerical experiments were carried out with a point source of cooling or a mixed-layer depth anomaly. The perturbations induced by a point of cooling source have a wave-train-like feature (Fig. 4.110). This picture suggests that perturbations are a combination of lower-order modes and higher-order modes, with the lower-order modes dominating. Such a complicated spatial structure is not inconsistent with the alternating signs shown in Figure 4.102.

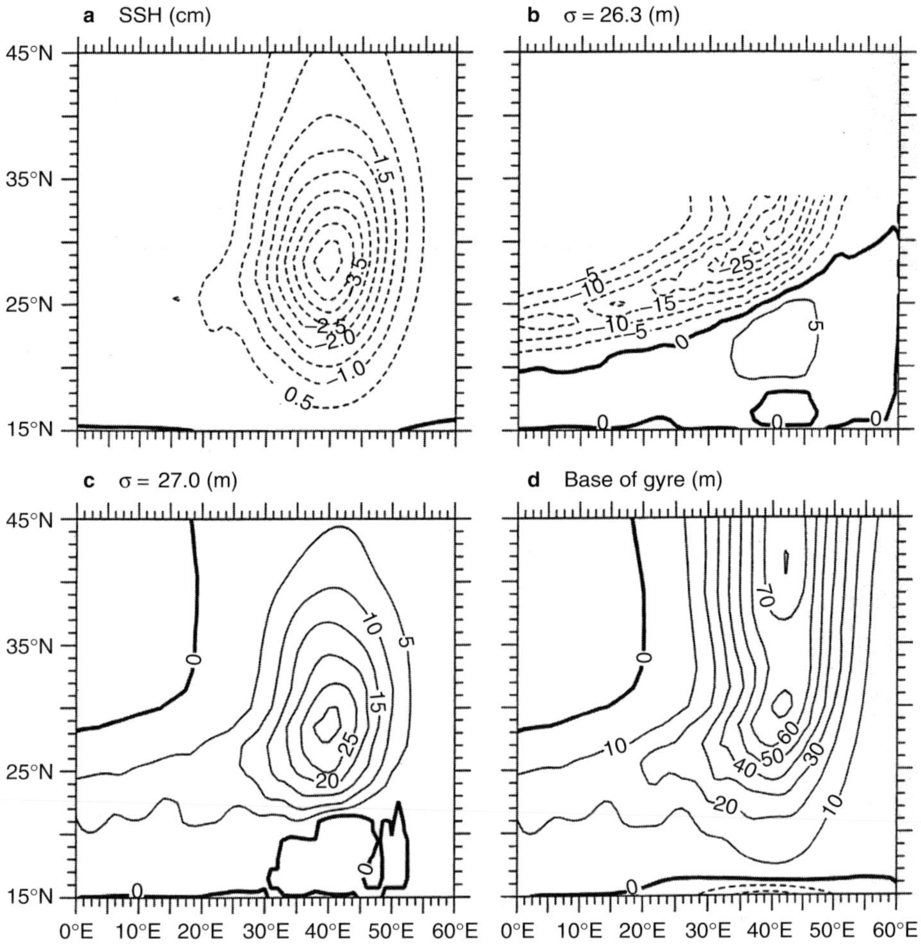

Fig. 4.108 Perturbations induced by a localized cooling: **a** sea surface elevation (cm); **b** depth of isopycnal surface $\sigma = 26.3\ \text{kg/m}^3$ (m); **c** depth of isopycnal surface $\sigma = 27\ \text{kg/m}^3$ (m); **d** depth of the wind-driven gyre (m) (Huang, 2000b).

Changes in subpolar mode water formation

The rate of subpolar mode water formation changes on time scales ranging from interannual to millennial. As a result, the background stratification, or the potential vorticity for the unventilated water, changes in response. We explored this issue using our model under the same forcing, except that we alternated the potential thickness for the unventilated thermocline, as indicated by the thick solid line in the right part of Figure 4.111.

With this change in the potential thickness of the unventilated thermocline, all ventilated and unventilated isopycnals move down in response. Results from our model are quite consistent with analyses based on historical data (Joyce *et al.*, 1999).

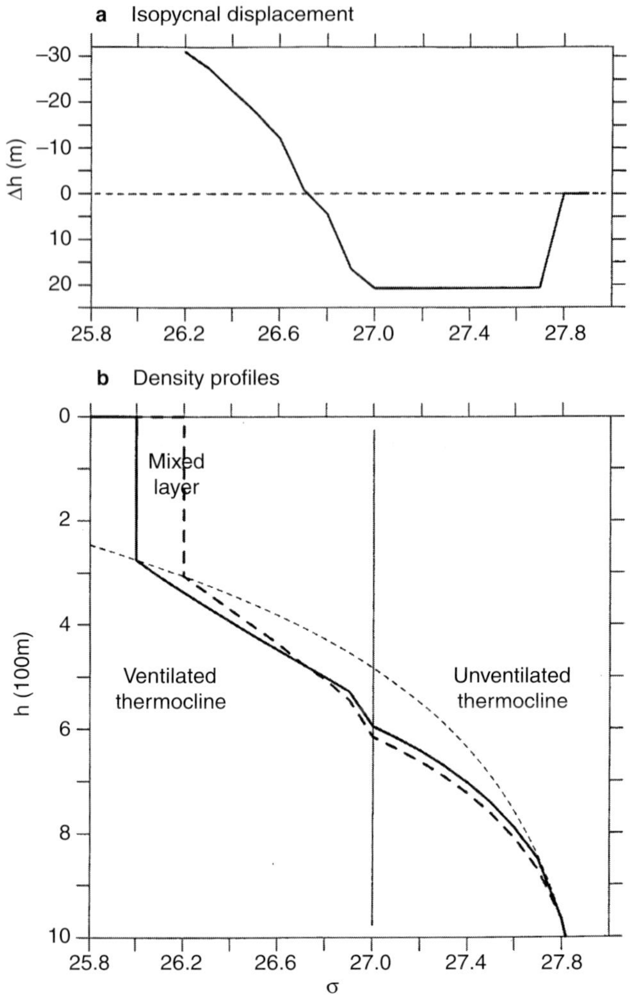

Fig. 4.109 A density profile at 30° N, 40° E: **a** the vertical displacement of isopycnals due to cooling; **b** the density profiles, with the *solid line* for the standard case and the *heavy dashed line* for the case with cooling. The *thin dashed line* indicates the background stratification (Huang, 2000b).

4.9.3 Decadal climate variability diagnosed from data and numerical models

Climate variability diagnosed from data

We owe the early development along these lines to Deser *et al.* (1996). By analyzing the climate data, they were able to show how the temperature anomaly was able to penetrate to the subsurface layers (Fig. 4.112). It can readily be seen that, on a decadal time scale, the temperature anomaly basically moves along isopycnal surfaces (Fig. 4.113).

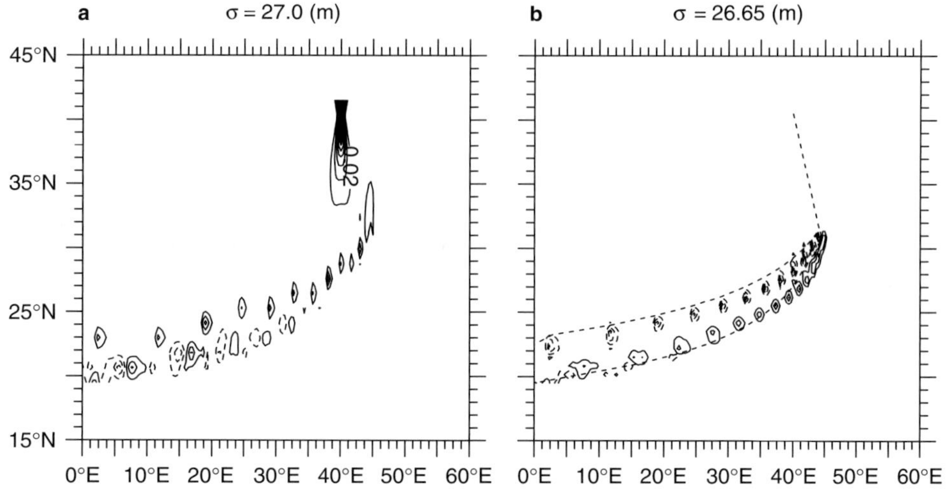

Fig. 4.110 Perturbation of the mean fields from a model with 181 × 21 grid points and under a localized cooling along a single outcrop line, with $\delta y = -0.05°$, and a width of $\delta x = 0.2°$. The western edge of the perturbation zone is defined by the streamline on the isopycnal surface of the primary characteristic due to cooling; the eastern edge of the perturbation zone is defined by the envelope of the streamline on the sea surface (Huang, 2000b).

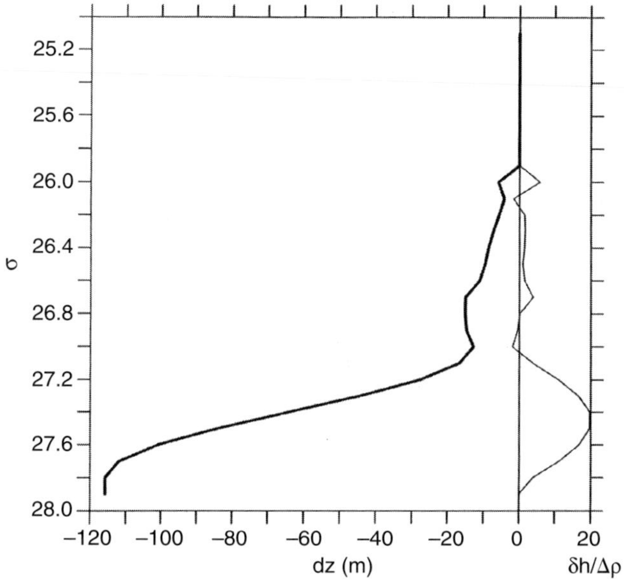

Fig. 4.111 Depth perturbation at a station (28.5° N, 10° E) in the model basin. The *thick solid line* indicates the vertical displacements on each isopycnal, and the *thin line* indicates thickness perturbation in each isopycnal layer ($\Delta\rho = 0.1\ \mathrm{kg/m^3}$) (Huang, 2000b).

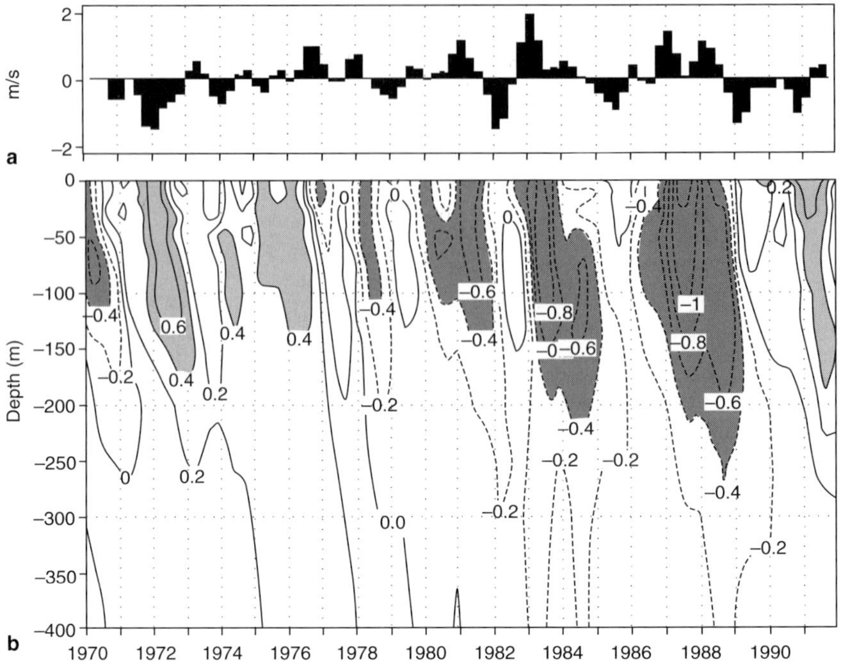

Fig. 4.112 Evolution of the anomalous temperature in the central North Pacific region (Deser *et al.*, 1996).

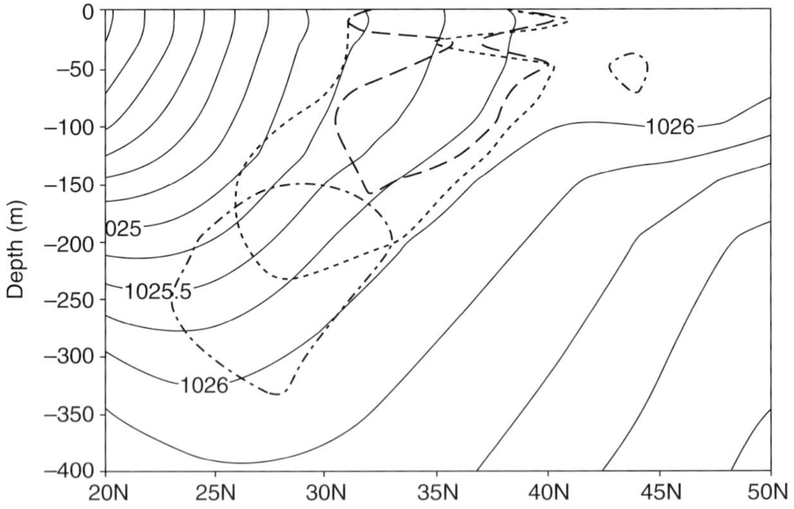

Fig. 4.113 The −3° C anomaly isotherms for 1977–81 (*dashed line*), 1982–6 (*dotted line*), and 1987–91 (*solid line*), superimposed upon the mean late-winter isopycnals (*thin solid lines labeled with numbers*, in kg/m^3) (Deser *et al.*, 1996).

From Figure 4.112, during the period from 1977 to 1989 the temperature anomaly moved down about 260 m, so the vertical velocity is slightly more than 20 m/year, which is on the same order as the vertical velocity in this area of ocean, as inferred from the Ekman pumping rate on the order of 30 m/year. Since Deser's work, there have been many papers published using this line of reasoning, and they all show a similar structure.

Reappearance of the temperature anomaly in the upper ocean

An interesting phenomenon identifiable from Figure 4.112 is the reappearance of the temperature anomaly in the upper ocean several years after the initial impact of a strong cooling event. Based on the ideas of ventilated thermocline theory, we can define an annual subduction depth $D_s = S_{ann}T$, where S_{ann} is the annual mean subduction rate, and $T = 1$ year is the time duration in defining the annual mean subduction rate. The subduction depth ratio is defined as $R_s = D_s/h_{max}$.

The distribution of this depth ratio for the North Pacific, inferred from an ideal-fluid thermocline model with continuous stratification, is shown in Figure 4.114. The inverse of this ratio is closely related to the time (in years) that a surface temperature anomaly can survive in the upper ocean, which can be seen by comparing this figure with the corresponding figures presented by Frankignoul and Reynolds (1983).

Using a simple one-dimensional model to analyze the thermal balance of the upper ocean heat content, it can readily be seen that it takes about $1/R_s$ years to renew the water property in the mixed layer. In addition, it is clear that the optimal season to form the anomaly is late winter (March 1) when the mixed layer is coldest and densest. In addition, only a cold

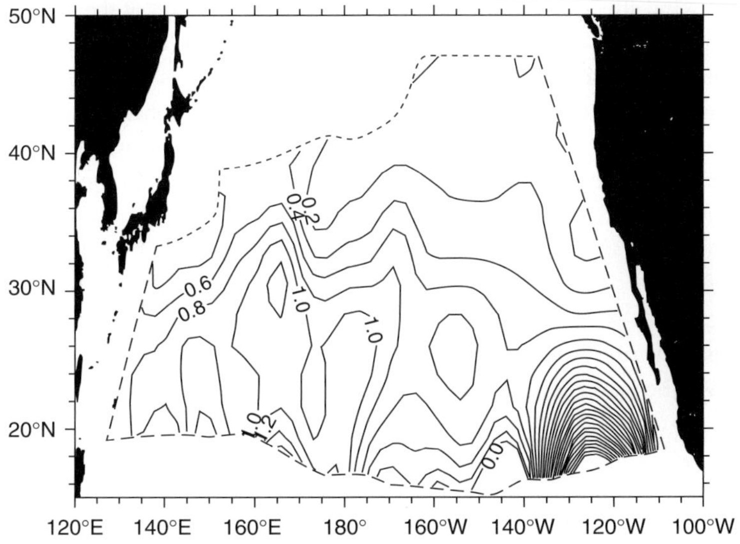

Fig. 4.114 Subduction depth ratio for the North Pacific, inferred from an ideal-fluid thermocline model (Huang and Russell, 1994).

anomaly is able to survive, because a warm anomaly does not penetrate deep enough and will be wiped out during the following late winter of a normal year, when the late-winter mixed layer is denser and deeper.

4.10 Inter-gyre communication due to regional climate variability

4.10.1 Introduction

Wind-driven circulation has been discussed in previous sections. However, in most cases, our discussions have focused on the steady circulation in individual gyres. For climate study, it is interesting to explore the climate variability of the circulation under different forcing.

A key element of decadal climate changes in the ocean is the possibility of inter-gyre communication. For example, Gu and Philander (1997) proposed an exchange between the subtropical and equatorial gyres through the subduction driven by Ekman pumping in the subtropical basin. This link has been actively pursued by many investigators.

In this section, we discuss a simple mechanism which can drive an inter-gyre communication over decadal time scales. It is well known that changes in wind-stress curl can lead to changes in the thermocline in a given gyre. However, a possible inter-gyre communication due to such a change can be studied as follows.

For a simple reduced-gravity model, the upper layer covers the whole surface, so the lower layer is isolated from air–sea interaction. The total volume of the lower layer is controlled by some rather slow processes such as deepwater formation and diapycnal mixing. We will make the further assumption that over a few decades these processes will not be affected by changes in the wind stress distribution. Therefore, the total amount of the lower layer water will remain unchanged over a relatively short decadal time scale. Because the total volume of the ocean is unchanged, the total volume of the upper layer water should remain unchanged. In order to conserve the volume of water masses, therefore, the global thermocline structure will change correspondingly. In this section, we use a simple steady-state reduced-gravity model to demonstrate the inter-gyre communication due to changes of forcing, such as wind stress and heating/cooling, in an individual basin.

4.10.2 Model formulation

The basic equations of a reduced-gravity model in spherical coordinates were discussed in Section 4.3, and the upper layer thickness obeys the following equation:

$$h^2 = h_e^2 + \frac{2a}{g'}\int_{\lambda}^{\lambda_e} P_r d\lambda \qquad (4.313)$$

where

$$P_r = -2a\omega \sin^2\theta w_e, \quad w_e = \frac{1}{2\omega\rho_0 a\sin\theta}\left[\frac{1}{\cos\theta}\tau_{\lambda}^{\theta} - \tau_{\theta}^{\lambda} + \frac{\tau^{\lambda}}{\sin\theta\cos\theta}\right] \qquad (4.314)$$

is the pumping rate, which is slightly different from the commonly used Ekman pumping rate. Although the meridional component of wind stress contributes an important part of the wind-stress curl, we will assume a purely zonal wind stress which is independent of longitudinal coordinates. It is straightforward to infer the effect due to changes in the meridional wind stress.

Although geostrophy is not valid right on the equator, it is a good approximation near the equator, except for a narrow zone of 100 km off the equator. Furthermore, layer thickness derived from Eqn. (4.313) remains finite approaching the equator, assuming that wind-stress curl is finite at the equator. Since the total amount of water in the upper layer should remain unchanged within the decadal time scale, the model is subject to the following constraint:

$$\iint_A dA \left(h_e^2 + \frac{2a}{g'} \int_{\lambda}^{\lambda_e} P_r d\lambda \right)^{1/2} = V_0 \tag{4.545}$$

where A is the whole basin in a single hemisphere, including the subpolar, subtropical, and equatorial gyres (Fig. 4.115). If the Ekman pumping rate in the subtropical basin declines, the thermocline bowl there is reduced (middle of Fig. 4.114).

The upward motion of the thermocline in the subtropical basin pushes the warm water above the main thermocline to the subpolar/equatorial basins. If the level of cold water along the eastern boundary remains unchanged, the volume of warm water in the subpolar/equatorial basins would remain the same as before. This is, of course, inconsistent with the upward motion in the subtropical basin. In order to compensate for the upward motion in the subtropical basin, the level of cold water in the subpolar/equatorial basins must move downward. The level of the cold water along the eastern boundary is adjusted in such a way that the total amount of warm water in the whole basin remains unchanged. For a given wind stress perturbation, the final level of the layer thickness along the eastern boundary,

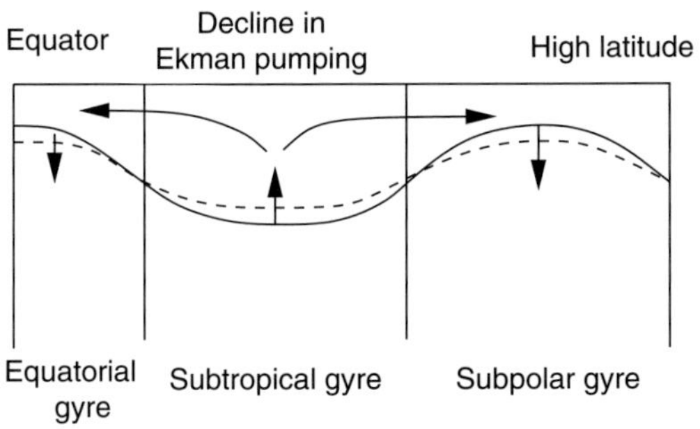

Fig. 4.115 Layer thickness adjustment due to weakening of the Ekman pumping in the subtropical basin.

$h_e + \delta h_e$, can be found by solving the following nonlinear equation:

$$\iint_A dA \left[(h_e + \delta h_e)^2 + \frac{2a}{g'} \int_{\lambda}^{\lambda_e} \left(P_r + P_r'\right) d\lambda \right]^{1/2} = V_0 \tag{4.546}$$

$$P_r' = -2a\omega \sin^2\theta w_e', \quad w_e' = \frac{1}{2\omega\rho_0 a \sin\theta} \left[-\tau_\theta^{\lambda'} + \frac{\tau^{\lambda'}}{\sin\theta\cos\theta} \right] \tag{4.547}$$

where $\tau\lambda'$ is the zonal wind stress perturbation. Assuming a small perturbation, one obtains

$$h_e \delta h_e \iint_A \frac{dA}{h(\lambda,\theta)} = -\frac{a}{g'} \iint_A \frac{\int_{\lambda}^{\lambda_e} P_r' d\lambda}{h(\lambda,\theta)} dA \tag{4.548}$$

where $h(\lambda,\theta) = \left(h_e^2 + \frac{2a}{g'}\int_{\lambda}^{\lambda_e} P_r d\lambda\right)^{1/2}$.

Thus, changes in the layer thickness along the eastern boundary are negatively correlated with changes in the basin-integrated pumping rate. For example, a reduction in upwelling rate in the subpolar basin should lead to a reduction of h_e. Similarly, stronger (weaker) Ekman pumping in the subtropical basin will lead to upward (downward) motion of the thermocline in both the equatorial and subpolar basins.

The model basin is 60° wide in the zonal direction, and extends from the equator to 70° N. The reduced gravity is chosen as $g' = 0.02\ \mathrm{m/s^2}$, and the zonal wind stress profile, in $\mathrm{N/m^2}$, is

$$\tau^{\lambda} = 0.02 - 0.08\sin(6\theta) - 0.05\left[1 - \tanh(10\theta)\right] - 0.05\left\{1 - \tanh\left[10\left(\frac{\pi}{2} - \theta\right)\right]\right\} \tag{4.549}$$

For a given upper layer thickness along the eastern boundary, the layer thickness clearly shows three gyres: an equatorial gyre, a subtropical gyre, and a subpolar gyre (Fig. 4.116).

First, for any given layer thickness along the eastern boundary, h_{e0}, we calculate the total volume of the upper layer water for the circulation driven by the standard wind stress, V_0. Second, we add on a small wind stress perturbation, in the form of a Gaussian profile:

$$\tau^{\lambda'} = \Delta\tau e^{(\theta-\theta_0)/\Delta\theta} \tag{4.550}$$

where $\Delta\tau = 0.03\ \mathrm{N/m^{-2}}$ is the amplitude, and $\Delta\theta = 10°$ is chosen for the wind stress perturbation profile.

Wind stress perturbations have been imposed on three locations; 70° N (on the northern boundary), 30° N (in the middle of the subtropical gyre), and 0° (at the equator). These perturbed wind stress profiles are shown in Figure 4.117a, b, c. In addition, the pumping rate and its perturbation, as defined in Eqns. (4.314) and (4.547), are included in these figures.

For the first case, the wind stress perturbation is imposed on the northern boundary, so the easterly near the northern boundary is weakened (Fig 4.117a). As a result, the amplitude of the upwelling rate in the subpolar basin declines (heavy dashed line in Fig. 4.117b).

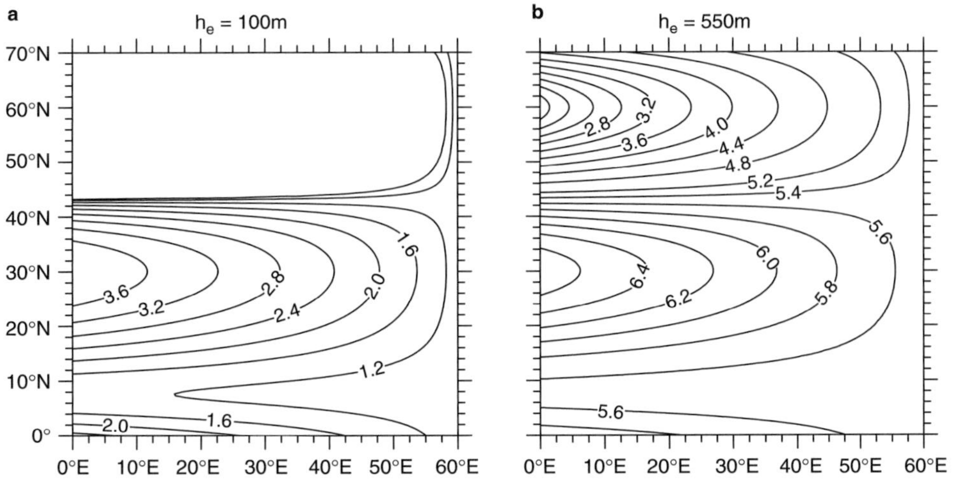

Fig. 4.116 Upper layer thickness maps (in 100 m) for the case in which the layer thickness along the eastern boundary is set to 100 m (**a**) and 550 m (**b**). The *open window* in the subpolar basin in *panel* **a** indicates the outcropping zone.

The thermocline dome in the subpolar basin is suppressed. The downward motion of the thermocline in the subpolar basin drives the adjustment in the whole basin. As a result of this adjustment, layer thickness along the eastern boundary is reduced, and layer thickness along the western boundary in both the subtropical basin and the equatorial basin is reduced (Fig. 4.118a). Note that when $h_{e0} < 550$ m, the upper layer outcrops within the subpolar basin. Results from such calculations are depicted by dashed lines in Figure 4.118a.

For example, when $h_{e0} = 550$ m, adding the wind stress perturbation leads to a decline of the mean layer thickness along the eastern boundary, $\delta h_e = -5.5$ m. Accordingly, layer thickness along the western boundary at the equator and in the subtropical gyre is reduced by 5.1 m and 4.5 m, respectively.

For the second case, the wind stress perturbation is imposed at 30° N, equivalent to a western wind anomaly in the middle of the subtropical gyre (Fig. 4.117c). This perturbation causes a slight southward shift and intensification of the pumping rate profile, which is mostly confined to the subtropical gyre (Fig. 4.117d). Note that the pumping rate is defined as the wind-stress curl multiplied by a factor of $\sin^2(\theta)$. As a result, the overall effect of this wind stress perturbation is a reduction in the total pumping rate.

Similar to the case just discussed above, the reduction in pumping rate leads to an upward motion of the thermocline in the subtropical gyre and drives a basin-wide downward motion of the thermocline interface. After the adjustment, the layer thicknesses along the eastern boundary and the western boundary become larger (Fig. 4.118b). When $h_{e0} < 550$ m, the upper layer outcrops in the subpolar basin, so the layer thickness minimum and its perturbation value there are always zero. Only if $h_{e0} \geq 550$ m does the minimum layer thickness in the subpolar gyre become non-zero; its changes due to wind stress perturbation

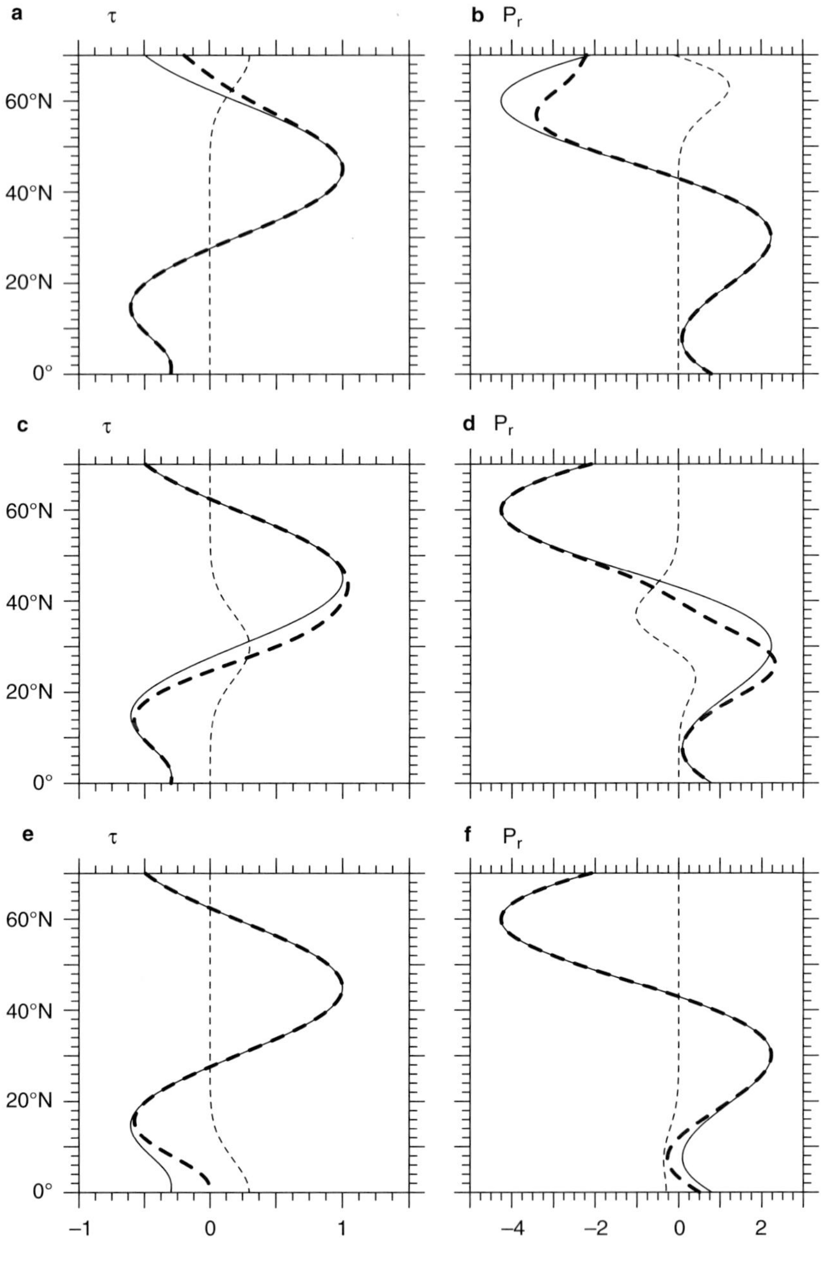

Fig. 4.117 Wind stress perturbation (*left panels* **a**, **c**, **e**, in 0.1 N/m^2) and the associated change in the pumping rate (*right panels* **b**, **d**, **f**, in 10^{-4} m^2/s^2) for three cases. The *thin solid lines* indicate the unperturbed profiles, the *thin dashed lines* indicate the perturbations, and the *heavy dashed lines* indicate the sums of the original and perturbed profiles.

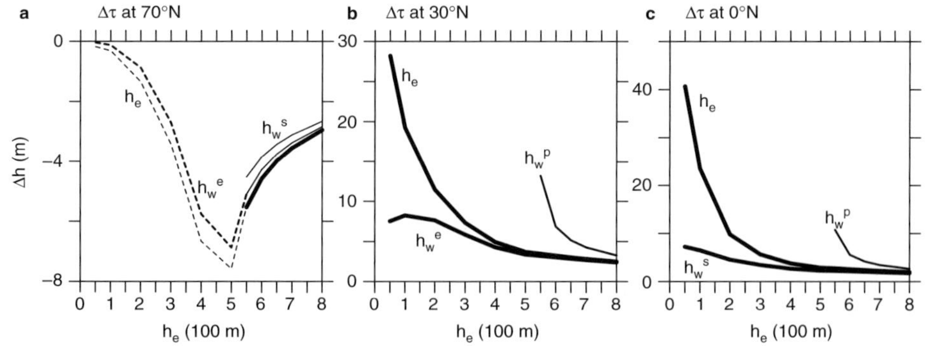

Fig. 4.118 Layer thickness adjustment in response to wind stress perturbations, described in Fig. 4.117. h_e is the original layer thickness along the eastern boundary. Curves labeled with h_e indicate changes in the layer thickness along the eastern boundary; h_w^e for changes of layer thickness at the western boundary and right at the equator; h_w^s for changes of maximum layer thickness along the western boundary of the subtropical gyre; and h_w^p for changes of maximum layer thickness along the western boundary of the subpolar gyre.

are depicted by the line labeled as h_w^p in Figure 4.118b. In a separate calculation, the model is forced by a stronger Ekman pumping rate over the subtropical basin. In response to such an anomalous forcing, the thermocline in the subtropical gyre interior moves downward. On the other hand, the thermocline moves upward in both the equatorial gyre and the subpolar gyre to compensate for the downward motion in the subtropical gyre.

For the third case, the wind stress perturbation is imposed at the equator. For the parameters we have chosen, this leads to exactly zero wind stress at the equator (Fig. 4.117e). Again, such a perturbation leads to a decline in pumping rate in the equatorial gyre, and a downward motion in the subtropical and subpolar basin (Fig. 4.118c).

For $h_{e0} \geq 550$ m, the lower layer is insulated from the air–sea interaction. Assuming that the balance of water mass sources and sinks for the lower layer does not change within a short period of 10–20 years, the total volume of the lower layer should remain unchanged. Since the total volume of the ocean remains the same, the total volume of the upper layer water should remain unchanged as well.

Under such an assumption, wind stress perturbation in either the subpolar gyre, the subtropical gyre, or the equatorial gyre can lead to vertical displacement of the thermocline in the whole basin. Depending on the original layer thickness along the eastern boundary, the vertical displacement is on the order of 5–20 m. If a very thin upper layer along the eastern boundary is chosen, say 100 m, the change due to this basin-wide adjustment can be more than 20 m.

This vertical movement of the thermocline is due to inter-gyre communication. In a departure from the traditional ways of studying the wind-driven circulation in each basin in isolation, a water mass conservation over the entire basin implies the dynamical consequence that changes in wind stress in an individual basin can lead to global changes in the thermocline. Such a global change will certainly give rise to new dynamics associated with the coupling of the thermocline with other components of the oceanic circulation.

As an example, wind stress changes at the mid latitudes can lead to substantial changes in the depth of the thermocline along the equator, which in turn will change the nature of the El Niño–Southern Oscillation (ENSO) cycle.

The discussion above is based on a simple reduced-gravity model, in which the density difference between the upper and lower layers is assumed to be uniform basin-wide. This assumption is a gross idealization. If an isopycnal surface is chosen as the interface, the density difference between water above and below this interface changes greatly from the subpolar basin to the equatorial basin. In order to include such a basin-wide density change, the reduced-gravity model can be converted into the so-called generalized reduced-gravity model (e.g., Huang, 1991b). The reduced gravity in the new model is now a function of horizontal coordinates, and the upper layer thickness can be calculated from a slightly modified equation:

$$h^2 = \frac{g_e'}{g'(\lambda, \theta)} h_e^2 + 2a \int_\lambda^{\lambda_e} \frac{P_r}{g'(\lambda, \theta)} d\lambda \tag{4.551}$$

where $g'(\lambda, \theta)$ is the reduced gravity, which can be calculated from the model, or it can be specified from data. The total volume of water in the upper layer is

$$\iint_A dA \left[\frac{g_e'}{g'(\lambda, \theta)} h_e^2 + 2a \int_\lambda^{\lambda_e} \frac{P_r}{g'(\lambda, \theta)} d\lambda \right]^{1/2} = V_0 \tag{4.552}$$

Parallel to the case with wind stress perturbation, a thermal perturbation gives rise to changes in the reduced gravity in a region of the basin. Within decadal time scales, the total amount of water in the lower layer remains unchanged. As a result, the volume of the upper-layer water remains unchanged. For simplicity, we will assume that reduced gravity in the unperturbed state depends on the latitude only, i.e., $g' = g'(\theta)$. Using the volume conservation constraint and assuming a small perturbation, we obtain

$$h_e \delta h_e \iint_A \frac{dA}{h(\lambda, \theta)} = a \iint_A \frac{\int_\lambda^{\lambda_e} P_r \delta g'(\lambda, \theta)\, d\lambda}{g'^2(\theta)\, h(\lambda, \theta)} dA \tag{4.553}$$

where $h(\lambda, \theta) = \left[h_e^2 + \frac{2a}{g'(\theta)} \int_\lambda^{\lambda_e} P_r d\lambda \right]^{1/2}$ and $\delta g'(\lambda, \theta)$ is a change in the reduced gravity specified.

The mechanism of this adjustment is very similar to that due to a wind stress anomaly. For example, if the subtropical ocean is cooled down, the upper layer density is increased, so $g'(\lambda, \theta) < 0$. According to Eqn. (4.553), $\delta h_e < 0$, i.e., cooling will lead to an upward motion of the thermocline in the whole basin. We recall that cooling leads to a smaller density difference between the upper and lower layers. According to the reduced-gravity model, a smaller density difference across the interface gives rise to a deeper thermocline at the given latitude. Thus, cooling behaves like an increase in Ekman pumping. Since the effect of cooling/heating is rather similar to that due to a wind stress anomaly, we do not show the corresponding numerical examples here.

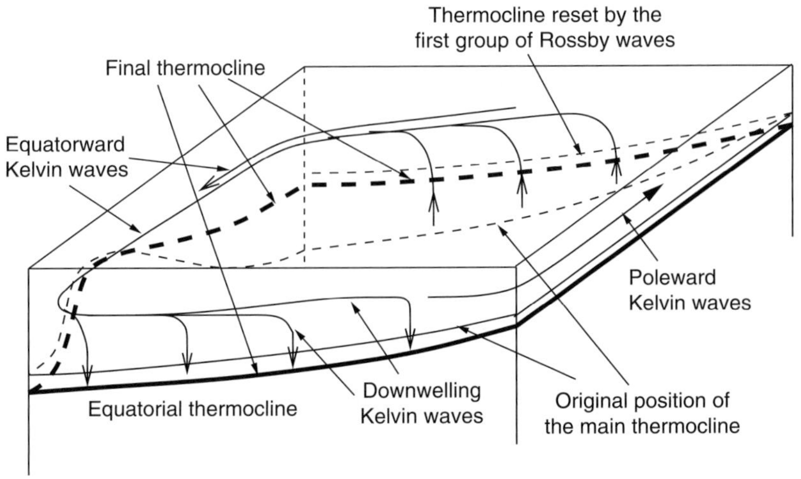

Fig. 4.119 Sketch of the adjustment in the subtropical-equatorial ocean induced by reduction of Ekman pumping rate in the subtropical basin interior.

In summary, both wind stress anomalies and thermal anomalies in a localized region can lead to global adjustment of the thermocline. As an example, stronger Ekman pumping or cooling in the subtropical basin forces a downward motion of the thermocline in the subtropical basin. Since the total volume of water mass is conserved, this leads to a global upward displacement of the thermocline in the whole basin. This global adjustment of the thermocline will interact with other processes in the ocean and produce complicated climate changes.

The calculations above are based on the equilibrium states of the model. In a time-dependent model, perturbations will propagate in the form of waves going through the whole basin. In particular, both Rossby waves and Kelvin waves play important roles in setting up the final solution. As shown schematically in Figure 4.119, the reduction in Ekman pumping rate in the subtropical basin excites the westward baroclinic Rossby waves. At this time the layer thickness along the eastern boundary remains unchanged; however, the thermocline moves upward after the Rossby waves pass. As the Rossby waves reach the western boundary, Kelvin waves are excited, which carry the signals equatorward. Due to the constraint of mass conservation, the Kelvin waves must carry a downwelling signal as they reach the equator and move eastward along the equatorial wave guide. At the eastern boundary the equatorial Kelvin waves split and turn poleward along the eastern boundary. As the waves pass through, the thermocline depth along the eastern boundary increases. Along the eastern boundary the Kelvin waves gradually lose their energy by sending out the westward Rossby waves. The final stage of the solution will be established after this wave loop repeats a couple of times and other physical processes in the ocean, such as dissipation, should also play some role. The connection between the mid-latitude wind stress perturbations and the equatorial thermocline and surface temperature anomaly can be explored in detail using a numerical model (Klinger *et al.*, 2002).

5
Thermohaline circulation

5.1 Water mass formation/erosion

The balance of water masses in the world's oceans consists of two major processes: water mass formation and erosion. Most water masses are formed near the upper surface and sink. Furthermore, through either transformation or erosion, water mass properties are continually transformed, so that a water mass gradually loses its identity. Therefore, some types of water mass are formed below the surface layers through the mixing of water masses originated from the sea surface; however, in this chapter, we primarily focus on formation/erosion of water masses in connection with surface processes.

According to the penetration depth, water-mass formation is generally separated into two major categories, those of deep water and mode water. The second category of water mass normally sinks to a relatively shallow part of the world's oceans. In this chapter, we first discuss deepwater formation and then mode water formation.

5.1.1 Sources of deep water in the world's oceans

In a broad sense, the balance of deep water in the world's oceans consists of two major opposing processes: the supply of newly formed water masses through deepwater formation and the removal of deep water through mixing and erosion. Deepwater formation is closely related to the downward branch of the vertical circulation, which continuously supplies the water masses, while deepwater erosion is closely related to the upward branch of the vertical circulation, which continuously removes the water masses. Both processes are essential for water mass balance and thermohaline circulation in the world's oceans. For example, mixing in the deep ocean or upwelling through strong fronts continuously removes the old-age deep water, thus making room for the newly formed deep water and maintaining the thermohaline circulation. Since continuous removal of deep water is closely related to vertical mixing in the stably stratified ocean (requiring a supply of external mechanical energy), without the continuous supply of this energy the thermohaline circulation would not be sustained.

The vertical circulation in the world's oceans is asymmetric in terms of upwelling and downwelling. In fact, the downward branch of the oceanic circulation is confined to a very few narrowly defined sites. Furthermore, the sites of the downwelling branch of the thermohaline circulation can be different from the major sites of buoyancy loss in the oceans.

In particular, the formation of deep water can take place near the edge of the horizontal gyre or within the boundary currents, instead of in the middle of the gyre.

In contrast, the upwelling branch of the circulation occurs on a much broader scale. Nevertheless, upwelling is also highly non-uniform in space. In fact, strong upwelling taking place along some of the narrow upwelling bands constitutes the major part of the upwelling in the world's oceans. One of such sites is located along the core of the Antarctic Circumpolar Current (ACC), where the strong westerly drives the strongest large-scale upwelling system in the world's oceans. In addition, strong upwelling along the equatorial band and coastal upwelling along the edges of the individual basins may constitute other major parts of the upwelling branch of the general circulation in the world's oceans. These relatively confined bands of strong upwelling constitute the major portion of water mass erosion in the world's oceans. Although water mass formation has been studied extensively, the opposite process, water mass erosion, has not received enough attention thus far. It is obvious that a complete theory of water mass balance in the world's oceans requires a more comprehensive study of both processes.

The discovery of cold deep water

Since surface water at low latitudes is rather warm, common sense might lead people to believe that water in the deep ocean is also warm. Thus, the discovery of cold water in the deep ocean at low latitudes was a great surprise. In 1751, Henry Ellis, captain of the British slave trading ship, the *Earl of Halifax*, described the low temperature observed at latitude $25°13'$ N, longitude $25°12'$ W from a depth of 5,346 ft (1,630 m):

> Upon the passage, I made several trials with the bucket sea-gage, in the latitude $25'13''$ north; longitude $25'12''$ west. I charged it and let it down to different depths, from 360 feet to 5346 feet; when I discovered, by a small thermometer of Fahrenheit, made by Mr. Bird, which went down in it, that the cold increased regularly, in proportion to the depths, till it descended to 3900 feet: from whence the mercury in the thermometer came up at 53 degrees; and tho' I afterwards sunk it to the depth of 5346 feet, that is a mile and 66 feet, it came up no lower. The warmth of the water on the surface, and that of the air, was at that time by the thermometer 84 degrees. I doubt not but that the water was a degree or two colder, when it enter'd the bucket, at the greatest depth, but in coming up had acquired some warmth (Warren, 1981).

Bottom water properties in the world's oceans

Deep water at low latitudes is much colder than the lowest temperature at the sea surface in winter-time; thus, such cold water cannot be formed locally, and the source of such water mass must be traced back to higher latitudes where cold winter conditions make the formation of water mass with such low temperature possible. The reasoning that cold water is formed at high latitudes and transported to low latitudes eventually led to the theories of thermohaline circulation in the world's oceans.

Subsequent observations through scientific expeditions established the fact that the bottom of the world's oceans is covered by cold water originating from a very few narrow sites at high latitudes, where severe winter conditions produce the coldest water in the

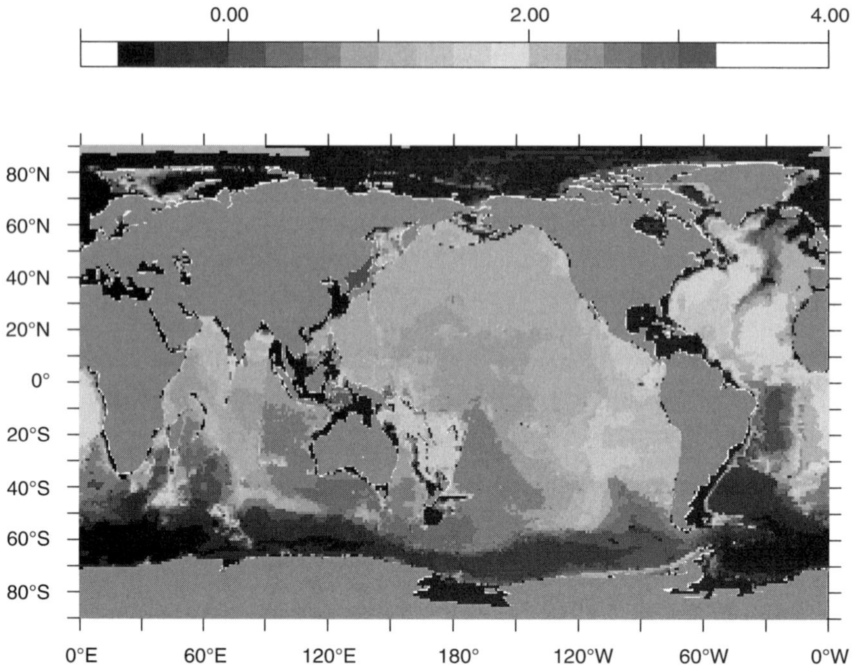

Fig. 5.1 Potential temperature at the bottom of the world's oceans based on Levitus *et al.*'s (1998) Climatology. Note that the sea floor is rather shallow along the mid-ocean ridge in the Atlantic Basin, so bottom water over the ridge is relatively warm. See color plate section.

oceans. Since oxygen solubility is high at low temperatures, high oxygen concentration in the abyssal oceans indicates the recently formed bottom/deepwater masses. Over the past century, extensive observation data over the world's oceans have been accumulated. Potential temperature distribution on the sea floor of the world's oceans, based on climatological data, is shown in Figure 5.1. As discussed in Chapter 2, potential temperature is a better tracer to use in the description of the oceanic environments because the effect of compression varying with depth is eliminated. From Figure 5.1 the following characteristics of bottom water temperature can readily be seen:

- Cold water on the bottom is formed around Antarctica, primarily in the Weddell Sea and Ross Sea. From these source regions, bottom water is carried northward and eastward by currents and eddies. The cold water mass formed around the edge of the Antarctic continent that sinks to the bottom of the world's oceans is called Antarctic Bottom Water (AABW).
- Cold bottom water spreads northward in each basin, and there is a tendency for the cold water to pile up in the western side of the basin. Bottom water temperature in the South Atlantic Ocean is the coldest among all basins.

In the South Atlantic Ocean, only the Brazil Basin receives AABW directly. The eastern basin, the Angola Basin, does not receive AABW from the south; it is a basin closed to cold bottom water from the south. In fact, AABW's effluence to this basin is through a narrow gap near the equator, where relatively cold water moves eastward and finally reaches the Angola Basin from the northern opening passage. The dynamical role of blocking and guiding by bottom topography will be discussed in later sections.

- At the northern end of the North Atlantic Basin there is a source of relatively cold water, which originates from the Norwegian and Greenland Seas. This water mass is called the North Atlantic Deep Water (NADW).

Over the past century, theories of thermohaline circulation have been developed in order to explain the general circulation related to the formation and spreading of bottom water and the connection with water in the upper ocean, where surface thermohaline forcing prevails. Our goal in this chapter is to explain the physical phenomena and the dynamical theories of the thermohaline circulation in the world's oceans.

Besides temperature the other major indicator of bottom water is the oxygen concentration. As an example, oxygen concentration along 165° W is shown in Figure 5.2. It is clear

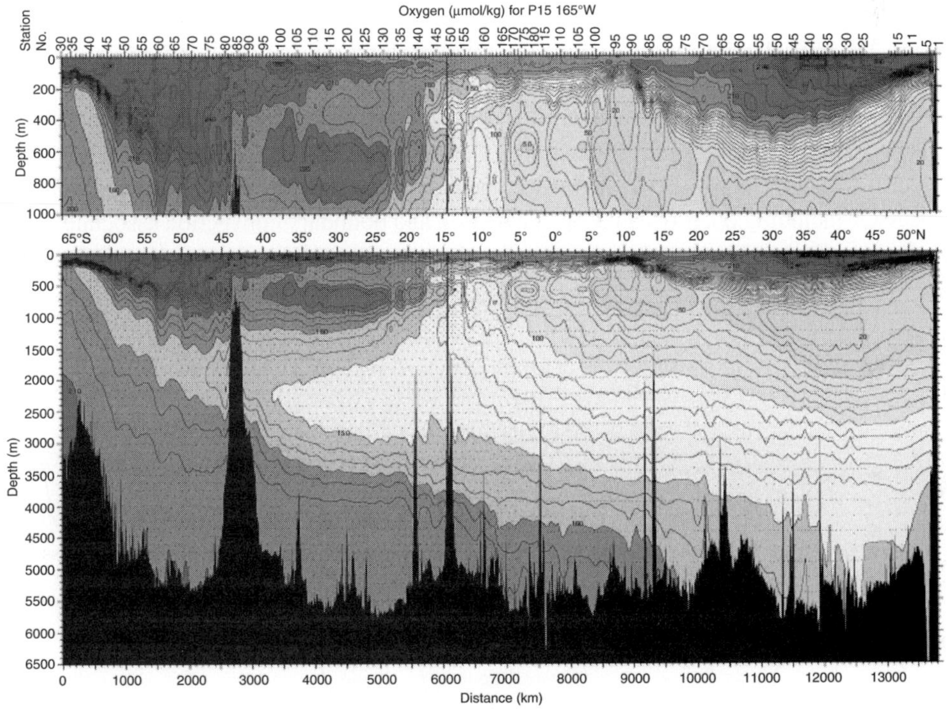

Fig. 5.2 Oxygen concentration P15 section (approximately along 165° W) from WOCE. The contour increment is 10 μmol/kg, and the concentration level between the yellow and light purple color is 150 μmol/kg. The concentration in the bottom layer in the Southern Hemisphere is above 190–200μmol/kg (Talley, 2007). See color plate section.

that, in the Pacific Basin, water of Antarctic origin with a high oxygen concentration fills up the lower part of the water column in the Southern Hemisphere. In contrast, water with very low oxygen concentration occupies the depths near 1 km in the high-latitude portion of the North Pacific Ocean, which indicates the poor ventilation of the water mass at these locations.

In addition, deep water with low oxygen concentration in the North Pacific Basin implies that there is no deepwater source in this basin. The lack of a deepwater source in the North Pacific Ocean is in great contrast to the abundant deepwater formation in the North Atlantic Ocean, and this contrast between the North Atlantic and North Pacific Oceans is one of the major features of the global thermohaline circulation under modern climate conditions.

Sources of deep/bottom water in the Atlantic Ocean

Deep water and bottom water in the Atlantic Ocean originate from marginal seas. There are primarily two sources: (1) along the edge of the Antarctic Continent, especially the Weddell Sea, and (2) Norwegian and Greenland Seas.

Water properties in these marginal seas and their modification during the outflow process are vitally important elements of the global thermohaline circulation. The circulation in the Atlantic Basin is a typical example, a two-dimensional sketch of which is shown in Figure 5.3. Note that the circulation is a complicated three-dimensional phenomenon; thus, the flow directions indicated in this diagram should not be interpreted as the actual flow path in the oceans. Some of the dynamical details related to this diagram will be discussed in later sections.

North Atlantic Deep Water is formed in the northern North Atlantic Ocean through two processes, including open-ocean deep convection and boundary convection associated with the horizontal gyre. Deep water formed in the Norwegian and Greenland Seas overflows the

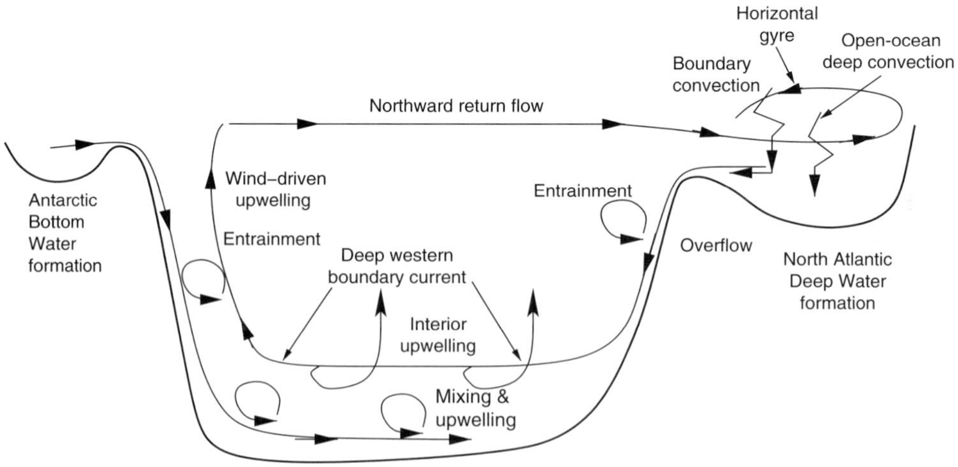

Fig. 5.3 Sketch of bottom/deep water formation and thermohaline circulation in the Atlantic Ocean.

Denmark Strait and enters the open North Atlantic Ocean. During the process of overflow, substantial entrainments take place, enhancing the total volumetric flux of the deep water. In the open North Atlantic Ocean, deep water appears as the deep western boundary current moving southward along the eastern coast of the American continent, gradually sending its water mass to the oceanic interior.

Although NADW may lose its mass through upwelling in the interior of the Atlantic Basin, one of the major pathways of NADW is through the wind-driven upwelling associated with the southern westerly and the subsequent northern return flow in the form of the Ekman flux. In addition, in the Southern Ocean, deep water from the northern source meets the bottom water from the southern source (AABW). Mixing between these two major sources of water masses in the deep ocean dictates the circulation in the abyss of the world's oceans.

Regimes of deep/bottom water formation in the world's oceans

Deep/bottom water is formed at many sites in the world's oceans. Although the details of deep/bottom water formation vary greatly from site to site, these sites can be classified into the two categories described below (Killworth, 1983a). An updated distribution of deep/bottom water sources in the world's oceans is shown in Figure 5.4.

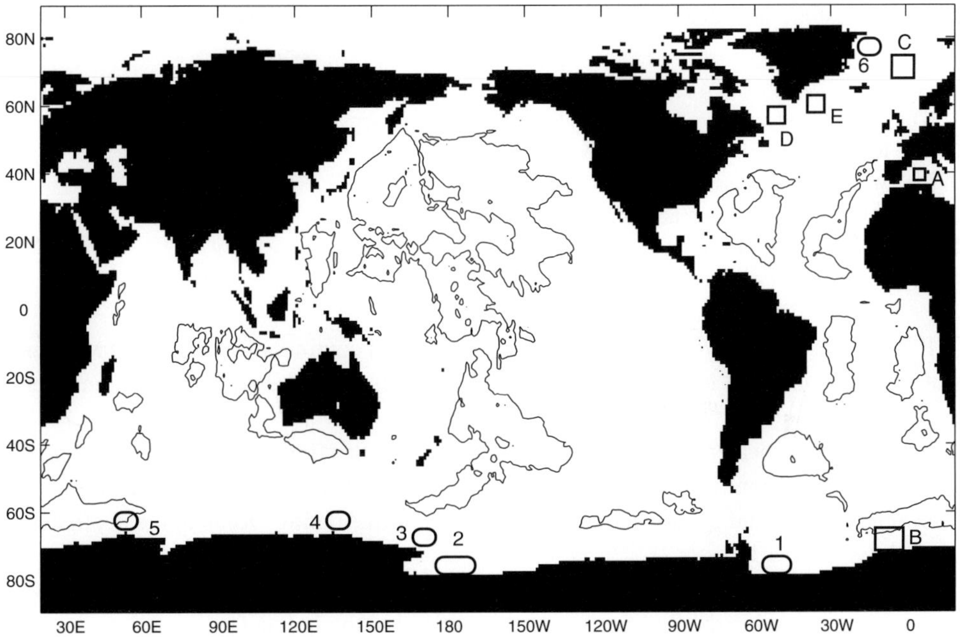

Fig. 5.4 Deep water formation sites in the world's oceans, with the 5 km depth contours; *squares* indicate water mass formation in the open ocean, and *circles* indicate formation in marginal seas.

Formation in the open ocean

Deep water can be formed through the so-called chimney formation. Water formed in the open ocean directly sinks to the bottom, and spreads to other parts of the oceans, at sites shown by the lettered squares in Figure 5.4. The following list was compiled with the help of B. Warren.

A. Gulf of Lions in the Mediterranean Sea (MEDOC Group, 1970);
B. The Weddell chimney (Gordon, 1978); the Weddell polynya (Martinson *et al.*, 1981; Gordon, 1982);
C. The Norwegian/Greenland Seas;
D. The Labrador Sea (Lazier, 1973; Clarke and Gascard, 1983; Pickart *et al.*, 2002);
E. The Irminger Sea (Pickart *et al.*, 2003).

In addition, there is deepwater formation in the Bransfield Strait; however, the deep water formed in this location may not be exported to other part of the world's oceans (Gordon and Nowlin, 1978).

Formation along the margins of the sea

Strong cooling along the edge of a continent creates favorable conditions for dense water mass formation in the marginal sea. Water formed in this way moves down along the continental slope and eventually reaches the bottom of the oceans (numbered circles in Fig. 5.4), including:

1. The western and southwestern Weddell Sea (Foster and Carmack, 1976);
2. The Ross Sea (Jacobs *et al.*, 1970; Warren, 1981);
3. The Wilkes Land (Carmack and Killworth, 1978);
4. The Adelie coast (Gordon and Tchernia, 1972);
5. The Enderly Land (Jacobs and Georgi, 1977);
6. The eastern coast of Greenland.

It is notable that deep/bottom water formation is not a continuous process. In fact, dense water formation tends to happen episodically, depending on the anomalous atmospheric conditions.

Although dense water is formed in the Mediterranean Sea during winter-time, under current climate conditions it cannot sink to the deep ocean. Instead, through strong entrainment, it becomes lighter and eventually spreads into the North Atlantic Ocean at the depth range of slightly below 1 km.

There is also dense water formation in the Red Sea due to excessive evaporation. However, dense water formed in the Red Sea is mostly confined to the Indian Ocean, so it will not be discussed here.

5.1.2 Bottom/deepwater formation

Antarctic Bottom Water (AABW) formation

The bottom layer of the world's oceans is filled with a thick layer of very cold water with potential temperature lower than 2°C, as shown in Figure 5.5.

The origin of this water mass is clearly from the edge of Antarctic (thus, it is called AABW), whose origin can be traced back to a few sites along the continental margin of Antarctica where cold water is formed during the Southern Hemisphere winter. The cold temperature of this water mass is directly linked to the strong cooling in winter-time when very cold wind from glacial ice on Antarctica blows over the coastal ocean adjacent to the ice edge, driving sea ice away from the coasts and thus creating coastal polynyas (small open water areas surrounded by sea ice). Owing to the increase in concentration of oxygen at low temperature, the newly formed bottom water is normally associated with very high oxygen concentration, about 200 μmol/kg (Fig. 5.2).

Strong cooling over polynyas produces more sea ice, and salt rejection during sea ice formation creates cold dense water with higher salinity. This dense water overflows the continental slope. The offshore transport of the newly formed bottom water is compensated by the onshore flow of water in the subsurface layer. However, flow pathways in the ocean are

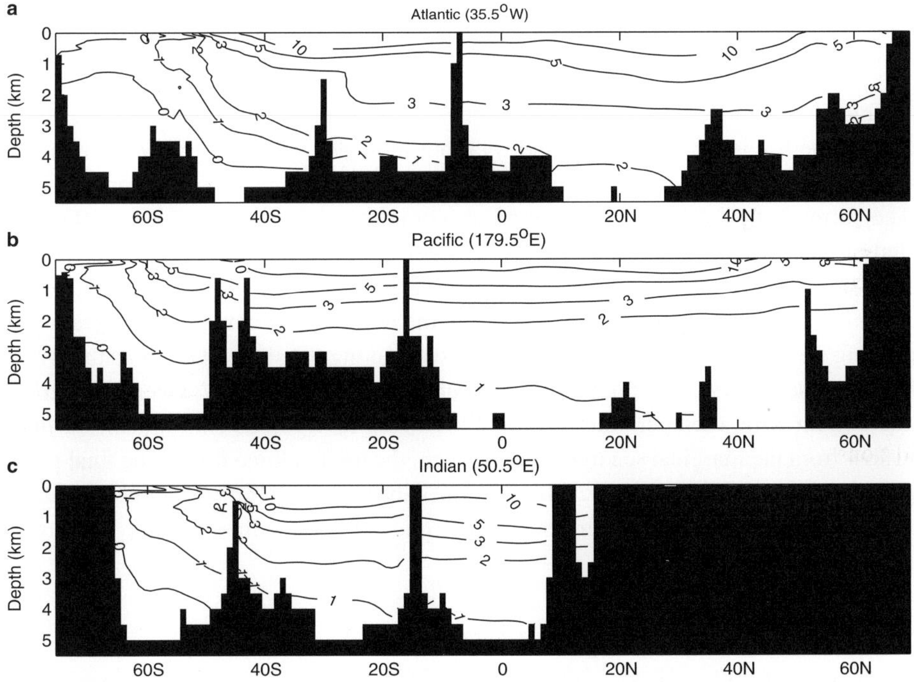

Fig. 5.5 Meridional distribution of potential temperature in **a** the Atlantic Ocean; **b** the Pacific Ocean; **c** the Indian Ocean.

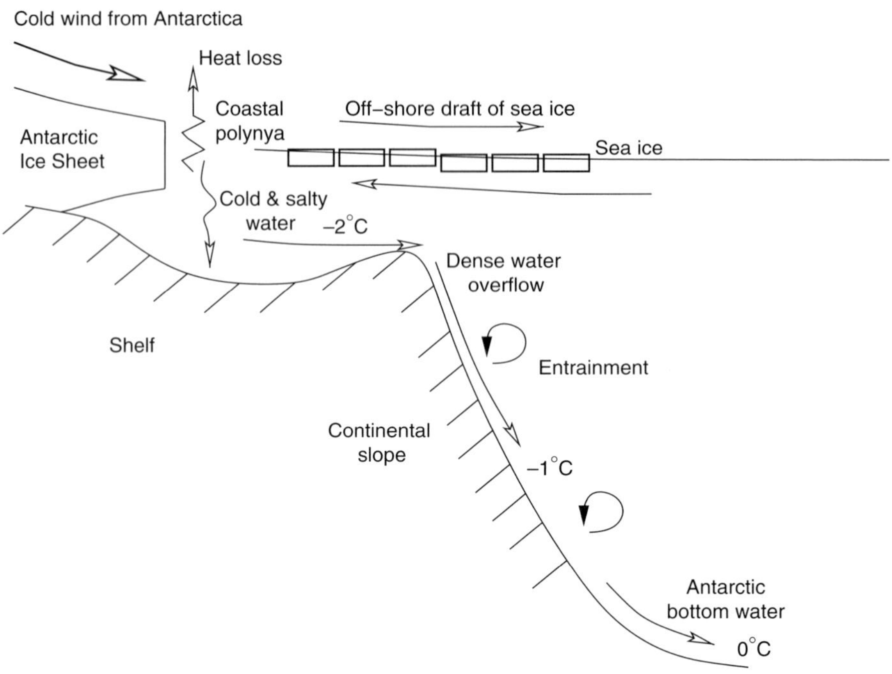

Fig. 5.6 Formation of AABW (redrawn from Gordon, 2002).

much more complicated due to many dynamical factors, including wind stress and surface thermohaline forcing, stratification and rotation; thus, the arrows in the two-dimensional sketch (Fig. 5.3) should not be taken as being the real trajectories of water parcels.

The formation of AABW involves many complicated physical processes (Fig. 5.6), including the formation of dense and salty water within the coastal polynyas, the transport of this water by the gyre circulation within the coastal area, the overflow from the marginal sea to the open ocean, and the entrainment during the descent of the gravity current along the continental slope. During the descent along the slope, it entrains the water in the environment; thus, it is slightly warmed up from $-2°C$ to $-1°C$. Eventually, it sinks to the bottom with a temperature of nearly $0°C$. Due to vigorous entrainment during the overflow from the marginal sea to the open ocean, the total volume flux of the final product is greatly increased (Gordon, 2002). In addition, cabbeling may further increase the density of the newly formed bottom water; thus, it may play a vital role in setting the properties of the final product.

Deep convection

Another form of bottom/deep water formation in the oceans is the deep convection taking place in the open ocean (Fig. 5.7). The major sites of deep convection include the northwestern Mediterranean, the Labrador Sea, and the Greenland Sea.

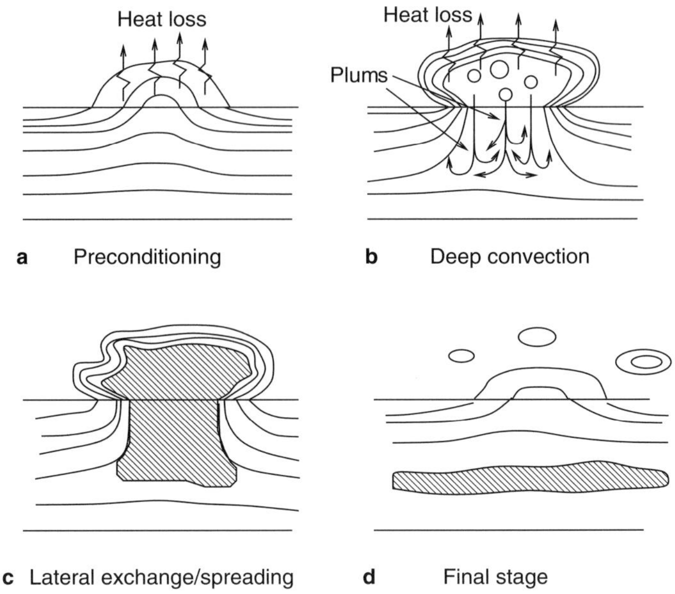

Fig. 5.7 Sketch of open ocean deep convection: **a** preconditioning, **b** deep convection, **c** lateral exchange/spreading, **d** final stage (redrawn from Marshall and Schott, 1999).

Basic processes

Deep convection in the open ocean involves dynamical processes of rather broad spectra in both spatial and temporal scales. As a concise description, it can be roughly classified into the following major processes (Marshall and Schott, 1999).

- **Preconditioning:** Strong cyclonic wind stress curl in early winter enhances the Ekman upwelling in the center of the cyclonic gyre, leading to a dome-shaped isopycnal structure. Within the center of the cyclonic gyre, stratification is very weak, and this weak stratification can facilitate deep convection (Fig. 5.7a).
- **Deep convection:** Strong buoyancy loss due to cooling and evaporation further reduces the stratification in the upper ocean within the central regime of the cyclonic gyre. Further cooling eventually sets in the deep convection, which consists of clusters of small-scale downward plumes (with a horizontal scale of 1 km or less) and eddies (with a horizontal scale of 10 km). Water in the small plumes moves downward with vertical velocity on the order of 0.1 m/s (Fig. 5.7b). Plumes and eddies form the mixed patch (this is also called a "chimney" in some early articles) with a horizontal scale on the order of 100 km.
- **Lateral exchange and spreading:** A few days after the onset of cooling, the dominating mode of heat exchange is shifted from vertical to horizontal through eddy activities on the geostrophic scale (Fig. 5.7c).
- **Final stage:** The chimney-like density structure associated with deep convection is gradually closed up and leaves behind a dome-shaped isopycnal structure, with a layer of cold water which settles at depth (Fig. 5.7d).

Two basic parameters play crucial roles in the establishment of deep convection. First, the buoyancy (Brunt–Väisälä frequency, $N^2 = -\frac{g\partial\rho_\Theta}{\rho_0\partial z}$), which is also a measure of the frequency of internal gravity waves. The ocean is normally stably stratified, i.e., $N^2 > 0$; however, owing to strong buoyancy forcing, there are small areas in the upper ocean where the stratification may become temporarily unstable, i.e., $N^2 < 0$, and convection ensues.

The second parameter is the Rossby deformation radius, defined as $R_d = NH/f_0$, where H is the thickness of the convective layer. Since the buoyancy frequency can be rewritten as $N = \sqrt{g'/H}$, the speed of gravity waves is $c_0 = \sqrt{g'H}$; thus, $R_d = c_0/f_0$ is a measure of how far gravity waves can travel over an inertial period. For the mid-latitude ocean, the typical scale of the Rossby deformation radius is on the order of 30 km. However, at high latitudes, weak stratification gives rise to a much smaller deformation radius, on the order of 10 km. Owing to strong cooling in winter-time it can be further reduced to a few kilometers. For horizontal scales on the order of the deformation radius or larger, geostrophic and hydrostatic balance dominate; however, for horizontal scales much smaller than the deformation radius, geostrophic and hydrostatic balances break down (Marshall *et al.*, 1997).

Scales of the plumes

Dimensional analysis can be used to predict the basic scales involved in deep convection (Marshall and Schott, 1999). Assume the surface buoyancy flux is B_0 and there is a layer of homogenized fluid with a depth of h. During the initial stage of the onset of convection, $t \ll 1/f$, rotation is unimportant; thus, both B_0 and t work as the only parameters controlling the formation of plumes. With these two parameters, dimensional analysis gives rise to the following scales for the horizontal length, velocity, and buoyancy of the plumes

$$l \sim \left(\mathsf{B}_0 t^3\right)^{1/2} \tag{5.1a}$$

$$u \sim w \sim (\mathsf{B}_0 t)^{1/2} \tag{5.1b}$$

$$b \sim (\mathsf{B}_0/t)^{1/2} \tag{5.1c}$$

For time scales that are long enough, the plumes evolve and reach to the bottom of the mixed layer. As the time scale approaches $1/f$, the role of rotation becomes dominating, and the corresponding scales are

$$l_{rot} \sim \left(\mathsf{B}_0/f^3\right)^{1/2} \tag{5.2a}$$

$$u_{rot} \sim w_{rot} \sim (\mathsf{B}_0/f)^{1/2} \tag{5.2b}$$

$$b_{rot} \sim (\mathsf{B}_0 f)^{1/2} \tag{5.2c}$$

Assume the flux of heat loss is 500 W/m^2, the corresponding buoyancy flux is $\mathsf{B}_0 = 2.25 \times 10^{-7}\text{m}^2/\text{s}^3$, and the scales of plumes are: $l_{rot} \sim 0.47$ km, $u_{rot} = w_{rot} \sim 0.05$ m/s.

North Atlantic Deep Water formation

North Atlantic Deep Water primarily consists of two parts: the overflow from the Norwegian Sea, and deep water formed in the Labrador Sea.

Deep water overflow from the Norwegian Sea may come from two sources (Mauritzen, 1996). The classical theory has been the following: winter deep convection in the Iceland and Greenland Seas produces cold and dense water that sinks to the deep part of the Norwegian Basin. As deep water accumulates and fills up to a level higher than the sills connecting the Norwegian Basin with the open northern North Atlantic Ocean, deep water overflows the sills and becomes the source of NADW.

However, such a scenario of deepwater formation in the North Atlantic Ocean has some potential problems. First, the existing estimates of deepwater formation rate are much smaller than the estimates of dense water overflows through the Greenland–Scotland Ridge. Second, this scenario implies that the overflow rate may have noticeable seasonal and interannual cycles. For example, observations indicate that the production of deep water in the Greenland Sea was greatly reduced in the 1980s (Schlosser *et al.*, 1991). However, there are no clear indications that the overflow rate changes much over such time scales.

Another source of deepwater overflow from the Norwegian Sea originates from the cooling-induced convection in the boundary currents flowing around the edge of the Norwegian Sea. In fact, Atlantic Water in the northward flowing Norwegian Atlantic Current becomes gradually denser due to heat loss, filling up the shallow and intermediate depths along the rim of the basin. This water mass flows over the sills and becomes the source of the NADW.

On the other hand, although cooling in the Norwegian Sea can produce Deep Norwegian Sea Water, this water mass is too cold and is denser than the overflow water. Tritium concentration analysis indicates that overflow water should come from a depth shallower than 1,000 m. As a matter of fact, three sills connecting the Norwegian Sea to the North Atlantic Ocean are relatively shallow: the Faroe–Shetland Channel (850 m), the Denmark Strait (600 m), and the Iceland–Faroe Ridge (500 m). Therefore, deep water overflowing these sills should be primarily from the relatively shallow sources along the rim of the basin. Thus, cooling-induced convection along the rim current in the Norwegian–Greenland Sea may be the major source of NADW.

Similarly, deep convection in the Labrador Sea contributes very little to the overall meridional overturning in the North Atlantic Ocean, and the most important pathway of water mass formation in the Labrador Sea is the gradual transition of water properties within the boundary current around the basin (Pickart and Spall, 2007).

5.1.3 Overflow of deep water

Topographic control of deep flow

The world's oceans are characterized by many basins separated by major ridge systems. Due to the existence of these ridges, bottom water movement and distribution of water properties are strongly confined by complicated dynamical laws. Basically, deepwater flows from one basin to the others must go over sills which exist as the lowest passages in these otherwise tall topographic barriers.

When water moves over a sill, it behaves very much like a deep waterfall. In many cases, a deep waterfall involves a volume flux on the order of 1–10 Sv (10^6–10^7 m^3/s), and an elevation change on the order of several hundreds of meters. Many people have visited the Niagara Falls, one of the largest land-based waterfalls in the world, which has a maximal volume flux of 3,000 m^3/s and elevation drop of 56 m. In comparison, deep waterfalls are much more powerful than any land-based waterfall on Earth, with a volume flux of more than 1,000 times and elevation drop of more than 10 times that of the Niagara Falls.

The meaning of rotating hydraulics

An essential feature common to the overflow associated with deepwater formation is that the overflow goes through the transition from subcritical to critical, and then to supercritical. Whether a flow is subcritical or not is defined in terms of the Froude number. Using the waterfall as an example, the concept of hydraulics for a non-rotating fluid can be explained as follows. For water flowing in an open channel, signals propagate with the velocity of the surface wave, i.e., $c = \sqrt{gh}$, where h is the depth of water. The Froude number is defined as $F = U/c$, where U is the horizontal velocity of the fluid. For most cases of flow through a channel, $F < 1$, so the fluid motion is subcritical, i.e., fluid travels at a speed slower than the speed of signals.

If the mean slope of the channel is gradually increased, fluid speed increases. At a critical value of slope, water travels so fast that its speed exactly matches the speed of surface waves. Generally, the bottom of the channel is not flat, and there is a place in the channel where the depth is the shallowest. This is called a sill, and depth both upstream and downstream from the sill is greater, as shown in Figure 5.8.

Assume that water motion upstream from the sill is subcritical and the depth of the sill is gradually reduced. The Froude number at the sill gradually increases, while flow in the

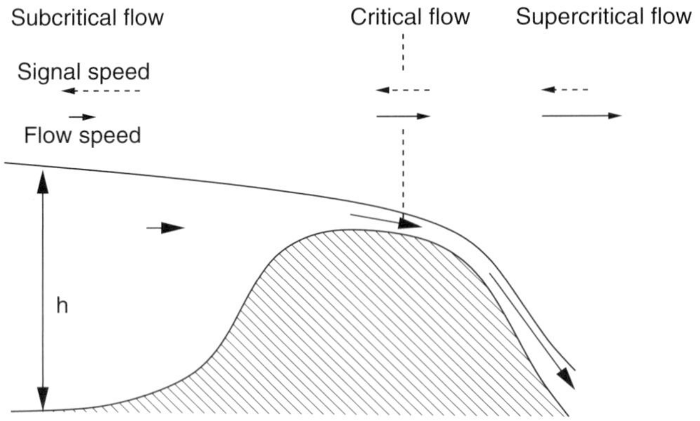

Fig. 5.8 Sketch of overflow from a marginal sea to the open ocean, involving rotating hydraulics.

whole channel remains subcritical. When the sill depth reaches a critical value, fluid motion at the sill becomes exactly critical, i.e., $U = c$ at this section. Although fluid motion upstream remains subcritical, fluid motion downstream becomes supercritical, i.e., fluid velocity is larger than the signal velocity. One of the major differences in supercritical and subcritical flows is that, in a supercritical flow, field signals cannot propagate upstream because the speed of the signals is slower than the flow speed.

If we stand by a waterfall, we realize that no matter how hard we disturb water in the fall, nothing happens upstream – signals cannot go upstream. What happens in the oceans is more complicated because we have to deal with stratification and rotation; thus, the study of overflow is called rotating hydraulics. As a first step in this direction, we can treat the oceanic flow with continuous stratification as a system with two density layers. In addition, we can assume that the upper layer moves so slowly that it can be assumed to be stagnant. Under such assumptions, the problem is reduced to the framework of the inverse reduced-gravity model discussed previously. The equivalent signal speed is $c = \sqrt{g'h}$, where $g' = g\Delta\rho/\rho$ is the reduced gravity and h is the thickness of the moving layer. Thus, the corresponding Froude number is defined as $F = u/\sqrt{g'h}$.

Similar to the case of the hydraulic problem in an open channel, the supercritical flow downstream from the sill is not very stable, and hydraulic jump-like phenomena and mixing with the environmental fluid ensue.

Another crucial phenomenon associated with deepwater overflow is that the newly formed cold and dense overflow water is piled up on the right-side bank of the channel (if we look in the downstream direction) because of the Coriolis force (Fig. 5.9). Of course, if the channel

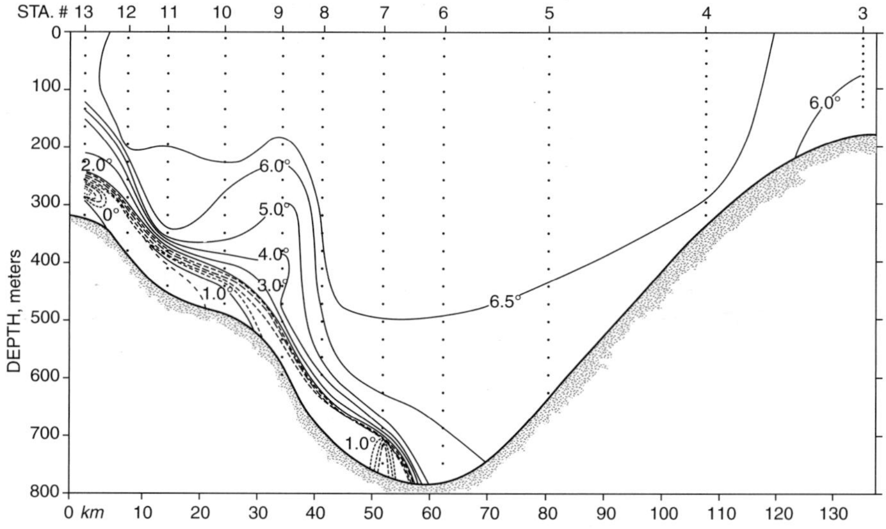

Fig. 5.9 Temperature section across the Denmark Strait in latitudes 65–66° N, illustrating the southward flow of cold water from the Norwegian Sea (Worthington, 1969).

is narrow enough, the overflow will basically fill up the whole channel. The word "narrow" is defined in terms of the first baroclinic radius of deformation. Due to the relatively weak stratification and shallow water depth, the first baroclinic radius of deformation associated with overflows is quite small, on the order of a few kilometers. The Denmark Strait is very wide in this sense, so the overflow must appear in the form of a current confined to the right-side bank of the channel.

The book by Pratt and Whitehead (2007) is highly recommended for the reader who is excited about deep waterfalls in the oceans. For the general reader who needs an introduction to hydraulics, the papers by Gill (1977) and Pratt and Helfrich (2003) are a good place to start.

Deep waterfalls in the world's oceans

There are numerous deep waterfalls in the world's oceans. Typical deep waterfalls include the water exchange through the Strait of Gibraltar, the overflow through the Denmark Strait, and many others. Deep waterfalls are regulated by rotating hydraulics. Flows through these falls play a critically important role in regulating water mass transportation and transformation in the oceans. As an example, the positions of these deep waterfalls in the Atlantic Basin are shown in Figure 5.10.

Spreading of AABW in the Brazil Basin

The spreading of AABW in the Brazil Basin is a good example for illustrating the movement/transition of bottom water in the world's oceans. Here the coldest water with potential temperature as low as $-0.4°C$ can be identified by the clusters of pink color in the middle of the basin in Figure 5.11. It is obvious that such cold water must come from the south, because there is otherwise no local source of such cold water in the basin. Note that the Vema Channel through which AABW enters the Brazil Basin from the southern edge is so narrow that it cannot be clearly shown in Figure 5.11.

From Figure 5.11 it can readily be seen that no AABW can escape through lateral boundaries of the basin; thus, all AABW entering through Vema and Hunter Channels in the south has to be removed from above through diapycnal mixing. The water mass budget in this basin requires a diapycnal diffusivity of $1\text{–}5(\times 10^{-4}\ \mathrm{m^2/s})$ (Morris *et al.*, 2001).

The flow of bottom water through the channel can be seen clearly through a meridional section along the coast, located around 30° W over the northern half of the basin (Fig. 5.12). Four neutral density surfaces, $\gamma = 28.27, 28.205, 28.16$, and $28.133\ \mathrm{kg/m^3}$, are also included as colored lines. The overflow can be readily seen on the bottom of the seafloor in Figure 5.12.

Thermodynamics of overflows

In many cases, dense water is formed in marginal seas and flows to the open ocean as outflow. Major outflows have several essential components in common (Price and Baringer, 1994):

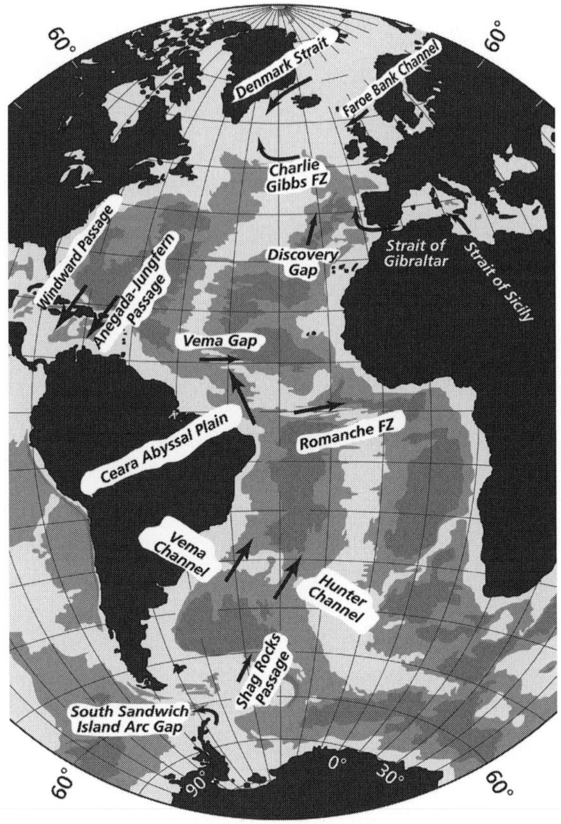

Fig. 5.10 Positions of deep waterfalls in the Atlantic Basin (courtesy of J. Whitehead).

- *Air–sea exchange* that produces dense water due to a combination of heat and freshwater fluxes from the ocean to atmosphere. In addition, saline rejection due to sea-ice formation can also contribute to the formation of dense water.
- *Marginal sea and open ocean exchange* that allows dense water formed in semi-closed marginal seas to flow into the open ocean.
- *Descent and entrainment* that modify the properties of the outflow water. In general, the volume of the overflow water increases more than 100%.

For the world's oceans there are four major sources of overflow: (1) Mediterranean Sea, (2) Denmark Strait, (3) Faroe Bank Channel, and (4) Filchner Ice Shelf (on the outer edge of sea ice near Antarctica).

The properties of these outflows are listed in Table 5.1. Note that although at the site of their original formation, the density of the Mediterranean outflow is the greatest of these four sources, at depths greater than 2 km, water from the Mediterranean outflow is the lightest. On the other hand, although at sea level the water mass formed off the Filchner Ice

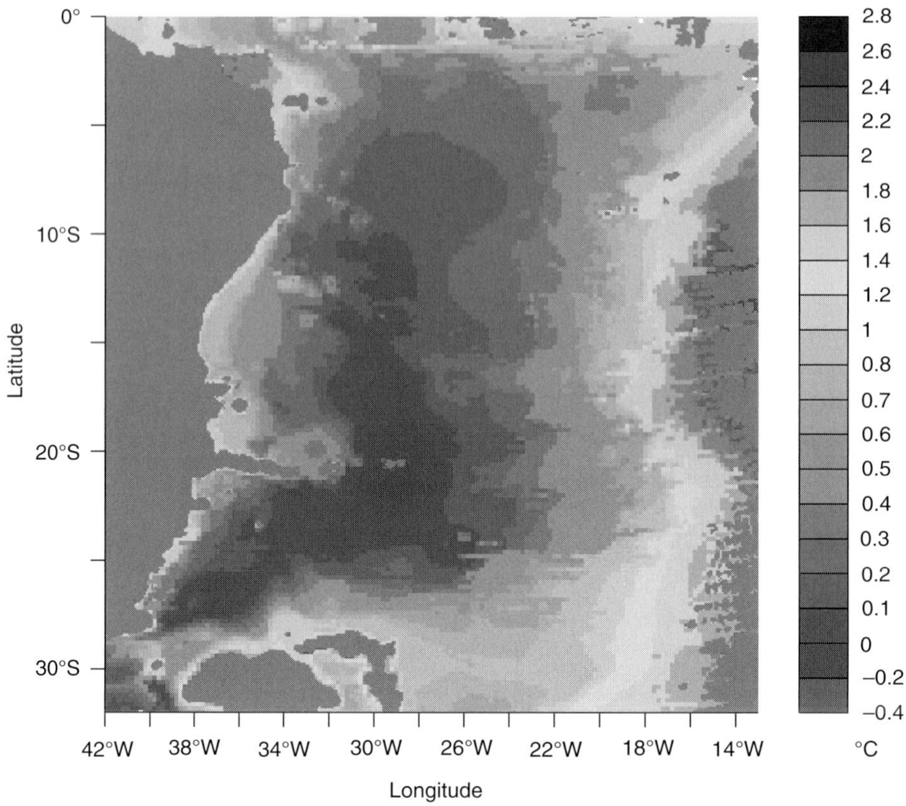

Fig. 5.11 Potential temperature on the sea floor in the Brazil Basin (Morris *et al.*, 2001). See color plate section.

Shelf is the lightest among these four, at great depth it is the densest (Fig. 5.13). As a result, the Mediterranean outflow cannot sink to the sea floor of the world's oceans; instead, the cold water mass formed off the Filchner Ice Shelf is the water mass that covers the bottom layer in the world's oceans.

There are several reasons why the density of outflow is reordered: the thermobaric effect and mixing, which are controlled by the density difference between the outflow and the environment, plus the topographic slope. As discussed in Section 2.4.9, the thermobaric effect is due to the nonlinearity of the equation of state: the compressibility of seawater strongly depends on temperature. In particular, cold seawater is more compressible than warm seawater. The effect of mixing during the descent of the overflow downstream of the sill will be discussed below.

The thermobaric effect can be seen clearly from Table 5.1 and Figure 5.13. Although the Mediterranean outflow is the heaviest near the sea surface, it is much warmer than other deep water from other sources. Since warm water is less compressible, when the *in situ*

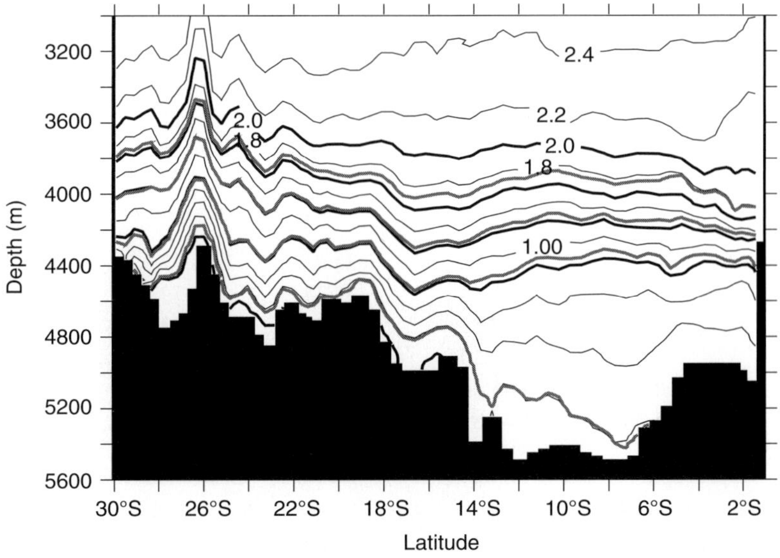

Fig. 5.12 A vertical section of potential temperature (*black*) and neutral density (*color*) from the western Brazil Basin, indicating the downward flow of water (over the seafloor from left to right) through the channel (Morris *et al.*, 2001). See color plate section.

pressure increases, its density increases much less than that of cold water. As a result, the density of these four deep waters is reorganized gradually with the increase in depth. In fact, below the depth of 1,000–1,300 m, the Mediterranean outflow is no longer the heaviest water among this group of four, even without mixing (Fig. 5.13).

In fact, the Mediterranean outflow spreads in the North Atlantic Ocean at a depth range of roughly 1–1.5 km in the upper ocean. As can be seen from Figure 5.14, the subsurface core of salinity maximum is a clear sign of the Mediterranean outflow.

The influence of the outflow can also be identified from the horizontal distribution of salinity at the depth of 1.2 km (Fig. 5.15). At this level, the core of high salinity extends all the way to the western part of the basin; thus, the outflow seems to play a vital role in the maintenance of the salinity balance in the North Atlantic Ocean.

However, it is important to note that this core does not necessarily indicate the direction of the mean flow. In fact, the existence of high salinity at this level in the western part of the basin may be largely due to salt diffusion induced by meso-scale eddies. Observations indicate that special types of eddy, with high salinity signature from the Mediterranean outflow, drift westward in this latitude band, and they may be the major contributor to the salinity tongue as seen from Figures 5.14 and 5.15.

An interesting question is whether the Mediterranean outflow could become the source water for the bottom circulation in the world's oceans. Assuming that the water mass properties of deepwater sources from the other three sites remain unchanged, an increase of

Table 5.1. *Density variation with depth for four sites of deepwater formation*

		Mediterranean	Denmark Strait	Faroe Bank Channel	Filchner Ice Shelf
T		13.4	0.0	−0.5	−2.1
S		37.80	34.9	34.92	34.67
Density at depth (db)	0	28.48	28.03	28.07	27.92
	1,000	32.85	32.74	32.79	32.69
	2,000	37.12	37.34	37.41	37.37
	3,000	41.30	41.84	41.92	41.93
	4,000	45.38	46.23	46.33	46.38
	4,000*	44.81	46.04	45.95	46.24
$\Delta\sigma_4^*$		0.57	0.19	0.38	0.14

The line marked by * indicates the potential density for the final product after mixing (Price and Baringer, 1994). $\Delta\sigma_4^*$ is the density decline due to mixing.

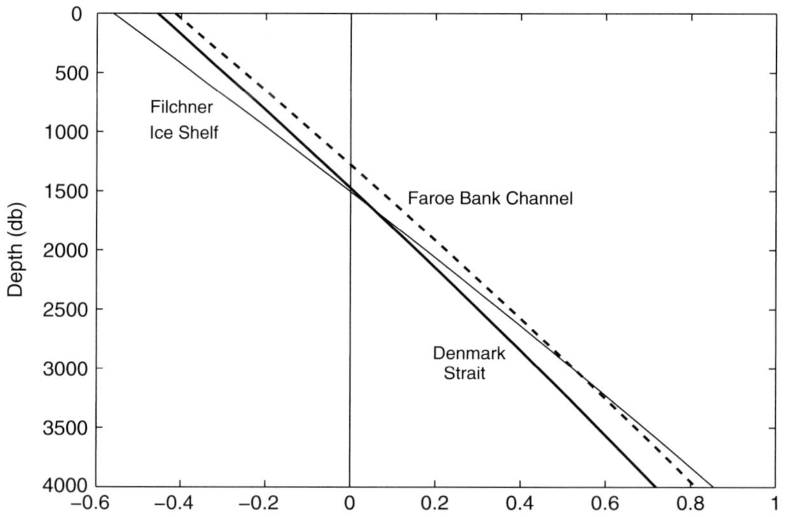

Fig. 5.13 Difference in the *in situ* density between the Mediterranean outflow and the other three sources of deep water, in kg/m^3.

salinity $\Delta S = 1.83$ or a decline in temperature $\Delta T = -4.82°\text{C}$ can produce a water mass that is the densest among all four sites of deepwater formation, so that the Mediterranean outflow would flow to the bottom of the world's oceans (Table 5.2).

This calculation does not include the effect of mixing. Were the effect of mixing taken into the consideration, the change in either the temperature or salinity required to make

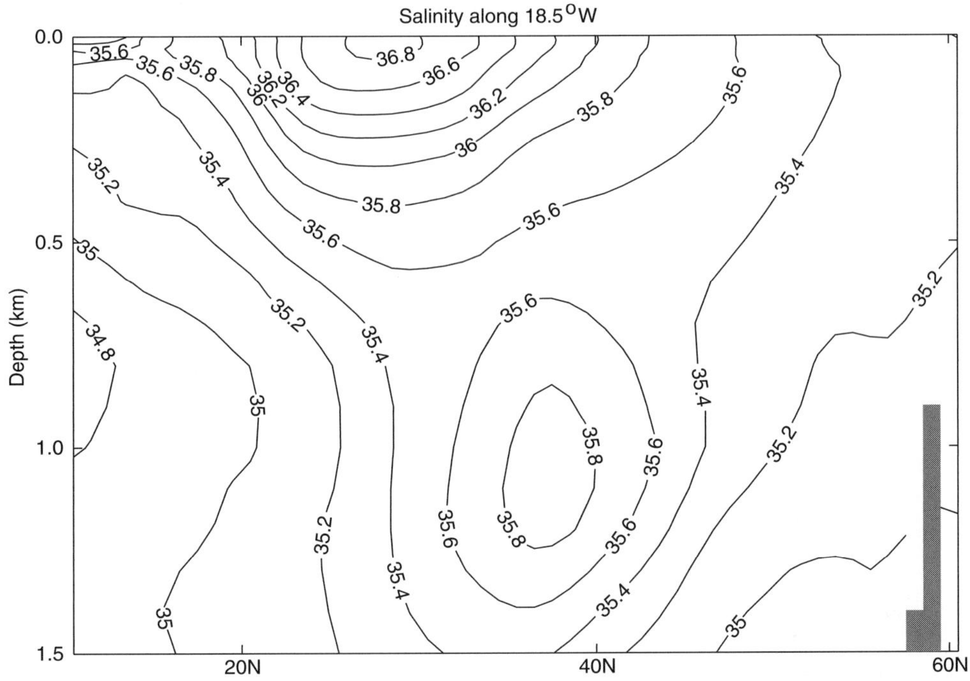

Fig. 5.14 Salinity distribution at 18.5° W in the Atlantic Ocean.

the water dense enough to sink to the bottom would be even larger. During the geological past, the evaporation rate over the Mediterranean Sea might have been much higher than the present day. Whether such a possibility could occur under such circumstances remains unclear.

Dynamics of overflows

Overflow from a marginal sea to the open ocean can be described in two steps (Fig. 5.16) (Price and Yang, 1998). First, the mass flux over the sill, a relatively narrow gap, is controlled by hydraulic processes. Second, the gravity current moves downstream from the sill.

The flow through the control section is regulated by the following constraints on mass fluxes, temperature and salinity:

$$M_{in} - M_{out} = (E - P)\,\rho_f A \tag{5.3}$$

$$T_{in}M_{in} - T_{out}M_{out} = QA/c_p \tag{5.4}$$

$$S_{in}M_{in} - S_{out}M_{out} = 0 \tag{5.5}$$

where M_{in} and M_{out} are the mass flux in and out of the marginal sea, $E - P$ is the rate of evaporation minus precipitation (including river run-off), ρ_f is the density of freshwater associated with evaporation and precipitation, A is the area of the marginal sea, $T_{in}, T_{out}, S_{in},$

Table 5.2. *Density difference between the Mediterranean outflow and other sites of deepwater formation, assuming changes in the deepwater properties of the Mediterranean outflow*

	Depth (m)	Denmark Strait	Faroe Bank Channel	Filchner Ice Shelf
Case A:	0	−1.88	−1.84	−1.98
$\delta T = 0, \delta S = 1.83$	5,000	−0.21	−0.10	−0.0002
Case B:	0	−1.36	−1.32	−1.46
$\delta T = -4.82, \delta S = 0$	5,000	−0.21	−0.10	−0.002

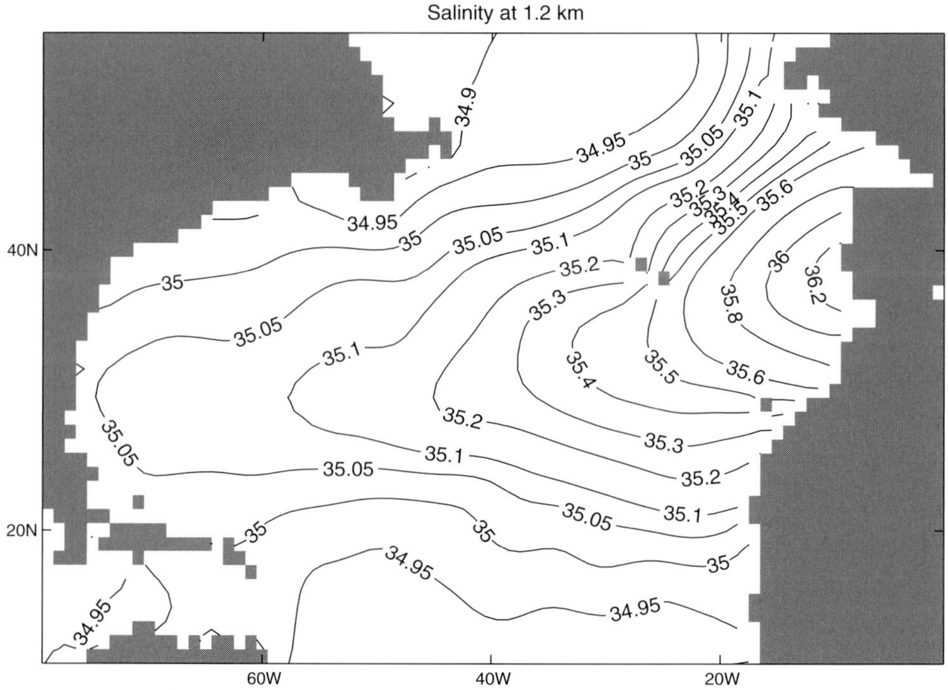

Fig. 5.15 Salinity distribution at 1.2 km in the Atlantic Ocean.

and S_{out} are the temperature and salinity for the inflow and outflow waters, Q is the rate of heat loss to the atmosphere in the marginal sea, and c_p is the specific heat of seawater.

Combining mass conservation Eqn. (5.3) and salt conservation Eqn. (5.4) leads to

$$M_{out} = \frac{S_{in}}{S_{out} - S_{in}} (E - P) \rho_f A \tag{5.6}$$

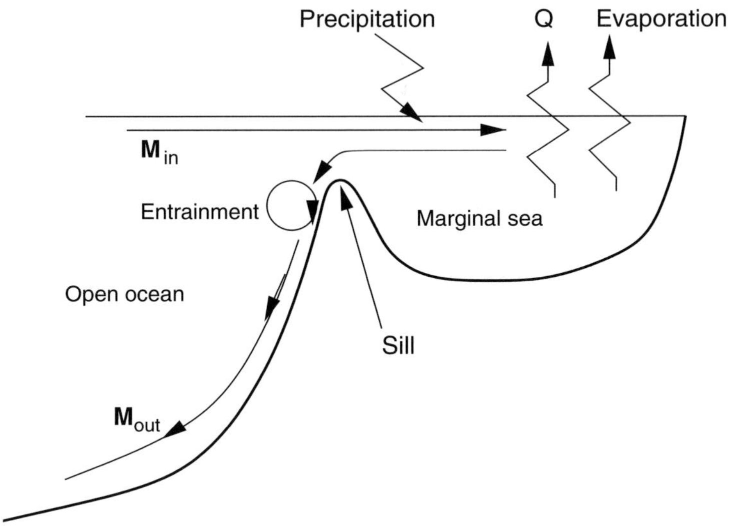

Fig. 5.16 Sketch of the overflow from a marginal sea.

Specifying $E - P, S_{in}$, and S_{out} would give rise to the solution of the overflow; however, water mass properties, such as salinity and temperature, are unknown before we solve the whole problem. As an alternative, the rate of overflow can be determined through the following approaches.

Two types of overflow

Flows through channels are complicated phenomena; however, they can be simulated with a simple two-layer model. The specific choice of the model depends on the relative width of the channel (Price and Yang, 1998).

Overflow through a narrow channel

For a channel with width smaller than the radius of deformation, which is on the order of 20–30 km for mid latitudes, and 10 km or less for high latitudes, the effect of rotation can be ignored. The Strait of Gibraltar is a good example. Thus, for steady and frictionless flows through a strait with width W and depth $H = h_1 + h_2$, the momentum and continuity equation for a two-layer model are

$$U_1 \frac{\partial u_1}{\partial x} + g \frac{\partial \zeta}{\partial x} = 0 \tag{5.7}$$

$$\rho_2 U_2 \frac{\partial u_2}{\partial x} + \rho_1 g \frac{\partial \zeta}{\partial x} + g (\rho_2 - \rho_1) \frac{\partial \zeta_{in}}{\partial x} = 0 \tag{5.8}$$

$$U_1 \frac{\partial}{\partial x}(\zeta - \zeta_{in}) + h_1 \frac{\partial u_1}{\partial x} = 0 \tag{5.9}$$

$$U_2 \frac{\partial \zeta_{in}}{\partial x} + h_2 \frac{\partial u_2}{\partial x} = 0 \tag{5.10}$$

where ζ is the free surface elevation, ζ_{in} is the interface perturbation, and advection terms $u_1\partial/\partial x$, $u_2\partial/\partial x$ are neglected in comparison with the advection with the mean flow terms $U_1\partial/\partial x$, $U_2\partial/\partial x$. This is a set of four equations for four unknown gradient terms. The resonance condition for this system can be obtained by setting the determinant of this system to zero, and this leads to

$$\frac{U_1^2}{g'h_1} + \frac{U_2^2}{g'h_2}\left(1 - \frac{U_1^2}{gh_1}\right) = 1 \tag{5.11}$$

where $g' = g\frac{\rho_2 - \rho_1}{\rho_2} \ll g$ is the reduced gravity. It can readily be seen that $U_1^2/gh_1 \ll U_1^2/g'h_1 \leq 1$; thus, this equation is reduced to

$$F^2 = F_1^2 + F_2^2 = \frac{U_1^2}{g'h_1} + \frac{U_2^2}{g'h_2} = 1 \tag{5.12}$$

where F is the Froude number. For a narrow strait such as the Strait of Gibraltar, there is a so-called over-mixed solution (Bryden and Stommel, 1984), which is obtained when density and salinity difference between the inflow and outflow are minimized, and the inflow and outflow, Eqn. (5.6), are maximized. In such a solution, the depth and rate of inflow and overflow are nearly equal, $h_1 \sim h_2 = H/2$, and the overflow rate is

$$M_{out} = 0.25\sqrt{g'H}HW \tag{5.13}$$

Including more realistic geometry of the channel, with a triangular section and the separation of the sill section and the narrowest section, leads to a formula with a smaller coefficient (Bryden and Kinder, 1991)

$$M_{out} = 0.075\sqrt{g'H}HW \tag{5.14}$$

Overflow through a wide channel

When the channel is wider than the radius of deformation, the inflow is free of the geometric constraint. For the ideal fluid, energy and potential vorticity are conserved along a streamline ψ

$$\left(u^2 + v^2\right)/2 + p/\rho = G(\psi) \tag{5.15}$$

$$\left(v_x - u_y + f\right)/h = F(\psi) \tag{5.16}$$

As discussed in Section 4.1.2, functions F and G are not independent, and they satisfy the following relation $dG = d\psi = F$. Assuming that the flow is straight and in the y-direction, so that the u component is negligible, these equations are reduced to

$$v^2/2 + p/\rho = G(\psi) \tag{5.17}$$

$$(v_x + f)/h = F(\psi) \tag{5.18}$$

When the strait is very wide, flow in the upper layer is quite slow, and to a good approximation, the pressure gradient in the upper layer is negligible; thus the flow in the lower layer can be treated in terms of the inverse reduced-gravity model. At the upstream location, flow in the lower layer is also negligible, the interface height relative to the sill is h_u, and the Bernoulli function of all streamlines should be the same: $G(\psi) = p/\rho = g'h_u$, where $g' = g(\rho_2 - \rho_1)/\rho_2$ is the reduced gravity. Thus, the corresponding function F should be $F = dG/d\psi = 0$, i.e., for dynamical consistency the flow from upstream should have a zero potential vorticity. Since $F = f/(h_u + H_0)$, where H_0 is the layer thickness below the sill depth, a zero potential vorticity implies that the layer depth below the sill should be infinite in the upstream. Geostrophy gives rise to an equation

$$fv = g'h_x \tag{5.19}$$

When the strait is wide enough, the interface intersects the bottom of the sill at $x = -b$ (Fig. 5.17a); the integration of Eqn. (5.19) gives the rate of overflow

$$Q = \int_{-b}^{0} hv dx = g'h_0^2/2f \tag{5.20}$$

where h_0 is the layer thickness at the side wall. As h_0 must be smaller than h_u, the interface height relative to the sill, the maximal overflow rate is $Q_{\max} = g'h_u^2/2f$.

When the strait is not wide enough, the layer interface intersects the other side wall, and the corresponding overflow rate is

$$Q = \left(\frac{2}{3}\right)^{3/2} L\sqrt{g'}\left(h_u - \frac{f^2L^2}{8g'}\right)^{3/2} \tag{5.21}$$

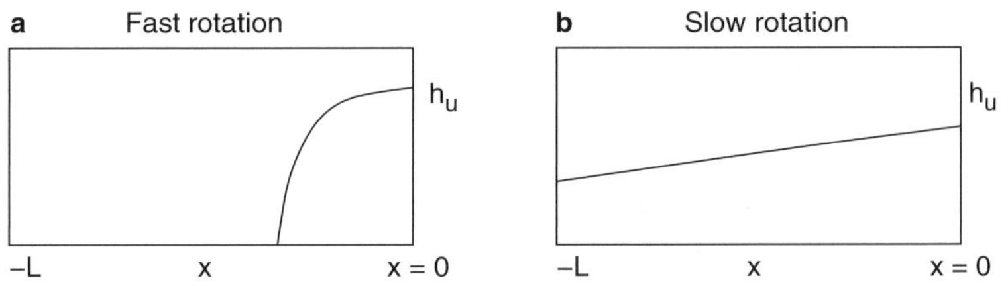

Fig. 5.17 Sketch of overflow through an idealized rectangular strait with width L in a rotating frame: **a** fast rotation, **b** slow rotation.

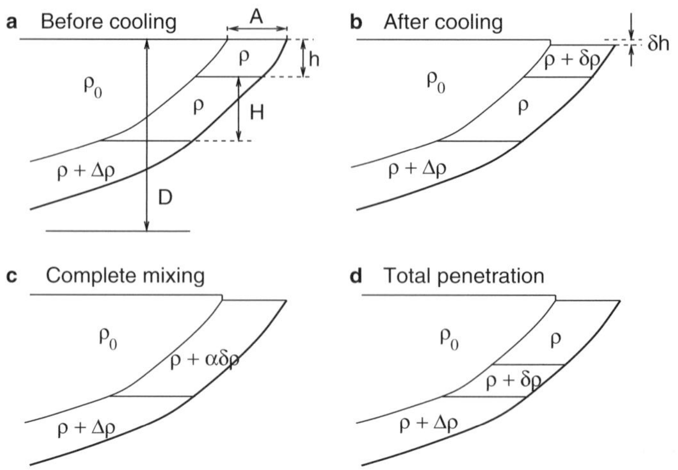

Fig. 5.18 Sketch of density structure related to overflow, **a** before cooling and **b** after cooling, with the *heavy curve* on the lower-right side indicating the sloping topography; **c** (complete mixing) and **d** (total penetration) depict two extreme scenarios of mixing between the newly formed deep water and the environment.

Dense water formation and overflow as a sink of GPE

At first glance, dense water formation and the overflow and gravity current flowing downslope may appear to provide the energy driving the circulation. However, this is not the case. As shown in Figure 5.18, dense water formation, the associated mixing, and movement are associated with the loss of gravitational potential energy (GPE) for the mean state. It is true that dense water formation and movement afterward can partly transform GPE lost from the mean state into kinetic energy and thus drive the circulation. On the other hand, since mechanical energy of the mean state is reduced owing to dense water formation and its subsequent overflow, another source of mechanical energy is required in order to maintain the circulation in a steady state.

Assume that the depth of the ocean is D, which can be used as the reference level for the definition of gravitational potential energy; however, it is more convenient to use the base of the second water parcel, i.e., depth $H + h$, as the reference level (Fig. 5.18a). Note that, for a mass-conserving model, the choice of reference level does not affect the loss/gain of gravitational potential energy. However, for models based on the traditional Boussinesq approximations, the choice of reference level can affect the loss/gain of gravitational potential energy because of the artificial source/sink of mass existing in such models. The water parcel near the sea surface is labeled with a subscript 1, and the water parcel below it is labeled with a subscript 2. The total amount of GPE and its changes before and after the deepwater formation and descent can be calculated as follows.

1. In the initial state before cooling, the total GPE of these two water parcels are

$$\chi_1^0 = \rho g A h \left(H + h/2\right),$$
$$\chi_2^0 = \rho g A H^2/2;$$

2. GPE loss of parcel 1 due to cooling is $\Delta\chi_1 = \rho g A h \delta h/2$;
3. GPE loss due to complete mixing with parcel 2 is $\Delta\chi_{1,2} = \rho g A H \delta h/2$;
4. GPE loss due to complete penetration with no mixing is $\delta\chi_{1,2} = \rho g A H \delta h$.

Thus, cooling and sinking of the dense water do not create mechanical energy. Instead, cooling reduces the GPE of water parcel 1, and subsequent sinking further reduces the total GPE of the mean state, with or without mixing. This loss of GPE of the mean state is partially converted into kinetic energy of the mean state and turbulence, and thus creates a gravity current in the ocean, which is a part of the thermohaline circulation. This is a concrete example demonstrating the basic theory of mechanical energy balance sustaining the thermohaline circulation in the world's oceans, as discussed in Chapter 3.

A modified streamtube model

The movement of deep water after flowing over the sill is an essential component of the deep-water formation and transportation in the oceans. An accurate simulation of such movement remains a challenge even for the complicated oceanic general circulation models (OGCM). The following discussion is based on a simple streamtube model, which can provide elegant solutions with clear physical insight (Price and Baringer, 1994).

Basic assumptions

For simplicity, we make the following assumptions:

- Outflows are density-driven bottom currents;
- A one-dimensional tube model is used;
- Flow is steady;
- Environments are stagnant and infinite, with both potential temperature and salinity profiles $\Theta_{ocn}(z)$ and $S_{ocn}(z)$ specified. Note that potential temperature is used here because density currents can move over a large depth range, so it is more accurate to use potential temperature as the independent variable.

Basic equations for the 1-D model

The structure of a streamtube model is shown in Figure 5.19. The horizontal momentum equations are

$$\vec{U} \cdot \nabla \vec{U} + \vec{f} \times \vec{U} = \frac{g \delta\rho \nabla D}{\rho_r} - \frac{\tau_b}{\rho_r H} - \frac{E\vec{U}}{H} \tag{5.22}$$

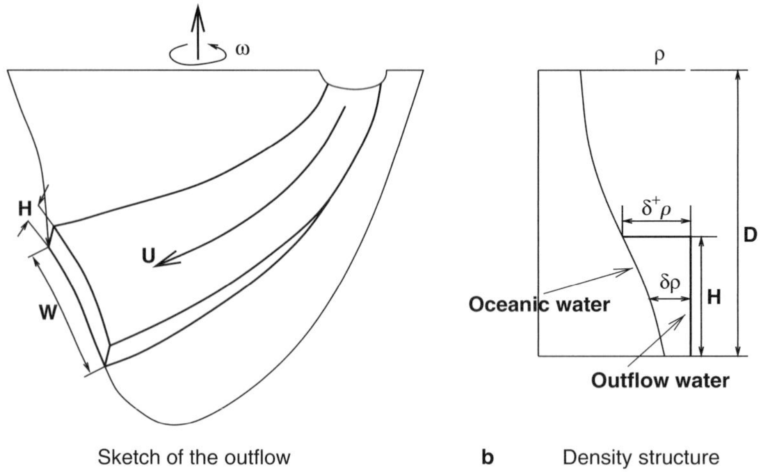

Fig. 5.19 A sketch of the streamtube model, where $\mathbf{U} = \vec{U}$ is the horizontal velocity vector; **a** sketch of the outflow, **b** density structure (modified from Price and Baringer, 1994).

where $\vec{U}$ is the velocity vector; $\rho_r = const.$ is the reference density; $\delta\rho = \rho - \rho_{ocn}(D - H/2)$ is the density difference calculated at mid-depth of the outflow and the oceanic water; the density difference calculated at the top of the outflow and the oceanic water will be denoted as $\delta^+\rho = \rho - \rho_{ocn}(D - H)$; the differences in potential temperature and salinity, $\delta^+\Theta$ and $\delta^+ S$, calculated at the top of the outflow and the oceanic water, are defined accordingly; $\tau_b = \rho_r C_d U \vec{U}$ is the bottom friction; and $C_d = 3 \times 10^{-3}$ is the bottom friction coefficient.

The entrainment rate E is parameterized in terms of the Richardson number $Ri = \frac{g\delta^+\rho H}{\rho_r U^2}$:

$$E = \begin{cases} \dfrac{U\,(0.08 - 0.1Ri)}{1 + 5Ri}, & \text{if } Ri \le 0.8 \\ 0, & \text{otherwise} \end{cases} \tag{5.23}$$

The heat and salt conservation equations are

$$\vec{U} \cdot \nabla\Theta = -\frac{E\delta^+\Theta}{H} \tag{5.24a}$$

$$\vec{U} \cdot \nabla S = -\frac{E\delta^+ S}{H} \tag{5.24b}$$

Note that the buoyancy of the outflow is calculated using the *in situ* density difference $\delta\rho = \rho\,(S, \Theta, P) - \rho_{ocn}\,(S_{ocn}(P), \Theta_{ocn}(P), P)$. Thus, the *in situ* density difference can change along the path

$$\frac{d\delta\rho}{dt} = \frac{\partial\rho}{\partial\Theta}\frac{d\Theta}{dt} + \frac{\partial\rho}{\partial S}\frac{dS}{dt} - \frac{dP}{dt}\left(\frac{\partial\Theta_{ocn}}{\partial P}\frac{\partial\rho_{ocn}}{\partial\Theta} + \frac{\partial S_{ocn}}{\partial P}\frac{\partial\rho_{ocn}}{\partial S}\right) + \frac{dP}{dt}\left(\frac{\partial\rho}{\partial P} - \frac{\partial\rho_{ocn}}{\partial P}\right) \tag{5.25}$$

The continuity equation is

$$\vec{U}\cdot\nabla H = E - H\vec{U}\cdot\frac{\nabla W}{W} - H\nabla\cdot\vec{U} \tag{5.26}$$

where the rate of increase of the width is specified by

$$\frac{\vec{U}\cdot\nabla W}{U} = \beta \tag{5.27}$$

If the outflow is in a channel, β can be specified *a priori*; however, it can be calculated in the general case by

$$\beta = 2K \tag{5.28}$$

where $K = \frac{\tau_b/\rho_r H}{fU}$ is the Ekman number; the ratio of bottom drag to the Coriolis force.

Given the initial values of $\vec{U}, \Theta, S, H$, and W, Eqns. (5.22), (5.24a), (5.24b), (5.26), and (5.27) can be used to calculate the evolution of the streamtube downstream.

Entraining density currents with rotation: an end-point model

We now discuss the simplified relations between the initial conditions and the final product of the outflow. In order to do so, we make three assumptions:

1. The currents are in geostrophy, so $U_{geo} = g'\alpha/f$, where $g' = g\delta\rho/\rho$ is the reduced gravity and α is the topographic slope.
2. The effect of bottom stress on width is retained through

$$W(x) = W_{src} + 2W_{geo}x \tag{5.29}$$

   using the continuity equation and neglecting entrainment,

$$H_{geo} = H_{src}U_{src}/U_{geo}\left(1 + 2K_{geo}x/W_{src}\right) \tag{5.30}$$

   where the subscript *src* indicates the properties of source water.
3. Mixing is treated as an 'entrainment event' where the Froude number is greater than one. Under these assumptions, the product water density of an outflow is

$$\rho_{prd} = \begin{cases} \rho_{src}, & \text{if } F_{geo} \le 1 \\ \rho_{src} - (\rho_{src} - \rho_{ocn})\left(1 - F_{geo}^{-2/3}\right), & \text{otherwise} \end{cases} \tag{5.31}$$

   where $F_{geo} = \frac{U_{geo}}{\sqrt{H_{geo}g'_{src}}}$ is the geostrophic Froude number.

Based on these relations, we can calculate the product density from the initial density and environmental density. Note that the larger the difference is, the smaller the product density will be. This seemingly strange result is due to the strong mixing induced by the large density difference that drives the Froude number supercritical.

Application of the streamtube model to the major sources of deep water in the world's oceans, as discussed above, gives rise to rough estimates for the amount of mixing during the process of descent and entrainment, listed as the last two rows in Table 5.1. From this table, it is clear that mixing for the Mediterranean outflow is much stronger than for the other three outflows. This is one of the major factors that control the depth where the spreading of the Mediterranean outflow is observed in the Atlantic Ocean under the current climate.

In conclusion, the selection of bottom water that can reach and fill up the bottom layer of the world's oceans is controlled by two competing processes: the thermobaric effect and entrainment during descent after overflow over the sill. Calculating the relative contributions due to mixing and entrainment remains a great challenge for theory and numerical modeling.

5.1.4 Mode water formation/erosion

Deepwater formation and bottom water formation provide the sources for the deep ocean, and thus establish the upstream conditions for the deep circulation. However, a complete picture of the meridional overturning circulation in the oceans also involves formation of water masses that sink to a relatively shallower depth. These water masses are also very important parts of the global thermohaline circulation.

Water masses with lighter density are formed primarily in the interiors of subpolar and subtropical basins, and are called mode water. The name "mode water" reflects the fact that these sources of water mass are not uniformly distributed in the temperature–salinity space; instead, owing to specific sea surface conditions favorable for the formation of these water masses, they appear in clusters in the parameter space.

Mode waters in the world's oceans

The historical development of the theories about mode water started with the identification of Eighteen Degree Water associated with the Gulf Stream recirculation in the North Atlantic Ocean (Worthington, 1959). The term "Mode Water" was first introduced by Masuzawa (1969) in the description of Subtropical Mode Water in the Kuroshio Extension in the North Pacific Ocean. McCartney (1977) extended the definition of mode water to the region north of the Subantarctic Front in the Southern Ocean and introduced the term "Subantarctic Mode Water." The term "Subpolar Mode Water" was introduced by McCartney and Talley (1982).

Subtropical mode waters exist at the equator side of the strong separated western boundary currents in the Northern Hemisphere and the strong ACC in the Southern Hemisphere. Three types of mode water formation and their primary sites have been discussed by Hanawa and Talley (2001), including the Subtropical Mode Water (and the Eastern Subtropical Mode Water), and the Subpolar Mode Water.

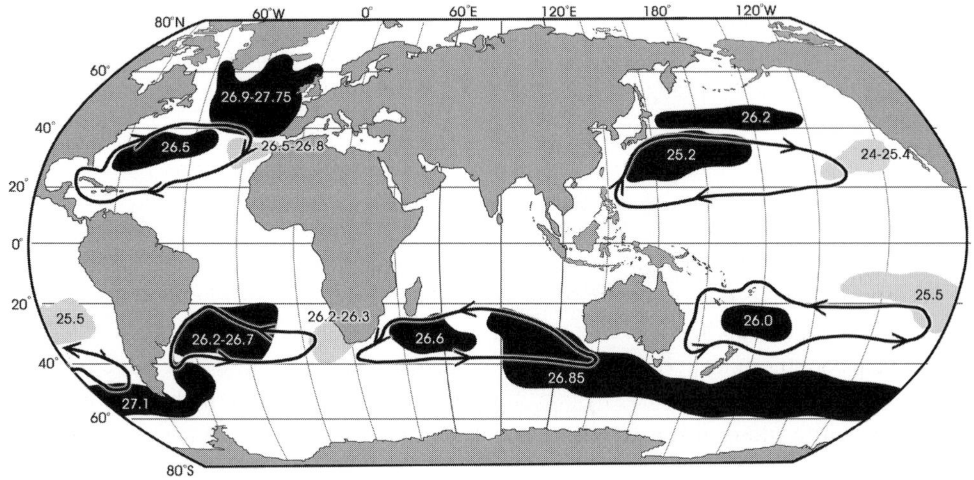

Fig. 5.20 Mode water formation sites in the world's oceans, with the *numbers* indicating the nominal density of the mode water oceans (Hanawa and Talley, 2001).

In general, mode water is used to describe some special types of water masses in the world's oceans which appear as local maxima of distribution density in the (T, S) space. Water identified as mode water shares similar temperature and salinity properties, and often appears as the local minimum in potential vorticity. Some of the most well-known types of mode water in the world's oceans are included in Figure 5.20, including Subtropical Mode Water, Eastern Subtropical Mode Water, Subpolar Mode Water, and Subantarctic Mode Water (in the Southern Ocean).

Mode water formation commonly occurs through subduction taking place in the upper ocean. Subduction of mode water from the late-winter mixed layer into the permanent thermocline of the subtropical basin interior is realized through the combined effects of vertical pumping and lateral induction. Vertical pumping is related to the Ekman pumping produced by the surface wind stress, and lateral induction is due to the horizontal advection of the wind-driven gyre and the horizontal gradient of the late-winter mixed layer depth. In fact, lateral induction is a dominant player in subtropical mode water formation.

Essential ingredients of subtropical mode water formation

As illustrated in Figure 5.21, the basic elements of subtropical mode water formation include the following.

- **Background circulation.** This transports newly formed mode water away from the formation site, and brings in new water from upstream, preparing the formation site for the next cycle of formation (lower panel of Fig. 5.21).
- **Strong seasonal cycle** of the mixed layer depth. This is induced by strong cooling due to cold and dry continental air blowing over the relatively warm water in the recirculation regime on the equator side of the separated western boundary current (upper left corner in lower panel of Figure 5.21).

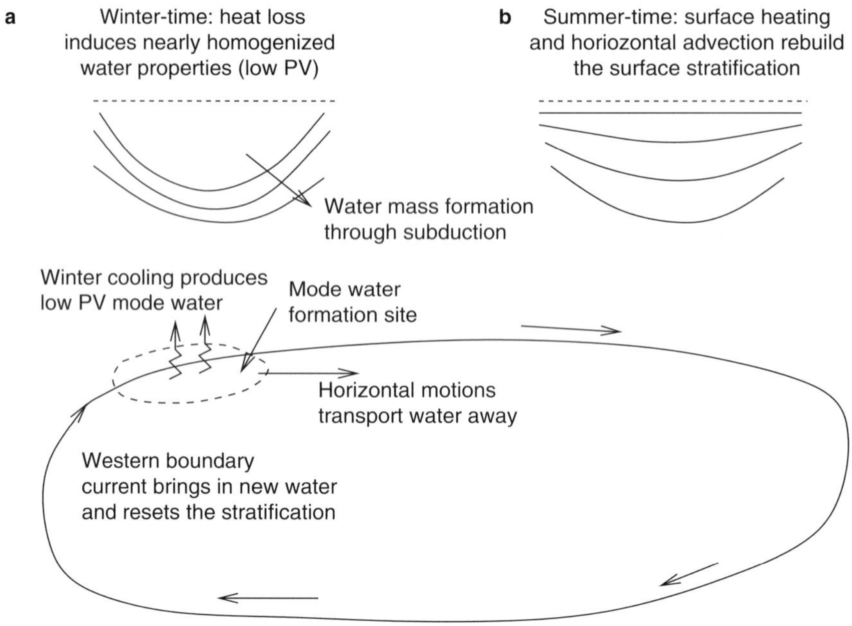

Fig. 5.21 Sketch of subtropical mode water mass formation through subduction: **a** winter-time, **b** summer-time.

During late winter, a large volume of mode water with nearly homogeneous properties, such as temperature and salinity (implying low potential vorticity) is formed (Fig. 5.21a). The rapid retreat of mixed layer depth in early spring leaves the nearly homogenized mode water behind and seals it with a shallow, strongly stratified layer on top, thus completing the formation phase of mode water (Fig. 5.21b).

- **Large horizontal gradient** of winter mixed layer depth. This combines with strong horizontal advection of the wind-driven gyre, giving rise to a strong lateral induction (Fig. 5.21a). This is explained in detail shortly.

As an example, the climatological mean temperature structure in the subtropical North Atlantic Ocean is shown in Figure 5.22. In winter-time, cold and dry continental air flowing over the warm water carried by the Gulf Stream forms a major site of heat loss in the ocean, as shown in the annual mean heat flux from oceans to atmosphere discussed in Section 1.1.1.

The strong cooling produces a large drop in near-surface temperature and thus a nearly homogeneous pool of water in the upper ocean, as indicated by the temperature structure in Figure 5.22a. Along the meridional section, late-winter cooling produces nearly vertical isothermals in the upper ocean (Fig. 5.23a). Since the contribution of salinity to density structure is relatively small, this implies a nearly homogenized density structure in the upper ocean, i.e., a very deep mixed layer and a pool of low potential vorticity ($f\,\Delta\rho/\Delta h$, where $\Delta\rho$ is the density jump across the layer interface and Δh is the layer thickness; thus a thick layer means low potential vorticity).

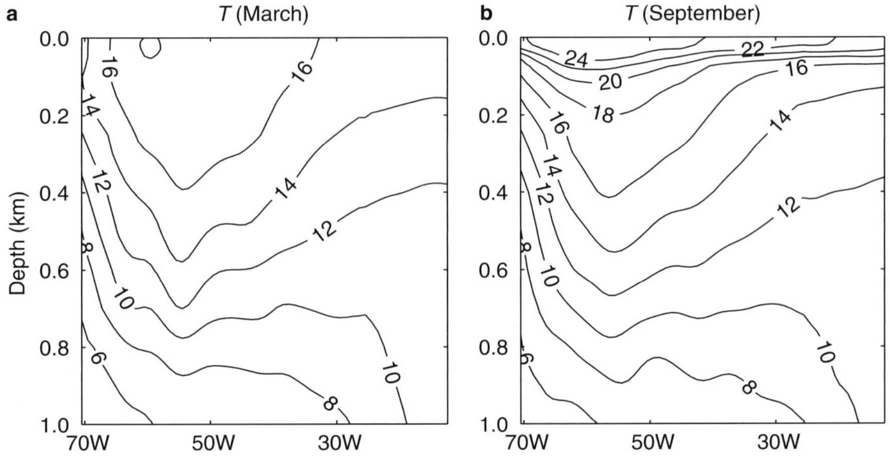

Fig. 5.22 Climatological temperature structure in the upper ocean along 38.5° N; **a** March, **b** September.

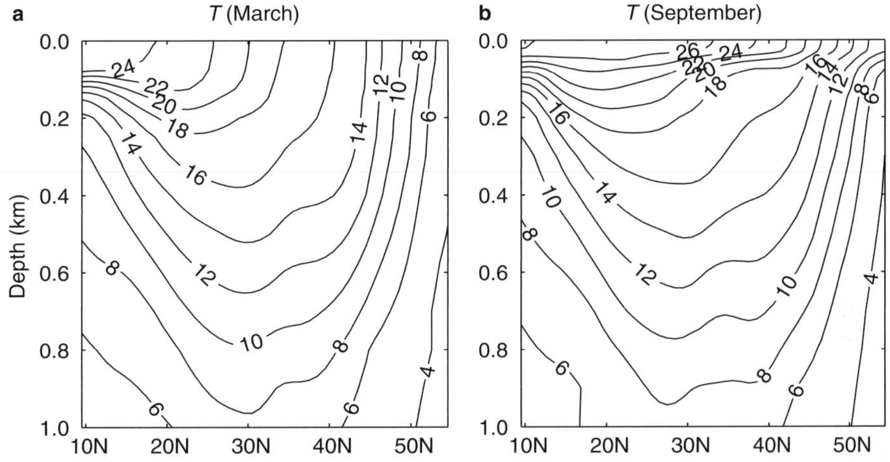

Fig. 5.23 Climatological temperature structure in the upper ocean along 39.5° W; **a** March, **b** September.

When spring comes, this pool of nearly homogenized water is covered by the strong stratification in the upper ocean built up by a rapid shoaling of the mixed layer. In addition, the horizontal advection associated with the subtropical gyre plays two important roles simultaneously. First, it transports the newly formed mode water into the permanent thermocline in the subtropical gyre interior. Second, it brings in new water from upstream, thus preparing the site for the next cycle of mode water formation, as shown in Figures 5.22b and 5.23b.

Mode water formation through subduction is a crucial component of the water mass balance in the world's oceans. In fact, sites of mode water formation are critical windows for the communication of atmospheric signals and tracer input into the oceans. The rate of mode water formation is a good index for climate variability in the oceans. In addition, mode water formation sites play a crucial role in resetting the potential vorticity of water masses in the oceans.

5.1.5 Subduction and obduction

Introduction

The basic ideas of mode water formation were discussed in the previous section. We now focus on the complex dynamical details of mode water formation. In particular, the rate of mode water formation is defined as the annual mean subduction rate, which is a vital index for climate study.

For the balance of water masses in the world's oceans, if there is water mass formation, there should be a process going on in the other direction; this is called water mass erosion. This process may also be called water mass transformation; however, the most suitable terminology has not yet been generally accepted. The rate of water mass erosion through processes in the upper ocean is called obduction, which will also be discussed in this section.

Iselin's model

As discussed in Section 4.1.5, one of the major conceptual difficulties in understanding how subsurface layers are set in motion is that they are not in direct contact with the local atmospheric forcing. However, most isopycnals are in contact with the atmosphere, primarily at high latitudes. Iselin (1939) postulated the preliminary framework for water mass formation through a link between the T–S relation found in a vertical section and the winter-time mixed layer at higher latitudes. His schematic picture for this ventilation and water mass formation process is shown in Figure 4.26. The arrows indicate the speculated motions. In modern terminology, the basic idea is that within the subtropical gyre a water mass is formed at the sea surface in late winter, and is pushed downward into the thermocline by Ekman pumping. Afterward, it downwells along isopycnals, continuing its equatorward motion induced by Sverdrup dynamics. The motion of the particles after their ejection from the base of the mixed layer is confined within the corresponding isopycnal surfaces because mixing is relatively weak within the main thermocline.

Iselin's model was the first prototype for water mass formation in the oceans; however, it was incomplete in two major regards. First, Iselin ignored the mixed layer, which plays a vitally important role in water mass formation. Second, since mixed layer depth and density change greatly from season to season, it was not clear how to make the link between water mass properties and winter-time mixed layer properties, as he postulated.

How do we calculate the water mass formation rate?

According to Iselin's model, the Ekman pumping rate might serve as the water mass formation rate. Although this seemingly simple concept had dominated for a long time, it turned out that the Ekman pumping rate is not exactly the rate of water mass formation. A better way is to calculate the mass flux across the base of the mixed layer. Mixed-layer models have been developed, and can provide an accurate description of the seasonal cycle of the entrainment/detrainment rate across the base of the mixed layer. Can we use the annually integrated detrainment rate as the local water mass formation rate? The answer is "No." Water leaving the mixed layer may not enter the permanent pycnocline; instead, some of the water detrained from the mixed layer at one location may be re-entrained into the mixed layer downstream. Similarly, the simple annually integrated rate of entrainment cannot be used as the rate of water mass erosion. This is due to the fact that water entrained into the mixed layer may not come from the permanent pycnocline; instead it may be water temporarily detrained upstream.

Subduction/obduction rate

A modified conceptual model is shown in Figure 5.24. The upper ocean is divided into four layers: the Ekman layer, the mixed layer, the seasonal pycnocline, and the permanent pycnocline. The Ekman layer plays the role of collecting the horizontal volume transport which is driven by surface wind stress and produces the convergence/divergence. In the subtropical basin, the convergence gives rise to Ekman pumping, and in the subpolar basin the divergence gives rise to Ekman sucking (upwelling). The mass exchange between the mixed layer and the seasonal pycnocline is called entrainment/detrainment, while the mass exchange between the seasonal pycnocline and the permanent pycnocline is called subduction/obduction. Accordingly, the annual mean subduction rate is defined as the total

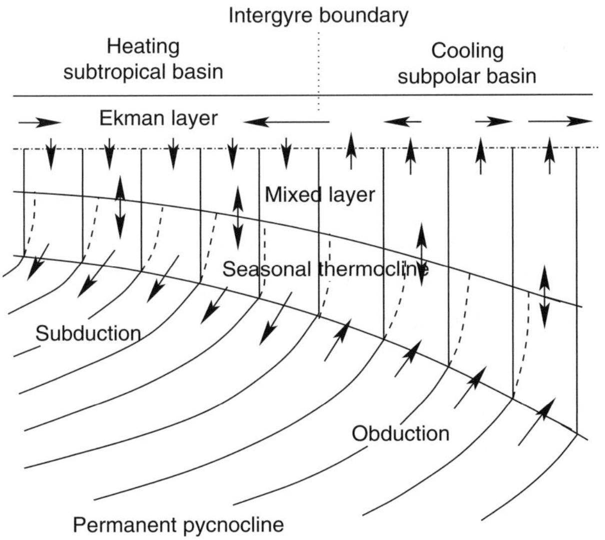

Fig. 5.24 Water mass formation and erosion through subduction and obduction processes. The *vertical two-way arrows* indicate a continuous mass exchange between the mixed layer and the seasonal thermocline.

amount of water going from the mixed layer, passing through the seasonal pycnocline, to the permanent pycnocline irreversibly in one year. This definition excludes the contribution due to the so-called temporal detrainment, i.e., the detrainment which re-enters the mixed layer downstream. Similarly, the annual mean obduction rate is defined as the total amount of water going from the permanent pycnocline, passing through the seasonal pycnocline, to the mixed layer irreversibly in one year.

The Stommel demon

A major technical difficulty in calculating the water mass formation rate is the complicated seasonal cycle in the mixed layer. Both water properties and mixed layer depth vary considerably within the seasonal cycle. By carefully analyzing the processes involved, Stommel (1979) was able to show that a process is at work that selects only the late-winter water for actual subduction into the permanent pycnocline (Fig. 5.25). This mechanism is now called the "Stommel demon."

As discussed in Section 4.1.7, the Stommel demon has become the backbone of the modern theory of wind-driven circulation. Similarly, the theory of mode water formation/erosion through subduction/obduction is also based on the Stommel demon. As discussed later, the effective detrainment period is marked by the Lagrangian trajectories of water particles released from the base of the mixed layer. For simplicity, we assume that the vertical velocity is nearly constant, equal to w_e, in the upper ocean, and the mixed layer depth is horizontally uniform. Such simplifications will be replaced by more accurate statements in the discussion below.

The basic mechanism is as follows. The mixed layer reaches its annual maximum density and depth in late winter, so there is a very thick layer of almost vertically homogenized water. When spring comes, the mixed layer shoals very quickly (as indicated by the sharp turning of the mixed layer depth in the upper and lower panels of Figure 5.25) and leaves the homogenized water behind, so that the water subducted has properties very close to those of the late-winter mixed layer. It can readily be seen that if the time evolution of the mixed layer depth is approximately a δ function, i.e., $\Delta T \to 0$, the subducted water would have the properties of late-winter water.

In the present case, both the vertical velocity and the annual maximal depth of the mixed layer are assumed to be constant everywhere. As a result, the annual mean subduction rate is equal to w_e.

Note that the subduction process in the oceans is a very complicated process involving the seasonal cycle. In fact, now the challenging problem is to calculate the annual mean subduction rate including the seasonal cycle. In one way, this mean can be considered as some kind of weighted average of the instantaneous detrainment rate. Choosing the late-winter properties is equivalent to using a δ function as the weight function. Stommel's suggestion has been used extensively in almost all theoretical models of the ventilated pycnocline. What Stommel suggested yields an elegant solution to this rather intricate problem. This can be used as the lowest-order solution. The next step is to find out a weight

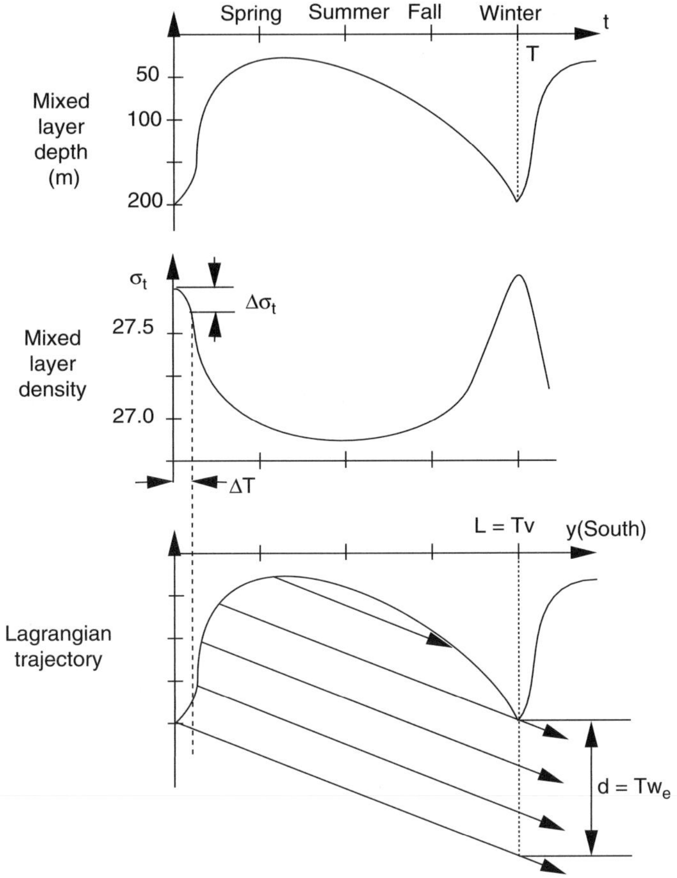

Fig. 5.25 The Stommel demon: the mixed layer properties at late winter are selected through the subduction process; the *horizontal axis* represents both time and distance along the 1-year trajectory.

function that is better than the δ function. In other words, we would like to know the next-order correction to the subduction rate calculated according to the Stommel formula. As such a correction must include the seasonal cycle, it is not an easy problem to solve.

Subduction

We begin with a layered model without a seasonal cycle. In such a model the ventilation/subduction process can be divided into two steps. First, ventilation occurs when water flows downward from the mixed layer into a layer below. Second, water in each density layer follows an equatorward motion induced by the Sverdrup dynamics. Thus, water in a dense layer will move underneath the next layer with lighter density. This process of submersion of a denser layer under a lighter one is called subduction. The term subduction has been used in geology to describe a similar process during the movement of tectonic plates. According

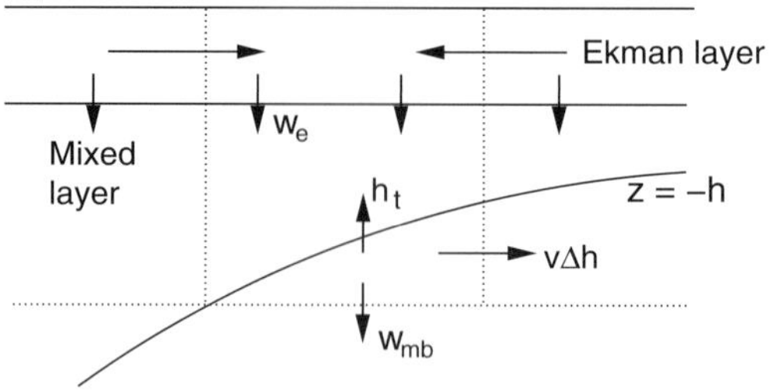

Fig. 5.26 The definition of instantaneous detrainment rate.

to this strict classification, as the number of layers increases, the first stage (ventilation) becomes shorter and shorter. It can readily be seen that, for a continuously stratified ocean, these two stages will merge; we use the term subduction, reserving the term ventilation for the general case of either subduction or obduction.

The seasonal cycle plays one of the most important roles in the upper ocean dynamics, so we must include the seasonal cycle in our subduction model. The most crucial parameter describing the subduction/ventilation process is the subduction rate. The instantaneous detrainment rate is defined as the volume flux of water leaving the base of the mixed layer per unit horizontal area (Cushman-Roisin, 1987):

$$D = -(w_{mb} + \vec{v}_{mb} \cdot \nabla h_m + \partial h_m/\partial t) \tag{5.32}$$

where $w_{mb} = w_e - \frac{\beta}{f}\int_{-h}^{0} v dz$ and $\vec{v}_{mb}$ are the vertical and horizontal velocity at the base of the mixed layer, and h_m is the mixed layer depth (Fig. 5.26). The first term on the right-hand side is the contribution due to vertical pumping at the base of the mixed layer, which is slightly smaller than the Ekman pumping rate due to the geostrophic flow in the mixed layer. The second term is due to the lateral induction. The third term is due to the temporal change of the mixed layer depth.

If there were no seasonal cycle, the subduction rate should equal the detrainment rate

$$S = D = -(w_{mb} + \vec{v}_{mb} \cdot \nabla h_m) \tag{5.33}$$

Thus, this equation can be used for calculating the subduction rate, if there is no seasonal cycle; however, the strong seasonal cycle in the oceans makes the calculation of the subduction rate much more complicated.

Another parameter commonly used in the description of tracer ventilation is the so-called ventilation rate of an individual isopycnal or water mass, defined as

$$V_r = \frac{(water\ mass)\ Volume}{S} \tag{5.34}$$

where S is the subduction rate defined above. Physically, the ventilation rate determines the average time (in years) it takes to renew the entire water mass through the ventilation process, or the average time that water particles remain in a water mass category (Jenkins, 1987).

If we ignore the mixed layer, i.e., set its thickness to zero, then the only term contributing to subduction is the vertical pumping, which is the same as the Ekman pumping, since the mixed layer thickness is zero. Such an oversimplified model for the subduction calculation can lead to the misconception that the subduction rate is the same as the Ekman pumping rate. Since the mixed layer depth is non-zero and it varies with time and location, each term on the right-hand side of Eqn. (5.32) contributes differently.

First, because the mixed layer has a finite thickness, the vertical velocity at the base of the mixed layer is slightly smaller than the Ekman pumping velocity. Second, the lateral induction term actually contributes to subduction substantially. In the North Atlantic Ocean, the winter-time mixed layer depth varies greatly. Within 3,000 km it increases northward from 100 m to about 400 m, so that the slope of mixed layer depth is about 0.0001. The meridional velocity in the mixed layer is about 0.01 m/s, so the lateral induction term is 10^{-6} m/s, which is of the same order as the vertical pumping term. According to a more accurate calculation for the North Atlantic Ocean, the contribution from the vertical pumping amounts to 12.1 Sv, and that from the lateral induction is about 12.7 Sv (Huang, 1990a).

When the time-dependent term is non-zero, the situation becomes even more complicated. There are two prominent cycles in the mixed layer, i.e., the diurnal cycle and the annual cycle. For simplicity, here we discuss the seasonal cycle only. First of all, there is the seasonal pycnocline between the mixed layer and the permanent pycnocline. Thus, a complete picture must consist of four layers, as shown in Figure 5.24. The seasonal pycnocline plays the role of a buffer, i.e., the mass exchange between the mixed layer and the permanent pycnocline must go through the seasonal pycnocline.

As stated above, the mass flux from the seasonal pycnocline to the permanent pycnocline is called subduction. Since we have assumed that the flow in the permanent pycnocline is time-independent, the subduction across the base of the seasonal pycnocline does not vary with time. The mass exchange between the mixed layer and the seasonal pycnocline has a prominent seasonal cycle, and the corresponding exchange rate is called the detrainment/entrainment rate. The subduction rate is, therefore, different from the detrainment rate because they represent different processes. Owing to the existence of the annual cycle, the commonly used subduction (obduction) rate discussed in this section is defined as the annual mean of the corresponding rates.

Between late winter and early fall, detrainment is activated owing to the Ekman pumping and mixed layer shoaling. This period can be further divided into two sub-phases. From late winter to early spring, water entering the seasonal pycnocline from the mixed layer will eventually reach the permanent pycnocline; this process is called the effective detrainment. From early spring to early fall, water entering the seasonal pycnocline will be re-taken by the rapid mixed layer deepening during the winter season, resulting in temporary (ineffective) detrainment (Fig. 5.27). From early fall to late winter, the mixed layer deepens rapidly – the entrainment phase. It appears that the temporal and spatial inhomogeneity of motions in the

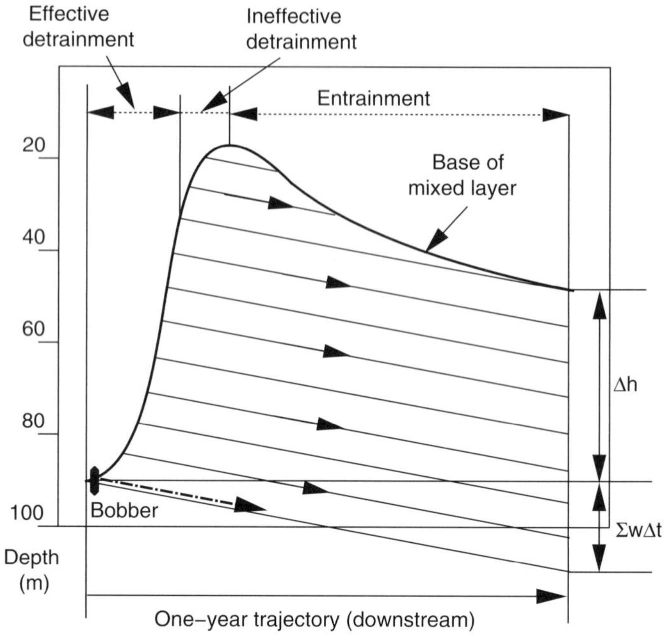

Fig. 5.27 The annual mean subduction rate defined in Lagrangian coordinates; the *horizontal axis* represents both spatial and temporal coordinates along the 1-year trajectory.

mixed layer can create a fairly complicated detrainment/entrainment process, and a comprehensive understanding of the intermittent and sporadic nature of detrainment/entrainment and its contribution to subduction is yet to come through observations and theoretical investigations.

Subduction rate defined as an integral property

Were the mixed layer to overlay a stagnant ocean, the subduction rate would be a purely local property. In the oceans, the mixed layer overrides currents in the seasonal/permanent pycnocline. As soon as water particles are left behind the mixed layer, they are carried downstream by currents. There is no chance that water particles could be overtaken at the same location where they first left the mixed layer. This situation is very similar to a human breathing air. A person confined in a small box may breathe the same air again and again; but a jogger can never inhale the same air that he exhales.

In the oceans, the mixed layer can overtake only the water pumped down from upstream, but not the water pumped down at the same location. Although a person sitting at a recording station may know the local rate of mixed layer entrainment/detrainment as a function of time, that person cannot be sure how much of this water actually reaches the permanent pycnocline. To obtain the correct answer, one has to check at stations downstream, because subduction is a non-local process.

The annual mean subduction rate can be defined in different ways depending on the coordinates used. First, it can be defined in Lagrangian coordinates (Woods and Barkmann, 1986):

$$S_L = -\left(\frac{1}{T}\int_{-T}^{0} w_{tr}dt + \frac{\Delta h_{m,L}}{T}\right) \tag{5.35}$$

where T is the time period over which the average is taken, assumed to be 1 year because of the seasonal cycle; w_{tr} is the vertical velocity along the 1-year trajectory; and $\Delta h_{m,L}$ represents the mixed layer depth change accumulated over a 1-year trajectory in Lagrangian coordinates. Thus, this definition includes both the temporal average and the spatial average over a 1-year trajectory. The schematic diagram in Figure 5.27 illustrates this definition for a two-dimensional case. An instrument called a Bobber is released in late winter, when effective detrainment starts at a station. This instrument can be checked continuously by means of acoustic signals. If we were following the instrument, we would see that during the first part of the trajectory effective detrainment takes place, i.e., the mixed layer retreats and leaves stratified water behind. During the second half of the trajectory, mixed layer entrainment takes place and re-takes part of the water which entered the seasonal pycnocline (at an earlier time). Thus, the seasonal cycle can be divided into three phases; (1) the effective detrainment phase, during which water that left the mixed layer flows geostrophically into the seasonal pycnocline and enters the permanent pycnocline irreversibly; (2) the ineffective detrainment phase, during which water that entered the seasonal pycnocline will be re-taken later (at a downstream location); and (3) the entrainment phase. Calculation of the subduction rate requires accurate information about the kinematic structure of the mixed layer and the velocity field in the pycnocline. Because such detailed information is very difficult to obtain from oceanic climatology, a simplified formula is

$$S_L = -\frac{\int_0^T w_{tr}dt + \Delta h_{m,L}}{T} = -\frac{(d_{tr,1} - d_{tr,0}) - (h_{m,1} - h_{m,0})}{T} \tag{5.36}$$

where $d_{tr,0}$ and $d_{tr,1}$ are the depth of the trajectory at the beginning and end of 1 year, $h_{m,0}$ and $h_{m,1}$ are the depth of the mixed layer at the beginning and end of 1 year, and $T = 1$ year is the duration of the motion. In Figure 5.27 we have assumed a simple case for the northern part of the subtropical basin, where the winter mixed layer depth increases northward. ***By definition, the subduction rate should be non-negative; therefore, a negative value calculated from this definition should be interpreted as a zero subduction rate.***

The annual mean subduction rate can also be defined in Eulerian coordinates. In this case, we stand at a fixed station. In order to calculate the annual mean subduction rate, we monitor the local mixed layer detrainment and entrainment. In addition, we have to follow the trajectories of particles released from this station to see whether these particles will eventually enter the permanent pycnocline (overtaken by the mixed layer entrainment) downstream.

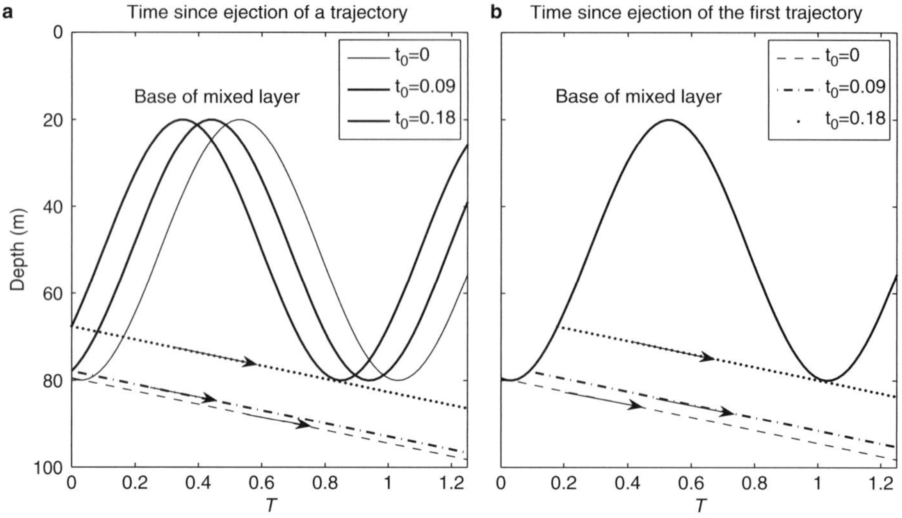

Fig. 5.28 Finding the critical trajectory that defines the end of the effective detrainment period by tracing trajectories released at a fixed station and regular time intervals: **a** time since ejection of a trajectory; **b** time since ejection of the first trajectory.

For example, let us assume that the seasonal cycle of mixed layer depth is a simple sinusoidal function of time which is independent of the geographic location. Effective detrainment takes place very slightly before the mixed layer depth reaches the annual maximum, and we will choose this as the zero point of time axis in Figure 5.28. At this station, at regular time intervals, one water parcel is released at the base of the mixed layer and its trajectory is monitored over 1 year. The trajectory of a water parcel released at the beginning of the effective detrainment period is called the first trajectory. By our choice of time axis, the effective detrainment first starts at time $t_0 = 0$, and continues to $t_0 = 0.18$ because this is the last trajectory of the water parcel that barely escapes the overtaking of a mixed layer downstream (Fig. 5.28a). Although the trajectories shown in Figure 5.28a started from the same position, they began at different times, as shown in Figure 5.28b. Since the mixed layer depth is assumed to be a function of time only, the local mixed layer depth appears to be the same for all trajectories. If the mixed layer depth is also a function of geographic location, the corresponding mixed layer depth for individual trajectories should be different as well.

Similar to the case in Lagrangian coordinates shown in Figure 5.27, the seasonal cycle in the mixed layer at a given station can be divided into three phases; effective detrainment, ineffective detrainment, and entrainment. The annual mean subduction rate is defined as

$$S_E = \frac{1}{T} \int_{T_S}^{T_E} D dt \tag{5.37}$$

where T_S and T_E indicate the starting and ending times of the effective detrainment.

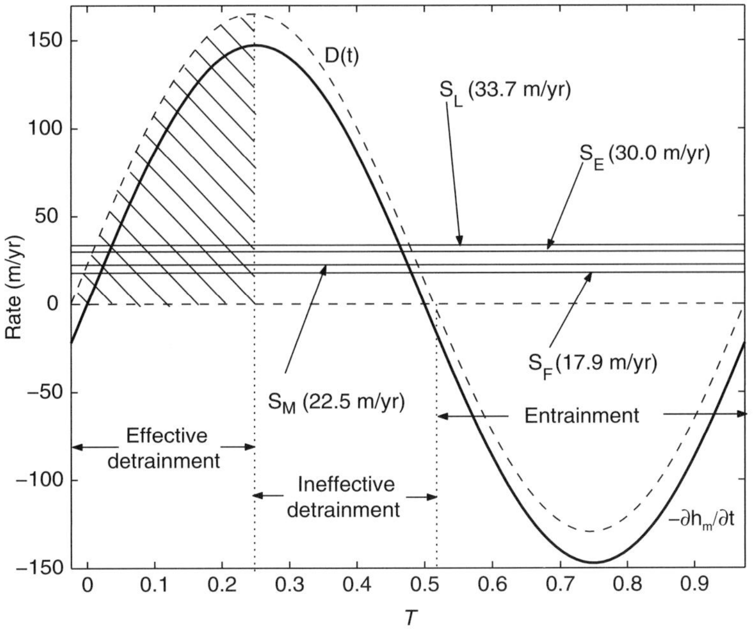

Fig. 5.29 An example of the seasonal cycle of the detrainment/ entrainment and different annual mean subduction rates.

Subduction rates calculated from these definitions for an idealized model are shown in Figure 5.29. The model is set for the northern part of the subtropical basin, where the mixed layer depth increases northward and the Ekman pumping velocity increases southward. In addition, the seasonal cycle is assumed to be a simple sinusoidal function of time. For comparison, two additional terms are introduced below

$$S_M = -\left(w_{mb} + v_{mb}\frac{d}{dy}h_{m,\max}\right) \tag{5.38}$$

where $h_{m,\max}$ is the annual maximal depth of the mixed layer:

$$S_F = -\left(w_{mb} + v_{mb}\frac{d}{dy}\overline{h}_m\right) \tag{5.39}$$

where $\overline{h}_m$ is the annual mean mixed layer depth. Note that both of these definitions treat the subduction in the local sense, so these two subduction rates do not include the average over the trajectory downstream. As a result, the rates calculated from these two equations are smaller than the rates calculated from the above-defined S_E and S_L, both of which include the contribution resulting from the spatial variance of the Ekman pumping velocity and mixed layer depth (Fig. 5.29). Thus, one should not use a simple annual mean for calculating the so-called annual mean circulation; there are always some nonlinear and non-local effects, which must be investigated carefully.

In this example, detrainment rate is controlled by the time-dependent term of the mixed layer depth, $-\partial h_m/\partial t$; however, due to the contribution of the vertical pumping, detrainment starts before $-\partial h_m/\partial t$ changes its sign from negative to positive. Although the detrainment rate is positive for about half of the cycle, only the first quarter of this positive detainment rate, indicated by the shading in Figure 5.29, contributes to the effective entrainment, so this is the part that really contributes to the annual mean subduction rate.

Potential vorticity in the ventilated thermocline

A key parameter for mode water formation is the potential vorticity of the newly formed water mass. For simplicity, the setting of potential vorticity of mode water through subduction can be illustrated in a two-dimensional sketch for an idealized case of steady circulation (see Fig. 5.30).

Using density conservation, we have

$$Q = \frac{f\Delta\rho}{\rho_0\Delta_z} = \frac{f}{\rho_0}\frac{\overline{\vec{u}_{ml}\cdot\nabla\rho_m}}{\overline{w_{tr}+\vec{u}_{tr}\cdot\nabla h_m}} \tag{5.40}$$

where the overbars indicate the mean over the 1-year trajectory, $\overrightarrow{u}_{ml}$ indicates the horizontal velocity in the mixed layer, and the subscript *tr* indicates the trajectory. The reason for emphasizing the mean over a 1-year trajectory stems from the definition of annual mean subduction rate. For example, the horizontal velocity near the Gulf Stream can be on the order of 0.1 m/s; thus, over 1 year, the trajectory can cover a distance of 2,000–3,000 km. The along-trajectory mean over such a long distance can be considerably different from the local term. Equation 5.40 states that potential vorticity in the ventilated thermocline is linearly proportional to the meridional density gradient of the mixed layer density, and inversely proportional to the sum of the vertical velocity at the base of the mixed layer plus

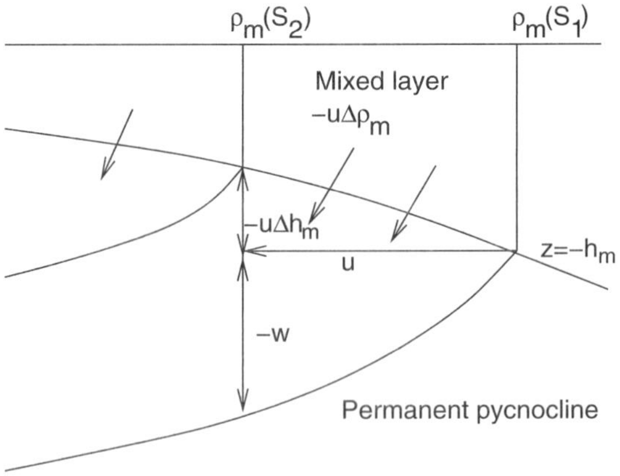

Fig. 5.30 Relation between mixed layer properties and the potential vorticity formed during subduction.

the horizontal increment of the mixed layer depth. Therefore, low potential vorticity water is formed when there is:

- low meridional gradient of mixed layer density
- strong Ekman pumping (implying a strong vertical velocity)
- a large horizontal gradient of late-winter mixed layer depth and large horizontal velocity.

Obduction

Upwelling/entrainment prevails in subpolar basins. When a water parcel moves from the permanent pycnocline to the mixed layer in the upper ocean, it loses its original identity, such as temperature and salinity. Thus, water mass is eliminated through erosion.

Similarly to what happens during the ineffective detrainment period, water entrained into the mixed layer may not actually come from the permanent pycnocline; instead, it may come from the seasonal pycnocline whose water was detrained from the mixed layer previously (Woods, 1985; Cushman-Roisin, 1987).

In order to clarify the physical processes involved in entrainment, we use the term obduction. Obduction has been used in geology for describing the process of upward thrusting of a crustal plate over the margin of an adjacent plate. Here obduction is borrowed to describe the process in which water from the permanent pycnocline upwells into the mixed layer and flows over the adjacent layers of water. Although obduction is basically a continuous process between the permanent pycnocline and the seasonal pycnocline, effective entrainment from the seasonal pycnocline to the mixed layer occurs only during part of the entrainment period (Fig. 5.31). During the effective entrainment phase, water from the permanent pycnocline, which has not been exposed to surface processes, is entrained into the mixed layer through the seasonal pycnocline. During the rest of the entrainment period (i.e., the ineffective entrainment period), water that has been exposed to air–sea interactions within the past year enters the mixed layer from the seasonal pycnocline, as indicated by the top five lines in Figure 5.31.

Using the special term "obduction" helps us to clarify the irreversible mass flux from the permanent pycnocline to the mixed layer. For example, although mixed-layer entrainment takes place in subtropical basins during a seasonal cycle, in most places water entrained into the mixed layer actually comes from the seasonal pycnocline, so there is no obduction. In fact, obduction takes place only within the subpolar basins and the subtropical–subpolar boundary regions, as will be shown in the following discussion.

The obduction rate can be defined in a way similar to the subduction rate. Though obduction is an antonym of subduction, obduction cannot be calculated as subduction with the opposite sign. There are two major differences between the two terms.

First, the physical processes involved in subduction and obduction are different. Subduction takes place in the subtropical basins, where water geostrophically flows down into the permanent pycnocline. As a result, water subducted to the permanent pycnocline carries late-winter mixed layer properties. In comparison, obduction takes place in the subpolar basin, where water from the permanent pycnocline below flows geostrophically upward into

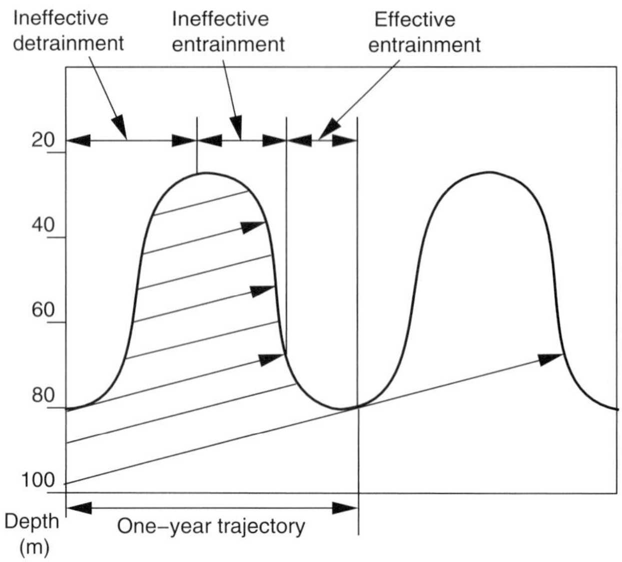

Fig. 5.31 An example of obduction, upwelling with a uniform upwelling of 18 m/year. The mixed layer depth has a simple sinusoidal cycle.

the seasonal pycnocline and eventually enters the mixed layer. As water enters the mixed layer, it quickly loses its identity as a result of strong mixing, and it is impossible to keep track of the trajectory of an individual water parcel afterwards.

The difference in the physics is reflected in the mathematical formulation of the suitable boundary value problems for the pycnocline structure in the subtropical and subpolar basins. The pycnocline equation is a nonlinear hyperbolic equation (Huang, 1988a, 1988b). In the subtropical basin, density is specified as an upper boundary condition because the upper surface is the upstream boundary. In the subpolar basin, the upper surface density cannot be specified. In fact, the mixed layer density is determined by the dynamics of the permanent pycnocline and is part of the solution.

Second, the times when effective detrainment and effective entrainment take place are different. It is well known that effective detrainment takes place after late winter, when the mixed layer reaches its annual maximum depth and density and starts to retreat. Effective entrainment takes place between late fall and early winter when the mixed layer deepens quickly, but before it reaches its annual maximum depth and density.

In calculating the obduction rate, it is important to trace back to the origin of the entrained water. To make the presentation clearer, assume $t = 0$ at late winter, say March 1. We begin with a simple case where the mixed layer depth has a simple sinusoidal cycle, and its amplitude is spatially uniform. The upwelling rate is 18 m/yr, and it is also uniform along the trajectory. Because the mixed layer is relatively shallow, we assume that the vertical velocity is approximately the same as the Ekman sucking rate everywhere along the trajectory. Below the base of the mixed layer, mixing is negligible, so the identity of water parcels is preserved. As a result, the particle trajectory can be used to trace the origin of the water before it is entrained into the mixed layer. As soon as water parcels enter the mixed layer, they lose their identity because of the strong vertical mixing within the mixed layer.

During spring and early summer the mixed layer retreats and leaves the stratified water behind, so this is the period of detrainment, although it is only a temporary detrainment in the present case. Beginning in early fall, the mixed layer deepens and entrainment takes place. The water entering the mixed layer during the first period does not really come from the permanent pycnocline. In fact, these water parcels were detrained into the seasonal pycnocline at an earlier time and at some upstream locations, as indicated by the top five lines in Figure 5.31. Such water was "contaminated" in the mixed layer during the past year, so this is not genuine effective entrainment. It is only within the second phase of entrainment that water from the permanent pycnocline enters the mixed layer, as indicated by the lower trajectories.

For the situation shown in Figure 5.31 a simple calculation shows that for the case with a uniform upwelling of 18 m/yr, the effective entrainment starts at $T_S = 0.8762$ and ends at $T_E = 1.0088$. In an ordinary year, the effective entrainment starts at January 18 and ends at March 4, yielding a duration of about 44 days.

The contrasts between subduction and obduction are shown in Table 5.3.

Obduction rate defined as an integral quantity

The annual mean obduction rate can be defined slightly differently, depending on the coordinates used. First, the obduction rate can be defined with the Lagrangian coordinates (Woods, 1985). Accordingly, the annual mean obduction rate is defined as

$$O_L = \frac{1}{T}\int_{-T}^{0} w_{tr}dt + \frac{\Delta h_{m,L}}{T} \tag{5.41}$$

where T is the time duration over which the average is taken, which is assumed to be 1 year because of the seasonal cycle; the subscript *tr* indicates the critical trajectory in Figure 5.31 (which marks the end of obduction); and $\Delta h_{m,L}$ indicates the mixed layer depth change accumulated over a 1-year Lagrangian trajectory. Note that both terms include the temporal average over the past year and the spatial average over the 1-year trajectory. ***By definition, an obduction rate should be non-negative, and a negative value calculated from this definition should be interpreted as a zero obduction rate.***

Second, an instant entrainment rate in Eulerian coordinates can be defined as

$$E = w_{mb} + \vec{u}_{mb} \cdot \nabla h_{mb} + \partial h_m/\partial t. \tag{5.42}$$

where the subscript *mb* indicates the base of the mixed layer. The instantaneous entrainment rate fluctuates greatly during one seasonal cycle. In addition, some of the entrained water during the phase of ineffective entrainment does not contribute to obduction. Thus, the

Table 5.3. *Ventilation: subduction vs. obduction*

	Subduction	Obduction
Mass flux	From the mixed layer to the permanent pycnocline	From the permanent pycnocline to the mixed layer
Water mass	Formation	Erosion
Time	Spring	Winter
Atmospheric forcing	Heating	Cooling
Mixed layer	Shoaling	Deepening
Trajectory tracking	Downstream	Upstream

entrainment cannot be used as an index for the water mass conversion rate. Similar to the definition of the annual-mean Eulerian subduction rate, it is more meaningful to use the annual mean obduction rate defined as

$$O_E = \frac{1}{T}\int_{T_S}^{T_E} E dt \tag{5.43}$$

where T_S and T_E are the times when the effective entrainment started and ended, and E is the instantaneous entrainment rate defined in Eqn. (5.42).

Some incorrect definitions for water mass erosion rate have been used in previous studies. In some cases, people have simply used "upwelling" as the term to describe the process opposite to subduction. As discussed above, upwelling is only part of the obduction term, and the other part is associated with lateral induction, which can be the dominant term in the subpolar basin.

Another potential pitfall in calculating the annual mean obduction rate arises from using a simple Eulerian mean at a given station over the whole year

$$\bar{O}_E = \frac{1}{T}\int_{-T}^{0} E dt \tag{5.44}$$

where substituting Eqn. (5.42) would lead to a wrong estimation of $\overline{O} = \overline{W_{mb}} + \overline{\vec{u}_{mb} \cdot \nabla h_m}$. In general, this substitution tends to underestimate the annual mean obduction rate.

The major differences between these two definitions of obduction rate are as follows. In Eulerian coordinates, water parcels entrained into the mixed layer at one station are monitored, and trajectories of parcels are traced back upstream for 1 year to determine whether the water comes from the permanent pycnocline or not. Accordingly, the beginning and the end of the effective entrainment are determined, and the annual obduction rate can be calculated by using Eqn. (5.43). In Lagrangian coordinates, water parcels entrained into the mixed layer at a station are monitored to determine the critical trajectory that marks the end of obduction. Given this trajectory, one can trace back along the trajectory for 1 year and calculate the obduction rate by using Eqn. (5.41).

Calculating the annual mean obduction rate according to these two definitions requires accurate information on the spatial and temporal evolution of the mixed layer, and such detailed information is almost impossible to obtain from any climatic data. Thus, a simple definition of the annual mean Lagrangian obduction rate can be used

$$O_L = \frac{(d_{tr,-1} - d_{tr,0}) - (h_{m,-1} - h_{m,0})}{T} \tag{5.45}$$

where d and h are the depth of the trajectory and the mixed layer base, and subscripts 0 and -1 indicate the fact that these quantities are calculated at the beginning of the year and the previous location by upstream-tracing of the critical trajectory for 1 year.

Although obduction rates calculated according to these definitions are slightly different, their difference is quite small compared with the errors existing in present-day climatology data. Therefore, the definition in Eqn. (5.45) can serve as a convenient and practical tool for calculating the obduction rate from climatological data with errors comparable to those of other processes.

Balancing the rate of water mass formation/erosion

Water mass formation and erosion take place in the oceans through the subduction and obduction processes discussed above. Along with the mixed layer and the Stommel demon, these processes can be schematically viewed in Figure 5.32. The ventilation/obduction rate has two components, the lateral induction and vertical pumping terms: $S = S_{li} + S_{vp}$, and $O = O_{li} + O_{vp}$.

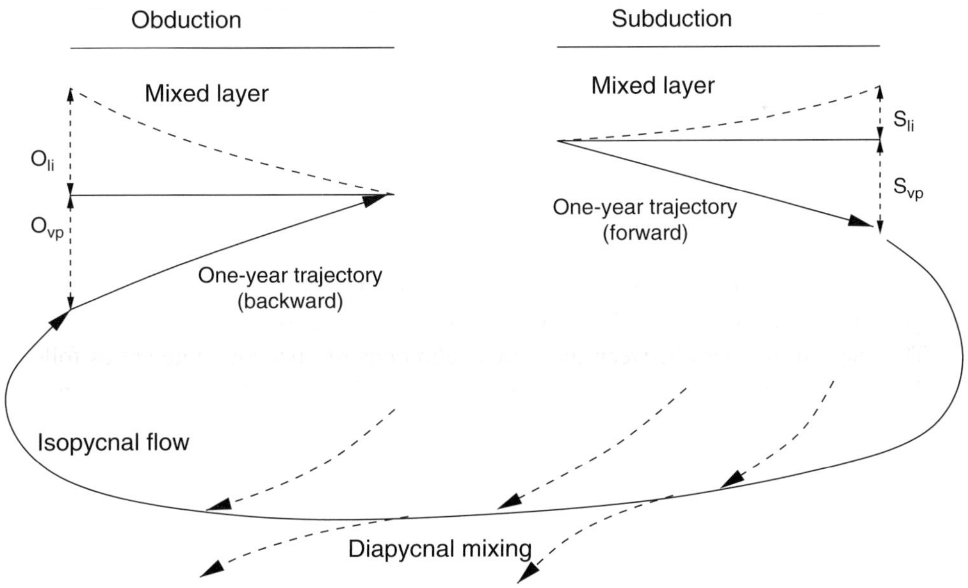

Fig. 5.32 Definition of the subduction/obduction rate in Lagrangian coordinates.

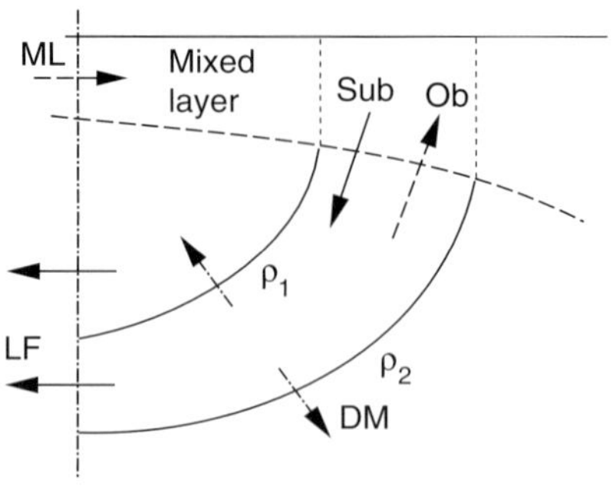

Fig. 5.33 Water mass balance, where *Sub* indicates subduction, *Ob* indicates obduction, *DM* indicates diapycnal mixing flux, and *LF* indicates lateral flux through the side boundary.

In a steady state, the water mass volume between two density surfaces must be constant with time; thus there is a balance (Fig. 5.33):

$$M_{Sub} - M_{Ob} + M_{DM} - M_{LF} = 0 \tag{5.46}$$

where M_{xx} denotes total mass flux associated with the process identified by the subscript xx. For a basin with open boundaries, the contribution of the last term should be balanced by the mass flux through the lateral boundaries of mixed layer, i.e.,

$$\iint_S m_{LF} ds = \iint_S m_{ML} dS \tag{5.47}$$

where S denotes the lateral boundaries of the basin and m_{xx} denotes the mass flux associated with process xx. The contribution due to the diapycnal mixing integrated over the entire density range also vanishes because this term can transform a water mass between different density categories. Thus, for a basin with open boundaries (Fig. 5.33) the total amount of subduction should equal the total amount of obduction plus the mixed layer inflow from the side boundary

$$\iint_A m_{Sub} dA = \iint_A m_{Ob} dA + \iint_S m_{ML} dS. \tag{5.48}$$

where A denotes the horizontal area of the basin. For a closed basin, there is no mixed layer inflow through the side boundaries; thus, the total amount of subduction and obduction should be exactly balanced:

$$\iint_A m_{Sub} dA = \iint_A m_{Ob} dA. \tag{5.49}$$

Equation (5.48) is one of the essential constraints for the water mass formation rates in an open basin and Eqn. (5.49) is for a closed basin, and this constraint should be satisfied for the global oceans.

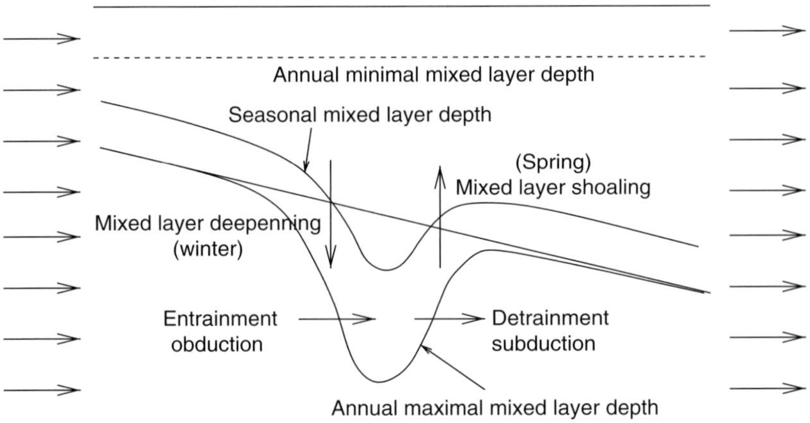

Fig. 5.34 A sketch for subduction/obduction at the Kuroshio Extension. Trajectories of water particles are depicted by *horizontal arrows*.

Examples of subduction/obduction

As an example, here we study the ventilation rate at the place where the Kuroshio (or the Gulf Stream) separates from the western boundary. A prominent maximum in the late-winter mixed layer depth exists off the coast of Japan, where there is strong cooling by cold, dry polar air from the continent. For simplicity, we assume that the along-current-path distribution of the late-winter mixed layer depth can be represented by the profile shown in Figure 5.34. The diagram illustrates the fact that, due to the local maximal winter mixed layer depth, both subduction and obduction can take place at the same location. The strong seasonal cycle can lead to the onset of obduction and subduction at different seasons, although they can also co-exist during the time of annual maximal mixed layer depth.

The seasonal cycle of the mixed layer depth, $f(t)$, has the typical profile shown in the lower panel of Figure 5.35. The mixed layer depth is a function of space and time:

$$h_m(x,t) = h_{\min} + \left[100 \times (1 + 0.001x) + 100e^{-(x/\Delta x)^2} - h_{\min}\right] f(t) \tag{5.50}$$

where $h_{min} = 40$ m is the mixed layer depth minimum, and Δx is the width of the Gaussian profile.

Where the Kuroshio separates from the coast, water travels relatively quickly. Combining with the strong mixed layer depth gradient, it makes a large contribution to the ventilation rate. In comparison, the contribution due to Ekman pumping is negligible because this is close to the inter-gyre boundary. To illustrate the basic ideas, we will neglect the vertical pumping and assume that water parcels travel 1,000 km in 1 year.

Our focus is on the station at the center of the winter mixed layer trough, $x = 0$. The instantaneous entrainment/detrainment rate calculated by Eqn. (5.32) is shown in Figure 5.35. Note that the time axis $t = 0$ corresponds to March 1. Effective detrainment begins at $t = 0$ when the mixed layer starts to shoal rapidly. The end of the effective detrainment,

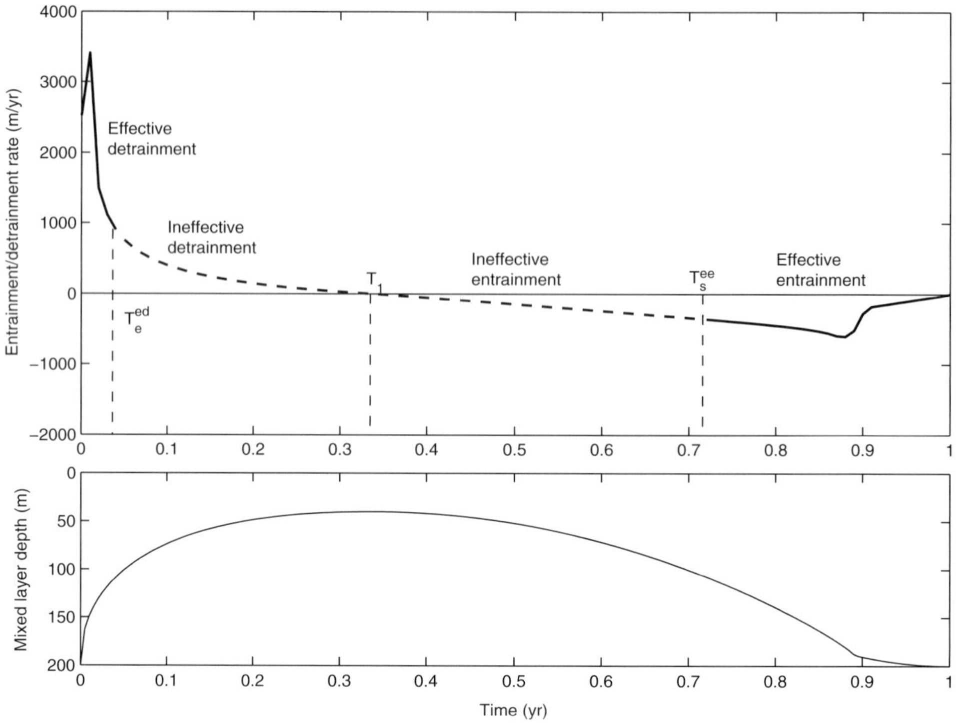

Fig. 5.35 Instantaneous entrainment/detrainment rate at the place of the maximal winter mixed layer depth. The *thin solid line* indicates the seasonal cycle of the local mixed layer depth.

$T_e^{ed} = 0.037$, is determined by tracing water particles released from this station at each time interval of $\Delta t = 0.001$. The ineffective detrainment period is between T_e^{ed} and $T_1 = 0.336$, because the water subducted during this period is entrained into the downstream mixed layer at a later time.

Entrainment starts at T_1. The beginning of the effective entrainment is determined by tracing water particles upstream for 1 year. It occurs in this case at $T_s^{ee} = 0.716$. Note that although detrainment takes place for one-third of a year, the period of effective detrainment is rather short, about 13 days. Similarly, entrainment takes place during two-thirds of the year, but only the second half of this period (about 103 days) is the period of effective entrainment.

At $x = 0$, as shown in Figure 5.35, effective entrainment takes place in winter. Near the end of winter, the entrainment mode slows down and eventually switches to the detrainment mode. The transition between these two modes is so quick that water which entered the mixed layer remains close to the location where it leaves the permanent pycnocline and the same water is left behind and geostrophically flows into the permanent pycnocline thereafter. Using this example, we have shown that obduction and subduction can take place at the

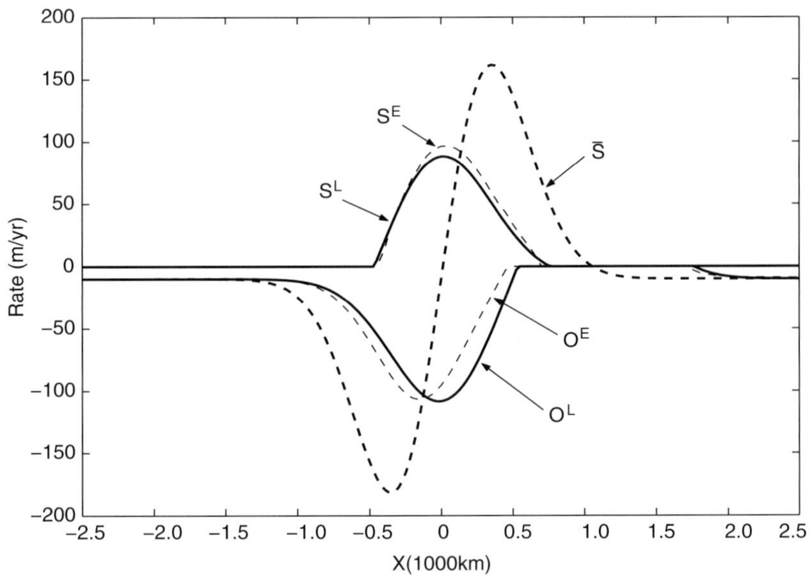

Fig. 5.36 Annual mean subduction (positive) and obduction (negative) rates: *heavy solid lines* indicate the Lagrangian rates, *thin dashed line* indicates the Eulerian rates, and the *heavy dashed line* is the rate calculated using the local winter mixed layer slope as defined in Eqn. (5.51).

same location, and the water masses involved may come through effective detrainment and entrainment during different phases of the seasonal cycle.

The spatial distribution of the annual mean subduction/obduction rates is shown in Figure 5.36, in which the subduction rate is defined as positive and the obduction rate is defined as negative. The solid lines denote the rates calculated in Lagrangian coordinates, while the thin dashed lines denote those calculated in Eulerian coordinates. Here, simplified versions of Eqns. (5.36, 5.45), i.e., $S_L = (h_{m,0} - h_{m,1})/T$ and $O_L = (h_{m,0} - h_{m,-1})/T$, have been used in computing the Lagrangian rates. The subduction/obduction rates calculated using the Lagrangian/Eulerian formulae agree well both in magnitude and in spatial distribution. The slight difference between them is due to the different weighting on the annual means assumed in these two definitions.

Four dynamically distinct regions appear in Figure 5.36. In the upstream direction, there is a region of pure obduction because the mixed layer depth increases monotonically downstream. In the middle of the mixed layer depth trough there is an ambiductive region, where both subduction and obduction take place and compensate for each other. Within the ambiductive region there is a local conversion from the obducted water to the subducted water, and the corresponding rate of water mass conversion is equal to the minimum of (S, O). There is a narrow band downstream where only subduction occurs. Farther downstream, an insulated region exists where there is neither effective subduction nor effective obduction.

For comparison, the subduction/obduction rate calculated using just the annual mean velocity and winter mixed layer depth

$$\overline{S} = -\overline{\overline{u}} \cdot \nabla h_{winter} \tag{5.51}$$

is shown in Figure 5.36 by the heavy dashed line. Such a formula is similar to that used by Marshall *et al.* (1993a) in their data analysis on subduction in the North Atlantic Ocean and that used in the analytical study of the subtropical North Pacific Ocean by Huang and Russell (1994). Although this definition is much easier to apply, it inflates the subduction/obduction rate near the mixed layer depth fronts. In addition, such a definition eliminates the possible overlapping and exclusion of subduction and obduction because it neglects the seasonal cycle and the along-trajectory changes.

Note that calculating the annual mean subduction/obduction rate in Eulerian coordinates requires detailed information about the seasonal cycle. On the other hand, the simplified formula from Eqns. (5.36, 5.45) in Lagrangian coordinates requires the annual mean velocity and winter mixed layer properties only. Such a calculation can provide much more accurate information about the ventilation rates, although care must be taken to set the rate to zero whenever it is negative.

Ventilation of the North Atlantic Ocean and North Pacific Ocean

One of the most crucial assumptions of the Stommel demon is that subduction takes place during a relatively short time period. In addition, March 1 has been widely used as the date for calculating the subduction rate in the Northern Hemisphere. In reality, the time of maximal mixed layer depth can be different from March 1, and the effective subduction in

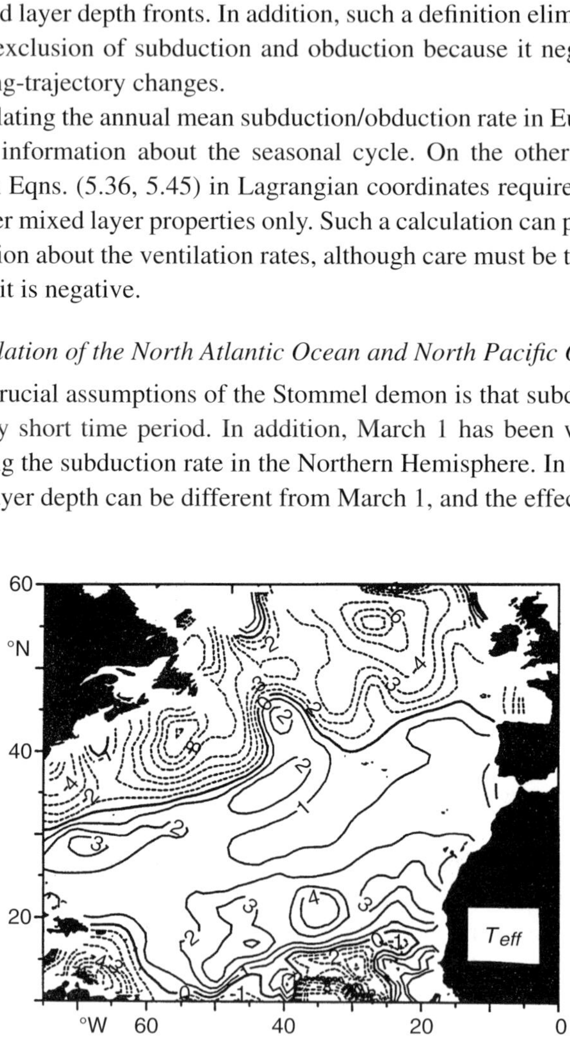

Fig. 5.37 Duration of the effective detrainment and entrainment (in months), or the period of subduction and obduction, diagnosed from a numerical model with a seasonal cycle of buoyancy forcing for the North Atlantic Ocean (Marshall *et al.*, 1993a).

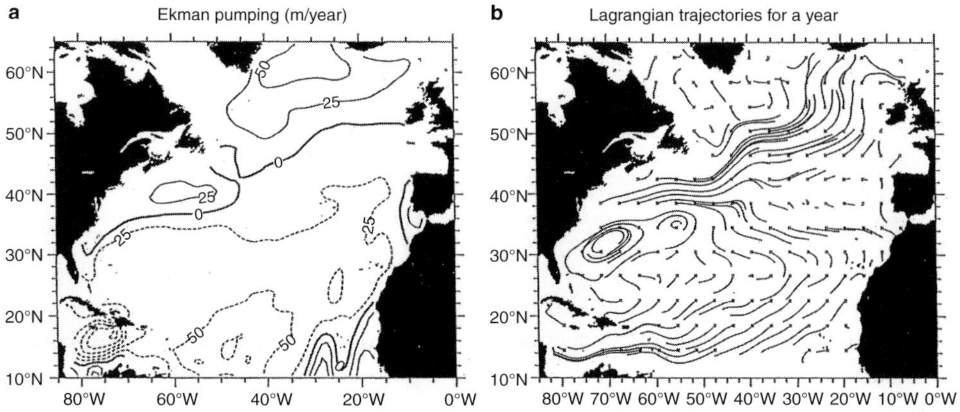

Fig. 5.38 **a** Ekman pumping rate in the North Atlantic Ocean, in m/yr; **b** 1-year trajectories of water particles released from the base of the mixed layer in March, with the along-isopycnal geostrophic velocity calculated using the 2,000-m depth as the reference level (Qiu and Huang, 1995).

the oceans happens during a finite duration of time. For example, Marshall *et al.* (1993a) calculated the duration of the effective subduction and obduction in the North Atlantic Ocean (Fig. 5.37). Since they defined obduction as the case with a negative value of subduction, the corresponding effective period of obduction is defined as negative. It is clear that the period of effective subduction and obduction in the oceans is relatively short, so that the Stommel demon does work properly. In addition, the duration of effective subduction or obduction also varies slightly in the basin.

Ventilation in the North Atlantic Ocean and North Pacific Ocean was examined by analyzing the Levitus (1982) climatological data and the Hellerman and Rosenstein (1983) wind stress data (Fig. 5.38a). The calculation was based on the geostrophic velocity, using the 2,000-m level as the reference, and the 1-year trajectories are shown in Figure 5.38b. During a 1-year period, water particles can travel over a great distance on the order of 1,000 km; thus, using the local gradient of the mixed layer depth, as in Eqn. (5.38), is inaccurate for calculating subduction/obduction.

Ventilation in the North Atlantic Ocean and North Pacific Ocean can be classified into four different types of region: the subductive regions, the obductive regions, the ambiductive regions where both subduction (in spring) and obduction (in winter) take place, and the insulated regions where neither subduction nor obduction happen (Fig. 5.39). Although the total subduction rates in these two oceans are comparable, the obduction rates are considerably different (Table 5.4). In the North Atlantic Ocean, obduction is strong (23.45 Sv), which is consistent with the notion of the fast thermohaline circulation and the short renewal time of the subpolar water masses in the Atlantic Basin. Obduction is weak in the North Pacific Ocean (7.83 Sv), consistent with the sluggish thermohaline circulation and the slower renewal process of the subpolar water masses there.

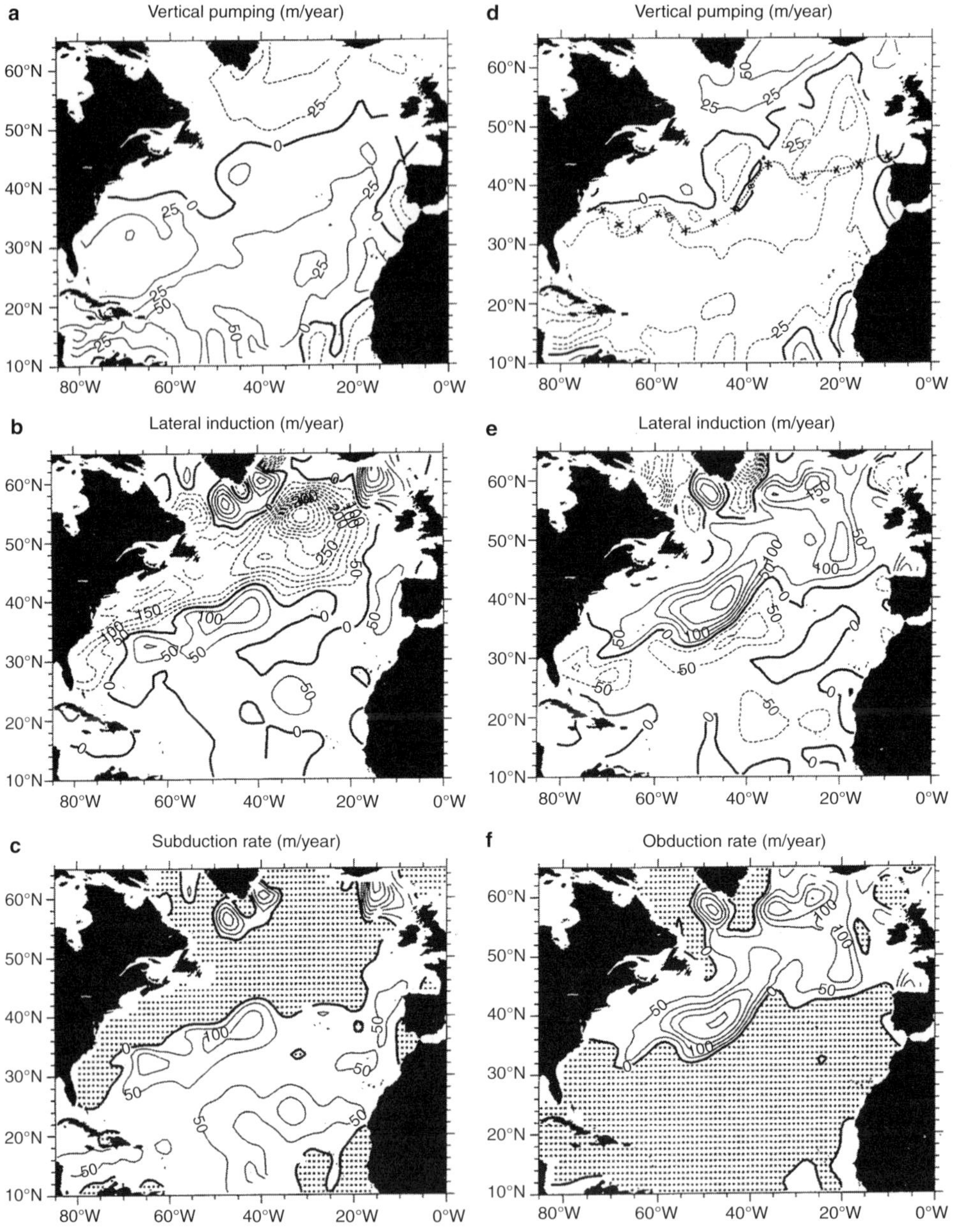

Fig. 5.39 Annual-mean ventilation rate for the North Atlantic Ocean. Subduction rate and its two components are shown in **a** the vertical pumping rate, **b** the lateral induction term, and **c** the subduction rate. *Stippled regions* in **c** indicate zero subduction rate. Obduction rate and its two components are shown in **d** the vertical pumping term, **e** the lateral induction term, and **f** the obduction rate. *Stippled regions* in **f** indicate zero obduction rate. The *dotted line with crosses* in **d** indicates the southern limit of the obduction zone (Qiu and Huang, 1995).

Table 5.4. *Ventilation rates for the North Atlantic Ocean and the North Pacific Ocean (Sv)*

	North Atlantic Ocean	North Pacific Ocean	Sum
Subduction			
Ekman pumping	−22.2	−30.8	−53.0
Vertical pumping	17.5	25.1	42.6
Lateral induction	9.5	10.1	19.6
Total subduction	27.0 + 3.1*	35.2	65.3
Obduction			
Ekman pumping	2.8	3.6	6.4
Vertical pumping	−0.62	3.1	2.5
Lateral induction	24.1	4.7	28.8
Total obduction	23.5	7.8	31.3
Local conversion	4.0	3.5	7.5

* 3.1 Sv is the sum of the localized subduction south of Iceland (1.9 Sv) and in the Labrador Sea (1.2 Sv).

The most interesting features of these maps are the ambiductive regions in the oceans, where the local water mass conversion rate can reach 80 m/yr. These local conversion rate maxima are very closely related to the heat flux maxima from the ocean to the atmosphere, as indicated by the dashed lines in Figure 5.40.

The water mass formation (erosion) rate, computed as the sum of the subduction (obduction) rate integrated over the corresponding outcropping density range, is plotted in Figure 5.41. The peaks of the subduction rate correspond to the subtropical mode water in the North Atlantic Ocean and North Pacific Ocean. A second peak in the North Atlantic Ocean indicates the subpolar mode water, which has no corresponding part in the North Pacific Ocean because shallow marginal seas, such as the Sea of Okhotsk, were not included in the calculation.

One of the most interesting features from Table 5.4 is that the basin-integrated obduction rate in the North Atlantic Ocean is 10 times larger than the total Ekman sucking rate. This indicates the critical dynamical role of the sloping mixed layer depth. Thus, it is clear that using the term "Ekman upwelling" to describe water mass erosion in the ocean can be very inaccurate and misleading.

Final remarks

In this section we discussed the formation of mode water through subduction and water mass erosion through obduction. In contrast to the wide interest in understanding water mass formation, there have been only a very few studies related to water mass erosion. It

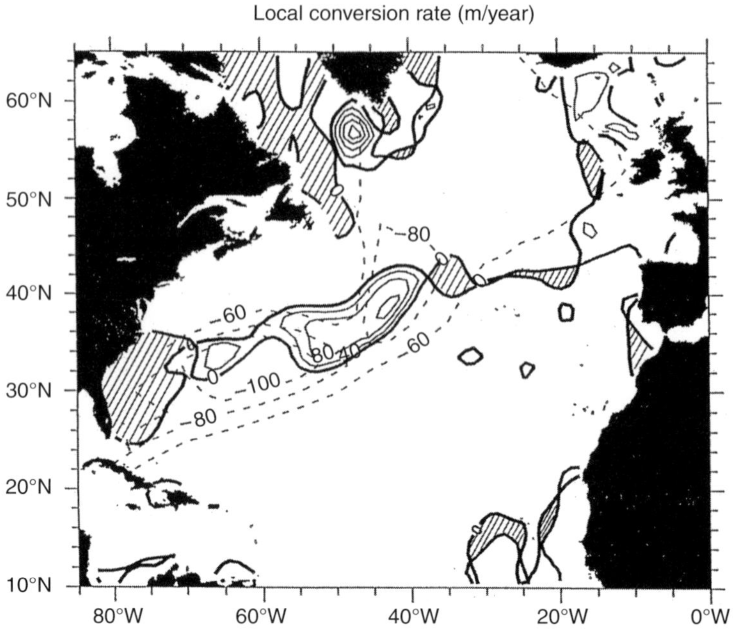

Fig. 5.40 Local mass conversion rate in the North Atlantic Ocean, defined within the ambiductive region, with a contour interval of 40 m/yr. The *shaded areas* are insulated regions where neither subduction nor obduction takes place. The *dashed lines* indicate the annual heat loss from the ocean to the atmosphere, in W/m^2, adapted from Hsiung (1985) (Qiu and Huang, 1995).

is obvious that a complete theory of water mass balance in the world's oceans requires the more comprehensive study of both processes.

In particular, water mass erosion is often related to the upwelling branch of the circulation in the oceans. Upwelling in the ocean is highly non-uniform in space. For example, the strong austral westerlies drive the upwelling in the core of the ACC, which is the strongest large-scale upwelling system in the world's oceans. In addition, there are other narrow upwelling systems, such as the equatorial upwelling, and the strong coastal upwelling along the edges of the basins. These upwelling systems are the major contributors to water mass erosion in the world's oceans.

5.2 Deep circulation

5.2.1 Observations

Deep currents in the oceans

Circulation in the deep ocean is directly related to deepwater formation. The mean flow in the deep ocean interior is very slow, with the horizontal velocity on the order of 0.01 m/s or less and vertical velocity on the order of 10^{-7} m/s; thus, direct observation of deep

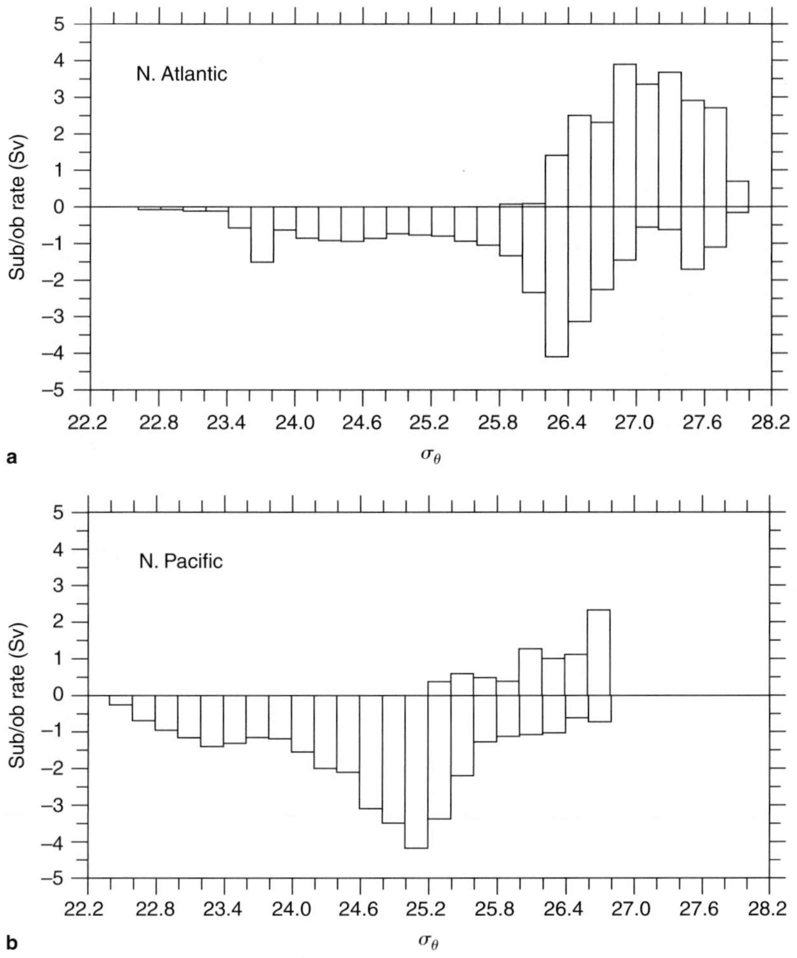

Fig. 5.41 Subduction/obduction rates per 0.2 σ_θ interval as a function of density for **a** the North Atlantic Ocean and **b** the North Pacific Ocean (Qiu and Huang, 1995).

circulation in the ocean interior is extremely difficult. However, the deep western boundary currents associated with the deepwater sources are the most remarkable features in the world's oceans. Since Swallow and Worthington (1957) made the first measurements of the deep western boundary current in the North Atlantic Ocean, the existence of deep boundary currents has been detected in many parts of the world's oceans.

Currents associated with deepwater formation in the North Atlantic Ocean

In the North Atlantic Ocean, deep water is formed in the Norwegian and Greenland Seas (Fig. 5.42). Worthington (1970) described the Norwegian Sea as a Mediterranean-like basin where the warm inflow water is turned into cold and dense deep water overflowing the sills

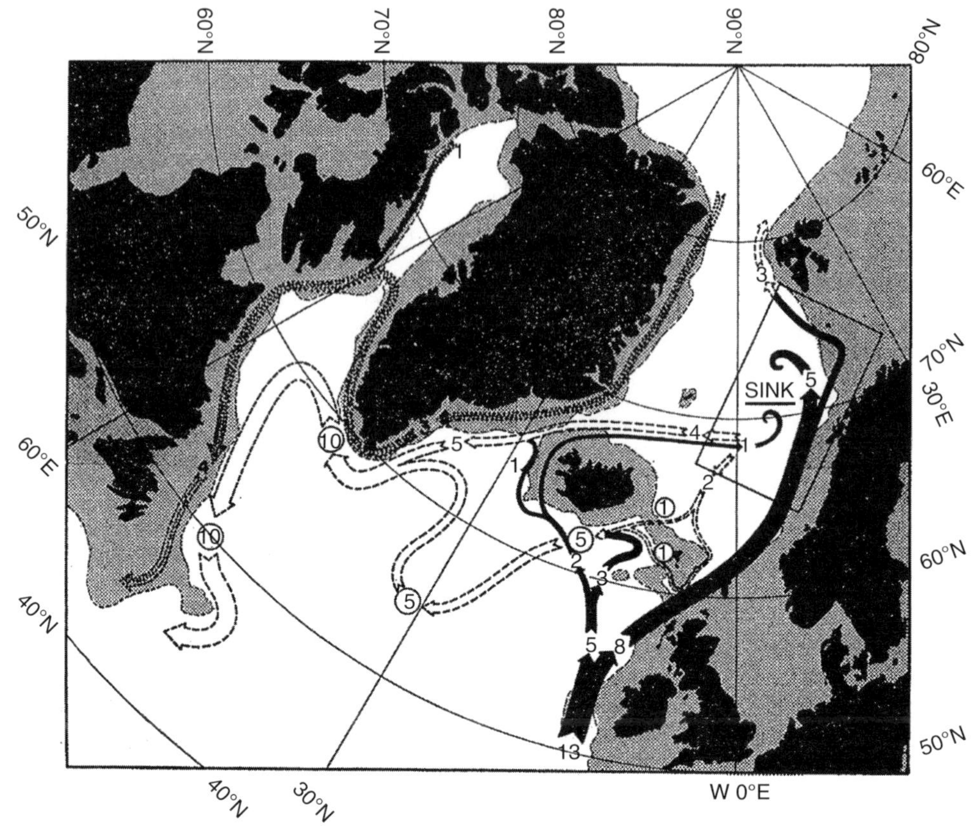

Fig. 5.42 Index map identifying the inflow and outflow currents for the Norwegian Sea and local places (Worthington, 1970).

separating the Norwegian and Greenland Seas. In Figure 5.42 the warm Atlantic inflow is depicted by solid arrows, and deep water formed at these locations overflows the Denmark Strait and the Faroe Bank Channel, forming the sources of the deep western boundary current (open arrows) which can be observed along the eastern coast of the North American Continent.

This deep western boundary current can be identified through the use of current meters and temperature distribution. For example, the low-temperature overflow water in a section through the Denmark Strait is shown in Figure 5.9. In addition, the deep western boundary current can be identified through the high oxygen content and other tracers of the newly formed deep water. The signature of North Atlantic Deep Water can be traced all the way southward along the eastern continental slopes of the North American and South American continents. Historically, the deep western boundary off the east coast of the North American continent was the first deep boundary current observed; it has been studied and monitored over the past 50 years.

In fact, there are many deep western boundary currents in the Atlantic Basin, including the southward-moving western boundary current associated with NADW and the northward-moving western boundary current associated with the AABW, although this latter current is less well defined due to the existence of complicated topography in the abyssal basins.

Note that, according to classical theory, NADW primarily originate from deep convection in the middle of the Norwegian Sea and Greenland Sea, as depicted in Figure 5.42. Recent studies (Mauritzen, 1996; Pickart and Spall, 2007) suggest that the primary source of the NADW is from the gradual cooling within the rim current in the Norwegian Sea and Greenland Sea. Thus, the index map shown in Figure 5.42 should be modified. However, for historical reasons, Figure 5.42 is used here to illustrate the basic pathway of the NADW, including its overflows through the three passages.

Overflows through these straits are regulated by rotating hydraulics, as discussed in Section 5.1.3. Since the channel is wide enough, overflow is pushed against the right bank of the strait (looking in the downstream direction). As shown in Figure 5.9, very cold water with temperature lower than 1°C is pushed against the bank of the channel at a depth of 300–500 m. On the left-hand side of the bank there is virtually no flow. The closeness of the isotherms and steep slope indicate a swift deep current through the channel at this section.

Deep western boundary currents in the South Atlantic Ocean

The situation in the South Atlantic Ocean is both complicated and interesting. From the potential temperature and salinity sections (Fig. 5.43), three western boundary currents can be identified. The most outstanding feature in these tracer sections is the southward deep western boundary current at a depth of 2 km, which is characterized by the core of relatively warm and salty water (with salinity greater than 34.96).

In the upper ocean there is a shallow southward western boundary current, which is associated with the wind-driven subtropical gyre in the South Atlantic Ocean. This current is closely linked to the narrow temperature and salinity fronts in the upper ocean.

There are clear signs of isothermal slope and isohaline slope inversion around the depth of 3 km, which suggests the inversion of flow direction of the deep western boundary current. In fact, below 4 km, there is an outstanding sign of cold (with temperature lower than 0°C) and relatively fresh water (with salinity less than 34.7) flowing northward along the bottom topography.

Deep western boundary currents can easily be identified from density sections. Since potential density with reference either to the sea surface or 2 km depth is valid only for the depth range close to the reference pressure, we use the global pressure-corrected density σ_g, which is a better choice for analysis over the whole depth of the ocean. Because the deep western boundary current moves much faster than the water above it, to a good approximation we can assume that the water above it is nearly stagnant, so that the horizontal pressure gradient at mid-level, say around 2.5 km depth, is negligible. In the Northern Hemisphere and below this reference level, dense water piles up along the western wall (Fig. 5.44a); integrating the hydrostatic pressure gives rise to a horizontal pressure gradient force pointing east. The deep western boundary current is subject to geostrophy, so that in

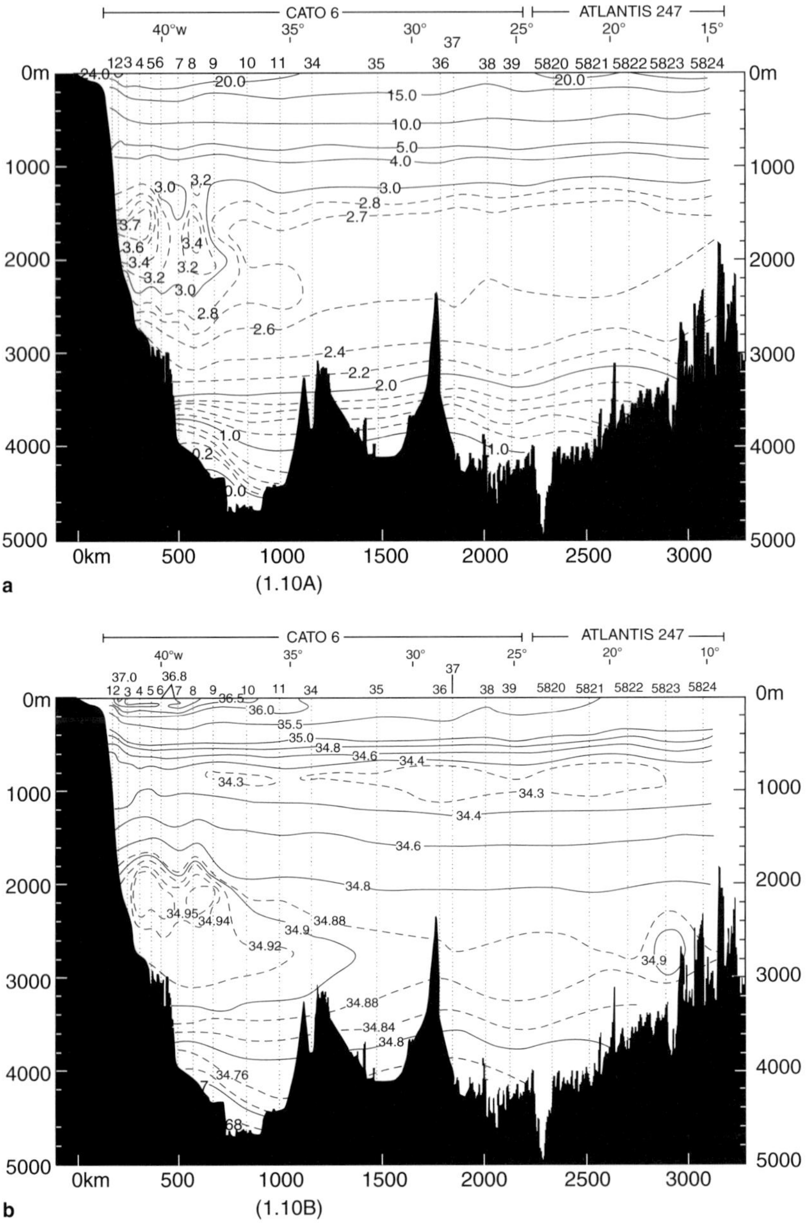

Fig. 5.43 Section of **a** potential temperature (°C), and **b** salinity along roughly 30° S from South America (*left*) to the Mid-Atlantic Ridge, illustrating the two deep western boundary currents of the South Atlantic Ocean; namely, the northward-flowing Antarctic Bottom Water and the southward-flowing North Atlantic Deep Water above (Reid *et al.*, 1977).

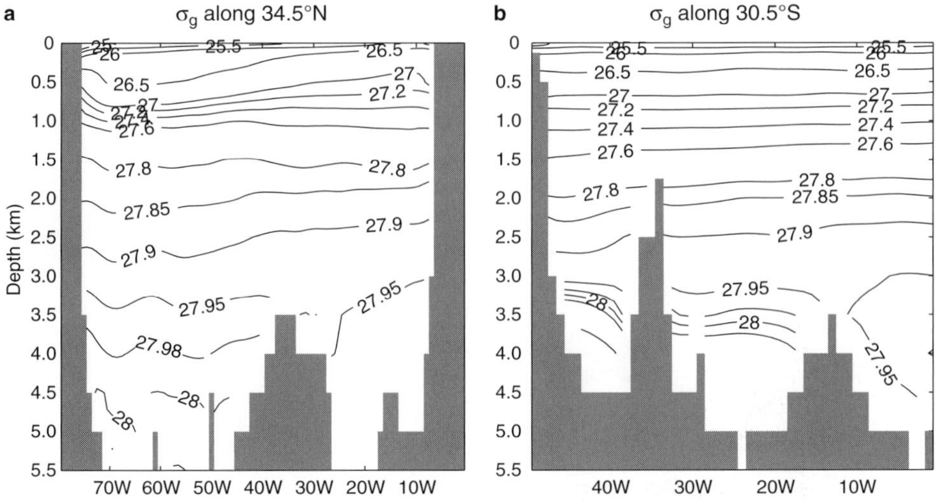

Fig. 5.44 Zonal sections of σ_g (kg/m^3) in the Atlantic Ocean, illustrating the deep western boundary currents in **a** the North Atlantic Ocean and **b** the South Atlantic Ocean.

order to balance this eastward pressure gradient force, flow must be equatorward. Similarly, the cold dense water piling up against the western wall below 3 km (Fig. 5.44b) indicates a strong equatorward deep western boundary current.

Deep boundary currents in the world's oceans

Although deep western boundary currents have been observed in many places, the task of finding these deep currents is far from being complete. It is more appropriate to call them deep boundary currents, because some of them are not along a western boundary. Figure 5.45 shows an up-to-date map of the observed deep boundary currents. For most of the Pacific Ocean, there is no sharp bottom topography to guide the deep boundary currents, so we may still not really know where the deep boundary currents are.

The following list was compiled with the help of B. Warren. The numbers in Figure 5.45 indicate the relevant papers where these boundary currents are described, based on *in situ* observations. 1: Owens and Warren (2001); 2: Whitworth *et al.* (1999); 3: Roemmich *et al.* (1996); 4: Warren (1973); 5: Whitworth *et al.* (1991); 6: Hogg and Owens (1999); 7: Warren and Speer (1991) and Mercier *et al.* (1994); 8: Dickson *et al.* (1990); 9: Saunders (1990); 10: Pickart (1992); 11: Speer and Forbes (1994); 12: Fieux *et al.* (1986); 13: Johnson *et al.* (1991a); 14: Toole and Warren (1993); 15: Johnson *et al.* (1991b). Note that flow across the mid-ocean ridge in the Atlantic Ocean takes place near the equator, and this consists primarily of the AABW flowing over the Romanche fracture zone. However, this overflow water also includes a substantial contribution from NADW through diapycnal mixing, as discussed by Polzin *et al.* (1996).

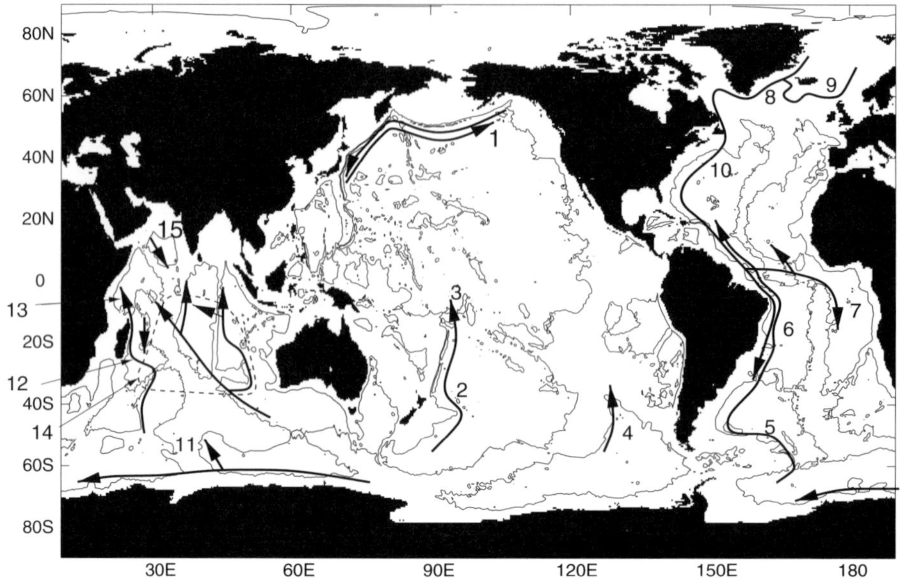

Fig. 5.45 Deep boundary currents observed in the world's oceans, with 4 km depth contours (modified from Warren, 1981).

5.2.2 Simple theory of the deep circulation

Inverse reduced-gravity model

For the circulation in the deep basin, flow near the bottom is much faster than flow above, so that as an approximation we can assume that the pressure gradient in the water above can be neglected; thus, an inverse reduced-gravity model can be formulated. Such $1\frac{1}{2}$-layer models have been used in the study of abyssal circulation (e.g., Stommel and Arons, 1960a). In addition, the amount of cold water is finite. If the amount of cold water is not enough to cover the whole basin, the interface between the cold and warm water can intersect the seafloor. Such a phenomenon is called grounding, which is equivalent to the layer outcropping in the upper ocean. Thus, the grounding phenomenon can be studied by using the same concept first developed by Parsons (1969). As an example, this technique was used to study bottom water circulation by Speer and McCartney (1992).

Following the analysis in Section 4.1.1, the pressure gradient term in an inverse gravity model can be derived as follows. Using the pressure expression derived for the case with a rigid lid, and assuming that the upper layer is very thick and motionless, i.e., $\nabla p_1 = 0$, leads to an expression of the gradient of the equivalent atmospheric pressure $\nabla p_a = 0$. Substituting these relations into the pressure gradient expression in the second layer (Fig. 5.46), we obtain

$$\nabla p_2 = -g(\rho_2 - \rho_1)\nabla(H - h) \tag{5.52}$$

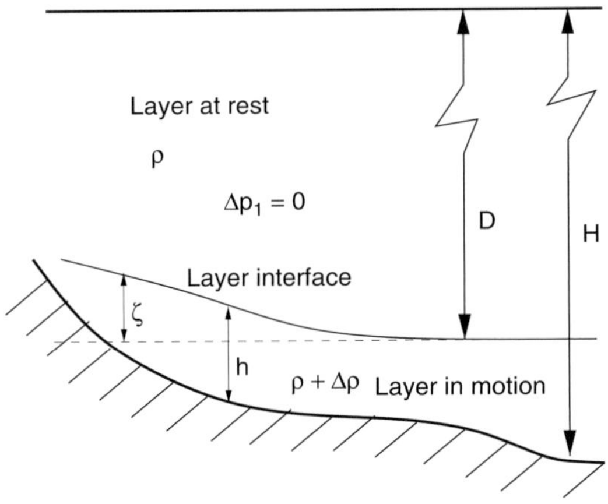

Fig. 5.46 An inverse reduced-gravity model.

Assuming geostrophy, the momentum equations for an inverse reduced-gravity model are

$$-fv = g'(H-h)_x \tag{5.53a}$$

$$fu = g'(H-h)_y \tag{5.53b}$$

If we denote the sea floor as $d = H$, Eqn. (5.53a, b) can be rewritten as

$$-fv = g'(h+d)_x \tag{5.54a}$$

$$fu = g'(h+d)_y \tag{5.54b}$$

Using the interfacial displacement from the mean position at depth D, $\zeta = D - (H-h)$, Eqn. (5.53a, b) can be also be rewritten as

$$-fv = -g'\zeta_x \tag{5.55a}$$

$$fu = -g'\zeta_y \tag{5.55b}$$

The continuity equation is

$$h_t + (hu)_x + (hv)_y = -w^* \tag{5.53}$$

where w^* is the vertical velocity leaving the upper surface of the lower layer due to entrainment to the layer above.

For example, along the western boundaries of the oceans, there are deep western boundary currents. These boundary currents are closely associated with the dense water originating from deepwater formation sites at high latitudes. The shoaling of the isopycnal surface near the western boundary (Fig. 5.46) suggests that the deep western boundary current flows equatorward.

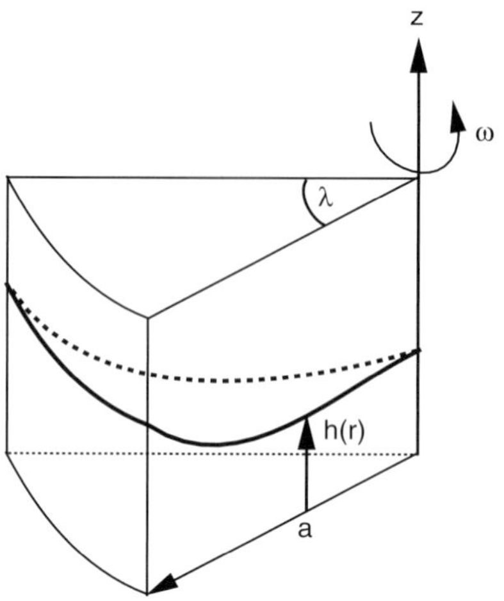

Fig. 5.47 Sketch of the laboratory equipment for the rotating pie experiment.

Deep circulation simulated by the rotating sector experiments

The theory of deep circulation was first developed in the late 1950s and early 1960s. In order to explore the possible patterns of circulation, a very simple and elegant pie-shaped experimental tool was designed (Stommel *et al.*, 1958). The system rotated uniformly, and the circulation was observed with the release of dye (Fig. 5.47).

The parabolic-shaped upper surface of the water in the rotating experiments produces a dynamical effect equivalent to the β-effect in the world's oceans. For "large-scale" motions in such pie-shaped experiments, the barotropic potential vorticity is f/h. Since h is large near the rim of the sector, potential vorticity is low. For an ocean with uniform depth on the surface of Earth, potential vorticity f/h is low at low latitudes; therefore, the rim of the rotating pie corresponds to the low-latitude ocean on the Earth.

By placing the source and sink at different locations in the model, different patterns of circulation can be observed (Fig. 5.48).

From the potential vorticity argument, the system allows three types of motion only:

1. Geostrophic flow along circles of constant radius.
2. Radial flow in the interior is possible only if there is a source or a sink.
3. A western boundary current going northward or southward. Similar to the wind-driven circulation discussed in Chapter 4, balancing potential vorticity in the whole model basin rules out the possibility of closing the circulation by an eastern boundary current.

Note that an increase (decrease) in water level corresponds to an upwelling leaving the abyssal layer, so that it is a uniform sink (source). As will be explained shortly, a source/sink implies a stretching which must be balanced by radial flow as required by the linear vorticity balance.

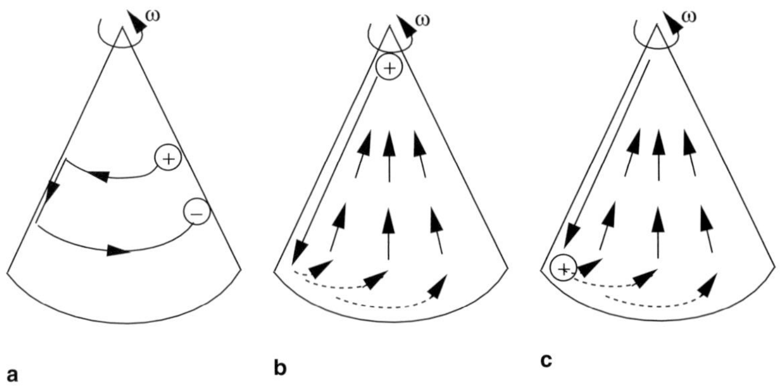

Fig. 5.48 **a** Diagram of circulation induced in a rotating sector by source (+) and sink (−); **b** sketch of flow pattern expected with source (+) at apex of sector, surface of fluid rising uniformly; **c** sketch of flow pattern expected with source (+) at the western edge of rim, with surface of fluid rising uniformly (redrawn from Stommel *et al.*, 1958).

Dynamical analysis

The projection of the centrifugal force and gravity force onto the tangent of the free surface should be in balance:

$$\omega^2 r \cos\alpha = g \sin\alpha \tag{5.57}$$

where $\tan\alpha$ is the slope of the free surface. This relation can be rewritten as

$$\omega^2 r/g = \tan\alpha = dh/dr \tag{5.58}$$

which integration leads to the shape of the free surface

$$h = h_0 \left(1 + \frac{\omega^2}{2gh_0} r^2\right) \tag{5.59}$$

The basic equations include geostrophy, hydrostatic equilibrium, and continuity:

$$2\omega v_\lambda = g \frac{\partial \zeta}{\partial r} \tag{5.60a}$$

$$-2\omega v_r = \frac{g}{r} \frac{\partial \zeta}{\partial \lambda} \tag{5.60b}$$

$$\frac{\partial}{\partial r}(hrv_r) + \frac{\partial}{\partial \lambda}(hv_\lambda) = -r\dot{\zeta} \tag{5.61}$$

where $\dot{\zeta} = \frac{\partial \zeta}{\partial t}$ is the time rate of the free surface change due to the source/sink. This term is assumed to be uniform basin-wide. (Note that the uniformity of the source term or the upwelling rate is a technical assumption, and there is no reason, *a priori*, why it should be uniform at all. In fact, upwelling of the abyssal ocean is not uniform, and the consequence of the non-uniform upwelling will be discussed later.)

Substituting Eqns. (5.60a) and (5.60b) into (5.61) leads to the vorticity equation

$$v_r \frac{dh}{dr} = -\dot{\zeta} \tag{5.62}$$

This equation corresponds to the Sverdrup relation in the wind-driven circulation theory. Here the slope of the parabolic free surface, dh/dr, plays a role equivalent to the β-effect in the ocean. Thus, the meridional velocity satisfies

$$v_r = -\frac{g\dot{\zeta}}{\omega^2 r} \tag{5.63}$$

Two cases with no net source/sink of water in the interior

Because there is no net source/sink, the free surface does not vary with time; thus, $\dot{\zeta} = 0$, so that $\frac{\partial(rv_r)}{\partial r} = 0$, and $\frac{\partial v_\lambda}{\partial \lambda} = -\frac{\partial(rv_r)}{\partial r} = 0$. There are two cases. First, if a source/sink is placed near the eastern boundary, there are zonal jets crossing the model basin, as shown in Figure 5.48a. Second, if there is no source/sink at a given "latitude" r, there should be no zonal flow. This is because the eastern wall is a rigid boundary, $v_\lambda = 0$; thus, $v_\lambda \equiv 0$ everywhere. There is an exception to this rule. Near the southern wall, friction is not negligible, so geostrophy breaks down. Water particles leave the southern boundary and enter the interior, as shown in Figure 5.48b, c.

The case with a net source of water

For this case, $\zeta > 0$, so that $v_r < 0$, which means that water in the interior moves northward! For the case with point sources/sinks, the free surface elevation increases with time $\dot{\zeta} = \frac{2}{\lambda_0 a^2} S$, where λ_0 is the angle of the pie-shaped equipment, and S is the sum of the source(defined positive) and sink (defined negative). Thus, the meridional velocity is

$$v_r = -\frac{2gS}{\lambda_0 \omega^2 a^2 r} \tag{5.64}$$

Note that this is independent of the location of the source; even if the source is put at the pole, water in the interior still moves toward the pole – toward the source. Of course, there must be a place where this dynamical constraint breaks down. As in many other cases of oceanic circulation, the western boundary layer plays a major role in fulfilling the requirement of mass balance and potential vorticity balance for a closed circulation.

Mass balance for a sector

Mass must be balanced in a steady state, and this balance gives rise to the surprising fact that the western boundary current required for mass balance in the sector may flow toward the source. The mass flux across the southern boundary of the controlled volume

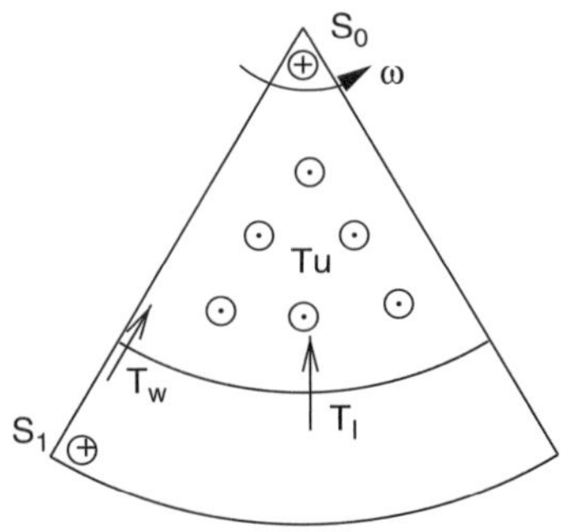

Fig. 5.49 Mass balance in a sector (redrawn from Stommel *et al.*, 1958).

(Fig. 5.49) is

$$T_I = -hr\lambda_0 v_r = \frac{S}{A}\left(1 + A\frac{r^2}{a^2}\right), \quad A = \frac{\omega^2 a^2}{2gh_0} \tag{5.65}$$

The mass flux due to surface elevation change is

$$T_u = -\frac{\lambda_0 r^2}{2}\dot{\zeta} = -S\frac{r^2}{a^2} \tag{5.66}$$

Thus, mass balance in the sector leads to the mass flux in the western boundary current

$$T_w = -T_I - T_u - S_0 = -S_0\left(1 + \frac{1 + S_1/S_0}{A}\right) \tag{5.67}$$

There are two interesting cases.

1. *A single source at the pole:* $S_1 = 0$, $S_0 > 0$. The mass flux in the western boundary layer is $T_w = -S_0(1 + 1/A)$. If $A = 1$, then $T_w = -2S_0$, i.e., there is a constant mass flux going southward, half of which is just the recirculating water.
2. *A source at the southwestern corner:* $S_0 = 0, S_1 > 0$. The mass flux in the western boundary layer is $T_w = -S_1/A$, i.e., there is a southward flux going toward the source.

Singular nature of the SAF experiments

Physically, every source has a finite dimension, i.e., it is a continuously distributed source. In fact, flow patterns obtained by releasing dye from the eastern boundary, as depicted in Figure 5.48a, are not in the form of a few clearly marked streamlines; instead, the flows appear in the form of clouds. These clouds may be interpreted as mini-gyres driven by the source and sink. Using the principle of vorticity balance, the local source/sink should induce the mini-gyres (Fig. 5.50). The net transportation of mass from the source to the sink is accomplished by the "equatorward" western boundary current along the left edge of the sector.

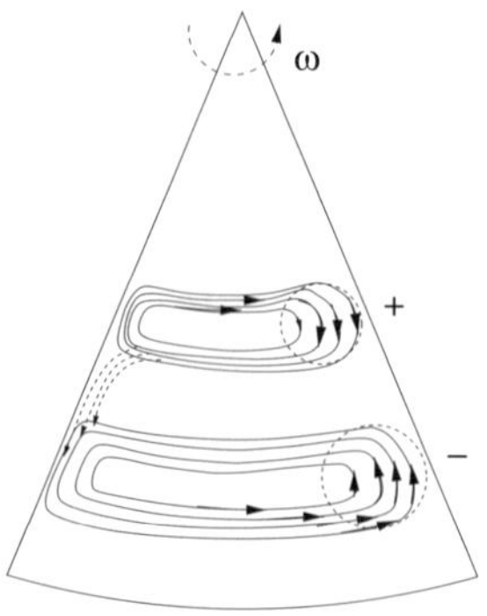

Fig. 5.50 Sketch of the mini-gyres driven by distributed source (+) and sink (−).

Source-sink driven flow on a sphere

The source-sink driven flow on a sphere was discussed by Stommel and Arons (1960a). The model consists of a single homogeneous layer of depth h covering the surface of the Earth, and the free surface elevation is ζ. The basic equations for a steady circulation are

$$-2\omega \sin\theta v = -\frac{g}{a\cos\theta}\frac{\partial\zeta}{\partial\lambda} \tag{5.68}$$

$$2\omega \sin\theta_u = -\frac{g}{a}\frac{\partial\zeta}{\partial\theta} \tag{5.69}$$

$$\frac{\partial}{\partial\lambda}(hu) + \frac{\partial}{\partial\theta}(hv\cos\theta) = aQ\cos\theta \tag{5.70}$$

The notation used in their original paper is different from the current common notations in oceanography; thus, we will adopt the latter, i.e., $0 \leq \lambda \leq 2\pi$ is the longitude and $-\pi/2 \leq \theta \leq \pi/2$ is the latitude. In addition, Q is defined as positive (negative) for distributed source (sink) of deep water. In the following analysis, we will assume that there is no bottom topography and the free surface satisfies the following relation, $\zeta \ll h$, so that h can be treated as approximately constant. By cross-differentiating Eqns. (5.68, 5.69), subtracting and using Eqn. (5.70), we obtain

$$v = -\tan\theta\frac{Qa}{h} \tag{5.71}$$

which corresponds to the Sverdrup relation in the wind-driven circulation. For example, a distributed source with $Q > 0$ drives a southward flow in the interior. Substituting Eqn. (5.71) into Eqn. (5.68) gives

$$\frac{\partial\zeta}{\partial\lambda} = -\frac{2\omega a^2\sin^2\theta}{gh}Q \tag{5.72}$$

From Eqn. (5.70) we obtain

$$\frac{\partial u}{\partial \lambda} = \frac{a}{h}\left(\sin\theta\frac{\partial Q}{\partial \theta} + 2Q\cos\theta\right) \tag{5.73}$$

Note that for the steady case, mass conservation requires the source distribution to satisfy

$$\iint Qa\cos\theta d\theta d\lambda = 0 \tag{5.74}$$

Evaporation–precipitation hemispheres with no boundaries

Assume a simple pattern of evaporation and precipitation

$$Q = -Q_0 \sin\lambda\cos\theta \tag{5.75}$$

The solution is

$$v = \frac{aQ_0}{h}\sin\lambda\sin\theta \tag{5.76}$$

$$u = -\frac{Q_0 a}{h}\left(3\sin^2\theta - 2\right)\cos\lambda - \frac{g}{2a\omega\sin\theta}\frac{dG}{d\theta} \tag{5.77}$$

$$\zeta = -\frac{2\omega a^2 Q_0}{gh}\sin^2\theta\cos\theta\cos\lambda + G(\theta) \tag{5.78}$$

where $G(\theta)$ is an arbitrary function. The circulation pattern is shown in Figure 5.51, with $G(\theta) = 0$. The basic parameters are $h = 5{,}000$ m, and $Q_0 = 1$ m/yr.

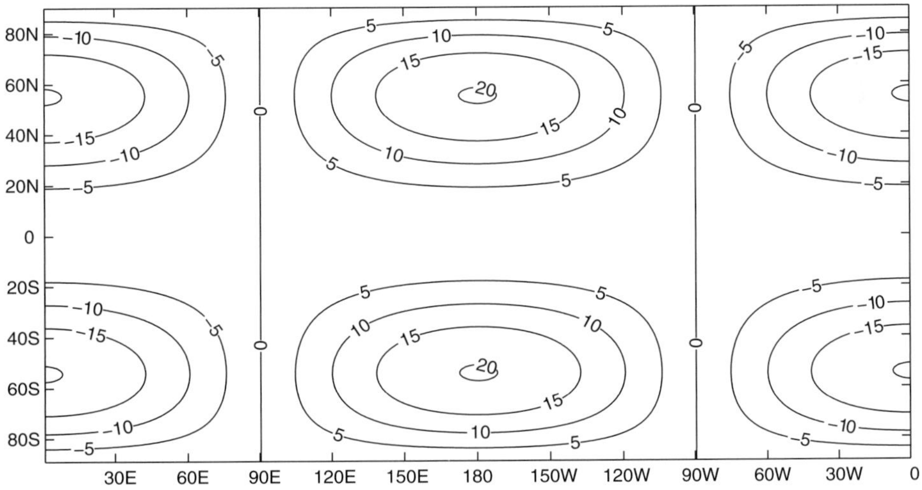

Fig. 5.51 Free surface elevation for the case with evaporation in the eastern hemisphere and precipitation in the western hemisphere.

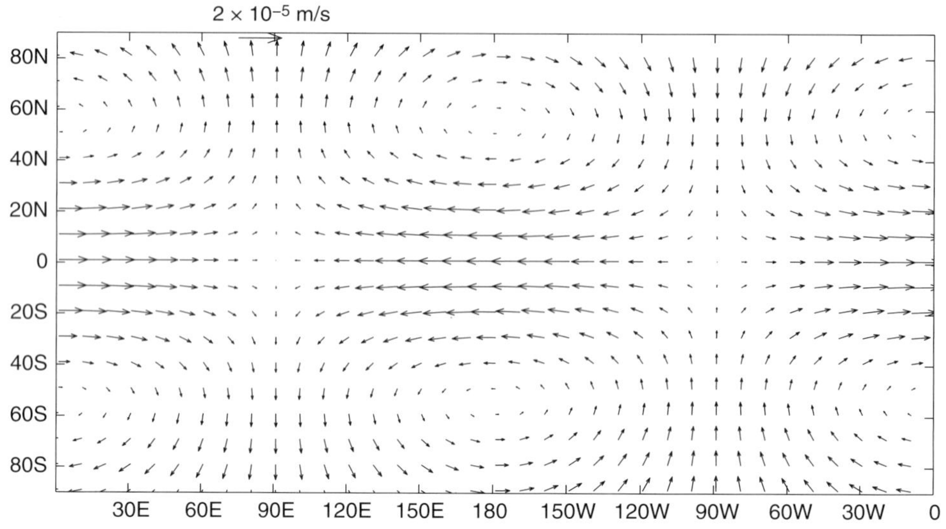

Fig. 5.52 Horizontal velocity for $Q = -Q_0 \sin\lambda \cos\theta$ with no meridional boundaries.

The corresponding velocity map is shown in Figure 5.52. There is clearly poleward flow in the region of evaporation (resembling the Ekman suction) in the eastern hemisphere, and equatorward flow in the western hemisphere. In addition, geostrophic flow is divergent because $f = 2\omega \sin\theta$ changes with θ.

Basins bounded by meridians

The circulation driven by a point source/sink in a basin confined by two meridians is particularly interesting because the western boundary current induced by source-induced interior flow can be very non-intuitive. We start from the case with a point source S_0 at the North Pole and uniform sink, i.e., $Q = -Q_0 < 0$ in the interior, which satisfies

$$S_0 = Q_0 a^2 (\lambda_2 - \lambda_1) \tag{5.79}$$

where λ_2 and λ_1 are the meridional boundaries of the basin. The solution is

$$v = \frac{Q_0 a}{h} \tan\theta \tag{5.80}$$

$$u = \frac{2aQ_0}{h} \cos\theta (\lambda_2 - \lambda) \tag{5.81}$$

$$\zeta = -\frac{2\omega a^2 Q_0}{gh} \sin^2\theta \, (\lambda_2 - \lambda) \tag{5.82}$$

There is a western boundary current whose transport can be determined by the mass balance within a sector bounded by the meridians λ_1 and λ_2, plus the equatorial boundary of θ. The total upwelling rate within this sector is

$$U_p = \int_{\lambda_1}^{\lambda^2} \int_{\theta}^{\pi/2} Q_0 a^2 \cos\theta d\lambda d\theta = Q_0 a^2 (1 - \sin\theta)(\lambda_2 - \lambda_1) \quad (5.83)$$

The poleward transport in the ocean interior across the latitude circle of θ is

$$I_p \int_{\lambda_1}^{\lambda_2} hva \cos\theta d\lambda = Q_0 a^2 \sin\theta (\lambda_2 - \lambda_2) \quad (5.84)$$

The mass balance within this sector leads to the transport of the western boundary current

$$T_w = U_p - S_0 - I_p = -2S_0 \sin\theta \quad (5.85)$$

Thus, the western boundary current should carry an equatorward transport, as shown in Figure 5.53a. In the basin interior there is a cyclonic poleward flow, and there is an equatorward western boundary current. This case is very similar to the result of Stommel *et al.* (1958) but, due to the spherical geometry, returning flow always exists along the western boundary.

For the general case of a point source placed at the western boundary at latitude θ_0, the solution is as follows. There is an interior upwelling driven by the source, so that there is always the same cyclonic flow in the basin interior. However, the transport of the western boundary current may vary depending on the exact location of the source (Fig. 5.54).

If the source is at the equator, the western boundary current is poleward south of 30° N; however, north of this critical latitude, the western boundary current is equatorward, as

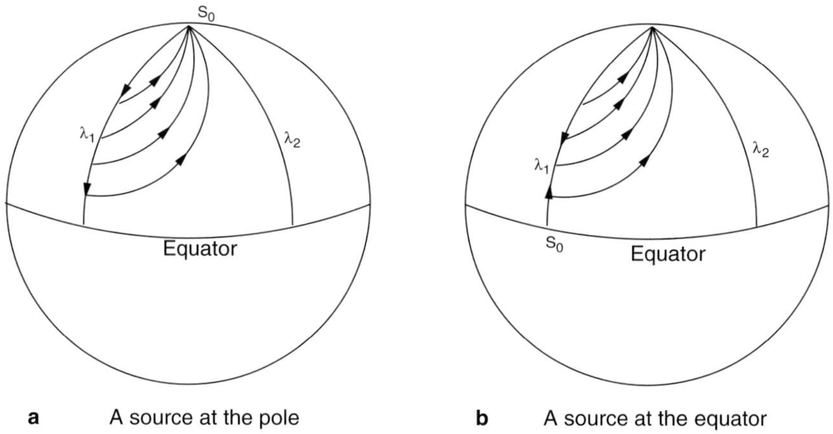

Fig. 5.53 Source-driven flow on a sphere: **a** source at the pole; **b** source at the equator (redrawn from Stommel and Arons, 1960a).

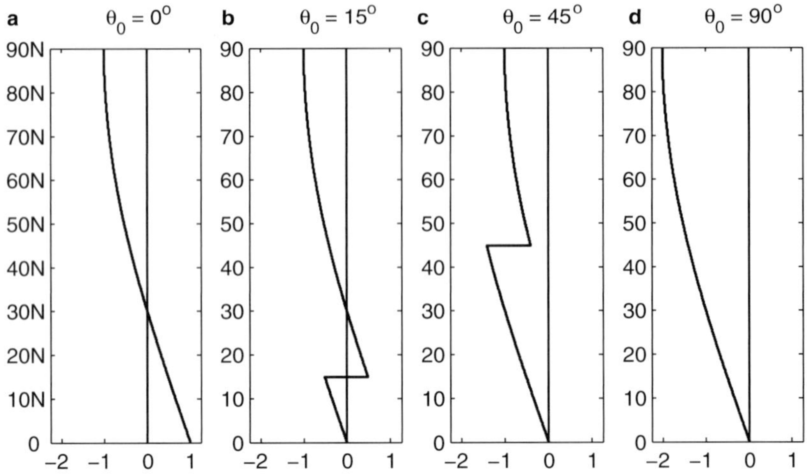

Fig. 5.54 **a–d** Transport of the western boundary current driven by a point source specified at location θ_0, in units of the total source S_0.

shown in Figures 5.53b and 5.54a. If the source is south of 30° N, the western boundary current is equatorward south of the source and poleward in the vicinity north of the source; however, north of 30° N, the western boundary current is equatorward (Fig. 5.54b). If the source is north of 30° N, the western boundary current is always equatorward. In particular, if the source is placed at the North Pole, the transport of the western boundary current is twice the strength of the source. Therefore, the western boundary current transport is quite non-intuitive, owing to mass balance of the model and the special nature of spherical coordinates.

If a sink is placed at the western boundary, the circulation is opposite to that produced by a source; the corresponding solutions are shown in Figure 5.55. It is interesting to note that if the sink is placed at 15° N, the western boundary current transport in the vicinity of the sink is toward the sink (Fig. 5.55b). Such a flow pattern is quite similar to the case of a non-rotating fluid; however, other than this case, the flow pattern driven by a point source/sink on a rotating sphere can be quite different from the corresponding pattern of flow produced by a point source/sink in the classical non-rotating fluid.

Based on theoretical reasoning, Stommel (1958) postulated a framework of deep circulation in the world's oceans (Fig. 5.56). Accordingly, deep circulation in the world's oceans is driven by point sources of deep water in the northern North Atlantic Ocean and the Weddell Sea. With this distribution of the deepwater source, there is a southward western boundary current in the Atlantic Basin which transports deep water away from the source. On its southward passage, it gradually loses its mass to the ocean interior. At the southern edge of the world's oceans, there is a simple circumpolar current around the edge of Antarctica and deep western boundary currents moving northward in individual basins which transport deep water from the southern source to the interior of the world's oceans. In the

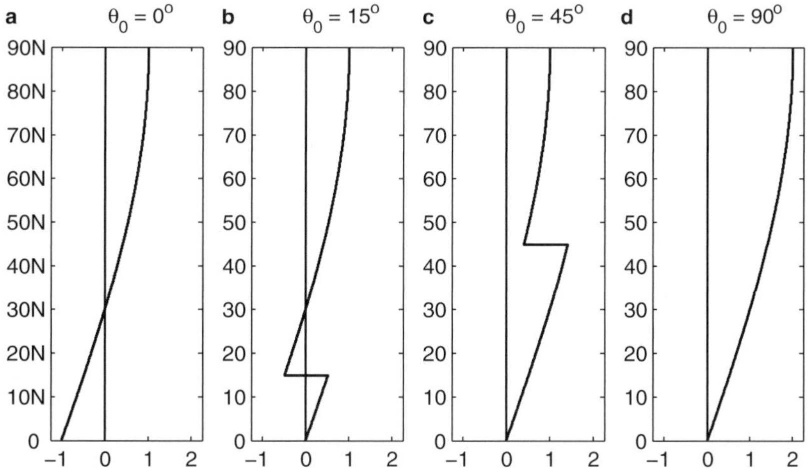

Fig. 5.55 **a–d** Transport of the western boundary current driven by a point sink specified at location θ_0, in units of the total sink S_0.

Fig. 5.56 Deep circulation in the world ocean postulated by Stommel (1958).

interior of each basin water uniformly moves upward, as assumed in the model, and this model-assumed uniform upwelling drives poleward flow, as dictated by the linear potential vorticity balance.

5.2.3 Generalized theories of deep circulation

Since Stommel and his colleagues proposed a theoretical framework for the deep circulation in the world's oceans in the 1960s, their theory has dominated the field of deep circulation in the world's oceans. It was only in the 1980s that people started to realize the limitations of the classical theory of deep circulation.

Stommel's theory is based on a very simple and solid theoretical foundation, including the steadiness of the circulation, the assumption of no bottom topography, and the assumption of uniform upwelling and point sources of deep water. Under such assumptions, the simple solution obtained from the model is the only logical result from the fundamentals of fluid dynamics.

It was very clear that direct observational confirmation of the poleward flow in the deep ocean interior predicted by the theory was very difficult because it is extremely slow. Thus, for a long time the relatively fast western boundary currents observed in the world's oceans have been used as the most concrete evidence in support of the theory. However, it eventually became clear that the simple assumptions made in the theory have rather serious limitations if one wants to describe the deep circulation in the world's oceans more accurately.

Modification of the Stommel and Arons theory

With the progress of *in situ* observations and further scrutiny of the physics involved, the discrepancy between the classical Stommel theory and the circulation in the deep ocean became clearer, and more realistic features have been added to the model in order to describe different physical features of the deep circulation. The most essential issues include the following.

Eastern boundary currents due to topographic β-effect

The bottom slope along the eastern boundary of the basin may be so steep that the corresponding topographic β-effect overpowers the planetary β-effect. As a result, a strong deep boundary current can appear along the eastern boundary of the basin.

Non-uniform upwelling

Although upwelling is the term used in many studies, strictly speaking "diapycnal velocity" is the correct term. In general, upwelling indicates a positive vertical velocity, which can be induced by diapycnal mixing; it can also be induced by adiabatic vertical motions due to flow over topography. In addition, strong upwelling can also be associated with the divergence of Ekman transport, such as the coastal upwelling and the strong upwelling in the Southern Ocean. In the following discussion, upward motion due to diapycnal mixing is referred to as upwelling.

For the idealized basin with a flat bottom, upwelling velocity is not necessarily uniform in the whole basin. Simple scaling suggests that the vertical velocity is related to the thermocline depth $w = \kappa/h$, where κ is the vertical diffusivity and h is the scale depth. Since the thermocline is shallow along the eastern boundary, it is expected that upwelling there can be much larger than in the western part of the basin.

Kawase (1987) postulated that the upwelling through the interface of a two-layer model is linearly proportional to the deviation from a prescribed reference position of the interface

$$w^* = \Gamma\,[h_0(x, y) - h(t, x, y)] \tag{5.86}$$

where Γ is the relaxation factor, which is the inverse of the relaxation time, and $h_0(x, y)$ and $h(t, x, y)$ are the depth of the reference state and the current state of the interface. As an alternative, the interface upwelling rate can be determined as a part of the solution through coupling the upwelling with the surface thermohaline forcing (Huang, 1993a).

Recent field observations indicate that diapycnal mixing is greatly enhanced near rough bottom topography, such as the mid-ocean ridge and seamounts (Ledwell *et al.*, 2000). Thus, the corresponding upwelling specified in the inverse reduced-gravity model should be non-uniform. Indeed, the dynamical consequence of non-uniform mixing/upwelling is a research frontier in abyssal circulation; a topic which is discussed in detail later in this chapter.

Baroclinic circulation in a stratified abyssal ocean

The classical Stommel and Arons theory is based on the assumption that density is uniform in the abyss, and that it predicts the barotropic circulation in the abyssal ocean. In a stratified ocean, density stratification gives rise to baroclinic circulation. The meridional velocity is governed by the Sverdrup relation, i.e., the linear vorticity balance $\beta v = f \partial w / \partial z$. In a zonal section, density along the eastern boundary is relatively light, and this density anomaly propagates westward by stationary diffusive Rossby waves; thus, baroclinic structure in velocity and density exists in the abyss (Pedlosky, 1992).

Hypsometry

In general, the bottom of an individual basin is not flat; instead of a flat bottom, most basins can be characterized in terms of the shape of a Chinese wok. Due to the decrease in horizontal area with depth, the circulation can be quite different from the classical theory of Stommel.

Assume that the horizontal area of the basin is $A(z)$ and the total amount of deepwater source for the abyss below level z is $S(z)$. Thus, the vertical velocity across level z is $w = S(z)/A(z)$. For the interior ocean, the linear potential vorticity balance is $\beta v = f \frac{dw}{dz} = f \frac{d}{dz}\left(\frac{S(z)}{A(z)}\right)$. Different combinations of $S(z)$ and $A(z)$ can give rise to dramatically different meridional velocity patterns at different levels (Rhines and McCready, 1989). For example, if $dw/dz < 0$, meridional velocity in the abyssal interior must move toward lower latitudes, opposite to the direction predicted by the classical theory.

Geothermal heating

Although the thermal isolation condition has been used in most ocean models, there is a heat flux through the sea floor. Geothermal heat flux released from major eruptions of volcanic activity may induce large plumes over the mid-ocean ridge, as discussed by Stommel (1982). However, for the global thermohaline circulation, the contribution from geothermal heat flux is small, and thus negligible. On the other hand, if we are interested in the abyssal circulation, geothermal heat flux can be a major contributor to the abyssal stratification, as demonstrated by Thompson and Johnson (1996). Adcroft *et al.* (2001) carried out numerical experiments using an oceanic general circulation model. By including a uniformly distributed geothermal heat flux of 50 mW/m^2, the bottom temperature increases by $0.1 - 0.3°C$, compared with the case without geothermal heat flux. Such a change in bottom water temperature is substantial; thus, it is clear that if we want to simulate bottom water properties and circulation accurately, geothermal heat flux must be included.

Grounding

Since the value of bottom water flux is finite, part of the bottom of the deep basin may not be covered by bottom water. The finiteness of bottom water makes the problem quite different from the cases discussed in the Stommel theory (Speer and McCartney, 1992). If the strength of the bottom water source is not great enough, say

$$\textbf{Source} < w^* \cdot (\textbf{Basin area}) \tag{5.87}$$

a phenomenon known as "grounding" occurs (w^* is the specified upwelling velocity through the upper surface of the bottom water); an example is shown in Figure 5.57.

This phenomenon is a mirror image of the outcropping phenomenon discussed in Section 4.1.4. The essential difference between the models for outcropping and for grounding is in the integral constraint. For the generalized Parsons model, the total amount of water in the upper layer must be constant, while in the grounding model, the total amount of upwelling should equal the source of bottom water.

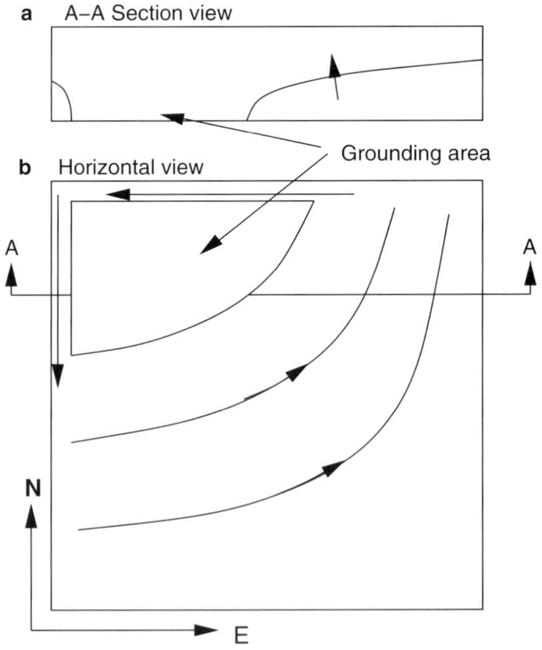

Fig. 5.57 Sketch of a model for bottom water circulation with the grounding phenomenon; **a** A–A section view, **b** horizontal view (adapted from Speer and McCartney, 1992).

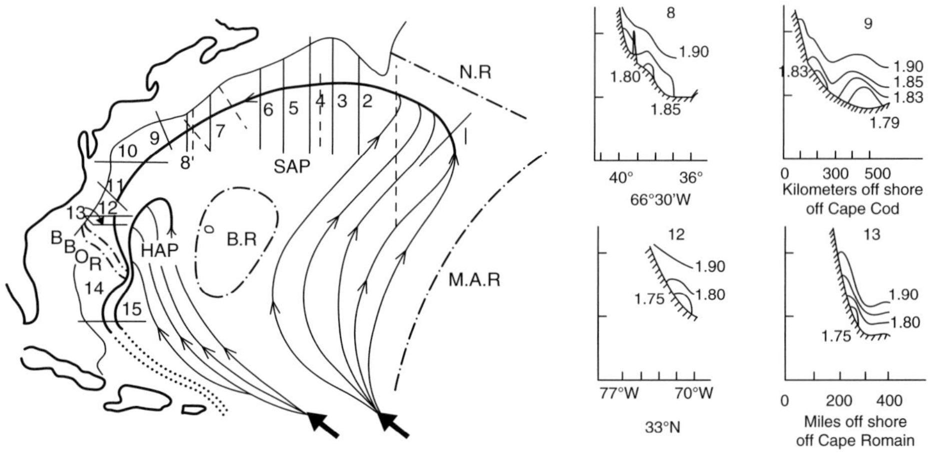

Fig. 5.58 *Left panel*: sketch of the North Atlantic Basin showing the locations of sections; the Mid-Atlantic Ridge (MAR), Newfoundland Ridge (NR), Bermuda Ridge (BR), and Blake–Bahama Outer Ridge (BBOR) are indicated by *dot-dashed lines*; HAP and SAP represent Hatteras Abyssal Plain and Sohm Abyssal Plain, respectively; *right panels*: potential temperature distribution for several sections (adapted from Weatherly and Kelley, 1985).

When grounding appears, bottom water is confined to the eastern boundary. Although it looks like an eastern boundary current, for the case of a flat bottom it is not a strong current. As a result of grounding, there will be an isolated northern boundary current and an isolated western boundary current whose structure has been discussed by Huang and Flierl (1987). These currents are separated from the main body of bottom water by a vast region where no bottom water exists.

Weatherly and Kelley (1985) have found a nearly continuous filament of cold water which flows equatorward at 40° N and 62° W, left panel in Figure 5.58. From the water mass properties, it was speculated that this cold filament consists of AABW, as shown by the potential temperature contours in the right panels of Figure 5.58. Note that when grounding takes place, the bottom water piles up along the edge of the deep basin, including the eastern part of the continental slope. Although this may look like an eastern boundary current, it is a relatively slow current; thus, this current is quite different from the strong eastern boundary current due to the topographic β-effect. As shown by Speer and McCartney (1992), the inverse reduced-gravity model can give a fairly good description of this flow pattern.

Topography and grounding in a simple inverse reduced-gravity model

The bottom water circulation can be simulated in terms of an inverse gravity model. For example, Speer *et al.* (1993) discussed numerical experiments of an inverse reduced-gravity model aimed at understanding the bottom water circulation in a single-hemisphere ocean.

The basic equations are obtained by slight modification from Eqn. (5.53a)

$$\frac{Ru}{h} - fv = -g'(h+d)_x \tag{5.88a}$$

$$\frac{Ru}{h} - fu = -g'(h+d)_y \tag{5.88b}$$

where R is the bottom friction parameter. The continuity equation is

$$h_t + \Delta \cdot (h\vec{u}) = -w^* \tag{5.89}$$

where w^* is the upwelling velocity specified as constant in their study. The corresponding potential vorticity ($q = f/h$) equation can be written as

$$\frac{d \ln q}{dt} = \frac{w^*}{h} - \frac{R}{f}\left[(v/h)_x - (x/h)_y\right] \tag{5.90}$$

Topographic β-effect

In the interior ocean friction is negligible, so the potential vorticity equation is reduced to

$$\beta_{eff}\, v = \frac{fw^*}{h} \tag{5.91}$$

where $\beta_{eff} = \beta - fh_y/h$ includes the topographic β term, which is proportional to the depth gradient h_y and inversely proportional to the layer depth.

Speer *et al.* (1993) carried out a numerical experiment with a model basin which has a sloping bottom in the equatorial part and a flat bottom elsewhere. In their model the topographic term reaches the maximum at the southwestern corner, where the layer thickness is minimal. The velocity pattern indicates an eastern boundary current in the southern basin. The flow pattern in the northern part of the basin is basically the same as predicted by the classical Stommel and Arons theory.

An example of bottom water circulation with grounding phenomenon obtained from a reduced-gravity model with a flat bottom is shown in Figure 5.59. The model is set up on a β plane with the central latitude at 30° N; there is an influx of 10 Sv specified along the southern boundary. Upwelling is assumed to be uniform in the basin interior, $w^* = 0.6 \times 10^{-6}$ m/s; wherever the layer grounds, the corresponding upwelling is set to zero.

As shown in Figure 5.59, bottom water is mostly confined to the eastern part of the basin, with a large "window" in the northwestern part of the basin. Although water seems to pile up in the eastern basin, there is no strong boundary current, as shown in the meridional/zonal sections. If we use a non-uniform upwelling rate and non-flat bottom, the circulation may become more complicated.

Dynamical role of topography and non-uniform upwelling in the world's oceans

The most outstanding features in the world's oceans include the mid-ocean ridge systems, seamounts and trenches (Fig. 5.60). Mid-ocean ridge systems in the world's oceans are the

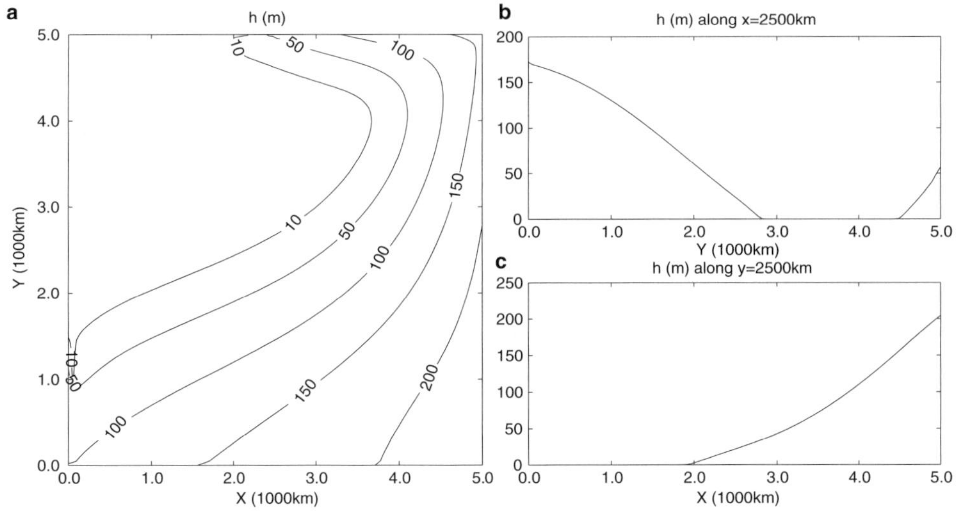

Fig. 5.59 **a–c** Bottom water circulation based on a numerical experiment of an inverse reduced-gravity model.

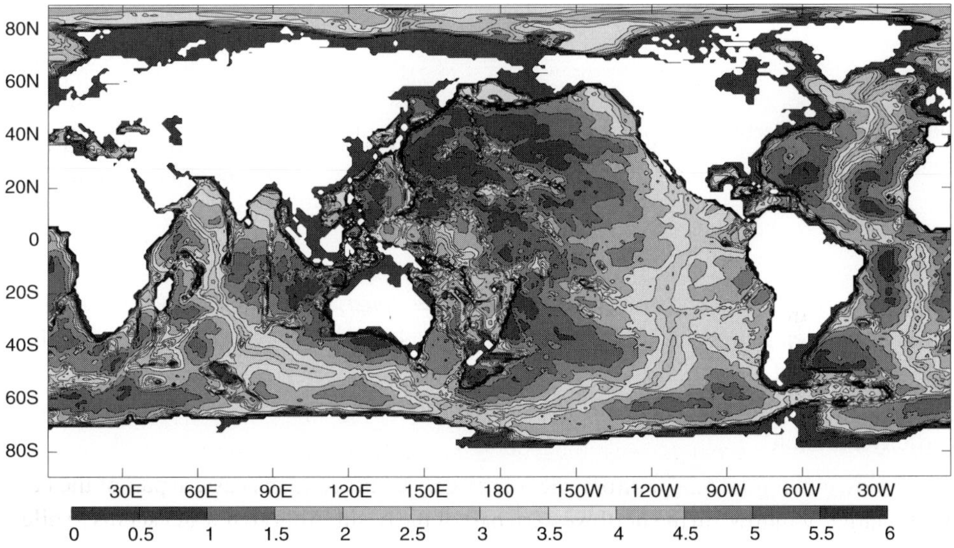

Fig. 5.60 Seafloor topography (in km) based on NOAA topographic dataset. See color plate section.

most conspicuous planetary-scale topography on the surface of the Earth, and they play a crucial role in the oceanic circulation. The Atlantic Basin is a young basin on a geologic time scale. As such, the mid-ocean ridge separates the basin into the eastern and western basin. On the other hand, the Pacific Basin is an old basin, and its mid-ocean ridge is pushed toward the eastern boundary. In addition, there are many deep trenches in the northern and western edge of the Pacific Basin. Because of the complicated features of these topographies, bottom circulation is quite different from that predicted from the classical Stommel–Arons theory based on the assumption of a flat bottom.

Owing to the shallow depth and rough sea floor of these topographic features, strong tidal mixing drives strong upwelling in their vicinity. In this section we explore the dynamical consequences of these topographic features and the associated strong upwelling, using the inverse reduced-gravity model. Apparently, flow in the middle depth of the ocean may not be very slow, so that the basic assumption of the inverse reduced-gravity model is not strictly valid; nevertheless, such models can be used to explore the abyssal circulation in simple theoretical studies and highlight the vitally important dynamical processes.

Model formulation

For simplicity, the β plane is used here with $f = f_0 + \beta(y - y_o), f_0 = \omega \sin 30°, \beta = 2\omega \cos 30°/a$, where $\omega = 7.27 \times 10^{-5}$/s and $a = 6,371$ km. Assuming that flow in the layer above moves much more slowly than the abyssal layer, the basic equations controlling the abyssal circulation are geostrophy

$$-fv = -g'\zeta_x \tag{5.92a}$$

$$fu = -g'\zeta_y \tag{5.92b}$$

where ζ is the interfacial displacement, and g' is the reduced gravity. In a steady state, the continuity equation is

$$(hu)_x + (hv)_y = -w^* \tag{5.93}$$

where $h = H - b + \zeta$ is the layer thickness; $H = const.$ is the unperturbed thickness of the abyssal layer; $b = b(x, y)$ is the bottom topography, and w^* is the interfacial upwelling velocity specified *a priori*.

Cross-differentiating Eqns. (5.92a) and (5.92b) and using Eqn. (5.93) leads to the potential vorticity equation

$$uq_x + vq_y = fw^*/h^2 \tag{5.94}$$

where $q = f/h$ is potential vorticity. Using geostrophy, this equation can be rewritten as

$$q_y\zeta_x - q_x\zeta_y = f^2w^*/g'h^2 \tag{5.95}$$

which is a first-order partial differential equation for ζ. The potential vorticity appears as the characteristic for this equation, and the equation can be solved by integration along characteristics. Since integration along characteristics may produce results in a non-uniform grid, it is convenient to project the characteristic equation onto an equal grid in the x-coordinate; the corresponding equation is

$$\frac{D\zeta}{Dx} = \frac{f^2 w^*}{g' h^2 q_y}, \text{ along } q = const. \tag{5.96}$$

In general, potential vorticity includes the contribution of the interfacial displacement, so this equation is a nonlinear equation. When the interfacial displacement is much smaller than the topographic height, its contribution to layer thickness can be neglected, and Eqn. (5.96) is reduced to a linearized equation in terms of potential vorticity coordinates, which is determined from the topography alone.

Flow over a one-dimensional ridge

For simplicity, we focus on the case with a one-dimensional mid-ocean ridge symmetric with respect to the ridge axis ($x = 0$), and the ridge topography is a sinusoidal function

$$b = \begin{cases} B[0.5 + 0.5\cos(x\pi/L)], & |x| \le L \\ 0, & |x| > L \end{cases} \tag{5.97}$$

where B is the maximal height of the ridge. The upwelling rate at the interface is specified as

$$w^* = w_{00} + w_0 b/B. \tag{5.98}$$

where w_{00} is the uniform upwelling rate over the whole basin, and w_0 is the additional maximal upwelling rate specified on the top of the topography. Since the topography is independent of y, $q_y = (f/h)_y = \beta/h$, and Eqn. (5.96) is reduced to

$$\frac{D\zeta}{Dx} = \frac{fq}{g'\beta} w^* \tag{5.99}$$

where q is conserved along each characteristic. This equation can be integrated from the eastern boundary with the boundary condition of $\zeta = 0$ along the eastern boundary.

Over the mid-ocean ridge, potential vorticity contours are bent equatorward owing to topographic stretching (Fig. 5.61).

At the eastern boundary on the q_0 contour, $f = f_0$; on the q_1 contour,

$$f = f_1 = f_0 + \beta \Delta y. \tag{5.100}$$

As potential vorticity is constant along q_1, we have

$$\frac{f_0 + \beta \Delta y}{H} = \frac{f_0}{H - B} \tag{5.101}$$

thus, the meridional separation between these two contours is

$$\Delta y = \frac{f_0 B}{\beta (H - B)} \tag{5.102}$$

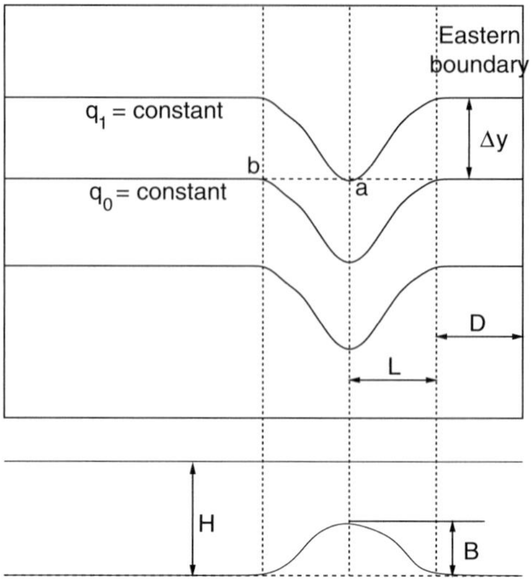

Fig. 5.61 Potential vorticity contours (*upper panel*) in a basin with a mid-ocean ridge (*lower panel*). The mean depth of the abyssal layer is H and the maximal height of the ridge is B.

The interfacial displacement at point $a(x = 0)$ is obtained by integrating Eqn. (5.99) backward along the characteristic $q_1 = const$. Along this characteristic

$$f = \begin{cases} q_1 H(1 - b/H), & 0 \leq x \leq L \\ q_1 H, & L \leq x \leq L + D \end{cases} \tag{5.103}$$

$$w^* = \begin{cases} w_{00} + w_0 b/B, & 0 \leq x \leq L \\ w_{00}, & L \leq x \leq L + D \end{cases} \tag{5.104}$$

thus, we have

$$\zeta_a = -\int_0^L \frac{fq_1}{g'\beta} w^* dx - \int_L^{L+D} \frac{fq_1}{g'\beta} w^* dx = -\frac{f_1^2}{g'\beta H} \left[w_{00} D + I \right],$$
$$I = \int_0^L (1 - b/H)(w_{00} + w_0 b/B)\, dx \tag{5.105}$$

Similarly, the interfacial displacement at point b is

$$\zeta_b = -\frac{f_0^2}{g'\beta H} [w_{00} D + 2I] \tag{5.106}$$

If $\zeta_a > \zeta_b$, geostrophic constraint requires that the meridional flow integrated over the western flank of the mid-ocean ridge should move poleward, as predicted by the classical Stommel–Arons theory; this pattern of circulation is called subcritical.

On the other hand, if $\zeta_a < \zeta_b$, the meridional flow integrated over the western flank of the mid-ocean ridge should move equatorward, opposite to the classical Stommel–Arons theory; the circulation and the topography are called supercritical.

For the present case, the corresponding constraint is

$$f_0^2(w_{00}D + 2I) < f_1^2(w_{00}D + I) \tag{5.107}$$

Since $w_{00} \ll w_0$, omitting the contribution due to the weak background upwelling reduces this relation to $\sqrt{2}f_0 < f_1$.

Using Eqn. (5.101), we obtain the critical condition

$$B > (1 - \sqrt{2}/2)H \approx 0.293H \tag{5.108}$$

For the general case, Eqn. (5.107) is reduced to a cubic equation for the critical height ratio $x = B/H$.

According to the classical theory of abyssal circulation of Stommel and Arons (1960a), under the assumptions of uniform upwelling and with no bottom topography, flow in the ocean interior is poleward, as shown in Figure 5.62a. A western boundary current is required in order to close the circulation; however, this is omitted from our discussion here.

For the case of enhanced upwelling over a meridional strip in the middle of the ocean and with no bottom topography, the circulation becomes stronger, but its pattern remains similar (Fig. 5.62c).

For the case of a ridge taller than the critical height under uniform upwelling, water over the western flank of the mid-ocean ridge moves equatorward, although water over the eastern flank or west of the mid-ocean ridge still moves poleward (Fig. 5.62b).

For the case with supercritical topography and enhanced upwelling over the mid-ocean ridge there is a strong circulation over the ridge (Fig. 5.62d).

This example demonstrates that topographic stretching is the primary force driving the model toward an equatorward circulation in the western flank of the ridge, opposite to the poleward flow predicted by the classical Stommel–Arons theory.

Flow over a seamount

As another example, we study a sinusoidal-shaped seamount with the mean depth of the abyssal layer $H = 3,000$ m. The seamount topography and upwelling are in the following forms

$$b = \begin{cases} 0.5B[1 + \cos(\pi r/r_0)], & \text{if } r = \sqrt{[(x - x_0)^2 + (y - y_0)^2]} \leq r_0 \\ 0, & \text{otherwise} \end{cases} \tag{5.109}$$

$$w^* = w_{00} + w_0 b/B \tag{5.110}$$

where $w_{00} = 10^{-7} m/s$, $w_0 = 5 \times 10^{-7} m/s$.

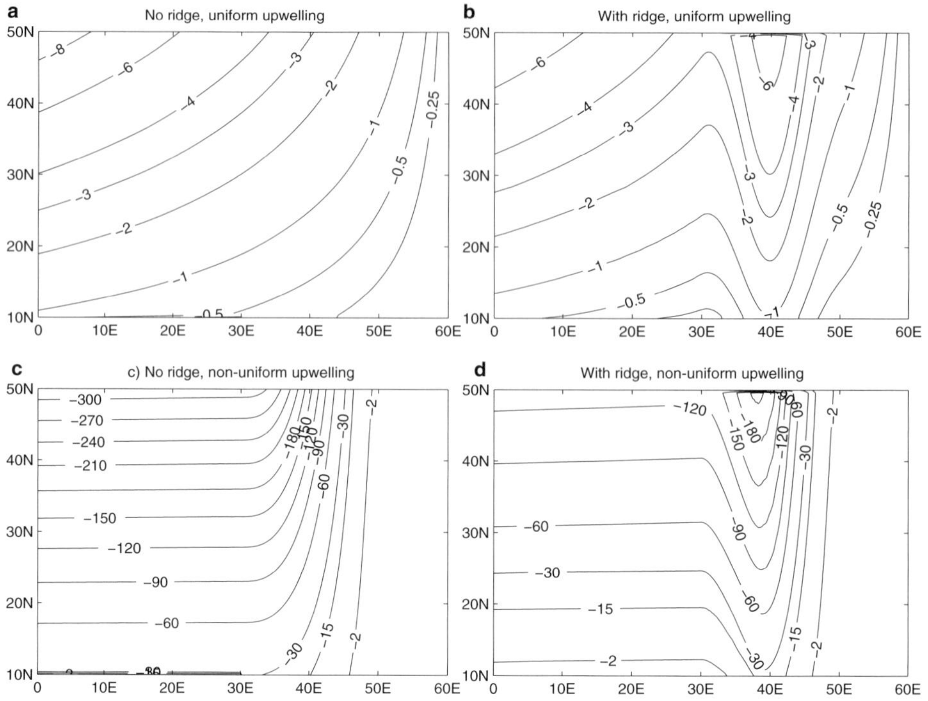

Fig. 5.62 Interfacial displacement (in units of m), the thickness of the abyssal layer is $H = 3,000$ m, the uniform upwelling rate is $w_{00} = 10^{-7}$ m/s, the amplitude of the enhanced upwelling rate over the mid-ocean ridge is $w_0 = 2 \times 10^{-5}$ m/s, and the maximum height of the mid-ocean ridge is $B = 1,500$ m (supercritical).

This case is different from the previous case of a one-dimensional mid-ocean ridge. In the present case, the topography is two-dimensional, and typical potential vorticity contours are shown in Figure 5.63. When closed potential vorticity contours exist, we cannot obtain the solution by simple integration along characteristics because some of the characteristics are closed, and solutions within such closed characteristics have to be found by invoking higher-order dynamics, such as bottom friction and inertial terms, which are omitted in the simple model discussed here. More elaborate numerical approaches are required for such cases; examples are given by Katsman (2006). Our discussion here is limited to the case of no closed geostrophic contours, and we will call such topography "sub-critical topography."

Flow in the abyss with subcritical topography (maximal height of 400 m) is shown in Figures 5.64 and 5.65. This case again shows that the enhanced upwelling near the topography alone does not change the fundamental structure of the abyssal circulation. Although the localized enhancement of upwelling can alter the zonal velocity near the latitude of the seamount, it does not change the direction of the meridional flow, as shown in Figures 5.64c and 5.65c.

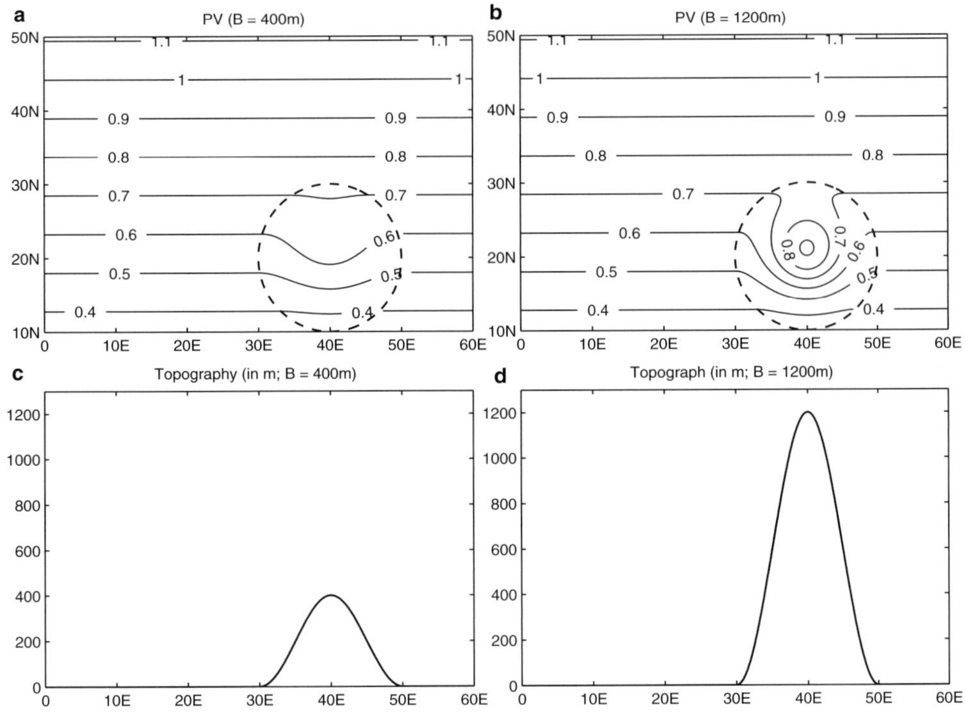

Fig. 5.63 **a–d** Seamount topography (**c**, **d**) and the associated normalized potential vorticity (**a**, **b**), $q = fH/h$ (in units of 10^{-4}/s). The maximal height of the seamount is 400 m (**a**, **c**) and 1,000 m (**b** , **d**).

It is interesting to see that perturbations induced by locally enhanced upwelling associated with an axis-symmetric seamount are nearly symmetric in the meridional direction, and the minor asymmetry is due to the increase of the Coriolis parameter in the meridional direction.

On the other hand, the topographic stretching can induce noticeable change in the meridional velocity pattern near the seamount. In fact, flow near the western slope of the seamount is equatorward (Fig. 5.65b), similar to the case of a mid-ocean ridge discussed above. Perturbations induced by topography are highly asymmetric in the meridional direction, as shown in Figure 5.64b. The negative anomaly in the interfacial displacement has a shape like a golf club.

Therefore, topographic stretching associated with both the mid-ocean ridge and an isolated seamount can induce equatorward flow on the western flank of the topography, opposite to the poleward flow predicted by the Stommel–Arons classical theory.

Flow over isolated steep bottom topography

In the world's oceans, there are many instances of isolated steep bottom topography, including seamounts and trenches. In terms of an inverse gravity model, most potential vorticity

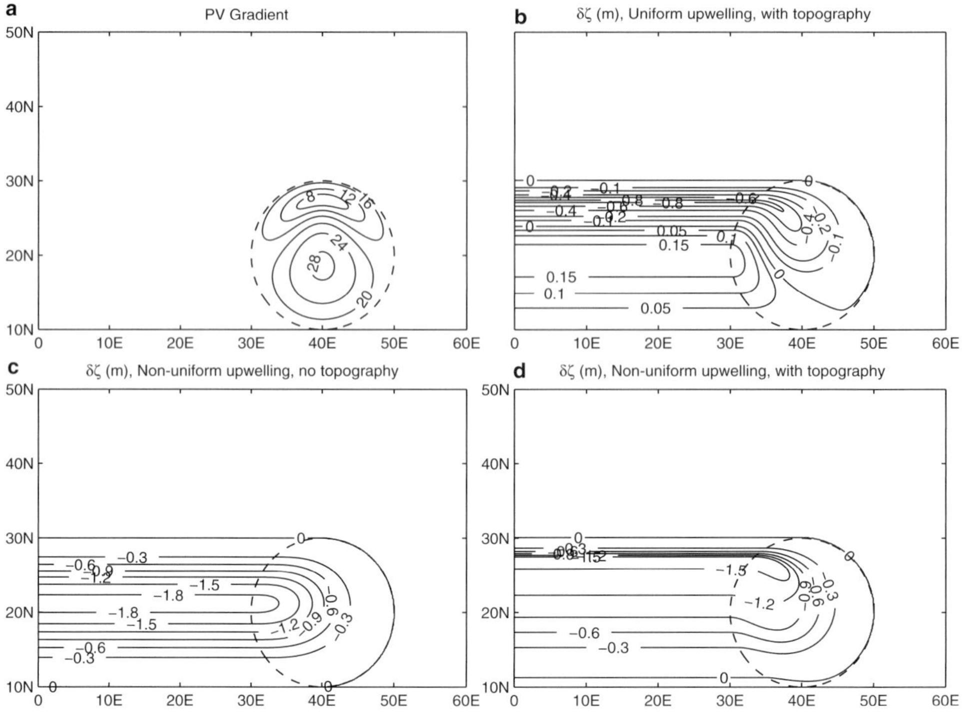

Fig. 5.64 Flow structure due to interfacial upwelling and a seamount (**a**); **b**, **c**, and **d** show changes in the interface displacement (in units of m), compared with the standard case of uniform upwelling and no topography.

isolines over these topographies are in the form of closed contours, and the circulation is in the regime of strongly nonlinear dynamics.

Since flow over steep topography can be quite strong, we reformulate the basic equations introduced in the subsection entitled "Inverse reduced-gravity model" in Section 5.2.2, including the nonlinear advection terms and the bottom friction terms. Assuming that the flow is steady, the momentum and continuity equations for an inverse reduced-gravity model are

$$uu_x + vu_y - fv = -g'\zeta_x - Ru/h_0. \tag{5.111a}$$

$$uu_x + vu_y - fu = -g'\zeta_y - Ru/h_0. \tag{5.111b}$$

$$(hu)_x + (hv)_y = -w^*. \tag{5.111c}$$

where h and h_0 are the layer thickness and its mean value, and $R(u, v)/h_0$ is the bottom friction term. The specific form of bottom drag is used here in order to find a simple analytical expression of the solution, as will be discussed shortly. Cross-differentiating and subtracting

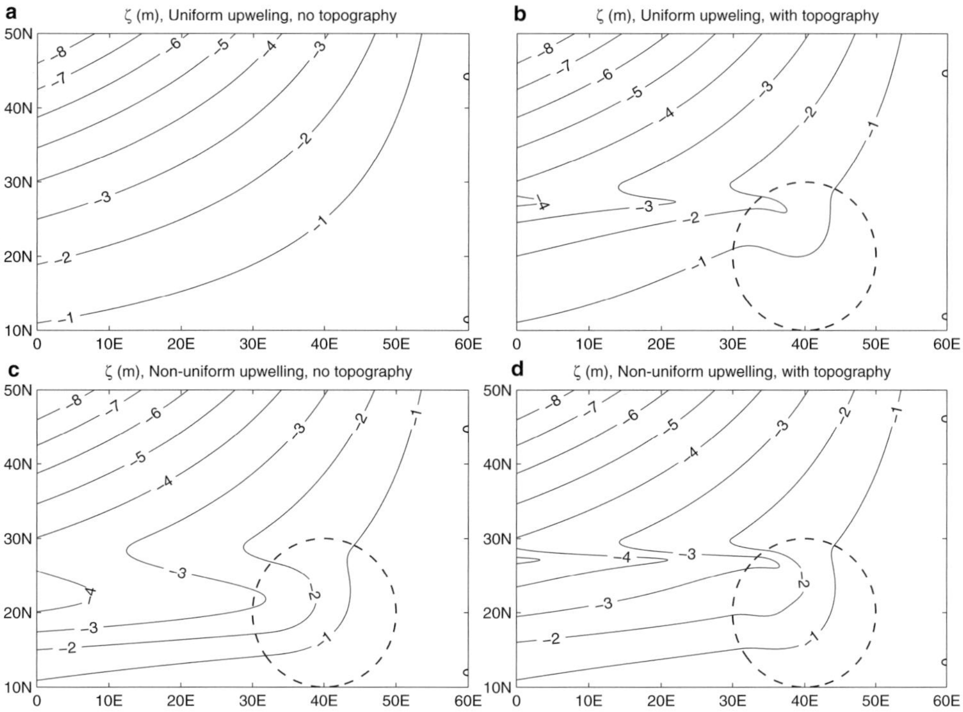

Fig. 5.65 **a–d** Interface displacement (in units of m) due to the combination of a sinusoidal seamount and enhanced upwelling $w_0 = 5 \times 10^{-7}$ m/s.

Eqns. (5.111a) and (5.111b) leads to a concise form of the potential vorticity equation

$$\nabla_h (h\vec{u}Q) = -R\left(v_x - u_y\right)/h_0 \tag{5.112}$$

where $Q = (f + v_x - u_y)/h$ is potential vorticity, including the contribution due to relative vorticity.

For the case with closed potential vorticity contours, we can calculate the integration of Eqn. (5.112) over the area A_Q within a closed potential vorticity contour C_Q. Note that, in the present case, there is a source/sink driving horizontal velocity; thus, there is no closed streamline as in the case discussed in Section 4.1.3. In fact, there is a mass flux crossing the potential vorticity contour C_Q and entering area A_Q. From continuity, the total mass flux across the boundary is equal to the total amount of upwelling inside area A_Q. Furthermore, this is a model with friction, so that potential vorticity is not conservation along streamlines. Thus, strictly speaking, potential vorticity contours are different from the so-called geostrophic contours used in a similar situation with closed contours.

Integrating the left-hand side term of Eqn. (5.112) leads to

$$\iint_{A_Q} \nabla_h(h\vec{u}Q)dA = Qc \oint_{C_Q} h\vec{u}\cdot\vec{n}dl = -Q_C \iint_{A_Q} w^* dA \tag{5.113}$$

In deriving this equation, we used the continuity equation (5.111c) and the fact that Q is constant along C_Q. Thus, the integration of Eqn. (5.112) is

$$\left(\frac{f + v_x - u_y}{h}\right)_c \iint_{A_Q} w^* dA = \frac{R}{h_0} \oint_{C_Q} \vec{u}\cdot d\vec{l}. \tag{5.114}$$

Equation (5.114) regulates flows within closed potential vorticity contours. First, assuming that there is a net upwelling inside a closed contour C_Q, the circulation must be cyclonic, as required by Eqn. (5.114), and this is true for both steep seamounts and trenches (Kawase and Straub, 1991; Johnson, 1998). Second, the circulation rate is inversely proportional to the frictional parameter R and the linearly proportional to total upwelling rate inside the closed contour. Third, since the relative vorticity is positive, $v_x - u_y > 0$, for cyclonic flow, the nonlinear term of the balance can further enhance the circulation.

Flow near deep trenches is one of the most interesting phenomena in the deep ocean because trenches are rather elongated features, with length scaling up to thousands of kilometers. Observations have indicated that there are indeed strong cyclonic circulations over steep trenches in the oceans. For example, there are strong deep boundary currents over the deep trenches in the northern and northwestern North Pacific Ocean, as shown in Figure 5.45 (Owens and Warren, 2001). Observations for other sites in the world's oceans were summarized by Johnson (1998).

Deep inertial western boundary currents

In the discussion above the existence of the western boundary currents is required by mass conservation, but we have not discussed any dynamical constraints on the deep western boundary currents. To explore the dynamical structure of these currents we need a dynamical framework. We have discussed the inertial western boundary currents in Chapter 4. According to simple inertial theory, the width of the surface inertial western boundary currents is on the order of 30–50 km. However, observations indicate that the deep western boundary currents can be very wide, on the order of 500 km. To explain this phenomenon, Stommel and Arons (1972) postulated that the essential ingredient of such a broad western boundary current is the sloping bottom over which the inertial western boundary currents flow.

Consider an inverse reduced-gravity model with a sloping bottom (Fig. 5.66).

The current is semi-geostrophic

$$-fv = -g'\eta_x = -g'(h+b)_x \tag{5.115a}$$

$$uv_x + vv_y + fu = -g'(h+b)_y \tag{5.115b}$$

$$(hu)_x + (hv)_y = 0 \tag{5.115c}$$

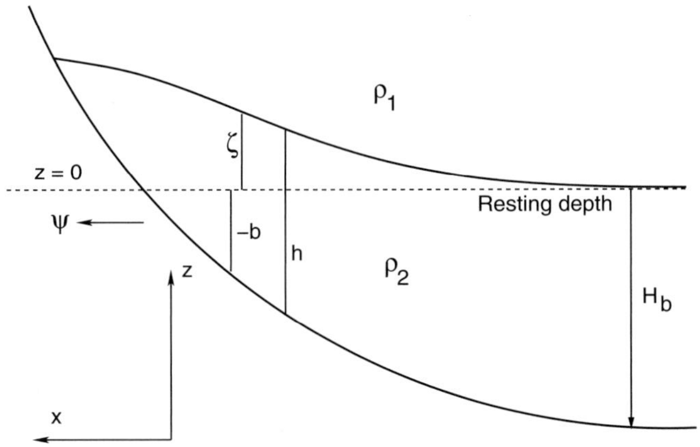

Fig. 5.66 Sketch of the model with exponential bottom topography (Pickart and Huang, 1995).

From these equations, we obtain the potential vorticity conservation and the Bernoulli conservation laws

$$\frac{f + v_x}{h} = Q(\psi) \tag{5.116}$$

$$\frac{v^2}{2} + g'(h + b) = B(\psi) \tag{5.117}$$

where ψ is the streamfunction, and $Q(\psi) = \frac{dB(\psi)}{d\psi}$.

Stommel and Arons discussed the case of uniform potential vorticity. In this case, the solution for the inertial deep western boundary currents can be represented in terms of exponential functions. However, it is much easier to solve this problem using the streamfunction coordinates transform, as we discussed for the surface western boundary currents in Section 4.1.

Introducing the nondimensional variables, $x = Lx'$, $h = Hh'$, $f = f_0 f'$, $v = Vv'$, and $\psi = \Psi\psi'$, then the basic equations (dropping the primes) are the following:

$$v_\psi^2 = \frac{2}{\varepsilon}\left(b\frac{h-f}{h}\right) \tag{5.118}$$

$$h = \psi - \frac{\varepsilon}{2}v^2 - sx \tag{5.119}$$

$$x_\psi = \frac{1}{hv} \tag{5.120}$$

where $s = \alpha L/H$ is the nondimensional bottom slope, $\varepsilon = (\lambda_D/L)^2$ is the Burger number, and $\lambda_D = (g'H)^{1/2}/f_0$ is the internal radius of deformation.

Stommel and Arons (1972) showed that there are two types of boundary condition for the onshore side of the current: vanishing layer thickness or zero velocity. In the second case, there is a stagnant region between the continental slope and the edge of the deep western boundary current. For the North Atlantic deep western boundary current, the second condition applies, although at low latitude the zero-thickness condition eventually takes over.

A very interesting phenomenon is that, as the deep western boundary current flows toward the equator, it moves up along the continental slope. Pickart and Huang (1995) performed the calculation for a case with uniform potential vorticity mimicking the deep western boundary current in the North Atlantic Ocean, and the solution is shown in Figure 5.67. Note that at low latitude the current becomes wider and the maximum velocity becomes larger.

Pickart and Huang (1995) also discussed the case with non-uniform potential vorticity and exponential type of continental slope. The calculations were carried out in the streamfunction coordinates. By using the potential vorticity profile diagnosed from data, the model simulation of the deep western boundary current has been improved (Fig. 5.68).

5.2.4 *Mixing-enhanced deep circulation*

Abyssal flow induced by bottom-intensified mixing over topography

In the Stommel and Arons theory, it is assumed that deep circulation is driven by uniform upwelling fed by point sources of deep water that are confined to several isolated places in

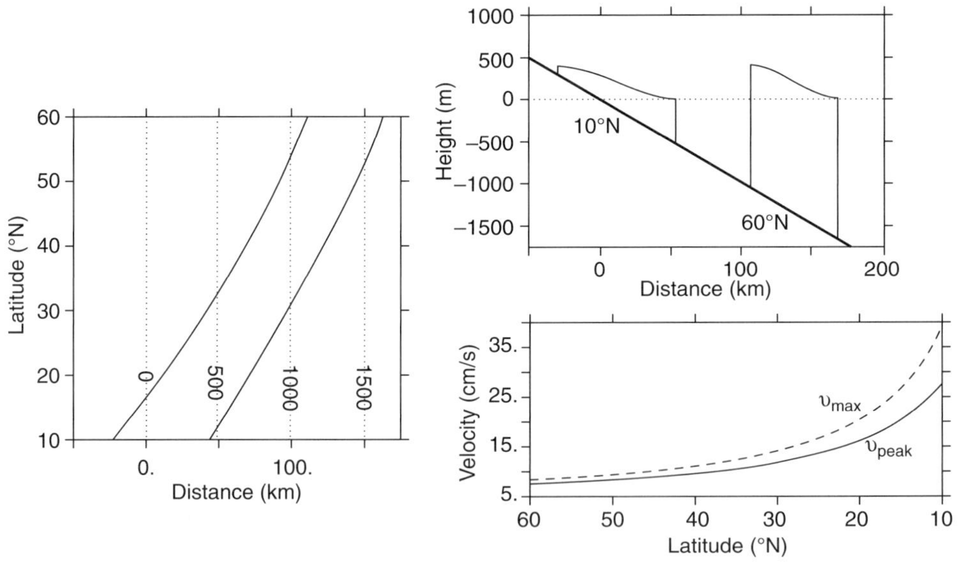

Fig. 5.67 Along-stream evolution of the uniform potential vorticity current with parameters $g' = 0.001$, $\alpha = 0.01$, transport = 5 Sv: **a** along-stream path of the current; the topography is indicated by the *dotted lines*; **b** cross-section of the interface revealing the change in structure from 60° N to 10° N; **c** along-stream trend in peak velocity of the current, where the maximum possible value is indicated by the *dashed line* (Pickart and Huang, 1995).

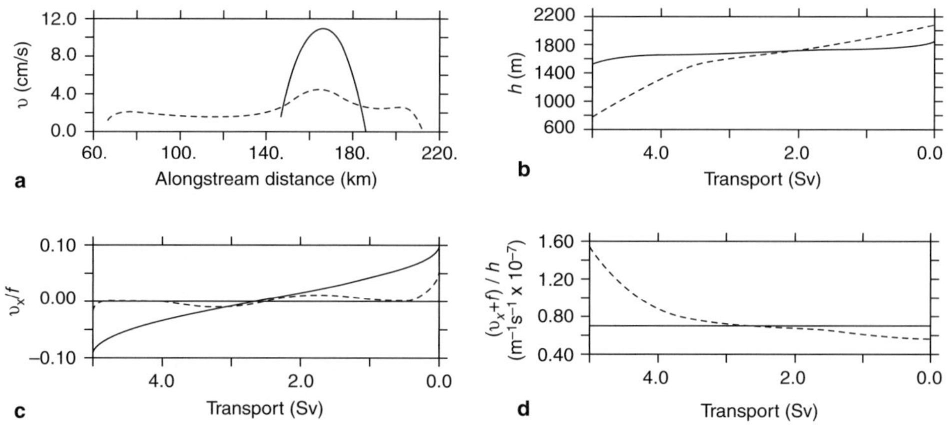

Fig. 5.68 Cross-stream structure of the realistic Q DWBC (*dashed line*) compared with the uniform Q solution (*solid line*) at 55° N: **a** downstream velocity, **b** layer thickness, **c** vorticity, **d** potential vorticity (Pickart and Huang, 1995).

the world's oceans. Since their model includes no mixing, no dissipation, and no bottom topography, the circulation can be readily derived from a simple linear vorticity balance. This leads to a simple circulation pattern, including poleward flow in the ocean interior which obeys the linear vorticity equation, plus the western boundary currents which are constructed from mass balance at each latitude.

In a similar way, if one assumes some simple forms of vorticity dissipation, the deep circulation in the whole basin, including the meridional boundaries, can be determined from the vorticity balance. As an example, Yang and Price (2000) extended the classical theory of Stommel and Arons by invoking a simple form of bottom drag and vertical velocity field.

Potential vorticity balance in the deep ocean is intimately related to mixing. Since mixing is highly non-uniform in space, the dynamical consequences of mixing is an important research frontier in deep circulation. Flow induced by bottom-intensified mixing over sloping boundaries has been studied extensively over the past several decades. Phillips (1970) and Wunsch (1970) pointed out that the thermal insulation boundary condition applied to a sloping bottom requires that isotherms must be perpendicular to the local slope. In a rotating fluid the density gradient constraint near the bottom boundary thus induces an along-slope flow (the primary circulation) and an uphill flow (the secondary circulation) in the bottom boundary layer. In addition, Phillips *et al.* (1986) carried out laboratory experiments and showed that a tertiary flow perpendicular to the slope might exist due to the convergence and divergence of the secondary circulation. A comprehensive review of the boundary layer induced by bottom-intensified mixing can be found in papers by Garrett (1991) and Garrett *et al.* (1993).

The fundamental physics of flow induced by bottom-intensified mixing is as follows. The thermal isolation boundary condition over the topographic slope requires that isothermals

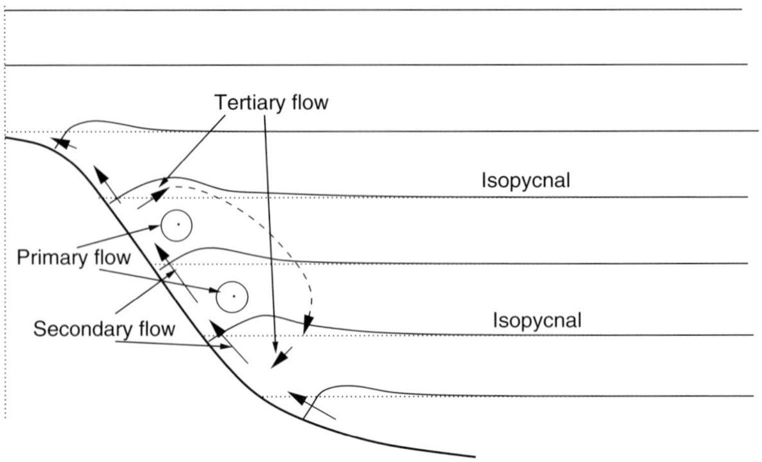

Fig. 5.69 Sketch of bottom circulation induced by bottom-intensified mixing.

must be perpendicular to the slope; thus, there is a thermal boundary layer near the topography (Fig. 5.69). The isopycnal surfaces near the topography are deformed in the following way. Away from topography, isopycnal surfaces move upward; thus, on the same geopotential level, density is higher than the ambient value, and this induces a geostrophic flow leaving the page and pointing toward the reader. In the vicinity of topography, density is lower than its background value, and such a density field induces a geostrophic flow leaving the reader and going into the page. These geostrophic currents are the primary flows induced by boundary mixing in the rotating flow.

On the other hand, close to the topography the horizontal pressure force is pointed toward the topography; thus, there is an upslope pressure force in the immediate vicinity of topography, which drives an upslope secondary flow within a thin boundary layer adjacent to topography. Over different parts of the slope, the mass flux in this boundary layer changes, and the corresponding convergence/divergence gives rise to the tertiary flow (Fig. 5.69).

If mixing is limited to the low level of molecular diffusion, the boundary layer is thin and the associated mass flux is very small and thus may be totally negligible for large-scale oceanic circulation. However, due to strong mixing induced by tidal dissipation over rough topography, mixing near rough bottom topography may induce strong flow in the vicinity of topography, which constitutes a major component of the large-scale oceanic circulation in the deep ocean.

Deep circulation induced by bottom-enhanced mixing over idealized topography

As an example, flow induced by bottom-enhanced mixing over a seamount calculated from a sigma-coordinate model is shown in Figure 5.70. The seamount has a radius of 15 km, a

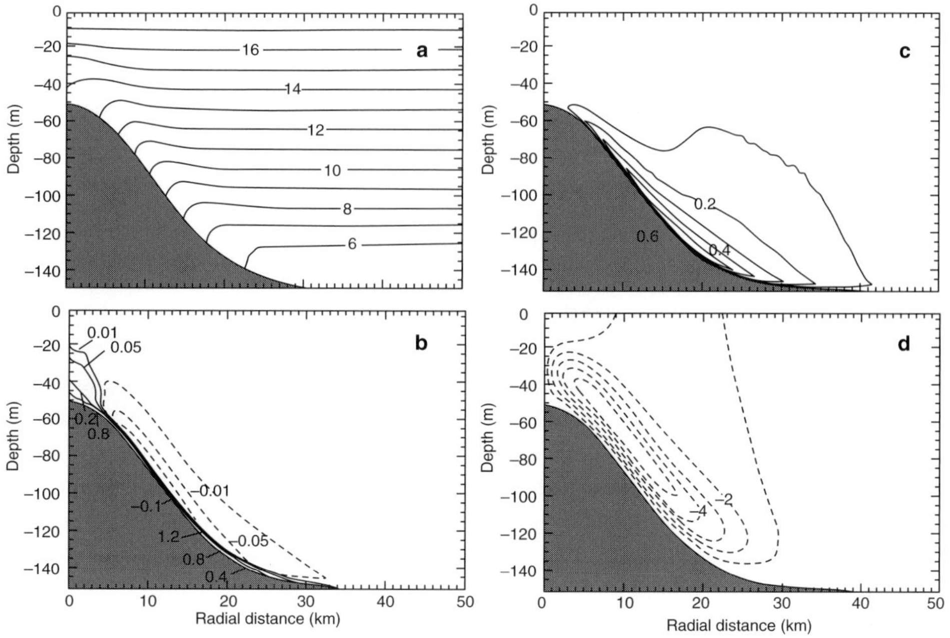

Fig. 5.70 Circulation induced by bottom-intensified mixing over a seamount; **a** temperature, in units of °C; **b** vertical velocity, in units of m/day; **c** overturning streamfunction, in units of $0.1\text{m}^2/\text{s}$; **d** azimuthal velocity, in units of 0.01 m/s (Cummins and Foreman, 1998).

height of 100 m, and both the vertical viscosity and diffusivity decay exponentially away from the bottom according to

$$(\kappa_v(z), A_v(z)) = (\kappa_0, A_0) \exp[-(H + z)/h] \tag{5.121}$$

where z is the vertical coordinate, with $z = -H$ set at the bottom topography, $h = 10$ m is the e-folding depth, $\kappa_0 = A_0 = 10^{-4}$ m^2/s. The flow field discussed here is entirely driven by bottom-enhanced mixing, without any external source/sink. As discussed in the previous section, the thermal boundary condition requires that in the vicinity of the sea floor, isothermals are perpendicular to the slope (Fig. 5.70a). Vertical velocity is positive within a very thin boundary layer adjacent to the bottom boundary and on the top of the seamount; however, away from this thin boundary, vertical velocity is negative everywhere, except near the top of the seamount (Fig. 5.70b). This vertical velocity field corresponds to a radial overturning circulation (Fig. 5.70c), with rising upslope motion close to the bottom and downward motion away from the bottom. The corresponding azimuthal velocity is anticyclonic (negative) over most of the depth, except for a very thin layer adjacent to the bottom boundary (Fig. 5.70d).

Deep circulation in the South Atlantic Ocean induced by bottom-intensified mixing over the mid-ocean ridge

Deep flow over topography induced by bottom-intensified mixing can be rather complicated, owing to the combination of mixing and topography. The deep circulation in the South Atlantic Ocean is a good example.

Model formulation

The numerical model is in z-coordinates based on the GFDL MOM 2 code, including 110 layers each of uniform thickness, 50 m (Huang and Jin, 2002a). The major feature of the model is the inclusion of strong diapycnal mixing over both sides of the mid-ocean ridge. Field observations (Polzin *et al.*, 1997; Ledwell *et al.*, 2000) indicate that mixing is very strong in the eastern half of the Brazil Basin, but it is very weak in the western half of the basin. Such strong bottom-intensified mixing over the western slope of the mid-ocean ridge is apparently related to the strong turbulence and internal wave breaking induced by tidal flow over rough topography of the mid-ocean ridge. Similar bottom-intensified mixing should exist for the eastern slope of the mid-ocean ridge as well; so mixing in the model is idealized as follows,

$$\kappa = \kappa_0 + \Delta\kappa_1 \cdot Exp\left[-\left(\frac{z+h}{\Delta h}\right)^2\right] \tag{5.122}$$

where $\kappa_0 = 10^{-5}$ m^2/s is the low background mixing coefficient; $\Delta\kappa_1 = 10^{-3}$ m^2/s is the strong bottom-intensified mixing on both sides of the mid-ocean ridge, which is set to zero for the western half of the Brazil Basin and the eastern half of the Angola Basin; h is the depth of the ocean; and $\Delta h = 600$ m is the e-folding distance of the bottom-intensified mixing.

The model is forced by the monthly-mean wind stress data of Hellerman and Rosenstein (1983) and sea surface temperature and salinity relaxation to the monthly mean climatology. The northern (southern) boundary is set along 6° N (41° S) where a sponge layer is used in which the model's thermodynamic variables are relaxed back to the climatology.

Owing to the low horizontal resolution, the Vema Channel cannot be fully resolved, so that an artificial channel (4,600 m, around 30° S and 39° W) is dug in the model to ensure that Antarctic Bottom Water (AABW) can flow into the Brazil Basin at the right depth. Similarly, the Romanche fracture zone is treated with caution by digging a channel 4,350 m deep (around 19° W near the equator); thus, the communication of AABW through the southern end of the Brazil Basin and its further communication to the Angola Basin are simulated by channels of the appropriate depth. The topography is shown in Figure 5.71a.

Several numerical experiments were carried out. In Case A, bottom-intensified diapycnal mixing was applied to both sides of the mid-ocean ridge. The boundary between strong mixing and background weak mixing was set along a "meridional" line that follows the maximum depth of the sub-basins on both sides of the mid-ocean ridge. Such a distribution of mixing gives rise to strong vertically averaged mixing over the mid-ocean ridge, while

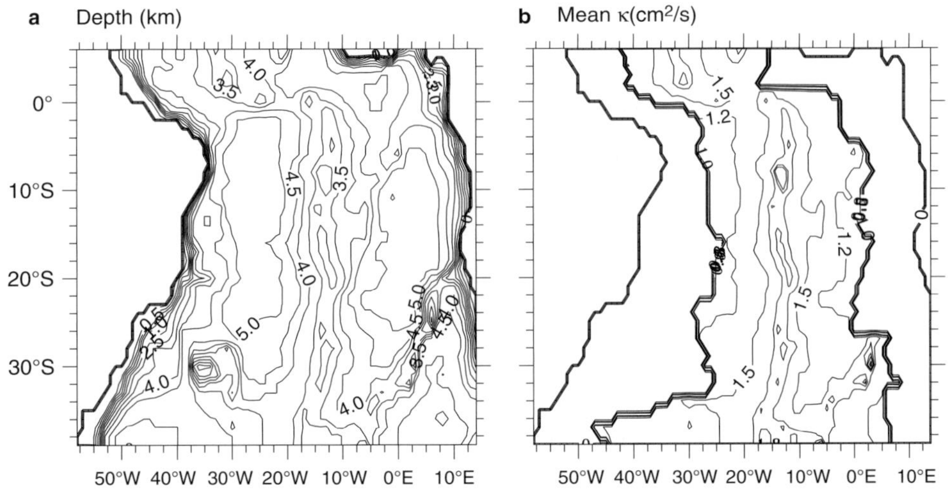

Fig. 5.71 The geometry of a model for the South Atlantic Ocean: **a** topography, in units of km; **b** vertically averaged diapycnal mixing coefficient, in units of 10^{-4} m^2/s (Huang and Jin, 2002a).

the vertically averaged mixing coefficient is constant for the eastern half of the Angola Basin and the western half of the Brazil Basin (Fig. 5.71b).

For comparison, there were two additional experiments, with a uniform mixing coefficient: In Case B, the mixing coefficient was set to the background value of $\kappa = 10^{-5}$ m^2/s. In Case C, the mixing coefficient was set to $\overline{\kappa} = 0.8541 \times 10^{-4}$m^2/s, equal to the basin-mean mixing coefficient for Case A. In comparison, the basin-average diapycnal mixing coefficient for AABW is estimated as $\overline{\kappa} \simeq (1-5) \times 10^{-4}$m^2/s.

Numerical results

All numerical experiments began from the initial state of an ocean at rest and integrated for 250 years. The volume flux of the AABW into the Brazil Basin is estimated as 4 Sv (Hogg *et al.*, 1982), so the renewal time for the water mass is about 100 years. Thus, under the relaxation conditions along both the northern and southern boundaries, 250 years should be long enough for the model to reach a quasi-equilibrium.

The horizontal velocity field

The most prominent feature of the circulation is the strong bottom current near the mid-ocean ridge in Case A (Fig. 5.72). In the Brazil Basin, there is a strong equatorward bottom current along the western slope of the mid-ocean ridge, which is almost as strong as the deep western boundary current along the western boundary of the basin (Fig. 5.72c); thus, it is opposite to the circulation inferred from the Stommel–Arons theory.

In the Brazil Basin there are three western boundary currents flowing along the eastern coast of South America, as seen in the horizontal velocity diagram (Fig. 5.72). Over the

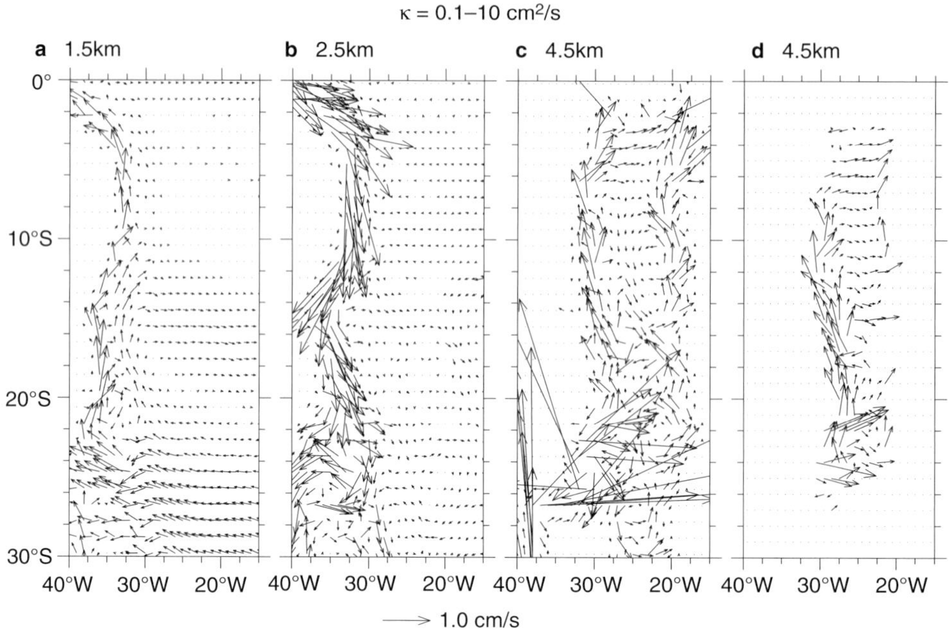

Fig. 5.72 Horizontal velocity within the Brazil Basin at different levels (**a** 1.5 km, **b** 2.5 km, **c** 4.5 km, **d** 5.0 km) for Case A (Huang and Jin, 2002a).

upper 1.5 km, the surface western boundary current flows northward (Fig. 5.72a). This surface western boundary current is due to the contribution from both the wind-driven circulation in the upper layer and the northward branch of the meridional thermohaline overturning cell.

For the depth range of 2–3.5 km, the western boundary current flows southward, carrying North Atlantic Intermediate Water and NADW (Fig. 5.72b). Around the depth of 4 km, the direction of the western boundary current switches to northward, carrying the AABW into the South and North Atlantic Oceans (Fig. 5.72c, d).

The dynamical effect of bottom-intensified mixing can be clearly seen at 4.5 km and 5 km depths (Fig. 5.72c, d). In fact, at the 4.5 km level, the eastern boundary current and the western boundary current in the Brazil Basin have roughly the same strength, with horizontal velocity on the order of 0.01 m/s. Although such an equatorward flow along the western flank of the ridge is consistent with the simple theoretical analysis in Section 5.2.3, *in situ* observations in the South Atlantic Ocean show a different pattern of circulation. In fact, owing to rough topography of the mid-ocean ridge, such a continuously equatorward flow near the mid-ocean ridge is interrupted and replaced by strong tidal flow within the narrow valleys (with a horizontal scale of tens of kilometers) which is perpendicular to the main axis of the mid-ocean ridge (Thurnherr and Speer, 2003).

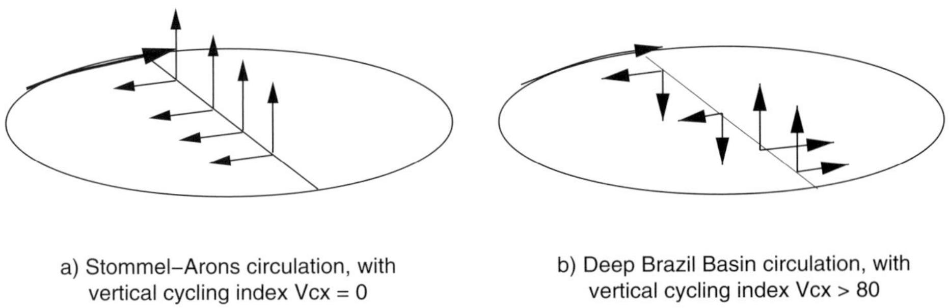

Fig. 5.73 **a, b** Two types of circulation pattern in a deep basin in the Southern Hemisphere.

This departure from the Stommel–Arons theory is expected, because the circulation in the deep basin is strongly constrained by both the topography and the bottom-intensified mixing, which were absent in the original Stommel–Arons theory.

The non-uniform upwelling

A fundamental assumption made by Stommel and Arons (1960a) was that upwelling is uniform basin-wide. However, such an assumption is not appropriate for the deep circulation. As Phillips *et al.* (1986) pointed out, boundary mixing can induce not only uphill flow, but also a tertiary flow that is perpendicular to the slope. In the present case, the tertiary flow appears in the form of upwelling near the mid-ocean ridge and downwelling in the basin interior. The prototype deep circulation predicted by Stommel and Arons and the deep circulation calculated from our numerical experiments are quite different, as shown in the sketches in Figure 5.73.

In order to illustrate the departure from the classic theory based on a uniform upwelling, we introduce a vertical cycling index, which is defined as

$$V_{cx}(z) = \text{sign}(W_p + W_n) \frac{\min(W_p, |W_n|)}{\max(W_p, |W_n|)} \times 100\% \tag{5.123}$$

where W_p is the basin-integrated upwelling rate, and W_n is the basin-integrated downwelling rate.

Stommel and Arons assumed a uniform upwelling, so $W_n = 0$, thus $V_{cx} = 0$ for their model. On the other hand, $V_{cx} = 100$ corresponds to the case with pure vertical recycling, with no net flux through a given level; such a case is the theoretical limit of the non-Stommel–Arons circulation. In the oceans both upwelling and downwelling exist at any given depth, so V_{cx} is normally non-zero. In fact, for most cases V_{cx} is far from being zero. For example, in the Brazil Basin, V_{cx} is close to either 100 or -100, indicating that there is a strong recirculation in the vertical direction (Fig. 5.74). As an example, at the depth of 4 km in Case A, the upwelling branch carries 5.39 Sv, and the downwelling branch carries 4.62 Sv, so the net upwelling rate is 0.77 Sv, which is fed from the AABW flux through the Vema Channel. There is also a large amount of mass flux through the side boundaries.

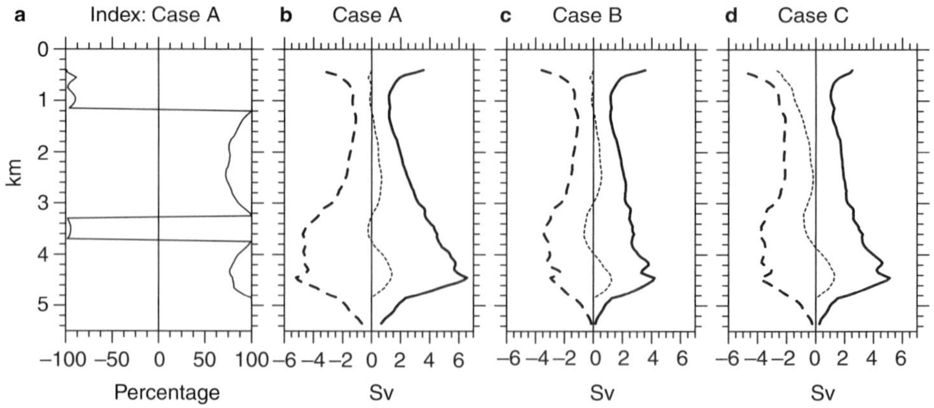

Fig. 5.74 Vertical mass flux profiles in the Brazil Basin: the *solid line* is the total upwelling, the *dashed line* is the total downwelling, and the *thin dotted line* is the net vertical mass flux: **a** the vertical cycling index. The fluxes are shown in the other panels: **b** Case A, **c** Case B, **d** Case C (Huang and Jin, 2002a).

At depths below 4 km, the net upwelling rate is primarily controlled by the amount of mass flux of the AABW through the Vema Channel, so it is not very sensitive to the choice of diapycnal mixing coefficient; however, the strength of upwelling and downwelling depends on the choice of diapycnal mixing coefficient rather sensitively, as shown in Figure 5.74. In fact, the upwelling branch in Case A is the strongest, indicating a vigorous vertical cycling of water in the Brazil Basin. In comparison, the vertical cycling of water in Case B is the weakest.

The choice of diapycnal mixing affects not only the deep circulation, as discussed above; it also affects the vertical upwelling rate over the whole depth range. This can be seen from the net upwelling rate in the Brazil Basin (Fig. 5.74). For example, there is a net upwelling at the depth range of 2–3 km for both Case A and Case B; however, there is a net downwelling for Case C at this depth range.

The difference in vertical net upwelling rate is directly related to the horizontal velocity field, indicating that the Ekman pumping effect due to the surface wind stress curl in the subtropical basin penetrates quite deeply for Case C. In comparison, the penetration of the Ekman pumping is much shallower for both Cases A and B. This difference is reflected in the depth of the main thermocline and the depth of the subtropical meridional cell. It is well known that a model with such a strong diapycnal mixing coefficient gives rise to a solution with a thermocline too deep and too diffusive; thus, the circulation in the upper ocean obtained from Case C is not realistic.

Properties on zonal sections

The structure of the deep circulation can be closely examined through temperature and velocity sections (below 2 km) for different zonal latitudes. (The salinity gradient is rather weak for the deep basin, so maps of salinity are less informative and are therefore not

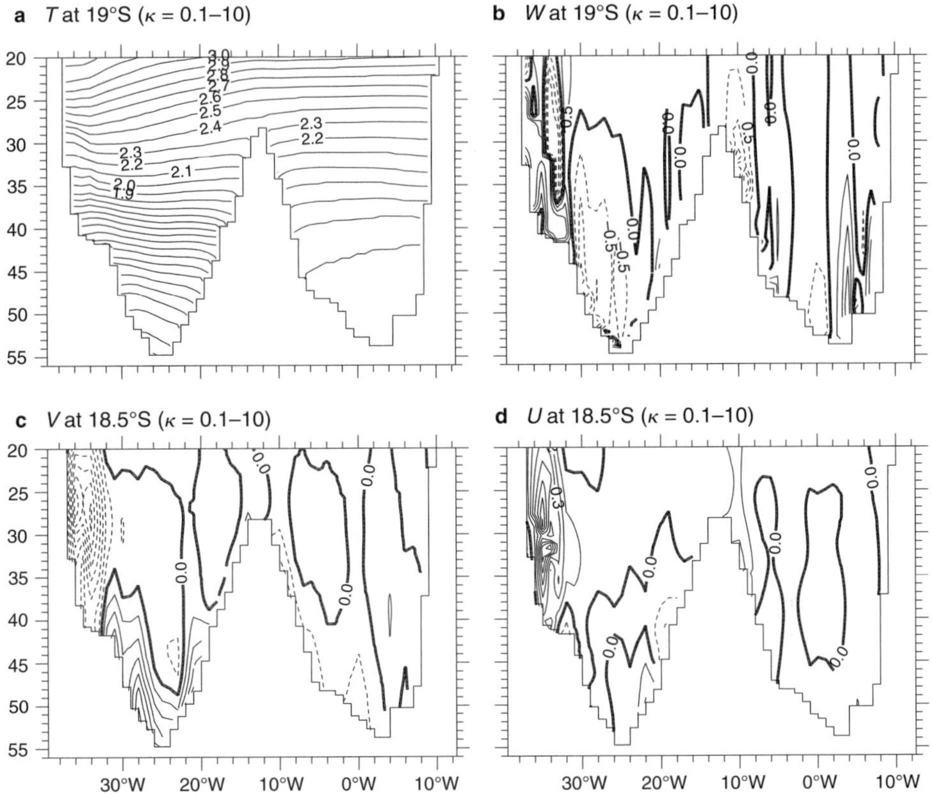

Fig. 5.75 Circulation at 19° S for Case A, depth in km: **a** potential temperature, in °C; **b** vertical velocity, in 10^{-6} m/s; **c** meridional velocity, in 0.01 m/s, contour interval 0.001 m/s; **d** zonal velocity, in 0.01 m/s (Huang and Jin, 2002a).

included.) First, the structure of the circulation obtained for Case A at 19° S is shown in Figure 5.75. The isothermal contours obtained from Case A have many essential features.

- In the Brazil Basin, the abyssal thermocline can easily be identified around a potential temperature range of 0.8–2.0°C.
- On both sides of the mid-ocean ridge, isotherms dive toward the ridge and seem more "perpendicular" to the local slope, consistent with the theory of a bottom boundary layer with an isolating condition (e.g., Phillips, 1970; Wunsch, 1970).
- Isotherms in the Brazil Basin interior slope down to the east.

In the Brazil Basin, water flows eastward below 4.5 km (Fig. 5.75b, d); this eastward and uphill flow over the bottom topography is primarily induced by the strong bottom-intensified mixing over the ridge. However, the flow direction reverses above 4.5 km: a generally westward flow occurs between 3.7 and 4.5 km, which is not inconsistent with the circulation pattern inferred from deep floats (Hogg and Owens, 1999). In the Angola Basin, there are actually two zonal cells rotating in opposite directions. In the eastern part, there is an anticlockwise circulation, with a mass flux of more than 0.3 Sv. This cell consists of eastward flow over the sea floor and the westward return flow at a depth of 4–4.5 km. In the western part, there is a weak clockwise circulation, with a slow uphill flow over the eastern slope of the mid-ocean ridge, apparently induced by the strong bottom-intensified mixing there. However, the strength of the upward motion along the eastern slope is much weaker than the corresponding upward motion along the western slope of the mid-ocean ridge. Thus, the circulation in the zonal section is similar to the case discussed by Cummins and Foreman (1998) for the flow induced by bottom-intensified mixing over a seamount.

The relatively steep isothermal contours indicate relatively strong bottom boundary currents along both sides of the mid-ocean ridge, as shown in Figure 5.75c. There the associated zonal and vertical velocities are also stronger, compared with Cases B and C (figures not included). Note that the bottom-intensified mixing coefficient induces upwelling over the western slope of the mid-ocean ridge and downwelling over the lower part of the ridge (Fig. 5.75b). This is similar to the flow pattern inferred from the tracer distribution observations in the deep Brazil Basin (St Laurent *et al.*, 2001). This can be explained as follows. Using a one-dimensional model as a crude approximation for density conservation, the basic balance is

$$w = \kappa_z + \kappa \frac{\rho_{zz}}{\rho_z} \tag{5.124}$$

For the model with bottom-intensified mixing in Case A, the first term on the right-hand side is on the order of $\kappa_z > -2 \times 10^{-6}$ m/s, so it is a substantial contribution to the downwelling near the outer part of the mid-ocean ridge. Owing to the continuity conservation, this downwelling should enhance the upwelling near the axis of the mid-ocean ridge, where the second term on the right-hand side of Eqn. (5.124) is positive owing to an almost zero density gradient near the sea floor. In comparison, for Cases B and C the κ_z term is identically equal to zero, so there is no downwelling near the bottom of the mid-ocean ridge.

Remarks

Numerical experiments based on a z-coordinates model indicate that with realistic bottom topography the abyssal circulation deviates from the classical Stommel–Arons circulation pattern noticeably, even for the case with uniformly weak mixing. For the case with strong bottom-intensified mixing over the mid-ocean ridge, the circulation departs from the classical Stommel–Arons circulation substantially. These results are consistent with many previous studies, such as the boundary mixing theory summarized by Garrett (1991), the numerical study by Cummins and Foreman (1998), and the theoretical study by Spall (2001). In summary, these results demonstrate the vitally important dynamical role of bottom topography in controlling the deep circulation. In order to simulate the deep circulation and water mass properties, the following issues are of critical importance.

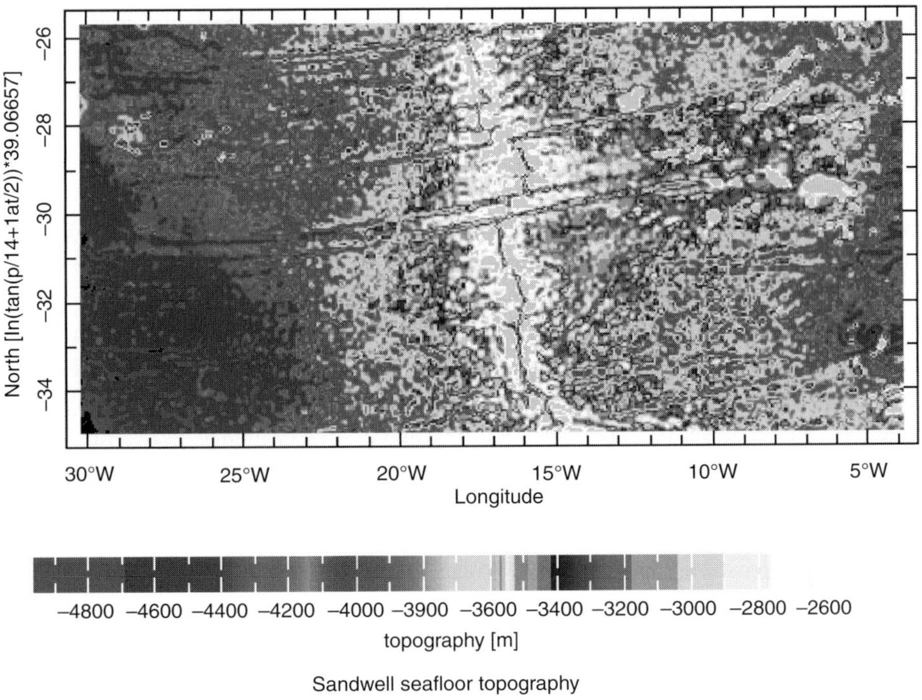

Fig. 5.76 Fine structure of seafloor topography (based on Smith and Sandwell, 1997). See color plate section.

Treating the bottom topography accurately

It is of vital importance to have the deep channels between different sub-basins connected for low-resolution models. Although a low-resolution model can provide useful information about deep circulation in the world's oceans, results from such numerical simulations can be rather sensitive to the horizontal resolution of the model. As seen from a fine-resolution map of the sea floor topography near the mid-ocean ridge (Fig. 5.76), flow can be very complicated due to the rough topography associated with the newly formed seafloor on both sides of the mid-ocean ridge. In particular, due to the existence of deep valleys perpendicular to the axis of the mid-ocean ridge, the geostrophic flow parallel to the axis of the mid-ocean ridge may be interrupted near the bottom topography. For example, the equatorward geostrophic flow shown in Figure 5.72 was not observed in the field experiments involving a tracer released near the mid-ocean ridge. Instead, deep floats deployed in the Brazil Basin indicated that flows at the mid-depth are dominated by alternating zonal jets, as discussed in the next section.

Using the best estimate for the mixing coefficient

A spatially varying mixing coefficient is clearly one of the most critical parts of ocean models for simulating the deep circulation. In particular, extremely strong localized mixing associated with overflow through sills connecting different basins has been omitted in most previous basin-scale simulations; however, such strong localized mixing may be crucial in simulating the deep circulation accurately.

5.2.5 Mid-depth circulation

According to the classical theory of Stommel and Arons (1960a) circulation in the deep ocean interior is dominated by broad eastward and poleward flows driven by the uniform upwelling (Fig. 5.56). Deep western boundary currents are required for completing the mass balance in individual basins. The deep western boundary currents predicted by their theory are probably the most robust feature in the ocean, and these have been confirmed through observations in the world's oceans. On the other hand, the broad interior flows predicted by the theory have never been observed.

Much effort has been made to explore the deep circulation through observation. For example, a large-scale field observation program, the Deep Basin Experiment, was specially designed to observe the deep circulation in the Brazil Basin. A large number of neutrally buoyant floats were released within the Brazil Basin during the 1990s in an attempt to measure directly the circulation in the deep ocean interior (Hogg and Owens, 1999).

It appears that the flow in the deep Brazil Basin is unlike the prediction from the classical theory. Although the deep western boundary currents observed confirm the theory, the interior flow inferred from the neutrally buoyant floats is dominated by zonal flows with unexpected small meridional scales, as shown in Figure 5.77.

It is not surprising that zonal flows dominate the deep circulation because potential vorticity contours in the deep ocean are primarily zonally oriented (O'Dwyer and Williams, 1997). As an example, a potential vorticity map for the abyssal ocean is shown in Figure 5.78. Weak diapycnal mixing in the basin interior gives rise to small vertical velocity divergence; thus, the corresponding meridional velocity is small and flows should be primarily zonal.

The study of the deep circulation may require basin-scale high-resolution data, which are currently not available from observations; thus, our discussion here is limited to results from eddy-permitting numerical models. According to a study by Nakano and Hasumi (2005), the zonal currents in the sub-surface oceans can be classified into two categories. First, there are broad-scale zonal flows which have a poleward slanting pattern in the meridional section. Second, there are fine-scale zonal jets, which have a meridional scale of $3°-5°$, formed in each broad zonal flow.

There are many possible origins for the zonal flows. First, baroclinic instability may induce such flows. Treguier *et al.* (2003) showed that although the mean flow in the Brazil Basin is baroclinically unstable, the corresponding growth rates are small; thus, this process is unlikely to be the sole source of the zonal jets observed there. Second, the topographic feature may also be responsible for the existence of zonal jets. However, the primary mechanism for the zonally alternating jets may be the response to wind stress. Using numerical experiments for the Pacific Ocean, Nakano and Suginohara (2002)

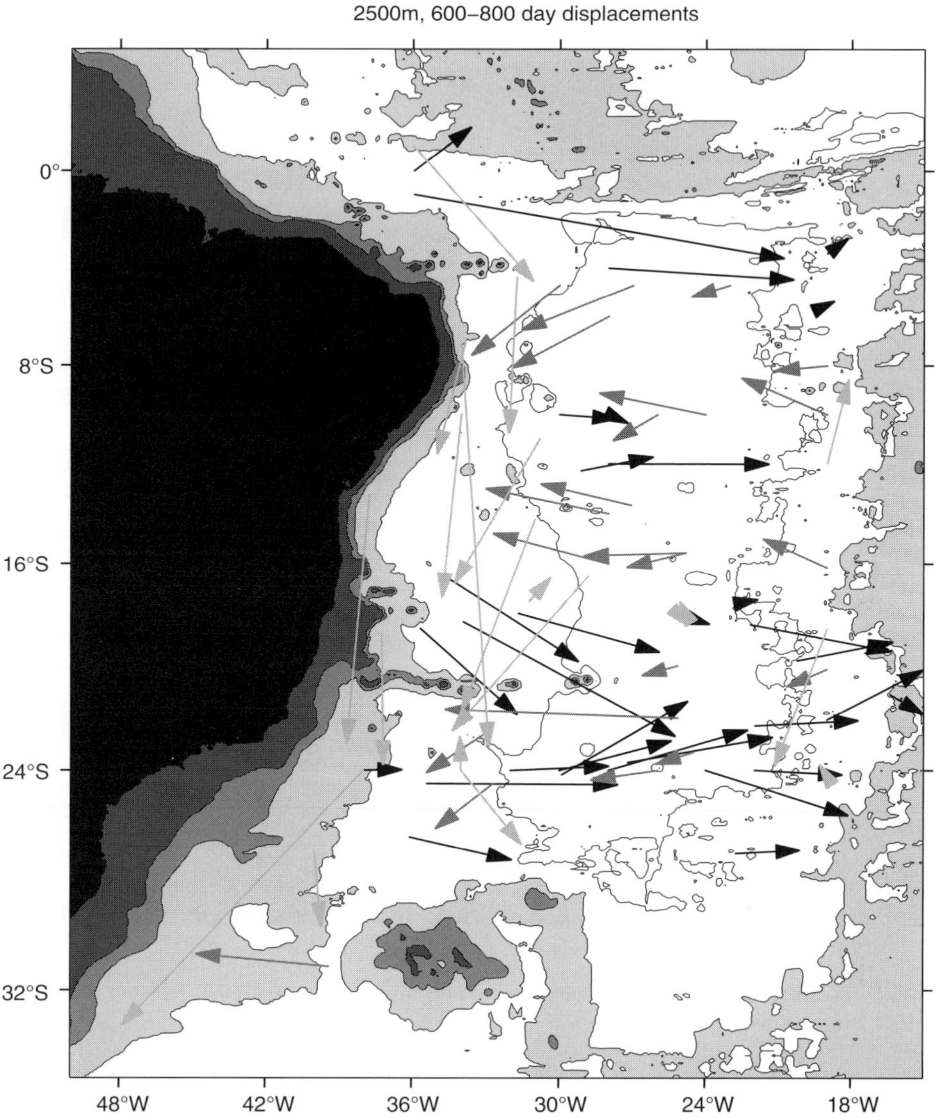

Fig. 5.77 600–800 days displacement of floats at 2,500 m in the Brazil Basin (Hogg and Owens, 1999).

demonstrated zonal flows driven by wind. The basic idea is that wind-driven circulation in the ocean is established through Rossby waves. The first few vertical modes move across the basin within a few decades and establish the barotropic and first baroclinic components of the wind-driven circulation. Higher modes take a much longer time to move across the basin. Due to dissipation, the Rossby waves of high modes can never complete their

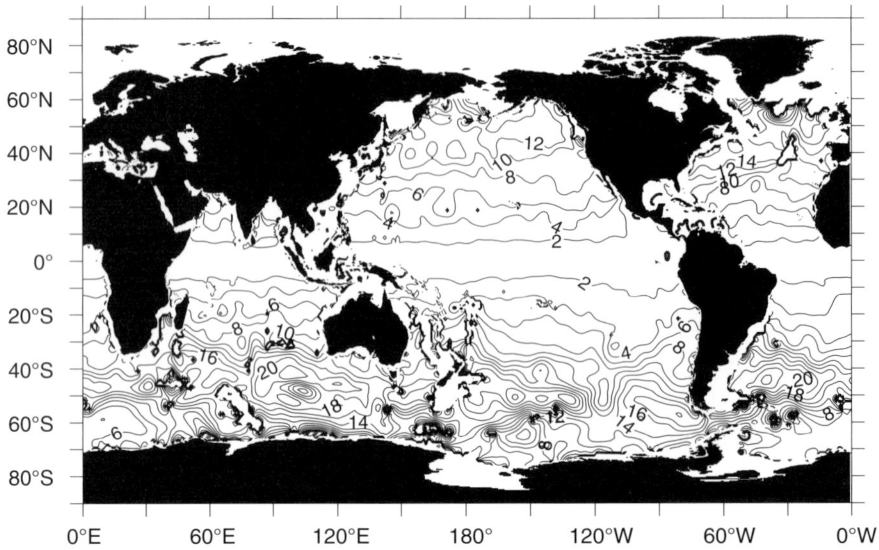

Fig. 5.78 Potential vorticity (in units of 10^{-12} m/s) at a depth of 2,750 m in the world's oceans, based on Levitus *et al.*'s (1998) Climatology.

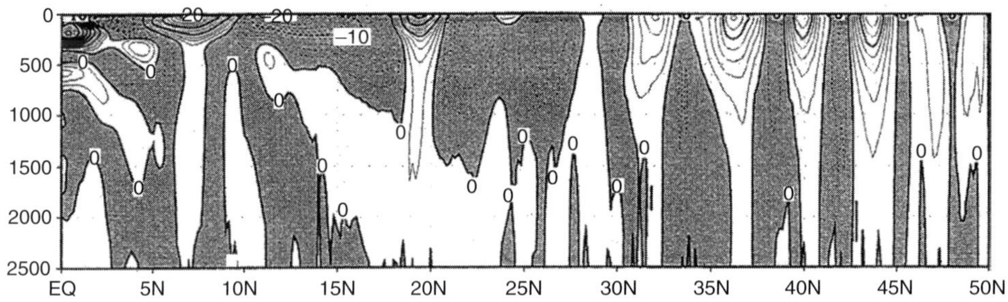

Fig. 5.79 A zonal velocity section along 180°, from a high-resolution ($^1/_4°$ × 1/6°) model with contour intervals of 2 cm/s; *shaded areas* indicate westward velocity bands (Nakano and Hasumi, 2005).

journey to the western coast, thus leaving a steady zonal flow with the characteristic meridional and vertical structure of the equatorial waves. Numerical experiments for the Brazil Basin produced results that seem quite consistent with the Lagrangian float data (Treguier *et al.*, 2003).

Numerical experiments with high resolution produced both types of current in the North Pacific Ocean (Fig. 5.79). Apparently, baroclinic instability plays the major role in producing the fine-scale zonal jets embedded in the broad-scale zonal flows. In fact, in both the atmosphere and oceans, there are zonal jet-like features which resemble the flow pattern observed from other planets, such as Jupiter and Saturn.

5.3 Haline circulation

5.3.1 Hydrological cycle and poleward heat flux

Poleward heat flux in the climate system has been one of the main focuses of climate study. According to traditional classification, the atmospheric component of the poleward heat flux dominates the total poleward heat flux at high latitudes. For example, at 35° latitude, where the poleward heat flux is maximal, the atmospheric transport accounts for 78% of the total in the Northern Hemisphere and 92% in the Southern Hemisphere. The oceans seem unimportant for mid-latitude and high-latitude climate, in terms of carrying the heat flux that is much needed for warming up high latitudes. But is that true? In particular, the Southern Hemisphere is covered primarily by oceans, especially between 35° S and 70° S. Therefore, it is almost impossible to believe that the oceans play such a minor role in poleward heat flux.

Definition of poleward heat flux

Oceanic heat flux has been discussed in many papers and textbooks. As both instruments and numerical models improve, poleward heat fluxes are becoming better diagnosed. For up-to-date information, the reader is referred to the review by Bryden and Imawaki (2001).

The main point is that although heat flux data may be further improved, there seems to be a fundamental problem in the definition of poleward heat fluxes that gives rise to misconceptions and similar mistakes by many people. From thermodynamics, it is well known that heat flux cannot be well defined for a system that has net mass gain (or loss) through the lateral boundary of the system. For example, Trenberth and Caron (2001) made the point that the poleward heat fluxes in the South Pacific and Indian Oceans are not well defined because of the existence of the Indonesian Throughflow.

The same caution should apply to both the oceans and atmosphere, because neither the ocean nor the atmosphere is a closed system in terms of mass – there is water exchange between them. In fact, the hydrological cycle of evaporation and precipitation is an essential ingredient for heat flux in the atmosphere – the latent heat flux. The mass flux associated with evaporation and precipitation has been traditionally ignored in heat flux calculations by oceanographers, for the following reasons.

First, such a small mass flux is rather hard to identify from the classical dynamical calculation based on either a reference level or other means. Second, most oceanic general circulation models have been based on the Boussinesq approximations; thus, the dynamical effects associated with evaporation and precipitation are simulated in terms of the virtual salt flux condition on the upper surface, while the mass flux associated with evaporation and precipitation is totally ignored. Improvements in oceanic general circulation models include a new generation of models formulated under the natural boundary condition (Huang, 1993b). As a result, the mass flux associated with evaporation and precipitation is exactly accounted for. A more accurate definition of heat flux is therefore necessary. A simple definition of heat flux is found in the Appendix to this chapter, which can be used for an oceanic section with a net mass flux exchange.

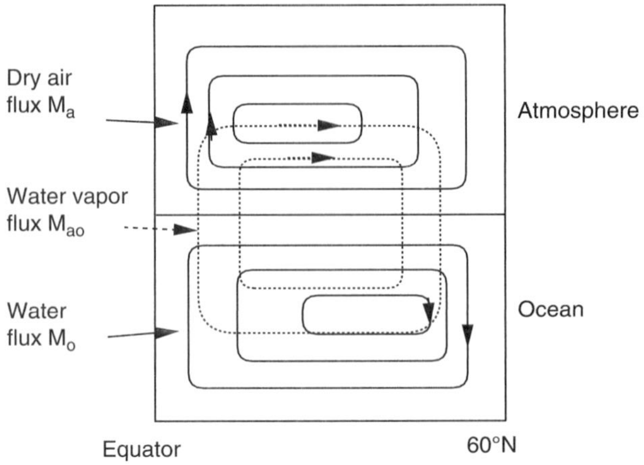

Fig. 5.80 Sketch of the atmosphere–ocean coupled climate system (Huang, 2005b).

Strictly speaking, the poleward heat flux in the climate system should be defined as three terms: sensible heat flux in the oceans, sensible heat flux in the atmosphere, and latent heat flux in the atmosphere–ocean coupled system, as shown in Figure 5.80. Over the latitudinal range from equator to pole there is a continuous exchange of heat between these three loops.

As an example, we calculate the freshwater flux through the air–sea interface. We use the evaporation and precipitation rates over the world's oceans as reported by Da Silva *et al.* (1994) (Fig. 5.81). This dataset has taken river run-off into consideration, so the globally integrated evaporation and precipitation rates are balanced. Near the equator, precipitation dominates, especially between 0° N and 10° N. However, evaporation dominates over the subtropics and thus produces a net water vapor flux that is transported by the atmosphere to high latitudes. At high latitudes (roughly beyond 40° off the equator) precipitation dominates.

However, this dataset seems to give rise to a poleward freshwater flux that is too large in the Southern Hemisphere. As an alternative, we use the water vapor flux calculated from atmospheric circulation models (Gaffen *et al.*, 1997). The freshwater flux reported in their study is based on the average of 25 atmospheric general circulation models. This poleward water vapor flux, M_w, is the major mechanism of poleward heat flux in the climate system (Fig. 5.82).

The poleward heat flux associated with the water vapor cycle is closely related to the latent heat content of water vapor

$$H_f = (L_h - h_w) M_w \tag{5.125}$$

where $L_h = 2,500$ J/g is the latent heat content for water vapor, and h_w is the enthalpy of the return water in the ocean, which is much smaller than L_h and is neglected in the

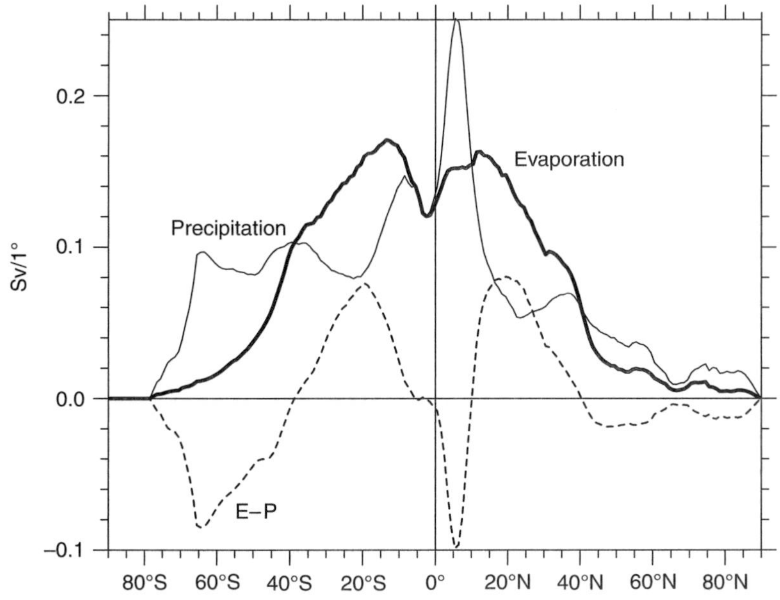

Fig. 5.81 Water vapor source and sink due to evaporation and precipitation, integrated zonally (Huang, 2005b).

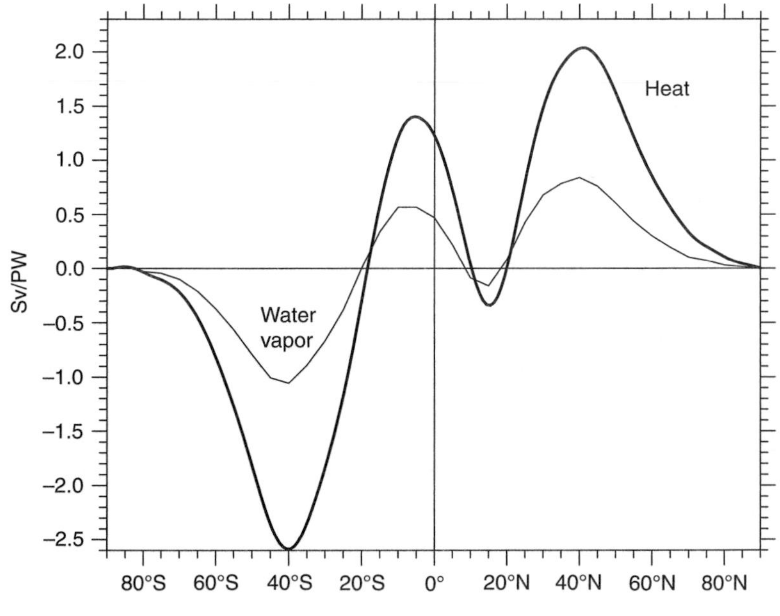

Fig. 5.82 Northward water vapor flux (Sv) and associated latent heat flux (PW) due to evaporation and precipitation, as calculated from atmospheric general circulation models (Huang, 2005b).

Table 5.5. *Mass transport maxima in the climate system*

Current systems	Gulf Stream	Kuroshio	ACC	Meridional cell in atmosphere	Jet Stream	Poleward water vapor
Mass flux (Sv)	150	130	134	200	~ 500	1

Sources: Gulf Stream (Hogg and Johns, 1995), Kuroshio (Wijffels *et al.*, 1998), ACC (Whitworth and Peterson, 1985), meridional overturning rate and the Jet Stream in the atmosphere (Peixoto and Oort, 1992), poleward water vapor flux (Gaffen *et al.*, 1997).

calculation. The heat flux associated with the water vapor flux should be attributed to the ocean–atmosphere–land coupled system.

However, the mass flux associated with the water vapor flux is two orders of magnitude smaller than other mass fluxes in the climate system (Table 5.5). This is one of the reasons why this mass flux has been overlooked in many papers discussing climate. On the other hand, this seemingly tiny mass flux is associated with a large heat flux. One kilogram of water vapor can deliver 2.5×10^6 J of heat. When one kilogram of water is cooled down by 10°C, the heat released is 4.18×10^4 J. Thus, water vapor is 60 times more efficient than water in transporting heat.

Here we use the Sverdrup as a unit for mass flux, as it has been widely used as a volume flux unit in oceanography: $1Sv = 10^6$ m^3/s. Since seawater density is very close to 10^3 kg/m^3, as a mass flux unit we have 1 Sv $= 10^9$ kg/s. As shown in Table 5.5, the Sverdrup is a very convenient unit for our climate system because mass fluxes in both the oceans and atmosphere have the same order of magnitude, although the density of water is one thousand times larger than that of air. In particular, using the Sverdrup as a mass flux unit is quite convenient for describing the flux of water vapor. For example, if we say that the water vapor flux across a certain latitude in the atmosphere is 0.5 Sv, then in the steady state of climate there should be 0.5 Sv of pure water going in the opposite direction, and there is no scope for confusion.

Another major advantage of using the Sverdrup as a mass flux unit is the application to planetary boundary layers in the climate system. According to Newton's Third Law, frictional forces in the atmospheric planetary boundary layer and the oceanic planetary boundary layer have the same magnitude but opposite signs. As a result, mass transport in both Ekman boundary layers should be the same, but going in opposite directions.

The mass flux in the Jet Stream is inferred from angular momentum. According to Peixoto and Oort (1992), the maximum angular momentum in the Northern Hemisphere is about 9.6×10^{25} kg/m^2/s. Assuming that the mean latitude is at 30° N, this gives a rough estimate of 500 Sv for the mass flux associated with the westerly in the Northern Hemisphere.

In the Northern Hemisphere the water vapor flux due to the oceanic process peaks at 40° N, with a mass flux of 0.84 Sv and a corresponding poleward heat flux of 2.1 PW.

In the Southern Hemisphere, the poleward water vapor flux reaches the maximum at 40° S, with a mass flux of 1.05 Sv and a corresponding poleward heat flux of 2.64 PW. It is clear that the latent heat flux associated with evaporation and precipitation over the world's oceans consists of a substantial portion of the poleward heat flux in the current climate system. From the hydrological cycle alone, we can estimate the poleward heat flux associated with latent heat; this is a major component of the poleward heat flux (e.g., Schmitt, 1995). Such a large poleward heat flux is, of course, intimately connected to the general circulation in the oceans. In the event of climate change, surface temperature and salinity change will certainly affect the latent heat flux, and thus the entire climate system.

After introducing the poleward heat flux associated with the water vapor cycle, the poleward heat fluxes in the current climate system are as shown in Figure 5.83. The sum of these three fluxes is the total poleward heat flux diagnosed from satellite measurements. The oceanic sensible heat flux is the oceanic heat flux from the NCEP (National Centers for Environmental Prediction) data (Trenberth and Caron, 2001). Although the atmospheric sensible heat flux is still the largest component in both hemispheres, it is not much larger than the other components.

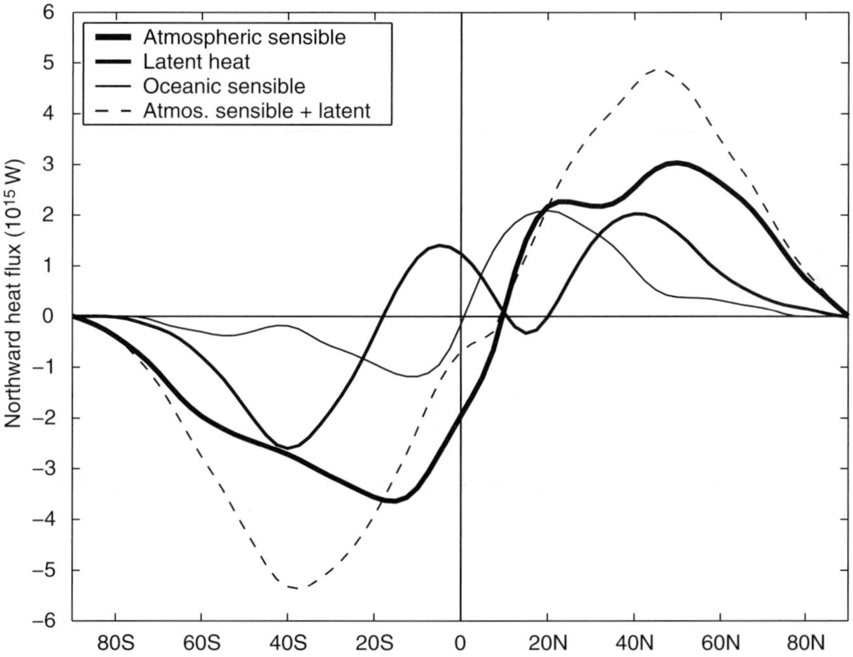

Fig. 5.83 Northward heat flux in PW, including atmospheric sensible heat flux, the oceanic sensible heat flux, and the latent heat flux associated with the hydrological cycle in the atmosphere–ocean coupled system; the *dashed line* indicates the sum of the atmospheric sensible and latent heat fluxes.

Heat flux divergence

Care must be taken in the interpretation of the poleward heat flux. Heat flux itself may not be very important. For each location what really matters is the divergence of the heat flux, because this is what affects the local heat balance. For example, at 65° S atmospheric sensible heat flux is stronger than the latent heat flux. However, a close examination reveals that the latent heat flux divergence is much stronger than that of the atmospheric sensible heat flux (Fig. 5.84). Similarly, in the Northern Hemisphere oceanic sensible heat flux divergence is the dominating source of heat south of 40° N; north of this latitude the divergence of latent heat flux becomes more important, and north of 50° N the atmospheric sensible heat flux dominates.

In both hemispheres, poleward heat flux is carried by three components working like a relay team. In the subtropics, oceanic sensible heat flux is the dominating contributor to the poleward heat flux divergence, and in mid latitudes the latent heat flux divergence is the dominating contributor. Finally, at high latitudes the atmospheric sensible heat flux divergence dominates (Fig. 5.84). The situation in the Southern Hemisphere is an excellent example, demonstrating the roles of these three components. The most noticeable term is the latent heat flux divergence that dominates over mid latitudes. The atmospheric sensible

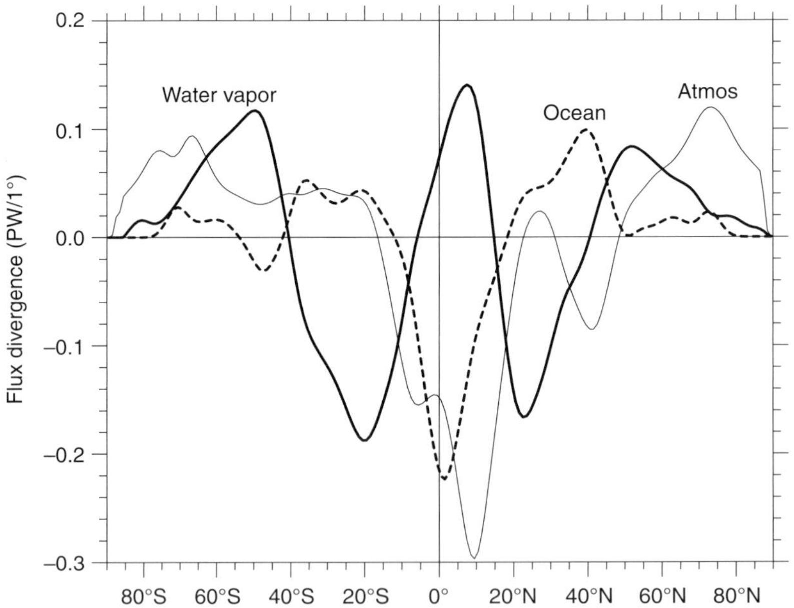

Fig. 5.84 Heat flux divergence: *Atmos* indicates the atmospheric sensible heat flux divergence, *Ocean* indicates the oceanic sensible heat flux divergence, and *Water vapor* indicates the latent heat flux divergence associated with the hydrological cycle in the atmosphere–ocean coupled system (Huang, 2005b).

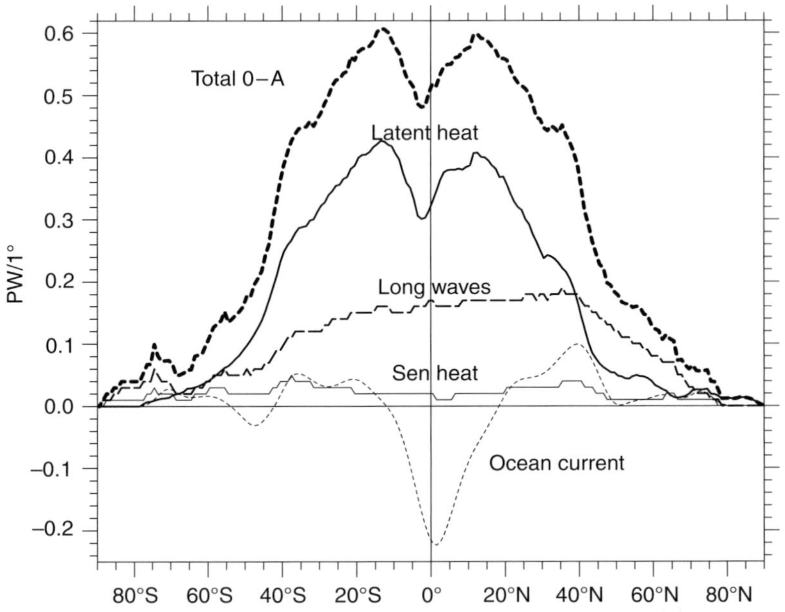

Fig. 5.85 Heat fluxes from the ocean to the atmosphere, integrated zonally, and the divergence of the oceanic sensible heat flux, labeled *Ocean current* (Huang, 2005b).

heat flux divergence plays a major role south of 70° S, where the Southern Ocean meets Antarctica.

The heat flux balance for a given water column is

$$\begin{aligned}&\textit{Net short wave radiation} + \textit{divergence of oceanic sensible heat flux}\\&\quad = \textit{latent heat} + \textit{net long wave radiation} + \textit{sensible heat}\end{aligned}$$

It is important to note that divergence of the oceanic sensible heat flux is a relatively small term in the oceanic heat flux balance at any given location (Fig. 5.85). In comparison, latent heat released to the atmosphere is much larger than the divergence of the oceanic sensible heat flux at any given location. The latent heat flux released to the atmosphere is thus a continuous source of heat for the atmospheric circulation. Although the atmosphere may play a major role in sending heat to higher latitudes, without the continuous heat support from the oceans, the warm climate we enjoy now might not be realized at all.

The hydrological cycle

In the climate system, the solar radiation is primarily absorbed by the ocean first. The atmosphere is primarily heated through evaporation and latent heat release in the atmosphere, plus the absorption of heat by so-called “greenhouse gases” in the atmosphere. As shown in Figure 5.85, the amount of latent heat flux from the ocean to the atmosphere is huge. The total amount of evaporation is 12.45 Sv, so the latent heat associated with evaporation is about 31.1 PW. As the oceanic circulation changes, the latent heat flux will be substantially modified; therefore, we should study how this is going to happen.

In previous studies, the crucial role of the atmosphere–ocean hydrological cycle associated with evaporation and precipitation has been overlooked in the oceanographic literature. In oceanic heat flux calculations, the poleward heat flux associated with the hydrological cycle has been neglected, and this was incorrectly attributed to the atmosphere alone. Thus, what most oceanographers have calculated in the past should be called the oceanic sensible heat flux.

On the other hand, meteorologists have claimed that the poleward heat flux associated with the hydrological cycle is entirely due to atmospheric processes, and has very little connection with the ocean circulation. Such an entire separation of heat fluxes is inconsistent with the physics. Without the oceanic currents to bring heat northward, there would be no high surface temperature in the Gulf Stream (or Kuroshio) region. As a result, there would be no strong latent heat release in these areas.

It is well known that the latent heat flux is the dominating form of poleward heat flux in the atmospheric circulation; however, this component has long been thought of as primarily an atmospheric process. The reason for such a misconception is, at least partially, due to the seemingly small amount of mass associated with this water vapor flux. The global integrated meridional water vapor flux is on the order of 1 Sv (10^9 kg/s), which is much smaller than the mass flux of the other components of the oceanic general circulation and the atmospheric circulation. Thus, the return branch of the global hydrological cycle, which takes place in the oceans, has been overlooked. This misconception is widespread and can be seen in many documents, papers, and books about climate.

First, although the main branch of the global hydrological cycle is through the oceans, many existing scientific programs related to the hydrological cycle completely ignore the oceanic component of the cycle. Since the oceans cover about 70% of the Earth's surface, the major component of the global hydrological cycle takes place over the oceans.

Second, the thermohaline circulation in many existing climate models is not realistically simulated. For example, the oceanic component in some climate models is represented in terms of the so-called swamp ocean, i.e., a very shallow layer of water (50 m or slightly deeper) with no current.

In such models the atmosphere takes up water vapor at low latitudes and transports it to high latitudes. After releasing the latent heat from the water vapor, the water has been discarded from the model, and only a few people have thought about the dynamical consequences of this seemingly small amount of water flux. Obviously, an ocean without current would be unable to transport the freshwater from high latitudes, where precipitation prevails, back to lower latitudes, where evaporation prevails. Thus, without the oceanic currents the subtropical ocean would eventually dry up, with all water piling up at high latitudes – an equilibrium state which is impossible to envisage. It is very unlikely that, when a quasi-steady state is reached, such models can reproduce the strong thermohaline circulation in the oceans. Thus, the sea surface conditions, such as surface temperature, surface salinity, and air–sea heat fluxes, from such a model may be quite different from the current climate.

Many of the existing climate models have an oceanic component based on the so-called Boussinesq approximations. In such models, the ocean is treated as an incompressible fluid environment, and the salinity balance in the model is controlled by the so-called virtual salt flux condition at the sea surface, or the traditional technique of requiring the surface salinity relaxation to the climatological mean surface salinity. Thus, changes in the water vapor cycle in such models have no dynamical consequences at all.

With improvements in the oceanic general circulation models, the dynamical role of freshwater will be simulated more realistically. Among many other features, evaporation and precipitation can drive the so-called Goldsbrough–Stommel circulation (Huang and Schmitt, 1993), which is a barotropic circulation with a horizontal circulation on the order of 1 Sv. This circulation is much smaller than the wind-driven or thermohaline circulation, so it has been excluded from most discussions about ocean circulation and climate. However, the seemingly small freshwater flux associated with evaporation and precipitation is responsible for driving the haline circulation in the oceans. In fact, given enough mechanical energy from wind stress and tides, evaporation and precipitation alone (without wind stress and heating) can drive a three-dimensional baroclinic circulation that is two orders of magnitude stronger than that of the barotropic Goldsbrough–Stommel circulation, i.e., the same order of strength as the wind-driven circulation or the thermally driven circulation (Huang, 1993b).

Thus, evaporation and precipitation comprise one of the major forces in regulating the oceanic general circulation. As the climate system shifts to a new state, the freshwater flux through the ocean–atmosphere system will change in response. As a result, the thermohaline circulation will be in a different state, and the sea surface temperature distribution will be different from the present-day distribution. Consequently, the surface air–sea heat fluxes shown in Figure 5.85 may be dramatically different. To achieve a comprehensive understanding will require much effort in studying the ocean–atmosphere coupled system.

The implications of a swamp ocean with no current

The role of the oceans in the ocean–atmosphere coupled system is illustrated in Figure 5.86. In the current climate, there are three components in this climate heat conveyor, and they work together like a relay team. Although the atmosphere is responsible for sending heat to high latitudes, it might be misleading to say that the atmospheric heat transport alone is responsible for the high-latitude poleward heat flux.

If there were no water vapor cycle closely related to the thermohaline circulation in the oceans, there would be no strong latent heat flux in the climate system. Thus, the whole Earth would cool down substantially. Since there would be no evaporation and precipitation, there would be no strong salinity gradient in the oceans. Due to less heat flux transport from low latitudes, the temperature at high latitudes would be lower than the present climatic conditions. Due to the greater-than-present meridional temperature difference and almost-zero salinity difference, the oceanic sensible heat flux would be intensified, which might

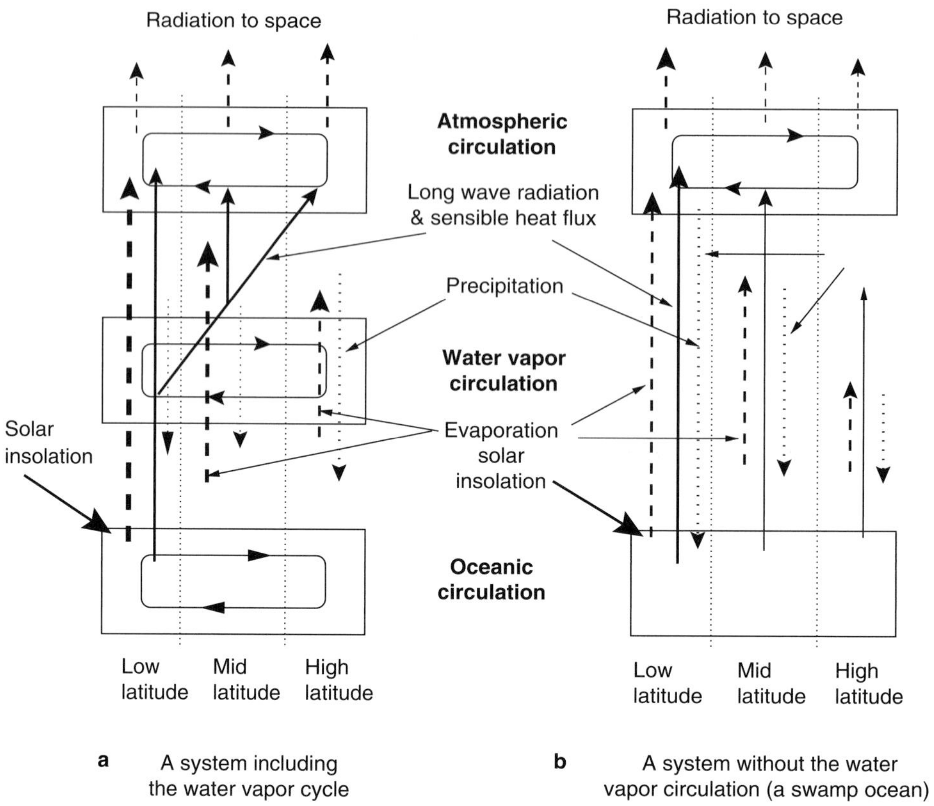

Fig. 5.86 Sketches for **a** a climate model with current and **b** a climate model with a swamp ocean.

partially compensate for the loss of poleward heat flux due to lack of the hydrological cycle, as shown in Figure 5.86b.

In some climate models, the oceanic component has been treated as a swamp, i.e., without the current. Although it is clear that such a model would not allow heat flux transport associated with current, a far stronger constraint on the climate system has not been clearly spelled out. A close examination reveals the following constraints for such a model:

- There are neither currents nor meso-scale eddies in the ocean, so there is no horizontal heat flux because molecular heat diffusion is negligible.
- Furthermore, there is no current to transport freshwater horizontally.
- As a result, evaporation and precipitation must be locally balanced.
- A major and somewhat unexpected consequence is that there will be no latent heat flux in the atmosphere.

The consequences of such assumptions are as follows. Although the ocean can still perform the role of absorbing short-wave radiation from the sun and heating up the atmosphere through latent heat release associated with evaporation, long-wave radiation, and sensible heat flux, there is neither oceanic sensible heat flux nor latent heat flux in the atmosphere, and the only mechanism to transport heat is the dry atmosphere circulation.

In summary, it is more appropriate to separate the poleward heat flux into three components: the atmospheric sensible heat flux, the oceanic sensible heat flux, and the atmosphere–ocean coupled latent heat flux. Although poleward heat flux can provide essential information about the global heat balance, for heat balance and climate at each location it is important to examine the divergence of these fluxes and their interaction. Furthermore, it is of great significance to explore how changes in the oceanic circulation affect the surface conditions, such as surface temperature and the air–sea heat fluxes, and the global climate conditions on Earth.

Contribution of salinity to stratification and horizontal pressure gradient in the oceans

Density is a nonlinear function of temperature, salinity, and pressure. In particular, the thermal expansion coefficient strongly depends on temperature. As a result, density at warm temperatures is primarily controlled by temperature; however, at high latitudes the thermal expansion coefficient is so small (Fig. 5.87) that density is primarily controlled by salinity.

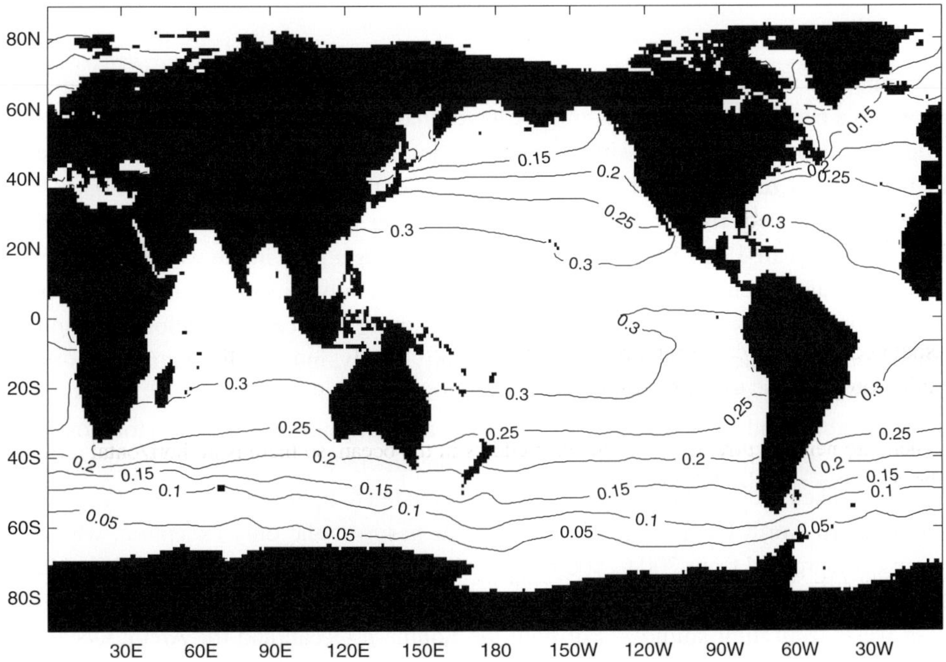

Fig. 5.87 Thermal expansion coefficient at the sea surface in the world ocean, in $10^{-3}/°C$.

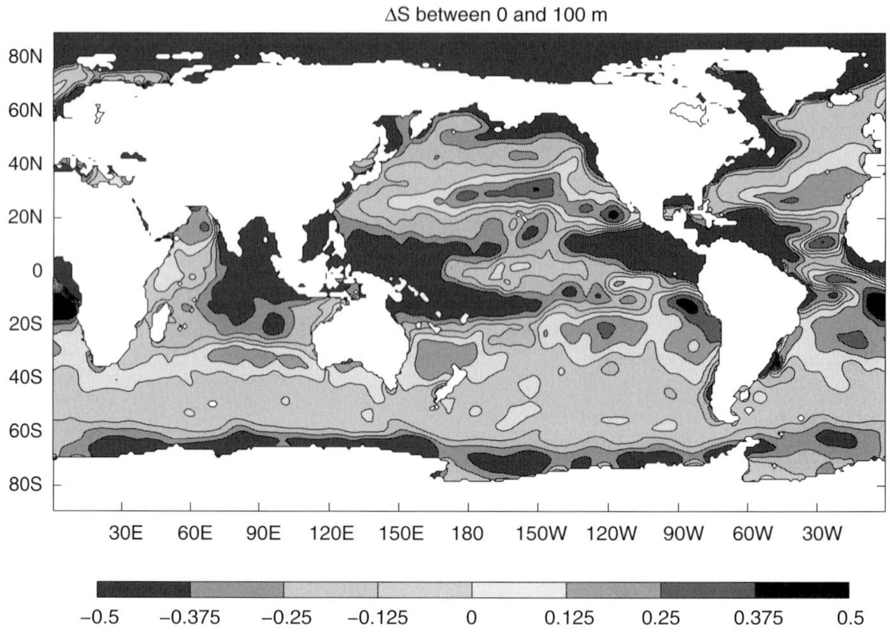

Fig. 5.88 Annual mean salinity difference between sea surface and 100 m depth. See color plate section.

Salinity is a major contributor to the stratification in the upper ocean. In fact, for most of the world's oceans, sea surface salinity is lower than that at 100 m depth, i.e., the vertical salinity gradient tends to intensify the stratification in the upper 100 m. In order to show the boundary of zero salinity difference, values lower than -0.5 are cut off in Figure 5.88. Therefore, relatively fresher water in the surface layer is a stabilizer for the stratification in the upper oceans. This is particularly true for the warm pool in the western Pacific and eastern Indian Oceans, plus high-latitude oceans. The only exceptions are the centers of the subtropical gyres in both hemispheres, where strong surface evaporation leads to high salinity on the surface and a relatively low salinity below, indicated in Figure 5.88 by the areas with red and yellow colors.

On the other hand, in the main thermocline, i.e., a depth range of 200–500 m for most of the world's oceans, salinity is gradually reduced in the downward direction, so its contribution tends to weaken the stratification, as indicated by the red to light green colors in Figure 5.89. The northern North Pacific Ocean (the subpolar basin) is the only exception, where the salinity is increased from 200 to 500 m. In addition, there is a tongue of negative value in the eastern subtropical basin, which is apparently created by the subduction of relatively fresh water from high latitudes. For both the Southern Ocean and the Arctic Ocean, the surface salinity gradient is always negative, indicating a strong halocline in these areas, as shown in Figures 5.88 and 5.89.

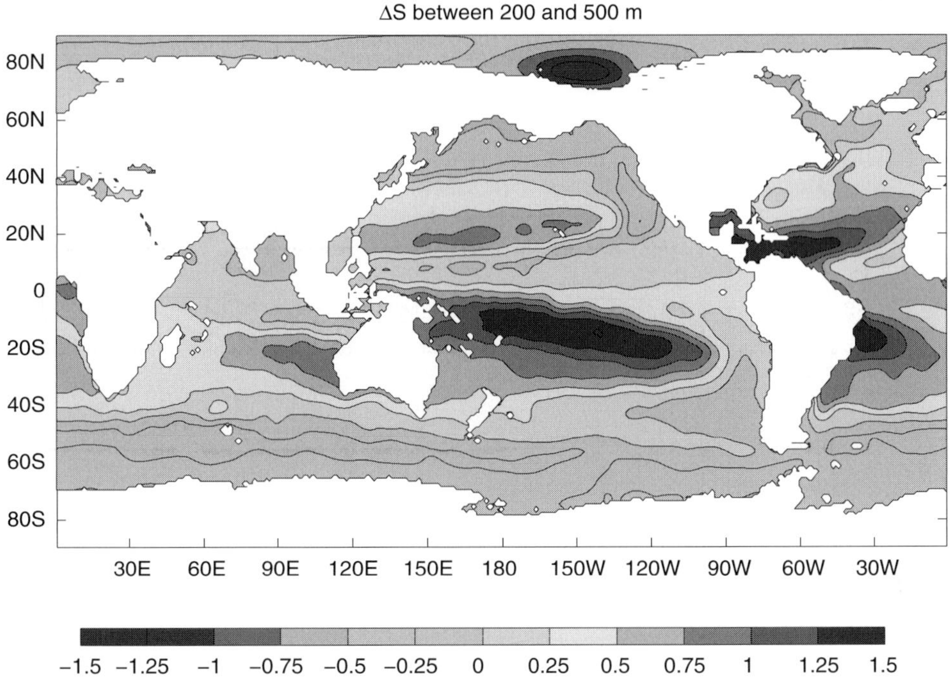

Fig. 5.89 Annual mean salinity difference between 200 and 500 m depth. See color plate section.

At low temperatures, salinity can make a crucial contribution to the stratification in the ocean; this can be illustrated through the following diagnosis based on the *World Ocean Atlas 2001* (WOA-01; Conkright *et al.*, 2002) climatological mean T, S properties in the world's oceans. We first examine how much salinity contributes to the vertical stratification in the water column. As an example, we show the stratification ratio, which is defined as follows.

The commonly used stratification can be extended as

$$N^2 = -\frac{g}{\rho_0}\frac{d\rho_\Theta}{dz}, \quad N_T^2 = -\frac{g}{\rho_0}\frac{d\rho_\Theta(T, S_0)}{dz}, \quad N_S^2 = N^2 - N_T^2 \tag{5.126}$$

where N_T^2 is defined as the equivalent stratification due to the vertical distribution of temperature only, with salinity set to a constant value of $S_0 = 35$. The difference between N^2 and N_T^2 is defined as N_S^2, which indicates the contribution of salinity to the stratification. For example, the places where $|N_S^2/N^2| \ll 1$ indicate that the contribution of the vertical salinity gradient to the local stratification is quite small. As shown in Figure 5.90a, in the upper ocean stratification is primarily due to the density gradient associated with temperature profile, while the salinity gradient tends to weaken the stratification. At nearly 1.5 km below sea level in the Southern Hemisphere, the stratification is primarily controlled by the relatively fresh water of Antarctic Intermediate Water (Fig. 5.90b, c).

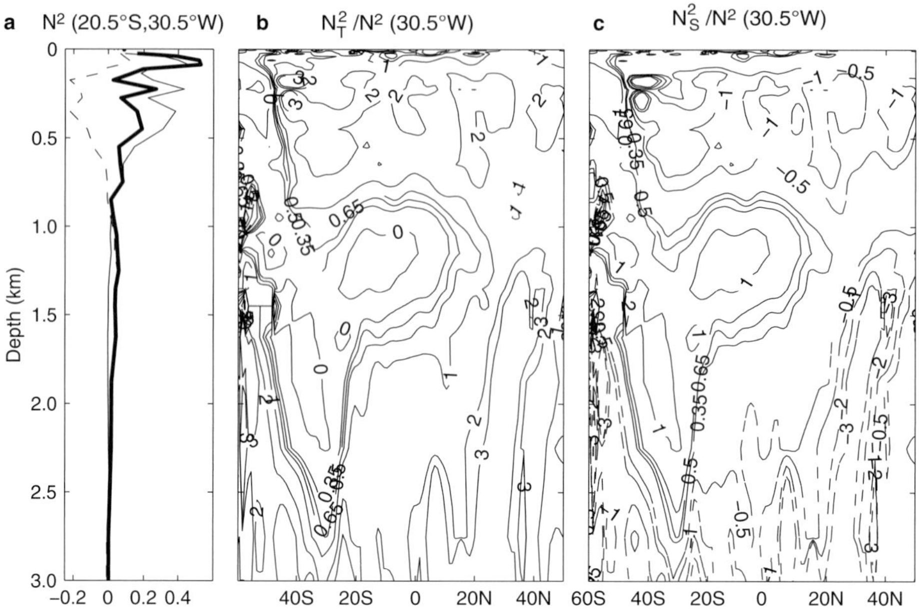

Fig. 5.90 Contribution of temperature and salinity to stratification along 30.5° W, inferred from climatology: **a** N^2(*heavy line*), N^2 due to temperature only (*thin line*), and N^2 due to salinity only (*dashed line*), in $10^{-4}/s^2$; **b** stratification contribution due to temperature; **c** stratification contribution due to salinity.

In the Atlantic Ocean, salinity's control of the stratification can be clearly identified from depth below the upper kilometer (Fig. 5.90b). In particular, the salinity gradient associated with both Antarctic Intermediate Water and North Atlantic Deep Water clearly plays a major role in shaping the stratification at mid depth.

On the other hand, although the temperature contribution dominates the stratification for the upper ocean in the Pacific Ocean, the contribution of salinity to the stratification below the top 1 km is very important (Fig. 5.91). In particular, the contribution of salinity is very large for latitudes higher than 20° off the equator.

Another way to illustrate the role of salinity is through its contribution to the horizontal pressure gradient. Since the sea-surface elevation is unknown, we will omit the potential contribution due to difference in sea-surface elevation and focus on the baroclinic pressure gradient associated with the T, S distribution only. Furthermore, we will also subtract the horizontal mean pressure, and present the deviation from the horizontal mean pressure at each level only. Similar to the discussion above, we separate the pressure into two components as follows:

$$p = \int_z^0 \rho dz, \quad p_T(T, S_0) = \int_z^0 \rho(T, S_0)\, dz, \quad p_S = p - p_T \tag{5.127}$$

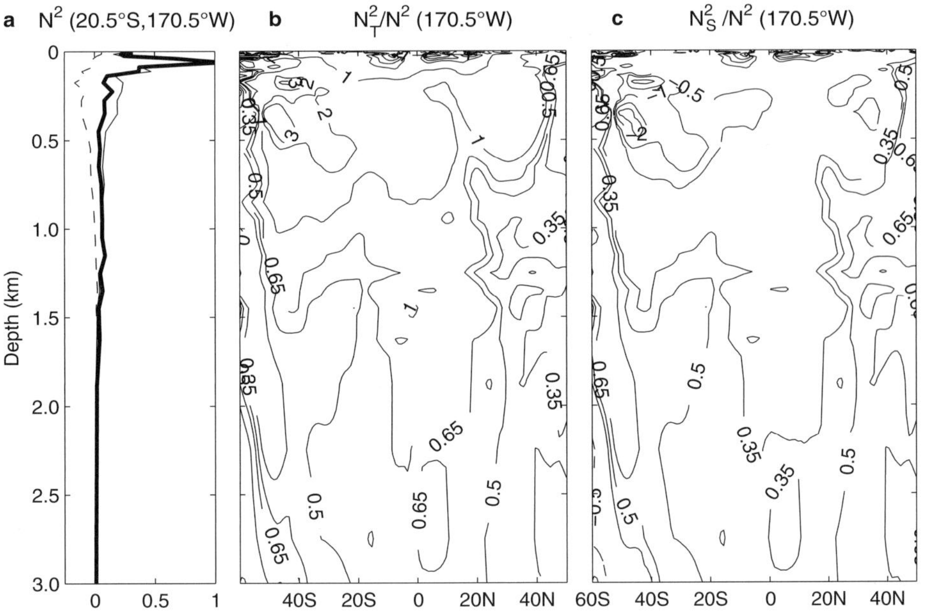

Fig. 5.91 Contribution of temperature and salinity to stratification along 169.5° W, inferred from climatology: **a** N^2(*heavy line*), N^2 due to temperature only (*thin line*), and N^2 due to salinity only (*dashed line*), in $10^{-4}/s^2$; **b** stratification contribution due to temperature; **c** stratification contribution due to salinity.

Thus, p is the commonly used pressure, p_T is the pressure due to temperature distribution in the water column only, and p_S is the pressure due to salinity distribution only. Since density is a nonlinear function of temperature, salinity, and pressure, the above definition does not provide an exact partition between temperature and salinity.

As shown in Figure 5.92a, at the middle level there is a southward pressure gradient force in the Atlantic Basin, which is responsible for maintaining the southward flow at the middle level as the return component of the present meridional overturning in the Atlantic Ocean. By examining the pressure distribution in the North Atlantic Ocean, it can readily be seen that temperature and salinity contributions have opposite signs at middle depth. In fact, the southward pressure force at the middle level is mostly provided by the salinity distribution (Fig. 5.92d), rather than the temperature (Fig. 5.92c). This suggests that meridional overturning in the Atlantic Ocean may be rather sensitive to changes in salinity, which is in turn related to the hydrological cycle. This connection will be explored shortly in the discussion about the thermohaline catastrophe associated with freshening of the subpolar North Atlantic Ocean.

On the other hand, the strong fronts associated with the ACC are primarily due to the temperature front, with a relatively small contribution from the salinity distribution.

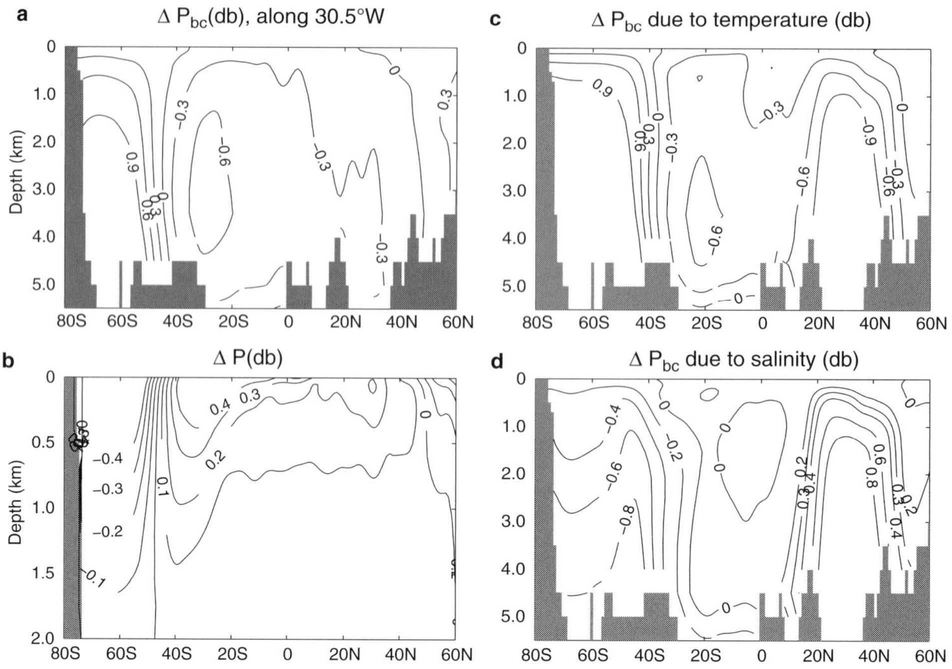

Fig. 5.92 **a–d** Baroclinic pressure deviation from the horizontal mean, averaged over 3° zonally with the central line taken along 30.5° W. Panel **b** indicates the pressure difference, using 2,000 m as the reference level, i.e., assuming no horizontal pressure gradient at this level.

The situation in the Pacific Ocean is different from that in the Atlantic Ocean. In the Pacific Basin, the pressure distribution at upper and middle levels is roughly symmetric with respect to the equator, except south of 45° S (Fig. 5.93a, b). The symmetric baroclinic pressure implies no cross-equator flow in the Pacific Ocean. This symmetric pattern in baroclinic pressure is primarily due to the temperature distribution, as shown in the upper and lower panels in Figure 5.93. Salinity distribution implies a northward pressure gradient force at the mid and deep levels, which seems to be linked to the northward movement of water at these levels.

The feedback between the hydrological cycle and the meridional thermal circulation

One of the possible reasons for ignoring the freshwater flux in heat flux analysis could be the misconception that freshwater flux in the ocean is so small that it seems to play a minuscule role in controlling the heat flux. As will be shown shortly, however, the freshwater flux through the oceans plays a major role in controlling the strength of the meridional overturning and poleward heat flux.

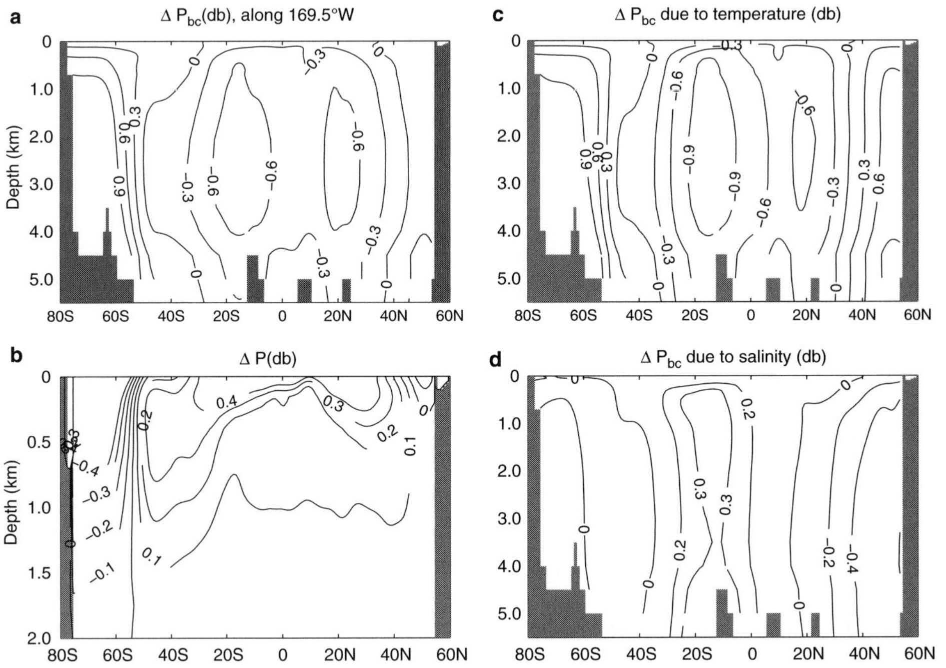

Fig. 5.93 **a–d** Baroclinic pressure deviation from the horizontal mean, averaged over 3° zonally with the central line taken along 169.5° W.

Due to evaporation at low latitudes and precipitation at high latitudes, there is a meridional salinity gradient at the sea surface. As shown in the previous section, the meridional density gradient due to the salinity difference is opposite to the meridional density gradient due to the thermal forcing in the upper oceans. Thus, in the present climate setting, the hydrological cycle is a brake for the poleward transport of heat transportation carried by the meridional thermally-forced circulation in the oceans. In other words, the poleward latent heat flux associated with the water vapor cycle has a negative feedback on the oceanic sensible heat flux.

If there were no hydrological cycle, the meridional overturning cell and the associated poleward heat flux in all five basins would be intensified. As an example, in the North Atlantic Ocean the surface density difference between the equator and high latitudes is about $\Delta\sigma = 3.39$ kg/m^3, based on the climatological data (Fig. 5.94a). If there were no evaporation and precipitation, there would be very little salinity difference in the oceans, so we set salinity to a constant value of 35. Assuming that the surface temperature distribution remains unchanged – a very bold assumption that is unlikely to be true in the real world – the corresponding surface density difference would be increased to $\Delta\sigma = 5.39$ kg/m^3 (Fig. 5.94b). Due to the increase in the north–south density difference, both the meridional overturning rate and poleward heat flux should increase.

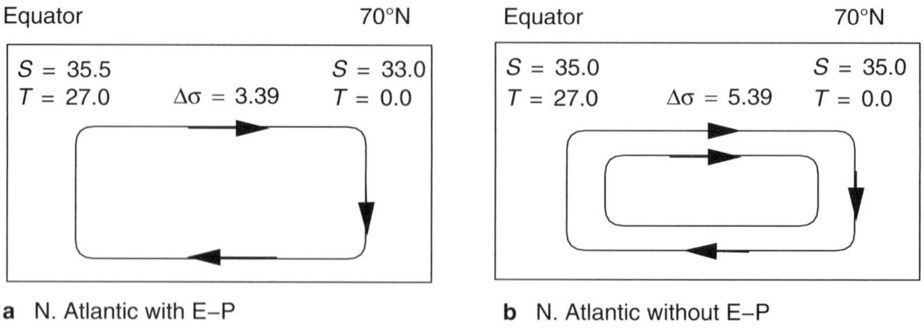

Fig. 5.94 Two models for the North Atlantic: **a** the standard model, forced by observed sea surface temperature and salinity; **b** a conceptual model with uniform salinity.

Four numerical experiments for the Atlantic Ocean were carried out in order to explore the dynamical role of freshwater flux in the meridional circulation. The MOM 2 (Pacanofsky, 1995) was used with a horizontal resolution of 2.5° × 2° and 15 levels vertically, with a constant diapycnal diffusivity of 0.4×10^{-4} m^2/s. The model's temperature and salinity were initialized with the climatologic data, and in each experiment the model was run for 1,000 years to reach a quasi-equilibrium.

In experiment A, both the surface temperature and salinity were relaxed toward the monthly mean climatology, along with the Hellermann and Rosenstein (1983) monthly mean wind stress data. In experiment B, salinity was set to a constant value of 35, but all other parameters remained the same as in experiment A.

In experiments C and D, the density effect due to temperature was ignored. Thus, the salinity difference was the only driving force for the circulation. In order to calculate the poleward heat flux, temperature was treated as a passive tracer, with an upper boundary condition such that the sea surface was relaxed toward the monthly mean climatology. In experiment C, the surface salinity was subject to the surface relaxation condition, assuming that the climatological monthly mean sea surface salinity remained unchanged. In experiment D, the model's salinity was subject to the virtual salt flux diagnosed from the steady circulation obtained from experiment A.

The meridional overturning circulation (MOC) in experiment B was 68% higher than in experiment A (Table 5.6), which is quite close to our scaling analysis. The poleward heat flux (PHF) in the second experiment was 21% higher than for experiment A, which is somewhat lower than the prediction from the simple scaling.

In experiments C and D, the meridional overturning cell reversed, and the associated meridional heat flux was equatorward, as shown in Table 5.6. The amount of equatorward heat flux was about 0.23 PW for the case with virtual salt flux. Thus, these numerical experiments demonstrated that the haline circulation is an essential factor controlling the poleward heat flux.

It is much more important to examine the poleward heat flux profile and the associated heat flux divergence. It is clear that, in experiment B, the ocean transported much more heat to high latitudes. In particular, the poleward heat flux divergence was much larger for

Table 5.6. *Four numerical experiments for the model Atlantic Ocean*

	A	B	C	D
Temperature	SST relaxation	SST relaxation	Treated as a dynamically passive tracer	Treated as a dynamically passive tracer
Salinity	SSS relaxation	Set to $S = 35$	SSS relaxation	Virtual salt flux
MOC (Sv)	12.58	21.10(+68%)	−13.27	−18.70
Poleward heat flux (PW)	0.72	0.87(+21%)	−0.07	−0.23

latitudes higher than 30° N (Fig. 5.95). Thus, in the case without a hydrological cycle, the oceanic current will transport more heat poleward and release it to higher latitudes, partially compensating for the decline in poleward heat flux caused by the lack of latent heat flux associated with the hydrological cycle.

Other experiments for the global oceans were also run with a low horizontal resolution of 4° × 4° and 15 layers. The Indonesian Throughflow is included in these low-resolution simulations. The model's temperature and salinity were initialized with the Levitus and Boyer (1994) data in each experiment. The model was run for 1,000 years to reach a quasi-equilibrium.

In the standard experiment, both the surface temperature and salinity were relaxed toward the monthly mean climatology, using the Hellermann and Rosenstein (1983) monthly mean wind stress data. In the second experiment, salinity was set to a constant value of 35, but everything else remained the same as in the standard experiment. The poleward heat flux was calculated, using the definition discussed in the Appendix, so the heat flux in both the South Pacific and Indian Oceans was uniquely defined.

It is clear that by turning off the salinity effect, the total northward heat flux in the Northern Hemisphere increases, while it declines in the Southern Hemisphere (Fig. 5.96). Most interestingly, the northward heat flux in the North Pacific Ocean increases substantially. This increase in heat flux at mid latitudes is probably associated with the newly created thermal mode of the meridional circulation in the North Pacific Ocean, caused by the lack of the salinity effect which opposes the thermal mode. The northward heat flux in the Atlantic Ocean declines due to the diminishing effect of the global conveyor belt that is driven, at least partially, by the salinity difference between the Pacific and Atlantic Oceans. As a result of the decline of the global conveyor belt, the poleward heat flux in the South Indian Ocean declines. Note that the heat flux in the South Atlantic Ocean is now southward, as required by the thermal mode in this basin.

A major assumption that we made use of is that the vertical mixing coefficient and wind stress remain the same for all these experiments. This is a highly idealized assumption. Since thermohaline circulation is so closely related to climate, changes in the circulation

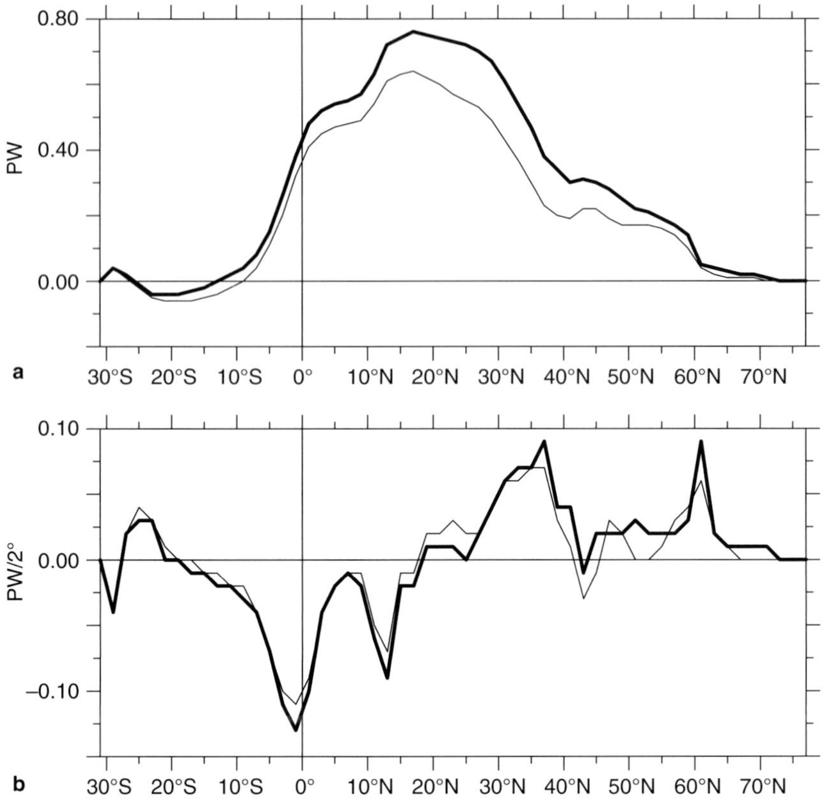

Fig. 5.95 Poleward heat flux (**a**) and its divergence (**b**) (due to advection) diagnosed from the standard model with both temperature and salinity relaxed (*thin line*) and the model with uniform salinity ($S = 35$, *heavy line*) (Huang, 2005b).

must affect both the wind stress and mixing coefficient; however, a more comprehensive study is difficult, and this is left for readers interested in pursuing this line of thought.

5.3.2 Surface boundary conditions for salinity

Salinity balance is one of the most critical components of the oceanic circulation. The suitable upper boundary conditions for salinity balance have evolved gradually over the past several decades. Early in its development, two types of salinity boundary conditions were used, i.e., the relaxation condition and the virtual salt flux condition. As our understanding of oceanic circulation physics deepens, the natural boundary condition is now used in more and more model studies. This section is devoted to the examination of salinity boundary conditions in the upper ocean.

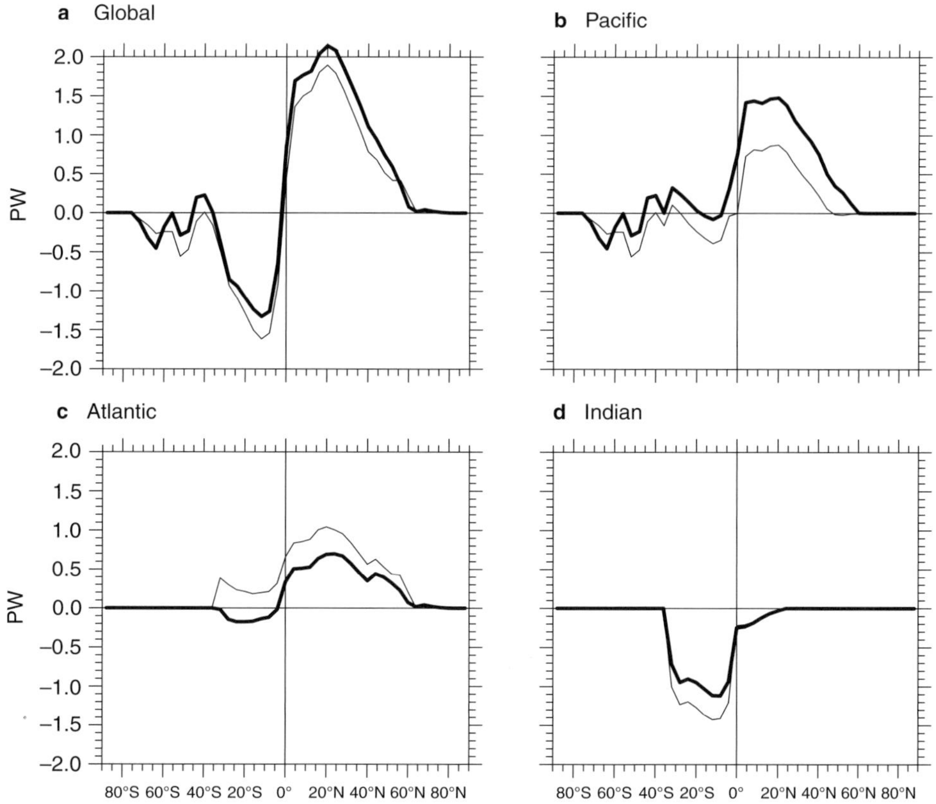

Fig. 5.96 **a–d** Poleward heat flux in the world's oceans, diagnosed from a model based on observed SST and SSS (*thin lines*) and a model based on relaxation to observed SST and with constant salinity (*heavy lines*) (Huang, 2005b).

Upper boundary conditions for salinity balance

The salinity boundary condition on the sea surface is directly linked to the air–sea freshwater flux, so that its formulation depends on how the free surface is handled. We begin with the formulation for models based on the rigid-lid application because the boundary conditions can be explained quite clearly in terms of the relevant physics. Then we will discuss the corresponding formulation for models with a free surface.

The rigid-lid approximation

Early in model development, the rigid-lid approximation was widely used. As explained in Section 4.1.1, the basic idea was to replace the free surface by the flat surface $z = 0$, so that the complexity of dealing with the moving free boundary $z = \zeta$ is reduced to a fixed and flat boundary $z = 0$. The suitable boundary condition for the vertical velocity under the rigid lid is as follows.

The kinematic boundary condition on the free surface is

$$w = \frac{D\zeta}{Dt} + (e - p), \text{ at } z = \zeta \tag{5.128}$$

where w is the vertical velocity, $D/Dt = \partial/\partial t + u\partial/\partial x + v\partial/\partial y$ is the total derivative, ζ is the free surface elevation, and $e - p$ is evaporation minus precipitation. For motions in the ocean interior, we introduce the following scaling

$$(x, y) = L(x', y'), \quad (u, v) = U(u', v') \tag{5.129}$$

$$t = Tt', \quad \zeta = \frac{fLU}{g}\zeta' \tag{5.130}$$

$$w = \delta U w', \quad e - p = \lambda\delta U\,(e - p)' \tag{5.131}$$

where $\delta = H/L \ll 1$ is the aspect ratio, and $\lambda \simeq 0.01 \ll 1$. It is assumed that the time scale is determined by the horizontal advection. Substituting these relations into the boundary condition, Eqn. (5.128), and dropping the primes leads to

$$w = \varepsilon_T F\frac{\partial\zeta}{\partial t} + \varepsilon F\vec{u}\cdot\nabla\zeta + \lambda(e - p), \text{ at } z = \varepsilon F\zeta \tag{5.132}$$

where $\varepsilon_T = 1/fT \ll 1$ is the Kibel number, $\varepsilon = U/fL \ll 1$ is the Rossby number, and $F = f^2L^2/gH \approx O(1)$. Since the time scale is set by the advection time scale L/U, the first two terms are of the same magnitude. For motions with a horizontal scale of 1,000 km, the nondimensional number εF is approximately 10^{-4}, which is much smaller than λ. Thus, the upper boundary condition can be linearized as

$$w = e - p, \ \ at\ z = 0 \tag{5.133}$$

The rigid-lid approximation is not accurate for motions with a time scale shorter than decadal, motions on the planetary scale, or motions in shallow seas. For example, for motions of planetary scale, $L = 10^7$ m, for wave motion on seasonal time scales or shorter, $T \leq 10^7$ s, so $\varepsilon_T F \approx 0.02$. Thus, the vertical velocity due to wave motions of the free surface elevation has the same magnitude as that of evaporation minus precipitation. Under such circumstances, the rigid-lid approximation, in which the free surface motion is neglected, is inaccurate.

In fact, the rigid-lid approximation has been used in many models with non-zero vertical velocity at $z = 0$. For example, in many quasi-geostrophic models the dynamical effect of wind stress curl is simulated by imposing the equivalent Ekman pumping velocity at the upper surface. In general, the Ekman pumping velocity is about 30 times larger than the vertical velocity associated with precipitation and evaporation.

Three boundary conditions for salinity under the rigid-lid approximation

For simplicity, we discuss different types of boundary condition for the haline circulation in the $x - z$ plane of local Cartesian coordinates. The salinity balance in a surface box

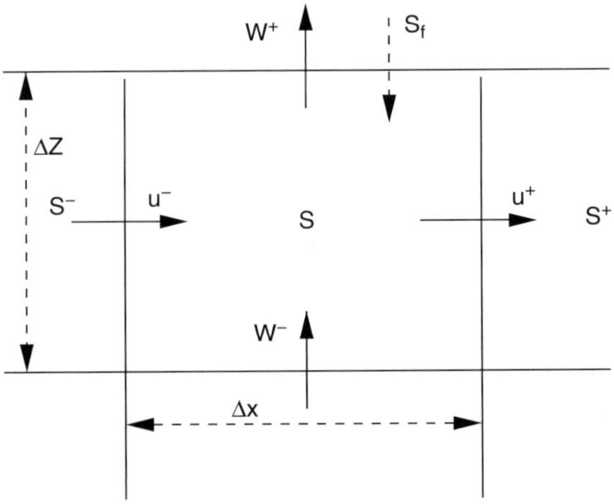

Fig. 5.97 A finite difference grid box in the $x - z$ plane.

(Fig. 5.97), can be written as

$$\frac{\partial S}{\partial t} + \frac{1}{\Delta x}\left[(uS)^+ - (uS)^-\right] + \frac{1}{\Delta z}\left[(wS)^+ - (wS)^-\right] = \frac{\kappa H}{\Delta x}\left(S_x^+ - S_x^-\right) + \frac{1}{\Delta z}\left(S_f - \kappa_V S_z^-\right). \quad (5.134)$$

where $(uS)^+$ and $(uS)^-$ are the salt fluxes advected across the right and left boundaries, and S_x^+ and S_x^- are the horizontal salinity gradients at the right and left boundaries; similarly, $(wS)^+$ and $(wS)^-$ are the salt fluxes advected through the upper and lower boundaries, and S_f and $(wS)^-$ are the vertical salinity fluxes at the upper and lower boundaries.

In addition, vertical velocity is prescribed on the upper surface

$$w = w^+, \text{ at } z = 0 \quad (5.135)$$

where w^+ is the vertical velocity and its value depends on the choice of model.

Relaxation condition

For this case, the rigid lid is interpreted as a solid boundary with no mass flux across it, so that the velocity boundary condition is

$$\text{(a) } w^+ = 0, \text{ so } (wS)^+ \text{vanishes} \quad (5.136)$$

As discussed above, however, the rigid-lid approximation does not necessarily require the vertical velocity to be zero at the upper surface. The salinity boundary condition is

$$\text{(b) } S_f = \Gamma(S^* - S) \quad (5.137)$$

where Γ is the relaxation constant, and S^* is specified according to the climatological mean surface salinity. This boundary condition can be traced back to the relaxation boundary condition for the sea surface temperature (Haney, 1971). His original equation is based on a detailed analysis of the heat flux through the air–sea interface, including solar insolation, latent heat flux, sensible heat flux, and the turbulent heat flux:

$$\kappa T_z = \Gamma(T^* - T_s) \tag{5.138}$$

where T^* is the reference temperature, or the equivalent atmospheric temperature, which should be calculated by taking into consideration all heat flux terms; T_s is the sea surface temperature. In many studies the climatological mean SST is used as T^*. It is clear that such an approach may introduce certain errors because of the difference between T^* and the climatological SST.

Virtual salt flux condition

For this case, the rigid lid is also interpreted as a solid boundary with no mass flux across it, so that the velocity boundary condition is

$$\text{(a) } w^+ = 0, \text{ so } (wS)^+ \text{ vanishes} \tag{5.139}$$

however, the salinity boundary condition is a flux condition

$$\text{(b) } S_f = (e - p)S \tag{5.140}$$

where $e - p$ is evaporation minus precipitation, and S is the salinity in the box. This is the virtual salt flux required for the salinity balance. However, using $(e - p)S$ to force a model may cause the total salt in a basin to increase infinitely. This can be explained as follows. For a steady climate, the global precipitation and evaporation should be nearly balanced:

$$\iint (e - p)dA = 0. \tag{5.141}$$

However, the global integration of the virtual salt flux is larger than zero:

$$\iint (e - p)SdA > 0. \tag{5.142}$$

This is due to a positive correlation between evaporation minus precipitation and salinity, as shown in Figure 1.2.3. For example, evaporation is very strong in the subtropical North Atlantic Ocean, giving rise to a salinity of more than 37; meanwhile, excess precipitation in the subpolar Pacific Ocean gives rise to a surface salinity lower than 32. Thus, to avoid the salinity explosion one must use the virtual salt flux defined by

$$S_f = (e - p)\overline{S}_s \tag{5.143}$$

where $\overline{S}_s$ is the mean sea surface salinity averaged over the whole model domain. This formulation can introduce a systematic bias which we will address later.

Natural boundary condition

As discussed above, for a climatic time scale the velocity boundary condition on the upper surface is

$$\text{(a) } w^+ = e - p \tag{5.144}$$

and this is the freshwater flux controlling the salinity balance; the corresponding boundary condition for the salinity balance is

$$\text{(b) } (wS)^+ - \kappa_V S_z^+ = 0 \tag{5.145}$$

Note that within the water column there is always turbulent salt flux $\kappa_V S_z$ and advective salt flux wS. However, in the air there is no salt, so both these terms are identically zero, although w is not zero. At the air–sea interface $\kappa_v S_z = (e - p)S$, i.e., the turbulent flux exactly cancels the advective flux so that there is no salt flux across the air–sea interface, as required by the physics. Thus, we may call this turbulent flux $S_f = \kappa_v S_z$ an anti-advective salt flux, which is the same as the virtual salt flux discussed above, $S_f = (e-p)S$. From our discussion, we can see that there is no need for any salt flux across the air–sea interface if we use the natural boundary condition, because these two fluxes exactly cancel each other; however, if we interpret the rigid-lid approximation as a zero vertical velocity condition, the virtual salt flux is needed in order to simulate the effect of freshwater flux through the air–sea interface. Accordingly, under the natural boundary condition, the salinity balance for a surface box is reduced to

$$\frac{\partial S}{\partial t} + \frac{1}{\Delta x}\left[(uS)^+ - (uS)^-\right] - \frac{1}{\Delta z}(wS)^- = \frac{\kappa_H}{\Delta x}\left(S_x^+ - S_x^-\right) - \frac{\kappa_V}{\Delta z} S_z^- \tag{5.146}$$

The natural boundary condition, in combination with the continuity equation, is a statement of salt conservation

$$\frac{D}{Dt}\iiint_V \rho S dv = 0 \tag{5.147}$$

This equation means that the total salt in the world's oceans is conserved. The local salinity change is due to freshwater dilution/concentration, and there is no need for the virtual salt flux. Although our discussion above and Figure 5.97 are based on the case with a rigid lid, the same argument can be applied for the case with free surface, as discussed later.

A virtual natural boundary condition

Evaporation and precipitation data are difficult to collect, and there are no reliable historical data. As a compromise, one can use the sea surface salinity as a forcing field to reconstruct the salinity balance in the past, using the virtual natural boundary condition as follows.

First, the equivalent evaporation and precipitation field can be inferred from the historical sea surface salinity data from the model as

$$e - p = \Gamma(S^* - S)/\overline{S}_s \tag{5.148}$$

where S^* and S are the climatological mean salinity and the observed salinity at a given time in the past, and $\bar{S}_s$ is the global mean sea surface salinity. Second, this equivalent evaporation and precipitation field can be used as the freshwater flux through the air–sea interface in the model.

Salinity condition for models with free surface

The natural boundary condition discussed above is based on the rigid-lid approximation, which was used extensively in the past. However, in the new generation of numerical models, the free surface is the preferred choice. The suitable boundary condition for salinity balance is a simple mathematical statement that $p-e$ is a source of freshwater at the upper surface of the ocean, with no salinity flux through the air–sea interface. We note that the upper surface of the ocean is neither a Eulerian surface nor a Lagrangian surface, because freshwater moves across this surface, as discussed in Chapter 3.

As an example, we discuss the suitable upper boundary condition for salinity used in mass-conserving models. A convenient choice of the vertical coordinate for a mass-conserving numerical model is the pressure coordinate discussed in Section 2.8. Since both the surface and bottom pressure change with time, the pressure-η coordinates can be used. The concept of the η-coordinate system was introduced by Mesinger and Janjic (1985). In a pressure-η coordinate system (Huang and Jin, 2007), the vertical coordinate is defined as

$$\eta = (p - p_t)/r_p, \quad r_p = p_{bt}/p_B, \quad p_{bt} = p_b - p_t \tag{5.149}$$

where $p_b = p_b(x, y, t)$ is the bottom pressure, $p_t = p_t(x, y, t)$ is the hydrostatic pressure at the upper surface of the water column, and $p_B = p_B(x, y)$ is the time-invariant reference bottom pressure, which is calculated from the basin-averaged stratification prescribed in the initial state.

Since p_t is the specified pressure at the upper boundary, owing to evaporation and precipitation the increment in hydrostatic pressure is $\delta(p - p_t) = -\rho_f g \delta Q_{E-P}$, where ρ_f is the density of freshwater and Q_{E-P} is the freshwater flux across the air–sea interface associated with evaporation and precipitation. Thus, the upper boundary condition is

$$r_p \dot{\eta} = -\rho_f g Q_{E-P}/p_B \tag{5.150}$$

where $\dot{\eta} = d\eta/dt$ is the virtual vertical velocity, which has a dimension different from the vertical velocity used in the traditional z-coordinates.

The bottom pressure is prognostically calculated from the bottom pressure tendency equation (Huang *et al.*, 2001). We begin with the continuity equation in pressure-η coordinates

$$\frac{\partial r_p}{\partial t} + \left(\frac{\partial r_p u}{\partial x} + \frac{\partial r_p v}{\partial y} \right) + \frac{\partial r_p \dot{n}}{\partial \eta} = 0 \tag{5.151}$$

Integrating Eqn. (5.151) from $\eta = 0$ (sea surface) to $\eta = p_B$ (bottom) and applying the corresponding boundary condition at the sea surface, we obtain the bottom pressure tendency equation

$$\frac{\partial p_{bt}}{\partial t} + \nabla_h \cdot \left(p_{bt} \vec{V}_{baro} \right) = -\rho_f g Q_{E-P} \tag{5.152}$$

where $\vec{V}_{baro}$ is the vertically integrated horizontal velocity and ∇_h is the horizontal divergence operator. Thus, precipitation minus evaporation can directly affect the bottom pressure due to the adding of mass. The contribution of the air–sea freshwater flux regulates the salinity distribution through the dilution of seawater by the mean of mass continuity.

The corresponding salinity condition at the upper surface is that the net salt flux due to salt advection and vertical salt diffusion exactly cancel each other, i.e.,

$$S_f = S_{f,adv} + S_{f,diffu} = 0, \quad \text{at the surface} \tag{5.153}$$

Note that, in the pressure coordinate, the sea surface elevation is a diagnostic variable calculated by integrating the hydrostatic relation

$$\zeta = z_b - \frac{1}{p_B(x,y)} \int_{p_B}^{0} \frac{p_{bt}}{\rho g} d\eta \tag{5.154}$$

In the traditional z-coordinate, the corresponding salinity condition at the upper surface is the same as Eqn. (5.153), i.e., there is no net salt flux across the air–sea interface. The effect of air–sea freshwater flux in the system is reflected in terms of a mass source in the free surface elevation. In the z-coordinate model, the vertical velocity at the sea surface ζ satisfies

$$w = \frac{\partial \zeta}{\partial t} + \vec{u}_h \cdot \nabla_h \eta + (e - p)\rho_f / \rho_s \tag{5.155}$$

where $\vec{u}_h$ is the horizontal velocity, and ρ_f and ρ_s are the freshwater density and sea surface density. Vertical integration of this equation leads to a prognostic equation for the free surface

$$\frac{\partial \zeta}{\partial t} \left(\frac{\partial}{\partial x} \int_{-H}^{\zeta} u dz + \frac{\partial}{\partial y} \int_{-H}^{\zeta} v dz \right) + (p - e)\rho_f / \rho_s + RT \tag{5.156}$$

where RT means rest terms associated with thermohaline processes in the water column. Thus, precipitation tends to increase the sea level. However, the local sea level is also closely related to the horizontal convergence/divergence of the vertically integrated velocity field (Huang and Jin, 2002b).

The pitfalls of the relaxation and virtual salt flux conditions

The relaxation condition implies a strong negative feedback on the surface salinity. As a result, the solutions obtained under this boundary condition match the observed surface salinity; in addition, the solutions are stable for most cases. These features can be advantageous for simulating the present climate. However, this boundary condition may not be suitable for simulating the oceanic circulation for general situations.

First, although the relaxation condition, Eqn. (5.138), for sea surface temperature is based on sound physical reasoning, the salt relaxation condition, Eqn. (5.137), lacks physical background. In addition, using such a large relaxation time seems unreasonable.

Second, the salinity relaxation condition is not suitable for climate study or forecasting because the reference salinity is unknown for climate conditions different from the present-day climate.

The virtual salt flux condition is also unphysical. First, to balance the salt in the oceans, a huge virtual salt flux through the air–sea interface and the atmosphere is required for taking up the salt from the subpolar basin, transporting it equatorward, and dumping it there. The virtual salt flux required in the North Atlantic Ocean can be estimated as

$$\sum (E - P)S > 10^9 \text{ kg/s} \times \frac{35}{1000} > 3.5 \times 10^7 \text{ kg/s} \tag{5.157}$$

It is obvious that such a huge virtual salt flux is not real and should be avoided.

Note that the original definition of virtual salt flux, Eqn. (5.140), includes a weakly positive feedback because of the fact that S tends to be high wherever evaporation overpowers precipitation. However, the virtual salt flux is based on the basin-mean salinity $\overline{S}_x$; thus, it has nothing to do with the local salinity. Even if the virtual salt flux can be accepted as a parameterization, the physics associated with the virtual salt flux is distorted.

In addition, this constraint can introduce large errors wherever the local salinity is much different from the basin or global mean. When the local salinity is very low, this formulation exaggerates the equivalent haline forcing and could give rise to negative salinity due to the large exaggerated virtual salt flux. For example, near the mouth of the Amazon River, sea surface salinity is very low. As a result, the model's salinity near the mouth of the river may become negative. For the world's oceans the virtual salt flux condition can also introduce non-negligible errors. Surface salinity in the North Pacific Ocean can be lower than 33, and the surface salinity in the North Atlantic Ocean can be higher than 37. Supposing that $\overline{S}_s$ is defined as 35 for a world ocean circulation model, a systematic bias of 10% will be introduced through the upper boundary condition for the salinity.

The nonlinear nature of the surface buoyancy boundary conditions

Differences in the surface thermohaline forcing conditions

The most important differences between thermal and haline boundary conditions at the sea surface are the following. First, thermal forcing in the upper ocean is a flux of internal energy, while the surface haline forcing is a freshwater flux, which is a mass flux plus a small amount of gravitational potential energy. According to Eqn. (3.5.31), the amount of GPE due to precipitation is $\iint g\rho\zeta\omega dA$, where ζ is the free surface elevation and ω is the rate of precipitation.

Second, sea surface thermal forcing is associated with a strong negative feedback between the surface temperature and air–sea heat flux. As a result, the e-folding decay time scale for thermal anomalies tends to be short, i.e., a thermal anomaly cannot survive for a long time. On the other hand, there is no direct feedback between the local salinity and evaporation/precipitation. Thus, salinity anomalies tend to last much longer than thermal anomalies.

Survival of temperature and salinity anomalies in the surface layer

Two aspects of physics affect the survival characteristics of surface temperature and salinity anomalies. First, the equation of state is nonlinear. In particular, the thermal expansion coefficient is large at high temperatures, but is very small near the freezing point. As a result, density structure and thus the circulation at high latitudes are primarily controlled by salinity rather than temperature. On the other hand, density and circulation at low latitudes are primarily controlled by temperature rather than salinity, with exceptions near the mouths of rivers where the large salinity difference may play a dominating role.

Second, surface thermal anomaly is subject to a rather strong negative feedback; thus, thermal anomalies can be dissipated rather quickly owing to the strong air–sea heat flux feedback. On the other hand, salinity is not directly linked to the evaporation and precipitation rate across the air–sea interface.

The following two figures (Figs. 5.98 and 5.99) illustrate the difference between temperature and salinity anomalies at low and high latitudes in the North Atlantic Ocean. The heavy solid lines indicate temperature, salinity, and density profiles at two stations, near 29° N and 69° N. For the climatic mean state, a stable stratification is maintained in the following way. At low latitudes, stratification is characterized by warm and salty water lying over cold and fresh water. On the other hand, at high latitudes, relatively cold and fresh water lies over warm and salty water. Assume that there are surface temperature/salinity perturbations in the upper ocean, with linear profiles in the upper 75 m and the maximum

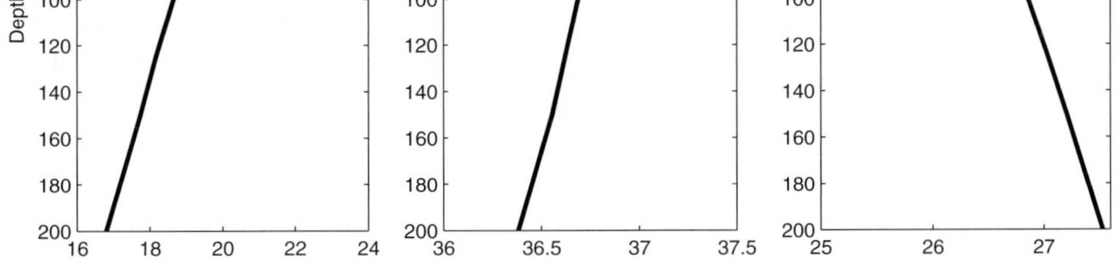

Fig. 5.98 **a** Temperature, **b** salinity, and **c** density profiles at a station in the North Atlantic (20.5° W, 29.5° N). *Thin solid lines* indicate perturbations due to temperature anomaly in the upper 75 m, *thin dashed lines* indicate perturbations due to salinity anomaly in the upper 75 m.

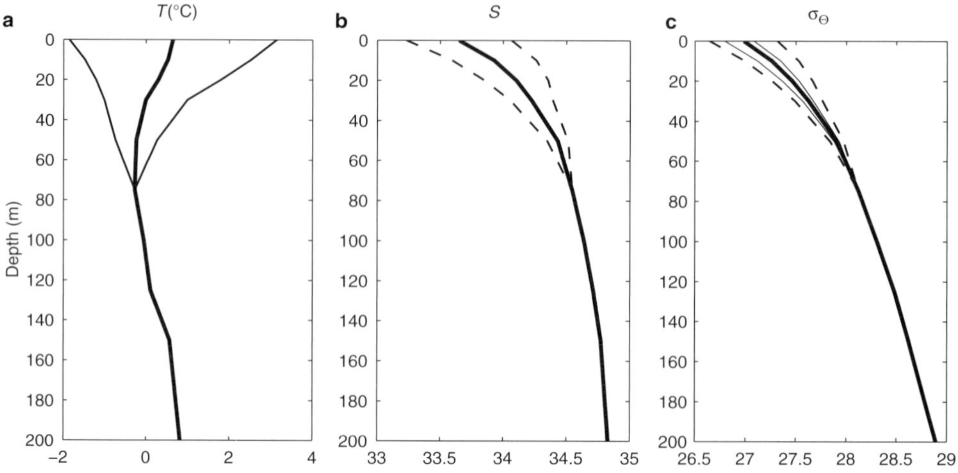

Fig. 5.99 **a** Temperature, **b** salinity, and **c** density profiles at a station in the North Atlantic (20.5° W, 69.5° N). *Thin solid lines* indicate perturbations due to temperature anomaly in the upper 75 m, *thin dashed lines* indicate perturbations due to salinity anomaly in the upper 75 m.

values of 3°C and 0.5 at the surface. It is clear that, at low latitudes, the density anomaly is primarily due to temperature (Fig. 5.98c); however, at high latitudes, it is primarily due to the salinity anomaly (Fig. 5.99c).

Note that a negative salinity anomaly (or freshwater anomaly) in the upper ocean can survive much longer. This is due to the fact that a freshwater anomaly in the upper ocean lowers the surface density, thus forming a strong halocline which is rather stable to perturbations. The lack of deep convection can lead to a further decline of heat loss to the atmosphere and a smaller evaporation rate. These physical processes work together to promote the longevity of surface freshwater anomalies in the high-latitude oceans. These stable freshwater layers on the top of the ocean can cause the halocline catastrophe, which will be discussed later.

Freshwater transport in a Boussinesq model

Most numerical models currently used in simulating ocean circulation and climate are based on the Boussinesq approximations. These models use the volume conservation to replace the physically more accurate mass conservation. As a result, virtual salt fluxes across the air–sea interface and meridional sections appear in these models. Such salt fluxes are artifacts of these models; so the meaning of such salt fluxes is unclear when the model is in a state of transition.

However, when the model reaches a quasi-steady state, the virtual salt fluxes diagnosed from the model may be interpreted as follows. When the model in a closed basin reaches a quasi-steady state, the meridional transport of salt across a latitude circle should vanish

$$0 = \iint \rho v S dx dz = \iint \rho v \overline{S} dx dz + \iint \rho v (S - \overline{S}) dx dz \tag{5.158}$$

where $\overline{S}$ is the basin mean salinity; thus, the freshwater transport through this section is

$$\iint \rho v dxdz = \iint \rho v(1 - S/\overline{S})dxdz \tag{5.159}$$

Since seawater density is nearly constant, as a good approximation the freshwater volumetric transport in these models can be defined as

$$F_{fw} = \iint v(1 - S/\overline{S})dxdz \tag{5.160}$$

Because the meridional volume transport for a Boussinesq model in a closed basin vanishes in a quasi-steady state, this equation can be rewritten as

$$F_{fw} = -\frac{1}{\overline{S}} \iint vSdxdz \tag{5.161}$$

i.e., the equivalent freshwater flux through a meridional section is equal to the meridional salt flux diagnosed from the model, multiplied by a negative sign and divided by the basin mean salinity (Bryan, 1969; Huang, 1993b).

It is worth noting that the definition of equivalent freshwater flux includes the basin mean salinity. Thus, the equivalent freshwater fluxes defined for different regions may not be comparable, because they are based on different mean salinity. This inconsistency is, unfortunately, intrinsic to the definition itself. It is worth emphasizing that the formula in Eqn. (5.161) is applicable for a closed basin only. At the latitudes of ACC, it cannot be used to infer the equivalent meridional freshwater transport for individual sectors, such as the Atlantic or Pacific sectors, because the meridional volumetric transport for individual sectors is non-zero.

5.3.3 Haline circulation induced by evaporation and precipitation

Historically, evaporation/precipitation was the first mechanism explored as the driving force for the oceanic general circulation. However, the physics involved in the haline circulation in the ocean has been overlooked for many decades. In fact, in most previous oceanic circulation models, the haline circulation was driven by either the salt relaxation condition or the virtual salt flux condition; thus, the real physics of haline circulation remained obscure for a long time. Before 1990, there were only a very few papers published in which the haline component of the oceanic circulation was forced by evaporation and precipitation.

In Chapter 3 we discussed in great detail why surface thermal forcing alone cannot drive or maintain a meridional overturning circulation. The situation in relation to surface freshwater flux is quite similar. The major difference between surface heat flux and freshwater flux is as follows. Surface freshwater flux is always associated with a mass transport across the air–sea interface. In general, freshwater is taken up from the surface at low latitudes in the form of moisture. Through the meridional circulation in the atmosphere, the moisture is transported to high latitudes; where it is put back into the ocean. If there were no other forcing, such as wind stress, heat flux, and tidal dissipation, the air–sea freshwater flux alone could drive a barotropic circulation in the ocean, which is called the Goldsbrough–Stommel circulation.

As will be discussed below, the barotropic circulation in the world's oceans driven by evaporation and precipitation is quite slow, with the total transport being on the order of 1 Sv. In the ocean, however, the haline component of the thermohaline circulation is one order of magnitude stronger. Surface freshwater flux alone cannot provide the mechanical energy required for sustaining such a strong circulation against friction. Thus, evaporation and precipitation alone cannot drive the baroclinic component of the haline circulation energetically. However, in the following discussion we will keep the use of the word "drive" for the presentation of the classical results, and the question of what really drives the thermal and haline circulation in the ocean will be discussed in detail in Section 5.4.

Classical circulations driven by evaporation and precipitation

Hough's solution

Evaporation and precipitation can drive oceanic circulation, and this was first explored by Hough (1897) in his tidal paper. Hough originally assumed the $P-E$ pattern to be a second Legendre polynomial:

$$P - E = P_2(x) = (3x^2 - 1)/2, x = \sin\theta \tag{5.162}$$

i.e., precipitation over high latitudes and evaporation over low latitudes.

Hough did not include any figures in his paper; his solution became much clearer after it was illustrated graphically by Stommel (1957). For simplicity, Stommel assumed a simpler pattern of precipitation over the Northern Hemisphere and evaporation over the Southern Hemisphere (Fig. 5.100).

Hough did not know how to parameterize friction, so there was no friction in his model. In fact, his solution may be valid only for the initial stage of the problem, when the solution can be treated as infinitesimal, so that the nonlinear effect can be neglected. In the beginning, the ocean covers the solid Earth with a uniform depth of water. Water from precipitation enters the sea surface in the Northern Hemisphere and leaves the sea surface in the Southern Hemisphere. Owing to the small difference in the meridional sea level, there is a meridional velocity pointed to the south (Fig. 5.100a).

As time progresses, more water is piled up in the Northern Hemisphere and less water is left in the Southern Hemisphere; this leads to an asymmetric shape of the water sphere. Owing to the Coriolis force, a zonal velocity gradually builds up, which is westward in the Northern Hemisphere and eastward in the Southern Hemisphere (Fig. 5.100b).

Since there is no friction in the model, it is not suitable for describing the long-term evolution of the solution. As time goes on, the free surface elevation and the zonal velocity would be unbounded, so the solution is not valid. An accurate solution can be found through more rigorous theory and modeling.

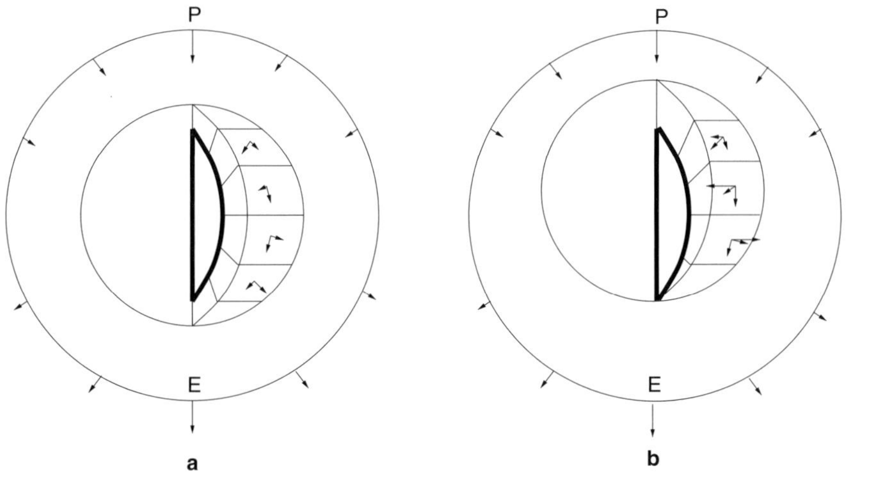

Fig. 5.100 Two successive stages (**a** and **b**) of the Hough-type circulation pattern, driven by precipitation (P) distributed over the Northern Hemisphere and evaporation (E) distributed over the Southern Hemisphere. The *hovering arrows* indicate the distribution of precipitation–evaporation. The *arrows drawn on the surface of the spheres* are velocity components. The zonal currents grow with time, as can be seen by comparing **a** with **b**. The *heavy lines in the center* indicate the solid portion of the Earth (redrawn from Stommel, 1957).

Goldsbrough's solution

For a long time, Hough's solution remained unnoticed by the oceanography community. Through an entirely independent approach, Goldsbrough (1933) discussed the case of the steady flow driven by evaporation and precipitation. He must have realized that in order to have a closed circulation driven by evaporation and precipitation, the pattern of evaporation and precipitation must satisfy a zero net flux condition in the zonal direction for a closed basin. By choosing a specific pattern of $E - P$, he was able to construct a steady solution without involving friction terms. Most importantly, he first derived the vorticity constraint for large-scale circulation:

$$\beta hv = f\,(E - P) \tag{5.163}$$

This is now called the Goldsbrough relation; according to this relation, evaporation and precipitation work as "pumping" in driving the large-scale oceanic circulation.

According to Eqn. (5.163), precipitation drives equatorward flow. If we want to close the circulation with freshwater flux alone, evaporation is needed at each latitude circle to bring the water back poleward. In order to balance the mass at each latitude circle, the zonal integration of evaporation minus precipitation must be zero. Therefore, the trick in finding the Goldsbrough circulation is to assume a pattern of precipitation in the eastern basin and evaporation in the western basin, so that the zonal integration of $E - P$ is zero along any latitude circle.

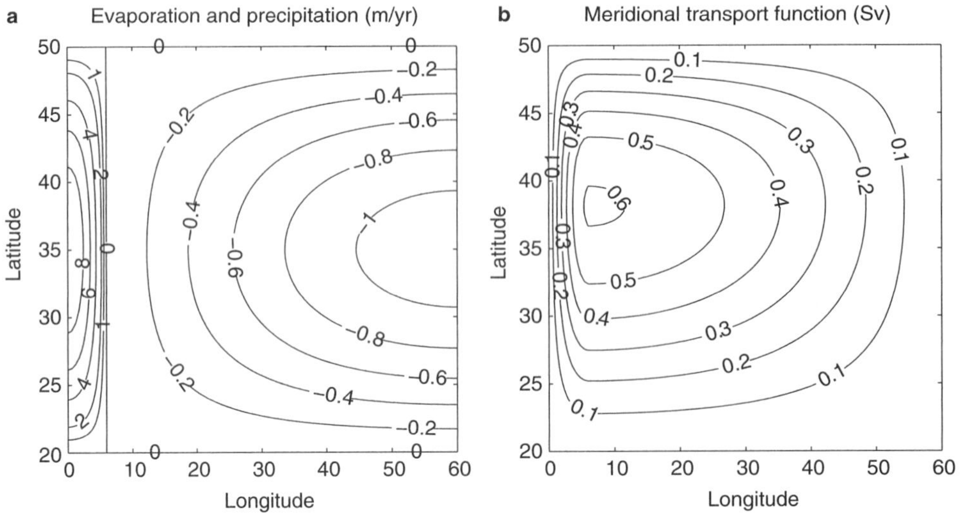

Fig. 5.101 A Goldsbrough-type solution for a closed basin, with precipitation and evaporation balanced for each latitudinal circle: **a** evaporation and precipitation (m/yr), **b** meridional transport function (Sv).

To illustrate this type of solution, we choose the following evaporation and precipitation pattern for a basin 60° by 30° (Fig. 5.101):

$$E - P = (e - p)\sin\left[\frac{\pi\,(y - y_n)}{(y_n - y_s)}\right], \quad e - p = \begin{cases} \frac{1}{\Delta}\cos\left(\frac{\pi x}{2\Delta}\right), & 0 \le x < \Delta \\ \frac{1}{1-\Delta}\cos\left(\frac{\pi x}{2(1-\Delta)}\right), & \Delta \le x \le 1 \end{cases} \tag{5.164}$$

where $e - p$ is in units of m/yr, and x is the nondimensional coordinate for each zonal circle, $\Delta = 0.06$. Therefore, the zonal sum of evaporation and precipitation is zero; an essential constraint implied in Goldsbrough's original model. The meridional velocity can be calculated from Eqn. (5.163), and the corresponding meridional transport function can be calculated through a zonal integration of the meridional velocity, starting from either the eastern boundary or the western boundary. For this model, the meridional transport function can also be calculated by integration from the western boundary, because the solution itself is entirely frictionless and there is no frictional boundary layer in the model; thus, integrating the meridional velocity from either the eastern or western boundaries gives the same meridional transport function pattern.

This freshwater flux induces a circulation that is closed within the basin. Although there is a relatively narrow western boundary layer in this solution, neither friction nor inertial terms are included in the model, so that both friction force and inertial term are neglected in the solution shown in Figure 5.101. In addition, it is important to note that barotropic circulation induced by evaporation and precipitation is on the order of 1 Sv, which is too small compared with the circulation observed in the oceans; thus, this circulation cannot be used to explain the much stronger circulation observed in the oceans.

Stommel's solution

Goldsbrough's main contribution was to derive the vorticity constraint for the ocean interior. He did not know how to treat friction; therefore, in order to find a steady solution in a closed basin, he used a distribution of evaporation minus precipitation that is quite unrealistic. The general solution driven by a much more realistic pattern of freshwater flux, with evaporation at low latitudes and precipitation at high latitudes, was solved by Stommel (1957, 1984a). Stommel pointed out that, in principle, circulation driven by evaporation minus precipitation in the ocean interior can be closed by western boundary currents. Therefore, the freshwater-driven circulation should be called the Goldsbrough–Stommel circulation. Figure 5.102 shows Stommel's generalization of Goldsbrough's idea for a single-hemisphere basin.

Goldsbrough–Stommel circulation in the world's oceans

The solution discussed by Stommel represents an idealized situation. The application of this idea to the oceans was carried out much later. Using the existing data sets of evaporation minus precipitation, Huang and Schmitt (1993) calculated the Goldsbrough–Stommel circulation in the world's oceans (Fig. 5.103). The solution is based on zonally uniform

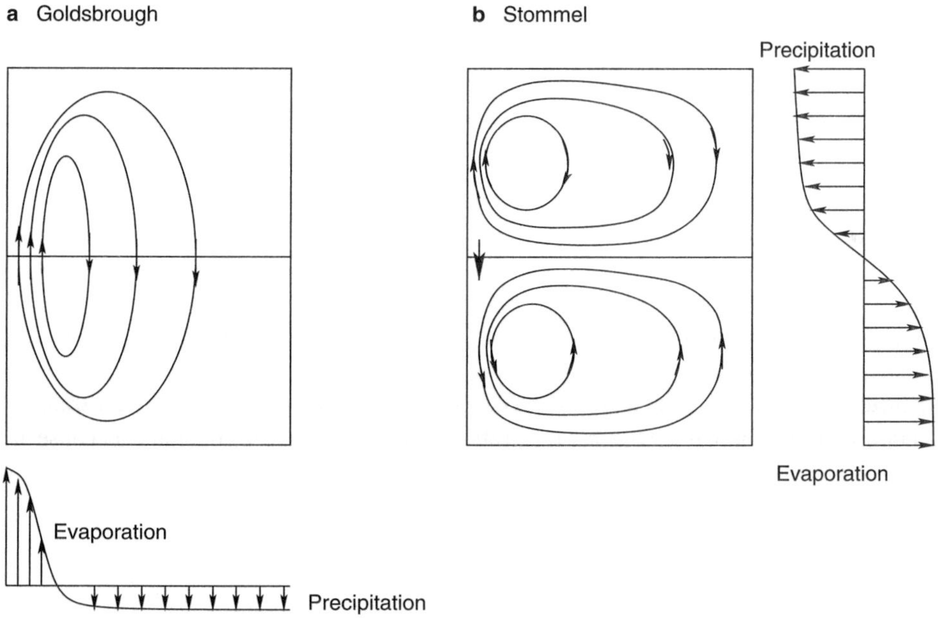

Fig. 5.102 **a** The Goldsbrough gyre driven by evaporation–precipitation and presented by him in 1933 as a model of the North Atlantic; **b** Stommel's (1957, 1984a) idea of using the western boundary currents to close a circulation forced by a more realistic distribution of evaporation–precipitation.

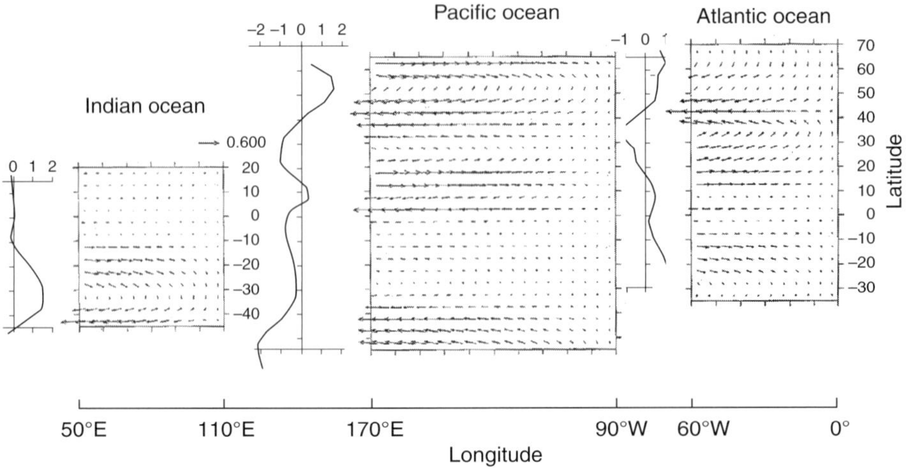

Fig. 5.103 The Goldsbrough–Stommel circulation of the world's oceans, neglecting the inter-basin transport. Each *arrow* indicates the horizontal mass flux integrated over a 5° × 5° box, in Sv; along the western boundary of each basin, there is a *curve* indicating the northward mass flux (in 10^6 m^3/s) within the western boundary, which is required to close the circulation (Huang and Schmitt, 1993).

evaporation and precipitation. The western boundary currents that perform the role of closing the meridional mass flux at each latitude in each basin are then attached to the interior solution.

Baroclinic haline circulation in the oceans

The circulation discussed above is only the barotropic component of the saline circulation, which is tiny compared with the wind-driven circulation and the thermal circulation in the oceans. As discussed above, the Goldsbrough–Stommel circulation has volume flux on the order of 1 Sv, but wind-driven circulation is on the order of tens of Sverdrups.

For a long time, the Goldsbrough–Stommel circulation was treated as an abstract theoretical idea not directly linked to the oceanic general circulation and thus not particularly useful for practical application. A close examination reveals, however, that the Goldsbrough–Stommel circulation is only one aspect of the circulation related to freshwater flux across the air–sea interface. It turns out that the Goldsbrough–Stommel circulation is only the barotropic component of the circulation induced by the air–sea freshwater flux. If there were no salt in the ocean, the Goldsbrough–Stommel circulation would be the only circulation induced by the freshwater flux. Within another theoretical limit of no external mechanical energy (either from wind stress or tidal flow) available for sustaining diapycnal mixing, there would be no baroclinic component of the haline circulation; thus, the Goldsbrough–Stommel circulation would be the only possible circulation induced by the air–sea freshwater flux.

There is no salinity involved in the dynamical analysis presented above. However, if salinity and salinity mixing driven by external sources of mechanical energy are included, the entire picture will be totally different, because baroclinic circulation associated with salt mixing and transportation as strong as the wind-driven circulation or the thermal circulation will be developed, as discussed in the next section.

To understand the baroclinic haline circulation, we begin with the circulation in an estuary. Estuaries are the interface between the freshwater-dominated river flow and the saltwater-dominated oceanic circulation. Water from the river upstream provides the freshwater input, and the open ocean provides the downstream condition.

Freshwater-flux-induced circulation in a salty estuary

The circulation in an estuary depends on many factors, such as the amount of river run-off, the mean salinity, and tidal mixing. In the estuary, freshwater from river run-off overlies the salty water from the open ocean; thus, there is a strong stratification, which is mostly due to the salinity difference. As discussed in Chapter 3, diapycnal mixing in such a strongly stratified environment requires an external source of mechanical energy, from tides and wind.

If there were no external mechanical energy available for sustaining vertical mixing, the freshwater from river run-off would flow over the salty water in the estuary. As a result, the only circulation would be movement of the top layer, and the lower layer below would be stagnant. Since there is no mixing, the volume flux in the upper layer remains constant over the whole path through the estuary. There is no salt in the upper layer, so water there remains fresh; the salinity of the lower layer remains the same as in the open ocean, which we take as 35 (Fig. 5.104a). However, with tidal mixing, a small amount of river run-off can induce a huge return flow in an estuary (Fig. 5.104b). In the discussion in this section we will assume that there is always an energy source available for mixing, such as the barotropic and internal tides, internal waves, wind stress, and other sources; however, the exact nature of the energy source is not our concern here.

The North Atlantic Ocean as an estuary

Similar to the case discussed above, the North Atlantic Ocean can be treated as a huge estuary, with evaporation at low latitudes and precipitation at high latitudes that exactly balance each other. First, let us assume that there is no external mechanical energy available for sustaining mixing; thus diapycnal diffusivity is zero. At time $t = 0$, evaporation starts at low latitudes and rain starts to come into the subpolar basin. Precipitation at high latitudes builds up the free surface, and water starts to flow toward low latitudes (rotation would modify the path). At low latitudes, in the beginning, evaporation would make some water saltier and sink to depth, and this would give rise to motion in the salty water. However, as freshwater arrives at the low latitudes and gradually covers up the entire upper surface of the basin, evaporation can affect only the freshwater, but not the salty water. As the residual motions in the deep water gradually lose their kinetic energy, the only motion remaining

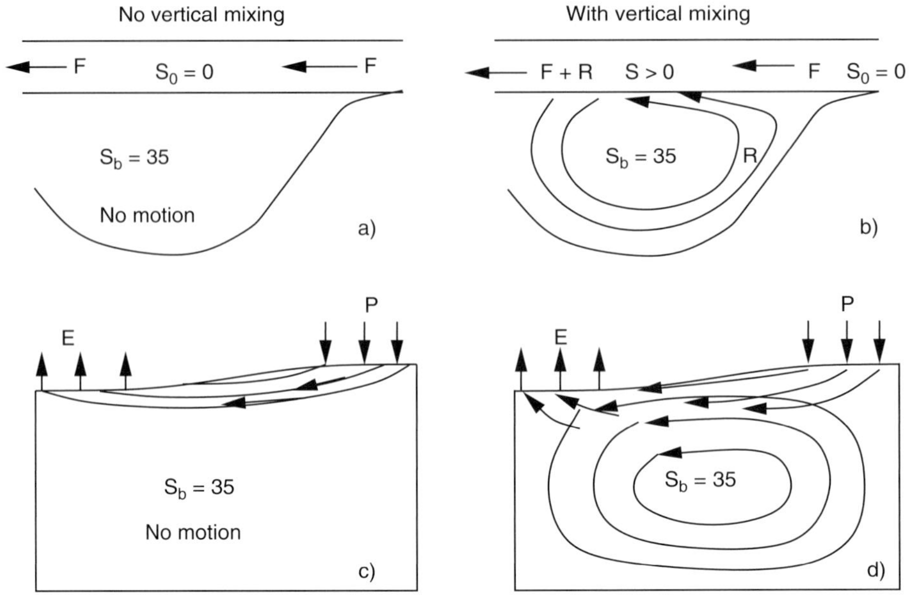

Fig. 5.104 Sketch of models with freshwater-driven circulation in an estuary and an open ocean: **a**, **c** model without vertical mixing; **b**, **d** model with vertical mixing.

will be the equatorward flow of freshwater on top of the stagnant deep and salty water (Fig. 5.104c).

Second, if there is external mechanical energy available for sustaining vertical mixing, there will be a very strong return flow induced by vertical mixing (Fig. 5.104d). Through salt conservation, the overturning rate is related to the salinity difference between the upper and lower layers of the ocean

$$R = \frac{S_0}{\Delta S} F \gg F \tag{5.165}$$

Because ΔS is much smaller than S_0, a small amount of precipitation can induce a strong meridional circulation.

Haline circulation under freshwater forcing

Haline circulation induced by freshwater flux is a complicated system, and the most convenient way to examine such circulation is to use a numerical model. The model is a mass-conserving model with a free surface, with $2^\circ \times 2^\circ$ resolution (4° N–64° N, $\Delta\theta = 60^\circ$) and $\Delta\lambda = 60^\circ$ wide in the zonal direction, subjected to a "linear" profile of $E - P$, as shown in Figure 5.105,

$$E - P = w_0 \left(1 - 2\frac{\theta - \theta_s}{\Delta\theta}\right) / \cos\theta, \quad w_0 = 1 \text{ m/yr} \tag{5.166}$$

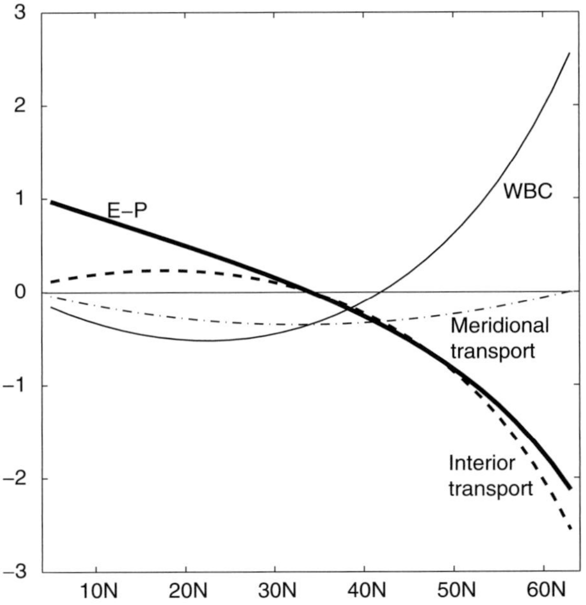

Fig. 5.105 Meridional distribution of mass fluxes: $E - P$ is the evaporation minus precipitation (in m/yr), *Interior* is the poleward mass flux in the interior, *WBC* is the mass flux of the western boundary current, *Meridional transport* is the total poleward mass flux; all these mass fluxes are in Sv.

The corresponding zonally integrated volume fluxes in the ocean interior, the whole basin, and the western boundary regime are

$$V_{int} = \frac{f}{\beta}(E - P)\, r \cos\theta \Delta\lambda \tag{5.167}$$

$$V_{total} = -\int_{64°}^{\theta} (E - P)\, r^2 \cos\theta \Delta\lambda d\theta \tag{5.168}$$

$$V_{wbc} = V_{total} - V_{int} \tag{5.169}$$

Note that the model's only surface forcing is the freshwater flux through the upper surface. The total meridional mass flux (indicated by the dot-dashed line) associated with $E - P$ is about 0.3 Sv; this is really very small compared with the wind-driven circulation.

The vertical mixing in the model is set to a constant value of 0.3×10^{-4} m^2/s. After 5,000 years of integration, the model reaches a state of aperiodic oscillation. The following discussion is based on the mean circulation averaged over several hundred years; details of the time evolution will be discussed in Section 5.4. Owing to precipitation at high latitudes, sea level there is increased. In fact, a center of sea-level high exists in the middle of the northern half of the basin, and a sea-level low exists at low latitudes near the eastern boundary (Fig. 5.106).

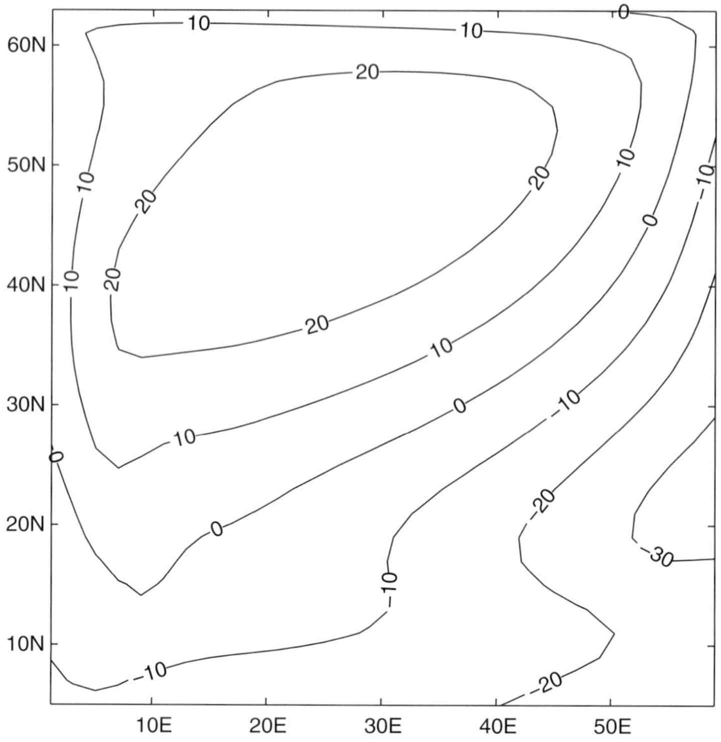

Fig. 5.106 Sea surface elevation, in cm, associated with the haline circulation.

The sea-level high is closely related to the barotropic anticyclonic circulation in the northern half of the basin (Fig. 5.107). There is a rather weak barotropic cyclonic circulation in the southern half of the basin. In addition, there are clearly western boundary currents required for closing the circulation in the basin (Fig. 5.107). This shows the barotropic meridional transport function obtained by zonally integrating the meridional volume flux over the whole depth. Readily seen are the southward flow driven by precipitation in the subpolar basin, the poleward flow driven by evaporation in the subtropical basin, and the western boundary currents required for closing the circulation, with the volume flux discussed above. The barotropic circulation is predicted by the Goldsbrough–Stommel theory. As discussed above, the barotropic circulation exists, independently of whether there is salt in the ocean or whether there is enough external mechanical energy to sustain the vertical mixing in the ocean.

The major new feature different from the Goldsbrough–Stommel theory is the existence of strong three-dimensional baroclinic circulation. The most important component is the meridional overturning cell, with an overturning rate more than 50 Sv, as shown in Figure 5.108. Water in the southern half of the basin is much saltier due to evaporation, and it sinks to the sea floor. Away from the southern boundary, the salty water gradually upwells

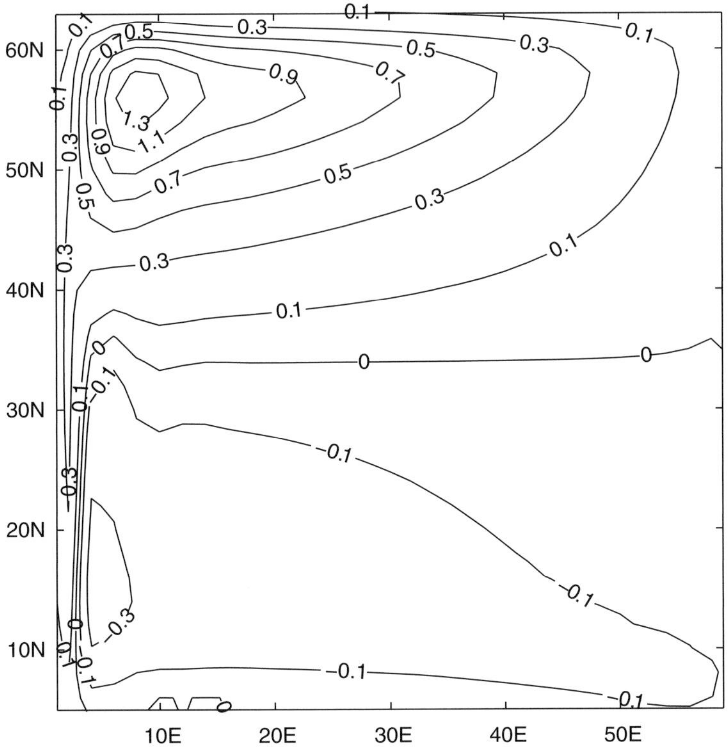

Fig. 5.107 Barotropic meridional transport function, in Sv.

and mixes with the relatively fresh water above. As discussed in Chapter 3, mixing in such a stratified fluid requires an external source of mechanical energy, which comes from both tidal dissipation and surface wind stress. Without such an energy source, the freshwater flux through the air–sea interface cannot drive strong circulation, as shown in Figure 5.108.

Precipitation tends to be very irregular in nature. Thus, in the mind's eye, there seems to be no clear connection between the tiny raindrops of precipitation and the large-scale organized haline circulation in the oceans. However, precipitation and evaporation are responsible for creating the salinity difference in the oceans, and thus regulating the haline component of the oceanic general circulation. In this model, precipitation at high latitudes reduces the salinity in the north, and evaporation at low latitudes increases salinity in the south; this is what really drives the circulation (Fig. 5.109). The salinity maximum is located in the southeastern corner, roughly at the location of the sinking branch of the haline circulation.

Under the freshwater forcing, there is a complicated three-dimensional haline circulation; thus, one should not take the barotropic circulation as the circulation pattern in the ocean. In fact, the strong subsurface upwelling in the middle of the basin works like an "Ekman compressor." This "compressor" moves upward and squeezes the water column in the upper

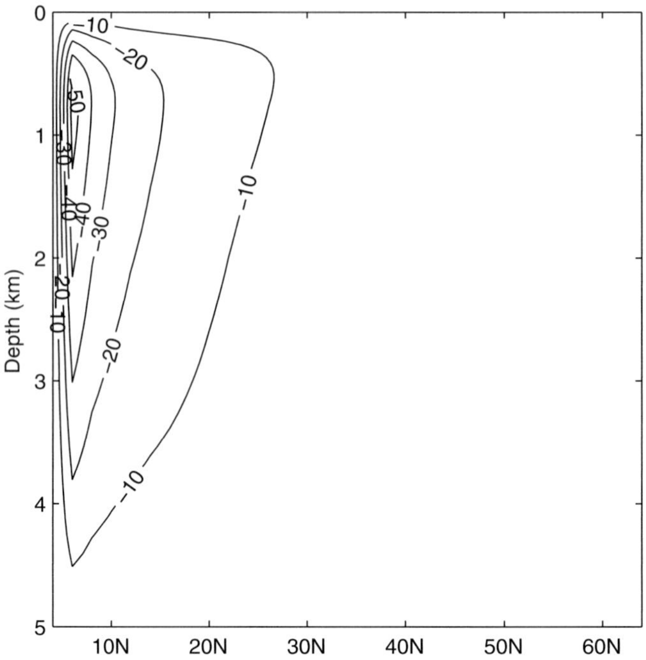

Fig. 5.108 Time-mean meridional streamfunction in Sv.

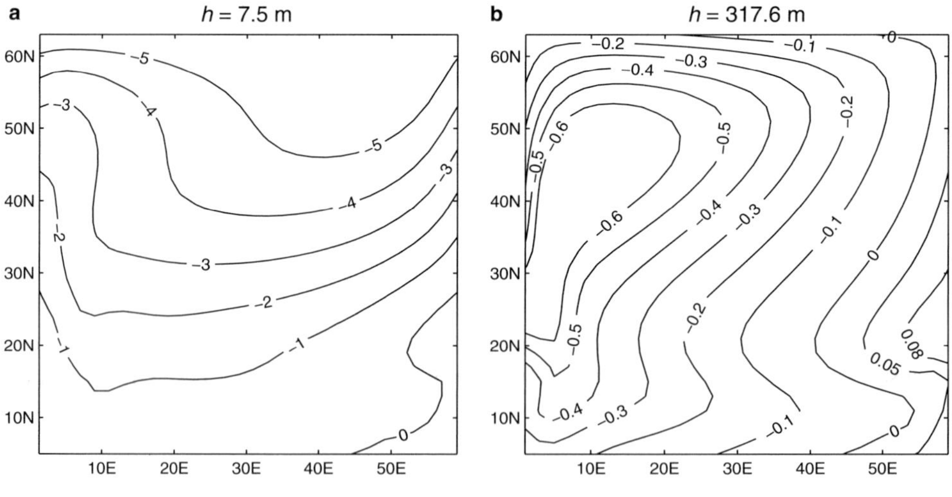

Fig. 5.109 Salinity (deviation from the mean) distribution at **a** 7.5 m and **b** 317.6 m.

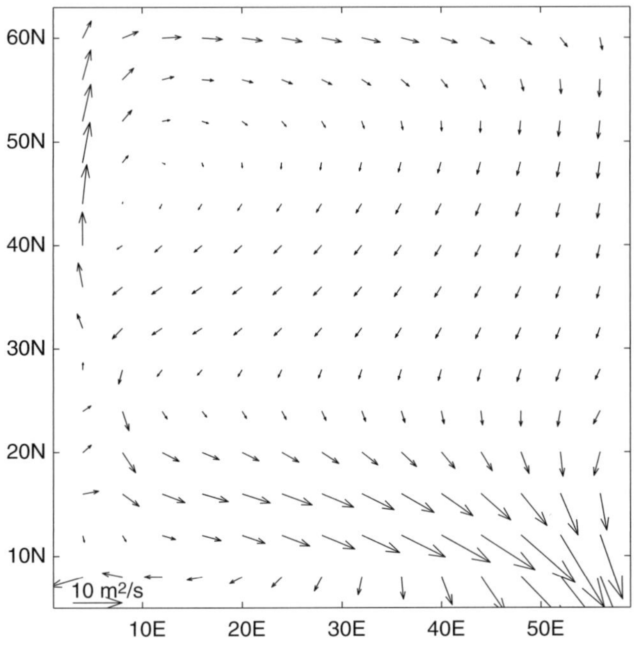

Fig. 5.110 Horizontal volume flux integrated for the upper 980 m.

ocean. In order to conserve potential vorticity $f/\Delta h$, the water column moves equatorward, very similarly to the case of a wind-driven subtropical gyre in the upper ocean. As a result, there is an anticyclonic gyre in the upper ocean, as shown in the northern part of the model basin in Figure 5.110. Below this level, there is another circulation which rotates in the opposite direction. In this model, deep water is formed near the southeastern corner of the model basin and sinks to the deep ocean, as indicated by the large velocity convergence in the lower right corner in Figure 5.110.

The baroclinic circulation associated with the haline circulation driven by the combination of strong vertical mixing and surface freshwater flux is one order of magnitude stronger than the barotropic circulation predicted by the Goldsbrough–Stommel theory. The comparison of these two types of circulation is schematically shown in Figure 5.111. The baroclinic circulation in an ocean with salt and mixing consists of a clockwise circulation in the upper ocean, driven by the basin-wise upwelling; however, there is also a deep anticlockwise circulation. In addition, there are a large meridional overturning cell and a zonal overturning cell. The strength of each of these cells or gyres is at least one order of magnitude larger than the so-called Goldsbrough–Stommel circulation. It is very important to note that the strength of the three-dimensional circulation strongly depends on the mean salinity and the amount of mechanical energy available for mixing, as revealed by the scaling law of the saline circulation discussed in Section 5.5.1.

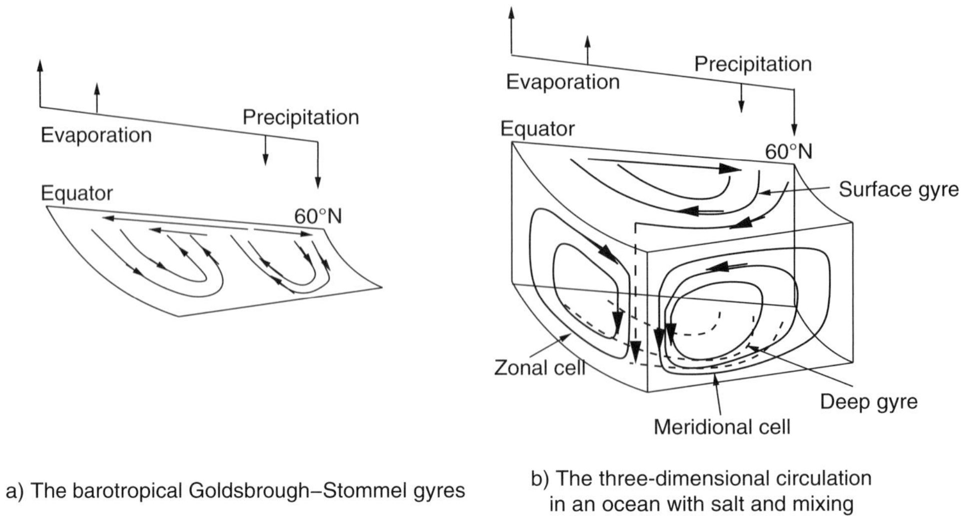

Fig. 5.111 **a**, **b** Schematic structure of the saline circulation driven by precipitation at high latitudes and evaporation at low latitudes.

In some previous studies, the salt flux diagnosed from models driven by the virtual salt flux has been misinterpreted as real salt flux in the oceans. In order to demonstrate the meaning of salt flux diagnosed from numerical models, we present the results obtained from two rigid-lid models. As shown in Figure 5.112, the net meridional salt flux approaches a quasi-equilibrium, as the circulation itself approaches the final equilibrium (Fig. 5.112a). As discussed in the previous section, the large advective salt flux diagnosed from models based on the virtual salt flux condition should be reinterpreted, and after such a reinterpretation the meridional salt flux diagnosed from models based on the virtual salt flux should also reach a quasi-equilibrium. ***Note that for a time-dependent problem, the so-called salt flux diagnosed from models based on the virtual salt flux condition is meaningless.***

5.3.4 Double diffusion

Seawater can be treated as a two-component system, with water and salt; and the most important thermodynamic variables are temperature and salinity. Owing to spatial inhomogeneity, diffusion of temperature and salt can take place simultaneously; this is called double diffusion. Double diffusion has been discussed in many papers and books, such as a special issue of *Progress in Oceanography* (Volume 56, 2003). Since saltwater (or freshwater) diffusion is associated with the exchange of molecules, at the molecular level, thermal diffusivity is two orders of magnitude faster than that of salt diffusivity. However, on scales much larger than molecular, instability takes place and gives rise to a different ratio between the equivalent diffusivity of temperature and salt. Among many intricate phenomena, two

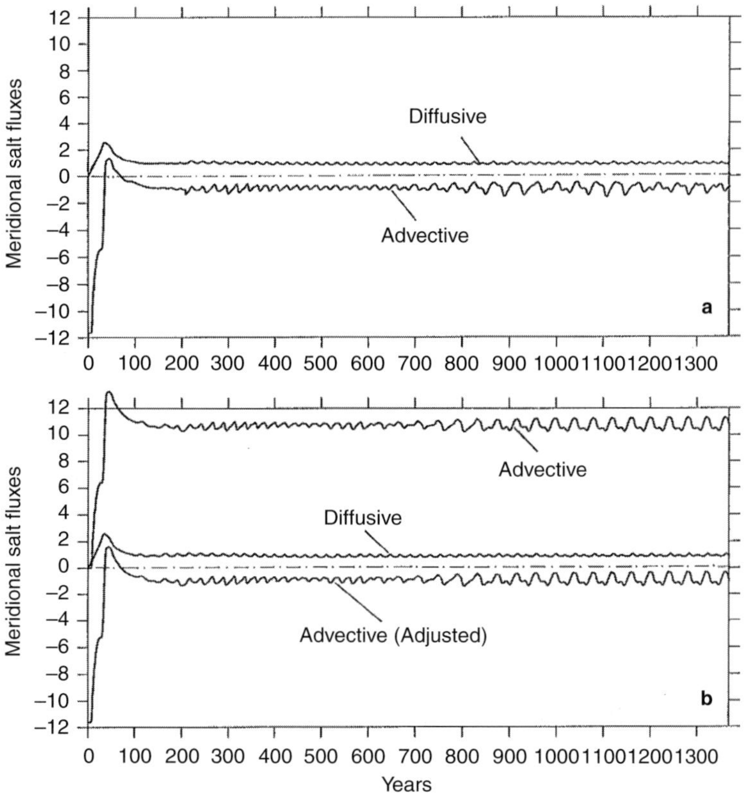

Fig. 5.112 Time evolution of the meridional salt fluxes at 28° N (in 10^6 kg/s): **a** from a rigid-lid model under the natural boundary condition; **b** from a rigid-lid model under the virtual salt flux condition (Huang, 1993b).

types of double diffusion are of great interest: (1) salt fingering, which happens when warm and salty water lies over relatively cold and fresh water; and (2) diffusive layering, which takes place when cold and fresh water lies over relatively warm and salty water. A convenient parameter widely used in double diffusion studies is the density ratio, which is defined as $R_\rho = \alpha \bar{T}_z / \beta \bar{S}_z$, where the overbar indicates that the vertical derivatives are taken over a depth range larger than the typical scale of the so-called thermohaline staircase. Since the background stratification for each water column should be stable, salt fingering takes place in the domain of $R_\rho > 1$, and diffusive layering takes place in the domain of $R_\rho < 1$.

For large-scale oceanic general circulation, salt fingers may play a major role in the maintenance of the thermohaline circulation. Salt fingering primarily happens in subtropical basins, where strong solar insolation gives rise to warm surface temperature and strong evaporation leads to high surface salinity. Although each water column is stably stratified, it can become unstable for the double diffusive process. As shown in Figure 5.113, if a

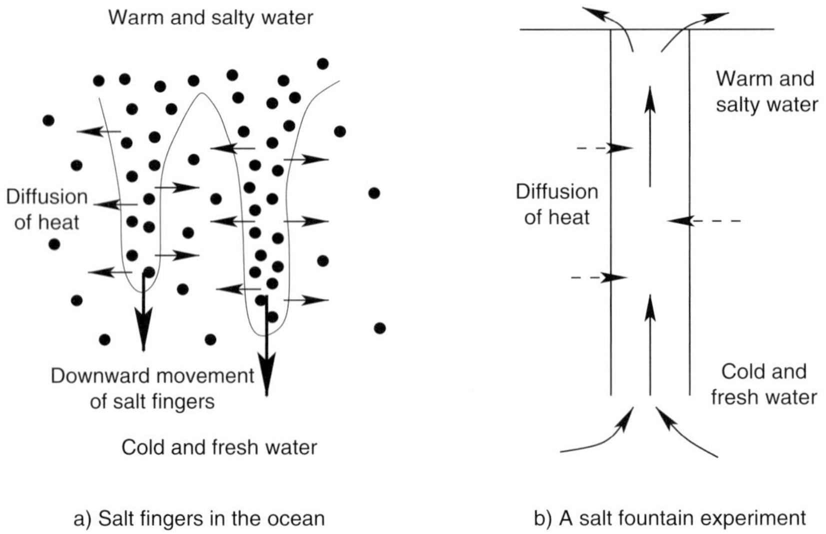

Fig. 5.113 Sketch of the salt fingering phenomenon: **a** salt fingers in the ocean; **b** a salt fountain experiment.

finger-shaped water parcel from the upper layer intrudes into the lower layer, rapid heat exchange between the finger and the environment leads to the loss of heat but not the salt content, because the thermal diffusivity is 100 times larger than the salt diffusivity. As a result, the finger-shaped parcel becomes heavier and its downward movement is accelerated. This kind of perturbation releases gravitational potential energy from the mean state, and this energy release continuously provides mechanical energy for the growth of the instability.

The dynamical consequences of the difference between heat and salt diffusion can be demonstrated by a laboratory experiment as follows. In such an experiment, a pipe is put into a stratified water container, with warm and salty water lying on top of the relatively cold and fresh water. The thin wall of the pipe allows heat exchange across the wall of the pipe, but not salt exchange. Cold water from the deep part of the container is warmed up by heat transfer through the thin wall of the pipe. Since salt diffusion through the wall is not permitted, the water inside the pipe becomes more buoyant; under the buoyancy force, water upwells through the pipe, thus creating a salt fountain (Fig. 5.113b).

A typical case of salt fingering is the subtropical North Atlantic Ocean. This area is characterized by high temperature and salinity in the upper ocean induced by strong heating and excessive evaporation (Fig. 5.114). The density ratio is around 1.5 for the depth range of 300–500 m, which favors salt fingering. Salt fingering in the oceans is organized as a thermohaline staircase. As shown in Figure 5.114, there are clear signs of a thermohaline staircase in the depth range 300–500 m.

Salt fingering involves complicated small-scale dynamical processes which are difficult to resolve in a numerical model for the large-scale circulation. Thus, such processes have to

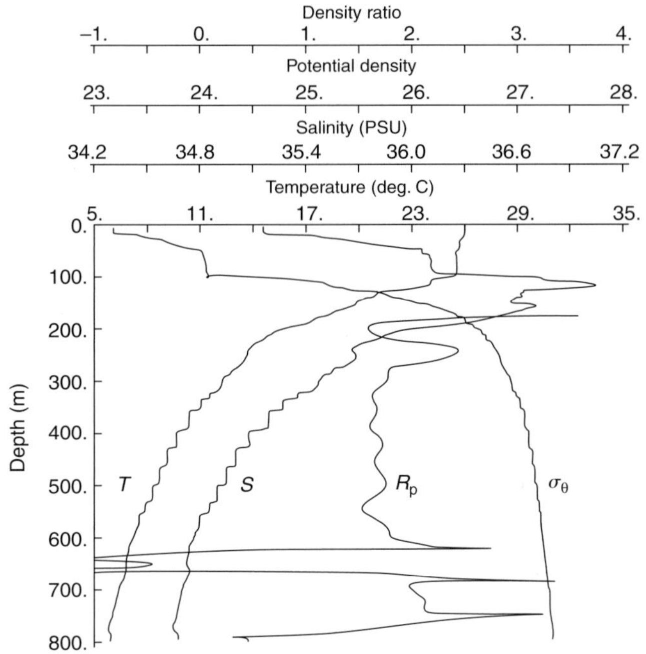

Fig. 5.114 Thermohaline staircase from *in situ* observation, including temperature (T), salinity (S), potential density (σ_θ), and density ratio (R_ρ, calculated over a sliding 40 m vertical interval) (Schmitt *et al.*, 1987).

be parameterized in terms of eddy diffusivity of heat and salt. For example, the following conditions were used to identify the oceanic regime where salt fingering takes place (Zhang *et al.*, 1998):

$$\Theta_z > 0, \quad S_z > 0, \quad 1 < R_\rho < \kappa_T/\kappa_s, \quad |\Theta_z| > \Theta_{z,c} \tag{5.170}$$

where Θ_z is the vertical gradient of potential temperature; $R_\rho = \alpha\Theta_z/\beta S_z$ is the density ratio; $\Theta_{z,c} = 2.5 \times 10^{-4}$ °C/m is the critical vertical gradient of temperature for double diffusion; S_z is the vertical gradient of salinity; and κ_s and κ_Γ are the molecular heat and salt diffusivity, $\kappa_s \approx 1.1 \times 10^{-9}$m^2/s, $\kappa_T \approx 1.4 \times 10^{-7}$m^2/s; thus, we have set $\kappa_T/\kappa_s = 100$ in the formula above.

The diapycnal eddy diffusivities for salinity and temperature are, respectively:

$$K_s = \frac{K^*}{1 + (R_\rho/R_c)^n} + K^\infty \tag{5.171}$$

$$K_T = \frac{0.7K^*}{R_\rho[1 + (R_\rho/R_c)^n]} + K^\infty \tag{5.172}$$

where $K^* = 10^{-4}$ m^2/s is the maximum diapycnal diffusivity for salt fingering; K^∞ is the constant diapycnal diffusivity due to other mixing processes unrelated to double diffusion. $K^\infty = 0.3 \times 10^{-4}$ m^2/s was used in many studies (e.g., Zhang *et al.*, 1998). $R_c = 1.6$ is the critical density ratio above which the diapycnal mixing due to salt fingering drops dramatically, due to the absence of staircases; $n = 6$ is an index to control the decay of K_Θ, K_S with increasing R_ρ.

For the case of diffusive layering, the eddy diffusivities can be parameterized as

$$K_T = CR_a^{1/3}\kappa_T + K^\infty \tag{5.173}$$

$$K_\Theta = R_F R_\rho (K_T - K^\infty) + K^\infty \tag{5.174}$$

where $C = 0.0032 \exp\left\{4.8R_\rho^{0.72}\right\}$, $R_F = \frac{1/R_\rho + 1.4(1/R_\rho - 1)^{3/2}}{1 + 14(1/R_\rho - 1)^{3/2}}$, $R_a = 0.25 \times 10^9 R_\rho^{-1.1}$. Accordingly, the dependence of eddy diffusivity of heat and salt is shown in Figure 5.115.

During salt fingering, density flux induced by vertical mixing is downward, i.e., the center of mass for a water column moves downward in the gravity field. As a result, the total amount of gravitational potential energy of the water column is reduced. The release

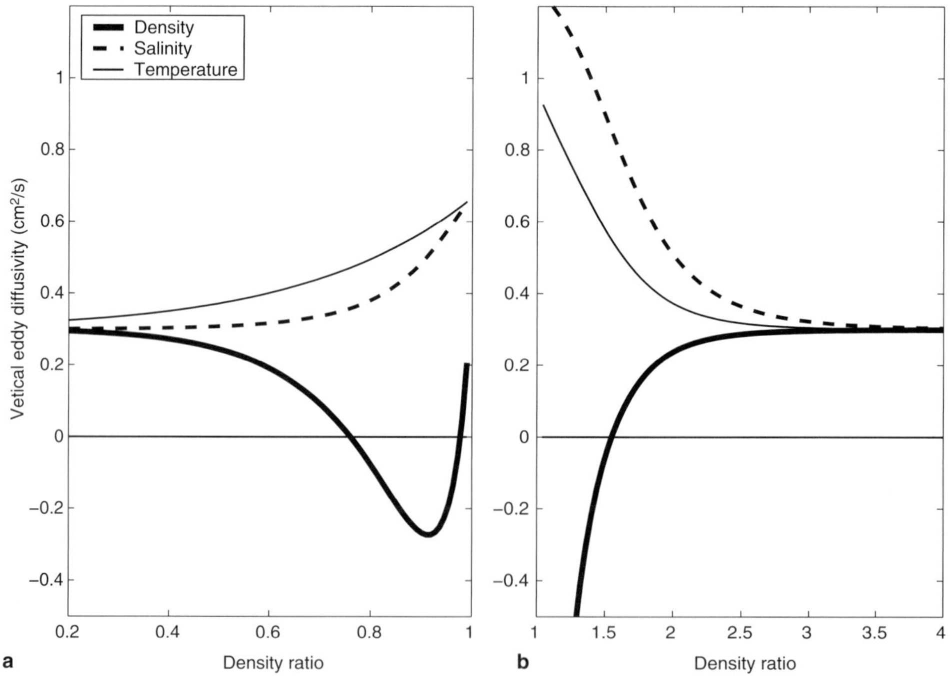

Fig. 5.115 **a**, **b** Diapycnal diffusivity of temperature (*solid line*), salinity (*dashed line*), and density (*dotted line*) as function of density ratio R_ρ. Note that salt fingering is represented by the domain of the right panel (**b**) where $R_\rho > 1$; while diffusive layering corresponds to the domain of $R_\rho < 1$ (Zhang *et al.*, 1998).

of gravitational potential energy associated with salt fingering in the world's oceans can be estimated from Eqns. (5.171) and (5.172). Assuming that the background value of $K^{\infty} = 0$, the total amount of gravitational potential energy released in the global oceans is estimated as 8 GW. The horizontal distribution of this energy conversion was included in Figure 3.18. This is much smaller than the energy associated with wind stress input and tidal dissipation; however, this energy is most concentrated in the subtropical basin interior, where it is much larger than the corresponding part of tidal dissipation. Therefore, energy release from salt fingering may play a dominant role in regulating the structure of the main thermocline and circulation in the subtropical gyre interior.

We emphasize that salt fingering itself does not create mechanical energy; instead, the double diffusive process can only release the gravitational potential energy stored in the mean state of the circulation system, which in turn is created by the external sources of mechanical energy from wind stress and tidal dissipation. In addition, the separation of water from seawater implies a certain amount of equivalent mechanical energy and the removal of entropy produced through salty and fresh water mixing. The energetics related to the maintenance of double diffusion in connection with global-scale circulation remains unclear.

5.4 Theories for the thermohaline circulation

5.4.1 Conceptual models for the thermohaline circulation

Thermal overturning circulation driven by sea-surface differential heating

How is the thermal circulation set up? The major physical processes involved are schematically shown in Figure 5.116. Assume that the ocean was initially of uniform temperature T_0 and motionless, so that no north–south differences in sea level or bottom pressure would

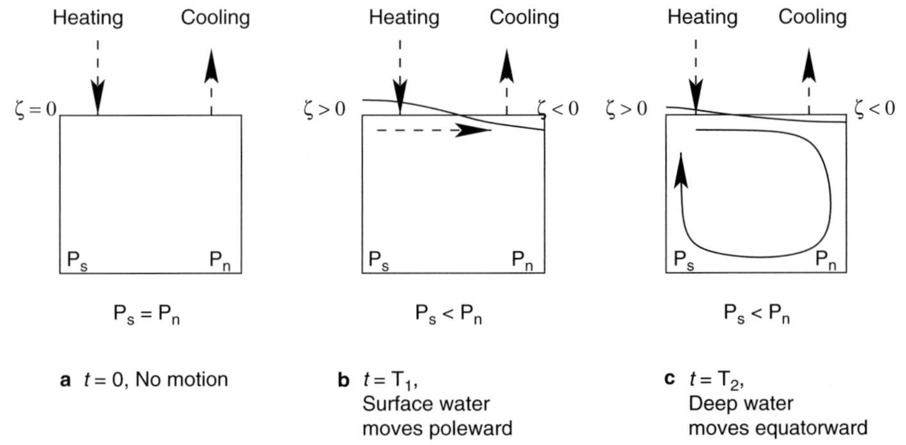

Fig. 5.116 **a–c** The initial stages of the meridional overturning circulation induced by surface heating/cooling.

exist (Fig. 5.116a). At time $t > 0$, heating at low latitudes and cooling at high latitudes apply to the sea surface. As a result, water at low latitudes expands and the sea surface uplifts, while water at high latitudes contracts and sea surface level declines. The difference in sea level creates a poleward pressure gradient in the upper ocean; however, there is no pressure gradient in the deep ocean at this time. The surface pressure gradient drives surface water poleward (Fig. 5.116b). The existence of Earth's rotation makes the poleward movement more complicated; however, as long as there is a meridional boundary, the zonal-mean meridional difference in sea surface level will eventually push water to high latitudes. Due to the piling up of water at high latitudes, a meridional pressure gradient develops in the deep ocean, which drives an equatorward flow of deep water (Fig. 5.116c). Here again, the rotation of the Earth makes deep water movement more complicated than that for the case without rotation. Therefore, a thermal overturning cell, consisting of a poleward flow in the upper ocean and an equatorward flow in the deep ocean, is established.

Although heating and cooling do drive a circulation, it is unclear whether such a circulation can be maintained and how strong the steady circulation is. Surface heating and cooling work quite differently. Surface heating at the southern end (low latitudes) of the model ocean creates a thin warm layer on the top of the ocean, resulting in a slightly higher sea-surface elevation, and the ocean is in a state of stable stratification. As shown in the left part of Figure 5.117a, the thin vertical line represents the uniform density profile before surface heating, and the heavy solid curve represents the thin layer of strong and stable stratification after heating.

On the other hand, surface cooling at the northern end of the model ocean creates a slab of dense water sitting on top of the water column with relatively low temperature and

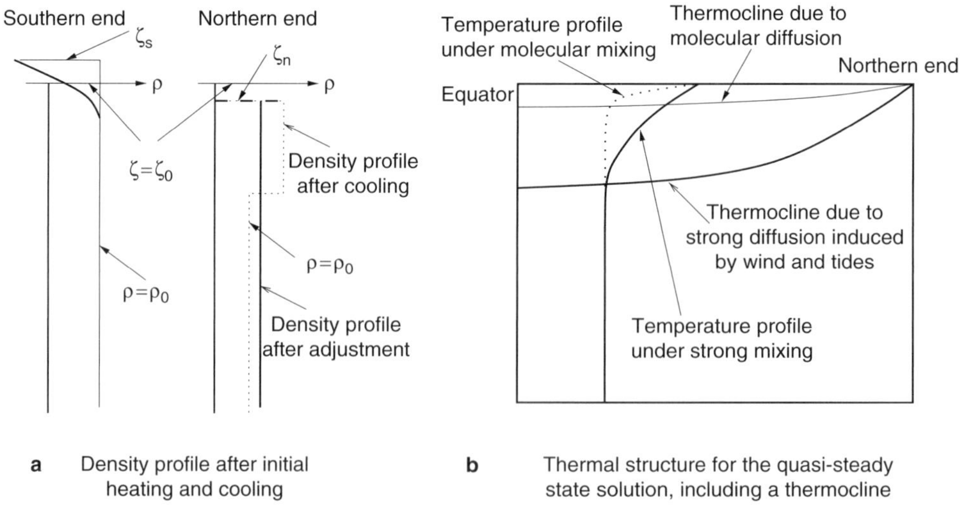

Fig. 5.117 A sketch for a Northern Hemisphere basin: **a** the density structure in two stations after initial heating/cooling and the ensuing adjustment; **b** a meridional-section view.

high density, as shown by the dotted line in the right part of Figure 5.117a. Free surface elevation declines after cooling; thus, the gravitational potential energy (GPE) of the mean state is reduced. Owing to this unstable stratification, convective adjustment ensues. GPE of the mean state is further reduced during the convective adjustment, as discussed in Section 5.1.3. After the convective adjustment, the whole water column has a uniform density (solid line in the right part of Fig. 5.117a) and the free surface elevation (heavy dashed line) is lower than the original.

The key point here is that in the northern part of the model ocean, cooling creates an unstable stratification and convective adjustment ensues, during which GPE in the mean state is released and converted into kinetic energy which drives the mean flow, turbulence, and internal waves, as discussed in Chapter 3. Therefore, we can see that cooling the surface creates no GPE; instead, it releases the GPE originally stored in the ocean.

On the other hand, heating creates a stable stratification in the southern part of the model ocean. When no source of mechanical energy is available, the stable stratification will remain. If the penetration of radiation into the ocean is neglected, GPE created by heating the skin of the surface is very small and negligible. Therefore, heating and cooling the upper surface do not generate GPE (neglecting GPE generated by molecular diffusion); instead, GPE is lost through convective adjustment associated with cooling.

To maintain a circulation, a source of GPE is required, and in the case of pure thermal circulation, such a source is generated through vertical mixing. Vertical mixing in a stably stratified fluid pushes light (heavy) fluid downward (upward), raising the center of mass. As a result, mixing increases GPE and maintains the thermal circulation connected to surface thermal forcing. Therefore, in order to maintain a circulation that is strong enough to be observable, external sources of mechanical energy are required.

In the case where an external source of mechanical energy to sustain mixing in the ocean interior is lacking, cold water will gradually pile up in the ocean. Eventually, the whole basin is filled up with the cold water, except for a very thin layer on the top of the ocean where the surface heating maintains a very sharp thermal boundary layer (Fig. 5.117b). Therefore, virtually no circulation in the ocean interior exists in this case due to the absence of an external source of mechanical energy. The very weak flow within the thin boundary layer on the sea surface is driven by the small amount of mechanical energy converted from internal energy through molecular diffusion. In contrast, the corresponding deep main thermocline in the ocean is depicted as the heavy curve. There is a strong meridional overturning circulation in connection with the strong main thermocline, the maintenance of which requires external sources of mechanical energy.

Concepts of thermohaline circulation

Dense deep water formed at high latitudes has long been thought to be the driving force for the thermohaline circulation in the oceans. Such a viewpoint seems to have arisen from people's understanding of what happens in the atmosphere, where solar insolation heats up the low and middle parts of the atmosphere, and convection takes place, while outer space works as the cooling source for this heat engine. Although the atmospheric thermal engine is of very low efficiency, of the order of only 0.8%, it is a heat engine indeed. The question relevant to oceanography is whether surface heating/cooling can work as a driving force for the thermohaline circulation.

Classical view of thermohaline circulation

Ever since the early stage of theoretical development, the surface thermal forcing has been identified as the driving force for the thermohaline circulation, and the classical picture of thermohaline circulation postulated by Wyrtki (1961) is illustrated in Figure 5.118.

By reasoning that "The circulation must take place in a meridional plane between the region of heating in low latitudes and the region of cooling in high latitudes," Wyrtki (1961) postulated a circulation system consisting of four principal processes: heating of the surface layer and the poleward flow at the surface; sinking of the heaviest water in the highest latitudes; spreading toward the equator in the deep layer; and ascension of the deep water through the thermocline into the surface layer.

Note that water mass transformation in this framework takes place in the surface layer only. Accordingly, in the circulation described by Wyrtki, the surface layer constitutes the essential ingredient of the water cycle, where water mass properties are changed by air–sea buoyancy flux. His discussion primarily focused on the air–sea heat flux in water mass transformation within the surface layer, and no internal mixing was required. In this framework, the wind-driven circulation was completely separated from the thermohaline circulation, and the potential role of wind stress in setting up the thermohaline circulation was excluded. Thus, in some sense, the circulation discussed in this framework is a "pure" thermohaline circulation.

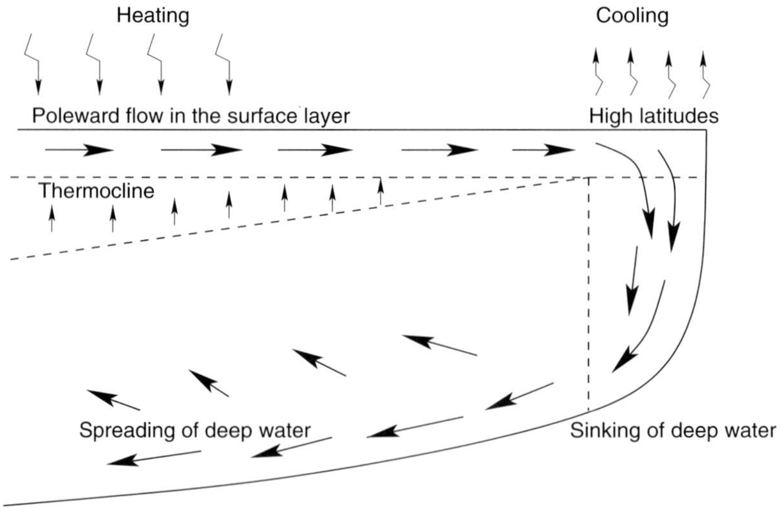

Fig. 5.118 A sketch of the classical view of thermohaline circulation driven by surface cooling/heating (modified from Wyrtki, 1961).

Later on, this framework was modified as follows: mixing in the surface layer was replaced by diapycnal mixing at the mid depth of the ocean interior. Accordingly, the main balance in the oceanic interior is between the vertical upwelling and the downward heat diffusion, as discussed by Munk (1966). In this modified framework, there is no need for wind stress to function because the poleward flow of the surface layer is presumably driven by the pressure gradient generated by the thermohaline circulation itself. Therefore, the circulation is literally a pure thermohaline circulation. Similar to the previous framework, the circulation system consists of four segments: cold and dense water formed at high latitudes sinks to great depth; dense deep water spreads to the whole basin; deep water upwells through the base of the main thermocline and gradually warms up; surface water turns poleward and completes the cycle.

Three schools of the thermohaline circulation

Two theories, or schools, have been proposed for explaining the thermohaline circulation in the oceans. The first one, which has dominated our thinking about thermohaline circulation, postulates that thermohaline circulation is driven by deepwater formation at high latitudes, as shown in Figure 5.119a. This will be called the "school of pushing."

The second theory postulates that a thermohaline circulation needs mechanical energy to overcome the friction; thus, the thermohaline circulation is driven by external sources of mechanical energy, such as tidal dissipation and wind stress. We will call this the "school of pulling." The school of pulling is further separated into two sub-schools, as explained below.

1. **School of pushing:** Deepwater formation pushes the deep current and thus maintains the thermohaline circulation.

In this old school of thought, the thermohaline circulation is assumed to be driven by surface thermohaline forcing, in particular the surface cooling/heating. Surface cooling produces dense water that sinks to a great depth. The high-latitude ocean is filled up with cold and dense water from surface to bottom. Combining with the warm and light water in the upper ocean at low latitudes, this creates a pressure force in the abyssal ocean which causes

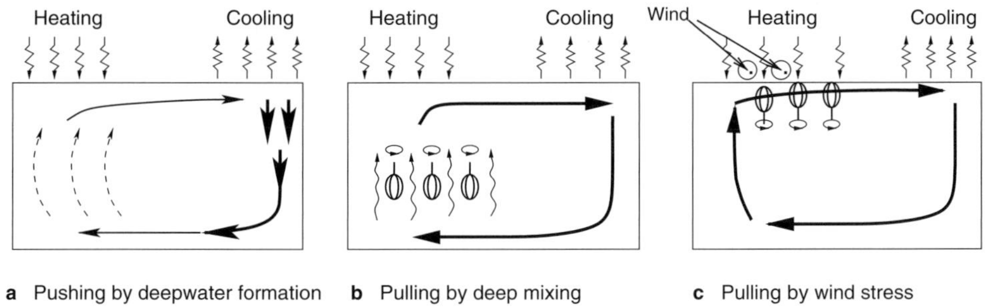

Fig. 5.119 **a–c** Three schools of theory for the thermohaline circulation.

cold bottom water to move toward low latitudes, thus pushing the meridional circulation. Owing to the Coriolis force, flow does not simply move in the down-pressure direction; nevertheless, this argument provides a connection between meridional overturning circulation and the pressure difference induced by surface thermohaline forcing and deepwater formation.

As will be discussed shortly, Stommel's (1961) classical two-box model belongs to this category, because he assumed that the circulation rate is proportional to the north–south pressure difference. Furthermore, the proportional constant relating the pressure difference and overturning rate is assumed to be invariant under different climate conditions.

The school of pushing views the thermohaline circulation as being controlled by the pressure gradient. Although the school of pushing may not work for the steady circulation, many people believe that the strong circulation induced by sudden cooling may well be a strong support for this theory. According to the energetic theory discussed in Chapter 3, however, the strong circulation after the onset of sudden cooling is due to the release of a large amount of GPE during the cooling-induced convective adjustment. In fact, the total amount of mechanical energy for the mean state is greatly reduced during such sudden cooling. Furthermore, the circulation would die out later if there were no continuous supply of an external source of mechanical energy.

2. **School of pulling by deep mixing:** Deep mixing removes cold water from the abyss, pulls and maintains the circulation.

The major problem of the school of pushing is that, due to the lack of an external source of mechanical energy to sustain mixing, cold water piles up in the ocean, and the solution is eventually reduced to a very weak circulation. To maintain a sizable circulation, the cold water in the abyss should be removed. Deep mixing sustained by tidal mixing can transform cold water into warm water in the deep ocean, creating room for newly formed deepwater and thus pulling the thermohaline circulation. Substantial arguments supporting this school have been presented by Munk and Wunsch (1998) and Huang (1999).

3. **School of pulling by wind stress:** The Southern westerly pulls cold water from the deep ocean and thus maintains the global thermohaline circulation.

Under present-day climatic conditions and modern geographic and topographic distribution, a strong Ekman upwelling around the latitude band of 50–60° S exists, which is closely related to the ACC. Due to this strong upwelling, North Atlantic Deep Water (NADW) is pulled up to the upper ocean, where water properties are gradually modified in the surface mixed layer on the way northward. In the upper ocean, wind stress continues to input mechanical energy through Ekman layer and surface waves. Theoretically, wind-driven upwelling and mixing in the upper ocean can build up the major parts of the thermohaline circulation and water mass transformation in the world's oceans, while the remaining parts of the water mass transformation in the deep ocean are relatively small and can be accomplished by external mechanical energy from tidal mixing. The circulation and sources

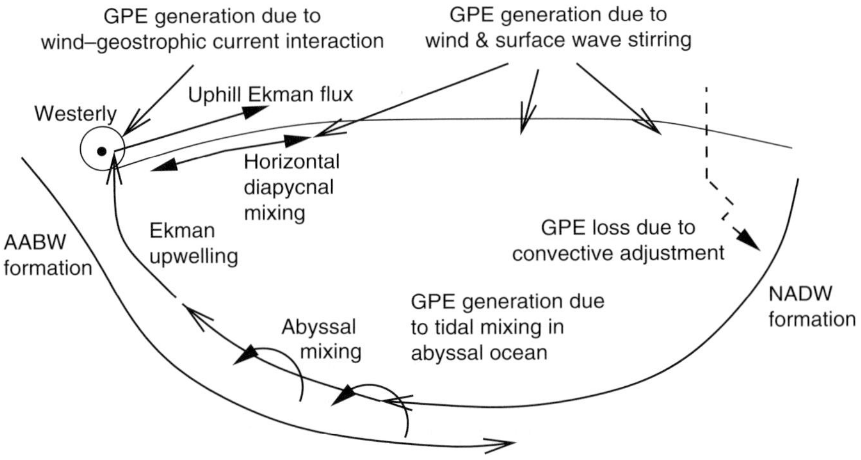

Fig. 5.120 A sketch of the different roles of mixing and upwelling in the Atlantic Ocean.

of external mechanical energy in the Atlantic sector according to this mechanism can be sketched out, as shown in Figure 5.120.

The basic ideas of this school can be traced back to early work by Toggweiler and Samuels (1995). Through a series of numerical experiments pertaining to the world's oceans, they demonstrated that in the limit of zero vertical mixing in the subsurface ocean, there is a sizable meridional overturning circulation and northward heat flux. This is also consistent with the fact that wind stress energy input to the ocean is much larger than that due to tidal mixing. In addition, this argument can be demonstrated by simple scaling, as discussed in Section 5.5.1.

4. Combination of schools of pulling

According to the school of pulling by deep mixing, the total amount of mechanical energy required for sustaining global upwelling is estimated as 2 TW, including the contribution of tidal dissipation in the deep ocean and wind stress energy input to surface geostrophic currents (Munk and Wunsch, 1998). However, wind stress energy input to surface geostrophic currents may be directly converted into the geopotential energy of the large-scale circulation; this is exactly what is claimed by the school of pulling by wind stress. By separating the energy sustaining upwelling of the North Atlantic Deep Water, the net energy required for sustaining the upwelling of Antarctic Bottom Water is reduced to 0.6 TW (Webb and Suginohara, 2001), which is much smaller than the 2 TW energy requirement postulated by Munk and Wunsch. Therefore, a more realistic picture is that upwelling in the world's oceans is sustained by both wind-stress-driven Ekman transport and deep mixing (Kuhlbrodt *et al.*, 2007).

Three paradigms

Thermohaline circulation in the ocean is a very complex system; thus, it can be studied from quite different angles. For example, one can study the momentum balance of the system, and this is close to the spirit of the school of pushing. One can also study the circulation from the point of view of mechanical energy balance, falling into the framework of the school of pulling. Therefore, there are currently three paradigms for the thermohaline circulation: their basic frameworks can be summarized as follows.

Old thermal paradigm

Thermohaline circulation is driven by horizontal buoyancy difference or deepwater formation. In simple box models, the circulation rate is assumed to be linearly proportional to the north–south density difference. In many oceanic general circulation models, and other numerical models used in climate study, the common practice has been to treat the diapycnal diffusivity as an external parameter for the model, which is chosen by tuning the model to reproduce an overturning rate that matches present-day observations.

The salient points of this paradigm are: the diffusivity is considered as fixed parameters intrinsic to the model, which can be chosen through tuning the model to fit observed circulations. The same diffusivity is used in the model to simulate circulation under different climate conditions, such as the circulation during the last glacial maximum, or circulation under the global warming scenario for the next 100 years.

New mechanical energy paradigm

Thermal forcing in the upper surface cannot generate the mechanical energy required to overcome friction and dissipation associated with circulation. In order to maintain the circulation against loss through friction and dissipation, therefore, external sources of mechanical energy are needed. The thermohaline circulation is a mechanical conveyor transporting mass, heat, and freshwater fluxes, and it should be directly driven by the external source of mechanical energy. Diapycnal mixing is subject to the energy constraint. Since the sources of mechanical energy change with the atmospheric wind conditions and tidal dissipation, diapycnal diffusivity should change in response to changes in climate and tides. Thus, the thermohaline circulation is energy-controlled; however, the circulation rate is not necessarily linearly proportional to the amount of mechanical energy sustaining the circulation.

New entropy paradigm

Entropy balance is one of the fundamental thermodynamic laws governing the universe; thus, we can also study the circulation from the point of view of entropy balance. Examining the thermohaline circulation from the balance of entropy may provide us with some new insights on the subject; however, this approach has not been pursued, with a few exceptions. A preliminary balance of entropy in the world's oceans was discussed in Section 3.8.

Table 5.7. *Three paradigms for the thermohaline circulation*

Paradigm	Old (Thermal)	New (Mechanical energy)	New (Entropy)
Nature of the circulation	A thermal engine maintained by surface thermohaline forcing	A conveyor maintained by external mechanical stirring	An orderly dissipation system maintained by negative entropy flux
What regulates the circulation?	Horizontal density difference set up by surface thermohaline forcing	Vertical pulling (upwelling and mixing) induced by wind and tides	External source/sink of entropy and internal entropy production in the system
Key parameters of the model	Vertical diffusivity	External sources of mechanical energy	External and internal sources/sinks of entropy
How to use the key parameter	The diffusivity is tuned to produce a circulation with a rate matching the present-day value	The energy constraint can be used to parameterize diapycnal mixing in the model	To be developed
Does the key parameter change?	The diffusivity is intrinsic to each model, and the same diffusivity is used for all climate conditions	Diffusivity of the model changes with the availability of external sources of mechanical energy	To be explored

The essential differences existing in these three paradigms reflect different views of the fundamental physics regulating the circulation and the ways in which to parameterize sub-grid scale processes in the models. The major points are summarized in Table 5.7.

It is important to emphasize that the mechanical energy paradigm is relatively young, and the relevant theories and parameterization will take a long time and much effort to be developed. With recent interest in this new paradigm, it will become more mature and competent in the near future. The entropy paradigm is completely new. Many of its aspects remain obscure for the time being; however, exploration along these lines may eventually be very fruitful in deepening our understanding of the oceanic general circulation.

5.4.2 Thermohaline circulation based on box models

Introduction

Thermohaline circulation in the oceans is controlled by thermal forcing and freshwater flux through the air–sea interface. The physical processes involved in these two types of forcing are quite different, although their differences were not fully recognized for quite a long time. Traditionally, both temperature and salinity are treated similarly in oceanic circulation models. For example, the same diffusion coefficient is used for both constituents, and the same kind of Rayleigh condition is used as the upper boundary condition for both temperature and salinity.

Simply based on the natural feeling that the oceans are extremely large and the circulation must potentially have a large inertia, a picture of quasi-steady thermohaline circulation appears to be the logical choice. Although there is much evidence that the thermohaline circulation does change on centennial and millennial time scales, most physical oceanographers have been busy working with the circulations under current climate conditions, with much less attention being paid to potential circulation patterns under different climate conditions.

Under strong relaxation conditions, models tend to reproduce solutions with SST (sea surface temperature) and SSS (sea surface salinity) that are comparable with climatological measurements, and modelers have been happy with this approach. People have tried other types of boundary condition for salinity; however, the model's solution always drifted away from the present circulation. The solution's drift signified that there were multiple solutions for thermohaline circulations; however, people failed to pay due attention to the possibility of multiple solution associated with such a drift.

Stommel (1961) seems to have been the first to recognize the possibility of multiple solutions for the thermohaline circulation. The sudden change associated with transition from one state of thermohaline circulation to another might be related to catastrophic change in the climate. However, Stommel's two-box model did not receive much attention for nearly 20 years after its publication. As climate scientists started to search for possible mechanisms for climate conditions different from the current state, attention was turned toward a possible role of the oceanic thermohaline circulation. In a sense, Stommel's model was rediscovered and the possibility of the thermohaline circulation being in the opposite direction from overturning became an issue of intense interest.

Rooth (1982) raised the issue of multiple solutions again. In an article about the crucial role of the hydrological cycle, he extended Stommel's two-box model into a three-box model, which includes both hemispheres. He speculated that the symmetric mode with equator-to-pole motion in both hemispheres may be unstable to small perturbations. After much effort, F. Bryan (at that time a PhD student in the Geophysical Fluid Dynamical Laboratory at Princeton) was able to find the multiple solutions and thus confirmed Rooth's hypothesis.

In the following, we discuss simple models based on the classical conceptions that the thermohaline circulation is driven by meridional differences in buoyancy. Recently, quite a few exciting new insights have been reported based on model runs under the energy constraint. This is still a new research frontier; we will touch on these results only briefly.

Multiple solutions in a two-box model

Stommel (1961) used a two-box model to illustrate his basic idea: The ocean is idealized as two boxes that are connected by an open channel in the upper level and a pipe at the bottom (Fig. 5.121). The upper channel represents the communication through the ocean interior, and the pipe below is equivalent to the deep western boundary current in the ocean. The

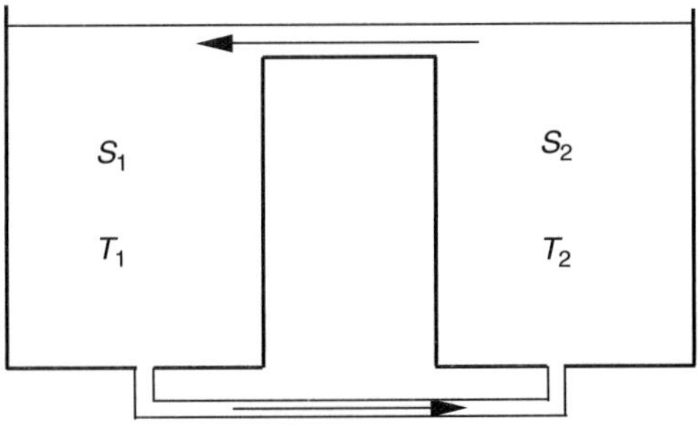

Fig. 5.121 A sketch of the two-box model (Stommel, 1961) for the thermohaline circulation.

formulation of this two-box model is discussed in detail in the following section devoted to the 2 × 2 box model.

One of the essential assumptions made by Stommel is that the circulation rate q is proportional to the north–south density difference

$$q = (\rho_2 - \rho_1)/k \tag{5.175}$$

where k is a constant, and ρ_1 and ρ_2 density in each box are assumed to be linear functions of temperature and salinity:

$$\rho_i = \rho_0(1 - \alpha T_i + \beta S_i), \quad i = 1, 2 \tag{5.176}$$

Introducing new variables $T = T_1 - T_2$ and $S = S_1 - S_2$, the control equations for the model are reduced to one set of differential equations for these new variables. Under the relaxation condition for both temperature and salinity, the balance equations are:

$$\frac{dT}{dt} = \Gamma_T(T^* - T) - |2q|T \tag{5.177}$$

$$\frac{dS}{dt} = \Gamma_S(S^* - S) - |2q|S \tag{5.178}$$

where d/dt is the time derivative , Γ_T and Γ_S are the inverse of relaxation time for temperature and salinity, and T^* and S^* are reference temperature and salinity. Although the common practice was to assume that $\Gamma_T = \Gamma_S$, Stommel realized that the relaxation time for salinity should be much longer than that for temperature, i.e., $\Gamma_T \gg \Gamma_S$.

Two stages of adjustment

Owing to the difference in the relaxation time, the system possesses two time scales for the adjustment toward a quasi-steady-state solution. On the short time scale, the process is temperature-controlled, i.e., temperature goes through a large amplitude change, which dominates changes in density and circulation. On the long time scale, the process is salinity-controlled. Owing to the weak relaxation condition applied to salinity, changes in salinity evolve over a much longer time, during which the system slowly adjusts its salinity, density, and circulation. The dramatic differences in the temperature adjustment and salinity adjustment play a critically important role in the thermohaline circulation, in particular in the process of slow or rapid transition from one state to another. We explore this in detail in this chapter.

Multiple solutions from the two-box model

Stommel demonstrated that if the ratio of these two relaxation times is larger than a critical value, there can be three possible steady states under the same forcing. The solutions can be plotted in a nondimensional T–S diagram as shown in Figure 5.122. Using this pair of new variables, T and S, the steady states of the model can be found analytically. The solutions can be graphically displayed in the traditional T–S diagram. For typical parameters, the solutions include:

- *A stable state that is thermally controlled and with a relatively fast circulation.* This is represented by point a in Figure 5.122. Because of the strong relaxation condition applied to temperature, the temperature difference is almost the same as the reference temperature difference. For a fast circulation, the salinity difference is very small, and its contribution to density is secondary, compared with that due to temperature; thus, circulation is temperature-controlled.
- *A stable state that is salinity-controlled and with quite a slow circulation.* This is represented by point c in Figure 5.122. The relatively slow circulation in the haline mode can be explained as follows. The haline mode is dominated by a salinity difference that is large enough to overcome the

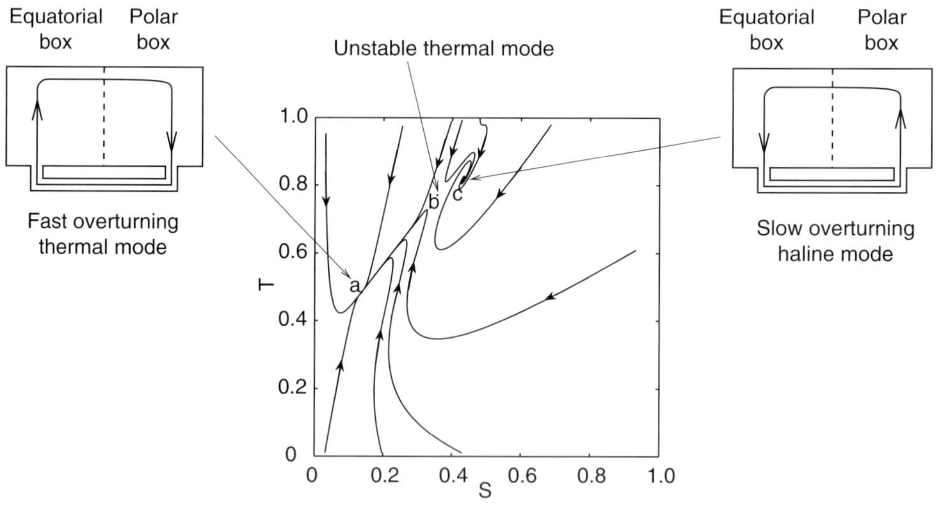

Fig. 5.122 Three equilibriums for a two-box model of Stommel (1961): the stable note (indicated by a), the saddle (indicated by b), and the stable spiral (indicated by c); the *arrows* indicate the behavior of the model's solutions in the T–S diagram.

density difference due to the temperature component. Such a large salinity difference is possible for a slow circulation only because of the long relaxation time scale.

- *An unstable state that is thermally controlled*, represented by the saddle point b in Figure 5.122.

Three-box model by Rooth (1982)

Stommel's idealized two-box model unfortunately went unnoticed for 20 years after its publication. The potential role of wind stress in the thermohaline circulation was also investigated by Stommel and Rooth (1968) in a similar two-box model; however, the role of wind stress was again overlooked for an even longer time. In fact, this paper has seldom been cited to date. Like many of the simple models postulated by Stommel, this two-box model seemed so completely different from the real world that for a long time nobody paid much attention to such an apparently simple model.

Stommel's original model is formulated for the thermohaline circulation in a single hemisphere, i.e., between the equator and high latitudes. Rooth (1982) made the second step by formulating a three-box model for the thermohaline circulation (Fig. 5.123), which consists of one equatorial box and two hemisphere boxes.

The system possesses four possible modes, including two symmetric modes and two asymmetric modes (Fig. 5.124). The two symmetric modes include: one mode with upwelling in the equatorial box and sinking at high-latitude boxes (top row), so that this mode is composed of two thermal modes in both hemispheres; another mode consists of upwelling in high-latitude boxes and sinking in the equatorial box (second row), so this mode is composed of two haline modes in both hemispheres.

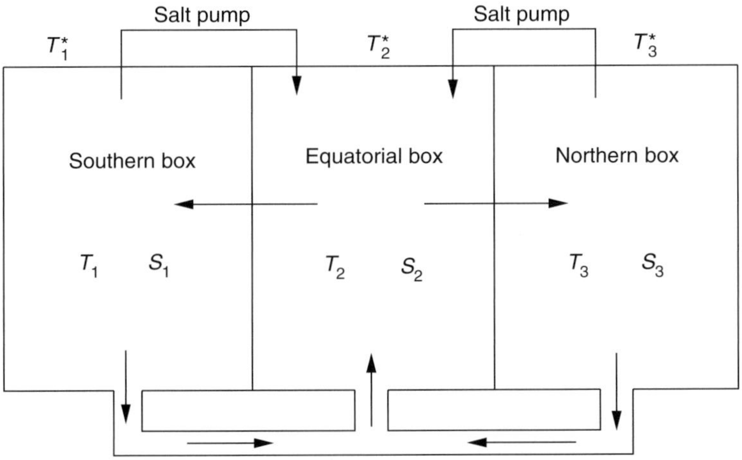

Fig. 5.123 A three-box model for a two-hemisphere ocean, where the temperature in each box is maintained by a hot/cold bath of T^*, and two ion pumps move salt from two side boxes into the central box.

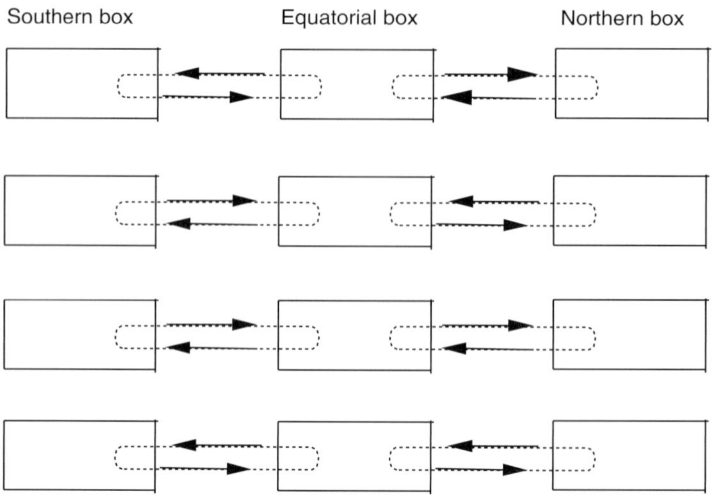

Fig. 5.124 Three-box model for a two-hemisphere ocean.

The two asymmetric modes consist of a thermal mode in one hemisphere and a haline mode in the other hemisphere, and these can be called pole-to-pole modes, as shown in the third and fourth rows in Figure 5.124.

An important postulation made by Rooth is that a symmetric solution might be unstable and could drift toward a pole-to-pole mode. Such a drift may cause a dramatic change in the oceanic circulation and the climate of the Earth.

Therefore, Stommel's (1961) pioneer work laid down the foundations for revealing multiple states for the thermohaline circulation; however, his work did not explore the transition between the thermal mode and the haline mode. Rooth's (1982) work set out the possibility of a transition in the state of the thermohaline circulation. The potential abrupt changes in thermohaline circulation or the so-called "thermohaline catastrophe" finally emerged in the 1980s, becoming one of the most exciting research frontiers for oceanic general circulation and climate.

Instability of the symmetric thermal modes

A more accurate study about the instability associated with the symmetric mode was carried out by Walin (1985). Assume that the surface thermohaline forcing, including both thermal forcing and freshwater forcing, is symmetric with respect to the equator (Fig. 5.125). The initial steady state of the thermohaline circulation is symmetric with respect to the equator, with sinking at high latitudes due to strong cooling and upwelling in the equatorial band (Fig. 5.125).

Similar to Stommel's model, the volume transport in each hemisphere is assumed to be linearly proportional to the pole-equator density difference

$$m = cHL[\alpha(T_0 - T) - \beta(S_0 - S)] \tag{5.179}$$

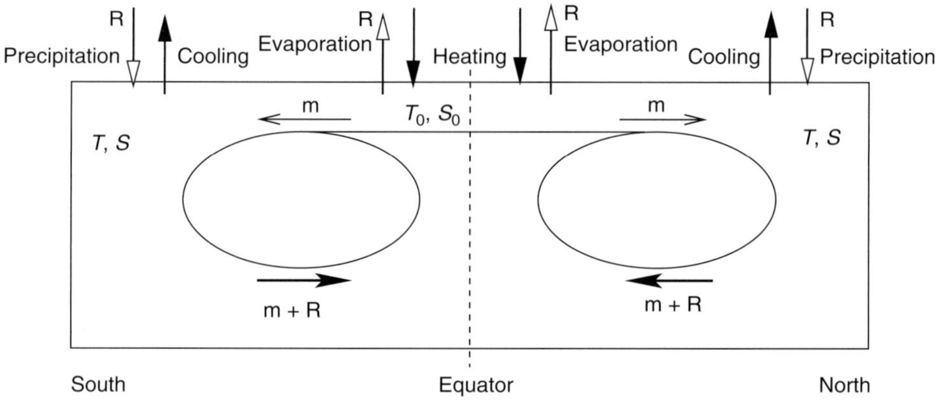

Fig. 5.125 Thermohaline circulation in a two-hemisphere basin (modified from Walin, 1985).

where T, S are temperature and salinity at high latitudes and T_0, S_0 are temperature and salinity at the equator. H and L are the height and width of the model ocean, and c is a constant. The salinity balance is

$$m_0 - (m + R)S = 0 \tag{5.180}$$

Because temperature is subject to a strong relaxation condition, the temperature perturbation T' is negligible. On the other hand, salinity perturbations parasitic to the thermal mode may not be negligible. In fact, we will show that such salinity perturbations may grow. Starting from Eqn. (5.180), the salinity perturbation satisfies

$$\frac{\partial S'}{\partial t} \propto (m + m')S_0(m + m' + R)(S + S') \simeq -\{cHL[\alpha(T_0 - T) - 2\beta(S_0 - S)] + R\}S' \tag{5.181}$$

If $R_\rho = \frac{\alpha(T_0 - T)}{\beta(S_0 - S)} < 2$, a perturbation to the saline component parasitic to the thermal mode is unstable and will grow. Since the total amount of salt in the ocean is constant, salinity perturbations in the two hemispheres must have opposite signs. As a result, the system evolves in the following way. First, the thermohaline circulation becomes asymmetric with respect to the equator, and there is cross-equator flow of the thermohaline circulation, i.e., the pole–pole mode grows as the result of this instability. In fact, the meridional circulation in both the Atlantic and Pacific Oceans includes a strong component in the pole–pole mode. Second, the density ratio, R_ρ, is driven toward a value greater than 2. Note that Stommel (1993) proposed a salt regulator theory of the oceanic mixed layer. According to his theory, owing to the stochastic forcing by rainstorms, the density ratio in the ocean is driven toward a value of 2.

A 2 × 2 box model

The model formulation

A straightforward extension of Stommel's two-box model is a 2 × 2 box model (Huang *et al.*, 1992) (Fig. 5.126). Assume that properties are uniform within each box, and the density is calculated through a linear equation of state

$$\rho_i = \rho_0(1 - \alpha T_i + \beta S_i), \ i = 1, 2, 3, 4 \tag{5.182}$$

The pressure is calculated from the hydrostatic relation and the corresponding horizontal differences of pressure in the upper and lower layers are

$$P_1 - P_2 = \Delta P_a + \frac{gH}{2}(\rho_1 - \rho_2) \tag{5.183}$$

$$P_3 - P_4 = \Delta P_a + gH(\rho_1 - \rho_2) + \frac{g\delta H}{2}(\rho_3 - \rho_4) \tag{5.184}$$

where ΔP_a is the unknown atmospheric pressure difference, which can be eliminated using the continuity equation.

This box model is based on the traditional Boussinesq approximations, so mass conservation is replaced by volume conservation. Due to volume conservation, the meridional

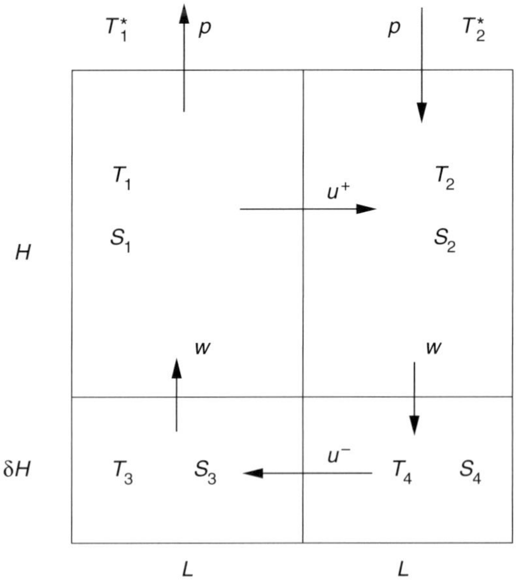

Fig. 5.126 A 2 × 2 box model, including two boxes (Boxes 1 and 3) at low latitudes, and two boxes (Boxes 2 and 4) at high latitudes. The height of the upper (lower) box is H (δH), and the box width is L in both the meridional and zonal directions. The model is also subject to reference air temperature T_i^*, amplitude of precipitation (evaporation) p, temperature (salinity) of boxes T_i (S_i), and velocities u^+, w, u^-.

velocity in the lower layers is inversely proportional to layer thickness δH (where δ is the thickness ratio of the lower box to the upper box; it is a parameter of the model), so the total mass flux in the lower layer $\delta H u^-$ remains finite when $\delta \to 0$. A finite volume flux associated with a very thin lower layer is a rough parameterization of the deep western boundary currents observed in the ocean.

Following Stommel's postulation, it is assumed that the meridional velocity is proportional to the horizontal pressure gradient

$$u^+ = c\frac{P_1 - P_2}{L}, \quad u^- = -c\frac{P_3 - P_4}{\delta L} \tag{5.185}$$

where the model parameter c is assumed to be invariant under different climate conditions.

Volume conservation requires

$$w = u^- \delta H / L \tag{5.186}$$

$$u^+ H + pL = u^- \delta H \tag{5.187}$$

From Eqns. (5.183) to (5.187), the corresponding velocity expression is

$$u^+ = -\frac{L}{2H}p - \frac{cgH}{4L}\left[\rho_1 - \rho_2 + \delta\left(\rho_3 - \rho_4\right)\right] \tag{5.188}$$

Introducing the nondimensional variables $\left(u^+, u^-\right) = \frac{L\Gamma}{H}\left(u^{+'}, u^{-'}\right)$, $w = \Gamma w'$, $p = \Gamma p'$, $T = T_0^* T'$, $S = S_0^* S'$ over these new variables (after dropping primes), this equation is reduced to

$$u^+ = -\frac{p}{2} + \left[A\left(T_1 - T_2 + \delta T_3 - \delta T_4\right) - B\left(S_1 - S_2 + \delta S_3 - \delta S_4\right)\right] \tag{5.189}$$

where

$$A = C\alpha T_0^*, \; B = C\beta S_0^*, \; C = \frac{cgH^2\rho_0}{4\Gamma L^2} \tag{5.190}$$

The fundamental assumption made in this model is Eqn. (5.185); namely, the circulation is frictionally controlled; in addition, the rotational effect is ignored, since it is troublesome to include rotation in such a highly truncated model. It is rather surprising to find that such a frictional model is able to simulate the meridional overturning quite successfully; the success of the model also implies that the meridional circulation is primarily controlled by frictional processes within the western boundary currents.

The upper boundary condition for temperature is a Rayleigh condition

$$H_f = \rho_0 c_p \Gamma\left(T^* - T\right) \tag{5.191}$$

where $\Gamma \simeq 8.1 \times 10^{-6} m/s$ (Haney, 1971). The upper boundary condition for the salinity is a specified flux of freshwater at each upper-layer box: the algebraic sum of the applied freshwater fluxes is zero.

Temperature balance in boxes 1 and 3 obeys the following equations:

$$\rho_0 c_p HL^2 \frac{dT_1}{dt} = -\rho_0 c_p HLu^+ T_1 + \rho_0 c_p L^2 w T_3 + \rho_0 c_p L^2 \Gamma (T_1^* - T_1) \quad (5.192)$$

$$\rho_0 c_p \delta HL^2 \frac{dT_3}{dt} = -\rho_0 c_p L^2 w T_3 + \rho_0 c_p L \delta H u^- T_4 \quad (5.193)$$

Using the nondimensional variables introduced above, plus a nondimensional time $t = t'H/\Gamma$, these equations and similar equations for the temperature and salinity balances in other boxes can be made nondimensional. The nondimensional equations for the heat and salt balances in upper boxes are:

$$\frac{dT_1}{dt} = -u^+ T_1 + wT_3 + 1 - T_1 \quad (5.194)$$

$$\frac{dT_2}{dt} = u^+ T_1 - wT_2 - T_2 \quad (5.195)$$

$$\frac{dS_1}{dt} = -u^+ S_1 + wS_3 \quad (5.196)$$

$$\frac{dS_2}{dt} = u^+ S_1 - wS_2 \quad (5.197)$$

The equations for lower boxes are:

$$\frac{dT_3}{dt} = \frac{u^-}{\delta}(T_4 - T_3) \quad (5.198)$$

$$\frac{dT_4}{dt} = \frac{u^-}{\delta}(T_2 - T_4) \quad (5.199)$$

$$\frac{dS_3}{dt} = \frac{u^-}{\delta}(S_4 - S_3) \quad (5.200)$$

$$\frac{dS_4}{dt} = \frac{u^-}{\delta}(S_2 - S_4) \quad (5.201)$$

where the velocities satisfy the continuity constraints

$$w = u^+ + p, \quad w = \delta u^- \quad (5.202)$$

These equations are based on the so-called upwind scheme, i.e., property transport is defined as velocity multiplied by property in the upwind box. These equations are valid for the circulation pattern indicated in Figure 5.126 only. If the circulation reverses its direction, the corresponding equations will be slightly different, yet they can be derived accordingly. In modeling, parameter $C = 0.05$ was adopted to give a vertical velocity comparable to observations, and the precipitation rate p was about 0.38×10^{-7} m/s.

Three types of mixing exist in the model. First, properties advected into each box are homogenized. Second, numerical mixing exists owing to the upwind scheme. Third, there is mixing associated with convective overturning, the so-called convective adjustment in many numerical models. This is a crude parameterization of the physics: A gravitationally unstable situation may occur due to cooling at the surface, and water within the unstable column will mix vertically until a stable stratification is obtained. However, there is no other source of mixing.

Reduction to a two-box model

As $\delta \to 0$, the model is reduced to the classic two-box model of Stommel (1961), with a minor difference, as the present model is based on a freshwater flux condition on the upper surface, instead of the salt relaxation condition used in Stommel's model. At the limit of infinitesimal δH, the meridional volume flux through boxes 3 and 4 remains finite. Essentially, boxes 3 and 4 shrink and become a "pipe" linking boxes 1 and 2.

For the thermal mode, $u^+ > 0$,

$$T_3 = T_4 = T_2,\ S_3 = S_4 = S_2 \tag{5.203}$$

Therefore, the equations are reduced to

$$\frac{dT_1}{dt} = -u^+ T_1 + wT_2 + 1 - T_1 \tag{5.204}$$

$$\frac{dT_2}{dt} = u^+ T_1 - wT_2 - T_2 \tag{5.205}$$

$$\frac{dS_1}{dt} = -u^+ S_1 + wS_2 \tag{5.206}$$

$$\frac{dS_2}{dt} = u^+ S_1 - wS_2 \tag{5.207}$$

The velocities are

$$u^+ = -\frac{p}{2} + C(T - S),\ w = u^+ + p \tag{5.208}$$

where

$$T = \alpha T_0(T_1 - T_2),\ S = \beta S_0(S_1 - S_2) \tag{5.209}$$

There are three zones in the (S, T) phase space: *region I* – where $u^+ \geq 0$, $u^- \geq 0$; *region II* – where $u^+ < 0$, $u^- \leq 0$; and *region III* – where $u^+ u^- < 0$ (Fig. 5.127). Among the three, region III is unphysical because salt would be wiped out from half of the model basin when both velocities are in the same direction.

Now let us look at the steady states that can be characterized by equilibrium curves:

Heat equilibrium curves within region I and II:

$$S = \frac{1}{2C} + T - \frac{1 + p}{2C}\frac{T_0}{T} \qquad \text{Region I} \tag{5.210}$$

$$S = -\frac{1}{2C} + T + \frac{1 + p}{2C}\frac{T_0}{T} \qquad \text{Region II} \tag{5.211}$$

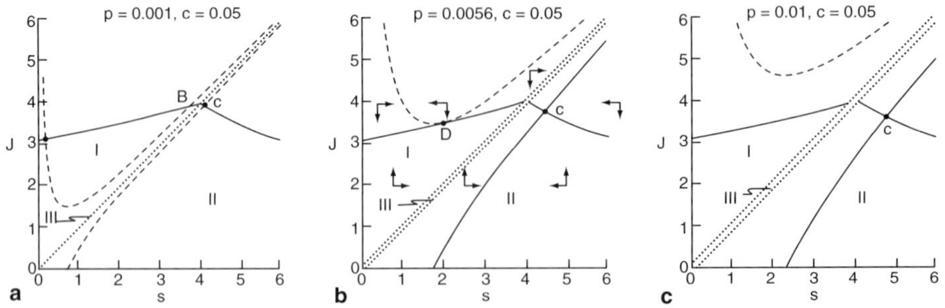

Fig. 5.127 S–T phase diagram for the 2 × 1 box ($\delta \to 0$). Panels **a–c** are for different values of p (nondimensional). The phase space consists of three regions: I, II, and III, characterizing different velocity patterns: the *solid curve* represents heat equilibrium, and the *dashed curve* represents salt equilibrium. Intersections of these two sets of curves define equilibrium states. The intersection at point A is a stable thermally driven equilibrium; point B is unstable; point C is a stable salinity-dominated equilibrium.; point D represents the coalescence of points A and B at the critical $p_c = 0.0056$. In panel **b** there are *pairs of arrows* to show the phase velocity of S_i and T_i (because away from the equilibrium points, they both change with time) in a region bounded by the equilibrium curves (Huang *et al.*, 1992).

Salt equilibrium curves within regions I and II:

$$T = S + \frac{pS_0}{CS} \qquad \text{Region I} \tag{5.212}$$

$$T = S - \frac{pS_0}{CS} \qquad \text{Region II} \tag{5.213}$$

The intersections of these two curves are steady solutions for the model (Fig. 5.127).

We note that there is a critical value of precipitation p_c in the following sense. When $p < p_c$ three solutions exist: one stable thermal-mode solution, one unstable thermal-mode solution, and one unstable haline-mode solution. As $p \to p_c$, the two thermal-mode solutions coalesce; this is a critical situation where a "catastrophe" would occur if p was slightly increased. When p is greater than p_c, only the haline mode exists. A noteworthy feature of this model is that there is always a stable haline mode for whatever value of precipitation p; in contrast, however, thermal modes exist for the domain of $p \le p_c$ only.

Transition from a thermal mode to a saline mode

As stressed above, in this model a stable haline mode exists for any value of precipitation; however, the thermal modes exist only for $p \le p_c$. Assume that the system is initially in the thermal-mode state. When p is increased beyond the critical value p_c, the thermal mode will collapse and the system eventually settles down into a stable haline mode.

In the neighborhood of the critical value p_c, a small increase in p would induce a catastrophic transition from thermal mode to haline mode. The transition can be subdivided into three stages: (1) a searching stage, (2) a catastrophic stage, and (3) an adjustment stage during which the deep boxes adjust their temperature and salinity to the new equilibrium values (Fig. 5.128). In this figure, the phase speed is defined as $\phi = \sqrt{\sum_{i=1}^{4}(T_{it}^2 + S_{it}^2)}$. Near year 312, the phase speed reaches a maximum, indicating the rapid changes of temperature and salinity in the model. As shown in this experiment, the final adjustment of deep water properties takes a long time. Even at the end of a 904-year run, the salinity and temperature of the bottom boxes is far from approaching the final equilibrium state. Although the results were obtained from a simple box model, they are valuable for understanding the much more complicated physics of catastrophic processes in oceanic general circulation models.

The system exhibits a sort of hysteric behavior as the strength of evaporation minus precipitation gradually increases or decreases (Fig. 5.129). As the precipitation increases, the system remains in the thermal mode while the meridional circulation decreases gradually. As p approaches p_c, the system can no longer remain in the thermal mode; however, owing to the existence of a "forbidden region," indicated by the narrow shaded wedge in Figure 5.129a, the transition to the haline mode must be catastrophic.

The system remains in the haline mode after the transition, even if the precipitation is gradually reduced, as indicated by the solid arrows in Figure 5.129. The system returns to the thermal mode through another catastrophic change when precipitation is reduced to zero. Therefore, the behavior of the system appears as the hysteric loop (Fig. 5.129).

We can now see that the essential ingredient for thermohaline catastrophe is the difference in upper boundary conditions for temperature and salinity. The temperature is subject to a relaxation condition with a short relaxation time, but the salinity is subject to either a relaxation condition with a long relaxation time (Stommel, 1961) or a flux condition for the salinity (the present case).

Monte Carlo experiments

The idea of the 2×2 box model can be extended to other multiple-box models. An interesting case is a 3×2 box model mimicking the North Atlantic Ocean, including three boxes in the upper ocean and three boxes in the abyss. The model is subject to a linear profile of reference temperature $T^* = (25, 12.5, 0)$, plus evaporation in the equatorial box and precipitation in the high-latitude box $(-p, 0, p)$.

From the study of the 2×2 box model, we expect that this 3×2 box model could have multiple solutions. If a system possesses multiple solutions, its parameter space can be separated into the so-called "basins of attraction." The concept of a basin of attraction originates from drainage basins in the world: precipitation over land is collected by rivers which eventually converge into different drainage basins. Since the thermohaline circulation system has multiple solutions, one way of studying the size of the basin of attraction is to run Monte Carlo experiments. The basic idea of Monte Carlo experiments is to find the probability of the system falling into different steady states. For a model with many parameters, the search for such a probability distribution function over the entire parameter space based on using evenly distributed sampling grids is very time-consuming. Thus, a commonly used efficient method of finding such information is to run the model by starting

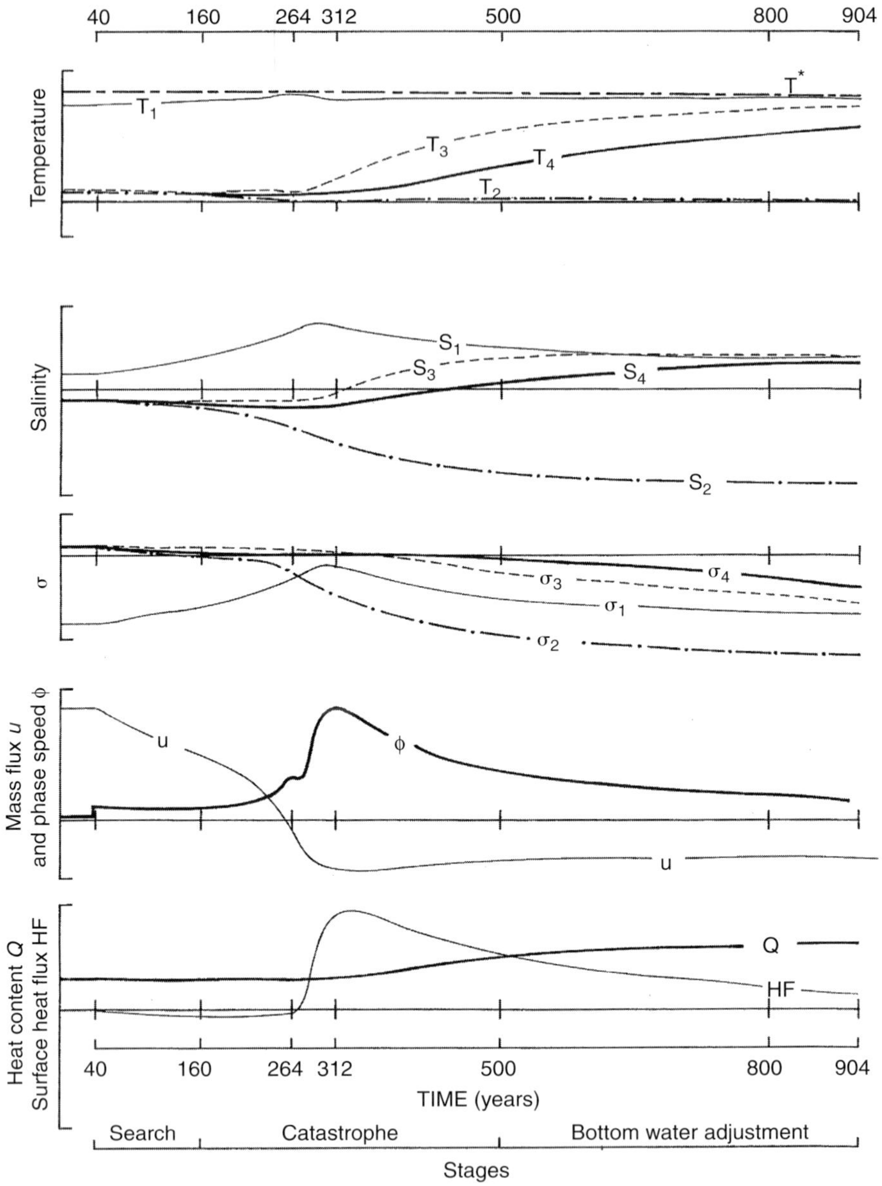

Fig. 5.128 Transition of a 2 × 2 ($\delta = 1$) system from a thermally controlled equilibrium state to a salinity-controlled equilibrium state, as caused by a discrete increase of p at time year 40, when p exceeds its critical value. The transition passes through three stages, as labeled at the bottom of the figure (Huang *et al.*, 1992).

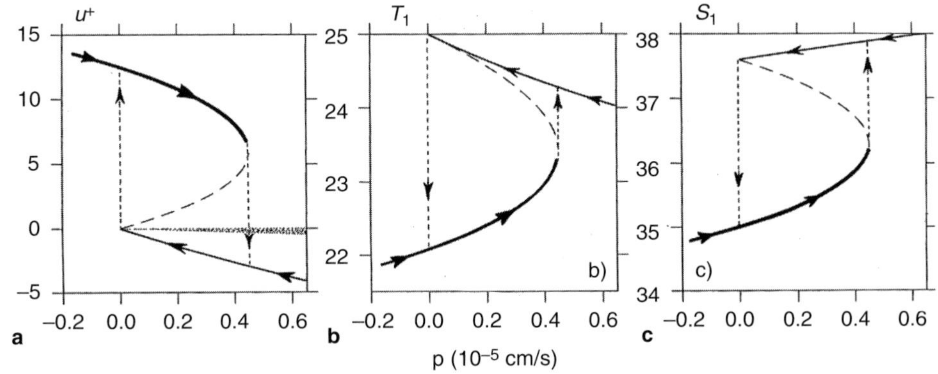

Fig. 5.129 Dependence of the equilibrium values of **a** u^+ (in 10^{-5} cm/s), **b** T_1 and **c** S_1 upon p, for the 2×2 ($\delta = 1$) box model: *heavy solid curve* for p increasing by small increments; *thin solid curve* for p decreasing (thus illustrating the hysterics of the system); *long-dashed lines* for an unstable state; the *short-dashed arrows* indicate abrupt transitions (Huang *et al.*, 1992).

from initial states with random temperature and salinity, letting the model run into its final equilibrium states, and saving the information for a final statistical summary.

For a given p, the model is integrated from an initial state of random temperatures in each box, between 0°C and 25°C, and random salinity between 31.5 and 38.5. The results of the Monte Carlo experiments indicate that the model has four stable modes: a thermal mode, a haline mode, and two intermediate modes. The probability of the results falling into each of the four stable modes is shown in Figure 5.130. In the calculation of the probability, solutions which are mirror images of each other are classified as the same mode in this figure. Triple numbers appearing in this figure represent the magnitude and sign of the vertical velocity in each of the three upper boxes; where 1 indicates strong upwelling, ε indicates very weak upwelling, and -1 indicates strong downwelling. According to the results of this model, the thermohaline circulation in the North Atlantic Ocean is close to a critical state: a small increase in precipitation may cause a catastrophic transition from the current thermal mode to the haline mode.

The results of the Monte Carlo experiment carried out for this 3×2 model demonstrate the existence of grand modes (the most probable modes). There are also many other possible modes; however, most of these modes have very low probability and are very unlikely to appear (Fig. 5.130).

Results of similar experiments for a 3×2 model forced by the observed distribution of evaporation minus precipitation show that for the current level of evaporation minus precipitation the most likely mode is an intermediate mode which is characterized by sinking at the mid-latitude box. The intermediate mode may have some connection with the intensification of the intermediate water formation which existed during the last glaciations.

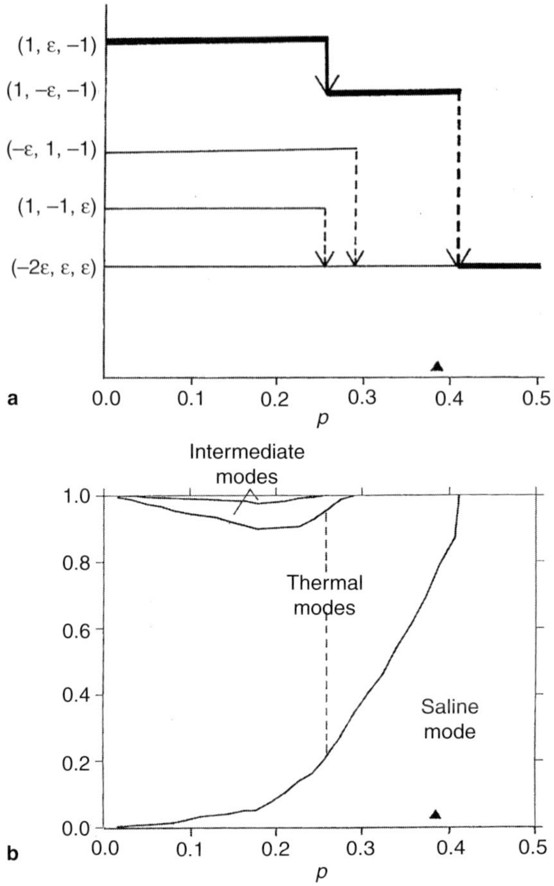

Fig. 5.130 Monte Carlo experiments for a 3×2 box model, p in 10^{-7} m/s. **a** The four possible steady states and the catastrophic pathways as p is gradually increased. The *horizontal axis* is the strength of precipitation/evaporation, and the *small triangle* indicates the amplitude of freshwater for the current climate. **b** The vertical space between the curves indicates the probability of each equilibrium state as a function of the freshwater flux (Huang *et al.*, 1992).

Extension of the two-box model: The energy constraint and wind-driven gyration

The wind-driven component is of critical significance in setting up the global thermohaline circulation; however, the role of wind-driven circulation is not included in most box models, with a few exceptions (e.g., Stommel and Rooth, 1968; Huang and Stommel, 1992). It is desirable to include this physical component in box models.

Model formulation

The dynamical effects of wind-driven circulation upon the thermohaline circulation can be explored in simple models (e.g., Pasquero and Tziperman, 2004; Longworth *et al.*, 2005).

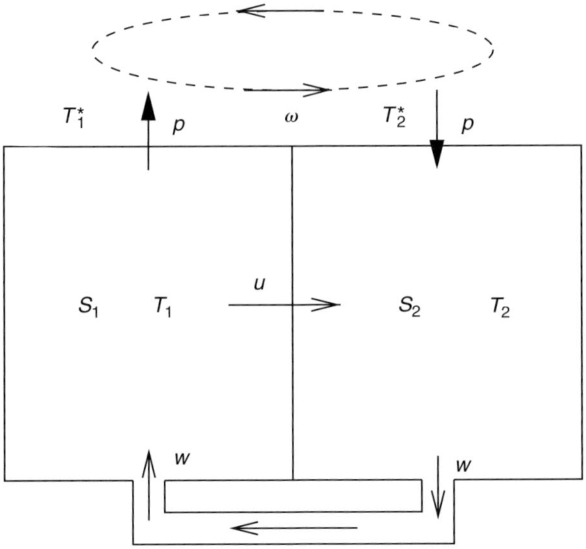

Fig. 5.131 A two-box model for the thermohaline circulation including the additional mass exchange due to a wind-driven gyre (the *elliptical dashed line*) (Guan and Huang, 2008).

The simplest choice is a two-box model for a single-hemisphere ocean, in which temperature is relaxed toward the specified reference temperature of $T_1^* = T_0$, $T_2^* = 0$ and an air–sea freshwater flux $[p, -p]$. To simulate the effect of the wind-driven gyre, a volume transport ω between the two boxes is added (Huang and Stommel, 1992). Following Guan and Huang (2008), this volume transport is prescribed *a priori*, independent of the energy for mixing and temperature/salinity (Fig. 5.131).

Assuming that the flow is poleward in the upper layer and equatorward in the abyssal ocean (the so-called thermal mode), the temperature balance in Box 1 obeys

$$HL^2\frac{dT_1}{dt} = -HLuT_1 + L^2wT_2 + L^2\Gamma(T_0 - T_1) - HL\omega(T_1 - T_2) \tag{5.214}$$

where L and H are the width and depth of each box, T_1 and T_2 are the temperature in each box, u and w are the horizontal and vertical velocity, ω is the strength of gyration, and Γ is the surface relaxation constant. Introducing the nondimensional variables through $u = \frac{\Gamma L}{H}u'$, $w = \Gamma w'$, $\omega = \frac{\Gamma L}{H}\omega'$, $t = \frac{H}{\Gamma}t'$, and dropping the primes, the temperature balance for Box 1 is reduced to

$$\frac{dT_1}{dt} = -uT_1 + wT_2 + T_0 - T_1 - \omega(T_1 - T_2) \tag{5.215}$$

Similarly, the temperature balance for Box 2 is

$$\frac{dT_2}{dt} = uT_1 - wT_2 - T_2 + \omega(T_1 - T_2) \tag{5.216}$$

The salinity balances for Boxes 1 and 2 are

$$\frac{dS_1}{dt} = -uS_1 + wS_2 - \omega(S_1 - S_2) \tag{5.217}$$

$$\frac{dS_2}{dt} = -uS_1 - wS_2 + \omega(S_1 - S_2) \tag{5.218}$$

The continuity equation is

$$w = u + p \tag{5.219}$$

Subtracting Eqn. (5.216) from Eqn. (5.215) and using Eqn. (5.219), we obtain

$$\frac{d}{dt}(T_1 - T_2) = -2\omega(T_1 - T_2) + 2pT_1 + T_0 - T_1 + T_2 - 2\omega(T_1 - T_2). \tag{5.220}$$

Integrating the sum of Eqns. (5.215) and (5.216) leads to a constraint: for any initial conditions, the sum of temperature in the two boxes converges exponentially, $T_1 + T_2 \to T_0$. Thus, the $2pT_1$ term in Eqn. (5.220) can be replaced by $p(T_0 + T_1 - T_2)$. Adding Eqns. (5.217) and (5.218) leads to the salt conservation

$$S_1 + S_2 = 2S_0 \tag{5.221}$$

where S_0 is the mean reference salinity of the model ocean. We now introduce new temperature and salinity variables: $\Delta T = T_1 - T_2$, $\Delta S = S_1 - S_2$.

Using Eqns. (5.219) and (5.220), we obtain a pair of equations:

$$\frac{d\Delta T}{dt} = -2w\Delta T + (p - 1 - 2\omega)\Delta T + (p + 1)T_0 \tag{5.222}$$

$$\frac{d\Delta S}{dt} = -2w\Delta S + (p - 2\omega)\Delta S + 2pS_0 \tag{5.223}$$

When the circulation is in the haline mode, the direction of mass transport between these two boxes is reversed, but the contribution due to wind-driven circulation stays unchanged. Thus, equations corresponding to Eqns. (5.222) and (5.223) are

$$\frac{d\Delta T}{dt} = -2w\Delta T - (p + 1 + 2\omega)\Delta T + (p + 1)T_0 \tag{5.224}$$

$$\frac{d\Delta S}{dt} = -2w\Delta S - (p + 2\omega)\Delta S + 2pS_0 \tag{5.225}$$

These sets of ordinary differential equations, (5.224) and (5.225), govern the thermohaline circulation, including the effect of wind-driven circulation. In order to calculate the circulation, we need one more constraint that determines the vertical velocity w.

Constraints regulating the circulation rate

The constraint regulating the meridional overturning rate is the remaining critical part of the model formulation. Stommel (1961) postulated a constraint which can be classified as a buoyancy constraint. Using Eqn. (5.175), the corresponding vertical velocity scale is as follows:

$$W_s = c(\rho_0 \alpha \Delta T - \rho_0 \beta \Delta S) \tag{5.226}$$

where c is a constant. This formulation implies that the circulation rate is regulated by the surface buoyancy difference. In other words, surface thermohaline forcing drives thermohaline circulation.

Stommel's assumption has been widely adopted in most existing box models. A careful examination of Stommel's classical assumption, namely Eqn. (5.226), suggests that the reasoning behind Stommel's choice of such a constraint is not obvious; in addition, it is not the only possible constraint and may not even be the best constraint for modeling thermohaline circulation under different climate conditions.

More suspiciously, this constant c has been treated as a constant intrinsic to each model, and it is assumed to be invariant under different climate conditions. From the most basic consideration of balance of mechanical energy, however, to maintain a steady circulation in the ocean, dense deep water must be brought upward through the thermocline, and heat absorbed in the upper ocean must be mixed downward against the upwelling of the cold and dense water through diapycnal mixing. Therefore, the balance of the external sources and the dissipation of mechanical energy are what really control the thermohaline circulation. As an alternative, a constraint based on mechanical energy sustaining diapycnal mixing can be formulated as follows. Assuming a one-dimensional balance of density in the vertical direction

$$w_e \rho_z = \kappa \rho_{zz} \tag{5.227}$$

where κ is the vertical (diapycnal) diffusivity, we obtain the following scale for the vertical velocity:

$$W_e = \kappa / D \tag{5.228}$$

where D is the depth scale of the main thermocline. The differences in temperature, salinity, and density between the surface and abyssal ocean are ΔT, ΔS, and $\Delta \rho$. Note that $\Delta \rho = \rho_0 (-\alpha \Delta T + \beta \Delta S) = \rho_1 - \rho_2 < 0$. In a two-layer model, the rate of GPE increase (per unit area) due to vertical mixing κ is $-g\kappa \Delta \rho$. Using the above notations, the rate of GPE created by mixing in the box model is

$$E_m = -g\kappa \Delta \rho L^2 = -gDL^2 W_e \Delta \rho = gDL^2 W_e (\rho_0 \alpha \Delta T - \rho_0 \beta \Delta S) \tag{5.229}$$

Therefore, the scale of W_e associated with meridional overturning satisfies

$$W_e = \frac{e}{\rho_0 \alpha \Delta T - \rho_0 \beta \Delta S} \tag{5.230}$$

where $e = \frac{E_m}{gDL^2} = \frac{E}{D}$, $E = \frac{E_m}{gL^2}$ (E is the rate, per unit area, of external mechanical energy supply) has a dimension of density multiplied by velocity and represents the strength of the external source of mechanical energy sustaining mixing. A noteworthy point here is that an external source of mechanical energy can vary with climate; thus, e in the energy-constraint model is treated as an external parameter which can change with climate.

Scaling laws

Some useful insights can be obtained through simple scaling. The continuity relation is scaled as

$$UD = WL \tag{5.231}$$

where U and W are horizontal and vertical velocity scales. The salinity balance associated with the meridional overturning cell obeys a simple scaling relation

$$(UDL + \Omega L^2)\Delta S = F\bar{S}L^2 \tag{5.232}$$

where F is the scale of freshwater flux, and $\Omega = \omega D/L$ is the scale of gyration. Therefore, the scale of salinity difference is

$$\Delta S = F\bar{S}/(W + \Omega) \tag{5.233}$$

When $\Omega = 0$, the vertical velocity scale under the energy constraint is reduced to

$$W_e = (\rho_0 \beta F\bar{S} + e)/\rho_0 \alpha \Delta T \tag{5.234}$$

Under the temperature relaxation condition, $\Delta T \sim T_0$ is a good approximation. The corresponding scale for the poleward heat flux is

$$H_f = c_p \rho_0 UDLT_0 = \frac{c_p L^2}{\alpha}(e + \rho_0 \beta F\bar{S}) \tag{5.235}$$

In other words, the scaling analysis implies that strong freshwater forcing enhances the upwelling rate, and thus both the meridional overturning rate and the poleward heat flux are enhanced.

In contrast, the Stommel-like model predicts that strong freshwater forcing reduces the meridional density difference, and thus the meridional overturning rate. Indeed, the assumptions of the fixed buoyancy constant c or the energy constant e represent two extreme conditions, and the real-world condition may be somewhere in between.

The role of mechanical energy

If $\omega = 0$ (i.e., no wind-driven gyre), it is well known that in a quite wide region of the parameter space there exist three steady states in the Stommel-like model (Fig. 5.132a): a stable thermal state, an unstable thermal state, and a stable haline state; whereas three

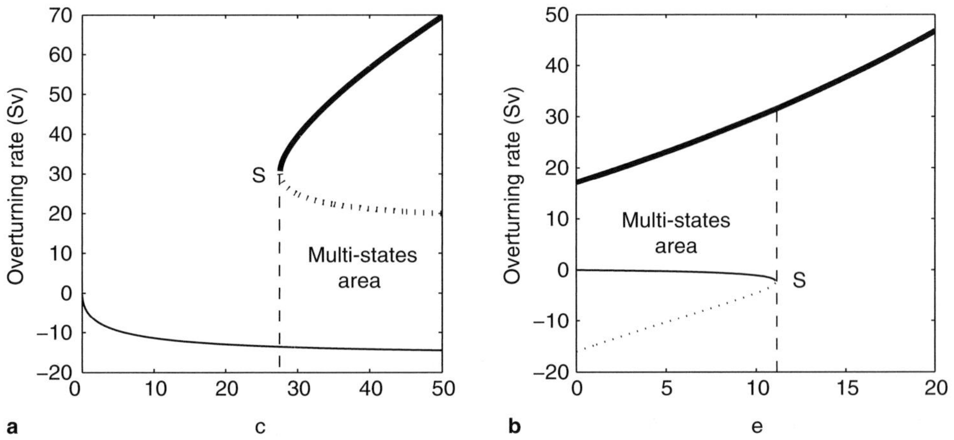

Fig. 5.132 Thermohaline circulation bifurcation for **a** the Stommel-like model, c in 10^{-7} m^4/kg/s; **b** the energy-constraint model, e in 10^{-7} kg /m^2/s,$T_0 = 15°C$, $S_0 = 35$, $p = 2$ m/yr. *Heavy lines* indicate the thermal modes, *thin lines* indicate the haline modes, *dotted lines* for unstable modes; S is the saddle-node bifurcation point (*dashed line*) (Guan and Huang, 2008).

steady states existing in the energy-constraint model (Fig. 5.132b) are a stable thermal state, a stable haline state, and an unstable haline state.

It can readily be seen that when $\omega = 0$ the bifurcation structure of the Stommel-like model is dramatically different from that of the energy-constraint model. In particular, the stable haline mode from the energy-constraint model is characterized by an extremely large salinity difference, i.e., salinity in the pole box is nearly zero. However, including the gyration dramatically reduces the salinity difference, ΔS, for the haline mode in the energy-constraint model; and it becomes much closer to the corresponding solution obtained from the Stommel-like model. The fact that the inclusion of wind-driven circulation can change the bifurcation structure of thermohaline circulation clearly demonstrates the importance of coupling the wind-driven gyre with thermohaline circulation.

The role of gyration

In this section we closely examine the role of gyration in detail. We adopt the following set of parameters for both the Stommel-like model and the energy-constraint model: the basin has a length scale of 4,000 km, is 4 km deep, has a reference temperature $T_0 = 15°C$, the mean salinity is $S_0 = 35$, and the other parameters are the same as in the above discussion. As shown in Figure 5.133, for a certain parameter range, changes in wind-driven circulation alone can induce a bifurcation of thermohaline circulation in both the Stommel-like model and the energy-constraint model.

For the Stommel-like model, an increasing wind-driven circulation enhances the meridional overturning rate in the thermal mode and reduces the meridional overturning rate

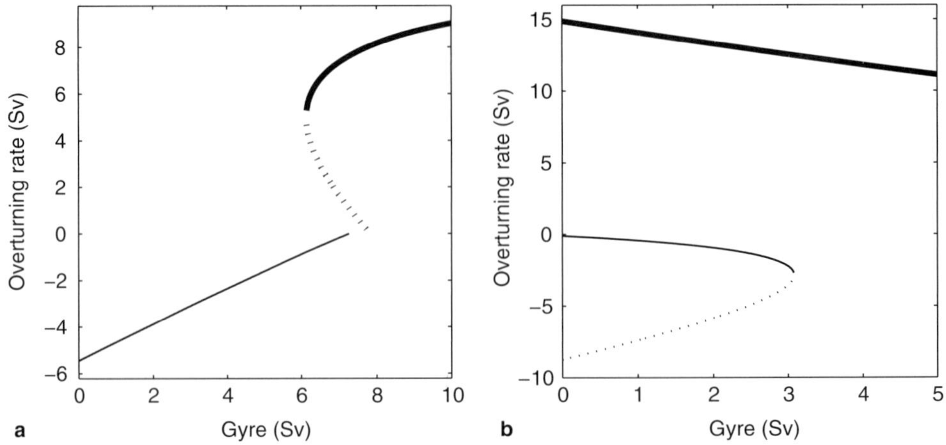

Fig. 5.133 Overturning rate as a function of strength of the wind-driven gyre, $p = 1$ m/yr, $T_0 = 15°C$, $S_0 = 35$: **a** the Stommel-like model, $c = 5 \times 10^{-7}$ m^4/kg/s; **b** the energy-constraint model, $e = 2.5 \times 10^{-7}$ kg/m^2/s. *Heavy lines* for the thermal modes, *thin lines* for the haline modes, and *dotted lines* for unstable modes (Guan and Huang, 2008).

in the haline mode (Fig. 5.133a). However, the effect of wind-driven circulation on the meridional overturning rate is reversed in the energy-constraint model (Fig. 5.133b).

Note that when gyration is smaller than a critical value, the energy-constraint model has three steady solutions: a stable thermal mode, a stable haline mode, and an unstable haline mode; beyond this critical value the model has only one stable solution – a thermal mode. For the current parameters: $p = 1$ m/yr and $e = 2.5 \times 10^{-7}$ kg/m^2/s, the critical value of the volumetric transport of the gyre is 3.1 Sv (Fig. 5.133b). Therefore, if the model starts from an initial state in the thermal mode and with a small wind-driven circulation, the solution will remain in the stable thermal mode. However, the story can be quite different if the model starts from an initial state of a stable haline mode with a weak wind-driven circulation. When wind-driven circulation is increased beyond the critical value, the haline mode is no longer a viable state and the system will be switched to the thermal mode through a catastrophic change.

Limitations of box models

In spite of the fact that models have been widely used in various applications, there are some inherent limitations of box models. Most importantly, the number of boxes in the model is often rather limited, so they cannot really simulate the details of the general circulation. Although box models with a large number of boxes have been constructed and used to study the thermohaline circulation, such complicated box models are not very popular because it is much easier to construct a model based on partial differential equations and finite difference. In addition, owing to the nature of the highly truncated boxes, large numerical diffusion due to the upwind scheme used in box models arises. Excessive numerical diffusion in box models prevents their application to cases where more accurate results are desirable.

Another major limitation of most box models is due to the exclusion of the Coriolis force. Lacking the effect of rotation makes the box model less useful as a tool for understanding the circulation. However, this is not an intrinsic limit for box models. Maas (1994) discussed a three-dimensional box model in which the dynamical effect of rotation is included. In fact, the effect of rotation, even the so-called β-effect, can be included in a box model. For example, a 3×3 barotropic box model based on the C-grid can include the Coriolis force and can be used to demonstrate the β-effect; i.e., western intensification can be demonstrated in the model. Two rows to the east in the model play the role of the ocean interior, while the western row plays the role of the western boundary layer. However, the model is so highly truncated that the parameters used in this kind of model are far from realistic.

5.4.3 Thermohaline circulation based on loop models

The water-wheel model

The water-wheel experiments were first carried out by Malkus and Howard at the Massachusetts Institute of Technology in the 1970s (Malkus, 1972). The experimental set-up is relatively simple; the simplest version is just a slightly tilted toy water-wheel with leaky paper cups suspended from its rim. Water is supplied from the tap on the top of the rim (Fig. 5.134).

The experiments show that when the water flow rate is slow, the top cups are not filled up enough to overcome the friction, since the cups are leaky. The wheel remains motionless, owing to friction. When the flow rate is large, the wheel starts to rotate, resulting in steady rotation in either direction. When the flow rate is even greater and larger than a certain

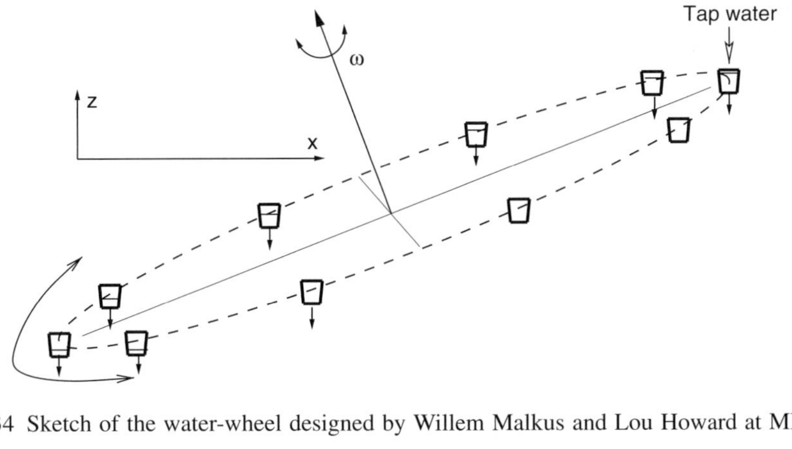

Fig. 5.134 Sketch of the water-wheel designed by Willem Malkus and Lou Howard at MIT in the 1970s.

threshold, the motion becomes chaotic, i.e., the wheel rotates clockwise and anticlockwise, with angular velocity changing with time in a chaotic way. The above experiments can be described in terms of the water-wheel equation, which will be detailed in connection with a so-called loop model in this section.

A loop model forced by evaporation and precipitation

Loop models have been studied and used extensively in many fields, including many engineering applications; thus, a large number of papers have been published along these lines. The thermohaline circulation in the oceans can have regular or irregular oscillations over a wide spectrum in both temporal and spatial space, and loop models can be used as a one-dimensional idealization. These models can be formulated for thermal circulation, haline circulation, or thermohaline circulation.

In this section, we discuss a simple loop model of the haline circulation. The system can be formulated under two slightly different boundary conditions: the natural boundary condition and the virtual salt flux condition. A model forced by the natural boundary condition is shown in Figure 5.135. The model based on the virtual salt flux condition can be reduced to an ordinary differential equation system, the water-wheel equation (Dewar and Huang, 1995; Huang and Dewar, 1996).

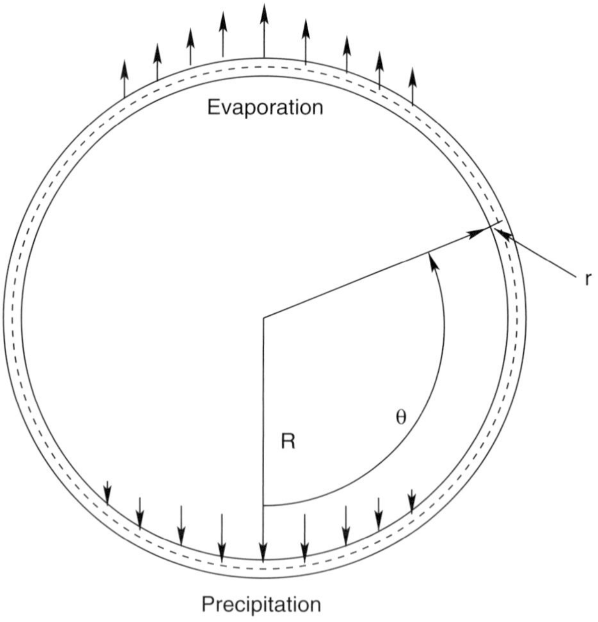

Fig. 5.135 A loop oscillator for the haline circulation – a tube filled with salty water. The fresh water is passed through the skin of the tube.

Model formulation

The model consists of a water tube loop filled with salty water (Fig. 5.135). The radius of the loop is R and the radius of the tube is r, where $R \gg r$. In the following analysis, we assume that the flow and water properties are uniform in each cross-section. At the surface of the tube, freshwater is exchanged with the environment as precipitation (p) and evaporation (e):

$$p - e = E \cos\theta \tag{5.236}$$

where E is the amplitude of the freshwater flux. We now set out the equations describing this loop system.

Continuity

Under the Boussinesq approximations, continuity requires that the velocity convergence equals the freshwater flux.

$$u_\theta = 2RE \cos\theta / r \tag{5.237}$$

In the following analysis, we use angular velocity, defined as

$$\omega = u/R \tag{5.238}$$

Thus, integrating Eqn. (5.237) leads to

$$\omega = \Omega + 2E \sin\theta / r \tag{5.239}$$

where Ω is the mean angular velocity of the water circulating the loop, and the second term on the right-hand side is due to the accumulated water caused by the freshwater flux through the surface of the tube.

Salt conservation

Similarly, the salt conservation equation is

$$S_t + (\omega S)_\theta = KS_{\theta\theta}/R^2 \tag{5.240}$$

where K is the coefficient of salt diffusivity. The contribution to salinity balance from evaporation and precipitation is included in the continuity equation, namely through the angular velocity relation, Eqn. (5.239); thus, such a model is under the natural boundary condition (NBC).

For the model subject to virtual salt flux (VSF) and relaxation boundary conditions, the corresponding salt conservation equations are

$$S_t + (\Omega S)_\theta = \frac{-2E}{r}\overline{S}\cos\theta + \frac{K}{R^2}S_{\theta\theta} \tag{5.241}$$

$$S_t + (\Omega S)_\theta = \frac{2\Gamma}{r}\left(S^* - S\right) + \frac{K}{R^2}S_{\theta\theta} \tag{5.242}$$

where Γ is a relaxation coefficient, $\overline{S}$ is the average salinity, and S^* is a specified reference salinity (traditionally chosen to be the observed mean salinity). Note that the net salt in the problem is guaranteed to be a constant if Eqns. (5.240) or (5.241) are used. No such guarantee exists if Eqn. (5.242) is used.

Momentum equation

The momentum balance for an infinitely small sector of water tube is

$$\rho_0 (u_t + uu_l) = -P_l - \rho g \sin\theta - \varepsilon\rho_0 u \tag{5.243}$$

where l is the arc element along the axis of the tube, P_l is the pressure gradient, and ε is the frictional coefficient. Integrating Eqn. (5.243) around the entire loop, all the gradient terms vanish and we obtain

$$\Omega_t = -\varepsilon\Omega - \frac{g}{2\pi R}\int_0^{2\pi} \frac{\rho}{\rho_0} \sin\theta d\theta \tag{5.244}$$

In this study we apply a linear equation of state

$$\rho = \rho_0 (1 + \beta S) \tag{5.245}$$

Thus, Eqn. (5.244) can be written as

$$\Omega_t = -\varepsilon\Omega - \frac{g\beta}{R} \langle S \sin\theta \rangle \tag{5.246}$$

where $\langle\rangle = \frac{1}{2\pi}\int_0^{2\pi} \cdot d\theta$ is an averaging operator.

Nondimensionalization

Equations (5.240) and (5.246) are nondimensionalized using

$$S = \overline{S}S', t = Tt', \omega = \omega'/T \tag{5.247}$$

where

$$T = \sqrt{R/g\beta\overline{S}} \tag{5.248}$$

After dropping the primes, we obtain

$$\Omega_t = -\alpha\Omega - \langle S \sin\theta \rangle \tag{5.249a}$$

$$S_t + [(\Omega + \lambda \sin\theta) S]_\theta = \kappa S_{\theta\theta} \tag{5.249b}$$

where

$$\alpha = \varepsilon T = \varepsilon \left(\frac{R}{g\beta\overline{S}}\right)^{1/2}, \lambda = \frac{2ET}{r} = \frac{2E}{r}\left(\frac{R}{g\beta\overline{S}}\right)^{1/2}, \kappa = \frac{KT}{R^2} = \frac{K}{R}\left(\frac{1}{gR\beta\overline{S}}\right)^{1/2} \tag{5.250}$$

are the nondimensional parameters of the model. The interpretation of T in Eqn. (5.248) is as a loop "flushing" time, as determined by buoyancy anomalies; thus, α represents the ratio of flushing time to viscous decay time, λ the ratio of flushing time to a tube "filling" time (due to freshwater flux), and κ the ratio of flushing time to salt diffusion time. In view of these interpretations, we are interested in the limit of small values of α, λ, and κ. Note that Eqn. (5.249b) implies a normalization condition for S, i.e., $\int_0^{2\pi} S d\theta = 2\pi$.

System behavior

We now consider the solutions of the coupled nonlinear, integro-differential equation set (5.249). For the given initial conditions and parameter values, this set of equations can be solved by a Fourier spectral approach and a fourth-order Runge–Kutta algorithm in time stepping. An advantage of the spectral technique is that the torque in the momentum equation appears naturally as the coefficient of the lowest Fourier mode. Solutions from this simple model can help us understand the oscillations of more complicated climate models; hence, we will focus on steady solutions and limit cycle oscillations. The nature of the stationary solutions depends critically on the relative significance of the parameters α, λ, and κ.

When $\kappa = 0$, the steady solution of Eqn. (5.249) is straightforward; i.e., for arbitrary α and λ:

$$S = (1 - \lambda/\Omega)^{1/2} (1 + \lambda \sin\theta/\Omega)^{-1} \tag{5.251}$$

where

$$\Omega = \pm\sqrt{\frac{\lambda}{\alpha(2 - \alpha\lambda)}} \tag{5.252}$$

is the mean angular velocity.

A much more interesting case is that of non-zero κ, because diffusion exists naturally and is also necessary for numerical stability. From the preceding discussion, the regime of most physical interest involves small values of α and λ, and we shall exploit this by assuming that $\alpha \approx \lambda \ll 1$. When κ is non-zero, the solutions for Ω and S can simply be obtained by employing a perturbation approach which makes use of these parametric restrictions. Indeed, a convenient classification of the solutions to Eqn. (5.249) can be found if we consider κ larger than, comparable to, and smaller than α and λ, respectively.

The solution of this loop model can be obtained by expanding the salinity in a Fourier series

$$S(\theta) = 1 + \sum_1^{\infty}(2a_n \sin n\theta + 2b_n \cos_n \theta) \tag{5.253}$$

If we are interested only in the long-term stability of the model, only the first two modes, $\sin\theta$ and $\cos\theta$, are required. For example, the corresponding equations for the

VSF loop are

$$
\begin{aligned}
\dot{\Omega} &= \alpha\Omega - a_1 \\
\dot{a}_1 &= \Omega b_1 - \kappa a_1 \\
\dot{b}_1 &= \Omega a_1 - \kappa b_1 - \lambda/2
\end{aligned}
\tag{5.254}
$$

Using a linear transformation, we can rewrite Eqn. (5.254) as a simple set of ordinary differential equations, the water-wheel equation (Huang and Dewar, 1996):

$$
\begin{aligned}
\dot{x} &= \alpha(y - x) \\
\dot{y} &= rx - \kappa y - xz \\
\dot{z} &= xy - \kappa z
\end{aligned}
\tag{5.255}
$$

This system is very similar to the well-known Lorenz equation:

$$
\begin{aligned}
\dot{x} &= \alpha(y - x) \\
\dot{y} &= rx - y - xz \\
\dot{z} &= xy - bz
\end{aligned}
\tag{5.256}
$$

If $\kappa = 1$, the water-wheel equation and the VSF loop model under the present simple forcing are equivalent to the Lorenz system with $b = 1$. The model's behavior is very similar to that of the well-known Lorenz model (Sparrow, 1982) (Figs. 5.136 and 5.137).

Both the water-wheel model and the loop model are idealizations of the much more complicated three-dimensional thermohaline circulation in the oceans. Thus, the above described chaotic behavior of such simplified one-dimensional models suggests that thermohaline circulation in the world's oceans is as complicated as the chaotic motions in the atmospheric circulation.

5.4.4 Two-dimensional thermohaline circulation

Thermal circulation driven by horizontal differential heating

The next level of complexity in the simulation of the thermohaline circulation is the thermally forced circulation in a two-dimensional rectangular tank. The laboratory study of thermal circulation in a quasi-two-dimensional tank has been discussed in Chapter 3 in connection with the energetics of the oceanic circulation. In this section, we focus on the numerical study of the two-dimensional thermohaline circulation.

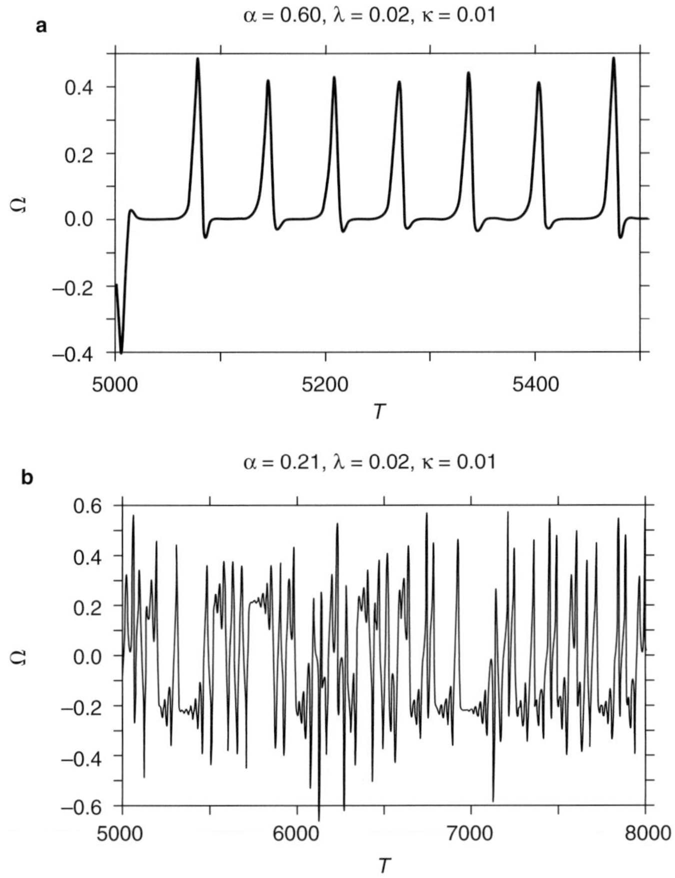

Fig. 5.136 **a**, **b** Chaotic solutions for the model under the natural boundary conditions (Huang and Dewar, 1996).

The model used is based on the Boussinesq approximations with the effect of rotation omitted (Sun and Sun, 2007). The basic equations are:

$$\vec{u}_t + \vec{u} \cdot \Delta \vec{u} = -\nabla p + \alpha g T \cdot \vec{k} + \nu \nabla^2 \vec{u} \tag{5.257}$$

$$\nabla \cdot \vec{u} = 0 \tag{5.258}$$

$$T_t + \vec{u} \cdot \nabla T = \kappa \nabla^2 T \tag{5.259}$$

where $\vec{u} = (u, w)$ is the velocity vector. Since the flow field is incompressible, the velocity is divergenceless, and a streamfunction exists

$$u = -\psi_z, \quad w = \psi_x \tag{5.260}$$

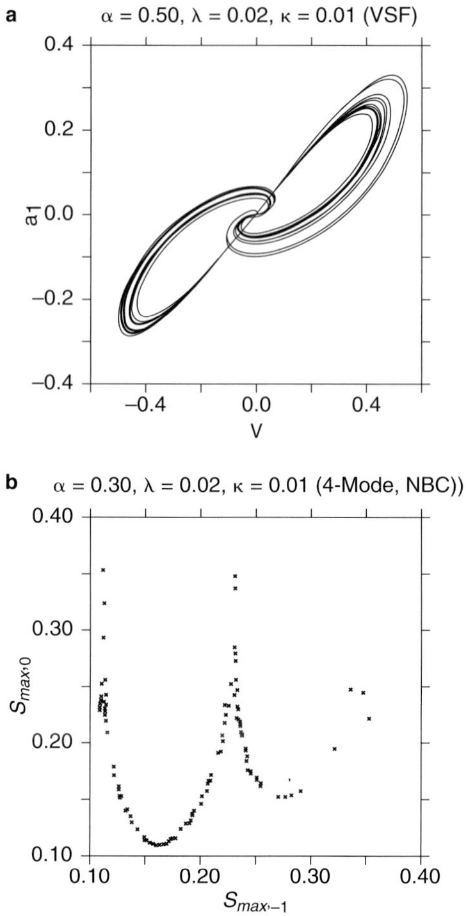

Fig. 5.137 **a** Butterfly map, amplitude of the $\sin\theta$ mode a_1 vs. angular velocity Ω, generated from the two-mode model based on VSF; **b** Poincaré return map for the four-mode model based on NBC (natural boundary condition), where $S_{max,-1}$ and $S_{max,0}$ are the salinity maxima of the previous and present quasi-cycles (Huang and Dewar, 1996).

For a tank of width L and depth H, we introduce the following nondimensional variables:

$$(x, z) = (Lx', Hz'), \quad T = \Delta T T', \quad t = (H^2/\kappa)t', \quad \psi = \kappa \psi' \tag{5.261}$$

where ΔT is the temperature scale of the boundary thermal forcing. The basic equations are reduced to the following nondimensional forms, after dropping primes:

$$\frac{\partial}{\partial t}\nabla^2\psi + J\left(\psi, \nabla^2\psi\right) = \Pr Ra \partial T/\partial x + \Pr \nabla^4 \psi \tag{5.262}$$

$$\frac{\partial T}{\partial t} + J(\psi, T) = \nabla^2 T \tag{5.263}$$

This problem has three nondimensional numbers, i.e., the Rayleigh number, the Prandtl number, and the aspect ratio:

$$Ra = \frac{\alpha g \Delta T H^3}{\kappa \nu}, \quad \Pr = \frac{\nu}{\kappa}, \quad \sigma = \frac{L}{H} \tag{5.264}$$

A typical thermal boundary condition is to specify the temperature, $T = \sin(\pi x/2)$, at the upper or lower boundary and a thermal insulation condition at other boundaries.

For seawater with temperature and salinity in the normal range, $\alpha \simeq 2 \times 10^{-4}/°C, \kappa \simeq 1.5 \times 10^{-7} m^2/s$, and $\nu \simeq 1.2 \times 10^{-6} m^2/s$; therefore, $\Pr \simeq 8$. For a tank with 0.2 m dimensions and a temperature difference of 10°C, the corresponding Rayleigh number is about 10^9. Although this set of equations looks relatively easy to solve, careful consideration reminds us that, in order to obtain an accurate steady solution, the grid used in calculating the numerical solution must be fine enough to resolve the turbulence. As shown in Figure 5.138, the number of grid points in each dimension should be linearly proportional to the actual length of the tank. If the resolution is not fine enough, some artificial numerical instability may appear. Physically, this indicates that the resolution of the numerical model must be fine enough to resolve turbulent motions of the smallest energetic scale.

For the range of Rayleigh numbers between 10^7 and 10^{10} and $\Pr = 8$, the overturning cell from the numerical solution appears as a partial cell, i.e., most of the flow is confined to a thin boundary layer adjacent to the boundary with thermal forcing (Fig. 5.139), which is consistent with the laboratory experimental results obtained by W. Wang and Huang (2005).

Numerical experiments indicate that the cell depth declines approximately in proportion to $Ra^{-0.2}$ (Fig. 5.140), consistent with the laboratory experiments by W. Wang and Huang (2005). Thus, for a high Rayleigh number, the overturning cell is much more confined to the boundary where thermal forcing is applied.

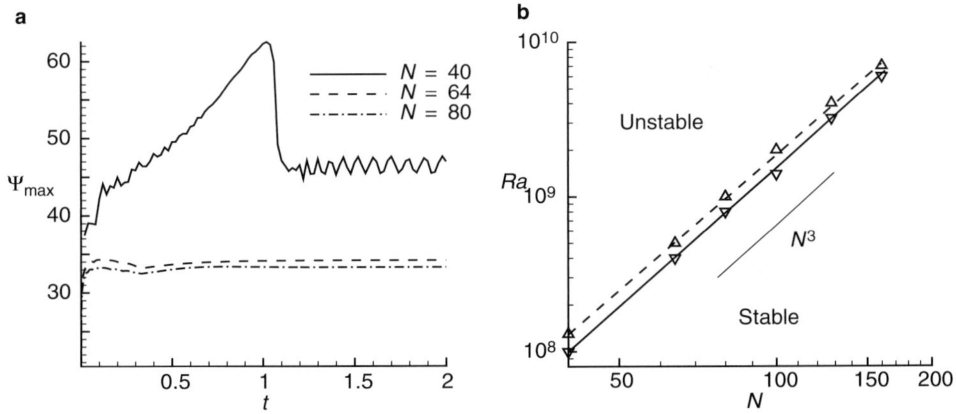

Fig. 5.138 **a, b** Relation between grid resolution and Rayleigh number, with the *circles* and *solid line* indicating the stable solutions, and the *squares* and *dashed line* indicating the unstable solutions; the *short line* indicates the relation $Ra \sim N^3$ (courtesy of Liang Sun).

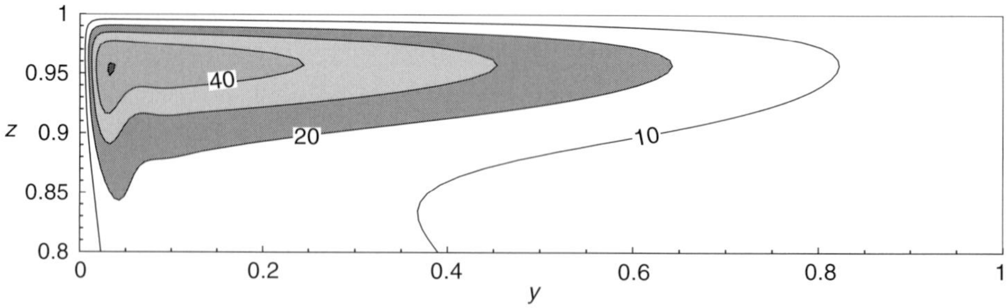

Fig. 5.139 Overturning circulation driven by horizontal differential heating applied to the upper boundary, Pr $= 8$, $Ra = 5 \times 10^8$, and aspect ratio $\sigma = 1$ (courtesy of Liang Sun).

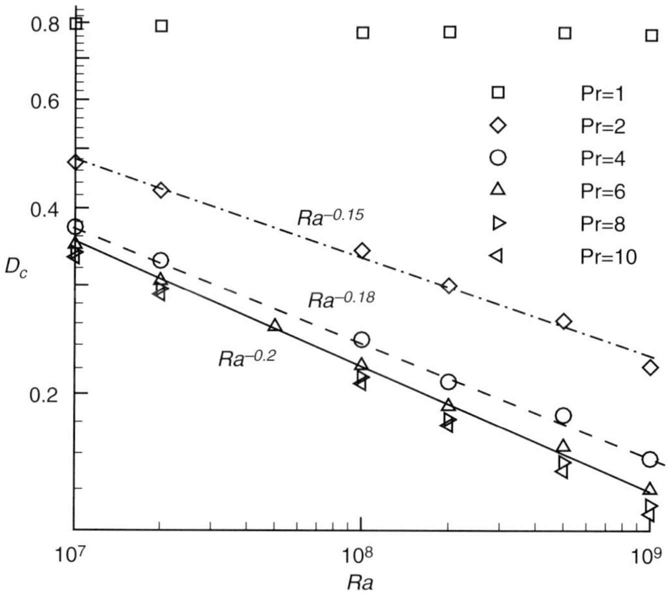

Fig. 5.140 The cell thickness as a function of the Rayleigh number (courtesy of Liang Sun).

The numerical study of flow forced by horizontal differential heating in such a two-dimensional tank is much easier than laboratory experiments, and can provide very useful insights into the physics of the thermal circulation. However, some essential differences exist between such a model and the circulation in the oceans.

First and foremost is the huge difference in Rayleigh number between the two. As the Rayleigh number increases, the flow may go through transitions from one dynamical regime to the next. For example, W. Wang and Huang (2005) found a regime transition around $Ra \simeq 5 \times 10^8$. Other transition regimes may exist at much higher Rayleigh numbers. For example, Roche *et al.* (2001) reported a transition in the classical Rayleigh–Benard convection through experiments with low-temperature liquid helium in the vicinity of $Ra \simeq 10^{12}$. Whether the flow under the horizontal differential heating has similar transition remains unknown. Assuming that the ocean basin has a horizontal dimension on the order of 10^7 m, the corresponding Rayleigh number based on the horizontal dimension is on the order of 10^{30}. Numerical simulation of turbulence with such a Rayleigh number remains a major challenge. Furthermore, even if we could carry out physical experiments with such a huge dimension, the character of the resulting flow might be completely different from what we have seen under the low range of Rayleigh number.

Second, many of the physical processes that exist in the oceans have been excluded from such a two-dimensional model, such as rotation and external sources of mechanical energy that support a diffusivity over 100 times stronger than the molecular diffusivity. Therefore, the results obtained from such two-dimensional model must be interpreted with caution.

Zonally averaged model for the thermohaline circulation

A zonally averaged model

The description of complicated three-dimensional circulation in the ocean can be greatly simplified and studied if a zonally averaged two-dimensional numerical model is utilized. Consider an ocean of width L and depth H; the steady circulation is represented by the following zonally averaged momentum equations:

$$-f\overline{v} = \frac{p_e - p_w}{\rho_0 L} + A_h \overline{u}_{zz} \tag{5.265a}$$

$$-f\overline{u} = -\frac{1}{\rho_0} + \overline{p}_y + A_h \overline{v}_{zz} \tag{5.265b}$$

where the overbars indicate the zonal mean and (p_e, p_w) are pressure at the eastern and western boundaries. Eliminating $\overline{u}$, we obtain an equation for $\overline{v}$:

$$f^2\overline{v} + A_h^2 \overline{v}_{zzzz} = \frac{f}{\rho_0 L}(p_e - p_w) + \frac{A_h}{\rho_0}\overline{p}_{yzz} \tag{5.266}$$

In order to solve this equation we need a parameterization of the eastern–western pressure term; Marotzke *et al.* (1988) suggested that this term should be parameterized in terms of the north–south pressure gradient, but they did not actually carry out such a parameterization. Instead, they took a short-cut by assuming a different balance:

$$A^* \overline{v}_{zz} = \frac{1}{\rho_0}\overline{p}_y \tag{5.267}$$

Note that this is equivalent to setting $f \equiv 0$, so the Coriolis force was excluded in their formulation. In the zonally averaged model, the continuity equation is

$$\overline{v}_y + \overline{w}_z = 0 \tag{5.268}$$

Thus, a streamfunction can be introduced

$$\psi_y = -\overline{w}, \ \psi_z = \overline{v} \tag{5.269}$$

Assuming the equation of state is linear

$$\rho = \rho_0 (1 - \alpha T + \beta S) \tag{5.270}$$

The hydrostatic relation is

$$\overline{p}_z = -\overline{\rho} g \tag{5.271}$$

By cross-differentiating Eqns. (5.267) and (5.271) twice and using Eqn. (5.270), we obtain a single equation for ψ:

$$\psi_{zzzz} = \frac{g}{A^*}(-\alpha T_y + \beta S_y) \tag{5.272}$$

There are two prognostic equations for the temperature and salinity

$$T_t - \psi_z T_y + \psi_y T_z = \kappa T_{zz} + C(T) \tag{5.273}$$

$$S_t - \psi_z S_y + \psi_y S_z = \kappa S_{zz} + C(S) \tag{5.274}$$

where $C(T)$ and $C(S)$ indicate the convective mixing due to overturning. The boundary conditions for this equation are

$$\begin{aligned} &\psi = 0, \ \psi_{zz} = 0, \ \text{at } z = 0 \\ &\psi = 0, \ \psi_z = 0, \ \text{at } z = -H \\ &\psi = 0, \ \text{at } y = \pm L \end{aligned} \tag{5.275}$$

These boundary conditions mean no mass flux through the bottom and side wall, no slip at the bottom, and no wind stress on the upper surface.

The boundary conditions for temperature and salinity are a no-flux condition at the solid boundaries

$$T_z = S_z = 0, \ at \ z = -H; \ T_y = S_y = 0, \ at \ y = \pm L \tag{5.276}$$

On the upper surface, the model is subject to the Dirichlet boundary conditions

$$T = T_0 \left[1 + \cos\left(\frac{y\pi}{L}\right)\right], \ S = S_0 \left[1 + \cos\left(\frac{y\pi}{L}\right)\right] \tag{5.277}$$

or the mixed boundary conditions

$$T = T_0 \left[1 + \cos\left(\frac{y\pi}{L}\right)\right], \ \kappa S_z = Q_s(y) \tag{5.278}$$

Early in the numerical modeling of the thermohaline circulation, both temperature and salinity were subject to strong relaxation conditions. Under such strong relaxation conditions, it was rather difficult to find multiple solutions. The breakthrough took place when people started to try different types of boundary condition.

F. Bryan's technique

Notable among the efforts made to search for multiple solutions and catastrophic change in the thermohaline circulation, F. Bryan developed a widely used technique, which can be summarized as follows:

First, run the model to a quasi-equilibrium state under the relaxation conditions for both temperature and salinity, $\frac{dT}{dt} = \Gamma(T^* - T)$, $\frac{dS}{dt} = \Gamma(S^* - S)$.

Second, when the system reaches the quasi-equilibrium, the equivalent salt flux required for maintaining the salt distribution is diagnosed, $S_f = \Gamma(S^* - S)$.

Third, this salt flux is used as the forcing for the salinity in the second stage of the numerical experiments. The model is restarted from the quasi-equilibrium state and run under a new set of boundary conditions for temperature and salinity, $\frac{dT}{dt} = \Gamma(T^* - T)$, $\frac{dS}{dt} = S_f$, i.e., a relaxation condition for temperature and a flux condition for salinity. Since the relaxation condition still applies to the temperature field, this new set of boundary conditions has been called the mixed boundary conditions. Note that the salt flux is now considered as fixed.

While the Bryan technique brought about quite a few enlightening results, there exist some potential problems to be dealt with in this kind of modeling.

- *Relaxation for salinity is unphysical*, so that the choice of this Γ coefficient is artificial anyway. Tziperman *et al.* (1994) suggested that a small value of Γ should be used. However, this does not really solve the problem.
- *The virtual salt flux used in the model is unphysical.* In particular, the salt flux diagnosed in the model has no physical meaning when the model is still in the process of reaching quasi-equilibrium. In fact, there is no salt flux across the air–sea interface; thus, it is more appropriate to use the natural boundary condition for salinity.

The basic steps in numerical experiments

To trigger the transition to the other states, sometimes an initial perturbation is needed. As has been found by trial and error, salinity perturbations at high latitudes are the most efficient way to initiate the transition. Occasionally, the symmetric state is unstable to very small perturbations, so the solution will drift away without ever adding on any initial perturbations. The most crucial results obtained from such a two-dimensional model are as follows.

- *A positive salinity anomaly* added at high latitudes can lead to a slow evolution toward the one-cell solution (pole–pole). The basic mechanism of this instability is the advective feedback discussed by Walin (1985).
- *A negative salinity anomaly* added at high latitudes can lead to a catastrophic change in the meridional circulation. Because water at high latitudes becomes too fresh to sink, the deepwater formation at one polar basin, where the negative salinity perturbations are added, is suddenly interrupted. As a result, the two-cell circulation pattern suddenly switches to the one-cell circulation. This mechanism is called convective feedback, which has a much shorter time scale.

Halocline catastrophe in a single-hemisphere model

Marotzke (1989) discussed a model with a single hemisphere. After switching into the mixed boundary condition and applying negative salinity perturbation at high latitudes, the deepwater formation was suddenly interrupted, and the polar halocline catastrophe set in. The model shifted to a mode that is saline-controlled. In fact, deep water is formed at the equator and sinks. This mode of circulation is quite slow compared with the thermal mode, with sinking at high latitudes.

The most interesting aspect of this mode is its instability in connection with the high-latitude convective overturning. After many thousands of years, deep water in the polar ocean becomes warm and salty owing to a slow mixing process. The cold and fresh water lying on top of the warm and salty water at high latitudes is potentially very unstable. At a certain stage, small perturbations can trigger catastrophic overturning, during which the deep water comes to the surface where it loses its heat constant quickly, but keeps its salinity unchanged. As a result, the water column becomes gravitationally unstable and strong overturning sets in. This is the most energetic phase of the circulation, which is called *flushing*. At the end of flushing, the polar halocline gradually sets in, and the whole cycle is repeated.

Note that the precondition for flushing is the existence of warm and salty deep water in the polar ocean, which definitely does not exist in the present-day oceans. Whether such a phenomenon could have happened in the past remains to be examined very carefully. Flushing will be discussed in detail in Section 5.4.7.

The modified zonally averaged model

As an improvement on the model by Marotzke *et al.* (1988), Wright and Stocker (1991) introduced a parameterization of the zonal pressure gradient term. In Cartesian coordinates, their argument is as follows. Assuming semi-geostrophy,

$$f\overline{v} = -\frac{p_e - p_w}{\rho_0 D} \tag{5.279a}$$

$$f\overline{u} = -\frac{1}{\rho_0}p_y - R\overline{v} \tag{5.279b}$$

Eliminating $\overline{v}$, one obtains

$$\frac{p_e - p_w}{D} = \frac{f}{R}\left(1 - \frac{\overline{u}}{\overline{u}_G}\right)\frac{\partial \overline{p}}{\partial y} = -\varepsilon_1 \frac{\partial \overline{p}}{\partial y} \tag{5.280}$$

where

$$\overline{u}_G = -\frac{1}{\rho_0 f}\frac{\partial \overline{p}}{\partial y} \tag{5.281}$$

is the zonally averaged geostrophic component of the total zonal velocity. Substituting Eqn. (5.280) back into Eqn. (5.279a), we obtain

$$\overline{v} = \frac{\varepsilon_1}{f \rho_0} \frac{\partial \overline{p}}{\partial y} \tag{5.282}$$

Note that this relation is similar to the assumptions used in the box model, namely that the meridional velocity is proportional to the meridional pressure gradient, but is different from the parameterization used by Marotzke *et al.* (1988). The coefficient ε_1 is diagnosed by averaging data collected from a three-dimensional oceanic general circulation model (OGCM); it varies from 0.1 to 0.3, depending on the exact formulation of the model. With this parameterization, their model can provide much more accurate solutions that look very much like the solutions obtained from a three-dimensional model by zonal averaging. Their model has been used to study the global ocean circulation and to simulate the Younger Dryas event, with very encouraging results.

The limitation of the 2-D models

Although two-dimensional models have the merits of simple formulation and easy application, one should be aware of their intrinsic limitations. First, it is difficult to include wind stress and the associated horizontal advection in such models. Second, velocity in the models is a diagnostic variable. Finally, a zonally averaged model lacks the capability of simulating the third dimension; thus, anything associated with the horizontal gyre is not explicitly simulated. As a result, the model cannot be used for simulating the decadal/interdecadal variability of the thermohaline circulation, which is essentially a three-dimensional phenomenon.

5.4.5 Thermal circulation in a three-dimensional basin

To understand the complex structure of three-dimensional thermohaline circulation in the ocean, one can conceptually separate it into the thermal and haline components. The quasi-steady-state solution of haline circulation is presented in Section 5.3.3. Since the haline contraction coefficient is nearly constant, the mean sea level does not change much during the spin-up of the haline circulation.

On the other hand, owing to the dependence of the thermal expansion coefficient on the temperature, the mean sea level changes greatly during the spin-up of the thermal circulation. As an example for the thermohaline circulation in the ocean, in this section we examine the thermal circulation in an idealized three-dimensional basin, using a mass-conserving numerical model. Since it conserves mass, it can keep tracking the time evolution of both the sea level and bottom pressure; thus, large changes in sea level during spin-up can be accurately simulated.

Model set-up

The model ocean is 5 km deep, with a modest horizontal resolution of $2^\circ \times 2^\circ$ and 30 vertical levels. It is started from an initial state of rest with the sea level at $z = 0$ and uniform temperature of either 25°C (warm start) or 0°C (cold start). The salinity is set to a uniform value of 35. The model is subject to a relaxation condition for temperature, with a reference temperature declining from 25°C at the southern boundary (4° N) to 0°C at the northern boundary (64° N). The vertical diffusivity is set to 0.3×10^{-4} m^2/s uniformly. Choosing the southern boundary away from the equatorial band avoids the special dynamics associated with the vanishing Coriolis parameter near the equator.

In both of the two cases (cold start and warm start), the basin-mean bottom pressure p_b remains unchanged because the model conserves mass. Owing to the difference in initial temperature, the bottom pressure of the model started from a cold initial state is higher than that started from a warm initial state, $\overline{p_{b,25}} < \overline{p_{b,0}}$. However, when the model runs to a quasi-equilibrium state, the solutions obtained from these two cases are virtually indistinguishable in terms of the nondimensional pressure coordinate $p' = p/\overline{p}_b$. For convenience, results are presented here in terms of a nominal z-coordinate, assuming the sea level is at $z = 0$.

Time evolution of the thermal circulation

The time evolution of the model is sensitively dependent on the initial state of the model; however, after a 5,000-year run, the model reaches a quasi-equilibrium state which is virtually independent of the initial temperature distribution. For the case of a warm start, over 5,000 years the water is gradually cooled down, as indicated by the thin line (Fig. 5.141a). As water is cooled down, its density increases, and the mean sea level declines. As a result, at the end of 5,000 years, the mean sea level is more than 16 m lower than that in the initial state (Fig. 5.141b). Decline in sea level does not occur at a uniform speed in the basin. In fact, mean sea surface height along the southern boundary declines more slowly than that along the northern boundary. Thus, there exists a large difference in mean sea surface height between the southern and northern boundaries (Fig. 5.141c). When the model reaches a quasi-steady state, this difference is about 0.57 m.

For the case of a warm start, the overturning circulation rate varies greatly during the spin-up of the circulation (Fig. 5.141d). Although a strong meridional overturning circulation is apparently directly linked to cooling at high latitudes, cooling itself does not generate the mechanical energy needed for sustaining the circulation. As discussed in Chapter 3, however, cooling actually reduces the total amount of mechanical energy in the mean state of the ocean. What really happens during cooling is that cooling increases the density of water parcels and reduces the sea surface height, as indicated by the large drop of the mean sea surface level in the model during the cooling. The downward migration of the center of mass leads to the release of a large amount of GPE, although only a small part of this energy released can be converted into the kinetic energy of the mean flow, leaving most of it lost to turbulence and internal waves.

The spin-up in the case of a cold start goes through a very different path compared with that of a warm start. All the major indexes of the circulation change rather slowly and smoothly, without the dramatic ups and downs that appear in the case of a warm start. The mean temperature in the basin increases quite slowly (heavy line in Fig. 5.141a), rather than the rapid decline of the basin mean temperature that appears in the case of a warm

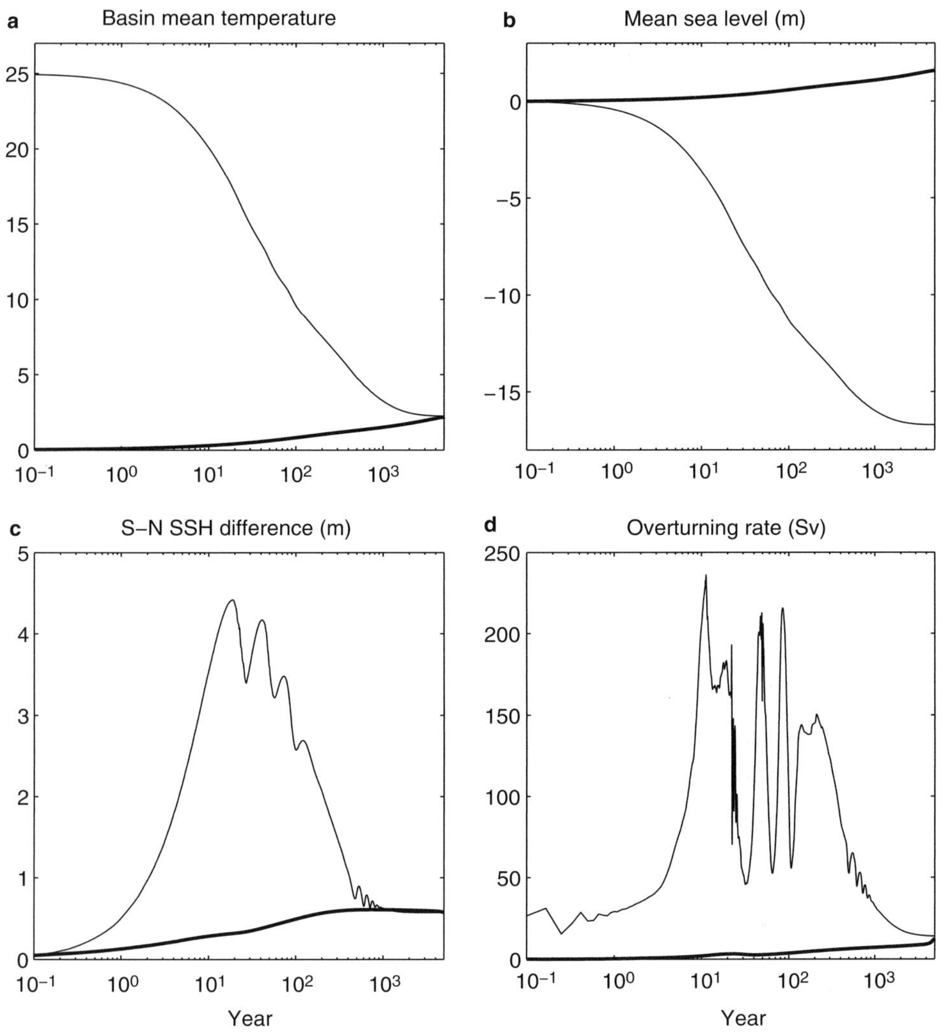

Fig. 5.141 **a–d** Spin-up of a thermal circulation, *heavy lines* for a cold start and *thin lines* for a warm start.

start. Over the same time period, the mean sea level increases slightly, less than 2 m (heavy line in Fig. 5.141b), which is much smaller than the sea level decline of more than 16 m for the case of a warm start. As for the difference in the north–south sea surface height and the meridional overturning rate in the case of a cold start, both increase slowly and monotonically, as shown by the heavy lines in Figure 5.141c, d.

As discussed in Section 3.7, in the case of a cold start there is virtually no convective adjustment or new deepwater formation at the beginning of the simulation because the initial temperature is set to 0°C. The system gradually accumulates its energy, which can be used to maintain the circulation primarily from GPE generated through vertical mixing. Under surface heating, the upper ocean water at low latitudes is warmed up, and sea level there becomes higher than that at high latitudes; through vertical mixing, the GPE of the system increases.

Driven by the meridional pressure gradient at sea level, warm surface water flows to high latitudes where cooling induces convective overturning, causing GPE to be partially converted into kinetic energy which drives the overturning circulation. However, this is a much less dramatic process, as compared with the sudden onset of cooling in the case of a warm start.

By comparing these two spin-up processes, the role of the mechanical energy balance in setting up the thermal circulation in the ocean is clearly demonstrated. We see that surface thermal forcing itself cannot create much mechanical energy. Instead, cooling at high latitudes constitutes a major sink of GPE, although the GPE lost through cooling can be partially converted into kinetic energy of the mean state, maintaining a strong circulation. However, the abnormally strong circulation during the sudden cooling cannot be sustained, and the circulation eventually turns to a more gentle state that can be sustained by the energy supplied by vertical mixing.

Three-dimensional structure of the thermal circulation

The difference in mean sea surface height between the southern and northern boundaries reflects the difference in temperature distribution in the upper ocean. Since the model is subject to a strong temperature relaxation condition on the upper surface, temperature in the surface layer is almost a linear function of latitude (Fig. 5.142a). The noticeable departure from the zonal distribution appears along the western boundary, which indicates a strong northward flow there.

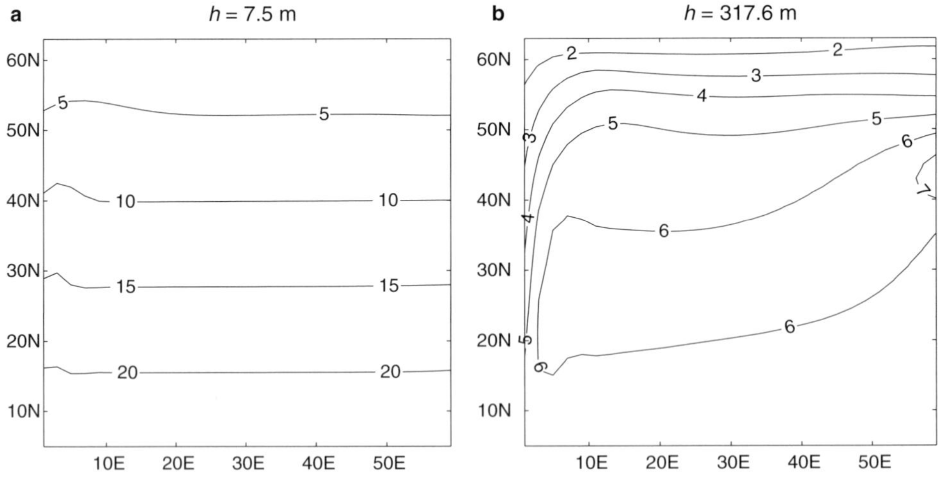

Fig. 5.142 Temperature distribution in the final steady state: **a** $h = 7.5$ m; **b** $h = 317.6$ m.

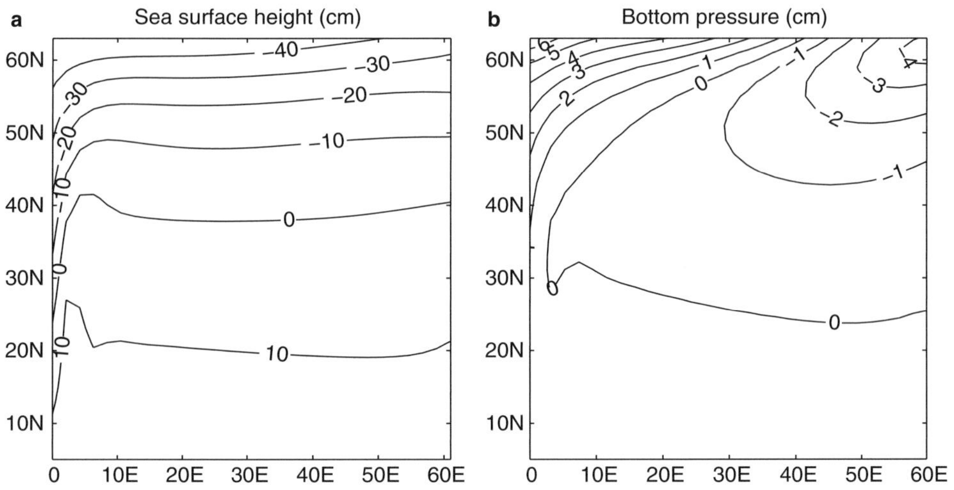

Fig. 5.143 Sea surface height (cm) anomaly (**a**) and bottom pressure anomaly (cm) (**b**).

Below the upper layers, temperature distribution is obviously non-zonal. At 317 m depth, a temperature maximum appears in the mid-latitude band, with lower temperatures appearing in both the northern and southern parts of the basin (Fig. 5.142b). The relatively low temperature in the southern part of the basin indicates the upwelling of cold water from the abyss.

In the final steady state, there is a band of low sea surface height along the northern boundary, with a global minimum in the northwest corner (Fig. 5.143a). The meridional difference in sea surface height is directly responsible for driving the poleward flow in the surface layer, as discussed in Section 5.4.1. The integrated volume flux in the upper 980 m indicates a very strong western boundary current moving northward that continues as a strong northern boundary current feeding the sinking along the northern boundary (Fig. 5.143a). In the basin interior, there is a relatively weak anticyclonic circulation, which is apparently due to the compression of cold water upwelling in the middle basin.

Although, in the northern half of the basin, the pattern of thermal circulation in the upper ocean is quite similar to that of haline circulation shown in Figure 5.111, the patterns of circulation in the southern half of the basin in these two cases are quite different. Strong sinking in the southeast corner of the basin in the case of haline circulation induces very strong current in that part of the basin, causing dramatically different surface flow patterns in these two cases.

On the sea floor, the pressure distribution has a pattern completely different from the sea surface height map (Fig. 5.143b). Since the bottom pressure has great absolute value, it is more illustrative to just show the pressure deviation from the mean bottom pressure, in units of equivalent water depth. As indicated in Figure 5.143b, along the western boundary there is a large meridional pressure gradient, which is a clear indication of the existence of a strong equatorward deep western boundary current. The bottom pressure pattern shown in Figure 5.143b suggests a cyclonic-like circulation in the abyssal interior.

As for the circulation with thermal forcing alone, no detectable barotropic circulation exists; this is also one of the major differences between the thermal circulation and the freshwater-forced circulation. Such differences primarily come from the fact that freshwater flux represents a source/sink of mass while the thermal forcing carries no surface source/sink of mass; thus, there is no barotropic flow in the model ocean.

The circulation in the upper ocean consists of two parts. First, a strong poleward western boundary current and an eastward northern boundary current exist, playing the role of transporting water toward the northeast corner of the model ocean. Second, a relatively slow clockwise circulation exists in the ocean interior, which is driven by the basin-wide upwelling (Fig. 5.144a).

In the meantime, the pattern integrated for the depth below the top kilometer shows itself to be exactly opposite to the pattern in the upper ocean. In particular, there are a strong westward northern boundary current and a southward western boundary current in the deep ocean, which transport water mass from the deep source in the northeast corner of the model ocean, as shown in Figure 5.144b. Nevertheless, the horizontal circulation below the surface layer is much weaker than that on the surface layer.

Perhaps the most noteworthy feature in the horizontal velocity map is the strong western boundary current. As discussed in Chapter 4, western boundary currents play a vital role in balancing mass, vorticity, and mechanical energy. Strong western boundary currents can be identified from the meridional map of the northward velocity along a section next to the western boundary (Fig. 5.145).

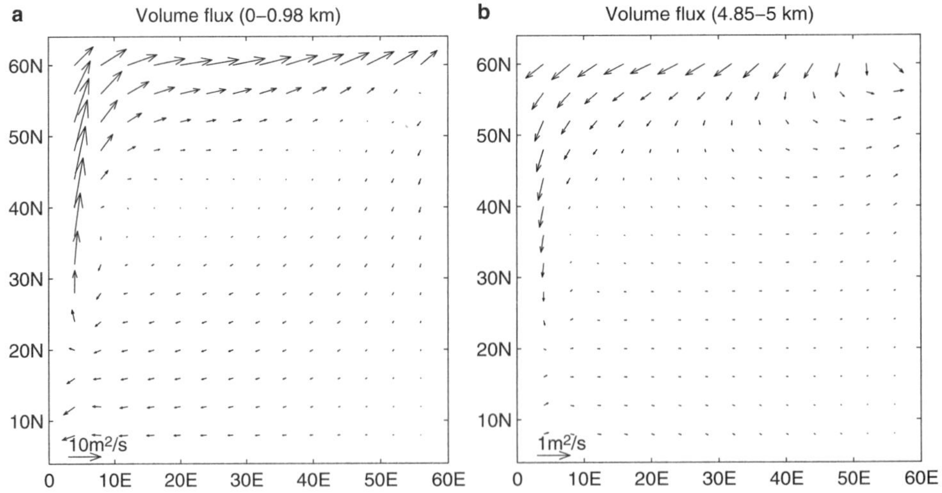

Fig. 5.144 Horizontal volume flux for **a** the upper ocean and **b** the bottom layer, in m^2/s.

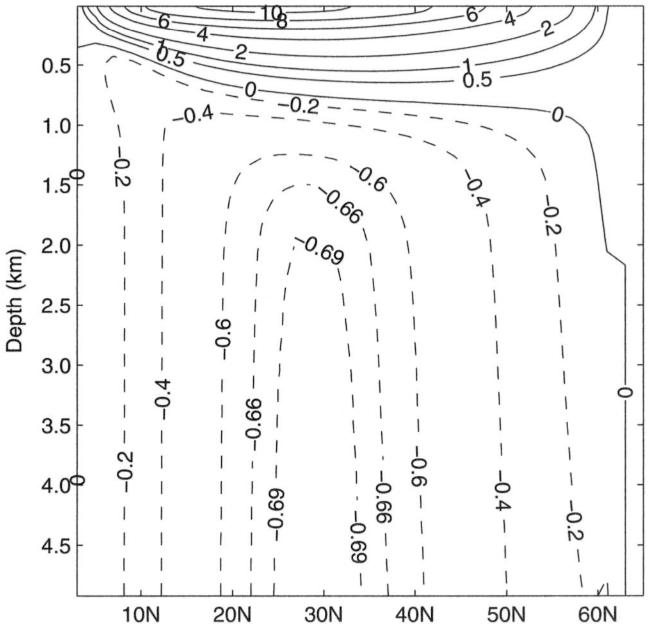

Fig. 5.145 Meridional velocity distribution along the western boundary of the model basin, in 0.01 m/s.

The meridional overturning circulation constitutes the most important component of the thermal circulation (Fig. 5.146). In turn, this consists of the following four branches:

1. a poleward flow of warm water in the upper ocean, as indicated by the relatively compact streamlines in the upper ocean;
2. a narrow sinking branch near the northern boundary;
3. a relatively broad and slow upwelling in the ocean interior;
4. a narrow deep western boundary current, as shown in Figures 5.144b and 5.145, which transports the cold bottom water equatorward.

The meridional overturning circulation plays the role of a conveyor bringing heat and other important tracers poleward and thus contributing to the machinery of the global circulation and climate. For the present case, the maximal poleward heat flux is about 0.12 PW (Fig. 5.147); a much lower value than that observed in the ocean, since no wind stress force is included in this model.

What mechanism is responsible for producing the main thermocline?

A crucial and long-debated issue relating to the existence of the main thermocline in the world's oceans is the key mechanism which produces the main thermocline. This problem can best be illustrated using a numerical model to simulate the thermal circulation in the ocean. To illustrate the point, we compare the results from two numerical experiments of

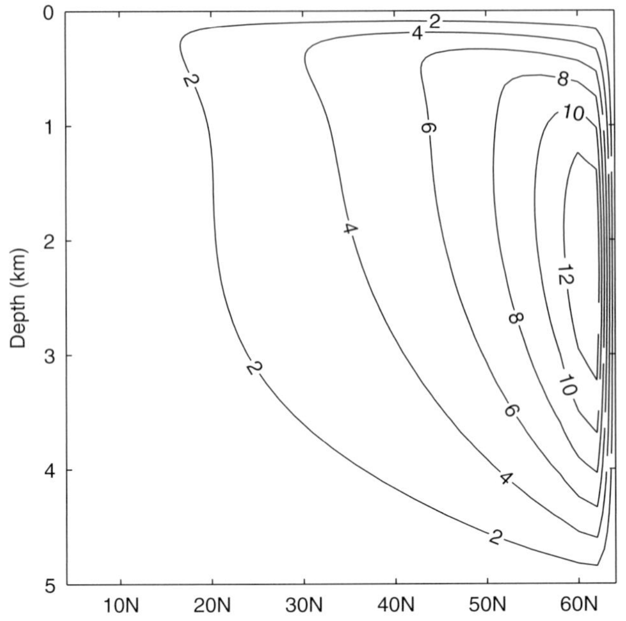

Fig. 5.146 Meridional overturning streamfunction, in Sv.

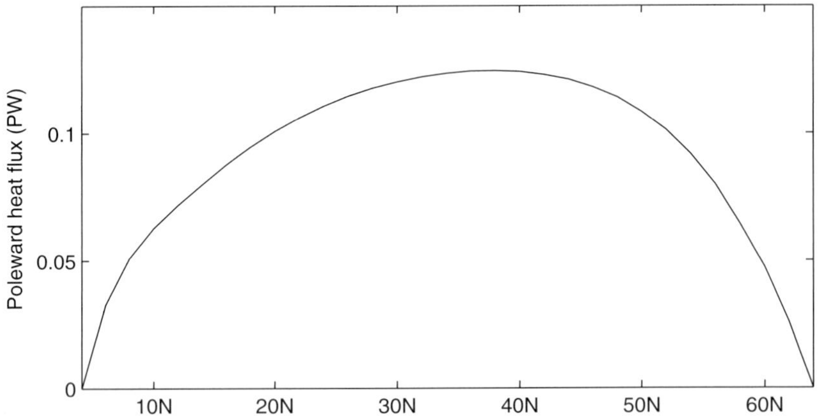

Fig. 5.147 Poleward heat flux associated with the meridional overturning circulation, as diagnosed from the numerical model.

thermal circulation under the same forcing conditions, with the only difference being that one is with wind stress and the other one with no wind stress.

As shown in Figure 5.148a, a purely thermal circulation, with uniform vertical mixing and a linear reference temperature profile applied at the upper surface, cannot produce a

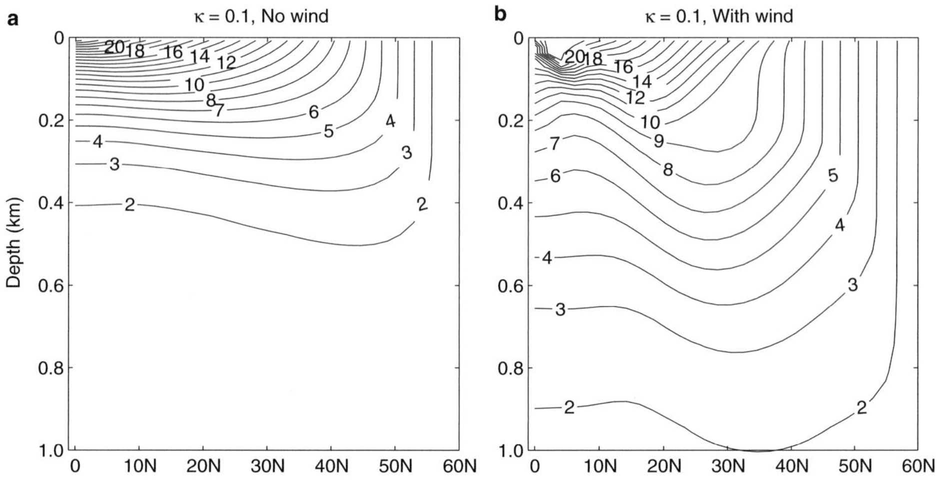

Fig. 5.148 Temperature distribution along the middle of the basin with a uniform vertical diffusivity of 0.1×10^{-4} m^2/s; **a** without wind stress; **b** with wind stress.

main thermocline in the subsurface region. In fact, the vertical stratification monotonically declines from the sea surface downward to the deep ocean. However, when a standard wind stress is applied, including equatorial easterlies, mid-latitude westerlies, and/or high-latitude easterlies, a main thermocline forms at the mid latitudes, around the depth of 400–500 m.

These two cases demonstrate that wind stress is the essential ingredient in the formation of the main thermocline in the subtropical ocean. The physical processes involved in the wind stress working are analyzed as follows. Wind stress drives the Ekman transport in the surface layer; the horizontal convergence of Ekman transport gives rise to Ekman pumping, which pushes the isopycnal layers in the upper ocean downward. Since isopycnal layers in the deep ocean are nearly flat, the bowl-shaped isopycnal layers in the subtropical ocean give rise to a strong subsurface maximum of vertical temperature gradient, i.e., the main thermocline.

5.4.6 Thermohaline circulation: multiple states and catastrophe

The first numerical confirmation of thermohaline catastrophe was reported by F. Bryan (1986). At that time it was remarkable to find a solution for the pole–pole circulation. Bryan started with a two-hemisphere model ocean with symmetric forcing and relaxation conditions for both temperature and salinity. A circulation symmetric with respect to the equator was established.

In order to find the asymmetric circulation, he tried many things to trigger the collapse of the symmetric circulation, including imposing a large positive salinity anomaly at high latitudes. It turned out that the symmetric circulation is insensitive to such perturbations. However, he found that if a relatively large negative salinity anomaly was imposed at the northern limit of the model ocean, a rapid response of the meridional circulation would take place. The collapse of the circulation is apparently caused by the interruption of deepwater formation at high northern latitudes due to the existence of the new halocline – a freshwater lid over the upper ocean. Since this kind of abrupt change is induced by the halocline, it is now termed the **halocline catastrophe**.

Thermohaline circulation under mixed boundary conditions

Early in the numerical modeling of the thermohaline circulation, both temperature and salinity were subject to strong relaxation conditions. Under such strong relaxation conditions, it was rather difficult to find multiple solutions. The breakthrough took place when people started to try different types of boundary condition.

As discussed above, a flux condition applied to the salinity balance is a major ingredient for multiple solutions and catastrophe. In addition, introducing the freshwater anomaly at high latitudes induces the thermohaline catastrophe. Although this newly formed layer of relatively fresher water may be subject to cooling, nevertheless its density cannot be reduced much because, at low temperature, seawater density is insensitive to changes in temperature. Consequently, deepwater formation at this pole is cut off; the interruption of deepwater formation at one pole sends a strong signal to the whole circulation system, and thermohaline catastrophe ensues.

With recent advances in numerical modeling, the halocline catastrophe has been reproduced through many different pathways, and in most numerical model experiments such initial impacts may no longer be needed since, in many cases, a model's solutions automatically drift away from the initial quasi-steady state and reach other states.

Halocline catastrophe according to F. Bryan

What is a halocline?

A halocline is defined as a layer where the vertical salinity gradient is locally maximal. Haloclines exist in the world's oceans, mostly as the interface between a layer of low-salinity water on the upper surface and the relatively salty water below. An outstanding permanent halocline exists in the subpolar gyre in the North Pacific Ocean and the Arctic Ocean (Fig. 5.149). Formation of the halocline in these locations is closely related to excess precipitation or local convergence of relatively fresh water in the upper ocean.

In addition, a non-permanent halocline can also be developed through various dynamical processes. For example, diurnal heating and rain events may substantially influence the turbulent motions in the upper ocean and lead to the formation of a halocline in the upper ocean (Soloviev and Lukas, 1996).

A freshwater cap may appear in the subpolar northern North Atlantic Ocean, which may be caused either by the import of freshwater from melting ice or by excessive precipitation over the subpolar basin. A halocline in the upper ocean can substantially cut down the air–sea heat flux, as discussed in Section 5.3.2. For example, a freshwater cap in the northern North Atlantic Ocean is considered as one of the major mechanisms capable of triggering a halocline catastrophe and an abrupt climate change in the Atlantic sector.

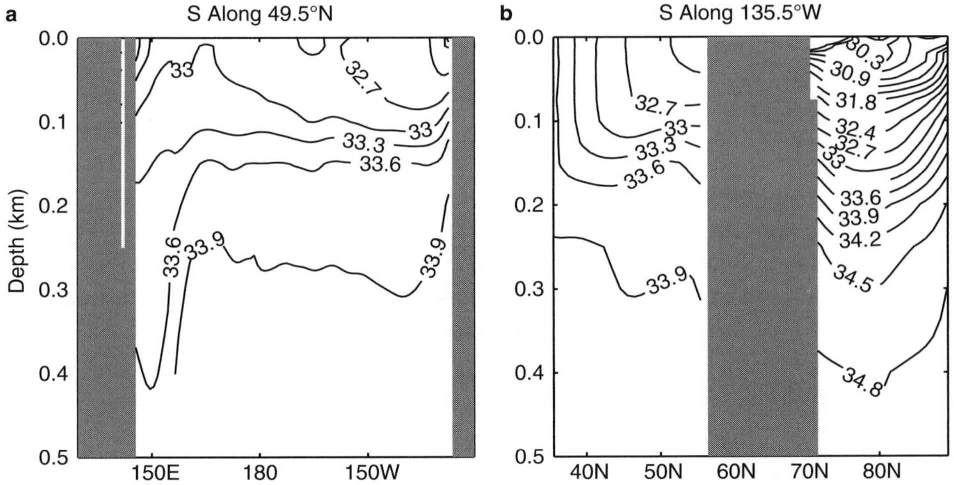

Fig. 5.149 **a**, **b** Halocline in the North Pacific Ocean based on climatology.

Halocline catastrophe from numerical experiments

Using a model for a two-hemisphere sector, including both poles, F. Bryan (1986) carried out the first set of successful numerical experiments demonstrating the idea of the halocline catastrophe. Under relaxation conditions for both temperature and salinity, the model was first run to a quasi-equilibrium of circulation which was symmetric with respect to the equator (Fig. 5.150a). However, when the model was restarted and run under the new mixed boundary conditions, such a symmetric circulation was potentially unstable. In fact, with the addition of negative salinity perturbation, the symmetric circulation became unstable and a polar halocline catastrophe set in. As a result, the symmetric mode of the thermohaline circulation collapsed and an asymmetric pole–pole circulation was established gradually (Fig. 5.150b). Thus, the entire solution changed accordingly, with the strength of the overturning cell in the Southern Hemisphere being nearly doubled.

The most important consequence of the halocline catastrophe is that the poleward heat flux is switched from a symmetric mode to an asymmetric mode, as shown in Figure 5.151. The specific asymmetric mode shown in Figure 5.151 suggests that after the halocline sets in, the poleward heat flux in the Northern Hemisphere nearly vanishes. Such a dramatic change in the oceanic poleward heat flux must bring about an abrupt and fundamental change in the climate conditions on Earth.

Implications of F. Bryan's numerical experiments

The thermohaline catastrophe has been reproduced in many numerical experiments based on the OGCM during the past two decades. The renewed interest in the thermohaline circulation is due to its close link with climate change. Oceans cover 70% of the Earth's

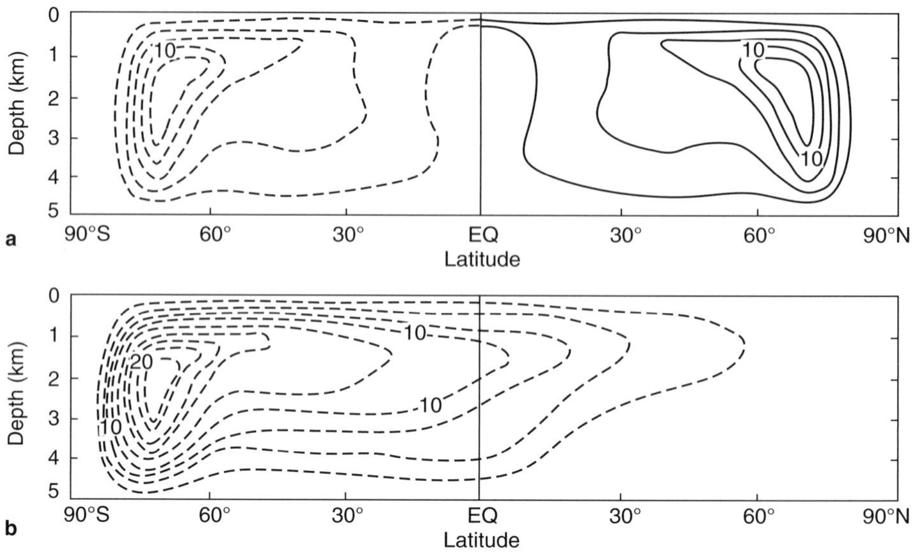

Fig. 5.150 Sketch of the two equilibrium states of an idealized North Atlantic Ocean under symmetric forcing for temperature and salinity (overturning rate in Sv): **a** the steady state symmetric with respect to the equator; **b** the asymmetric pole-to-pole mode set up after the halocline catastrophe (redrawn from F. Bryan, 1986).

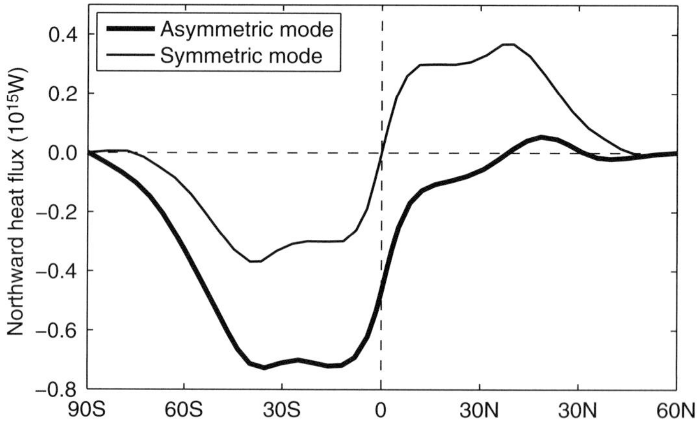

Fig. 5.151 Sketch of the northward heat flux associated with the two equilibrium states (redrawn from F. Bryan, 1986).

surface. Water has a much higher heat capacity than air; thus, the ocean is the dominating factor influencing the climate, and climate on a planet without oceans would be extremely different from that on a planet with oceans.

However, as shown in the numerical experiments by F. Bryan, this poleward heat flux can be changed substantially if the thermohaline circulation in the oceans collapses. The oceans are so immense that it is hard to imagine that the oceanic circulations could ever change. However, according to paleoclimate records, oceanic circulation has changed many times in the past.

The most remarkable example is the so-called Younger Dryas period, dated about 11,000 years ago. Abundant and strong evidence suggests that, during the last glaciations, the thermohaline circulation in the North Atlantic Ocean was interrupted due to the extensive ice cover over the northern North Atlantic Ocean. At the end of the last glaciations, the thermohaline circulation in the North Atlantic Ocean gradually returned to a situation rather similar to the present-day situation. However, during a very short period, the so-called Younger Dryas period, the thermohaline circulation in the North Atlantic Ocean was interrupted again. The interruption of the thermohaline circulation and the sudden cooling of the European and North American continents have been confirmed by many paleoclimate records.

It has been speculated that the large amount of freshwater from the ice melting flooded the subpolar basin of the North Atlantic Ocean. This cold and fresh water cap prevented evaporation and cut down the heat flux from the ocean to the atmosphere. As a result, deepwater formation was interrupted, and so was the thermohaline circulation. There have been many studies of the thermohaline circulation and halocline catastrophe.

At present, numerical models of the thermohaline circulation in the oceans cannot reproduce some of the basic structure of the circulation. For example, the thermocline reproduced in many numerical models is much too thick. The poleward heat flux in these models cannot match observations, and the strength of the thermohaline circulation is different from observations. Apparently, much work has still to be done before the numerical models can simulate the oceanic circulation properly.

The conveyor belt

The conveyor belt was first proposed by Broecker as a conceptual model of the global thermohaline circulation, in an article appearing in *Natural History* in 1987. This concept is so appealing that the conveyor has become a hot topic, and Broecker (1991) wrote a further article especially devoted to it. From the beginning, the concept of a "conveyor" has been the subject of fierce debate! A further modified version of the conveyor belt was postulated by Schmitz (1996a) (Fig. 5.152). It was assumed that this conveyor is driven by sinking in the northern North Atlantic Ocean. As discussed above, however, the new energy theory of the thermohaline circulation claims that such a conveyor must be maintained by mechanical energy from wind stress and tidal dissipation.

There are many ways of estimating the sinking rate of the NADW. For example, Broecker (1991) used a tracer balance method based on the fact that radiocarbon ^{14}C decays by 1% in 82.7 years, thereby the ^{14}C deficiency for the Atlantic Ocean suggests a residence time of 83–250 years, or an average age of 180 years. The total volume of the deep Atlantic

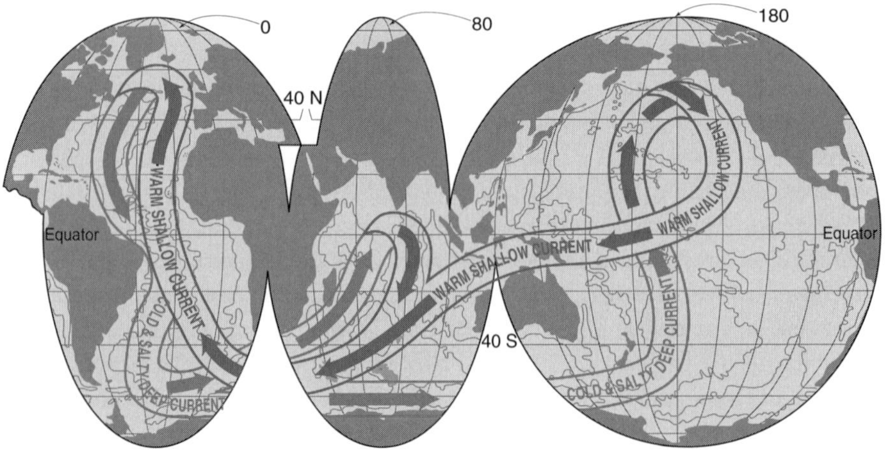

Fig. 5.152 A two-layer thermohaline conveyor belt originally proposed by Broecker (1991), modified and redrawn by Schmitz (1995) (adapted from Schmitz, 1996a). See color plate section.

Basin is $2{,}500 \times 6.2 \times 10^3 = 1.55 \times 10^7$ (m^3). Thus, the ventilation flux is $V/T = 8.6 \times 10^{14}$ m^2/yr ≈ 27 Sv.

The formation rate of the AABW is estimated to be 4 Sv. The original source of the deepwater formation rate in the Norwegian/Greenland Sea is only about 0.47 Sv. However, note that, along the path of the deep western boundary current, there is much more water entrained into the deep western boundary current. The flow segment along the eastern coast of Greenland for density greater than $\sigma_\Theta \geq 27.8$ kg/m^3 has a total volume flux of 10.7 Sv (Dickson *et al.*, 1990); when it reaches the southern tip of Greenland, the volume flux is increased to 13.3 Sv. The total volume flux of the deep western boundary current in the North Atlantic Ocean can be further enhanced owing to some contribution from the Labrador Sea; thus, Broecker's estimation is not inconsistent with other observations.

The conveyor belt plays a vital role in the global thermohaline circulation under the current climate condition because it transports heat, freshwater, and many other tracers throughout the world's oceans. The conveyor belt, as it was originally postulated and depicted (see Fig. 5.152), is clearly an extremely oversimplified concept. Because of its simplicity, many people have used this term as a tool to describe the thermohaline circulation in the world's oceans. However, the dynamical picture implied by this diagram is so oversimplified that sometimes it can be quite misleading. One of the most critical problems hidden in this diagram is the omission of the ACC, which plays the role of the global artery of the world ocean circulation. A better picture of the global circulation was proposed by Schmitz (1996b) (Fig. 5.153). According to this theory, a major part of the water mass transport has to go through the whole loop of the ACC before it can return to the northern North Atlantic Ocean and repeat the cycle again. During this process, wind stress forcing in the Southern Ocean plays a vital role in regulating this global conveyor.

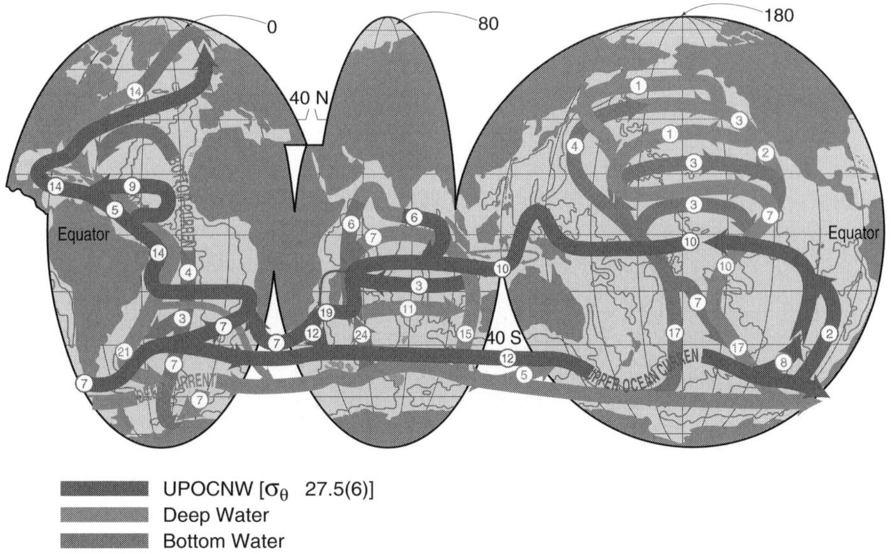

Fig. 5.153 Water mass transports (in Sv) in the world's oceans; the *blue lines* for bottom water, *green lines* for deep water, and *red* for the upper ocean (Schmitz, 1996b). See color plate section.

Multiple solutions for the global oceans

Results from an idealized oceanic model

Marotzke and Willebrand (1991) carried out a series of numerical experiments for the global oceans. The model consisted of a highly idealized geometry, with two ocean basins of equal width and meridional length (left panel of Fig. 5.154). The model was first spun up under relaxation conditions for both temperature and salinity. A global zonal mean SSS was used as the reference salinity (right panel of Fig. 5.154).

After the model reached quasi-equilibrium, the virtual salt flux was diagnosed and the model was restarted and run under the mixed boundary conditions. Since there are four sites where deep water can be formed, theoretically there can be $2^4 = 16$ possible patterns of deepwater formation. Several modes of the global circulations were discussed. For example, in Equilibrium 1, sinking took place in both the North Atlantic and Pacific Oceans, as shown in the left column of Figure 5.155. As a result, there was a strong poleward heat flux in the Northern Hemisphere. Similarly, in Equilibrium 4, sinking took place in the Southern Ocean, including both the Atlantic and Pacific Basins; there was a strong poleward heat flux in the Southern Hemisphere, and the poleward heat flux in the Northern Hemisphere was very small, as shown in the right column of Figure 5.155.

Another interesting mode mimics the conveyor belt observed under the current climatic conditions. This mode is characterized by a strong sinking in the North Atlantic Ocean, with a weak sinking in the Southern Pacific Ocean (Fig. 5.156a, b). The corresponding

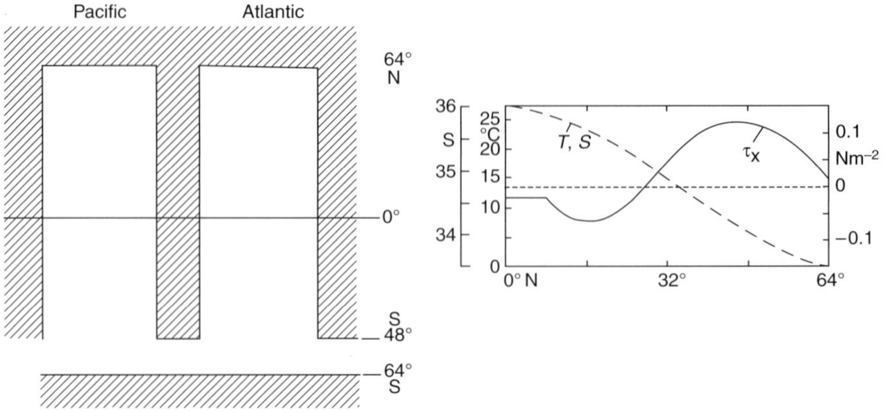

Fig. 5.154 A global model to simulate the global modes of the thermohaline circulation: *left panel*, the geometry of the model; *right panel*, forcing fields of the model (Marotzke and Willebrand, 1991).

poleward heat flux consists of a strong northward component in the North Atlantic Basin and a relatively weak southward component in the South Pacific Basin.

Note that in Marotzke and Willebrand's experiments the same global mean sea surface salinity measurements were used. independent of longitude; therefore their solutions for the Atlantic and Pacific Oceans were made symmetric, and their solutions did not take into consideration the vapor flux across Central America. Whether the mode with deepwater formation in the North Pacific Ocean can ever exist in reality is questionable.

Two stable equilibriums for the Atlantic Ocean in the air–sea coupled system

Manabe and Stouffer (1988) carried out numerical experiments based on the Princeton atmosphere–ocean coupled model. In their experiments, two equilibria were found. The first one, labeled EXP I, represents the present-day climate, while the second one, labeled EXP II, turns out to be quite different, with the North Atlantic Ocean being much colder and fresher (Fig. 5.157). The maximum temperature and salinity differences are 5°C and 3, respectively.

Implications for climate change

Three of the models discussed above, namely Stommel's two-box model, F. Bryan's two-hemispheric numerical model, and the Manabe–Stouffer coupled model, represented different degrees of idealization for the climate system. However, the multiple stable equilibrium states obtained from these models share a fundamentally similar structure. Paleoclimate records indicate that the meridional circulation in the North Atlantic Basin has gone through cycles of On/Off; thus, one of the most important research frontiers in oceanic circulation and climate study is the question of whether such an abrupt change in the North Atlantic circulation may happen in the near future. In particular, the main focus

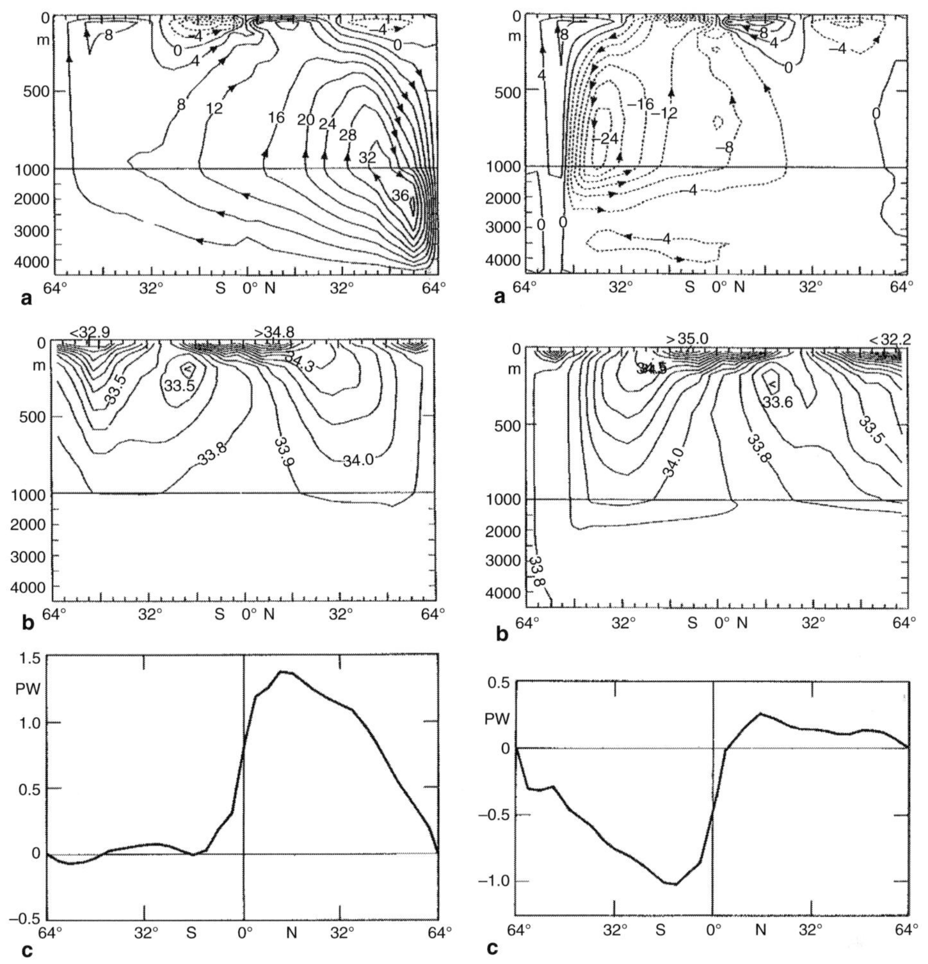

Fig. 5.155 *Left column*: Equilibrium 1: northern sinking. **a** Meridional streamfunction in Sv as the sum of both basins; **b** zonally averaged salinity; and **c** northward heat flux in PW for both basins combined. *Right column*: Equilibrium 4: southern sinking. **a** Meridional streamfunction in Sv as the sum of both basins; **b** zonally averaged salinity; and **c** northward heat flux in PW for both basins combined (Marotzke and Willebrand, 1991).

is the potential impact of changes of hydrological cycle and freshwater flux on the climate system:

- *Due to global warming*, the hydrological cycle may be intensified, with more evaporation at low latitudes and more precipitation at high latitudes. The meridional salinity gradient can be enhanced, producing a meridional pressure difference against that due to thermal forcing. As a result, the meridional overturning cell associated with the NADW may slow down and may even be interrupted.

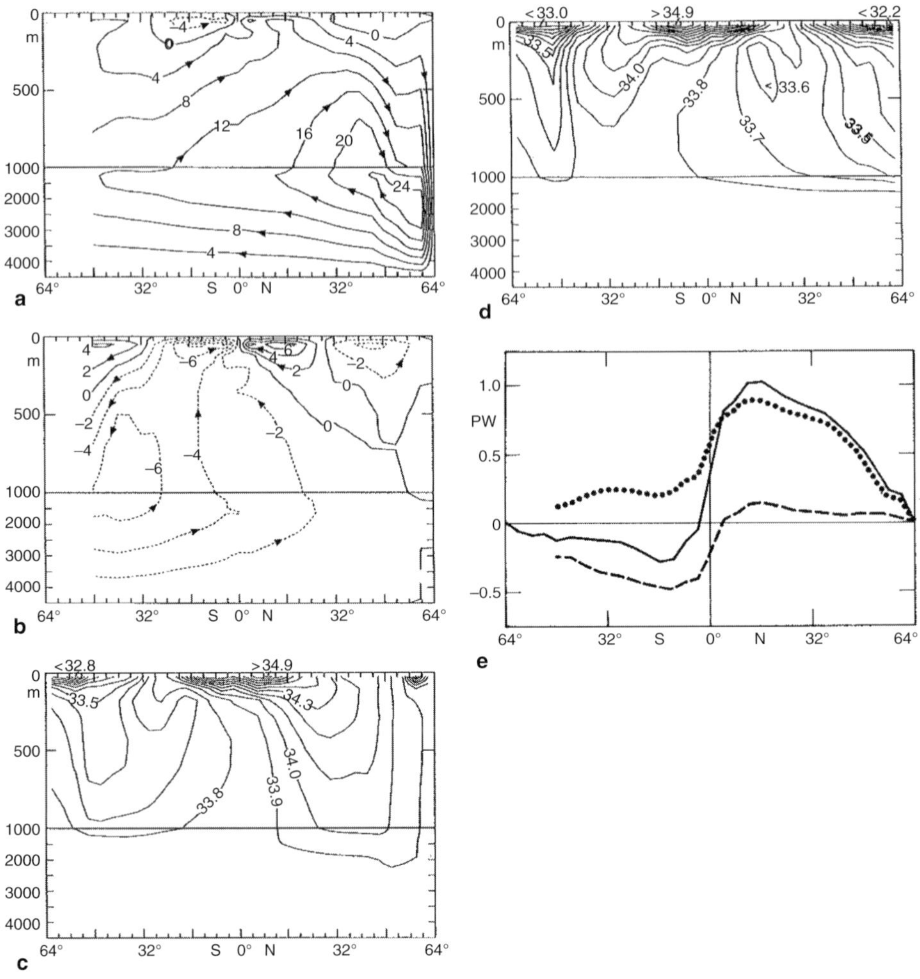

Fig. 5.156 Equilibrium 2: the conveyor belt: **a** Atlantic and **b** Pacific meridional streamfunction in Sv; **c** Atlantic and **d** Pacific zonally averaged salinity; **e** northward heat flux in PW for Atlantic (*dotted*), Pacific (*dashed*), and the two basins combined (*solid*) (Marotzke and Willebrand, 1991).

Accordingly, many of the state-of-art climate models predict that meridional overturning circulation in the Atlantic Basin will be substantially reduced over the next hundred years.

- *Due to global climate change*, within the next 30–50 years the Arctic Ocean may be ice-free in summer-time. Without the ice, the large amount of relatively fresh water in the Arctic Ocean may not be held within the Arctic Basin. If some water from this freshwater pool floods the North Atlantic Ocean, it may cause a halocline catastrophe, similar to that simulated by F. Bryan and others.

Note that the results obtained from these models were based on the assumption that diapycnal mixing coefficients are invariant under different climate conditions. From the

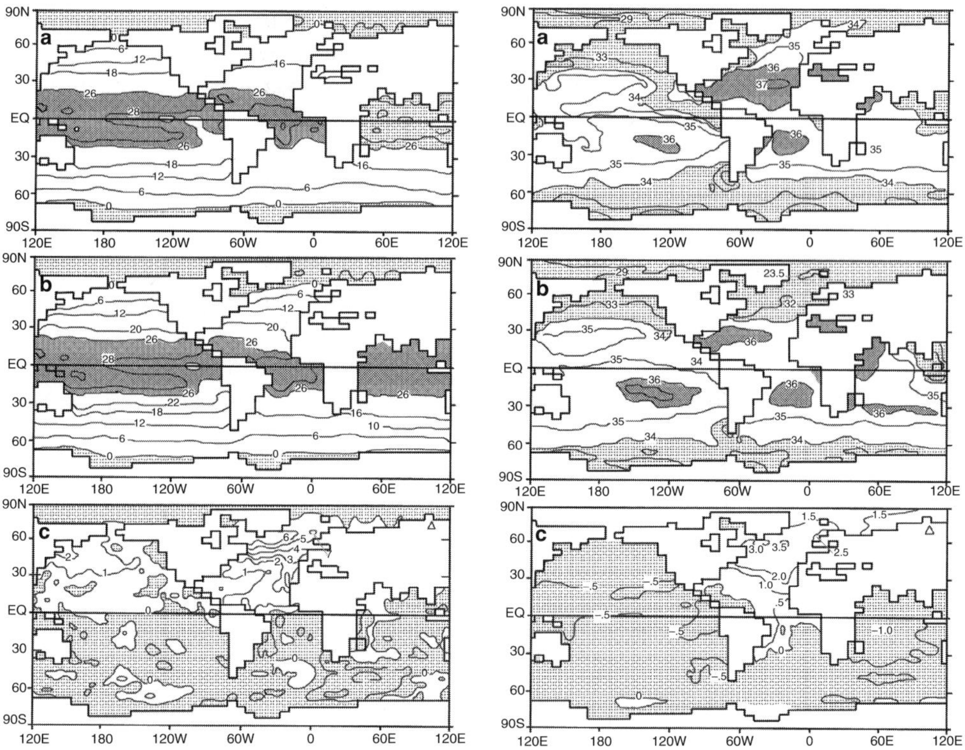

Fig. 5.157 *Left column*: **a** sea surface temperature from EXP I; **b** sea surface temperature from EXP II; **c** the difference between the two experiments (**a** minus **b**). *Right column*: **a** sea surface salinity from EXP I; **b** sea surface salinity from EXP II; **c** the difference between EXP I and EXP II (**a** minus **b**) (Manabe and Stouffer, 1988).

point of view of the new energy theory, wind stress – and, to a lesser degree, tidal dissipation – can be quite different under different climate conditions. Thus, the results from such model experiments remain questionable. It is hoped that, with the rapid progress being made toward unraveling the mystery of mixing and oceanic circulation, we will be able to simulate and predict the oceanic circulation and climate more accurately in the near future.

5.4.7 Thermohaline oscillations

Thermohaline circulation encompasses a wide spectrum of different phenomena. Although each of them has its own characteristics, most of these phenomena are closely linked to the salinity balance in the system, including the connection with the hydrological cycle in the climate system or the turbulent diffusion of salty (or fresh) water in the oceanic interior. In this section, we deal with a hierarchy of saline/thermohaline oscillations.

A heat–salt oscillator

First of all, let us introduce a prototype of a thermohaline oscillator which can be illustrated using a simple box model (Welander, 1982). The model consists of a well-mixed layer, forced from above by a heat flux $\kappa_T(T_A - T)$ and salt flux $\kappa_S(S_A - S)$ (Fig. 5.158).

The turbulent flux between the mixed layer and the water below is parameterized in terms of a transient mixing coefficient, $\kappa = \kappa(\rho)$ (where ρ is the density of the well-mixed layer), depending on the density difference (Fig. 5.159). Thus, if we set $T_0 = S_0 = \rho_0 = 0$, the basic equations are

$$\dot{T} = \kappa_T(T_A - T) - \kappa(\rho)T \tag{5.283}$$

$$\dot{S} = \kappa_S(S_A - S) - \kappa(\rho)S \tag{5.284}$$

$$\rho = -\alpha T + \beta S \tag{5.285}$$

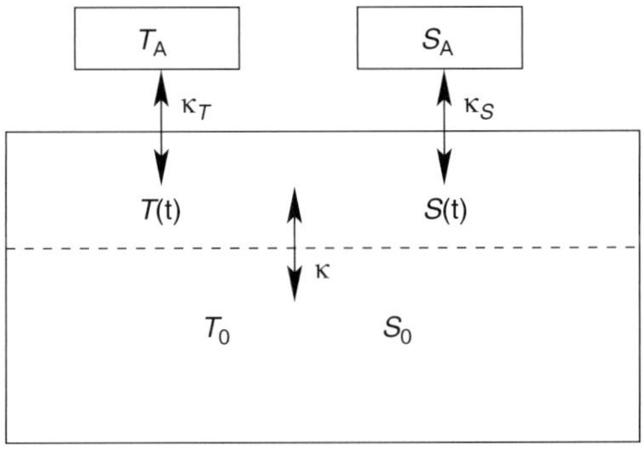

Fig. 5.158 Welander's box model. The values T_A and S_A are the effective temperature and salinity external forcing; T_0 and S_0 are the temperature and salinity of the deep reservoir; κ_T, κ_S, and κ are the coefficients in assumed Newtonian-type transfer laws.

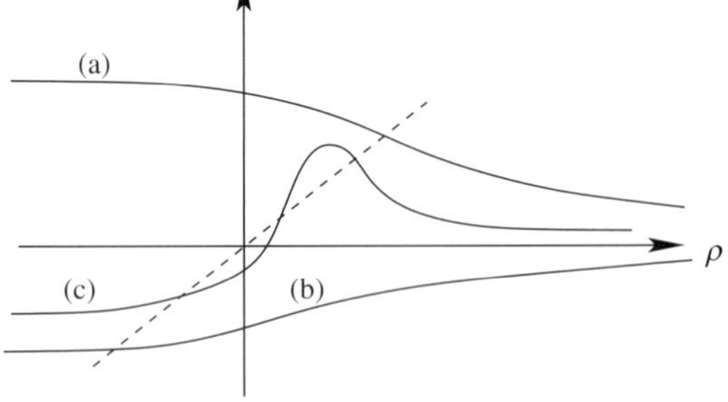

Fig. 5.159 Examples of possible forms for $F(\rho)$ in a case where $\kappa(\rho)$ increases monotonically (redrawn from Welander, 1982).

A crucial assumption made here is that $\kappa(\rho)$ is a positive function, monotonically increasing with ρ.

The steady state of the system is

$$\overline{T} = \frac{\kappa_T T_A}{\kappa_T + \kappa(\overline{\rho})}, \quad \overline{S} = \frac{\kappa_S S_A}{\kappa_S + \kappa(\overline{\rho})} \tag{5.286}$$

$$\overline{\rho} = -\alpha\overline{T} + \beta\overline{S} = \frac{\alpha\kappa_T T_A}{\kappa_T + \kappa(\overline{\rho})} + \frac{\beta\kappa_S S_A}{\kappa_S + \kappa(\overline{\rho})} = F\overline{\rho}) \tag{5.287}$$

The solutions to this problem can be found graphically as the intersections between the dashed straight line $\bar{\rho} = \rho$ and the curve $\bar{\rho} = F(\rho)$ (Fig. 5.159). As shown in this figure, in cases (a) and (b) there is one steady state; in case (c) there are three steady states.

A simple case of oscillator is obtained by assuming that the mixing coefficient κ takes two fixed values only:

$$\begin{aligned} \kappa &= \kappa_0 \quad \textit{for}\ \rho \leq -\varepsilon \\ \kappa &= \kappa_1, \quad \textit{for}\ \rho > -\varepsilon \end{aligned} \tag{5.288}$$

and κ_0 is smaller than κ_1. Thus, there exist two "attractors" in the phase plane, as shown in Figure 5.160. An attractor in the phase space means that all trajectories in the vicinity are "attracted" to the attractor, i.e., they flow toward the attractors, but can never reach them. As soon as the system crosses the diagonal, $\rho = 0$, in the phase plane, the system is attracted by the attractor on the other side of the diagonal (Fig. 5.160); thus, the system can never reach a steady state, instead there is a limit cycle (Fig. 5.161).

Introducing the following nondimensional variables:

$$T^* = T/T_A, S' = S/S_A, \rho' = \rho/\beta S_A, t' = \kappa_A t \tag{5.289}$$

Fig. 5.160 Schematic picture of the flip-flop model in a T–S plane. Points 1 and 2 represent the non-existing steady states which act as "attractors" when the solution point is in the half-plane $\rho \leq -\varepsilon$ and $\rho > -\varepsilon$, respectively. The slope dS/dT has a discontinuity at the lines $\rho = -\varepsilon$ (redrawn from Welander, 1982).

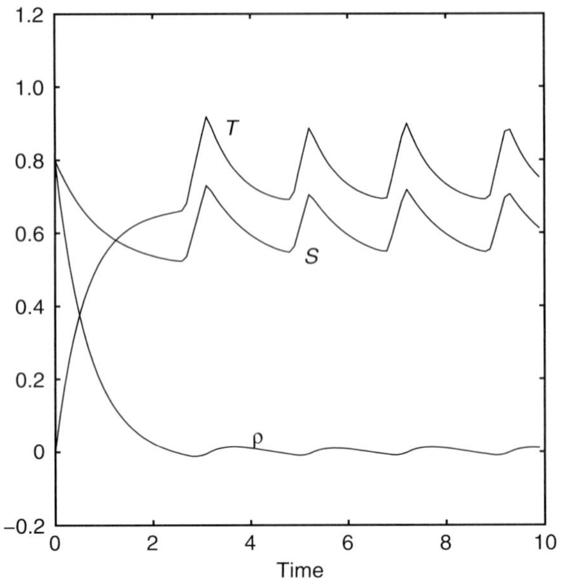

Fig. 5.161 The solution $T(t)$, $S(t)$, and ρ (t) through a few cycles.

Then Eqns. (5.283)–(5.285), after dropping the primes, are reduced to

$$\dot{T} = 1 - T - \kappa(\rho)T \tag{5.290}$$

$$\dot{S} = 0.5(1 - S) - \kappa(\rho)S \tag{5.291}$$

$$\rho = -0.8T + S$$

where the diffusivity function is chosen as

$$\kappa(\rho) = \frac{1}{\pi} \arctan\left(3000\rho + \frac{50}{3} + \frac{\phi}{2}\right) \tag{5.292}$$

The system oscillates as shown in Figure 5.161.

In fact, Welander's model is an idealization of the convective overturning in the more complicated three-dimensional OGCM, and convective adjustment is an important factor in OGCM. As is well known, the primary cause for any thermohaline variability is the nonlinear feedback between different components of the system. Since the system is very complicated, it is a great challenge to try and sort out the various types of feedback mechanism and understand the role of each of them in controlling the climate.

Stommel (1986) independently formulated a thermohaline oscillator. He arranged two boxes horizontally and assumed that the exchange rate between the boxes would become very large when the density between them was very small, which presumably would mimic isopycnal mixing. His solution posed a limit cycle rather similar to Welander's model.

Oscillations and chaotic behavior of haline circulation

As discussed in Section 5.4.3, haline circulation simulated by a loop model can appear in steady-state, oscillatory, and chaotic forms. In general, haline circulation simulated from different models shows a wealth of oscillatory and chaotic behavior. It turns out that the most crucial ingredient for oscillatory or chaotic behaviors is the flux condition applied to the model.

Between the two conditions applied in saline modeling: the relaxation condition and the flux condition, the latter is more natural and accurate, since in reality haline circulation in the ocean is forced by freshwater flux through the air–sea interface. A close examination of the relaxation condition $S_f = \Gamma(S^* - S)$ reveals two theoretical limitations of this constraint. First, when the relaxation constant is very large, or the corresponding relaxation time approaches zero, this condition is equivalent to specifying the surface salinity. Therefore, this is an extremely strong constraint on the surface salinity. Second, when $\Gamma \to 0$ but $\Gamma S^* \to$ constant, the relaxation constraint converges to a flux constraint. Compared with a model under strong relaxation conditions, both the relaxation constraint with a small relaxation constant and the flux constraint are weaker constraints on surface salinity, and saline oscillations can be developed quite easily under such constraints.

Haline oscillations stem from the negative feedback between the sea surface salinity and the overturning circulation as follows. A positive salinity anomaly at low latitudes enhances the north–south pressure difference at sea level. As a result, the meridional overturning circulation is intensified, bringing more freshwater from high latitudes to low latitudes and reducing the salinity anomaly there. On the other hand, the positive salinity anomaly, formed in the upper surface earlier, is advected into the subsurface ocean; from there it moves toward high latitudes and eventually arrives in the upper ocean at high latitudes. Thus, the meridional salinity perturbations give rise to a smaller meridional pressure difference in the upper ocean, reducing the meridional overturning circulation, and the next half-cycle of the oscillation ensues.

The oscillatory nature of the system depends sensitively on the formulation of the model and the parameters used in the numerical experiments. For example, Huang and Chou (1994) discussed the behavior of the haline circulation obtained from a model under the rigid-lid approximation and a freshwater flux forcing which is a linear function of latitude; there can be steady-state, oscillatory, and chaotic solutions, depending on the choice of parameters (Fig. 5.162).

As an example for the haline oscillation, the solution discussed in Section 5.3.3 is in a state of aperiodic oscillation, with a quasi-period of approximately 27 years. All the major indexes of the circulation go through aperiodic oscillations (Fig. 5.163).

Over a reasonably short time scale, the system behaves in a way close to a periodic oscillation, with a period of 26.8 years, i.e., all the major indexes of the circulation go through nearly periodic cycles, such as the meridional overturning rate (Fig. 5.164a). The peak in the meridional overturning circulation is also closely linked to the peak of the total kinetic energy (Fig. 5.164d). The oscillation in the overturning rate is directly linked

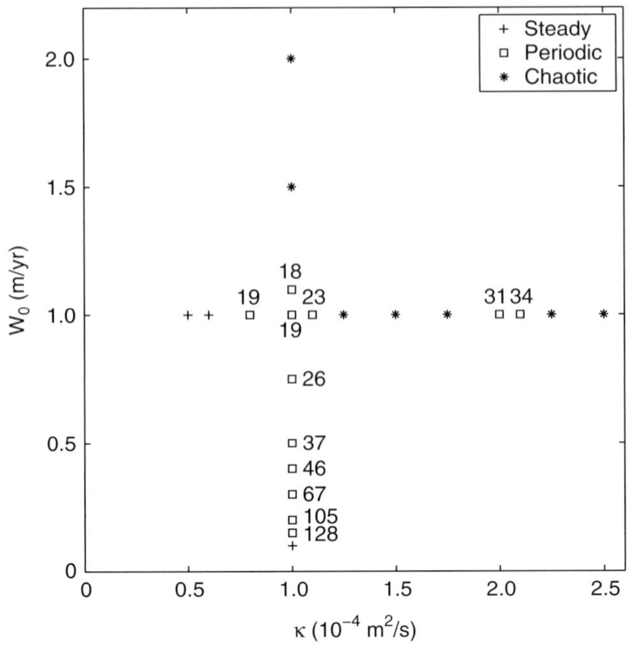

Fig. 5.162 Behavior of haline circulation in the phase space of the vertical diffusivity (κ) and fresh-water flux amplitude (W_0); *crosses* indicate steady solutions, *stars* indicate chaotic solutions, and *squares* indicate periodic solutions, with the *numbers* indicating the period in years.

to the north–south pressure difference at sea level, as indicated by the difference in the mean sea level at the northern and southern boundaries (Fig. 5.164b). In fact, the peak of the overturning rate is reached slightly after the peak time of the north–south sea-level difference. The difference in the north–south mean sea level is directly linked to the mean surface salinity at the northern and southern boundaries (Fig. 5.164c).

It is interesting to note that the peak of the overturning circulation corresponds to the minimum of the meridional difference in bottom pressure (Fig. 5.164e). This suggests that the abyssal flow lags behind the surface branch. In addition, the mean surface salinity reaches its minimum several years after the peak of the meridional overturning circulation (Fig. 5.164f).

Decadal/interdecadal variability in the oceanic general circulation model

Decadal/interdecadal variability of the thermohaline circulation is an extremely relevant topic for the study of climate because reliable instrumental records of the ocean are available only for the past several decades. In this respect, the loop oscillation with feedbacks between horizontal advection and surface salinity anomaly, similar to that discussed above, has been explored as one of the major mechanisms.

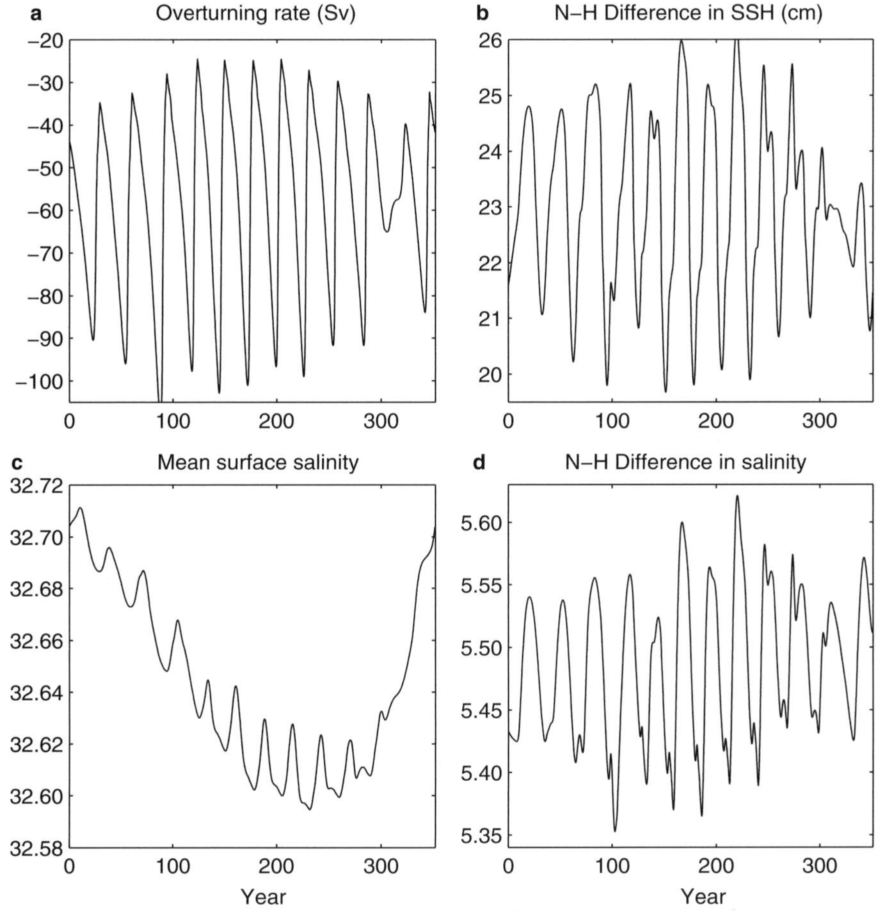

Fig. 5.163 **a–d** Time evolution of the haline oscillations over a time span of several hundred years.

One of the essential ingredients for decadal/interdecadal variability is the mixed boundary conditions, i.e., a relaxation condition for temperature and a flux condition for salinity. As discussed in the previous section, haline circulation under the freshwater flux condition alone can develop oscillatory solutions. Thus, it is easy to understand why a model under the mixed boundary conditions tends to develop oscillatory solutions.

In the ocean, the decadal/interdecadal oscillations may be due to a combination of SST (sea surface temperature) and salinity. A higher-than-normal SST can induce a positive salinity anomaly at lower latitudes. Advection brings the salinity anomaly to high latitudes. Since the thermal expansion coefficient is small at high latitudes, a positive salinity anomaly gives rise to a higher density and lower sea level there. The resulting stronger-than-normal meridional pressure gradient induces an enhancement of the meridional overturning circulation, and the system develops oscillatory behavior similar to that discussed above (Fig. 5.165).

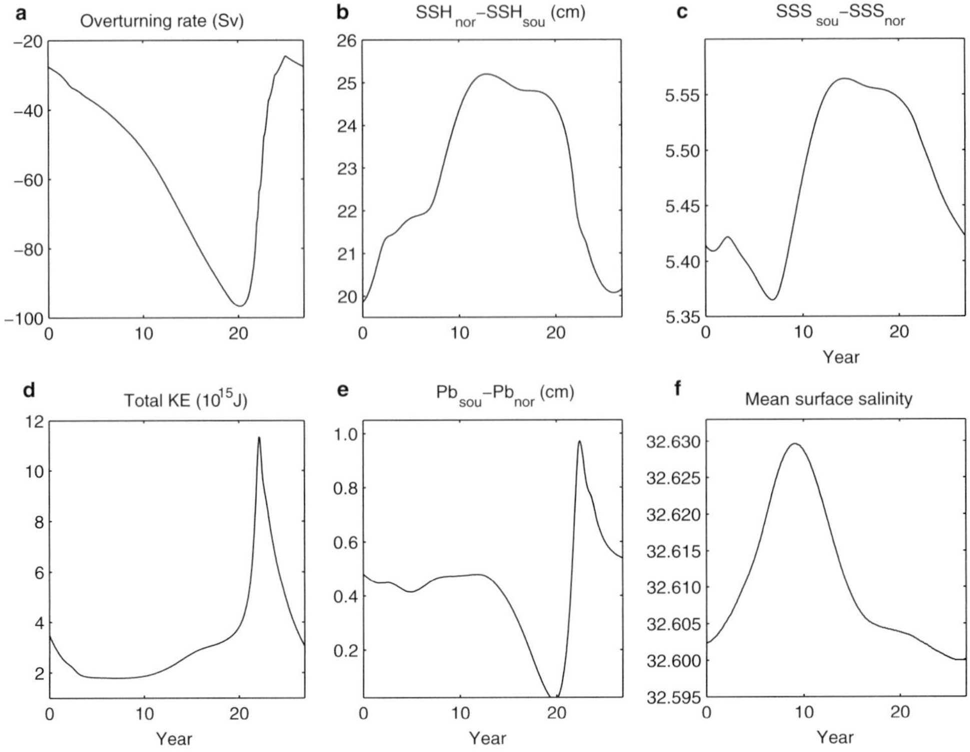

Fig. 5.164 **a–f** System behavior over a quasi-cycle, sampled from the freshwater-induced haline circulation with a quasi-periodic oscillation.

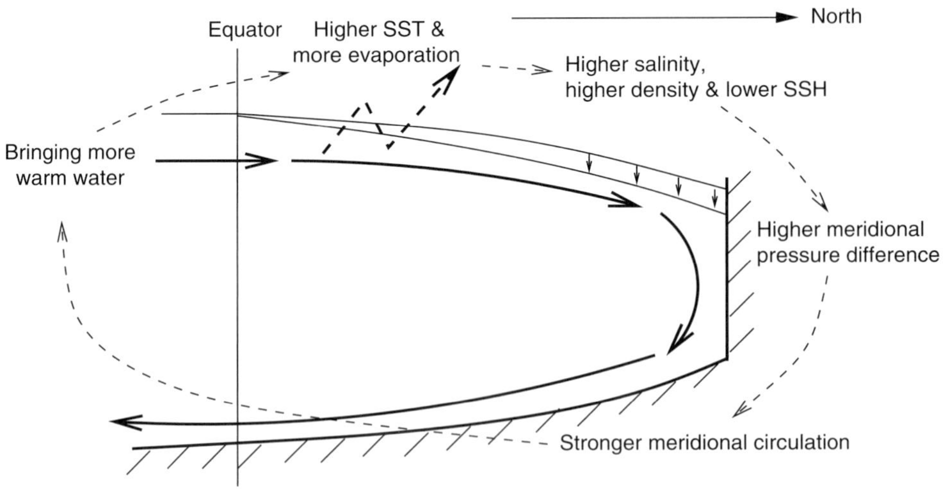

Fig. 5.165 A sketch of the oscillatory circulation in the meridional plane under the mixed boundary conditions.

Decadal/interdecadal variability of the oceanic circulation can be found through numerical modeling as follows. First, a numerical model for a closed basin is spun up, with relaxation conditions for both temperature and salinity. When the model's solution reaches a quasi-steady state, the equivalent virtual salt flux can be diagnosed from this quasi-steady solution. Next, the model is restarted from this quasi-steady state and run under mixed boundary conditions, i.e., the same relaxation condition for temperature, but salinity is now under a flux condition, specified according to the diagnosed virtual salt flux. In many cases, a wealth of decadal variability can be found through the numerical simulations. Of course, the exact characteristics of such decadal/interdecadal variability depend on the model's geometry and, above all, on the choice of forcing field and parameters.

As an example, the decadal variability was identified in a single-hemisphere ocean model forced by the mixed boundary condition (Weaver *et al.*, 1991). Similarly, the interdecadal variability in an idealized model of the North Atlantic Ocean was discussed by Weaver *et al.* (1994). Under the mixed boundary condition, the model possesses a limit cycle with a period of approximately 22 years. As shown in Figure 5.166, both the mean kinetic energy and the air–sea heat flux oscillate during each cycle. This limit cycle is associated with a great change of the thermohaline circulation in the model basin. For example, the poleward heat flux varied from a maximum of 0.8 PW at 39° N to a minimum of 0.5 PW. In addition, deepwater formation at the northern boundary of the Labrador Sea was shut off and turned on during each cycle. Therefore, the decadal/interdecadal cycle may play a crucial role in the global oceanic circulation and climate system.

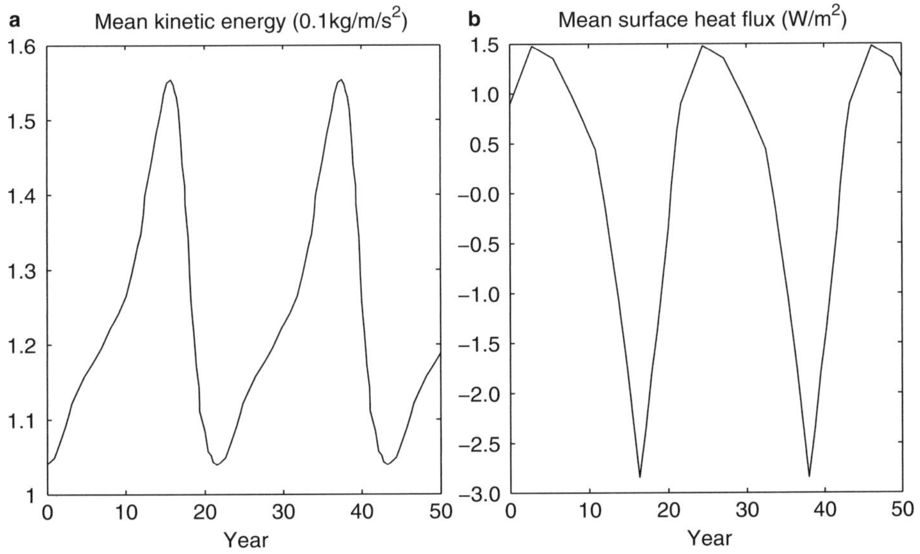

Fig. 5.166 Interdecadal variability in an idealized model of the North Atlantic: **a** mean kinetic energy; **b** surface heat flux (redrawn from Weaver *et al.*, 1994).

Variability on a diffusive time scale: flushing

Owing to the extremely different natures of the boundary conditions applied to temperature and salinity, there exists one more type of thermohaline variability called flushing (Marotzke, 1989), which is on rather a long diffusive time scale, i.e., on the centennial and millennial time scales. Flushing can be studied using a numerical model as follows. A model is first run to quasi-equilibrium under the relaxation condition for both salinity and temperature with no wind stress. After diagnosing the virtual salt flux, the model is continued, running under the mixed boundary conditions. By adding a small (1‰) fresh (negative salinity) perturbation at high latitudes, the polar halocline catastrophe sets in. The system slowly evolves into a state with very weak equatorial downwelling and upwelling elsewhere, i.e., it is in a haline mode, in reverse of the thermal mode. During a period of several thousand years the horizontal diffusive process produces warm and salty abyssal water in the whole basin.

At high latitudes, cold and fresh water overlays the warm and salty water there, with a relatively weak stratification which is potentially unstable. Imagine a small water parcel moving in a vertical direction, bringing warm and salty deep water to the surface. Near the surface the water parcel quickly loses its heat owing to the strong thermal relaxation. Under the flux condition, surface salinity evolves on a much longer time scale, so it remains virtually unchanged over a relatively short time scale.

As a result of surface cooling, the water parcel becomes heavier than the ambient density and sinks all the way down to the bottom. The GPE released in this process feeds energy to the convective overturning. Consequently, the system goes through a very violent phase, called flushing, until the potentially stored energy is consumed completely. The system is in a quasi-thermal mode during the flushing period, but it returns to the quasi-haline mode after flushing.

Gravitational potential energy balance is the key element in flushing. To demonstrate the connection between GPE and flushing, we analyze the evolution of GPE during the flushing. Assume a model ocean of two layers, with equal thickness of h and unit area (Fig. 5.167). The upper layer is subject to a relaxation temperature $T^* = T_1$. At the initial time, temperature and salinity are T_1 and S_1 in the upper layer, and T_2 and S_2 in the lower layer. Water density is a linear function of T and S: $\rho = \rho_0(1 - \alpha T + \beta S)$. We also assume, for simplicity, that there is no stratification at the initial time, i.e., $\rho_1 = \rho_2$, or $\alpha(T_2 - T_1) = \beta(S_2 - S_1)$.

If these two layers are flipped upside down (the second stage in Fig. 5.167), the total GPE remains unchanged. Cooling takes place afterward. After a time period, which is much longer than the relaxation time for surface temperature but much shorter than the time scale for surface salinity change, temperature in the upper layer is reduced to T_1, but the salinity there remains unchanged. Changes in GPE can be calculated according to the model used as follows.

If a Boussinesq model is used, the layer thickness remains unchanged after cooling. Due to the artificial source of mass in the model, the total mass of the upper layer increases $\delta m = h\rho_0\alpha(T_2 - T_1)$. Using the bottom as the reference level for GPE, the total GPE

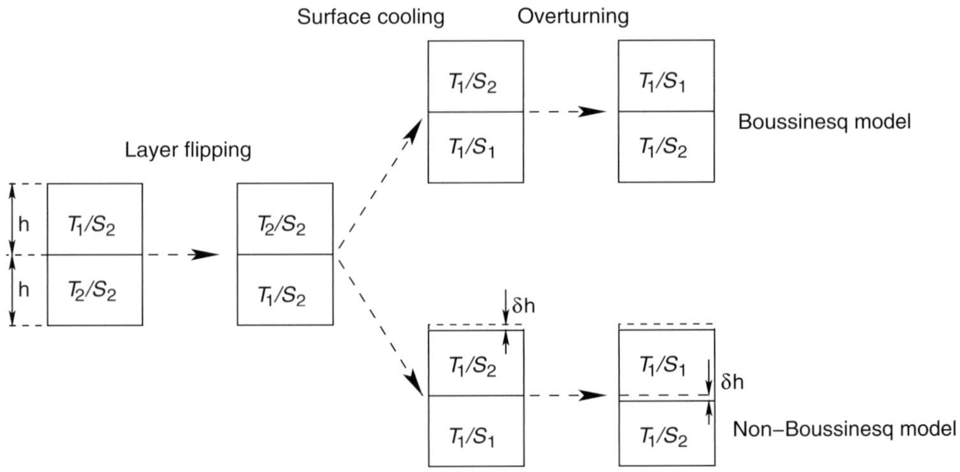

Fig. 5.167 Four conceptual stages of a two-layer ocean during the process of flushing, simulated in terms of the Boussinesq and non-Boussinesq models.

increase from stage 2 to stage 3 is

$$\chi_3 - \chi_2 = 1.5gh^2\rho_2\alpha(T_2 - T_1) > 0 \tag{5.293}$$

Obviously such a state is gravitationally unstable, and overturning takes place. With the heavy layer sinking to the bottom, from stage 3 to stage 4 the total GPE is reduced

$$\chi_4 - \chi_3 = -gh^2\rho_0\beta\,(S_2 - S_1) < 0 \tag{5.294}$$

Evolution of GPE during this process is shown in Figure 5.168. Note that GPE in stage 4 is slightly higher than in the initial stage 1,

$$\chi_4 - \chi_1 = 0.5gh^2\rho_0\beta\,(S_2 - S_1) = 0.5gh^2\rho_0\alpha\,(T_2 - T_1) > 0 \tag{5.295}$$

If a non-Boussinesq model is used, the total mass of each layer remains unchanged, but its thickness shrinks during cooling. Thus, from stage 2 to stage 3, the upper layer thickness is reduced to

$$h' \simeq [1 - \alpha(T_2 - T_1)]h \tag{5.296}$$

Owing to the downward movement of the center of mass of the upper layer, the total GPE is reduced

$$\chi_3 - \chi_2 = -0.5gh^2\rho_0\alpha\,(T_2 - T_1) < 0 \tag{5.297}$$

This stage is gravitationally unstable, and overturning takes place, during which the total GPE is reduced

$$\chi_4 - \chi_3 = -gh^2\rho_0\beta\,(S_2 - S_1) < 0 \tag{5.298}$$

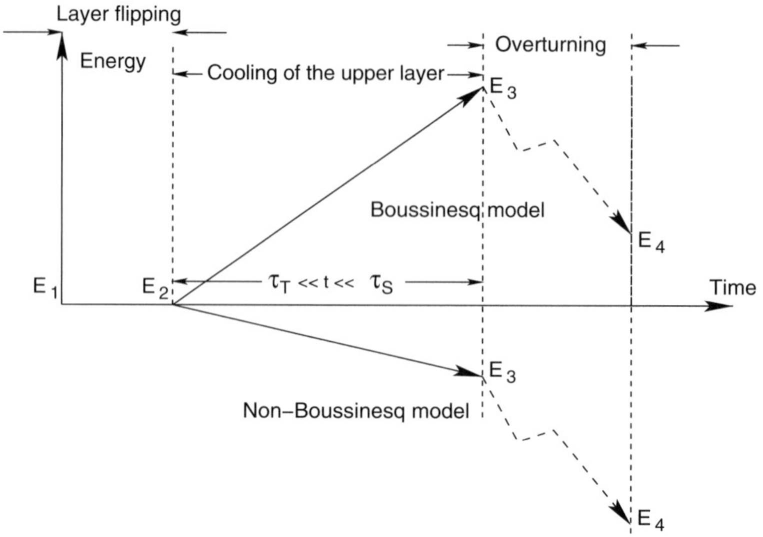

Fig. 5.168 Evolution of GPE in a two-layer ocean during the process of flushing, as simulated in terms of Boussinesq and non-Boussinesq models; assume that $\alpha = \beta = 1$ and T, S, and energy are in nondimensional units.

The evolution of GPE for the non-Boussinesq model is also shown in the lower part of Figure 5.168.

It is clear that only the non-Boussinesq model can capture the evolution of GPE accurately. In fact, GPE of the system is reduced during the steps shown in Figure 5.168. At the end of these steps, GPE of the system is at its lowest. To return to the initial stage, the cold water in the lower layer has to be warmed up. Expansion of the lower layer pushes its center, and thus the center of the layer above, upward.

On the other hand, GPE in the Boussinesq model is falsely increased during the transition from stage 2 to stage 3. Furthermore, the final stage 4 in the Boussinesq model has slightly more GPE than the initial stage; another artifact in the model due to the artificial source of mass introduced during cooling. Finally, to return to the initial stage 1, the model must go through the heating of the lower layer. During heating of the lower layer, an artificial sink of mass takes place, resulting in a reduced GPE of the lower layer, while the GPE of the upper layer remains unchanged. Therefore, only in the transition from stage 3 to stage 4 does the change in GPE stay the same in Boussinesq and non-Boussinesq models, since this transition does not involve a change in density of the water in both layers.

The flushing phenomenon also exists for the case with wind stress forcing. The period of flushing depends on the parameters of the model, but it is always on the order of thousands of years. The cycle of flushing can be clearly seen from many indexes of the circulation, such as the meridional overturning rate (Fig. 5.169a), the sea surface salinity (Fig. 5.169b), the basin mean temperature (Fig. 5.169c), and the heat loss (Fig. 5.169d) (Huang, 1994).

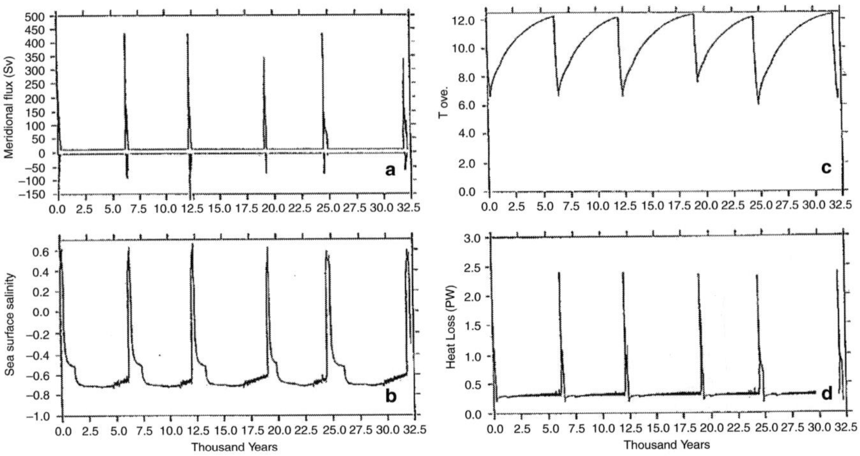

Fig. 5.169 Time evolution of thermohaline circulation in a single-hemisphere Boussinesq model ocean forced by wind stress, thermal relaxation, and freshwater flux, including: **a** meridional overturning rate (in Sv); **b** mean sea surface salinity (deviation from the mean salinity of 35); **c** basin-mean temperature (in °C); **d** heat flux to atmosphere (not counting the heat flux from atmosphere to the ocean, in PW) (Huang, 1994).

During flushing, meridional overturning is extremely strong, and relatively warm and salty water is brought to the upper ocean, making the sea surface salinity higher. At the same time, rapid heat loss to the atmosphere reduces the basin mean temperature, as shown in Figure 5.169c, d.

An interesting application of this idea is the deep ocean anoxia during the Late Permian era (Zhang *et al.*, 2001). As the models, including a three-box model and an oceanic general circulation model, have indicated, the oceanic circulation at that time may have switched between a long-time haline mode and a short burst of thermal mode, with a period of approximately 3,330 years.

During the phase of the thermal mode (Fig. 5.170), there may exist a superposition of the wind-driven circulation and the thermohaline circulation looping through the Tethys Sea. The meridional overturning cell is strong, with sinking near the South Pole, and it penetrates all the way to the sea floor (Fig. 5.170b). The circulation is much faster, with a large meridional heat flux, and is southward at all latitudes.

In comparison, during the phase of the haline mode, the meridional circulation is much slower and confined to the upper 1.5 km (Fig. 5.171). The meridional heat flux is much smaller and more symmetrical with respect to the equator.

5.5 Combining wind-driven and thermohaline circulation

5.5.1 Scaling of pycnocline and thermohaline circulation

The wind-driven circulation in the upper ocean can be clearly described in terms of the horizontal advection of water within the Ekman layer and the horizontal gyre below. We

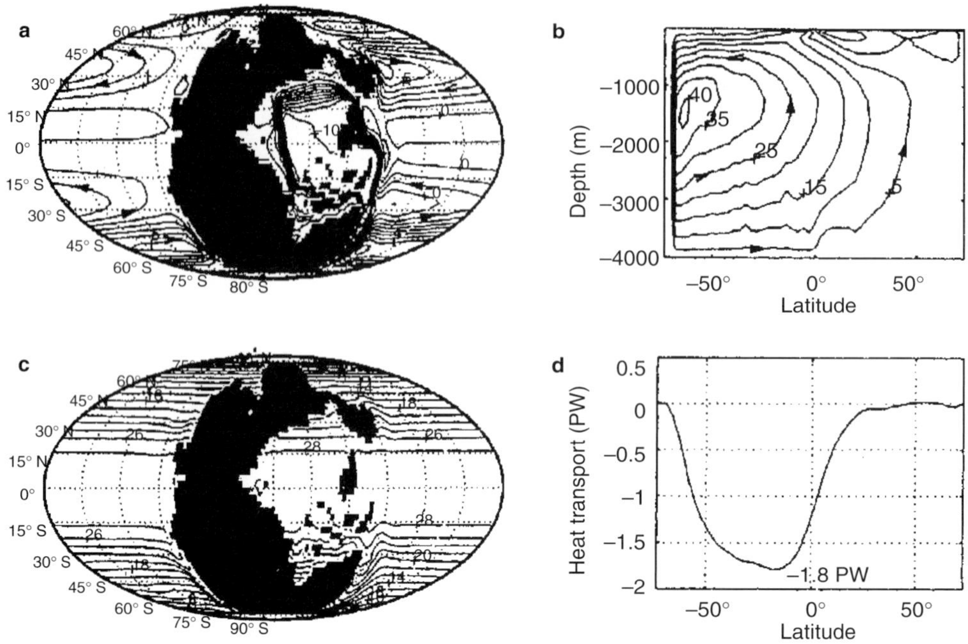

Fig. 5.170 Possible thermal mode of the Late Permian ocean circulation: **a** horizontal streamfunction (in Sv); **b** overturning streamfunction (in Sv); **c** sea surface temperature (in °C); **d** meridional oceanic heat transport (PW) (Zhang *et al.*, 2001).

have some beautiful theories and exact formulae to describe the strength of the circulation, including the Ekman transport and the Sverdrup transport.

For the thermohaline circulation, however, there is no simple formula which can be used to predict the strength of the thermohaline circulation, such as the meridional overturning rate and poleward heat flux. For the study of oceanic circulation and climate, it is desirable to use some basic parameters to predict these indexes. Since there are no exact formulae for the description of the circulation, scaling analysis has been used to describe the circulation in many studies (e.g., Bryan and Cox, 1967; Welander, 1971b; Bryan, 1987; Gnanadesikan, 1999). Although scaling analysis constitutes a very useful tool for understanding the roles of different forcing mechanisms, this approach remains empirical, and we will examine the details of this approach in this section.

Scaling of the meridional velocity

The meridional velocity can be linked to meridional pressure gradient as follows. The first step is using the thermal wind relation to connect the vertical shear of the zonal velocity with the meridional density gradient

$$fu_z = \frac{g\rho_y}{\rho_0} \tag{5.299}$$

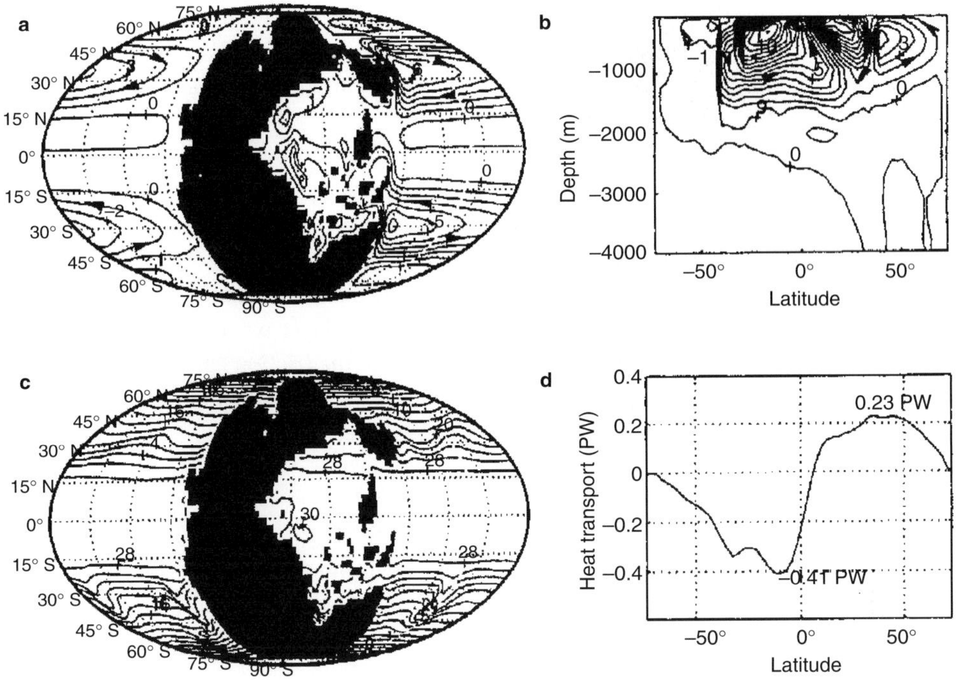

Fig. 5.171 The possible haline mode of the Late Permian ocean circulation: **a** horizontal streamfunction (in Sv); **b** overturning streamfunction (in Sv); **c** sea surface temperature (in °C); **d** meridional oceanic heat transport (in PW) (Zhang *et al.*, 2001).

This leads to a relation between the zonal velocity scale U and the scale depth of the thermocline D

$$U = \frac{g\Delta\rho}{f\rho_0 L_y} D = \frac{g'}{fL_y} D \tag{5.300}$$

where $\Delta\rho$ is the meridional density difference, $g' = g\Delta\rho/\rho_0$ is the reduced gravity, and L_y is the meridional scale of the gyre.

The next step is to create a link between the zonal geostrophic velocity scale U and the meridional velocity scale V. Physically speaking, there is no simple relation between these two components. This step is also the weakest part of the scaling analysis. We propose the following relation

$$V = cU = \frac{cg'}{fL_y} D \tag{5.301}$$

where c is an empirical constant, which is assumed to be 1 in many studies. There is no simple means available to determine what the suitable value of c is. In fact, the equivalent c for the North Pacific Ocean is negative because of the fact that the meridional volumetric flux above the main pycnocline is driven equatorward by the upwelling at high latitudes. In the North Atlantic Ocean, the meridional overturning rate can be positive, zero, or negative, despite the surface density at high latitudes always being higher than that at low latitudes.

The corresponding meridional overturning rate is

$$M = VDL_x = \frac{cg'L_x}{fL_y}D^2 \tag{5.302}$$

where L_x is the zonal scale of the gyre.

Another way of deriving the meridional velocity is by using the frictional boundary layer existing near the western boundary. Within this boundary layer we assume that the time-dependent term and the inertial terms are negligible, and the friction can be parameterized as lateral friction. Since the zonal velocity within the western boundary current is near zero, the meridional momentum equation is reduced to a two-term balance

$$A_h \frac{\partial^2 v}{\partial x^2} = \frac{1}{\rho_0}\frac{\partial p}{\partial y} \tag{5.303}$$

where A_h is the horizontal momentum diffusivity. According to Eqn. (4.1.2.44), the width of the Munk boundary layer is $\delta_M = (A_h/\beta)^{1/3}$. The meridional pressure gradient on the right-hand side of Eqn. (5.303) is scaled as

$$\frac{1}{\rho_0}\frac{\partial p}{\partial y} = \frac{g'D}{L_y} \tag{5.304}$$

Thus, the boundary balance gives rise to the meridional velocity scale V_{wbc} in the western boundary layer

$$A_h^{1/3}\beta^{2/3}V_{wbc} = \frac{g'D}{L_y} \tag{5.305}$$

The meridional velocity scale for the basin interior is linked to the meridional velocity in the western boundary layer through

$$V = c\frac{\delta_m}{L_x}V_{wbc} \tag{5.306}$$

where the constant c is introduced to take into account the dynamical effects of geometry and other factors in connection with these two velocity scales. Therefore, the meridional velocity scale for the basin interior is

$$V = \frac{cg'D}{\beta L_x L_y} \tag{5.307}$$

Finally, the meridional overturning rate is

$$M = VDL_x \frac{cg'D}{\beta L_y}D^2 \tag{5.308}$$

This scaling law is essentially the same as Eqn. (5.302) because we essentially have the relation $\beta \sim f/L_x$.

We can also use the Stommel boundary layer theory to derive a similar scaling law. In this case, the meridional momentum equation is reduced to a two-term balance

$$Rv = \frac{1}{\rho_0}\frac{\partial p}{\partial y} \tag{5.309}$$

where R is the bottom friction parameter, which should be interpreted as a crude parameterization of the baroclinic instability, as discussed in Section 4.1.3. Following the discussion above, we can derive equations similar to Eqns. (5.307) and (5.308).

The major difficulty in this scaling is how to link the meridional geostrophic velocity with the meridional pressure gradient. In a non-rotating fluid environment, it is well known that velocity is down pressure gradient and proportional to the pressure gradient. In a rotating fluid, geostrophy links the meridional pressure gradient with the zonal velocity through the thermal wind relation. Speaking tentatively, the postulation that meridional velocity is proportional to meridional pressure difference is probably not a bad assumption. As an example, let us examine the relation between bottom pressure difference and sea surface height difference between the northern and southern boundaries, diagnosed from a simple single-hemisphere model which is forced by surface differential heating and different diapycnal diffusivity only (Fig. 5.172). It can readily be seen that the meridional overturning rate and poleward heat flux are nearly proportional to the meridional pressure difference.

Circulation under a constant surface density condition

In many oceanic circulation models, strong relaxation conditions are imposed on the surface temperature and salinity; thus, $\Delta\rho$ can be treated as an external parameter. A close examination reveals that the surface density difference is slightly smaller than the imposed reference density, and this can lead to deviations from the scaling law discussed below; however, such deviations are very small and are negligible in practical applications.

From the upper surface, the volumetric flux due to wind-stress-induced Ekman pumping contributes

$$T_w = \frac{\tau}{f\rho_0}L_x \approx w_{e,0}L_xL_y \tag{5.310}$$

where τ is wind stress, and $w_{e,0}$ is the amplitude of Ekman pumping rate. From the base of the main pycnocline, the upwelling associated with thermohaline circulation drives a volumetric flux T_u. In order to estimate this volumetric flux we need to find the scale of the vertical velocity. Assuming that the vertical diffusion is balanced by vertical diffusion

$$w\rho_z = \kappa\rho_{zz} \tag{5.311}$$

where κ is the turbulent diapycnal diffusivity. This equation leads to the scale of the vertical velocity

$$W = \kappa/D \tag{5.312}$$

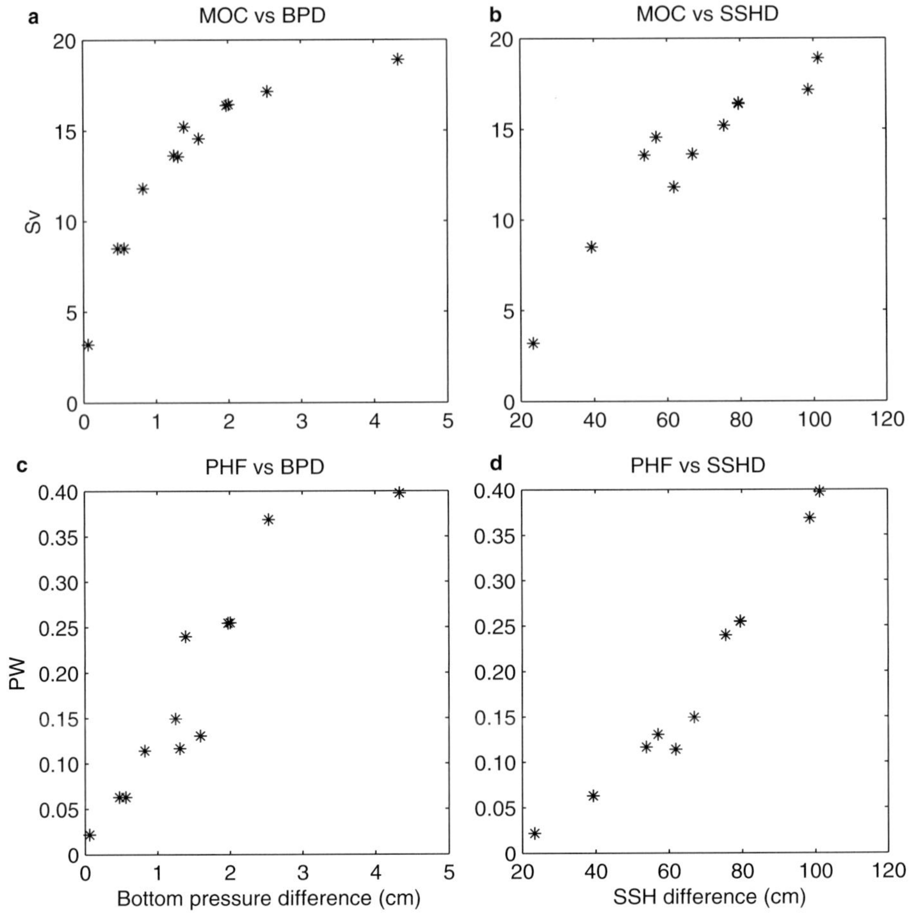

Fig. 5.172 The dependence of the meridional overturning rate (MOC) (**a**, **b**) and poleward heat flux (PHF) (**c**, **d**) on the bottom pressure difference (BPD) between the northern and southern boundaries (**a**, **c**, in cm of water column height) and the sea surface height difference (SSHD) between the northern and southern boundaries (**b**, **d**, in cm), diagnosed from a mass-conserving oceanic general circulation model (OGCM) with different diapycnal diffusivity and no wind forcing.

Thus, the total upwelling rate in a basin is

$$T_u = \frac{\kappa L_x L_y}{D} \tag{5.313}$$

As shown in Figure 5.173, the volumetric transport of the basin is

$$M = T_w + T_u \tag{5.314}$$

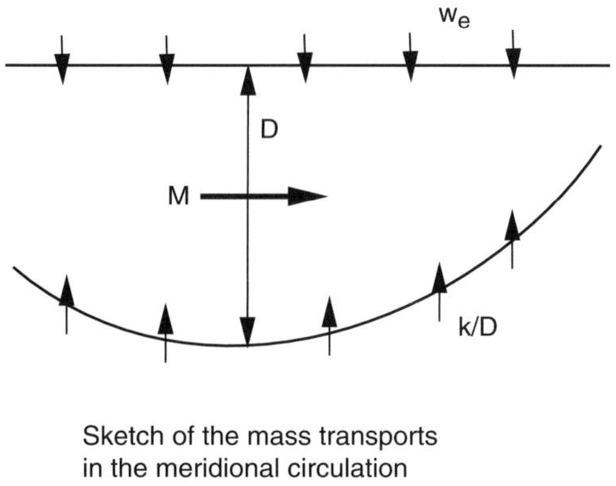

Fig. 5.173 Sketch of the volumetric transport in the meridional circulation.

Using Eqn. (5.312), this relation leads to an equation determining the depth of the pycnocline:

$$D^3 - \frac{w_{e,0} f L_y^2}{cg'} D = \frac{\kappa f L_y^2}{cg'} \tag{5.315}$$

According to dimensional analysis, each term of this equation should have the same dimension, $[L^3]$; thus, we can introduce two length scales and rewrite this equation as follows:

$$D^3 - d_w^2 D = d_\kappa^3 \tag{5.316}$$

where

$$d_w = \left(\frac{w_{e,0} f L_y^2}{cg'} \right)^{1/2} = a^{1/2} \tag{5.317}$$

$$d_\kappa = \left(\frac{\kappa f L_y^2}{cg'} \right)^{1/3} = b^{1/3} \tag{5.318}$$

are the scaling depths of the pycnocline due to Ekman pumping and diapycnal mixing.

For this cubic equation, there are three real roots, if the following condition is satisfied:

$$d_w \geq \left(\frac{27}{4} \right)^{1/6} d_\kappa \approx 1.37 d_\kappa \tag{5.319}$$

For a subtropical basin, we assume that $L_y = 3{,}000$ km, $g' = 0.015\ \mathrm{m/s^2}, f = 10^{-4}\ \mathrm{s^{-1}}$, $w_{e,0} = 10^{-6}$ m/s, $\kappa = 10^{-5}\ \mathrm{m^2/s}$, $c = 1$; so that the typical values are $d_w \approx 245$ m, $d_\kappa \approx 84$ m. Therefore, the depth of the pycnocline is primarily controlled by the Ekman pumping term, which is linearly proportional to wind stress τ. In general, there are two limits worth noticing:

The case of strong wind and weak diapycnal mixing, or $d_w \gg d_k$. Using d_w as the length scale and introducing the nondimensional thermocline depth, $d = D/d_w$, equation (5.316) is reduced to

$$d^3 - d = \varepsilon \tag{5.320}$$

where $\varepsilon = (d_k/d_w)^3 \ll 1$. This cubic equation has three real roots, $d = 1+O(\varepsilon), -1+O(\varepsilon)$, and $-\varepsilon$; but only the positive root is physically meaningful. Using the perturbation method, this positive root is approximately

$$d = 1 + \frac{\varepsilon}{2} - \frac{3}{8}\varepsilon^2 + \frac{1}{2}\varepsilon^3 + \cdots \tag{5.321}$$

In dimensional form, this is

$$D_w = d_w \left(1 + \frac{b}{2a^{3/2}} - \frac{3b^2}{8a^3} + \frac{b^3}{2a^{9/2}} + \cdots\right) \sim \left(\frac{w_{e,0} f L_y^2}{cg'}\right)^{1/2} \tag{5.322}$$

and the corresponding meridional overturning rate is

$$M_w = c\frac{g' L_x}{f L_y} d_w^2 \left(1 + \frac{b}{a^{3/2}} + \cdots\right) \approx w_{e,0} L_x L_y \tag{5.323}$$

Therefore the pycnocline depth is proportional to the square root of the wind stress (or the Ekman pumping rate), and the meridional overturning rate is linearly proportional to the wind stress. As shown in Eqn. (4.31) in Section 4.1.2, in the reduced-gravity model the depth of the main thermocline is proportional to the square root of the meridional gradient of wind stress (i.e., the Ekman pumping rate) and inversely proportional to the reduced gravity. In the present case, the diapycnal mixing is relatively weak, so it can only make a small linear correction to both the pycnocline depth and the meridional overturning rate.

The case of weak wind and strong diapycnal mixing, or $d_w \ll d_\kappa$. Using d_κ as the length scale and introducing a nondimensional layer depth, $d = D/d_\kappa$, Eqn. (5.315) is reduced to the following nondimensional form:

$$d^3 - \lambda d = 1 \tag{5.324}$$

where $\lambda = (d_w/d_\kappa)^2 \ll 1$. Note that this equation has three real roots, one positive and two negative, for $\lambda \geq 1.89$. However, for the case with $\lambda \ll 1$ this cubic equation has only one real root and it is positive. Using the perturbation method, this root is approximately

$$d = 1 + \frac{\lambda}{3} - \frac{\lambda^3}{81} + \cdots \tag{5.325}$$

or in dimensional form

$$D_\kappa = d_\kappa \left(1 + \frac{a}{3b^{2/3}} - \frac{a^3}{81b^2} + \cdots\right) \tag{5.326}$$

and the meridional overturning rate is

$$M_w = c\frac{g'L_x}{fL_y}d_\kappa^2(1 + \frac{2a}{3b^{2/3}} + \cdots) \approx \left(\frac{cg'\kappa^2L_x^3L_y}{f}\right)^{1/3} \quad (5.327)$$

Therefore, the pycnocline depth is proportional to the $\frac{1}{3}$ power of the diffusivity, and the meridional overturning rate is proportional to the $\frac{2}{3}$ power of the diffusivity. In this case the wind stress makes a small correction to both the pycnocline depth and the meridional overturning rate.

Scaling of the pycnocline thickness

For the case with wind-driven circulation in the upper ocean, the depth of the main pycnocline and the thickness of the main pycnocline are different, so we need to find the scaling law for these two length scales (Samelson and Vallis, 1997). As shown in Figure 5.174, the main pycnocline has a depth scale D and a thickness scale of δ.

Therefore, within the main pycnocline the thermal wind relation gives the following scaling relation:

$$\frac{fV}{\delta} \approx \frac{g\Delta\rho}{\rho_0 d} = \frac{g'}{d} \quad (5.328)$$

where δ is the scale thickness of the main thermocline and d is the width of the main pycnocline; the latter is linked to L_y, the meridional scale of the circulation, through the following relation:

$$d \approx \delta L_y/D \quad (5.329)$$

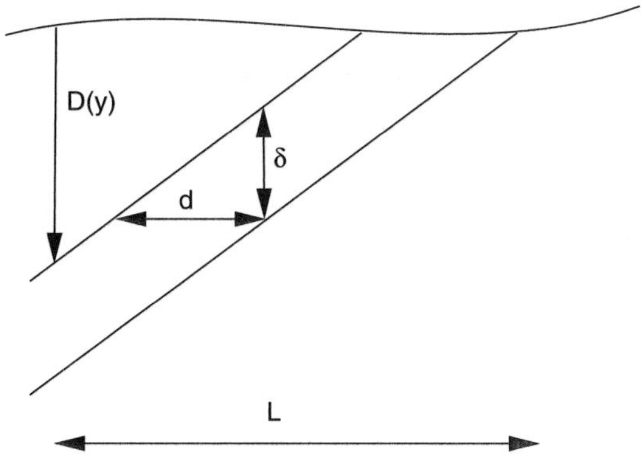

Fig. 5.174 Sketch of the main pycnocline, indicating the basis for the scaling analysis.

where D is the vertical scale depth of the pycnocline discussed above. Using the continuity and the balance between the vertical heat transport and vertical diffusion, we obtain

$$\frac{V}{L_y} \approx \frac{W}{\delta} \approx \frac{\kappa}{\delta^2} \tag{5.330}$$

From equations (5.328), (5.39) and (5.330), we obtain the scale thickness for the main pycnocline:

$$\delta \approx \left(\frac{f\kappa L_y^2}{g'D}\right)^{1/2} \tag{5.331}$$

There are two cases. First, for the case when diffusion dominates, the vertical scale of the pycnocline depth and the pycnocline thickness is the same; thus, Eqn. (5.331) gives rise to the same scale as Eqn. (5.318). Second, for the case with strong wind forcing and weak diapycnal mixing, the pycnocline depth obeys Eqn. (5.317), so the final expression of pycnocline thickness is

$$\delta \approx \left(\frac{f\kappa^2 L_y^2}{g'w_{e,0}}\right)^{1/4} \tag{5.332}$$

Note that the depth and thickness of the pycnocline depend on the relative strength of the Ekman pumping and the diffusion (Table 5.8). In the case with strong wind forcing, the thickness of the main pycnocline is linearly proportional to $\kappa^{1/2}$, but inversely proportional to the $\frac{1}{4}$ power of the Ekman pumping rate. This means that strong wind forcing produces a deep and thin main pycnocline.

Circulation forced by surface freshwater flux

A relaxation condition is not suitable for the haline circulation, and a more realistic way of simulating the haline circulation is to specify the freshwater flux at the surface. As discussed in Section 5.3, the salinity balance in the meridional cell is

$$VDL_x(\bar{S} - \Delta S) = \left(VDL_x - L_x L_y E\right)\bar{S} \tag{5.333}$$

where D is the depth of the halocline, $\bar{S}$ is basin mean salinity, ΔS is the meridional salinity difference, and E is the rate of evaporation minus precipitation. This leads to a simple relation between the scale of salinity difference and the evaporation rate:

$$VD\Delta S = \bar{S}EL_y \tag{5.334}$$

Accordingly, the meridional density difference is

$$\Delta\rho = \frac{\rho_0\beta_s\bar{S}EL_y}{VD} \tag{5.335}$$

Table 5.8. *Scales for the pycnocline and vertical velocity for the cases with Ekman pumping dominating and diffusion dominating*

Dominating forcing	Depth of the pycnocline	Thickness of the pycnocline	Vertical velocity within the main pycnocline	Meridional overturning rate
Ekman pumping	$D \simeq \left(\frac{fL_y^2 w_{e,0}}{cg'}\right)^{1/2}$	$\delta \simeq \left(\frac{f\kappa^2 L_y^2}{g' w_{e,0}}\right)^{1/4}$	$W \sim \frac{\kappa}{\delta} \sim \kappa^{1/2}$	$M = w_{e,0} L_x L_y$
Vertical diffusion	$D \simeq \delta \simeq \left(\frac{f\kappa L_y^2}{cg'}\right)^{1/3}$		$W \sim \frac{\kappa}{\delta} \sim \kappa^{2/3}$	$M = \left(\frac{cg'\kappa^2 L_x^3 L_y}{f}\right)^{1/3}$

where β_s is the salt contraction coefficient. Assuming that $V = cU$, Eqn. (5.300) is reduced to

$$V = \left(\frac{cg\beta_s \bar{S} E}{f}\right)^{1/2} \tag{5.336}$$

Thus, the meridional velocity scale is proportional to the square root of the mean salinity and the amplitude of freshwater flux, and is independent of wind stress and diffusivity. The corresponding meridional volumetric flux is

$$M = VDL_x \tag{5.337}$$

The volumetric balance equation (5.315) leads to

$$D^2 - h_w D - h_\kappa^2 = 0 \tag{5.338}$$

where

$$h_w = w_{e,0} L_y / V, \quad h_\kappa = \left(\frac{\kappa L_y}{V}\right)^{1/2} \tag{5.339}$$

are the halocline depth scales associated with wind stress and diffusion.

The solution of Eqn. (5.338) is

$$D = \frac{h_w + \sqrt{h_w^2 + 4h_\kappa^2}}{2} \tag{5.340}$$

There are two limit cases:

The case of strong wind stress and weak diapycnal mixing

Introduce a small parameter:

$$\varepsilon = h_\kappa / h_w = \frac{1}{w_{e,0}} \left(\frac{\kappa V}{L_y} \right)^{1/2} \ll 1 \tag{5.341}$$

so from Eqn. (5.340), the halocline scale depth is

$$D = h_w (1 + \varepsilon^2 - 2\varepsilon^4 + \cdots) = \frac{w_{e,0} L_y}{V} \left(1 + \varepsilon^2 + \cdots \right) \tag{5.342}$$

and the meridional overturning rate is

$$M = w_{e,0} L_x L_y \left(1 + \varepsilon^2 + \cdots \right) \tag{5.343}$$

therefore, both the halocline depth and the meridional overturning rate are linearly proportional to the Ekman pumping rate, and the diffusivity makes a small linear correction.

The case of weak wind stress and strong diapycnal mixing

Introduce a small parameter:

$$\lambda = h_w / h_\kappa = w_{e,0} \left(\frac{L_y}{\kappa V} \right)^{1/2} \ll 1 \tag{5.344}$$

so from Eqn. (5.340), the halocline scale depth is

$$D = h_\kappa \left(1 + \frac{\lambda}{2} + \frac{\lambda^2}{8} - \frac{\lambda^4}{64} + \cdots \right) = \left(\frac{\kappa L_y}{V} \right)^{1/2} \left(1 + \frac{\lambda}{2} + \cdots \right) \tag{5.345}$$

and the meridional overturning rate is

$$M = \left(\kappa V L_y \right)^{1/2} L_x \left(1 + \frac{\lambda}{2} + \cdots \right) \approx \left(\kappa L_y L_x^2 \right)^{1/2} \left(\frac{c g \beta_s \bar{S} E}{f} \right)^{1/4} \left(1 + \frac{\lambda}{2} \right) \tag{5.346}$$

thus, both the halocline depth and the meridional overturning rate are proportional to the $\frac{1}{2}$ power of the diffusivity, as discussed by Huang and Chou (1994), while the Ekman pumping rate makes a small linear correction. This displays a much stronger dependence on the diapycnal mixing than the $\frac{1}{3}$-power law for the case of the relaxation conditions discussed above.

In addition, the strength of the meridional circulation depends on the $\frac{1}{4}$ power of the precipitation amplitude and the mean salinity in the ocean. For example, if there were no salt in the ocean, the barotropic gyres described by Goldsbrough (1933) would be the only circulation driven by evaporation and precipitation, with no baroclinic return flow at all.

Circulation under mixed boundary conditions

The general case is circulation under mixed boundary conditions, i.e., the thermal balance is forced by a temperature relaxation toward a specified reference temperature, but the salt balance is driven by the specified air–sea freshwater flux. The meridional density difference corresponding to Eqn. (5.335) is now

$$\Delta\rho = -\rho_0\alpha\Delta T + \frac{\rho_0\beta_s\bar{S}EL_y}{VD} \tag{5.347}$$

The case of fixed diapycnal diffusivity

Assuming that the zonal velocity and the meridional velocity obey the same empirical relation, $V = cU$, the thermal wind relation, Eqn. (5.300), is reduced to

$$V^2 = -\frac{cg\alpha\Delta T}{fL_y}DV + \frac{cg\beta_s\bar{S}E}{f} \tag{5.348}$$

This equation includes the two limit cases corresponding to the cases discussed above, i.e., the case with fixed reference temperature and the case with a fixed rate of freshwater flux specified at the upper surface. The solution to this equation is

$$V = \frac{cg\alpha\Delta T}{2fL_y}D \pm \left[\left(\frac{cg\alpha\Delta T}{2fL_y}D\right)^2 + \frac{cg\beta_s\bar{S}E}{f}\right]^{1/2} \tag{5.349}$$

Substituting this relation into the meridional volumetric flux balance equation leads to a cubic equation, which includes a square root term in the cubic term. This equation can be solved using the same perturbation method (Zhang *et al.*, 1999).

The case of fixed rate of energy sustaining mixing

Following discussion in the subsection entitled "Extension of the two-box model: The energy constraint and wind-driven gyration" in Section 5.4.2, the total amount of external mechanical energy supporting diapycnal mixing $\dot{E}_m$ is fixed, which is used to support the diapycnal mixing, i.e.,

$$\dot{E}_m = -gHAW\Delta\rho = g\rho_0 HAW(\alpha\Delta T - \beta_s\Delta S) \tag{5.350}$$

where H and A are the mean depth and the total horizontal area of the ocean. The salinity balance associated with the meridional overturning cell gives rise to a simple relation

$$\Delta S = \bar{S}E/W \tag{5.351}$$

where E is the freshwater flux. Thus, Eqn. (5.350) is reduced to

$$W = \frac{e + \beta_s\bar{S}E}{\alpha\Delta T} \tag{5.352}$$

where $e = \dot{E}_m / g\rho_0 HA$ represents the strength of the external source of mechanical energy and, under the temperature relaxation condition $\Delta T \simeq \Delta T^*$, can be used as a good approximation (ΔT^* is the meridional difference of reference temperature).

The corresponding meridional density difference can be rewritten as

$$\Delta\rho = \rho_0 \left(\alpha \Delta T^* - \beta_s \bar{S} E / W \right) \tag{5.353}$$

Using the scaling relation between the meridional and vertical velocity $V = WL_y/D$, Eqn. (5.301) is reduced to the pycnocline depth scale

$$D = \left[\frac{f(e + \beta_s \bar{S} E)\left(1 + \beta_s \bar{S} E / e\right)}{cg} \right]^{1/2} \frac{L_y}{\alpha \Delta T^*} \tag{5.354}$$

The corresponding poleward heat flux is defined as

$$H_f = c_p \rho_0 W A \Delta T = c_p \rho_0 A (e + \beta_s \bar{S} E)/\alpha \tag{5.355}$$

The scaling analysis implies that strong freshwater forcing enhances the upwelling rate, and thus the meridional overturning rate and poleward heat flux. On the other hand, a large meridional temperature difference implies strong stratification, and thus it can suppress the overturning rate and poleward heat flux. This scaling law has been verified by some idealized model experiments (Huang, 1999; Nilsson *et al.*, 2003); however, model behavior under energy constraint may not exactly follow the scaling law discussed above, and the suitable scaling law remains unclear for the general cases.

A scaling law for the meridional circulation in the Atlantic Ocean

The pycnocline depth and the meridional overturning cell in the Atlantic Ocean are closely related. In order to consider a scaling law for these two aspects of the circulation, it is very important to include the crucial role played by the ACC and the associated wind stress forcing due to the southern westerly (Gnanadesikan, 1999). The basic methodology is a volumetric transport balance of all the relevant components of the circulation system, as depicted in Figure 5.175.

According to Eqn. (5.308), meridional volume flux associated with the northern sinking is

$$T_N = \frac{cg'}{\beta L_y^n} D^2 \tag{5.356}$$

where L_y^n is the meridional length scale of the North Atlantic Basin, and D is the scale depth of the main pycnocline.

A major physical process unique to the Southern Ocean is the strong eddy activity associated with the ACC. The eddy-induced flow gives rise to a reduction of the transport, indicated by the T_E term on the left-hand edge of Figure 5.175.

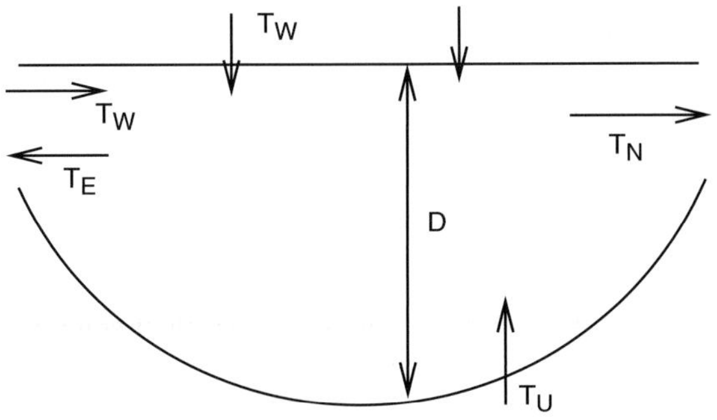

Fig. 5.175 Sketch of the pycnocline and meridional overturning rate for the Atlantic Ocean, including the contribution due to eddies.

Physically, eddy activity reduces the horizontal density gradient, leading to a decline of the isopycnal slope. Imagining the isopycnal surface as a rubber bow, then baroclinic instability tends to bend the southern edge of the rubber bow and makes it flatter, thus causing the water to leak out. This is a sink of water, which can be written as

$$T_E = -v_{eddy} D L_x \tag{5.357}$$

where L_x is the circumference of the Earth at the latitude of the Drake Passage, $v_{eddy} = A_I \partial S_l / \partial z = A_I / L_y^s$, where A_I is an eddy diffusion coefficient, $S_l = D/L_y^s$ is the slope of the isopycnals, and L_y^s is the meridional width of the pycnocline associated with the ACC.

The contribution due to wind stress is

$$T_W = \tau L_x / f \rho_0 \tag{5.358}$$

and the contribution due to upwelling is

$$T_U = \kappa L_x L_y / D \tag{5.359}$$

where L_y is the meridional scale of the wind-driven circulation. The volume flux balance for the whole basin is

$$T_U + T_W + T_E = T_N \tag{5.360}$$

which leads to a cubic equation for D. Once again, dimensional homogeneity of this equation suggests that we write it in the following form:

$$D^3 + d_e D^2 - d_w^2 D - d_\kappa^3 = 0 \tag{5.361}$$

where three pycnocline depth scales are introduced, i.e., the eddy-controlled pycnocline depth scale d_e, the wind stress-controlled pycnocline depth scale d_w, and the diffusion controlled pycnocline depth scale d_κ

$$d_e = \frac{\beta A_I L_x L_y^n}{cg' L_y^s} \tag{5.362}$$

$$d_w = \left(\frac{\tau \beta L_x L_y^n}{cg' f \rho_0}\right)^{1/2} \tag{5.363}$$

$$d_\kappa = \left(\frac{\kappa \beta L_x L_y L_y^n}{cg'}\right)^{1/3} \tag{5.364}$$

Assume the following values for the basic parameters of the circulation: $f = 10^{-4}$/s, $\beta = 2 \times 10^{-11}$/s/m, $L_x = 2.5 \times 10^7$ m, $L_y = 10^7$ m, $L_y^n = L_y^s = 1.5 \times 10^6$ m, $A_I = 1{,}000$ m^2/s, $c = 0.16$, $g' = 0.01$ m/s^2, $\kappa = 10^{-5}$ m^2/s and the typical values for these depth scales are: $d_e \simeq 313$m, $d_w \simeq 685$m, and $d_\kappa \simeq 368$m.

Comparing these three depth scales, it is clear that pycnocline depth associated with wind stress is the dominating scale; thus, it is convenient to use this depth scale and introduce the nondimensional depth $d = D/d_w$, for which the corresponding equation is

$$d^3 + \lambda d^2 - d - \varepsilon = 0 \tag{5.365}$$

where $\lambda = d_I/d_w \simeq 0.457 < 1$ and $\varepsilon = (d_\kappa/d_w)^3 \simeq 0.16 \ll 1$. Treating both λ and ε as small parameters, the series solution of this equation is

$$d = \left(1 - \frac{\lambda}{2} + \frac{\lambda^2}{8}\right) + \left(\frac{1}{2} + \frac{\lambda}{4} - \frac{\lambda^3}{32}\right)\varepsilon + \left(-\frac{3}{8} - \frac{\lambda}{2} - \frac{15\lambda^2}{64}\right)\varepsilon^2 + \cdots \tag{5.366}$$

or

$$d = 1 + \frac{\varepsilon - \lambda}{2} + \text{higher-order terms} \tag{5.367}$$

In the dimensional units, the pycnocline depth and the northern sinking rate are

$$D = \left(\frac{\tau \beta L_x L_y^n}{cg' f \rho_0}\right)^{1/2} \left(1 + \frac{\varepsilon - \lambda}{2}\right) \tag{5.368}$$

$$T_N = \frac{\tau L_x}{f \rho_0}(1 - \lambda + \varepsilon) \tag{5.369}$$

Thus, under the current climate conditions, North Atlantic Deep Water is primarily controlled by wind stress over the ACC, the eddy term is very important, and diapycnal mixing plays a relatively minor role.

In other words, to the lowest order, pycnocline depth and the meridional overturning rate are primarily controlled by the Ekman pumping rate. To the next order, eddy mixing tends to reduce the pycnocline depth and northern sinking rate, and this is consistent with the physical conception that baroclinic instability tends to flatten isopycnal surfaces and to reduce the northern sinking rate (i.e., it is a leakage in the circulation system). In addition, deep ocean upwelling enhances the pycnocline depth. However, the contribution of diapycnal mixing to the pycnocline depth and the meridional overturning rate is rather small, on the order of 16% only. Thus, under current climate conditions, diapycnal mixing in the deep ocean is not the most crucial controller for the meridional overturning cell associated with the NADW in the Atlantic Ocean (Toggweiler and Samuels, 1998).

The dominant controller is wind stress associated with the southern Jet Stream, and it appears in the lowest-order term and the "first-order" correction term associated with λ. Note that we have assumed a kind of eddy-mixing parameterization $A_I \simeq 1,000$ m^2/s in order to decide the depth scale of eddy mixing. The nondimensional parameter λ is about 0.46; thus, it is not really a smaller parameter. As a result, treating the eddy term as a small contribution term may not be valid in general.

In fact, eddy dynamics may play a critically important role in the Southern Ocean. Over the past couple of decades, the southern westerly has intensified (Yang *et al.*, 2007); however, both *in situ* observations and numerical experiments based on a fine-resolution model have indicated that the zonal transport of the ACC remains roughly the same. This phenomenon is called eddy saturation (Marshall *et al.*, 1993; Hallberg and Gnanadesikan, 2001). On the other hand, although results from numerical simulations with fine resolution indicate that strong wind stress can lead to stronger meridional circulation in the North Atlantic Ocean, there are no observations supporting this claim; thus, this remains controversial (Hallberg and Gnanadesikan, 2006). ***It is clear that the oceans behave in a much more complex manner than the simple scaling law discussed here would indicate.***

5.5.2 Interaction between wind-driven and deep circulations

Wind-driven horizontal gyre and thermohaline overturning cell

Combined wind-driven and thermohaline circulations

In the previous chapters, we discussed theories about wind-driven and thermohaline circulations, treating them separately. In reality, however, the oceanic general circulation is a combination of wind-driven circulation and thermohaline circulation. Conceptually, the ocean can be divided into three layers (Fig. 5.176). The top 30 m is the Ekman layer, where wind stress drives the horizontal Ekman transport, and its convergence and divergence give rise to Ekman pumping in the subtropical basin and Ekman upwelling in the subpolar basin. Within the depth range of 30–1,500 m the wind-driven subtropical, subpolar, and equatorial gyres are the dominating features. In addition to the linear Sverdrup gyre in the ocean interior, there is a strong recirculation in the northwestern corner of the subtropical gyre. Motions below 1,500 m are dominated by the thermohaline circulation, including the deepwater formation at high latitudes, equatorward deep western boundary currents, and the slow upwelling in the basin interior.

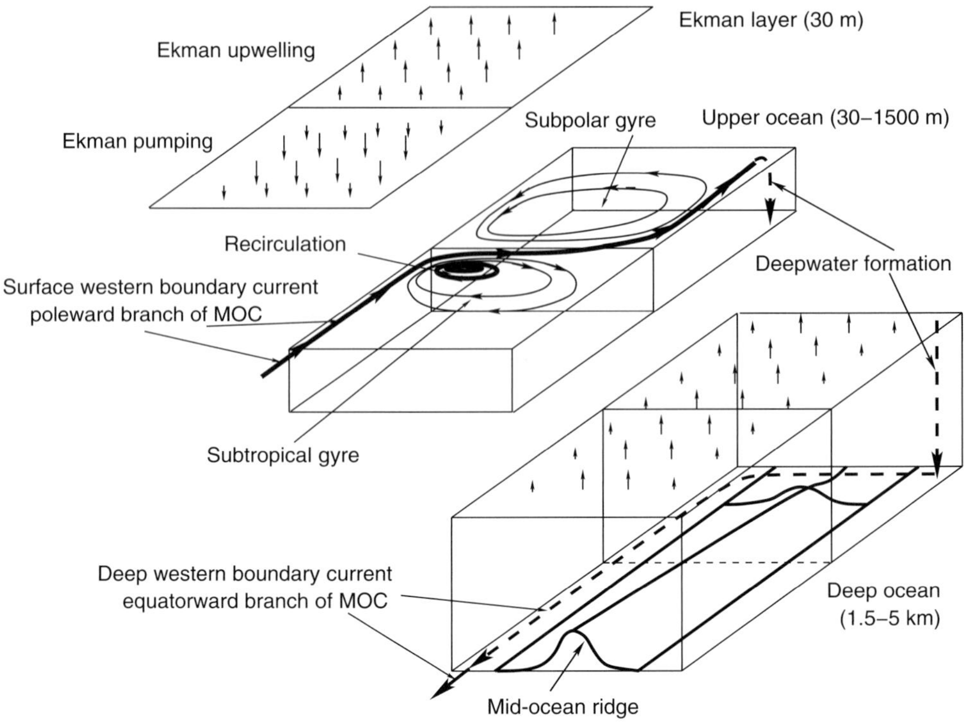

Fig. 5.176 Sketch of the interaction between the wind-driven gyre and the thermohaline circulation in a Northern Hemisphere basin.

Superimposed on the wind-driven circulation in the upper ocean, there is poleward mass flux induced by the thermohaline circulation, i.e., the upper branch of the meridional overturning cell (MOC), indicated by the arrow on the western side of the middle box in Figure 5.176. This branch of thermohaline circulation moves through the entire meridional length of the upper ocean dominated by the wind-driven gyre, and eventually leaves the upper ocean and sinks to the deep ocean through deepwater formation in the northeast corner of the basin.

In the deep ocean, the circulation is primarily of the thermohaline type. The deep circulation can be traced back to the deepwater source in the northeast corner of the basin. In addition, the recirculation regime is the place of strong cooling and mode water formation. Thus, this is also a branch of the oceanic general circulation, which links the wind-driven circulation with the shallow part of the thermohaline circulation. A comprehensive picture of the oceanic general circulation should include this branch; however, for simplicity, this component of the circulation is not included in the sketch.

Deep water gradually warms up and returns to the upper ocean as a slow and broad upwelling, as shown in the lowest part of Figure 5.176. Deepwater upwelling is not uniform

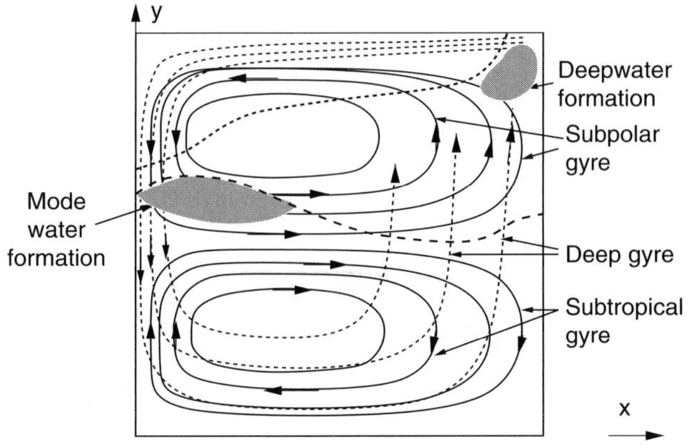

Fig. 5.177 Schematic structure of a two-gyre circulation in a multi-layer model.

throughout the ocean. In particular, upwelling is stronger wherever mechanical-energy-sustaining diapycnal mixing is abundant, such as near the mid-ocean ridge. It is necessary to emphasize that the strongest upwelling system in the world's oceans is associated with the westerlies over the ACC; however, this is not included in the diagram.

The most outstanding features in the framework of the oceanic general circulation are the western boundary current in the upper kilometer and its northeastward extension at mid/high latitudes. In addition, there is a strong deep western boundary current stemming from the deepwater formation site and the corresponding return flow in the upper ocean.

The connection between the wind-driven gyres in the upper ocean and the thermohaline circulation in the deep ocean is further illustrated in Figure 5.177, where the horizontal view of circulation in a $4\frac{1}{2}$-layer model is shown. This model ocean includes a mixed layer of a uniform depth of 50 m (layer 1), and three moving layers below (layers 2, 3 and 4), plus a bottom layer (layer 5) which is very thick and motionless (Huang, 1989b).

The model is forced by wind stress, and the temperature in the mixed layer is relaxed to a linear profile of reference temperature. In Figure 5.177 the heavy dashed lines depict the outcrop lines (between layer 2 and layer 3 and between layer 3 and layer 4); lines with arrows depict streamlines of the wind-driven circulation in the upper ocean; and dashed lines with arrows depict the slow deep circulation driven by deep mixing and upwelling.

Deep water formed at the northeastern corner flows westward and continues its movement as the deep western boundary current depicted by the dashed arrows in Figure 5.177. Deep water returns to the ocean interior and the upper ocean through upwelling driven by mechanical-energy-sustained mixing. Due to the basin-wide upwelling driven by the deepwater formation at the northeastern corner, there is cyclonic circulation in the deep ocean, depicted by the dashed arrows.

The downward branches of the thermohaline circulation are quite narrow. On the other hand, the upward branch of the thermohaline circulation is rather broad. Of course,

Figure 5.177 represents a highly idealized circulation in a single-hemisphere basin. Many dynamical processes, such as the flow across the equator, flow over bottom topography, and flow induced by non-uniform mixing in the deep ocean, make the dynamical picture much more complicated.

Meridional overturning circulation in different coordinates

Thermohaline circulation, in particular the poleward heat and freshwater transports associated with the meridional overturning circulation, are key components in the climate system on Earth. The meridional overturning circulation rate in geopotential height coordinates has been widely used as an index for the strength of the thermohaline circulation. For example, the zonally integrated meridional streamfunction in the Atlantic Ocean diagnosed from the annual mean circulation, including temperature, salinity and velocity, for the year 2007 from the SODA (Carton and Giese, 2008) data are shown in Figure 5.178a. One of the most important features in this figure is the two-cell-structure of circulation. The clockwise meridional circulation in the upper part of the water column is clearly related to the NADW, and this includes the northward transport of warm water in the upper kilometer, the formation of NADW at high latitudes, and the return flow at mid depth ($\sim$3.5 km). This relatively shallow cell is the dominating part of the meridional circulation; there is a secondary meridional cell rotating in an anticlockwise direction. This cell is located below 3.5 km and represents the deep overturning related to the northward transport of AABW and its modification due to mixing between AABW and NADW.

In many studies, the thermohaline circulation is characterized in terms of a meridional overturning circulation rate defined as the maximum of the meridional overturning streamfunction, which can be identified from Figure 5.178. Since there are at least two overturning cells, a single index, such as the meridional overturning circulation rate, is not enough for the description of the circulation.

Furthermore, thermohaline circulation is a complicated phenomenon in three-dimensional space. Poleward heat flux carried by the ocean is closely related to the meridional overturning circulation, horizontal gyration, and other aspects of the wind-driven and thermohaline circulation. Therefore, the traditional meridional overturning rate defined in the θ–z coordinates (θ is the latitude) may not be the best index for climate study. The role played by horizontal wind-driven gyres can be included through other diagnosis methods, as discussed in the second part of this section.

In this section, we introduce other coordinates which can be used to map out thermohaline circulation, such as the potential density coordinates, potential temperature, and salinity. Using these coordinates, the meridional overturning circulation in the Atlantic Ocean, diagnosed from SODA (Carton and Giese, 2008), can be calculated; the corresponding streamfunction maps are shown in Figure 5.178. The corresponding maximal overturning rates in the northern North Atlantic Ocean (30° N and higher) in these four coordinates are 15.0 Sv (in θ–z coordinates); 23.1 Sv (in θ–σ_2 coordinates, where σ_2 is potential density, using 2,000 db as the reference pressure); 29.6 Sv (in θ–Θ coordinates, where Θ is potential temperature); and 18.9 Sv (in θ–S coordinates, where S is salinity).

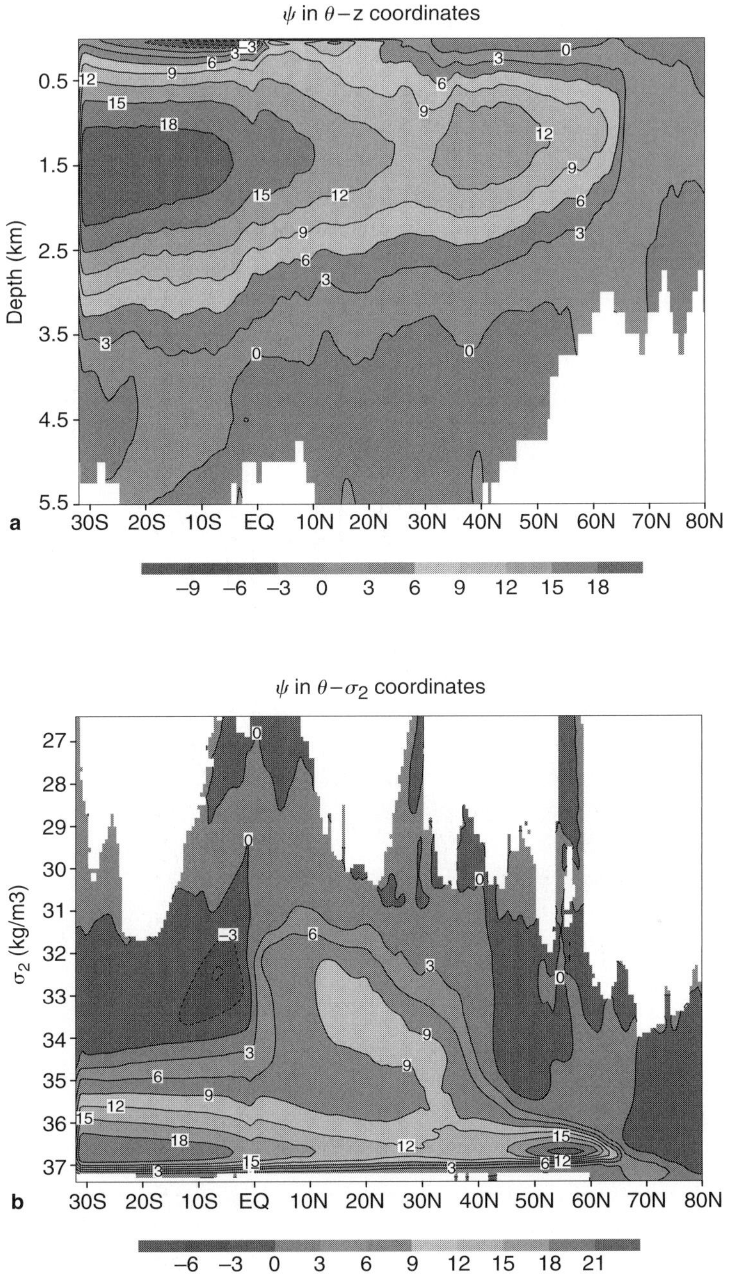

Fig. 5.178 **a** Annual-mean meridional overturning circulation of the Atlantic Ocean, θ–z coordinates, in Sv; **b** annual-mean meridional overturning circulation of the Atlantic Ocean, θ–σ coordinates, in Sv; **c** annual-mean meridional overturning circulation of the Atlantic Ocean, θ–Θ coordinates, in Sv (overleaf); **d** annual-mean meridional overturning circulation of the Atlantic Ocean, θ–S coordinates, in Sv (overleaf). See color plate section.

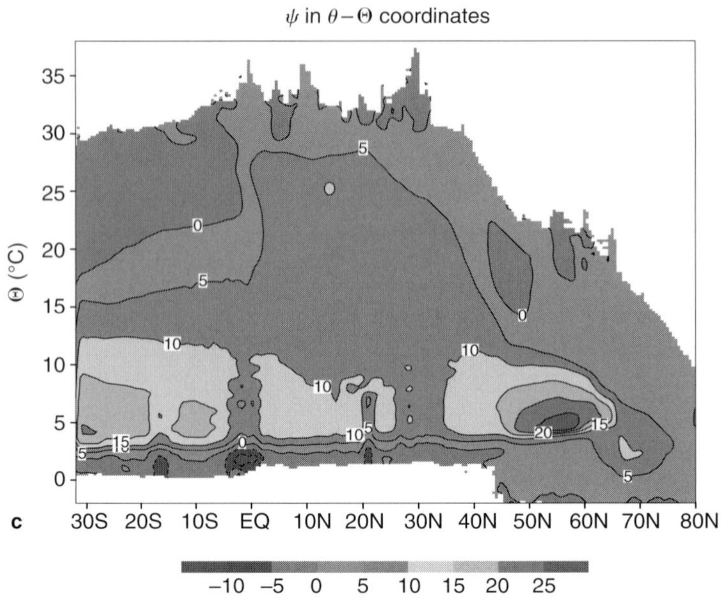

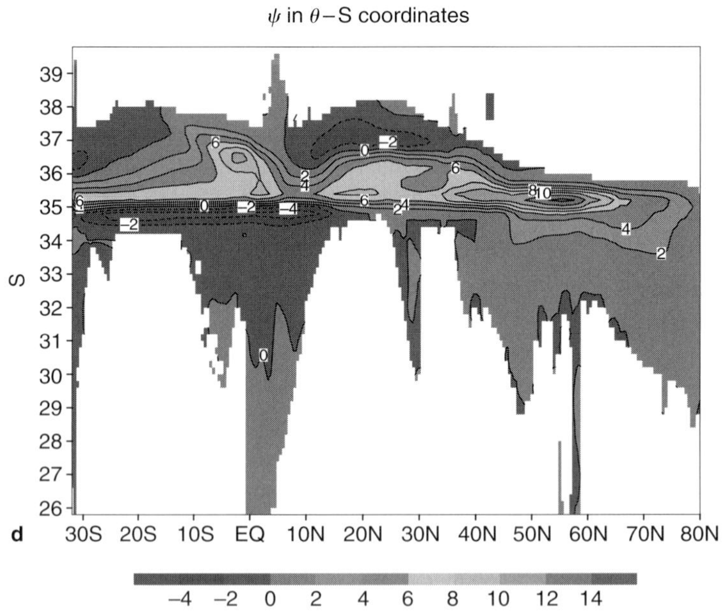

Fig. 5.178 Continued.

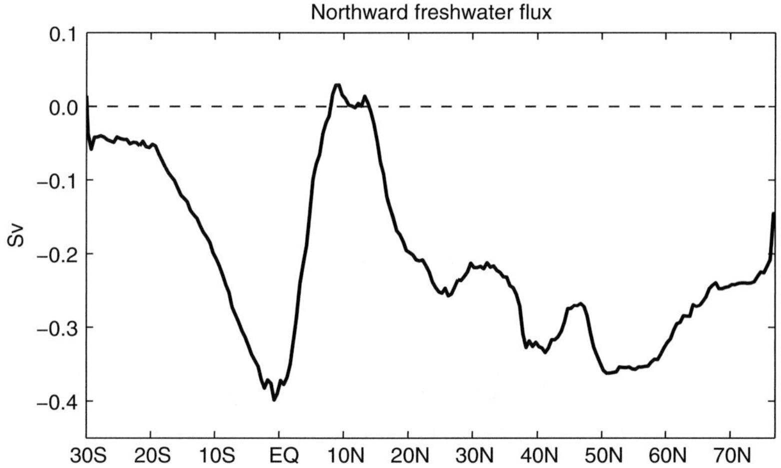

Fig. 5.179 Annual-mean northward freshwater transport in the Atlantic Ocean, in Sv.

In particular, the overturning streamfunction in potential temperature coordinates is closely related to poleward heat flux; therefore, for climate study, the overturning streamfunction in temperature coordinates may serve as the best diagnosis tool. The cores of the overturning streamfunction maps, labeled in red, represent the poleward transport of surface water and the return flow at depth.

Since the model used to generate the SODA data is a Boussinesq model, the time-dependent meridional salt transport diagnosed from the model has no clear physical meaning. As discussed in Section 5.3.2, however, the corresponding meridional salt flux in the steady-state circulation, such as the climate annual mean circulation shown in Figure 5.178d, can be interpreted as a freshwater transport in the opposite direction. Using Eqn. (5.161), the equivalent freshwater flux in the Atlantic Ocean can be diagnosed from the model for the mean circulation (Fig. 5.179).

The most outstanding feature revealed in other coordinates is the meridional circulation of relatively light, warm and salty water in the low-latitude upper ocean, as shown in Figure 5.178b, c, d.

The role of gyration in the oceanic general circulation

Oceans play critically important roles in transporting mass and heat fluxes meridionally through water mass formation and transformation processes, which are essential factors in climate change. The interpretation of the oceanic role in climate relies on the diagnosis of the three-dimensional oceanic circulation. The traditional diagnosis of thermohaline circulation is based on the meridional streamfunction map, obtained by integrating the meridional velocity in the zonal direction, as shown in Figure 5.178a.

This interpretation of oceanic transports is somewhat limited. First, the deep flow in the ocean interior is actually poleward, which is quite different from what is implied by the

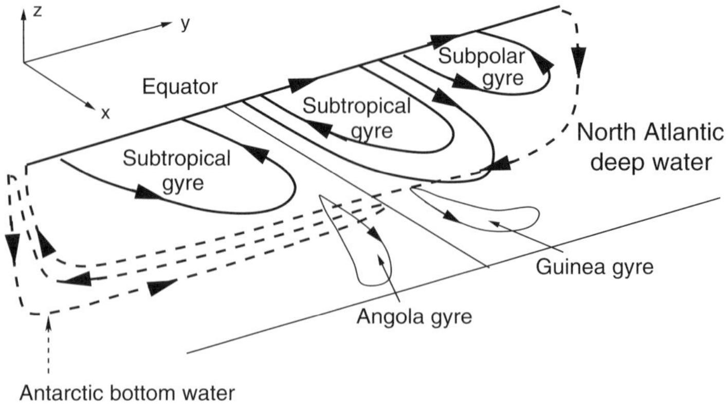

Fig. 5.180 Sketch of the horizontal gyres and meridional cells in the Atlantic Ocean.

zonally integrated meridional streamfunction map. Second, the location and the value of the maximum volume flux from the meridional streamfunction map provide an incomplete description of the circulation only. Third, the meridional streamfunction map does not provide much information about the three-dimensional structure of the circulation. In fact, the horizontal wind-driven circulation is excluded from such a map.

As shown in Figure 5.180, there are many wind-driven gyres in the upper ocean. However, they make no contribution to the zonally integrated overturning streamfunction. Although using the meridional overturning streamfunction defined in terms of potential temperature or potential density coordinates (described in Fig. 5.178) may provide better information about the oceanic circulation, such streamfunction maps have similar shortcomings, as discussed above.

To study the ocean's role in the climate system, a simple diagnostic tool for the meridional volume and heat transport by the horizontal wind-driven gyres can be used. Using this tool to analyze three-dimensional circulation data can provide a clear diagnosis of the vertical and horizontal locations of the wind-driven gyres, the meridional volume and heat transports of each individual gyre in the oceans. The application of this tool to time-dependent circulation data can also provide information on the time variability of these basic parameters of the oceanic circulation.

Diagnosing gyration in the circulation

The diagnosis is based on identifying closed loops of flow in each z-level. For a given grid point (θ_j, z_k) in the θ–z plane a zonal-accumulated meridional volume flux is defined as

$$\psi_k(\lambda, \theta_j) = \Delta z_k \int_{\lambda_k^e}^{\lambda} v(\lambda, \theta_j, z_k) a \cos\theta d\lambda \tag{5.370}$$

where Δz_k is the thickness of the given level k, and $\lambda_k^e = \lambda_k^e\left(\theta_j, z_k\right)$ is the eastern boundary.

The meridional throughflow volume flux for this grid point is

$$m_k^t = \psi_k(\lambda_k^w, \theta_j) \tag{5.371}$$

where $\lambda_k^w = \lambda_k^w\left(\theta_j, z_k\right)$ is the western boundary.

The next step is to search for the zero values of $\lambda_{k,i}$, where $\psi_{k,i}^0(\lambda_{k,i}) = 0, i = 1, 2, \ldots, N$. By definition, $\lambda_{k,0} = \lambda_k^w\left(\theta_j, z_k\right)$ and $\lambda_{k,N} \equiv \lambda_k^e$. In general, the zero crossing of ψ is not exactly a grid point, so it is calculated by a linear interpolation.

Within each interval $\lambda = [\lambda_{k,i}, \lambda_{k,i+1}], i = 0, 1, 2, \ldots, N-1$, the maximum (or minimum) is

$$\psi_{k,i}^m = \mathrm{Max}_{\lambda \in (\lambda_{k,i}, \lambda_{k,i+1})} \psi_k\left(\lambda, \theta_j\right) \geq 0 \tag{5.372}$$

$$\psi_{k,i}^n = \mathrm{Min}_{\lambda \in (\lambda_{k,i}, \lambda_{k,i+1})} \psi_k\left(\lambda, \theta_j\right) \leq 0 \tag{5.373}$$

Note that within the interval of each pair of zeros, if it is a maximum (minimum), the corresponding value of the minimum (maximum) is set to zero. By definition, the location where ψ reaches the non-trivial $\psi_{k,i}^m$ and $\psi_{k,i}^n$ must alternate. In addition, for each grid (θ_j, z_k) in the θ–z plane, if the throughflow is non-zero, a correction to the first value of maximum (or minimum) is needed

$$\psi_{k,1}^m = \psi_{k,1}^m - m_k^t, \quad \text{if } m_k^t > 0 \tag{5.374}$$

$$\psi_{k,1}^n = \psi_{k,1}^n - m_k^t, \quad \text{if } m_k^t < 0 \tag{5.375}$$

The total meridional volume transports due to the clockwise and anticlockwise circulation for grid (θ_j, z_k) are defined as the sum of individual local minimum and maximum of ψ

$$G_k^p(\theta_j) = \sum_i \psi_{k,i}^m, \quad G_k^n(\theta_j) = \sum_i \psi_{k,i}^n \tag{5.376}$$

This technique can be used to diagnose all gyres (or large eddies) at given levels separately.

At each latitude y the total contribution due to gyration is defined as

$$M_g(\theta) = \sum_{k=1}^{K} \left[|G_k^n| + G_k^p\right] \tag{5.377}$$

where K is the maximal number of levels at that latitude. The meridional throughflow rate M_t, i.e., the net contribution due to throughflow, is

$$M_t(\theta) = \sum_{k=1,\, m_k^t > 0}^{K} m_k^t \tag{5.378}$$

The total meridional circulation rate is defined as

$$M_c(\theta) = M_g(\theta) + M_t(\theta) \tag{5.379}$$

Since the vertical grids are uneven, in order to show the volume flux over a certain depth range, the volume flux in each layer is re-scaled as

$$G_k^{p\prime} = G_k^p \frac{h_0}{\Delta z_k}, \quad G_k^{n\prime} = G_k^n \frac{h_0}{\Delta z_k} \tag{5.380}$$

where $h_0 = 100$ m is the typical scale for most essential features associated with gyre-scale circulation.

In comparison, the commonly used MOC streamfunction is defined as the vertical integration

$$\psi_{MOC}(\theta, k) = \sum_{kk=1}^{k} m_{kk}^t \tag{5.381}$$

where $m_k^t = \psi_k(\lambda_k^w, \theta_j)$ is the meridional throughflow volume flux defined above. Thus, the MOC streamfunction includes both positive and negative contributions from m_{kk}^t. On the other hand, the meridional throughflow rate M_t accounts for the positive contribution term only.

The commonly used MOC rate is to define

$$\Psi(\theta) = \max_{-H<z<0} (\psi_{MOC}) \tag{5.382}$$

In addition, the maximum of this flow rate around 20–50° N is defined as the MOC rate,

$$\Psi_{\max} = \max(\Psi(\theta)) \tag{5.383}$$

The definitions introduced above are different from the commonly used terms. First, the meridional throughflow rate M_t is the vertical accumulation of all positive contributions from the meridional throughflow volume flux m_k^t at each vertical level. On the other hand, the commonly used meridional streamfunction $\psi_{MOC}(\theta, k)$ is the vertical integration of the meridional throughflow volume flux m_k^t; thus, it includes both positive and negative contributions from different levels. For the cases when multiple meridional overturning cells exist, M_t is larger than $\psi_{MOC}(\theta, k)$.

Second, the total meridional circulation rate M_c includes the contribution due to horizontal gyration, so it is much larger than both the meridional throughflow rate and the commonly used MOC rate.

Heat flux calculation

The poleward heat flux can be separated into two components: throughflow and gyration. The calculation consists of three steps.

First, calculate the poleward heat flux due to gyration within each pair of zeros of streamfunction ψ: $\psi_{k,i}^0(\lambda_{k,i}) = 0$, $i = 1, 2, \ldots, \mathrm{N}$.

Second, the total meridional heat flux due to the clockwise (counterclockwise) circulation for grid (θ_j, z_k) is defined as the sum of individual local minima (maxima) of ψ:

$$H_k^p(\theta_j) = \rho_0 c_p \sum_{i=1}^{N} \int_{P_i} v\Theta a \cos\theta d\lambda \Delta z_k \quad (5.384)$$

$$H_k^n(\theta_j) = \rho_0 c_p \sum_{j=1}^{M} \int_{N_j} v\Theta a \cos\theta d\lambda \Delta z_k \quad (5.385)$$

where P_i (N_j) is the subdomain between two sequential roots of the streamfunction ψ when the value ψ is positive (negative), and Θ is potential temperature.

Third, the segment next to the western boundary $\lambda = [\lambda_k^w, \lambda_{k,1}]$ needs special handling, and we will discuss the case when $\psi_k^w = \psi_k(\lambda_k^w, \theta_j) > 0$. There are two possible situations:

(a) $\psi_k^w < \psi_{k,1}^m$, where $\psi_{k,1}^m$ is the zonal accumulated meridional flux maximum within this subdomain, and the subdomain consists of two sub-segments, P_w and T_w, which are connected at λ_{ww}, the first point east of λ_1 that satisfies $\psi = \psi_w$. The heat fluxes due to throughflow and anticlockwise gyration are

$$H_k^{through} = \rho_0 c_p \int_{T_W} v\Theta a \cos\theta d\lambda \Delta z_k \quad (5.386)$$

$$\Delta H_k^p = \rho_0 c_p \int_{P_w} v\Theta a \cos\theta d\lambda \Delta z_k \quad (5.387)$$

(b) $\psi_k^w = \psi_{k,1}^m$. There is no anticlockwise flow in this segment, and the heat flux contribution to the throughflow can be calculated by Eqn. (5.386).

Gyres and their contribution to meridional fluxes in the Atlantic Ocean

These formulae were used to diagnose the circulation in the Atlantic Ocean, based on the SODA (Simple Ocean Data Assimilation) data (Carton *et al.*, 2000a, 2000b).

Horizontal gyres

As an example, the zonal accumulated meridional flux at 82.5 m depth in the Atlantic Ocean includes two subtropical gyres (G2, G5), the Subpolar Gyre (G1), the anticlockwise Guinea Gyre (G3) at the latitude of the ITCZ due to positive Ekman pumping, and the clockwise Angola Gyre (G4) near the western coast of Africa (Fig. 5.181). Our algorithm also provides the contribution to the meridional volume transport of each gyre (Fig. 5.182). For example, the North Atlantic Subtropical Gyre (G2) is located within the latitudinal band of 15° N–45° N.

Furthermore, the maximal strength of meridional volume and heat fluxes for each gyre are identified through searching within a particular latitudinal band. The meridional position of the center of a gyre is defined as the latitude of the maximal meridional volume transport

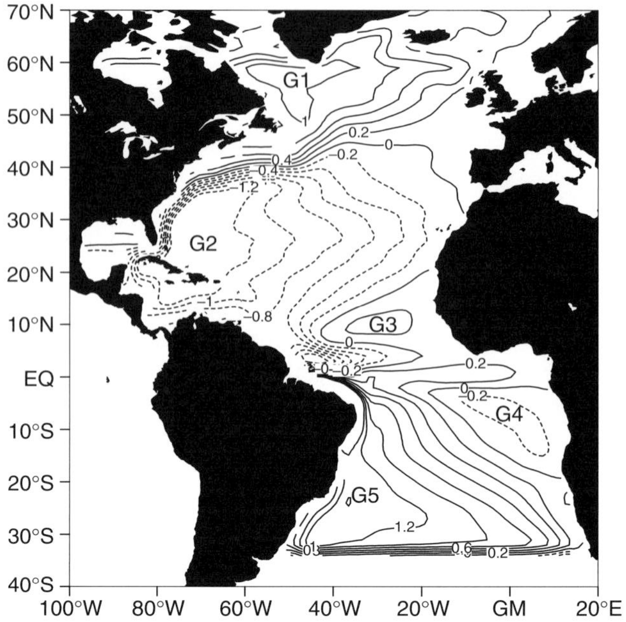

Fig. 5.181 Climatological-mean zonally integrated meridional volume transport (in Sv), defined in Eqn. (5.370), at depth 82.5 m, the thickness of the layer is 15 m. G1 is the Subpolar Gyre, G2 is the North Subtropical Gyre, G3 is the Guinea Gyre, G4 is the Angola Gyre, and G5 is the South Subtropical Gyre (Jiang *et al.*, 2008).

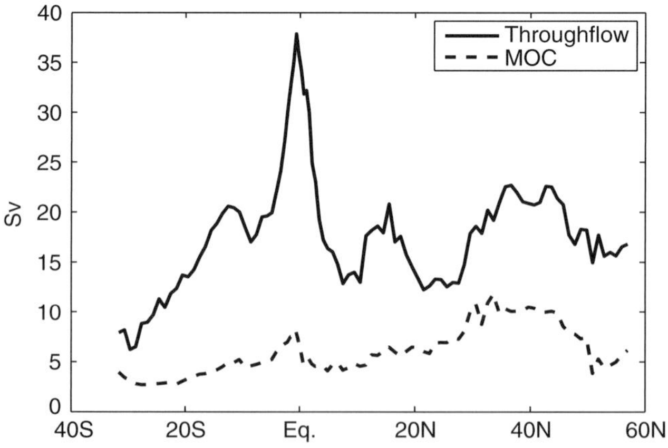

Fig. 5.182 Climatological-mean throughflow and MOC (Jiang *et al.*, 2008).

of the gyre. The vertical position of the center of an anticlockwise gyre can be defined as

$$C_z^p = \frac{\sum_{k=k_1}^{k_2} \sum_{j=j_1}^{j_2} G_k^p(\theta_j) \cdot z_k}{\sum_{k=k_1}^{k_2} \sum_{j=j_1}^{j_2} G_k^p(\theta_j)} \tag{5.388}$$

where (j_1, j_2) and (k_1, k_2) is the domain for the gyre in question, and these can be chosen suitably for each gyre.

Meridional volume fluxes

The zonally integrated meridional volume transport is dominated by the North Atlantic Deep Water (NADW) with southward transport of 18 Sv at most latitudes and Antarctic Bottom Water cells; the maximum overturning rate is estimated at 26 Sv at 59° N. The corresponding meridional throughflow rate diagnosed from the model is approximately 20 Sv and the commonly used term, MOC rate in Eqn. (5.383), is slightly larger than 10 Sv (Fig. 5.182). This large difference between observations and the model output is due to the low horizontal resolution of the numerical model used here.

On the other hand, the total meridional circulation rate reaches the maximum of 95 Sv at 33° N and a second maximum of 83 Sv near the equator (heavy solid line in Fig. 5.183). Thus, the total meridional circulation rate is much larger than the maximum meridional volume transport identified from the zonally integrated meridional streamfunction, as commonly used in diagnosing numerical model output. The essential difference is due to the contributions of multiple meridional cells in the vertical plane and gyres (or large eddies) in the horizontal plane, which may cancel each other in the calculation of the traditional overturning rate. As discussed above, different overturning streamfunction and overturning

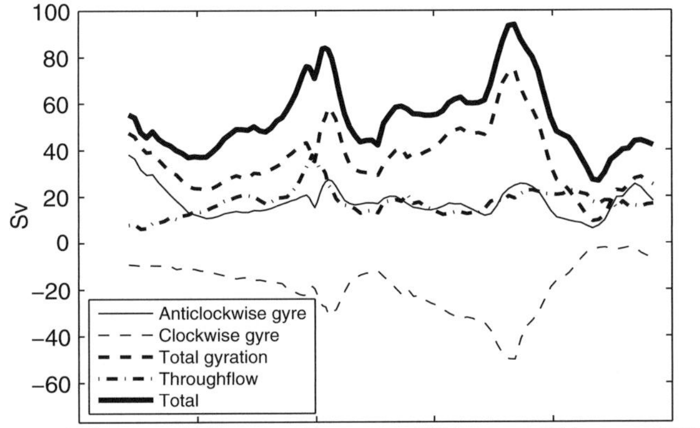

Fig. 5.183 Climatological-mean depth-integrated northward volume transports due to different components in the Atlantic Ocean, in Sv. The *dashed line below the horizontal axis* is the volume transport due to clockwise gyration (plotted as negative in this figure) (Jiang *et al.*, 2008).

rates, defined in θ–Θ or θ–σ_2 planes, can be used for diagnostic study. Although such overturning rates are larger than that defined in the traditional θ–z plane, they are smaller than the throughflow rate defined in this section and they cannot convey the complete information of the three-dimensional circulation.

The most important feature of this algorithm is that it allows a clear identification of the dynamical roles of the wind-driven gyres, including their contribution to both the meridional volume and heat fluxes. These gyres contribute strongly to the total meridional volume flux (heavy dashed line in Fig. 5.183). The total amount of meridional volume flux associated with gyration has a maximum of 75 Sv around 33° N. The maximal volume fluxes of clockwise and anticlockwise circulation are associated with the North Subtropical Gyre (50 Sv) and the South Subtropical Gyre (38 Sv).

In the North Atlantic Ocean, the maximal volume flux associated with the Gulf Stream recirculation system is on the order of 150 Sv (Hogg and Johns, 1995). Although the maximal volume transport associated with the recirculation is mostly zonally oriented, the maximal meridional volume transport can be quite large. Thus, the total meridional circulation rate (as defined above, in the North Atlantic Ocean, including the contribution of the meridional throughflow, horizontal wind-driven gyres, meso-scale eddies, and other deep recirculation gyres) may be more than 100 Sv.

Meridional heat flux

The maximal poleward heat flux in the Atlantic Basin is estimated as 0.8 PW. The total poleward heat flux in the ocean is separated into two major components. For the climatological state the flux associated with the meridional throughflow reaches a maximum of 0.65 PW around 28° N and the horizontal gyration reaches 0.2 PW around 30° N. Throughflow includes the contribution due to multiple cells in the vertical direction; thus, it is a better description of the circulation physics than the more commonly used MOC rate.

Interannual–decadal variability

Sverdrup function with time delay

For a steady circulation, the meridional volume transport of the wind-driven gyres obeys

$$M = -\frac{f}{\beta}\int_{\lambda}^{\lambda_e} w_e a \cos\theta \, d\lambda \tag{5.389}$$

$$w_e = \frac{1}{a\rho_0 2\omega \sin\theta}\left(\frac{1}{\cos\theta}\frac{\partial \tau^\theta}{\partial \lambda} - \frac{\partial \tau^\lambda}{\partial \theta} + \frac{\tau^\lambda}{\sin\theta\cos\theta}\right) \tag{5.390}$$

where w_e is the Ekman pumping rate, $f = 2\omega \sin\theta$ is the Coriolis parameter, $\beta = df/ad\theta$, a is the radius of the Earth, ω is the Earth's angular velocity, θ is the latitude, λ is the longitude, and τ^θ and τ^λ are the meridional/longitudinal components of wind stress.

Equation (5.389) applies to steady circulations only. As discussed in Section 4.8, when wind stress changes with time, the circulation at a station λ adjusts to wind stress changes east of this station with a time delay due to the finite speed of Rossby wave propagation. The circulation is completely established after all baroclinic Rossby waves reach the western boundary of the basin; however, the arrival of the first baroclinic Rossby waves initiated from the eastern boundary indicates the near completion of the adjustment at each station. Therefore, the corresponding formula is

$$M(t) = -\frac{f}{\beta}\int_{\lambda}^{\lambda_e} w_e[\lambda, \theta, t - a(\lambda_e - \lambda)\cos\theta / c_g] a\cos\theta d\lambda \tag{5.391}$$

where $c_g = \beta R_e^2$ is the group velocity of Rossby waves, and R_e is the first baroclinic radius of deformation.

Decadal variability of gyres in the North Atlantic Ocean

The maximal strength of meridional volume transport (defined as the meridional circulation rate between 20° N and 40° N) and the corresponding poleward heat flux associated with gyration and throughflow in the Atlantic Ocean vary greatly over a decadal time scale (Fig. 5.184).

The most outstanding feature was the dramatic intensification of gyration in the 1970s, apparently associated with the regime shift in the atmospheric circulation around 1975–6. Although the volume flux associated with throughflow remains virtually constant, the gyration rate went up more than 100% during this time period. In addition, there seems to be a gradual intensification of gyration over the past 50 years.

Poleward heat flux associated with throughflow and gyration has changed greatly over the past 50 years. There is a clear trend of decline in poleward heat flux associated with throughflow. In particular, there is a large amplitude change that took place in the 1970s and 1980s; the contributions due to both gyration and meridional throughflow went up, although the corresponding increase in the meridional throughflow was relatively small. This is may be due to large changes in the thermal structure in the upper ocean associated with the wind-driven circulation in response to the regime shift in the 1970s.

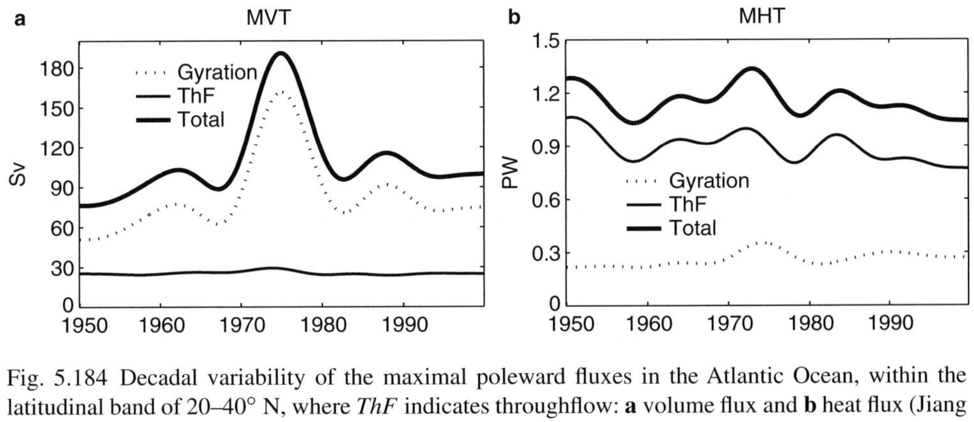

Fig. 5.184 Decadal variability of the maximal poleward fluxes in the Atlantic Ocean, within the latitudinal band of 20–40° N, where *ThF* indicates throughflow: **a** volume flux and **b** heat flux (Jiang *et al.*, 2008).

The poleward heat flux associated with gyration also changed. The regime shift seems to have had a noticeable impact on the poleward heat flux associated with gyration in the 1970s. In addition, the poleward heat flux associated with gyration has increased slightly over the past 50 years, although the overall trend of poleward heat flux declined slightly over the same period owing to the weakening of the contribution associated with the throughflow.

5.5.3 Global adjustment of the thermocline

Adjustment to a quasi-steady state

Introduction

Paleoproxy evidence indicates that the production of North Atlantic Deep Water (NADW) was much reduced or even shut down at times in the past that coincided with rapid changes in global climate (as recorded in ice cores and elsewhere; see, e.g. Broecker, 1998). Although the atmospheric response was mostly confined to the Northern Hemisphere, planetary wave propagation through the world's oceans may have brought about rapid global changes (Döscher *et al.*, 1994). The global adjustment process associated with the planetary waves can be examined by studying an idealized model in which the global circulation is represented in terms of a linear single mode.

One of the major assumptions in the classical theory of deep circulation by Stommel and Arons (1960a) is that the circulation is stationary. For the non-stationary circulation, planetary Kelvin waves along the coasts and in the equatorial waveguide, and Rossby waves, both play a major role in setting up the circulation in a closed basin by transporting water masses (Kawase, 1987). In this section we extend the theory to the multi-basin world oceans.

Model formulation

We begin with a linear shallow water model on an equatorial β plane (Kawase, 1987):

$$u_t - \beta y v = -g h_x - K u \tag{5.392}$$

$$v_t + \beta y u = -g h_y - K v \tag{5.393}$$

$$h_t + H\left(u_x + v_y\right) = -\lambda h + Q \tag{5.394}$$

Note that H is the equivalent depth of the shallow water equation, and its specification will be discussed later. Thus, the horizontal momentum equations are geostrophic balance plus the time-dependent term and a simple Rayleigh friction, Ku or Kv; the continuity equation includes the time-dependent term, the deepwater source distribution Q, and a simple Rayleigh damping term λh. The most crucial departure from the Stommel–Arons formulation in this model is the inclusion of the time-dependent terms and replacing the specified upwelling with the term λh. These equations can be nondimensionalized by introducing the following scales for velocity, length, depth, and time:

$$c = (KgH/\lambda)^{1/2}, \ L = (c/\beta)^{1/2}, \ H, \ T = (c\beta)^{-1/2} \tag{5.395}$$

For $\lambda = K$, this equation set is reduced to

$$u_t - yv = -h_x - ru \tag{5.396}$$

$$v_t + yu = -h_y - rv \tag{5.397}$$

$$h_t + (u_x + v_y) = -rh + Q' \tag{5.398}$$

where $r = KT$ is the new nondimensional friction parameter, and $Q' = TQ/H$.

Apart from the western boundary, the solution consists of equatorial Kelvin waves leaving the western boundary to propagate toward the eastern boundary. At the eastern boundary the equatorial Kelvin waves are reflected as westward Rossby waves, plus two poleward Kelvin waves. On their poleward paths, the coastal Kelvin waves send out Rossby waves propagating westward and establishing the circulation in the interior. Along their pathways, wave amplitude gradually declines due to dissipation. As the circulation approaches equilibrium, the time-dependent terms drop off, and the steady-state solution, which is the combination of the Kelvin waves and their reflection in Rossby waves, in an ocean bounded at $X = 0$ in the west and $X = L_B$ in the east, is (Cane, 1989):

$$h = AF, \quad F = (\cosh 2r\xi)^{1/2} e^{-y^2/2 \tanh 2r\xi} \tag{5.399}$$

where A is a constant for each basin to be determined by volumetric flux balance in the model, $F > e^{-r\xi y^2}$ for very small r and $F \equiv 1$ along the eastern boundary and the equator, and $\xi = (L_B - X)/L$ is the nondimensional zonal coordinate.

When r is very small, the volumetric communication rate M between two adjacent basins is primarily controlled by the semi-geostrophic current around the southern tip of the continent separating the basins. Denoting its latitude by y^s, the volumetric flux from one basin to the next is $M = \Delta h/y^s$, where Δh is the layer thickness change across the boundary current. For example, the volumetric balance in the Indian Ocean is the influx equal to the upwelling inside the basin plus the outflow

$$(A_A - A_I F_{Iw})/y^s_{A,I} = r \iint_{S_I} h dx dy + (A_I - A_P F_{Pw})/y^s_{I,P} \tag{5.400}$$

The subscripts A, I, and P indicate the Atlantic, Indian, and Pacific Oceans; S_I is the surface area of the Indian Ocean; $F_{Iw} \simeq e^{-r\xi_{Iw} y^s_{I,P} y^s_{I,P}}$; ξ_{Iw} is the western boundary of the Indian Ocean; and $y^s_{A,I}$ ($y^s_{I,P}$) represent the southern tip of the continent separating the Atlantic and Indian (Indian and Pacific) Oceans (Fig. 5.185).

For small friction, $r \iint_{S_I} h dx dy \simeq r A_I S_I$ to the first order in$r$. There is a similar relation for the Pacific Ocean

$$(A_I - A_P F_{Pw})/y^s_{I,P} = r \iint_{S_P} h dx dy + A_P/y^s_{P,A} \tag{5.401}$$

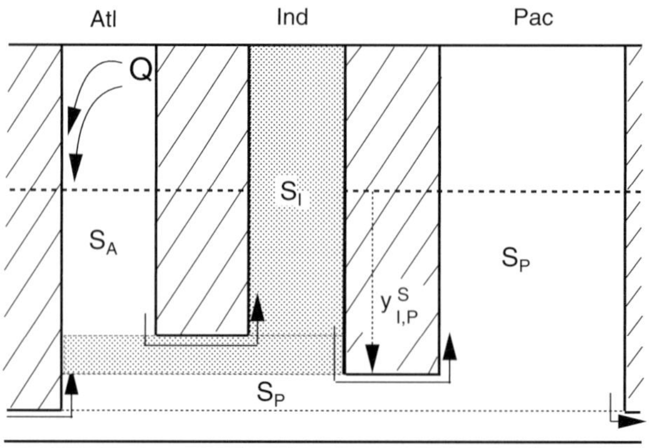

Fig. 5.185 Circulation in the idealized world's oceans driven by a point source of deep water Q. S_A, S_I, and S_P are the areas of each ocean, and $y^s_{I,P}$ is the southern tip of the continent separating the Indian and Pacific Oceans (Huang *et al.*, 2000).

In addition, we have the total volumetric balance in the world's oceans:

$$Q' = r\,(A_A S_A + A_I S_I + A_P S_P) \tag{5.402}$$

The solution can be found by combining Eqns. (5.400, 5.401, 5.402). When r is very small, the first term on the right-hand side of Eqns. (5.400, 5.401) can be neglected.

Numerical experiments

The numerical model is integrated using $\Delta x = 1°$ and $\Delta y = 0.5°$. In order to simulate the sea level and main thermocline appropriately, a combination of the first baroclinic and second baroclinic modes is used, and the equivalent depth scale is $H = 0.92$ m (Zebiak and Cane, 1987). The Rayleigh friction and damping coefficients are $K = \lambda = 4 \times 10^{-10}$/s, which corresponds to a nondimensional friction coefficient $r = 4.8 \times 10^5$. For a mean thermocline depth of 300 m, this corresponds to a diapycnal diffusivity of 0.36×10^{-4} m^2/s, which is larger than the low open-ocean mixing coefficient of 10^{-5} m^2/s, inferred from tracer release experiments; however, it is smaller than the $O(10^{-4}$ m^2/s) pivotal values inferred from the global tracer budget.

A source of 10 Sv, approximately the rate at which NADW currently crosses the equator from the North Atlantic Ocean, is uniformly distributed within a latitudinal band north of 50° N in the North Atlantic Ocean. After 500 years the solution seems to approach a quasi-steady state. Note that a slowdown in deepwater formation is equivalent to a source of upper layer water and a sink of lower layer water. Thus, the h field discussed below should be interpreted as the downward motion of the thermocline in response to a slowdown of deepwater formation, or its upward motion in response to increased deepwater formation.

Several numerical experiments were carried out, including cases in which the world's oceans are represented by the highly idealized rectangular basins (Fig. 5.185). All results from these numerical experiments fit the simple formulae very closely. As an example, experiments with a realistic coastline are shown in Figure 5.186.

The first case is the control case: the Indonesian Passage is closed, the corresponding southern tips of the African, Australian and American continents are 34° S, 44° S, and 55° S, respectively, and the amplitude of the thermocline depth perturbation in the world's oceans is $A_A : A_I : A_P = 105 : 87 : 46$ (m) (Fig. 5.186a). Thus, the shallow-water model predicts that a deepwater formation rate of 10 Sv induces an upward motion of the thermocline on the order of 50–100 meters. Assuming $r > 0$, Eqns. (5.400, 5.401, 5.402) are in error by less than 10%. Errors are due to the approximation in neglecting friction, additional dissipation in the numerical model for a realistic, jagged coastline, and the fact that the solution has not completely reached the steady state. It is interesting to note that there is a current of 3 Sv going through the Drake Passage. From Eqn. (5.402), it can readily be seen that if the friction/damping coefficient is reduced 10 times, this recirculation will increase 10 times because the amplitude of the perturbation is inverse to r.

In the second case, the Indonesian Passage is open, and the thermocline perturbation ratio is $A_A : A_I : A_P = 106 : 67 : 54$ (Fig. 5.186b). The effective southern tip of the boundary between the Indian and Pacific Oceans is moved up to 9.1° S, still far enough away from the equator so that the boundary current is geostrophically controlled. The deepwater flow takes a short-cut through the Indonesian Passage. This direct route gives rise to a slightly larger thermocline depth perturbation in the Pacific Ocean. The volumetric flux through the Indonesian Passage is 3.3 Sv. Here again, reducing r 10 times can give rise to a flux 10 times larger.

In the third case, both the Indonesian Passage and Drake Passage are closed, and the thermocline perturbation ratio is $A_A : A_I : A_P = 98 : 81 : 43$ (Fig. 5.186c). With the Drake Passage closed, Eqns. (5.400) and (5.401) need modification: the last term on the right-hand side of Eqn. (5.401) is eliminated and the area of the Pacific Basin is expanded to the Antarctic coast. It is interesting that in this simple model the thermocline adjustment in the global oceans is not sensitive to whether the Drake Passage is open or closed. However, the Drake Passage plays such a vitally important role in controlling the thermohaline and thermocline circulation in the world's oceans that the meaning of this result should be interpreted with caution.

The role of Kelvin/Rossby waves

Both Kelvin and Rossby waves establish the steady solutions discussed above. When NADW is formed, Kelvin waves are generated at the site of deepwater formation and move to the other parts of world's oceans, as shown in Figure 5.187.

The pathways of Kelvin waves are as follows. First, they move southward along the eastern coastline of North America in the form of coastal Kelvin waves (Leg 1). At the equator, these waves turn eastward and propagate along the equatorial waveguide (Leg 2). At the eastern boundary they bifurcate and become the poleward Kelvin waves (Legs 3 and 11). After turning the corner of Cape of Hope, they continue to move equatorward along the eastern coast of Africa (Leg 4). The rest of the passage (Legs 5, 6, 7, 8, 9, and 10) is similar.

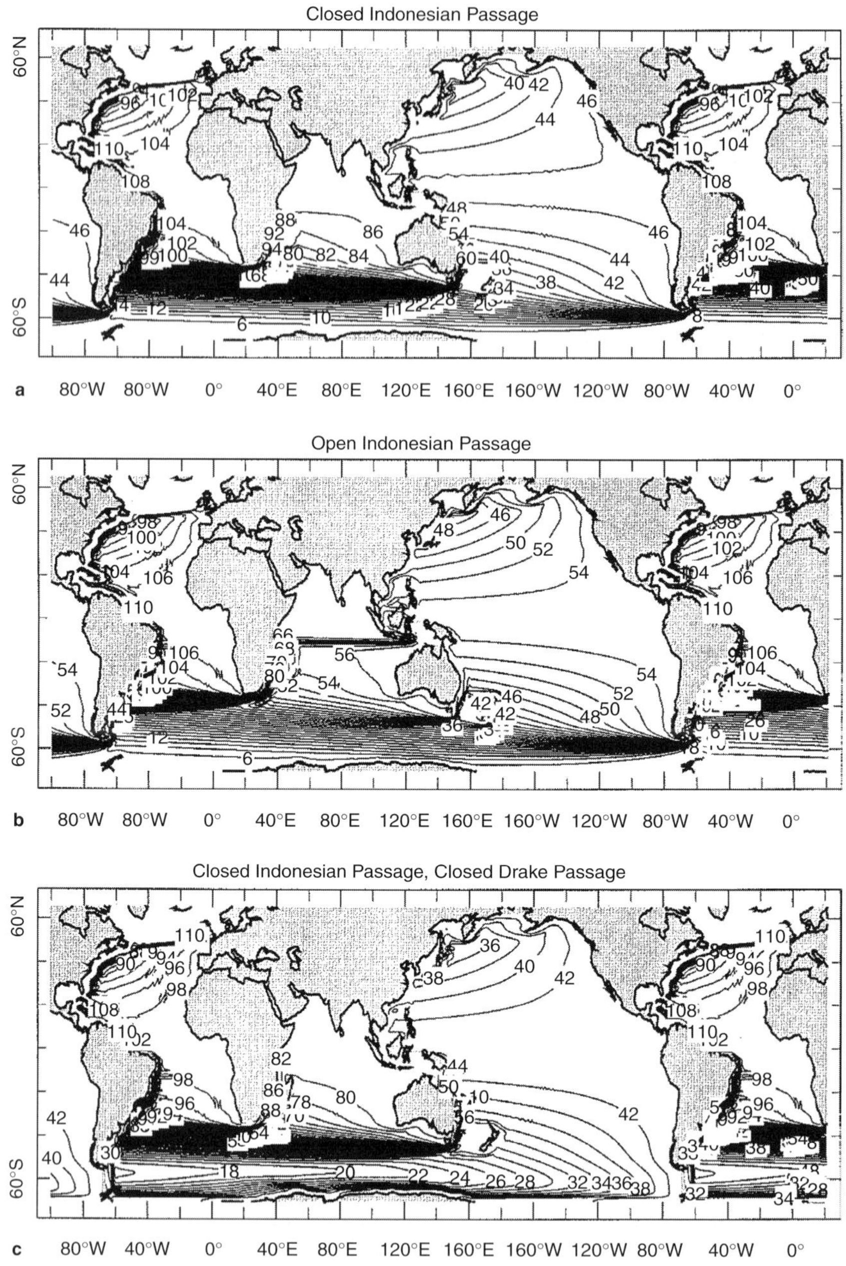

Fig. 5.186 The displacement of the thermocline (m) induced by NADW source of 10 Sv: **a** closed Indonesian Passage; **b** open Indonesian Passage; **c** closed Indonesian Passage, closed Drake Passage (Huang *et al.*, 2000).

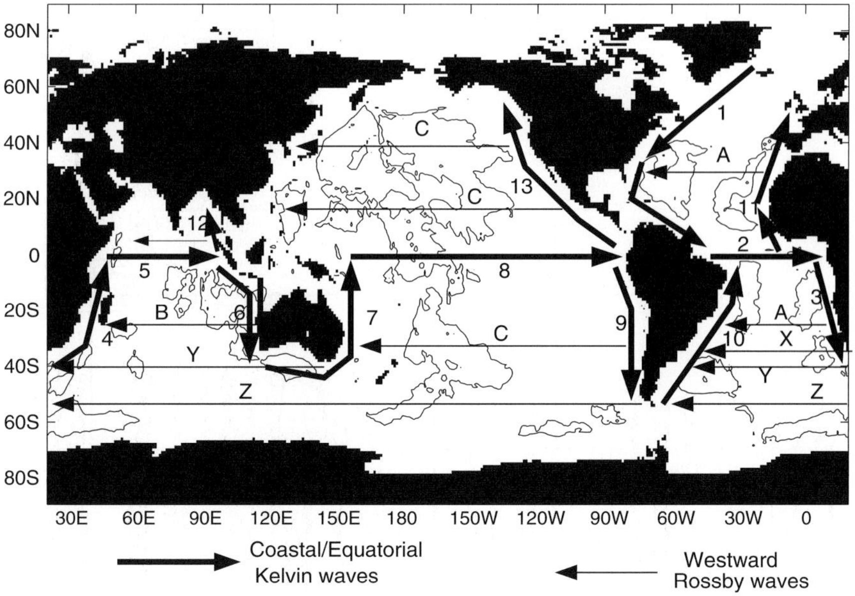

Fig. 5.187 Sketch of the propagation of signals induced by deepwater formation in the world's oceans in the form of Kelvin waves and Rossby waves: the Indonesian Passage is blocked by an artificial bridge near the northwestern corner of Australia.

As Kelvin waves move poleward along the eastern boundaries of each basin, they send westward Rossby waves, as indicated by thin arrows (labeled A, B, C) in Figure 5.187; these Rossby waves carry the signals to the ocean interior.

Kelvin waves travel very fast: 3 m/s for the first baroclinic mode. Within one month the signals can reach the western equatorial Atlantic Ocean (Fig. 5.188a). From here the Kelvin waves travel along the equator. Reaching the eastern boundary, the signal propagates poleward along the coast into both hemispheres, generating Rossby waves that propagate westward.

The first signal will reach the equatorial Indian Ocean within a year, but it takes about 5 years for the amplitude there to become appreciable, as shown by the dashed line in Figure 5.188a. From there it takes another 5 years to generate a significant signal in the equatorial Pacific Ocean (dot-dashed line in Fig. 5.188a), although the first sign of change will reach the equatorial Pacific Ocean within a year.

It takes a much longer time for the signals to reach the ocean interior, because locations away from the equator can be reached only by the Rossby waves (indicated by the thin lines with arrows in Figure 5.187) generated along the eastern coast of the individual basin. The higher the latitude, the longer it takes for the signal to cross back to the western boundary. For example, signals are visible at a station in the South Atlantic Ocean (20° S, 30° W) at

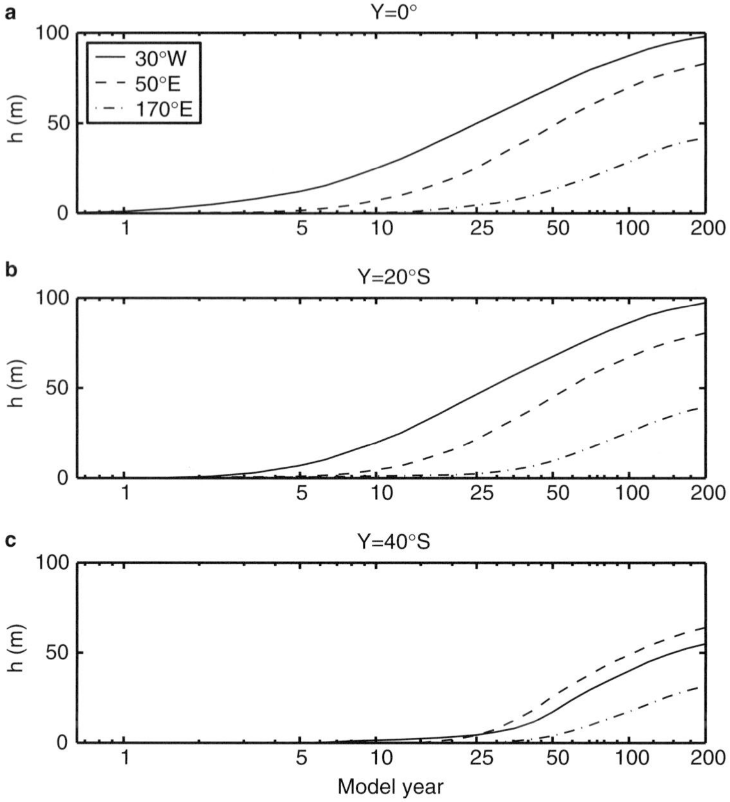

Fig. 5.188 **a–c** Time evolution of the depth anomalies at different locations in the world's oceans, indicating the time delay due to the propagation of Kelvin waves and Rossby waves.

year 3; they reach a station in the South Indian Ocean (20° S, 50° E) at year 5, and a station in the South Pacific Ocean (20° S, 170° E) at year 12.5 (Fig. 5.188b).

At 40° S the signals arrive with a much longer delay. They reach the corresponding longitude positions at years 10, 16, and 25 (Fig. 5.188c). The corresponding pathways of long Rossby waves are depicted by the thin lines with arrows X, Y, Z in Figure 5.187. It is very interesting to note that signals arrive at the station in the South Atlantic Ocean (40° S, 30° W) in two stages. For the first stage, a weak signal comes from the westward-propagating Rossby waves within the Atlantic Ocean. Much stronger signals arrive later when the Rossby waves in the Indian Ocean reach this location, but it takes 25 years for these signals to reach there.

Thus, the eastern coast of South America consists of three regimes (Fig. 5.185), controlled by Rossby waves generated along the southern coast of Africa (S_A), along the coast of Australia S_I (the shaded area), and along the western coast of South America (S_P). Given the much longer arrival times of the Rossby waves at high latitudes and over longer distances, the response of the thermocline to changes in deepwater formation in the North Atlantic Ocean is much delayed at these locations.

Adjustment on shorter time scales

The details of the global thermocline adjustment on time scales from months to decades by the baroclinic Kelvin waves and Rossby waves has been discussed by Johnson and Marshall (2002). As an example, assume that the ocean is at rest initially, and the thermocline has a constant depth of 500 m. At $t = 0$ a deepwater source of 10 Sv is switched on. The evolution of the thermocline depth on a time scale of months is shown in Figure 5.189. Since the first baroclinic Kelvin waves travel rather fast, it takes less than one month for them to travel from the northern boundary to the equator, and it takes less than a month to cross the equator.

Layer thickness along the meridional boundaries

The upper layer thickness along the eastern boundary is approximately constant, and it declines almost linearly with time (Fig. 5.190a).

- *The constant layer thickness* along the eastern boundary is due to the no-flux boundary condition along the eastern wall; thus, weak friction gives rise to geostrophy and a layer depth that is roughly independent of y.
- *The non-constant layer depth* along the western wall is due to friction associated with the western boundary current. There is also a potential contribution due to the inertial terms.

A time-delay equation predicting layer thickness along the eastern boundary

The adjustment in the basin interior is carried out through baroclinic Rossby waves

$$\frac{\partial h}{\partial t} - \frac{c}{a\cos\theta}\frac{\partial h}{\partial \lambda} = 0, \quad c = \frac{\beta g' H}{f^2} \tag{5.403}$$

Note that the speed of long Rossby waves depends on the latitude. Based on quasi-geostrophic theory, the speed of long Rossby waves can be readily calculated from a simple theoretical formula from the stratification (Chelton *et al.*, 1998). However, satellite observations indicate that the speed diagnosed from satellite observations is different from the value predicted from the eigenvalue problem of the normal mode in quasi-geostrophic theory. This discrepancy was first reported by Chelton and Schlax (1996), shown as the black dots (satellite data analysis) and black lines (theory) in Figure 5.191b. As more refined satellite data are now available, this issue has been explored in detail. A recent study by Chelton *et al.* (2007) revealed a complicated picture for the propagation of nearly linear large-scale eddies and the strongly nonlinear relatively small-scale eddies (Fig. 5.191a). It is clear that eddy dynamics requires much deeper study.

A major source of the discrepancy between the theoretical value and that diagnosed from observations is due to the existence of strong currents and eddies in the oceans. Along this line there have been many studies devoted to wave–mean-current interaction in the oceans;

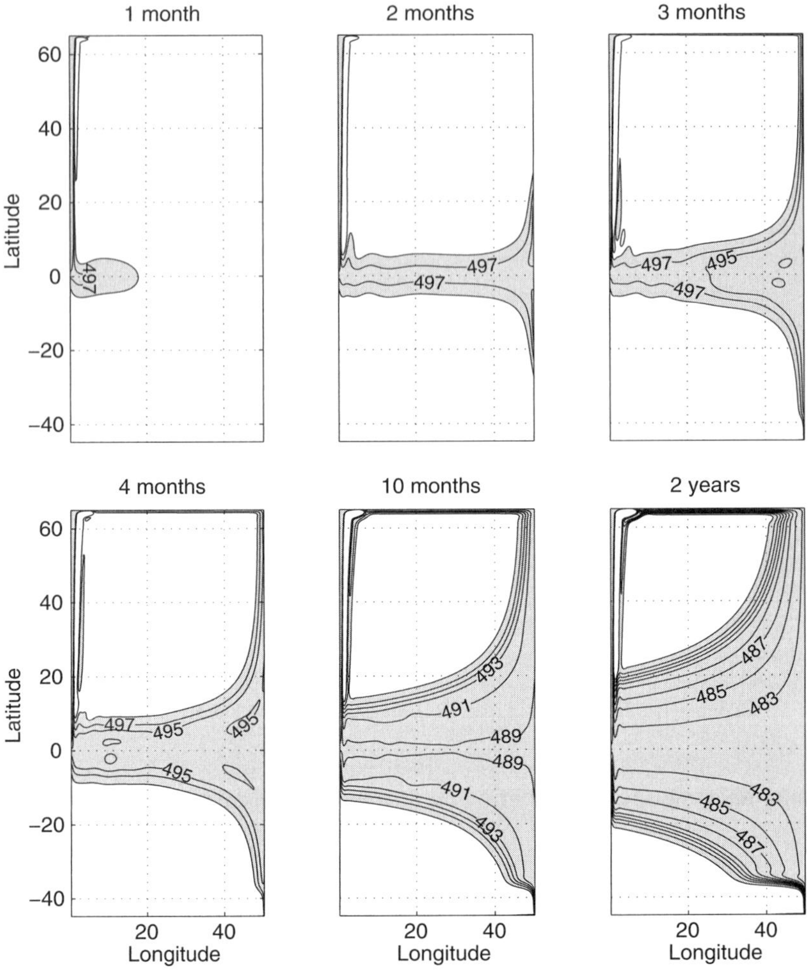

Fig. 5.189 Time evolution of surface layer thickness, after switching on a deepwater source of 10 Sv. There is a sponge layer along the southern boundary, where influence can clearly be seen (Johnson and Marshall, 2002).

however, a simple and practical solution for this dilemma is to diagnose the wave speed from satellite data as shown in Figure 5.192 by Qiu (2003). On the other hand, since the model discussed above is only a theoretical tool aiming at understanding the physics of the circulation, the simple formula in Eqn. (5.403) can be used.

Integrating this equation leads to

$$\frac{\partial}{\partial t}\int_{\lambda_w}^{\lambda_e} ha\cos\theta d\lambda = c\left[h_e(t) - h_b(\theta, t)\right] \tag{5.404}$$

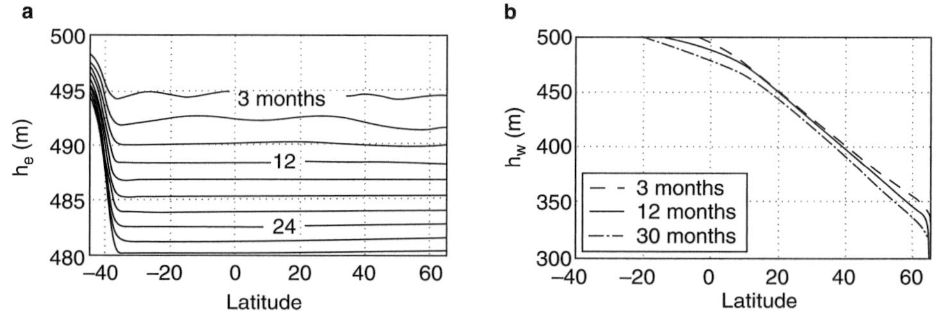

Fig. 5.190 Evolution of surface layer thickness along: **a** the eastern boundary, **b** the western boundary (Johnson and Marshall, 2002).

where h_b is the layer thickness at the outer edge of the western boundary and h_e at the eastern boundary. Since h_b is set by the Rossby waves from the eastern boundary, we have the following time-delay relation:

$$h_b(t) = h_e\left(t - \frac{L}{c}\right) \tag{5.405}$$

where $L/c = t^*(\theta)$ is the time for the Rossby waves to cross the basin. Integrating the continuity equation gives

$$\frac{\partial}{\partial t}\int_{\lambda_w}^{\lambda_e} ha\cos\theta d\lambda = -\frac{\partial T(\theta, t)}{a\partial\theta} \tag{5.406}$$

where $T(\theta, t)$ is the volumetric flux within the western boundary current

$$\frac{\partial}{a\partial\theta}T(\theta, t) = c\left[h_e(t - L/c) - h_e(t)\right] \tag{5.407}$$

Integrating Eqn. (5.407) over the latitudinal band of the basin leads to

$$h_e(t) = \frac{1}{\int c(\theta)ad\theta}\left\{\int h_e\left[t - \frac{L}{c(\theta)}\right]c(\theta)\,ad\theta - T_N + T_S\right\} \tag{5.408}$$

where $T_S = g'\left(h_e^2 - h_0^2\right)/2f_s$ and T_N (specified) are the volumetric flux at the southern and northern ends of the western boundary (Fig. 5.193). This is a time-delay equation that describes the time evolution of the thermocline thickness along the eastern boundary. As shown clearly, it takes one month for the signal to arrive at the equator. At this time, most of the water is carried out eastward by the equator-bounded Kelvin waves (Fig. 5.193a). The western boundary current is gradually established after the long Rossby waves radiated from the eastern boundary arrive at the western boundary, as shown in Figure 5.193b.

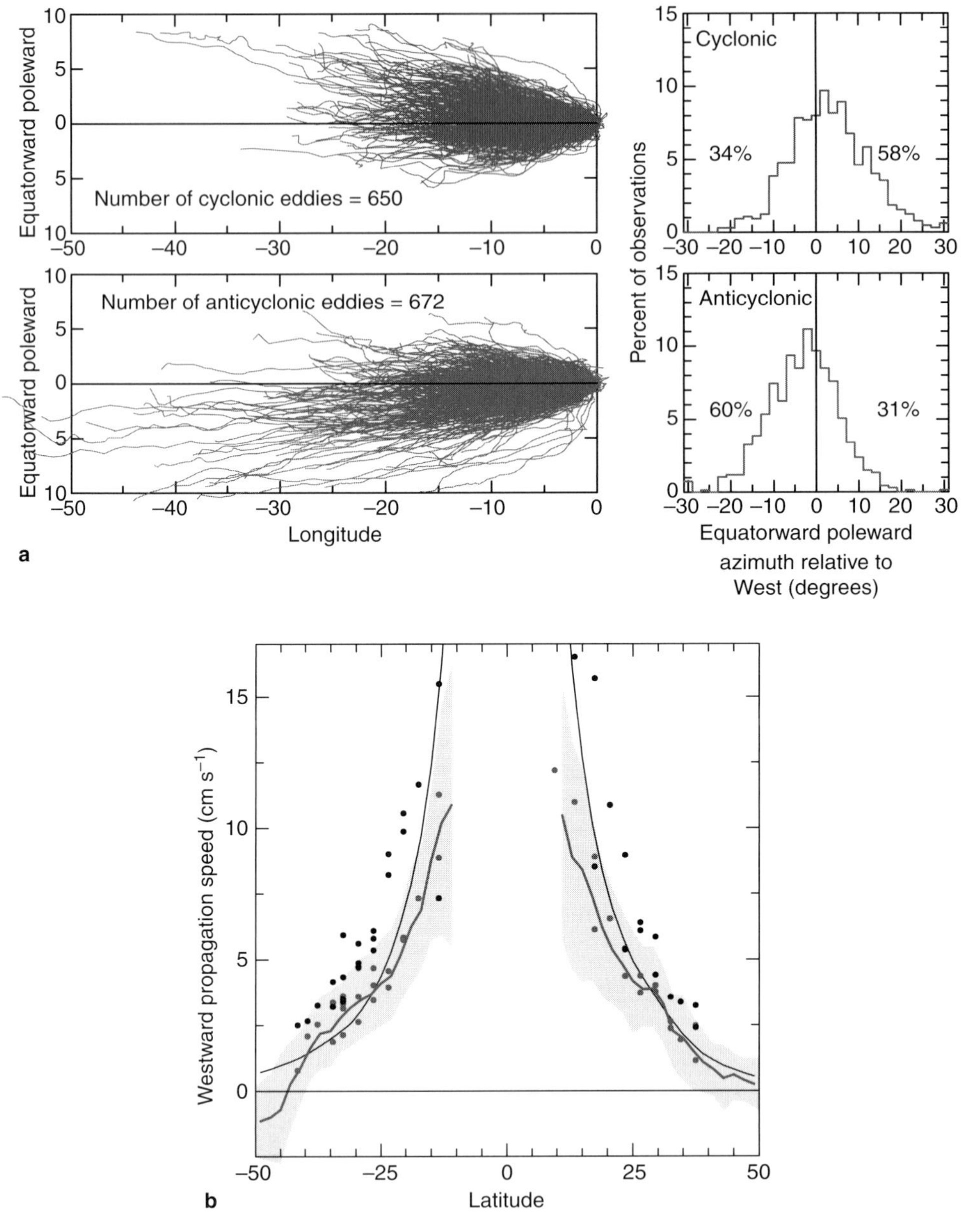

Fig. 5.191 **a** The global propagation of eddies with lifetimes ≥12 weeks; *left panels* show changes in relative position, *right panels* are histograms of the mean propagation angle relative to due west; **b** the latitudinal variation of the westward zonal propagation speeds of large-scale SSH (*black dots*) and small-scale eddies (*red dots*); the *red line* indicates the zonal average of the propagation speeds of all eddies with lifetimes ≥12 months, the *gray shading* indicates the central 68% of the distribution in each latitudinal band, and the *black line* indicates the propagation speed of the non-dispersive baroclinic Rossby waves (Chelton *et al.*, 2007). See color plate section.

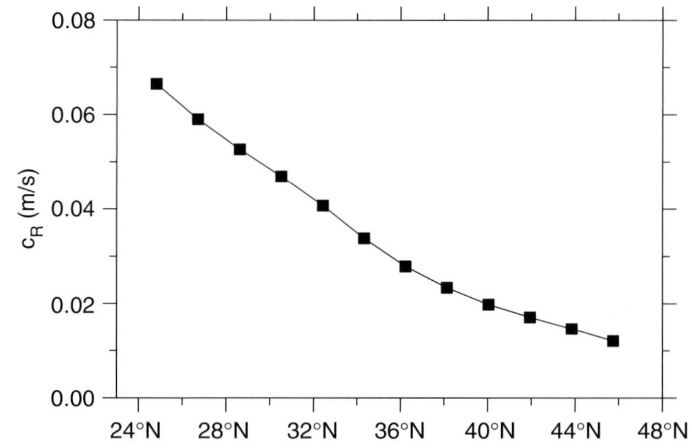

Fig. 5.192 Westward phase speed of the first baroclinic Rossby waves as a function of latitude, diagnosed from satellite observations (Qiu, 2003).

5.5.4 Dynamical role of the mixed layer in regulating meridional mass/heat fluxes

What really controls the meridional overturning circulation?

Most studies relating to the energetics of the oceanic circulation have primarily focused on the importance of subsurface diapycnal mixing in regulating the MOC and poleward heat flux. However, oceans receive a huge amount of mechanical energy both through surface waves and through ageostrophic currents in the Ekman layer. It is commonly believed that most of such energy is dissipated in the upper ocean, primarily through the turbulence in the oceanic mixed layer. There is, however, an unanswered question: Does the huge amount of mechanical energy input by the surface waves and ageostrophic currents somehow contribute to the maintenance and regulation of the MOC and poleward heat flux in the steady state of the oceanic circulation?

In this section we explore the role of the mixed layer depth distribution in regulating the MOC and poleward heat flux. In particular, we examine the connection between the mixed layer depth, the energy required to sustain the thermohaline circulation, the MOC, and poleward heat flux.

The oceanic mixed layer is one of the most important components of the oceanic general circulation, because it is the buffer between the ocean and the atmosphere. In general, the mixed layer depth is controlled by external mechanical energy input and surface buoyant forcing. When the mixed layer is shallow, its depth is controlled by wind-stress-induced turbulence; however, when the mixed layer is deep, surface turbulence cannot penetrate to the base of the mixed layer, so that mixed layer depth is primarily controlled by convective adjustment induced by surface cooling.

At high latitudes, the only way of producing effective perturbations in the mixed layer is through surface cooling. In a steady-state model with a linear equation of state, the mixed layer should penetrate to the sea floor, so that such perturbations cannot change

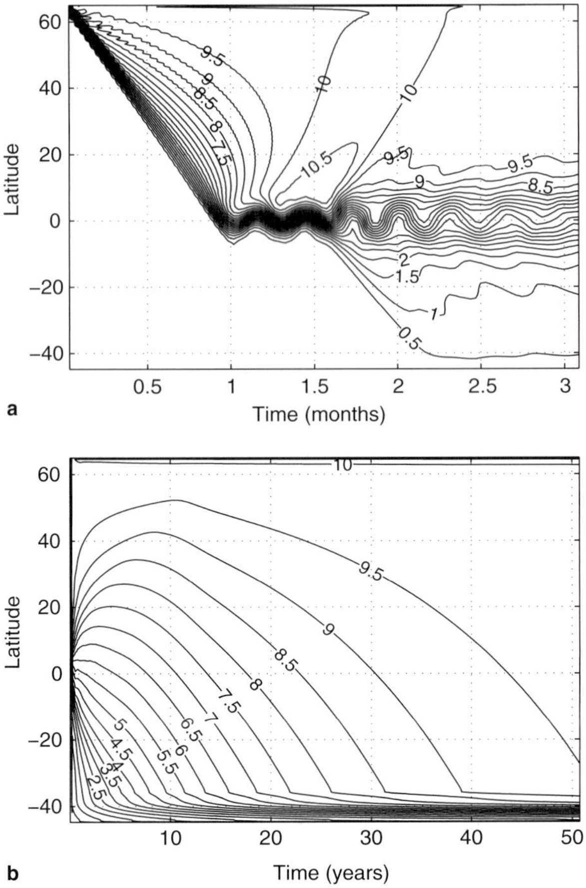

Fig. 5.193 Net northward volume transport (in Sv) as a function of latitude and time: **a** the initial Kelvin wave response, and **b** the longer time scale adjustment (Johnson and Marshall, 2002).

the meridional pressure difference. As a result, perturbing the mixed layer at high latitudes does not generate change in the meridional circulation and poleward mass/heat fluxes. Similarly, perturbing the mixed layer at mid latitudes can only slightly change the meridional overturning circulation and poleward mass/heat fluxes, as illustrated in Figure 5.194a, b.

On the other hand, perturbing the mixed layer at low latitudes may induce changes in the meridional overturning circulation and poleward mass/heat fluxes, as explained in the following discussion.

A simple two-dimensional model

Assume a two-dimensional model with width L, depth H, and a constant stratification below the mixed layer. The model ocean is in a steady state without a seasonal cycle. Assume a

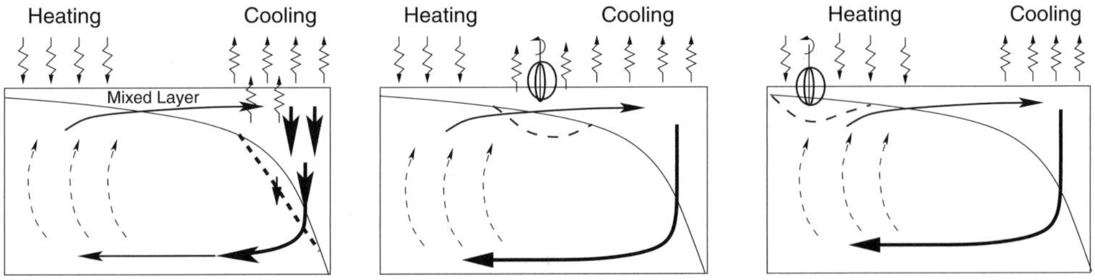

a High-latitude cooling
No change in N–S pressure difference MOC, Poleward heat flux

b Cooling and stirring at mid latitudes
Small changes in N–S pressure difference MOC, Poleward heat flux

c Stirring at low latitudes
Big changes in N–S pressure difference MOC, Poleward heat flux

Fig. 5.194 **a–c** Sketch of three types of mixed layer perturbations, including anomalies in surface thermal and mechanical forcing.

linear equation of state $\rho = \overline{\rho}(1 - \alpha T)$, where α is the thermal expansion coefficient, so that there is no cabbeling effect. Thus, at the northern boundary, the mixed layer depth should be equal to the depth H, and in the whole basin its distribution is assumed to be a linear function of latitude

$$h(y) = H(ay + 1 - a) \tag{5.409}$$

where $y = (\theta - \theta_0)/\Delta\theta$, $\Delta\theta$ is the meridional width of the model basin, and $a \leq 1$ is a constant in nondimensional units. Density in the mixed layer is vertically homogenized and is a linear function of latitude

$$\rho_m = \rho_0 + \Delta\rho y, \quad -h \leq z \leq 0, \quad 0 \leq y \leq 1 \tag{5.410}$$

Since stratification below the mixed layer is assumed to be constant, density in the ocean interior can be written as

$$\rho_i = \rho_0 + \Delta\rho - \frac{\Delta\rho}{aH}(H + z), \quad -H \leq z < -h \tag{5.411}$$

where ρ_0 is the mixed layer density at the southern boundary of the model basin, and $\Delta\rho$ is the mixed layer density difference between the southern and northern boundaries. Typical density profiles under three different mixed layer depth distributions are shown in Figure 5.195a.

If the mixed layer depth at low latitudes increases (Fig. 5.196a), stratification below the mixed layer (i.e., $\Delta\rho/aH$ in the above formula) in a steady state is actually enhanced.

Under the rigid-lid approximation, pressure in the mixed layer can be calculated by integrating from $z = 0$ downward

$$p_m = p_a - g\rho_m z = p_a - g(\rho_0 + \Delta\rho y)z \tag{5.412}$$

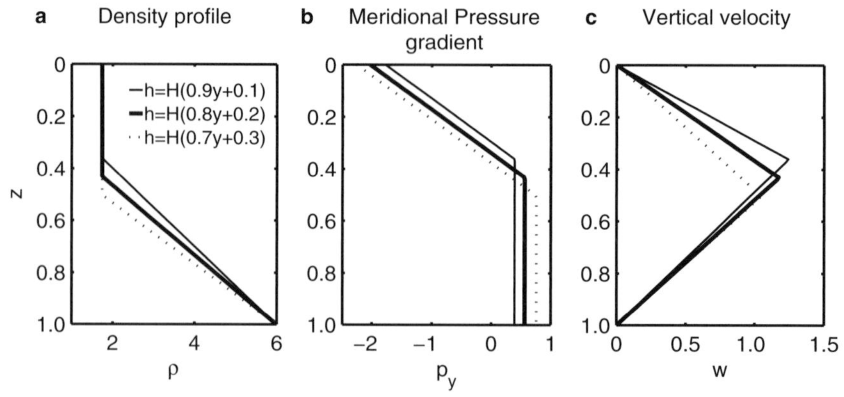

Fig. 5.195 Structure of the solution at a station $y = 0.3$ for three different mixed layer depth profiles: **a** density profile; **b** meridional pressure gradient; and **c** vertical velocity (all in nondimensional units) (Huang *et al.*, 2007).

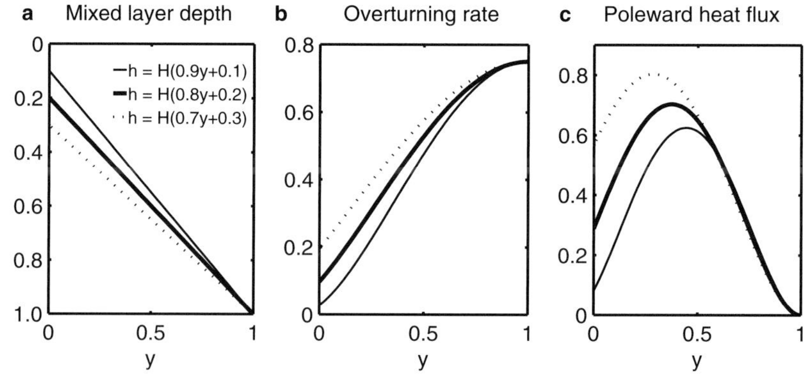

Fig. 5.196 Meridional structure of the solution for three different mixed layer depth profiles (in nondimensional units): **a** mixed layer depth; **b** meridional overturning rate; and **c** poleward heat flux (Huang *et al.*, 2007).

where $p_a(y)$ is the unknown "atmospheric pressure." Pressure in the ocean interior and below the mixed layer is

$$p_i = p_a + g\rho_m h + g\int_z^{-h} \rho_i dz \tag{5.413}$$

From these two relations, the meridional pressure gradients in these two regions are

$$p_{m,y} = p_{a,y} - g\Delta\rho z \tag{5.414a}$$

$$p_{i,y} = p_{a,y} + g\Delta\rho h \tag{5.414b}$$

where the second subscript y indicates the partial derivative with respect to y. The unknown atmospheric pressure gradient $p_{a,y}$ can be eliminated as follows.

First, we assume that the meridional velocity is proportional to the meridional pressure gradient

$$v = -cp_y \tag{5.415}$$

where c is a constant.

Second, under the Boussinesq approximations, continuity requires that the vertically integrated meridional volume flux crossing each meridional position y must be zero:

$$\int_{-H}^{0} p_y dz = 0 \tag{5.416}$$

Substituting Eqns. (5.414a, 5.414b) into Eqn. (5.416) leads to the relation determining the unknown atmospheric pressure gradient

$$p_{a,y} = -g\Delta\rho h\left(1 - \frac{h}{2H}\right) \tag{5.417}$$

Thus, the corresponding pressure gradients in the mixed layer and below are

$$p_{m,y} = -g\Delta\rho\left[h\left(1 - \frac{h}{2H}\right) + z\right] \tag{5.418a}$$

$$p_{i,y} = \frac{g\Delta\rho}{2H}h^2 \tag{5.418b}$$

Typical meridional pressure gradient profiles are shown in Figure 5.195b. It can readily be seen that if the mixed layer depth increases at a given station, meridional pressure gradients in the mixed layer and in the ocean interior are enhanced. Therefore, the MOC should strengthen because the meridional velocity is proportional to the meridional pressure gradient in the model.

Substituting Eqns. (5.418a, 5.418b) into Eqn. (5.415), the corresponding meridional velocity profiles in the mixed layer and in the interior below are

$$v_m = cg\Delta\rho\left[h\left(1 - \frac{h}{2H}\right) + z\right] \tag{5.419a}$$

$$v_i = -\frac{cg\Delta\rho}{2H}h^2 \tag{5.419b}$$

The continuity equation is

$$v_y + w_z = 0 \tag{5.420}$$

Since the vertical velocity at the surface or the bottom vanishes, it can be calculated from the vertical integration at each meridional location

$$w = -\int_{z}^{0} v_y dz \tag{5.421}$$

Typical vertical velocity profiles are shown in Figure 5.195c.

In order to find the boundary between the northward and southward flows at each station, we use Eqn. (5.419a) and set $v_m = 0$, leading to the vertical position of the zero-meridional velocity layer:

$$h_0 = -h(1 - h/2H) \tag{5.422}$$

By definition, the meridional velocity changes its sign at depth h_0; thus, the meridional overturning rate is

$$\Psi = -\int_{0}^{h_0} v_m dz = \frac{cg\Delta\rho}{2} h^2 (1 - h/2H)^2 \tag{5.423}$$

The poleward heat flux can be calculated as follows. The poleward heat flux is

$$H_f = c_p \int_{-H}^{0} \overline{\rho v T} dz = -\frac{c_p}{\alpha}\left(\int_{-h}^{0} v_m \rho_m dz + \int_{-H}^{-h} v_i \rho_i dz\right) = \frac{c_p}{\alpha}\frac{cg\Delta\rho^2}{4aH^2} h^2 (H - h)^2 \tag{5.424}$$

We assume that the density balance in the oceanic interior can be approximately treated in terms of a one-dimensional balance between the vertical advection and vertical diffusion:

$$w\frac{\partial \rho}{\partial z} = \kappa \frac{\partial^2 \rho}{\partial z^2} \tag{5.425}$$

Using this equation, the scale for the mixing coefficient is $\kappa = WD$, where W and D are the scales of vertical velocity and thickness of the stratified water. Therefore, the mechanical energy required for sustaining mixing in the ocean interior is estimated as

$$e_i = \kappa(\rho_0 + \Delta\rho - \rho_m) = W(H - h)(\Delta\rho - \Delta\rho y) = \frac{cg\Delta\rho^2}{H} h(H - h)^3 \tag{5.426}$$

The sensitivity of the MOC and poleward heat flux to changes in mixed layer properties in the model ocean can be explored through three cases with different linear mixed layer depth profiles (Fig. 5.196a). As stated in the previous section, the mixed layer depth in our simple model should be equal to the depth H of the ocean at the northern boundary, so that the differences between these three cases appear at lower latitudes. It can readily be seen that for these cases, if the mixed layer depth is larger at low and mid latitudes, the circulation rate is also larger there but remains unchanged at the northern boundary (Fig. 5.196b). At the same time, poleward heat flux is enhanced, and the position of maximum poleward heat

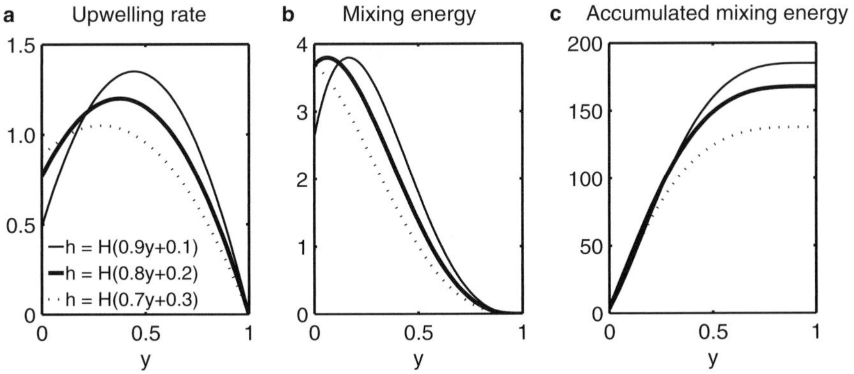

Fig. 5.197 Meridional structure of the solution for three different mixed layer depth profiles (in nondimensional units): **a** the upwelling rate; **b** energy required for supporting diapycnal mixing; **c** meridionally accumulated energy required for sustaining mixing (Huang *et al.*, 2007).

flux drifts to lower latitudes with the deepening of the mixed layer depth at lower latitudes (Fig. 5.196c).

On the other hand, the upwelling rate at the base of the mixed layer at mid and high latitudes for the case with the deepest mixed layer $h = H\ (0.7y + 0.3)$ is the lowest, although it is the highest near the southern boundary (Fig. 5.197a). Similarly, the energy required for sustaining the subsurface diapycnal mixing is the lowest at mid and high latitudes, and the corresponding meridionally accumulated energy for sustaining subsurface diapycnal mixing is the smallest among all three cases (Fig. 5.197b, c).

In contrast, for the case with a shallow mixed layer at lower latitudes, $h = H\ (0.9y + 0.1)$, the MOC transports less water and heat; at the same time the circulation may require more energy for sustaining the subsurface diapycnal mixing, as shown in Figures 5.196 and 5.197.

Thus, these three cases demonstrate that, with a deep mixed layer at lower latitudes, the meridional overturning cell can transport more water at low and mid latitudes, and more heat poleward. At the same time, the circulation may require less mechanical energy for sustaining diapycnal mixing in the ocean interior.

Results obtained from an oceanic general circulation model

The analytical model discussed above is for a highly-simplified two-dimensional model. Similar results can be obtained from numerical experiments based on an OGCM. In most existing models, wind energy input to the mixed layer is parameterized in terms of energy input to the turbulence using simple formulae. For example, wind energy input can be parameterized as

$$e = \rho m * u^{*3} \tag{5.427}$$

where $m*$ is an empirical constant in nondimensional units, and $u*$ is the (atmospheric-side) frictional velocity.

The dynamical impact of wind stress to the surface ocean is complicated. Obviously, the contribution of wind energy input to the ocean should include contributions from a wide spectrum in space and time; thus, a simple formula such as Eqn. (5.427) is unlikely to convey the complete information about energy transport from the atmosphere to the ocean.

In OGCMs m^* is usually set to 1.25. However, a quite different value of m^* has also been used in previous studies. Noh *et al.* (2004) estimated turbulent kinetic energy at the surface with $m^* = 1.40$; but m^* was taken to be 3.50 in the study of Craig and Banner (1994). (The air-side frictional velocity is used here, so that there is a coefficient of about 28.6 in the formula in our calculation, compared with that used in the studies using the water-side frictional velocity.) However, Stacey (1999) analyzed data from the Knight Inlet in southwestern Canada, and concluded that using $m^* = 5.25$ provided the best fit to the data.

In the following experiments, the model was spun up from a state of rest for 1,000 years, in which m^* was set to 1.25. Based on this run, five experiments were then run for 400 years, in which m^* was set to 0.4, 1.25, 3.75, 7.5, and 12.5. The experiment with $m^* = 1.25$ is taken as the standard case in this study. A higher value of m^* implies that a larger amount of wind energy input is used to deepen the mixed layer depth under the same wind stress.

The basic idea behind this approach is as follows. The contributions of wind stress to the oceanic circulation are rather complicated, but at least they can be separated into two categories: the large-scale mean wind stress, which drives the Ekman transport and the wind-driven gyre in the upper ocean, and the source for the turbulent kinetic energy in the upper ocean. Simply changing the "mean" wind stress profile in the ocean model cannot represent the complicated contribution from wind accurately. As a compromise in our conceptual study here, we choose to alternate this single parameter m^* to simulate contributions to surface turbulence in the upper ocean from the wind stress components with relatively high spatial and temporal resolution.

An increase in m^* has the profound effect of mixed layer deepening at low and mid latitudes, where mixed layer depth is primarily controlled by the turbulence kinetic energy input from the winds Fig. 5.198a). It is generally accepted that increasing m^* can increase the mixed layer depth in the bulk mixed-layer model (Gaspar, 1988). Since a bulk mixed-layer model is used in this study, choosing a high value of m^* may lead to a deep mixed layer depth.

An immediate consequence of a deeper mixed layer at low and mid latitudes is the increase in the meridional pressure gradient in the upper ocean; this leads to an intensification of the MOC. For example, when $m^* = 7.5$, the MOC increases to nearly 1.6 Sv, compared to the case with $m^* = 1.25$ for the latitudinal bands of 20–45° N (Fig. 5.198b).

Due to the increase in the meridional mass transport, the circulation can carry more heat poleward. In fact, the maximal poleward heat flux increases from 0.38 PW for the case of $m^* = 1.25$, to 0.57 PW for the case of $m^* = 7.5$ (Fig. 5.199a).

In terms of mechanical energy balance, diapycnal mixing in the ocean interior is controlled by mechanical energy input from the winds and tides (Munk and Wunsch, 1998; Huang, 1999). A larger amount of mechanical energy input can result in stronger diapycnal mixing in

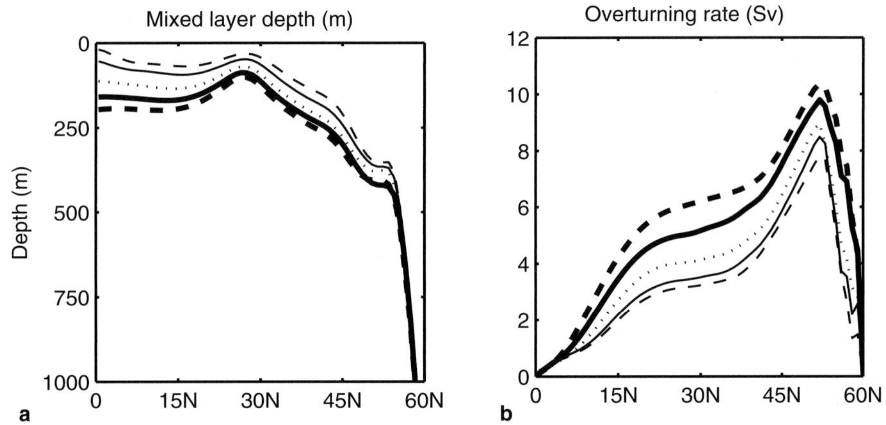

Fig. 5.198 **a** Zonal-mean mixed layer depth in the 3-D ocean model (in m); **b** perturbing rate obtained from a 3-D ocean model (in Sv) (Huang *et al.*, 2007).

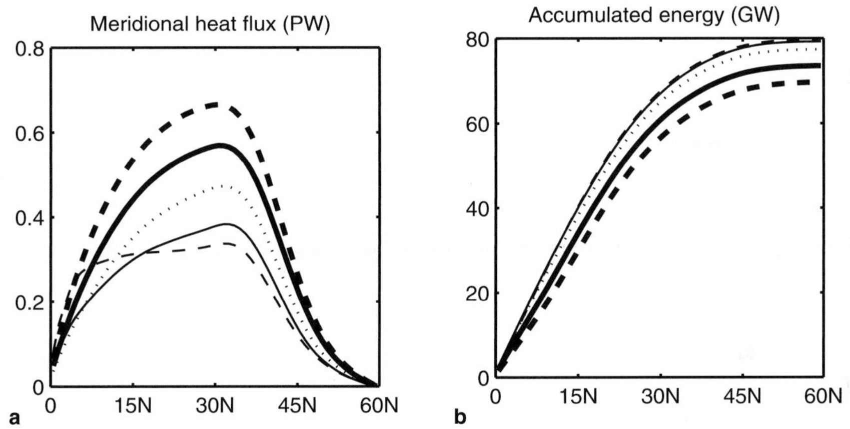

Fig. 5.199 **a** Poleward heat flux in the 3-D ocean model (in 10^{15} W); **b** meridionally accumulated energy required for sustaining subsurface diapycnal mixing in the 3-D ocean model (in 10^9 W) (Huang *et al.*, 2007).

the ocean interior, which may enhance the MOC and poleward heat flux. On the other hand, more mechanical energy input can increase the mixed layer depth. This process also plays an important role in regulating the MOC and poleward heat flux. For example, tropical cyclones can input a large amount of mechanical energy into the ocean, which not only enhances diapycnal mixing in the ocean interior but also deepens the mixed layer greatly (Price, 1981), so tropical cyclones play an essential role in driving the MOC and poleward heat flux (Emanuel, 2001). However, there is no simple linear relationship between the amount of mechanical energy input and the strength of the MOC and the related poleward heat flux. The complicated nature of such connections remains a critically important issue for further study.

Another interesting result is that the energy required for sustaining subsurface diapycnal mixing is actually reduced for the cases with a deeper mixed layer at lower latitudes (Fig. 5.199b). This is due to the fact that a deeper mixed layer implies a smaller density difference between the bottom water and the base of the mixed layer. Since the amount of energy required for sustaining subsurface diapycnal mixing is proportional to this density difference, it is reduced with this smaller density difference.

Appendix: Definition of the oceanic sensible heat flux

Traditionally, the poleward oceanic heat flux is defined as

$$H_o = \iint \rho c_p v \Theta a \cos\theta d\lambda dz \tag{5.A1}$$

where ρ is the *in situ* temperature, c_p is the specific heat under constant pressure, v is the meridional velocity, and Θ is the potential temperature. It is well known that if the system has a net mass flux through the section, the heat flux calculated from this formula may depend on the choice of temperature scale. In general, heat flux itself has no direct physical meaning, and the most essential quantity associated with the heat flux is its divergence. There are two criteria in defining heat flux: first, the divergence of heat flux should satisfy the local heat balance equation; second, the heat flux should depend on the local properties only.

The best-known example is the circulation in the South Pacific and Indian Oceans. Due to the Indonesian Throughflow, the poleward heat flux in the South Pacific and Indian Oceans is not uniquely defined. To overcome the complication due to the net mass flux through the sections, many different approaches have been used. For example, Zhang and Marotzke (1999) proposed a procedure in which the heat flux associated with the westward flow in the strait is used to evaluate the poleward heat flux. Although the heat flux definition they proposed satisfies the first criterion, it does not satisfy the second criterion. In fact, their definition includes a term that depends on the thermal condition in the cross-section of the Throughflow. If the condition within the strait changes, the heat flux evaluated will also change, even if the circulation and water properties at the given section south of the strait do not change.

To overcome such a problem, a much simpler definition is recommended:

$$H_o = \iint \rho c_p v(\Theta - \Theta_0) a \cos\theta d\lambda dz \tag{5.A2}$$

where Θ and Θ_0 are potential temperature and the reference potential temperature, which can be chosen arbitrarily; the most obvious choice is the global-mean potential temperature, about 2° C. This definition satisfies the two criteria listed above. We analyze the meaning of the heat flux defined in this way as follows.

In the oceans, there is indeed a net mass flux through any given section due to either evaporation/precipitation or throughflow:

$$H_o = H_{o,0} + H_{o,1} \tag{5.A3}$$

$$H_{o,0} = \iint \rho c_p v(\Theta - \Theta_0) a \cos\theta d\lambda dz \tag{5.A4}$$

$$H_{o,1} = \iint \overline{\rho v} c_p (\overline{\Theta} - \Theta_0) a \cos\theta d\lambda dz \tag{5.A5}$$

where $\overline{\rho v} = \iint \rho v a \cos\theta d\lambda dz / \iint a \cos\theta d\lambda dz$ is the mass flux averaged over the section. The first component, $H_{o,0}$, is the oceanic sensible heat flux through this section, carried by the circulation without the net mass flux. By definition, it is independent of the choice of reference temperature. On the other hand, the second component, $H_{o,1}$, clearly depends on the choice of reference temperature, so it should be isolated from the first component. In the oceans, the net mass flux through a given section is due either to evaporation/precipitation or to an inter-basin loop current, such as the Indonesian Throughflow. We will separate the net mass flux through each section into two parts:

$$\overline{\rho v} = \overline{\rho v}_{loop} + \overline{\rho v}_{emp} \tag{5.A6}$$

where $\overline{\rho v}_{loop}$ is the zonal-mean meridional mass flux associated with the loop current going through the whole basin, and the total amount of mass flux associated with this component should be constant at any zonal section, such as the Indonesian Throughflow; and $\overline{\rho v}_{emp}$ is the zonal-mean meridional mass flux associated with evaporation and precipitation. This component of the net mass flux through the section is associated with the water vapor cycle in the atmosphere, so it should not be counted as part of the oceanic sensible heat flux.

Accordingly, poleward sensible heat flux in the Atlantic Ocean can be defined as

$$H^A_{o,sen} = \iint_A \rho c_p v \left(\Theta - \Theta_0\right) a \cos\theta d\lambda dz - H^A_{o,emp} \tag{5.A7}$$

$$H^A_{o,emp} = c_p \iint_A \overline{\rho v}_{emp}(\theta) \left[\overline{\Theta}_A(\theta) - \Theta_0\right] a \cos\theta d\lambda dz = c_p M_{emp,A}(\theta) \left[\overline{\Theta}_A(\theta) - \Theta_0\right] \tag{5.A8}$$

where $\Theta_A(\theta)$ is the mean potential temperature at a given zonal section in the Atlantic Ocean, and $M_{emp,A}$ is the zonally integrated meridional mass flux due to evaporation/precipitation. Similarly, the poleward heat flux in the Pacific and Indian Oceans are defined as

$$H^P_{o,sen} = \iint_P \rho c_p v \left(\Theta - \Theta_0\right) a \cos\theta d\lambda dz - H^P_{o,emp},$$

$$H^P_{o,emp} = c_p M_{emp,P}(\theta) \left[\overline{\Theta}_P(\theta) - \Theta_0\right] \tag{5.A9}$$

$$H_{o,sen}^{I} = \iint_{I} \rho c_p v \left(\Theta - \Theta_0\right) a \cos\theta d\lambda dz - H_{o,emp}^{I},$$

$$H_{o,emp}^{I} = c_p M_{emp,I}(\theta) \left[\overline{\Theta}_I(\theta) - \Theta_0\right] \tag{5.A10}$$

The sum of water flux due to evaporation and precipitation, including the river run-off, in all oceans should equal the water vapor flux in the atmosphere. Thus, the sum of the evaporation and precipitation in three basins is

$$M_{emp} = M_{emp,A} + M_{emp,P} + M_{emp,I} \tag{5.A11}$$

The heat flux associated with this mass flux is

$$H_{o,emp} = c_p \left[M_{emp,A}\overline{\Theta}_A(\theta) + M_{emp,P}\overline{\Theta}_P(\theta) + M_{emp,I}\overline{\Theta}_I(\theta)\right] - c_p M_{emp}\Theta_0 \tag{5.A12}$$

This heat flux is independent of the choice of temperature scale.

This example demonstrates that a working definition of heat flux should be independent of the specific choice of temperature scale. In order to achieve this goal, one has to separate the net mass flux through a given section, and link this net mass flux with the corresponding return flow in the climate system, i.e., the water vapor flux in the atmosphere,

$$H_{emp}^{air-sea} = -M_{emp}\left\{L_h + c_p\left[\overline{\Theta}_{atmos}(\theta) - \Theta_0\right]\right\} + H_{o,emp} \tag{5.A13}$$

where the first term indicates the heat content flux in the atmospheric branch of the water vapor loop, and the second term is the heat content flux in the oceanic branch of the water cycle. This definition is, of course, independent of the choice of temperature scale.

References

Adcroft, A., Scott, J. R. and Marotzke, J. (2001). Impact of geothermal heating on the global ocean circulation. *Geophys. Res. Lett.*, **28**, 1735–1738.

Alford, M. H. (2003). Improved global maps and 54-year history of wind-work on ocean inertial motions. *Geophys. Res. Lett.*, **30**, 1424, doi:10.1029/2002GL016614.

Anderson, D. L. T. and Gill, A. (1975). Spin-up of a stratified ocean, with applications to upwelling. *Deep Sea Res.*, **22**, 583–596.

Anderson, D. L. T. and Moore, D. W. (1979). Cross-equatorial inertial jets with special relevance to very remote forcing of the Somali current. *Deep Sea Res.*, **26**, 1–22.

Blandford, R. R. (1965). Notes on the theory of the thermocline. *J. Mar. Res.*, **23**, 18–29.

Blanke, B. and Raynaud, S. (1997). Kinematics of the Pacific Equatorial Undercurrent: An Eulerian and Lagrangian approach from GCM results. *J. Phys. Oceanogr.*, **27**, 1038–1053.

Broecker, W. S. (1991). The great ocean conveyor. *Oceanogr.*, **4**, 79–89.

Broecker, W. S. (1998). Paleocean circulation during the last deglaciation: A bipolar seesaw? *Paleoceanography*, **13**, 119–121.

Bromley, L. A. (1968). Relative enthalpies of sea salt solutions at 25°C. *J. Chemi. Eng. Data*. **13**, 399–402.

Bryan, F. (1986). High-latitude salinity effects and interhemispheric thermohaline circulations. *Nature*, **323**, 301–304.

Bryan, F. (1987). Parameter sensitivity of primitive equation ocean general circulation models. *J. Phys. Oceanogr.*, **17**, 970–985.

Bryan, K. (1969). Climate and the ocean circulation. III: The ocean model. *Mon. Wea. Rev.*, **97**, 806–827.

Bryan, K. and Cox, M. (1967). A numerical investigation of the oceanic general circulation. *Tellus*, **19**, 54–80.

Bryden, H. L. and Brady, E. C. (1985). Diagnostic study of the three-dimensional circulation in the upper equatorial Pacific Ocean. *J. Phys. Oceanogr.*, **15**, 1255–1273.

Bryden, H. L. and Imawaki, S. (2001). Ocean heat transport. In *Ocean Circulation and Climate*, ed. G. Siedler, J. Church and J. Gould, International Geophysical Series, Academy Press, New York, pp. 455–474.

Bryden, H. L. and Kinder, T. H. (1991). Steady two-layer exchange through the Strait of Gibraltar. *Deep Sea Res.*, **38**, (Suppl. 1), S445–S463.

Bryden, H. L. and Stommel, H. M. (1984). Limiting processes that determine basic features of the circulation in the Mediterranean Sea. *Oceanol. Acta*, **7**, 289–296.

Cahn, A. (1945). An investigation of the free oscillations of a simple current system. *J. Meteorol.*, **2**, 113–119.

Cane, M. A. (1989). A mathematical note on Kawase's study of the deep-ocean circulation. *J. Phys. Oceanogr.*, **19**, 548–550.

Carmack, E. C. and Killworth, P. D. (1978). Formation and interleaving of abyssal water masses off Wilkes Land, Antarctica. *Deep Sea Res.*, **25**, 357–369.

Carton, J. A. and Giese, B. S. (2008). A reanalysis of ocean climate using Simple Ocean Data Assimilation (SODA). *Month. Wea. Rev.*, **136**, 2999–3017.

Carton, J. A., Chepurin, G., Cao, X. and Giese, B. S. (2000a). A Simple Ocean Data Assimilation analysis of the global upper ocean 1950–1995. Part 1: Methodology. *J. Phys. Oceanogr.*, **30**, 294–309.

Carton, J. A., Chepurin, G. and Cao, X. (2000b). A Simple Ocean Data Assimilation analysis of the global upper ocean 1950–1995. Part 2: Results, *J. Phys. Oceanogr.*, **30**, 311–326.

Cessi, P. (1988). A stratified model of the inertial recirculation. *J. Phys. Oceanogr.*, **18**, 662–682.

Cessi, P., Ierley, G. and Young, W. (1987). A model of the inertial recirculation driven by potential vorticity anomalies. *J. Phys. Oceanogr.*, **17**, 1640–1652.

Charney, J. G. (1955). The Gulf Stream as an inertial boundary layer. *Proc. Nat. Acad. Sci.*, **41**, 731–740.

Chelton, D. B. and Schlax, M. G. (1996). Global observations of oceanic Rossby waves. *Science*, **272**, 234–238.

Chelton, D. B., de Szoeke, R. A. and Schlax, M. G. (1998). Geographical variability of the first baroclinic Rossby radius of deformation. *J. Phys. Oceanogr.*, **28**, 433–460.

Chelton, D. B., Schlax, M. G., Samelson, R. M. and de Szoeke, R. A. (2007). Global observations of large oceanic eddies. *Geophys. Res. Lett.*, **34**, L15606, doi:10.1029/2007GL030812.

Chereskin, T. K. (1995). Direct evidence for an Ekman balance in the California Current. *J. Geophys. Res.*, **100**, 18,261–18,269.

Chereskin, T. K. and Price, J. F. (2001). Ekman transport and pumping. In *Encyclopedia of Ocean Sciences*, ed. J. H. Steele, K. K. Turekian and S. A. Thorpe, Academic Press, New York, pp. 809–815.

Christopher, E. and Pedlosky, J. (1995). The influence of distributed sources and upwelling on the baroclinic structure of the abyssal circulation. *J. Phys. Oceanogr.*, **25**, 2259–2284.

Chu, P.C. (1995). P-vector method for determining absolute velocity from hydrographic data. *Mar. Techn. Soc. J.*, **29**, 3–14.

Clarke, R. A. and Gascard, J. C. (1983). The formation of Labrador Sea Water, Part I: Large-scale processes, *J. Phys. Oceanogr.*, **13**, 1764–1778.

Conkright, M. E., Locarnini, R. A., Garcia, H. E., O'Brien, T. D., Boyer, T. P., Stephens, C. and Antonov, J. I. (2002). *World Ocean Atlas 2001: Objective Analysis, Data Statistics, and Figures*, CD-ROM Documentation. National Oceanographic Data Center, Silver Spring, MD, 17 pp.

Cox, M. D. (1985). An eddy-resolving numerical model of the ventilated thermocline. *J. Phys. Oceanogr.*, **15**, 1312–1324.

Cox, M. D. and Bryan, K. (1984). A numerical model of the ventilated thermocline. *J. Phys. Oceanogr.*, **14**, 674–687.

Craig, P. D. and Banner, M. L. (1994). Modeling wave-enhanced turbulence in the ocean surface layer. *J. Phys. Oceanogr.*, **24**, 2546–2559.

Cummins, P. F. and Foreman, M. G. G. (1998). A numerical study of circulation driven by mixing over a submarine bank. *Deep Sea Res.*, I, **45**, 745–769.

Cushman-Roisin, B. (1987). Subduction. In: Dynamics of the oceanic surface mixed layer. Proceedings Hawaiian Winter Workshop, 181–196.

Cushman-Roisin, B. (1994). *Introduction to Geophysical Fluid Dynamics*. Prentice Hall Press, 320 pp.

Da Silva, A. M., Young, C. C. and Levitus, S. (1994). Atlas of surface marine data 1994: Volume 1: Algorithms and procedures. US Dept. of Commerce. Washington, DC. NOAA Atlas NESDIS 6, 83 pp. [G1046 .C1 N33 v.1]

Defant, A. (1961). *Physical Oceanography*, Vol. 1, Elsevier, New York, 728 pp.

DeLaughter, J. E., Stein, C. A. and Stein, S. (2005). *Hotspots: A view from the swells*. Geological Society of America, Special Paper 388.

Deser, C., Alexander, M. A. and Timlin, M. S. (1996). Upper-ocean thermal variability in the North Pacific during 1970–1991. *J. Climate*, **9**, 1840–1855.

Dewar, W. K (1986). On the potential vorticity structure of weakly ventilated isopycnals: a theory of subtropical mode water maintenance. *J. Phys. Oceanogr.*, **16**, 1204–1216.

Dewar, W. K. and Huang, R. X. (1995). Fluid flow in loops driven by freshwater and heat fluxes. *J. Fluid Mech.*, **297**, 153–191.

Dewar, W. K. and Killworth, P. (1990). On the cylinder collapse problem, mixing, and the merger of isolated eddies. *J. Phys. Oceanogr.*, **20**, 1563–1575.

Dewar, W. K., Bingham, R. J., Iverson, R. L., Nowacek, D. P., St Laurent, L. C. and Wiebe, P. H. (2006). Does the marine biosphere mix the ocean? *J. Mar. Res.*, **64**, 541–561.

Dewar, W. K., Samelson, R. M. and Vallis, G. K. (2005). The ventilated pool: a model of Subtropical Mode Water. *J. Phys. Oceanogr.*, **35**, 137–150.

Dickson, R. R., Gmitrowicz, E. M. and Watson, A. J. (1990). Deep-water renewal in the Northern North Atlantic. *Nature*, **344**, No. 6269, 848–850.

Döscher, R., Boning, C. W. and Herrmann, P. (1994). Response of circulation and heat transport in the North Atlantic to changes in thermohaline forcing in northern latitudes: A model study. *J. Phys. Oceanogr.*, **24**, 2306–2320.

Eden, C. and Willebrand, J. (1999). Neutral density revisited, *Deep Sea Res. II*, **46**, 33–54.

Edwards, C. A. and Pedlosky, J. (1998). Dynamics of nonlinear cross-equatorial flow. Part I: Potential vorticity transformation. *J. Phys., Oceanogr.*, **28**, 2382–2406.

Edwards, N. R. (1996). Unsteady similarity solutions and oscillating ocean gyres. *J. Mar. Res.*, **54**, 793–826.

Egbert, G. D., Ray, R. D. and Bills, B. G. (2003). Numerical modelling of the global semidiurnal tide in the present day and in the last glacial maximum. *J. Geophys. Res.*, **109**, C03003, doi:10.1029/2003JC001973.

Ekman, V. W. (1905). On the influence of the earth's rotation on ocean currents. *Arch. Math. Astron. Phys.*, **2**, 1–52.

Emanuel, K. (2001). Contribution of tropical cyclones to meridional heat transport by the oceans. *J. Geophys. Res.*, **106**(D14), 14771–14781.

Faller, A. (1966). Sources of energy for the ocean circulation and a theory of the mixed layer. *Proceedings of the Fifth US National Congress of Applied Mechanics*, Minnesota, pp. 651–672.

Feistel, R. (1993). Equilibrium thermodynamics of seawater revisited. *Prog. Oceanogr.*, **31**, 101–179.

Feistel, R. (2003). A new extended Gibbs thermodynamic potential of seawater. *Prog. Oceanogr.*, **58**, 43–114.

Feistel, R. (2005). Numerical implementation and oceanographic application of the Gibbs thermodynamic potential of seawater. Ocean Science Discussion, 1 (2004), 1–19, www.ocean-science.net/osd/1/1, Ocean Science 1 (2005), 9–16, www.ocean-science.net/os/1/9

Feistel, R. and Hagen, E. (1995). On the Gibbs thermodynamic potential of seawater. *Prog. Oceanogr.*, **36**, 249–327.

Feng, C. Y., Wang, W. and Huang, R. X. (2006). Meso-scale available gravitational potential energy in the world oceans. *Oceanology Bulletin Sinica*, **25**, No.5, 1–13.

Ferrari, R. and Wunsch, C. (2009). Ocean circulation kinetic energy: Reservoirs, sources, and sinks. *Ann. Rev. Fluid Mech.*, **41**, 253–282.

Fieux, M., Schott, F. and Swallow, J. C. (1986). Deep boundary currents in the western Indian Ocean revisited. *Deep Sea Res.*, **33**, 415–426.

Filippov, U. G. (1968). Application of group analysis for the solution of certain ocean flow problems. *Meteorology and Hydraulics*, **9**, 53–62.

Fine, R., Reid, J. L. and Ostlund, H. G. (1981). Circulation of tritium in the Pacific Ocean. *J. Phys. Oceanogr.*, **11**, 3–14.

Fine, R. A., Peterson, W. H. and Ostlund, H. G. (1987). The penetration of tritium into the tropical Pacific. *J. Phys. Oceanogr.*, **17**, 553–564.

Flament, P. (2002). A state variable for characterizing water masses and their diffusive stability: Spiciness. *Progr. Oceanogr.*, **54**, 493–501.

Fofonoff, N. P. (1954). Steady flow in a frictionless homogeneous ocean. *J. Mar. Res.*, **13**, 254–262.

Fofonoff, N. P. (1962). Physical properties of seawater. In *The Sea: Ideas and Observations on Progress in the Study of the Seas, 1: Physical Oceanography*, ed. M. N. Hill, Wiley, Interscience, New York, pp. 323–395.

Foster, T. D. and Carmack, E. C. (1976). Temperature and salinity structure in the Weddell Sea. *J. Phys. Oceanogr.*, **6**, 36–44.

Fox-Kemper, B. and Pedlosky, J. (2004). Wind-driven barotropic gyre I: Circulation controlled by eddy vorticity fluxes to an enhanced removal region. *J. Mar. Res.* **62**, 169–193.

Frankignoul, C. and Reynolds, R. W. (1983). Testing a dynamical model for mid-latitude sea surface temperature anomalies. *J. Phys. Oceanogr.*, **13**, 1131–1145.

Fratantoni, D. M., Johns, W. E., Townsend, T. L. and Hurlburt, H. E. (2000). Low-latitude circulation and mass transport pathways in a model of the tropical Atlantic Ocean. *J. Phys. Oceanogr.*, **30**, 1944–1966.

Gaffen, D. J., Rosen, R. D., Salstein, D. A. and Boyle, J. S. (1997). Evaluation of tropospheric water vapor simulations from the atmospheric Model Intercomparison Project. *J. Climate*, **10**, 1648–1661.

Gan, Z., Yan, Y. and Qi, Y. (2007). Climatological mean distribution of specific entropy in the oceans. *Ocean Science Discussion*, **4**, 129–144.

Garrett, C. (1991). Marginal mixing theories. *Atmos. Ocean*, **29**, 313–339.

Garrett, C., MacCready, P. and Rhines, P. (1993). Boundary mixing and arrested Ekman layers: rotating stratified flow near a sloping boundary. *Annual Rev. Fluid Mech.*, **25**, 291–323.

Gaspar, P. (1988). Modeling the seasonal cycle of the upper ocean. *J. Phys. Oceanogr.*, **18**, 161–180.

Gent, P. R. and McWilliams, J. (1990). Isopycnal mixing in ocean circulation models. *J. Phys. Oceanogr.*, **20**, 150–155.

Gent, P. R., Willebrand, J., McDougall, T. J. and McWilliams, J. (1995). Parameterizing eddy-induced tracer transports in ocean circulation models. *J. Phys. Oceanogr.*, **25**, 463–474.

Gill, A. E. (1971). Ocean Models. *Phil. Trans. Roy. Soc. Lond. A*, **270**, 391–413.

Gill, A. E. (1977). The hydraulics of rotating-channel flow. *J. Fluid Mech.*, **80**, 641–671.

Gill, A. E., Green, J. S. A. and Simmons, A. J. (1974). Energy partition in the large-scale ocean circulation and the production of mid-ocean eddies. *Deep Sea Res.*, **21**, 499–528.

Gnanadesikan, A. (1999). A simple predictive model for the structure of the oceanic pycnocline. *Science*, **283**, 2077–2079.

Goldsbrough, G. (1933). Ocean currents produced by evaporation and precipitation. *Proc. Roy. Soc. Lond.* A, **141**, 512–517.

Godfrey, J. S. (1989). A Sverdrup model of the depth-integrated flow for the world ocean allowing for island circulations. *Geophys. Astrophys. Fluid Dyn.*, **45**, 89–112.

Godfrey, J. S. (1993). The effect of the Indonesian throughflow on ocean circulation and heat exchange with the atmosphere: A review. *J. Geophys. Res.*, **101**, 12,217–12,237.

Gordon, A. L. (1978). Deep Antarctic convection west of Maud Rise. *J. Phys. Oceanogr.*, **8**, 600–612.

Gordon, A. L. (1982). Weddell deep water variability. *J. Mar. Res.*, **40**, 199–217.

Gordon, A. L. (2002). Bottom water formation. In *Encyclopedia of ocean sciences*, ed. J. H. Steele, K. K. Turkrekian and S. A. Thorpe, Academic Press, New York, pp. 334–340.

Gordon, A. L. and Nowlin, W. D. (1978). Basin waters of Bransfield Strait, *J. Phys. Oceanogr.*, **8**, 258–264.

Gordon, A. L. and Tchernia, P. (1972). Waters of the continental margin off the Adelie Coast Antarctic. In *Antarctic Oceanology II: The Australian–New Zealand Section. Antarctic Res. Ser.*, **19**, ed. D. E. Haynes, *Amer. Geophys. Union*, 59–69.

Gouriou, Y. and Toole, J. (1993). Mean circulation of the upper layers of the western equatorial Pacific Ocean. *J. Geophys. Res.*, **98**, 22,495–22,520.

Greatbatch, R. J., Fanning, A. F., Goulding, A. D. and Levitus, S. (1991). A diagnosis of interpentadal circulation changes in the North Atlantic. *J. Geophys. Res.*, **96**, 22,009–22,023.

Gu, D. and Philander, S. G. (1997). Interdecadal climate fluctuations that depend on exchanges between the tropics and extratropics. *Science*, **275**, 805–807.

Guan, Y. P. and Huang, R. X. (2008). Stommel's box model of thermohaline circulation revisited – The role of mechanical energy supporting mixing and wind-driven gyration. *J. Phys. Oceanogr.*, **38**, 909–917.

Hallberg, R. and Gnanadesikan, A. (2001). An exploration of the role of transient eddies in determining the transport of a zonally reentrant current. *J. Phys. Oceanogr.*, **31**, 3312–3330.

Hallberg, R. and Gnanadesikan, A. (2006). The role of eddies in determining the structure and response of the wind-driven Southern Hemisphere overturning: Results from the modeling eddies in the Southern Ocean (MESO) project. *J. Phys. Oceanogr.*, **36**, 2232–2252.

Hanawa, K. and Talley, L. D. (2001). Mode waters. In *Ocean Circulation and Climate*, ed. G. Siedler, J. Church and J. Gould, International Geophysical Series, Academy Press, New York, pp. 373–386.

Haney, R. L. (1971). Surface thermal boundary condition for ocean circulation models. *J. Phys. Oceanogr.*, **1**, 241–248.

Hayes, S. P. (1982). A comparison of geostrophic and measured velocities in the Equatorial Undercurrent. *J. Mar. Res.*, **40**, (Suppl.), 219–229.

Helland-Hansen, B. (1912). The ocean waters, An introduction to physical oceanography. *Intern. Rev. d. Hydrobiol.* (Suppl.) Bd. III, Ser. 1, H. **2**, 1–84.

Hellerman, S. and Rosenstein, M. (1983). Normal monthly wind stress over the world ocean with error estimates. *J. Phys. Oceanogr.*, **13**, 1093–1104.

Hendershott, M. C. (1987). Single-layer models of the general circulation. In *General Circulation of the Ocean*, ed. H. D. A. Abarbanel and W. R. Young, Springer-Verlag, New York, pp. 202–267.

Hogg, N. G. and Johns, W. E. (1995). Western boundary currents. US National Report to International Union of Geodesy and Geophysics 1991–1994, *Rev. Geophys.*, (Suppl.), 1311–1334.

Hogg, N. G. and Owens, W. B. (1999). Direct measurement of the deep circulation within the Brazil Basin. *Deep Sea Res.*, II, **46**, 335–353.

Hogg, N., Biscaye, P., Gardner, W. and Schmitz, W. J., Jr. (1982). On the transport and modification of Antarctic Bottom Water in the Vema Channel. *J. Mar. Res.*, **40**, (Suppl.), 231–263.

Holton, J. R. (2004). *An Introduction to Dynamic Meteorology*, Volume 88, Fourth edition (International Geophysics), Academic Press, New York, 535 pp.

Hood, S. and Williams, R. G. (1996). On frontal and ventilated modes of the main thermocline. *J. Mar. Res.*, **54**, 211–238.

Hough, S. S. (1897). On the application of harmonic analysis to the dynamical theory of the tides. – Part I. On Laplace's "Oscillations of the First Species", and on the dynamics of ocean currents. *Phil. Trans. Roy. Soc. Lond.* A, **189**, 201–257.

Hsiung, J. (1985). Estimates of global oceanic meridional heat transport. *J. Phys. Oceanogr.*, **15**, 1405–1413.

Hua, B. L., Moore, D. and Le Gentil, S. (1997). Inertial nonlinear equilibration of the equatorial flows. *J. Fluid Mech.* **331**. 345–371.

Huang, B. and Liu, Z. (1999). Pacific subtropical-tropical thermocline water exchange in the National Centers for Environmental Prediction ocean model. *J. Geophys. Res.*, **104**, 11,065–11,076.

Huang, N. E. (1979). On surface drift currents in the ocean. *J. Fluid Mech.*, **91**, 191–208.

Huang, R. X. (1988a). On boundary value problems of the ideal-fluid thermocline. *J. Phys. Oceanogr.*, **18**, 619–641.

Huang, R. X. (1988b). Ideal-fluid thermocline with weakly convective adjustment in a subpolar basin. *J. Phys. Oceanogr.*, **18**, 642–651.

Huang, R. X. (1989a). The generalized eastern boundary conditions and the three-dimensional structure of the ideal fluid thermocline. *J. Geophys. Res.*, **94**, 4855–4865.

Huang, R. X. (1989b). Sensitivity of a multilayered oceanic general circulation model to the sea surface thermal boundary condition. *J. Geophys. Res.*, **94**, 18,011–18,021.

Huang, R. X. (1990a). On the three-dimensional structure of the wind-driven circulation in the North Atlantic. *Dyn. Atmos. Oceans*, **15**, 117–159.

Huang, R. X. (1990b). On the structure of inertial western boundary currents with two moving layers. *Tellus*, **42**A, 594–604.

Huang, R. X. (1990c). Matching a ventilated thermocline model with inertial western boundary currents. *J. Phys. Oceanogr.*, **20**, 1599–1607.

Huang, R. X. (1991a). The three-dimensional structure of wind-driven gyres: ventilation and subduction. US National Report to International Union of Geodesy and Geophysics 1987–1990, *Rev. Geophys.*, (Suppl.), 590–609.

Huang, R. X. (1991b). A note on combining wind and buoyancy forcing in a simple one-layer ocean model. *Dyn. Atmos. Oceans*, **15**, 535–540.

Huang, R. X. (1993a). A two-level model for the wind and buoyancy-forced circulation. *J. Phys. Oceanogr.*, **23**, 104–115.

Huang, R. X. (1993b). Real freshwater flux as a natural boundary condition for the salinity balance and thermohaline circulation forced by evaporation and precipitation. *J. Phys. Oceanogr.*, **23**, 2428–2446.

Huang, R. X. (1994). Thermohaline circulation: Energetics and variability in a single-hemisphere basin model. *J. Geophys. Res.*, **99**, 12,471–12,485.

Huang, R. X. (1998a). On the energy balance of the oceanic general circulation. *Chinese J. Atmos. Science*, **22**, 452–467.

Huang, R. X. (1998b). Mixing and available potential energy in a Boussinesq ocean. *J. Phys. Oceanogr.*, **28**, 669–678.

Huang, R. X. (1999). Mixing and energetics of the thermohaline circulation. *J. Phys. Oceanogr.*, **29**, 727–746.

Huang, R. X. (2000a). Parameter study of a continuously stratified model of the ideal-fluid thermocline. *J. Phys. Oceanogr.*, **30**, 1372–1388.

Huang, R. X. (2000b). Climate variability inferred from a continuously stratified model of the ideal-fluid thermocline. *J. Phys. Oceanogr.*, **30**, 1389–1406.

Huang, R. X. (2001). An analytical solution of the ideal-fluid thermocline. *J. Phys. Oceanogr.*, **31**, 2441–2457.

Huang, R. X. (2005a). Available potential energy in the world oceans. *J. Mar. Res.*, **63**, 141–158.

Huang, R. X. (2005b). Contribution of oceanic circulation to the poleward heat flux. *J. Ocean Uni. China*, **4**, 277–287.

Huang, R. X. and Bryan, K. (1987). A multi-layer model of the thermohaline and wind-driven ocean circulation. *J. Phys. Oceanogr.*, **17**, 1909–1924.

Huang, R. X. and Chou, R. L. (1994). Parameter sensitivity study of the saline circulation. *Climate Dyn.*, **9**, 391–409.

Huang, R. X. and Dewar, W. K. (1996). Haline circulation: Bifurcation and chaos. *J. Phys. Oceanogr.*, **26**, 2093–2106.

Huang, R. X. and Flierl, G. R. (1987). Two-layer models for the thermocline and current structure in subtropical/subpolar gyres. *J. Phys. Oceanogr.*, **17**, 872–884.

Huang, R. X. and Jin, F. F. (2003). The asymmetric nature of the equatorial undercurrent in the Pacific and Atlantic. *J. Phys. Oceanogr.*, **33**, 1083–1094.

Huang, R. X. and Jin, X.-Z. (2002a). Deep circulation in the South Atlantic induced by bottom-intensified mixing over the mid-ocean ridge. *J. Phys. Oceanogr.*, **32**, 1150–1164.

Huang, R. X. and Jin, X.-Z. (2002b). Sea surface elevation and bottom pressure anomalies due to thermohaline forcing: Part I: isolated perturbations. *J. Phys. Oceanogr.*, **32**, 2131–2150.

Huang R. X. and Jin, X.-Z. (2006). Gravitational potential energy balance for the thermal circulation in a model ocean. *J. Phys. Oceanogr*, **36**, 1420–1429.

Huang, R. X. and Jin, X.-Z. (2007). On the natural boundary conditions applied to the sea-ice coupled model. *J. Geophys. Res.*, **112**, doi:10.1029/2006JC003735.

Huang, R. X. and Pedlosky, J. (1999). Climate variability inferred from a layered model of the ventilated thermocline. *J. Phys. Oceanogr.*, **29**, 779–790.

Huang, R. X. and Pedlosky, J. (2000a). Climate variability of the equatorial thermocline inferred from a two-moving-layer model of the ventilated thermocline. *J. Phys. Oceanogr.*, **30**, 2610–2626.

Huang, R. X. and Pedlosky, J. (2000b). Climate variability induced by anomalous buoyancy forcing in a multilayer model of the ventilated thermocline. *J. Phys. Oceanogr.*, **30**, 3009–3021.

Huang, R. X. and Qiu, B. (1994). Three-dimensional structure of the wind-driven circulation in the subtropical North Pacific. *J. Phys. Oceanogr.*, **24**, 1608–1622.

Huang, R. X. and Russell, S. (1994). Ventilation of the subtropical North Pacific. *J. Phys. Oceanogr.*, **24**, 2589–2605.

Huang, R. X. and Schmitt, R. W. (1993). The Goldsbrough–Stommel circulation of the world oceans. *J. Phys. Oceanogr.*, **23**, 1277–1284.

Huang, R. X. and Stommel, H. M. (1990). Cross-sections of two-layer inertial Gulf Stream. *J. Phys. Oceanogr.*, **20**, 907–911.

Huang, R. X. and Stommel, H. M. (1992). Convective flow patterns in an eight-box cube driven by combined wind stress, thermal, and saline forcing. *J. Geophys. Res.*, **97**, 2347–2364.

Huang, R. X. and Wang, Q. (2001). Interior communication from the subtropical to the tropical oceans. *J. Phys. Oceanogr.*, **31**, 3538–3550.

Huang, R. X. and Wang, W. (2003). Gravitational potential energy sinks in the oceans: Near-boundary processes and their parameterization. Proceedings, Hawaii Winter Workshop, pp. 239–247.

Huang, R. X., Cane, M. A., Naik, N. and Goodman, P. (2000). Global adjustment of the thermocline in response to deepwater formation. *Geophys. Res. Lett.*, **27**, 759–762.

Huang, R. X., Huang, C. J. and Wang, W. (2007). Dynamical roles of mixed layer in regulating the meridional mass/heat fluxes. *J. Geophys. Res.*, **112**, C05036, doi:10.1029/2006JC004046.

Huang, R. X., Jin, X.-Z. and Zhang, X.-H. (2001). An ocean general circulation model in pressure coordinates. *Adv. Atmos. Sci.*, **18**, 1–22.

Huang, R. X., Luyten, J. M. and Stommel, H. M. (1992). Multiple equilibrium states in combined thermal and saline circulation. *J. Phys. Oceanogr.*, **22**, 231–246.

Huang, R. X., Wang, W. and Liu, L. L. (2006). Decadal variability of wind energy input to the world ocean, *Deep Sea Res II,* **53**, 31–41.

Huthnance, J. M. (1984). Slope currents and "JEBAR". *J. Phys. Oceanogr.*, **14**, 795–810.

Ierley, G. R. and Sheremet, V. A. (1995). Multiple solutions and advection-dominated flows in the wind-driven circulation. Part I: Slip. *J. Mar. Res.*, **53**, 703–737.

Ierley, G. R. and Young, W. R. (1983). Can the western boundary layer affect the potential vorticity distribution in the Sverdrup interior of a wind gyre? *J. Phys. Oceanogr.*, **13**, 1753–1763.

Iselin, C. O'D. (1939). The influence of vertical and lateral turbulence on the characteristics of the waters at mid-depths. *Trans. Amer. Geophys. Union*, **20**, 414–417.

Iselin, C. O'D. (1940). Preliminary report on long-period variations in the transport of the Gulf Stream. *Papers in Physical Oceanography and Meteorology*, **8**: 1, 40 pp.

Jackett, D. R. and McDougall, T. J. (1997). Neutral density variable of the world's oceans. *J. Phys. Oceanogr.*, **27**, 237–263.

Jacobs, S. S. and Georgi, D. T. (1977). Observations on the southwest Indian/Antarctic Ocean. In *A Voyage of Discovery*, George Deacon 70th Anniversary Volume, ed. M. Angel, Pergamon Press, New York, pp. 43–84.

Jacobs, S. S., Amos, A. F. and Bruchhause, P. M. (1970). Ross Sea oceanography and antarctic bottom water formation. *Deep Sea Res.*, **17**, 935–962.

Jayne, S. R., Hogg, N. G. and Malanotte-Rizzoli, P. (1996). Recirculation gyres forced by a beta-plane jet. *J. Phys. Oceanogr.*, **26**, 492–504.

Jeffreys, H. (1925). On fluid motions produced by differences of temperature and humidity. *Quart. Jour. R. Meteor. Soc.*, **51**, No. 216, 347–356.

Jenkins, A. (1987). Wind and wave induced currents in a rotating sea with depth-varying eddy viscosity. *J. Phys. Oceanogr.*, **17**, 938–951.

Jiang, H., Huang, R. X. and Wang, H. (2008). The role of gyration in the oceanic general circulation, Part 1: The Atlantic Ocean. *J. Geophys. Res.*, Vol. **113**, C03014, doi:10.1029/2007/JC004134.

Johnson, G. C. (1998). Deep water properties, velocity, and dynamics over ocean trenches. *J. Mar. Res.*, **56**, 329–347.

Johnson, G. C. and McPhaden, M. J. (1999). Interior pycnocline flow from the subtropical to the equatorial Pacific Ocean. *J. Phys. Oceanogr.*, **29**, 3073–3089.

Johnson, G. C., Warren, B. A. and Olson, D. B. (1991a). Flow of bottom water in the Somali Basin, *Deep Sea Res.*, **38**, 637–652.

Johnson, G. C., Warren, B. A. and Olson, D. B. (1991b). A deep boundary current in the Arabian Basin. *Deep Sea Res.*, **38**, 653–661.

Johnson, H. L. and Marshall, D. P. (2002). A theory for the surface Atlantic response to thermohaline variability. *J. Phys. Oceanogr.*, **32**, 1121–1132.

Joyce, T. M. (1988). On wind-driven cross-equatorial flow in the Pacific Ocean. *J. Phys. Oceanogr.*, **18**, 19–24.

Joyce, T. M., Pickart, R. S. and Millard, R. C. (1999). Long-term hydrographic changes at 52° and 66°W in the North Atlantic Subtropical Gyre and Caribbean. *Deep Sea Res. II*, **46**, 245–278.

Kagan, B. A. and Sundermann, J. (1996). Dissipation of tidal energy, paleotides, and evolution of the Earth–Moon system. *Advances in Geophysics*, **38**, 179–266.

Kasahara, A. (1974). Various vertical coordinate systems used for numerical weather prediction. *Mon. Wea. Rev.*, **102**, 509–522.

Katsman, C. A. (2006). Impacts of localized mixing and topography on the stationary abyssal circulation. *J. Phys. Oceanogr.*, **36**, 1660–1671.

Kawase, M. (1987). Establishment of deep ocean circulation driven by deep-water production. *J. Phys. Oceanogr.*, **17**, 2294–2317.

Kawase, M. and Straub, D. (1991). Spinup of source-driven circulation in an abyssal basin in the presence of bottom topography. *J. Phys. Oceanogr.*, **21**, 1501–1514.

Killworth, P. D. (1983a). Deep convection in the world ocean. *Rev. Geophys. Space Phys.*, **21**, 1–26.

Killworth, P. D. (1983b). Some thoughts on the thermocline equations. *Ocean Modelling.*, **48,** 1–5 (unpublished manuscript).

Killworth, P. D. (1991). Cross-equatorial geostrophic adjustment. *J. Phys. Oceanogr.*, **21,** 1581–1601.

Kistler, R., Kalnay, E., Collins, W., Saha, S., White, G., Woollen, J., Chelliah, M., Ebisuzaki, W., Kanamitsu, M., Kousky, V., van den Dool, H., Jenne, R. and Fiorino, M. (2001). The NCEP-NCAR 50-Year reanalysis: Monthly mean CD-ROM and documentation. *Bull. Amer. Meteo. Soc.*, **82**, 247–267.

Kittel, C. and Knoemer, H. (1980). *Thermal Physics*, Freeman, New York, 496 pp.

Klinger, B. A., McCreary, J. P., Jr. and Kleeman, R. (2002). The relationship between oscillating subtropical wind stress and equatorial temperature. *J. Phys. Oceanogr.*, **32**, 1507–1521.

Kuhlbrodt, T., Griesel, A., Montoya, M., Levermann, A., Hofmann, M. and Rahmstorf, S. (2007). On the driving processes of the Atlantic meridional overturning circulation. *Rev. Geophys.*, **45**, RG2001, doi:10.1029/2004RG000166.

Kundu, P. K. (1990). *Fluid Mechanics*, Academic Press, New York, 638 pp.

Laplace, P. S. (1775). Recherches sur plusieurs points du système du monde. *Memoires de l'Académie Royale des Sciences de Paris* **88**, 75–182. Reprinted in Oeuvres Completes de Laplace, Gauthier-Villars, Paris, 9 (1893).

Landau, L. D. and Lifshitz, E. M. (1959). *Fluid Mechanics*, Pergamon, London, 536 pp.

Lazier, J. R. N. (1973). Temporal changes in some fresh water temperature structure. *J. Phys. Oceanogr.*, **3**, 226–229.

Ledwell, J. R., Montgomery, E. T., Polzin, K. L., St Laurent, L. C., Schmitt, R. W. and Toole, J. M. (2000). Evidence for enhanced mixing over rough topography in the abyssal ocean. *Nature*, **403**, 179–182.

Ledwell, J. R., Watson, A. J. and Law, C. B. (1993). Evidence for slow mixing across the pycnocline from an open-ocean tracer-release experiment. *Nature*, **364**, 701–703.

Levitus, S. (1982). *Climatological Atlas of the World Ocean*, NOAA Prof. Paper No.13. US Dept. of Commerce, Washington, DC.

Levitus, S. and Boyer, T. (1994). *World Ocean Atlas*, vol. 4, *Temperature*, NOAA Atlas NESDIS 4, Natl. Oceanic and Atmos. Admin., US Dept. of Commerce, Washington, DC.

Levitus, S., Boyer, T., Conkright, M. E., O'Brien, T., Antonov, J., Stephens, C., Stathoplos, L., Johnson, D. and Gelfeld, R. (1998). *World Ocean Database 1998*. Volume 1: *Introduction*, NOAA Atlas NESDIS 18, Natl. Oceanic and Atmos. Admin., U.S. Dept. of Commerce, Washington, DC., 346 pp.

Lewis, G. N. and Randall, M. (1961). *Thermodynamics*. McGraw-Hill, New York, 723 pp.

Liu L. L., Wang, W. and Huang, R. X. (2008). Mechanical energy input to the ocean induced by tropical cyclones. *J. Phys. Oceanogr.*, **38**, 1253–1266.

Liu, Z. (1994). A simple model of the mass exchange between the subtropical and tropical ocean. *J. Phys. Oceanogr.*, **24**, 1153–1165.

Liu, Z. and Pedlosky, J. (1994). Thermocline forced by annual and decadal surface temperature variation. *J. Phys. Oceanogr.*, **24**, 587–608.

Liu, Z. and Philander, G. H. (2001). Tropical–extratropical oceanic exchange pathways. In *Ocean Circulation and Climate*, ed. G. Siedler, J. Church and J. Gould, International Geophysical Series, Academy Press, New York, pp. 247–257.

Longworth, H., Marotzke, J. and Stocker, T. F. (2005). Ocean gyres and abrupt change in the thermohaline circulation: A conceptual analysis. *J. Climate*, **18**, 2403–2416.

Lorenz, E. N. (1955). Available potential energy and the maintenance of the general circulation. *Tellus*, **7**, 157–167.

Lorenz, E. N. (1967). *The Nature and Theory of General Circulation of the Atmosphere*. World Meteorological Organization, Geneva, WMO No. **218**, T. P. 115, 161 pp.

Lu, P. and McCreary, J. P. (1995). Influence of the ITCZ on the flow of the thermocline water from the subtropical to the equatorial Pacific Ocean. *J. Phys. Oceanogr.*, **25**, 3076–3088.

Lu, P., McCreary, J. P. and Klinger, B. A. (1998). Meridional circulation cells and the source waters of the Pacific equatorial Undercurrent. *J. Phys. Oceanogr.*, **28**, 62–84.

Luyten, J. R. and Stommel, H. (1985). Upstream effects of the Gulf Stream on the structure of the mid-ocean thermocline. *Progr. Oceanogr.*, **14**, 387–399.
Luyten, J. R. and Stommel, H. (1986). Gyres driven by combined wind and buoyancy flux. *J. Phys. Oceanogr.*, **16**, 1551–1560.
Luyten, J. R., Pedlosky, J. and Stommel, H. M. (1983). The ventilated thermocline. *J. Phys. Oceanogr.*, **13**, 292–309.
Luyten, J., Stommel, H. and Wunsch, C. (1985). A diagnostic study of the Northern Atlantic subpolar gyre. *J. Phys. Oceanogr.*, **15**, 1344–1348.
McCalphin, J. D. (1995). Rossby wave generation by poleward-propagating Kelvin waves: The midlatitude quasigeostrophic approximation. *J. Phys. Oceanogr.*, **25**, 1415–1425.
McCartney, M. S. (1977). Subantarctic Mode Water. In *A Voyage of Discovery*, George Deacon 70th Anniversary Volume, *Deep Sea Res.*, (Suppl.), 103–119.
McCartney, M. S. (1982). The subtropical recirculation of mode water. *J. Mar. Res.*, **40**, 427–464.
McCartney, M. S. and Talley, L. D. (1982). The Subpolar Mode Water of the North Atlantic Ocean. *J. Phys. Oceanogr.*, **12**, 1169–1188.
McCreary, J. P. and Lu, P. (1994). Interaction between the subtropical and equatorial ocean circulation: the subtropical cell. *J. Phys. Oceanogr.*, **24**, 466–497.
McDowell, S., Rhines, P. and Keffer, T. (1982). North Atlantic potential vorticity and its relation to the general circulation. *J. Phys. Oceanogr.*, **12**, 1417–1436.
McPhaden, M. J. and Fine, R. A. (1988). A dynamical interpretation of the tritium maximum in the Central Equatorial Pacific. *J. Phys. Oceanogr.*, **18**, 1454–1457.
Maas, L. R. M. (1994). A simple model for the three-dimensional, thermally and wind-driven ocean circulation. *Tellus*, **46A**, 671–680.
Madsen, O. S. (1977). A realistic model of the wind-induced Ekman boundary layer. *J. Phys. Oceanogr.*, **7**, 248–255.
Malkus, W. R. (1972). Nonperiodic convection at high and low Prandtl number. *Mem. Soc. R. Sci. Liège, 6e Ser.*, **IV**, 125–128.
Manabe, S. and R. J. Stouffer (1988). Two stable equilibria of a coupled ocean–atmosphere model. *J. Clim.*, **1**, 841–866.
Margules, M. (1905). *Über die Energie der Sturme*, Wien: K. K. Hof- und Staatsdruckerei, 26 pp.
Marotzke, J. (1989). Instabilities and steady states of the thermohaline circulation. In *Ocean Circulation Models: Combining Data and Dynamics*, ed. D. L. T. Anderson and J. Willebrand, Kluwer, pp. 501–511.
Marotzke, J. (1990). Instabilities and multiple equilibria of the thermohaline circulation. Ph. D. thesis. Ber. Inst. Meeresk. Kiel, 194, 126 pp.
Marotzke, J. and Willebrand, J. (1991). Multiple equilibria of the global thermohaline circulation. *J. Phys. Oceanogr.*, **21**, 1372–1385.
Marotzke, J., Welander, P. and Willebrand, J. (1988). Instability and multiple equilibria in a meridional-plane model of the thermohaline circulation. *Tellus*, **40**A, 162–172.
Marshall, J. C. and Nurser, A. J. G. (1991). A continuously stratified thermocline model incorporating a mixed layer of variable thickness and density. *J. Phys. Oceanogr.*, **21**, 1780–1792.
Marshall, J. and Schott, F. (1999). Open-ocean convection: observations, theory and models. *Rev. Geophys.*, **37**, 1–64.
Marshall, J., Hill, C., Perelman, L. and Adcroft, A. (1997). Hydrostatic, quasi-hydrostatic, and nonhydrostatic ocean modeling. *J. Geophys. Res.*, **102**, 5733–5752.

Marshall, J. C., Nurser, A. J. G. and Williams, R. G. (1993a). Inferring the subduction rate and period over the North Atlantic. *J. Phys. Oceanogr.*, **23**, 1315–1329.
Marshall, J. C., Olbers, D., Wolf-Gladrow, D. and Ross, H. (1993b). Potential vorticity constraints on the hydrography and transport of the southern oceans. *J. Phys. Oceanogr.* **23**, 465–487
Martinson, D. S., Killworth, P. D. and Gordon, A. L. (1981). A convective model for the Weddell polynya. *J. Phys. Oceanogr.*, **11**, 466–488.
Masuzawa, J. (1969). Subtropical model water. *Deep Sea Res.*, **16**, 463–472.
Mauritzen, C. (1996). Production of dense overflow waters feeding the North Atlantic across the Greenland–Scotland ridge. Part I: Evidence for a revised circulation scheme, *Deep Sea Res.*, **43**, 769–806.
Mauritzen, C., Polzin, K. L., McCartney, M. S., Millard, R. C. and West-Mark, D. E. (2002). Evidence in hydrography and density fine structure for enhanced vertical mixing over the Mid-Atlantic Ridge in the western Atlantic. *J. Geophys. Res.*, **107**, 3147, doi:10.1029/2001JC001114.
MEDOC Group (1970). Observation of formation of deep water in the Mediterranean Sea, 1969. *Nature*, **227**, 1037–1040.
Mercier, H., Speer, K. G. and Honnorez, J. (1994). Flow pathways of bottom water through the Romanche and chain fracture-zones. *Deep Sea Res., I*, **41**, 1457–1477.
Mesinger, F. and Janjic, Z. I. (1985). Problems and numerical methods of the incorporation of mountains in the atmospheric models. *Lectures in Applied Mathematics*, **22**, 81–120
Mihaljan, J. M. (1963). An exact solution of the Rossby adjustment problem. *Tellus*, **15**, 150–154.
Miles, J. W. (1972). Kelvin waves on oceanic boundaries. *J. Fluid Mech.*, **55**, 113–127.
Millero, F. J. and Leung, W. H. (1976). The thermodynamics of seawater at one atmosphere. *Amer. J. Sci.*, **276**, 1035–1077.
Milliff, R. F. and McWilliams, J. C. (1994). The evolution of boundary pressure in ocean basin. *J. Phys. Oceanogr.*, **24**, 1317–1338.
Montgomery, R. B. (1937). A suggested method for representing gradient flow in isentropic surfaces. *Bull. Amer. Meteor. Soc.*, **18**, 210–212.
Morel, A. and Antoine, D. (1994). Heating rate within the upper ocean in relation to its bio-optical state. *J. Phys. Oceanogr.*, **24**, 1652–1665.
Morgan, G. W. (1956). On the wind-driven ocean circulation. *Tellus*, **8**, 301–320.
Morris, M. Y., Hall, M. M., St Laurent, L. C. and Hogg, N. G. (2001). Abyssal mixing in the Brazil Basin. *J. Phys. Oceanogr.*, **31**, 3331–3348.
Munk, W. H. (1950). On the wind-driven ocean circulation. *J. Meteor.*, **7**, 79–93.
Munk, W. H. (1966). Abyssal recipes. *Deep Sea Res.*, **13**, 707–730.
Munk, W. H. (1981). Internal waves and small-scale processes. In *Evolution of Physical Oceanography*, ed. B. A. Warren and C. Wunsch, MIT Press, pp. 264–291.
Munk, W. H. and Wunsch, C. (1998). Abyssal recipes II: Energetics of the tidal and wind mixing. *Deep Sea Res.*, I, **45**, 1977–2010.
Nakano, H. and Hasumi, H. (2005). A series of zonal jets embedded in the broad zonal flows in the Pacific obtained in eddy-permitting ocean general circulation models. *J. Phys. Oceanogr.*, **35**, 474–488.
Nakano, H. and Suginohara, N. (2002). A series of mid-depth zonal flows in the Pacific driven by winds. *J. Phys. Oceanogr.*, **32**, 161–176.
Needler, G. (1967). A model for the thermohaline circulation in an ocean of finite depth. *J. Mar. Res.*, **25**, 329–342.

Neumann, G. and Pierson, W. J. (1966). *Principles of physical oceanography*, Prentice-Hall, Englewood Cliffs, 545 pp.

Newton, I. (1687). *Philosophiae Naturalis Principia Mathematica.* See Newton's *Principia*. Cajori's 1946 revision of Motte's 1729 translation, University of California Press, Berkeley, 680 pp.

Niiler, P. P. (1966). On the theory of wind-driven ocean circulation. *Deep Sea Res.*, **13**, 597–606.

Nilsson, J., Brostrom, G. and Walin, G. (2003). The thermohaline circulation and vertical mixing: Does weaker density stratification give stronger overturning? *J. Phys. Oceanogr.*, **33**, 2781–2795.

Noh, Y., Min, H. S. and Raasch, S. (2004). Large eddy simulation of the ocean mixed layer: The effects of wave breaking and Langmuir circulation. *J. Phys. Oceanogr.*, **34**, 720–735.

Obukhov, A. M. (1949). On the question of the geostrophic wind (in Russian), *Izv. Akad. Nauk SSSR Ser. Geograf.-Geofiz.*, **13**(4), 281–306.

O'Dwyer, J. and Williams, R. G. (1997). The climatological distribution of potential vorticity over the abyssal ocean. *J. Phys. Oceanogr.*, **27**, 2488–2506.

Oliver, P. J. (1986). *Applications of Lie Groups to differential equations*, Springer-Verlag, New York, 497 pp.

Oort, A. H., Anderson, L. A. and Peixoto, J. P. (1994). Estimates of the energy cycle of the oceans. *J. Geophys. Res.*, **99**, 7665–7688.

Oort, A. H., Ascher, S. C., Levitus, S. and Peixoto, J. P. (1989). New estimates of the available potential energy in the world ocean. *J. Geophys. Res.*, **94**, 3187–3200.

Osborn, T. R. (1980). Estimates of the local rate of vertical diffusion from dissipation measurements. *J. Phys. Oceanogr.*, **10**, 83–89.

Owens, W. B. and Warren, B. A. (2001). Deep circulation in the northwest corner of the Pacific Ocean. *Deep Sea Res.* I, **48**, 959–993.

Pacanofsky, R. C. (1995). *MOM2 documentation user's guide and reference manual*, version 2. GFDL Ocean Group Tech. Rep. No.3, GFDL/Princeton University, 232 pp.

Paparella, F. and Young, W. R. (2002). Horizontal convection is non turbulent. *J. Fluid Mech.*, **205**, 466–474.

Parsons, A. T. (1969). A two-layer model of Gulf Stream separation. *J. Fluid Mech.*, **39**, 511–528.

Pasquero, C. and Tziperman, E. (2004). Effects of a wind-driven gyre on thermohaline circulation variability. *J. Phys. Oceanogr.*, **34**, 805–816.

Pedlosky, J. (1983). Eastern boundary ventilation and the structure of the thermocline. *J. Phys. Oceanogr.*, **13**, 2038–2044.

Pedlosky, J. (1984). Cross-gyre ventilation of the subtropical gyre: An internal mode in the ventilated thermocline. *J. Phys. Oceanogr.*, **14**, 1172–1178.

Pedlosky, J. (1986). The buoyancy and wind-driven ventilated thermocline. *J. Phys. Oceanogr.*, **16,** 1077–1087.

Pedlosky, J. (1987a). *Geophysical Fluid Dynamics.* Springer-Verlag, New York, 710 pp.

Pedlosky, J. (1987b). An inertial theory of the equatorial undercurrent. *J. Phys. Oceanogr.*, **17**, 1978–1985.

Pedlosky, J. (1992). The baroclinic structure of the abyssal circulation. *J. Phys. Oceanogr.*, **22**, 652–659.

Pedlosky, J. (1996). *Ocean Circulation Theory.* Springer-Verlag, Heidelberg, 453 pp.

Pedlosky, J. (2006). A history of thermocline theory. In *Physical Oceanography Developments Since 1950*, ed. M. Jochum and R. Murtugudde, Springer, New York, pp. 139–152.

Pedlosky, J. and Robbins, P. (1991). The role of finite mixed-layer thickness in the structure of the ventilated thermocline. *J. Phys. Oceanogr.*, **21**, 1018–1031.

Peixoto, J. P. and Oort, A. H. (1992). *Physics of Climate*, American Institute of Physics, New York, 520 pp.

Peltier, W. R. and Caulfield, C. P. (2003). Mixing efficiency in stratified shear flows. *Ann. Rev. Fluid Mech.*, **35**, 135–167.

Phillips, N. A. (1966). The equations of motion for a shallow rotating atmosphere and the "traditional approximation". *J. Atmos. Sci.*, **23**, 626–628.

Phillips, O. M. (1970). On flows induced by diffusion in a stably stratified fluid. *Deep Sea Res.*, **17**, 435–440.

Phillips, O. M., Shyu, J. and Salmun, H. (1986). An experiment on boundary mixing: mean circulation and transport rates. *J. Fluid Mech.*, **173**, 473–499.

Pickart, R. S. (1992). Water mass components of the North Atlantic deep western boundary current. *Deep Sea Res.*, **39**, 1553–1572.

Pickart, R. S. and Huang, R. X. (1995). Structure of an inertial deep western boundary current. *J. Mar. Res.*, **53**, 739–770.

Pickart, R. S. and Spall, M. A. (2007). Impact of Labrador Sea convection on the North Atlantic meridional overturning circulation. *J. Phys. Oceanogr.*, **37**, 2207–2227.

Pickart, R. S., Spall, M. A., Ribergaard, M. H., Moore, G. W. K. and Milliff, R. F. (2003). Deep convection in the Irminger Sea forced by the Greenland tip jet. *Nature*, **424**, No. 6945, 152–156.

Pickart, R. S., Torres, D. J. and Clarke, R. A. (2002). Hydrography of the Labrador Sea during active convection. *J. Phys. Oceanogr.*, **32**, 428–457.

Plueddemann, A. J. and Farrar, J. T. (2006). Observations and models of the energy flux from the wind to mixed-layer inertial currents. *Deep Sea Res.*, II, **53**, 5–30.

Polzin, K. L., Speer, K. G., Toole, J. M. and Schmitt, R. W. (1996). Intense mixing of Antarctic Bottom water in the equatorial Atlantic Ocean. *Nature*, **380**, 54–57.

Polzin, K. L., Toole, J. M., Ledwell, J. R. and Schmitt, R. W. (1997). Spatial variability of turbulent mixing in the abyssal ocean. *Science*, **276**, 93–96.

Pratt, L. J. and Helfrich, K. R. (2003). Rotating hydraulics and upstream basin circulation. *J. Phys. Oceanogr.*, **33**, 1651–1663.

Pratt, L. J. and Whitehead, J. (2007). *Rotating Hydraulics*, Springer, New York, 589 pp.

Price, J. F. (1981). Upper ocean response to a hurricane. *J. Phys. Oceanogr.*, **11**, 153–175.

Price, J. F. and Baringer, M. O. (1994). Outflows and deep water production by marginal seas. *Progr. Oceanogr.*, **33**, 161–200.

Price, J. F. and Sundermeyer, M. A. (1999). Stratified Ekman layers. *J. Geophys. Res.*, **104**, 20,467–20,494.

Price, J. F. and Yang, J. (1998). Marginal sea overflows for climate simulations. In: *Ocean Modelling and Parameterizations*, ed. E. P. Chassignet and J. Verron, Kluwer Academic Publishers, pp. 155–170.

Price, J. F., Weller, R. A. and Schudlich, R. R. (1987). Wind-driven ocean currents and Ekman transport. *Science*, **238**, 1534–1538.

Qiu, B. (2002a). The Kuroshio Extension system: Its large-scale variability and role in the midlatitude ocean–atmosphere interaction. *J. Oceanogr.*, **58**, 57–75.

Qiu, B. (2002b). Large-scale variability in the midlatitude subtropical and subpolar North Pacific Ocean: Observations and causes. *J. Phys. Oceanogr.*, **32**, 353–375.

Qiu, B. (2003). Kuroshio extension variability and forcing of the Pacific decadal oscillations: Responses and potential feedback. *J. Phys. Oceanogr.*, **33**, 2465–2482.

Qiu, B. and Huang, R. X. (1995). Ventilation of the North Atlantic and North Pacific: subduction versus obduction. *J. Phys. Oceanogr.*, **25**, 2374–2390.

Reid, J. L., Nowlin, W. D. and Patzert, W. C. (1977). On the characteristics and circulation of the southwestern Atlantic Ocean. *J. Phys. Oceanogr.*, **7**, 62–91.

Reid, R. O., Elliott, B. A. and Olson, D. B. (1981). Available potential energy: A clarification. *J. Phys. Oceanogr.*, **11**, 15–29.

Rhines, P. B. and MacCready, P. M. (1989). Boundary control over the large-scale circulation, *Proceedings of the 'Aha Huliko'a Symposium*, ed. P. Muller, Hawaii Inst. of Geophysics, pp. 75–97.

Rhines, P. B. and Young, W. R. (1982a). Homogenization of potential vorticity in planetary gyres. *J. Fluid Mech.*, **122**, 347–367.

Rhines, P. B. and Young, W. R. (1982b). A theory of wind-driven circulation: I. Mid-ocean gyres. *J. Mar. Res.,* **40** (Suppl.), 559–596.

Robinson, A. R. and Stommel, H. (1959). The oceanic thermocline and the associated thermohaline circulation. *Tellus*, **11**, 295–308.

Roche, P.-E., Castaing, B., Chabaud, B. and Hebral, B. (2001). Observation of the $^1/_2$ power law in Rayleigh–Bénard convection. *Phys. Rev. E*, **63**, DOI: 10.1103/PhysRevE.63.045303.

Roemmich, D., Hautala, S. and Rudnick, D. (1996). Northward abyssal transport through the Samoan passage and adjacent regions. *J. Geophys. Res.*, **101**, 14,039–14,055.

Rogers, C. and Ames, W. F. (1989). *Nonlinear Boundary Value Problems in Science and Engineering*, Academic Press, New York, 416 pp.

Rooth, C. (1982). Hydrology and ocean circulation. *Prog. Oceanogr.*, **11**, 131–149.

Rooth, C., Stommel, H. and Veronis, G. (1978). On motions in steady, layered, geostrophic models, *J. Oceanogr. Soc. Japan*, **34**, 265–267.

Rossby, C. G. (1938). On the mutual adjustment of pressure and velocity distributions in certain simple current systems, II. *J. Mar. Res.*, **1**, 239–263.

Rossby, T. (1965). On thermal convection driven by non-uniform heating from below; An experimental study. *Deep Sea Res.*, **12**, 9–10.

Rossby, T. (1998). Numerical experiments with a fluid heated non-uniformly from below. *Tellus*, **50**A, 242–257.

Salmon, R. (1990). The thermocline as an "internal boundary layer". *J. Mar. Res.*, **48**, 437–469.

Salmon, R. (1991). Similarity solutions of the thermocline equations. *J. Mar. Res.*, **49**, 249–280.

Salmon, R. and Hollerbach, R. (1991). Similarity solutions of the thermocline equations. *J. Mar. Res.*, **49**, 249–280.

Samelson, R. M. and Vallis, G. K. (1997). Large-scale circulation with small diapycnal diffusion: The two-thermocline limit. *J. Mar. Res.*, **55**, 223–275.

Saunders, P. M. (1990). Cold outflow from the Faroe Bank Channel. *J. Phys. Oceanogr.*, **20**, 29–43.

Sandstrom, J. W. (1908). Dynamicsche Versuche mit Meerwasser. *Annalen der Hydrographie under Maritimen Meteorologie*, **36**, 6–23.

Sandstrom, J. W. (1916). Meteorologische Studien im schwedischen Hochgebirge. *Goteborgs K. Vetenskaps-och Vitterhetssamhalles Handl.*, Ser. **4**, **22**, No. 2, 48 pp.

Schlosser, P., Bonisch, G., Rhein, M. and Bayer, R. (1991). Reduction of deepwater formation in the Greenland Sea during the 1980s: Evidence from tracer data. *Science*, **251**, 1054–1056.

Schmitt, R. W. (1995). The ocean component of the global water cycle. *US National Report to International Union of Geodesy and Geophysics 1991–1994. Rev. Geophys., (Suppl.)*, 1395–1409.

Schmitt, R. W., Ledwell, J. R., Montgomery, E. T., Polzin, K. L. and Toole, J. M. (2005). Enhanced diapycnal mixing by salt fingers in the thermocline of the tropical Atlantic. *Science*, **308** (5722), 685–688.

Schmitt, R. W., Perkins, H., Boyd, J. D. and Stalcup, M. C. (1987). C-SALT: An investigation of the thermohaline staircase in the western tropical North Atlantic. *Deep Sea Res.*, **34**, 1655–1665.

Schmitz, Jr., W. J. (1995). On the interbasin-scale thermohaline circulation, *Rev. Geophys.*, **33**, 151–173.

Schmitz, Jr., W. J. (1996a). On the world ocean circulation: Volume I, Some global features/North Atlantic circulation. Woods Hole Oceanographic Institution Technical Report WHOI-96–03, 148 pp.

Schmitz, Jr., W. J. (1996b). On the world ocean circulation: Volume II, The Pacific and Indian Oceans / A global update. Woods Hole Oceanographic Institution Technical Report WHOI-96–08, 241 pp.

Schott, F. and Stommel, H. (1978). Beta spirals and absolute velocities in different oceans. *Deep Sea Res.*, **25**, 961–1010.

Sedov, L. I. (1959). *Similarity and dimensional methods in mechanics*, Academic Press, New York, 363 pp.

Sen, A., Scott, R. B. and Arbic, B. K. (2008). Global energy dissipation rate of deep-ocean low-frequency flows by quadratic bottom boundary layer drag: Computations from current-meter data. *Geophys. Res. Lett.*, **35**, doi:10.1029/2008GL033407.

Smith, W. H. F. and Sandwell, D. T. (1997). Global seafloor topography from satellite altimetry and ship depth soundings. *Science*, **277**, 1956–1962.

Smolarkiewicz, P. K. (2006). Multidimensional positive definite advection transport algorithm: An overview. *Int. J. Num. Method in fluid*, **50**, 1123–1144.

Soloviev, A. and Lukas, R. (1996). Observation of spatial variability of diurnal thermocline and rain-formed halocline in the western Pacific warm pool. *J. Phys. Oceanogr.*, **26**, 2529–2538.

Spall, M. A. (1992). Cooling spirals and recirculation in the subtropical gyre. *J. Phys. Oceanogr.*, **22**, 564–571.

Spall, M. A. (2001). Large-scale circulations forced by localized mixing over a sloping bottom. *J. Phys. Oceanogr.*, **31**, 2369–2384.

Sparrow, C. (1982). The Lorenz equations: Bifurcations, chaos, and strange attractors. Springer-Verlag, New York, 269 pp.

Speer, K. G. and Forbes, A. (1994). A deep western boundary current in the South Indian basin. *Deep Sea Res., I*, **41**, 1289–1303.

Speer, K. G. and McCartney, M. S. (1992). Bottom water circulation in the western North Atlantic. *J. Phys. Oceanogr.*, **22**, 83–92.

Speer, K. G., Tziperman, E. and Feliks, Y. (1993). Topography and grounding in a simple bottom layer model. *J. Geophys. Res.*, **98**, 8547–8558.

St Laurent, L. C., Toole, J. M. and Schmitt, R. W. (2001). Buoyancy forcing by turbulence above rough topography in the abyssal Brazil Basin. *J. Phys. Oceanogr.*, **31**, 3476–3495.

Stacey, M. W. (1999). Simulations of the wind-forced near-surface circulation in Knight Inlet: A parameterization of the roughness length. *J. Phys. Oceanogr.*, **29**, 1363–1367.

Stein, C. A. and Stein, S. (1992). A model for the global variation in oceanic depth and heat flow with lithospheric age. *Nature*, **359**, 123–129.

Stein, C. A. and Stein, S. (1994). Comparison of plate and asthenospheric flow models for the thermal evolution of oceanic lithosphere. *Geophys. Res. Lett.*, **21**, 709–712.

Stommel, H. (1948). The westward intensification of wind-driven ocean currents. *Trans. Amer. Geophys. Union*, **29**, 202–206.

Stommel, H. (1954). Why do our ideas about the ocean circulation have such a peculiarly dream-like quality? In *Collected Works of Henry Stommel*, ed. N. G. Hogg and R. X. Huang, Amer. Meteor. Soc., Boston, Vol. 1, pp. 124–134.

Stommel, H. (1957). A survey of ocean current theory. *Deep Sea Res.*, **4**, 149–184.

Stommel, H. (1958). The abyssal circulation. Letter to the editors. *Deep Sea Res.*, **5**, 80–82.

Stommel, H. M. (1961). Thermohaline convection with two stable regimes of flow. *Tellus*, **13**, 224–230.

Stommel, H. M. (1979). Determination of water mass properties of water pumped down from the Ekman layer to the geostrophic flow below. *Proc. Natl. Acad. Sci. USA*, **76**, 3051–3055.

Stommel, H. (1982). Is the South Pacific helium-3 plume dynamically active? *Earth and Plan. Sci. Lett.*, **61**, 63–67.

Stommel, H. (1984a). The delicate interplay between wind-stress and buoyancy input in ocean circulation: The Goldsbrough variations. Crafoord prize Lecture presented at the Royal Swedish Academy of Sciences, Stockholm, on September 28, 1983. *Tellus*, **36**(A), 111–119.

Stommel, H. (1984b). The Sea of the Beholder. In *Collected Works of Henry Stommel*, ed. Hogg and R. X. Huang, Amer. Meteor. Soc., Boston, Vol. I, pp. 5–112.

Stommel, H. (1986). A thermohaline oscillator. *Ocean Modelling*, **72**, 5–6 (unpublished manuscript). In *Collected Works of Henry Stommel*, ed. N. G. Hogg and R. X. Huang, Amer. Meteor. Soc., Boston, Vol. II, pp. 648–649.

Stommel, H. M. (1993). A conjectural regulating mechanism for determining the thermohaline structure of the oceanic mixed layer. *J. Phys. Oceanogr.*, **23**, 142–148.

Stommel, H. and Arons, A. B. (1960a). On the abyssal circulation of the world ocean – I. Stationary planetary flow patterns on a sphere. *Deep Sea Res.*, **6**, 140–154.

Stommel, H. and Arons, A. B. (1960b). On the abyssal circulation of the world ocean – II. An idealized model of the circulation pattern and amplitude in oceanic basins. *Deep Sea Res.*, **6**, 217–233.

Stommel, H. and Arons, A. B. (1972). On the abyssal circulation of the world ocean – V. The influence of bottom slope on the broadening of inertial boundary currents. *Deep Sea Res.*, **19**, 707–718.

Stommel, H. and Rooth, C. (1968). On the interaction of gravitational and dynamic forcing in simple circulation models. *Deep Sea Res.*, **15**, 165–170.

Stommel, H. and Schott, F. (1977). The beta spiral and the determination of the absolute velocity field from hydrographic station data. *Deep Sea Res.*, **24,** 325–329.

Stommel, H., Arons, A. B. and Faller, A. J. (1958). Some examples of stationary planetary flow patterns in bounded basins. *Tellus*, **10**, 179–187.

Suginohara, N. (1973). An eastward flow at lower, middle latitudes derived from a three-layer model of a wind-driven ocean circulation. *J. Oceanogr. Soc. Japan*, **29**, 227–235.

Sun, L. and Sun, D. J. (2007). Numerical simulation of partial-penetrating flow in horizontal convection. In *New Trends in Fluid Mechanics Research*, Proceedings of the Fifth International Conference on Fluid Mechanics, Aug. 15–19 (2007), Shanghai, China. Tsinghua University Press Springer, pp. 391–394.

Sverdrup, H. U. (1947). Wind-driven currents in a baroclinic ocean, with application to the equatorial currents of the eastern Pacific. *Proc. Nat. Acad. Sci. USA*, **33**, 318–326.

Sverdrup, H. U., Johnson, M. W. and Fleming, R. H. (1942). *The oceans*. Prentice-Hall, 1987 pp.

Swallow, J. C. and Worthington, L. V. (1957). Measurements of deep currents in the western North Atlantic. *Nature*, **179**, 1183–1184.

Talley, L. D. (1985). Ventilation of the subtropical North Pacific: The shallow salinity minimum. *J. Phys. Oceanogr.*, **15**, 633–649.

Talley, L. D. (1988). Potential vorticity distribution in the North Pacific. *J. Phys. Oceanogr.*, **18**, 89–106.

Talley, L. D. (2007). *Hydrographic Atlas of the World Ocean Circulation Experiment* (WOCE: Volume 2: Pacific Ocean). ed. M. Sparrow, P. Chapman and J. Gould, International WOCE Project Office, Southampton, U.K., ISBN 0–904175-54–5.

Terray, E. A., Donelan, M. A., Agraval, Y. C., Drennan, W. M., Kahma, K. K., Williams III, A. J., Hwang, P. A. and Kitaigorodskii, S. A. (1996). Estimates of kinetic energy dissipation under breaking waves. *J. Phys. Oceanogr.*, **26**, 792–807.

Terray, E. A., Drennan, W. M. and Donelan, M. A. (1999). The vertical structure of shear and dissipation in the ocean surface layer. In *The Wind-Driven Air–Sea Interface*, ed. M. L. Banner, School of Mathematics, University of New South Wales, 239–245.

Thompson, L. and Johnson, G. C. (1996). Abyssal currents generated by diffusion and geothermal heating over rises. *Deep Sea Res.*, **43**, 903–211.

Thorpe, S. A. (2005). *The Turbulent Ocean*, Cambridge Univ. Press, 458 pp.

Thurnherr, A. M. and Speer, K. G. (2003). Boundary mixing and topographic blocking on the Mid-Atlantic Ridge in the South Atlantic. *J. Phys. Oceanogr.*, **33**, 848–862.

Toggweiler, J. R. and Samuels, B. (1993). Is the magnitude of the deep outflow from the Atlantic Ocean actually governed by Southern Hemisphere winds? In *The Global Carbon Cycle*, ed. M. Heimann, Springer, pp. 303–331.

Toggweiler, J. R. and Samuels, B. (1995). Effect of Drake Passage on the global thermohaline circulation. *Deep Sea Res. I,* **42**, 477–500.

Toggweiler, J. R. and Samuels, B. (1998). On the ocean's large-scale circulation near the limit of no vertical mixing. *J. Phys. Oceanogr.*, **28**, 1832–1852.

Toole, J. M. and Warren, B. A. (1993). A hydrographic section across the subtropical south Indian Ocean. *Deep Sea Res.*, I, **40**, 1973–2019.

Treguier, A. M., Hogg, N. G., Maltrud, M., Speer, K. and Thierry, V. (2003). The origin of deep zonal flows in the Brazil Basin. *J. Phys. Oceanogr.*, **33**, 580–599.

Trenberth, K. E. and Caron, J. M. (2001). Estimates of meridional atmosphere and ocean heat transport. *J. Climate*, **14**, 3433–3443.

Tsuchiya, M., Lukas, R., Fine, R. A., Firing, E. and Lindstrom, E. (1989). Source waters of the Pacific Equatorial Undercurrent. *Progr. Oceanogr.*, **23**, 101–147.

Turner, J. S. (1973). *Buoyancy Effects in Fluid*, Cambridge Press, 367 pp.

Tziperman, E., Toggweiler, J. R., Bryan, K. and Feliks, Y. (1994). Instability of the thermohaline circulation with respect to mixed boundary conditions: Is it really a problem for realistic models?, *J. Phys. Oceanogr.*, **24**, 217–232.

Uppala, S.M. et al. (2005). The ERA-40 reanalysis. *Quart. J. R. Meteorol. Soc.*, **131**, 2961–3012
Vallis, G. K. (2000). Large-scale circulation and production of stratification: effects of wind, geometry and diffusion. *J. Phys. Oceanogr.*, **30,** 933–954.
Veronis, G. (1966). Wind-driven ocean circulation – Part I. Linear theory and perturbation analysis. *Deep Sea Res.*, **13**, 17–29.
Veronis, G. (1969). On theoretical models of the thermocline circulation. *Deep Sea Res.*, **16** (Suppl.), 301–323.
Veronis, G. (1973). Model of world ocean circulation: I. Wind-driven, two layer. *J. Mar. Res.*, **31**, 228–288.
von Mises, R. (1927). Bemerkungen zur Hydrodynamik, *Z. Angnew. Math. Mech.*, **7**, 425–431.
Walin, G. (1985). The thermohaline circulation and the control of ice ages. *Palaeogeogr., Palaeoclimatol., Palaeoecol.*, **50**, 323–332.
Wang, Q. and Huang, R. X. (2005). Decadal variability of pycnocline flows from the subtropical to the equatorial Pacific. *J. Phys. Oceanogr.*, **35**, 1861–1875.
Wang, W. and Huang, R. X. (2004a). Wind energy input to the Ekman Layer. *J. Phys. Oceanogr.*, **34**, 1267–1275.
Wang, W. and Huang, R. X. (2004b). Wind energy input to the surface waves. *J. Phys. Oceanogr.*, **34**, 1276–1280.
Wang, W. and Huang, R. X. (2005). An experimental study on thermal circulation driven by horizontal differential heating. *J. Fluid Mech.*, **540**, 49–73.
Wang, W., Qian, C. C. and Huang, R. X. (2006). Mechanical energy input to the world oceans due to atmospheric loading. *Chinese Science Bulletin*, **51**, 327–330.
Warren, B. (1973). Transpacific hydrographic sections at Lats. 43°S and 28°S: the SCORPIO expedition – II. Deep water. *Deep Sea Res.*, **20**, 9–38.
Warren, B. A. (1981). Deep circulation of the world ocean. In *Evolution of Physical Oceanography*, ed. B. A. Warren and C. Wunsch, MIT Press, Cambridge, MA, pp. 6–41.
Warren, B. A. and Speer, K. G. (1991). Deep circulation in the eastern South Atlantic Ocean. *Deep Sea Res.* I, **38**, S281–S322.
Weatherly, G. L. and Kelley, E. A., Jr. (1985). Two views of the cold filament, *J. Phys. Oceanogr.*, **15**, 68–81.
Weaver, A. J. and Sarachik, E. S. (1991). The role of mixed boundary conditions in numerical models of the ocean's climate. *J. Phys. Oceanogr.*, **21**, 1470–1493.
Weaver, A. J., Aura, S. M. and Myers, P. G. (1994). Interdecadal variability in an idealized model of the North Atlantic. *J. Geophys. Res.*, **99**(C6), 12,423–12,441.
Weaver, A. J., Sarachik, E. S. and Marotzke, J. (1991). Freshwater flux forcing of decadal and interdecadal oceanic variability. *Nature*, **353**, 836–838.
Webb, D. J. and Suginohara, N. (2001). Vertical mixing in the ocean. *Nature*, **409**, 37.
Welander, P. (1959). An advective model of the ocean thermocline. *Tellus*, **11**, 309–318.
Welander, P. (1971a). Some exact solutions to the equation describing an ideal-fluid thermocline. *J. Mar. Res.*, **29**, 60–68.
Welander, P. (1971b). The thermocline problem. *Philos. Trans. R. Soc. London Ser.* A, **270**, 415–421.
Welander, P. (1982). A simple heat-salt oscillator. *Dyn. Atmos. Ocean*, **6**, 233–242.
Whitworth, T., III and Peterson, R. G. (1985). Volume transport of the Antarctic Circumpolar Current from bottom pressure measurements. *J. Phys. Oceanogr.*, **15**, 810–816.

Whitworth, T., Nowlin, W. D., Pillsbury, R. D., Moore, M. I. and Weiss, R. F. (1991). Observations of the Antarctic Circumpolar Current and deep boundary current in the southwest Atlantic. *J. Geophys. Res.*, **96**, 15105–15118.

Whitworth, T., Warren, B. A., Nowlin, W. D., Rutz, S. B., Pillsbury, R. D. and Moore, M. I. (1999). On the deep western-boundary current in the Southwest Pacific Basin. *Progr. Oceanogr.*, **43**, 1–54.

Wijffels, S. (1993). *Exchanges between hemisphere and gyres: A direct approach to the mean circulation of the equatorial Pacific*. Ph.D. thesis, MIT/WHOI, 267 pp.

Wijffels, S. E., Hall, M. M., Joyce, T., Torres, D. J., Hacker, P. and Firing, E. (1998). Multiple deep gyres of the western North Pacific: A WOCE section along 149°E. *J. Geophys. Res.*, **103**, 12,985–13,009.

Williams, R. G. (1989). Influence of air–sea interaction on the ventilated thermocline. *J. Phys. Oceanogr.*, **19**, 1255–1267.

Williams, R. G. (1991). The role of mixed layer in setting the potential vorticity of the main thermocline. *J. Phys. Oceanogr.*, **21**, 1803–1814.

Woods, I. R. (1968). Selective withdrawal from a steady stratified fluid. *J. Fluid Mech.*, **32**, 209–223.

Woods, J. D. (1985). The physics of pycnocline ventilation. In *Coupled Ocean–Atmosphere Models*, ed. J. C. Nihoul, Elsevier Sci. Pub., pp. 543–590.

Woods, J. D. and Barkmann, W. (1986). A Lagrangian mixed layer model of Atlantic 18°C water formation. *Nature*, **319**, 574–576.

Worthington, L. V. (1959). The 18° water in the Sargasso Sea. *Deep Sea Res.*, **5**, 297–305.

Worthington, L. V. (1969). An attempt to measure the volume transport of Norwegian Sea overflow water through the Denmark Strait. *Deep Sea Res.*, **16** (Suppl.), 421–432.

Worthington, L. V. (1970). The Norwegian Sea as a Mediterranean basin. *Deep Sea Res.*, **17**, 77–84.

Wright, D. G. and Stocker, T. F. (1991). A zonally averaged ocean model for the thermohaline circulation. Part I: Model development and flow dynamics. *J. Phys. Oceanogr.*, **21**, 1713–1724.

Wunsch, C. (1970). On oceanic boundary mixing. *Deep Sea Res.*, **17**, 293–301.

Wunsch, C. (1998). The work done by the wind on the oceanic general circulation. *J. Phys. Oceanogr.*, **28**, 2332–2342.

Wunsch, C. (2002). What is the thermohaline circulation? *Science*, **298**, 1179–1181.

Wunsch, C. and Ferrari, R. (2004). Vertical mixing, energy, and the general circulation of the oceans. *Ann. Rev. of Fluid Mech.*, **36**, 281–314.

Wunsch, C. and Stammer, D. (1997). Atmospheric loading and oceanic "inverted barometer" effect. *Rev. of Geophys.*, **35**, 79–107.

Wyrtki, K. (1961). The thermohaline circulation in relation to the general circulation in the oceans. *Deep Sea Res.*, **8**, 39–64.

Wyrtki, K. and Kilonsky, B. (1984). Mean water and current structure during the Hawaii–Tahiti Shuttle Experiment. *J. Phys. Oceanogr.*, **14**, 242–254.

Yan, Y., Gan, Z. and Qi, Y. (2004). Entropy budget of the ocean system. *Geophys. Res. Lett.*, **31**, L14311, doi:10.1029/2004GL019921.

Yang, J. and Price, J. (2000). Water-mass formation and potential vorticity balance in an abyssal ocean circulation. *J. Mar. Res.*, **58**, 789–808.

Yang, X.-Y., Huang, R. X. and Wang, D. X. (2007). Decadal changes of wind stress over the Southern Ocean associated with Antarctic ozone depletion. *J. Climate*, **20**, 3395–3410.

Yeh, T. C. (1957). On the formation of quasi-geostrophic motion in the atmosphere. *J. Meteor. Soc. Japan*, the 75th Anniversary Volume, 130–134.

Young, W. R. (1981). *The vertical structure of the wind-driven circulation*. Ph. D thesis, MIT/WHOI Joint Program in Oceanography, 215 pp.

Young, W. R. and Ierley, G. R. (1986). Eastern boundary conditions and weak solutions of the ideal thermocline equations. *J. Phys. Oceanogr.*, **16**, 1884–1900.

Zalesak, S. T. (1979). Fully multidimensional flux-corrected transport algorithms for fluids. *J. Comput. Phys.*, **31**, 335–362.

Zang, X. and Wunsch, C. (2001). Spectral description of low-frequency oceanic variability. *J. Phys. Oceanogr.*, **31**, 3073–3095.

Zebiak, S. E. and Cane, M. A. (1987). A model EL NINO – Southern Oscillation. *Month. Wea. Rev.*, **115**, 2262–2278.

Zhang, J., Schmitt, R. W. and Huang, R. X. (1998). Sensitivity of the GFDL Modular Ocean Model to parameterization of double-diffusive processes. *J. Phys. Oceanogr.*, **28**, 589–605.

Zhang, J., Schmitt, R. W. and Huang, R. X. (1999). The relative influence of diapycnal mixing and hydrologic forcing on the stability of the thermohaline circulation. *J. Phys. Oceanogr.*, **29**, 1096–1108.

Zhang, K. Q. and Marotzke, J. (1999). The importance of open-boundary estimation for an Indian Ocean GCM-data system. *J. Mar. Res.*, **57**, 305–334.

Zhang, R., Follows, M. J., Grotzinger, J. P. and Marshall, J. (2001). Could the late Permian deep ocean have been anoxic? *Paleoceanography*, **16**, 317–329.

Suggested reading

Chapter 1

Pickard, G. L, W. J. Emery, and L. D. Talley (2009). *Descriptive Physical Oceanography*, Elsevier.

Stommel, H. (1984). The Sea of the Beholder. In Hogg, N. G. and R. X. Huang (eds.) *Collected Works of Henry Stommel*, American Meteor. Soc., Boston, Vol. I, 5–112.

Chapter 2

Cushman-Roisin, B. (1994). *Introduction to Geophysical Fluid Dynamics*. Prentice Hall Press, 320 pp.

Feistel, R. and E. Hagen (1995). On the Gibbs thermodynamic potential of seawater. *Prog. Oceanogr.*, **36**, 249–327.

Fofonoff, N. P. (1992). Lecture Notes EPP-226, Harvard University, 66pp.

Gill, A. E. (1982). *Atmosphere–Ocean Dynamics*. Academic Press, New York, 662pp.

Holton, J. R. (2004). *An Introduction to Dynamic Meteorology*, Volume 88, Fourth edition (International Geophysics), Academic Press, New York, 535pp.

Sedov, L. I. (1959). *Similarity and Dimensional Methods in Mechanics*, Academic Press, New York, 363 pp.

Chapter 3

Faller, A. (1966). Sources of energy for the ocean circulation and a theory of the mixed layer. *Proceedings of the Fifth US National Congress of Applied Mechanics*, Minnesota, pp. 651–672.

Huang, R. X. (2004). Ocean, energy flow in. In C. J. Cleveland (ed.), *Encyclopedia of Energy*, Elsevier, Vol. 4, pp. 497–509.

Kuhlbrodt, T., A. Griesel, M. Montoya, A. Levermann, M. Hofmann, and S. Rahmstorf (2007). On the driving processes of the Atlantic meridional overturning circulation. *Rev. Geophys.*, **45**, RG2001, doi:10.1029/2004RG000166.

Munk, W. H. and C. Wunsch (1998). Abyssal recipes II: energetics of the tidal and wind mixing. *Deep Sea Res.*, I, **45**, 1977–2010.

Wunsch, C. and R. Ferrari (2004). Vertical mixing, energy, and the general circulation of the oceans. *Ann. Rev. of Fluid Mech.*, **36**, 281–314.

Chapter 4

Luyten, J., J. Pedlosky, and H. M. Stommel (1983). The ventilated thermocline. *J. Phys. Oceanogr.*, **13**, 292–309.

McCreary, J. P. and P. Lu (1994). Interaction between the subtropical and equatorial ocean circulation: the subtropical cell. *J. Phys. Oceanogr.*, **24**, 466–497.

Pedlosky, J (2006). A history of thermocline theory. In M. Jochum and R. Murtugudde (eds.), *Physical Oceanography Developments Since 1950*, Springer, New York, pp. 139–152.

Pedlosky, J. (1996). *Ocean Circulation Theory*. Springer-Verlag, Heidelberg, 453 pp.

Rhines, P. B. and W. R. Young (1982). A theory of the wind-driven circulation. I. Mid-ocean gyres. *J. Mar. Res.,* **40** (Suppl.), 559–596.

Stommel, H. (1948). The western intensification of wind-driven ocean currents. *Trans., Amer. Geophys. Union*, **29**, 202–206.

Stommel, H. (1984). The delicate interplay between wind-stress and buoyancy input in ocean circulation: the Goldsbrough variations. Crafoord prize Lecture presented at the Royal Swedish Academy of Sciences, Stockholm, on September 28, 1983. *Tellus*, **36**(A), 111–119.

Chapter 5

Bryden, H. L. and S. Imawaki (2001). Ocean heat transport. In G. Siedler, J. Church, and J. Gould (eds.): "*Ocean Circulation and Climate*", International Geophysical Series, Academy Press, New York, pp. 455–474.

Kuhlbrodt, T., A. Griesel, M. Montoya, A. Levermann, M. Hofmann, and S. Rahmstorf (2007). On the driving processes of the Atlantic meridional overturning circulation. *Rev. Geophys.*, **45**, RG2001, doi:10.1029/2004RG000166.

Marshall, J. and F. Schott (1999). Open-ocean convection: observations, theory and models. *Rev. Geophys.*, **37**, 1–64.

Schmitz, Jr., W. J. (1986). On the world ocean circulation: Volume I, Some global features/North Atlantic circulation. Woods Hole Oceanographic Institution Technical Report WHOI-96–03, 148pp.

Schmitz, Jr., W. J. (1986). On the world ocean circulation: Volume II, The Pacific and Indian Oceans / A global update. Woods Hole Oceanographic Institution Technical Report WHOI-96–08, 241pp.

Stommel, H. M. (1961). Thermohaline convection with two stable regimes of flow. *Tellus*, **13**, 224–230.

Toggweiler, J. R. and B. Samuels (1998). On the ocean's large-scale circulation near the limit of no vertical mixing. *J. Phys. Oceanogr.*, **9**, 1832–1852.

Warren, B. A. (1981). Deep circulation of the world ocean. In B. A. Warren and C. Wunsch (eds.), *Evolution of Physical Oceanography*, Massachusetts Institute of Technology Press, Cambridge, 623pp.

Whitehead, A. A. (1998). Topographic control of ocean flows in deep passages and straits. *Rev. Geophys.*, **36**, 423–440.

Woods, J. D. (1985). The physics of pycnocline ventilation. In J. C. Nihoul (ed.), *Coupled Ocean-Atmosphere Models*. Elsevier Sci. Pub., pp. 543–590.

Index